Annals of Mathematics Studies

Number 195

Asymptotic Differential Algebra and Model Theory of Transseries

Matthias Aschenbrenner
Lou van den Dries
Joris van der Hoeven

PRINCETON UNIVERSITY PRESS
PRINCETON AND OXFORD
2017

Published by Princeton University Press, 41 William Street,
Princeton, New Jersey 08540

In the United Kingdom: Princeton University Press, 6 Oxford Street,
Woodstock, Oxfordshire OX20 1TR

press.princeton.edu

Library of Congress Cataloging-in-Publication Data

Names: Aschenbrenner, Matthias, 1972– | van den Dries, Lou | Hoeven, J. van der (Joris)

Title: Asymptotic differential algebra and model theory of transseries / Matthias Aschenbrenner, Lou van den Dries, Joris van der Hoeven.

Description: Princeton : Princeton University Press, 2017. | Series: Annals of mathematics studies ; number 195 | Includes bibliographical references and index.

Identifiers: LCCN 2017005899 | ISBN 9780691175423 (hardcover : alk. paper) | ISBN 9780691175430 (pbk. : alk. paper)

Subjects: LCSH: Series, Arithmetic. | Divergent series. | Asymptotic expansions. | Differential algebra.

Classification: LCC QA295 .A87 2017 | DDC 512/.56–dc23 LC record available at https://lccn.loc.gov/2017005899

British Library Cataloging-in-Publication Data is available

The publisher would like to acknowledge the authors of this volume for providing the camera-ready copy from which this book was printed.

This book has been composed in LaTeX.

Printed on acid-free paper. ∞

10 9 8 7 6 5 4 3 2 1

Had the apparatus [of transseries and analyzable functions] been introduced for the sole purpose of solving Dulac's "conjecture," one might legitimately question the wisdom and cost-effectiveness of such massive investment in new machinery. However, [these notions] have many more applications, actual or potential, especially in the study of analytic singularities. But their chief attraction is perhaps that of giving concrete, if partial, shape to G. H. Hardy's dream of an *all-inclusive, maximally stable algebra of "totally formalizable functions."*

— Jean Écalle, *Six Lectures on Transseries, Analysable Functions and the Constructive Proof of Dulac's Conjecture.*

The virtue of model theory is its ability to organize succinctly the sort of tiresome algebraic details associated with elimination theory.

— Gerald Sacks, *The Differential Closure of a Differential Field.*

Les analystes p-adiques se fichent tout autant que les géomètres algébristes ..., des gammes à plus soif sur les valuations composées, les groupes ordonnés baroques, sous-groupes pleins desdits et que sais-je. Ces gammes méritent tout au plus d'enrichir les exercices de Bourbaki, tant que personne ne s'en sert.

— Alexander Grothendieck, letter to Serre dated October 31, 1961.

I don't like either writing or reading two-hundred page papers. It's not my idea of fun.

— John H. Conway, quoted in *Genius at Play: The Curious Mind of John Horton Conway* by Siobhan Roberts.

Contents

Preface

We develop here the algebra and model theory of the *differential field of transseries,* a fascinating mathematical structure obtained by iterating a construction going back more than a century to Levi-Civita and Hahn. It was introduced about thirty years ago as an exponential ordered field by Dahn and Göring in connection with Tarski's problem on the real field with exponentiation, and independently by Écalle in his proof of the Dulac Conjecture on plane analytic vector fields.

The analytic aspects of transseries have a precursor in Borel's summation of divergent series. Indeed, Écalle's theory of accelero-summation vastly extends Borel summation, and associates to each *accelero-summable* transseries an *analyzable* function. In this way many non-oscillating real-valued functions that arise naturally (for example, as solutions of algebraic differential equations) can be represented faithfully by transseries.

For about twenty years we have studied the differential field of transseries within the broader program of developing *asymptotic differential algebra.* We have recently obtained decisive positive results on its model theory, and we describe these results in an *Introduction and Overview.* That introduction assumes some rudimentary knowledge of differential fields, valued fields, and model theory, but no acquaintance with transseries. It is intended to familiarize readers with the main issues in this book and with the terminology that we frequently use.

Initially, Joris van der Hoeven in Paris and Matthias Aschenbrenner and Lou van den Dries in Urbana on the other side of the Atlantic worked independently, but around 2000 we decided to join forces. In 2011 we arrived at a rough outline for proving some precise conjectures: see our programmatic survey *Toward a model theory for transseries.* All the conjectures stated in that paper (with one minor change) did turn out to be true, even though some seemed to us at the time rather optimistic.

Why is this book so long? For one, several problems we faced had no short solutions. Also, we have chosen to work in a setting that is sufficiently flexible for further developments, as we plan to show in a later volume. Finally, we have tried to be reasonably self-contained by assuming only a working knowledge of basic algebra: groups, rings, modules, fields. Occasionally we refer to Lang's *Algebra.*

After the *Introduction and Overview* this book consists of 16 chapters and 2 appendices. Each chapter has an introduction and is divided into sections. Each section has subsections, the last one often consisting of (partly historical) notes and comments. Many chapters state in the beginning some assumptions—sometimes just notational in nature— that are in force throughout that chapter, and of course the reader should be aware of those in studying a particular chapter, since we do not repeat these assumptions when stating theorems, etc. The same holds for many sections and subsections. The end of the volume has a list of symbols and an index.

ACKNOWLEDGMENTS

Part of this work was carried out while some of the authors were in residence at various times at the Fields Institute (Toronto), the Institut des Hautes Études Scientifiques (Bures-sur-Yvette), the Isaac Newton Institute for Mathematical Sciences (Cambridge), and the Mathematical Sciences Research Institute (Berkeley). The support and hospitality of these institutions is gratefully acknowledged.

Aschenbrenner's work was partially supported by the National Science Foundation under grants DMS-0303618, DMS-0556197, and DMS-0969642. Visits by van der Hoeven to Los Angeles were partially supported by the UCLA Logic Center.

We thank the following copyright holders for permission to reproduce the text in the epigraphs in the front of this book: Springer Science and Business Media, New York, for the quote by Jean Écalle from [121]; the American Mathematical Society for the quote by Gerald Sacks from [376], © 1972 American Mathematical Society; Professor Jean-Pierre Serre for the quote by Alexander Grothendieck from [88]; and Siobhan Roberts for the quote by John H. Conway that appears in her book *Genius at Play: The Curious Mind of John Horton Conway* [344] © Siobhan Roberts, published by Bloomsbury Publishing, Inc., 2016.

We thank David Marker and Angus Macintyre for their interest and steadfast moral support over the years. To Santiago Camacho, Andrei Gabrielov, Tigran Hakobyan, Elliot Kaplan, Nigel Pynn-Coates, Chieu Minh Tran, and especially to Allen Gehret, we are indebted for numerous comments on and corrections to the manuscript. We are also grateful to Philip Ehrlich for setting us right on some historical points, and to the anonymous reviewers for useful suggestions and for spotting some errors. We are of course solely responsible for any remaining inadequacies.

Finally, we thank our editor, Vickie Kearn, and the other staff at Princeton University Press, notably Nathan Carr and Glenda Krupa, for helping us to bring this book into its final form.

Matthias Aschenbrenner, Los Angeles

Lou van den Dries, Urbana

Joris van der Hoeven, Paris

September 2015

Conventions and Notations

Throughout, m and n range over the set $\mathbb{N} = \{0, 1, 2, \dots\}$ of natural numbers. For sets X, Y we distinguish between $X \subseteq Y$, meaning that X is a subset of Y, and $X \subset Y$, meaning that X is a proper subset of Y.

For an (additively written) abelian group A we set $A^{\neq} := A \setminus \{0\}$. By *ring* we mean an associative but possibly non-commutative ring with identity 1. Let R be a ring. A *unit* of R is a $u \in R$ with a right-inverse (an $x \in R$ with $ux = 1$) and a left-inverse (an $x \in R$ with $xu = 1$). If u is a unit of R, then u has only one right-inverse and only one left-inverse, and these coincide. With respect to multiplication the units of R form a group $R^\times$ with identity 1. Thus the multiplicative group of a field K is $K^\times = K \setminus \{0\} = K^{\neq}$. Subrings and ring morphisms preserve 1.

A *domain* is a ring with $1 \neq 0$ such that for all x, y in the ring, if $xy = 0$, then $x = 0$ or $y = 0$. Usually domains are commutative, but not always. However, an *integral* domain is always commutative, that is, a subring of a field.

Let R be a ring. An R-module is a left R-module unless specified otherwise, and the scalar $1 \in R$ acts as the identity on any R-module. Let M be an R-module and $(x_i)_{i \in I}$ a family in M. A family $(r_i)_{i \in I}$ in R is *admissible* if $r_i = 0$ for all but finitely many $i \in I$. An *R-linear combination of* (x_i) is an $x \in M$ such that $x = \sum_i r_i x_i$ of M for some admissible family (r_i) in R. We say that (x_i) *generates* M if every element of M is an R-linear combination of (x_i). We say that (x_i) is *R-dependent* (or *linearly dependent over* R) if $\sum_i r_i x_i = 0$ for some admissible family $(r_i)_{i \in I}$ in R with $r_i \neq 0$ for some $i \in I$; for $I = \{1, \dots, n\}$ we also abuse language by expressing this as: $x_1, \dots, x_n$ *are R-dependent*. We say that (x_i) is *R-independent* (or *linearly independent over* R) if (x_i) is not R-dependent. We call M *free on* (x_i) (or (x_i) *a basis of* M) if (x_i) generates M and (x_i) is R-independent. Sometimes we use this terminology for sets $X \subseteq M$ to mean that for some (equivalently, for every) index set I and bijection $i \mapsto x_i \colon I \to X$ the family (x_i) has the corresponding property.

Let K be a commutative ring. A *K-algebra* is defined to be a ring A together with a ring morphism $\phi \colon K \to A$ that takes its values in the center of A; we then refer to ϕ as the structural morphism of the K-algebra A, and construe A as a K-module by $\lambda a := \phi(\lambda) a$ for $\lambda \in K$ and $a \in A$.

Given a field extension F of a field K and a family (x_i) in F we use the expressions (x_i) *is algebraically (in)dependent over* K and (x_i) *is a transcendence basis of* F *over* K in a way similar to the above linear analogues; likewise, a set $X \subseteq F$ can be referred to as being a transcendence basis of F over K.

When a vector space V over a field C is given, then *subspace of* V means *vector subspace of* V.

Leitfaden

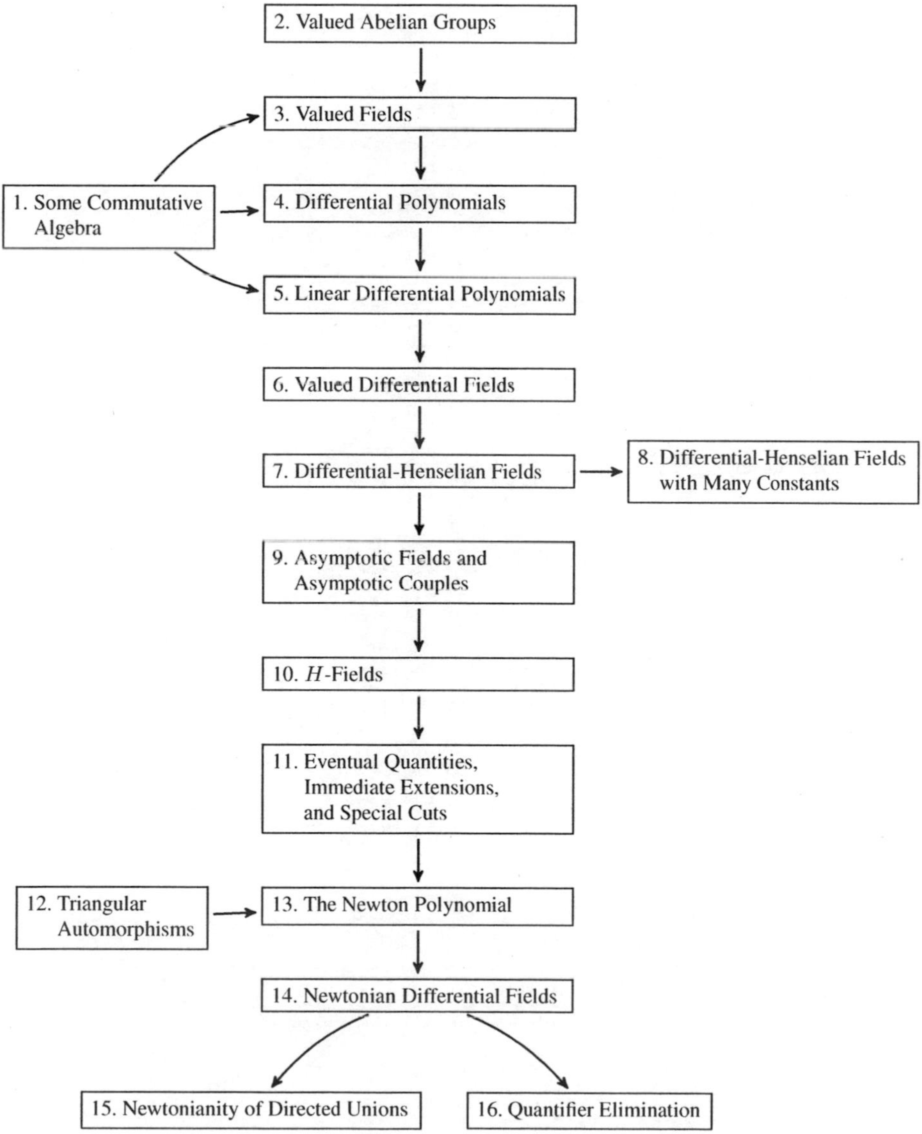
2. Valued Abelian Groups
3. Valued Fields
1. Some Commutative Algebra
4. Differential Polynomials
5. Linear Differential Polynomials
6. Valued Differential Fields
7. Differential-Henselian Fields
8. Differential-Henselian Fields with Many Constants
9. Asymptotic Fields and Asymptotic Couples
10. H-Fields
11. Eventual Quantities, Immediate Extensions, and Special Cuts
12. Triangular Automorphisms
13. The Newton Polynomial
14. Newtonian Differential Fields
15. Newtonianity of Directed Unions
16. Quantifier Elimination

Dramatis Personæ

We summarize here the definitions of some notions prominent in our work, together with a list of attributes that apply to them. We include the page number where each concept is first introduced. We let m, n, r range over $\mathbb{N} = \{0, 1, 2, \dots\}$. Below K is a field, possibly equipped with further structure. We let a, b, f, g, y, z range over elements of K, and we let Y be an indeterminate over K. If K comes equipped with a valuation, then we let $\mathcal{O}$ be the valuation ring of K, and we freely employ the dominance relations on K introduced in Section 3.1. If K comes equipped with a derivation ∂, then we also write $f', f'', \dots, f^{(n)}, \dots$ for $\partial f, \partial^2 f, \dots, \partial^n f, \dots$, and $f^\dagger = f'/f$ for the logarithmic derivative of any $f \neq 0$; in this case $C = \{f : f' = 0\}$ denotes the constant field of K, and c ranges over C. We let Γ be an ordered abelian group, and let α, β, γ range over Γ.

VALUED FIELDS

Let K be a *valued field,* that is, a field equipped with a valuation on it; p. 112.

Complete: every cauchy sequence in K has a limit in K; p. 84.

Spherically complete: every pseudocauchy sequence in K has a pseudolimit in K; p. 78. “Spherically complete” is equivalent to “maximal” as defined below.

Maximal: there is no proper immediate valued field extension of K; p. 129.

Algebraically maximal: there is no proper immediate algebraic valued field extension of K; p. 130.

Henselian: for every $P \in K[Y]$ with $P \preccurlyeq 1$, $P(0) \prec 1$, and $P'(0) \asymp 1$, there exists $y \prec 1$ with $P(y) = 0$; p. 136.

DIFFERENTIAL FIELDS

Let K be a *differential field,* that is, a field of characteristic zero equipped with a derivation on it; p. 200.

Linearly surjective: for all $a_0, \dots, a_r \in K$ with $a_r \neq 0$ there exists y such that

$$a_0 y + a_1 y' + \cdots + a_r y^{(r)} = 1; \qquad \text{p. 253.}$$

Linearly closed: for all $r \geqslant 1$ and $a_0, \dots, a_r \in K$ there are $b_0, \dots, b_{r-1}, b \in K$ with $a_0Y + a_1Y' + \dots + a_rY^{(r)} = b_0(Y' + bY) + b_1(Y' + bY)' + \dots + b_{r-1}(Y' + bY)^{(r-1)}$; p. 252.

Picard-Vessiot closed (or *pv-closed*): for all $r \geqslant 1$ and $a_0, \dots, a_r \in K$ with $a_r \neq 0$ there are C-linearly independent $y_1, \dots, y_r$ such that $a_0y_i + a_1y_i' + \dots + a_ry_i^{(r)} = 0$ for $i = 1, \dots, r$; p. 254.

Differentially closed: for all $P \in K[Y, \dots, Y^{(r)}]^{\neq}$ and $Q \in K[Y, \dots, Y^{(r-1)}]^{\neq}$ such that $\frac{\partial P}{\partial Y^{(r)}} \neq 0$ there is y with $P(y, y', \dots, y^{(r)}) = 0$ and $Q(y, y', \dots, y^{(r-1)}) \neq 0$; p. 237.

VALUED DIFFERENTIAL FIELDS

Let K be a *valued differential field*, that is, a differential field equipped with a valuation on it; p. 221.

Small derivation: $f \prec 1 \Rightarrow f' \prec 1$; p. 222.

Monotone: $f \prec 1 \Rightarrow f' \preccurlyeq f$; p. 226.

Few constants: $c \preccurlyeq 1$ for all c; p. 226.

Many constants: for every f there exists c with $f \asymp c$; p. 226.

Differential-henselian (or d-*henselian*): K has small derivation and:

(DH1) for all $a_0, \dots, a_r \preccurlyeq 1$ in K with $a_r \asymp 1$ there exists $y \asymp 1$ such that

$$a_0y + a_1y' + \dots + a_ry^{(r)} \sim 1;$$

(DH2) for every $P \in K[Y, Y', \dots, Y^{(r)}]$ with $P \preccurlyeq 1$, $P(0) \prec 1$, and $\frac{\partial P}{\partial Y^{(n)}}(0) \asymp 1$ for some n, there exists $y \prec 1$ such that $P(y, y', \dots, y^{(r)}) = 0$; p. 340.

ASYMPTOTIC FIELDS

Let K be an *asymptotic field*, that is, a valued differential field such that for all nonzero $f, g \prec 1$: $f \prec g \Longleftrightarrow f' \prec g'$; p. 379.

H-asymptotic (or *of H-type*): $0 \neq f \prec g \prec 1 \Rightarrow f^\dagger \succcurlyeq g^\dagger$; p. 379.

Differential-valued (or d-*valued*): for all $f \asymp 1$ there exists c with $f \sim c$; p. 379.

Grounded: there exists nonzero $f \not\asymp 1$ such that $g^\dagger \succcurlyeq f^\dagger$ for all nonzero $g \not\asymp 1$; p. 384.

Asymptotic integration: for all $f \neq 0$ there exists $g \not\asymp 1$ with $g' \asymp f$; p. 383.

Asymptotically maximal: K has no proper immediate asymptotic field extension; p. 380.

Asymptotically d-*algebraically maximal*: K has no proper immediate differential-algebraic asymptotic field extension; p. 380.

λ-free: H-asymptotic, ungrounded, and for all f there exists $g \succ 1$ with $f - g^{\dagger\dagger} \succcurlyeq g^{\dagger}$; p. 506.

ω-free: H-asymptotic, ungrounded, and for all f there is $g \succ 1$ with $f - \omega(g^{\dagger\dagger}) \succcurlyeq g^{\dagger}$, where $\omega(z) := -(2z' + z^2)$; p. 515.

Newtonian: H-asymptotic, ungrounded, and every $P \in K[Y, Y', \dots, Y^{(r)}]^{\neq}$ of Newton degree 1 has a zero in $\mathcal{O}$; p. 640. (See p. 480 for Newton degree.)

ORDERED VALUED DIFFERENTIAL FIELDS

Let K be an *ordered valued differential field*, that is, a valued differential field equipped with an ordering in the usual sense of *ordered field*; p. 378.

Pre-H-field: $\mathcal{O}$ is convex in the ordered field K, and for all f:

$$f > \mathcal{O} \Longrightarrow f' > 0; \qquad \text{p. 452.}$$

H-field: $\mathcal{O}$ is the convex hull of C in the ordered field K, and for all f:

$$f > C \Longrightarrow f' > 0, \quad f \asymp 1 \Longrightarrow \text{ there exists } c \text{ with } f \sim c; \qquad \text{p. 451.}$$

Liouville closed: K is a real closed H-field and for all f, g there exists $y \neq 0$ such that $y' + fy = g$; p. 460.

ASYMPTOTIC COUPLES

Let (Γ, ψ) be an *asymptotic couple*, that is, the ordered abelian group Γ is equipped with a map $\psi \colon \Gamma^{\neq} \to \Gamma$ such that for all $\alpha, \beta \neq 0$:

(AC1) $\alpha + \beta \neq 0 \Rightarrow \psi(\alpha + \beta) \geqslant \min(\psi(\alpha), \psi(\beta))$;

(AC2) $\psi(k\alpha) = \psi(\alpha)$ for all $k \in \mathbb{Z}^{\neq}$;

(AC3) $\alpha > 0 \Rightarrow \alpha' := \alpha + \psi(\alpha) > \psi(\beta)$;

p. 322. For $\gamma \neq 0$ we set $\gamma' := \gamma + \psi(\gamma)$.

H-asymptotic (or *of H-type*): $0 < \alpha < \beta \Rightarrow \psi(\alpha) \geqslant \psi(\beta)$; p. 323.

Grounded: $\Psi := \{\psi(\alpha) : \alpha \neq 0\}$ has a largest element; p. 388.

Small derivation: $\gamma > 0 \Rightarrow \gamma' > 0$; p. 388.

Asymptotic integration: for all α there exists $\beta \neq 0$ with $\alpha = \beta'$; p. 383.

Asymptotic Differential Algebra and Model Theory of Transseries

century. Hardy considered *logarithmico-exponential functions*: real-valued functions built up from constants and the variable x using addition, multiplication, division, exponentiation and taking logarithms. He showed that such a function, when defined on some interval $(a, +\infty)$, has eventually constant sign (no oscillation!), and so the germs at $+\infty$ of these functions form an ordered field H with derivation $\frac{d}{dx}$. Thus H is what Bourbaki [62] calls a *Hardy field*: a sub*field* K of the ring of germs at $+\infty$ of differentiable functions $f\colon (a, +\infty) \to \mathbb{R}$ with $a \in \mathbb{R}$, closed under taking derivatives; for more precision, see Section 9.1. Each Hardy field is naturally an ordered differential field. The Hardy field H is rather special: every $f \in H$ satisfies an algebraic differential equation over $\mathbb{R}$. But H lacks some closure properties that are desirable for a comprehensive theory. For instance, H has no antiderivative of e^{x^2} (by Liouville; see [361]), and the functional inverse of $(\log x)(\log\log x)$ doesn't lie in H, and is not even asymptotic to any element of H: [111, 190]; see also [333].

With $\mathbb{T}$ and transseries we go beyond H and logarithmico-exponential functions by admitting *infinite* sums. It is important to be aware, however, that by virtue of its inductive construction, $\mathbb{T}$ does not contain, for example, the series

$$x + \log x + \log\log x + \log\log\log x + \cdots,$$

which does make sense in a suitable extension of $\mathbb{T}$. Thus $\mathbb{T}$ allows only certain kinds of infinite sums. Nevertheless, it turns out that the differential field $\mathbb{T}$ enjoys many remarkable closure properties that H lacks. For instance, $\mathbb{T}$ is closed under natural operations of exponentiation, integration, composition, compositional inversion, and the resolution of *feasible* algebraic differential equations (where the meaning of *feasible* can be made explicit). This makes $\mathbb{T}$ of interest for different areas of mathematics:

Analysis

In connection with the Dulac Problem, $\mathbb{T}$ is sufficiently rich for modeling the asymptotic behavior of so-called Poincaré return maps. This analytically deep result is a crucial part of Écalle's solution of the Dulac Problem [119, 120, 121]. (At the end of this introduction we discuss this in more detail.)

Computer algebra

Many transseries are concrete enough to compute with them, in the sense of computer algebra [190, 402]. Moreover, many of the closure properties mentioned above can be made effective. This allows for the automation of an important part of asymptotic calculus for functions of one variable.

Logic

Given an o-minimal expansion of the real field, the germs at $+\infty$ of its definable one-variable functions form a Hardy field, which in many cases can be embedded into $\mathbb{T}$. This gives useful information about the possible asymptotic behavior of these definable functions; see [21, 292] for more about this connection.

Soon after the introduction of $\mathbb{T}$ in the 1980s it was suspected that $\mathbb{T}$ might well be a kind of *universal domain* for the differential algebra of Hardy fields and similar ordered differential fields, analogous to the role of the algebraically closed field $\mathbb{C}$ as a universal domain for algebraic geometry of characteristic 0 (Weil [461, Chapter X, §2]), and of $\mathbb{R}$, $\mathbb{Q}_p$, and $\mathbb{C}((t))$ in related ordered and valued settings. This is corroborated by the strong closure properties enjoyed by $\mathbb{T}$. See in particular p. 148 of Écalle's book [120] for eloquent expressions of this idea. The present volume and the next substantiate the *universal domain* nature of the differential field $\mathbb{T}$, using the language of *model theory*. The model-theoretic properties of the classical fields $\mathbb{C}$, $\mathbb{R}$, $\mathbb{Q}_p$ and $\mathbb{C}((t))$ are well established thanks to Tarski, Seidenberg, Robinson, Ax & Kochen, Eršov, Cohen, Macintyre, Denef, and others; see [443, 395, 350, 28, 29, 131, 84, 275, 100]. Our goal is to analyze likewise the differential field $\mathbb{T}$, which comes with a definable ordering and valuation, and in this book we achieve this goal.

The ordered and valued differential field $\mathbb{T}$

For what follows, it will be convenient to quickly survey some of the most distinctive features of $\mathbb{T}$. Appendix A contains precise definitions and further details.

Each transseries $f = f(x)$ can be uniquely decomposed as a sum

$$f = f_{\succ} + f_{\asymp} + f_{\prec},$$

where $f_{\succ}$ is the *infinite part of* f, $f_{\asymp}$ is its constant term (a real number), and $f_{\prec}$ is its *infinitesimal part*. In the example (1) above,

$$\begin{aligned}
\varphi_{\succ} &= -3\,\mathrm{e}^{\mathrm{e}^x} + \mathrm{e}^{\frac{\mathrm{e}^x}{\log x} + \frac{\mathrm{e}^x}{\log^2 x} + \frac{\mathrm{e}^x}{\log^3 x} + \cdots} - x^{11}, \\
\varphi_{\asymp} &= 7, \\
\varphi_{\prec} &= \frac{\pi}{x} + \frac{1}{x\log x} + \cdots .
\end{aligned}$$

In this example, $\varphi_{\succ}$ happens to be a finite sum, but this is not a necessary feature of transseries: take for example $f := \frac{\mathrm{e}^x}{\log x} + \frac{\mathrm{e}^x}{\log^2 x} + \frac{\mathrm{e}^x}{\log^3 x} + \cdots$, with $f_{\succ} = f$. Declaring a transseries to be positive iff its dominant (= leftmost) coefficient is positive turns $\mathbb{T}$ into an *ordered* field extension of $\mathbb{R}$ with $x > \mathbb{R}$. In our example (1), the dominant transmonomial of $\varphi(x)$ is $\mathrm{e}^{\mathrm{e}^x}$ and its dominant coefficient is -3, whence $\varphi(x)$ is negative; in fact, $\varphi(x) < \mathbb{R}$.

The inductive definition of $\mathbb{T}$ involves constructing a certain exponential operation $\exp\colon \mathbb{T} \to \mathbb{T}^{\times}$, with $\exp(f)$ also written as e^f, and

$$\exp(f) = \exp(f_{\succ}) \cdot \exp(f_{\asymp}) \cdot \exp(f_{\prec}) = \exp(f_{\succ}) \cdot \mathrm{e}^{f_{\asymp}} \cdot \sum_{n=0}^{\infty} \frac{f_{\prec}^n}{n!}$$

where the first factor $\exp(f_{\succ})$ is a transmonomial, the second factor $\mathrm{e}^{f_{\asymp}}$ is the real number obtained by exponentiating the real number $f_{\asymp}$ in the usual way, and the third factor $\exp(f_{\prec}) = \sum_{n=0}^{\infty} \frac{f_{\prec}^n}{n!}$ is expanded as a series in the usual way. Conversely, each transmonomial is of the form $\exp(f_{\succ})$ for some transseries f. Viewed as an exponential field, $\mathbb{T}$ is an *elementary* extension of the exponential field of real numbers; see [111]. In particular, $\mathbb{T}$ is real closed, and so its ordering is existentially definable (and universally definable) from its ring operations:

(2) $$f \geqslant 0 \iff f = g^2 \text{ for some } g.$$

However, as emphasized above, our main interest is in $\mathbb{T}$ as a *differential field,* with derivation $f \mapsto f'$ on $\mathbb{T}$ defined termwise, with $r' = 0$ for $r \in \mathbb{R}$, $x' = 1$, $(\mathrm{e}^f)' = f'\,\mathrm{e}^f$, and $(\log f)' = f'/f$ for $f > 0$. Let us fix here some notation and terminology in force throughout this volume: a *differential field* is a field K of characteristic 0 together with a single derivation $\partial\colon K \to K$; if ∂ is clear from the context we often write a' instead of $\partial(a)$, for $a \in K$. The *constant field* of a differential field K is the subfield

$$C_K \ := \ \{a \in K \,:\, a' = 0\}$$

of K, also denoted by C if K is clear from the context. The constant field of $\mathbb{T}$ turns out to be $\mathbb{R}$, that is,

$$\mathbb{R} \ = \ \{f \in \mathbb{T} :\ f' = 0\}.$$

By an *ordered differential field* we mean a differential field equipped with a total ordering on its underlying set making it an *ordered field* in the usual sense of that expression. So $\mathbb{T}$ is an *ordered differential field.* More important than the ordering is the *valuation* on $\mathbb{T}$ with valuation ring

$$\mathcal{O}_{\mathbb{T}} \ := \ \big\{f \in \mathbb{T} :\ |f| \leqslant r \text{ for some } r \in \mathbb{R}\big\} \ = \ \{f \in \mathbb{T} :\ f_{\succ} = 0\},$$

a convex subring of $\mathbb{T}$. The unique maximal ideal of $\mathcal{O}_{\mathbb{T}}$ is

$$\mathcal{o}_{\mathbb{T}} \ := \ \big\{f \in \mathbb{T} : |f| \leqslant r \text{ for all } r > 0 \text{ in } \mathbb{R}\big\} \ = \ \{f \in \mathbb{T} :\ f = f_{\prec}\}$$

and thus $\mathcal{O}_{\mathbb{T}} = \mathbb{R} + \mathcal{o}_{\mathbb{T}}$. Its very definition shows that $\mathcal{O}_{\mathbb{T}}$ is *existentially* definable in the differential field $\mathbb{T}$. However, $\mathcal{O}_{\mathbb{T}}$ is *not* universally definable in the differential field $\mathbb{T}$: Corollary 16.2.6. In light of the model completeness conjecture discussed below, it is therefore advisable to add the valuation as an extra primitive, and so in the rest of this introduction *we construe $\mathbb{T}$ as an ordered and valued differential field, with valuation given by $\mathcal{O}_{\mathbb{T}}$.* By a *valued differential field* we mean throughout a differential field K equipped with a valuation ring of K that contains the prime subfield $\mathbb{Q}$ of K.

Grid-based transseries

When referring to transseries we have in mind the *well-based transseries of finite logarithmic and exponential depth* of [190], also called *logarithmic-exponential series* in [112]. The construction of the field $\mathbb{T}$ in Appendix A allows variants, and we briefly comment on one of them.

Each transseries f is an infinite sum $f = \sum_{\mathfrak{m}} f_{\mathfrak{m}}\mathfrak{m}$ where each $\mathfrak{m}$ is a transmonomial and $f_{\mathfrak{m}} \in \mathbb{R}$. The *support* of such a transseries f is the set $\operatorname{supp}(f)$ of transmonomials $\mathfrak{m}$ for which the coefficient $f_{\mathfrak{m}}$ is nonzero. For instance, the transmonomials in the support of the transseries φ of example (1) are

$$\mathrm{e}^{\mathrm{e}^x},\ \mathrm{e}^{\frac{\mathrm{e}^x}{\log x}+\frac{\mathrm{e}^x}{(\log x)^2}+\frac{\mathrm{e}^x}{(\log x)^3}+\cdots},\ x^{11},$$
$$1,\ \frac{1}{x},\ \frac{1}{x\log x},\ \ldots,\ \frac{1}{x^2},\ \frac{1}{x^3},\ldots,\ \mathrm{e}^{-x},\ \mathrm{e}^{-x^2},\ \ldots.$$

By imposing various restrictions on the kinds of permissible supports, the construction from Appendix A yields various interesting differential subfields of $\mathbb{T}$.

To define multiplication on $\mathbb{T}$, supports should be *well-based*: every nonempty subset of the support of a transseries f should contain an asymptotically dominant element. So well-basedness is a minimal requirement on supports. A much stronger condition on $\operatorname{supp}(f)$ is as follows: there are transmonomials $\mathfrak{m}$ and $\mathfrak{n}_1,\ldots,\mathfrak{n}_k \in \mathcal{O}_{\mathbb{T}}$ $(k \in \mathbb{N})$ such that

$$\operatorname{supp} f \subseteq \left\{\mathfrak{m}\,\mathfrak{n}_1^{i_1}\cdots\mathfrak{n}_k^{i_k} : i_1,\ldots,i_k \in \mathbb{N}\right\}.$$

Supports of this kind are called *grid-based*. Imposing this constraint all along, the construction from Appendix A builds the differential subfield $\mathbb{T}_{\mathrm{g}}$ of *grid-based* transseries of $\mathbb{T}$. Other suitable restrictions on the support yield other interesting differential subfields of $\mathbb{T}$.

The differential field $\mathbb{T}_{\mathrm{g}}$ of grid-based transseries has been studied in detail in [194]. In particular, that book contains a kind of algorithm for solving *algebraic differential equations* over $\mathbb{T}_{\mathrm{g}}$. These equations are of the form

$$P(y,\ldots,y^{(r)}) = 0, \tag{3}$$

where $P \in \mathbb{T}_{\mathrm{g}}[Y,\ldots,Y^{(r)}]$ is a nonzero polynomial in Y and a finite number of its formal derivatives $Y',\ldots,Y^{(r)}$. We note here that by combining results from [194] and the present volume, any solution $y \in \mathbb{T}$ to (3) is actually grid-based. Thus transseries outside $\mathbb{T}_{\mathrm{g}}$ such as $\varphi(x)$ from (1) or $\zeta(x) = 1 + 2^{-x} + 3^{-x} + \cdots$ are differentially transcendental over $\mathbb{T}_{\mathrm{g}}$; see the *Notes and comments* to Section 16.2 for more details, and Grigor′ev-Singer [155] for an earlier result of this kind.

Model completeness

One reason that "geometric" fields like $\mathbb{C}$, $\mathbb{R}$, $\mathbb{Q}_p$ are more manageable than "arithmetic" fields like $\mathbb{Q}$ is that the former are *model complete*; see Appendix B for this and other basic model-theoretic notions used in this volume. A consequence of the model completeness of $\mathbb{R}$ is that any finite system of polynomial equations over $\mathbb{R}$ (in any number of unknowns) with a solution in an ordered field extension of $\mathbb{R}$, has a solution in $\mathbb{R}$ itself. By the $\mathbb{R}$-version of (2) we can also allow polynomial inequalities in such a system. (A related fact: if such a system has real algebraic coefficients, then it has a real algebraic solution.)

For a more geometric view of model completeness we first specify an algebraic subset of $\mathbb{R}^n$ to be the set of common zeros,

$$\{y = (y_1, \dots, y_n) \in \mathbb{R}^n : P_1(y) = \dots = P_k(y) = 0\},$$

of finitely many polynomials $P_1, \dots, P_k \in \mathbb{R}[Y_1, \dots, Y_n]$. Define a subset of $\mathbb{R}^m$ to be *subalgebraic* if it is the image of an algebraic set in $\mathbb{R}^n$ for some $n \geqslant m$ under the projection map

$$(y_1, \dots, y_n) \mapsto (y_1, \dots, y_m) : \mathbb{R}^n \to \mathbb{R}^m.$$

Then a consequence of the model completeness of $\mathbb{R}$ is that the complement in $\mathbb{R}^m$ of any subalgebraic set is again subalgebraic. Model completeness of $\mathbb{R}$ is a little stronger in that only polynomials with integer coefficients should be involved.

A nice analogy between $\mathbb{R}$ and $\mathbb{T}$ is the following intermediate value property, announced in [193] and established for $\mathbb{T}_g$ in [194]: Let $P(Y) = p(Y, \dots, Y^{(r)})$ be a differential polynomial over $\mathbb{T}$, that is, with coefficients in $\mathbb{T}$, and let f, h be transseries with $f < h$; then $P(g)$ takes on all values strictly between $P(f)$ and $P(h)$ for transseries g with $f < g < h$. Underlying this opulence of $\mathbb{T}$ is a more robust property that we call *newtonianity,* which is analogous to henselianity for valued fields. The fact that $\mathbb{T}$ is newtonian implies, for instance, that any differential equation $y' = Q(y, y', \dots, y^{(r)})$ with $Q \in x^{-2}\mathcal{O}_{\mathbb{T}}[Y, Y', \dots, Y^{(r)}]$ has an infinitesimal solution $y \in \mathcal{o}_{\mathbb{T}}$. The definition of "newtonian" is rather subtle, and is discussed later in this introduction.

Another way that $\mathbb{R}$ and $\mathbb{T}$ are similar concerns the factorization of linear differential operators: any linear differential operator $A = \partial^r + a_1\partial^{r-1} \cdots + a_r$ of order $r \geqslant 1$ with coefficients $a_1, \dots, a_r \in \mathbb{T}$, is a product of such operators of order one and order two, with coefficients in $\mathbb{T}$. Moreover, any linear differential equation $y^{(r)} + a_1 y^{(r-1)} + \dots + a_r y = b$ $(a_1, \dots, a_r, b \in \mathbb{T})$ has a solution $y \in \mathbb{T}$ (possibly $y = 0$). In particular, every transseries f has a transseries integral g, that is, $f = g'$. (It is noteworthy that a *convergent* transseries can very well have a *divergent* transseries as an integral; for example, the transmonomial $\frac{e^x}{x}$ has as an integral the divergent transseries $\sum_{n=0}^{\infty} n! \frac{e^x}{x^{n+1}}$. The analytic aspects of transseries are addressed by Écalle's theory of *analyzable functions* [120], where genuine functions are associated to transseries such as $\sum_{n=0}^{\infty} n! \frac{e^x}{x^{n+1}}$, using the process of accelero-summation, a far reaching generalization of Borel summation; these analytic issues are not addressed in the present volume.)

These strong closure properties make it plausible to conjecture that $\mathbb{T}$ is model complete, as a valued differential field. This and some other conjectures to be mentioned in this introduction go back some 20 years, and are proved in the present volume. To state model completeness of $\mathbb{T}$ geometrically we use the terms d-*algebraic* and d-*polynomial* to abbreviate *differential-algebraic* and *differential polynomial* and we define a d-*algebraic set* in $\mathbb{T}^n$ to be the set of common zeros,

$$\{f = (f_1, \dots, f_n) \in \mathbb{T}^n : P_1(f) = \dots = P_k(f) = 0\}$$

of some d-polynomials $P_1, \dots, P_k$ in differential indeterminates $Y_1, \dots, Y_n$,

$$P_i(Y_1, \dots, Y_n) = p_i(Y_1, \dots, Y_n, Y_1', \dots, Y_n', Y_1'', \dots, Y_n'', Y_1''', \dots, Y_n''', \dots)$$

over $\mathbb{T}$. We also define an *H-algebraic set* to be the intersection of a d-algebraic set with a set of the form

$$\{y = (y_1, \dots, y_n) \in \mathbb{T}^n : y_i \in o_{\mathbb{T}} \text{ for all } i \in I\} \quad \text{where } I \subseteq \{1, \dots, n\},$$

and we finally define a subset of $\mathbb{T}^m$ to be *sub-H-algebraic* if it is the image of an H-algebraic set in $\mathbb{T}^n$ for some $n \geqslant m$ under the projection map

$$(f_1, \dots, f_n) \mapsto (f_1, \dots, f_m) : \mathbb{T}^n \to \mathbb{T}^m.$$

It follows from the model completeness of $\mathbb{T}$ that the complement in $\mathbb{T}^m$ of any sub-H-algebraic set is again sub-H-algebraic, in analogy with Gabrielov's "theorem of the complement" for real subanalytic sets [145]. (The model completeness of $\mathbb{T}$ is a little stronger: it is equivalent to this "complement" formulation where the defining d-polynomials of the d-algebraic sets involved have integer coefficients.) A consequence is that for subsets of $\mathbb{T}^m$,

$$\text{sub-}H\text{-algebraic} \quad = \quad \text{definable in } \mathbb{T}.$$

The usual model-theoretic approach to establishing that a given structure is model complete consists of two steps. (There is also a preliminary choice to be made of *primitives*; our choice for $\mathbb{T}$: its ring operations, its derivation, its ordering, and its valuation.) The first step is to record the basic compatibilities between primitives; "basic" here means in practice that they are also satisfied by the *substructures* of the structure of interest. For the more familiar structure of the ordered field $\mathbb{R}$ of real numbers, these basic compatibilities are the ordered field axioms. The second and harder step is to find some closure properties satisfied by our structure that together with these basic compatibilities can be shown to imply *all* its elementary properties. In the model-theoretic treatment of $\mathbb{R}$, it turns out that this job is done by the closure properties defining *real closed fields*: every positive element has a square root, and every odd degree polynomial has a zero.

H-fields

For $\mathbb{T}$ we try to capture the first step of the axiomatization by the notion of an *H-field.* We chose the prefix H in honor of E. Borel, H. Hahn, G. H. Hardy, and F. Hausdorff, who pioneered our subject about a century ago [55, 162, 164, 171], and who share the initial H, except for Borel. To define H-fields, let K be an ordered differential field (with constant field C) and set

$$\mathcal{O} := \{a \in K : |a| \leqslant c \text{ for some } c > 0 \text{ in } C\} \qquad \text{(a convex subring of } K\text{)},$$
$$o := \{a \in K : |a| < c \text{ for all } c > 0 \text{ in } C\}.$$

These notations should remind the reader of Landau's big O and small o. The elements of o are thought of as *infinitesimal,* the elements of $\mathcal{O}$ as *bounded,* and those of $K \setminus \mathcal{O}$ as *infinite.* Note that $\mathcal{O}$ is definable in the ordered differential field K, and is a valuation ring of K with (unique) maximal ideal o. We define K to be an *H-field* if it satisfies the two conditions below:

(H1) for all $a \in K$, if $a > C$, then $a' > 0$,

(H2) $\mathcal{O} = C + \scriptstyle\mathcal{O}$.

By (H2) the constant field C can be identified canonically with the residue field $\mathcal{O}/\scriptstyle\mathcal{O}$ of $\mathcal{O}$. As we did with $\mathbb{T}$ we construe an H-field K as an *ordered valued differential field.* An H-field K is said to have *small derivation* if $\partial\scriptstyle\mathcal{O} \subseteq \scriptstyle\mathcal{O}$ (and thus $\partial\mathcal{O} \subseteq \scriptstyle\mathcal{O}$). If K is an H-field and $a \in K$, $a > 0$, then K with its derivation ∂ replaced by $a\partial$ is also an H-field. Such changes of derivation play a major role in our work.

Among H-fields with small derivation are $\mathbb{T}$ and its ordered differential subfields containing $\mathbb{R}$, and any Hardy field containing $\mathbb{R}$. Thus $\mathbb{R}(x)$, $\mathbb{R}(x, \mathrm{e}^x, \log x)$ as well as Hardy's larger field of logarithmico-exponential functions are H-fields.

Closure properties

Let $\mathrm{Th}(\boldsymbol{M})$ be the first-order theory of an $\mathcal{L}$-structure $\boldsymbol{M}$, that is, $\mathrm{Th}(\boldsymbol{M})$ is the set of $\mathcal{L}$-sentences that are true in $\boldsymbol{M}$; see Appendix B for details. In terms of H-fields, we can now make the model completeness conjecture more precise, as was done in [19]:

$$\mathrm{Th}(\mathbb{T}) \;=\; \text{model companion of the theory of } H\text{-fields with small derivation},$$

where $\mathbb{T}$ is construed as an ordered and valued differential field. This amounts to adding to the earlier model completeness of $\mathbb{T}$ the claim that any H-field with small derivation can be embedded as an ordered valued differential field into some ultrapower of $\mathbb{T}$. Among the consequences of this conjecture is that any finite system of algebraic differential equations over $\mathbb{T}$ (in several unknowns) has a solution in $\mathbb{T}$ whenever it has one in some H-field extension of $\mathbb{T}$. It means that the concept of "H-field" is intrinsic to the differential field $\mathbb{T}$. It also suggests studying systematically the extension theory of H-fields: A. Robinson taught us that for a theory to have a model companion at all—a rare phenomenon—is equivalent to certain embedding and extension properties of its class of models. Here it helps to know that H-fields fall under the so-called *differential-valued fields* (abbreviated as d-*valued fields* below) of Rosenlicht, who began a study of these valued differential fields and their extensions in the early 1980s; see [364]. (A d-*valued* field is defined to be a valued differential field such that $\mathcal{O} = C + \scriptstyle\mathcal{O}$, and $a'b \in b'\scriptstyle\mathcal{O}$ for all $a, b \in \scriptstyle\mathcal{O}$; here $\mathcal{O}$ is the valuation ring with maximal ideal $\scriptstyle\mathcal{O}$, and C is the constant field.) Most of our work is actually in the setting of valued differential fields where no field ordering is given, since even for H-fields the valuation is a more robust and useful feature than its field ordering.

Besides developing the extension theory of H-fields we need to isolate the relevant *closure properties* of $\mathbb{T}$. First, $\mathbb{T}$ is real closed, but that property does not involve the derivation. Next, $\mathbb{T}$ is closed under integration and, by its very construction, also under exponentiation. In terms of the derivation this gives two natural closure properties of $\mathbb{T}$:

$$\forall a \exists b\, (a = b'), \qquad \forall a \exists b\, (b \neq 0 \,\&\, ab = b').$$

An H-field K is said to be *Liouville closed* if it is real closed and satisfies these two sentences; cf. Liouville [260, 261]. So $\mathbb{T}$ is Liouville closed. It was shown in [19] that any H-field has a *Liouville closure,* that is, a minimal Liouville closed H-field extension. If K is a Hardy field containing $\mathbb{R}$ as a subfield, then it has a unique Hardy field extension that is also a Liouville closure of K, but it can happen that an H-field K has two Liouville closures that are not isomorphic over K; it cannot have more than two. Understanding this "fork in the road" and dealing with it is fundamental in our work. Useful notions in this connection are *comparability classes*, *groundedness*, and *asymptotic integration.* We discuss this briefly below for H-fields. (Parts of Chapters 9 and 11 treat these notions for a much larger class of valued differential fields.) Later in this introduction we encounter an important but rather hidden closure property, called ω-*freeness*, which rules over the fork in the road. Finally, there is the very powerful closure property of newtonianity that we already mentioned earlier.

Valuations and asymptotic relations

Let K be an H-field, let a, b range over K, and let $v\colon K \to \Gamma_\infty$ be the (Krull) valuation on K associated to $\mathcal{O}$, with value group $\Gamma = v(K^\times)$ and $\Gamma_\infty := \Gamma \cup \{\infty\}$ with $\Gamma < \infty$. Recall that Γ is an ordered abelian group, additively written as is customary in valuation theory. Then

$$va < vb \iff |a| > c|b| \text{ for all } c > 0 \text{ in } C.$$

Thinking of elements of K as germs of functions at $+\infty$, we also adopt Hardy's notations from asymptotic analysis:

$$a \succ b, \qquad a \succcurlyeq b, \qquad a \prec b, \qquad a \preccurlyeq b, \qquad a \asymp b, \qquad a \sim b$$

are defined to mean, respectively,

$$va < vb, \quad va \leqslant vb, \quad va > vb, \quad va \geqslant vb, \quad va = vb, \quad v(a-b) > va.$$

(Some of these notations from [165] actually go back to du Bois-Reymond [48].) Note that $a \succ 1$ means that a is infinite, that is, $|a| > C$, and $a \prec 1$ means that a is infinitesimal, that is, $a \in \smallo$. It is crucial that the asymptotic relations above can be differentiated, provided we restrict to nonzero a, b with $a \not\asymp 1$, $b \not\asymp 1$:

$$a \succ b \iff a' \succ b', \qquad a \asymp b \iff a' \asymp b', \qquad a \sim b \iff a' \sim b'.$$

For $a \neq 0$ we let $a^\dagger := a'/a$ be its logarithmic derivative, so $(ab)^\dagger = a^\dagger + b^\dagger$ for $a, b \neq 0$. Elements $a, b \succ 1$ are said to be *comparable* if $a^\dagger \asymp b^\dagger$; if K is a Hardy field containing $\mathbb{R}$ as subfield, or $K = \mathbb{T}$, this is equivalent to the existence of an $n \geqslant 1$ such that $|a| \leqslant |b|^n$ and $|b| \leqslant |a|^n$. Comparability is an equivalence relation on the set of infinite elements of K, and the comparability classes $\operatorname{Cl}(a)$ of K are totally ordered by $\operatorname{Cl}(a) \leqslant \operatorname{Cl}(b) :\iff a^\dagger \preccurlyeq b^\dagger$.

EXAMPLE. For $K = \mathbb{T}$, set $e_0 = x$ and $e_{n+1} = \exp(e_n)$. Then the sequence $(\operatorname{Cl}(e_n))$ is strictly increasing and cofinal in the set of comparability classes. More important are the ℓ_n defined recursively by $\ell_0 = x$, and $\ell_{n+1} = \log \ell_n$. Then the sequence $\operatorname{Cl}(\ell_0) >$

$\mathrm{Cl}(\ell_1) > \mathrm{Cl}(\ell_2) > \cdots > \mathrm{Cl}(\ell_n) > \cdots$ is coinitial in the set of comparability classes of $\mathbb{T}$. For later use it is worth noting at this point that

$$\ell_n^\dagger = \frac{1}{\ell_0 \cdots \ell_n}, \qquad -\ell_n^{\dagger\dagger} = \frac{1}{\ell_0} + \frac{1}{\ell_0\ell_1} + \cdots + \frac{1}{\ell_0\ell_1\cdots\ell_n}.$$

We call K *grounded* if K has a smallest comparability class. Thus $\mathbb{T}$ is not grounded. If $\Gamma^>$ contains an element α such that for every $\gamma \in \Gamma^>$ we have $n\gamma \geqslant \alpha$ for some $n \geqslant 1$, then K is grounded; this condition on Γ is in particular satisfied if $\Gamma \neq \{0\}$ and Γ has finite archimedean rank. If K is grounded, then K has only one Liouville closure (up to isomorphism over K).

The H-field K is said to have *asymptotic integration* if K satisfies $\forall a \exists b (a \asymp b')$, equivalently, $\{vb' : b \in K\} = \Gamma_\infty$. It is obvious that every Liouville closed H-field has asymptotic integration; in particular, $\mathbb{T}$ has asymptotic integration. In general, at most one $\gamma \in \Gamma$ lies outside $\{vb' : b \in K\}$; if K is grounded, then such a γ exists, by results in Section 9.2, and so K cannot have asymptotic integration.

STRATEGY AND MAIN RESULTS

Model completeness of $\mathbb{T}$ concerns finite systems of algebraic differential equations over $\mathbb{T}$ with asymptotic side conditions in several differential indeterminates.

Robinson's strategy for establishing model completeness applied to $\mathbb{T}$ requires us to move beyond $\mathbb{T}$ to consider H-fields and their extensions. If we are lucky—as we are in this case—it will suffice to consider extensions of H-fields by one element y at a time. This leads to equations $P(y) = 0$ with an asymptotic side condition $y \prec g$. Here $P \in K\{Y\}$ is a univariate differential polynomial with coefficients in an H-field K with $g \in K^\times$, and $K\{Y\} = K[Y, Y', Y'', \dots]$ is the differential domain of d-polynomials in the differential indeterminate Y over K. The key issue: when is there a solution in some H-field extension of K? A detailed study of such equations in the special case $K = \mathbb{T}_{\mathrm{g}}$ and where we only look for solutions in $\mathbb{T}_{\mathrm{g}}$ itself was undertaken in [194], using an assortment of techniques (for instance, various fixpoint theorems) heavily based on the particular structure of $\mathbb{T}_{\mathrm{g}}$. Generalizing these results to suitable H-fields is an important guideline in our work.

Differential Newton diagrams

Let K be an H-field, and consider a d-algebraic equation with asymptotic side condition,

$$(4) \qquad P(y) = 0, \qquad y \prec g,$$

where $P \in K\{Y\}$, $P \neq 0$, and $g \in K^\times$; we look for nonzero solutions in H-field extensions of K. For the sake of concreteness we take $K = \mathbb{T}_{\mathrm{g}}$ and look for nonzero solutions in $\mathbb{T}_{\mathrm{g}}$, focusing on the example below:

$$(5) \qquad \mathrm{e}^{-\mathrm{e}^x} y^2 y'' + y^2 - 2xyy' - 7\,\mathrm{e}^{-x} y' - 4 + \frac{1}{\log x} = 0, \qquad y \prec x.$$

We sketch briefly how [194] goes about solving (5). First of all, we need to find the possible *dominant terms* of solutions y. This is done by considering possible cancellations. For example, y^2 and -4 might be the terms of least valuation in the left side of (5), with all other terms having greater valuation, so negligible compared to y^2 and -4. This yields a cancellation $y^2 \sim 4$, so $y \sim 2$ or $y \sim -2$, giving 2 and -2 as potential dominant terms of a solution y.

Another case: $\mathrm{e}^{-\mathrm{e}^x} y^2 y''$ and y^2 are the terms of least valuation. Then we get a cancellation $\mathrm{e}^{-\mathrm{e}^x} y^2 y'' \sim -y^2$, that is, $y'' \sim -\mathrm{e}^{\mathrm{e}^x}$, which leads to $y \sim -\mathrm{e}^{\mathrm{e}^x}/\mathrm{e}^{2x}$. But this possibility is discarded, since (5) also requires $y \prec x$. (On the other hand, if the asymptotic condition in (5) had been $y \prec \mathrm{e}^{\mathrm{e}^x}$, we would have kept $-\mathrm{e}^{\mathrm{e}^x}/\mathrm{e}^{2x}$ as a potential dominant term of a solution y.)

What makes things work in these two cases is that the cancellations arise from terms of *different* degrees in $y, y', y'', \dots$. Such cancellations are reminiscent of the more familiar setting of algebraic equations where the dominant monomials of solutions can be read off from a *Newton diagram* and the corresponding dominant coefficients are zeros of the corresponding *Newton polynomials*; see Section 3.7. This method still works in our d-algebraic setting, for cancellations among terms of different degrees, but requires the construction of so-called *equalizers*.

A different situation arises for cancellations between terms of the *same* degree. Consider for example the case that y^2 and $-2xyy'$ have least valuation among the terms in the left side of (5), with all other terms of higher valuation. Then $y^2 \sim 2xyy'$, so $y^\dagger \sim \frac{1}{2x}$. Now $y^\dagger = \frac{1}{2x}$ gives $y = cx^{1/2}$ with $c \in \mathbb{R}^\times$, but the weaker condition $y^\dagger \sim \frac{1}{2x}$ only gives $y = ux^{1/2}$ with $u \neq 0$, $u^\dagger \prec x^{-1}$, that is, $|v(u)| < |v(x)|/n$ for all $n \geqslant 1$. Substituting $ux^{1/2}$ for y in (5) and considering u as the new unknown, the condition on $v(u)$ forces $u \asymp 1$, so after all we do get $y \sim cx^{1/2}$ with $c \in \mathbb{R}^\times$, giving $cx^{1/2}$ as a potential dominant term of a solution y. It is important to note that here an *integration constant* c gets introduced.

Manipulations as we just did are similar to rewriting an equation $H(y) = 0$ with *homogeneous* nonzero $H \in K\{Y\}$ of positive degree as a (Riccati) equation $R(y^\dagger) = 0$ with R of *lower* order than H.

This technique can be shown to work in general for cancellations among terms of the same degree, provided we are also allowed to transform the equation to an equivalent one by applying a suitable iteration of the *upward shift* $f(x) \mapsto f(\mathrm{e}^x)$. (For reasonable H-fields K one can apply instead compositional conjugation by positive active elements; see below for *compositional conjugation* and *active*.)

Having determined a possible dominant term $f = c\mathfrak{m}$ of a solution of (4), where $c \in \mathbb{R}^\times$ and $\mathfrak{m}$ is a transmonomial, we next perform a so-called *refinement*

$$(6) \qquad P(f + y) = 0, \qquad y \prec f$$

of (4). For instance, taking $f = 2$, the equation (5) transforms into

$$\begin{aligned} &\mathrm{e}^{-\mathrm{e}^x} y^2 y'' + y^2 - 2xyy' + 4\,\mathrm{e}^{-\mathrm{e}^x} yy'' \\ &+ 4y - (4x + 7\,\mathrm{e}^{-x})y' + 4\,\mathrm{e}^{-\mathrm{e}^x} y'' + \tfrac{1}{\log x} \;=\; 0, \qquad y \prec 2. \end{aligned}$$

Now apply the same procedure to this refinement, to find the "next" term.

Roughly speaking, this yields an infinite process to obtain all possible asymptotic expansions of solutions to any asymptotic equation. How do we make this into a finite process? For this, it is useful to introduce the *Newton degree* of (4). This notion is similar to the Weierstrass degree of a multivariate power series and corresponds to the degree of the asymptotically significant part of the equation. If the Newton degree is 0, then (4) has no solution. The Newton degree of (5) turns out to be 2: this has to do with the fact that $\mathrm{e}^{-\mathrm{e}^x} y^2 y'' \prec y^2$ whenever $y \prec x$. We shall return soon to the precise definition of Newton degree for differential polynomials over rather general H-fields. As to the resolution of asymptotic equations over $K = \mathbb{T}_{\mathrm{g}}$, the following key facts were established in [194]:

- The Newton degree stays the same or decreases under refinement.
- If the Newton degree of the refinement (6) equals that of (4), we employ so-called *unravelings*; these resemble the *Tschirnhaus transformations* that overcome similar obstacles in the algebraic setting. Combining unravelings with refinements as described above, we arrive after finitely many steps at an asymptotic equation of Newton degree 0 or 1.
- The H-field $\mathbb{T}_{\mathrm{g}}$ is newtonian, that is, any asymptotic equation over $\mathbb{T}_{\mathrm{g}}$ of Newton degree 1 has a solution in $\mathbb{T}_{\mathrm{g}}$.

All in all, we have for any given asymptotic equation over $\mathbb{T}_{\mathrm{g}}$ a more or less finite procedure for gaining an overview of the entire space of solutions in $\mathbb{T}_{\mathrm{g}}$.

To define the Newton degree of an asymptotic equation (4) over rather general H-fields, we first need to introduce the *dominant part* of P and then, based on a process called *compositional conjugation,* the *Newton polynomial* of P.

The dominant part

Let K be an H-field. We extend the valuation v of K to the integral domain $K\{Y\}$ by setting

$$vP \;=\; \min\{va : \, a \text{ is a coefficient of } P\},$$

and we extend the binary relations $\asymp$ and $\sim$ on K to $K\{Y\}$ accordingly. It is also convenient to fix a monomial set $\mathfrak{M}$ in K, that is, a subset $\mathfrak{M}$ of $K^{>}$ that is mapped bijectively by v onto the value group Γ of K. This allows us to define the *dominant part* $D_P(Y)$ of a nonzero d-polynomial $P(Y)$ over K to be the unique element of $C\{Y\} \subseteq K\{Y\}$ with $P \sim \mathfrak{d}_P D_P$, where $\mathfrak{d}_P \in \mathfrak{M}$ is the *dominant monomial* of P determined by $P \asymp \mathfrak{d}_P$. (Another choice of monomial set would just multiply D_P by some positive constant.) For $K = \mathbb{T}$ we always take the set of transmonomials as our monomial set.

EXAMPLE 1. Let $K = \mathbb{T}$. For $P = x^5 + (2 + \mathrm{e}^x)Y + (3\,\mathrm{e}^x + \log x)(Y')^2$, we have $\mathfrak{d}_P = \mathrm{e}^x$ and $D_P = Y + 3(Y')^2$. For $Q = Y^2 - 2xYY'$ we have $D_Q = -2YY'$.

For K with small derivation we can use D_P to get near the zeros $a \asymp 1$ of P: if $P(a) = 0$, $a \asymp 1$, then $D_P(c) = 0$ where c is the unique constant with $a \sim c$. We need to understand, however, the behavior of $P(a)$ not only for $a \asymp 1$, that is, $va = 0$, but also for "sufficiently flat" elements $a \in K$, that is, for va approaching $0 \in \Gamma$. For instance, in $\mathbb{T}$, the iterated logarithms

$$\ell_0 = x, \quad \ell_1 = \log x, \quad \ell_2 = \log\log x, \ \ldots$$

satisfy $v(\ell_n) \to 0$ in $\Gamma_{\mathbb{T}}$ and likewise $v(1/\ell_n) \to 0$. The *dominant term* $\mathfrak{d}_P D_P$ of P often provides a good approximation for P when evaluating at sufficiently flat elements, but not always: for $K = \mathbb{T}$ and Q as in Example 1 we note that for $y = \ell_2$ we have: $y^2 = \ell_2^2 \succ 2xyy' = 2\ell_2/\ell_1$, so $Q(y) \sim y^2 \not\sim (\mathfrak{d}_Q D_Q)(y)$.

In order to approximate $P(y)$ by $(\mathfrak{d}_P D_P)(y)$ for sufficiently flat y, we need one more ingredient: *compositional conjugation.* For $K = \mathbb{T}$ and Q as in Example 1, this amounts to a change of variables $x = \mathrm{e}^{\mathrm{e}^{\tilde{x}}}$, so that $Q(y) = y^2 - 2y(dy/d\tilde{x})\,\mathrm{e}^{-\tilde{x}}$ for $y \in \mathbb{T}$. With respect to this new variable $\tilde{x}$, the dominant term Y^2 of the adjusted d-polynomial $Y^2 - 2YY'\,\mathrm{e}^{-\tilde{x}}$ is then an adequate approximation of Q when evaluating at sufficiently flat elements of $\mathbb{T}$. Such changes of variable do not make sense for general H-fields, but as it turns out, compositional conjugation is a good substitute.

Compositional conjugation

We define this for an arbitrary differential field K. For $\phi \in K^\times$ we let K^ϕ be the differential field obtained from K by replacing its derivation ∂ by the multiple $\phi^{-1}\partial$. Then a differential polynomial $P(Y) \in K\{Y\}$ defines the same function on the common underlying set of K and K^ϕ as a certain differential polynomial $P^\phi(Y) \in K^\phi\{Y\}$: for $P = Y'$, we have $P^\phi(Y) = \phi Y'$ (since over K^ϕ we evaluate Y' according to the derivation $\phi^{-1}\partial$), for $P = Y''$ we have $P^\phi(Y) = \phi' Y' + \phi^2 Y''$ (with $\phi' = \partial\phi$), and so on. This yields a ring isomorphism

$$P \mapsto P^\phi \colon K\{Y\} \to K^\phi\{Y\}$$

that is the identity on the common subring $K[Y]$. It is also an automorphism of the common underlying K-algebra of $K\{Y\}$ and $K^\phi\{Y\}$, and studied as such in Chapter 12. We call K^ϕ the *compositional conjugate of K by ϕ,* and P^ϕ the *compositional conjugate of P by ϕ.* Note that K and K^ϕ have the same constant field C. If K is an H-field and $\phi \in K^>$, then so is K^ϕ. It pays to note how things change under compositional conjugation, and what remains invariant.

The Newton polynomial

Suppose now that K is an H-field with asymptotic integration. For $\phi \in K^>$ we say that ϕ is *active* (in K) if $\phi \succcurlyeq a^\dagger$ for some nonzero $a \not\asymp 1$ in K; equivalently, the derivation $\phi^{-1}\partial$ of K^ϕ is small. Let $\phi \in K^>$ range over the active elements of K in what follows, fix a monomial set $\mathfrak{M} \subseteq K^>$ of K, and let $P \in K\{Y\}$, $P \neq 0$. The dominant part D_{P^ϕ} of P^ϕ lies in $C\{Y\}$, and we show in Section 13.1 that it eventually stabilizes as ϕ varies: there is a differential polynomial $N_P \in C\{Y\}$ and an active $\phi_0 \in K^>$ such that for all $\phi \preccurlyeq \phi_0$,

$$D_{P^\phi} = c_\phi N_P, \qquad c_\phi \in C^>.$$

We call N_P the *Newton polynomial* of P. It is of course only determined up to a factor from $C^>$, but this ambiguity is harmless. The (total) degree of N_P is called the *Newton degree* of P.

EXAMPLE 2. Let $K = \mathbb{T}$. Then $f \in K^>$ is active iff $f \succcurlyeq \ell_n^\dagger = \frac{1}{\ell_0 \ell_1 \cdots \ell_n}$ for some n. If P is as in Example 1, then for each ϕ,

$$P^\phi = x^5 + (2 + \mathrm{e}^x)Y + \phi^2(3\,\mathrm{e}^x + \log x)(Y')^2,$$

so $D_{P^\phi} = Y$ if $\phi \prec 1$. This yields $N_P = Y$, so P has Newton degree 1. It is an easy exercise to show that for $Q = Y^2 - 2xYY'$ we have $N_Q = Y^2$.

A crucial result in [194] (Theorem 8.6) says that if $K = \mathbb{T}_{\mathrm{g}}$, then $N_P \in \mathbb{R}[Y](Y')^{\mathbb{N}}$. A major step in our work was to isolate a robust class of H-fields K with asymptotic integration for which likewise $N_P \in C[Y](Y')^{\mathbb{N}}$ for all nonzero $P \in K\{Y\}$. This required several completely new tools to be discussed below.

The special cuts γ, λ and ω

Recall that ℓ_n denotes the nth iterated logarithm of x in $\mathbb{T}$, so $\ell_0 = x$ and $\ell_{n+1} = \log \ell_n$. We introduce the elements

$$\begin{aligned}
\gamma_n &= \ell_n^\dagger &&= \frac{1}{\ell_0 \cdots \ell_n} \\
\lambda_n &= -\gamma_n^\dagger &&= \frac{1}{\ell_0} + \frac{1}{\ell_0 \ell_1} + \cdots + \frac{1}{\ell_0 \ell_1 \cdots \ell_n} \\
\omega_n &= -2\lambda_n' - \lambda_n^2 &&= \frac{1}{\ell_0^2} + \frac{1}{\ell_0^2 \ell_1^2} + \cdots + \frac{1}{\ell_0^2 \ell_1^2 \cdots \ell_n^2}
\end{aligned}$$

of $\mathbb{T}$. As $n \to \infty$ these elements approach their formal limits

$$\begin{aligned}
\gamma_{\mathbb{T}} &= \frac{1}{\ell_0 \ell_1 \ell_2 \cdots} \\
\lambda_{\mathbb{T}} &= \frac{1}{\ell_0} + \frac{1}{\ell_0 \ell_1} + \frac{1}{\ell_0 \ell_1 \ell_2} + \cdots \\
\omega_{\mathbb{T}} &= \frac{1}{\ell_0^2} + \frac{1}{\ell_0^2 \ell_1^2} + \frac{1}{\ell_0^2 \ell_1^2 \ell_2^2} + \cdots,
\end{aligned}$$

which for now are just suggestive expressions. Indeed, our field $\mathbb{T}$ of transseries *of finite logarithmic and exponential depth* does not contain any pseudolimit of the pseudocauchy sequence (λ_n), nor of the pseudocauchy sequence (ω_n). There are, however, immediate H-field extensions of $\mathbb{T}$ where such pseudolimits exist, and if we let $\lambda_{\mathbb{T}}$ be such a pseudolimit of (λ_n), then in some further H-field extension we have an element suggestively denoted by $\exp(\int -\lambda_{\mathbb{T}})$ that can play the role of $\gamma_{\mathbb{T}}$.

Even though $\gamma_{\mathbb{T}}$, $\lambda_{\mathbb{T}}$ and $\omega_{\mathbb{T}}$ are not in $\mathbb{T}$, we can take them as elements of some H-field extension of $\mathbb{T}$, as indicated above, and so we obtain sets

$$\begin{aligned} \Gamma(\mathbb{T}) &= \{a \in \mathbb{T} : a > \gamma_{\mathbb{T}}\} \\ \Lambda(\mathbb{T}) &= \{a \in \mathbb{T} : a < \lambda_{\mathbb{T}}\} \\ \Omega(\mathbb{T}) &= \{a \in \mathbb{T} : a < \omega_{\mathbb{T}}\} \end{aligned}$$

that can be shown to be definable in $\mathbb{T}$. For instance,

$$\begin{aligned} \Gamma(\mathbb{T}) &= \{a \in \mathbb{T} : \forall b \in \mathbb{T}\, (b \succ 1 \Rightarrow a \neq b^\dagger)\} \\ &= \{-a' : a \in \mathbb{T},\ a \succcurlyeq 0\}. \end{aligned}$$

In other words, $\gamma_{\mathbb{T}}$, $\lambda_{\mathbb{T}}$ and $\omega_{\mathbb{T}}$ realize *definable cuts* in $\mathbb{T}$.

For any ungrounded H-field $K \neq C$ we can build a sequence (ℓ_ρ) of elements $\ell_\rho \succ 1$, indexed by the ordinals ρ less than some infinite limit ordinal, such that

$$\sigma > \rho \Rightarrow \ell_\sigma^\dagger \prec \ell_\rho^\dagger, \qquad v(\ell_\rho) \to 0 \text{ in } \Gamma.$$

These ℓ_ρ play in K the role that the iterated logarithms ℓ_n play in $\mathbb{T}$. In analogy with $\mathbb{T}$ they yield the elements

$$\gamma_\rho := \ell_\rho^\dagger, \qquad \lambda_\rho := -\gamma_\rho^\dagger, \qquad \omega_\rho := -2\lambda_n' - \lambda_n^2,$$

of K, and (λ_ρ) and (ω_ρ) are pseudocauchy sequences. As with $\mathbb{T}$ this gives rise to definable sets $\Gamma(K)$, $\Lambda(K)$ and $\Omega(K)$ in K. The fact mentioned earlier that $\mathbb{T}$ does not contain $\gamma_{\mathbb{T}}$, $\lambda_{\mathbb{T}}$ or $\omega_{\mathbb{T}}$ turns out to be very significant: in general, we have

$$\gamma_K \in K \quad \Rightarrow \quad \lambda_K \in K \quad \Rightarrow \quad \omega_K \in K$$

and each of the four mutually exclusive cases

$$\gamma_K \in K, \qquad \gamma_K \notin K \;\&\; \lambda_K \in K, \qquad \lambda_K \notin K \;\&\; \omega_K \in K, \qquad \omega_K \notin K$$

can occur; see Section 13.9. Here we temporarily abuse notations, since we should explain what we mean by $\gamma_K \in K$ and the like; see the next subsections.

On gaps and forks in the road

Let K be an H-field. We say that an element $\gamma \in K$ is a *gap* in K if for all $a \in K$ with $a \succ 1$ we have

$$a^\dagger \succ \gamma \succ (1/a)'.$$

The existence of such a gap is the formal counterpart to the informal statement that $\gamma_K \in K$. If K has a gap γ, then γ has no primitive in K, so K is not closed under integration. If K has trivial derivation (that is, $K = C$), then K has a gap $\gamma = 1$.

There are also K with $K \neq C$ (even Hardy fields) that have a gap. Not having a gap is equivalent to being grounded or having asymptotic integration.

We already mentioned the result from [19] that K may have two Liouville closures that are not isomorphic over K (but fortunately not more than two). Indeed, if K has a gap γ, then in one Liouville closure all primitives of γ are infinitely large, whereas in the other γ has an infinitesimal primitive. Even if K has no gap, the above fork in the road can arise more indirectly: Assume that K has asymptotic integration and $\lambda \in K$ is such that for all $a \in K^\times$ with $a \prec 1$,

$$a'^{\dagger} < -\lambda < a^{\dagger\dagger}.$$

Then K has no element $\gamma \neq 0$ with $\lambda = -\gamma^\dagger$, but K has an H-field extension $K\langle\gamma\rangle$ generated by an element γ with $\lambda = -\gamma^\dagger$, and any such γ is a gap in $K\langle\gamma\rangle$. It follows again that K has two Liouville closures that are not K-isomorphic.

For real closed K with asymptotic integration, the existence of such an element λ corresponds to the informal statement that $\gamma_K \notin K$ & $\lambda_K \in K$. We define K to be *λ-free* if K has asymptotic integration and satisfies the sentence

$$\forall a \exists b \big[b \succ 1 \ \&\ a - b^{\dagger\dagger} \succcurlyeq b^\dagger \big].$$

It can be shown that for real closed K with asymptotic integration, λ-freeness is equivalent to the nonexistence of an element λ as above. More generally, K is λ-free iff K has asymptotic integration and (λ_ρ) has no pseudolimit in K.

The property of ω-freeness

Even λ-freeness might not prevent a fork in the road for some d-algebraic extension. Let K be an H-field, and define

$$\omega = \omega_K \colon K \to K, \qquad \omega(z) := -2z' - z^2.$$

Assume that K is λ-free and $\omega \in K$ is such that for all $b \succ 1$ in K,

$$\omega - \omega(b^{\dagger\dagger}) \prec (b^\dagger)^2.$$

Then the first-order differential equation $\omega(z) = \omega$ admits no solution in K, but K has an H-field extension $K\langle\lambda\rangle$ generated by a solution $z = \lambda$ to $\omega(z) = \omega$ such that $K\langle\lambda\rangle$ is no longer λ-free (and with a fork in its road towards Liouville closure).

For λ-free K the existence of an element ω as above corresponds to the informal statement that $\lambda_K \notin K$ & $\omega_K \in K$. We say that K is *ω-free* if no such ω exists, more precisely, K has asymptotic integration and satisfies the sentence

$$\forall a \exists b \big[b \succ 1 \ \&\ a - \omega(b^{\dagger\dagger}) \succcurlyeq (b^\dagger)^2 \big].$$

(It is easy to show that if K is ω-free, then it is λ-free.) For K with asymptotic integration, ω-freeness is equivalent to the pseudocauchy sequence (ω_ρ) not having a pseudolimit in K. Thus $\mathbb{T}$ is ω-free. More generally, if K has asymptotic integration and is a union of grounded H-subfields, then K is ω-free by Corollary 11.7.15.

Much deeper and very useful is that if K is an ω-free H-field and L is a d-algebraic H-field extension of K, then L is also ω-free and has no comparability class smaller than all those of K; this is part of Theorem 13.6.1. Thus the property of ω-freeness is very robust: if K is ω-free, then forks in the road towards Liouville closure no longer occur, even for d-algebraic H-field extensions of K (Corollary 13.6.2). There are, however, Liouville closed H-fields that are not ω-free; see [22].

Another important consequence of ω-freeness is that Newton polynomials of differential polynomials then take the same simple shape as those over $\mathbb{T}_g$:

THEOREM 1. *If K is ω-free and $P \in K\{Y\}$, $P \neq 0$, then $N_P \in C[Y](Y')^{\mathbb{N}}$.*

The proof in Chapter 13 depends heavily on Chapter 12, where we determine the invariants of certain automorphism groups of polynomial algebras in infinitely many variables $Y_0, Y_1, Y_2, \ldots$ over a field of characteristic zero.

The function ω and the notion of ω-freeness are closely related to second order linear differential equations over K. More precisely (Riccati), for $y \in K^\times$, $4y'' + fy = 0$ is equivalent to $\omega(z) = f$ with $z := 2y^\dagger$; so the second-order linear differential equation $4y'' + fy = 0$ reduces in a way to a first-order (but non-linear) differential equation $\omega(z) = f$. (The factor 4 is just for convenience, to get simpler expressions below.)

EXAMPLE. The differential equation $y'' = -y$ has no solution $y \in \mathbb{T}^\times$, whereas the Airy equation $y'' = xy$ has two $\mathbb{R}$-linearly independent solutions in $\mathbb{T}$ [308, Chapter 11, (1.07)]. Indeed, in Sections 11.7 and 11.8 we show that for $f \in \mathbb{T}$, the differential equation $4y'' + fy = 0$ has a solution $y \in \mathbb{T}^\times$ if and only if $f < \omega_{\mathbb{T}}$, that is, $f < \omega_n = \frac{1}{\ell_0^2} + \frac{1}{\ell_0^2\ell_1^2} + \cdots + \frac{1}{\ell_0^2\ell_1^2\cdots\ell_n^2}$ for some n. This fact reflects classical results [167, 184] on the question: for which logarithmico-exponential functions f (in Hardy's sense) does the equation $4y'' + fy = 0$ have a non-oscillating real-valued solution (more precisely, a nonzero solution in a Hardy field)?

Newtonianity

This is the most consequential elementary property of $\mathbb{T}$. An ω-free H-field K is said to be *newtonian* if every d-polynomial $P(Y)$ over K of Newton degree 1 has a zero in $\mathcal{O}$. This turns out to be the correct analogue for valued differential fields like $\mathbb{T}$ of the property of being henselian for a valued field. We chose the adjective *newtonian* since it is this property that allows us to develop in Chapter 13 a Newton diagram method for differential polynomials. It is good to keep in mind that the role of newtonianity in the results of Chapters 14, 15, and 16 is more or less analogous to that of henselianity in the theory of valued fields and as the key condition in the Ax-Kochen-Eršov results.

We already mentioned the result from [194] that $\mathbb{T}_g$ is newtonian. That $\mathbb{T}$ is newtonian is a consequence of the following analogue in Chapter 15 of the familiar valuation-theoretic fact that spherically complete valued fields are henselian:

THEOREM 2. *If K is an H-field, $\partial K = K$, and K is a directed union of spherically complete grounded H-subfields, then K is* (*ω-free and*) *newtonian.*

EXAMPLE. Let $K = \mathbb{T}$ and consider for $\alpha \in \mathbb{R}$ the differential polynomial

$$P(Y) = Y'' - 2Y^3 - xY - \alpha \in \mathbb{T}\{Y\}.$$

For $\phi \in \mathbb{T}^{\times}$ we have $(Y'')^{\phi} = \phi^2 Y'' + \phi' Y'$ for $\phi \in \mathbb{T}^{\times}$, so

$$P^{\phi} = \phi^2 Y'' + \phi' Y' - 2Y^3 - xY - \alpha.$$

Now $\phi^2, \phi' \prec 1 \prec x$ for active $\phi \prec 1$ in $\mathbb{T}^{>}$. Hence $N_P \in \mathbb{R}^{\times} Y$, so P has Newton degree 1. Thus the Painlevé II equation $y'' = 2y^3 + xy + \alpha$ has a solution $y \in \mathcal{O}_{\mathbb{T}}$. (It is known that P has a zero $y \preccurlyeq 1$ in the differential subfield $\mathbb{R}(x)$ of $\mathbb{T}$ iff $\alpha \in \mathbb{Z}$; see for example [156, Theorem 20.2].)

The main results of Chapter 14 amount for H-fields to the following:

THEOREM 3. *If K is a newtonian ω-free H-field with divisible value group, then K has no proper immediate* d*-algebraic H-field extension.*

COROLLARY 1. *Let K be a real closed newtonian ω-free H-field, and let $K^{\mathrm{a}} = K[\mathrm{i}]$ (where $\mathrm{i}^2 = -1$) be its algebraic closure. Then:*

(i) *each* d*-polynomial in $K^{\mathrm{a}}\{Y\}$ of positive degree has a zero in K^{a};*

(ii) *each linear differential operator in $K^{\mathrm{a}}[\partial]$ of positive order is a composition of such operators of order* 1*;*

(iii) *each* d*-polynomial in $K\{Y\}$ of odd degree has a zero in K; and*

(iv) *each linear differential operator in $K[\partial]$ of positive order is a composition of such operators of order* 1 *and order* 2.

THEOREM 4. *If K is an ω-free H-field with divisible value group, then K has an immediate* d*-algebraic newtonian H-field extension, and any such extension embeds over K into every ω-free newtonian H-field extension of K.*

An extension of K as in Theorem 4 is minimal over K and thus unique up to isomorphism over K. We call such an extension a *newtonization* of K.

THEOREM 5. *If K is an ω-free H-field, then K has a* d*-algebraic newtonian Liouville closed H-field extension that embeds over K into every ω-free newtonian Liouville closed H-field extension of K.*

An extension of K as in Theorem 5 is minimal over K and thus unique up to isomorphism over K. We call such an extension a *Newton-Liouville closure* of K.

The main theorems

We now come to the results in Chapter 16, which in our view justify this volume. First, the various elementary conditions we have discussed axiomatize a model complete theory. To be precise, construe H-fields in the natural way as $\mathcal{L}$-structures where $\mathcal{L} := \{0, 1, +, -, \cdot, \partial, \leqslant, \preccurlyeq\}$, and let T^{nl} be the $\mathcal{L}$-theory whose models are the newtonian ω-free Liouville closed H-fields.

THEOREM 6. *T^{nl} is model complete.*

The theory T^{nl} is not complete and has exactly two completions, namely $T^{\mathrm{nl}}_{\mathrm{small}}$ (small derivation) and $T^{\mathrm{nl}}_{\mathrm{large}}$ (large derivation). Thus newtonian ω-free Liouville closed H-fields with small derivation have the same elementary properties as $\mathbb{T}$.

Every H-field with small derivation can be embedded into a model of $T^{\mathrm{nl}}_{\mathrm{small}}$; thus Theorem 6 yields the strong version of the model completeness conjecture from [19] stated earlier in this introduction. As $T^{\mathrm{nl}}_{\mathrm{small}}$ is complete and effectively axiomatized, it is decidable. In particular, there is an algorithm which, for any given d-polynomials $P_1, \dots, P_m$ in indeterminates $Y_1, \dots, Y_n$ with coefficients from $\mathbb{Z}[x]$, decides whether there is a tuple $y \in \mathbb{T}^n$ such that $P_1(y) = \cdots = P_m(y) = 0$. Such an algorithm with $\mathbb{T}$ replaced by its differential subring $\mathbb{R}[[x^{-1}]]$ is due to Denef and Lipshitz [101], but no such algorithm can exist with $\mathbb{T}$ replaced by $\mathbb{R}((x^{-1}))$ or by any of various other natural H-subfields of $\mathbb{T}$ [20, 155].

Theorem 6 is the main step towards an elimination of quantifiers, in a slightly extended language: Let $\mathcal{L}^{\iota}_{\Lambda,\Omega}$ be $\mathcal{L}$ augmented by the unary function symbol ι and the unary predicates Λ, Ω, and extend T^{nl} to the $\mathcal{L}^{\iota}_{\Lambda,\Omega}$-theory $T^{\mathrm{nl},\iota}_{\Lambda,\Omega}$ by adding as defining axioms for these new symbols the universal closures of

$$\begin{aligned} &\big[a \neq 0 \longrightarrow a \cdot \iota(a) = 1\big] \;\&\; \big[a = 0 \longrightarrow \iota(a) = 0\big], \\ &\Lambda(a) \;\longleftrightarrow\; \exists y\big[y \succ 1 \;\&\; a = -y^{\dagger\dagger}\big], \\ &\Omega(a) \;\longleftrightarrow\; \exists y\big[y \neq 0 \;\&\; 4y'' + ay = 0\big]. \end{aligned}$$

For a model K of T^{nl} this makes the sets $\Lambda(K)$ and $\Omega(K)$ downward closed with respect to the ordering of K. For example, for $f \in \mathbb{T}$,

$$\begin{aligned} f \in \Lambda(\mathbb{T}) &\iff f < \lambda_n = \frac{1}{\ell_0} + \frac{1}{\ell_0\ell_1} + \cdots + \frac{1}{\ell_0\ell_1\cdots\ell_n} \quad \text{for some } n, \\ f \in \Omega(\mathbb{T}) &\iff f < \omega_n = \frac{1}{\ell_0^2} + \frac{1}{\ell_0^2\ell_1^2} + \cdots + \frac{1}{\ell_0^2\ell_1^2\cdots\ell_n^2} \quad \text{for some } n, \end{aligned}$$

that is, $\Lambda(\mathbb{T})$ and $\Omega(\mathbb{T})$ are the cuts in $\mathbb{T}$ determined by $\lambda_{\mathbb{T}}$, $\omega_{\mathbb{T}}$ introduced earlier. We can now state what we view as the main result of this volume:

THEOREM 7. *The theory $T^{\mathrm{nl},\iota}_{\Lambda,\Omega}$ admits elimination of quantifiers.*

We cannot omit here either Λ or Ω. In Chapter 16 we do include for convenience one more unary predicate I in $\mathcal{L}^{\iota}_{\Lambda,\Omega}$: for a model K of T^{nl} and $a \in K$,

$$\mathrm{I}(a) \;\longleftrightarrow\; \exists y\big[a \preccurlyeq y' \;\&\; y \preccurlyeq 1\big] \;\longleftrightarrow\; a = 0 \vee \big[a \neq 0 \;\&\; \neg\Lambda(-a^{\dagger})\big],$$

where the first equivalence is the defining axiom for I, and the second shows that I is superfluous in Theorem 7. We note here that this predicate I governs the solvability of first-order linear differential equations *with asymptotic side condition.* More precisely, for K as above and $f \in K$, $g, h \in K^{\times}$, the following are equivalent:

(a) there exists $y \in K$ with $y' = fy + g$ and $y \prec h$;

(b) $\big[(f - h^{\dagger}) \in \mathrm{I}(K) \text{ and } (g/h) \in \mathrm{I}(K)\big]$ or $\big[(f - h^{\dagger}) \notin \mathrm{I}(K) \text{ and } (g/h) \prec f - h^{\dagger}\big]$.

This equivalence is part of Corollary 11.8.12 and exemplifies Theorem 7 (but is not derived from that theorem, nor used in its proof).

In the proof of Theorem 7, and throughout the construction of suitable H-field extensions, the predicates I, Λ and Ω act as switchmen. Whenever a fork in the road occurs due to the presence of a gap γ, then $\mathrm{I}(\gamma)$ tells us to take the branch where $\int\gamma \preccurlyeq 1$, while $\neg\mathrm{I}(\gamma)$ forces $\int\gamma \succ 1$. Likewise, the predicates Λ and Ω control what happens when adjoining elements γ and λ with $\gamma^\dagger = -\lambda$ and $\omega(\lambda) = \omega$.

From the above defining axioms for Λ and Ω it is clear that these predicates are (uniformly) existentially definable in models of T^{nl}. By model completeness of T^{nl} they are also uniformly *universally* definable in these models; Section 16.5 deals with such algebraic-linguistic issues.

Next we list some more intrinsic consequences of our elimination theory.

COROLLARY 2. *Let K be a newtonian ω-free Liouville closed H-field, and suppose the set $X \subseteq K^n$ is definable. Then X has empty interior in K^n (with respect to the order topology on K and the product topology on K^n) if and only if for some nonzero $P \in K\{Y_1, \dots, Y_n\}$ we have $X \subseteq \{y \in K^n : P(y) = 0\}$.*

In (i) below the intervals are in the sense of the ordered field K.

COROLLARY 3. *Let K be a newtonian ω-free Liouville closed H-field. Then:*

(i) *K is o-minimal at infinity: if $X \subseteq K$ is definable in K, then for some $a \in K$, either $(a, +\infty) \subseteq X$, or $(a, +\infty) \cap X = \emptyset$;*

(ii) *if $X \subseteq K^n$ is definable in K, then $X \cap C^n$ is semialgebraic in the sense of the real closed constant field C of K;*

(iii) *K has* NIP. (*See Appendix B for this very robust property.*)

It is hard to imagine obtaining these results for $K = \mathbb{T}$ without Theorem 7. Item (i) relates to classical bounds on solutions of algebraic differential equations over Hardy fields; see [20, Section 3]. To illustrate item (ii) of Corollary 3, we note that the set of real parameters $(\lambda_0, \dots, \lambda_n) \in \mathbb{R}^{n+1}$ for which the system

$$\lambda_0 y + \lambda_1 y' + \cdots + \lambda_n y^{(n)} = 0, \qquad 0 \neq y \prec 1$$

has a solution in $\mathbb{T}$ is a semialgebraic subset of $\mathbb{R}^{n+1}$; in fact, it agrees with the set of all $(\lambda_0, \dots, \lambda_n) \in \mathbb{R}^{n+1}$ such that the polynomial $\lambda_0 + \lambda_1 Y + \cdots + \lambda_n Y^n \in \mathbb{R}[Y]$ has a negative zero in $\mathbb{R}$; see Corollary 11.8.26. To illustrate item (iii), let $Y = (Y_1, \dots, Y_n)$ be a tuple of distinct differential indeterminates; for an m-tuple $\sigma = (\sigma_1, \dots, \sigma_m)$ of elements of $\{\prec, \asymp, \succ\}$ we say that $P_1, \dots, P_m \in \mathbb{T}\{Y\}$ *realize* σ if there exists $a \in \mathbb{T}^n$ such that $P_i(a)\,\sigma_i\,1$ holds for $i = 1, \dots, m$. Then a special case of (iii) says that for fixed $d, n, r \in \mathbb{N}$, the number of tuples $\sigma \in \{\prec, \asymp, \succ\}^m$ realized by some $P_1, \dots, P_m \in \mathbb{T}\{Y\}$ of degree at most d and order at most r grows only polynomially with m, even though the total number of tuples is 3^m. These manifestations of (ii) and (iii), though instructive, are perhaps a bit misleading, since they can be obtained without appealing to (ii) and (iii).

In the course of proving Theorem 6 we also get:

THEOREM 8. *If K is a newtonian ω-free Liouville closed H-field, then K has no proper* d*-algebraic H-field extension with the same constant field.*

For $K = \mathbb{T}_{\mathrm{g}}$ this yields: every $f \in \mathbb{T} \setminus \mathbb{T}_{\mathrm{g}}$ is d-transcendental over $\mathbb{T}_{\mathrm{g}}$.

We can also enlarge $\mathbb{T}$. For example, the series $\sum_{n=0}^{\infty} e_n^{-1}$, with e_n the nth iterated exponential of x, does not lie in $\mathbb{T}$ but does lie in a certain completion $\mathbb{T}^{\mathrm{c}}$ of $\mathbb{T}$. This completion $\mathbb{T}^{\mathrm{c}}$ is naturally an ordered valued differential field extension of $\mathbb{T}$, and by Corollary 14.1.6 we have $\mathbb{T} \preccurlyeq \mathbb{T}^{\mathrm{c}}$.

ORGANIZATION

Here we discuss the somewhat elaborate organization of this volume into chapters, some technical ingredients not mentioned so far, and some material that goes beyond the setting of H-fields. Indeed, the supporting algebraic theory deserves to be developed in a broad way, and there are more notions to keep track of than one might expect.

Background chapters

To make our work more accessible and self-contained, we provide in the first five chapters background on commutative algebra, valued abelian groups, valued fields, differential fields, and linear differential operators. This material has many sources, and we thought it would be convenient to have it available all in one place. In addition we have an appendix with the construction of $\mathbb{T}$, and an appendix exposing the (small) part of model theory that we need.

The basic facts on Hahn products, pseudocauchy sequences and spherical completeness in these early chapters are used throughout the volume. Some readers might prefer to skip in a first reading cauchy sequences, completeness (for valued abelian groups and valued fields) and step-completeness, which are not needed for the main results in this volume (but see Corollary 14.1.6). Some parts, like Sections 2.3 and 5.4, fit naturally where we put them, but are mainly intended for use in the next volume. On the other hand, Section 5.7 on compositional conjugation is elementary and frequently referred to in subsequent chapters, but this material seems virtually absent from the literature.

Valued differential fields

We also profited from examining arbitrary valued differential fields K with small derivation, that is, $\partial\smallo \subseteq \smallo$ for the maximal ideal $\smallo$ of the valuation ring $\mathcal{O}$ of K. This yields the continuity of the derivation ∂ with respect to the valuation topology and gives $\partial\mathcal{O} \subseteq \mathcal{O}$, and so induces a derivation on the residue field. To our surprise, we could establish in Chapters 6 and 7 some useful facts in this very general setting when the induced derivation on the residue field is nontrivial, for example the Equalizer Theorem 6.0.1. We need this result in deriving an "eventual" version of it for ω-free H-fields in Chapter 13, which in turn is crucial in obtaining our main results, via its role in constructing an appropriate Newton diagram for d-polynomials.

Asymptotic couples

A useful gadget is the *asymptotic couple* of an H-field K. This is the value group Γ of K equipped with the map $\gamma \mapsto \gamma^\dagger \colon \Gamma^{\neq} \to \Gamma$ defined by: if $\gamma = vf$, $f \in K^\times$, then $\gamma^\dagger = v(f^\dagger)$. This map is a valuation on Γ, and we extend it to a map $\Gamma \to \Gamma_\infty$ by setting $0^\dagger := \infty$. Two key facts are that $\alpha^\dagger < \beta + \beta^\dagger$ for all $\alpha, \beta > 0$ in Γ, and $\alpha^\dagger \geqslant \beta^\dagger$ whenever $0 < \alpha \leqslant \beta$ in Γ. The condition on an H-field of having small derivation can be expressed in terms of its asymptotic couple; the same holds for having a gap, for being grounded, and for having asymptotic integration, but not for being ω-free.

Asymptotic couples were introduced by Rosenlicht [364] for d-valued fields. In Chapter 6 we assign to *any* valued differential field with small derivation an asymptotic couple, with good effect. Asymptotic couples play also an important role in Chapters 9, 10, 11, 13, and 16.

Differential-henselian fields

Valued differential fields with small derivation include the so-called *monotone* differential fields defined by the condition $a' \preccurlyeq a$. In analogy with the notion of a *henselian* valued field, Scanlon [382] introduced *differential-henselian* monotone differential fields. Using the Equalizer Theorem we extend this notion and basic facts about it to arbitrary valued differential fields with small derivation in Chapter 7. (We abbreviate *differential-henselian* to d-*henselian*.) This material plays a role in Chapter 14, using the following relation between d-*henselian* and *newtonian*: an ω-free H-field K is newtonian iff for every active $\phi \in K^>$ the compositional conjugate K^ϕ is d-henselian, with the valuation v on K^ϕ replaced by the coarser valuation $\pi \circ v$ where $\pi \colon v(K^\times) = \Gamma \to \Gamma/\Delta$ is the canonical map to the quotient of Γ by its convex subgroup

$$\Delta := \{\gamma \in \Gamma : \gamma^\dagger > v\phi\}.$$

We pay particular attention to two special cases: $v(C^\times) = \{0\}$ (few constants), and $v(C^\times) = \Gamma$ (many constants). The first case is relevant for newtonianity, the second case is considered in a short Chapter 8, where we present Scanlon's extension of the Ax-Kochen-Eršov theorems to d-henselian valued fields with many constants, and add some things on definability.

While d-henselianity is defined in terms of solving differential equations in one unknown, it implies the solvability of suitably non-singular systems of n differential equations in n unknowns: this is shown at the end of Chapter 7, and has a nice consequence for newtonianity: Proposition 14.5.7.

Asymptotic differential fields

To keep things simple we confined most of the exposition above to H-fields, but this setting is a bit too narrow for various technical reasons. For example, a differential subfield of an H-field with the induced ordering is not always an H-field, and passing to an algebraic closure like $\mathbb{T}[i]$ destroys the ordering, though $\mathbb{T}[i]$ is still a d-valued field. On occasion we also wish to change the valuation of an H-field or d-valued field by coarsening. For all these reasons we introduce in Chapter 9 the class of *asymptotic differential fields*, which is larger and more flexible than Rosenlicht's class of d-valued fields. Many basic facts about H-fields and d-valued fields do have good analogues for asymptotic differential fields. This is shown in Chapter 9, which also contains a lot of basic material on asymptotic couples. Chapter 10 deals more specifically with H-fields.

Immediate extensions

Indispensable for attaining our main results is the fact that every H-field with divisible value group and with asymptotic integration has a spherically complete immediate H-field extension. This is part of Theorem 11.4.1, and proving it about five years ago removed a bottleneck. It provides the only way known to us of extending every H-field to an ω-free H-field. Possibly more important than Theorem 11.4.1 itself are the tools involved in its proof. In view of the theorem's content, it is ironic that models of T^{nl} are never spherically complete, in contrast to all prior positive results on elementary theories of valued fields with or without extra structure, cf. [28, 29, 41, 131, 382].

The differential Newton diagram method

Chapters 13 and 14 present the differential Newton diagram method in the general context of asymptotic fields that satisfy suitable technical conditions, such as ω-freeness. Before tackling these chapters, the reader may profit from first studying our exposition of the Newton diagram method for ordinary one-variable polynomials over henselian valued fields of equicharacteristic zero in Section 3.7. Some of the issues encountered there (for example, the *unraveling* technique) appear again, albeit in more intricate form, in the differential context of these chapters. In the proofs of a few crucial facts about the special cuts λ and ω in Chapter 13 we use some results from the preceding Chapter 12 on triangular automorphisms. Chapter 12 is a bit special in being essentially independent of the earlier chapters.

Proving newtonianity

Chapter 15 contains the proof of Theorem 2, and thus establishes that $\mathbb{T}$ is a model of our theory $T^{\text{nl}}_{\text{small}}$. This theorem is also useful in other contexts: In [43], Berarducci and Mantova construct a derivation on Conway's field **No** of surreal numbers [92, 150] turning it into a Liouville closed H-field with constant field $\mathbb{R}$. From Theorem 2 and the completeness of T^{nl} it follows that **No** with this derivation and $\mathbb{T}$ are elementarily equivalent, as we show in [24].

Quantifier elimination

In Chapter 16 we first prove Theorem 6 on model completeness, next we consider H-fields equipped with a $\Lambda\Omega$-structure, and then deduce Theorem 7 about quantifier elimination with various interesting consequences, such as Corollaries 2 and 3. The introduction to this chapter gives an overview of the proof and the role of various embedding and extension results in it.

THE NEXT VOLUME

The present volume focuses on achieving quantifier elimination (Theorem 7), and so we left out various things we did since 1995 that were not needed for that. In a second volume we intend to cover these things as required for developing our work further. Let us briefly survey some highlights of what is to come.

Linear differential equations

We plan to consider linear differential equations in much greater detail, comprising the corresponding differential Galois theory, in connection with constructing the linear surjective closure of a differential field, factoring linear differential operators over suitable algebraically closed d-valued fields, and explicitly constructing the Picard-Vessiot extension of such an operator. Concerning the latter, the complexification $\mathbb{T}[i]$ of $\mathbb{T}$ is no longer closed under exponential integration, since oscillatory "transmonomials" such as e^{ix} are not in $\mathbb{T}[i]$. Adjoining these oscillatory transmonomials to $\mathbb{T}[i]$ yields a d-valued field that contains a Picard-Vessiot extension of $\mathbb{T}$ for each operator in $\mathbb{T}[\partial]$.

Hardy fields

We also wish to pay more attention to Hardy fields, and this will bring up analytic issues. For example, every Hardy field containing $\mathbb{R}$ can be shown to extend to an ω-free Hardy field. Using methods from [195], we also hope to prove that it always extends to a newtonian ω-free Hardy field. Indeed, that paper proves among other things the following pertinent result (formulated here with our present terminology): Let $\mathbb{T}_{\mathrm{g}}^{\mathrm{da}}$ consist of the grid-based transseries that are d-algebraic over $\mathbb{R}$. Then $\mathbb{T}_{\mathrm{g}}^{\mathrm{da}}$ is a newtonian ω-free Liouville closed H-subfield of $\mathbb{T}_{\mathrm{g}}$ and is isomorphic over $\mathbb{R}$ to a Hardy field containing $\mathbb{R}$.

Embedding into fields of transseries

Another natural question we expect to deal with is whether every H-field can be given some kind of transserial structure. This can be made more precise in terms of the axiomatic definition of a *field of transseries* in terms of a transmonomial group $\mathfrak{M}$ in Schmeling's thesis [388]. For instance, one axiom there is that for all $\mathfrak{m} \in \mathfrak{M}$ we have $\operatorname{supp} \log \mathfrak{m} \subseteq \mathfrak{M}^{\succ}$. We hope that any H-field can be embedded into such a field of transseries. This would be a natural counterpart of Kaplansky's theorem [209] embedding certain valued fields into Hahn fields, and would make it possible to think of H-field elements as generalized transseries.

More on the model theory of $\mathbb{T}$

In the second volume we hope to deal with further issues around $\mathbb{T}$ of a model-theoretic nature: for example, identifying the induced structure on its value group (conjectured to be given by its H-couple, as specified in [18]); and determining the definable closure of a subset of a model of T^{nl}, in order to get a handle on what functions are definable in $\mathbb{T}$.

A by-product of the present volume is a full description of several important 1-types over a given model of T^{nl}, but the entire space of such 1-types remains to be surveyed. Theorem 8 suggests that the model-theoretic notions of *non-orthogonality to C* or *C-internality* may be significant for models of T^{nl}; see also [25].

FUTURE CHALLENGES

We now discuss a few more open-ended avenues of inquiry.

Differentiation and exponentiation

The restriction to $\mathcal{O}_{\mathbb{T}}$ of the exponential function on $\mathbb{T}$ is easily seen to be definable in $\mathbb{T}$, but by part (ii) of Corollary 3, the restriction to $\mathbb{R}$ of this exponential function is not definable in $\mathbb{T}$. This raises the question whether our results can be extended to the differential field $\mathbb{T}$ *with exponentiation*, or with some other extra o-minimal structure on it.

Logarithmic transseries

A transseries is *logarithmic* if all transmonomials in it are of the form $\ell_0^{r_0}\cdots\ell_n^{r_n}$ with $r_0,\dots,r_n\in\mathbb{R}$. (See Appendix A.) The logarithmic transseries make up an ω-free newtonian H-subfield $\mathbb{T}_{\log}$ of $\mathbb{T}$ that is not Liouville closed. We conjecture that $\mathbb{T}_{\log}$ as a valued differential field is model complete. The asymptotic couple of $\mathbb{T}_{\log}$ has been successfully analyzed by Gehret [146], and turns out to be model-theoretically tame, in particular, has NIP [147]. (There is also the notion of a transseries being *exponential.* The exponential transseries form a real closed H-subfield $\mathbb{T}_{\exp}$ of $\mathbb{T}$ in which the set $\mathbb{Z}$ is existentially definable, see [20]. It follows that the differential field $\mathbb{T}_{\exp}$ does not have a reasonable model theory: it is as complicated as so-called *second-order arithmetic.*)

Accelero-summable transseries

The paper [195] on transserial Hardy fields yields on the one hand a method to associate a genuine function to a suitable formal transseries, and in the other direction also provides means to associate concrete asymptotic expansions to elements of Hardy fields. We expect that more can be done in this direction.

Écalle's theory of analyzable functions has a more canonical procedure that associates a function to an *accelero-summable* transseries. These transseries make up an H-subfield $\mathbb{T}_{\text{as}}$ of $\mathbb{T}$. This procedure has the advantage that it does not only preserve the ordered field structure, but also composition, functional inversion, and several other operations. In its full generality, however, Écalle's theory requires sophisticated analytic tools, and is beyond the scope of this volume. It is clear that $\mathbb{T}_{\text{as}}$ is analytically more important than $\mathbb{T}$, but the latter might help in understanding the former. The H-subfield $\mathbb{T}_{\text{as}}$ of $\mathbb{T}$ contains $\mathbb{R}$, is ω-free and Liouville closed. Is it newtonian? In view of Theorem 8, a positive answer would confirm Écalle's belief [120, p. 148] that any solution in $\mathbb{T}$ of an algebraic differential equation $P(Y) = 0$ over $\mathbb{T}_{\text{as}}$ with $P \neq 0$ lies in $\mathbb{T}_{\text{as}}$.

Beyond H-fields

The derivation of a differentially closed field K cannot be continuous with respect to a nontrivial valuation on K; see Section 10.7. This sets a limit for the study of valued differential fields with a reasonable interaction between valuation and derivation. However, one may close off the d-valued field $\mathbb{T}[i]$ under exponential integration, by adding oscillatory transmonomials recursively. This results in valued differential fields of *complex transseries* over which a version of the Newton diagram method for $\mathbb{T}_{\text{g}}$ goes through; see [192]. It would be interesting to find out more about the model theory of these rich valued differential fields.

A HISTORICAL NOTE ON TRANSSERIES

The differential field of transseries was first defined and extensively used in Écalle's solution of Dulac's problem, which is about plane analytic vector fields. Its solution shows in particular that a plane polynomial vector field admits only a finite number of limit cycles. (A *limit cycle* of a planar vector field is a periodic trajectory with an annular neighborhood not containing any other periodic trajectory.) It was long believed that in 1923 Dulac [115] had given a proof of this finiteness statement, until Il′yashenko [197] found a gap in 1981: Dulac was operating with asymptotic expansions of germs of functions as if they faithfully represented these germs. To justify this in Dulac's case is not easy: it was done independently by Écalle [120] and Il′yashenko [198], and required fundamental new ideas (and hundreds of pages). We briefly sketch here the role of transseries in Écalle's approach.

Suppose towards a contradiction that some polynomial vector field on $\mathbb{R}^2$ has infinitely many limit cycles. Classical facts about planar vector fields such as the Poincaré-Bendixson Theorem allow us to reduce to the case where infinitely many of these limit cycles accumulate at a so-called *polycycle* of the vector field; see [199, Theorem 24.22] for details. Such a polycycle σ consists of finitely many trajectories

$$S_1 \to S_2, \ \ldots \ , S_{r-1} \to S_r, \ S_r \to S_1 \qquad \text{(the edges)}$$

between singularities $S_1, \ldots, S_r$ (the vertices) of the vector field; see Figure 0.1 where $r = 3$. Draw lines $\ell_1, \ldots, \ell_r$ that cross these edges $S_r \to S_1$, $S_1 \to S_2, \ldots,$ $S_{r-1} \to S_r$ transversally at points $O_1, \ldots, O_r$. For any trajectory φ of the vector field that is sufficiently close to σ we consider the successive points where φ meets

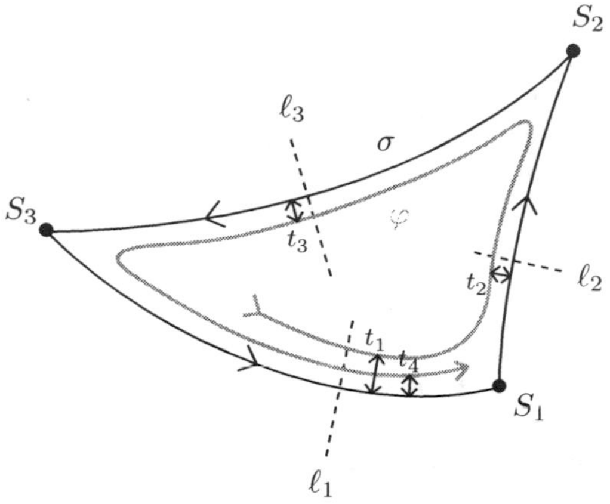

Figure 0.1: A polycycle σ and a close trajectory φ.

$\ell_1, \ell_2, \ldots, \ell_r, \ell_1$ and denote their distances to $O_1, \ldots, O_r, O_1$ by $t_1, t_2, \ldots, t_r, t_{r+1}$. The behavior of the vector field near S_i yields for some $\varepsilon_i > 0$ a real analytic function $g_i \colon (0, \varepsilon_i) \to (0, \infty)$ such that $t_{i+1} = g_i(t_i)$. We have $g_i(t) \to 0$ as $t \to 0^+$ but g_i does not necessarily extend analytically to 0.

The composition $f := g_r \circ \cdots \circ g_1$ is defined on some interval $(0, \varepsilon)$ and is called the *Poincaré return map* of the polycycle σ (relative to our choice of $\ell_1, \ldots, \ell_r$). We have $t_{r+1} = f(t_1)$, for trajectories close enough to σ. Thus $f(t) = t$ corresponds to a periodic trajectory, and so it suffices to show that either f is the identity, or $f(t) \neq t$ for all sufficiently small $t > 0$. One can even ask whether this non-oscillation property holds for Poincaré return maps of polycycles of plane *analytic* (not necessarily polynomial) vector fields; this is *Dulac's problem*.

It is convenient to work at infinity by setting $x = t^{-1}$ and replace these functions $f, g_1, \ldots, g_r$ of t by functions $F, G_1, \ldots, G_r$ of x with $F = G_r \circ \cdots \circ G_1$. Dulac [115] provides formal series expansions $\widetilde{G_i}$ of the G_i, which are rather simple transseries, usually divergent, and which by formal composition yields an often complicated transseries expansion $\widetilde{F} = \widetilde{G_r} \circ \cdots \circ \widetilde{G_1}$ of F.

Écalle is able to reconstitute the germs G_i and F from their formal counterparts $\widetilde{G_i}$ and $\widetilde{F}$ by developing a delicate analytic machinery of *accelero-summation*. More precisely, he constructs an (accelero-summation) operator $\widetilde{G}(x) \mapsto G$ whose domain of definition is a certain differential subfield $\mathbb{T}_{\text{as}}$ of $\mathbb{T}$ and whose values are germs of real analytic functions at $+\infty$; it assigns to each $\widetilde{G_i}$ the germ G_i. Moreover, $\mathbb{T}_{\text{as}}$ is closed under composition, and accelero-summation preserves real constants, addition, multiplication, differentiation, composition, and the (total) field ordering: if $\widetilde{G}(x) \in \mathbb{T}_{\text{as}}$, $\widetilde{G}(x) > 0$, then $G(x) > 0$ for all sufficiently large real x; here the x in $\widetilde{G}(x)$ is an indeterminate, while in $G(x)$ it ranges over real numbers. Applying this operator to

The Zariski topology

Let $f \in R$ and $\mathfrak{p} \in \operatorname{Spec}(R)$, and think of the residue class $f + \mathfrak{p} \in R/\mathfrak{p}$ as the *value of f at the point* $\mathfrak{p}$. Then "f vanishes at $\mathfrak{p}$" just means that $f \in \mathfrak{p}$. For a set $S \subseteq R$, the subset

$$\mathrm{Z}(S) \;:=\; \{\mathfrak{p} \in \operatorname{Spec}(R) : \mathfrak{p} \supseteq S\}$$

of $\operatorname{Spec}(R)$ is then the set of points in $\operatorname{Spec}(R)$ at which all $f \in S$ vanish: the set of common zeros of the "functions" $f \in S$. Thus $\mathrm{Z}(\emptyset) = \operatorname{Spec}(R)$, and $\mathrm{Z}(R) = \emptyset$. For $f_1, \dots, f_n \in R$ we also write $\mathrm{Z}(f_1, \dots, f_n)$ instead of $\mathrm{Z}(\{f_1, \dots, f_n\})$. If I is the ideal of R generated by $S \subseteq R$, then $\mathrm{Z}(S) = \mathrm{Z}(I)$. The sets $\mathrm{Z}(S)$ with $S \subseteq R$ are the closed sets of a topology on $\operatorname{Spec}(R)$, called its **Zariski topology**: use that

$$\textstyle\bigcap_{\lambda\in\Lambda} \mathrm{Z}(S_\lambda) \;=\; \mathrm{Z}\left(\bigcup_{\lambda\in\Lambda} S_\lambda\right)$$

for any family $(S_\lambda)_{\lambda\in\Lambda}$ of subsets of R, and that for $S_1, S_2 \subseteq R$,

$$\mathrm{Z}(S_1) \cup \mathrm{Z}(S_2) \;=\; \mathrm{Z}(S_1 S_2), \qquad S_1 S_2 \;:=\; \{f_1 f_2 : f_1 \in S_1,\ f_2 \in S_2\}.$$

The sets $\mathrm{D}(f) := \operatorname{Spec}(R) \setminus \mathrm{Z}(f)$ with $f \in R$ form a base of open sets for this Zariski topology. The Zariski topology on $\operatorname{Spec}(R)$ is not in general hausdorff.

Let $\varphi\colon R \to S$ be a morphism of commutative rings. Then for each $\mathfrak{q} \in \operatorname{Spec}(S)$ we have $\varphi^{-1}(\mathfrak{q}) \in \operatorname{Spec}(R)$, giving rise to an inclusion-preserving map

$$\varphi^*\colon \operatorname{Spec}(S) \to \operatorname{Spec}(R), \qquad \varphi^*(\mathfrak{q}) := \varphi^{-1}(\mathfrak{q})$$

with $(\varphi^*)^{-1}(\mathrm{Z}(I)) = \mathrm{Z}(\varphi(I))$ for ideals I of R. So φ^* is continuous, and if φ is surjective with kernel I, then φ^* is a homeomorphism onto its image $\mathrm{Z}(I)$.

Radical ideals

Let I be an ideal of R. The **radical** of I (more precisely, the nilradical of I) is the ideal

$$\sqrt{I} \;:=\; \{f \in R :\ f^n \in I \text{ for some } n\}$$

of R. Note that $\sqrt{I} \supseteq I$, and $\sqrt{I} = R$ iff $I = R$. One says that I is **radical** if $I = \sqrt{I}$. Every prime ideal of R is radical, the radical of an ideal of R is radical, and the intersection $\bigcap_{\lambda\in\Lambda} I_\lambda$ of a family $(I_\lambda)_{\lambda\in\Lambda}$ of radical ideals of R is radical. Moreover, if I is radical and S is a subset of R, then

$$(I : S) \;:=\; \{f \in R :\ fS \subseteq I\}$$

is a radical ideal of R containing I. We call

$$\operatorname{nil}(R) \;:=\; \sqrt{\{0\}} \;=\; \{a \in R :\ f^n = 0 \text{ for some } n\}$$

the **nilradical** of R. If $\operatorname{nil}(R)$ is finitely generated, then, as is easily verified, $\operatorname{nil}(R)^n = \{0\}$ for some $n \geqslant 1$. One says that R is **reduced** if $\operatorname{nil}(R) = \{0\}$, equivalently, $\{0\}$ is a radical ideal of R.

For $X \subseteq \mathrm{Spec}(R)$, the radical ideal of all $f \in R$ that vanish at each point of X is

$$\mathrm{I}(X) := \bigcap_{\mathfrak{p} \in X} \mathfrak{p},$$

which by convention is R for $X = \emptyset$. Note that $\mathrm{Z}\big(\mathrm{I}(X)\big)$ is the closure of X in the space $\mathrm{Spec}(R)$. Here is an abstract version of Hilbert's Nullstellensatz:

PROPOSITION 1.1.1. *For each ideal I of R we have* $\mathrm{I}\big(\mathrm{Z}(I)\big) = \sqrt{I}$.

Towards the proof we first define a **multiplicative subset of** R to be a set $S \subseteq R$ such that $1 \in S$, and $fg \in S$ for all $f, g \in S$.

LEMMA 1.1.2 (Krull). *Let S be a multiplicative subset of R. Let I be an ideal of R disjoint from S and maximal with these properties. Then I is prime.*

PROOF. Let $f, g \in R$ be such that $fg \in I$. Assume towards a contradiction that $f, g \notin I$. Then $Rf + I$ and $Rg + I$ are ideals of R properly containing I, so we have $s, t \in S$ with $s \in Rf + I$, $t \in Rg + I$. Then we have a contradiction:

$$st \in S \cap (Rf + I)(Rg + I) \subseteq S \cap I = \emptyset. \qquad \square$$

Using also Zorn, this lemma applied to $S = \{1\}$ gives the well-known fact that every ideal $I \neq R$ of R is contained in a prime ideal of R.

PROOF OF PROPOSITION 1.1.1. As already noted, $\mathrm{I}\big(\mathrm{Z}(I)\big)$ is radical, which in view of $\mathrm{I}\big(\mathrm{Z}(I)\big) \supseteq I$ gives $\mathrm{I}\big(\mathrm{Z}(I)\big) \supseteq \sqrt{I}$. Conversely, suppose $f \in R$, $f \notin \sqrt{I}$. Then $f \notin \mathrm{I}\big(\mathrm{Z}(I)\big)$, since Zorn and Lemma 1.1.2 yield a prime ideal $\mathfrak{p} \supseteq I$ of R disjoint from the multiplicative subset $\{f^n : n = 0, 1, 2, \dots\}$ of R, so $f \notin \mathfrak{p} \in \mathrm{Z}(I)$. $\square$

COROLLARY 1.1.3. *The map $I \mapsto \mathrm{Z}(I)$ is an inclusion-reversing bijection from the set of radical ideals of R onto the set of closed subsets of* $\mathrm{Spec}(R)$, *with inverse given by $X \mapsto \mathrm{I}(X)$.*

COROLLARY 1.1.4. $\mathrm{nil}(R) = \bigcap_{\mathfrak{p} \in \mathrm{Spec}(R)} \mathfrak{p}$.

We now continue with some purely topological considerations that will help in coming to grips with non-traditional spaces like $\mathrm{Spec}(R)$.

Irreducibility

Let X be a topological space. Call X **irreducible** if $X \neq \emptyset$ and X is not the union of two proper closed subsets; note that then X is connected and each nonempty open subset of X is irreducible and dense in X. A subset of X is called irreducible if it is irreducible as a subspace of X. It is easy to see that if $X \neq \emptyset$, then

X is irreducible $\iff$ any two nonempty open subsets of X have a nonempty intersection

$\iff$ every nonempty open subset of X is dense in X.

Thus if X is a subspace of Y, then X is irreducible iff its closure $\mathrm{cl}(X)$ in Y is irreducible. In particular, for any $x \in X$ the subspace $\mathrm{cl}(\{x\})$ of X is irreducible. If $f\colon X \to Y$ is continuous and X is irreducible, then $f(X)$ is irreducible.

DEFINITION 1.1.5. An **irreducible component** of X is a maximal irreducible subset of X.

Thus the irreducible components of X are closed. One-point spaces are irreducible, so every topological space is the union of its irreducible components, in view of:

PROPOSITION 1.1.6. *Every irreducible subset of X is contained in an irreducible component of X.*

PROOF. By Zorn, it suffices to verify that the union Y of a family $\{Y_\lambda\}_{\lambda\in\Lambda}$ of irreducible subsets of X, linearly ordered by inclusion, is irreducible. For this, let $U_i \subseteq X$ be open with $U_i \cap Y \neq \emptyset$, $i = 1, 2$. For $i = 1, 2$, take $\lambda_i \in \Lambda$ with $U_i \cap Y_{\lambda_i} \neq \emptyset$. Assuming $Y_{\lambda_1} \subseteq Y_{\lambda_2}$, we get $U_1 \cap U_2 \cap Y_{\lambda_2} \neq \emptyset$, since Y_{λ_2} is irreducible, and thus $U_1 \cap U_2 \cap Y \neq \emptyset$. □

COROLLARY 1.1.7. *Suppose X is a finite union of irreducible subsets. Then X has only finitely many irreducible components and no irreducible component of X is contained in the union of the others.*

PROOF. By Proposition 1.1.6 we have $X = X_1 \cup \cdots \cup X_m$ where $X_1, \ldots, X_m$ are irreducible components of X. It remains to note that if Y is any irreducible subset of X, then $Y \subseteq X_i$ for some i, using $Y = \bigcup_i (Y \cap X_i)$. □

We can now identify the irreducible closed subsets of $\mathrm{Spec}(R)$:

LEMMA 1.1.8. *Let X be a closed subset of $\mathrm{Spec}(R)$. Then X is irreducible if and only if $\mathrm{I}(X)$ is a prime ideal of R.*

PROOF. Suppose X is irreducible, and $f, g \in R$, $fg \in \mathrm{I}(X)$. Then for all $\mathfrak{p} \in X$ we have $f \in \mathfrak{p}$ or $g \in \mathfrak{p}$, so $X = \big(X \cap \mathrm{Z}(f)\big) \cup \big(X \cap \mathrm{Z}(g)\big)$, hence $X \subseteq \mathrm{Z}(f)$ or $X \subseteq \mathrm{Z}(g)$, and thus $f \in \mathrm{I}(X)$ or $g \in \mathrm{I}(X)$. Conversely, assume $\mathrm{I}(X)$ is prime, and $X = X_1 \cup X_2$ with closed X_1, X_2. Then $\mathrm{I}(X) = \mathrm{I}(X_1 \cup X_2) = \mathrm{I}(X_1) \cap \mathrm{I}(X_2)$, hence $\mathrm{I}(X) = \mathrm{I}(X_1)$ or $\mathrm{I}(X) = \mathrm{I}(X_2)$, so $X = X_1$ or $X = X_2$. □

Thus $X \mapsto \mathrm{I}(X)$ is an inclusion-reversing bijection from the set of closed irreducible subsets of $\mathrm{Spec}(R)$ onto the set of prime ideals of R.

Let I be an ideal of R. A **minimal prime divisor of** I is a point $\mathfrak{p} \in \mathrm{Z}(I)$ such that no $\mathfrak{q} \in \mathrm{Z}(I)$ is strictly contained (as a set) in $\mathfrak{p}$. For $I = \{0\}$, the minimal prime divisors of I are exactly the minimal prime ideals of R (with respect to inclusion). The above bijection $X \mapsto \mathrm{I}(X)$ maps the set of irreducible components of the closed subset $\mathrm{Z}(I)$ of $\mathrm{Spec}(R)$ onto the set of minimal prime divisors of I. Thus by Proposition 1.1.6, every $\mathfrak{p} \in \mathrm{Z}(I)$ contains a minimal prime divisor of I.

Prime avoidance

For use in Sections 1.5 and 1.6 we show:

LEMMA 1.1.9. *Let* $I, J_1, \ldots, J_n$ *be ideals of* R *with* $n \geqslant 2$ *and* $I \subseteq J_1 \cup \cdots \cup J_n$. *Assume* $J_1, \ldots, J_{n-2}$ *are prime. Then* $I \subseteq J_j$ *for some* $j \in \{1, \ldots, n\}$.

PROOF. We argue by induction on $n = 2, 3, \ldots$. Let $n = 2$, and assume that neither $I \subseteq J_1$ nor $I \subseteq J_2$. Take $a_1 \in I \setminus J_2$, $a_2 \in I \setminus J_1$. Then $a_1 \in J_1$, $a_2 \in J_2$, hence $a_1 + a_2 \in I \setminus (J_1 \cup J_2)$. Next, let $n \geqslant 3$ and let j, k range over $\{1, \ldots, n\}$. It suffices to find k with $I \subseteq \bigcup_{j \neq k} J_j$, since then we are done by inductive hypothesis. Towards a contradiction, assume that $I \not\subseteq \bigcup_{j \neq k} J_j$ for all k. For each k, take $a_k \in I \setminus \left(\bigcup_{j \neq k} J_j\right)$. Then $a_k \in J_k$ for each k, hence $a := \prod_{k \neq 1} a_k \in I \cap \bigcap_{k \neq 1} J_k$; but $a \notin J_1$, since J_1 is prime. Thus $a_1 + a \in I$ and $a_1 + a \notin J_k$ for all k, a contradiction. □

COROLLARY 1.1.10. *Let* I, J *be ideals of* R *such that* $I \not\subseteq J$, *and let* $\mathfrak{p}_1, \ldots, \mathfrak{p}_n \in \operatorname{Spec}(R)$ *be such that* $I \setminus J \subseteq \mathfrak{p}_1 \cup \cdots \cup \mathfrak{p}_n$. *Then* $I \subseteq \mathfrak{p}_j$ *for some* $j \in \{1, \ldots, n\}$.

PROOF. Use $I \subseteq \mathfrak{p}_1 \cup \cdots \cup \mathfrak{p}_n \cup J$ and Lemma 1.1.9. □

Chain conditions

In this subsection the set S is *partially* ordered by $\leqslant$, with

$$s < s' \;:\Longleftrightarrow\; s \leqslant s' \text{ and } s \neq s',$$

for $s, s' \in S$. A **maximal** element of a set $X \subseteq S$ is an $x \in X$ such that there is no $x' \in X$ with $x < x'$. The following are equivalent:

(1) Every nonempty subset of S has a maximal element;

(2) there is no sequence $s_1 < s_2 < \cdots < s_n < s_{n+1} < \cdots$ in S.

We say that S satisfies the **ascending chain condition** (or has **acc**) if S fulfills one of these equivalent conditions. We say that S satisfies the **descending chain condition** (or has **dcc**) if S with the reversed partial ordering $\geqslant$ has acc. Thus if the ordering of S is total, then S has dcc iff S is well-ordered.

Noetherian rings

Call R **noetherian** if the set of its ideals, partially ordered by inclusion, has acc. If R is noetherian, then so is its image under any ring morphism.

LEMMA 1.1.11. R *is noetherian* $\Longleftrightarrow$ *every ideal of* R *is finitely generated.*

PROOF. If the ideal I of R is not finitely generated, then we obtain inductively a sequence (r_n) in I with $r_{n+1} \notin (r_0, \ldots, r_n)$ for all n, giving a strictly increasing infinite sequence $(r_0) \subset (r_0, r_1) \subset (r_0, r_1, r_2) \subset \cdots$ of ideals of R. Conversely, if every ideal of R is finitely generated, it follows easily that there cannot exist a strictly increasing infinite sequence of ideals of R. □

Fields and, more generally, principal ideal domains (like the ring of integers) are noetherian. In Chapter 3 we study valuation rings, which are only noetherian in a very special (but important) case, as we now explain. We define here a **valuation ring** to be an integral domain R such that for all $a, b \in R$: $a \in bR$ or $b \in aR$.

Suppose that R is a valuation ring. For any $a_1, \dots, a_n \in R$, $n \geqslant 1$, we have $(a_1, \dots, a_n)R = a_j R$ for some j; in particular, every finitely generated ideal of R is principal. The set $R \setminus R^\times$ of nonunits of R is clearly the largest proper ideal of R, and is thus the unique maximal ideal $\mathfrak{m}_R$ of R.

Fields are valuation rings, but are viewed as trivial within the class of valuation rings. Next in complication are *discrete valuation rings*: a **discrete valuation ring** (or **DVR**) is an integral domain R with an element t such that $t \neq 0$, $t \notin R^\times$, and $R^{\neq} = R^\times t^{\mathbb{N}}$. Note that every DVR is indeed a valuation ring. The power series ring $\boldsymbol{k}[[t]]$ in an indeterminate t over a field $\boldsymbol{k}$ is clearly a DVR.

Suppose R is a DVR. Then R is a PID (principal ideal domain): with t as in the definition of *discrete valuation ring*, any ideal $I \neq \{0\}$ of R is clearly generated by t^m where m is minimal such that $t^m \in I$. In particular, a DVR is noetherian:

COROLLARY 1.1.12. *Let R be a valuation ring that is not a field. Then*

$$R \text{ is noetherian} \iff R \text{ is a PID} \iff R \text{ is a DVR.}$$

PROOF. Suppose R is a PID. Take $t \in R$ such that $\mathfrak{m}_R = Rt$. Then t is prime in R. If t' is also prime in R, then, R being a valuation ring, $t = at'$ or $t' = at$ with $a \in R$, and in either case, $a \in R^\times$. Thus, R being factorial, every $r \in R^{\neq}$ has the form $r = ut^n$ with $u \in R^\times$. So R is a DVR. The rest is clear. □

Many rings of natural origin are noetherian, by a famous result:

PROPOSITION 1.1.13 (Hilbert's Basis Theorem). *If R is noetherian, then so is the ring $R[X]$ of polynomials in the indeterminate X over R.*

PROOF. Suppose I is an ideal of $R[X]$ that is not finitely generated. Set $f_0 := 0$, and with $f_0, f_1, \dots, f_n \in I$, take $f_{n+1} \in I \setminus (f_1, \dots, f_n)$ of minimal degree. For $n \geqslant 1$, let $r_n \in R^{\neq}$ and $d_n \in \mathbb{N}$ be such that $f_n = r_n X^{d_n} +$ lower degree terms. Then $d_1 \leqslant d_2 \leqslant \cdots$. Also, $(r_1, \dots, r_n) \neq (r_1, \dots, r_n, r_{n+1})$: otherwise $r_{n+1} = \sum_{i=1}^n a_i r_i$ with all $a_i \in R$, so $f_{n+1} - \sum_{i=1}^n a_i X^{d_{n+1} - d_i} f_i$ has smaller degree than f_{n+1}, contradicting the choice of f_{n+1}. Hence R is not noetherian. □

Thus if R is noetherian, then so is every finitely generated commutative R-algebra.

Noetherian spaces

Let X, Y be topological spaces. Call X **noetherian** if its collection of closed sets satisfies the descending chain condition: there is no strictly descending infinite sequence $X_0 \supset X_1 \supset \cdots$ of closed subsets of X; equivalently, each nonempty collection of closed subsets of X has a minimal element with respect to inclusion. If R is noetherian, then the space $\operatorname{Spec}(R)$ is noetherian.

REMARKS. A noetherian space is quasicompact (every covering by open subsets has a finite subcovering), but in the absence of being hausdorff this is less useful than some other facts:

(1) each subspace of a noetherian space is noetherian;

(2) if X is noetherian and $f\colon X \to Y$ is continuous, then $f(X) \subseteq Y$ is noetherian;

(3) if X is covered by finitely many noetherian subspaces, then X is noetherian.

Suppose X is noetherian. Then X is a finite union of irreducible closed subsets: if not, take a minimal closed subset Y of X that is not a finite union of irreducible closed subsets. Then $Y \neq \emptyset$ and Y is not irreducible, so $Y = Y_1 \cup Y_2$ with Y_1, Y_2 proper closed subsets of Y. Each Y_i is a finite union of irreducible closed subsets, and so is Y, a contradiction. Thus by Corollary 1.1.7:

COROLLARY 1.1.14. *If* $\operatorname{Spec}(R)$ *is noetherian (in particular, if R is noetherian) and I is an ideal of R, then I has only finitely many minimal prime divisors.*

So if $\operatorname{Spec}(R)$ is noetherian, then R has only finitely many minimal prime ideals. By a **zero divisor** of R we mean an $a \in R$ such that $ab = 0$ for some $b \in R^{\neq}$.

COROLLARY 1.1.15. *Suppose* $\operatorname{Spec}(R)$ *is noetherian and $\mathfrak{p}_1, \dots, \mathfrak{p}_n$ are the minimal prime ideals of R. Then* $\operatorname{nil}(R) = \mathfrak{p}_1 \cap \cdots \cap \mathfrak{p}_n$, *the elements of $\mathfrak{p}_1 \cup \cdots \cup \mathfrak{p}_n$ are zero divisors of R, and if R is reduced, $\mathfrak{p}_1 \cup \cdots \cup \mathfrak{p}_n$ is the set of zero divisors of R.*

PROOF. The first statement holds by Corollary 1.1.4. The second statement holds clearly for $n = 0, 1$, so assume $n \geqslant 2$ and $\mathfrak{p}_1, \dots, \mathfrak{p}_n$ are distinct. Let i, j range over $\{1, \dots, n\}$, and let $r \in \mathfrak{p}_j$. For $i \neq j$, take $s_i \in \mathfrak{p}_i \setminus \mathfrak{p}_j$. Then $s := \prod_{i \neq j} s_i \in \bigcap_{i \neq j} \mathfrak{p}_i$, $s \notin \mathfrak{p}_j$. Then $rs \in \bigcap_i \mathfrak{p}_i$, so we have $m \geqslant 1$ with $(rs)^m = 0$. Since $s \notin \mathfrak{p}_j$, we have $s^m \neq 0$, so we get $i \in \mathbb{N}$ with $i < m$ and $r^i s^m \neq 0$, $r^{i+1} s^m = 0$. Thus r is a zero divisor of R. For the third statement, assume that R is reduced, that is, $\mathfrak{p}_1 \cap \cdots \cap \mathfrak{p}_n = \{0\}$. Let $r \in R$ be a zero divisor. Take $s \in R^{\neq}$ with $rs = 0$, and then take $j \in \{1, \dots, n\}$ with $s \notin \mathfrak{p}_j$. From $rs = 0 \in \mathfrak{p}_j$ we obtain $r \in \mathfrak{p}_j$. □

Krull dimension

This is a notion of dimension suitable for noetherian spaces. Let X, Y be topological spaces. In this subsection we take suprema and infima in $\mathbb{N} \cup \{-\infty, +\infty\}$. We define the (Krull) **dimension** of X to be the supremum of the set of n for which there is a strictly increasing sequence $X_0 \subset X_1 \subset \cdots \subset X_n$ of irreducible closed subsets of X. In particular, the dimension of X is $-\infty$ iff $X = \emptyset$. We denote the dimension of X by $\dim(X)$. (If X is a nonempty hausdorff space, then $\dim(X) = 0$, so Krull dimension is of no interest for hausdorff spaces.) If X is a subspace of Y, then $\dim(X) \leqslant \dim(Y)$. (Use that the closure in Y of an irreducible subset is irreducible.) Moreover,

$$\dim(X) = \sup\bigl\{\dim(Y) : Y \text{ is an irreducible component of } X\bigr\}.$$

Some special prime ideals

Let K be a field, and let a family $X = (X_\lambda)_{\lambda \in \Lambda}$ of distinct indeterminates be given. If Λ is finite, then the ring $K[X]$ is noetherian by Proposition 1.1.13. If Λ is infinite, then $K[X]$ is not noetherian. However, such rings appear later as rings of differential polynomials, and in Section 4.6 we use:

LEMMA 1.1.16. *Let I be an ideal of $K[X]$ generated by homogeneous polynomials of degree 1. Then I is a prime ideal of $K[X]$.*

PROOF. Take a K-independent set $H \subseteq K[X]$ of homogeneous polynomials of degree 1 that generates the ideal I. Then H is part of a basis B of the K-linear subspace of $K[X]$ generated by the X_λ. Take a bijection $\lambda \mapsto b_\lambda \colon \Lambda \to B$. Then $P(X) \mapsto P\big((b_\lambda)\big)$ is an automorphism of the K-algebra $K[X]$. Replacing I by the inverse image of I under this automorphism, we arrange that $H = \{X_\lambda : \lambda \in \Lambda_0\}$, with $\Lambda_0 \subseteq \Lambda$. Hence $K[X]/I$ is isomorphic as a K-algebra to $K[X_\lambda : \lambda \in \Lambda \setminus \Lambda_0]$. $\square$

Notes and comments

The notion of *noetherian ring* and the basic facts about it are due to E. Noether [304]. (The short proof of Hilbert's Basis Theorem [180] given above is from [380].) Krull's Lemma 1.1.2 is from [227, p. 732]. Antecedents for the Zariski topology on $\operatorname{Spec}(R)$ include Stone [435, 436], Jacobson [203], and Zariski [467]. Irreducible and noetherian spaces come from Serre [399].

1.2 RINGS AND MODULES OF FINITE LENGTH

The algebra in this section is not just commutative: *R is a ring, possibly not commutative, and M, N range over R-modules.*

Composition series

An **M-series** (of length m) is a strictly increasing sequence

$$\{0\} = M_0 \subset M_1 \subset \cdots \subset M_m = M \tag{1.2.1}$$

of submodules of M. A **refinement** of an M-series (1.2.1) is an M-series

$$\{0\} = N_0 \subset N_1 \subset \cdots \subset N_n = M$$

such that $\{M_0, \ldots, M_m\} \subseteq \{N_0, \ldots, N_n\}$ (so $m \leqslant n$). Two M-series

$$\{0\} = M_0 \subset M_1 \subset \cdots \subset M_m = M, \quad \{0\} = N_0 \subset N_1 \subset \cdots \subset N_n = M$$

of M are said to be **equivalent** if $m = n$ and there is a permutation $i \mapsto i'$ of $\{1, \ldots, m\}$ such that $M_i/M_{i-1} \cong N_{i'}/N_{i'-1}$, as R-modules, for $i = 1, \ldots, m$. The next result is known as the Schreier Refinement Theorem.

THEOREM 1.2.1. *Any two M-series have equivalent refinements.*

See [249, Chapter I, §3] for a proof of an analogue of this theorem for groups which adapts in a straightforward way to modules.

We call M **simple** (or **irreducible**) if $M \neq \{0\}$ and M has no submodules other than $\{0\}$ and M. If M is simple, then $M = Re$ for each $e \in M^{\neq}$. Hence M is simple iff $M \cong R/\mathfrak{m}$ for some maximal left ideal $\mathfrak{m}$ of R. It is clear that if M and N are simple R-modules, then every R-linear map $\varphi\colon M \to N$ with $\varphi \neq 0$ is an isomorphism of R-modules (Schur's Lemma).

A **composition series** of M is an M-series (1.2.1) such that M_{i+1}/M_i is simple for $i = 0, \dots, m-1$. Every M-series equivalent to a composition series of M is itself a composition series of M, and a composition series of M cannot be refined to a strictly longer M-series. Thus Theorem 1.2.1 yields the Jordan-Hölder Theorem:

COROLLARY 1.2.2. *Any two composition series of M are equivalent.*

Euler-Poincaré maps

Let A be an abelian group. Suppose that for certain R-modules M there is defined a quantity $\chi(M) \in A$, such that:

(1) $\chi(\{0\})$ is defined and equal to $0 \in A$,

(2) if $0 \to K \to M \to N \to 0$ is an exact sequence of R-modules, then $\chi(M)$ is defined if and only if both $\chi(K)$ and $\chi(N)$ are defined, and in this case, $\chi(M) = \chi(K) + \chi(N)$.

(Such an assignment $M \mapsto \chi(M)$ is called an **Euler-Poincaré map** on R-modules.) Clearly if $M \cong N$, then $\chi(M)$ is defined iff $\chi(N)$ is defined, and in this case $\chi(M) = \chi(N)$. For example, if $R = \mathbb{Z}$, then setting $\chi(M) := \log|M|$ for a finite abelian group M defines an Euler-Poincaré map on abelian groups with values in the additive group $A = \mathbb{R}$.

LEMMA 1.2.3. *Let an exact sequence*

$$0 \longrightarrow M_0 \xrightarrow{\varphi_0} M_1 \xrightarrow{\varphi_1} \cdots \longrightarrow M_n \longrightarrow 0$$

of R-modules be given such that $\chi(M_i)$ is defined for $i = 0, \dots, n$. Then

$$\sum_{i=0}^{n} (-1)^i \chi(M_i) = 0.$$

PROOF. By induction on n. The cases $n = 0, 1, 2$ are obvious. Suppose $n \geqslant 3$, and put $N := \operatorname{im}\varphi_1 = \ker\varphi_2$. Then we have exact sequences

$$0 \longrightarrow M_0 \xrightarrow{\varphi_0} M_1 \xrightarrow{\varphi_1} N \longrightarrow 0, \qquad 0 \longrightarrow N \xrightarrow{\subseteq} M_2 \longrightarrow \cdots \longrightarrow M_n \longrightarrow 0.$$

By the first sequence $\chi(N)$ is defined and $\chi(M_0) - \chi(M_1) + \chi(N) = 0$. Now use the inductive hypothesis on the second sequence. □

Modules of finite length

If M has a composition series, then all composition series of M have a common length, called the **length** of M, denoted by $\ell(M)$. If M does not have a composition series, we put $\ell(M) := \infty$. Note: $\ell(M) = 0$ iff $M = \{0\}$. If N is a submodule of M, then M has finite length if and only if N and M/N have finite length, in which case $\ell(M) = \ell(N) + \ell(M/N)$. Hence $M \mapsto \ell(M)$, defined for M of finite length, is a $\mathbb{Z}$-valued Euler-Poincaré map on R-modules.

EXAMPLE. Suppose $R = K$ is a field. Then $\ell(M) = \dim_K M$. In this way Lemma 1.2.3 contains Lemma 2.3.19 as a special case.

Call M **noetherian** if the set of submodules of M, partially ordered by inclusion, has acc. Call M **artinian** if this partially ordered set has dcc. As in the proof of Lemma 1.1.11, M is noetherian iff all its submodules are finitely generated.

LEMMA 1.2.4. *M is noetherian and artinian if and only if M has finite length. Thus if M has finite length, then M is finitely generated.*

PROOF. Previous remarks make it clear that if $\ell(M) < \infty$, then M is noetherian and artinian. Conversely, suppose M is artinian. If $M \neq \{0\}$, let M_1 be a minimal nonzero submodule of M; if $M \neq M_1$, let M_2 be a submodule of M which is minimal among the submodules of M properly containing M_1, and so on. This yields a strictly increasing sequence $\{0\} = M_0 \subset M_1 \subset M_2 \subset \cdots$ of submodules of M. If M is also noetherian, this construction stops with $M_m = M$ for some m, and then we have a composition series of M. □

Rings of finite length

In this subsection R is commutative. In the phrases "R has finite length" and "R is artinian" we view R as an R-module in the usual way.

LEMMA 1.2.5. *Suppose R is artinian. Then every prime ideal of R is maximal.*

PROOF. Let $\mathfrak{p}$ be a prime ideal of R; then $R/\mathfrak{p}$ is an artinian integral domain, and replacing R by $R/\mathfrak{p}$ we may assume that R is an integral domain, and need to show that R is a field. Let $r \in R^{\neq}$. Considering the chain $Rr \supseteq Rr^2 \supseteq \cdots$, we obtain $n \geqslant 1$ and $s \in R$ such that $r^n = r^{n+1}s$; hence $1 = rs$. □

PROPOSITION 1.2.6. *The following conditions on R are equivalent:*

(i) *R is noetherian and every prime ideal of R is maximal;*

(ii) *every finitely generated R-module has finite length;*

(iii) *R has finite length;*

(iv) *R is noetherian and artinian.*

PROOF. If $R = \{0\}$, then all four conditions are trivially satisfied, so let $R \neq \{0\}$. Assume (i) holds, and let $\mathfrak{m}_1, \dots, \mathfrak{m}_n$ be the minimal prime ideals of R; so each $\mathfrak{m}_i$ is maximal. Put $\mathfrak{n} := \mathfrak{m}_1 \cdots \mathfrak{m}_n$, so $\mathfrak{n} \subseteq \mathfrak{m}_1 \cap \cdots \cap \mathfrak{m}_n = \mathrm{nil}(R)$, giving $m \geqslant 1$ with $\mathfrak{n}^m = \{0\}$. Let M be a finitely generated R-module. Consider the chain

$$M \supseteq \mathfrak{m}_1 M \supseteq \mathfrak{m}_1\mathfrak{m}_2 M \supseteq \cdots \supseteq \mathfrak{m}_1 \cdots \mathfrak{m}_n M = \mathfrak{n}M$$

of submodules of M. For $i = 0, \dots, n-1$, the quotient $\mathfrak{m}_1 \cdots \mathfrak{m}_i M / \mathfrak{m}_1 \cdots \mathfrak{m}_{i+1} M$ is a finitely generated vector space over the field $R/\mathfrak{m}_{i+1}$, hence has finite length as an R-module. So $M/\mathfrak{n}M$ has finite length. Likewise, $\mathfrak{n}M/\mathfrak{n}^2 M$ has finite length, and so $M/\mathfrak{n}^2 M$ has finite length. Proceeding this way we see that $M/\mathfrak{n}^m M$ has finite length. As $\mathfrak{n}^m = \{0\}$, so does M, showing (ii). The implication (ii) $\Rightarrow$ (iii) is obvious, (iii) $\Longleftrightarrow$ (iv) is Lemma 1.2.4, and Lemma 1.2.5 gives (iv) $\Rightarrow$ (i). □

Notes and comments

The Schreier Refinement Theorem is from [391], and the Jordan-Hölder Theorem from [206, 187]. In connection with Proposition 1.2.6 we mention that every artinian commutative ring is automatically noetherian: theorem of Akizuki [7]; see also [288, Theorem 3.2].

1.3 INTEGRAL EXTENSIONS AND INTEGRALLY CLOSED DOMAINS

In this section we establish some facts for use in Chapter 3. But first a reminder on matrices. Let R be a commutative ring, $n \geqslant 1$, and let $R^{n \times n}$ be the R-algebra of $n \times n$ matrices over R, with multiplicative identity I_n, the $n \times n$ identity matrix.

Let a matrix $T = (T_{ij}) \in R^{n \times n}$ be given. It has determinant $\det T \in R$, and transpose $T^{\mathrm{t}} \in R^{n \times n}$. For $i, j = 1, \dots, n$, let T^{ij} be the determinant of the $(n-1) \times (n-1)$ matrix obtained from T by removing the ith row and the jth column from T; by convention this determinant equals 1 for $n = 1$. The matrix

$$T^* := \left((-1)^{i+j} T^{ij}\right)^{\mathrm{t}} \in R^{n \times n}$$

is called the **adjoint of** T and satisfies

$$TT^* = T^*T = (\det T)\,\mathrm{I}_n \qquad \text{(Cramer's Rule, Laplace expansion)}.$$

Integral extensions

In this subsection A is a subring of the commutative ring B. An element $b \in B$ is said to be **integral over** A if there are $n \geqslant 1$ and $a_1, \dots, a_n \in A$ such that $b^n + a_1 b^{n-1} + \cdots + a_{n-1} b + a_n = 0$.

LEMMA 1.3.1. *Let $b \in B$. The following are equivalent:*

(i) *b is integral over A;*

(ii) *the submodule $A[b]$ of the A-module B is finitely generated;*

(iii) *some subring of B contains $A[b]$ and is finitely generated as an A-module.*

PROOF. If $b^n + a_1 b^{n-1} + \cdots + a_{n-1}b + a_n = 0$ with $n \geqslant 1$ and $a_1, \ldots, a_n \in A$, then clearly $A[b] = A + Ab + \cdots + Ab^{n-1}$. This gives (i) $\Rightarrow$ (ii), and (ii) $\Rightarrow$ (iii) is trivial. To show (iii) $\Rightarrow$ (i), let $C \supseteq A[b]$ be a subring of B that is finitely generated as an A-module. Take $c_1, \ldots, c_n \in C$ such that $C = Ac_1 + \cdots + Ac_n$, $n \geqslant 1$. Then

$$\begin{aligned} bc_1 &= a_{11}c_1 + \cdots + a_{1n}c_n \\ \vdots \quad & \quad \vdots \qquad \vdots \quad \vdots \\ bc_n &= a_{n1}c_1 + \cdots + a_{nn}c_n \end{aligned}$$

for certain $a_{ij} \in A$. Then for the $n \times n$ matrix $T = b\,\mathrm{I}_n - (a_{ij})$ we have $Tc = 0$, where c is the column vector $(c_1, \ldots, c_n)^{\mathrm{t}} \in C^n$ and likewise, $0 = (0, \ldots, 0)^{\mathrm{t}}$ in C^n. Multiplying $Tc = 0$ on the left by T^* we obtain $\det(T) = 0$. This gives an equality $b^n + a_1 b^{n-1} + \cdots + a_{n-1}b + a_n = 0$ with $a_1, \ldots, a_n \in A$. □

We say that B is **integral over** A if every element of B is integral over A.

COROLLARY 1.3.2. *Let $b_1, \ldots, b_m \in B$. Then the following are equivalent:*

(i) *each b_i is integral over A;*

(ii) *the submodule $A[b_1, \ldots, b_m]$ of the A-module B is finitely generated;*

(iii) *$A[b_1, \ldots, b_m]$ is integral over A.*

PROOF. Suppose $P_i(b_i) = 0$ for monic $P_i \in A[X]$ of degree $d_i \geqslant 1$, $i = 1, \ldots, m$. Then $A[b_1, \ldots, b_m]$ is generated as an A-module by the products $b_1^{j_1} \cdots b_m^{j_m}$ with $0 \leqslant j_1 < d_1, \ldots, 0 \leqslant j_m < d_m$. This gives (i) $\Rightarrow$ (ii). The implication (ii) $\Rightarrow$ (iii) follows from (iii) $\Rightarrow$ (i) in Lemma 1.3.1, and (iii) $\Rightarrow$ (i) is obvious. □

COROLLARY 1.3.3. *Let c be an element in a commutative ring extension of B. Suppose c is integral over B and B is integral over A. Then c is integral over A.*

PROOF. Take $b_1, \ldots, b_n \in B$ such that c is integral over the ring $A[b_1, \ldots, b_n]$; the latter is a finitely generated A-module, and $A[b_1, \ldots, b_n, c]$ is finitely generated as an $A[b_1, \ldots, b_n]$-module, hence also as an A-module. So c is integral over A. □

We say that A is **integrally closed in** B if every $b \in B$ that is integral over A already lies in A. The set of elements of B that are integral over A is called the **integral closure of** A **in** B. By the previous two corollaries, the integral closure of A in B is a subring of B that contains A, is integral over A, and integrally closed in B.

Prime ideals under integral extensions

In this subsection A is a subring of the commutative ring B, and B is integral over A.

LEMMA 1.3.4. *Assume $1 \neq 0$ in B, and let J be an ideal of B containing an element that is not a zero divisor of B. Then $J \cap A \neq \{0\}$.*

PROOF. Suppose $b \in J$ is not a zero divisor of B. Take $n \geqslant 1$ minimal such that there exist $a_1, \ldots, a_n \in A$ with $b^n + a_1 b^{n-1} + \cdots + a_n = 0$. For such $a_1, \ldots, a_n$ we have $0 \neq a_n \in bB \cap A \subseteq J \cap A$. □

COROLLARY 1.3.5. *Let $\mathfrak{q}, \mathfrak{q}' \in \operatorname{Spec}(B)$, and suppose $\mathfrak{q} \subseteq \mathfrak{q}'$ and $\mathfrak{q} \cap A = \mathfrak{q}' \cap A$. Then $\mathfrak{q} = \mathfrak{q}'$.*

PROOF. Let $\mathfrak{p} := \mathfrak{q} \cap A = \mathfrak{q}' \cap A \in \operatorname{Spec}(A)$, and let $\overline{A}$ be the image of $A/\mathfrak{p}$ in $\overline{B} := B/\mathfrak{q}$ under the natural embedding $A/\mathfrak{p} \to B/\mathfrak{q}$. Then the domain $\overline{B}$ is integral over its subring $\overline{A}$, and $\mathfrak{q}'/\mathfrak{q}$ is a prime ideal of $\overline{B}$ that intersects $\overline{A}$ trivially. By Lemma 1.3.4 this yields $\mathfrak{q} = \mathfrak{q}'$. □

LEMMA 1.3.6. *Let I be an ideal of A. Then $IB \cap A \subseteq \sqrt{I}$.*

PROOF. Let $b \in IB \cap A$. Take a finitely generated subalgebra C of the A-algebra B with $b \in IC$; then C is also finitely generated as A-module. Proceeding as in (iii) $\Rightarrow$ (i) in the proof of Lemma 1.3.1, and using the notations there, we get all $a_{ij} \in I$, and so obtain $b^n + a_1 b^{n-1} + \cdots + a_n = 0$ where $n \geqslant 1$ and $a_1, \ldots, a_n \in I$, so $b^n \in I$. □

COROLLARY 1.3.7. *For each $\mathfrak{p} \in \operatorname{Spec}(A)$ there is a $\mathfrak{q} \in \operatorname{Spec}(B)$ with $\mathfrak{p} = \mathfrak{q} \cap A$.*

PROOF. Let $\mathfrak{p} \in \operatorname{Spec}(A)$, and set $S := A \setminus \mathfrak{p}$. Then $\mathfrak{p}B \cap S = \emptyset$ by Lemma 1.3.6, so Lemma 1.1.2 gives $\mathfrak{q} \in \operatorname{Spec}(B)$ with $\mathfrak{p}B \subseteq \mathfrak{q}$ and $\mathfrak{q} \cap S = \emptyset$, hence $\mathfrak{q} \cap A = \mathfrak{p}$. □

COROLLARY 1.3.8. *Let $\mathfrak{p}, \mathfrak{p}' \in \operatorname{Spec}(A)$ and $\mathfrak{q} \in \operatorname{Spec}(B)$ be such that $\mathfrak{p} \subseteq \mathfrak{p}'$ and $\mathfrak{p} = \mathfrak{q} \cap A$. Then there exists $\mathfrak{q}' \in \operatorname{Spec}(B)$ such that $\mathfrak{q} \subseteq \mathfrak{q}'$ and $\mathfrak{p}' = \mathfrak{q}' \cap A$.*

PROOF. Apply Corollary 1.3.7 to the prime ideal $\mathfrak{p}'/\mathfrak{p}$ of $A/\mathfrak{p}$ and the natural embedding $A/\mathfrak{p} \to B/\mathfrak{q}$. □

COROLLARY 1.3.9. *Let $\mathfrak{q} \in \operatorname{Spec}(B)$ and $\mathfrak{p} = \mathfrak{q} \cap A$. Then:*

$$\mathfrak{p} \textit{ is a maximal ideal of } A \iff \mathfrak{q} \textit{ is a maximal ideal of } B.$$

PROOF. Corollary 1.3.5 gives "$\Rightarrow$" and Corollary 1.3.8 yields "$\Leftarrow$." □

Thus any maximal ideal $\mathfrak{p}$ of A is contained in some prime ideal $\mathfrak{q}$ of B; for any such $\mathfrak{p}, \mathfrak{q}$ we have: $\mathfrak{p} = \mathfrak{q} \cap A$, $\mathfrak{q}$ is a maximal ideal of B, and the field $B/\mathfrak{q}$ is algebraic over the field $A/\mathfrak{p}$ (after identifying the latter with its natural image in $B/\mathfrak{q}$).

An application

In Section 4.6 we shall need the following basic fact about integral domains that are finitely generated over an infinite field:

LEMMA 1.3.10. *Let the infinite field K be a subring of the integral domain B. Assume that B is finitely generated as a K-algebra and $x \in B$ is transcendental over K. Then $x - c \in B^\times$ for only finitely many $c \in K$.*

PROOF. Let F be the fraction field of B. Take $x_1, \dots, x_n \in B$ such that $B = K[x_1, \dots, x_n]$, $x = x_1$, and $x_1, \dots, x_r$ is a transcendence basis of F over K, where $1 \leqslant r \leqslant n$. Take $g \in K[x_1, \dots, x_r]^{\neq}$ such that $x_{r+1}, \dots, x_n \in F$ are integral over the subring $A := K[x_1, \dots, x_r, g^{-1}]$ of F. Then $B[g^{-1}] \subseteq F$ is integral over A, so every maximal ideal of A extends to a maximal ideal of $B[g^{-1}]$. Thus every K-algebra morphism $A \to K$ extends to a K-algebra morphism $B[g^{-1}] \to K^{\mathrm{a}}$, where K^{a} is an algebraic closure of K. Treating $x_1, x_2, \dots, x_r$ as indeterminates over K and using that K is infinite, we have $c_2, \dots, c_r \in K$ such that $g(x_1, c_2, \dots, c_r) \neq 0$ in $K[x_1]$. Let $c \in K$ be such that $g(c, c_2, \dots, c_r) \neq 0$. All but finitely many elements of K satisfy this condition, so it only remains to show that $x - c \notin A^\times$. Taking $c_1 := c$ we get a K-algebra morphism $B \to K$ sending x_i to c_i for $i = 1, \dots, r$. It extends to a K-algebra morphism $A[g^{-1}] \to K^{\mathrm{a}}$ sending $x - c$ to 0, so $x - c \notin A^\times$. □

Integrally closed domains

We define an **integrally closed domain** to be an integral domain that is integrally closed in its fraction field. The ring of integers, and polynomial rings (in any set of variables) over fields are integrally closed domains, since they are factorial domains (also called *unique factorization domains*):

LEMMA 1.3.11. *Factorial domains are integrally closed domains.*

PROOF. Let A be a factorial domain with fraction field K, and suppose $f \in K^\times$ is integral over A. Then $f^n + a_1 f^{n-1} + \cdots + a_{n-1} f + a_n = 0$ with coefficients $a_1, \dots, a_n \in A$. We arrange $f = a/b$ where $a, b \in A^{\neq}$ and no irreducible element of A divides both a and b in A. Then $a^n + a_1 a^{n-1} b + \cdots + a_{n-1} a b^{n-1} + a_n b^n = 0$, so b divides a^n, which forces $b \in A^\times$, and thus $f \in A$. □

Next we characterize the integral closure of an integrally closed domain in a field extension of its fraction field:

LEMMA 1.3.12. *Suppose A is an integrally closed domain with fraction field K and $L \supseteq K$ is a field extension. An element of L is integral over A if and only if it is algebraic over K and its minimum polynomial over K has its coefficients in A.*

PROOF. Suppose b is integral over A; then clearly b is algebraic over K. Let $P \in K[X]$ be the minimum polynomial of b over K, say of degree n. Let K^{a} be an algebraic closure of K. Then in $K^{\mathrm{a}}[X]$ we have $P = (X - b_1) \cdots (X - b_n)$ where each $b_i \in K^{\mathrm{a}}$ is a conjugate of b, that is, of the form $\sigma(b)$ for some embedding $\sigma \colon K(b) \to K^{\mathrm{a}}$

over K. Hence every b_i is integral over A, so the coefficients of P are integral over A, and thus lie in A, since A is integrally closed. □

LEMMA 1.3.13. *Suppose that A is an integrally closed domain with fraction field K, $L \supseteq K$ is a separable field extension of finite degree, and B is the integral closure of A in L. Then $L = K(x)$ for some $x \in B$. For any such x with minimum polynomial P over K we have $B \subseteq P'(x)^{-1}A[x]$.*

PROOF. The Primitive Element Theorem [249, Chapter V, Theorem 4.6] gives $x \in L$ with $L = K(x)$. Multiplying x by a suitable element of $A^{\neq}$ we get $x \in B$. Let P be the minimum polynomial of x over K; then $P \in A[X]$ by Lemma 1.3.12. Take a field extension M of L such that $M|K$ is a Galois extension of finite degree. Set $G := \mathrm{Aut}(M|K)$, $H := \mathrm{Aut}(M|L)$, and take a coset decomposition

$$G = \sigma_1 H \cup \cdots \cup \sigma_n H, \qquad n = [L : K] = \deg P.$$

So $\sigma_1|L, \ldots, \sigma_n|L$ are the distinct K-embeddings of L into M. We take $\sigma_1 = \mathrm{id}_M$. Note that $P = \prod_{i=1}^n \big(X - \sigma_i(x)\big)$. For $i = 1, \ldots, n$ put $Q_i := P/\big(X - \sigma_i(x)\big) \in M[X]$. Since $x \in B$, we have $b_0, \ldots, b_{n-2} \in B$ such that

$$\begin{aligned} Q_1 &= P/(X - x) = X^{n-1} + b_{n-2}X^{n-2} + \cdots + b_0 \quad \text{hence} \\ Q_i &= \sigma_i(Q_1) = X^{n-1} + \sigma_i(b_{n-2})X^{n-2} + \cdots + \sigma_i(b_0). \end{aligned}$$

Also note that $Q_1(x) = P'(x)$ and $Q_i(x) = 0$ for $i \geqslant 2$. Let $b \in B$. Then

$$\begin{aligned} Q_1(x)b &= \sum_{i=1}^n Q_i(x)\sigma_i(b) = \sum_{i=1}^n \big(x^{n-1} + \sigma_i(b_{n-2})x^{n-2} + \cdots + \sigma_i(b_0)\big)\sigma_i(b) \\ &= \Big(\sum_i \sigma_i(b)\Big) x^{n-1} + \Big(\sum_i \sigma_i(b_{n-2}b)\Big) x^{n-2} + \cdots + \sum_i \sigma_i(b_0 b). \end{aligned}$$

The coefficients $\sum_i \sigma_i(b), \sum_i \sigma_i(b_{n-2}b), \ldots, \sum_i \sigma_i(b_0 b)$ are in K since they are traces of elements of L, and they are integral over A, hence they are in A. Therefore $P'(x)b = Q_1(x)b \in A[x]$, so $b \in P'(x)^{-1}A[x]$. □

Notes and comments

The notions of *integral element* and *integral closure* arose from that of an *algebraic integer* and the *ring of integers* of an algebraic number field (19th century: Gauss, Dirichlet, Kummer, Dedekind). The general theory was developed by E. Noether [306], Krull [229], and Cohen-Seidenberg [82].

1.4 LOCAL RINGS

Throughout this section A is a commutative ring. Call A **local** if A has exactly one maximal ideal. Thus valuation rings are local rings. Often we denote a local ring A by $(A, \mathfrak{m})$, thus indicating its maximal ideal $\mathfrak{m}$. A local ring $(A, \mathfrak{m})$ has **residue field**

$\boldsymbol{k} := A/\mathfrak{m}$ with a surjective ring morphism $a \mapsto \overline{a} := a+\mathfrak{m} : A \to \boldsymbol{k}$. Local rings have rather good properties compared to arbitrary commutative rings. This is exemplified by the generation of modules over local rings, the first topic of this section. Next we describe the maximal ideals of certain integral extensions of a local ring. We then discuss the process of localization, which can often simplify a problem by reduction to a local ring issue. Finally, we consider regular sequences, as needed in the study of regular local rings in Section 1.6.

Nakayama's Lemma

In this subsection M is a finitely generated A-module. Thus any set of generators of M has a finite subset generating M. We denote by $\mu(M)$ the least m such that M is generated by a subset of size m. We say that $G \subseteq M$ is a **minimal set of generators** of M if G generates M, but no proper subset of G does. Any set of generators of M contains a minimal set of generators of M, and every set of generators of M of size $\mu(M)$ is minimal.

LEMMA 1.4.1 (Nakayama). *Suppose the ideal I of A is contained in every maximal ideal of A, and $IM = M$. Then $M = \{0\}$.*

PROOF. Towards a contradiction, assume $n = \mu(M) \geqslant 1$. Take $x_1, \dots, x_n \in M$ with $M = Ax_1 + \cdots + Ax_n$. Since $IM = M$, we can take $a_1, \dots, a_n \in I$ such that $x_n = a_1x_1 + \cdots + a_nx_n$. Then $(1-a_n)x_n \in Ax_1 + \cdots + Ax_{n-1}$ where $1 - a_n \in A^{\times}$. Hence $M = Ax_1 + \cdots + Ax_{n-1}$, a contradiction. □

In the rest of this subsection $(A, \mathfrak{m})$ is a local ring with residue field $\boldsymbol{k} = A/\mathfrak{m}$. We view $\overline{M} := M/\mathfrak{m}M$ as a $\boldsymbol{k}$-linear space in the natural way, and for $x \in M$ we set $\overline{x} := x + \mathfrak{m}M \in \overline{M}$.

COROLLARY 1.4.2. *Let N be a submodule of M. If $M = N + \mathfrak{m}M$, then $M = N$.*

PROOF. Apply Lemma 1.4.1 to $\mathfrak{m}$ and M/N in place of I and M. □

COROLLARY 1.4.3. *Let $x_1, \dots, x_n \in M$. Then*

$$M = Ax_1 + \cdots + Ax_n \qquad \Longleftrightarrow \qquad \overline{M} = \boldsymbol{k}\,\overline{x_1} + \cdots + \boldsymbol{k}\,\overline{x_n}.$$

PROOF. Apply the previous corollary to $N = Ax_1 + \cdots + Ax_n$. □

Familiar properties of bases of $\boldsymbol{k}$-vector spaces and the preceding corollary yield:

COROLLARY 1.4.4. *Let $x_1, \dots, x_m \in M$ be distinct. Then:*

(i) *$\{x_1, \dots, x_m\}$ is a minimal set of generators of M if and only if $\overline{x_1}, \dots, \overline{x_m}$ is a basis of the $\boldsymbol{k}$-linear space $\overline{M}$; thus $\mu(M) = \dim_{\boldsymbol{k}} \overline{M}$;*

(ii) *$\{x_1, \dots, x_m\}$ is contained in a minimal set of generators of M if and only if $\overline{x_1}, \dots, \overline{x_m}$ are $\boldsymbol{k}$-linearly independent.*

Let $x \in \mathfrak{m} \setminus \mathfrak{m}^2$; then $A^* := A/Ax$ is a local ring with maximal ideal $\mathfrak{m}^* := \mathfrak{m}/Ax$. For $a \in A$, put $a^* := a + Ax \in A^*$.

COROLLARY 1.4.5. *Let $x_1, \dots, x_n \in \mathfrak{m}$, and suppose $x_1^*, \dots, x_n^*$ are distinct, and $\{x_1^*, \dots, x_n^*\}$ is a minimal set of generators of $\mathfrak{m}^*$. Then $x \notin \{x_1, \dots, x_n\}$, and $\{x_1, \dots, x_n, x\}$ is a minimal set of generators of $\mathfrak{m}$.*

PROOF. Clearly we have $x \notin \{x_1, \dots, x_n\}$, and $x_1, \dots, x_n, x$ generate $\mathfrak{m}$. For the set $\{x_1, \dots, x_n, x\}$ of generators of $\mathfrak{m}$ to be minimal, it is enough by Corollary 1.4.4 that the residue classes $x_1 + \mathfrak{m}^2, \dots, x_n + \mathfrak{m}^2, x + \mathfrak{m}^2 \in \mathfrak{m}/\mathfrak{m}^2$ are $\boldsymbol{k}$-linearly independent. Thus, let $a_1, \dots, a_n, a \in A$ be such that

$$a_1 x_1 + \dots + a_n x_n + ax \in \mathfrak{m}^2.$$

Then in A^* we have

$$a_1^* x_1^* + \dots + a_n^* x_n^* \in (\mathfrak{m}^*)^2$$

and hence $a_1^*, \dots, a_n^* \in \mathfrak{m}^*$ by hypothesis and Corollary 1.4.4, so $a_1, \dots, a_n \in \mathfrak{m}$. This yields $ax \in \mathfrak{m}^2$ and thus $a \in \mathfrak{m}$, since $x \notin \mathfrak{m}^2$. □

COROLLARY 1.4.6. *Suppose $\mathfrak{m}$ is finitely generated. Then either:*

(i) $\mathfrak{m}^n \neq \mathfrak{m}^{n+1}$ *for all n, or*

(ii) $\mathfrak{m}^n = \{0\}$ *for some n, in which case* $\operatorname{Spec}(A) = \{\mathfrak{m}\}$.

PROOF. Suppose $\mathfrak{m}^n = \mathfrak{m}^{n+1}$. Then $\mathfrak{m}^n = \{0\}$ by Nakayama's Lemma, and so for $\mathfrak{p} \in \operatorname{Spec}(A)$ we have $\mathfrak{m}^n \subseteq \mathfrak{p}$, hence $\mathfrak{m} \subseteq \mathfrak{p}$. □

Here is a characterization of DVRs among local rings:

LEMMA 1.4.7. *Suppose A is noetherian, $t \in A$ is not a zero divisor of A, and $\mathfrak{m} = tA$. Then A is a* DVR.

PROOF. Using that t is not a zero divisor we get $\mathfrak{m}I = I$ for $I := \bigcap_n \mathfrak{m}^n = \bigcap_n t^n A$, so $I = \{0\}$ by Nakayama's Lemma. Therefore, given $a \in A^{\neq}$, we have n with $a \in t^n A \setminus t^{n+1} A$, and then $a = t^n u$, $u \in A^\times$. Thus A is a DVR. □

The maximal ideals of an integral extension of a local ring

Let A be a local ring with maximal ideal $\mathfrak{m}$ and X an indeterminate. We extend the surjective ring morphism $a \mapsto \overline{a} := a + \mathfrak{m} \colon A \to \boldsymbol{k} := A/\mathfrak{m}$ to a surjective ring morphism $P \mapsto \overline{P} \colon A[X] \to \boldsymbol{k}[X]$ that sends X to X. With these notations we have:

LEMMA 1.4.8. *Let $P \in A[X]$ be monic of positive degree, and $A[x] := A[X]/(P)$ with $x := X + (P)$. Suppose $\overline{P} \in \boldsymbol{k}[X]$ factors as*

$$\overline{P} = \overline{P_1}^{e_1} \cdots \overline{P_n}^{e_n}$$

where each $e_i \geqslant 1$, *each* $P_i \in A[X]$ *is monic, and* $\overline{P_1}, \ldots, \overline{P_n} \in \boldsymbol{k}[X]$ *are irreducible and distinct. Then* $\mathfrak{m}_1, \ldots, \mathfrak{m}_n$ *with*

$$\mathfrak{m}_i \ := \ \mathfrak{m}A[x] + P_i(x)A[x]$$

are the distinct maximal ideals of $A[x]$.

PROOF. Consider for each i the composite of the natural surjections

$$A[x] = A[X]/(P) \to \boldsymbol{k}[X]/(\overline{P}) \to \boldsymbol{k}[X]/(\overline{P_i}).$$

It is easy to see that the composite map has kernel $\mathfrak{m}_i$; since $\boldsymbol{k}[X]/(\overline{P_i})$ is a field, $\mathfrak{m}_i$ is a maximal ideal of $A[x]$. If $1 \leqslant i < j \leqslant n$, then $\overline{P_i} \not\equiv 0 \mod \overline{P_j}$ in $\boldsymbol{k}[X]$, so $\mathfrak{m}_i \neq \mathfrak{m}_j$. It remains to show that the $\mathfrak{m}_i$ are the only maximal ideals of $A[x]$. First note that $A[x]$ is integral over A, so each maximal ideal of $A[x]$ contains $\mathfrak{m}$. The polynomial $P - \prod_i P_i^{e_i}$ is in $\mathfrak{m}A[X]$, and $P(x) = 0$, so $\prod_i P_i(x)^{e_i} \in \mathfrak{m}A[x]$. Thus each maximal ideal of $A[x]$ contains some $P_i(x)$, and thus equals $\mathfrak{m}_i$ for some i. □

Localization

Let S be a multiplicative subset of A. Recall the construction of the localization $S^{-1}A$ of A at S: this is the ring whose elements are the equivalence classes of the equivalence relation $\sim$ on $A \times S$ defined by

$$(a_1, s_1) \sim (a_2, s_2) \quad :\Longleftrightarrow \quad a_1 s_2 s = a_2 s_1 s \text{ for some } s \in S,$$

with the equivalence class of $(a, s) \in A \times S$ denoted by $\frac{a}{s}$ or a/s, and with addition and multiplication given by

$$\frac{a_1}{s_1} + \frac{a_2}{s_2} = \frac{a_1 s_2 + a_2 s_1}{s_1 s_2}, \qquad \frac{a_1}{s_1} \cdot \frac{a_2}{s_2} = \frac{a_1 a_2}{s_1 s_2}.$$

Then $S^{-1}A$ has zero element $0/1$ and identity $1/1$, and $S^{-1}A = \{0\} \Longleftrightarrow 0 \in S$. The map $\iota = \iota_A^S \colon A \to S^{-1}A$ sending $a \in A$ to $a/1$ is a ring morphism with

$$\iota(S) \subseteq (S^{-1}A)^\times, \qquad \ker \iota \ = \ \{a \in A : \ as = 0 \text{ for some } s \in S\}.$$

We recall the key universal property of $\iota \colon A \to S^{-1}A$: for every ring morphism $\phi \colon A \to B$ into a commutative ring B with $\phi(S) \subseteq B^\times$ there is a unique ring morphism $\phi' \colon S^{-1}A \to B$ such that $\phi = \phi' \circ \iota$.

Let I be an ideal of A. Then $S^{-1}I := \left\{\frac{a}{s} : a \in I,\ s \in S\right\}$ is the ideal of $S^{-1}A$ generated by $\iota(I)$. The ideal

$$\iota^{-1}(S^{-1}I) \ = \ \{a \in A : \ as \in I \text{ for some } s \in S\}$$

of A contains I, with $\iota^{-1}(S^{-1}I) = A$ iff $I \cap S \neq \emptyset$. If J is an ideal of $S^{-1}A$ and $I = \iota^{-1}(J)$, then $J = S^{-1}I$. Hence if A is noetherian, then so is $S^{-1}A$. An ideal J of $S^{-1}A$ is prime iff the ideal $\iota^{-1}(J)$ of A is prime and disjoint from S. Thus:

COROLLARY 1.4.9. *The map* $\iota^*: \operatorname{Spec}(S^{-1}A) \to \operatorname{Spec}(A)$ *is a homeomorphism onto its image* $\{\mathfrak{p} \in \operatorname{Spec}(A) : \mathfrak{p} \cap S = \emptyset\}$, *with inverse given by* $\mathfrak{p} \mapsto S^{-1}\mathfrak{p}$.

EXAMPLE. For $S = \{1, s, s^2, \dots\}$ with $s \in A$ we denote $S^{-1}A$ also by A_s, and the image of ι^* in this case is $\mathrm{D}(s) = \{\mathfrak{p} \in \operatorname{Spec}(A) : s \notin \mathfrak{p}\}$.

Localization produces local rings: Let $\mathfrak{p} \in \operatorname{Spec}(A)$; then $S = A \setminus \mathfrak{p}$ is a multiplicative subset of A with $0 \notin S$, and the localization $S^{-1}A$ of A at S is then denoted by $A_{\mathfrak{p}}$ and called the localization of A at $\mathfrak{p}$. The ring $A_{\mathfrak{p}}$ is indeed local, with maximal ideal $\mathfrak{p}A_{\mathfrak{p}}$: by the preceding corollary ι^* maps $\operatorname{Spec} A_{\mathfrak{p}}$ bijectively onto

$$\{\mathfrak{q} \in \operatorname{Spec}(A) : \mathfrak{q} \subseteq \mathfrak{p}\}.$$

The morphism $\iota: A \to A_{\mathfrak{p}}$ induces an embedding $A/\mathfrak{p} \to A_{\mathfrak{p}}/\mathfrak{p}A_{\mathfrak{p}}$ of the domain $A/\mathfrak{p}$ into the field $F := A_{\mathfrak{p}}/\mathfrak{p}A_{\mathfrak{p}}$, and F is the fraction field of the image of this embedding in the sense that every $f \in F$ equals $s^{-1}a$ for some a, s in the image of this embedding, $s \neq 0$.

Localization in integral domains

Suppose A is an integral domain and S is a multiplicative subset of A with $0 \notin S$. Then we have the ring embedding

$$a/s \mapsto s^{-1}a \; : \; S^{-1}A \to \operatorname{Frac} A \qquad (a \in A,\ s \in S),$$

via which, throughout this volume, we identify $S^{-1}A$ with a subring of the fraction field $\operatorname{Frac} A$ of A. (Thus $S^{-1}A = \operatorname{Frac} A$ for $S = A^{\neq}$.) Note that if A is a PID, then so is $S^{-1}A$. For $\mathfrak{p} \in \operatorname{Spec}(A)$ and $S = A \setminus \mathfrak{p}$ we obtain the local domain $A_{\mathfrak{p}}$. Thus if A is a PID and $\{0\} \neq \mathfrak{p} \in \operatorname{Spec}(A)$, then $A_{\mathfrak{p}}$ is a DVR by Lemma 1.4.7.

LEMMA 1.4.10. *Let* $\mathfrak{m}$ *be a maximal ideal of the integral domain* A, *and let* B *be a local ring such that* $A \subseteq B \subseteq A_{\mathfrak{m}}$. *Then* $B = A_{\mathfrak{m}}$.

PROOF. Let $\mathfrak{n}$ be the maximal ideal of B and $s \in A \setminus \mathfrak{m}$; it is enough to show that then $s \in B^{\times}$, that is, $s \notin \mathfrak{n}$. Take $t \in A$ with $st \equiv 1 \bmod \mathfrak{m}$, so $st - 1 \in \mathfrak{m}$. Since $\mathfrak{m} \subseteq \mathfrak{m}A_{\mathfrak{m}} \cap B \subseteq \mathfrak{n}$, this yields $st - 1 \in \mathfrak{n}$, so $s \notin \mathfrak{n}$. □

We can now complete Lemma 1.3.13 as follows:

COROLLARY 1.4.11. *Let* A *be an integrally closed domain with fraction field* K, *let* $L \supseteq K$ *be a separable field extension of finite degree. Let* B *be the integral closure of* A *in* L, *let* $x \in B$ *with* $L = K(x)$ *have minimum polynomial* P *over* K, *and let* $\mathfrak{m}$ *and* $\mathfrak{n}$ *be maximal ideals of* $A[x]$ *and* B, *respectively, such that* $P'(x) \notin \mathfrak{m} = \mathfrak{n} \cap A[x]$. *Then we have* $A[x]_{\mathfrak{m}} = B_{\mathfrak{n}}$.

PROOF. From $P'(x) \notin \mathfrak{m}$ and Lemma 1.3.13 we get $B \subseteq A[x]_{\mathfrak{m}} \subseteq B_{\mathfrak{n}}$. Now Lemma 1.4.10 applied to B, $A[x]_{\mathfrak{m}}$, $\mathfrak{n}$ in place of A, B, $\mathfrak{m}$, yields $A[x]_{\mathfrak{m}} = B_{\mathfrak{n}}$. □

In the next four lemmas A is an integral domain and S is a multiplicative subset of A with $0 \notin S$. So for $\mathfrak{p} \in \operatorname{Spec}(A)$ with $\mathfrak{p} \cap S = \emptyset$ we have $S^{-1}\mathfrak{p} \in \operatorname{Spec}(S^{-1}A)$. We set $K := \operatorname{Frac}(A)$.

LEMMA 1.4.12. *If $\mathfrak{p} \in \operatorname{Spec}(A)$ and $\mathfrak{p} \cap S = \emptyset$, then $(S^{-1}A)_{S^{-1}\mathfrak{p}} = A_{\mathfrak{p}}$ in K.*

Let X be an indeterminate over A in the next two lemmas.

LEMMA 1.4.13. *We have $(S^{-1}A)[X] = S^{-1}\big(A[X]\big)$ inside $K(X)$.*

LEMMA 1.4.14. *Let $\mathfrak{P} \in \operatorname{Spec}\big(A[X]\big)$. Then $\mathfrak{p} := \mathfrak{P} \cap A \in \operatorname{Spec}(A)$, $\mathfrak{P}$ generates a prime ideal $\mathfrak{Q}$ in $A_{\mathfrak{p}}[X] \subseteq K[X]$, and $A[X]_{\mathfrak{P}} = A_{\mathfrak{p}}[X]_{\mathfrak{Q}}$ inside $K(X)$.*

PROOF. The set $S := A \setminus \mathfrak{p}$ is a multiplicative subset of A as well as of $A[X]$, and $\mathfrak{P} \cap S = \emptyset$. So $\mathfrak{P}$ does generate a prime ideal $\mathfrak{Q} = S^{-1}\mathfrak{P}$ in $S^{-1}\big(A[X]\big) = A_{\mathfrak{p}}[X]$. Now apply Lemma 1.4.12 to $A[X]$ and $\mathfrak{P}$ in the role of A and $\mathfrak{p}$. □

In Section 5.9 we use:

LEMMA 1.4.15. *Let I be an ideal of A and $\mathfrak{p} \in Z(I)$. Then the ideal*

$$\iota^{-1}(IA_{\mathfrak{p}}) \;=\; \{a \in A : \; as \in I \textit{ for some } s \in A \setminus \mathfrak{p}\}$$

of A satisfies $I \subseteq \iota^{-1}(IA_{\mathfrak{p}}) \subseteq \mathfrak{p}$ and is contained in every prime ideal $\mathfrak{q}$ of A with $I \subseteq \mathfrak{q} \subseteq \mathfrak{p}$. If $\mathfrak{q} := \iota^{-1}(IA_{\mathfrak{p}})$ itself is prime, then $\mathfrak{q}A_{\mathfrak{q}} = IA_{\mathfrak{q}}$.

PROOF. The first statement is clear. Suppose $\mathfrak{q} := \iota^{-1}(IA_{\mathfrak{p}})$ is prime. We have $\mathfrak{q}A_{\mathfrak{p}} = IA_{\mathfrak{p}}$, and applying the ring morphism $A_{\mathfrak{p}} \to A_{\mathfrak{q}}$ yields $\mathfrak{q}A_{\mathfrak{q}} = IA_{\mathfrak{q}}$. □

Localization of modules

Let S be a multiplicative subset of A. Every A-module M gives rise to the $S^{-1}A$-module $S^{-1}M$, whose elements are the formal fractions x/s ($x \in M$, $s \in S$), with

$$x/s \;=\; 0 \iff tx = 0 \text{ for some } t \in S.$$

Addition of these fractions is given by

$$(x_1/s_1) + (x_2/s_2) \;=\; (s_2x_1 + s_1x_1)/s_1s_2 \qquad (x_1, x_2 \in M,\; s_1, s_2 \in S)$$

and their multiplication by scalars from $S^{-1}A$ is given by

$$(a/s)(x/t) \;=\; ax/st \qquad (a \in A,\; s, t \in S,\; x \in M).$$

Any $S^{-1}A$-module is construed as an A-module via $\iota\colon A \to S^{-1}A$. The map $\iota_M = \iota_M^S\colon M \to S^{-1}M$ defined by $\iota_M(x) = x/1$ is A-linear, and for any A-linear map $f\colon M \to N$ into an $S^{-1}A$-module N there is a unique $S^{-1}A$-linear map $f'\colon S^{-1}M \to N$ such that $f = f' \circ \iota_M$.

Regular sequences

Let M be an A-module. Call $a \in A$ a **zero divisor on** M if $ax = 0$ for some $x \in M^{\neq}$. Thus $a \in A$ is not a zero divisor on M iff the A-linear map $x \mapsto ax\colon M \to M$ is injective. Let $\vec{r} = (r_1, \ldots, r_n) \in A^n$, $n \geqslant 1$.

DEFINITION 1.4.16. The sequence $\vec{r}$ is called **regular on** M, if

(1) $M \neq r_1 M + \cdots + r_n M$; and

(2) r_j is not a zero divisor on $M/(r_1 M + \cdots + r_{j-1} M)$, for $j = 1, \ldots, n$.

This notion, applied to the A module A, is motivated by:

EXAMPLE. Let K be a commutative ring with $0 \neq 1$, let $X_1, \ldots, X_n$ be distinct indeterminates, $n \geqslant 1$, $A = K[X_1, \ldots, X_n]$. Then $(X_1, \ldots, X_n)$ is regular on A.

It is easy to see that for $1 \leqslant m < n$,

$$\text{(1.4.1)} \qquad \vec{r} \text{ is regular on } M \iff \begin{cases} (r_1, \ldots, r_m) \text{ is regular on } M \\ \text{and } (r_{m+1}, \ldots, r_n) \text{ is regular on} \\ M/(r_1 M + \cdots + r_m M). \end{cases}$$

The following are also easy to verify:

LEMMA 1.4.17. *If $\vec{r}$ is regular on A and X is an indeterminate, then $\vec{r}$ is regular on the $A[X]$-module $A[X]$.*

LEMMA 1.4.18. *If $\vec{r} = (r_1, \ldots, r_n)$ is regular on A and $r_1, \ldots, r_n \in \mathfrak{p}$ where $\mathfrak{p} \in \operatorname{Spec}(A)$, then $(\frac{r_1}{1}, \ldots, \frac{r_n}{1})$ is regular on the $A_{\mathfrak{p}}$-module $A_{\mathfrak{p}}$.*

In the next lemma and its corollary X is an indeterminate and I is an ideal of $A[X]$ such that $\mathfrak{m} := I \cap A$ is a maximal ideal of A.

LEMMA 1.4.19. *If $I \neq \mathfrak{m}A[X]$, then $I = \mathfrak{m}A[X] + PA[X]$ where $P \in A[X]$ is not a zero divisor on the $A[X]$-module $A[X]/\mathfrak{m}A[X]$.*

PROOF. Extend the canonical map $A \to A/\mathfrak{m}$ to the ring morphism $A[X] \to (A/\mathfrak{m})[X]$ by sending X to X. Since the kernel of this extension is $\mathfrak{m}A[X]$, we obtain a ring isomorphism $A[X]/\mathfrak{m}A[X] \xrightarrow{\cong} (A/\mathfrak{m})[X]$. Now use that the polynomial ring $(A/\mathfrak{m})[X]$ over the field $A/\mathfrak{m}$ is a domain and that its ideals are principal. □

From Lemmas 1.4.17 and 1.4.19 we obtain:

COROLLARY 1.4.20. *If $\vec{r} = (r_1, \ldots, r_n)$ is regular on A and $\mathfrak{m} = r_1 A + \cdots + r_n A$, and $I \neq \mathfrak{m}A[X]$, then there exists $P \in A[X]$ such that $I = r_1 A[X] + \cdots + r_n A[X] + PA[X]$ and $(r_1, \ldots, r_n, P)$ is regular on the $A[X]$-module $A[X]$.*

Notes and comments

Nakayama's Lemma in its various forms is due to Krull, Azumaya, and Nakayama; see the discussion in [301, pp. 212–213]. Early studies of local rings are Krull [230] and Chevalley [76]. Lemma 1.4.8 is essentially due to Kummer [243] and Dedekind [97]. Localization goes back to Grell [154], and in the generality above, to Chevalley [77] and Uzkov [452].

1.5 KRULL'S PRINCIPAL IDEAL THEOREM

Throughout this section R is a commutative ring. The **height** of a prime ideal $\mathfrak{p}$ of R, denoted by $\mathrm{ht}(\mathfrak{p})$, is the supremum, in $\mathbb{N} \cup \{\infty\}$, of the lengths n of strictly increasing sequences $\mathfrak{p}_0 \subset \mathfrak{p}_1 \subset \cdots \subset \mathfrak{p}_n = \mathfrak{p}$ of prime ideals of R. The (Krull) **dimension** of R is

$$\dim R := \sup \{\mathrm{ht}(\mathfrak{p}) : \mathfrak{p} \in \mathrm{Spec}(R)\}$$

with supremum in $\mathbb{N} \cup \{-\infty, +\infty\}$. So $\dim R = \dim \mathrm{Spec}(R)$ by Lemma 1.1.8, with $\dim \mathrm{Spec}(R)$ defined as at the end of Section 1.1. More generally, for each ideal I of R, we have $\dim R/I = \dim \mathrm{Z}(I)$. Note:

$$\dim R = -\infty \iff 1 = 0 \text{ in } R \iff \mathrm{Spec}(R) = \emptyset.$$

EXAMPLES. A prime ideal of R has height 0 if and only if it is minimal. The ring R has dimension 0 if and only if $R \neq \{0\}$ and every prime ideal of R is maximal; thus an integral domain has dimension 0 if and only if it is a field. See Proposition 1.2.6 for a characterization of 0-dimensional noetherian commutative rings.

Clearly $\mathrm{ht}(\mathfrak{p}) + \dim(R/\mathfrak{p}) \leqslant \dim R$. If R is local with maximal ideal $\mathfrak{m}$, then $\mathrm{ht}(\mathfrak{m}) = \dim R$. Therefore, if $\mathfrak{p}$ is a prime ideal of R, then $\mathrm{ht}(\mathfrak{p}) = \mathrm{ht}(\mathfrak{p}R_{\mathfrak{p}}) = \dim R_{\mathfrak{p}}$.

The following is clear from the definitions:

LEMMA 1.5.1. *Let $\varphi\colon R \to S$ be a surjective ring morphism and $\mathfrak{p}$ be a prime ideal of R with $\mathfrak{p} \supseteq \ker \varphi$. Then $\varphi(\mathfrak{p})$ is a prime ideal of S, and* $\mathrm{ht}(\mathfrak{p}) \geqslant \mathrm{ht}(\varphi(\mathfrak{p}))$.

From Corollaries 1.3.5, 1.3.7, and 1.3.8 we obtain:

COROLLARY 1.5.2. *If R is a subring of the commutative ring S, and S is integral over R, then* $\dim(R) = \dim(S)$.

In the rest of this section we assume that R is noetherian. The following theorem is a key result about Krull dimension in this noetherian setting:

THEOREM 1.5.3 (Principal Ideal Theorem). *If $x \in R$ and $\mathfrak{p}$ is a minimal prime divisor of Rx, then* $\mathrm{ht}(\mathfrak{p}) \leqslant 1$.

Since $\mathrm{ht}(\mathfrak{p}) = \dim R_{\mathfrak{p}}$ for $\mathfrak{p} \in \mathrm{Spec}(R)$, this immediately follows from the case where R is a local ring and $\mathfrak{p}$ is its maximal ideal, treated in the next lemma:

LEMMA 1.5.4. *Suppose $(R, \mathfrak{m})$ is a local ring, $x \in R$, and $\mathfrak{m}$ is a minimal prime divisor of Rx. Then* $\dim R \leqslant 1$.

PROOF. Let $\mathfrak{q} \neq \mathfrak{m}$ be a prime ideal of R; we need to show that $\mathrm{ht}(\mathfrak{q}) = 0$. Let $\iota: R \to R_{\mathfrak{q}}$ be the morphism given by $\iota(r) = \frac{r}{1}$. Set $\mathfrak{q}^{(n)} := \iota^{-1}(\mathfrak{q}^n R_{\mathfrak{q}})$ (the *nth symbolic power* of $\mathfrak{q}$) and consider the chain

$$\mathfrak{q}^{(1)} + Rx \supseteq \mathfrak{q}^{(2)} + Rx \supseteq \cdots$$

of ideals of R. The ring $\overline{R} := R/Rx$ has exactly one prime ideal (namely $\mathfrak{m}/Rx$), so $\dim \overline{R} = 0$. Then Proposition 1.2.6 gives n such that $\mathfrak{q}^{(n)} + Rx = \mathfrak{q}^{(n+1)} + Rx$.

CLAIM: $\mathfrak{q}^{(n)} = \mathfrak{q}^{(n+1)} + \mathfrak{m}\mathfrak{q}^{(n)}$.

To prove this claim, let $a \in \mathfrak{q}^{(n)}$. Then $a = b + rx$ where $b \in \mathfrak{q}^{(n+1)}$, $r \in R$. Since $rx \in \mathfrak{q}^{(n)}$ and $x \notin \mathfrak{q}$, we obtain $r \in \mathfrak{q}^{(n)}$ and hence $a = b + rx \in \mathfrak{q}^{(n+1)} + \mathfrak{m}\mathfrak{q}^{(n)}$.

The claim and Corollary 1.4.2 yields $\mathfrak{q}^{(n)} = \mathfrak{q}^{(n+1)}$ and hence $\mathfrak{q}^n R_{\mathfrak{q}} = \mathfrak{q}^{n+1} R_{\mathfrak{q}}$. Thus $\mathrm{ht}(\mathfrak{q}) = \dim R_{\mathfrak{q}} = 0$ by Corollary 1.4.6, as required. □

Combining Theorem 1.5.3 with Corollary 1.1.15 gives:

COROLLARY 1.5.5. *If $x \in R$ is not a zero divisor and $\mathfrak{p}$ is a minimal prime divisor of Rx, then* $\mathrm{ht}(\mathfrak{p}) = 1$.

Below we extend Theorem 1.5.3 to non-principal ideals. First a lemma:

LEMMA 1.5.6. *Let $\mathfrak{p}, \mathfrak{q}_1, \ldots, \mathfrak{q}_m$ be prime ideals of R such that $\mathfrak{p} \not\subseteq \bigcup_{i=1}^m \mathfrak{q}_i$, and $\mathrm{ht}(\mathfrak{p}) \geqslant n \geqslant 1$. Then there is a strictly increasing sequence $\mathfrak{p}_0 \subset \mathfrak{p}_1 \subset \cdots \subset \mathfrak{p}_n$ of prime ideals of R with $\mathfrak{p}_n = \mathfrak{p}$ and $\mathfrak{p}_j \not\subseteq \bigcup_{i=1}^m \mathfrak{q}_i$ for $j = 1, \ldots, n$.*

PROOF. By induction on n. The case $n = 1$ being trivial, suppose $n \geqslant 2$. Since $\mathrm{ht}(\mathfrak{p}) \geqslant n \geqslant 2$, we can take a prime ideal $\mathfrak{q} \subseteq \mathfrak{p}$ of R with $\mathrm{ht}(\mathfrak{p}/\mathfrak{q}) \geqslant 2$ in $R/\mathfrak{q}$ and $\mathrm{ht}(\mathfrak{q}) \geqslant n - 2$. Lemma 1.1.9 provides $x \in \mathfrak{p} \setminus (\mathfrak{q} \cup \bigcup_{i=1}^m \mathfrak{q}_i)$. Take a minimal prime divisor $\mathfrak{p}' \subseteq \mathfrak{p}$ of $xR + \mathfrak{q}$. Since $\mathrm{ht}(\mathfrak{p}'/\mathfrak{q}) \leqslant 1$ by Theorem 1.5.3, we have $\mathfrak{p}' \neq \mathfrak{p}$. From $x \in \mathfrak{p}'$ we get $\mathfrak{p}' \not\subseteq \bigcup_{i=1}^m \mathfrak{q}_i$. Now apply the inductive hypothesis to $\mathfrak{p}'$. □

THEOREM 1.5.7 (Generalized Principal Ideal Theorem). *If $\mathfrak{p}$ is a minimal prime divisor of an ideal of R generated by n elements, then* $\mathrm{ht}(\mathfrak{p}) \leqslant n$.

PROOF. By induction on n. The case $n = 0$ is trivial. Let $n \geqslant 1$, and let $\mathfrak{p}$ be a minimal prime divisor of $I = Rx_1 + \cdots + Rx_n$, $x_1, \ldots, x_n \in R$. Let $\mathfrak{q}_1, \ldots, \mathfrak{q}_m$ be the minimal prime divisors of $J := Rx_1 + \cdots + Rx_{n-1}$; thus $\mathrm{ht}(\mathfrak{q}_i) \leqslant n - 1$ for $i = 1, \ldots, m$ by inductive assumption. If $\mathfrak{p} \subseteq \bigcup_{i=1}^m \mathfrak{q}_i$, then for some $i \in \{1, \ldots, m\}$ we have $\mathfrak{p} \subseteq \mathfrak{q}_i$ and hence $\mathrm{ht}(\mathfrak{p}) \leqslant \mathrm{ht}(\mathfrak{q}_i) \leqslant n - 1$, and we are done. So assume $\mathfrak{p} \not\subseteq \bigcup_{i=1}^m \mathfrak{q}_i$. Towards a contradiction, suppose that $\mathrm{ht}(\mathfrak{p}) \geqslant n + 1$. Lemma 1.5.6 yields a strictly increasing sequence $\mathfrak{p}_0 \subset \mathfrak{p}_1 \subset \cdots \subset \mathfrak{p}_{n+1}$ of prime ideals of R with $\mathfrak{p}_{n+1} = \mathfrak{p}$ and $\mathfrak{p}_j \not\subseteq \bigcup_{i=1}^m \mathfrak{q}_i$ for $j = 1, \ldots, n$. Now consider the ring $\overline{R} := R/J$, whose minimal prime ideals are the $\overline{\mathfrak{q}_i} := \mathfrak{q}_i/J$ $(i = 1, \ldots, m)$. Setting $\overline{\mathfrak{p}} := \mathfrak{p}/J$, we have $\mathrm{ht}(\overline{\mathfrak{p}}) \leqslant 1$ by Theorem 1.5.3. Therefore $\overline{\mathfrak{p}}$ is a minimal prime divisor of the ideal $(\mathfrak{p}_1 + J)/J$ of $\overline{R}$, since $\mathfrak{p}_1 \not\subseteq \bigcup_{i=1}^m \mathfrak{q}_i$. Hence $\mathfrak{p}$ is a minimal prime divisor of $\mathfrak{p}_1 + J$. Thus the prime ideal $\mathfrak{p}/\mathfrak{p}_1$ of $R/\mathfrak{p}_1$ has height $\geqslant n$ and is a minimal prime divisor of the ideal generated by $x_1 + \mathfrak{p}_1, \ldots, x_{n-1} + \mathfrak{p}_1$ in $R/\mathfrak{p}_1$, in contradiction to the inductive hypothesis. □

Important consequences of Theorem 1.5.7 are the following:

COROLLARY 1.5.8. *Every prime ideal of* R *has finite height.*

COROLLARY 1.5.9. *If* R *is local with maximal ideal* $\mathfrak{m}$, *then* $\dim(R) \leqslant \mu(\mathfrak{m})$.

Here is a generalization of Corollary 1.5.5:

COROLLARY 1.5.10. *If* $m \geqslant 1$, *the sequence* $(r_1, \dots, r_m) \in R^m$ *is regular on* R, *and* $\mathfrak{p}$ *is a minimal prime divisor of* $Rr_1 + \cdots + Rr_m$, *then* $\mathrm{ht}(\mathfrak{p}) = m$.

This follows from Theorem 1.5.7 and an easy induction on m, using:

LEMMA 1.5.11. *Let* $\mathfrak{p} \in \mathrm{Spec}(R)$, *suppose* $x \in \mathfrak{p}$ *is not a zero divisor, and set* $\overline{R} := R/Rx$, $\overline{\mathfrak{p}} := \mathfrak{p}/Rx \in \mathrm{Spec}(\overline{R})$. *Then* $\mathrm{ht}(\mathfrak{p}) \geqslant \mathrm{ht}(\overline{\mathfrak{p}}) + 1$.

PROOF. For $n = \mathrm{ht}(\overline{\mathfrak{p}})$ we get a strictly increasing sequence $\mathfrak{p}_0 \subset \mathfrak{p}_1 \subset \cdots \subset \mathfrak{p}_n = \mathfrak{p}$ of prime ideals of R with $Rx \subseteq \mathfrak{p}_0$. Corollary 1.1.15 and $x \in \mathfrak{p}_0$ give that $\mathfrak{p}_0$ is not a minimal prime ideal of R. Thus $\mathrm{ht}(\mathfrak{p}) \geqslant n + 1$. □

The Generalized Principal Ideal Theorem has a converse:

PROPOSITION 1.5.12. *Let* $\mathfrak{p}$ *be a prime ideal of* R *and* $n = \mathrm{ht}(\mathfrak{p})$. *Then there are* $x_1, \dots, x_n \in \mathfrak{p}$ *such that* $\mathfrak{p}$ *is a minimal prime divisor of* $Rx_1 + \cdots + Rx_n$.

PROOF. By induction on n. The case $n = 0$ being trivial, suppose $n \geqslant 1$. Let $\mathfrak{q}_1, \dots, \mathfrak{q}_m$ be the minimal prime ideals of R; then $\mathfrak{p} \not\subseteq \mathfrak{q}_j$ for $j = 1, \dots, m$. By Lemma 1.1.9 we get $x_1 \in \mathfrak{p}$ with $x_1 \notin \bigcup_{j=1}^{m} \mathfrak{q}_j$. Put $\overline{R} := R/Rx_1$ and $\overline{\mathfrak{p}} := \mathfrak{p}/Rx_1$. Then $\mathrm{ht}(\overline{\mathfrak{p}}) \leqslant n - 1$ by the choice of x_1. So inductively we have $x_2, \dots, x_n \in R$ with $\overline{\mathfrak{p}}$ a minimal prime divisor of the ideal of $\overline{R}$ generated by the cosets $x_j + Rx_1$, $j = 2, \dots, n$. Thus $\mathfrak{p}$ is a minimal prime divisor of $Rx_1 + \cdots + Rx_n$. □

COROLLARY 1.5.13. *Let* $\mathfrak{p}$ *be a prime ideal of* R *of height* n. *Let* $x \in \mathfrak{p}$ *and* $\overline{R} := R/Rx$, $\overline{\mathfrak{p}} := \mathfrak{p}/Rx$. *Then* $\mathrm{ht}(\overline{\mathfrak{p}}) = n$ *or* $\mathrm{ht}(\overline{\mathfrak{p}}) = n - 1$.

PROOF. By Lemma 1.5.1 it remains to show that $\mathrm{ht}(\overline{\mathfrak{p}}) \geqslant n-1$. Suppose that $\mathrm{ht}(\overline{\mathfrak{p}}) \leqslant n - 2$. Then $n \geqslant 2$, and Proposition 1.5.12 gives $x_1, \dots, x_{n-2} \in \mathfrak{p}$ such that $\overline{\mathfrak{p}}$ is a minimal prime divisor of the ideal of $\overline{R}$ generated by the cosets $x_j + Rx$, $j = 1, \dots, n-2$. Then $\mathfrak{p}$ is a minimal prime divisor of $Rx + Rx_1 + \cdots + Rx_{n-2}$, hence $\mathrm{ht}(\mathfrak{p}) \leqslant n - 1$ by Theorem 1.5.7, a contradiction. □

Notes and comments

The notion of Krull dimension and the Principal Ideal Theorem are due to Krull [226, 228]. It seems that Kronecker [225, p. 80] already knew the Generalized Principal Ideal Theorem for polynomial rings over a field. Not every noetherian ring has finite Krull dimension; see [301, Appendix A1].

1.6 REGULAR LOCAL RINGS

In this section $(R, \mathfrak{m})$ *is a noetherian local ring with residue field* $\boldsymbol{k} = R/\mathfrak{m}$.

Definition and basic properties

The **embedding dimension** of R is defined as $\operatorname{edim}(R) := \mu(\mathfrak{m})$, the minimal number of generators of the maximal ideal $\mathfrak{m}$ of R. Corollary 1.4.4 gives $\operatorname{edim}(R) = \dim_{\boldsymbol{k}} \mathfrak{m}/\mathfrak{m}^2$, and Corollary 1.5.9 says that $\dim(R) \leqslant \operatorname{edim}(R)$. We call $(R, \mathfrak{m})$ a **regular** local ring if $\dim(R) = \operatorname{edim}(R)$. If $\dim R = 0$, then R is regular iff R is a field.

PROPOSITION 1.6.1. *Every regular local ring is an integral domain.*

In the proof of this proposition we use:

LEMMA 1.6.2. *Suppose $(R, \mathfrak{m})$ is a regular local ring, and $x \in \mathfrak{m} \setminus \mathfrak{m}^2$. Then $\overline{R} := R/Rx$ is a regular local ring with $\dim R = 1 + \dim \overline{R}$.*

PROOF. We have $\dim R = \operatorname{edim} R = 1 + \operatorname{edim} \overline{R} \geqslant 1 + \dim \overline{R}$, using Corollary 1.4.5 for the second equality. Corollary 1.5.13 gives $\dim R \leqslant 1 + \dim \overline{R}$. □

PROOF OF PROPOSITION 1.6.1. Let $(R, \mathfrak{m})$ be a regular local ring; we show by induction on the dimension n of R that R is a domain. This is clear if $\dim R = 0$, so assume $n \geqslant 1$. Let $\mathfrak{p}_1, \ldots, \mathfrak{p}_m$ be the minimal prime ideals of R. If $\mathfrak{m} \setminus \mathfrak{m}^2 \subseteq \bigcup_{i=1}^m \mathfrak{p}_i$, then Corollary 1.1.10 gives $\mathfrak{m} \subseteq \mathfrak{p}_i$ for some $i \in \{1, \ldots, m\}$, which contradicts $\operatorname{ht}(\mathfrak{m}) = n \geqslant 1$. Take $x \in \mathfrak{m} \setminus \mathfrak{m}^2$ with $x \notin \bigcup_{i=1}^m \mathfrak{p}_i$. By Lemma 1.6.2, $\overline{R} := R/Rx$ is a regular local ring of dimension $n - 1$, hence an integral domain by inductive hypothesis. So Rx is a prime ideal of R, and we can take $i \in \{1, \ldots, m\}$ with $\mathfrak{p}_i \subseteq Rx$. We claim that $\mathfrak{p}_i = \{0\}$ (and hence R is an integral domain). To see this, let $a \in \mathfrak{p}_i$, and take $r \in R$ such that $a = rx$. Since $x \notin \mathfrak{p}_i$, we have $r \in \mathfrak{p}_i$, so $a = rx \in \mathfrak{p}_i\mathfrak{m}$. This yields $\mathfrak{p}_i = \mathfrak{m}\mathfrak{p}_i$, hence $\mathfrak{p}_i = \{0\}$ by Nakayama's Lemma. □

COROLLARY 1.6.3. *Suppose $(R, \mathfrak{m})$ is a regular local ring. Let $I \neq R$ be an ideal of R. Then the following are equivalent:*

(i) *there is a minimal set G of generators of $\mathfrak{m}$ with distinct $x_1, \ldots, x_m \in G$ such that $I = Rx_1 + \cdots + Rx_m$;*

(ii) *R/I is a regular local ring (and thus I is a prime ideal of R).*

In that case and with m as in (i), we have $\dim R = m + \dim R/I$.

PROOF. Let m be as in (i). Then induction on m and Lemma 1.6.2 gives (i) ⇒ (ii) and $\dim R = m + \dim R/I$. To show (ii) ⇒ (i), suppose R/I is a regular local ring. Let $x_1, \ldots, x_d \in \mathfrak{m}$, where $d := \dim R/I = \operatorname{edim} R/I$, be such that $x_1 + I, \ldots, x_d + I$ are the distinct elements of a minimal set of generators of the maximal ideal $\mathfrak{m}/I$ of R/I. The natural exact sequence of $\boldsymbol{k}$-linear maps

$$0 \longrightarrow I/I \cap \mathfrak{m}^2 \longrightarrow \mathfrak{m}/\mathfrak{m}^2 \longrightarrow (\mathfrak{m}/I)/(\mathfrak{m}/I)^2 \longrightarrow 0$$

and Corollary 1.4.4 yield $x_{d+1}, \ldots, x_n \in I$ such that $x_1, \ldots, x_n$ are the distinct elements of a minimal set of generators of $\mathfrak{m}$. Put $I' := Rx_{d+1} + \cdots + Rx_n$. By what we showed before, $R' := R/I'$ is a regular local ring of dimension d and I' is a prime ideal of R. Since $I' \subseteq I$, this yields $I' = I$. □

Next, a useful characterization of regularity:

PROPOSITION 1.6.4. *Assume R is not a field. Then the local ring $(R, \mathfrak{m})$ is regular if and only if there is an $n \geqslant 1$ and a sequence $\vec{r} = (r_1, \ldots, r_n) \in R^n$ such that $\vec{r}$ is regular on R and $\mathfrak{m} = Rr_1 + \cdots + Rr_n$. Moreover, if $r_1, \ldots, r_n$ are the distinct elements of a minimal set of generators of $\mathfrak{m}$ and $\dim R = n$, then the sequence $(r_1, \ldots, r_n)$ is regular on R.*

PROOF. Suppose $n \geqslant 1$, the sequence $(r_1, \ldots, r_n) \in R^n$ is regular on R, and $\mathfrak{m} = Rr_1 + \cdots + Rr_n$. Then by Corollaries 1.5.9 and 1.5.10

$$\dim(R) = \operatorname{ht}(\mathfrak{m}) = n \geqslant \operatorname{edim}(R) \geqslant \dim(R),$$

so $\dim(R) = \operatorname{edim}(R)$. Conversely, suppose that $(R, \mathfrak{m})$ is a regular local ring and $r_1, \ldots, r_n$ are the distinct elements of a minimal set of generators of $\mathfrak{m}$. We show by induction on n that then the sequence $(r_1, \ldots, r_n)$ is regular on R. Since R is an integral domain, r_1 is not a zero divisor on R. Also, $r_1 \in \mathfrak{m} \setminus \mathfrak{m}^2$, so $\overline{R} := R/Rr_1$ is regular of dimension $n-1$, by Lemma 1.6.2. We are done if $n = 1$, so let $n \geqslant 2$. Put $\overline{r} := r + Rr_1$ for $r \in R$. Then $\overline{r_2}, \ldots, \overline{r_n}$ are the distinct elements of a minimal set of generators of the maximal ideal $\mathfrak{m}/Rr_1$ of $\overline{R}$. Hence the sequence $(\overline{r_2}, \ldots, \overline{r_n})$ is regular on $\overline{R}$, by inductive hypothesis. Thus $(r_1, \ldots, r_n)$ is regular on R, by (1.4.1). □

Examples of regular local rings

First the 1-dimensional case:

LEMMA 1.6.5. *R is regular and $\dim(R) = 1 \iff R$ is a DVR.*

PROOF. The direction $\Leftarrow$ is clear. Assume R is regular (hence an integral domain) and $\dim(R) = 1$. Then $\mathfrak{m} = tR$ with $0 \neq t \in R$, so R is a DVR by Lemma 1.4.7. □

PROPOSITION 1.6.6. *Let R be a regular local ring, X an indeterminate over R and $\mathfrak{P}$ a prime ideal of $R[X]$ with $\mathfrak{P} \cap R = \mathfrak{m}$. Then $R[X]_{\mathfrak{P}}$ is a regular local ring.*

PROOF. Set $n := \dim R$. If $n = 0$, then R is a field, and so $R[X]_{\mathfrak{P}}$ is either a field (iff $\mathfrak{P} = \{0\}$) or a DVR. Assume $n \geqslant 1$. Then Proposition 1.6.4 gives a sequence $(r_1, \ldots, r_n) \in R^n$ that is regular on R with $\mathfrak{m} = Rr_1 + \cdots + Rr_n$. Then the desired result follows from Corollary 1.4.20, Lemma 1.4.18, and Proposition 1.6.4. □

The following consequence will be used in Section 5.9:

COROLLARY 1.6.7. *Let K be a PID, $X_1, \ldots, X_n$ distinct indeterminates, and $\mathfrak{P}$ a prime ideal of $K[X_1, \ldots, X_n]$. Then $K[X_1, \ldots, X_n]_{\mathfrak{P}}$ is a regular local ring.*

PROOF. By induction on n. The case $n = 0$ reduces to Lemma 1.6.5. Assume $n \geqslant 1$, put $A := K[X_1, \ldots, X_{n-1}]$ and $\mathfrak{p} := A \cap \mathfrak{P}$. Inductively, $A_{\mathfrak{p}}$ is a regular local ring. Also, $K[X_1, \ldots, X_n]_{\mathfrak{P}} = A_{\mathfrak{p}}[X_n]_{\mathfrak{Q}}$ for some $\mathfrak{Q} \in \operatorname{Spec}(A_{\mathfrak{p}}[X_n])$, by Corollary 1.4.14. Hence $K[X_1, \ldots, X_n]_{\mathfrak{P}}$ is a regular local ring by Proposition 1.6.6. □

Notes and comments

Regular local rings were studied by Krull [230], which contains Proposition 1.6.1. The terminology comes from [76]. The algebraic-geometric significance of regular local rings is largely due to Zariski [466].

1.7 MODULES AND DERIVATIONS

The rest of this chapter is dominated by the general idea of linearization, and by the useful dualities that come with it. Thus we consider tensor products of modules, derivations, and (dually) differentials. *Throughout this section we fix a commutative ring R and let M, N range over R-modules.*

Modules of morphisms

Let A, B, C be R-modules. We denote by $\mathrm{Hom}_R(A, M)$ (or $\mathrm{Hom}(A, M)$ if R is clear from the context) the set of R-linear maps $A \to M$, made into an R-module in the natural way. Let $\phi\colon A \to B$ be R-linear. Then

$$\beta \mapsto \beta \circ \phi\colon \mathrm{Hom}(B, M) \to \mathrm{Hom}(A, M)$$

is an R-linear map, denoted by ϕ^* or $\mathrm{Hom}(\phi, M)$. If $\psi\colon B \to C$ is also R-linear, then $(\psi \circ \phi)^* = \phi^* \circ \psi^*$.

LEMMA 1.7.1. *The following are equivalent for ϕ and R-linear $\psi\colon B \to C$:*

(i) *the sequence $A \xrightarrow{\phi} B \xrightarrow{\psi} C \longrightarrow 0$ is exact;*

(ii) *for each M the sequence*

$$\mathrm{Hom}(A, M) \xleftarrow{\phi^*} \mathrm{Hom}(B, M) \xleftarrow{\psi^*} \mathrm{Hom}(C, M) \longleftarrow 0$$

of induced morphisms is exact.

PROOF. The direction (i) $\Rightarrow$ (ii) is entirely routine. Assume (ii). Exactness at C in (i) means surjectivity of ψ. Let $\alpha\colon C \to C/\operatorname{im}\psi$ be the canonical map. Since $\alpha \circ \psi = 0$, exactness in (ii) at $\mathrm{Hom}(C, M)$ for $M = C/\operatorname{im}\psi$ gives $\alpha = 0$, so $\operatorname{im}\psi = C$. Next, applying $(\psi \circ \phi)^* = 0$ to id_C gives $\psi \circ \phi = 0$. It only remains to show that $\operatorname{im}\phi \supseteq \ker\psi$. Let $\beta\colon B \to M := B/\operatorname{im}\phi$ be the canonical map. Then $\beta \circ \phi = 0$, so $\beta = \gamma \circ \psi$ for some $\gamma \in \mathrm{Hom}(C, M)$, so $\ker\psi \subseteq \ker\beta = \operatorname{im}\phi$. □

Tensor products

The tensor product $M \otimes_R N$ of the R-modules M and N is an R-module with an R-bilinear map

$$(x, y) \mapsto x \otimes_R y \,:\, M \times N \to M \otimes_R N$$

that is universal: for any R-module B and R-bilinear map $\beta\colon M \times N \to B$ there is a unique R-linear $b\colon M \otimes_R N \to H$ such that $b(x \otimes y) = \beta(x, y)$ for all $x \in M$,

$y \in N$. For use below we recall the construction of $M \otimes N$: Let x, x' range over M, y, y' over N, and r over R. Let F be the free R-module with basis $M \times N$, and let G be the submodule of F generated by all elements of the form

$$\begin{aligned} (x+x',y)-(x,y)-(x',y), &\qquad (x,y+y')-(x,y)-(x,y'), \\ (rx,y)-r(x,y), &\qquad (x,ry)-r(x,y). \end{aligned}$$

Put $M \otimes_R N := F/G$; then

$$(x,y) \mapsto x \otimes y := (x,y)+G : \ M \times N \to M \otimes N$$

is an R-bilinear map having the desired universal property. We drop the subscript R in expressions like $M \otimes_R N$ and $x \otimes_R y$ when R is clear from the context. We shall need the following variant of the universal property of $M \otimes N = M \otimes_R N$:

LEMMA 1.7.2. *Let $\phi\colon M \times N \to Z$ be a biadditive map into an abelian group Z such that $\phi(rx,y) = \phi(x,ry)$ for all $r \in R$, $x \in M$, $y \in N$. Then there is a unique additive map $f\colon M \otimes N \to Z$ such that $f(x \otimes y) = \phi(x,y)$ for all $x \in M$, $y \in N$.*

PROOF. With the notations in the above construction of $M \otimes N$, we have the internal direct sum decomposition $F = \bigoplus_{(x,y)} R \cdot (x,y)$, and for given x, y, the map $r \mapsto \phi(rx,y)\colon R \to Z$ is additive. Thus we have an additive map $\widehat{\phi}\colon F \to Z$ with $\widehat{\phi}\big(r(x,y)\big) = \phi(rx,y)$ for all r, x, y. Clearly $G \subseteq \ker\widehat{\phi}$, so $\widehat{\phi}$ induces an additive map $f\colon M \otimes N = F/G \to Z$ such that $f(x \otimes y) = \phi(x,y)$ for all x, y. There can be at most one such map f, since every element of $M \otimes N$ is a finite sum of elements of the form $x \otimes y$. □

We use this variant to construct derivations. A **derivation** on R is a map $\partial\colon R \to R$ such that $\partial(a+b) = \partial(a) + \partial(b)$ and $\partial(ab) = \partial(a)b + a\partial(b)$ for all $a, b \in R$. The map $R \to R$ sending every element of R to 0 is a derivation on R, called the **trivial** derivation (on R). If ∂ is a derivation on R and $a \in R$, then $a\partial$ is also a derivation on R. Below ∂ is a derivation on R. It gives a subring $\{a \in R : \partial(a) = 0\}$ of R. If a set $X \subseteq R$ generates a subring R_0 of R with $\partial(X) \subseteq R_0$, then $\partial(R_0) \subseteq R_0$. A **$\partial$-compatible derivation** on M is an additive map $d\colon M \to M$ such that $d(rx) = \partial(r)x + rd(x)$ for all $r \in R$ and $x \in M$.

COROLLARY 1.7.3. *Let d_M and d_N be ∂-compatible derivations on M and N. Then there is a unique ∂-compatible derivation δ on $M \otimes N$ such that*

$$\delta(x \otimes y) = d_M(x) \otimes y + x \otimes d_N(y) \quad \textit{for all } x \in M,\ y \in N. \tag{1.7.1}$$

PROOF. The map

$$(x,y) \mapsto \phi(x,y) := d_M(x) \otimes y + x \otimes d_N(y) : \ M \times N \to M \otimes N$$

is biadditive and satisfies $\phi(rx,y) = \phi(x,ry)$ for all $r \in R$, $x \in M$, $y \in N$. Hence Lemma 1.7.2 yields an additive map $\delta\colon M \otimes N \to M \otimes N$ satisfying (1.7.1), and this map is easily checked to be a ∂-compatible derivation on $M \otimes N$. □

Now let A be a commutative R-algebra. Then $A \otimes_R M$ is not only an R-module, but even an A-module inducing the given R-module structure, with

$$a \cdot (b \otimes x) \;=\; ab \otimes x \qquad (a, b \in A,\ x \in M).$$

Let (x_i) be a family of elements in M. If the R-module M is free on (x_i), then the A-module $A \otimes_R M$ is free on $(1 \otimes x_i)$; the converse holds if the structural morphism $R \to A$ is injective.

COROLLARY 1.7.4. *Let ∂_A be a derivation on A such that $\partial_A(r1_A) = \partial(r)1_A$ for all $r \in R$. Then ∂_A is a ∂-compatible derivation on the R-module A, and if d is a ∂-compatible derivation on M, then the ∂-compatible derivation δ on $A \otimes_R M$ with*

$$\delta(a \otimes x) \;=\; \partial_A(a) \otimes x + a \otimes d(x) \quad \textit{for all } a \in A,\ x \in M$$

is ∂_A-compatible.

Let B be a second commutative R-algebra. Then the A-module $A \otimes_R B$ is even a (commutative) A-algebra with multiplicative identity $1_A \otimes 1_B$ and multiplication given by

$$(a_1 \otimes b_1) \cdot (a_2 \otimes b_2) \;=\; a_1a_2 \otimes b_1b_2 \qquad (a_1, a_2 \in A,\ b_1, b_2 \in B).$$

We have ring morphisms $a \mapsto a \otimes 1_B\colon A \to A \otimes B$ and $b \mapsto 1_A \otimes b\colon B \to A \otimes B$, which are injective if R is a field and $1_A \neq 0$, $1_B \neq 0$.

COROLLARY 1.7.5. *Let ∂_A and ∂_B be derivations on the rings A and B, respectively, such that $\partial_A(r1_A) = \partial(r)1_A$ and $\partial_B(r1_B) = \partial(r)1_B$ for all $r \in R$. Then there is a unique derivation δ on the ring $A \otimes_R B$ such that*

$$\delta(a \otimes b) \;=\; \partial_A(a) \otimes b + a \otimes \partial_B(b) \quad \textit{for all } a \in A,\ b \in B.$$

Some useful isomorphisms

Let A be a commutative R-algebra. Let E be an A-module, viewed as an R-module via the structural morphism $R \to A$. For an R-linear map $\phi\colon M \to E$ and $a \in A$ we define the R-linear map $a\phi\colon M \to E$ by $a\phi(x) := a \cdot \phi(x)$ $(x \in M)$. This makes $\mathrm{Hom}_R(M, E)$ an A-module inducing its given R-module structure. For any R-linear $\phi\colon M \to E$ we have the A-linear map

$$\phi_A : A \otimes_R M \to E, \qquad \phi_A(a \otimes x) = a\phi(x) \quad (a \in A,\ x \in M),$$

and it is routine to check that the resulting map

$$(1.7.2) \qquad \phi \mapsto \phi_A : \mathrm{Hom}_R(M, E) \to \mathrm{Hom}_A(A \otimes_R M, E)$$

is an isomorphism of A-modules.

Let I be an ideal of R and consider $A := R/I$ as an R-algebra in the obvious way. Then M/IM is naturally an A-module, and we have an A-linear map

$$(1.7.3) \qquad x + IM \mapsto 1_A \otimes x : M/IM \to A \otimes_R M \qquad (x \in M).$$

This map (1.7.3) is an isomorphism of A-modules: its inverse is the A-linear map $A \otimes_R M \to M/IM$ induced by the R-bilinear map $A \times M \to M/IM$ that sends $(r + I, x)$ to $rx + IM$ for $r \in R$, $x \in M$.

Let S be a multiplicative subset of R, and consider $A := S^{-1}R$ as an R-algebra via the canonical map $\iota\colon R \to S^{-1}R = A$. Then we have an isomorphism

$$(1.7.4) \qquad A \otimes_R M \to S^{-1}M$$

of A-modules that sends $(r/s) \otimes x$ to $(rx)/s$ for $r \in R$, $s \in S$, $x \in M$.

Let $h\colon A \to B$ be a morphism between commutative R-algebras A and B. We have the A-module $A \otimes_R M$, and so the B-module $B \otimes_A (A \otimes_R M)$, by construing B as an A-algebra via h. We also have a B-linear map

$$(1.7.5) \quad B \otimes_R M \to B \otimes_A (A \otimes_R M), \qquad b \otimes_R x \mapsto b \otimes_A (1 \otimes_R x) \text{ for } x \in M.$$

It is easy to check (by explicit construction of an inverse) that this map is in fact an isomorphism of B-modules: *transitivity of base change*.

Rational rank

Let M be an abelian group, in other words, a $\mathbb{Z}$-module. Then we define the **rational rank** $\operatorname{rank}_{\mathbb{Q}} M$ of M by

$$\operatorname{rank}_{\mathbb{Q}} M := \dim_{\mathbb{Q}} \mathbb{Q} \otimes_{\mathbb{Z}} M,$$

if the $\mathbb{Q}$-linear space $\mathbb{Q} \otimes_{\mathbb{Z}} N$ has finite dimension, and set $\operatorname{rank}_{\mathbb{Q}} M := \infty$ otherwise. Using for example (1.7.4) it follows that this rational rank is the largest m for which there are $\mathbb{Z}$-independent $x_1, \dots, x_m \in M$, if such a largest m exists, and is ∞ otherwise. In particular, the free abelian group $\mathbb{Z}^m$ has rational rank m. It is easy to check that the rational rank, when restricted to abelian groups of finite rational rank, yields an Euler-Poincaré map with values in $\mathbb{Z}$.

Independence at a prime

Let $f_i \in M$ for $i \in I$. Given $\mathfrak{p} \in \operatorname{Spec}(R)$, the family (f_i) is said to be **independent at** $\mathfrak{p}$ if the family $(f_i + \mathfrak{p}M)$ in the $R/\mathfrak{p}$-module $M/\mathfrak{p}M$ is linearly independent over $R/\mathfrak{p}$. The fact below is used in Section 5.9:

LEMMA 1.7.6. *Suppose the R-module M is free and $\mathfrak{p}, \mathfrak{q} \in \operatorname{Spec}(R)$, $\mathfrak{p} \supseteq \mathfrak{q}$. If (f_i) is independent at $\mathfrak{p}$, then (f_i) is independent at $\mathfrak{q}$.*

PROOF. Let $f_1, \dots, f_m \in M$ with $m \geqslant 1$ be such that $f_1 + \mathfrak{p}M, \dots, f_m + \mathfrak{p}M$ are linearly independent over $R/\mathfrak{p}$. It suffices to show that then $f_1 + \mathfrak{q}M, \dots, f_m + \mathfrak{q}M$ are linearly independent over $R/\mathfrak{q}$. Take a basis of M and take distinct $e_1, \dots, e_n$ from that basis such that for $i = 1, \dots, m$ we have $f_i = f_{i1}e_1 + \cdots + f_{in}e_n$ with all $f_{ij} \in R$. The vectors $e_1 + \mathfrak{p}M, \dots, e_n + \mathfrak{p}M$ of $M/\mathfrak{p}M$ are linearly independent over $R/\mathfrak{p}$, so $m \leqslant n$, and some $m \times m$ submatrix of (f_{ij}) has determinant $D \notin \mathfrak{p}$. Then $D \notin \mathfrak{q}$, so $f_1 + \mathfrak{q}M, \dots, f_m + \mathfrak{q}M$ are linearly independent over $R/\mathfrak{q}$. □

Notes and comments

Lemma 1.7.6 is taken from [205].

1.8 DIFFERENTIALS

Below we define Kähler differentials and prove some basic facts about them, as needed in Section 5.9. *Throughout, K is a commutative ring, A is a commutative K-algebra, and M is an A-module.* (Later we only use the case where K is a field of characteristic zero, but that doesn't simplify things.) We let a, b range over A.

K-derivations

A map $\Delta\colon A \to M$ is said to be a **K-derivation** if Δ is K-linear and $\Delta(ab) = a\Delta(b) + b\Delta(a)$ for all a, b. A K-derivation $A \to A$ is called a **K-derivation on A.** So every K-derivation on A is a derivation on A in the sense of Section 1.7, and a derivation on a commutative ring R is the same thing as a $\mathbb{Z}$-derivation on R, where R is viewed as a $\mathbb{Z}$-algebra.

In the rest of this subsection $\Delta\colon A \to M$ is a K-derivation, so $\Delta(\kappa \cdot 1) = 0$ for all $\kappa \in K$, and $A^{\Delta} := \ker \Delta$ is a subalgebra of A. The rules below for computing with Δ are easy to verify:

LEMMA 1.8.1. *For all $n \geqslant 1$ and $a_1, \dots, a_n \in A$,*

$$\Delta(a_1 \cdots a_n) = \sum_{j=1}^{n} a_1 \cdots a_{j-1} a_{j+1} \cdots a_n \,\Delta(a_j).$$

In particular, $\Delta(a^n) = na^{n-1}\Delta(a)$ for $n \geqslant 1$, and if I is an ideal of A and $n \geqslant 1$, then $\Delta(I^n) \subseteq I^{n-1}M$.

Let $X = (X_1, \dots, X_n)$ be a tuple of distinct indeterminates, $n \geqslant 1$. Consider a polynomial $P = \sum_{\boldsymbol{i}} P_{\boldsymbol{i}} X^{\boldsymbol{i}} \in A[X]$ with $\boldsymbol{i} = (i_1, \dots, i_n)$ ranging over a finite subset of $\mathbb{N}^n$ and with $P_{\boldsymbol{i}} \in A$ and $X^{\boldsymbol{i}} := X_1^{i_1} \cdots X_n^{i_n}$ for all $\boldsymbol{i}$.

LEMMA 1.8.2. *For $\boldsymbol{a} = (a_1, \dots, a_n) \in A^n$, and setting $\boldsymbol{a}^{\boldsymbol{i}} := a_1^{i_1} \cdots a_n^{i_n}$,*

$$\Delta(P(\boldsymbol{a})) = \sum_{\boldsymbol{i}} \boldsymbol{a}^{\boldsymbol{i}} \Delta(P_{\boldsymbol{i}}) + \sum_{j=1}^{n} \frac{\partial P}{\partial X_j}(\boldsymbol{a}) \Delta(a_j).$$

In particular, for polynomials over K:

COROLLARY 1.8.3. *Given $F \in K[X]$ and $\boldsymbol{a} = (a_1, \dots, a_n) \in A^n$, we have*

$$\Delta(F(\boldsymbol{a})) = \sum_{j=1}^{n} \frac{\partial F}{\partial X_j}(\boldsymbol{a})\, \Delta(a_j).$$

If A is a field, then A^{Δ} is a subfield of A:

The universal K-derivation

A **universal** K-derivation of A is an A-module Ω together with a K-derivation $\mathrm{d}\colon A \to \Omega$ such that for any A-module M and K-derivation $\Delta\colon A \to M$ there is a unique A-linear map $\phi\colon \Omega \to M$ with $\Delta = \phi \circ \mathrm{d}$:

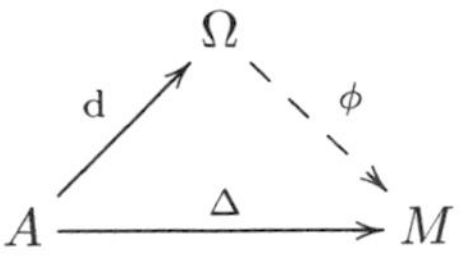

If $\mathrm{d}_1\colon A \to \Omega_1$ and $\mathrm{d}_2\colon A \to \Omega_2$ are universal K-derivations of A, then the unique A-linear map $\phi\colon \Omega_1 \to \Omega_2$ with $\mathrm{d}_2 = \phi \circ \mathrm{d}_1$ is an isomorphism of A-modules, by the usual argument. So there is at most one universal K-derivation of A up to canonical isomorphism. To get a universal K-derivation of A, take some bijection $d\colon A \to d(A)$ and let F be the free A-module on the set $d(A)$. Let G be the A-submodule of F generated by the elements

$$d(ab) - ad(b) - bd(a), \quad d(a+b) - d(a) - d(b), \quad d(\kappa a) - \kappa d(a)\ (\kappa \in K).$$

Let Ω be the quotient module F/G, and let $\mathrm{d}\colon A \to \Omega$ be given by $\mathrm{d}(a) := d(a) + G$. It is clear that then Ω with $\mathrm{d}\colon A \to \Omega$ is a universal K-derivation of A. For definiteness, we set $\Omega_{A|K} := \Omega$ and let $\mathrm{d}\colon A \to \Omega_{A|K}$ be the universal K-derivation just defined. Note that the A-module $\Omega_{A|K}$ is generated by its elements $\mathrm{d}(a)$. Instead of $\mathrm{d}(a)$, we also write $\mathrm{d}\,a$, or $\mathrm{d}_{A|K}(a)$ if we need to indicate the dependence on A and K. The A-module $\Omega_{A|K}$ is called the module of (Kähler) **differentials** of the K-algebra A, and $\mathrm{d}\,a$ is called the **differential** of a.

It may help to think of the elements of A as K-valued functions on a space, with $\Omega_{A|K}$ as a kind of cotangent bundle for this space. In any case, the K-derivation $\mathrm{d}\colon A \to \Omega_{A|K}$ can be useful in linearizing problems involving A.

Note that if A is generated as a K-algebra by its elements a_i with $i \in I$, then by Corollary 1.8.3 the A-module $\Omega_{A|K}$ is generated by the differentials $\mathrm{d}\,a_i$ of these elements. Using both Corollary 1.8.3 and the Quotient Rule 1.8.4 we get:

COROLLARY 1.8.10. *If K is a field and $L = K(x_1, \dots, x_n)$ a field extension of K, then the vector space $\Omega_{L|K}$ over L is generated by $\mathrm{d}\,x_1, \dots, \mathrm{d}\,x_n$, so $\dim_L \Omega_{L|K} \leqslant n$.*

Each A-module M gives an isomorphism of A-modules

$$\text{(1.8.2)} \qquad \mathrm{Hom}_A(\Omega_{A|K}, M) \xrightarrow{\cong} \mathrm{der}_K(A, M), \qquad \phi \mapsto \phi \circ \mathrm{d}\,.$$

In particular we have an isomorphism of A-modules

$$\text{(1.8.3)} \qquad \Omega^*_{A|K} := \mathrm{Hom}_A(\Omega_{A|K}, A) \xrightarrow{\cong} \mathrm{der}_K(A), \qquad \phi \mapsto \phi \circ \mathrm{d},$$

so we have an A-bilinear map

$$\langle\ ,\ \rangle : \Omega_{A|K} \times \mathrm{der}_K(A) \to A$$

given by $\langle \omega, \Delta \rangle := \phi(\omega)$ for $\omega \in \Omega_{A|K}$, $\Delta \in \mathrm{der}_K(A)$, where $\phi\colon \Omega_{A|K} \to A$ is the A-linear map such that $\phi \circ \mathrm{d} = \Delta$. Thus for $\Delta \in \mathrm{der}_K(A)$,

$$\langle \mathrm{d}\, a, \Delta \rangle = \Delta(a). \tag{1.8.4}$$

LEMMA 1.8.11. *Suppose $A = K[X]$ where $X = (X_i)_{i \in I}$ is a family of distinct indeterminates over K. Then the A-module $\Omega_{A|K}$ is free with basis $(\mathrm{d}\, X_i)_{i \in I}$, and for $P \in A$ we have $\frac{\partial P}{\partial X_i} = 0$ for all but finitely many $i \in I$ and*

$$\mathrm{d}\, P = \sum_{i \in I} \frac{\partial P}{\partial X_i}\, \mathrm{d}\, X_i. \tag{1.8.5}$$

PROOF. Let i, j range over I. Corollary 1.8.3 yields (1.8.5) for each $P \in A$. Thus the A-module $\Omega_{A|K}$ is generated by the $\mathrm{d}\, X_i$. Next, $\partial/\partial X_j \in \mathrm{der}_K(A)$, and by (1.8.4) we have $\langle \mathrm{d}\, X_i, \partial/\partial X_j \rangle = \delta_{ij}$ (Kronecker delta). This "orthogonality" shows that the A-module $\Omega_{A|K}$ is free on $(\mathrm{d}\, X_i)_{i \in I}$. □

Let S be a multiplicative subset of A. The K-derivation $\mathrm{d}\colon A \to \Omega := \Omega_{A|K}$ yields by Lemma 1.8.5 a K-derivation

$$S^{-1}\mathrm{d} : S^{-1}A \to S^{-1}\Omega$$

into the $S^{-1}A$-module $S^{-1}\Omega$.

LEMMA 1.8.12. *The K-derivation $S^{-1}\mathrm{d}\colon S^{-1}A \to S^{-1}\Omega$ is a universal K-derivation of $S^{-1}A$. In particular, $\Omega_{S^{-1}A|K} \cong S^{-1}\Omega_{A|K}$ as $S^{-1}A$-modules.*

PROOF. Let N be an $S^{-1}A$-module and $D \in \mathrm{der}_K\big(S^{-1}A, N\big)$. The $S^{-1}A$-module $S^{-1}\Omega$ is generated by the $S^{-1}\mathrm{d}(a/1) = (\mathrm{d}\, a)/1$, so there is at most one $S^{-1}A$-linear map $\psi\colon S^{-1}\Omega \to N$ with $\psi \circ S^{-1}\mathrm{d} = D$; our job is to find such ψ. The K-derivation $\Delta := D \circ \iota\colon A \to N$ yields an A-linear $\phi\colon \Omega \to N$ with $\Delta := \phi \circ \mathrm{d}$. From ϕ we get an $S^{-1}A$-linear $\psi\colon S^{-1}\Omega \to N$ with $\psi(\omega/1) = \phi(\omega)$ for all $\omega \in \Omega$. This map ψ has the desired property. □

Next we give a useful alternative construction of the module of Kähler differentials of A. Consider the K-algebra $A \otimes_K A$, with identity $1 \otimes 1$ and multiplication given by

$$(a_1 \otimes a_2) \cdot (b_1 \otimes b_2) = a_1 b_1 \otimes a_2 b_2.$$

Then $A \otimes_K A$ is an A-algebra with structural morphism

$$a \mapsto a \otimes 1\colon A \to A \otimes_K A \quad (\text{so } a(b_1 \otimes b_2) = ab_1 \otimes b_2)$$

which induces its given K-algebra structure. We have an A-algebra morphism

$$\mu\colon A \otimes_K A \to A, \qquad \mu(a \otimes b) = ab.$$

Set $J := \ker \mu$. Then $1 \otimes a - a \otimes 1 \in J$ for all $a \in A$, and we get a K-linear map

$$d_A\colon A \to J/J^2, \qquad d_A(a) := (1 \otimes a - a \otimes 1) + J^2,$$

which is easily verified to be a K-derivation, using

$$1 \otimes ab - b \otimes a - a \otimes b + ab \otimes 1 \;=\; (1 \otimes a - a \otimes 1) \cdot (1 \otimes b - b \otimes 1) \;\in\; J^2.$$

The universal property of d thus yields an A-linear map

$$\Omega_{A|K} \to J/J^2 : \;\; \mathrm{d}\,a \mapsto d_A(a) \;=\; (1 \otimes a - a \otimes 1) + J^2.$$

LEMMA 1.8.13. *This map $\Omega_{A|K} \to J/J^2$ is an isomorphism of A-modules. (Thus the K-derivation d_A is universal.)*

PROOF. First, J is generated as an A-module by the $1 \otimes a - a \otimes 1$: let $s \in J$ and take $a_1, \ldots, a_n, b_1, \ldots, b_n \in A$ with $s = \sum_i a_i \otimes b_i$; then $\mu(s) = \sum_i a_i b_i = 0$, so

$$s \;=\; \sum_i a_i(1 \otimes b_i - b_i \otimes 1),$$

which proves the generation claim. Therefore, as an A-module, J/J^2 is generated by the $d_A(a)$. It remains to find an A-linear map $\phi\colon J/J^2 \to \Omega_{A|K}$ with $\phi \circ d_A = \mathrm{d}$. The K-bilinear map $(a, b) \to a\,\mathrm{d}\,b\colon A \times A \to \Omega_{A|K}$ yields a K-linear map $\psi\colon A \otimes_K A \to \Omega_{A|K}$ with $\psi(a \otimes b) = a\,\mathrm{d}\,b$ for all a, b. Then ψ is actually A-linear, $\psi(1 \otimes a - a \otimes 1) = \mathrm{d}\,a$ for all a, and ψ sends each product $(1 \otimes a - a \otimes 1) \cdot (1 \otimes b - b \otimes 1)$ to 0, and so ψ induces an A-linear map $\phi\colon J/J^2 \to \Omega_{A|K}$ with $\phi \circ d_A = \mathrm{d}$. □

COROLLARY 1.8.14. *Let ∂ be a derivation on K and ∂_A a derivation on A such that $\partial_A(\kappa 1) = \partial(\kappa)1$ for all $\kappa \in K$. Then there is a unique ∂_A-compatible derivation ∂^* on the A-module $\Omega_{A|K}$ such that $\partial^*(\mathrm{d}\,a) = \mathrm{d}\,\partial_A(a)$ for all a.*

PROOF. Since $\Omega_{A|K}$ is generated as an A-module by the $\mathrm{d}\,a$, there can be at most one such ∂^*. Corollary 1.7.5 yields a derivation δ on the ring $A \otimes_K A$ such that

$$\delta(a \otimes b) \;=\; \partial_A(a) \otimes b + a \otimes \partial_A(b) \qquad \text{for all } a,\, b.$$

Then $\delta(J) \subseteq J$, so $\delta(J^2) \subseteq J^2$, hence δ induces a ∂_A-compatible derivation ∂^* on J/J^2 with $\partial^*(d_A(a)) = d_A(\partial_A(a))$ for all a. Now use Lemma 1.8.13. □

The fundamental exact sequences

Let L be a commutative ring, B be a commutative L-algebra, and suppose we are given a commutative diagram

$$\begin{array}{ccc} A & \xrightarrow{\;h\;} & B \\ \uparrow & & \uparrow \\ K & \longrightarrow & L \end{array}$$

of ring morphisms, where the vertical arrows are the structural morphisms.

LEMMA 1.8.15. *With the B-module $\Omega_{B|L}$ construed as an A-module via h, there is exactly one A-linear map $\Omega_h \colon \Omega_{A|K} \to \Omega_{B|L}$ such that the diagram below commutes:*

$$\begin{array}{ccc} \Omega_{A|K} & \xrightarrow{\Omega_h} & \Omega_{B|L} \\ {\scriptstyle \mathrm{d}_{A|K}}\uparrow & & \uparrow{\scriptstyle \mathrm{d}_{B|L}} \\ A & \xrightarrow{h} & B \end{array}$$

PROOF. The map $\mathrm{d}_{B|L} \circ h$ is a K-derivation of A; so the existence and uniqueness of Ω_h follow from the universal property of $\mathrm{d}_{A|K}$. □

Now let B be a commutative K-algebra and $h\colon A \to B$ a K-algebra morphism. Then we are in the situation above with $L = K$ and the identity map $K \to L$, so Lemma 1.8.15 yields an A-linear map

$$\Omega_h : \Omega_{A|K} \to \Omega_{B|K}, \quad \Omega_h(\mathrm{d}\,a) = \mathrm{d}\,h(a),$$

which in turn (see previous section) yields a B-linear map

$$\alpha := (\Omega_h)_B \colon B \otimes_A \Omega_{A|K} \to \Omega_{B|K}, \quad \alpha(1 \otimes \mathrm{d}\,a) = \mathrm{d}\,h(a) \ \text{ for all } a.$$

Applying Lemma 1.8.15 to another diagram yields a B-linear map

$$\beta : \Omega_{B|K} \to \Omega_{B|A}, \qquad \beta(\mathrm{d}_{B|K}\, f) = \mathrm{d}_{B|A}\, f \ \text{ for } f \in B.$$

PROPOSITION 1.8.16 (First fundamental exact sequence). *The following sequence of B-modules and B-linear maps is exact:*

$$B \otimes_A \Omega_{A|K} \xrightarrow{\alpha} \Omega_{B|K} \xrightarrow{\beta} \Omega_{B|A} \longrightarrow 0$$

PROOF. Let E be a B-module. Then (1.7.2) yields an isomorphism

$$\mathrm{Hom}_A\big(\Omega_{A|K}, E\big) \xrightarrow{\cong} \mathrm{Hom}_B\big(B \otimes_A \Omega_{A|K}, E\big)$$

of B-modules. Identifying these two B-modules via this isomorphism, we have a commutative diagram of A-linear maps:

$$\begin{array}{ccccccc} \mathrm{Hom}_A(\Omega_{A|K}, E) & \xleftarrow{\alpha^*} & \mathrm{Hom}_B(\Omega_{B|K}, E) & \xleftarrow{\beta^*} & \mathrm{Hom}_B(\Omega_{B|A}, E) & \longleftarrow & 0 \\ \downarrow{\scriptstyle \cong} & & \downarrow{\scriptstyle \cong} & & \downarrow{\scriptstyle \cong} & & \\ \mathrm{der}_K(A, E) & \xleftarrow{\mathrm{der}_K(h,E)} & \mathrm{der}_K(B, E) & \xleftarrow{\supseteq} & \mathrm{der}_A(B, E) & \longleftarrow & 0 \end{array}$$

where the vertical arrows are isomorphisms given by (1.8.2) and the bottom row is exact by Lemma 1.8.8. Hence the top row is also exact. Now use Lemma 1.7.1. □

Next, let I be an ideal of A, set $B := A/I$, and let $h\colon A \to B$ be the canonical map. To determine $\ker \alpha$ in this case, first note that $\mathrm{d}_{A|K}(I^2) \subseteq I\Omega_{A|K}$. Composing the natural surjection $\Omega_{A|K} \to \Omega_{A|K}/I\Omega_{A|K}$ with the restriction of $\mathrm{d}_{A|K}$ to I yields an A-linear map $I \to \Omega_{A|K}/I\Omega_{A|K}$ whose kernel contains I^2, hence in turn induces a B-linear map $\gamma\colon I/I^2 \to \Omega_{A|K}/I\Omega_{A|K}$. Identifying the B-modules $\Omega_{A|K}/I\Omega_{A|K}$ and $B \otimes_A \Omega_{A|K}$ as in (1.7.3), we get

$$\gamma : I/I^2 \to B \otimes_A \Omega_{A|K}, \quad \gamma(a + I^2) = 1_B \otimes \mathrm{d}\, a \quad \text{for } a \in I.$$

PROPOSITION 1.8.17 (Second fundamental exact sequence). *The following sequence of B-modules and B-linear maps is exact:*

$$I/I^2 \xrightarrow{\gamma} B \otimes_A \Omega_{A|K} \xrightarrow{\alpha} \Omega_{B|K} \longrightarrow 0$$

PROOF. Let E be a B-module. As in the proof of Proposition 1.8.16 we get a commutative diagram of A-linear maps:

$$\begin{array}{ccccccc}
\mathrm{Hom}_B(I/I^2, E) & \xleftarrow{\gamma^*} & \mathrm{Hom}_A(\Omega_{A|K}, E) & \xleftarrow{\alpha^*} & \mathrm{Hom}_B(\Omega_{B|K}, E) & \longleftarrow & 0 \\
\Big\downarrow = & & \Big\downarrow \cong & & \Big\downarrow \cong & & \\
\mathrm{Hom}_B(I/I^2, E) & \xleftarrow{\overline{\Delta} \hookleftarrow \Delta} & \mathrm{der}_K(A, E) & \xleftarrow{\mathrm{der}_K(h,E)} & \mathrm{der}_K(B, E) & \longleftarrow & 0
\end{array}$$

with exact bottom row by Lemma 1.8.9. Now use Lemma 1.7.1. □

Let R be a commutative K-algebra, $\mathfrak{p} \in \mathrm{Spec}(R)$, and set $A = R_{\mathfrak{p}}$. Then A is a K-algebra and a local ring with maximal ideal $\mathfrak{m} = \mathfrak{p}A$. Let $F := A/\mathfrak{m}$ be its residue field. Then Lemma 1.8.12 and the isomorphisms (1.7.4) and (1.7.5) give

$$F \otimes_A \Omega_{A|K} \cong F \otimes_A (A \otimes_R \Omega_{R|K}) \cong F \otimes_R \Omega_{R|K} \qquad (F\text{-linear isomorphisms})$$

with $1 \otimes \mathrm{d}(r/1) \in F \otimes_A \Omega_{A|K}$ corresponding to $1 \otimes \mathrm{d}\, r \in F \otimes_R \Omega_{R|K}$ for $r \in R$. In view of Proposition 1.8.17 this leads to an exact sequence

$$\mathfrak{m}/\mathfrak{m}^2 \xrightarrow{\gamma_0} F \otimes_R \Omega_{R|K} \xrightarrow{\alpha_0} \Omega_{F|K} \longrightarrow 0 \tag{1.8.6}$$

of vector spaces over F and F-linear maps, with $\gamma_0\big((r/1) + \mathfrak{m}^2\big) = 1 \otimes \mathrm{d}\, r$ for $r \in \mathfrak{p}$ and $\alpha_0(1 \otimes \mathrm{d}\, r) = \mathrm{d}\, h(r/1)$ for $r \in R$, where $h\colon A \to F$ is the canonical map.

Notes and comments

The module $\Omega_{A|K}$ of differentials was introduced by Kähler [207]. The description of $\Omega_{A|K}$ via Lemma 1.8.13 is due to Cartier [71]; see [245] for the history. For more information on $\Omega_{A|K}$, see [244]. Corollary 1.8.14 is from [204]. If K is a field of characteristic zero, then the F-linear map γ_0 from (1.8.6) is injective; see [244, Corollary 6.5 (a)] or [288, Theorems 25.2 and 26.9]. See Chapter 12 below for K-derivations on non-commutative K-algebras.

1.9 DERIVATIONS ON FIELD EXTENSIONS

Let K be a field and L a field extension of K. Below we indicate ways to extend a derivation on K to a derivation on L. In the case that $\operatorname{char}(K) = 0$ we also relate bases for the vector space $\Omega_{L|K}$ over L to transcendence bases of L over K.

Extending derivations

In this subsection we fix a derivation ∂ on K. Let $X = (X_1, \dots, X_n)$ be a tuple of distinct indeterminates and let $P = \sum_{\boldsymbol{i}} P_{\boldsymbol{i}} X^{\boldsymbol{i}} \in K[X]$ be a polynomial: the $\boldsymbol{i} = (i_1, \dots, i_n)$ range over a finite subset of $\mathbb{N}^n$ and $P_{\boldsymbol{i}} \in K$, $X^{\boldsymbol{i}} := X_1^{i_1} \cdots X_n^{i_n}$ for all $\boldsymbol{i}$. Set $P^{\partial} := \sum_{\boldsymbol{i}} \partial(P_{\boldsymbol{i}}) X^{\boldsymbol{i}} \in K[X]$. It is easy to verify that $P \mapsto P^{\partial}$ is a derivation on $K[X]$ extending ∂. The following identity is a special case of Lemma 1.8.2 (but we allow $n = 0$ here) and is used frequently:

LEMMA 1.9.1. *For all $a = (a_1, \dots, a_n) \in K^n$,*

$$\partial(P(a)) \;=\; P^{\partial}(a) + \sum_{j=1}^{n} \frac{\partial P}{\partial X_j}(a) \cdot \partial(a_j).$$

LEMMA 1.9.2. *Suppose that L is separably algebraic over K. Then ∂ extends uniquely to a derivation on L.*

PROOF. Let ∂ be extended to a derivation on L. Let us denote this extended derivation also by ∂. For any $a \in L$ with minimum polynomial $P \in K[X]$ over K we have $P(a) = 0$ and $P'(a) \neq 0$, and hence by 1.9.1,

$$\partial(a) \;=\; \frac{-P^{\partial}(a)}{P'(a)}. \tag{1.9.1}$$

So there is at most one such derivation on L. To construct such a derivation, let $a \in L$; it suffices to get a derivation on $K[a] = K(a)$ extending ∂. Let $P \in K[X]$ be the minimum polynomial of a over K, so the ring morphism

$$Q \mapsto Q(a) \,:\, K[X] \to K[a]$$

has kernel $PK[X]$. Consider the additive map

$$d\colon K[X] \to K[a] = K(a), \qquad d(Q) \;=\; Q^{\partial}(a) + Q'(a)\frac{-P^{\partial}(a)}{P'(a)}.$$

An easy computation shows that $d(Q) = 0$ for $Q \in PK[X]$, so d induces an additive map $K[a] \to K[a]$ sending $Q(a)$ to $Q^{\partial}(a) + Q'(a)\frac{-P^{\partial}(a)}{P'(a)}$, for all $Q \in K[X]$. This last map is easily checked to be a derivation on $K[a]$ extending ∂. □

Let L be as in Lemma 1.9.2, and let ∂ be extended to a derivation on L. Then any subfield of L containing K is closed under this derivation of L, by (1.9.1).

LEMMA 1.9.3. *Let $x_1, \dots, x_n, y_1, \dots, y_n \in L$ be such that $x_1, \dots, x_n$ are algebraically independent over K, set $x = (x_1, \dots, x_n)$, and consider L as a $K[x]$-module via the inclusion $K[x] \to L$. Then there is a unique $\mathbb{Z}$-derivation $K[x] \to L$ that extends ∂ and sends x_j to y_j for $j = 1, \dots, n$.*

PROOF. By Lemma 1.8.2 there is at most one such extension. The map

$$P(x) \mapsto P^{\partial}(x) + \sum_{j=1}^{n} \frac{\partial P}{\partial X_j}(x) y_j \, : \, K[x] \to L$$

is easily verified to be an extension as required. □

Here is an easy consequence of Lemma 1.9.3 and Corollary 1.8.6:

COROLLARY 1.9.4. *Suppose that $L = K(x)$ where $x = (x_i)_{i \in I}$ is a family in L that is algebraically independent over K. Then there is for each family $(y_i)_{i \in I}$ in L a unique extension of ∂ to a derivation on L with $\partial(x_i) = y_i$ for all $i \in I$.*

LEMMA 1.9.5. *Suppose that K has characteristic $p > 0$ and $L = K(a)$ where $a^p = c \in K \setminus K^p$ and $\partial(c) = 0$. Then there is for each $b \in L$ a unique extension of ∂ to a derivation on L with $\partial(a) = b$.*

PROOF. Let $b \in L$. The minimum polynomial of a over K is $X^p - c$, so the ring morphism $K[X] \to K[a] \colon Q \mapsto Q(a)$ is surjective with kernel $(X^p - c)K[X]$. View $K[a] = K(a) = L$ as a $K[X]$-algebra via this morphism. Then the map

$$d \, : \, K[X] \to K[a], \qquad d(Q) := Q^{\partial}(a) + Q'(a)b,$$

is a $\mathbb{Z}$-derivation, and $d(X^p - c) = -\partial(c) = 0$, hence d induces a derivation on $K[a]$ sending a to b and extending ∂. Uniqueness is clear by Lemma 1.9.1. □

COROLLARY 1.9.6. *There exists an extension of ∂ to a derivation on L. There is a unique such extension iff L is separably algebraic over K.*

PROOF. Let B be a transcendence basis for $L|K$, and let F be the subfield of L consisting of the elements of L that are separably algebraic over $K(B)$. We can extend ∂ to a derivation on $K(B)$ by Corollary 1.9.4, in more than one way if $B \neq \emptyset$. Any such extension then extends further to a derivation on F by Lemma 1.9.2. If $F \neq L$, then any derivation on F extends in more than one way to a derivation on the purely inseparable extension L of F, by Lemma 1.9.5. □

The next result will be used in the proof of Proposition 3.3.16. For an n-tuple $P = (P_1, \dots, P_n) \in K[X]^n$ we define the $n \times n$ matrix

$$P' := \left(\frac{\partial P_i}{\partial X_j} \right)_{i,j=1,\dots,n} \qquad \text{(entries in } K[X]\text{)}.$$

Evaluating its entries at a point $a \in L^n$ gives the $n \times n$ matrix $P'(a)$ over L.

COROLLARY 1.9.7. *Let* $P = (P_1, \dots, P_n) \in K[X]^n$, $a = (a_1, \dots, a_n) \in L^n$ *be such that* $P(a) = 0$ *and* $\det P'(a) \neq 0$. *Then* $a_1, \dots, a_n$ *are separably algebraic over* K.

PROOF. Let $\Delta \in \operatorname{der}_K K(a)$. Then Corollary 1.8.3 and the assumptions $P(a) = 0$ and $\det P'(a) \neq 0$ give $\Delta(a_1) = \cdots = \Delta(a_n) = 0$, so $\Delta = 0$. Hence $K(a)|K$ is separably algebraic by Corollary 1.9.6 applied to the trivial derivation on K. □

Kähler differentials for field extensions

If L is separably algebraic over K, then $\Omega_{L|K} = \{0\}$ by Lemma 1.9.2 and the isomorphism (1.8.3). Below we generalize this fact for K of characteristic zero.

LEMMA 1.9.8. *Let* $(x_i)_{i \in I}$ *be a family in L. If the family* (x_i) *is algebraically independent over* K, *then the family* $(\mathrm{d}\, x_i)$ *in* $\Omega_{L|K}$ *is linearly independent over L. The converse holds when* $\operatorname{char}(K) = 0$.

PROOF. Suppose (x_i) is algebraically independent over K. Corollary 1.9.6 and its proof yield for each $i \in I$ a K-derivation ∂_i on L such that $\partial_i(x_j) = \delta_{ij}$ (Kronecker delta) for all $j \in I$. Thus $(\mathrm{d}\, x_i)$ is linearly independent over L by (1.8.4).

Next, let $\operatorname{char}(K) = 0$ and assume $x_1, \dots, x_n \in L$ are algebraically dependent over K. Set $x := (x_1, \dots, x_n)$ and let $P \in K[X_1, \dots, X_n]$ be a nonzero polynomial such that $P(x) = 0$, and of minimal degree with these properties. This minimality gives i with $\frac{\partial P}{\partial X_i} \neq 0$ and so $\frac{\partial P}{\partial X_i}(x) \neq 0$. Then by Corollary 1.8.3,

$$\sum_{i=1}^{n} \frac{\partial P}{\partial X_i}(x)\, \mathrm{d}\, x_i \;=\; \mathrm{d}\, P(x) \;=\; 0 \text{ in } \Omega_{L|K},$$

so $\mathrm{d}\, x_1, \dots, \mathrm{d}\, x_n$ are linearly dependent over L. □

In particular, if $\operatorname{char}(K) = 0$, then a family $(x_i)_{i \in I}$ in L is a transcendence basis for $L|K$ iff $(\mathrm{d}\, x_i)$ is a basis of the vector space $\Omega_{L|K}$ over L, and so

$$\dim_L \Omega_{L|K} \;=\; \operatorname{trdeg}(L|K).$$

In Section 5.9 we prove an analogue of this for extensions of differential fields.

Notes and comments

This section stems from A. Weil [461, Chapter I, §5].

Chapter Two

Valued Abelian Groups

Our main objects of interest are fields like $\mathbb{T}$ that are equipped with a compatible valuation and derivation. To analyze these objects we need valuation theory, and so we include two chapters containing the purely valuation-theoretic tools. Since these tools come from a variety of sources scattered over the literature and some are new, we include all but the most routine proofs.

After introducing some terminology concerning ordered sets in Section 2.1, this chapter treats valued abelian groups (Sections 2.2, 2.3) and ordered abelian groups (Section 2.4); the latter will occur as value groups of valued fields. Valued abelian groups arise in our work because the logarithmic derivative map on a valued differential field like $\mathbb{T}$ induces a valuation on the value group that turns out to be very useful. Moreover, the notion of a *pseudocauchy sequence* makes perfect sense in the general setting of valued abelian groups, and the basic facts about these sequences yield a natural proof of a generalized Hahn Embedding Theorem which can serve as a model for later proofs of several much deeper embedding theorems.

2.1 ORDERED SETS

This section serves mainly to fix notations and terminology. By convention *ordered set* means *totally ordered set* unless specified otherwise. This agrees with the usual meaning of *ordered abelian group* and *ordered field.*

Let S be an ordered set. We denote the ordering on S by $\leqslant$, the corresponding strict ordering by $<$, and the reversals of $\leqslant$ and $<$ by $\geqslant$ and $>$, respectively. We extend these notations to sets $A, B \subseteq S$ by $A \leqslant B :\Longleftrightarrow a \leqslant b$ for all $(a,b) \in A \times B$, and $A < B :\Longleftrightarrow a < b$ for all $(a,b) \in A \times B$. Also for $b \in S$, $A \leqslant b :\Longleftrightarrow A \leqslant \{b\}$, and so on. We view a subset of S as ordered by the induced ordering. We put $S_\infty := S \cup \{\infty\}$, $\infty \notin S$, with the ordering on S extended to a (total) ordering on S_∞ by $S < \infty$. Occasionally, we even take two distinct elements $-\infty, \infty \notin S$, and extend the ordering on S to an ordering on $S \cup \{-\infty, \infty\}$ by $-\infty < S < \infty$. For $A \subseteq S$ and $b \in S$ we set $A^{\leqslant b} := \{a \in A : a \leqslant b\}$; similarly for $<$, $\geqslant$, and $>$ instead of $\leqslant$. For $a, b \in S$ we put

$$[a,b] = [a,b]_S := \{x \in S : a \leqslant x \leqslant b\}.$$

A subset C of S is said to be **convex** in S if for all $a, b \in C$ we have $[a,b] \subseteq C$. For $A \subseteq S$ we let

$$\operatorname{conv}(A) := \{x \in S : a \leqslant x \leqslant b \text{ for some } a, b \in A\}$$

be the **convex hull of** A in S, that is, the smallest convex subset of S containing A. For $-\infty \leqslant a < b \leqslant \infty$ we also set

$$(a, b) = (a, b)_S := \{x \in S : a < x < b\}.$$

The sets of the form (a, b) are called **intervals** in S. The intervals in S form a basis for a hausdorff topology on S, the **interval topology** or **order topology** on S.

The ordered set S is said to be **dense** if for all $a < b$ in S we have $(a, b) \neq \emptyset$, and **without endpoints** if for all $a, b \in S$ we have $(a, \infty) \neq \emptyset$ and $(-\infty, b) \neq \emptyset$.

Let S' be also an ordered set and $f: S \to S'$ a map. Then f is said to be

(1) **increasing** if for all $a, b \in S$: $a \leqslant b \Rightarrow f(a) \leqslant f(b)$;

(2) **strictly increasing** if for all $a, b \in S$: $a < b \Rightarrow f(a) < f(b)$;

(3) **decreasing** if for all $a, b \in S$: $a \leqslant b \Rightarrow f(a) \geqslant f(b)$;

(4) **strictly decreasing** if for all $a, b \in S$: $a < b \Rightarrow f(a) > f(b)$.

An increasing bijection $S \to S'$ is called an **isomorphism** of ordered sets. We say that f has the **intermediate value property** if for all $a < b$ in S,

$$f(a) < f(b) \implies \big(f(a), f(b)\big)_{S'} \subseteq f\big((a, b)_S\big),$$
$$f(a) > f(b) \implies \big(f(b), f(a)\big)_{S'} \subseteq f\big((a, b)_S\big).$$

The ordered set S is said to be **well-ordered** if every nonempty subset of S has a smallest element; equivalently, there is no infinite sequence

$$a_0 > a_1 > \cdots > a_n > \cdots$$

in S. We say that two ordered sets **have the same order type** if they are isomorphic. If S is well-ordered then there is a unique ordinal number, denoted by $\operatorname{ot}(S)$, with the same order type as S. (As usual, an ordinal is construed here in von Neumann's sense as the set of all smaller ordinals.) If S is well-ordered and S' is an ordered subset of S, then S' is well-ordered with $\operatorname{ot}(S') \leqslant \operatorname{ot}(S)$.

LEMMA 2.1.1. *Let S' be an ordered set and $f: S \to S'$ increasing and surjective. If S is well-ordered, then so is S', with* $\operatorname{ot}(S) \geqslant \operatorname{ot}(S')$.

PROOF. For each $s' \in S'$ pick $g(s') \in f^{-1}(s')$. Then the map $g: S' \to S$ is strictly increasing, and the claims follow. □

A subset A of S is said to be a **cut** in S, or **downward closed** in S, if for all $a \in A$ and $s \in S$ we have $s < a \Rightarrow s \in A$. The empty subset of S and S itself are cuts in S; these are called the **trivial** cuts in S. The collection of cuts in an ordered set is totally ordered by inclusion (with largest element S and smallest element $\emptyset$), and the union and intersection of any set of cuts are also cuts. We say that an element x of an ordered set extending S **realizes** the cut A in S if $A < x < S \setminus A$.

Dually, $A \subseteq S$ is **upward closed** in S if for all $a \in A$ and $s \in S$ we have $a < s \Rightarrow s \in A$. (So $A \subseteq S$ is upward closed in S iff $S \setminus A$ is downward closed in S.) For $A \subseteq S$ we put

$$A^{\downarrow} := \{s \in S : s \leqslant a \text{ for some } a \in A\},$$
$$A^{\uparrow} := \{s \in S : a \leqslant s \text{ for some } a \in A\};$$

clearly $A^{\downarrow}$ is the smallest downward closed subset of S containing A, and $A^{\uparrow}$ is the smallest upward closed subset of S containing A. A subset A of S is said to be **cofinal** in S if $A^{\downarrow} = S$, i.e., if for each $s \in S$ there is $a \in A$ with $s \leqslant a$, and A is said to be **coinitial** in S if $A^{\uparrow} = S$, that is, if for each $s \in S$ there is $a \in A$ with $s \geqslant a$. The relation of cofinality is transitive: if $A \subseteq S$ is cofinal in S and $B \subseteq A$ is cofinal in A then B is cofinal in S. A map $f\colon S \to S'$ into an ordered set S' is said to be **cofinal** if $f(S)$ is cofinal in S'.

As a consequence of the Axiom of Choice, there always exists a well-ordered cofinal subset of S. The **cofinality** of S, denoted by $\operatorname{cf}(S)$, is defined to be the smallest ordinal λ such that there exists a cofinal subset of S of order type λ. This cofinality λ is actually a cardinal; here we identify as usual an ordinal α with the set of ordinals $< \alpha$, and a cardinal κ with the least ordinal of cardinality κ. Dually, the **coinitiality** of S, denoted by $\operatorname{ci}(S)$, is the cofinality of S equipped with the reversed ordering, that is, the smallest ordinal λ such that there exists a coinitial subset of S of reversed order type λ. Thus $\operatorname{cf}(\alpha) \leqslant \alpha$ for every ordinal α. An ordinal α such that $\operatorname{cf}(\alpha) = \alpha$ is called **regular.** Using transitivity of cofinality one easily sees that $\operatorname{cf}(S)$ is regular.

LEMMA 2.1.2. *Let S' be a cofinal subset of S. Then* $\operatorname{cf}(S') = \operatorname{cf}(S)$.

PROOF. Transitivity of cofinality gives $\operatorname{cf}(S') \geqslant \operatorname{cf}(S)$. To get $\operatorname{cf}(S) \geqslant \operatorname{cf}(S')$ we can replace S' by a well-ordered cofinal subset of order type $\operatorname{cf}(S')$ and assume that S' is well-ordered. Take a strictly increasing and cofinal sequence $\big(s(\gamma)\big)$ in S indexed by the ordinals $\gamma < \lambda := \operatorname{cf}(S)$. Define the sequence $\big(s'(\gamma)\big)$ in S' indexed by the same ordinals by

$$s'(\gamma) := \min\big\{s' \in S' : s' \geqslant s(\gamma)\big\}.$$

The sequence $\big(s'(\gamma)\big)$ is increasing, hence by Lemma 2.1.1 the order type of its image $s'(\lambda)$ satisfies $\lambda \geqslant \operatorname{ot}\big(s'(\lambda)\big)$. Since $s'(\lambda)$ is well-ordered and cofinal in S', we also have $\operatorname{ot}\big(s'(\lambda)\big) \geqslant \operatorname{cf}(S')$, and thus $\operatorname{cf}(S) = \lambda \geqslant \operatorname{cf}(S')$. □

COROLLARY 2.1.3. *Let α be a regular ordinal. Then every subset of α of order type α is cofinal in α.*

PROOF. Let β be an ordinal and S' a cofinal subset of β with $\operatorname{ot}(S') = \alpha$. Then $\alpha = \operatorname{cf}(\alpha) = \operatorname{cf}(S') = \operatorname{cf}(\beta) \leqslant \beta$, where the third equality holds by Lemma 2.1.2. □

COROLLARY 2.1.4. *Let S' be an ordered set without a largest element, and let $f\colon S \to S'$ be an increasing and cofinal map. Then* $\operatorname{cf}(S) = \operatorname{cf}(S')$.

PROOF. Define an equivalence relation $\sim$ on S by $x \sim y :\Longleftrightarrow f(x) = f(y)$. Pick a set R of representatives for this equivalence relation, and observe that f restricts to an isomorphism $R \to f(R) = f(S)$ of ordered sets and R is cofinal in S. Hence $\mathrm{cf}(S) = \mathrm{cf}(R) = \mathrm{cf}(f(S)) = \mathrm{cf}(S')$, by Lemma 2.1.2. □

A **well-indexed sequence** is a sequence (a_ρ) whose terms a_ρ are indexed by the elements ρ of an infinite well-ordered set without a last element. Restricting a well-indexed sequence (a_ρ) to a cofinal subset of its index set yields a well-indexed sequence; it is what we call a **cofinal subsequence** of (a_ρ). Given a set A, a **well-indexed sequence in** A is a well-indexed sequence whose terms are all in A.

Notes and comments

The material in this section is standard fare from the theory of ordered sets. For more on ordered sets see [168]. The proof of Corollary 2.1.4 is from the proof of the "Fact" on p. 22 of [106].

2.2 VALUED ABELIAN GROUPS

We introduce the basic notions concerning valued abelian groups, treat pseudocauchy and cauchy sequences in this context, and prove a fixpoint theorem.

Valued abelian groups

Let G be an abelian (additively written) group. A **valuation** on G is a function $v\colon G \to S_\infty$ where S is an ordered set, such that for all $x, y \in G$ the following conditions are satisfied:

(VA1) $v(x) = \infty \Longleftrightarrow x = 0$;

(VA2) $v(-x) = v(x)$;

(VA3) $v(x+y) \geqslant \min\big(v(x), v(y)\big)$.

Let $v\colon G \to S_\infty$ be a valuation on G. Note that if $x, y \in G$ and $vx < vy$, then $v(x+y) = vx$. (Use that $vx = v(x+y-y) \geqslant \min\big(v(x+y), vy\big)$.) More generally, if $x_1, \dots, x_n \in G$, $n \geqslant 1$, and $v(x_1) < v(x_2), \dots, v(x_n)$, then $v(x_1 + \cdots + x_n) = v(x_1)$. Another elementary fact that is often used is that the condition $v(x-y) > vx$ for $x, y \in G$ (which implies $x, y \neq 0$) defines an equivalence relation on $G^{\neq}$.

Assume now that the valuation $v\colon G \to S_\infty$ is surjective. For $s \in S$ and $a \in G$ we define the **open ball** $B_a(s)$ and the **closed ball** $\overline{B}_a(s)$ by

$$\begin{aligned} B_a(s) &:= \big\{x \in G : v(x-a) > s\big\}, \\ \overline{B}_a(s) &:= \big\{x \in G : v(x-a) \geqslant s\big\}. \end{aligned}$$

We refer to $B_a(s)$ as the **open ball centered at** a **with valuation radius** s, and to $\overline{B}_a(s)$ as the **closed ball centered at** a **with valuation radius** s. Note that $B(s) := B_0(s)$ is a subgroup of G, and the $B_a(s) = a + B(s)$ are its cosets. Likewise, $\overline{B}(s) := \overline{B}_0(s)$ is a subgroup of G, and the $\overline{B}_a(s) = a + \overline{B}(s)$ are its cosets. Viewing balls in this way as cosets, it is clear that if D, E are balls (of any kind) with nonempty intersection, then $D \subseteq E$ or $E \subseteq D$. Likewise, any point in a ball can serve as a center of that ball.

The open balls form a basis for a topology on G, the **v-topology**. (If the valuation v is understood from the context we also speak of the **valuation topology** on G.) It is easy to see that balls (of any kind) are both open and closed in the v-topology, and that G with the v-topology is a hausdorff topological group. If S has a largest element, then the v-topology on G is discrete. This is in particular the case if v is **trivial**, that is, S is a singleton. If $S = \emptyset$, then of course $G = \{0\}$.

We have $B(s) \subseteq \overline{B}(s)$, and we let $G(s) := \overline{B}(s)/B(s)$ be the corresponding quotient group, which is nontrivial. Here is a useful and suggestive cardinality bound on G in terms of the cardinalities of these quotients:

LEMMA 2.2.1. $|G| \leqslant \prod_{s\in S} |G(s)|$.

PROOF. Let $s \in S$, and let $\mathcal{B}_s$ be the set of all open balls with valuation radius s. Each $B \in \mathcal{B}_s$ is contained in a unique closed ball $\overline{B}$ of valuation radius s, and $\overline{B}$ is the disjoint union of exactly $|G(s)|$-many $E \in \mathcal{B}_s$. Thus we have a map $f_s \colon \mathcal{B}_s \to G(s)$ such that for all $D, E \in \mathcal{B}_s$, if $\overline{D} = \overline{E}$ and $D \neq E$, then $f_s(D) \neq f_s(E)$. We now associate to each $a \in G$ the element $\widetilde{a}$ of the cartesian product set $\prod_{s\in S} G(s)$ such that $\widetilde{a}(s) = f_s\big(B_a(s)\big)$ for all $s \in S$. Then the map

$$a \mapsto \widetilde{a} \, : \, G \to \prod_{s\in S} G(s)$$

is injective: if $a, b \in G$ and $a \neq b$, then for $s := v(a-b)$ we have $\widetilde{a}(s) \neq \widetilde{b}(s)$ since a and b are in the same closed ball of valuation radius s but in different open balls of valuation radius s. □

EXAMPLE (Hahn products). Let $(G_s)_{s\in S}$ be a family of nontrivial abelian groups indexed by an ordered set S. For any element $g = (g_s)$ of the product group $\prod_s G_s$ we define its support by

$$\operatorname{supp} g \ := \ \{s \in S : \ g_s \neq 0\}.$$

We define the **Hahn product** of the family (G_s) to be the subgroup $H\big[(G_s)\big]$ of $\prod_s G_s$ consisting of the $g \in \prod_s G_s$ with well-ordered support. The (surjective) valuation on this Hahn product $G := H\big[(G_s)\big]$ given by

$$v \, : \, G \to S_\infty, \qquad v(g) \ := \ \min(\operatorname{supp} g) \text{ for } g \neq 0,$$

is called the **Hahn valuation** of G. Let G be equipped with its Hahn valuation. Given $\sigma \in S$ we have the obvious projection map

$$(g_s) \mapsto g_\sigma \, : \, G \to G_\sigma,$$

which restricts to a surjective group morphism $\overline{B}(\sigma) \to G_\sigma$ with kernel $B(\sigma)$, thus inducing a group isomorphism $G(\sigma) \to G_\sigma$. If all G_s are equal, say $G_s = A$ for each $s \in S$, where A is an abelian group, then we set $H[S, A] := H\big[(G_s)\big]$.

We define a **valued abelian group** to be a triple (G, S, v) with G an abelian group, S an ordered set, and $v\colon G \to S_\infty$ a surjective valuation. We call the ordered set S the **value set** of the valued abelian group (G, S, v). A Hahn product is considered as a valued abelian group by equipping it with its Hahn valuation. When an ambient valued abelian group (G, S, v) is given, then we set for $x, y \in G$,

$$\begin{aligned} x \preccurlyeq y &:\Longleftrightarrow v(x) \geqslant v(y), & x \prec y &:\Longleftrightarrow v(x) > v(y), \\ x \succcurlyeq y &:\Longleftrightarrow v(x) \leqslant v(y), & x \succ y &:\Longleftrightarrow v(x) < v(y), \\ x \asymp y &:\Longleftrightarrow v(x) = v(y), & x \sim y &:\Longleftrightarrow v(x-y) > v(x). \end{aligned}$$

These relational notations are shorter and often more suggestive than notations using v. In particular, $\sim$ is an equivalence relation on $G^{\neq}$.

Let (G, S, v) and (G', S', v') be valued abelian groups. Then we say that (G, S, v) is a **valued subgroup** of (G', S', v'), or (G', S', v') **extends** (G, S, v), or (G', S', v') is an **extension** of (G, S, v), if G is a subgroup of G', S is an ordered subset of S', and $v(x) = v'(x)$ for all $x \in G$; notation: $(G, S, v) \subseteq (G', S', v')$.

Assume $(G, S, v) \subseteq (G', S', v')$. Then we have for each $s \in S$ a natural group embedding $G(s) \to G'(s)$ sending, for $a \in \overline{B}(s)$, the coset $a + B(s)$ of the open ball $B(s)$ in G to the coset $a + B'(s)$ of the open ball $B'(s) = \{x \in G' : v'(x) > s\}$ in G'. This extension of valued abelian groups is said to be **immediate** if $S = S'$ and for each $s \in S$ the group embedding $G(s) \to G'(s)$ is bijective; equivalently, for each $0 \neq x' \in G'$ there is $x \in G$ with $x \sim x'$. For example, if G is dense in G' in the v'-topology, then (G', S', v') is an immediate extension of (G, S, v). Such immediate extensions, in the setting of asymptotic differential fields, will play an important role in what follows. A tool for coming to grips with immediate extensions is the notion of pseudoconvergence, to which we turn in the next subsection. At this stage we can make one useful definition: Call a valued abelian group (G, S, v) **maximal** if it has no proper immediate valued abelian group extension. Hahn products are maximal: this is Corollary 2.2.7 below.

COROLLARY 2.2.2. *Every valued abelian group has an immediate valued abelian group extension that is maximal.*

PROOF. Use Zorn and Lemma 2.2.1. □

By imposing extra structure on our valued abelian groups, we can add to this existence result a corresponding uniqueness property; see Corollaries 2.3.2 and 2.3.3.

Pseudoconvergence

Fix a valued abelian group (G, S, v). Let (a_ρ) be a well-indexed sequence in G, and $a \in G$. Then (a_ρ) is said to **pseudoconverge to** a (notation: $a_\rho \rightsquigarrow a$), if $\big(v(a - a_\rho)\big)$ is eventually strictly increasing, that is, for some index ρ_0 we have $a - a_\sigma \prec a - a_\rho$ whenever $\sigma > \rho > \rho_0$. We also say in that case that a **is a pseudolimit of** (a_ρ). Note that if $a_\rho \rightsquigarrow a$, then $a_\rho + b \rightsquigarrow a + b$ for each $b \in G$, and that

$$a_\rho \rightsquigarrow 0 \iff (va_\rho) \text{ is eventually strictly increasing.}$$

LEMMA 2.2.3. *Let (a_ρ) be a well-indexed sequence in G such that $a_\rho \rightsquigarrow a$ where $a \in G$. With $s_\rho := v(a - a_\rho)$, we have:*

(i) *either $a \prec a_\rho$ eventually, or $a \sim a_\rho$ eventually;*

(ii) *(va_ρ) is either eventually strictly increasing, or eventually constant;*

(iii) *for each $b \in G$: $\quad a_\rho \rightsquigarrow b \iff v(a-b) > s_\rho$ eventually.*

PROOF. Let ρ_0 be as in the definition of "$a_\rho \rightsquigarrow a$." Suppose $a_\rho \preccurlyeq a$, where $\rho > \rho_0$. Then for $\sigma > \rho$ we have $a - a_\sigma \prec a - a_\rho \preccurlyeq a$, so $a \sim a_\sigma$. This proves (i). Now (ii) follows from (i) by noting that if $a \prec a_\rho$, then $va_\rho = v(a - a_\rho)$, and if $a \sim a_\rho$, then $va_\rho = va$. We leave (iii) to the reader. □

If (a_ρ) is a well-indexed sequence in G and $a \in G'$ where (G', S', v') is a valued abelian group extending (G, S, v), then "$a_\rho \rightsquigarrow a$" is to be interpreted in this valued extension, that is, by considering the sequence (a_ρ) in G as a sequence in G'.

LEMMA 2.2.4. *Suppose (G', S', v') is an immediate valued abelian group extension of (G, S, v), and let $a \in G' \setminus G$. Then there is a well-indexed sequence (a_ρ) in G such that $a_\rho \rightsquigarrow a$ and (a_ρ) has no pseudolimit in G.*

PROOF. We claim that the subset $\{v'(a - x) : x \in G\}$ of S has no largest element. To see this, let $x \in G$; we shall find $y \in G$ such that $v'(a - y) > v'(a - x)$. We have $s := v'(a - x) \in S$ and $G(s) = G'(s)$, so we can take $b \in G$ with $a - x \in b + B'(s)$, hence $v'(a - y) > s$ for $y := x + b$, as claimed. It follows that we can take a well-indexed sequence (a_ρ) in G such that the sequence $\big(v'(a - a_\rho)\big)$ is strictly increasing and cofinal in $\{v'(a - x) : x \in G\}$. Thus $a_\rho \rightsquigarrow a$. If $a_\rho \rightsquigarrow g \in G$, then $v'(a - g) > v'(a - a_\rho)$ for all ρ by Lemma 2.2.3(iii), hence $v'(a - g) > v'(a - x)$ for all $x \in G$, a contradiction. □

Pseudocauchy sequences

As before, fix a valued abelian group (G, S, v). To capture within (G, S, v) that a well-indexed sequence (a_ρ) in G has a pseudolimit in some valued abelian group extension we make the following definition.

A **pseudocauchy sequence in** G, more precisely, in (G, S, v), is a well-indexed sequence (a_ρ) in G such that for some index ρ_0 we have

$$\tau > \sigma > \rho > \rho_0 \implies a_\tau - a_\sigma \prec a_\sigma - a_\rho.$$

We also write *pc-sequence* for *pseudocauchy sequence*. A cofinal subsequence of a pc-sequence in G is a pc-sequence in G.

LEMMA 2.2.5. *Let (a_ρ) be a well-indexed sequence in G. Then (a_ρ) is a pc-sequence in G if and only if (a_ρ) has a pseudolimit in some valued abelian group extension of (G, S, v). In that case, (a_ρ) has even a pseudolimit in some elementary extension of the two-sorted structure (G, S, v).* (See B.5 for *elementary extension*.)

PROOF. Suppose (a_ρ) is a pc-sequence in G, and let ρ_0 be as in the definition of *pseudocauchy sequence.* We refer to B.9 for the notion of type. Consider the type in the variable x consisting of the formulas

$$x - a_\sigma \prec x - a_\rho \qquad (\sigma > \rho > \rho_0).$$

Every finite subset of this type is realized by a_τ for any sufficiently large τ. Thus we can realize this type by a suitable $a \in G'$ for some elementary extension (G', S', v') of (G, S, v), and then $a_\rho \rightsquigarrow a$.

For the converse, suppose $a_\rho \rightsquigarrow a$ where $a \in G'$ and (G', S', v') is a valued abelian group extension of (G, S, v). Let ρ_0 be as in the definition of pseudolimit, and let $\sigma > \rho > \rho_0$. Then $a_\sigma - a_\rho = (a_\sigma - a) - (a_\rho - a)$, so $a_\sigma - a_\rho \asymp a - a_\rho$. So if in addition $\tau > \sigma$, then $a_\tau - a_\sigma \asymp a - a_\sigma \prec a - a_\rho \asymp a_\sigma - a_\rho$. □

It will be relevant to us that the last part of Lemma 2.2.5 goes through with the same proof and the corresponding notion of elementary extension when (G, S, v) has extra (first-order) structure. For example, we shall apply it to valued differential fields: fields equipped with both a valuation and a derivation, as defined later.

COROLLARY 2.2.6. *If every pc-sequence in G has a pseudolimit in G, then the valued abelian group (G, S, v) is maximal.*

PROOF. Immediate from Lemmas 2.2.4 and 2.2.5. □

The converse of Corollary 2.2.6 holds when our valued abelian group has suitable extra structure; see Corollary 2.3.2. It is easy to check that every pc-sequence in a Hahn product pseudoconverges in it. Thus by Corollary 2.2.6:

COROLLARY 2.2.7. *Any Hahn product is maximal as a valued abelian group.*

Lemma 2.2.5 and part (ii) of Lemma 2.2.3 yield:

COROLLARY 2.2.8. *If (a_ρ) is a pc-sequence in G, then (va_ρ) is either eventually strictly increasing (so $a_\rho \rightsquigarrow 0$), or eventually constant (so $a_\rho \not\rightsquigarrow 0$).*

Let (a_ρ) be a pc-sequence in G, pick ρ_0 as above, and put

$$s_\rho := v(a_{\rho'} - a_\rho) \in S \qquad \text{for } \rho' > \rho > \rho_0;$$

this depends only on ρ as the notation suggests. Then $(s_\rho)_{\rho > \rho_0}$ is strictly increasing.

LEMMA 2.2.9. *Let $a \in G$. Then the following are equivalent:*

(i) $a_\rho \rightsquigarrow a$;

(ii) $v(a - a_\rho) = s_\rho$ *for all* $\rho > \rho_0$;

(iii) $v(a - a_\rho) \geqslant s_\rho$ *for all* $\rho > \rho_0$;

(iv) $v(a - a_\rho) \geqslant s_\rho$ *eventually.*

PROOF. Suppose $a_\rho \rightsquigarrow a$, and let $\rho > \rho_0$. Since $v(a - a_\sigma)$ is eventually strictly increasing, $v(a - a_\sigma) \notin \{v(a - a_\rho), s_\rho\}$ for sufficiently large $\sigma > \rho$; for such σ,

$$s_\rho = v(a_\sigma - a_\rho) = \min\big(v(a - a_\rho), v(a - a_\sigma)\big) = v(a - a_\rho).$$

This shows (i) $\Rightarrow$ (ii), and (ii) $\Rightarrow$ (iii) $\Rightarrow$ (iv) are trivial. For (iv) $\Rightarrow$ (i), suppose $\rho_1 \geqslant \rho_0$ is such that $v(a - a_\rho) \geqslant s_\rho$ for all $\rho > \rho_1$. It suffices to show that then $v(a - a_\rho) = s_\rho$ for $\rho > \rho_1$. Suppose towards a contradiction that $\rho > \rho_1$ is such that $v(a - a_\rho) > s_\rho$. Take any $\sigma > \rho$; then $v(a - a_\sigma) > s_\rho$, so

$$s_\rho \;=\; v(a_\sigma - a_\rho) \;=\; v\big((a - a_\rho) - (a - a_\sigma)\big) \;>\; s_\rho,$$

a contradiction. □

The **width** of (a_ρ) is an upward closed subset of S_∞, namely

$$\{s \in S_\infty : \ s > s_\rho \text{ for all } \rho > \rho_0\} \qquad \text{(independent of the choice of } \rho_0).$$

Its significance is that if $a, b \in G$ and $a_\rho \rightsquigarrow a$, then by Lemma 2.2.3,

$$a_\rho \rightsquigarrow b \iff v(a - b) \text{ is in the width of } (a_\rho).$$

Fixpoint theorem

As before, (G, S, v) is a valued abelian group. To visualize the property that every pc-sequence has a pseudolimit, define a **nest of balls** in G to be a collection of balls in G any two of which meet. So a nest of balls in G is (totally) ordered by inclusion. We call (G, S, v) **spherically complete** if every nonempty nest of closed balls in G has a point in its intersection. (For example, if S is well-ordered under the reversed ordering, then trivially (G, S, v) is spherically complete.)

LEMMA 2.2.10. *(G, S, v) is spherically complete if and only if each pc-sequence in G has a pseudolimit in G.*

PROOF. Let (a_ρ) be a pc-sequence in G, and let ρ_0 be as in the definition of pc-sequence. Then $\mathcal{B} = \big\{\overline{B}_{a_\rho}(s_\rho) : \ \rho > \rho_0\big\}$ is a nonempty nest of closed balls in G, and $\bigcap \mathcal{B} = \{a \in G : a_\rho \rightsquigarrow a\}$ by Lemma 2.2.9. Thus if (G, S, v) is spherically complete, then each pc-sequence in G has a pseudolimit in G. Conversely, suppose the latter condition holds, and let $\mathcal{B}$ be a nonempty nest of closed balls in G; we need to show that $\bigcap \mathcal{B} \neq \emptyset$. Eliminating a trivial case, assume that $\mathcal{B}$ has no smallest ball in it. Replacing $\mathcal{B}$ by a coinitial subset (under inclusion), we arrange that $\mathcal{B} = \{B_\rho : \ \rho < \lambda\}$ for some infinite limit ordinal λ, with B_ρ strictly containing B_σ whenever $\rho < \sigma < \lambda$. For each $\rho < \lambda$ we take $a_\rho \in B_\rho \setminus B_{\rho+1}$. It follows that (a_ρ) is a pc-sequence. Take $a \in G$ such that $a_\rho \rightsquigarrow a$. Then $a \in \bigcap \mathcal{B}$. □

A routine argument shows:

LEMMA 2.2.11. *If $X \subseteq G$ and $B_x(s) \subseteq X$ whenever $x, y \in X$, $x \neq y$, $s = v(x - y)$, then any pseudolimit in G of any pc-sequence in X lies in X.*

For example, any ball (open or closed) in G satisfies the hypothesis of Lemma 2.2.11, and so does, trivially, $X = G$. A map $f\colon X \to X$ with $X \subseteq G$ is said to be **contractive** if for all distinct $x, y \in X$ we have $f(x) - f(y) \prec x - y$.

THEOREM 2.2.12. *Let $f\colon X \to X$ be a contractive map, where $\emptyset \neq X \subseteq G$ and each pc-sequence in X has a pseudolimit in X. Then f has a unique fixpoint.*

PROOF. It is clear that f has at most one fixpoint. Take any point $x_0 \in X$ and make it the initial term of a sequence $(x_\lambda)_{\lambda<\nu}$ in X indexed by the ordinals less than an ordinal $\nu > 0$, such that

(1) $x_\lambda \neq x_\mu$ whenever $\lambda < \mu < \nu$;

(2) $x_{\lambda+1} = f(x_\lambda)$ whenever $\lambda < \lambda + 1 < \nu$;

(3) $x_{\lambda''} - x_{\lambda'} \prec x_{\lambda'} - x_\lambda$ whenever $\lambda < \lambda' < \lambda'' < \nu$.

If $\nu = \mu + 1$ is a successor ordinal, and x_μ is not yet a fixpoint of f, then we set $x_\nu := f(x_\mu)$, and then the extended sequence $(x_\lambda)_{\lambda<\nu+1}$ satisfies (1)–(3) with $\nu + 1$ instead of ν. If ν is a limit ordinal, then we let x_ν be a pseudolimit in X of the pc-sequence $(x_\lambda)_{\lambda<\nu}$, and then the extended sequence $(x_\lambda)_{\lambda<\nu+1}$ satisfies again (1)–(3) with $\nu + 1$ instead of ν. This building process must come to a halt by producing a fixpoint of f. □

COROLLARY 2.2.13. *Suppose (G, S, v) is spherically complete and $e\colon G \to G$ is a contractive map. Then $\mathrm{id}_G + e\colon G \to G$ is bijective.*

PROOF. Injectivity is immediate from e being contractive. To get surjectivity, let $a \in G$, and define $f\colon G \to G$ by $f(x) = a - e(x)$. Then f is contractive. Let x be the fixpoint of f. Then $a = x + e(x)$. □

The variant below is also useful. Let $\emptyset \neq P \subseteq S$, $\emptyset \neq X \subseteq G$, and let a map $f\colon X \to X$ be given. A **P-fixpoint** of f is a point $x \in X$ such that $v(f(x) - x) \geqslant s$ for some $s \in P$. We call f **contractive up to** P if for all $x, y \in X$ with $v(x - y) < P$ we have $f(x) - f(y) \prec x - y$.

LEMMA 2.2.14. *Suppose every pc-sequence in X has a pseudolimit in X, and f is contractive up to P. Then f has a P-fixpoint.*

PROOF. Take any point $x_0 \in X$ and make it the initial term of a sequence $(x_\lambda)_{\lambda<\nu}$ in X indexed by the ordinals less than an ordinal $\nu > 0$, such that

(1) $v(x_\mu - x_\lambda) < P$ whenever $\lambda < \mu < \nu$;

(2) $x_{\lambda+1} = f(x_\lambda)$ whenever $\lambda < \lambda + 1 < \nu$;

(3) $x_{\lambda''} - x_{\lambda'} \prec x_{\lambda'} - x_\lambda$ whenever $\lambda < \lambda' < \lambda'' < \nu$.

We now continue as in the proof of Theorem 2.2.12, replacing "fixpoint" by "P-fixpoint" throughout. □

Generalizing closed balls

Let (G, S, v) be a valued abelian group. We define a **union-closed ball** in G to be the union of a nonempty nest of closed balls in G. We can assume the closed balls in such a nest to have a common center: Let $\{B_i : i \in I\}$ be a nest of closed balls in G where I is a nonempty index set and $B_i = \overline{B}_{a_i}(s_i)$, with $a_i \in G$ and $s_i \in S$ for $i \in I$; then $\bigcup_i B_i = \bigcup_i \overline{B}_a(s_i)$ for any $a \in \bigcup_i B_i$. Using this fact, one easily shows:

LEMMA 2.2.15. *Let B_1 and B_2 be union-closed balls in G such that $B_1 \cap B_2 \neq \emptyset$. Then $B_1 \subseteq B_2$ or $B_2 \subseteq B_1$. If in addition B_1 is properly contained in B_2, then there is a closed ball B in G such that $B_1 \subseteq B \subseteq B_2$.*

Note that if S has no largest element, then every open ball in G is a union-closed ball. A **nest of union-closed balls** in G is a collection of union-closed balls in G any two of which meet. Such a nest is (totally) ordered by inclusion.

COROLLARY 2.2.16. *If (G, S, v) is spherically complete, then any nonempty nest of union-closed balls in G has a point in its intersection.*

PROOF. Let (B_i) a family of union-closed balls in G indexed by a nonempty well-ordered set I such that I has no largest element and B_i properly contains B_j whenever $i < j$ in I. For each i, let $s(i)$ be the immediate successor of i in I, and use Lemma 2.2.15 to get a closed ball D_i in G such that $B_{s(i)} \subseteq D_i \subseteq B_i$. Then $\{D_i : i \in I\}$ is a nest of closed balls, and $\bigcap_i D_i = \bigcap_i B_i$.

The corollary is an easy consequence of this construction. □

Equivalence of pc-sequences

Let (G, S, v) be a valued abelian group.

LEMMA 2.2.17. *Given pc-sequences (a_ρ) and (b_σ) in G (with possibly different index sets), the following conditions are equivalent:*

(i) *(a_ρ) and (b_σ) have the same pseudolimits in every valued abelian group extension of (G, S, v);*

(ii) *(a_ρ) and (b_σ) have the same width, and have a common pseudolimit in some valued abelian group extension of (G, S, v);*

(iii) *there are arbitrarily large ρ and σ such that for all $\rho' > \rho$ and $\sigma' > \sigma$ we have $a_{\rho'} - b_{\sigma'} \prec a_{\rho'} - a_\rho$, and there are arbitrarily large ρ and σ such that for all $\rho' > \rho$ and $\sigma' > \sigma$ we have $a_{\rho'} - b_{\sigma'} \prec b_{\sigma'} - b_\sigma$.*

PROOF. Suppose $s \in S$ is in the width of (a_ρ) but not in the width of (b_σ). Take $x \in G$ with $vx = s$ and take a in some valued abelian group extension of (G, S, v) such that $a_\rho \rightsquigarrow a$. Then also $a_\rho \rightsquigarrow a + x$, but a and $a + x$ cannot both be pseudolimits of (b_σ). This argument proves (i) ⇒ (ii). To prove (ii) ⇒ (iii), assume (ii), and take a in some valued abelian group extension of (G, S, v) such that $a_\rho \rightsquigarrow a$ and $b_\sigma \rightsquigarrow a$. Let ρ_0 be so large that $v(a - a_\rho)$ is strictly increasing for $\rho > \rho_0$ and $a - a_\rho \asymp a_{\rho'} - a_\rho$

for all $\rho' > \rho > \rho_0$. Likewise, let σ_0 be so large that $v(a - b_\sigma)$ is strictly increasing for $\sigma > \sigma_0$, and $a - b_\sigma \asymp b_{\sigma'} - b_\sigma$ for all $\sigma' > \sigma > \sigma_0$. Take any $\rho > \rho_0$ and then take $\sigma > \sigma_0$ such that $a - b_\sigma \prec a - a_\rho$. Then for $\rho' > \rho$ and $\sigma' > \sigma$ we have $a_{\rho'} - b_{\sigma'} = (a_{\rho'} - a) + (a - b_{\sigma'})$, with

$$a_{\rho'} - a \prec a - a_\rho \asymp a_{\rho'} - a_\rho, \qquad a - b_{\sigma'} \prec a - b_\sigma \prec a_{\rho'} - a_\rho,$$

so $a_{\rho'} - b_{\sigma'} \prec a_{\rho'} - a_\rho$. Likewise, we obtain the second part of (iii).

For (iii) ⇒ (i), assume (iii), and let a be a pseudolimit of (a_ρ) in a valued abelian group extension of (G, S, v). Take ρ_0 and σ_0 such that $v(a - a_\rho)$ is strictly increasing for $\rho > \rho_0$, and $s_\sigma := v(b_{\sigma'} - b_\sigma)$ for $\sigma' > \sigma > \sigma_0$ depends only on σ (not on σ') and is strictly increasing as a function of $\sigma > \sigma_0$. Let any $\sigma > \sigma_0$ be given. Then $a - b_\sigma = (a - a_\rho) + (a_\rho - b_\sigma)$ and

$$a_\rho - b_\sigma = (a_\rho - b_{\sigma'}) + (b_{\sigma'} - b_\sigma) \asymp b_{\sigma'} - b_\sigma, \text{ and } v(b_{\sigma'} - b_\sigma) = s_\sigma$$

for all sufficiently large $\rho > \rho_0$ and $\sigma' > \sigma$. As $v(a - a_\rho)$ is strictly increasing for $\rho > \rho_0$, and $v(a - b_\sigma)$ and $v(a_\rho - b_\sigma) = s_\sigma$ depend only on σ (for big enough ρ) it follows that $v(a - b_\sigma) = s_\sigma$. Thus $b_\sigma \rightsquigarrow a$. □

REMARK. Lemma 2.2.17 goes through with the same proof when the valued abelian group (G, S, v) is equipped with extra first-order structure and in (i) and (ii) the extensions are required to be elementary extensions of the expanded structure.

Two pc-sequences (a_ρ) and (b_σ) in G are said to be **equivalent** if the conditions of Lemma 2.2.17 are satisfied. This equivalence relation on the class of pc-sequences in G is relative to the ambient valued abelian group (G, S, v). However, condition (iii) of Lemma 2.2.17 shows that for pc-sequences (a_ρ) and (b_σ) in G and any valued abelian group extension (G', S', v') of (G, S, v) we have: (a_ρ) and (b_σ) are equivalent with respect to (G, S, v) iff (a_ρ) and (b_σ) are equivalent with respect to (G', S', v'). Note that any cofinal subsequence of a pc-sequence (a_ρ) in G is equivalent to (a_ρ).

Let a be an element of a valued abelian group extension of (G, S, v) with $a \notin G$. For convenience, denote the valuation of that extension also by v, and set

$$v(a - G) := \{v(a - g) : g \in G\},$$

a nonempty subset of the value set of that extension. The next lemmas collect some basic facts about this situation. The easy proofs are left to the reader.

LEMMA 2.2.18. *The following are equivalent:*

(i) $v(a - G)$ *has no largest element;*

(ii) $a_\rho \rightsquigarrow a$ *for some pc-sequence* (a_ρ) *in* G *without pseudolimit in* G.

If a *lies in an immediate valued abelian group extension of* (G, S, v), *then* (i) *holds. If* (i) *holds, then* $v(a - G)$ *is a downward closed subset of* S.

LEMMA 2.2.19. *Let (a_ρ) be a well-indexed sequence in G such that $v(a-a_\rho)$ is strictly increasing as a function of ρ. Then $a_\rho \rightsquigarrow a$, and we have:*

$$\big(v(a-a_\rho)\big) \text{ is cofinal in } v(a-G) \iff (a_\rho) \text{ has no pseudolimit in } G.$$

A **divergent** pc-sequence in G is a pc-sequence in G without pseudolimit in G.

COROLLARY 2.2.20. *Any two divergent pc-sequences in G with a common pseudolimit in an immediate valued abelian group extension of (G, S, v) are equivalent.*

Coarsenings

Let surjective valuations $v\colon G \to S_\infty$ and $v'\colon G \to S'_\infty$ on the abelian group G be given. We say that v' is a **coarsening** of v (or v' is **coarser** than v, or v is **finer** than v') if for all $a, b \in G$ we have $v(a) \leqslant v(b) \Rightarrow v'(a) \leqslant v'(b)$; equivalently, there exists an increasing surjection $i\colon S_\infty \to S'_\infty$ such that $v' = i \circ v$. Note that for such a map i we have $i(\infty) = \infty$ and $i(S) = S'$, since $v'g \neq \infty$ for $g \in G^{\neq}$. We call v and v' **equivalent** if v is coarser than v' and v' is coarser than v; equivalently, there exists an isomorphism $i\colon S_\infty \to S'_\infty$ of ordered sets such that $v' = i \circ v$ (and such i is then uniquely determined). If v and v' are equivalent, then the v-topology agrees with the v'-topology. If v' is coarser than v and S' has no largest element, then the v-topology also agrees with the v'-topology.

LEMMA 2.2.21. *Suppose v' is coarser than v, and (a_ρ) is a pc-sequence in G with respect to v'. Then (a_ρ) is a pc-sequence with respect to v, and for all $a \in G$,*

$$a_\rho \rightsquigarrow a \text{ with respect to } v \iff a_\rho \rightsquigarrow a \text{ with respect to } v'.$$

To verify $\Rightarrow$, use for example Lemma 2.2.9.

COROLLARY 2.2.22. *If (G, S, v) is spherically complete and v' is coarser than v, then (G, S', v') is spherically complete.*

Cauchy sequences

Let (G, S, v) be a valued abelian group and (a_ρ) a well-indexed sequence in G. We say that (a_ρ) is a **cauchy sequence** (or a **c-sequence**) in G if for every $s \in S$ there is ρ_0 such that $v(a_\rho - a_{\rho'}) > s$ for all $\rho, \rho' > \rho_0$. Thus if S has a largest element, then (a_ρ) is a c-sequence iff a_ρ is eventually constant. A c-sequence (a_ρ) in G remains a c-sequence in every extension (G', S', v') of (G, S, v) with S cofinal in S'. For $a \in G$ we say that (a_ρ) **converges to** a if for each $s \in S$ there is some ρ_0 such that $v(a - a_\rho) > s$ for all $\rho > \rho_0$; in symbols: $a_\rho \to a$. We say that (a_ρ) **converges in** G if $a_\rho \to a$ for some $a \in G$. Note that if (a_ρ) converges in some extension of G then (a_ρ) is a c-sequence. There is at most one $a \in G$ with $a_\rho \to a$, and if there is such an a we call it the **limit of** (a_ρ). Clearly each pc-sequence of width $\{\infty\}$ is a c-sequence. (For a partial converse of this statement see Lemma 2.2.35 below.) If (a_ρ) is a pc-sequence of width $\{\infty\}$ and $a \in G$, then we have the equivalence $a_\rho \rightsquigarrow a \iff a_\rho \to a$. The following is obvious.

LEMMA 2.2.23. *Let* (a_ρ), (b_ρ) *be c-sequences in* G *with the same index set, and* $a, b \in G$. *Then*

(i) $(-a_\rho)$ *is a c-sequence, and if* $a_\rho \to a$ *then* $-a_\rho \to -a$;

(ii) $(a_\rho + b_\rho)$ *is a c-sequence, and if* $a_\rho \to a$ *and* $b_\rho \to b$ *then* $a_\rho + b_\rho \to a + b$.

We also have an analogue of Corollary 2.2.8 for c-sequences:

LEMMA 2.2.24. *Let* (a_ρ) *be a c-sequence in* G. *Then either for every* $s \in S$ *there is a* ρ_0 *such that* $va_\rho > s$ *for all* $\rho > \rho_0$ *(that is,* $a_\rho \to 0$*), or* va_ρ *takes an eventually constant value in* S *(so* $a_\rho \not\to 0$*).*

PROOF. Suppose $a_\rho \not\to 0$. Take $s \in S$ such that for every ρ there is some $\rho' > \rho$ with $va_{\rho'} \leqslant s$. Also, (a_ρ) being a c-sequence, take ρ_0 such that $v(a_\rho - a_{\rho'}) > s$ for all $\rho, \rho' > \rho_0$. Let $\rho' > \rho > \rho_0$; we claim that $va_\rho = va_{\rho'}$. To see this take $\rho'' > \rho'$ such that $va_{\rho''} \leqslant s$. Then $v(a_\rho - a_{\rho''}) > s$ and hence $va_\rho = va_{\rho''}$, and similarly $va_{\rho'} = va_{\rho''}$, whence the claim. □

For every c-sequence (a_ρ) in G with $a_\rho \not\to 0$ there is an $s \in S$ with $va_\rho = s$ eventually, by Lemma 2.2.24; we call s the eventual valuation of (a_ρ), and if $a_\rho \to 0$ we define the eventual valuation of (a_ρ) to be ∞. Note that if a is an element in an extension of G with $a_\rho \to a$ then the eventual valuation of (a_ρ) is va.

A cofinal subsequence (b_σ) of a c-sequence (a_ρ) is a c-sequence, with $a_\rho \to a$ iff $b_\sigma \to a$, for all $a \in G$. The next lemma allows us to restrict our attention to c-sequences indexed by the ordinals less than $\mathrm{cf}(S)$:

LEMMA 2.2.25. *Suppose* (a_ρ) *is a c-sequence that is not eventually constant. Then the index set of* (a_ρ) *has cofinality* $\mathrm{cf}(S)$.

PROOF. The assumption implies that $S \neq \emptyset$ and S has no largest element. Choose a sequence (s_γ) in S, indexed by the ordinals $\gamma < \mathrm{cf}(S)$, which is strictly increasing and cofinal in S. For each γ, define

$$I_\gamma := \{\rho : \ v(a_{\rho_1} - a_{\rho_2}) > s_\gamma \text{ for all } \rho_1, \rho_2 \geqslant \rho\},$$

so $I_\gamma \neq \emptyset$ and I_γ is upward closed in the set of indices ρ. Set $\rho(\gamma) := \min I_\gamma$. The well-indexed sequence $(\rho(\gamma))$ is increasing, since $I_{\gamma'} \subseteq I_\gamma$ for $\gamma' \geqslant \gamma$. Also $\bigcap_\gamma I_\gamma = \emptyset$, since (a_ρ) is not ultimately constant. Therefore, given any index ρ we can choose γ with $\rho \notin I_\gamma$, and then $\rho(\gamma) > \rho$. Hence the sequence $(\rho(\gamma))$ is cofinal in the set of indices ρ, and thus by Corollary 2.1.4 the set of indices ρ has cofinality $\mathrm{cf}(\mathrm{cf}(S)) = \mathrm{cf}(S)$. □

We call an extension $(G', S', v') \supseteq (G, S, v)$ of valued abelian groups **dense** if G is dense in G' (in the v'-topology on G'). Every dense extension of valued abelian groups is immediate. Hence by Lemma 2.2.4 and its proof:

LEMMA 2.2.26. *Given a dense extension* $(G', S', v') \supseteq (G, S, v)$ *and* $a \in G' \setminus G$, *there is a divergent pc-sequence* (a_ρ) *of width* $\{\infty\}$ *in* G, *indexed by the ordinals* $\rho < \mathrm{cf}(S)$, *such that* $a_\rho \rightsquigarrow a$.

One says that G is **complete** if every c-sequence in G converges in G. Thus if S has a largest element, then G is automatically complete. If v' is a coarsening of the valuation v of G and the value set S' of v' has no largest element, then Lemma 2.2.21 goes through with c-sequences instead of pc-sequences, and $a_\rho \to a$ instead of $a_\rho \rightsquigarrow a$. Therefore:

COROLLARY 2.2.27. *If* G *is complete with respect to* v, *then* G *is complete with respect to any coarsening of* v.

By Lemma 2.2.26, if G is complete, then G has no proper dense extension. On the other hand, every c-sequence in G converges in some dense extension of G:

THEOREM 2.2.28. *Every valued abelian group has a dense complete extension.*

PROOF. If $S = \emptyset$ (so $G = \{0\}$), or S has a largest element, then G is complete and $G^{\mathrm{c}} := G$ has the required properties. For the rest of the proof we assume S is nonempty and has no largest element (so $\mathrm{cf}(S)$ is infinite). Let G^{cs} be the set of all c-sequences (a_ρ) in G indexed by the ordinals $\rho < \mathrm{cf}(S)$. By Lemma 2.2.23, G^{cs} is an abelian group under componentwise addition of sequences, and

$$N := \big\{(a_\rho) \in G^{\mathrm{cs}} : a_\rho \to 0\big\}$$

is a subgroup of G^{cs}; we let $G^{\mathrm{c}} := G^{\mathrm{cs}}/N$ be the quotient group. For $(a_\rho), (b_\rho) \in G^{\mathrm{cs}}$ with $b_\rho \to 0$ the eventual valuations of (a_ρ) and $(a_\rho + b_\rho)$ are the same, hence we can define a function

$$v^{\mathrm{c}} \colon G^{\mathrm{c}} \to S_\infty, \qquad v^{\mathrm{c}}\big((a_\rho) + N\big) = \text{eventual valuation of } (a_\rho).$$

It is easy to check that v^{c} is a valuation of the abelian group G^{c}. The map which associates to each element $a \in G$ the coset of N containing the constant sequence $(a) \in G^{\mathrm{cs}}$ is an embedding $G \to G^{\mathrm{c}}$ of groups, and identifying G with its image under this embedding, the valued group $(G^{\mathrm{c}}, v^{\mathrm{c}}, S)$ is an extension of (G, v, S).

Let $(a_\rho) \in G^{\mathrm{cs}}$ and $s \in S$. Since (a_ρ) is a c-sequence, we can take ρ_0 such that $v(a_\rho - a_{\rho_0}) > s$ for all $\rho > \rho_0$; thus the eventual value of the c-sequence $(a_\rho - a_{\rho_0})$ is larger than s, hence $a_{\rho_0} \in G \cap B_{(a_\rho)+N}(s)$. Thus G is dense in G^{c}. By a similar argument, $a_\rho \to (a_\rho) + N$ in G^{c}. Therefore, by Lemma 2.2.25, *every* c-sequence in G converges in G^{c}. It now follows that G^{c} is complete: Let (b_σ) be a c-sequence in G^{c}; we need to show that (b_σ) has a limit in G^{c}. By what we have shown, applied to G^{c} in place of G, (b_σ) has a limit b' in some dense valued abelian group extension G' of G^{c}. Since $G' \supseteq G$ is also dense, by Lemma 2.2.26 there is a c-sequence (a_ρ) in G with $a_\rho \to b'$ in G'. This c-sequence converges in G^{c}, thus $b' \in G^{\mathrm{c}}$. □

COROLLARY 2.2.29. *G is complete iff G has no proper dense extension.*

So "spherically complete $\Rightarrow$ maximal $\Rightarrow$ complete" by Corollaries 2.2.10 and 2.2.6. Thus any Hahn product is complete, by Corollary 2.2.7.

Corollary 2.2.33 below says that any two dense complete extensions of G are isomorphic over G by a unique isomorphism. We obtain this from a more general observation (Lemma 2.2.31) about the extension of continuous maps.

Let (a_ρ), (b_σ) be c-sequences in G (with possibly different index sets). We call (a_ρ) and (b_σ) equivalent (in symbols: $(a_\rho) \sim (b_\sigma)$) if for each $s \in S$ there are ρ_0, σ_0 such that $v(a_\rho - b_\sigma) > s$ for all $\rho > \rho_0$, $\sigma > \sigma_0$. So if (a_ρ) and (b_σ) have the same limit in some abelian valued group extension of G, then $(a_\rho) \sim (b_\sigma)$; conversely, if $(a_\rho) \sim (b_\sigma)$ then (a_ρ) and (b_σ) have the same limit in some dense abelian valued group extension of G (by Theorem 2.2.28). The relation $\sim$ is an equivalence relation on the class of c-sequences in G. If (b_σ) is a cofinal subsequence of (a_ρ) then $(a_\rho) \sim (b_\sigma)$. Moreover, given $a \in G$ we have $a_\rho \to a$ iff $(a_\rho) \sim (a)$, where (a) denotes an arbitrary constant c-sequence all of whose terms equal a.

Let $X \subseteq G$; by a c-sequence in X we mean a c-sequence (a_ρ) in G such that $a_\rho \in X$ for each ρ. In the next two lemmas we consider a map $f\colon X \to G_1$ where (G_1, S_1, v_1) is a valued abelian group. We say that f is **uniformly continuous** if for each $s_1 \in S_1$ there is an $s \in S$ such that for all $a, b \in X$, if $v(a-b) > s$, then $v\big(f(a) - f(b)\big) > s_1$. We say that f is **cauchy-continuous** (for short: **c-continuous**) if for every c-sequence (a_ρ) in X, the image sequence $\big(f(a_\rho)\big)$ is a c-sequence in G_1. Clearly we have the implications

$$\text{uniformly continuous} \Rightarrow \text{c-continuous} \Rightarrow \text{continuous}.$$

Conversely, if G is complete, X is closed, and f is continuous, then f is c-continuous.

LEMMA 2.2.30. *Suppose f is c-continuous, (a_ρ) and (b_σ) are c-sequences in X, and $(a_\rho) \sim (b_\sigma)$. Then $\big(f(a_\rho)\big) \sim \big(f(b_\sigma)\big)$.*

PROOF. By passing to cofinal subsequences of (a_ρ), (b_σ), we arrange that the two sequences have the same ordered index set, by Lemma 2.2.25. Order the set of pairs (ρ, i), where $i = 0, 1$, by $(\rho, i) < (\rho', i')$ iff $\rho < \rho'$ or $\rho = \rho'$ and $i = 0$, $i' = 1$. Then the sequence $(c_{(\rho,i)})$ with $c_{(\rho,0)} = a_\rho$ and $c_{(\rho,1)} = b_\rho$ is a c-sequence in X equivalent to both (a_ρ) and (b_ρ). Since f is c-continuous, $\big(f(c_{(\rho,i)})\big)$ is a c-sequence in G_1, and $\big(f(c_{(\rho,i)})\big)$ is equivalent to both $\big(f(a_\rho)\big)$ and $\big(f(b_\sigma)\big)$. $\square$

LEMMA 2.2.31. *Suppose that for each c-sequence (a_ρ) in X, the sequence $\big(f(a_\rho)\big)$ converges in G_1. Then f has a unique extension to a continuous map $X' \to G_1$, where X' is the closure of X in G in the v-topology. This extension is uniformly continuous if f is uniformly continuous.*

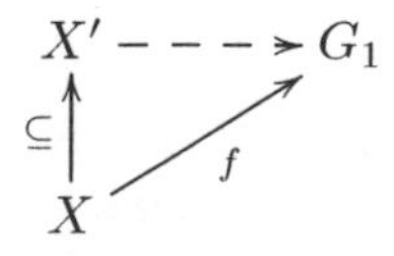

PROOF. We only show existence of such an extension, since uniqueness then follows easily. Let $a' \in X'$ be given; choose a c-sequence (a_ρ) in X such that $a_\rho \to a'$, and let $f'(a')$ be the limit of $\big(f(a_\rho)\big)$ in G_1. Note that $f'(a')$ does not depend on our choice of the sequence (a_ρ): if (b_σ) is another c-sequence in X such that $b_\sigma \to a'$, then $(a_\rho) \sim (b_\sigma)$ and hence $\big(f(a_\rho)\big) \sim \big(f(b_\sigma)\big)$ by the previous lemma, so $\big(f(a_\rho)\big)$ and $\big(f(b_\rho)\big)$ have the same limit in G_1. The map $f' \colon X' \to G_1$ so defined clearly extends f and is continuous. It is easy to verify that if f is uniformly continuous, then so is f'. □

Let (G', S', v') be a valued abelian group. A **morphism** $(G, S, v) \to (G', S', v')$ of valued abelian groups is a pair (i, j), where i is a group morphism $G \to G'$ and j is an increasing map $S_\infty \to S'_\infty$ such that $j\big(v(x)\big) = v'\big(i(x)\big)$ for all $x \in G$ (so $j(\infty) = \infty$). Such a morphism (i, j) is an **embedding** if j is injective (in which case i is also injective). If $(G, S, v) \subseteq (G', S', v')$, then we have an embedding

$$(G, S, v) \to (G', S', v')$$

of valued abelian groups whose components are the natural inclusions $G \to G'$ and $S_\infty \to S'_\infty$. A morphism $(i, j) \colon (G, S, v) \to (G', S', v')$ of valued abelian groups is said to be an **isomorphism** if $i \colon G \to G'$ and $j \colon S_\infty \to S'_\infty$ are bijections; note that then $(i^{-1}, j^{-1}) \colon (G', S', v') \to (G, S, v)$ is also an isomorphism (of valued abelian groups). An embedding $(i, j) \colon (G, S, v) \to (G', S', v')$ determines an isomorphism $(i, j) \colon (G, S, v) \to (iG, jS, v'|_{iG})$ onto a valued subgroup of (G', S', v').

COROLLARY 2.2.32. *Let $(G, S, v) \xrightarrow{(i,j)} (G', S', v')$ be a morphism of valued abelian groups where i is continuous. Let (G_1, S_1, v_1) be a dense valued abelian group extension of (G, S, v) and (G'_1, S'_1, v'_1) a valued abelian group extension of (G', S', v') such that S' is cofinal in S'_1 and every c-sequence in G' converges in G'_1.*

$$\begin{array}{ccc}
(G_1, S_1, v_1) & \dashrightarrow & (G'_1, S'_1, v'_1) \\
\uparrow & & \uparrow \\
(G, S, v) & \xrightarrow{(i,j)} & (G', S', v')
\end{array}$$

Then there is a unique extension of (i, j) to a morphism $(G_1, S_1, v_1) \to (G'_1, S'_1, v'_1)$ whose valued group component is continuous.

PROOF. Since i is actually uniformly continuous and S' is cofinal in S'_1, the group morphism $G \xrightarrow{i} G' \xrightarrow{\subseteq} G'_1$ is uniformly continuous. Hence it extends uniquely to a continuous map $i_1 \colon G_1 \to G'_1$. Then i_1 is a group morphism since it restricts to a group morphism $G \to G'$ on the dense subgroup G of G_1. Moreover, for every c-sequence (a_ρ) in G with limit $a \in G_1$ we have $i(a_\rho) \to i_1(a)$ in G'_1, hence

$$\begin{aligned}
j\big(v_1(a)\big) &= j\big(\text{eventual valuation of } (a_\rho)\big) \\
&= \text{eventual valuation of } i(a_\rho) \;=\; v'_1\big(i_1(a)\big).
\end{aligned}$$

Thus (i_1, j) is a morphism of valued abelian groups as desired. □

This gives the uniqueness of dense complete extensions:

COROLLARY 2.2.33. *Let G' be a dense complete extension of G and G_1 a dense extension of G. Then there is a unique embedding $G_1 \to G'$ over G, and if G_1 is in addition complete, this embedding is an isomorphism.*

Thanks to this corollary we may refer to a dense complete extension of G as the **completion** of G, to be denoted by G^c. By Corollary 2.2.32, taking completions is functorial in the following way:

COROLLARY 2.2.34. *If (G', S', v') is a valued abelian group extension of (G, S, v) and S is cofinal in S', then the inclusion $G \to G'$ extends uniquely to a valued abelian group embedding $G^c \to G'^c$.*

Next we clarify the relationship between pc-sequences and c-sequences:

LEMMA 2.2.35. *Each c-sequence in G that is not eventually constant has a cofinal subsequence which is a pc-sequence of width $\{\infty\}$. (Hence G is complete iff every pc-sequence of width $\{\infty\}$ has a pseudolimit in G.)*

PROOF. Let (a_ρ) be a c-sequence in G that is not eventually constant. Thus $\operatorname{cf}(S)$ is infinite. Take $a \in G^c$ such that $a_\rho \to a$. By passing to a suitable cofinal subsequence we arrange that $a_\rho \neq a$ for all ρ. By Lemma 2.2.25 we also arrange that the ordered index set of (a_ρ) is the set of all ordinals $\rho < \operatorname{cf}(S)$. Then the subset

$$\{v(a - a_\rho) : \rho < \operatorname{cf}(S)\}$$

of S is cofinal in S. An obvious transfinite recursion yields an ordinal λ and a strictly increasing sequence $\big(\rho(\gamma)\big)_{\gamma<\lambda}$ in the set of indices $\rho < \operatorname{cf}(S)$, such that $v\big(a - a_{\rho(\gamma)}\big)$ as a function of $\gamma < \lambda$ is strictly increasing and cofinal in S. Then $\operatorname{cf}(\lambda) = \operatorname{cf}(S)$ by Corollary 2.1.4. By passing to a cofinal subsequence and reindexing we replace λ by $\operatorname{cf}(\lambda)$, and then $\lambda = \operatorname{cf}(S)$. It remains to note that then $\{\rho(\gamma) : \gamma < \lambda\}$ is cofinal in $\{\rho : \rho < \operatorname{cf}(S)\}$ by Corollary 2.1.3. □

For use in Section 3.2 we also note:

LEMMA 2.2.36. *Suppose (G, S, v) is complete and (G', S', v') is a valued abelian group extension of (G, S, v). Let $a' \in G' \setminus G$ be such that $v'(a' - G)$ (a subset of S') has no largest element. Then $v'(a' - G) \subseteq S$, and $v'(a' - G)$ is not cofinal in S.*

PROOF. First, given $g \in G$, take $a \in G$ such that $v'(a' - g) < v'(a' - a)$. Then $v'(a' - g) = v(a - g)$. Thus $v'(a' - G) \subseteq S$. Towards a contradiction, assume that $v'(a' - G)$ is cofinal in S. Take a well-indexed sequence (a_ρ) in G such that $\big(v'(a' - a_\rho)\big)$ is strictly increasing and cofinal in $v'(a' - G)$, and hence in S. Then (a_ρ) is a pc-sequence of width $\{\infty\}$ in G, and thus $a_\rho \rightsquigarrow a \in G$, so $v'(a' - a) > v'(a' - G)$, a contradiction. □

Direct product of valued abelian groups

Let (G_i, S_i, v_i) be a valued abelian group for $i = 1, \ldots, n$, where $n \geqslant 1$. Suppose S is an ordered set containing each S_i as an ordered subset, with $S = S_1 \cup \cdots \cup S_n$. Set $G := G_1 \times \cdots \times G_n$ (direct product of abelian groups), and for each $g = (g_1, \ldots, g_n) \in G$, set

$$vg := \min\big(v_1(g_1), \ldots, v_n(g_n)\big).$$

Then $v\colon G \to S_\infty$ is a valuation on G. We call the valued abelian group (G, S, v) the **direct product** of the valued abelian groups (G_i, S_i, v_i). Given a closed ball $B = \overline{B}_g(s)$ in (G, S, v), with $g = (g_1, \ldots, g_n) \in G$, $s \in S$, we have

$$\begin{aligned} B &= B_1 \times \cdots \times B_n \quad \text{where } B_i = \big\{x \in G_i : v_i(x - g_i) \geqslant s\big\}, \text{ so} \\ B_i &= \bigcup \big\{\overline{B}_{g_i}(s_i) :\ s_i \in S_i,\ s_i \geqslant s\big\} \quad \text{if there is an } s_i \geqslant s \text{ in } S_i, \text{ and} \\ B_i &= \{g_i\} \quad \text{if there is no } s_i \geqslant s \text{ in } S_i. \end{aligned}$$

LEMMA 2.2.37. *If each (G_i, S_i, v_i) is spherically complete, then so is (G, S, v).*

PROOF. Let $\mathcal{B} \neq \emptyset$ be a nest of closed balls in (G, S, v). For each $B \in \mathcal{B}$, let $B_1, \ldots, B_n$ be as above. Let $i \in \{1, \ldots, n\}$ be given. If B_i is a singleton for some $B \in \mathcal{B}$, then $\bigcap\{B_i : B \in \mathcal{B}\}$ is a singleton. Otherwise, by the remarks preceding the lemma, $\{B_i : B \in \mathcal{B}\}$ is a nonempty nest of union-closed balls in (G_i, S_i, v_i). If each (G_i, S_i, v_i) is spherically complete, then Corollary 2.2.16 yields for each i a point $g_i \in \bigcap\{B_i : B \in \mathcal{B}\}$, and thus a point $g = (g_1, \ldots, g_n) \in \bigcap \mathcal{B}$. □

LEMMA 2.2.38. *Suppose each S_i is cofinal in S. Then:*

$$(G, S, v) \textit{ is complete} \iff \textit{each } (G_i, S_i, v_i) \textit{ is complete.}$$

PROOF. Let (g_ρ) be a well-indexed sequence in G, with $g_\rho = (g_{\rho 1}, \ldots, g_{\rho n})$ and $g_{\rho i} \in G_i$ for each i and ρ. By the cofinality assumption, (g_ρ) is a c-sequence in G iff $(g_{\rho i})$ is for every i a c-sequence in G_i. This yields the desired equivalence. □

We now consider the case where all (G_i, S_i, v_i) are equal, say $(G_i, S_i, v_i) = (H, S, v)$ for $i = 1, \ldots, n$. Then the valuation topology on G with respect to the valuation $v\colon G \to S_\infty$ on G defined above coincides with the product topology on $G = H^n$ with respect to the valuation topology of H. By the previous lemma, the dense extension $(H^{\mathrm{c}})^n$ of $G = H^n$ is complete, so we may take $(H^{\mathrm{c}})^n$ as the completion G^{c} of G. Note also that by Lemma 2.2.31:

COROLLARY 2.2.39. *Every c-continuous map $X \to H$ ($X \subseteq H^n$) has a unique continuous extension $X' \to H^{\mathrm{c}}$ to the closure X' of X in $(H^{\mathrm{c}})^n$.*

Notes and comments

The material in this section is classical valuation theory, some of it generalized from valued fields to valued abelian groups. An early explicit mention of valued abelian groups is in [144, Chapter IV, §4]. Lemma 2.2.1 is due to Krull [229]; the short proof given here was found by Gravett [153]. The concept of pseudocauchy sequence and its basic properties, for rank 1 valued fields, are due to Ostrowski [314]. (See also Section 3.2 below.) The Fixpoint Theorem 2.2.12 is due to Prieß-Crampe [324] (in a more general setting of ultrametric spaces).

Equivalence of pc-sequences will show up in several later chapters. This notion of equivalence seems to have been used first in [382], and the flexibility it provides was exploited heavily in [41] in connection with valued difference fields.

The completion of a valued abelian group G can also be introduced as the completion of the hausdorff topological group G (with respect to the valuation topology) equipped with the natural extension of the valuation of G to this completion by continuity (see, e.g., [323, §II.4]); we chose the approach via c-sequences in order to stress the analogy with pc-sequences and spherical completeness. Lemma 2.2.35 appears in [323, II, §4, Lemma 11], with a corrected proof (different from the one here) given in [375, Theorem 5.7].

2.3 VALUED VECTOR SPACES

In this section we define valued vector spaces, characterize spherically complete valued vector spaces, and then prove a version of the Hahn Embedding Theorem for valued vector spaces. We pay special attention to Hahn spaces, which are particularly well-behaved valued vector spaces.

Let C be a field. A **valued vector space over** C is a vector space G over C equipped with a surjective valuation $v\colon G \to S_\infty$ on its underlying additive group such that $v(\lambda a) = v(a)$ for all $\lambda \in C^\times$ and $a \in G$. Note that then for each $s \in S$ the subgroups $B(s) = B_0(s)$ and $\overline{B}(s) = \overline{B}_0(s)$ of G are subspaces of C, and hence $G(s) = \overline{B}(s)/B(s)$ is in a natural way a vector space over C.

EXAMPLE (Hahn products). Let S be an ordered set and $(G_s)_{s\in S}$ a family of vector spaces $G_s \neq \{0\}$ over C. Then the Hahn product $H = H\big[(G_s)\big]$ of (G_s) is a subspace of the vector space $\prod_s G_s$ over C, and H with its Hahn valuation is a valued vector space over C. For each $s \in S$ the natural group isomorphism $H(s) \to G_s$ is C-linear.

For the remainder of this section (G,S,v) and (G',S',v') are valued vector spaces over C. We call (G,S,v) a **valued subspace** of (G',S',v'), or (G',S',v') an **extension** of (G,S,v), if, as valued abelian groups, (G,S,v) is a valued subgroup of (G,S,v), and G is a subspace (not merely a subgroup) of G'.

An **embedding** of valued vector spaces over C is an embedding

$$(i,j)\ :\ (G,S,v) \to (G',S',v')$$

of valued abelian groups as defined in Section 2.2 such that $i\colon G \to G'$ is C-linear. Such an embedding $(i,j)\colon (G,S,v) \to (G',S',v')$ is said to be an **isomorphism** (of valued vector spaces over C) if $i\colon G \to G'$ and $j\colon S_\infty \to S'_\infty$ are bijections; note that then $(i^{-1}, j^{-1})\colon (G',S',v') \to (G,S,v)$ is also an isomorphism. Such an embedding $(i,j)\colon (G,S,v) \to (G',S',v')$ determines an isomorphism

$$(i,j) \;:\; (G,S,v) \to (iG, jS, v'|iG)$$

onto a valued subspace of (G',S',v').

We consider C as the 1-dimensional valued vector space over C whose valuation is trivial. Note that any 1-dimensional valued vector space over C is isomorphic to the valued vector space C.

Below (G,S,v) is usually referred to as *the valued vector space* G, in particular when S and v are clear from the context. Also, when $S = S'$, as ordered sets, then by an embedding $i\colon G \to G'$ of valued vector spaces over C we mean an embedding $(i, \mathrm{id}_{S_\infty})\colon (G,S,v) \to (G',S,v')$ of valued vector spaces over C.

Maximal valued vector spaces

Note that if (a_ρ) is a pc-sequence in G and $a_\rho \rightsquigarrow a \in G$, then for $g \in G$ and $\lambda \in C^\times$ we have a pc-sequence $(g + \lambda a_\rho)$ in G with $g + \lambda a_\rho \rightsquigarrow g + \lambda a$.

LEMMA 2.3.1. *Let (a_ρ) be a divergent pc-sequence in G. Then G has an immediate extension $G \oplus Ca$ with $a_\rho \rightsquigarrow a$, such that any extension $G \oplus Cb$ of G with $a_\rho \rightsquigarrow b$ gives an isomorphism $G \oplus Ca \to G \oplus Cb$ of valued vector spaces over C that is the identity on G and sends a to b.*

PROOF. Take an extension $G \oplus Ca$ of G such that $a_\rho \rightsquigarrow a$. (This is possible by Lemma 2.2.5 and the remark following its proof.) We claim that this extension is immediate. Indeed, we have $s \in S$ such that

$$v(a - a_\rho) \;>\; va \;=\; va_\rho \;=\; s, \text{ eventually,}$$

in particular, $va \in S$, and, eventually, $a + B(s) = a_\rho + B(s)$ in $G \oplus Ca$. Instead of a, this also works for each of its affine transforms $g + \lambda a$ with $g \in G$ and $\lambda \in C^\times$, since $(g + \lambda a_\rho)$ is a divergent pc-sequence in G with $g + \lambda a_\rho \rightsquigarrow g + \lambda a$, for such g, λ. This proves the immediacy claim. If $G \oplus Cb$ is also an extension with $a_\rho \rightsquigarrow b$, then by the above, for any $g \in G$ and $\lambda \in C^\times$,

$$g + \lambda a \;\asymp\; g + \lambda a_\rho \;\asymp\; g + \lambda b, \text{ eventually,}$$

and so we have an isomorphism $G \oplus Ca \to G \oplus Cb$ of valued vector spaces over C that is the identity on G and sends a to b. □

Call G **maximal** if it has no proper immediate (valued vector space) extension. From Corollary 2.2.6 and Lemmas 2.2.10, 2.3.1 we obtain:

COROLLARY 2.3.2. *The following are equivalent:*

(i) *G is maximal as a valued abelian group;*

(ii) *G is maximal;*

(iii) *each pc-sequence in G has a pseudolimit in G;*

(iv) *G is spherically complete.*

Using Zorn, we obtain from Lemmas 2.2.1, 2.2.4, and 2.3.1:

COROLLARY 2.3.3. *The valued vector space G has a maximal immediate extension, and all maximal immediate extensions of G are isomorphic over G.*

The next result is a form of the Hahn Embedding Theorem and identifies the maximal immediate extension of G from Corollary 2.3.3.

PROPOSITION 2.3.4. *There exists an embedding $i\colon G \to H\big[(G(s))\big]$ of valued vector spaces over C such that for all $g \in G^{\neq}$,*

$$i(g)_s \;=\; g + B(s) \in G(s) \quad \textit{where } s = v(g).$$

Note that then $H\big[(G(s))\big]$ is an immediate extension of its valued subspace iG.

PROOF. Set $H := H\big[(G(s))\big]$, and let H' be the subspace of H consisting of all $g \in \prod_s C(s)$ with *finite* support. The valuation of H restricts to a surjective valuation $H' \to S_\infty$, making H an immediate extension of its valued subspace H'.

Choose for each $s \in S$ a C-linear right-inverse $f_s\colon G(s) \to \overline{B}(s)$ of the natural projection $\overline{B}(s) \to \overline{B}(s)/B(s) = G(s)$, and let f be the C-linear map $H' \to G$ such that $f((g_s)) = \sum_s f_s(g_s)$. Then G is an immediate extension of its valued subspace $G' := f(H')$, and we have an isomorphism $i'\colon G' \to H'$ of valued vector spaces over C given by $i'(g) = h$ iff $f(h) = g$, for all $g \in G', h \in H'$.

Take a maximal immediate extension $\widetilde{G}$ of G. This is also a maximal immediate extension of G'. Since H is a maximal immediate extension of H', it follows from Corollary 2.3.3 that i' extends to an isomorphism $\widetilde{G} \to H$. Then the restriction of this isomorphism to G given an embedding $i\colon G \to H$ as in the proposition. □

In Section 2.2 we defined the completion G^c of G as a valued abelian group. We make G^c into a vector space over C as follows. Given $\lambda \in C$, the map

$$x \mapsto \lambda x \,:\, G \to G$$

is uniformly continuous, so extends uniquely to a continuous map $G^c \to G^c$ by Lemma 2.2.32. Denoting the image of $x \in G^c$ under this map also by λx and varying λ we have an operation

$$(\lambda, x) \mapsto \lambda x \,:\, C \times G^c \to G^c$$

that makes G^c a valued vector space over C extending G.

Valuation bases

Let G_0 be a subspace of G, and let $\mathcal{B} \subseteq G$. We say that $\mathcal{B}$ is **valuation-independent over** G_0 if $0 \notin \mathcal{B}$, and for every family $(\lambda_b)_{b\in\mathcal{B}}$ of elements of C, with $\lambda_b \neq 0$ for only finitely many $b \in \mathcal{B}$, and every $g_0 \in G_0$,

$$v\left(\sum_{b\in\mathcal{B}} \lambda_b b + g_0\right) = \min\left(\{vb : b \in \mathcal{B}, \lambda_b \neq 0\} \cup \{v(g_0)\}\right).$$

This gives a linearly independent family $(b + G_0)_{b\in\mathcal{B}}$ in the C-vector space G/G_0. For a set $X \subseteq G$ we let $\langle X\rangle$ be the C-linear subspace of G generated by X. If $\mathcal{B}$ is valuation-independent over G_0 and $\langle\mathcal{B} \cup G_0\rangle = G$, then we call $\mathcal{B}$ a **valuation basis** of G over G_0. If $\mathcal{B}$ is valuation-independent over $\{0\}$, then we just say that $\mathcal{B}$ is valuation-independent, and if $\mathcal{B}$ is a valuation basis of G over $\{0\}$, then we just say that $\mathcal{B}$ is a valuation basis of G. Thus:

LEMMA 2.3.5. *Let $\mathcal{B}, \mathcal{B}' \subseteq G$ be disjoint. Then $\mathcal{B} \cup \mathcal{B}'$ is valuation-independent over G_0 if and only if $\mathcal{B}$ is valuation-independent over G_0 and $\mathcal{B}'$ is valuation-independent over $\langle\mathcal{B} \cup G_0\rangle$.*

Every valuation basis of G over G_0 is clearly maximal among the subsets of G that are valuation-independent over G_0. It is also easy to see that an increasing union of subsets of G, each valuation-independent over G_0, is valuation-independent over G_0; hence by Zorn, there exists a maximal (possibly empty) subset of G which is valuation-independent over G_0.

LEMMA 2.3.6. *The extension $G_0 \subseteq G$ is immediate if and only if G has no nonempty subset that is valuation-independent over G_0.*

PROOF. Suppose $G_0 \subseteq G$ is immediate, and let $g \in G^{\neq}$. Take $g_0 \in G_0$ such that $g \sim g_0$. Then $v(g - g_0) \neq \min(vg, vg_0)$, so $\{g\}$ is not valuation-independent over G_0. On the other hand, suppose $G_0 \subseteq G$ is not immediate. Then we can take $g \in G^{\neq}$ such that for each $g_0 \in G_0$ we have $v(g - g_0) \leqslant vg$ and hence $v(g - g_0) = \min(vg, vg_0)$; then $\{g\}$ is valuation-independent over G_0. □

COROLLARY 2.3.7. *There exists a subspace $G_1 \supseteq G_0$ of G such that G_1 admits a valuation basis over G_0, and $G_1 \subseteq G$ is immediate.*

PROOF. Let $\mathcal{B} \subseteq G$ be maximal valuation-independent over G_0 and $G_1 := \langle\mathcal{B} \cup G_0\rangle$. Then G_1 admits a valuation basis over G_0. Moreover, if $\emptyset \neq \mathcal{B}' \subseteq G$ is valuation-independent over G_1, then $\mathcal{B} \cup \mathcal{B}'$ is valuation-independent over G_0, contradicting maximality of $\mathcal{B}$. Hence by Lemma 2.3.6, $G_1 \subseteq G$ is immediate. □

COROLLARY 2.3.8. *Suppose the set $v(G \setminus G_0)$ is reverse well-ordered. Then G has a valuation basis over G_0.*

PROOF. Take G_1 as in the previous lemma. For a contradiction assume $G \neq G_1$. Then the nonempty subset $v(G \setminus G_1)$ of $v(G \setminus G_0)$ is also reverse well-ordered, so we can take $g \in G \setminus G_1$ such that $vg = \max v(G \setminus G_1)$. Since $G_1 \subseteq G$ is immediate there is some $g_1 \in G_1$ with $v(g - g_1) > vg$, and $g - g_1 \in G \setminus G_1$, a contradiction. □

We denote the dimension of a vector space V over C by $\dim_C V$.

LEMMA 2.3.9. $\dim_C G/G_0 \geqslant |v(G) \setminus v(G_0)|$.

PROOF. If $\mathcal{B}$ is a subset of G such that $vb \notin v(G_0)$ for each $b \in \mathcal{B}$ and $b \not\asymp b'$ for all $b \neq b'$ in $\mathcal{B}$, then $\mathcal{B}$ is C-linearly independent over G_0. □

Note that by Corollary 2.3.8 and Lemma 2.3.9, every finite-dimensional valued vector space over C has a valuation basis. More generally:

COROLLARY 2.3.10. *Every valued vector space over C of countable dimension has a valuation basis.*

PROOF. Suppose G has dimension $\aleph_0$. Then $G = \bigcup_n G_n$ where

$$G_0 = \{0\}, \quad G_n \subseteq G_{n+1}, \quad \dim_C G_{n+1}/G_n = 1 \qquad \text{(for every } n\text{)}.$$

Then each G_n is finite-dimensional, so $v(G_n)$ is finite by Lemma 2.3.9. Thus G_n as a valued subspace of G is spherically complete and hence maximal (Corollary 2.3.2), so $G_n \subseteq G_{n+1}$ is not immediate. By Corollary 2.3.7 and since $\dim_C G_{n+1}/G_n = 1$, we can take $b_n \in G_{n+1}$ such that $\{b_n\}$ is a valuation basis for G_{n+1} over G_n. Then $\{b_n : n \geqslant 0\}$ is a valuation basis of G, by Lemma 2.3.5. □

In Section 7.6 we shall need "good" right-inverses to certain linear maps:

LEMMA 2.3.11. *Suppose G is maximal and $A \colon G \to G$ is a C-linear map such that $g \succcurlyeq A(g)$ for all $g \in G$, and for all $h \in G$ there is $g \in G$ with $A(g) = h$ and $g \asymp h$. Then there exists a C-linear map $B \colon G \to G$ such that $A \circ B = \mathrm{id}_G$ and $B(h) \asymp h$ for all $h \in G$.*

PROOF. Assume $H_0 \neq G$ is a C-linear subspace of G and $B_0 \colon H_0 \to G$ is a C-linear map such that $A \circ B_0$ is the inclusion $H_0 \to G$ and $B_0(h) \asymp h$ for all $h \in H_0$.

CLAIM: *There exists $h_1 \in G \setminus H_0$ such that B_0 extends to a C-linear map*

$$B_1 : H_1 \to G, \qquad H_1 := H_0 + Ch_1,$$

for which $A \circ B_1$ is the inclusion $H_1 \to G$ and $B_1(h) \asymp h$ for all $h \in H_1$.

It is clear that Lemma 2.3.11 follows from this claim. To prove the claim we first pick an element $b \in G \setminus H_0$ and distinguish two cases:

CASE 1: *$H_0 + Cb$ is not an immediate extension of H_0 (where both are viewed as valued subspaces of G).* Then we set $H_1 := H_0 + Cb$. In view of Lemma 2.3.6 we can take $h_1 \in H_1 \setminus H_0$ such that $\{h_1\}$ is a valuation basis of H_1 over H_0. Take any $g \in G$

with $A(g) = h_1$ and $g \asymp h_1$, and let $B_1 \colon H_1 \to G$ be the C-linear extension of B_0 with $B_1(h_1) = g$. Then B_1 has the claimed property.

CASE 2: *$H_0 + Cb$ is an immediate extension of H_0.* Take a divergent pc-sequence (b_ρ) in H_0 such that $b_\rho \leadsto b$. Then $(a_\rho) := \big(B_0(b_\rho)\big)$ is a divergent pc-sequence in $B_0(H_0)$, and has a pseudolimit a in G. Then $A(a_\rho) = b_\rho \leadsto h_1 := A(a)$ by Lemma 2.2.9, so $h_1 \notin H_0$. Set $H_1 := H_0 + Ch_1$ and let $B_1 \colon H_1 \to G$ be the C-linear extension of B_0 with $B_1(h_1) = a$. Then B_1 has the claimed property. To see this, consider any $h = h_0 + \lambda h_1 \in H_1$ with $h_0 \in H_0$ and $\lambda \in C^\times$. Then $\big(B_0(h_0) + \lambda a_\rho\big)$ is a divergent pc-sequence in $B_0(H_0)$ and $B_0(h_0) + \lambda a_\rho \leadsto B_0(h_0) + \lambda a$, so

$$B_1(h) \;=\; B_0(h_0) + \lambda a \;\asymp\; B_0(h_0) + \lambda a_\rho, \quad \text{eventually}$$

by Lemma 2.2.3. Applying A to $B_0(h_0) + \lambda a_\rho$ we get $h_0 + \lambda b_\rho \leadsto h_0 + \lambda h_1 = h$, and so $h \asymp h_0 + \lambda b_\rho$, eventually. Since $B_0(h_0) + \lambda a_\rho \asymp h_0 + \lambda b_\rho$ for all ρ, we get $B_1(h) \asymp h$, as required. □

The rest of this section is mainly intended for use in a later volume.

Hahn spaces

If for all $a, b \in G^{\neq}$ with $a \asymp b$ there exists $\lambda \in C$ such that $a \sim \lambda b$, then we call G a **Hahn space over** C. Equivalently, G is a Hahn space over C iff all vector spaces $G(s)$ have dimension 1. In particular, an immediate extension of a Hahn space is a Hahn space, and a valued subspace of a Hahn space is again a Hahn space. Given a family (G_s) of 1-dimensional vector spaces over C, indexed by the elements s of an ordered set S, the Hahn product $H\big[(G_s)\big]$ of (G_s) is a Hahn space over C. By Proposition 2.3.4, every Hahn space embeds into a Hahn product of this kind:

COROLLARY 2.3.12. *If G is a Hahn space over C, then there is an embedding $i \colon G \to H[S, C]$ of valued vector spaces over C such that $H[S, C]$ is an immediate extension of its valued subspace iG.*

Suppose G is a Hahn space over C and $\mathcal{B} \subseteq G$ with $0 \notin \mathcal{B}$. Then $\mathcal{B}$ is valuation-independent iff $b \not\asymp b'$ for all $b \neq b'$ in $\mathcal{B}$, and $\mathcal{B}$ is maximal valuation-independent iff it is valuation-independent and $v(\mathcal{B}) = S$. Thus if G admits a valuation basis, then $\dim_C G = |S|$, and G is determined (up to isomorphism of valued vector spaces) by the ordered set S. By Corollary 2.3.10:

LEMMA 2.3.13. *Suppose G is a Hahn space of countable dimension as a vector space over C. Then G has a basis $\mathcal{B}$ with $v(\mathcal{B}) = S$ and $b \not\asymp b'$ for all $b \neq b'$ in $\mathcal{B}$.*

Scalar extension

In this subsection G is a Hahn space over C, and D is an extension field of C. We construe $G_D := D \otimes_C G$ as a vector space over D in the usual way. Identifying $g \in G$ with $1 \otimes g \in G_D$ makes G a C-linear subspace of G_D.

LEMMA 2.3.14. *Let $h \in G_D$. There are $d_i \in D^{\neq}$, $g_i \in G^{\neq}$, $i = 1, \dots, m$, such that*

$$h = \sum_{i=1}^{m} d_i g_i, \qquad g_1 \succ \cdots \succ g_m. \tag{2.3.1}$$

PROOF. We have $h = \sum_{j=1}^{n} e_j h_j$ ($e_j \in D$, $h_j \in G$). We use induction on n to get h in the form (2.3.1). If $n = 1$ we can do this with $m = 0$ or $m = 1$, so suppose $n > 1$. We can assume $h_1 \succcurlyeq \cdots \succcurlyeq h_n \neq 0$. Take $c_j \in C$ with $h_j - c_j h_1 \prec h_1$ for $j = 2, \dots, n$. Then

$$h = \left(e_1 + \sum_{j=2}^{n} c_j e_j\right) h_1 + \sum_{j=2}^{n} e_j (h_j - c_j h_1).$$

It remains to apply the inductive hypothesis to $\sum_{j=2}^{n} e_j (h_j - c_j h_1)$. □

COROLLARY 2.3.15. *There is a unique valuation w on the abelian group G_D that extends the valuation v of G and makes G_D a valued vector space over D. It makes G_D a Hahn space over D with $w(G_D) = v(G)$.*

PROOF. For such a valuation w and $h \in G_D^{\neq}$ as in (2.3.1) we have $w(h) = v(g_1)$. As to existence: Corollary 2.3.12 gives an embedding $G \to H[S, C]$ of valued vector spaces over C; it extends to a D-linear injective map $G_D \to H[S, D]$. The Hahn valuation of $H[S, D]$ yields in this way a valuation w on G_D as claimed. □

Fredholm operators

In order to discuss some properties of linear operators on Hahn spaces, we begin with some generalities about Fredholm operators. Let V and V' be vector spaces over C and $A\colon V \to V'$ be a C-linear map, with kernel $\ker A := \{v \in V : A(v) = 0\}$ and cokernel $\operatorname{coker} A := V'/A(V)$. One says that A is a **Fredholm operator** if both C-vector spaces $\ker A$ and $\operatorname{coker} A$ are finite-dimensional, and in that case we define the **index** of A by

$$\operatorname{index} A := \dim_C \ker A - \dim_C \operatorname{coker} A \qquad \text{(an integer)}.$$

If both V and V' are finite-dimensional, then every C-linear map $V \to V'$ is a Fredholm operator of index $\dim_C V - \dim_C V'$.

PROPOSITION 2.3.16. *If $A\colon V \to V'$ and $B\colon V' \to V''$ are Fredholm operators, then so is $B \circ A\colon V \to V''$, and* $\operatorname{index} B \circ A = \operatorname{index} B + \operatorname{index} A$.

Towards the proof of this proposition, we first note an easy lemma:

LEMMA 2.3.17. *Let $A\colon V \to V'$ and $B\colon W \to W'$ be C-linear. Then A and B are Fredholm operators iff $A \oplus B\colon V \oplus W \to V' \oplus W'$ is one; in that case*

$$\operatorname{index}(A \oplus B) = \operatorname{index} A + \operatorname{index} B.$$

We also need two small items from linear algebra; for the first one, see for example [249, Lemma III.9.1]; the second one is a special case of Lemma 1.2.3.

LEMMA 2.3.18 (Snake Lemma). *Assume that the following diagram of vector spaces over C and C-linear maps is commutative with exact rows:*

$$\begin{array}{ccccccc} & & V' & \xrightarrow{i} & V & \xrightarrow{p} & V'' & \longrightarrow & 0 \\ & & \downarrow{\scriptstyle F'} & & \downarrow{\scriptstyle F} & & \downarrow{\scriptstyle F''} & & \\ 0 & \longrightarrow & W' & \xrightarrow{j} & W & \xrightarrow{q} & W'' & & \end{array}$$

Then we have an exact sequence of C-linear maps

$$\ker F' \longrightarrow \ker F \longrightarrow \ker F'' \xrightarrow{\delta} \operatorname{coker} F' \longrightarrow \operatorname{coker} F \longrightarrow \operatorname{coker} F''$$

where the maps besides the "connecting" map δ are the natural ones.

LEMMA 2.3.19. *For each exact sequence*

$$0 \longrightarrow V_1 \longrightarrow V_2 \longrightarrow \cdots \longrightarrow V_n \longrightarrow 0$$

of finite-dimensional vector spaces over C we have

$$\sum_{i=1}^{n} (-1)^i \dim_C V_i = 0.$$

Suppose in Lemma 2.3.18 the map i is injective, and the map q is surjective. Then by Lemmas 2.3.18 and 2.3.19, if two of the three maps F, F' and F'' are Fredholm operators, then so is the third, and $\operatorname{index} F = \operatorname{index} F' + \operatorname{index} F''$.

PROOF OF PROPOSITION 2.3.16. Let $A\colon V \to V'$ and $B\colon V' \to V''$ be Fredholm operators. Let

$$i\colon V \to V \oplus V', \quad p\colon V \oplus V' \to V', \quad j\colon V' \to V'' \oplus V', \quad q\colon V'' \oplus V' \to V'$$

be the C-linear maps given by

$$\begin{aligned} i(x) &= \big(x, A(x)\big), & p(x,x') &= A(x) - x', \\ j(x') &= \big(B(x'), x'\big), & q(x'',x') &= x'' - B(x') \qquad (x \in V, x' \in V', x'' \in V''). \end{aligned}$$

Then the diagram

$$\begin{array}{ccccccccc} 0 & \longrightarrow & V & \xrightarrow{i} & V \oplus V' & \xrightarrow{p} & V' & \longrightarrow & 0 \\ & & \downarrow{\scriptstyle A} & & \downarrow{\scriptstyle (B\circ A)\oplus \mathrm{id}_{V'}} & & \downarrow{\scriptstyle B} & & \\ 0 & \longrightarrow & V' & \xrightarrow{j} & V'' \oplus V' & \xrightarrow{q} & V'' & \longrightarrow & 0 \end{array}$$

commutes and has exact rows. It now remains to use the remark preceding this proof and Lemma 2.3.17. □

Linear operators on Hahn spaces

In this subsection G is a Hahn space over C, and $A\colon G \to G$ is a C-linear operator. We set $\ker^{\neq} A := (\ker A)^{\neq}$.

LEMMA 2.3.20. *Suppose that* $\ker A$ *is finite-dimensional. Then there exists a C-linear subspace M of G such that*

$$G = M \oplus \ker A, \qquad v(M) \cap v(\ker^{\neq} A) = \emptyset.$$

PROOF. By Corollary 2.3.12 we may assume that G is a valued subspace of the Hahn product $H[S, C]$. Let $y_1, \dots, y_d \in \ker^{\neq} A$ be such that $vy_1 > \cdots > vy_d$ are the distinct elements of $v(\ker^{\neq} A)$. Then

$$M := \{a \in G : \operatorname{supp}(a) \cap \{vy_1, \dots, vy_d\} = \emptyset\}$$

is a C-linear subspace of G such that that $v(M) \cap v(\ker^{\neq} A) = \emptyset$ (in particular $M \cap \ker A = \{0\}$). Given $a \in G$ we let $i(a)$ denote the largest $i \in \{1, \dots, d\}$ such that $vy_i \in \operatorname{supp}(a)$, if such an i exists, and $i(a) := 0$ otherwise. We show that for every $a \in G$ we have $a \in M + \ker A$ by induction on $i(a)$. If $i(a) = 0$, then $a \in M$. If $i(a) = i > 0$, then there exists $\lambda \in C^{\times}$ such that $vy_i \notin \operatorname{supp}(a - \lambda y_i)$, hence $i(a - \lambda y_i) < i$ and therefore $a = (a - \lambda y_i) + \lambda y_i \in M + \ker A$ by inductive hypothesis. Thus M has the required properties. □

COROLLARY 2.3.21. *If* $\ker A$ *is finite-dimensional, then*

$$v(A(G)) = \{v(A(y)) : y \in G,\ vy \notin v(\ker^{\neq} A)\}.$$

The set $\mathcal{F}_G$ of Fredholm operators $G \to G$ is a monoid under composition (by Proposition 2.3.16), with identity element id_G. In the lemma below we let $\mathcal{S}_G$ be the submonoid of $\mathcal{F}_G$ consisting of all surjective Fredholm operators $G \to G$.

LEMMA 2.3.22. *There exists a map $A \mapsto A^{-1}\colon \mathcal{S}_G \to \mathcal{F}_G$ such that for all $A \in \mathcal{S}_G$:*

$$A \circ A^{-1} = \mathrm{id}_G, \quad v(A^{-1}(G)) \cap v(\ker^{\neq} A) = \emptyset,$$

and if $B\colon G \to G$ is a bijective C-linear operator, then

$$(B \circ A)^{-1} = A^{-1} \circ B^{-1}.$$

PROOF. As in the proof of Lemma 2.3.20 we may assume that G is a valued subspace of $H[S, C]$. For each $A \in \mathcal{F}_G$ we define the subspace $M_A = M$ as in that proof. Let $A \in \mathcal{S}_G$. For $g \in G$ we let $A^{-1}(g)$ be the unique $f \in M_A$ with $A(f) = g$. This yields a C-linear operator $A^{-1}\colon G \to G$ with $A(A^{-1}(g)) = g$ and $v(A^{-1}(g)) \notin v(\ker^{\neq} A)$ for all $g \in G$. Note that A^{-1} is injective and $A^{-1}(G) = M_A$; in particular, $\ker A^{-1} = \{0\}$ and $\operatorname{coker} A^{-1} = G/M_A \cong \ker A$ are both finite-dimensional, hence $A^{-1} \in \mathcal{F}_G$. Let in addition $B\colon G \to G$ be a bijective C-linear operator; then $\ker B \circ A = \ker A$, hence $M_{B \circ A} = M_A$. So if $g \in G$, then $f := A^{-1}(B^{-1}(g))$ satisfies $f \in M_{B \circ A}$ and $(B \circ A)(f) = g$, hence $f = (B \circ A)^{-1}(g)$. Thus $(B \circ A)^{-1} = A^{-1} \circ B^{-1}$. □

Notes and comments

Corollaries 2.3.2 and 2.3.3 are in Gravett [152]. (For valued abelian *groups,* the equivalence of maximality and spherical completeness remains true, whereas the uniqueness in Corollary 2.3.3 does not go through; see [140].) Proposition 2.3.4 is from Conrad [91]; it generalizes the classical embedding theorem for ordered abelian groups (Corollary 2.4.19 below) of Hahn [162]. The proof given here follows [152]. (See [124] for a detailed discussion of Hahn's work and its various spinoffs.) Corollary 2.3.10 is due to Brown [65]. Our presentation follows [238]. Our notion of Hahn space generalizes that of [18] where Hahn spaces are a special kind of ordered vector spaces; see Section 2.4 below. Lemma 2.3.14 is implicit in the proof of [364, Theorem 3] and is a variant of [18, Lemma 2.1]. The concept of Fredholm operator and Proposition 2.3.16 are well-known in operator theory; see [379].

2.4 ORDERED ABELIAN GROUPS

An **ordered abelian group** is an abelian group G (usually written additively) equipped with a (total) ordering $\leqslant$ on its underlying set such that for all $x, y, z \in G$,

$$x \leqslant y \quad \Rightarrow \quad x + z \leqslant y + z.$$

Each ordered abelian group is a hausdorff topological group with respect to its order topology. *In the rest of this section G is an ordered abelian group*, and we let a, b, c range over G. We put $|a| := \max(a, -a)$. We set $G^< := G^{<0}$ and $G^> := G^{>0}$, and similarly with $\leqslant$ and $\geqslant$ in place of $<$ and $>$, respectively. We extend the addition on G to G_∞ by $a + \infty = \infty + a = \infty + \infty = \infty$.

EXAMPLE (Hahn products). Let $(G_s)_{s \in S}$ be a family of ordered abelian groups $G_s \neq \{0\}$ indexed by an ordered set S. Then there is a unique ordering making the Hahn product $H\big[(G_s)\big]$ of (G_s) an ordered abelian group such that

$$0 < g \quad \Longleftrightarrow \quad 0 < g_{vg},$$

for all nonzero $g = (g_s) \in H\big[(G_s)\big]$; call it the **Hahn ordering** of $H\big[(G_s)\big]$.

Every ordered abelian group is torsion-free. We shall consider each torsion-free abelian group A as a subgroup of the divisible abelian group $\mathbb{Q}A = \mathbb{Q} \otimes_{\mathbb{Z}} A$ (the divisible hull of A) via the embedding $x \mapsto 1 \otimes x$. We also equip $\mathbb{Q}G$ with the unique ordering that makes it an ordered abelian group containing G as an ordered subgroup. Any torsion-free abelian group A embeds as a group into a Hahn product $H[S, \mathbb{Q}]$ for some ordered set S: picking a basis S of the vector space $\mathbb{Q}A$ over $\mathbb{Q}$ and an order on the set S yields group inclusions and a group isomorphism

$$A \subseteq \mathbb{Q}A = \bigoplus_{s \in S} \mathbb{Q}s \subseteq H[(\mathbb{Q}s)] \cong H[S, \mathbb{Q}].$$

Thus by the example on Hahn products above, every torsion-free abelian group can be expanded to an ordered abelian group.

Convex valuations and archimedean classes

Let $v\colon G \to S_\infty$ be a surjective valuation on G. We say that v is **compatible** with the ordering of G or a **convex valuation** of G if one of the following equivalent conditions is satisfied:

(C1) for all a, b, if $0 < a < b$, then $va \geqslant vb$;

(C2) for every $s \in S$ the subgroup $B(s) = \{a : va > s\}$ of G is convex;

(C3) for every $s \in S$ the subgroup $\overline{B}(s) = \{a : va \geqslant s\}$ of G is convex.

Note that then $va \leqslant vb$ whenever $n|a| \geqslant |b|$, and $v(ka) = va$ when $k \in \mathbb{Z}^{\neq}$. Hence if v is convex, then v has a unique extension to a convex valuation $\mathbb{Q}G \to S_\infty$.

The **archimedean class** of a is defined by

$$[a] := \big\{g \in G : |a| \leqslant n|g| \text{ and } |g| \leqslant n|a| \text{ for some } n \geqslant 1\big\}.$$

The archimedean classes partition G. Each archimedean class $[a]$ with $a \neq 0$ is the disjoint union of the two convex sets $[a] \cap G^{<}$ and $[a] \cap G^{>}$. We order the set

$$[G] := \big\{[a] : a \in G\big\}$$

of archimedean classes by

$$[a] < [b] \quad :\Longleftrightarrow \quad n|a| < |b| \text{ for all } n \geqslant 1.$$

We also write $a = o(b)$ instead of $[a] < [b]$. We have $[0] < [a]$ for all $a \in G^{\neq}$, and

$$[a] \leqslant [b] \quad \Longleftrightarrow \quad |a| \leqslant n|b| \text{ for some } n \geqslant 1.$$

We also write $a = O(b)$ instead of $[a] \leqslant [b]$. Equipping $[G]$ with the reversed ordering on $[G]$ so as to make $[0]$ its largest element, the map $x \mapsto [x]$ becomes a convex valuation of G, called the **standard valuation** of G. The surjective convex valuations on G are precisely the coarsenings of the standard valuation of G.

LEMMA 2.4.1. *Let v be a convex valuation on G such that $S = v(G^{\neq})$ has no largest element. Then the v-topology on G agrees with the order topology on G.*

PROOF. If $a > 0$, then $B(va) \subseteq (-a, a)$. If $s \in S$, then $(-b, b) \subseteq B(s)$ for any b with $vb > s$. □

Let v be a convex valuation on G and (a_ρ) a pc-sequence in G with respect to v. We wish to view the pseudolimits of (a_ρ) in terms of the ordering of G. For simplicity, assume $v(a_\tau - a_\sigma) > v(a_\sigma - a_\rho)$ for all indices $\tau > \sigma > \rho$, in particular, $a_\rho \neq a_\sigma$ for all $\rho \neq \sigma$. For each ρ, let $s(\rho)$ be the immediate successor of ρ in the set of indices, and set $d_\rho := a_{s(\rho)} - a_\rho$. We divide the set of indices into two disjoint subsets:

$$L := \{\lambda : a_\lambda < a_{s(\lambda)}\}, \qquad R := \{\rho : a_{s(\rho)} < a_\rho\}.$$

Then $a_\lambda < a_\sigma < a_\lambda + 2d_\lambda$ whenever $\lambda \in L$ and $\lambda < \sigma$, and $a_\rho + 2d_\rho < a_\sigma < a_\rho$ whenever $\rho \in R$ and $\rho < \sigma$. Thus $a_\lambda < a_\rho$ for all $\lambda \in L$ and $\rho \in R$. Set

$$P := \{a_\lambda : \lambda \in L\} \cup \{a_\rho + 2d_\rho : \rho \in R\},$$
$$Q := \{a_\rho : \rho \in R\} \cup \{a_\lambda + 2d_\lambda : \lambda \in L\}.$$

LEMMA 2.4.2. *We have $P < Q$, and for every $a \in G$,*

$$a_\rho \rightsquigarrow a \iff P < a < Q.$$

In particular, if G has no least positive element, and κ is a cardinal such that the set of indices ρ has cardinality $< \kappa$ and G is κ-saturated as an ordered set, then (a_ρ) has a pseudolimit in G. (See B.9 for the notion of κ-saturation.)

PROOF. We get $P < Q$ from the inequalities already stated. Suppose $a_\rho \rightsquigarrow a \in G$. Then for each ρ we have $v(a - a_{s(\rho)}) > vd_\rho$, so $a_\rho < a < a_\rho + 2d_\rho$ when $\rho \in L$, and $a_\rho + 2d_\rho < a < a_\rho$ when $\rho \in R$. This gives $P < a < Q$. Next, assume $P < a < Q$. Then $v(a - a_\rho) \geqslant vd_\rho$ for all ρ, and so $a_\rho \rightsquigarrow a$ by Lemma 2.2.9. □

We say that G is **archimedean** if $[G^{\neq}] := [G] \setminus \{[0]\}$ is a singleton, that is, $G \neq \{0\}$, and for all nonzero a and b there is some n such that $|a| \leqslant n|b|$. Thus if G is archimedean, then there is no nontrivial convex valuation on G. Moreover:

LEMMA 2.4.3 (Hölder). *If G is archimedean and $e \in G^{>}$, then there is a unique embedding of G into the ordered additive group of $\mathbb{R}$ sending e to 1.*

We leave the easy proof of this fact to the reader.

EXAMPLE. Let $(G_s)_{s \in S}$ be a family of ordered abelian groups $G_s \neq \{0\}$ indexed by an ordered set S, and $H = H\big[(G_s)\big]$ the Hahn product of (G_s). Then the valuation v of H is convex. If every ordered abelian group G_s is archimedean, then v is equivalent to the standard valuation of H.

Let G_0 be a subgroup of G. We view G_0 as an ordered subgroup, and $[G_0]$ is accordingly identified with an ordered subset of $[G]$ in the obvious way. Then the standard valuation of G restricts to the standard valuation of G_0, and if G_0 is dense in G (in the order topology on G), then $[G_0] = [G]$. Also $[\mathbb{Q}G] = [G]$.

LEMMA 2.4.4. *Let G_0 be a subgroup of G and $a \in G$. Then:*

(i) $[a] \notin [G_0] \Rightarrow [G_0 + \mathbb{Z}a] = [G_0] \cup \big\{[a]\big\}, [a] \leqslant [g]$ *for all* $g \in (G_0 + \mathbb{Z}a) \setminus G_0$*;*

(ii) $[G_0 + \mathbb{Z}a] \neq [G_0] \Rightarrow [G_0 + \mathbb{Z}a] = [G_0] \cup \big\{[g]\big\}$ *for some* $g \in G_0 + \mathbb{Z}a$.

We leave the proof of this lemma to the reader; (ii) follows easily from (i).

LEMMA 2.4.5. *Let C be a cut in the ordered set $[G^{\neq}]$. Then there is an ordered group extension $G + \mathbb{Z}x$ of G such that:*

(i) $x > 0$ *and* $[x]$ *realizes the cut* C*;*

(ii) *for any ordered abelian group extension H of G and $y \in H^{>}$ such that $y > 0$ and $[y]$ realizes C there is a unique ordered group embedding $G + \mathbb{Z}x \to H$ that is the identity on G and sends x to y.*

PROOF. Consider G in the usual way as a subgroup of an abelian group $G \oplus \mathbb{Z}x$ with $nx \neq 0$ for all $n \geqslant 1$. The ordering of G extends uniquely to an abelian group ordering on $G \oplus \mathbb{Z}x$ such that for all $g \in G$ and $n \geqslant 1$,

$$g + nx > 0 \Leftrightarrow g \geqslant 0 \text{ or } [g] \in C, \qquad g - nx > 0 \Leftrightarrow g > 0 \text{ and } [g] > C.$$

Then $G \oplus \mathbb{Z}x = G + \mathbb{Z}x$ with its element x has the desired properties. □

Convex subgroups and rank

It is easy to see that for a subgroup H of G, the following are equivalent:

(1) H is convex;

(2) for all $g, h \in G$, if $0 \leqslant |g| \leqslant |h|$ and $h \in H$, then $g \in H$;

(3) for all $g, h \in G$, if $[g] \leqslant [h]$ and $h \in H$, then $g \in H$.

As a consequence, the map $H \mapsto [H]$ is an inclusion-preserving bijection from the set of convex subgroups of G onto the set of nonempty cuts in $[G]$. (In particular, the set of convex subgroups of G is totally ordered by inclusion.) The **rank** of G, denoted by $\operatorname{rank}(G)$, is defined to be n if there are exactly n nontrivial convex subgroups of G, and defined to be ∞ if there are infinitely many convex subgroups of G. Thus G has rank 1 iff G is archimedean. If G_0 is an ordered subgroup of G, then clearly $\operatorname{rank}(G_0) \leqslant \operatorname{rank}(G)$.

Given a subgroup H of G, the convex hull $\operatorname{conv}(H)$ of H in G is a convex subgroup of G containing H; in fact $\operatorname{conv}(H)$ is the smallest convex subgroup of G containing H. In particular, for $g \in G$,

$$\operatorname{conv}(\mathbb{Z}g) = \{x \in G : [x] \leqslant [g]\}$$

is the smallest convex subgroup of G containing g. Convex subgroups of G of the form $\operatorname{conv}(g\mathbb{Z})$, where $g \in G$, are said to be **principal.** Note that if $\operatorname{rank}(G) < \infty$, then every convex subgroup of G is principal. Thus

$$\operatorname{rank}(G) < \infty \iff [G] \text{ is finite,}$$

and if $[G]$ is finite, then $\operatorname{rank}(G) = |[G^{\neq}]|$.

Recall from Section 1.7 that the **rational rank** of G, denoted by $\operatorname{rank}_{\mathbb{Q}}(G)$, is the dimension of $\mathbb{Q}G$ as a vector space over $\mathbb{Q}$ (which by convention is ∞ if this vector space is not finitely generated). Clearly, if $(g_i)_{i \in I}$ is a family in $G^{\neq}$ and $[g_i] \neq [g_j]$ for all $i \neq j$, then (g_i) is $\mathbb{Q}$-linearly independent in $\mathbb{Q}G$. Thus $\operatorname{rank}(G) \leqslant \operatorname{rank}_{\mathbb{Q}}(G)$.

Sometimes we use the following easy observation:

LEMMA 2.4.6. *Let* (s_ρ) *be a strictly increasing well-indexed sequence in* $G^>$. *Then the following are equivalent:*

(i) (s_ρ) *is cofinal in some convex subgroup of* G;

(ii) *for each* ρ *there is* $\sigma > \rho$ *with* $s_\sigma \geqslant 2s_\rho$.

Ordered quotient groups

Let H be a convex subgroup of G with quotient group $\dot{G} := G/H$ and canonical surjective group morphism

$$x \mapsto \dot{x} := x + H : G \to \dot{G}.$$

We equip $\dot{G}$ with the unique ordering making $\dot{G}$ an ordered abelian group such that for all a, b, if $a \leqslant b$, then $\dot{a} \leqslant \dot{b}$; this is done by declaring $\dot{a} > \dot{0}$ iff $a > H$. It is easy to verify that for $a, b \notin H$ we have $[a] \leqslant [b] \iff [\dot{a}] \leqslant [\dot{b}]$, hence $x \mapsto \dot{x}$ induces an isomorphism $[G] \setminus [H] \to [\dot{G}^{\neq}]$ of ordered sets. Hence

$$\operatorname{rank}(G) = \operatorname{rank}(H) + \operatorname{rank}(\dot{G}).$$

EXAMPLE. Suppose H, H' are subgroups of G with H convex in G and $G = H \oplus H'$ (internal direct sum of subgroups of G). Then G is ordered by the reverse lexicographic ordering: for $h \in H$, $h' \in H'$, $h + h' \geqslant 0$ iff $h' > 0$ or $h' = 0$, $h \geqslant 0$. The restriction of the morphism $G \to \dot{G} = G/H$ to H' is an isomorphism $H' \xrightarrow{\cong} \dot{G}$ of ordered abelian groups, and so $\operatorname{rank}(G) = \operatorname{rank}(H) + \operatorname{rank}(H')$.

EXAMPLE. Suppose G is divisible. Then every convex subgroup of G is also divisible. Hence if H is a convex subgroup of G, then we can take a (divisible) subgroup H' of G such that $G = H \oplus H'$ (internal direct sum of subgroups of G), and G is ordered by the reverse lexicographic ordering as described in the previous example.

LEMMA 2.4.7. *Let* $v \colon G \to S_\infty$ *be a convex valuation on* G *and let* (G', S, v') *be an immediate extension of the valued abelian group* (G, S, v). *Then just one ordering of* G' *makes* G' *an ordered abelian group extension of* G *and* v' *a convex valuation.*

PROOF. Let $a' \in (G')^{\neq}$, and take $a \in G$ such that $a' \sim a$. For an ordering of G' as in the lemma, we would have $a' > 0 \iff a > 0$. Conversely, this equivalence determines an ordering as claimed. □

LEMMA 2.4.8. *Let* $v \colon G \to S_\infty$ *be the standard valuation, so* $S_\infty = [G]$ *with the reversed ordering. For* $s \in S$ *we have convex subgroups* $B(s) \subseteq \overline{B}(s)$ *of* G, *and* $G(s) = \overline{B}(s)/B(s)$ *with the quotient ordering is archimedean.*

Steady functions and slow functions

For use in Section 9.2 we establish here a useful condition for functions on ordered abelian groups to have the intermediate value property. Let $v\colon G \to S_\infty$ be a convex valuation on G. Thus for all $x, y \in G$, if $vx > vy$, then $x = o(y)$. For $y \in G$ we let $o_v(y)$ stand for any element $x \in G$ with $vx > vy$, and accordingly, for $x \in G$ we use $x = o_v(y)$ as a suggestive notation for $vx > vy$, sometimes preferred over $x \prec y$, which has the same meaning.

Let U be a nonempty convex subset of G. A function $i\colon U \to G$ is said to be v-**steady** if i has the intermediate value property and $i(x) - i(y) = x - y + o_v(x-y)$ for all distinct $x, y \in U$. (Note that then i is strictly increasing.) In particular, a v-steady function $G \to G$ is bijective. If $a \in U$ and the restrictions of $i\colon U \to G$ to $U^{\leqslant a}$ and $U^{\geqslant a}$ are v-steady, then i is v-steady. A function $\eta\colon U \to G$ is said to be v-**slow on the right** if for all $x, y, z \in U$,

(s1) $\eta(x) - \eta(y) = o_v(x-y)$ if $x \neq y$;

(s2) $\eta(y) = \eta(z)$ if $x < y < z$ and $z - y = o_v(z-x)$.

We define $\eta\colon U \to G$ to be v-**slow on the left** in the same way except that in clause (s2) we replace "$x < y < z$" by "$x > y > z$."

LEMMA 2.4.9. *Suppose $i\colon U \to G$ is v-steady and $\eta\colon U \to G$ is v-slow on the left or on the right. Then $i + \eta\colon U \to G$ is v-steady.*

PROOF. It is clear that $(i+\eta)(x) - (i+\eta)(y) = x - y + o_v(x-y)$ for all distinct $x, y \in U$, and thus $i + \eta$ is strictly increasing. To prove that $i + \eta$ has the intermediate value property, assume that η is v-slow on the right, let $a, b \in U$ with $a < b$, put $c := b - a$ and define $i_1, \eta_1\colon [0, c] \to G$ by

$$i_1(x) = i(a+x) - i(a), \qquad \eta_1(x) = \eta(a+x) - \eta(a).$$

Then i_1 is v-steady and η_1 is v-slow on the right, and it suffices to prove the intermediate value property for $x \mapsto i_1(x) + \eta_1(x)$. So we can assume $U = [0, c]$ and $i(0) = \eta(0) = 0$. Let $0 < y < i(c) + \eta(c)$; it suffices to get $x \in (0, c)$ with $i(x) + \eta(x) = y$. Note: $i(c) = c + o_v(c)$, $\eta(c) = o_v(c)$. We distinguish two cases:

CASE 1: $c - y = o_v(c)$. Then $0 < y - \eta(c) < i(c)$, so $y - \eta(c) = i(x)$ with $0 < x < c$. It follows easily that $c - x = o_v(c)$, so $\eta(x) = \eta(c)$ by (s2), and thus $i(x) + \eta(x) = y$.

CASE 2: $c - y \neq o_v(c)$. Then $0 < y < c$ and $0 < y - \eta(y) < i(c)$, so $y - \eta(y) = i(x)$ with $0 < x < c$. It follows easily that $v(x) = v(y)$, $x - y = o_v(x)$, and $x - y = o_v(y)$ so $\eta(x) = \eta(y)$ by (s2), and thus $i(x) + \eta(x) = y$. $\square$

LEMMA 2.4.10. *Suppose the functions $\eta_1, \eta_2\colon U \to G$ are v-slow on the right. Then $-\eta_1, \eta_1 + \eta_2, \min(\eta_1, \eta_2)$ are also v-slow on the right.*

PROOF. It is clear that η is v-slow on the right for $\eta = -\eta_1$ and for $\eta = \eta_1 + \eta_2$. Put $\eta := \min(\eta_1, \eta_2)$. Then (s2) holds. Let $a, b \in U$, $a < b$. If for some $i \in \{1, 2\}$, we have $\eta(a) = \eta_i(a)$ and $\eta(b) = \eta_i(b)$, then $\eta(a) - \eta(b) = \eta_i(a) - \eta_i(b) = o_v(a - b)$. Suppose $\eta_1(a) \leqslant \eta_2(a)$ and $\eta_2(b) \leqslant \eta_1(b)$. If $\eta(a) \geqslant \eta(b)$, this yields

$$0 \leqslant \eta(a) - \eta(b) = \eta_1(a) - \eta_2(b) \leqslant \eta_2(a) - \eta_2(b) = o_v(a - b),$$

and if $\eta(a) \leqslant \eta(b)$, then

$$0 \leqslant \eta(b) - \eta(a) = \eta_2(b) - \eta_1(a) \leqslant \eta_1(b) - \eta_1(a) = o_v(a - b).$$

The remaining case $\eta_2(a) \leqslant \eta_1(a)$ and $\eta_1(b) \leqslant \eta_2(b)$ follows by symmetry. □

Completeness

In this subsection $G^>$ has no smallest element. Note that this condition holds for divisible ordered abelian groups. An easy consequence of this assumption on G is that for each $\varepsilon \in G^>$ there exists a $\delta \in G^>$ such that $2\delta < \varepsilon$. Below we often use this fact without comment.

Let (a_ρ) be a well-indexed sequence in G. We say that (a_ρ) is a **cauchy sequence** (or a **c-sequence**) in G if for every $\varepsilon \in G^>$ there is ρ_0 such that $|a_\rho - a_{\rho'}| < \varepsilon$ for all $\rho, \rho' > \rho_0$. For $a \in G$ we say that (a_ρ) **converges to** a if for each $\varepsilon \in G^>$ there is an index ρ_0 such that $|a - a_\rho| < \varepsilon$ for all $\rho > \rho_0$; in symbols: $a_\rho \to a$. We say that (a_ρ) **converges in** G if $a_\rho \to a$ for some $a \in G$. There is at most one $a \in G$ with $a_\rho \to a$. If (a_ρ) converges in some ordered abelian group extension G' of G such that $G^>$ is coinitial in $(G')^>$, then (a_ρ) is a c-sequence in G. If (a_ρ) is a c-sequence in G with its standard valuation, then (a_ρ) is a c-sequence in the ordered abelian group G; similarly, for $a \in G$, if $a_\rho \to a$ in G with its standard valuation, then $a_\rho \to a$ in the ordered abelian group G. The converses of these implications hold if $[G^{\neq}]$ with its natural ordering has no smallest element.

It is easy to check that Lemma 2.2.23 goes through for this notion of c-sequence in an ordered abelian group. We also have an analogue of Lemma 2.2.25 with the same proof except for trivial rewordings:

LEMMA 2.4.11. *Let (a_ρ) be a c-sequence in G which is not eventually constant. Then the index set of (a_ρ) has cofinality* $\operatorname{ci}(G^>)$.

A **dense** extension of G is an extension $G' \supseteq G$ of ordered abelian groups such that G is dense in G' in the order topology; then $G^>$ is coinitial in $(G')^>$. Thus:

LEMMA 2.4.12. *Given a dense extension $G' \supseteq G$ and $a \in G' \setminus G$, there is a c-sequence (a_ρ) in G, indexed by the ordinals $\rho < \operatorname{ci}(G^>)$, such that $a_\rho \to a'$.*

We say that G is **complete** if every c-sequence in G converges in G. A function $f\colon X \to G$ with $X \subseteq G$ is said to be **uniformly continuous** if for each $\varepsilon \in G^>$ there exists a $\delta \in G^>$ such that for all $x, y \in X$, if $|x - y| < \delta$, then $|f(x) - f(y)| < \varepsilon$. Completeness and uniform continuity are related in the usual way:

LEMMA 2.4.13. *Suppose G is complete and $f\colon X \to G$ with $X \subseteq G$ is uniformly continuous. Then f extends uniquely to a continuous map from the topological closure of X in G into G, and this extension is also uniformly continuous.*

Here is the analogue of Theorem 2.2.28:

THEOREM 2.4.14. *G has a dense complete extension. (Hence G is complete iff G has no proper dense extension.)*

PROOF. We mimic the construction in the proof of Theorem 2.2.28. Let G^{cs} be the set of c-sequences (a_ρ) in G indexed by the ordinals $\rho < \mathrm{ci}(G^>)$, made into an abelian group via the componentwise addition of such sequences. Then

$$N := \{(a_\rho) \in G^{\mathrm{cs}} : a_\rho \to 0\}$$

is a subgroup of G^{cs}, and we let $G^{\mathrm{d}} := G^{\mathrm{cs}}/N$. If $(a_\rho) \in G^{\mathrm{cs}} \setminus N$, then for some $\varepsilon \in G^>$ we have $a_\rho > \varepsilon$ eventually, or $a_\rho < -\varepsilon$ eventually. Using this fact we make G^{d} into an ordered abelian group such that for all $(a_\rho) \in G^{\mathrm{cs}}$,

$$(a_\rho) + N > 0 \quad \Longleftrightarrow \quad \text{for some } \varepsilon \in G^>,\ a_\rho > \varepsilon \text{ eventually.}$$

The map which associates to each $a \in G$ the coset $(a) + N$, where $(a) \in G^{\mathrm{cs}}$ is the constant sequence with value a, is an embedding $G \to G^{\mathrm{d}}$ of ordered abelian groups; we identify G with an ordered subgroup of G^{d} via this embedding.

It is routine to show that G is dense in G^{d}, and that for $(a_\rho) \in G^{\mathrm{cs}}$ we have $a_\rho \to a := (a_\rho) + N$ in G^{d}. Thus by Lemma 2.4.11, every c-sequence in G converges in G^{d}. As in the proof of Theorem 2.2.28 one now argues that G^{d} is complete, using the previous lemma in place of Lemma 2.2.26. □

Below G^{d} is a dense complete extension of G (not necessarily the one constructed in the proof of Theorem 2.4.14). Using Lemma 2.4.12, we have:

COROLLARY 2.4.15. *For any dense extension G' of G there exists a unique ordered abelian group embedding $G' \to G^{\mathrm{d}}$ over G.*

It follows that the ordered abelian group extension G^{d} of G is determined uniquely (up-to-unique-isomorphism-over G) by the property of being a dense complete extension of G. So there is no harm in referring to it as the **completion of G.**

Ordered vector spaces

An **ordered field** is a field C with a (total) ordering $\leqslant$ of C such that for all $x, y, z \in C$,

$$x \leqslant y \Rightarrow x + z \leqslant y + z, \qquad x \leqslant y \;\&\; z \geqslant 0 \Rightarrow xz \leqslant yz.$$

(Thus the ordering $\leqslant$ makes the additive group of C an ordered abelian group.) Just one ordering makes the field $\mathbb{Q}$ of rational numbers into an ordered field, and it is given by $q \geqslant 0 \Longleftrightarrow q = m/n$ for some m, n with $n \neq 0$. This is how we view $\mathbb{Q}$ as an ordered field. If C is an ordered field, then $x^2 \geqslant 0$ for all $x \in C$, $\mathrm{char}(C) = 0$, and we always take $\mathbb{Q}$ as an ordered subfield of C. Also just one ordering makes the field $\mathbb{R}$ of real numbers into an ordered field, and it is given by $r \geqslant 0 \Longleftrightarrow r = x^2$ for some $x \in \mathbb{R}$; this is how $\mathbb{R}$ is considered as an ordered field. If C is an ordered field, then C is a topological field with respect to the order topology, that is, the addition and multiplication maps $C \times C \to C$ (with the product topology on $C \times C$), and the map $x \mapsto x^{-1} \colon C^\times \to C$ are continuous.

Let C be an ordered field, and let our ordered abelian group G come equipped with an operation $(\lambda, x) \mapsto \lambda x \colon C \times G \to G$ making G a vector space over C. Then G is said to be an **ordered vector space over** C if for all $\lambda \in C$ and $x \in G$:

$$\lambda > 0 \,\&\, x > 0 \quad \Rightarrow \quad \lambda x > 0.$$

Note that then $|\lambda x| = |\lambda| \cdot |x|$ for all $\lambda \in C$ and $x \in G$. Every divisible ordered abelian group, viewed as a vector space over $\mathbb{Q}$, is an ordered vector space over the ordered field $\mathbb{Q}$. We view C as an ordered vector space over C in the obvious way. If G is a 1-dimensional ordered vector space over C and $e \in G^>$, then $\lambda \mapsto \lambda e \colon C \to G$ is an isomorphism of ordered vector spaces over C.

Let G be an ordered vector space over C, and let P be a cut in G, that is, P is a downward closed subset of G. We set $Q := G \setminus P$, so $P < Q$. Take a vector space extension $G + Cb \supseteq G$ over C with $b \notin G$. There is a unique ordering on $G + Cb$ that extends the ordering of G and makes $G+Cb$ into an ordered vector space over C, such that $P < b < Q$: in this ordering we have for $\lambda \in C^>$ and $g, h \in G$,

$$g + \lambda b < h \Leftrightarrow b < \lambda^{-1}(h - g) \Leftrightarrow \lambda^{-1}(h-g) \in Q.$$

In the next two lemmas we consider $G + Cb = G \oplus Cb$ as an ordered vector space over C in this way. The first lemma is obvious:

LEMMA 2.4.16. *For any vector b' in any ordered vector space extension $G' \supseteq G$ over C with $P < b' < Q$, there is a unique embedding $G + Cb \to G'$ of ordered vector spaces over C that is the identity on G and sends b to b'.*

LEMMA 2.4.17. *G is dense in $G + Cb$ if and only if*

(i) *for each $\varepsilon \in G^>$ there are $p \in P$, $q \in Q$ with $q - p < \varepsilon$, and*

(ii) *P has no largest element and Q has no least element.*

PROOF. It is clear that if G is dense in $G + Cb$, then (i) and (ii) hold. Assume (i) and (ii). For each $a \in G$ there is by (ii) an $\varepsilon \in G^>$ with $|a - b| > \varepsilon$. For each $\varepsilon \in G^>$ there is by (i) an $a \in G$ with $|a - b| < \varepsilon$. These properties of b are inherited by every affine transform $a + \lambda b$ with $a \in G$ and $\lambda \in C^\times$, so there is for every $\delta \in (G + Cb)^>$ an $\varepsilon \in G^>$ with $\delta > \varepsilon$. Thus every neighborhood of b in $G + Cb$ contains a vector from G; this property of b is inherited by every vector in $G + C^\times b$ and trivially holds for vectors in G. Thus G is dense in $G + Cb$. □

Suppose G is an ordered vector space over C with $G \neq \{0\}$; so $G^>$ has no least element. Given any $\lambda \in C$, the map $x \mapsto \lambda x \colon G \to G$ is uniformly continuous, so extends uniquely to a continuous map $G^{\mathrm{d}} \to G^{\mathrm{d}}$ by Lemma 2.4.13. Denoting the image of any $x \in G^{\mathrm{d}}$ under this map also by λx and varying λ we have an operation

$$(\lambda, x) \mapsto \lambda x \;:\; C \times G^{\mathrm{d}} \to G^{\mathrm{d}}$$

that makes G^{d} into an ordered vector space over C that extends G as an ordered vector space over C. In particular, the completion of any nontrivial divisible ordered abelian group is divisible.

Valued ordered vector spaces

Let C be an ordered field. A **valued ordered vector space over** C is an ordered vector space G over C with a convex valuation $v\colon G \to S_\infty$ such that (G, S, v) is a valued vector space over C. Note that then for $s \in S$ we have the convex subspaces $B(s) \subseteq \overline{B}(s)$ of G, and so $G(s) = \overline{B}(s)/B(s)$ with the quotient ordering is an ordered vector space over C.

Let (G_s) be a family of ordered vector spaces $G_s \neq \{0\}$ over C, indexed by the elements s of an ordered set S. Then the Hahn product $H\big[(G_s)\big]$ with its Hahn ordering and its Hahn valuation is a valued ordered vector space over C, and this is how such a Hahn product will be construed as a valued ordered vector space over C. In this context we can add something to Proposition 2.3.4:

PROPOSITION 2.4.18. *Let G be a valued ordered vector space over C, and let*

$$i : G \to H := H\big[(G(s))\big]$$

be as in Proposition 2.3.4. Then i preserves order: if $g \in G^{>}$, then $i(g) \in H^{>}$.

With this, the embedding of an ordered abelian group into its divisible hull and Lemmas 2.4.3 and 2.4.8 yield the classical Hahn Embedding Theorem:

COROLLARY 2.4.19. *Any ordered abelian group G embeds into the Hahn ordered group $H[S, \mathbb{R}]$ where $S = [G^{\neq}]$ with reversed ordering.*

Here is a complement to Lemma 2.4.7:

COROLLARY 2.4.20. *Let G be a valued ordered vector space over C and $G' \supseteq G$ an immediate extension of valued vector spaces, with valuation v'. Then just one ordering on G' makes G' an ordered abelian group extension of G with convex v'. With v' and this ordering, G' is a valued ordered vector space over C.*

This follows easily by inspecting the proof of Lemma 2.4.7.

The C-valuation of an ordered vector space over C

Let G be an ordered vector space over the ordered field C. The **C-archimedean class** of $a \in G$ is

$$[a]_C := \big\{g \in G : |a| \leqslant \lambda|g| \text{ and } |g| \leqslant \lambda|a| \text{ for some } \lambda \in C^{>}\big\}.$$

For G as an ordered vector space over the ordered subfield $\mathbb{Q}$ of C we get $[a]_{\mathbb{Q}} = [a]$ for $a \in G$, where $[a]$ is the archimedean class of a as defined earlier in this section.

Let $[G]_C$ be the set of C-archimedean classes. Then $[G]_C$ is a partition of G, and we linearly order $[G]_C$ by

$$\begin{aligned} [a]_C < [b]_C \quad &:\Longleftrightarrow \quad \lambda|a| < |b| \text{ for all } \lambda \in C^{>} \\ &\Longleftrightarrow \quad [a]_C \neq [b]_C \text{ and } |a| < |b|. \end{aligned}$$

Thus $[0]_C = \{0\}$ is the smallest C-archimedean class. The map

$$x \mapsto [x]_C \; : \; G \to [G]_C$$

with the reversed ordering on $[G]_C$ (so with $[0]_C$ as the largest element) is a convex valuation of G making G a valued ordered vector space over C. This valuation is called the **C-valuation** of the ordered vector space G, and is a coarsening of the standard valuation of the ordered additive group of G defined in Section 2.4. A subspace G_0 of G is considered as an ordered vector space over C by restricting the ordering to G_0, and $[G_0]_C$ then becomes an ordered subset of $[G]_C$ via the obvious identification; the C-valuation of G then restricts to the C-valuation of G_0.

We say that G is **C-archimedean** if $[G^{\neq}]_C := [G]_C \setminus \{[0]\}$ is a singleton. Viewing G as a valued ordered vector space over C by the C-valuation, and given $s \in [G^{\neq}]_C$, the ordered vector space $G(s) = \overline{B}(s)/B(s)$ over C is C-archimedean.

EXAMPLE. Let (G_s) be a family of C-archimedean ordered vector spaces over C indexed by the elements of an ordered set S. Then the Hahn valuation of the Hahn product $H := H\big[(G_s)\big]$ is equivalent to the C-valuation of H.

For use in the proof of Lemma 10.4.5 we record the following:

LEMMA 2.4.21. *Let (a_ρ) be a divergent pc-sequence in G with respect to the C-valuation of G. Then G has an extension $G \oplus Ca$ as an ordered vector space over C such that $[G \oplus Ca]_C = [G]_C$, and $a_\rho \rightsquigarrow a$ with respect to the C-valuation of $G \oplus Ca$.*

PROOF. Lemma 2.3.1 and Corollary 2.4.20 provide an immediate valued ordered vector space extension $G \oplus Ca$ of G with $a_\rho \rightsquigarrow a$. It follows easily that the valuation on $G \oplus Ca$ is the C-valuation. □

Let $G \oplus Ca$ be as in Lemma 2.4.21, and let $G \oplus Cb$ also be as in that lemma, with b instead of a. Then there is an isomorphism $G \oplus Ca \to G \oplus Cb$ of ordered vector spaces over C which is the identity on G and sends a to b.

Completion of valued ordered vector spaces

Let G be a valued ordered vector space over an ordered field C such that the value set $S = v(G^{\neq})$ is not empty and has no largest element. Then the valuation topology and the order topology on G coincide, by Lemma 2.4.1. Thus G is complete as a valued abelian group if and only if G is complete as an ordered abelian group.

Recall that G^{c} is the completion of G as a valued abelian group. We construe it as a valued vector space over C as in Section 2.3. Corollary 2.4.20 then gives a unique ordering on G^{c} making it a valued ordered vector space over C extending the valued order vector space G. With this ordering G^{c} is also complete as an ordered abelian group, and G is dense in G^{c} in the order topology.

On the other hand, G^{d} is the completion of G as an ordered abelian group, and we construe it here as an ordered vector space over the ordered field C as indicated at the end of the subsection on ordered vector spaces. Thus by the uniqueness property of G^{d} we have:

COROLLARY 2.4.22. *There is a unique isomorphism $G^{\mathrm{c}} \to G^{\mathrm{d}}$ of ordered vector spaces over C which is the identity on G.*

Hahn spaces over ordered fields

Let C be an ordered field and G an ordered vector space over C. Call G a **Hahn space** over C if G with its C-valuation is a Hahn space over C as defined in Section 2.3, that is, for all $a, b \in G^{\neq}$ with $[a]_C = [b]_C$ there exists $\lambda \in C^\times$ such that $[a - \lambda b]_C < [a]_C$.

EXAMPLES.

(1) Any 1-dimensional ordered vector space over C is a Hahn space over C.

(2) The ordered vector space $\mathbb{Q} + \mathbb{Q}\sqrt{2} \subseteq \mathbb{R}$ over $\mathbb{Q}$ is not a Hahn space over $\mathbb{Q}$.

(3) Any ordered vector space over the ordered field $\mathbb{R}$ is a Hahn space over $\mathbb{R}$.

(4) If (G_s) is a family of Hahn spaces $G_s \neq \{0\}$ over C indexed by the elements s of an ordered set, then the Hahn product $H\big[(G_s)\big]$ as a vector space over C with its Hahn ordering is a Hahn space over C.

Using Proposition 2.4.18 we obtain easily:

COROLLARY 2.4.23. *If G is a Hahn space over C equipped with its C-valuation, then there is an embedding $i\colon G \to H[S, C]$ of valued ordered vector spaces over C such that $H[S, C]$ is an immediate extension of its valued subspace iG.*

Notes and comments

The fact that torsion-free abelian groups are orderable is in [256]. Lemma 2.4.1 is in [287], Lemma 2.4.3 is from [188]. The archimedean property was isolated by Stolz [434]. (See [125, 139].) The material on steady and slow functions extends a result in [18]. The completion of an ordered abelian group can also be obtained by realizing certain kinds of Dedekind cuts; see [83]. (The particular cuts required were already considered by Veronese [457].) Completing valued abelian groups and ordered abelian group are special cases of completing a uniform space; see [60, Section II.3.7] and [323, Lemmas II.4.8 and III.4.5]. The notion of "Hahn space over an ordered field" is from [18]. For more on ordered algebraic structures, see [144, 323].

Chapter Three

Valued Fields

In this chapter we assume familiarity with basic field theory, including the rudiments of the theory of ordered fields (which are summarized, without proofs, in Section 3.5 below). In Section 3.1 we take up valued fields and establish their basic properties, focusing in particular on extensions of valued fields. Next, in Section 3.2 we study pseudocauchy sequences in valued fields; these sequences will be needed later in constructing solutions of algebraic differential equations in immediate extensions of suitable valued differential fields. In Section 3.3 we consider *henselian* valued fields. (In Chapter 7 we study for valued differential fields the analogous notion of *differential-henselian*.) Whenever it simplifies matters we restrict our attention in this section to valued fields whose residue field is of characteristic zero, since this is the only case that arises later. In Section 3.4 we show how to decompose a valuation on a field into simpler ones. This leads to a study of various special types of pseudocauchy sequences. The valuation of $\mathbb{T}$ is compatible with its natural ordering. Consequently, Section 3.5 of this chapter contains basic facts about fields with compatible ordering and valuation. In Section 3.6 we review some basic model theory of valued fields. In the final Section 3.7 we consider the Newton diagram and Newton tree of a polynomial over a valued field.

3.1 VALUATIONS ON FIELDS

Let A be an integral domain. A **valuation** on A is a map $v\colon A^{\neq} \to \Gamma$, where Γ is an ordered abelian group, such that for all $x, y \in A^{\neq}$:

(V1) $v(xy) \;=\; v(x) + v(y)$;

(V2) $v(x+y) \;\geqslant\; \min(v(x), v(y))$ when $x + y \neq 0$.

Let $v\colon A^{\neq} \to \Gamma$ be a valuation on A. Then $v1 = v(-1) = 0$, since v restricted to the group $A^{\times}$ of units of A is a group morphism. Hence $vx = v(-x)$ for all $x \in A^{\neq}$. By convention we extend v to $v\colon A \to \Gamma_{\infty}$ by $v(0) := \infty$, which makes v a valuation on the additive group of A as defined in Section 2.2. We can extend v uniquely to a valuation $v\colon K^{\times} \to \Gamma$ on the fraction field K of A, by

$$v(x/y) \;=\; vx - vy \qquad (x, y \in A^{\neq}).$$

Thus $v(K^{\times})$ is a subgroup of Γ. When we refer to a valuation $v\colon K^{\times} \to \Gamma$ on a field K, we assume from now on that $v(K^{\times}) = \Gamma$, unless specified otherwise. We call $\Gamma = v(K^{\times})$ the **value group** of v. The valuation v on K is trivial iff $\Gamma = \{0\}$.

Let $v\colon K^\times \to \Gamma$ be a valuation on the field K. We set

$$\mathcal{O} := \{x \in K : vx \geqslant 0\}, \qquad \mathcal{o} := \{x \in K : vx > 0\}.$$

If we need to indicate the dependence on v, we write $\mathcal{O}_v$ and $\mathcal{o}_v$ instead of $\mathcal{O}$ and $\mathcal{o}$. Thus $\mathcal{O}$ is a subring of K and $\mathcal{o}$ is an ideal of $\mathcal{O}$. In fact, $vx = 0 \iff v(x^{-1}) = 0$, for $x \in K^\times$, so $\mathcal{O}^\times = \mathcal{O} \setminus \mathcal{o}$. Note that for $x, y \in \mathcal{O}$ we have $x \in \mathcal{O}y$ iff $vx \geqslant vy$. Therefore $\mathcal{O}$ is a valuation ring as defined in Section 1.1, and in particular a local ring with maximal ideal $\mathcal{o}$. We call $\mathcal{O}$ the **valuation ring** of v. The **residue field** of v is the field $\boldsymbol{k}_v := \mathcal{O}/\mathcal{o}$. With the notation of Section 2.2 and working in the valued abelian group (K, Γ, v) we have $\mathcal{O} = \overline{B}(0)$, $\mathcal{o} = B(0)$, $K(0) = \mathcal{O}/\mathcal{o}$, and for each $a \in K^\times$ the group isomorphism

$$x \mapsto xa : \overline{B}(0) \to \overline{B}(va)$$

induces a group isomorphism $K(0) \to K(va)$. The v-topology on K makes K into a hausdorff topological field.

Local rings

Recall that a commutative ring is said to be **local** if it has exactly one maximal ideal. Thus a commutative ring A is local iff $A \setminus A^\times$ is an ideal of A, and in this case, $\mathfrak{m}_A := A \setminus A^\times$ is the (unique) maximal ideal of A. We let $\boldsymbol{k}_A := A/\mathfrak{m}_A$ be the **residue field** of the local ring A. Given local rings A, B, we say that B **lies over** A if A is a subring of B and $\mathfrak{m}_A \subseteq \mathfrak{m}_B$ (and hence $\mathfrak{m}_B \cap A = \mathfrak{m}_A$). In this case we have an induced embedding $\boldsymbol{k}_A = A/\mathfrak{m}_A \to B/\mathfrak{m}_B = \boldsymbol{k}_B$ of residue fields, by means of which we identify $\boldsymbol{k}_A$ with a subfield of $\boldsymbol{k}_B$. We then have a commutative diagram

$$\begin{array}{ccc} B & \longrightarrow & \boldsymbol{k}_B \\ \subseteq\uparrow & & \subseteq\uparrow \\ A & \longrightarrow & \boldsymbol{k}_A \end{array}$$

where the horizontal arrows are the residue morphisms.

A **valuation ring of a field** K is a subring A of K such that for each $x \in K^\times$ we have $x \in A$ or $x^{-1} \in A$. Each valuation ring of a field is clearly a valuation ring as defined in Section 1.1. Let $K \subseteq L$ be a field extension, and let A, B be valuation rings of K and L, respectively. Then $B \cap K$ is a valuation ring of K, and

$$B \text{ lies over } A \iff A = B \cap K.$$

Correspondence between valuation rings and valuations

Let K be a field. The valuation ring of a valuation on K is a valuation ring of K. Conversely, to a valuation ring A of K we associate a valuation on K as follows: Consider the (abelian) quotient group $\Gamma_A = K^\times/A^\times$, written additively. The binary relation $\geqslant$ on Γ_A defined by

$$xA^\times \geqslant yA^\times :\iff x/y \in A \qquad (x, y \in K^\times)$$

makes Γ_A into an ordered abelian group, and the natural map

$$v_A\colon K^\times \to \Gamma_A, \qquad v_A x := xA^\times,$$

is a valuation on K. The valuation ring of v_A is A, and every valuation on K with valuation ring A is equivalent to v_A, as defined in Section 2.2. In fact, for valuations $v\colon K^\times \to \Gamma = v(K^\times)$ and $v'\colon K^\times \to \Gamma' = v'(K^\times)$ on the field K we have: v and v' are equivalent as defined in Section 2.2, if and only if v and v' have the same valuation ring, and also if and only if there is an isomorphism $i\colon \Gamma \to \Gamma'$ of ordered abelian groups such that $v' = i \circ v$.

Let $K \subseteq L$ be a field extension, A a valuation ring of K, and B be a valuation ring of L lying over A. Then $\mathfrak{m}_A = \mathfrak{m}_B \cap K$, so we have an induced embedding $\boldsymbol{k}_A = A/\mathfrak{m}_A \to B/\mathfrak{m}_B = \boldsymbol{k}_B$ of residue fields, by means of which we identify $\boldsymbol{k}_A$ with a subfield of $\boldsymbol{k}_B$. Also, $B^\times \cap K = A^\times$, so we have an induced embedding $v_A(x) \mapsto v_B(x)\colon \Gamma_A \to \Gamma_B$ ($x \in K^\times$) of ordered abelian groups, by means of which Γ_A is identified with an ordered subgroup of Γ_B. Then $v_A(x) = v_B(x)$ for all $x \in K$.

A **valued field** is just a field equipped with one of its valuation rings. Let K be a valued field. Unless we specify otherwise we let $\mathcal{O}$ be the distinguished valuation ring of K, with v the corresponding valuation on K, and σ the maximal ideal of $\mathcal{O}$. The **value group** of K is $\Gamma = v(K^\times)$. If we need to indicate the dependence on K we attach a subscript K, so $\mathcal{O} = \mathcal{O}_K$, $v = v_K$, and so on. Sometimes (especially when K is an asymptotic differential field, see Section 9.1) we prefer more relational notation: define the binary relations $\preccurlyeq$ and $\prec$ on K by

$$a \preccurlyeq b :\Longleftrightarrow va \geqslant vb, \qquad a \prec b :\Longleftrightarrow va > vb,$$

just as we already did for for valued abelian groups. Accordingly,

$$K^{\preccurlyeq 1} := \{a \in K : a \preccurlyeq 1\}$$

is then the valuation ring of K, and $K^{\prec 1} := \{a \in K : a \prec 1\}$ is the maximal ideal of $K^{\preccurlyeq 1}$. Likewise, $K^{\succ 1} := \{a \in K : a \succ 1\}$. The **residue field** of K is

$$\operatorname{res}(K) := \boldsymbol{k}_{\mathcal{O}} = \mathcal{O}/\sigma = K^{\preccurlyeq 1}/K^{\prec 1},$$

and for $a \in \mathcal{O}$ we let $\operatorname{res}(a)$, or sometimes $\overline{a}$, be the residue class $a + \sigma \in \operatorname{res}(K)$. The residue fields of the valued fields of interest in our work (such as $\mathbb{T}$) are of characteristic zero. One says that K is of **equicharacteristic zero** if $\operatorname{res}(K)$ has characteristic zero. Note that then K also has characteristic zero, and the valuation ring $\mathcal{O}$ of K contains a subfield (i.e., a subring that happens to be a field): the unique ring morphism $\mathbb{Z} \to \mathcal{O}$ is injective, and identifying $\mathbb{Z}$ with its image in $\mathcal{O}$ under this morphism, each nonzero integer is a unit in $\mathcal{O}$, so the fraction field $\mathbb{Q} \subseteq K$ of $\mathbb{Z}$ is contained in $\mathcal{O}$. Note that if $\mathcal{O}$ contains a subfield C, then K as a vector space over C with the valuation v is a valued vector space over C (as defined in Section 2.3).

Let K be a valued field. A **valued field extension** of K is a field extension L of K equipped with a valuation ring of L that lies over $\mathcal{O}$; in this situation we also call K a **valued subfield** of L. Thus a valued field extension $K \subseteq L$ gives rise to a field extension $\operatorname{res}(K) \subseteq \operatorname{res}(L)$ and to an ordered group extension $\Gamma \subseteq \Gamma_L$.

Correspondence between dominance relations and valuations

The binary relation $\preccurlyeq$ introduced above is an example of a dominance relation:

DEFINITION 3.1.1. A **dominance relation** on a field K is a binary relation $\preccurlyeq$ on K such that for all $f, g, h \in K$:

(D1) $1 \not\preccurlyeq 0$;

(D2) $f \preccurlyeq f$;

(D3) $f \preccurlyeq g$ and $g \preccurlyeq h \Rightarrow f \preccurlyeq h$;

(D4) $f \preccurlyeq g$ or $g \preccurlyeq f$;

(D5) $f \preccurlyeq g \Leftrightarrow fh \preccurlyeq gh$, provided $h \neq 0$;

(D6) $f \preccurlyeq h$ and $g \preccurlyeq h \Rightarrow f + g \preccurlyeq h$.

If $f \preccurlyeq g$, we say that f is **dominated** by g, or g **dominates** f.

Thus, if v is a valuation on K with valuation ring $K^{\preccurlyeq 1}$, we obtain a dominance relation on K by setting, for $f, g \in K$:

(3.1.1) $$f \preccurlyeq g \;:\Longleftrightarrow\; vf \geqslant vg \;\Longleftrightarrow\; f = gh \text{ for some } h \in K^{\preccurlyeq 1}.$$

Conversely, if $\preccurlyeq$ is a dominance relation on K, then clearly $K^{\preccurlyeq 1} := \{f \in K : f \preccurlyeq 1\}$ is a valuation ring of K, and if v denotes the corresponding valuation on K, then the equivalence (3.1.1) holds, for all $f, g \in K$. We call v the valuation **associated** to the dominance relation $\preccurlyeq$. This yields a one-to-one correspondence between dominance relations on K and valuation rings of K. (That is why the reverse of a dominance relation is sometimes called a *valuation divisibility*.) We shall use the valuation and dominance terminologies interchangeably and switch between them without further comment.

NOTATIONS. Let K be a field with dominance relation $\preccurlyeq$ on K, let $v\colon K^\times \to \Gamma$ be the associated valuation, and let f, g denote elements of K. We define, just as we did for valued abelian groups in Section 2.2:

$$f \prec g \quad :\Longleftrightarrow \quad vf > vg \quad \Longleftrightarrow \quad f \preccurlyeq g \text{ and } g \not\preccurlyeq f.$$

If $f \prec g$, we say that f is **strictly dominated** by g. If $f \preccurlyeq g$ and $g \preccurlyeq f$, then we say that f and g are **asymptotic**, written as $f \asymp g$, and if $f - g \prec f$, then f and g are said to be **equivalent**, written as $f \sim g$. The relations $\asymp$ and $\sim$ are equivalence relations on K and $K^\times$, respectively. Note that

$$\begin{aligned} f \asymp g \quad &\Longleftrightarrow \quad vf = vg, \\ f \sim g \quad &\Longleftrightarrow \quad v(f - g) > vf. \end{aligned}$$

In particular, $f \sim g \Rightarrow f \asymp g$. Define

$$\begin{aligned} f \preccurlyeq 1 \quad &:\Longleftrightarrow \quad f \text{ is } \textbf{bounded}, \\ f \prec 1 \quad &:\Longleftrightarrow \quad f \text{ is } \textbf{infinitesimal}, \\ f \succ 1 \quad &:\Longleftrightarrow \quad f \text{ is } \textbf{infinite}. \end{aligned}$$

The elements of the maximal ideal $K^{\prec 1}$ of the valuation ring $K^{\preccurlyeq 1}$ are exactly the infinitesimals of K.

Well-based series

Let $\mathfrak{M}$ be an ordered set whose ordering is $\preccurlyeq$. We think of the elements of $\mathfrak{M}$ as *monomials* and accordingly denote its elements by $\mathfrak{m}$, $\mathfrak{n}$, and so on. A set $\mathfrak{S} \subseteq \mathfrak{M}$ is said to be **well-based** if it is well-ordered for the reverse ordering $\succcurlyeq$, that is, there is no strictly increasing infinite sequence $\mathfrak{m}_0 \prec \mathfrak{m}_1 \prec \mathfrak{m}_2 \prec \cdots$ in $\mathfrak{S}$. Clearly the union of two well-based subsets of S is well-based.

Next, let C be an (additive) abelian group whose elements are to be thought of as coefficients. In this spirit, a function $f\colon \mathfrak{M} \to C$ will be denoted as a series $\sum_{\mathfrak{m}\in\mathfrak{M}} f_{\mathfrak{m}}\mathfrak{m}$, with $f_{\mathfrak{m}} = f(\mathfrak{m})$, with **support** $\operatorname{supp} f := \{\mathfrak{m} \in \mathfrak{M} : f_{\mathfrak{m}} \neq 0\}$. Then

$$C[[\mathfrak{M}]] \ := \ \{f\colon \mathfrak{M} \to C : \ \operatorname{supp} f \subseteq \mathfrak{M} \text{ is well-based}\}$$

is a subgroup of the additive group of all functions $\mathfrak{M} \to C$ with pointwise addition. Indeed, $C[[\mathfrak{M}]]$ is just the Hahn product $H[\mathfrak{M}, C]$ of Section 2.2, with respect to the reverse ordering of $\mathfrak{M}$. For nonzero $f \in C[[\mathfrak{M}]]$ we define

$$\mathfrak{d}(f) \ := \ \max_{\preccurlyeq} \operatorname{supp} f,$$

the **dominant monomial** of f.

In the rest of this subsection $\mathfrak{M}$ is an ordered abelian group and C is a (coefficient) field. Thus $C[[\mathfrak{M}]]$ is a subspace of the C-vector space of all functions $\mathfrak{M} \to C$. We take $\mathfrak{M}$ as a *multiplicative* group, in view of the role of its elements as monomials. We leave the proof of the next result to the reader.

LEMMA 3.1.2. *Let $\mathfrak{S}_1, \mathfrak{S}_2 \subseteq \mathfrak{M}$ be well-based. Then for each $\mathfrak{m} \in \mathfrak{M}$ there are only finitely many $(\mathfrak{m}_1, \mathfrak{m}_2) \in \mathfrak{S}_1 \times \mathfrak{S}_2$ such that $\mathfrak{m} = \mathfrak{m}_1 \cdot \mathfrak{m}_2$, and the set*

$$\mathfrak{S}_1 \cdot \mathfrak{S}_2 \ := \ \{\mathfrak{m}_1 \cdot \mathfrak{m}_2 : \ \mathfrak{m}_1 \in \mathfrak{S}_1, \ \mathfrak{m}_2 \in \mathfrak{S}_2\}$$

is well-based.

Lemma 3.1.2 gives a binary operation on $C[[\mathfrak{M}]]$ by

$$f \cdot g \ := \ \sum_{\mathfrak{m}\in\mathfrak{M}} \left(\sum_{\mathfrak{n}_1\cdot\mathfrak{n}_2=\mathfrak{m}} f_{\mathfrak{n}_1} g_{\mathfrak{n}_2} \right) \mathfrak{m}.$$

With this operation as multiplication, $C[[\mathfrak{M}]]$ is a domain, with subfield C via the identification $c \mapsto f$ with $f_1 = c$ and $f_\mathfrak{m} = 0$ for all $\mathfrak{m} \neq 1$. We identify the group $\mathfrak{M}$ with a subgroup of $C[[\mathfrak{M}]]^\times$ via $\mathfrak{m} \mapsto f$, with $f_\mathfrak{m} = 1$ and $f_\mathfrak{n} = 0$ for all $\mathfrak{n} \neq \mathfrak{m}$.

Let Γ be an additive copy of the group $\mathfrak{M}$, with group isomorphism

$$\mathfrak{m} \mapsto v\mathfrak{m} : \mathfrak{M} \to \Gamma,$$

and equip Γ with the ordering $\leqslant$ such that for all $\mathfrak{m}, \mathfrak{n} \in \mathfrak{M}$: $\mathfrak{m} \succcurlyeq \mathfrak{n} \iff v\mathfrak{m} \leqslant v\mathfrak{n}$. Then the map

$$v\colon C[[\mathfrak{M}]] \to \Gamma_\infty, \qquad vf = \begin{cases} v(\mathfrak{d}(f)) & \text{if } f \neq 0 \\ \infty & \text{if } f = 0 \end{cases}$$

is a valuation on the C-vector space $C[[\mathfrak{M}]]$, making $C[[\mathfrak{M}]]$ a spherically complete Hahn space over C. Moreover, v is a valuation on the domain $C[[\mathfrak{M}]]$. The binary relation $\preccurlyeq$ on $C[[\mathfrak{M}]]$ associated to v satisfies, for nonzero $f, g \in C[[\mathfrak{M}]]$:

$$f \preccurlyeq g \quad \iff \quad \mathfrak{d}(f) \preccurlyeq \mathfrak{d}(g) \text{ (in the ordered set } \mathfrak{M}).$$

LEMMA 3.1.3. *$C[[\mathfrak{M}]]$ is a field.*

PROOF. Let $f \in C[[\mathfrak{M}]]$, $f \neq 0$; to get $g \in C[[\mathfrak{M}]]$ with $fg = 1$, divide f by $f_{\mathfrak{d}(f)}\mathfrak{d}(f)$ to arrange $f = 1 + \varepsilon$ with $\varepsilon \prec 1$. The map $\Phi\colon C[[\mathfrak{M}]] \to C[[\mathfrak{M}]]$ given by $\Phi(x) = 1 - \varepsilon x$ is contractive: if $x, y \in C[[\mathfrak{M}]]$, $x \neq y$, then $\Phi(x) - \Phi(y) = \varepsilon(y - x) \prec x - y$. Since the valued additive group $C[[\mathfrak{M}]]$ is spherically complete, Theorem 2.2.12 gives $g \in C[[\mathfrak{M}]]$ with $1 - \varepsilon g = \Phi(g) = g$, and so $fg = 1$. □

The valuation ring of the valued field $K = C[[\mathfrak{M}]]$ is

$$\mathcal{O} = \{f \in K : \operatorname{supp}(f) \subseteq \mathfrak{M}^{\preccurlyeq 1}\}.$$

The map sending $f \in \mathcal{O}$ to its constant term f_1 is a surjective ring morphism $\mathcal{O} \to C$ with kernel

$$\smallfrown\!\!\!\!\!\mathcal{o} = \{f \in K : \operatorname{supp}(f) \subseteq \mathfrak{M}^{\prec 1}\},$$

hence induces an isomorphism $\operatorname{res}(K) = \mathcal{O}/\smallfrown\!\!\!\!\!\mathcal{o} \to C$. We call K the valued field of **well-based series with coefficients in C and monomials from $\mathfrak{M}$.** A valued field of the form $C[[\mathfrak{M}]]$ is also referred to as a **Hahn field over C.**

Often we prefer an alternative notation for Hahn fields, as follows. Let C be a field and Γ an additive ordered abelian group. Then $C((t^\Gamma))$ is the field consisting of the formal series $f = \sum_\gamma c_\gamma t^\gamma$ (summation over all $\gamma \in \Gamma$) with all coefficients $c_\gamma \in C$, such that $\{\gamma : c_\gamma \neq 0\}$ is a well-ordered subset of Γ, and with the obvious addition, and multiplication according to $t^\alpha t^\beta = t^{\alpha+\beta}$ for $\alpha, \beta \in \Gamma$. Note that $C((t^\Gamma))$ is $C[[x^\Gamma]]$ in our original notation, with $x^\gamma := t^{-\gamma}$ for $\gamma \in \Gamma$, and with dominance relation $\preccurlyeq$ on the multiplicative group $x^{\Gamma} = t^{\Gamma}$ given by: $x^{\alpha} \preccurlyeq x^{\beta} \iff \alpha \leqslant \beta$. This yields the Hahn field $C((t^{\Gamma}))$ over C with valuation $v\colon C((t^{\Gamma}))^{\times} \to \Gamma$ given by

$$v(f) = \min\{\gamma \in \Gamma : c_{\gamma} \neq 0\}$$

for nonzero $f = \sum_{\gamma} c_{\gamma} t^{\gamma}$ as above. When Γ is the ordered group $\mathbb{Z}$ of integers, this gives the valued field $C((t^{\mathbb{Z}})) = C((t)) = C[[x^{\mathbb{Z}}]]$ of Laurent series in $t := t^1 = x^{-1}$ over C, with valuation ring $C[[t]]$ in conventional notation.

The overrings of a valuation ring

Let A be a domain with fraction field K. Recall from Section 1.4 that, given a prime ideal $\mathfrak{p}$ of A, the localization of A with respect to $\mathfrak{p}$ is the subring

$$A_{\mathfrak{p}} = \{a/s : a, s \in A, s \notin \mathfrak{p}\}$$

of K, that $A_{\mathfrak{p}}$ is a local ring with maximal ideal $\mathfrak{p}A_{\mathfrak{p}}$ generated by $\mathfrak{p}$, and that $\mathfrak{p}A_{\mathfrak{p}} \cap A = \mathfrak{p}$. Note also that for prime ideals $\mathfrak{p}$, $\mathfrak{p}'$ of A we have:

$$A_{\mathfrak{p}} \subseteq A_{\mathfrak{p}'} \iff \mathfrak{p} \supseteq \mathfrak{p}'.$$

In the rest of this subsection A is a valuation ring of K with associated valuation $v\colon K^{\times} \to \Gamma$ where $\Gamma = \Gamma_A$. The following is easy to verify:

LEMMA 3.1.4. *If B is a subring of K containing A, then B is a valuation ring of K, $\mathfrak{m}_B \subseteq \mathfrak{m}_A$, and $B = A_{\mathfrak{p}}$ for a unique prime ideal $\mathfrak{p}$ of A, namely $\mathfrak{p} = \mathfrak{m}_B$.*

Combining the previous lemma with the next lemma shows that the collection of overrings of A in its fraction field is totally ordered by inclusion. The proof of this lemma is also a routine verification left to the reader.

LEMMA 3.1.5. *We have an inclusion-reversing bijection from the set of convex subgroups of Γ onto the set of prime ideals of A given by*

$$\Delta \mapsto \mathfrak{p}_{\Delta} := \{x \in K : vx > \Delta\},$$

with inverse

$$\mathfrak{p} \mapsto \Delta_{\mathfrak{p}} := \{\gamma \in \Gamma : |\gamma| < vx \text{ for all } x \in \mathfrak{p}\}.$$

The rank of the valuation ring A is defined to be the rank of the ordered abelian group Γ, and similarly, by the rank of a valued field we mean the rank of its value group. (We use analogous terminology for rational rank in place of rank.) By the previous lemma, if A has finite rank r, then K has exactly $r+1$ subrings containing A and they form a tower: $A = B_0 \subseteq B_1 \subseteq \cdots \subseteq B_r = K$.

COROLLARY 3.1.6. *The following conditions on Γ and A are equivalent:*

(i) *Γ is archimedean;*

(ii) *A is proper subring of K and is maximal with respect to inclusion among the proper subrings of K.*

Let $\mathfrak{p}$ be a prime ideal of A and $B = A_{\mathfrak{p}}$, with associated valuation $v_B \colon K^\times \to \Gamma_B$, and let $\Delta = \Delta_{\mathfrak{p}}$ with ordered quotient group $\dot{\Gamma} = \Gamma/\Delta$. The inclusion $A \subseteq B$ gives rise to an ordered group morphism $\Gamma = K^\times/A^\times \to K^\times/B^\times = \Gamma_B$ with kernel Δ, so we have an isomorphism $\Gamma_B \to \dot{\Gamma}$ (of ordered groups) which fits into the commutative diagram

$$\begin{array}{ccc} K^\times & \xrightarrow{\ v\ } & \Gamma \\ {\scriptstyle v_B}\downarrow & & \downarrow \\ \Gamma_B & \xrightarrow{\ \cong\ } & \dot{\Gamma} \end{array}$$

where the arrow on the right is the natural surjection $\Gamma \to \dot{\Gamma}$.

Degree, residue degree, and ramification index

Let $K \subseteq L$ be a field extension. Then $[L : K]$ is its **degree**, that is, the dimension of L as a vector space over K, with the convention that $[L : K] = \infty$ if this dimension is infinite. Likewise, given an extension $\Gamma \subseteq \Gamma'$ of abelian groups we let $[\Gamma' : \Gamma]$ be its index, which by convention is ∞ if Γ'/Γ is infinite. We also have the transcendence degree $\operatorname{trdeg}(L|K)$ of L over K, set equal to ∞ if $\operatorname{trdeg}(L|K)$ is not finite. *Below in this subsection $K \subseteq L$ is a valued field extension.* We call $[\operatorname{res}(L) : \operatorname{res}(K)]$ the **residue degree** of L over K and $[\Gamma_L : \Gamma]$ the **ramification index** of L over K.

PROPOSITION 3.1.7. *Let $b_1, \dots, b_m \in \mathcal{O}_L$ be such that $\overline{b_1}, \dots, \overline{b_m} \in \operatorname{res}(L)$ are linearly independent over $\operatorname{res}(K)$, $m \geqslant 1$. Likewise, let $c_1, \dots, c_n \in L^\times$ be such that $vc_1, \dots, vc_n \in \Gamma_L$ lie in distinct cosets of Γ, $n \geqslant 1$. Then*

$$v\left(\sum_{i,j} a_{ij} b_i c_j\right) = \min_{i,j} v(a_{ij} c_j) \qquad (\textit{all } a_{ij} \in K). \tag{3.1.2}$$

In particular, the family $(b_i c_j)$ is K-linearly independent.

PROOF. First we show that $v(a_1 b_1 + \cdots + a_m b_m) = \min_i v(a_i)$ for $a_1, \dots, a_m \in K$. We can assume that some $a_i \neq 0$, and then dividing by an a_i of minimum valuation, we can reduce to the case that $v(a_i) \geqslant 0$ for all i and $v(a_i) = 0$ for some i. We must show that then $v(a_1 b_1 + \cdots + a_m b_m) = 0$. This follows by reduction mod o_L using the linear independence of $\overline{b_1}, \dots, \overline{b_m}$ over $\operatorname{res}(K)$. To show (3.1.2), we can assume that for each j there is an i with $a_{ij} \neq 0$. So for any j,

$$\gamma_j := v\left(\sum_i a_{ij} b_i c_j\right) = v\left(\left(\sum_i a_{ij} b_i\right) c_j\right) = v\left(\sum_i a_{ij} b_i\right) + vc_j = \min_i v(a_{ij}) + vc_j \in \Gamma + vc_j.$$

Now for $j \neq j'$ we have $\Gamma + vc_j \neq \Gamma + vc_{j'}$, so $\gamma_j \neq \gamma_{j'}$. Hence

$$v\left(\sum_{i,j} a_{ij} b_i c_j\right) = \min_j \gamma_j = \min_j\left(\min_i v(a_{ij}) + v(c_j)\right) = \min_{i,j} v(a_{ij} c_j). \qquad \square$$

COROLLARY 3.1.8. $[L:K] \geqslant \big[\operatorname{res}(L):\operatorname{res}(K)\big]\cdot[\Gamma_L:\Gamma]$.

Under suitable extra assumptions on K, the inequality in Corollary 3.1.8 is an equality; see Corollary 3.3.49. For algebraic extensions the previous corollary yields:

COROLLARY 3.1.9. *If* $[L:K]=n$, *then* $\big[\operatorname{res}(L):\operatorname{res}(K)\big]\leqslant n$ *(so* $\operatorname{res}(L)$ *is algebraic over* $\operatorname{res}(K)$*), and* $[\Gamma_L:\Gamma]\leqslant n$ *(so* $m\Gamma_L\subseteq\Gamma$ *for some* $m\in\{1,\dots,n\}$*).*

REMARK. Degree, residue degree and ramification index are multiplicative: for valued field extensions $K\subseteq L\subseteq M$ with $[M:K]<\infty$,

$$\begin{aligned}
[M:K] &= [M:L]\cdot[L:K],\\
\big[\operatorname{res}(M):\operatorname{res}(K)\big] &= \big[\operatorname{res}(M):\operatorname{res}(L)\big]\cdot\big[\operatorname{res}(L):\operatorname{res}(K)\big],\\
\big[\Gamma_M:\Gamma\big] &= [\Gamma_M:\Gamma_L]\cdot[\Gamma_L:\Gamma].
\end{aligned}$$

Corollary 3.1.8 also has an analogue for the transcendence degree of valued field extensions. To see this, note:

LEMMA 3.1.10. *Let* $x_1,\dots,x_m\in\mathcal{O}_L$ *be such that the residue classes* $\overline{x_1},\dots,\overline{x_m}\in\operatorname{res}(L)$ *are algebraically independent over* $\boldsymbol{k}=\operatorname{res}(K)$, *and let* $y_1,\dots,y_n\in L^\times$ *be such that the cosets* $vy_1+\Gamma,\dots,vy_n+\Gamma\in\Gamma_L/\Gamma$ *are* $\mathbb{Z}$*-linearly independent. Then* $x_1,\dots,x_m,y_1,\dots,y_n$ *are algebraically independent over* K.

This lemma is immediate from Proposition 3.1.7, and in turn now entails what is sometimes called the *Zariski-Abhyankar Inequality*:

COROLLARY 3.1.11. $\operatorname{trdeg}(L|K)\geqslant\operatorname{rank}_{\mathbb{Q}}(\Gamma_L/\Gamma)+\operatorname{trdeg}\big(\operatorname{res}(L)|\operatorname{res}(K)\big)$.

The following consequence of this inequality is used in Section 3.6:

COROLLARY 3.1.12. *Let* $E\subseteq F$ *be a field extension with* $\operatorname{trdeg}(F|E)=1$ *and let* A *be a valuation ring of* F *such that* $E\subseteq A\neq F$. *Then* A *is maximal among the proper subrings of* F.

PROOF. By Corollary 3.1.11 applied to $K:=E$ with the trivial valuation, and $L:=F$ with the valuation given by A, the value group Γ_L has $\operatorname{rank}_{\mathbb{Q}}(\Gamma_L)=1$, and so is archimedean. It remains to use Corollary 3.1.6. □

Integral closure and valuations

Next we relate integrality to valuations. *In this subsection we fix a field* K *and a local subring* A *of* K *with maximal ideal* $\mathfrak{m}=\mathfrak{m}_A$. The first proposition and Zorn imply that there is always a valuation ring of K lying over A. (See the beginning of this section for the meaning of *lying over* for local rings and in particular for valuation rings.)

PROPOSITION 3.1.13. *Consider the class of all local subrings of* K *lying over* A, *partially ordered by* $B\leqslant B'$ $:\Longleftrightarrow$ B' *lies over* B. *Any maximal element of this class is a valuation ring of* K.

In the proof of this proposition we use the following lemma.

LEMMA 3.1.14. *Suppose $x \in K^\times$ is such that $1 \in \mathfrak{m}A[x^{-1}] + x^{-1}A[x^{-1}]$. Then x is integral over A.*

PROOF. We have $1 = a_n x^{-n} + \cdots + a_1 x^{-1} + a_0$ where $a_1, \ldots, a_n \in A$ and $a_0 \in \mathfrak{m}$. Multiplying both sides by x^n yields:

$$x^n(1 - a_0) + (\text{terms of lower degree in } x) = 0.$$

Since $1 - a_0$ is a unit in A, it follows that x is integral over A. □

PROOF OF PROPOSITION 3.1.13. Replacing A by a maximal element of the class of local subrings of K lying over A, we arrange that A is the only local subring of K lying over A; we need to show that then A is a valuation ring of K. For this, let $x \in K^\times$. Suppose first that x is integral over A. Then $A[x]$ is integral over A, so Corollaries 1.3.5 and 1.3.7 give a maximal ideal $\mathfrak{n}$ of $A[x]$ with $\mathfrak{n} \cap A = \mathfrak{m}$, and then $A[x]_\mathfrak{n}$ is a local subring of K which lies over A. By maximality of A this gives $x \in A$. Next, suppose x is not integral over A. Then by the lemma above we have $1 \notin \mathfrak{m}A[x^{-1}] + x^{-1}A[x^{-1}]$, so we have a maximal ideal $\mathfrak{n}$ of $A[x^{-1}]$ such that $\mathfrak{n} \supseteq \mathfrak{m}$, and thus $\mathfrak{n} \cap A = \mathfrak{m}$. The local subring $A[x^{-1}]_\mathfrak{n}$ of K lies over A, and maximality of A yields $x^{-1} \in A$. □

COROLLARY 3.1.15. *Let $L \supseteq K$ be a field extension. For each valuation ring of K there exists a valuation ring of L which lies over it.*

LEMMA 3.1.16. *Each valuation ring is integrally closed.*

PROOF. Suppose that A is a valuation ring of K, let $v = v_A$ be the associated valuation, and let $x \in K$ satisfy $x^n + a_{n-1}x^{n-1} + \cdots + a_0 = 0$ (all $a_i \in A$). If $vx < 0$, then $v(x^n) = nvx < ivx + v(a_i) = v(a_i x^i)$ for $i = 0, \ldots, n-1$ and hence $v(x^n + \cdots + a_0) = nvx \neq \infty$, a contradiction. □

From Lemma 3.1.16 we obtain:

COROLLARY 3.1.17. *If K is algebraically closed and A is a valuation ring of K, then the field $\boldsymbol{k}_A = A/\mathfrak{m}$ is algebraically closed and the abelian group Γ_A is divisible.*

In view of Corollary 3.1.9 this gives:

COROLLARY 3.1.18. *Let L be an algebraic closure of K. Let A, B be valuation rings of K, L, respectively, such that B lies over A. (So (K, A) is a valued subfield of (L, B).) Then $\boldsymbol{k}_B$ is an algebraic closure of $\boldsymbol{k}_A$, and Γ_B is a divisible hull of Γ_A.*

PROPOSITION 3.1.19. *The integral closure of A in K equals the intersection of the valuation rings of K lying over A.*

PROOF. Any $x \in K$ integral over A lies in every valuation ring of K containing A as a subring. Next, suppose $x \in K^\times$ is not integral over A. Then, by the lemma above, $\mathfrak{m}A[x^{-1}] + x^{-1}A[x^{-1}]$ is a proper ideal of $A[x^{-1}]$. So we can take a maximal ideal $\mathfrak{n} \supseteq \mathfrak{m}$ of $A[x^{-1}]$ that contains x^{-1}. This yields a local subring $A[x^{-1}]_\mathfrak{n}$ of K that lies over A. Let V be a maximal element of the class of local subrings of K lying over $A[x^{-1}]_\mathfrak{n}$. Then V is a valuation ring of K, by Proposition 3.1.13, V lies over A, and $x^{-1} \in \mathfrak{m}_V$, so $x \notin V$. □

Valuations and algebraic field extensions

In this subsection K is a field and A is a valuation ring of K with maximal ideal $\mathfrak{m} = \mathfrak{m}_A$. Also, L is an algebraic field extension of K, and B is the integral closure of A in L.

PROPOSITION 3.1.20. *The valuation rings of L lying over A are exactly the $B_{\mathfrak{q}}$ with $\mathfrak{q}$ a maximal ideal of B.*

PROOF. Let V be a valuation ring of L lying over A. Valuation rings are integrally closed, so $B \subseteq V$. Set $\mathfrak{q} := \mathfrak{m}_V \cap B$, so $\mathfrak{q} \cap A = \mathfrak{m}$, hence $\mathfrak{q}$ is a maximal ideal of B by Corollary 1.3.5.

CLAIM: $V = B_{\mathfrak{q}}$.

To prove this claim, note first that $B \setminus \mathfrak{q} \subseteq V \setminus \mathfrak{m}_V$, hence $B_{\mathfrak{q}} \subseteq V$. For the other inclusion let $x \in V^{\neq}$. We have a relation $a_n x^n + \cdots + a_0 = 0$ where $a_0, \ldots, a_n \in A$, $a_n \neq 0$. Take $s \in \{0, \ldots, n\}$ maximal such that $v_A(a_s) = \min_i v_A(a_i)$, and put $b_i := a_i/a_s$. Dividing by $a_s x^s$ yields

$$\underbrace{(b_n x^{n-s} + \cdots + b_{s+1}x + 1)}_{y} + x^{-1}\underbrace{(b_{s-1} + \cdots + b_0/x^{s-1})}_{z} = 0.$$

So $x = -y^{-1}z$. Since $b_i \in \mathfrak{m}$ for $i = s+1, \ldots, n$ we have $y \in V^{\times}$, thus $y \notin \mathfrak{q}$. Hence to get $x \in B_{\mathfrak{q}}$ it suffices to show that $y, z \in B$. This will follow from Proposition 3.1.19 if we show that y, z lie in every valuation ring of L lying over A. If such a ring contains x, it also contains y, hence it contains $z = -yx$. If such a ring does not contain x, then it contains x^{-1}, and thus $z = b_0/x^{s-1} + \cdots$. This finishes the proof of our claim.

Conversely, let $\mathfrak{q}$ be a maximal ideal of B. It is clear that $B_{\mathfrak{q}}$ lies over A, and it remains to show that $B_{\mathfrak{q}}$ is a valuation ring of L. Proposition 3.1.13 gives a valuation ring V of L that lies over $B_{\mathfrak{q}}$, and so $V = B_{\mathfrak{p}}$ where $\mathfrak{p}$ is a maximal ideal of B. As $B_{\mathfrak{q}} \subseteq V$, this gives $\mathfrak{p} \subseteq \mathfrak{q}$, so $\mathfrak{p} = \mathfrak{q}$, and thus $B_{\mathfrak{q}} = V$. □

Proposition 3.1.20 yields a bijection $\mathfrak{q} \mapsto B_{\mathfrak{q}}$ from the set of maximal ideals of B onto the set of valuation rings of L lying over A. In particular, distinct valuation rings of L lying over A are incomparable with respect to inclusion. The next proposition concerns the behavior of valuations under *normal* field extensions:

PROPOSITION 3.1.21. *Suppose $L \supseteq K$ is normal. Then for any valuation rings V, V' of L lying over A there is some $\sigma \in \mathrm{Aut}(L|K)$ such that $\sigma(V) = V'$.*

In the proof we use the Chinese Remainder Theorem [249, II, §2]: *Let $I_1, \ldots, I_n$ $(n \geqslant 1)$ be ideals in the commutative ring R such that $I_i + I_j = R$ for all i, j with $1 \leqslant i < j \leqslant n$, and let $a_1, \ldots, a_n \in R$. Then there exists an $x \in R$ such that $x \equiv a_i \mod I_i$ for $i = 1, \ldots, n$.*

PROOF OF PROPOSITION 3.1.21. We assume $[L : K] < \infty$ below, since the proposition follows from its validity in that special case.

Let $\mathfrak{q}$, $\mathfrak{q}'$ be maximal ideals of B with $\mathfrak{q} \cap A = \mathfrak{q}' \cap A = \mathfrak{m}$; by Proposition 3.1.20 it suffices to show that there is $\sigma \in G := \mathrm{Aut}(L|K)$ with $\sigma(\mathfrak{q}) = \mathfrak{q}'$. Assume towards a contradiction that $\sigma(\mathfrak{q}) \neq \mathfrak{q}'$ for all $\sigma \in G$. Thus the two sets of maximal ideals of B, $\{\sigma(\mathfrak{q}) : \sigma \in G\}$ and $\{\sigma(\mathfrak{q}') : \sigma \in G\}$, are disjoint; since $[L : K] < \infty$, these two sets are also finite. By the Chinese Remainder Theorem we get $x \in B$ such that

$$x \equiv 0 \bmod \sigma(\mathfrak{q}), \quad x \equiv 1 \bmod \sigma(\mathfrak{q}') \qquad \text{for all } \sigma \in G.$$

Hence $\sigma x \in \mathfrak{q} \setminus \mathfrak{q}'$ for all $\sigma \in G$. Recall that

$$\mathrm{N}_{L|K}(x) := \Big(\prod_{\sigma \in G} \sigma x\Big)^{\ell}$$

lies in K, where $\ell = 1$ if $\operatorname{char} K = 0$, and $\ell = p^e$ for some $e \in \mathbb{N}$ if $\operatorname{char} K = p > 0$. Each σx is integral over A, so $\mathrm{N}_{L|K}(x) \in A$ since A is integrally closed. Also $\mathrm{N}_{L|K}(x) \in \mathfrak{q} \setminus \mathfrak{q}'$ because $\mathfrak{q}$ and $\mathfrak{q}'$ are prime ideals, hence $\mathrm{N}_{L|K}(x) \in A \cap \mathfrak{q} = \mathfrak{m} \subseteq \mathfrak{q}'$, so $\mathrm{N}_{L|K}(x) \in \mathfrak{q}'$, a contradiction. □

COROLLARY 3.1.22. *Suppose $[L : K] < \infty$. There are only finitely many valuation rings of L lying over A.*

PROOF. Take a field extension $L' \supseteq L$ such that $[L' : K] < \infty$ and $L' \supseteq K$ is normal, apply Proposition 3.1.21 to L', and use that $\mathrm{Aut}(L'|K)$ is finite. □

In the next result K is the valued field with valuation ring A.

COROLLARY 3.1.23. *Let K^{a} be an algebraic closure of K equipped with a valuation ring of K^{a} lying over A. Then any valued field embedding $K \to F$, where F is an algebraically closed valued field, extends to a valued field embedding $K^{\mathrm{a}} \to F$.*

PROOF. Let $K \to F$ be a valued field embedding, and let $j \colon K^{\mathrm{a}} \to F$ be a field embedding that extends the field embedding $K \to F$. Let V be the valuation ring of F. Then $j^{-1}(V)$ is a valuation ring of K^{a} lying over A, and so we have a field automorphism σ of K^{a} over K such that $\sigma^{-1}j^{-1}(V)$ is the valuation ring of K^{a}. Then $j\sigma$ is a valued field embedding $K^{\mathrm{a}} \to F$ as desired. □

Although for our purpose we need to consider only valued fields of characteristic zero, we note for the sake of completeness:

LEMMA 3.1.24. *Suppose $\operatorname{char}(K) = p > 0$ and $L \supseteq K$ is purely inseparable. Then there is a unique valuation ring of L that lies over A.*

PROOF. This unique valuation ring is $\{x \in L : x^{p^n} \in A \text{ for some } n\}$. □

LEMMA 3.1.25. *Let V be a valuation ring of L lying over A, with v as its associated valuation. Suppose $\sigma \in \mathrm{Aut}(L|K)$ is such that $\sigma(V) = V$. Then $v \circ \sigma = v$.*

PROOF. Otherwise we have $x \in K^\times$ with $v(\sigma(x)) < v(x)$, that is, $\sigma(x)/x \notin V$. By induction we get $\sigma^n(x)/\sigma^{n-1}(x) \notin V$ for all $n \geqslant 1$, and thus $v(\sigma^n(x)) < v(x)$ for all $n \geqslant 1$. Taking $n \geqslant 1$ such that $\sigma^n(x) = x$ we get a contradiction. □

COROLLARY 3.1.26. *Let L be a normal extension of K and V a valuation ring of L lying over A, with associated valuation v. Suppose V is the only valuation ring of L lying over A. Then for $x \in L^\times$ with minimum polynomial*

$$a_0 + a_1X + \cdots + a_{n-1}X^{n-1} + X^n \qquad (a_0, \dots, a_{n-1} \in K,\ n \geqslant 1)$$

over K we have $v(x) = \frac{1}{n}v(a_0)$.

PROOF. Use Lemma 3.1.25 and the fact that a_0 or $-a_0$ is a product $x_1 \cdots x_n$ of conjugates x_i of x. □

Recall that a Galois extension of a field E is an algebraic field extension of E that is both normal and separable over E. Let L be a Galois extension of K and $G := \mathrm{Aut}(L|K)$. Let V be a valuation ring of L lying over A. Then the subgroup

$$G^{\mathrm{d}} := \{\sigma \in G : \sigma(V) = V\}$$

of G is called the **decomposition group** of V over K, and the fixed field L^{d} of G^{d} is called the **decomposition field** of V over K.

COROLLARY 3.1.27. *Suppose L is a Galois extension of K. If $V' \neq V$ is another valuation ring of L lying over A, then $V \cap L^{\mathrm{d}} \neq V' \cap L^{\mathrm{d}}$. Moreover, L^{d} is the smallest subfield of L containing K and having this property.*

PROOF. If V' is a valuation ring of L lying over A with $V \cap L^{\mathrm{d}} = V' \cap L^{\mathrm{d}}$, then by Proposition 3.1.21 there is some $\sigma \in \mathrm{Aut}(L|L^{\mathrm{d}})$ such that $\sigma(V) = V'$; but $\mathrm{Aut}(L|L^{\mathrm{d}}) = G^{\mathrm{d}}$, hence for such σ we have $\sigma(V) = V$ and thus $V = V'$. Suppose L' is any subfield of L with $K \subseteq L'$ which also has the indicated property. Then $\mathrm{Aut}(L|L') \subseteq G^{\mathrm{d}}$: if $\sigma \in \mathrm{Aut}(L|L')$ then both $\sigma(V)$ and V are valuation rings of L lying over $V \cap L'$, hence $\sigma(V) = V$, and thus $\sigma \in G^{\mathrm{d}}$. Therefore $L^{\mathrm{d}} \subseteq L'$. □

Adjoining roots

Let K be a valued field with residue field $\boldsymbol{k} = \mathrm{res}(K)$. For use later in this section and in Section 3.3 we show:

LEMMA 3.1.28. *Let p be a prime number, and x an element in a field extension of K such that $x^p = a \in K^\times$ where $va \notin p\Gamma$. Then $X^p - a$ is the minimum polynomial of x over K, and v extends uniquely to a valuation $w\colon K(x)^\times \to \Delta$ with $\Delta \subseteq \mathbb{Q}\Gamma$ (as ordered groups). The residue field of w remains $\boldsymbol{k}$, and $[\Delta : \Gamma] = p$, with*

$$\Delta = \bigcup_{i=0}^{p-1} \Gamma + iw(x) \qquad (\textit{disjoint union}).$$

PROOF. Let $w\colon K(x)^\times \to \Delta$ with $\Delta \subseteq \mathbb{Q}\Gamma$ be a valuation extending v. (By Corollaries 3.1.15 and 3.1.9 there is such an extension.) Since $va \notin p\Gamma$, the elements

$$w(x^0) = 0, \quad w(x^1) = \frac{va}{p}, \ldots, w(x^{p-1}) = \frac{(p-1)va}{p}$$

of Δ are in distinct cosets of Γ, so $1, x, \ldots, x^{p-1}$ are K-linearly independent, and thus $X^p - a$ is the minimum polynomial of x over K. Also, for an arbitrary nonzero element $b = b_0 + b_1 x + \cdots + b_{p-1}x^{p-1}$ of $K(x)$ (all $b_i \in K$),

$$w(b) = \min\left\{v(b_i) + \frac{iva}{p} : i = 0, \ldots, p-1\right\},$$

showing the uniqueness of w. This also proves the claims made by the lemma about Δ. The residue field of w remains $\boldsymbol{k}$ by Corollary 3.1.8. □

Simple transcendental extensions

Let K be a valued field with value group Γ and residue field $\mathrm{res}(K)$. So far we mainly considered algebraic extensions, but in the rest of this section $L = K(x)$ is a field extension with x transcendental over K. We shall indicate some valuations of L that extend the given valuation v of K.

LEMMA 3.1.29. *Let Δ be an ordered abelian group extension of Γ and $\delta \in \Delta$. Then v extends uniquely to a valuation $v\colon L^\times \to \Delta$ of L such that*

$$va = \min_i \big(v(a_i) + i\delta\big) \qquad \text{for } a = \sum_i a_i x^i \in K[x]^{\neq} \text{ (all } a_i \in K). \tag{3.1.3}$$

Here we do not require $v\colon L^\times \to \Delta$ to be surjective.

PROOF. There is clearly at most one such extension as in the lemma. To prove existence, define $v\colon K[x]^{\neq} \to \Delta$ by (3.1.3). It suffices to show that v is a valuation on $K[x]$ (hence extends to a valuation on L). It is clear that (V2) is satisfied. To show (V1), let $a = \sum_i a_i x^i$, $b = \sum_j b_j x^j \in K[x]^{\neq}$ ($a_i, b_j \in K$). Then

$$ab = \sum_n c_n x^n \qquad \text{where } c_n = \sum_{i+j=n} a_i b_j,$$

and for all n,

$$v(c_n) + n\delta \geqslant \min_{i+j=n} \big((v(a_i) + i\delta) + (v(b_j) + j\delta)\big) \geqslant va + vb.$$

Take i_0 minimal such that $v(a_{i_0}) + i_0\delta = \min_i \big(v(a_i) + i\delta\big)$ and take j_0 minimal such that $v(b_{j_0}) + j_0\delta = \min_j \big(v(b_j) + j\delta\big)$. Now set $n_0 = i_0 + j_0$ and consider

$$c_{n_0} x^{n_0} = a_{i_0} b_{j_0} x^{n_0} + \underbrace{\text{terms } a_i b_j x^{n_0} \text{ with } i + j = n_0, \text{ and } i < i_0 \text{ or } j < j_0}_{\text{each has valuation} > v(a_{i_0}) + v(b_{j_0}) + n_0\delta}.$$

Thus $v(c_{n_0}) + n_0\delta = \big(v(a_{i_0}) + i_0\delta\big) + \big(v(b_{j_0}) + j_0\delta\big) = va + vb$. □

Next we consider in more detail two special cases.

LEMMA 3.1.30. *Let Δ be an ordered abelian group extension of Γ, and $\delta \in \Delta$ such that $n\delta \notin \Gamma$ for all $n \geqslant 1$. Then v extends uniquely to a (not necessarily surjective) valuation $v\colon L^\times \to \Delta$ of L with $vx = \delta$. The value group of this valuation is the internal direct sum $\Gamma \oplus \mathbb{Z}\delta$ in Δ, and its residue field is $\boldsymbol{k} := \operatorname{res}(K)$.*

PROOF. If v is an extension as in the lemma, then

$$v(1) = 0, \quad v(x) = \delta, \quad v(x^2) = 2\delta, \ \dots$$

lie in different cosets of Γ, so by Proposition 3.1.7, v satisfies (3.1.3), and thus v must be the extension of Lemma 3.1.29. Let v be the extension described in Lemma 3.1.29. Then $vx = \delta$, and $v(L^\times) = \Gamma \oplus \mathbb{Z}\delta$. To see that the residue field is still $\boldsymbol{k}$, let $a \in L$ with $va = 0$, so $a = b/c$ where $b, c \in K[x]^{\neq}$. Then $b = dx^m(1+r)$, $c = ex^n(1+s)$ where $d, e \in K^\times$ and $r, s \prec 1$ in L. Since $vd + m\delta = vb = vc = ve + n\delta$ we have $vd = ve$ and $m = n$. Thus $a = (d/e)(1+t)$ where $t \prec 1$ in L, so $\overline{a} = \overline{d/e} \in \boldsymbol{k}$. □

The valuation on L described in the next lemma is called the **gaussian extension** of v to L (with respect to x).

LEMMA 3.1.31. *There is a unique valuation ring of L that makes L a valued field extension of K with $x \preccurlyeq 1$ and $\overline{x}$ transcendental over $\boldsymbol{k} := \operatorname{res}(K)$. Equipping L with this valuation ring we have $\Gamma_L = \Gamma$ and $\operatorname{res}(L) = \boldsymbol{k}(\overline{x})$, and*

$$v_L(a) = \min_i v(a_i) \qquad \text{for } a = \sum_i a_i x^i \in K[x]\ (a_i \in K). \tag{3.1.4}$$

PROOF. If L is equipped with a valuation ring as in the lemma with associated valuation v_L, then $1, \overline{x}, \overline{x}^2, \dots$ are linearly independent over $\boldsymbol{k}$, so (3.1.4) holds by Proposition 3.1.7, which determines v_L uniquely, with $\Gamma_L = \Gamma$. As to existence, apply Lemma 3.1.29 with $\Delta = \Gamma$ and $\delta = 0$ to equip L with the valuation ring whose associated valuation v_L satisfies (3.1.4). Then clearly $v_L(x) = 0$, and $\overline{x}$ is transcendental over $\boldsymbol{k}$. To get $\operatorname{res}(L) = \boldsymbol{k}(\overline{x})$, let $a \in L$ with $v_L(a) = 0$; we claim that $\overline{a} \in \boldsymbol{k}(\overline{x})$. Now $a = b/c$ where $b, c \in K[x]^{\neq}$ with $v_L(b) = v_L(c) = 0$. Then $\overline{b} \neq 0$, $\overline{c} \neq 0$, and $\overline{b}, \overline{c} \in \boldsymbol{k}[\overline{x}]$, and so $\overline{a} = \overline{b}/\overline{c} \in \boldsymbol{k}(\overline{x})$ as desired. □

Let X be an indeterminate, and equip the field $K(X)$ with the gaussian extension of v with respect to X. A polynomial $P \in K[X]$ is said to be **primitive** if $vP = 0$. The proof of the next lemma is obvious.

LEMMA 3.1.32. *If $P \in K[X] \setminus K$ is primitive and $P = P_1 \cdots P_n$ with $P_i \in K[X]$ for $i = 1, \dots, n$, then there are $a_1, \dots, a_n \in K^\times$ such that each polynomial $Q_i = a_i P_i$ is primitive and $P = Q_1 \cdots Q_n$.*

Note that $\mathcal{O}[X]^\times = \mathcal{O}^\times$, since $\mathcal{O}$ is a domain.

LEMMA 3.1.33. *Suppose $P \in \mathcal{O}[X]$, $P \notin \mathcal{O}$. Then*

$$P \text{ is irreducible in } \mathcal{O}[X] \iff P \text{ is primitive and irreducible in } K[X].$$

PROOF. Take $a \in \mathcal{O}$ with $va = vP$. Then $P = a(a^{-1}P)$ and $a^{-1}P$ is primitive, so if P is irreducible in $\mathcal{O}[X]$, then P is primitive, hence irreducible in $K[X]$ by the previous lemma. The direction $\Leftarrow$ is obvious. □

Prescribing value group and residue field extensions

Let K be a valued field with value group Γ and residue field $\boldsymbol{k} = \mathrm{res}(K)$. We finish this section with the following existence result:

PROPOSITION 3.1.34. *Let Γ' be an ordered abelian group extension of Γ and $\boldsymbol{k}'$ a field extension of $\boldsymbol{k} = \mathrm{res}(K)$. Then there is a valued field extension K' of K with value group Γ' and with residue field isomorphic to $\boldsymbol{k}'$ over $\boldsymbol{k}$.*

This follows from Lemmas 3.1.28 and 3.1.30 (for the value group extension) and 3.1.31 and the following fact (for the residue field extension).

LEMMA 3.1.35. *Let ξ lie in an algebraic closure of $\boldsymbol{k}$, let $P \in \mathcal{O}[X]$ be monic such that $\overline{P} \in \boldsymbol{k}[X]$ is the minimum polynomial of ξ over $\boldsymbol{k}$, and let a be a zero of P in an algebraic closure of K. Then v extends uniquely to a (not necessarily surjective) valuation $v' \colon K(a)^\times \to \mathbb{Q}\Gamma$ on $K(a)$. The residue field $\boldsymbol{k}'$ of v' is $\boldsymbol{k}$-isomorphic to $\boldsymbol{k}(\xi)$, $v'(K(a)^\times) = \Gamma$, and $[K(a) : K] = [\boldsymbol{k}' : \boldsymbol{k}]$.*

PROOF. By Lemma 3.1.33, P is the minimum polynomial of a over K, hence

$$[K(a) : K] \;=\; \deg P \;=\; \deg \overline{P} \;=\; [\boldsymbol{k}(\xi) : \boldsymbol{k}].$$

Let $v' \colon K(a)^\times \to \mathbb{Q}\Gamma$ be a valuation on $K(a)$ extending v, with residue field $\boldsymbol{k}' \supseteq \boldsymbol{k}$. Then $v'a \geqslant 0$ since a is integral over $\mathcal{O}$. Now $\overline{P}(\overline{a}) = 0$ in $\boldsymbol{k}'$, hence there exists a $\boldsymbol{k}$-isomorphism $\boldsymbol{k}(\overline{a}) \to \boldsymbol{k}(\xi)$; in particular, $[K(a) : K] = [\boldsymbol{k}(\overline{a}) : \boldsymbol{k}]$ and thus $\boldsymbol{k}' = \boldsymbol{k}(\overline{a})$ and $v'(K(a)^\times) = \Gamma$, by Corollary 3.1.8. The uniqueness of v' follows from Proposition 3.1.7. □

Notes and comments

Everything in this section is classical valuation theory. The notations $\prec$ and $\sim$ were introduced by du Bois-Reymond [48, 49] in asymptotic analysis. (See [125, 126, 139] for discussions of his work.)

The valued fields $C[[\mathfrak{M}]]$ occur in Hahn [162]; a variant with $\mathfrak{M} = x^{\mathbb{R}}$ is in Levi-Civita [257]; generalizations of Hahn fields were considered by Mal′cev [282], Neumann [302], and Higman [179]; see [124] for some history.

As to our notation $C[[\mathfrak{M}]]$ for Hahn fields, another popular notation is $C((\mathfrak{M}))$, used for example in [112]. But here we only use the latter for $\mathfrak{M} = t^\Gamma$ as specified at the end of the subsection on well-based series.

Zariski [468, Chapter VI, §10] contains a special case of Corollary 3.1.11, and Abhyankar [3] a more general version for noetherian local rings. Propositions 3.1.13 and 3.1.19–3.1.21 and their corollaries are due to Krull [229], building on work by Deuring [102] and Ostrowski [314] for rank 1 valuations. In that setting Ostrowski defined the ramification index and residue degree, and proved Corollary 3.1.8 and

Lemma 3.1.24. Lemma 3.1.29 stems from Ostrowski [314] and Rella [332], and Proposition 3.1.34 from Mac Lane [277]. For the history of valuation theory before Krull's [229], see Roquette [360].

3.2 PSEUDOCONVERGENCE IN VALUED FIELDS

Let K be a valued field. Viewing the additive group of K as a valued abelian group, the material on pseudoconvergent and pseudocauchy sequences in Section 2.2 applies. Let (a_ρ) be a well-indexed sequence in K, and $a \in K$. Recall that (a_ρ) is said to pseudoconverge to a (notation: $a_\rho \rightsquigarrow a$) if $v(a - a_\rho)$ is eventually strictly increasing. We also say in that case that a is a pseudolimit of (a_ρ). Note that if $a_\rho \rightsquigarrow a$ and $b \in K$, then $a_\rho + b \rightsquigarrow a + b$, and $a_\rho b \rightsquigarrow ab$ if $b \neq 0$. More generally:

PROPOSITION 3.2.1. *If $a_\rho \rightsquigarrow a$, and $P \in K[X]$, $P \notin K$, then $P(a_\rho) \rightsquigarrow P(a)$.*

The proof of this proposition is based on Taylor expansion of polynomials: for a polynomial $P \in K[X]$ of degree at most d there are unique polynomials $P_{(i)} \in K[X]$ such that in the ring $K[X, Y]$ of polynomials over K in the distinct indeterminates X, Y, the identity

$$P(X+Y) = \sum_{i=0}^{d} P_{(i)}(X) \cdot Y^i \tag{3.2.1}$$

holds. For convenience we also set $P_{(i)} = 0$ for $i > d$, so $P_{(0)} = P$ and $P_{(1)} = P'$ (the formal derivative of P). If $\operatorname{char}(K) = 0$ then $P_{(i)} = \frac{1}{i!} P^{(i)}$ where $P^{(i)}$ is the usual ith formal derivative of P. Although this will not be used until the next section, we already note here a useful identity:

LEMMA 3.2.2. $P_{(i)(j)} = \binom{i+j}{i} P_{(i+j)}$ $(i, j \in \mathbb{N})$.

We also use a fact on ordered abelian groups, with the easy proof left to the reader:

LEMMA 3.2.3. *For each i in a finite nonempty set I let $\beta_i \in \Gamma$ and $n_i \in \mathbb{N}^{\geqslant 1}$, and let $\lambda_i \colon \Gamma \to \Gamma$ be the linear function given by $\lambda_i(\gamma) = \beta_i + n_i\gamma$. Assume that $n_i \neq n_j$ for all distinct $i, j \in I$. Let $\rho \mapsto \gamma_\rho$ be a strictly increasing function from an infinite linearly ordered set without largest element into Γ. Then there is an $i_0 \in I$ such that if $i \in I$ and $i \neq i_0$, then $\lambda_{i_0}(\gamma_\rho) < \lambda_i(\gamma_\rho)$ eventually.*

PROOF OF PROPOSITION 3.2.1. Assume $a_\rho \rightsquigarrow a$, and $P \in K[X]$ with $P \notin K$ has degree at most d. Substituting a for X and $a_\rho - a$ for Y in (3.2.1) yields

$$P(a_\rho) - P(a) = \sum_{i=1}^{d} P_{(i)}(a)(a_\rho - a)^i, \quad \text{and for } i = 1, \dots, d,$$

$$v\big(P_{(i)}(a)(a_\rho - a)^i\big) = \beta_i + i\gamma_\rho \quad \text{where } \beta_i := v\big(P_{(i)}(a)\big) \text{ and } \gamma_\rho := v(a_\rho - a).$$

Since $P = P(a) + \sum_{i=1}^{d} P_{(i)}(a)(X - a)^i$ and P is not constant, there is an $i \in \{1, \dots, d\}$ with $P_{(i)}(a) \neq 0$. Since (γ_ρ) is eventually strictly increasing, Lemma 3.2.3

yields $i_0 \in \{1, \dots, d\}$ with $P_{(i_0)}(a) \neq 0$ such that for every $i \in \{1, \dots, d\}$ with $i \neq i_0$ we have $\beta_{i_0} + i_0\gamma_\rho < \beta_i + i\gamma_\rho$ eventually. Then $v\big(P(a_\rho) - P(a)\big) = \beta_{i_0} + i_0\gamma_\rho$ eventually; in particular, the sequence $\big(v\big(P(a_\rho) - P(a)\big)\big)$ is eventually strictly increasing, that is, $P(a_\rho) \rightsquigarrow P(a)$. □

The sequence (a_ρ) is pc in K if and only if (a_ρ) has a pseudolimit in some valued field extension of K, and in that case, (a_ρ) has even a pseudolimit in some elementary extension of the valued field K. (See the remark following Lemma 2.2.5.) Together with the previous proposition this immediately yields:

COROLLARY 3.2.4. *Suppose (a_ρ) is a pc-sequence in K, and $P \in K[X]$ is nonconstant. Then $\big(P(a_\rho)\big)$ is a pc-sequence in K.*

If (a_ρ) is a pc-sequence, then $(v(a_\rho))$ is either eventually strictly increasing, or eventually constant. Hence:

COROLLARY 3.2.5. *Let (a_ρ) and (b_ρ) be pc-sequences in K. Then:*

$$a_\rho b_\rho \rightsquigarrow 0 \quad \Longleftrightarrow \quad a_\rho \rightsquigarrow 0 \text{ or } b_\rho \rightsquigarrow 0.$$

A valued field extension $L \supseteq K$ is said to be **immediate** if $\operatorname{res}(L) = \operatorname{res}(K)$ and $\Gamma_L = \Gamma$ (equivalently, the extension $L \supseteq K$ of valued additive groups is immediate as defined in Section 2.2).

In the rest of this subsection we assume that (a_ρ) is a pc-sequence in K, and we shall prove that then (a_ρ) has a pseudolimit in an immediate valued field extension of K. To prepare for that, let $P \in K[X]$ be nonconstant. Then by Corollaries 2.2.8 and 3.2.4 there are two possibilities:

either $\big(v(P(a_\rho))\big)$ is eventually strictly increasing (equivalently, $P(a_\rho) \rightsquigarrow 0$),
or $\big(v(P(a_\rho))\big)$ is eventually constant (equivalently, $P(a_\rho) \not\rightsquigarrow 0$).

We say that (a_ρ) is of **algebraic type over** K if the first possibility is realized for some nonconstant $P \in K[X]$, and then such a P of least degree is called a **minimal polynomial of** (a_ρ) **over** K. By Corollary 3.2.5 such a minimal polynomial of (a_ρ) over K is irreducible. We say that (a_ρ) is of **transcendental type over** K if the second possibility is realized for all nonconstant $P \in K[X]$. Note that for $a \in K$ we have $a_\rho \rightsquigarrow a$ iff $P(a_\rho) \rightsquigarrow 0$ for $P = X - a$; in particular, if (a_ρ) is of algebraic type over K and diverges in K, then a minimal polynomial of (a_ρ) over K has degree at least 2, and if (a_ρ) is of transcendental type over K, then (a_ρ) diverges in K. A pc-sequence in K of transcendental type over K determines an essentially unique immediate extension:

LEMMA 3.2.6. *Suppose (a_ρ) is of transcendental type over K. The valuation v on K extends uniquely to a valuation $v\colon K(X)^\times \to \Gamma$ such that*

$$vP = \textit{eventual value of } v\big(P(a_\rho)\big) \qquad \textit{for each } P \in K[X]. \tag{3.2.2}$$

With this valuation $K(X)$ is an immediate valued field extension of K in which $a_\rho \rightsquigarrow X$. Moreover, if $a_\rho \rightsquigarrow a$ in a valued field extension of K, then there is a valued field isomorphism $K(X) \to K(a)$ over K that sends X to a.

PROOF. It is clear that defining vP for $P \in K[X]$ as in (3.2.2), we have a valuation on $K[X]$ and thus on $K(X)$. The value group of this valuation is still Γ, and one checks easily that $a_\rho \rightsquigarrow X$. To verify that the residue field of $K(X)$ is the residue field of K we first note that because the value groups are equal, each $R \in K(X)$ with $vR = 0$ has the form $R = P/Q$ where $P, Q \in K[X]$ with $vP = vQ = 0$. So it is enough to consider a nonconstant $P \in K[X]$ with $vP = 0$, and find $b \in K$ with $v(P - b) > 0$. We have $0 = vP = v\big(P(a_\rho)\big)$ eventually, and $\big(v(P - P(a_\rho))\big)$ is eventually strictly increasing, so $v\big(P - P(a_\rho)\big) > 0$, eventually. Thus $b = P(a_\rho)$ for big enough ρ will do the job.

Finally, suppose $a_\rho \rightsquigarrow a$ with a in a valued field extension of K (whose valuation we continue to denote by v as usual). For nonconstant $P \in K[X]$ we have $P(a_\rho) \rightsquigarrow P(a)$ and thus $v\big(P(a)\big) = v\big(P(a_\rho)\big)$, eventually; in particular, $P(a) \neq 0$ and $v\big(P(a)\big) = vP \in \Gamma$. Thus a is transcendental over K and the field isomorphism $K(X) \to K(a)$ over K that sends X to a is even a valued field isomorphism. □

Here is an analogue for pc-sequences of algebraic type:

LEMMA 3.2.7. *Suppose (a_ρ) is divergent of algebraic type over K. Let $\mu(X)$ be a minimal polynomial of (a_ρ) over K, and a a zero of μ in an extension field of K. Then v extends uniquely to a valuation $v\colon K(a)^\times \to \Gamma$ such that*

$$v\big(P(a)\big) = \textit{eventual value of } v\big(P(a_\rho)\big) \qquad \textit{for each } P \in K[X] \textit{ of degree} < \deg \mu.$$

With this valuation $K(a)$ is an immediate valued field extension of K, and $a_\rho \rightsquigarrow a$. Moreover, if $\mu(b) = 0$ and $a_\rho \rightsquigarrow b$ in a valued field extension of K, then there is a valued field isomorphism $K(a) \to K(b)$ over K that sends a to b.

PROOF. Much of the proof duplicates the proof for the case of transcendental type. A difference is in how we obtain the multiplicative law for $v\colon K(a)^\times \to \Gamma$ as defined above. Let $s, t \in K(a)^\times$. Then $s = S(a)$, $t = T(a)$ with nonzero $S, T \in K[X]$ of degree less than $\deg \mu$, and $ST = Q\mu + R$ with $Q, R \in K[X]$ and $\deg R < \deg \mu$, so $st = R(a)$, and thus eventually

$$vs = v\big(S(a_\rho)\big), \quad vt = v\big(T(a_\rho)\big), \quad v(st) = v\big(R(a_\rho)\big).$$

Also

$$vs + vt = v\big(S(a_\rho)T(a_\rho)\big) = v\big(Q(a_\rho)\mu(a_\rho) + R(a_\rho)\big), \qquad \text{eventually.}$$

Since $\big(v\big(Q(a_\rho)\mu(a_\rho)\big)\big)$ is either eventually strictly increasing or eventually ∞, this forces $vs + vt = v(R(a_\rho))$, eventually, so $vs + vt = v(st)$. □

The previous two lemmas together now immediately imply:

COROLLARY 3.2.8. *Every pc-sequence in K has a pseudolimit in an immediate valued field extension of K.*

A minimal polynomial of a pc-sequence in K of algebraic type over K is in general non-unique, even if we require it to be monic. For example, adding to the constant term of a minimal polynomial $\mu \in K[X]$ of (a_ρ) over K any element of K whose valuation lies in the width of the pc-sequence $\big(\mu(a_\rho)\big)$ does not change its status as a minimal polynomial of (a_ρ) over K.

Maximal and algebraically maximal valued fields

A valued field is said to be **maximal** if it has no proper immediate valued field extension. By Lemma 2.2.1 and Zorn, every valued field has an immediate extension which is maximal.

COROLLARY 3.2.9. *The following are equivalent:*

(i) *K is maximal;*

(ii) *the valued additive group of K is maximal;*

(iii) *K is spherically complete;*

(iv) *every pc-sequence in K has a pseudolimit in K.*

PROOF. The implication (i) $\Rightarrow$ (iv) follows from Corollary 3.2.8, (iii) $\Leftrightarrow$ (iv) from Lemma 2.2.10, (iv) $\Rightarrow$ (ii) from Corollary 2.2.6, and (ii) $\Rightarrow$ (i) is trivial. □

For any field C and ordered abelian group $\mathfrak{M}$, the valued field $C[[\mathfrak{M}]]$ is maximal, by Corollaries 2.2.7 and 3.2.9. Thus by Corollaries 3.1.9 and 3.1.17:

COROLLARY 3.2.10. *Let C be a field and $\mathfrak{M}$ be an ordered abelian group. The field $C[[\mathfrak{M}]]$ is algebraically closed iff C is algebraically closed and $\mathfrak{M}$ is divisible.*

By Lemma 2.2.4, if $a \notin K$ is an element in an immediate valued field extension of K, then there is a divergent pc-sequence (a_ρ) in K such that $a_\rho \rightsquigarrow a$. The following lemma complements this result:

LEMMA 3.2.11. *Let a in an immediate valued field extension of K be algebraic over K, and $a \notin K$. Then there is a divergent pc-sequence (a_ρ) in K of algebraic type over K such that $a_\rho \rightsquigarrow a$.*

PROOF. Take a divergent pc-sequence (a_ρ) in K such that $a_\rho \rightsquigarrow a$. Let $P \in K[X]$ be the minimum polynomial of a over K. By the Taylor identity (3.2.1) we have

$$P(a_\rho) = P(a_\rho) - P(a) = (a_\rho - a) \cdot Q(a_\rho), \qquad Q \in K(a)[X]$$

and thus

$$v\big(P(a_\rho)\big) = v(a_\rho - a) + v\big(Q(a_\rho)\big).$$

Since $\big(v(a_\rho - a)\big)$ is eventually strictly increasing and $\big(v(Q(a_\rho))\big)$ is eventually strictly increasing or eventually constant, $\big(v(P(a_\rho))\big)$ is eventually strictly increasing, so (a_ρ) is of algebraic type over K. □

A valued field is **algebraically maximal** if it has no immediate proper algebraic valued field extension. Thus every maximal valued field is algebraically maximal, and every algebraically closed valued field is algebraically maximal. Lemmas 3.2.7 and 3.2.11 yield:

COROLLARY 3.2.12. *K is algebraically maximal if and only if each pc-sequence in K of algebraic type over K has a pseudolimit in K.*

By Zorn, K has an immediate valued field extension that is algebraically maximal and algebraic over K. By Corollary 3.3.4 and Theorem 3.3.29 such an extension is unique up to unique isomorphism over K, if K has equicharacteristic zero.

Completion of valued fields

Let K be a valued field. We say that a valued field extension $L \supseteq K$ is **dense** if K is dense in L in the valuation topology on L. Every dense extension of valued fields is immediate. The following complements Theorem 2.2.28:

THEOREM 3.2.13. *There is a dense valued field extension $K^c \supseteq K$ such that any dense valued field extension $L \supseteq K$ embeds uniquely over K into K^c.*

The proof of this theorem uses the following routine lemma:

LEMMA 3.2.14. *Let (a_ρ), (b_ρ) be c-sequences in K with the same index set. Then:*

(i) *$(a_\rho \cdot b_\rho)$ is a c-sequence;*

(ii) *if $a, b \in K$ and $a_\rho \to a$, $b_\rho \to b$, then $a_\rho \cdot b_\rho \to a \cdot b$;*

(iii) *if $a_\rho \neq 0$ for all ρ and $a_\rho \not\to 0$, then $(1/a_\rho)$ is a c-sequence.*

COROLLARY 3.2.15. *Suppose $L = K(a_i : i \in I)$ is a valued field extension of K such that Γ is cofinal in Γ_L and each generator a_i is the limit in L of a c-sequence in K. Then $L \supseteq K$ is dense.*

PROOF OF THEOREM 3.2.13. Let K^c be the completion of the valued additive group of K. (See Section 2.2.) By Lemmas 2.2.39 and 3.2.14, the multiplication $(x, y) \mapsto x \cdot y$ and inversion $x \mapsto 1/x$ $(x \neq 0)$ on K have unique extensions to continuous maps $K^c \times K^c \to K^c$ and $(K^c)^{\neq} \to K^c$, with the product topology on $K^c \times K^c$, and K^c is a field extension of K with the first map as multiplication and the second map as inversion. By Lemmas 2.2.24 and 3.2.14, the valuation of K^c is a valuation on the field K^c. Given a dense valued field extension L of K there is a unique embedding $L \to K^c$ of valued additive groups over K, by Corollary 2.2.33, and by continuity, this embedding is a valued field embedding. □

The properties of the valued field extension K^c of K postulated in Theorem 3.2.13 determine K^c up to a unique (valued field) isomorphism over K. We call K^c the **completion of** K. Note that by construction, K^c is indeed complete: every c-sequence in K^c converges in K^c. Hence:

COROLLARY 3.2.16. *A valued field is complete iff it has no proper dense valued field extension. (So every maximal valued field is complete.)*

For Hahn fields we make the above more concrete. Consider a (valued) Hahn field $C[[\mathfrak{M}]]$ over the field C with $\mathfrak{M} \neq \{1\}$. Since $C[[\mathfrak{M}]]$ is spherically complete, it is complete. For $S \subseteq C[[\mathfrak{M}]]$, set $\operatorname{supp} S := \bigcup_{f \in S} \operatorname{supp} f$, and let $\operatorname{cl}(S)$ be the closure of S in $C[[\mathfrak{M}]]$ with respect to its valuation topology. For $f = \sum_{\mathfrak{m}} f_{\mathfrak{m}} \mathfrak{m}$ in $C[[\mathfrak{M}]]$ and $\mathfrak{n} \in \mathfrak{M}$ we define the **truncation $f_{|\mathfrak{n}}$ of f at $\mathfrak{n}$** by

$$f_{|\mathfrak{n}} := \sum_{\mathfrak{m} \succ \mathfrak{n}} f_{\mathfrak{m}} \mathfrak{m},$$

so $f \in \operatorname{cl}\big(\{f_{|\mathfrak{n}} : \mathfrak{n} \in \mathfrak{M}\}\big)$. A set $S \subseteq C[[\mathfrak{M}]]$ is said to be **truncation closed** if for all $f \in S$ and $\mathfrak{n} \in \mathfrak{M}$ we have $f_{|\mathfrak{n}} \in S$. Note that if S is a truncation closed C-linear subspace of $C[[\mathfrak{M}]]$, then $\operatorname{supp} S \subseteq S$. We do not use this here, but it is worth mentioning that many subsets of $C[[\mathfrak{M}]]$ of a natural origin are truncation closed; see [109]. From the observations above we get:

LEMMA 3.2.17. *If S is a truncation closed subset of $C[[\mathfrak{M}]]$, then*

$$\operatorname{cl}(S) = \big\{f \in C[[\mathfrak{M}]] : f_{|\mathfrak{n}} \in S \text{ for every } \mathfrak{n} \in \mathfrak{M}\big\},$$

and so $\operatorname{cl}(S)$ is also truncation closed.

Let K be a (valued) subfield of $C[[\mathfrak{M}]]$. Then $\operatorname{cl}(K)$ is also a subfield of $C[[\mathfrak{M}]]$. Assume further that $\mathfrak{M} \subseteq K$. Then the valuation topology of $C[[\mathfrak{M}]]$ induces on K the valuation topology of the valued field K, and likewise with $\operatorname{cl}(K)$ instead of K. Thus the valued field extension $\operatorname{cl}(K) \supseteq K$ is dense. As a valued abelian group $\operatorname{cl}(K)$ is complete by Lemma 2.2.35, so the unique valued field embedding $\operatorname{cl}(K) \to K^{\mathrm{c}}$ over K is an isomorphism. Therefore it is reasonable to call $\operatorname{cl}(K)$ the **completion of K in $C[[\mathfrak{M}]]$**.

COROLLARY 3.2.18. *Suppose K is a truncation closed subfield of $C[[\mathfrak{M}]]$ such that $\mathfrak{M} \subseteq K$. Then the completion of K in $C[[\mathfrak{M}]]$ is*

$$\big\{f \in C[[\mathfrak{M}]] : f_{|\mathfrak{n}} \in K \text{ for every } \mathfrak{n} \in \mathfrak{M}\big\}.$$

EXAMPLE 3.2.19. Let $L := \big\{f \in C[[\mathfrak{M}]] : \operatorname{supp} f_{|\mathfrak{n}} \text{ is finite for all } \mathfrak{n} \in \mathfrak{M}\big\}$. Then $C[\mathfrak{M}] \subseteq L$, $C[\mathfrak{M}]$ is dense in L (for the valuation topology on $C[[\mathfrak{M}]]$), and L is a truncation closed subalgebra of the C-algebra $C[[\mathfrak{M}]]$. Hence $\operatorname{cl}(C[\mathfrak{M}]) = \operatorname{cl}(L) = L$ by Lemma 3.2.17; in particular, L is complete as a valued abelian group.

Suppose now that $\mathfrak{M}$ is archimedean. Then $(1-\varepsilon)^{-1} = 1 + \varepsilon + \varepsilon^2 + \cdots \in L$ for $\varepsilon \in L$ with $\varepsilon \prec 1$, and so L is a subfield of $C[[\mathfrak{M}]]$. Hence $C(\mathfrak{M}) \subseteq L$, and thus L is the completion of $C(\mathfrak{M})$ in $C[[\mathfrak{M}]]$.

Let $(\mathfrak{M}_i)_{i \in I}$ be a family of ordered subgroups of $\mathfrak{M}$ such that for all $i, j \in I$ there is a $k \in I$ with $\mathfrak{M}_i, \mathfrak{M}_j \subseteq \mathfrak{M}_k$ (automatic if the $\mathfrak{M}_i$ are convex in $\mathfrak{M}$) and $\mathfrak{M} = \bigcup_i \mathfrak{M}_i$. Then the hypothesis of Corollary 3.2.18 holds for $K := \bigcup_i C[[\mathfrak{M}_i]]$. If in addition the $\mathfrak{M}_i$ are convex in $\mathfrak{M}$, then $\operatorname{cl}(K) = C[[\mathfrak{M}]]$.

The valued field K is said to be **discrete** if $\Gamma \cong \mathbb{Z}$ (as ordered abelian groups), equivalently, the valuation ring $\mathcal{O}$ is a DVR as defined in Section 1.1. In that case, every pc-sequence in K has width $\{\infty\}$. Hence for discrete valued fields, the properties *complete*, *spherically complete*, and *maximal* are all equivalent.

From Corollary 2.2.34 we obtain:

COROLLARY 3.2.20. *If $L \supseteq K$ is a valued field extension and Γ is cofinal in Γ_L, then the inclusion $K \to L$ extends uniquely to a valued field embedding $K^c \to L^c$.*

Next we show the continuity of roots of a polynomial in the valuation topology. First an elementary fact: given $n \geqslant 1$ and $a_1, \dots, a_n, b_1, \dots, b_n \in K$ (not necessarily distinct), there are permutations σ and τ of $\{1, \dots, n\}$ such that for $i = 1, \dots, n$,

$$v\big(a_{\sigma(i)} - b_{\tau(i)}\big) = \max_{i \leqslant j \leqslant n} v\big(a_{\sigma(i)} - b_{\tau(j)}\big).$$

To see this, pick $i_1, j_1 \in \{1, \dots, n\}$ such that

$$v(a_{i_1} - b_{j_1}) = \max\big\{v(a_i - b_j) : 1 \leqslant i, j \leqslant n\big\},$$

and set $\sigma(1) = i_1$, $\tau(1) = j_1$. Now continue inductively with the a_i with $i \neq i_1$ and the b_j with $j \neq j_1$. In the next lemma $K[X]$ is equipped with the gaussian extension of the valuation of K.

LEMMA 3.2.21. *Let $P = \prod_{i=1}^n (X - a_i)$ with $n \geqslant 1$ and $a_1, \dots, a_n \in K$. Then for each $\gamma \in \Gamma$ there is a $\beta \in \Gamma$ with the property that if $Q = \prod_{i=1}^n (X - b_i)$ with $b_1, \dots, b_n \in K$ and $v(P - Q) > \beta$, and $v(a_i - b_i) = \max_{i \leqslant j \leqslant n} v(a_i - b_j)$ for each i, then $v(a_i - b_i) > \gamma$ for each i.*

PROOF. We proceed by induction on n, the case $n = 1$ being trivial. Suppose $n > 1$ and let $\gamma \in \Gamma$ be given. Set $\widetilde{P} := P/(X - a_1)$, and by inductive hypothesis choose $\widetilde{\beta} \in \Gamma$ such that if $\widetilde{Q} = \prod_{i=2}^n (X - b_i)$ (all $b_i \in K$) with $v(\widetilde{P} - \widetilde{Q}) > \widetilde{\beta}$, and $v(a_i - b_i) \geqslant v(a_i - b_j)$ for $2 \leqslant i \leqslant j \leqslant n$, then $v(a_i - b_i) > \gamma$ for each $i \geqslant 2$.

Let now $b_1, \dots, b_n \in K$ with $v(a_i - b_i) \geqslant v(a_i - b_j)$ for $i \leqslant j \leqslant n$, and $Q = \prod_{i=1}^n (X - b_i)$. Euclidean Division in $K[X]$ shows that the coefficients of $\widetilde{P}$ are polynomials (with integer coefficients) in a_1 and the coefficients P_i of P, and the same polynomials, with a_1 and P_i replaced by b_1 and Q_i, respectively, yield the coefficients of $\widetilde{Q} := \prod_{i=2}^n (X - b_i)$. Since polynomial functions are continuous in the valuation topology, we can therefore choose $\beta_0 \geqslant \gamma$ in Γ such that if $v(a_1 - b_1) > \beta_0$ and $v(P - Q) > \beta_0$, then $v(\widetilde{P} - \widetilde{Q}) > \widetilde{\beta}$. We have

$$v\big(Q(a_1)\big) = \sum_{i=1}^n v(a_1 - b_i) \leqslant nv(a_1 - b_1).$$

Using $P(a_1) = 0$ we get

$$v\big(Q(a_1)\big) = v\big((P - Q)(a_1)\big) \geqslant v(P - Q) + \min\big(0, nv(a_1)\big).$$

Hence if $v(P - Q) > \beta_1 := n\beta_0 - \min\big(0, nv(a_1)\big)$, then $v(a_1 - b_1) > \beta_0$. Thus $\beta := \max(\beta_0, \beta_1)$ has the required property. □

COROLLARY 3.2.22. *If K is algebraically closed, then so is its completion K^c.*

PROOF. Assume K is algebraically closed. Let L be an algebraic closure of the completion K^c of K, equipped with a valuation ring of L lying over the valuation ring of K^c. To get K^c algebraically closed, it is enough to show that K is dense in L, since then $K^c = L$. So let $a \in L$, and let $\gamma \in \Gamma$ be given. Let $P \in K^c[X]$ be the minimum polynomial of a over K^c. Choose $\beta \in \Gamma$ as in Lemma 3.2.21 applied to L in place of K. As K is dense in K^c we can take a monic $Q \in K[X]$ of the same degree as P with $v(P - Q) > \beta$. Since K is algebraically closed, it now follows from Lemma 3.2.21 that there is $b \in K$ with $Q(b) = 0$ and $v(a - b) > \gamma$. □

Valued vector spaces over valued fields

To study extensions of complete valued fields of finite degree, we temporarily move to the setting of valued vector spaces over valued fields. *Below K is a valued field.*

DEFINITION 3.2.23. A **valued vector space over** K is a vector space G over K, together with a surjective valuation $v\colon G \to S_\infty$ on the additive group of G, and an action of the value group Γ of K on the value set S,

$$(\gamma, s) \mapsto \gamma + s : \Gamma \times S \to S,$$

such that for all $\alpha, \beta \in \Gamma$, $s, s' \in S$, and $a \in K^\times$, $g \in G^{\neq}$,

(VS1) $\alpha \leqslant \beta \iff \alpha + s \leqslant \beta + s$;

(VS2) $s \leqslant s' \Rightarrow \alpha + s \leqslant \alpha + s'$;

(VS3) $v(ag) = va + vg$.

Note that here the same symbol v denotes the valuation of the valued abelian group G and of the valued field K. It is convenient to extend the action of Γ on S to a map $\Gamma_\infty \times S_\infty \to S_\infty$ by $\infty + s = \gamma + \infty = \infty$ for all $\gamma \in \Gamma_\infty$, $s \in S_\infty$. Then (VS1), (VS2), and (VS3) hold for all $\alpha, \beta \in \Gamma_\infty$, $s, s' \in S_\infty$, and $a \in K$, $g \in G$.

Below we denote such a valued vector space by (G, S, v), or simply by G. For a (vector) subspace H of G we have $\Gamma + v(H^{\neq}) \subseteq v(H^{\neq})$, and so H is a valued vector space by restricting the valuation of G to H and the action of Γ to $v(H^{\neq})$.

EXAMPLES.

(1) If the valuation of K is trivial, so $\Gamma = \{0\}$, then a valued vector space over K is the same as a valued vector space (G, S, v) over the *field* K as defined in Section 2.3, together with the trivial action of Γ on S.

(2) The valued field K is in a natural way a valued vector space (K, Γ, v) over itself, with the action being the addition on Γ. More generally, if $L \supseteq K$ is a valued field extension, with the valuation of L denoted by v, then (L, Γ_L, v), with the action of Γ on Γ_L by addition, is a valued vector space over K.

(3) Given a vector space $G \neq \{0\}$ over K and a basis $\mathcal{B}$ of G, we turn G into a valued vector space $(G, \Gamma, v_{\mathcal{B}})$ over K by setting $v_{\mathcal{B}}g = \min\{va_b : b \in \mathcal{B}\}$ for $g = \sum_{b \in \mathcal{B}} a_b b$, where $a_b \in K$ for all $b \in \mathcal{B}$, $a_b = 0$ for all but finitely many $b \in \mathcal{B}$, with the action of Γ on Γ given by addition.

Let (G, S, v) be a valued vector space over K. The notions "valuation-independent" and "valuation basis" from Section 2.3 extend to this more general setting: $\mathcal{B} \subseteq G$ is called **valuation-independent** if $0 \notin \mathcal{B}$, and for every family $(a_b)_{b \in \mathcal{B}}$ in K, with $a_b = 0$ for all but finitely many $b \in \mathcal{B}$, we have

$$v\left(\sum_{b \in \mathcal{B}} a_b b\right) = \min\big(\{v(a_b b) : b \in \mathcal{B}\} \cup \{\infty\}\big).$$

Every valuation-independent subset of G is K-linearly independent. (For example, the set $\{b_i c_j : 1 \leqslant i \leqslant m,\ 1 \leqslant j \leqslant n\}$ in Proposition 3.1.7 is valuation-independent in L as valued vector space over K.)

LEMMA 3.2.24. *Let $\mathcal{B} \subseteq G$ be valuation-independent with span H, and $g \in G \setminus H$. Then $\mathcal{B} \cup \{g\}$ is valuation-independent iff $vg \geqslant v(h+g)$ for all $h \in H$.*

PROOF. The forward direction being obvious, suppose that $vg \geqslant v(h+g)$ for all $h \in H$. It suffices to show that for all $h \in H$ and $a \in K^\times$ we have $v(h+ag) = \min(vh, va+vg)$. For this, after dividing by a, we may assume $a = 1$. If $vh \neq vg$, then $v(h+g) = \min(vh, vg)$, and if $vh = vg$, then $\min(vh, vg) \leqslant v(h+g) \leqslant vg$ by assumption, hence $v(h+g) = vg = \min(vh, vg)$. □

A **valuation basis** of G is a valuation-independent vector space basis of G. So in Example (3) above, $\mathcal{B}$ is a valuation basis of $(G, \Gamma, v_{\mathcal{B}})$.

Let $\mathcal{B}$ be a valuation basis of G and suppose $|\mathcal{B}| = n \geqslant 1$ is finite, say $\mathcal{B} = \{b_1, \dots, b_n\}$. Then (G, S, v) is isomorphic as a valued abelian group to the direct product of the valued abelian groups (K, S_i, v_i), $i = 1, \dots, n$, where $S_i = \Gamma + vb_i$ as an ordered subset of S, $v_i a = va + vb_i$ for $a \in K$. Each valued abelian group (K, S_i, v_i) in turn is isomorphic to the valued abelian group (K, Γ, v). Thus, by Lemma 2.2.37:

LEMMA 3.2.25. *If G has a finite valuation basis and K is spherically complete, then G is spherically complete.*

COROLLARY 3.2.26. *Suppose G has finite dimension as a K-vector space, and K is spherically complete. Then G has a valuation basis, hence is spherically complete.*

PROOF. Let $\mathcal{B} \subseteq G$ be valuation-independent, with span H. Then H as a valued vector space over K is spherically complete by the lemma above, so we are done if $G = H$. Suppose $G \neq H$ and take $g \in G \setminus H$. Then $v(g - H) \subseteq S$ has a largest element $v(g - h_0)$, $h_0 \in H$, by Lemma 2.2.18. Replacing g by $g - h_0$ we arrange that $vg \geqslant v(g-h)$ for all $h \in H$. Then by Lemma 3.2.24, the set $\mathcal{B} \cup \{g\}$ is valuation-independent. Continuing this way we build a valuation basis of G. □

With completeness instead of spherical completeness, we have:

PROPOSITION 3.2.27. *Suppose K is complete, G is finite-dimensional as a vector space over K, and $\Gamma + s$ is cofinal in S for all $s \in S$. Then (G, S, v) is complete.*

PROOF. Let $\mathcal{B}$ be a basis for G, where $n = |\mathcal{B}| \geqslant 1$, say $\mathcal{B} = \{b_1, \dots, b_n\}$. Then $\mathcal{B}$ is a valuation basis of $(G, \Gamma, v_{\mathcal{B}})$. As K is complete, $(G, \Gamma, v_{\mathcal{B}})$ is complete by Lemma 2.2.38 and the considerations preceding Lemma 3.2.25.

CLAIM: *$v_{\mathcal{B}}$ is* equivalent *to v, in the sense that there exist $s, t \in S$ with*

$$v_{\mathcal{B}}x + s \leqslant vx \leqslant v_{\mathcal{B}}x + t \qquad \textit{for all } x \in G.$$

(It follows that (G, S, v) is complete: if (x_ρ) is a c-sequence in (G, S, v), then (x_ρ) is a c-sequence in $(G, \Gamma, v_{\mathcal{B}})$, which gives $x \in G$ with $x_\rho \to x$ in $(G, \Gamma, v_{\mathcal{B}})$, and then $x_\rho \to x$ in (G, S, v).) To prove the claim, we set $s := \min(vb_1, \dots, vb_n)$, and then $vx \geqslant v_{\mathcal{B}}x + s$ for all $x \in G$. We show by induction on n that there exists $t \in S$ with $vx \leqslant v_{\mathcal{B}}x + t$ for all $x \in G$. For $n = 1$ this holds for $t := s$, so assume $n > 1$. For $i = 1, \dots, n$, let G_i be the span of $\mathcal{B} \setminus \{b_i\}$. Inductively, we can assume that G_i is complete as a valued subspace of G, so by Lemma 2.2.36 we can take $t_i \in S$ such that $v(b_i + g) \leqslant t_i$ for all $g \in G_i$. Then for $x = \sum_i a_i b_i \in G$ (all $a_i \in K$), we get $vx \leqslant va_i + t_i$ for all i, so $vx \leqslant v_{\mathcal{B}}x + t$ where $t = \max_i t_i$. □

In the next two corollaries $L \supseteq K$ is a valued field extension with $[L : K] < \infty$. Note that then Γ is cofinal in Γ_L, by Corollary 3.1.9. Thus by Corollary 3.2.26 and the previous proposition:

COROLLARY 3.2.28. *If K is spherically complete, then L is spherically complete. If K is complete, then L is complete.*

By Corollary 3.2.20 we have a unique valued field embedding $K^c \to L^c$ extending the inclusion $K \to L$. We view K^c as a valued subfield of L^c via this embedding, and this gives us the subfield K^cL of L^c.

COROLLARY 3.2.29. *We have $L^c = K^cL$ and $\mathcal{O}_{L^c} = \mathcal{O}_{K^c}\mathcal{O}_L$.*

PROOF. Since $[K^cL : K^c] \leqslant [L : K] < \infty$, the valued subfield LK^c of L^c is complete by Corollary 3.2.28. As L is dense in L^c, so is K^cL, and thus $K^cL = L^c$.

Next, take a basis of the vector space L over K that is contained in $\mathcal{O}_L$ and extract from it a basis $\mathcal{B} = \{b_1, \dots, b_n\} \subseteq \mathcal{O}_L$ of L^c as a vector space over K^c, where $[L^c : K^c] = n$. This gives us the valuation $v_{\mathcal{B}}$ on the vector space L^c over K^c. The proof of Proposition 3.2.27 gives $\alpha \in \Gamma_L$ such that $v_{\mathcal{B}}x \geqslant vx + \alpha$ for all $x \in L^c$. Let any $a \in \mathcal{O}_{L^c}$ be given. By density we get $b \in \mathcal{O}_L$ with $v(a - b) \geqslant -\alpha$, so $v_{\mathcal{B}}(a - b) \geqslant 0$, that is, $a = b + \sum_{i=1}^n a_i b_i$ with all $a_i \in \mathcal{O}_{K^c}$, and so $a \in \mathcal{O}_{K^c}\mathcal{O}_L$, as promised. □

Notes and comments

Proposition 3.2.1 for rank 1 is in Ostrowski [314]. The classification of pc-sequences into those of algebraic and transcendental type is due to Kaplansky [209], who also proved Lemmas 3.2.6 and 3.2.7 as well as (ii) $\Leftrightarrow$ (iii) in Corollary 3.2.9. Corollary 3.2.10 is in Mac Lane [278]. The completion of $\mathbb{R}(t^{\mathbb{R}})$ in $\mathbb{R}(\!(t^{\mathbb{R}})\!)$ (see Example 3.2.19) was first considered by Levi-Civita [257, 258], and later used for restricted versions of "non-standard analysis" by Laugwitz [253] and Lightstone and Robinson [259, 356]. Corollary 3.2.22 for rank 1 is from the early days of valuation theory [247, 374]. Corollary 3.2.26 is due to Krull [229]. The proof of Corollary 3.2.29 via Proposition 3.2.27 stems from [359] and is credited there to E. Artin. The notion of valuation-independence was introduced by Baur [38] (with different terminology).

3.3 HENSELIAN VALUED FIELDS

In this section K is a valued field with residue field $\boldsymbol{k} = \operatorname{res}(K)$. Recall that for a polynomial $P \in \mathcal{O}[X]$ we let $\overline{P}$ be the polynomial with coefficients in $\boldsymbol{k}$ obtained by replacing each coefficient of P by its residue class. The definition of henselianity isolates a key algebraic property of maximal valued fields (see Corollaries 3.3.4 and 3.3.21): given $P \in \mathcal{O}[X]$, every non-singular zero of $\overline{P}$ in $\boldsymbol{k}$ can be lifted to a zero of the original polynomial P in $\mathcal{O}$. More precisely:

DEFINITION 3.3.1. We call K **henselian** if for every polynomial $P \in \mathcal{O}[X]$ and $\alpha \in \boldsymbol{k}$ with $\overline{P}(\alpha) = 0$ and $\overline{P}'(\alpha) \neq 0$ there is $a \in \mathcal{O}$ with $P(a) = 0$ and $\overline{a} = \alpha$.

By the next lemma, the a in this definition is unique. In Chapter 7 we introduce a version of henselianity for valued differential fields lacking such uniqueness.

LEMMA 3.3.2. *Let $P \in \mathcal{O}[X]$ and $\alpha \in \boldsymbol{k}$ be such that $\overline{P}(\alpha) = 0$ and $\overline{P}'(\alpha) \neq 0$. Then there is at most one $a \in \mathcal{O}$ with $P(a) = 0$ and $\overline{a} = \alpha$.*

PROOF. Suppose $a \in \mathcal{O}$ satisfies $P(a) = 0$ and $\overline{a} = \alpha$. Then $\overline{P}'(\overline{a}) = \overline{P}'(\alpha) \neq 0$, hence $P'(a) \in \mathcal{O}^\times$. Taylor expansion in $x \in \mathcal{o}$ around a gives

$$\begin{aligned} P(a+x) &= P(a) + P'(a)x + bx^2 = P'(a)x + bx^2 \qquad (b \in \mathcal{O}) \\ &= P'(a)x\left(1 + P'(a)^{-1}bx\right). \end{aligned}$$

Since $P'(a)\left(1 + P'(a)^{-1}bx\right) \in \mathcal{O}^\times$ this gives $P(a+x) = 0$ iff $x = 0$. □

Next we show that algebraically maximal valued fields are henselian:

LEMMA 3.3.3. *Let $P \in \mathcal{O}[X]$ and $a \in \mathcal{O}$ be such that $P(a) \prec 1$, $P'(a) \asymp 1$, and P has no zero in $a + \mathcal{o}$. Then there is a pc-sequence (a_ρ) in K such that $P(a_\rho) \rightsquigarrow 0$, $a_\rho \equiv a \bmod \mathcal{o}$ for each ρ, and (a_ρ) has no pseudolimit in K.*

PROOF. Starting with a, one step of the classical Newton process for approximating the zeros of polynomials yields an element b of $\mathcal{O}$ such that $v(b-a) = v\big(P(a)\big) > 0$

and $v(P(b)) \geqslant 2v(P(a)) > 0$: use Taylor expansion to write

$$\begin{aligned} P(a+x) &= P(a) + P'(a)x + \text{terms of higher degree in } x \\ &= P'(a)\left(P'(a)^{-1}P(a) + x + \text{terms of higher degree in } x\right); \end{aligned}$$

setting $x = -P'(a)^{-1}P(a)$ then yields

$$P(a+x) = P'(a)\left(\text{multiple of } P(a)^2\right),$$

hence $b = a + x$ has the required property. Note that $b \equiv a \bmod \circ$ and thus $P'(b) \equiv P'(a) \bmod \circ$, in particular $P'(b) \asymp 1$. Hence the hypothesis on a in the statement of the lemma also applies to b. We now iterate this process. More precisely, let λ be a nonzero ordinal and (a_ρ) a sequence in $a + \circ$ indexed by the ordinals $\rho < \lambda$ such that $a_0 = a$, and $v(a_\sigma - a_\rho) = v(P(a_\rho))$ and $v(P(a_\sigma)) \geqslant 2v(P(a_\rho))$ whenever $\lambda > \sigma > \rho$. (For $\lambda = 2$ we have such a sequence with $a_0 = a$, $a_1 = b$.) If $\lambda = \mu+1$ is a successor ordinal, then we construct the next term $a_\lambda \in a+\circ$ by Newton approximation so that $v(a_\lambda - a_\mu) = v(P(a_\mu))$ and $v(P(a_\lambda)) \geqslant 2v(P(a_\mu))$.

Suppose now that λ is a limit ordinal. Then (a_ρ) is clearly a pc-sequence and $P(a_\rho) \rightsquigarrow 0$. If (a_ρ) has no pseudolimit in K, then we are done. If (a_ρ) has a pseudolimit in K, let a_λ be such a pseudolimit. Then $P(a_\rho) \rightsquigarrow P(a_\lambda)$. Since $(v(P(a_\rho)))$ is strictly increasing, this yields $v(P(a_\lambda)) \geqslant v(P(a_{\rho+1})) \geqslant 2v(P(a_\rho))$ for each index $\rho < \lambda$. It is also clear that $v(a_\lambda - a_\rho) = v(a_{\rho+1} - a_\rho) = v(P(a_\rho))$ for each $\rho < \lambda$, in particular, $a_\lambda \in a + \circ$. Thus we have extended our sequence by one more term. This building process must come to an end. □

COROLLARY 3.3.4. *Each algebraically maximal valued field is henselian. (Thus each maximal valued field and each algebraically closed valued field is henselian.)*

This follows from Corollary 3.2.12 and the previous lemma. For equicharacteristic zero valued fields, see Corollary 3.3.21 for a converse.

Let $P \in \mathcal{O}[X]$ and $a \in \mathcal{O}$ satisfy the hypothesis of Lemma 3.3.3, let (a_ρ) be the pc-sequence constructed in the proof of that lemma, and consider the strictly increasing sequence (γ_ρ), where $\gamma_\rho = v(a_\rho - a_\sigma)$ with $\sigma > \rho$. Then by construction of (a_ρ) we have $\gamma_\sigma \geqslant 2\gamma_\rho$ for $\sigma > \rho$; thus (γ_ρ) is cofinal in a convex subgroup $\neq \{0\}$ of Γ (Lemma 2.4.6). So if $\operatorname{rank}(\Gamma) = 1$, then (a_ρ) has width $\{\infty\}$. Thus:

COROLLARY 3.3.5. *Each complete valued field of rank* 1 *is henselian.*

Corollary 3.3.5 is commonly known as *Hensel's Lemma.*

LEMMA 3.3.6. *Let $P \in \mathcal{O}[X]$ and $a \in \mathcal{O}$ be such that $P(a) \prec 1$, $P'(a) \asymp 1$, and let x in a valued field extension L of K be a zero of P with $v(x - a) > 0$. Then $v(P(b)) = v(x - b)$ for all $b \in L$ with $v(a - b) > 0$.*

PROOF. Taylor expansion yields

$$\begin{aligned} P(x+Y) &= P(x) + P'(x)\cdot Y + \text{terms of higher degree in } Y \\ &= P'(x)\cdot Y \cdot (1+Q) \qquad \text{where } Q \in \mathcal{O}_L[Y],\ Q(0) = 0. \end{aligned}$$

Let $b \in L$, $v(a-b) > 0$. Substituting $b-x$ for Y yields

$$P(b) \;=\; P'(x)\cdot(b-x)\cdot\big(1+Q(b-x)\big) \qquad \text{with } Q(b-x) \prec 1,$$

which gives the desired conclusion. □

COROLLARY 3.3.7. *The following conditions on K are equivalent:*

(i) *the completion K^{c} of K is henselian;*

(ii) *for every polynomial $P \in \mathcal{O}[X]$ and $a \in \mathcal{O}$ with $P(a) \prec 1$ and $P'(a) \asymp 1$ and every $\gamma \in \Gamma^{>}$ there exists $b \in \mathcal{O}$ with $v\big(P(b)\big) > \gamma$ and $\overline{a} = \overline{b}$.*

In particular, if K is henselian, then so is K^{c}.

PROOF. Suppose K^{c} is henselian, $P \in \mathcal{O}[X]$, $a \in \mathcal{O}$, $P(a) \prec 1$, $P'(a) \asymp 1$, and $\gamma \in \Gamma^{>}$. Take $x \in K^{\mathrm{c}}$ with $x \preccurlyeq 1$, $P(x) = 0$, and $\overline{x} = \overline{a}$. Next, take $b \in K$ such that $v(b-x) > \gamma$. Then by Lemma 3.3.6 applied to $L = K^{\mathrm{c}}$ we obtain $v\big(P(b)\big) = v(x-b) > \gamma$. Thus (i) $\Rightarrow$ (ii). Assume (ii), and let $Q \in \mathcal{O}_{K^{\mathrm{c}}}[X]$ and $\alpha \in \boldsymbol{k}$ be such that $\overline{Q}(\alpha) = 0$ and $\overline{Q}'(\alpha) \neq 0$. We extend the valuation of K^{c} to its algebraic closure and use Corollary 3.3.4 to get a zero x of Q in this algebraic closure such that $\overline{x} = \alpha$. Let $\gamma \in \Gamma^{>}$. Since K is dense in K^{c}, we can take $P \in K[X]$ with $v(P-Q) > \gamma$. Then $P \in \mathcal{O}[X]$ and $\overline{P} = \overline{Q}$. By (ii) we have $b \in \mathcal{O}$ with $v\big(P(b)\big) > \gamma$ and $\overline{b} = \alpha$. Hence by Lemma 3.3.6 again we have $v(x-b) > \gamma$. By Corollary 3.2.15 and 3.2.16 this yields $x \in K^{\mathrm{c}}$, so (i) holds. □

Lifting the residue field

Suppose C is a subfield of the valuation ring $\mathcal{O}$ of K. Then C is mapped onto a subfield $\overline{C}$ of $\boldsymbol{k} = \operatorname{res}(K)$ under the residue map

$$x \mapsto \overline{x} : \mathcal{O} \to \mathcal{O}/\smallo = \boldsymbol{k}.$$

In case $\overline{C} = \boldsymbol{k}$ we call C a **lift** of $\boldsymbol{k}$ (in $\mathcal{O}$). For example, the subfield C of a Hahn field $C[[\mathfrak{M}]]$ is a lift of the residue field of this Hahn field. We say that the residue field of K can be lifted if there is a lift of $\boldsymbol{k}$ in $\mathcal{O}$.

PROPOSITION 3.3.8. *Suppose K is henselian of equicharacteristic zero. Then the residue field of K can be lifted; in fact, every maximal subfield of $\mathcal{O}$ is a lift of $\boldsymbol{k}$.*

In the proof we use:

LEMMA 3.3.9. *Let C be a maximal subfield of $\mathcal{O}$. Then C is algebraically closed in K, and the field extension $\boldsymbol{k} \supseteq \overline{C}$ is algebraic.*

PROOF. Since $\mathcal{O}$ is integrally closed in K, it contains the algebraic closure of C in K; hence C is algebraically closed in K. Let $\xi \in \boldsymbol{k}$ be transcendental over $\overline{C}$, and take $x \in \mathcal{O}$ such that $\overline{x} = \xi$. For $P \in C[X]$, $P \neq 0$, we have $\overline{P} \in \overline{C}[X]$, $\overline{P} \neq 0$, hence $\overline{P(x)} = \overline{P}(\xi) \neq 0$, so $P(x) \in \mathcal{O}^{\times}$. In particular, the subring $C[x]$ of $\mathcal{O}$ is mapped isomorphically onto the subring $\overline{C}[\xi]$ of $\boldsymbol{k}$ by the residue map. Thus $C[x]$

is a domain with fraction field $C(x)$ inside $\mathcal{O}$, and $C(x)$ is mapped isomorphically onto the subfield $\overline{C}(\xi)$ of $\boldsymbol{k}$. Thus $C(x)$ is a subfield of $\mathcal{O}$ that strictly contains C, a contradiction. □

PROOF OF PROPOSITION 3.3.8. Since $\operatorname{char} \boldsymbol{k} = 0$, the valuation ring $\mathcal{O}$ of K contains a subfield. By Zorn there is a maximal subfield C of $\mathcal{O}$. Suppose $x \in \mathcal{O}$ and $\overline{x} \in \boldsymbol{k} \setminus \overline{C}$. By the previous lemma, $\xi := \overline{x}$ is algebraic over $\overline{C}$. Let $P \in C[X]$ be a monic polynomial such that its image $\overline{P} \in \overline{C}[X]$ is the minimum polynomial of ξ over $\overline{C}$. Then P is irreducible since $\overline{P}$ is. Since $\operatorname{char}(\boldsymbol{k}) = 0$, the irreducible polynomial $\overline{P} \in \overline{C}[X]$ is separable, so $\overline{P}'(\xi) \neq 0$. As K is henselian, this gives $\varepsilon \in \smallo$ such that $P(x + \varepsilon) = 0$. Then $C[x + \varepsilon]$ is a subfield of $\mathcal{O}$, contradicting the maximality of C. Thus $\boldsymbol{k} = \overline{C}$, so C is a lift of $\boldsymbol{k}$. □

Characterizations of henselianity

The following elementary lemma contains useful reformulations of the henselianity condition:

LEMMA 3.3.10. *The following are equivalent:*

(i) *K is henselian;*

(ii) *each polynomial*

$$1 + X + P_2X^2 + \cdots + P_nX^n \qquad \text{where } n \geqslant 2,\ P_2, \dots, P_n \in \smallo$$

has a zero in $\mathcal{O}$ (of course, such a zero must lie in $-1 + \smallo$);

(iii) *each polynomial*

$$Y^n + Y^{n-1} + Q_{n-2}Y^{n-2} + \cdots + Q_0 \quad \text{where } n \geqslant 2,\ Q_0, \dots, Q_{n-2} \in \smallo$$

has a zero in $\mathcal{O}^\times$;

(iv) *given a polynomial $P \in \mathcal{O}[X]$ and $a \in \mathcal{O}$ such that $P(a) \prec P'(a)^2$, there is $b \in \mathcal{O}$ such that $P(b) = 0$ and $b - a \prec P'(a)$. (Newton version)*

PROOF. Assume K is henselian and let $P = 1 + X + P_2X^2 + \cdots + P_nX^n$ with $n \geqslant 2$ and $P_2, \dots, P_n \in \smallo$. Then for $a = -1$ we have $P(a) \prec 1$ and $P'(a) \asymp 1$. Thus P has a zero in $\mathcal{O}$. This shows (i) ⇒ (ii). For (ii) ⇔ (iii), use the substitution $X = 1/Y$. Suppose now that (ii) holds, let P, a be as in the hypothesis of (iv). Note that $P'(a) \neq 0$ and set $c := P(a)/P'(a)^2$ (so $c \prec 1$). Let $x \in \mathcal{O}$ and consider the expansion:

$$\begin{aligned} P(a + x) &= P(a) + P'(a)x + \sum_{i \geqslant 2} P_{(i)}(a)x^i \\ &= cP'(a)^2 + P'(a)x + \sum_{i \geqslant 2} P_{(i)}(a)x^i. \end{aligned}$$

Set $x = cP'(a)y$ where $y \in \mathcal{O}$. Then

$$P(a+x) = cP'(a)^2\left(1 + y + \sum_{i\geqslant 2} a_i y^i\right)$$

where the $a_i \in \mathcal{o}$ do not depend on y. By (ii) choose $y \in \mathcal{O}$ such that

$$1 + y + \sum_{i\geqslant 2} a_i y^i = 0.$$

This yields an element $b = a + x = a + cP'(a)y$ as required. This shows (ii) $\Rightarrow$ (iv), and (iv) $\Rightarrow$ (i) is clear. □

Given an algebraic closure K^{a} of K, the valuation $v\colon K^\times \to \Gamma$ extends to a valuation $(K^{\mathrm{a}})^\times \to \mathbb{Q}\Gamma$. The following proposition shows among other things that uniqueness of such an extension is equivalent to K being henselian:

PROPOSITION 3.3.11. *The following are equivalent:*

(i) *K is henselian;*

(ii) *for each algebraic field extension $L \supseteq K$ there is a unique valuation ring of L lying over $\mathcal{O}$;*

(iii) *for each monic polynomial $P \in \mathcal{O}[X]$ which is irreducible in $K[X]$ there is some $m \geqslant 1$ and a monic $Q \in \mathcal{O}[X]$ with $\overline{Q}$ is irreducible in $\boldsymbol{k}[X]$ and $\overline{P} = \overline{Q}^m$;*

(iv) *given monic $P, Q, R \in \mathcal{O}[X]$ with $\overline{P} = \overline{Q}\cdot\overline{R}$ and $\overline{Q}$, $\overline{R}$ relatively prime in $\boldsymbol{k}[X]$, there are monic $Q^*, R^* \in \mathcal{O}[X]$ with $P = Q^*R^*$ and $\overline{Q^*} = \overline{Q}$, $\overline{R^*} = \overline{R}$.*

PROOF. Suppose (ii) fails. Then we have a field extension $L \supseteq K$ with $[L : K] < \infty$ and more than one valuation ring of L lying over $\mathcal{O}$. By Lemma 3.1.24 we can replace L by the separable closure of K in L and arrange that L is separable over K. Replacing L by its normal closure over K (in an algebraic closure of K), we can also assume that L is a Galois extension of K. Let $\mathcal{O}_L$ be a valuation ring of L lying over $\mathcal{O}$, and let G^{d} be the decomposition group and L^{d} the decomposition field of $\mathcal{O}_L$ over K; see Corollary 3.1.27. Let $\mathcal{O}_1, \dots, \mathcal{O}_m$ be the distinct valuation rings of L lying over $\mathcal{O}$, with $\mathcal{O}_1 = \mathcal{O}_L$, and let $\mathcal{o}_i$ be the maximal ideal of $\mathcal{O}_i$. Then $m > 1$ and $\mathcal{O}_1 \cap L^{\mathrm{d}} \neq \mathcal{O}_i \cap L^{\mathrm{d}}$ for $i = 2, \dots, m$. Let B be the integral closure of $\mathcal{O}$ in L^{d}. Then $\mathcal{o}_1 \cap B, \dots, \mathcal{o}_m \cap B$ are maximal ideals of B, by Proposition 3.1.20, and $\mathcal{o}_1 \cap B \neq \mathcal{o}_i \cap B$ for $i = 2, \dots, m$. Next, the Chinese Remainder Theorem provides an element $x \in B$ such that $x \in \mathcal{o}_1$ and $x \notin \mathcal{o}_i$ for $i = 2, \dots, m$. Let

$$P := X^n + a_{n-1}X^{n-1} + \cdots + a_0 \qquad (a_i \in K)$$

be the minimum polynomial of x over K. From $m > 1$ we get $x \notin K$, so $n \geqslant 2$ and P does not have a zero in K. By Lemma 1.3.12 we have $P \in \mathcal{O}[X]$; we claim that $a_0 \in \mathcal{o}$ and $a_1 \notin \mathcal{o}$. To see this, let $x = x_1, x_2, \dots, x_n$ be the distinct conjugates of x under the action of $\mathrm{Aut}(L|K)$. Now $a_0 = (-1)^n x(x_2 \cdots x_n)$ and $x_2, \dots, x_n$ are integral over $\mathcal{O}$, so $x_2 \cdots x_n \in B$, and thus $a_0 \in xB \cap \mathcal{O} \subseteq \mathcal{o}$. Let $j \in \{2, \dots, n\}$ and take $\sigma \in \mathrm{Aut}(L|K)$ with $x_j = \sigma(x)$; since $x_j \neq x$ and $x \in L^{\mathrm{d}}$ we have $\sigma \notin G^{\mathrm{d}}$, hence $\sigma^{-1}(\mathcal{O}_1) = \mathcal{O}_i$ where $i \in \{2, \dots, m\}$. For such i we have $x \notin \mathcal{o}_i = \sigma^{-1}(\mathcal{o}_1)$, so $x_j = \sigma(x) \notin \mathcal{o}_1$. Therefore the sum $\sum_{j=1}^n x_1 \cdots x_{j-1} x_{j+1} \cdots x_n$ in

$$a_1 = (-1)^{n-1} \sum_{j=1}^{n} x_1 \cdots x_{j-1} x_{j+1} \cdots x_n,$$

has precisely one term, namely $x_2 \cdots x_n$, that misses the factor $x_1 = x$, and so this is the only term in the sum not in $\mathcal{o}_1$; thus $a_1 \notin \mathcal{o}$. It follows that K is not henselian. This proves (i) $\Rightarrow$ (ii), since we have established its contrapositive.

Suppose now that (ii) holds. Let K^{a} be an algebraic closure of K. Then by Proposition 3.1.20, the integral closure B of $\mathcal{O}$ in K^{a} is local. Let $P \in \mathcal{O}[X]$ be monic and irreducible in $K[X]$, and let $x \in K^{\mathrm{a}}$ be a zero of P. Then the subring $\mathcal{O}[x]$ of B is local, since distinct maximal ideals of $\mathcal{O}[x]$ would extend to distinct maximal ideals of B. Now P is monic and irreducible, so $\mathcal{O}[x] \cong \mathcal{O}[X]/(P)$ as ring extensions of $\mathcal{O}$. By Lemma 1.4.8, $\overline{P} = \overline{Q}^m$ for some $m \geqslant 1$ and monic irreducible polynomial $\overline{Q} \in \boldsymbol{k}[X]$. This shows (ii) $\Rightarrow$ (iii).

Suppose (iii) holds; let $P, Q, R \in \mathcal{O}[X]$ be monic such that $\overline{P} = \overline{Q} \cdot \overline{R}$ and $\overline{Q}, \overline{R}$ are relatively prime in $\boldsymbol{k}[X]$. We have $P = P_1 \cdots P_m$ where each $P_i \in K[X]$ is monic and irreducible. The coefficients of P_i are elementary symmetric functions in the zeros of P_i, and these zeros are among the zeros of P and hence integral over $\mathcal{O}$; thus $P_i \in \mathcal{O}[X]$ for each i. By (iii), $\overline{P_i}$ is a power of an irreducible polynomial in $\boldsymbol{k}[X]$. Since $\overline{Q}$, $\overline{R}$ are relatively prime, either $\overline{P_i}$ divides $\overline{Q}$ or $\overline{P_i}$ divides $\overline{R}$. Let Q^* be the product of those P_i such that $\overline{P_i}$ divides $\overline{Q}$, and let R^* be the product of the remaining P_i. Then Q^*, R^* have the required properties, showing (iv).

Assume (iv); to derive (i), it suffices, by the equivalence of (i) and (iii) in Lemma 3.3.10, that each polynomial $P = X^n + X^{n-1} + a_{n-2}X^{n-2} + \cdots + a_0$ with $n \geqslant 2$ and $a_0, \dots, a_{n-2} \in \mathcal{o}$ has a zero in $\mathcal{O}^\times$. Now $\overline{P}$ factors as $(X+1)X^{n-1}$. By (iv) take monic $Q^*, R^* \in \mathcal{O}[X]$ with $P = Q^* R^*$, $\overline{Q^*} = \overline{Q} = X + 1$, $\overline{R^*} = X^{n-1}$. Then $Q^* = X - a$ with $a \in \mathcal{O}$, so $\overline{a} = -1$, hence $a \in \mathcal{O}^\times$ and $P(a) = 0$. $\square$

The equivalence of (i) and (ii) in this proposition has an important consequence:

COROLLARY 3.3.12. *If K is henselian and L is an algebraic valued field extension of K, then L is henselian and $\mathcal{O}_L$ is the integral closure of $\mathcal{O}$ in L.*

To get a variant of property (iii) in Proposition 3.3.11 for not necessarily monic polynomials we use the following lemma:

LEMMA 3.3.13. *Suppose K is henselian, and let $P \in \mathcal{O}[X]$ be irreducible in $K[X]$ such that* $\deg \overline{P} \geqslant 1$. *Then* $\deg(\overline{P}) = \deg(P)$.

PROOF. Let K^{a} be an algebraic closure of K and $\mathcal{O}^{\mathrm{a}}$ the unique valuation ring of K^{a} lying over $\mathcal{O}$; we continue to denote the associated valuation $(K^{\mathrm{a}})^\times \to \mathbb{Q}\Gamma$ on K^{a}

by v. Then $v \circ \sigma = v$ for all $\sigma \in \mathrm{Aut}(K^{\mathrm{a}}|K)$, by Lemma 3.1.25. In $K^{\mathrm{a}}[X]$,

$$(3.3.1) \qquad P = a\prod_{j=1}^{n}(X - x_j) \qquad \text{where } a \in \mathcal{O},\ a \neq 0, x_1, \dots, x_n \in K^{\mathrm{a}}.$$

Since for all i, j there is $\sigma \in \mathrm{Aut}(K^{\mathrm{a}}|K)$ with $\sigma(x_i) = x_j$, we have $\gamma \in \mathbb{Q}\Gamma$ with $v(x_i) = \gamma$ for all i. We claim that $\gamma \geqslant 0$. Suppose $\gamma < 0$, and let

$$\prod_{j=1}^{n}(X - x_j) = X^n + b_1X^{n-1} + \dots + b_{n-1}X + b_n, \qquad b_1, \dots, b_n \in K.$$

Then $v(b_n) = n\gamma$, and $v(b_i) \geqslant i\gamma > n\gamma = v(b_n)$ for $i = 1, \dots, n-1$, and this holds also for $b_0 := 1$. But this contradicts $\deg \overline{P} \geqslant 1$, and thus $\gamma \geqslant 0$, as claimed. Applying the residue morphism to both sides in (3.3.1) and using $\overline{P} \neq 0$ now yields $\overline{a} \neq 0$, that is, $\deg(\overline{P}) = n = \deg(P)$. □

COROLLARY 3.3.14. *Suppose K is henselian, $P \in \mathcal{O}[X]$ is irreducible in $K[X]$, and $\deg \overline{P} \geqslant 1$. Then there exist $m \geqslant 1$, $a \in \mathcal{O}^\times$, and monic $Q \in \mathcal{O}[X]$ such that $\overline{Q}$ is irreducible in $\boldsymbol{k}[X]$ and $\overline{P} = \overline{a} \cdot \overline{Q}^m$.*

PROOF. The leading coefficient a of P lies in $\mathcal{O}^\times$ by the previous lemma, and so the desired result follows from Proposition 3.3.11(iii) applied to the monic polynomial $P/a \in \mathcal{O}[X]$. □

We also note a consequence of Corollary 3.1.26 and Proposition 3.3.11:

COROLLARY 3.3.15. *If K is henselian and x an element of an algebraic valued field extension of K, with minimum polynomial*

$$a_0 + a_1X + \dots + a_{n-1}X^{n-1} + X^n \qquad (a_0, \dots, a_{n-1} \in K,\ n \geqslant 1)$$

over K, then $v(x) = \frac{1}{n}v(a_0)$.

The henselian axiom concerns polynomials over $\mathcal{O}$ in a single variable, but implies an analogue for multivariate polynomials. To discuss this, let $n \geqslant 1$, and let $P = (P_1, \dots, P_n)$ be an n-tuple of polynomials $P_i \in K[Y_1, \dots, Y_n]$. For $a \in K^n$ we have $P(a) := \big(P_1(a), \dots, P_n(a)\big) \in K^n$, and we recall from Section 1.9 that

$$P'(a) := \left(\frac{\partial P_i}{\partial Y_j}(a)\right)_{i,j=1,\dots,n} \qquad \text{(an } n \times n\text{-matrix with entries in } K).$$

We equip the additive abelian group K^n with the valuation

$$v\colon K^n \to \Gamma_\infty, \qquad v(a) := \min_i v(a_i) \quad \text{for } a = (a_1, \dots, a_n) \in K^n.$$

We consider here the elements of K^n as column vectors, that is, as $n \times 1$-matrices with entries in K. In particular, this holds for the above $a \in K^n$ and $P(a) \in K^n$.

PROPOSITION 3.3.16. *Let K be henselian, let $P_1, \dots, P_n \in \mathcal{O}[Y_1, \dots, Y_n]$, $n \geqslant 1$, and let $a \in \mathcal{O}^n$. Then:*

(i) *if $v(P(a)) > 0$ and $v(\det P'(a)) = 0$, then there is a unique $b \in \mathcal{O}^n$ with $P(b) = 0$ and $v(a - b) > 0$;*

(ii) *Newton version: if $v(P(a)) > 2v(\det P'(a))$, then there is a unique $b \in \mathcal{O}^n$ with $P(b) = 0$ and $v(a - b) > v(\det P'(a))$.*

PROOF. To prove (i), let $v(P(a)) > 0$ and $v(\det P'(a)) = 0$. The square matrix $J := P'(a)$ has entries in $\mathcal{O}$ with $\det J \in \mathcal{O}^\times$, so has inverse J^{-1} with entries in $\mathcal{O}$. For $y \in \smallo^n$ we have by Taylor expansion

$$P(a + y) = P(a) + J \cdot y + Q(y), \qquad Q = (Q_1, \dots, Q_n),$$

where the $Q_i \in \mathcal{O}[Y_1, \dots, Y_n]$ are independent of y and contain only monomials of degree $\geqslant 2$. Thus for $y \in \smallo^n$,

$$J^{-1} \cdot P(a + y) = J^{-1} \cdot P(a) + y + R(y) \qquad R = (R_1, \dots, R_n),$$

where the $R_i \in \mathcal{O}[Y_1, \dots, Y_n]$ are independent of y and contain only monomials of degree $\geqslant 2$. Thus with $c := J^{-1} \cdot P(a)$ we have $c \in \smallo^n$, and it remains to show that there is exactly one $y \in \smallo^n$ with $c + y + R(y) = 0$, that is, $-c - R(y) = y$. Consider the map $f \colon \smallo^n \to \smallo^n$ given by $f(y) = -c - R(y)$. For $x, y \in \smallo^n$ we have $f(x) - f(y) = R(y) - R(x)$, and so $v(f(x) - f(y)) > v(x - y)$ if $x \neq y$. If K is maximal as a valued field, then the valued abelian group K^n is spherically complete, by Lemma 2.2.37 and Corollary 3.2.9, and so f has a unique fixpoint by Theorem 2.2.12, and (i) holds. Suppose K is not maximal. Then we take some algebraically closed maximal valued field extension L of K and obtain a unique $y \in \smallo_L^n$ with $P(a+y) = 0$. Then the entries y_i of the vector y are separably algebraic over K by Corollary 1.9.7.

Let K^{a} be the algebraic closure of K in L, so $\mathcal{O}^{\mathrm{a}} := \mathcal{O}_L \cap K^{\mathrm{a}}$ is the unique valuation ring of K^{a} lying over $\mathcal{O}$. Let $\smallo^{\mathrm{a}}$ be the maximal ideal of $\mathcal{O}^{\mathrm{a}}$, so $y \in (\smallo^{\mathrm{a}})^n$. Then for each $\sigma \in \operatorname{Aut}(K^{\mathrm{a}}|K)$,

$$P(a + \sigma(y)) = \sigma(P(a + y)) = 0, \quad \sigma(y) \in (\sigma\smallo^{\mathrm{a}})^n = (\smallo^{\mathrm{a}})^n,$$

so $\sigma(y) = y$ by uniqueness of y. Therefore $y \in \smallo^n$. This proves (i).

For (ii), set $J := P'(a)$, and put $d := \det J$, and assume $v(P(a)) > 2v(d)$. As in the proof of (i) we have $Q_1, \dots, Q_n \in \mathcal{O}[Y_1, \dots, Y_n]$, each having only monomials of degree $\geqslant 2$, such that for all $y \in K^n$,

$$P(a + y) = P(a) + J \cdot y + Q(y), \qquad Q = (Q_1, \dots, Q_n).$$

Note that $Q_i(dY_1, \dots, dY_n) = d^2 R_i(Y_1, \dots, Y_n)$ with $R_i \in \mathcal{O}[Y_1, \dots, Y_n]$. Then

$$P(a + dy) = P(a) + J \cdot dy + d^2 R(y), \qquad R = (R_1, \dots, R_n),$$

for all $y \in K^n$. Now J^{-1} has all its entries in $d^{-1}\mathcal{O}$, so for all $y \in K^n$,

$$J^{-1} \cdot P(a + dy) = J^{-1} \cdot P(a) + dy + dS(y), \qquad S = (S_1, \dots, S_n)$$

with $S_i \in \mathcal{O}[Y_1,\dots,Y_n]$ independent of y and all its monomials of degree $\geqslant 2$. So

$$d^{-1}J^{-1}\cdot P(a+dy) \;=\; d^{-1}J^{-1}\cdot P(a)+y+S(y),$$

for all $y \in K^n$, and $v\big(d^{-1}J^{-1}\cdot P(a)\big) > 0$. Thus (i) gives a unique $y \in o^n$ with $P(a+dy)=0$, which is the conclusion of (ii). □

In the corollary below we let $X_1,\dots,X_m,Y_1,\dots,Y_n$ be distinct indeterminates and set $X=(X_1,\dots,X_m)$, $Y=(Y_1,\dots,Y_n)$. Given an n-tuple $P=(P_1,\dots,P_n)$ of polynomials $P_i \in K[X,Y]$ and $a\in K^m$, $b \in K^n$ we define the $n\times n$-matrix

$$\frac{\partial P}{\partial Y}(a,b) \;:=\; \left(\frac{\partial P_i}{\partial Y_j}(a,b)\right)_{i,j=1,\dots,n}$$

with entries in K.

COROLLARY 3.3.17 (Implicit Function Theorem). *Suppose K is henselian. Let $P=(P_1,\dots,P_n)$ where $P_i \in \mathcal{O}[X,Y]$, $m,n\geqslant 1$. Let $a\in\mathcal{O}^m$, $b\in\mathcal{O}^n$ be such that*

$$P(a,b) \;=\; 0, \qquad \det\frac{\partial P}{\partial Y}(a,b) \;\neq\; 0,$$

and set $\delta := v\left(\det \frac{\partial P}{\partial Y}(a,b)\right) \in \Gamma$. Then for all $x\in\mathcal{O}^m$ with $v(x-a)>2\delta$ there is a unique $y\in\mathcal{O}^n$ such that $P(x,y)=0$ and $v(y-b)>\delta$.

PROOF. Let $x\in\mathcal{O}^m$, $v(x-a)>2\delta$. Taylor expansion yields

$$P(x,b) \;=\; P(a,b)+Q(x-a) \;=\; Q(x-a), \qquad Q=(Q_1,\dots,Q_n)$$

where the $Q_i \in \mathcal{O}[X_1,\dots,X_m]$ contain only monomials of degree $\geqslant 1$, and so $v\big(P(x,b)\big)\geqslant v(x-a) > 2\delta$. Similarly,

$$\det\frac{\partial P}{\partial Y}(x,b) \;=\; \det\frac{\partial P}{\partial Y}(a,b)+R(x-a) \quad \text{where } R\in\mathcal{O}[X],\ R(0)=0,$$

so $v\left(\det\frac{\partial P}{\partial Y}(x,b)\right) \;=\; v\left(\det\frac{\partial P}{\partial Y}(a,b)\right) \;=\; \delta$. Now apply Proposition 3.3.16(ii) to $\big(P_1(x,Y),\dots,P_n(x,Y)\big)$ and b in place of P and a. □

Hensel configuration and algebraic maximality

For K to be henselian is a condition on polynomials over its valuation ring $\mathcal{O}$. It is convenient to have an equivalent condition in terms of polynomials over K. Let $P\in K[X]$ be of degree at most d, and let $a\in K$, so

$$P_{+a}(X) \;:=\; P(a+X) \;=\; \sum_{i=0}^{d} P_{(i)}(a)X^i \;=\; P(a)+P'(a)X+\sum_{i=2}^{d}P_{(i)}(a)X^i.$$

We say that P is in **hensel configuration** at a if $P'(a)\neq 0$, and either $P(a)=0$ or $P(a)\neq 0$ and $\gamma := v\big(P(a)\big)-v\big(P'(a)\big)$ satisfies

$$v\big(P(a)\big) \;<\; v\big(P_{(i)}(a)\big)+i\gamma \quad \text{for } i=2,\dots,d.$$

LEMMA 3.3.18. *Suppose K is henselian and P as above is in hensel configuration at $a \in K$. Set $\gamma := v(P(a)) - v(P'(a)) \in \Gamma_\infty$. Then there is a unique $b \in K$ such that $P(a+b) = 0$ and $v(b) \geqslant \gamma$; this b satisfies $v(b) = \gamma$.*

PROOF. If $P(a) = 0$, then $b := 0$ works, so assume $P(a) \neq 0$. Take $g \in K$ such that $vg = \gamma$, and set $Q := P_{+a}$, $h := P(a)$, and

$$\widetilde{Q}(X) := Q(gX)/h = 1 + \underbrace{(P'(a)g/h)}_{\asymp 1} X + \sum_{i=2}^{d} \underbrace{(P_{(i)}(a)g^i/h)}_{\prec 1} X^i.$$

Since K is henselian and the image of $\widetilde{Q}$ under the natural surjection $\mathcal{O}[X] \to \boldsymbol{k}[X]$ has degree 1, the polynomial $\widetilde{Q}$ has a unique zero $u \in \mathcal{O}$. We have $u \asymp 1$ and $P(a+ug) = 0$, hence $b := ug$ has the required properties. □

PROPOSITION 3.3.19. *Suppose K is of equicharacteristic zero. Let (a_ρ) be a pc-sequence in K with $a_\rho \rightsquigarrow a \in K$, and set $\gamma_\rho := v(a - a_\rho)$. Let $P \in K[X]$ of degree $\leqslant d$ be such that $P(a_\rho) \rightsquigarrow 0$ and $P_{(i)}(a_\rho) \not\rightsquigarrow 0$ for $i = 1, \dots, d$. Then for $i = 2, \dots, d$, we have, eventually,*

$$v(P(a_\rho)) = v(P(a_\rho) - P(a)) = v(P'(a)) + \gamma_\rho < v(P_{(i)}(a)) + i\gamma_\rho.$$

PROOF. The proof of Proposition 3.2.1 yields a unique $i_0 \geqslant 1$ such that for each $i \geqslant 1$ with $i \neq i_0$,

$$v(P(a_\rho) - P(a)) = v(P_{(i_0)}(a)) + i_0\gamma_\rho < v(P_{(i)}(a)) + i\gamma_\rho, \quad \text{eventually.}$$

Now $P(a_\rho) \rightsquigarrow 0$, so $v(P(a_\rho)) = v(P(a_\rho) - P(a))$, eventually, and for $i \geqslant 1$, $i \neq i_0$:

$$v(P(a_\rho)) = v(P_{(i_0)}(a)) + i_0\gamma_\rho < v(P_{(i)}(a)) + i\gamma_\rho, \quad \text{eventually.}$$

We claim that $i_0 = 1$. Let $i > 1$; our claim will then follow by deriving

$$v(P'(a)) + \gamma_\rho \leqslant v(P_{(i)}(a)) + i\gamma_\rho, \quad \text{eventually.}$$

Now the proof of Proposition 3.2.1 applied to P' instead of P also yields

$$v(P'(a_\rho) - P'(a)) \leqslant v(P'_{(j)}(a)) + j\gamma_\rho, \quad \text{eventually}$$

for all $j \geqslant 1$. Since $v(P'(a_\rho)) = v(P'(a))$ eventually, this yields

$$v(P'(a)) \leqslant v(P'_{(j)}(a)) + j\gamma_\rho, \quad \text{eventually}$$

for all $j \geqslant 1$. Using $P'_{(j)} = (1+j)P_{(1+j)}$ (Lemma 3.2.2), this gives for $j \geqslant 1$:

$$v(P'(a)) \leqslant v(P_{(1+j)}(a)) + j\gamma_\rho, \quad \text{eventually.}$$

For $j = i - 1$, this yields

$$v(P'(a)) + \gamma_\rho \leqslant v(P_{(i)}(a)) + i\gamma_\rho, \quad \text{eventually.}$$

Thus $i_0 = 1$ as claimed. □

COROLLARY 3.3.20. *Suppose K is of equicharacteristic zero. Let (a_ρ) be a pc-sequence in K, and let $P \in K[X]$ be such that $P(a_\rho) \rightsquigarrow 0$ and $P_{(i)}(a_\rho) \not\rightsquigarrow 0$ for all $i \geqslant 1$. Then P is in hensel configuration at a_ρ, eventually, and in any henselian valued field extension of K there is a unique b such that $a_\rho \rightsquigarrow b$ and $P(b) = 0$.*

PROOF. Let a be a pseudolimit of (a_ρ) in some valued field extension of K (whose valuation we continue to denote by v), and set $\gamma_\rho := v(a - a_\rho)$. Since for each $i \geqslant 1$ we have $v\big(P_{(i)}(a_\rho)\big) = v\big(P_{(i)}(a)\big)$, eventually, Proposition 3.3.19 shows that P is in hensel configuration at a_ρ, eventually. Let K' be a henselian valued field extension of K. After deleting an initial segment of the sequence (a_ρ) we can assume that (γ_ρ) is strictly increasing, and that $v(a_\sigma - a_\rho) = \gamma_\rho$ whenever $\sigma > \rho$, and that $P'(a_\rho) \neq 0$ for all ρ. Likewise, by Lemma 3.3.18 and Proposition 3.3.19 we can assume that for every ρ there is a unique $b_\rho \in K'$ such that $P(b_\rho) = 0$ and $v(a_\rho - b_\rho) = v\big(P(a_\rho)\big) - v\big(P'(a_\rho)\big)$. Proposition 3.3.19 also shows that we can assume that $v\big(P(a_\rho)\big) - v\big(P'(a_\rho)\big) = \gamma_\rho$ for all ρ, so $v(a_\rho - b_\rho) = \gamma_\rho$ for all ρ. The uniqueness of b_ρ yields $b_\sigma = b_\rho$ whenever $\sigma > \rho$. Thus all b_σ are equal to a single b, which has the desired properties. □

COROLLARY 3.3.21. *Suppose K is of equicharacteristic zero. Then K is henselian if and only if K is algebraically maximal.*

PROOF. We already know that algebraically maximal valued fields are henselian (Corollary 3.3.4). Assume K is henselian, and let (a_ρ) be a pc-sequence in K of algebraic type over K. Take a monic minimal polynomial P of (a_ρ) over K. Then $P(a_\rho) \rightsquigarrow 0$ and $P_{(i)}(a_\rho) \not\rightsquigarrow 0$ for each $i \geqslant 1$, so P is in hensel configuration at a_ρ, eventually, and thus Lemma 3.3.18 gives $a \in K$ such that $P(a) = 0$. Since P is irreducible, we obtain $P = X - a$, so $a_\rho \rightsquigarrow a$. □

This leads to a partial converse of some earlier results (see 3.1.17 and 3.3.4):

COROLLARY 3.3.22. *Suppose K is of equicharacteristic zero. Then K is algebraically closed if and only if K is henselian,* $\operatorname{res}(K)$ *is algebraically closed, and Γ is divisible.*

EXAMPLE 3.3.23 (from Laurent series to Puiseux series). Let C be a field. The elements of the valued subfield

$$\mathrm{P}(C) := \bigcup_{n \geqslant 1} C(\!(t^{\frac{1}{n}\mathbb{Z}})\!)$$

of the Hahn field $C(\!(t^{\mathbb{Q}})\!)$ are called **Puiseux series** over C (in t). Note that $\mathrm{P}(C)$ has the same residue field (isomorphic to C) as its extension $C(\!(t^{\mathbb{Q}})\!)$ and the same value group $\mathbb{Q}$, so $C(\!(t^{\mathbb{Q}})\!)$ is an immediate extension of $\mathrm{P}(C)$. For $n \geqslant 1$ we have

$$C(\!(t^{\frac{1}{n}\mathbb{Z}})\!) = C(\!(t)\!) + C(\!(t)\!)a + \cdots + C(\!(t)\!)a^{n-1} = C(\!(t)\!)[a], \qquad a := t^{\frac{1}{n}},$$

so $C(\!(t^{\frac{1}{n}\mathbb{Z}})\!)$ is algebraic of degree n over $C(\!(t)\!)$. It follows that $\mathrm{P}(C)$ is algebraic over the Laurent series field $C(\!(t)\!) = C(\!(t^{\mathbb{Z}})\!)$. Now $C(\!(t^{\frac{1}{n}\mathbb{Z}})\!)$ is henselian for every $n \geqslant 1$, by Corollary 3.3.4, so $\mathrm{P}(C)$ is henselian. Hence by Corollary 3.3.22, if C is algebraically closed of characteristic zero, then $\mathrm{P}(C)$ is algebraically closed, and thus $\mathrm{P}(C)$ is the algebraic closure of $C((t))$ in $C((t^{\mathbb{Q}}))$.

COROLLARY 3.3.24. *Suppose K is henselian of equicharacteristic zero. Let $L = K(y)$ be a valued field extension of K with $nv(y) \notin \Gamma$ for all $n \geqslant 1$. Then no divergent pc-sequence in K has a pseudolimit in L.*

PROOF. Let (a_ρ) be a divergent pc-sequence in K with $a_\rho \rightsquigarrow a \in L$. Then (a_ρ) is of transcendental type over K, by Corollaries 3.2.12 and 3.3.21, so $K(a)$ is an immediate extension of K by Lemma 3.2.6. But a is transcendental over K and so y is algebraic over $K(a)$, and thus $nv(y) \in \Gamma_{K(a)} = \Gamma$ for some $n \geqslant 1$, a contradiction. □

Henselization

Let K be a valued field. A **henselization of** K is a henselian valued field extension K^{h} of K such that any valued field embedding $K \to L$ into a henselian valued field L extends uniquely to an embedding $K^{\mathrm{h}} \to L$.

PROPOSITION 3.3.25. *Every valued field has a henselization.*

PROOF. Fix an algebraic closure K^{a} of the underlying field of K and pick a valuation ring $\mathcal{O}^{\mathrm{a}}$ of K^{a} that lies over $\mathcal{O}$. Then $(K^{\mathrm{a}}, \mathcal{O}^{\mathrm{a}})$ is a henselian valued field extension of K, and the intersection E of the subfields $F \supseteq K$ of K^a such that $(F, \mathcal{O}^a \cap F)$ is henselian is itself a henselian valued field extension of K with valuation ring $\mathcal{O}_E := \mathcal{O}^a \cap E$. We claim that E is a henselization of K. To see this, let a valued field embedding $i \colon K \to L$ into a henselian valued field L be given. We have to show that i extends uniquely to a valued field embedding $E \to L$. We can arrange that K is a valued subfield of L with i the inclusion map. Replacing L by the relative algebraic closure of K in L we also reduce to the case that L is algebraic over K. Take an algebraic closure L^a of L and take a valuation ring V of L^a that lies over $\mathcal{O}_L$. Then Proposition 3.1.21 gives a valued field isomorphism $\sigma \colon (K^{\mathrm{a}}, \mathcal{O}^{\mathrm{a}}) \to (L^{\mathrm{a}}, V)$ over K, and so σ maps E onto the smallest henselian valued subfield of (L^a, V) containing K, and thus $\sigma(E) \subseteq L$. So we have a valued field embedding $\sigma|E \colon E \to L$ over K. Let $j \colon E \to L$ also be a valued field embedding over K; it is enough to show that then $j = \sigma|E$. Set $F := \{a \in E : j(a) = \sigma(a)\}$. Then $F \supseteq K$ is a subfield of E, and we take it as a valued subfield of E. Then F is henselian: let $P \in \mathcal{O}_F[X]$ and $a \in \mathcal{O}_F$ be such that $P(a) \prec 1$ and $P'(a) \asymp 1$. Then P has a unique zero $b \in E$ with $a - b \prec 1$, and likewise, the image of P under j in $L[X]$ has a unique zero c in L with $j(a) - c \prec 1$, and so $j(b) = c$. For the same reason, $\sigma(b) = c$, and so $b \in F$. We have now shown that F is henselian. Hence $E = F$ by the minimality of E, and thus $j = \sigma|E$. □

If K_1 and K_2 are henselizations of K, then the unique embedding $K_1 \to K_2$ over K is an isomorphism; thus there is no harm in referring to *the* henselization of K. Of course, if K is henselian, then it is its own henselization. We already know that K has an immediate algebraically maximal (and thus henselian) valued field extension that is algebraic over K. Hence:

COROLLARY 3.3.26. *The henselization of K is an immediate valued field extension of K and is algebraic over K.*

If K' is any henselian valued field extension of K, then by the **henselization of K in K'** we mean the henselization K^{h} of K with $K \subseteq K^{\mathrm{h}} \subseteq K'$ as valued fields.

We can now describe the absolute Galois group of the henselization of K:

LEMMA 3.3.27. *Let K^{a} be an algebraic closure of K and $\mathcal{O}^{\mathrm{a}}$ a valuation ring of K^{a} lying over $\mathcal{O}$. Let K^{h} be the henselization of K in $(K^{\mathrm{a}}, \mathcal{O}^{\mathrm{a}})$. Then*

$$\mathrm{Aut}(K^{\mathrm{a}}|K^{\mathrm{h}}) = \{\sigma \in \mathrm{Aut}(K^{\mathrm{a}}|K) : \sigma(\mathcal{O}^{\mathrm{a}}) = \mathcal{O}^{\mathrm{a}}\}.$$

PROOF. Let $\sigma \in \mathrm{Aut}(K^{\mathrm{a}}|K^{\mathrm{h}})$. Then $\sigma(\mathcal{O}^{\mathrm{a}})$ and $\mathcal{O}^{\mathrm{a}}$ are valuation rings of K^{a} lying over the valuation ring of K^{h}, hence they are equal by Proposition 3.3.11. Conversely, let $\sigma \in \mathrm{Aut}(K^{\mathrm{a}}|K)$ with $\sigma(\mathcal{O}^{\mathrm{a}}) = \mathcal{O}^{\mathrm{a}}$. Then σ is an automorphism of the valued field $(K^{\mathrm{a}}, \mathcal{O}^{\mathrm{a}})$, so its valued subfields K^{h} and $\sigma(K^{\mathrm{h}})$ are both henselizations of K in $(K^{\mathrm{a}}, \mathcal{O}^{\mathrm{a}})$, with isomorphism $\sigma|K^{\mathrm{h}}$ between them. Therefore $K^{\mathrm{h}} = \sigma(K^{\mathrm{h}})$ and $\sigma|K^{\mathrm{h}} = \mathrm{id}$. □

This leads to the following top-down construction of K^{h} when $\mathrm{char}(K) = 0$:

COROLLARY 3.3.28. *Assume $\mathrm{char}(K) = 0$, and let K^{a}, $\mathcal{O}^{\mathrm{a}}$, and K^{h} be as in the lemma above. Then K^{h} is the decomposition field of $\mathcal{O}^{\mathrm{a}}$ over K.*

THEOREM 3.3.29. *Suppose K is of equicharacteristic zero and L is an immediate henselian valued field extension of K and L is algebraic over K. Then L is a henselization of K.*

PROOF. By Corollary 3.3.21 the henselization K^{h} of K in L is algebraically maximal, and L is an immediate algebraic extension of K^{h}, so $K^{\mathrm{h}} = L$. □

We can now prove the following useful fact:

LEMMA 3.3.30. *Suppose K has equicharacteristic zero, and $L \supseteq K$ is an immediate valued field extension with $[L : K] < \infty$. Then $L = K(x)$ for some $x \prec 1$ in L whose minimum polynomial over K has the form*

$$X^n + a_{n-1}X^{n-1} + \cdots + a_1 X + a_0 \qquad \textit{with all } a_i \preccurlyeq 1,\ a_1 \asymp 1,\ a_0 \prec 1.$$

PROOF. Let K^{a} be an algebraic closure of K containing L as a subfield and take a valuation ring $\mathcal{O}^{\mathrm{a}}$ of K^{a} lying above $\mathcal{O}_L$. Let L^{h} be the henselization of L in $(K^{\mathrm{a}}, \mathcal{O}^{\mathrm{a}})$; then L^{h} is also the henselization of K in $(K^{\mathrm{a}}, \mathcal{O}^{\mathrm{a}})$, by Theorem 3.3.29. Let N be a Galois extension of K inside K^{a} which contains L with $[N : K] < \infty$. Then $\mathcal{O}_N = \mathcal{O}^{\mathrm{a}} \cap N$ is a valuation ring of N. Let $G := \mathrm{Aut}(N|K)$ and let

$$G^{\mathrm{d}} = \{\sigma \in G : \sigma(\mathcal{O}_N) = \mathcal{O}_N\}$$

be the decomposition group of $\mathcal{O}_N$ over K, with fixed field N^{d}. By Corollary 3.3.28 we have $N^{\mathrm{d}} = L^{\mathrm{h}} \cap N \supseteq L$; in particular, $G^{\mathrm{d}} \subseteq \mathrm{Aut}(N|L)$. Let A be the integral closure of $\mathcal{O}$ in L and B the integral closure of $\mathcal{O}$ in N, and $\mathfrak{q}$ the maximal ideal of B such that $B_{\mathfrak{q}} = \mathcal{O}_N$.

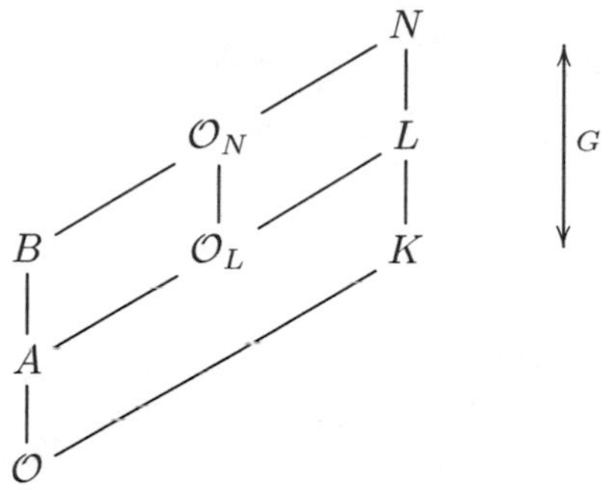

CLAIM: *Suppose* $\sigma \in G$ *and* $\sigma(\mathfrak{q}) \cap A = \mathfrak{q} \cap A$. *Then* $\sigma \in \mathrm{Aut}(N|L)$.

PROOF OF CLAIM. From $\mathcal{O}_N = B_{\mathfrak{q}}$ we get $\sigma(\mathcal{O}_N) = B_{\sigma(\mathfrak{q})}$, so $\mathcal{O}_N$ and $\sigma(\mathcal{O}_N)$ lie over $A_{\mathfrak{q}\cap A} = \mathcal{O}_L$. Since $N \supseteq L$ is normal, Proposition 3.1.21 gives $\tau \in \mathrm{Aut}(N|L)$ with $\tau(\mathcal{O}_N) = \sigma(\mathcal{O}_N)$, so $\tau^{-1}\sigma \in G^{\mathrm{d}} \subseteq \mathrm{Aut}(N|L)$, thus $\sigma \in \mathrm{Aut}(N|L)$. □

By the claim, the (finite) sets

$$\{\sigma(\mathfrak{q}) \cap A : \sigma \in \mathrm{Aut}(N|L)\}, \qquad \{\sigma(\mathfrak{q}) \cap A : \sigma \in G \setminus \mathrm{Aut}(N|L)\}$$

of maximal ideals of A are disjoint. Hence the Chinese Remainder Theorem (below Proposition 3.1.21) gives $x \in A$ such that

$$x \in \sigma(\mathfrak{q}) \text{ for all } \sigma \in \mathrm{Aut}(N|L), \qquad x \notin \sigma(\mathfrak{q}) \text{ for all } \sigma \in G \setminus \mathrm{Aut}(N|L).$$

We show that x has the required properties. Let H be the subgroup of G whose fixed field in N is $K(x)$. From $K(x) \subseteq L$ we get $\mathrm{Aut}(N|L) \subseteq \mathrm{Aut}(N|K(x)) = H$. On the other hand, $x \in \mathfrak{q}$ gives $x = \sigma(x) \in \sigma(\mathfrak{q})$ for all $\sigma \in H$, so $H \subseteq \mathrm{Aut}(N|L)$; thus $H = \mathrm{Aut}(N|L)$ and $L = K(x)$.

Now fix a coset decomposition $G = \sigma_1 H \cup \cdots \cup \sigma_n H$ of G with respect to its subgroup H, where $\sigma_1 = \mathrm{id}$ and $n = [G : H] = [L : K]$. Then $\sigma_1|L, \dots, \sigma_n|L$ are the distinct K-embeddings of L into N. The minimum polynomial of x over K is

$$P(X) = \prod_{i=1}^{n} (X - \sigma_i(x)) = X^n + a_{n-1}X^{n-1} + \cdots + a_1 X + a_0 \qquad (\text{all } a_j \in K).$$

An argument similar to the one at the end of the proof of the implication (i) ⇒ (ii) in Proposition 3.3.11 now shows that $a_j \preccurlyeq 1$ for all j and $a_0 \prec 1$, $a_1 \asymp 1$. □

The next lemma describes the valuation ring $\mathcal{O}_L$ of L in Lemma 3.3.30:

LEMMA 3.3.31. *Let $L = K(x)$ be a separable algebraic field extension of K where $x \prec 1$ and the minimum polynomial of x over K has the form*

$$X^n + a_{n-1}X^{n-1} + \cdots + a_1 X + a_0 \qquad \text{with all } a_i \preccurlyeq 1,\ a_1 \asymp 1,\ a_0 \prec 1.$$

Then $\scriptstyle\mathcal{O}(x) := \scriptstyle\mathcal{O}\mathcal{O}[x] + x\mathcal{O}[x]$ is a maximal ideal of $\mathcal{O}[x]$ and $\mathcal{O}_L = \mathcal{O}[x]_{\scriptstyle\mathcal{O}(x)}$.

PROOF. By Lemma 1.4.8, $\scriptstyle\mathcal{O}(x)$ is a maximal ideal of $\mathcal{O}[x]$. Let B be the integral closure of $\mathcal{O}$ in L. Then $\mathcal{O}_L = B_{\mathfrak{n}}$ for a maximal ideal $\mathfrak{n}$ of B, by Proposition 3.1.20. Hence $x \in \mathfrak{n}$, so $\scriptstyle\mathcal{O}(x) = \mathfrak{n} \cap \mathcal{O}[x]$, and thus $\mathcal{O}[x]_{\scriptstyle\mathcal{O}(x)} = B_{\mathfrak{n}}$ by Corollary 1.4.11. □

Monomial groups

A **monomial group** of the valued field K is a subgroup $\mathfrak{M}$ of $K^\times$ that is mapped bijectively onto Γ by v. This notion is equivalent to that of a **cross-section** s of the valued field K, that is, a group morphism $s\colon \Gamma \to K^\times$ such that $v(s\gamma) = \gamma$ for every $\gamma \in \Gamma$. Indeed, for such a cross-section s, the set $s(\Gamma)$ is a monomial group, and every monomial group of K is of this form for a unique cross-section s of K. For example, if C is a field and $\mathfrak{M}$ an ordered abelian multiplicative group and $K = C[[\mathfrak{M}]]$ the associated Hahn field, then the group $\mathfrak{M}$, identified with a subgroup of $K^\times$ in the natural way, is a monomial group of K. Other examples for monomial groups are provided by the following lemma:

LEMMA 3.3.32. *Let G be a divisible subgroup of $K^\times$ with $v(G) = \Gamma$. Then there exists a cross-section s of K with $s(\Gamma) \subseteq G$.*

PROOF. The inclusion $\mathcal{O}^\times \cap G \to G$ and the restriction of the valuation $v\colon K^\times \to \Gamma$ to G yield an exact sequence

$$1 \to \mathcal{O}^\times \cap G \to G \to \Gamma \to 0$$

of abelian groups. Since G is divisible, so is $\mathcal{O}^\times \cap G$, hence this exact sequence splits, which gives a section s as claimed. □

Thus algebraically closed valued fields have monomial groups. Likewise, if K is real closed (see Section 3.5 below), then K has a monomial group contained in $K^>$. It is also easy to see that if K is henselian with residue field $\boldsymbol{k} = \operatorname{res}(K)$ of characteristic zero, and the abelian groups $\boldsymbol{k}^\times$ and Γ are divisible, then $K^\times$ is divisible, hence K has a monomial group.

LEMMA 3.3.33. *Let K be an algebraically closed valued field and $\mathfrak{M}_0$ a monomial group of a valued subfield K_0 of K over which K is algebraic. Then $\mathfrak{M}_0$ extends to a monomial group $\mathfrak{M}$ of K.*

PROOF. Let $s_0\colon \Gamma_0 = v(K_0^\times) \to K_0^\times$ be the cross-section of K_0 with $s_0(\Gamma_0) = \mathfrak{M}_0$. Since the abelian group $K^\times$ is divisible, s_0 extends to a group morphism

$$s : \Gamma = v(K^\times) \to K^\times.$$

Using $\Gamma = \mathbb{Q}\Gamma_0$ it follows easily that $\mathfrak{M} := s(\Gamma)$ is a monomial group of K. □

Next we want to show that every valued field (possibly equipped with additional structure) has an elementary extension with a monomial group. This requires a digression on abelian groups. Let A, B be (additively written) abelian groups. Call A a **pure subgroup** of B if A is a subgroup of B such that $A \cap nB = nA$ for all $n \geqslant 1$. If A is a subgroup of B and B/A is torsion-free, then A is a pure subgroup of B. In case B is itself torsion-free, then

$$A \text{ is a pure subgroup of } B \iff B/A \text{ is torsion-free.}$$

Also, if A is a direct summand of B (that is, A is a subgroup of B and $B = A \oplus B'$, internally, for some subgroup B' of B), then A is a pure subgroup of B.

LEMMA 3.3.34. *Suppose A is a pure subgroup of B, and the group B/A is finitely generated. Then $B = A \oplus B'$, internally, for some subgroup B' of B.*

PROOF. By the fundamental theorem on finitely generated abelian groups we have

$$B/A \;=\; \mathbb{Z}(b_1 + A) \oplus \cdots \oplus \mathbb{Z}(b_m + A)$$

for suitable $b_1, \ldots, b_m \in B$. If $b_i + A$ has finite order $n_i \geqslant 1$ in B/A, then $n_i b_i \in A$, so $n_i b_i = n_i a_i$ with $a_i \in A$, and replacing b_i by $b_i - a_i$ we arrange $n_i b_i = 0$. With this adjustment of the b_is one checks easily that $B = A \oplus B'$ where

$$B' \;=\; \mathbb{Z}b_1 + \cdots + \mathbb{Z}b_m \;=\; \mathbb{Z}b_1 \oplus \cdots \oplus \mathbb{Z}b_m. \qquad \square$$

COROLLARY 3.3.35. *Suppose A is a pure subgroup of B, $e_{ij} \in \mathbb{Z}$ for $1 \leqslant i \leqslant m$ and $1 \leqslant j \leqslant n$, and $a_1, \ldots, a_m \in A$. Suppose the system of equations*

$$\begin{aligned} e_{11}x_1 + \cdots + e_{1n}x_n &= a_1 \\ \vdots \qquad \vdots \qquad \vdots \quad &\quad \vdots \\ e_{m1}x_1 + \cdots + e_{mn}x_n &= a_m \end{aligned}$$

has a solution in B, that is, there are $x_1, \ldots, x_n \in B$ for which the above equations hold. Then this system has a solution in A.

LEMMA 3.3.36. *Suppose A is a pure subgroup of B, and $b \in B$. Then there is a pure subgroup A' of B that contains A and b such that A'/A is countable.*

PROOF. By the Downward Löwenheim-Skolem Theorem (see B.5.10) we can take a subgroup A' of B that contains A and b such that A'/A is countable and $A'/A \preccurlyeq B/A$. It follows easily that A' is a pure subgroup of B. $\square$

PROPOSITION 3.3.37. *Let $h\colon A \to U$ be a group morphism into an $\aleph_1$-saturated (additive) abelian group U and suppose A is a pure subgroup of B. Then h extends to a group morphism $B \to U$.*

PROOF. By the previous lemma we reduce to the case that B/A is countable. (This reduction does not use that U is $\aleph_1$-saturated.) Let $b_0, b_1, b_2, \ldots$ generate B over A. If we can find elements $u_0, u_1, u_2, \ldots \in U$ such that $\sum_{i=0}^{n} e_i u_i = h(a)$ whenever $\sum_{i=0}^{n} e_i b_i = a$ (all $e_i \in \mathbb{Z}$, $a \in A$), then we can extend h as desired by sending b_n to u_n for each n. Note that each finite subset of this countable set of constraints on $(u_0, u_1, u_2, \ldots)$ can be satisfied, by Corollary 3.3.35. The desired result follows. □

COROLLARY 3.3.38. *Let U be an $\aleph_1$-saturated (additive) abelian group and a pure subgroup of B. Then U is a direct summand of B.*

PROOF. Apply Proposition 3.3.37 to the identity map $U \to U$. □

For our valued field K, construed as a field with valuation ring, we get:

LEMMA 3.3.39. *If K is $\aleph_1$-saturated, then it has a cross-section.*

PROOF. With $U = \mathcal{O}^\times$, the natural inclusion $U \to K^\times$ and the map $v \colon K^\times \to \Gamma$ yield the exact sequence of abelian groups

$$1 \to U \to K^\times \to \Gamma \to 0.$$

Since Γ is torsion-free, U is a pure subgroup of $K^\times$. If K is $\aleph_1$-saturated, then so is the group U, and thus the above exact sequence splits by Corollary 3.3.38. □

The following variant of this lemma is also sometimes useful:

LEMMA 3.3.40. *Let K be $\aleph_1$-saturated, and let E be a valued subfield of K such that Γ_E is pure in Γ and $\aleph_1$-saturated. Let s_E be a cross-section of E. Then s_E extends to a cross-section of K.*

PROOF. By Lemma 3.3.39 we have a cross-section s of K. By Corollary 3.3.38 we have an internal direct sum decomposition $\Gamma = \Gamma_E \oplus \Delta$ with Δ a subgroup of Γ. This gives a cross-section of K that coincides with s_E on Γ_E and with s on Δ. □

Uniqueness of maximal immediate extensions

In this subsection we assume that K is a valued field of equicharacteristic zero. By Corollaries 2.3.3 and 3.2.9 any two maximal immediate valued field extensions of K are isomorphic over K *as valued abelian additive groups*; we can now improve on this:

COROLLARY 3.3.41. *Any two maximal immediate valued field extensions of K are isomorphic over K as valued fields.*

PROOF. Let K_1, K_2 be maximal immediate valued field extensions of K. Below, each subfield of K_i ($i = 1, 2$) is viewed as a valued field by taking as valuation ring the intersection of the valuation ring of K_i with the subfield. Consider fields L_1, L_2 with $K \subseteq L_i \subseteq K_i$ for $i = 1, 2$, and suppose that we have a valued field isomorphism $L_1 \cong L_2$. Note that $L_1 = K_1$ iff $L_2 = K_2$.

Suppose $L_1 \neq K_1$ (and hence $L_2 \neq K_2$). It suffices to show that then we can extend the isomorphism $L_1 \cong L_2$ to a valued field isomorphism $L_1' \cong L_2'$ where L_i' is a field with $L_i \subseteq L_i' \subseteq K_i$ and $L_i' \neq L_i$, for $i = 1, 2$. If L_1 is not henselian, then we can take for L_i' the henselization of L_i in K_i. Thus, suppose L_1 is henselian. Take $b \in K_1 \setminus L_1$ and take a divergent pc-sequence (a_ρ) in L_1 such that $a_\rho \rightsquigarrow b$. Since L_1 is algebraically maximal by Corollary 3.3.21, (a_ρ) is of transcendental type over L_1. The image of (a_ρ) under our isomorphism $L_1 \cong L_2$ has a pseudolimit $c \in K_2$. By Lemma 3.2.6 we can extend this isomorphism to an isomorphism $L_1(b) \cong L_2(c)$. □

Assuming also that K has a monomial group we can identify the maximal immediate extension of K with a Hahn field:

COROLLARY 3.3.42. *Suppose K is maximal and $\mathfrak{M}$ is a monomial group of K. Then there is a valued field isomorphism $K \cong \boldsymbol{k}[[\mathfrak{M}]]$ that induces the identity on $\boldsymbol{k} = \operatorname{res}(K)$ and on $\mathfrak{M}$.*

PROOF. By Corollary 3.3.4, K is henselian. Thus by Proposition 3.3.8 there is a lift C of the residue field $\boldsymbol{k}$ in $\mathcal{O}$. Let i be the field isomorphism $c \mapsto \bar{c}\colon C \to \boldsymbol{k}$. Then i extends uniquely to a valued field embedding of the valued subfield $C(\mathfrak{M})$ of K into $\boldsymbol{k}[[\mathfrak{M}]]$ that is the identity on $\mathfrak{M}$. Clearly K is an immediate extension of $C(\mathfrak{M})$. By the previous corollary, this embedding extends to a valued field isomorphism $K \cong \boldsymbol{k}[[\mathfrak{M}]]$. □

The proof of Corollary 3.3.41 also yield a useful embedding property of the maximal immediate extension of K. To formulate this, set $\Gamma := v(K^\times)$ and let L be a valued field extension of K. We say that L is **Γ-maximal** if L is henselian and every pc-sequence in L of length $\leqslant \operatorname{cf}(S)$ for some $S \subseteq \Gamma$ pseudoconverges in L. In particular, if L is maximal, then L is Γ-maximal. Also, if L is henselian and $|\Gamma|^+$-saturated, then L is Γ-maximal.

COROLLARY 3.3.43. *If M is a maximal immediate valued field extension of K, then M embeds as a valued field over K into any Γ-maximal valued field extension of K.*

Figure 3.1 shows various valued field extensions of K discussed in this section. Here, $K^{\max}$ denotes a maximal immediate extension of K, and K^{h} a henselization of K in $K^{\max}$. We introduce $(K^{\mathrm{h}})^{\mathrm{unr}}$ in the following subsection.

Unramified extensions

In this subsection K is henselian. We fix an algebraic closure K^{a} of K, and equip it with the unique valuation ring $\mathcal{O}^{\mathrm{a}}$ making K^{a} a valued field extension of K. Any algebraic field extension of K we mention is assumed to be a subfield of K^{a}, and regarded as a *valued* subfield of K^{a}.

Let L be an algebraic field extension of K. If $[L : K] < \infty$, then we say that L is **unramified over** K if the associated extension $\operatorname{res}(L) \supseteq \operatorname{res}(K)$ of residue fields is separable and $[L : K] = [\operatorname{res}(L) : \operatorname{res}(K)]$ (and thus $\Gamma_L = \Gamma$, by Corollary 3.1.8).

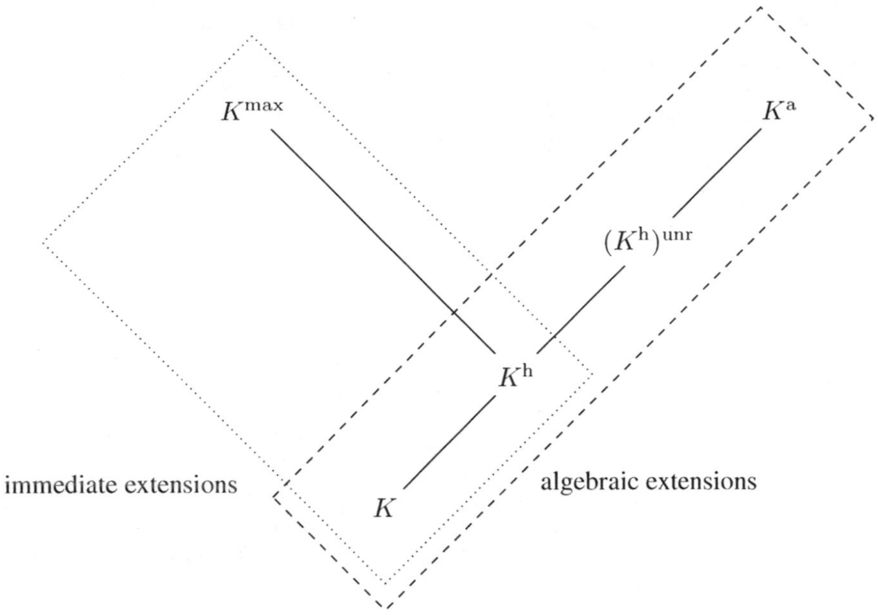

Figure 3.1: Inclusion diagram of some valued field extensions of a valued field K of equicharacteristic zero.

Suppose $[L:K] < \infty$ and E is an intermediate field: K is a subfield of E and E is a subfield of L. Then by the multiplicativity of degrees and properties of separability,

$$L \text{ is unramified over } K \iff L \text{ is unramified over } E \text{ and } E \text{ is unramified over } K.$$

Without assuming $[L:K] < \infty$ we call L **unramified over** K if every intermediate field E with $[E:K] < \infty$ is unramified over K (and thus $\Gamma_L = \Gamma$).

LEMMA 3.3.44. *Let $L' \supseteq K$ be an algebraic field extension, and L, K' subfields of L' containing K, such that $L' = LK'$ and L is unramified over K.*

$$\begin{array}{ccc} L & \longrightarrow & L' \\ \uparrow & & \uparrow \\ K & \longrightarrow & K' \end{array}$$

Then L' is unramified over K'.

PROOF. We may assume $[L:K] < \infty$. Take $x \in \mathcal{O}_L$ such that $\operatorname{res}(L) = \operatorname{res}(K)(\overline{x})$, and let $P \in K[X]$ be the minimum polynomial of x over K. Then $P \in \mathcal{O}[X]$ and

$$\begin{aligned} \big[\operatorname{res}(L) : \operatorname{res}(K)\big] &\leqslant \deg \overline{P} = \deg P \\ &= \big[K(x):K\big] \leqslant [L:K] = \big[\operatorname{res}(L) : \operatorname{res}(K)\big], \end{aligned}$$

hence $L = K(x)$ (and thus $L' = K'(x)$), and $\overline{P}$ is the minimum polynomial of $\overline{x}$ over $\operatorname{res}(K)$. Let $\mathcal{O}'$ be the valuation ring of K'; so $\mathcal{O}'$ is the integral closure of $\mathcal{O}$ in K'. Let $Q \in K'[X]$ be the minimum polynomial of x over K'. Then $Q \in \mathcal{O}'[X]$ and $P = QR$ with monic $R \in \mathcal{O}'[X]$, so $\overline{Q}$ divides $\overline{P}$ in $\operatorname{res}(K')[X]$, and hence $\overline{Q} \in \operatorname{res}(K')[X]$ is separable. Thus $\overline{Q} \in \operatorname{res}(K')[X]$ is irreducible by the irreducibility of $Q \in K'[X]$ and the lifting property of Proposition 3.3.11(iv). Therefore,

$$\begin{aligned}\big[\operatorname{res}(L') : \operatorname{res}(K')\big] \leqslant [L' : K'] &= \deg Q = \deg \overline{Q} \\ &= \big[\operatorname{res}(K')(\overline{x}) : \operatorname{res}(K')\big] \leqslant \big[\operatorname{res}(L') : \operatorname{res}(K')\big].\end{aligned}$$

It follows that $\operatorname{res}(L') = \operatorname{res}(K')(\overline{x})$ and $[L' : K'] = \big[\operatorname{res}(L') : \operatorname{res}(K')\big]$, and so L' is unramified over K'. □

COROLLARY 3.3.45. *Let $L \supseteq K$ be an algebraic field extension which is unramified over K. Let E be an intermediate field. Then L is unramified over E.*

PROOF. Lemma 3.3.44 applied to $K' = E$ shows that L is unramified over E. □

COROLLARY 3.3.46. *Suppose $L, L' \supseteq K$ are algebraic field extensions unramified over K. Then the compositum LL' is also unramified over K.*

PROOF. We can reduce to the case that $[L : K] < \infty$ and $[L' : K] < \infty$. It follows from Lemma 3.3.44 that LL' is unramified over L'. Then the equivalence just before Lemma 3.3.44 gives that LL' is unramified over K. □

Let L be an algebraic field extension of K. By the previous corollary, the composite of the intermediate fields that are unramified over K is itself unramified over K, and is called the **maximal unramified extension of K in L.**

LEMMA 3.3.47. *Let M be the maximal unramified extension of K in L. Then* $\operatorname{res}(M)$ *is the separable algebraic closure of* $\operatorname{res}(K)$ *in* $\operatorname{res}(L)$.

PROOF. Let $\xi \in \operatorname{res}(L)$ be separable over $\operatorname{res}(K)$; we need to show $\xi \in \operatorname{res}(M)$. Let $P \in \mathcal{O}[X]$ be monic such that $\overline{P}$ is the minimum polynomial of ξ over $\operatorname{res}(K)$. Then $P \in K[X]$ is irreducible. As L is henselian, P has a zero $x \in \mathcal{O}_L$ with $\overline{x} = \xi$. It follows that

$$\big[K(x) : K\big] = \big[\operatorname{res}(K)(\xi) : \operatorname{res}(K)\big],$$

hence $K(x)$ is unramified over K, so $K(x) \subseteq M$, and thus $\xi = \overline{x} \in \operatorname{res}(M)$. □

We let K^{unr} be the maximal unramified extension of K in K^{a}. By the previous lemma, the residue field of K^{unr} is the separable algebraic closure of $\operatorname{res}(K)$ in its algebraic closure $\operatorname{res}(K^{\mathrm{a}})$. Note that if $L \supseteq K$ is any algebraic field extension, then $K^{\mathrm{unr}} \cap L$ is the maximal unramified extension of K in L.

PROPOSITION 3.3.48. *Suppose K has equicharacteristic zero and $L \supseteq K$ is a field extension with $[L:K] < \infty$ and $\operatorname{res}(K) = \operatorname{res}(L)$. Then there is a chain of subfields*

$$K = K_0 \subseteq K_1 \subseteq \cdots \subseteq K_n = L$$

of L such that for $i = 1, \dots, n$ we have $K_i = K_{i-1}\big(b^{1/p}\big)$ for some prime number p and some $b \in K_{i-1}^{\times}$ with $vb \notin p\Gamma_{K_{i-1}}$. Moreover, $[L:K] = [\Gamma_L : \Gamma]$.

PROOF. The quotient group Γ_L/Γ is finite, so we have a chain of subgroups

$$\Gamma = \Gamma_0 \subseteq \Gamma_1 \subseteq \cdots \subseteq \Gamma_n = \Gamma_L$$

of Γ_L such that $[\Gamma_i : \Gamma_{i-1}]$ is prime for $i = 1, \dots, n$. Set $K_0 = K$, let $i \in \{1, \dots, n\}$ and suppose K_{i-1} is an intermediate field with $\Gamma_{K_{i-1}} = \Gamma_{i-1}$. With $[\Gamma_i : \Gamma_{i-1}] = p$, take $\alpha \in \Gamma_i$ such that $\alpha + \Gamma_{i-1}$ has order p in Γ_i/Γ_{i-1}. Then $\alpha = va$ where $a \in L^{\times}$, and $p\alpha = vb$ with $b \in K_{i-1}^{\times}$. We arrange $a^p = b$ as follows: since $\operatorname{res}(L) = \operatorname{res}(K)$ we have $a^p/b = cu$ where $c \in K$ and $u \in L$ with $c \asymp 1$ and $u \sim 1$. As L is henselian, the polynomial $X^p - u \in \mathcal{O}_L[X]$ has a zero x in $\mathcal{O}_L$ with $x \sim 1$; replacing a by a/x and b by bc we have $a^p = b$. With $K_i := K_{i-1}(a)$, this yields $\Gamma_{K_i} = \Gamma_i$ by Lemma 3.1.28. In this way we construct a chain $K_0 \subseteq K_1 \subseteq \cdots \subseteq K_n$ of subfields of L. Then the extension $L \supseteq K_n$ is immediate, so $L = K_n$ by Corollary 3.3.21. Lemma 3.1.28 also gives $[L:K] = [\Gamma_L : \Gamma]$ by multiplicativity. □

Proposition 3.3.48 together with Lemma 3.3.47 yield a key fact:

COROLLARY 3.3.49. *Suppose K has equicharacteristic zero and $L \supseteq K$ is a field extension with $[L:K] < \infty$. Then*

$$[L:K] = [\Gamma_L : \Gamma] \cdot \big[\operatorname{res}(L) : \operatorname{res}(K)\big].$$

PROOF. Let M be the maximal unramified extension of K in L. Then by definition $\big[\operatorname{res}(M) : \operatorname{res}(K)\big] = [M : K]$. Next, $\operatorname{res}(M) = \operatorname{res}(L)$ by Lemma 3.3.47, so $[L:M] = [\Gamma_L : \Gamma_M]$ by Proposition 3.3.48. It remains to note that $\Gamma_M = \Gamma$. □

Let $L \supseteq K$ be a field extension with $[L:K] < \infty$. One says that L is **purely ramified over** K if $[L:K] = [\Gamma_L : \Gamma]$. Thus by Corollaries 3.1.8 and 3.3.49, if K has equicharacteristic zero, then

$$L \text{ is unramified over } K \iff \Gamma_L = \Gamma,$$
$$L \text{ is purely ramified over } K \iff \operatorname{res}(L) = \operatorname{res}(K).$$

Notes and comments

Hensel [175] introduced the valued field of p-adic numbers and showed it to be henselian. Its generalization stated in Corollary 3.3.5 is due to Rychlík [374]. The characterizations of henselianity collected in Lemma 3.3.10 and Proposition 3.3.11 are due to Nagata [300], Rayner [331], and Rim [339]. The notion of henselization and the existence of the henselization of a valued field is also due to Nagata. (The henselization of rank 1 valued fields had been constructed already by Ostrowski [314].)

Puiseux series (see Example 3.3.23) stem from Newton [303] and Puiseux [326]. It is well-known that for a field C of characteristic $p > 0$, the Puiseux series field $\mathrm{P}(C)$ is not algebraically closed in its extension $C(\!(t^{\mathbb{Q}})\!)$: the polynomial $X^p - X - x$ with $x = t^{-1}$ has a zero $x^{1/p} + x^{1/p^2} + \cdots \in C(\!(t^{\mathbb{Q}})\!) \setminus \mathrm{P}(C)$. (This example is from Chevalley [78, p. 64]. See [212, 213] for more on this issue.)

For Lemma 3.3.30, see [382, Proposition 1.1]. The facts on abelian groups in 3.3.34–3.3.38 are taken from [74, Chapter V, §5] where this material is treated for modules over any ring. Lemma 3.3.39 is also in [74]. Cross-sections are from [28]. The notion of pure subgroup and basic facts about it are due to Prüfer [325]. Proposition 3.3.8 goes back to [169] for discrete valuations, and for general valuations to [209, 277]. Corollaries 3.3.41 and 3.3.42 are due to Kaplansky [209]. The maximal unramified extension plays a key role in the Galois theory of valuations, of which we only needed a few facts here; see [129, 236] for more.

Our use of "henselian" for valued fields is common in valuation theory, but clashes a bit with its use in the larger area of commutative algebra: there it would be the *valuation ring* of the valued field that is henselian, according to Azumaya [30] who defined a local ring $\mathcal{O}$ with residue field $\boldsymbol{k}$ to be henselian if it satisfies the condition in Definition 3.3.1. Some results in this section generalize to this local ring setting, for example Proposition 3.3.16, but in that case the proof is more difficult: [158, (18.5.11), (b)]; see also [14, (1.9)], [293, Theorem 4.2, (d′)].

3.4 DECOMPOSING VALUATIONS

A major theme in our work is constructing zeros of differential polynomials over valued differential fields. We approximate/construct these zeros by pc-sequences that have special properties. In this section we introduce these properties in a purely valuation-theoretic setting. A basic tool for analyzing such pc-sequences is the two-pronged process of coarsening and specializing a valuation, which we treat first. *Throughout this section K is a valued field with valuation ring $\mathcal{O}$, valuation $v\colon K^\times \to \Gamma$ and dominance relation $\preccurlyeq$.*

Coarsening and specialization

Let Δ be a convex subgroup of $\Gamma = v(K^\times)$. Then we have the ordered quotient group $\dot{\Gamma} = \Gamma/\Delta$ and the valuation

$$\dot{v} = v_\Delta \colon K^\times \to \dot{\Gamma}$$

on the field K, defined by $\dot{v}(f) := v(f) + \Delta$. We call $\dot{v}$ the **coarsening of v by Δ**, or the **Δ-coarsening of** v; the valued field $(K, \dot{v})$ whose valuation is $\dot{v}$ instead of v is called the **Δ-coarsening of** K. Note that $\dot{v}$ is indeed a coarsening of v as defined in Section 2.2. Thus if $\Delta \neq \Gamma$, then the v-topology agrees with the $\dot{v}$-topology. The dominance relation on K corresponding to the coarsened valuation $\dot{v}$ is denoted by $\preccurlyeq_\Delta$, or by $\dot{\preccurlyeq}$ if Δ is understood from the context, so $f \preccurlyeq g \Rightarrow f \mathrel{\dot{\preccurlyeq}} g$ for $f, g \in K$. The valuation ring of $\dot{v}$ is

$$\dot{\mathcal{O}} := \{a \in K : va \geqslant \delta \text{ for some } \delta \in \Delta\},$$

which has $\mathcal{O}$ as a subring and has maximal ideal

$$\dot{o} := \{a \in K : va > \Delta\} \subseteq o.$$

Put $\dot{K} := \dot{\mathcal{O}}/\dot{o}$, the residue field of $\dot{\mathcal{O}}$, and put $\dot{a} := a + \dot{o} \in \dot{K}$ for $a \in \dot{\mathcal{O}}$. Then for $a \in \dot{\mathcal{O}}$ with $a \notin \dot{o}$ the value va depends only on the residue class $\dot{a} \in \dot{K}$, which gives the valuation

$$v \colon \dot{K}^\times \to \Delta, \qquad v\dot{a} := va \tag{3.4.1}$$

with valuation ring $\mathcal{O}_{\dot{K}} = \{\dot{a} : a \in \mathcal{O}\}$. The maximal ideal of $\mathcal{O}_{\dot{K}}$ is

$$o_{\dot{K}} = \{\dot{a} : a \in o\}.$$

Throughout $\dot{K}$ stands for the valued field $(\dot{K}, \mathcal{O}_{\dot{K}})$. The composed map

$$\mathcal{O} \to \mathcal{O}_{\dot{K}} \to \operatorname{res}(\dot{K}) = \mathcal{O}_{\dot{K}}/o_{\dot{K}}$$

has kernel o, and thus induces a field isomorphism $\operatorname{res}(K) \xrightarrow{\cong} \operatorname{res}(\dot{K})$, and we identify $\operatorname{res}(K)$ and $\operatorname{res}(\dot{K})$ via this map. We call $\dot{K}$ a **specialization** of the valued field K. The following is now straightforward.

LEMMA 3.4.1. *Let (a_ρ) be a well-indexed sequence in $\dot{\mathcal{O}}$ such that $(\dot{a}_\rho)$ is a pc-sequence in $\dot{K}$. Then (a_ρ) is a pc-sequence in K, and for all $a \in \dot{\mathcal{O}}$,*

$$a_\rho \leadsto a \quad \text{in } K \iff \dot{a}_\rho \leadsto \dot{a} \quad \text{in } \dot{K}.$$

We picture the valuation v on K as follows:

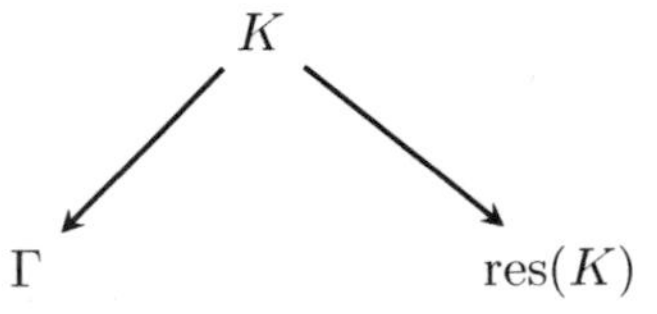

The greater the angle between the arrows, the larger the valuation ring. Given our convex subgroup Δ of Γ, this gives rise to the following diagram which displays the original valuation v, its coarsening by Δ, and the corresponding specialization:

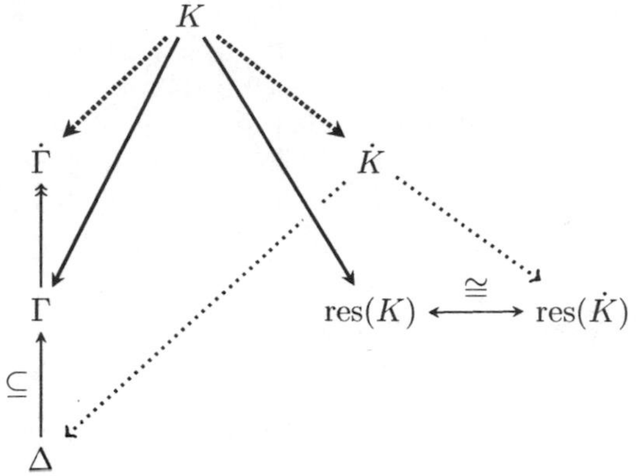

The original valuation $v\colon K^\times \to \Gamma$ can often be understood in terms of the two (usually simpler) valuations $\dot{v}\colon K^\times \to \dot{\Gamma}$ and $v\colon \dot{K}^\times \to \Delta$. Here is an example:

LEMMA 3.4.2. *K is henselian $\iff$ $(K, \dot{\mathcal{O}})$ and $\dot{K}$ are henselian.*

PROOF. We extend the natural surjection $x \mapsto \dot{x}\colon \dot{\mathcal{O}} \to \dot{\mathcal{O}}/\dot{\mathcal{o}} = \dot{K}$ to the ring morphism $Q \mapsto \dot{Q}\colon \dot{\mathcal{O}}[X] \to \dot{K}[X]$ which maps the indeterminate X to itself.

Suppose first that K is henselian. Then $(K, \dot{\mathcal{O}})$ clearly is henselian, using $\dot{\mathcal{o}} \subseteq \mathcal{o}$, $\mathcal{O} \subseteq \dot{\mathcal{O}}$, and the equivalence of (i) and (ii) in Lemma 3.3.10. To show that $\dot{K}$ is also henselian, we again use condition (ii) for henselianity from Lemma 3.3.10. Thus, let $P = 1+X+P_2X^2+\cdots+P_nX^n \in \mathcal{O}_{\dot{K}}[X]$ where $n \geqslant 2$ and $P_i \in \mathcal{o}_{\dot{K}}$ for $i = 2, \dots, n$; we have to show that P has a zero in $\mathcal{O}_{\dot{K}}$. Take $Q = 1 + X + Q_2X^2 + \cdots + Q_nX^n \in \dot{\mathcal{O}}[X]$ with $\dot{Q} = P$. Then $Q_i \in \mathcal{o}$ for $i = 2, \dots, n$, hence Q has a zero $x \in \mathcal{O}$, and then $\dot{x} \in \mathcal{O}_{\dot{K}}$ is a zero of P.

Conversely, suppose $(K, \dot{\mathcal{O}})$ and $\dot{K}$ are henselian. Let $P \in \mathcal{O}[X]$ and $a \in \mathcal{O}$ be such that $P(a) \in \mathcal{o}$ and $P'(a) \in \mathcal{O}^\times$; we need to show that P has a zero $b \in \mathcal{O}$ with $b \equiv a \bmod \mathcal{o}$. Since $\dot{K}$ is henselian, we can take $c \in \mathcal{O}$ (so $\dot{c} \in \mathcal{O}_{\dot{K}}$) such that $\dot{P}(\dot{c}) = 0$ and $\dot{c} \equiv \dot{a} \bmod \mathcal{o}_{\dot{K}}$. Then $P(c) \in \dot{\mathcal{o}}$ and $c \equiv a \bmod \mathcal{o}$, so $P'(c) \in \mathcal{O}^\times \subseteq \dot{\mathcal{O}}^\times$. Since $(K, \dot{\mathcal{O}})$ is henselian, this gives $b \in \dot{\mathcal{O}}$ with $P(b) = 0$ and $b \equiv c \bmod \dot{\mathcal{o}}$. Then b is a zero of P as required. □

Coarsening and valued field extensions

Let Δ be a convex subgroup of Γ with ordered quotient group $\dot{\Gamma} = \Gamma/\Delta$. As described above, this gives rise to the coarsened valuation $\dot{v} = v_\Delta\colon K^\times \to \dot{\Gamma}$ with residue field $\dot{K} = \dot{\mathcal{O}}/\dot{\mathcal{o}}$, the latter equipped with the valuation $v\colon \dot{K}^\times \to \Delta$ such that $v\dot{a} = va$ for all $a \in \dot{\mathcal{O}} \setminus \dot{\mathcal{o}}$.

Next, let L be a valued field extension of K, and let Δ_L be the convex hull of Δ in Γ_L. Then the natural inclusion $\Gamma \to \Gamma_L$ induces an embedding

$$\dot{\Gamma} = \Gamma/\Delta \to \dot{\Gamma}_L := \Gamma_L/\Delta_L$$

of ordered abelian groups, and identifying $\dot{\Gamma}$ with its image under this embedding, the coarsening $\dot{v} = v_{\Delta_L}$ of the valuation v of L by Δ_L extends the coarsening $\dot{v} = v_\Delta$ of the valuation of K by Δ, so we have a valued field extension $(L, \dot{\mathcal{O}}_L)$ of $(K, \dot{\mathcal{O}})$. As usual, we identify $\operatorname{res}(K)$ with a subfield of $\operatorname{res}(L)$ and the residue field $\dot{K}$ of v_Δ with a subfield of the residue field $\dot{L}$ of v_{Δ_L}. Then the valuation $v\colon \dot{L}^\times \to \Delta_L$ with $v\dot{a} = va$ for all $a \in \dot{\mathcal{O}}_L \setminus \dot{o}_L$ restricts to the valuation $v\colon \dot{K}^\times \to \Delta$ on $\dot{K}$. Thus the diagrams

$$\begin{array}{ccc} L^\times & \xrightarrow{v_{\Delta_L}} & \dot{\Gamma}_L \\ \uparrow & & \uparrow \\ K^\times & \xrightarrow{v_\Delta} & \dot{\Gamma} \end{array} \qquad \begin{array}{ccc} \dot{L}^\times & \xrightarrow{v} & \Delta_L \\ \uparrow & & \uparrow \\ \dot{K}^\times & \xrightarrow{v} & \Delta \end{array} \qquad \begin{array}{ccc} \operatorname{res}(L) & \xrightarrow{\cong} & \operatorname{res}(\dot{L}) \\ \uparrow & & \uparrow \\ \operatorname{res}(K) & \xrightarrow{\cong} & \operatorname{res}(\dot{K}) \end{array}$$

commute. (Here the vertical arrows are the natural inclusions.)

LEMMA 3.4.3. *Suppose* $[L : K] < \infty$. *Then* $[\dot{L} : \dot{K}] \leqslant [L : K] < \infty$. *Thus if* $\dot{K}$ *is spherically complete (respectively, complete), then so is* $\dot{L}$.

PROOF. The first statement follows from Corollary 3.1.9. The second statement follows from the first and Corollary 3.2.28. □

If $\Gamma_L = \Gamma$, then $\Delta_L = \Delta$, and if $\operatorname{res}(L) = \operatorname{res}(K)$, then $\operatorname{res}(\dot{L}) = \operatorname{res}(\dot{K})$. In particular, if L (with valuation ring $\mathcal{O}_L$) is an immediate extension of K, then $\dot{L}$ is an immediate extension of $\dot{K}$.

LEMMA 3.4.4. *Suppose* L *is a completion of the valued field* K *and* $\Delta \neq \Gamma$. *Then the coarsening* $(L, \dot{\mathcal{O}}_L)$ *is a completion of* $(K, \dot{\mathcal{O}})$.

PROOF. Note: L is an immediate extension of K. Lemma 2.2.27 gives completeness of $(L, \dot{\mathcal{O}}_L)$. Use $\Delta \neq \Gamma$ to get K dense in L with respect to $\dot{v}$. □

Conversely, suppose $(L, \dot{\mathcal{O}}_L)$ is an *immediate* valued field extension of $(K, \dot{\mathcal{O}})$, and let $\dot{v}$ denote both the valuation of $(K, \dot{\mathcal{O}})$ and of $(L, \dot{\mathcal{O}}_L)$. Thus $\dot{K} = \dot{L}$ after the usual identification, where $\dot{L}$ is the residue field of $(L, \dot{\mathcal{O}}_L)$.

We define a map $v\colon L^\times \to \Gamma$ extending the valuation $v\colon K^\times \to \Gamma$ as follows. For $f \in L^\times$, take $g \in K^\times$ and $u \in L^\times$ such that $f = gu$ and $\dot{v}(u) = 0$; then $\dot{u} \in \dot{L}^\times = \dot{K}^\times$, so $v(\dot{u}) \in \Delta$; it is easy to check that $v(g) + v(\dot{u}) \in \Gamma$ depends only on f and not on the choice of g, u; now put $v(f) := v(g) + v(\dot{u})$.

LEMMA 3.4.5. *The map* $v\colon L^\times \to \Gamma$ *is a valuation on* L, *its coarsening by* Δ *is exactly the given valuation* $\dot{v}$ *on* L, *and* L *with this valuation* v *is an immediate extension of the valued field* K.

PROOF. It is easy to check that $v\colon L^\times \to \Gamma$ is a group morphism extending v on $K^\times$, and that $\dot{v}(f) = v(f) + \Delta \in \dot{\Gamma}$ for $f \in L^\times$. Also, if $f \in L^\times$ and $\dot{v}f > 0$, then $vf > 0$ and $v(1+f) = 0$. Next, for $f_1, f_2 \in L^\times$ with $f_1 + f_2 \neq 0$ one shows that $v(f_1 + f_2) \geqslant \min\{vf_1, vf_2\}$ by distinguishing the cases $\dot{v}f_1 = \dot{v}f_2$ and $\dot{v}f_1 < \dot{v}f_2$. The rest of the lemma follows easily. □

COROLLARY 3.4.6. *K is maximal $\iff$ $(K,\dot{\mathcal{O}})$ and $\dot{K}$ are maximal.*

PROOF. Assume K is maximal. Then $(K,\dot{\mathcal{O}})$ is maximal by Corollary 2.2.22, and $\dot{K}$ is maximal by Lemma 3.4.1. Conversely, assume $(K,\dot{\mathcal{O}})$ and $\dot{K}$ are maximal, and let L be an immediate valued field extension of K. Then with the notations above, $\dot{L}$ is an immediate extension of $\dot{K}$, so $\dot{L}=\dot{K}$. It follows that $(L,\dot{\mathcal{O}}_L)$ is an immediate extension of $(K,\dot{\mathcal{O}})$, and so $K=L$. □

Step-completeness

Let (a_ρ) be a pc-sequence in K, and set

$$s(\rho) \;:=\; \text{immediate successor of } \rho.$$

Then $\gamma_\rho := v(a_{s(\rho)}-a_\rho)\in\Gamma$ for all sufficiently large ρ, and (γ_ρ) is eventually strictly increasing. Also, if $a_\rho \rightsquigarrow a\in L$, where L is a valued field extension of K, then $\gamma_\rho = v(a-a_\rho)$ for all sufficiently large ρ. Given a nontrivial convex subgroup Δ of Γ we say that (a_ρ) is **Δ-special** if (γ_ρ) is eventually cofinal in Δ, that is, $\gamma_\rho\in\Delta$ for all sufficiently large ρ, and for each $\delta\in\Delta$ we have $\gamma_\rho>\delta$ for all sufficiently large ρ. Thus (a_ρ) is Δ-special iff the width of (a_ρ) is $\{\gamma\in\Gamma_\infty:\gamma>\Delta\}$; in particular, if (a_ρ) is Δ-special, then so is every pc-sequence in K equivalent to (a_ρ). We call (a_ρ) **special** if it is Δ-special for some nontrivial convex subgroup Δ of Γ. We say that K is **step-complete** if each special pc-sequence in K has a pseudolimit in K. By Lemma 2.2.35, this is the same as requiring that for each convex subgroup $\Delta\neq\{0\}$ of Γ the residue field $\dot{K}$ of the coarsened valuation $\dot{v}=v_\Delta$ is complete with respect to the valuation v defined in (3.4.1). Therefore:

COROLLARY 3.4.7. *Let Δ be a convex subgroup of Γ, with ordered quotient group $\dot{\Gamma}=\Gamma/\Delta$, corresponding coarsening $\dot{v}=v_\Delta\colon K^\times\to\dot{\Gamma}$ of v, and residue field $\dot{K}$ of $\dot{v}$ with valuation $v\colon \dot{K}^\times\to\Delta$ as above. Then*

$$K \text{ is step-complete} \iff (K,\dot{\mathcal{O}}) \text{ and } \dot{K} \text{ are step-complete.}$$

We have the following implications for our valued field K:

$$\text{spherically complete} \implies \text{step-complete} \implies \text{complete.}$$

If K has rank 1, then the second arrow may be reversed. If K is discrete, then both arrows may be reversed. Thus by Corollaries 3.4.6 and 3.4.7 and induction on n:

LEMMA 3.4.8. *If K is step-complete and Γ is isomorphic to the lexicographically ordered group $\mathbb{Z}^n$, then K is spherically complete.*

We note a few more basic facts about step-complete valued fields:

COROLLARY 3.4.9. *If K is step-complete, then K is henselian.*

PROOF. Let $P\in\mathcal{O}[X]$ and $a\in\mathcal{O}$ be such that $P(a)\prec 1$, $P'(a)\asymp 1$ and P has no zero in $a+\smallo$. Then by the discussion preceding Corollary 3.3.5, there is a special divergent pc-sequence in K, so K is not step-complete. □

COROLLARY 3.4.10. *Suppose K is step-complete and L is a valued field extension of K with $[L:K]<\infty$. Then L is step-complete.*

PROOF. By Corollary 3.1.9, every convex subgroup of Γ_L is the convex hull in Γ_L of a convex subgroup of Γ. Now apply Lemma 3.4.3. □

Every valued field has a step-complete valued field extension; in the case of equicharacteristic zero there is such an extension with a semiuniversal property:

PROPOSITION 3.4.11. *Suppose K has equicharacteristic zero. Then there exists a step-complete valued field extension K^{sc} of K which embeds over K into any step-complete valued field extension of K.*

PROOF. Fix a maximal immediate extension M of K; for a valued subfield L of M we denote by L^{h} the henselization of L inside M. Starting with $K_0 = K^{\text{h}}$, we construct an increasing sequence (K_ρ) of henselian valued subfields of M, indexed by ordinals ρ, as follows. Let an ordinal $\mu > 0$ and an increasing sequence $(K_\rho)_{\rho<\mu}$ of henselian valued subfields of M be given. If μ is a limit ordinal, we set $K_\mu := \bigcup_{\rho<\mu} K_\rho$. Suppose μ is a successor ordinal, say $\mu = \lambda + 1$. If K_λ is step-complete, then we are done and set $K^{\text{sc}} := K_\lambda$. So assume K_λ is not step-complete. Then we take a convex subgroup $\Delta \neq \{0\}$ of Γ and a divergent Δ-special pc-sequence (a_ρ) in K_λ; note that (a_ρ) is of transcendental type, by Corollary 3.3.21. Take $a \in M$ with $a_\rho \leadsto a$ and set $K_\mu := K_\lambda(a)^{\text{h}}$. This construction has to terminate eventually. By Lemmas 3.2.6 and 3.4.9 and the universal property of henselization, K^{sc} has the required property. □

Suppose K has equicharacteristic zero. Call a valued field extension K^{sc} of K as in Proposition 3.4.11 a **step-completion** of K. Each step-completion of K is an immediate extension of K, since every maximal immediate extension of K is step-complete. Does K have up to isomorphism of valued fields over K a unique step-completion? (The answer is "yes" if K has rank 1, and also if $\Gamma = \mathbb{Z}^n$, ordered lexicographically, by Corollary 3.3.41 and Lemma 3.4.8.)

Fluent and jammed pc-sequences

Let (a_ρ) be a pc-sequence in K. Set

$$\gamma_\rho := v(a_{s(\rho)} - a_\rho) \quad \text{where } s(\rho) := \text{immediate successor of } \rho.$$

So $\gamma_\rho \in \Gamma$ for all sufficiently large ρ, and (γ_ρ) is eventually strictly increasing.

Let Δ be a convex subgroup of Γ. We say that (a_ρ) is **Δ-fluent** if (a_ρ) remains a pc-sequence for the coarsened valuation v_Δ; equivalently, for some index ρ_0,

$$\gamma_{\rho'} - \gamma_\rho > \Delta \quad \text{for all } \rho' > \rho > \rho_0.$$

The following is almost immediate but good to keep in mind.

LEMMA 3.4.12. *Suppose (a_ρ) is Δ-fluent and $a \in K$. Then*

$$a_\rho \leadsto a \text{ in } K \iff a_\rho \leadsto a \text{ in the } \Delta\text{-coarsening of } K.$$

We say that (a_ρ) is **Δ-jammed** if for some index ρ_0,

$$\gamma_{\rho'} - \gamma_\rho \in \Delta \quad \text{for all } \rho' > \rho > \rho_0.$$

It is easily checked that if (a_ρ) is not Δ-jammed, then it has a cofinal Δ-fluent subsequence. Let $W \subseteq \Gamma_\infty$ be the width of (a_ρ). Then (a_ρ) is Δ-jammed iff for some $\gamma \in \Gamma$ with $\gamma < W$ we have $\gamma' - \gamma \in \Delta$ for all γ' with $\gamma < \gamma' < W$. Thus being Δ-jammed depends only on the width. Therefore, if (a_ρ) is Δ-jammed, then so is every equivalent pc-sequence in K.

We say that (a_ρ) is **fluent** if it is Δ-fluent for some nontrivial convex subgroup Δ of Γ, and we say that (a_ρ) is **jammed** if it is Δ-jammed for every nontrivial convex subgroup Δ of Γ. If (a_ρ) is not jammed, then it has a fluent cofinal subsequence.

Let us say that K is **fluent** if every fluent pc-sequence in K has a pseudolimit in K; equivalently, every divergent pc-sequence in K is jammed; equivalently, every coarsening of K by a nontrivial convex subgroup of Γ is maximal as a valued field. Every maximal valued field is fluent. If K is fluent, then so is the residue field (with its induced valuation) of any coarsening of K. The notions introduced above are only relevant when $[\Gamma^{\neq}]$ has no smallest element.

LEMMA 3.4.13. *Suppose $[\Gamma^{\neq}]$ has no smallest element. Then each special pc-sequence in K has a cofinal fluent subsequence.*

PROOF. Let (a_ρ) be a Δ'-special pc-sequence in K where Δ' is a nontrivial convex subgroup of Γ. Take a nontrivial proper convex subgroup Δ of Δ'. It is enough to observe that (a_ρ) is not Δ-jammed, and so has a cofinal Δ-fluent subsequence. □

COROLLARY 3.4.14. *Suppose $[\Gamma^{\neq}]$ has no smallest element and K is fluent. Then K is step-complete, and thus henselian.*

PROOF. By Lemma 3.4.13 every special pc-sequence in K has a pseudolimit in K. It remains to appeal to Lemma 3.4.9. □

COROLLARY 3.4.15. *Suppose K is fluent and L is a valued field extension of K with $[L : K] < \infty$. Then L is fluent.*

PROOF. Let Δ' be a nontrivial convex subgroup of Γ_L; we show that the Δ'-coarsening of L is maximal. It follows from Corollary 3.1.9 that $\Delta := \Delta' \cap \Gamma$ is a nontrivial convex subgroup of Γ and $\Delta' = \Delta_L$. As the Δ-coarsening of K is maximal, it remains to appeal to Corollary 3.2.28. □

Δ-immediate extensions

In this subsection *extension* means *valued field extension*. Let Δ be a convex subgroup of Γ. A **Δ-immediate extension of** K is an immediate extension L of K such that each $a \in L \setminus K$ is a pseudolimit of a divergent Δ-fluent pc-sequence in K; equivalently, it is an immediate extension L of K such that no $a \in L \setminus K$ is a pseudolimit of a divergent Δ-jammed pc-sequence in K. If L is a Δ-immediate extension of K and M

is a Δ-immediate extension of L, then M is a Δ-immediate extension of K: this uses the characterization of "Δ-immediate extension" in terms of Δ-jammed pc-sequences. We now consider still another useful characterization in terms of coarsening.

Let $(K, \dot{\mathcal{O}})$ be the field K with the coarsened valuation $\dot{v} = v_\Delta$ and its valuation ring $\dot{\mathcal{O}}$. Given an immediate extension L of K and using again Δ to coarsen, we obtain a valued field extension $(L, \dot{\mathcal{O}}_L)$ of $(K, \dot{\mathcal{O}})$.

LEMMA 3.4.16. *Let L be an immediate extension of K. Then L is a Δ-immediate extension of K iff the extension $(L, \dot{\mathcal{O}}_L)$ of $(K, \dot{\mathcal{O}})$ is immediate.*

PROOF. Assume $L|K$ is Δ-immediate. Let $a \in \dot{\mathcal{O}}_L$; it is enough to find $b \in \dot{\mathcal{O}}$ such that $v(a - b) > \Delta$. For this we can assume $a \notin K$. Take a divergent Δ-fluent pc-sequence (a_ρ) in K such that $a_\rho \rightsquigarrow a$ and $\gamma_\rho := v(a_{s(\rho)} - a_\rho) = v(a - a_\rho)$ for all ρ. Take ρ such that $a \sim a_\rho$ (so $\gamma_\rho \geqslant \delta$ for some $\delta \in \Delta$), and $\gamma_{s(\rho)} - \gamma_\rho > \Delta$. Then for $b := a_{s(\rho)}$ we have $v(a - b) = \gamma_{s(\rho)} > \Delta + \gamma_\rho$, so $v(a - b) > \Delta$.

For the converse, assume the extension $(L, \dot{\mathcal{O}}_L)$ of $(K, \dot{\mathcal{O}})$ is immediate. Let $a \in L \setminus K$, and take a divergent pc-sequence (a_ρ) in $(K, \dot{\mathcal{O}})$ such that $a_\rho \rightsquigarrow a$, in $(L, \dot{\mathcal{O}}_L)$. Then (a_ρ) is a Δ-fluent pc-sequence with $a_\rho \rightsquigarrow a$ in L. If $b \in K$ were such that $a_\rho \rightsquigarrow b$ in K, then also $a_\rho \rightsquigarrow b$ in $(K, \dot{\mathcal{O}})$, and so no such b exists. $\square$

COROLLARY 3.4.17. *Let L be a Δ-immediate extension of K. Then L is also a Δ-immediate extension of any valued field F with $K \subseteq F \subseteq L$.*

COROLLARY 3.4.18. *The following conditions on K are equivalent:*

(i) *K has no proper Δ-immediate extension;*

(ii) *every Δ-fluent pc-sequence in K pseudoconverges in K;*

(iii) *$(K, \dot{\mathcal{O}})$ is a maximal valued field.*

If K has any of these properties, then $(K, \dot{\mathcal{O}})$ is henselian.

PROOF. We have (ii) $\Longleftrightarrow$ (iii) by Corollary 3.2.9 applied to the Δ-coarsening of K in place of K, and (iii) $\Rightarrow$ (i) is clear. Let $(L, \dot{\mathcal{O}}_L)$ be an immediate extension of $(K, \dot{\mathcal{O}})$. Then by Lemma 3.4.5 the valuation of K extends to a valuation $v\colon L \to \Gamma_\infty$ making L an immediate extension of K and $\dot{\mathcal{O}}_L$ the valuation ring of the coarsening of this valuation by Δ. Equipped with this valuation, L is a Δ-immediate extension of K. Thus (i) $\Rightarrow$ $(K, \dot{\mathcal{O}}) = (L, \dot{\mathcal{O}}_L)$. This shows (i) $\Rightarrow$ (iii). If any of the conditions (i), (ii), (iii) holds, then $(K, \dot{\mathcal{O}})$ is henselian by Corollary 3.3.4. $\square$

By Zorn there exists a Δ-immediate extension of K that has no proper Δ-immediate extension. Such an extension of K is called a **maximal Δ-immediate extension** of K; under certain conditions it is unique up to isomorphism over K. To discuss this, let us fix some maximal Δ-immediate extension K^Δ of K. Then K^Δ is also a maximal Δ-immediate extension of any intermediate valued field L with $K \subseteq L \subseteq K^\Delta$, and K^Δ is henselian with respect to its Δ-coarsened valuation.

PROPOSITION 3.4.19. *Suppose K has equicharacteristic zero. Then every maximal Δ-immediate extension of K is isomorphic over K to the valued field K^Δ.*

PROOF. Let L be a maximal Δ-immediate extension of K. Then both the Δ-coarsening of L and the Δ-coarsening of K^Δ are maximal immediate extensions of the Δ-coarsening of K. Then Corollary 3.3.41 gives an isomorphism $L \to K^\Delta$ of these Δ-coarsenings that is the identity on K. Since for each $b \in L^\times$ we have $b = a(1+\varepsilon)$ with $a \in K^\times$ and $\varepsilon \in \dot{o}_L$, this isomorphism is also an isomorphism with respect to the valuations of L and K^Δ, respectively. □

The following lemma is obvious, and is used in the next subsection.

LEMMA 3.4.20. *If L is a Δ-immediate extension of K, then any maximal Δ-immediate extension of L is also a maximal Δ-immediate extension of K.*

Fluent completion

In this subsection we assume that K has equicharacteristic zero, $\Gamma \neq \{0\}$ and $[\Gamma^{\neq}]$ has no smallest element. By *extension* we mean *valued field extension*, and *embedding* means *valued field embedding*.

Let L be an extension of K. A pc-sequence (a_ρ) in L is said to be K-**fluent** if there is a nontrivial convex subgroup Δ of Γ such that (a_ρ) has length $\leqslant \operatorname{cf}(S)$ for some $S \subseteq \Gamma/\Delta$, and (a_ρ) is Δ_L-fluent. We say that L is K-fluent if L is henselian and every K-fluent pc-sequence in L pseudoconverges in L. Note that if L is fluent, then L is K-fluent. Also, if L is Γ-maximal, then L is K-fluent. The following is a variant of Proposition 3.4.19.

LEMMA 3.4.21. *Let Δ be a nontrivial convex subgroup of Γ and let K^Δ be as above. Then K^Δ embeds over K into any K-fluent extension of K.*

PROOF. Let L be an intermediate valued field: $K \subseteq L \subseteq K^\Delta$, assume $L \neq K^\Delta$, and let $i\colon L \to F$ be an embedding over K into a K-fluent extension F of K. It suffices to show that i can be extended to an embedding into F of some intermediate valued field that strictly contains L. Note that i remains an embedding when replacing L by its Δ-coarsening, and F by its Δ_F-coarsening. Since the Δ-coarsening of K^Δ is henselian and the Δ_F-coarsening of F is henselian, we can extend i and arrange that the Δ-coarsening of L is henselian. Take $a \in K^\Delta \setminus L$. Take a divergent Δ-fluent pc-sequence (a_ρ) in L such that $a_\rho \rightsquigarrow a$. By passing to a cofinal subsequence we arrange that (a_ρ) has length $\leqslant \operatorname{cf}(S)$ for some $S \subseteq \Gamma/\Delta$. Then (a_ρ) is of transcendental type over the Δ-coarsening of L. In view of Corollary 3.2.4 and Lemma 3.4.12, (a_ρ) is of transcendental type over L. Also $\big(i(a_\rho)\big)$ is a K-fluent pc-sequence in F, so $i(a_\rho) \rightsquigarrow b$ with $b \in F$, hence we can extend i to an embedding $L(a) \to F$ sending a to b. □

PROPOSITION 3.4.22. *K has an immediate fluent extension that embeds over K into any K-fluent extension of K.*

We call an extension of K as in this theorem a **fluent completion** of K.

PROOF. Fix a decreasing coinitial sequence (Δ_α) of nontrivial convex subgroups of Γ indexed by the ordinals $\alpha < \lambda$ for some infinite limit ordinal λ, where "coinitial" means that every nontrivial convex subgroup Δ of Γ includes Δ_α for some α (and thus for all sufficiently large α). For each α we pick a maximal Δ_α-immediate extension K^{Δ_α} of K. We arrange this so that K^{Δ_α} is a valued subfield of K^{Δ_β} whenever $\alpha < \beta < \lambda$: by transfinite recursion and using Lemma 3.4.20, take $K^{\Delta_{\alpha+1}}$ to be a maximal $\Delta_{\alpha+1}$-immediate extension of K^{Δ_α} for $\alpha < \lambda$, and if $\beta < \lambda$ is an infinite limit ordinal, take K^{Δ_β} to be a maximal Δ_β-immediate extension of the Δ_β-immediate extension $\bigcup_{\alpha<\beta} K^{\Delta_\alpha}$ of K. Put

$$K^{\mathrm{f}} := \bigcup_{\alpha<\lambda} K^{\Delta_\alpha}.$$

Then K^{f} is an immediate extension of K with the following properties:

(1) K^{f} is henselian;

(2) every fluent pc-sequence in K pseudoconverges in K^{f};

(3) every $a \in K^{\mathrm{f}} \setminus K$ is a pseudolimit of some divergent fluent pc-sequence in K;

(4) K^{f} embeds over K into any K-fluent extension of K.

To get (1), let $n \geqslant 2$, $a_2, \dots, a_n \in K^{\mathrm{f}}$, $a_2, \dots, a_n \prec 1$; our job is to show that then the polynomial $1 + X + a_2X^2 + \cdots + a_nX^n$ has a zero in the valuation ring of K^{f}. Take $\alpha < \lambda$ so large that $a_2, \dots, a_n \in K^{\Delta_\alpha}$ and $v(a_2), \dots, v(a_n) > \Delta_\alpha$. (This is possible since $[\Gamma^{\neq}]$ has no smallest element.) Since K^{Δ_α} is henselian with respect to its Δ_α-coarsened valuation, our polynomial does have a zero $b \in K^{\Delta_\alpha}$ with $v(1+b) > \Delta_\alpha$, and so $v(b) \geqslant 0$. This proves (1).

To get (4), let L be any K-fluent extension of K. Now use Lemma 3.4.21 and transfinite recursion to obtain for each $\alpha < \lambda$ an embedding $i_\alpha \colon K^{\Delta_\alpha} \to L$ such that i_β extends i_α whenever $\alpha < \beta < \lambda$.

We define a *semifluent completion* of K to be an immediate extension of K with the properties (1)–(4) of K^{f}. So we have shown that K has a semifluent completion. Let any ordinal $\nu > 0$ be given. We now build an increasing sequence (K_μ) of immediate extensions of K, indexed by the ordinals $\mu < \nu$, such that $K_0 = K$, $K_{\mu+1}$ is a semifluent completion of K_μ whenever $\mu < \mu+1 < \nu$, and $K_\mu = \bigcup_{\alpha<\mu} K_\alpha$ whenever $\mu < \nu$ is an infinite limit ordinal. With ν large enough, we obtain $\mu < \mu + 1 < \nu$ such that $K_\mu = K_{\mu+1}$, and it is easy to check that then K_μ is a fluent completion of K. $\square$

Are all fluent completions of K isomorphic over K, for every K? We do not know the answer, but we can proceed without. Here is an important property of fluent completions, used in Sections 11.6 and 11.7.

LEMMA 3.4.23. *Suppose (a_ρ) is a jammed pc-sequence in K with a pseudolimit in a fluent completion of K. Then (a_ρ) has a pseudolimit in K.*

PROOF. Every fluent completion of K embeds over K into the fluent completion K_μ from the proof of Proposition 3.4.22. By transfinite induction this gives a reduction to the case that (a_ρ) has a pseudolimit in a semifluent completion of K as defined in that proof. Since every pc-sequence in K equivalent to (a_ρ) is also jammed, it remains to use property (3) of semifluent completions and Corollary 2.2.20. □

Approximation by special pc-sequences

The main goal of this subsection is the result below about approximating elements in the henselization K^{h} of the valued field K. It generalizes the fact (immediate from Corollary 3.3.5) that for K of rank 1 every element in K^{h} is the limit of a c-sequence in K.

PROPOSITION 3.4.24 (F.-V. Kuhlmann). *For each $x \in K^{\mathrm{h}} \setminus K$ there is a divergent special pc-sequence (x_ρ) in K and some $a \in K^\times$ such that $x_\rho \rightsquigarrow x/a$.*

We give the proof after a number of auxiliary results. *In the rest of this subsection, Δ is a nontrivial convex subgroup of Γ.* We begin with a reformulation of the conclusion of this proposition:

LEMMA 3.4.25. *Let x be an element of a valued field extension of K with $x \notin K$, and let $a \in K^\times$. Then the following are equivalent:*

(i) *there is a divergent Δ-special pc-sequence (x_ρ) in K such that $x_\rho \rightsquigarrow x/a$;*

(ii) *the coset $\alpha + \Delta$, where $\alpha = va$, is a cofinal subset of $v(x - K)$.*

PROOF. Let (x_ρ) be a divergent Δ-special pc-sequence in K such that $x_\rho \rightsquigarrow x/a$. We may assume that $v(x/a - x_\rho) \in \Delta$ for all ρ, so $v(x - ax_\rho) \in \alpha + \Delta$ for each ρ, and $\big(v(x - ax_\rho)\big)$ is cofinal in $\alpha + \Delta$. Thus $\alpha + \Delta \subseteq v(x - K)$ since $v(x - K)$ is downward closed. Since the pc-sequence (ax_ρ) in K diverges and $ax_\rho \rightsquigarrow x$, the sequence $\big(v(x - ax_\rho)\big)$ is cofinal in $v(x - K)$. Hence $\alpha + \Delta$ is cofinal in $v(x - K)$. Conversely, suppose $\alpha + \Delta \subseteq v(x - K)$ and $\alpha + \Delta$ is cofinal in $v(x - K)$. Since $\Delta \neq \{0\}$, the set $\alpha + \Delta$, and hence also the set $v(x - K)$, does not have a largest element. Choose a well-indexed sequence (y_ρ) in K such that $v(x - y_\rho) \in \alpha + \Delta$ for each ρ and $\big(v(x - y_\rho)\big)$ is strictly increasing and cofinal in $\alpha + \Delta$, and hence in $v(x-K)$. Then (x_ρ) where $x_\rho := y_\rho/a$ for each ρ is a divergent Δ-special pc-sequence in K such that $x_\rho \rightsquigarrow x/a$. □

Let x be an element of a valued field extension of K. We say that x is **almost Δ-special over** K if $x \notin K$ and for some $a \in K^\times$ the equivalent conditions (i) and (ii) in the previous lemma hold. We say that x is **Δ-special over** K if $x \notin K$ and conditions (i) and (ii) hold for $a = 1$. So if x is almost Δ-special over K, then some $K^\times$-multiple of x is Δ-special over K. Moreover, for $a, b \in K$, $a \neq 0$,

$$v(ax + b - K) = va + v(x - K),$$

hence $ax + b$ is almost Δ-special over K iff x is. We call x **almost special over** K if x is almost Δ-special over K for some Δ, and similarly we say that x is **special over** K if x is Δ-special over K for some Δ.

In the next lemmas we consider valued field extensions L of K. For such L we let Δ_L be the convex hull of Δ in Γ_L, and $\dot{K}$ is the valued residue field of the coarsening v_Δ of the valuation v in K by Δ, viewed as a valued subfield of the valued residue field $\dot{L}$ of the coarsening v_{Δ_L} of the valuation on L by Δ_L as usual.

LEMMA 3.4.26. *Let $x \in L$ and $vx \in \Delta_L$. Then x is Δ-special over K iff $\dot{x} \notin \dot{K}$ and $\dot{x}$ is the limit in $\dot{L}$ of a c-sequence in $\dot{K}$.*

PROOF. Let (x_ρ) be a divergent Δ-special pc-sequence in K with $x_\rho \rightsquigarrow x$. Then $vx = v(x_\rho)$ for sufficiently large ρ, and after passing to a cofinal subsequence, we may assume that this holds for all ρ. Then $(\dot{x}_\rho)$ is a c-sequence in $\dot{K}$ and $\dot{x}_\rho \to \dot{x}$. Also, $\dot{x} \notin \dot{K}$: otherwise, $\dot{x} = \dot{a}$, $a \in \dot{\mathcal{O}}$, and then $x_\rho \rightsquigarrow a$ by Lemma 2.2.9, a contradiction. Conversely, suppose (x_ρ) is a well-indexed sequence in $\dot{\mathcal{O}}$ such that $(\dot{x}_\rho)$ is a c-sequence in $\dot{K}$ and $\dot{x}_\rho \to \dot{x} \notin \dot{K}$. After passing to a cofinal subsequence, we may assume that $(\dot{x}_\rho)$ is a pc-sequence of width $\{\infty\}$ in $\dot{K}$, by Lemma 2.2.35; thus (x_ρ) is a Δ-special pc-sequence in K with $x_\rho \rightsquigarrow x$. If $a \in K$ is a pseudolimit of (x_ρ), then $va = vx_\rho$ eventually, in particular $va \in \Delta$, and $\dot{x}_\rho \to \dot{a} = \dot{x}$, a contradiction. Hence (x_ρ) does not have a pseudolimit in K, and so x is Δ-special over K. □

Note that if $x \in L$ is Δ-special over K with $vx \in \Delta_L$, then by the previous lemma the valued field extension $\dot{K}(\dot{x}) \supseteq \dot{K}$ is dense, and hence every $y \in L$ with $vy \in \Delta_L$ and $\dot{y} \in \dot{K}(\dot{x}) \setminus \dot{K}$ is Δ-special over K.

The following lemma indicates a source of special elements over K.

LEMMA 3.4.27. *Let $P \in \mathcal{O}[X]$ and $a \in \mathcal{O}$ be such that $P(a) \prec 1$, $P'(a) \asymp 1$, and P has no zero in $a + \smallo$. Let x in a valued field extension L of K be a zero of P with $x - a \prec 1$. Then $vx \in \Delta$ and x is Δ-special, for some Δ.*

PROOF. By the discussion preceding Corollary 3.3.5, we have Δ and a Δ-special divergent pc-sequence (a_ρ) in K such that $P(a_\rho) \rightsquigarrow 0$ and $a_\rho \equiv a \bmod \smallo$ for all ρ. Taylor expansion around $x \in L$ yields

$$\begin{aligned} P(x+Y) &= P(x) + P'(x)\cdot Y + \text{terms of higher degree in } Y \\ &= P'(x)\cdot Y \cdot (1+Q) \qquad \text{where } Q \in \mathcal{O}_L[Y],\ Q(0)=0. \end{aligned}$$

Substituting $a_\rho - x$ for Y yields

$$P(a_\rho) = P'(x)\cdot(a_\rho - x)\cdot\big(1 + Q(a_\rho - x)\big) \qquad \text{with } Q(a_\rho - x) \prec 1.$$

Thus $v\big(P(a_\rho)\big) = v(a_\rho - x)$ for all ρ, hence $a_\rho \rightsquigarrow x$, and so x is Δ-special over K. Also, $vx = va_\rho < v(x - a_\rho) \in \Delta$, eventually, so $vx \in \Delta$. □

We say that a valued field extension L of K is **almost Δ-special** if every element of $L \setminus K$ is almost Δ-special over K. A valued field extension L of K is **almost special** if every element of $L \setminus K$ is almost special over K. (So Proposition 3.4.24 says that the henselization of K is an almost special extension of K.) Thus every almost special valued field extension is immediate. The next lemma shows that if $M \supseteq L$ and $L \supseteq K$ are almost special valued field extensions, then so is $M \supseteq K$:

LEMMA 3.4.28. *Let L be an almost special valued field extension of K and x an element of a valued field extension of L such that x is almost special over L. Then x is almost special over K.*

PROOF. Clearly $v(x-K) \subseteq v(x-L)$, and if equality holds, then x is almost special over K. So suppose $v(x-K) \neq v(x-L)$. We can take $y \in L$ with $v(x-K) < v(x-y)$; then $y \notin K$, so y is almost special over K. For each $a \in K$ we have $v(y-a) = v\big((y-x)+(x-a)\big) = v(x-a)$, hence $v(y-K) = v(x-K)$. Thus x is almost special over K. □

We now define a condition on an element x of a valued field extension of K which ensures that $K(x) \supseteq K$ is almost special: call an element x in a valued field extension L of K **very Δ-special over** K if

(VS1) $vx \in \Delta_L$ and x is Δ-special over K;

(VS2) for all $n \geqslant 1$, if $1, x, \dots, x^n$ are K-linearly independent, then $1, \dot{x}, \dots, \dot{x}^n$ in $\dot{L}$ are $\dot{K}$-linearly independent.

Here $\dot{K}$ is the valued residue field of the coarsening v_Δ of the valuation v in K by Δ, viewed as a valued subfield of the valued residue field $\dot{L}$ of the coarsening v_{Δ_L} of the valuation on $L = K(x)$ by the convex hull Δ_L of Δ in Γ_L. Note that given (VS1), condition (VS2) expresses that $[K(x):K] = [\dot{K}(\dot{x}):\dot{K}] \in \mathbb{N} \cup \{\infty\}$. Thus a very Δ-special x over K is transcendental over K iff $\dot{x}$ is transcendental over $\dot{K}$.

LEMMA 3.4.29. *Let x be very Δ-special over K. Then $K(x) \supseteq K$ is almost Δ-special.*

PROOF. Let $y \in K(x)\setminus K$; we need to show that y is almost Δ-special over K. We first assume that x is algebraic over K. Then $y = P(x)$ with $P \in K[X]$ of degree m with $1 \leqslant m < [K(x):K]$. Replacing y by $ay+b$ and P by $aP+b$, for suitable $a, b \in K$, $a \neq 0$, we may assume that $vP = 0$ and $P(0) = 0$. Thus $P \in \dot{\mathcal{O}}[X]$ with $\deg \dot{P} \geqslant 1$. By (VS2), $1, \dot{x}, \dots, \dot{x}^m$ are $\dot{K}$-linearly independent, hence $\dot{y} = \dot{P}(\dot{x}) \notin \dot{K}$. Thus by (VS1) and the remark following Lemma 3.4.26, $y = P(x)$ is Δ-special over K.

Now suppose that x is transcendental over K; then $\dot{x}$ is transcendental over $\dot{K}$ by (VS2). So $y = P(x)/Q(x)$ with $P, Q \in K[X]$, $Q \neq 0$. After multiplying P, Q and y by suitable elements of $K^\times$ we may assume that $vP = vQ = 0$. Now $P = \sum_i P_i X^i$, $Q = \sum_j Q_j X^j$ with $P_i, Q_j \in \mathcal{O}$ for all i, j; set $n = \deg \dot{Q}$. Then $Q_n \neq 0$, so with $b := P_n/Q_n$ we have $P_n - bQ_n = 0$. Take $a \in K^\times$ such that $v(P-bQ) = va$. Then with $R := a^{-1}(P-bQ) \in \mathcal{O}[X]$, we have $v(R) = 0$ and $R_n = 0$, so $\dot{R}(\dot{x})/\dot{Q}(\dot{x}) \notin \dot{K}$. By (VS1) and the remark after Lemma 3.4.26, $R(x)/Q(x)$, and hence also $y = b + aR(x)/Q(x)$, are Δ-special over K. □

An element of a valued field extension of K is said to **very special over** K if it is very Δ-special over K for some Δ. Together with Lemma 3.4.28, the previous lemma implies:

Corollary 3.4.30. *Let L be an almost special valued field extension of K and x an element of a valued field extension of L such that x is very special over L. Then $L(x) \supseteq K$ is almost special.*

Here is another consequence of Lemma 3.4.29:

Corollary 3.4.31. *Suppose K is henselian of equicharacteristic zero. Let (a_ρ) be a divergent Δ-special pc-sequence in K, and let x be a pseudolimit of (a_ρ) in a valued field extension of K. Then $x - a$ is very Δ-special over K, for some $a \in K$; in particular, $K(x) \supseteq K$ is almost Δ-special.*

Proof. Take ρ_0 such that $v(x - a_\rho) = v(a_\rho - a_\sigma) \in \Delta$ for $\rho_0 \leqslant \rho < \sigma$, and set $a := a_{\rho_0}$; then $(b_\rho)_{\rho \geqslant \rho_0} := (a_\rho - a)_{\rho \geqslant \rho_0}$ is a divergent Δ-special pc-sequence in K with pseudolimit $y := x - a$, and $vy \in \Delta$ and $v(b_\rho) \in \Delta$ for $\rho_0 \leqslant \rho$. The sequence $(\dot{b}_\rho)$ in $\dot{K}$ is a divergent pc-sequence in $\dot{K}$ (of width $\{\infty\}$) with pseudolimit $\dot{y}$. Since K is henselian of equicharacteristic zero, so is $\dot{K}$, by Lemma 3.4.2. Thus (b_ρ) is of transcendental type over K and $(\dot{b}_\rho)$ is of transcendental type over $\dot{K}$, by Lemma 3.2.7 and Corollary 3.3.21. Hence y is transcendental over K and $\dot{y}$ is transcendental over $\dot{K}$ by Lemma 3.2.6. Thus y is very Δ-special over K, and so $K(x) = K(y) \supseteq K$ is almost Δ-special, by Lemma 3.4.29. □

In analogy to Lemma 3.4.27, we have:

Lemma 3.4.32. *Let $P \in \mathcal{O}[X]$ be monic and let $a \in \mathcal{O}$ be such that $P(a) \prec 1$, $P'(a) \asymp 1$, and P has no zero in $a + \smallo$. Suppose that for each Δ, either $\dot{P} \in \mathcal{O}_{\dot{K}}[X]$ is irreducible over $\dot{K}$ or has a zero in $\dot{a} + \smallo_{\dot{K}}$. Then each zero x of P in any valued field extension of K with $x - a \prec 1$ is very special over K.*

Proof. Let x be an element in a valued field extension of K with $P(x) = 0$ and $x - a \prec 1$. By Lemma 3.4.27 we can take Δ such that x is Δ-special over K and $vx \in \Delta$. Then $\dot{P}(\dot{a}) \prec 1$, $\dot{P}'(\dot{a}) \asymp 1$, as well as $\dot{P}(\dot{x}) = 0$, $\dot{x} - \dot{a} \prec 1$, and $\dot{x} \notin \dot{K}$ (by Lemma 3.4.26). Thus, by Lemma 3.3.2, $\dot{P}$ has no zero in $\dot{a} + \smallo_{\dot{K}}$, so by hypothesis, $\dot{P}$ is irreducible. It follows that $\dot{P}$ is the minimum polynomial of $\dot{x}$ over $\dot{K}$. With $L := K(x)$, we now have

$$[L : K] \leqslant \deg P = \deg \dot{P} = \big[\dot{K}(\dot{x}) : \dot{K}\big] \leqslant [\dot{L} : \dot{K}] \leqslant [L : K]$$

and thus $[L : K] = \big[\dot{K}(\dot{x}) : \dot{K}\big]$. Hence x is very Δ-special over K. □

We can now give the proof of Proposition 3.4.24:

Proof of Proposition 3.4.24. Let L be a valued subfield of K^{h} containing K such that $L \supseteq K$ is almost special. If $L = K^{\mathrm{h}}$, then we are done, so suppose otherwise; then L is not henselian, by the minimality of henselizations. Take a monic polynomial $P \in \mathcal{O}_L[X]$ of minimal degree such that for some $a \in \mathcal{O}_L$ we have $P(a) \prec 1$ and $P'(a) \asymp 1$, and P has no zero in $a + \smallo_L$. Fix such an a. We claim that for every Δ either $\dot{P} \in \dot{L}[X]$ is irreducible, or $\dot{P}$ has a zero in $\dot{a} + \smallo_{\dot{L}}$. To prove this claim, suppose towards a contradiction that Δ is such that $\dot{P} \in \dot{L}[X]$ is reducible and has no zero in $\dot{a} + \dot{o}_{\dot{L}}$. Then we have monic $Q, R \in \mathcal{O}_L[X]$ of degree $\geqslant 1$ such that $\dot{P} = \dot{Q}\dot{R}$. Then $\dot{Q}(\dot{a}) \prec 1$ or $\dot{R}(\dot{a}) \prec 1$, say $\dot{Q}(\dot{a}) \prec 1$. Then also $\dot{Q}'(\dot{a}) \asymp 1$ and $\dot{Q}$ has no zero in $\dot{a} + o_{\dot{L}}$. Hence $Q(a) \prec 1$, $Q'(a) \asymp 1$, and Q has no zero in $a + o_L$, and $\deg Q < \deg P$, contradicting the minimality of $\deg P$. This proves the claim. Take a zero $x \in K^{\mathrm{h}}$ of P with $x - a \prec 1$. By Lemma 3.4.32, x is very special over L, and so by Lemma 3.4.29, the valued field extension $L(x) \supseteq L$ is almost special. Hence $L(x) \supseteq K$ is almost special, by Lemma 3.4.28. It now remains to appeal to Zorn. □

We use Proposition 3.4.24 to refine Proposition 3.4.11:

COROLLARY 3.4.33. *Let K have equicharacteristic zero. Then any step-completion of K is almost special over K.*

PROOF. For a cardinal κ we let κ^+ be the next bigger cardinal. Fix a maximal immediate valued field extension M of K. In the proof of Proposition 3.4.11 we constructed a step-completion K^{sc} of K as the union of an increasing sequence $(K_\lambda)_{\lambda<\nu}$ of henselian valued subfields of M, indexed by all ordinals λ less than some ordinal $\nu < |K|^+$, as follows: $K_0 = K^{\mathrm{h}}$, and for $0 < \mu < \nu$, if μ is a limit ordinal, then $K_\mu = \bigcup_{\lambda<\mu} K_\lambda$, and if μ is a successor ordinal, $\mu = \lambda + 1$, then $K_\mu = K_\lambda(a)^{\mathrm{h}}$ where $a \in M$ is a pseudolimit of a special divergent pc-sequence in K_λ. (Here the superscript h refers to henselization in M.)

Now transfinite induction using Proposition 3.4.24 and Corollary 3.4.31 implies that $K_\mu \supseteq K$ is almost special for each $\mu < \nu$. So K^{sc} is almost special over K. Any step-completion of K embeds into K^{sc} over K, and is therefore almost special over K as well. □

Notes and comments

Lemma 3.4.2 is in Nagata [300]. Step-completeness was defined by Krull [229, p. 177] in an ideal-theoretic way; Ribenboim [334] has the connection to special pc-sequences, see also [336]. Lemma 3.4.7 is from [334], Lemma 3.4.8 is in Mac Lane [277], Lemma 3.4.9 in Krull [229], and Lemma 3.4.10 in Ribenboim [335]. The proofs of these facts given here follow [458]. According to [277] the question posed after Proposition 3.4.11 goes back to Krull. It is discussed, in the context of valued ordered fields, in [450]. Proposition 3.4.24 is proved in [234].

3.5 VALUED ORDERED FIELDS

The basic facts on ordered and real closed fields, due to Artin and Schreier, are summarized without proof in Theorems 3.5.4 and 3.5.9 below. Next, we focus on the relevant valuations on ordered fields. Indeed, the more robust features of a non-archimedean ordered field are better described in terms of a valuation than in terms of the ordering. Throughout this section K is (at least) a field.

Ordered fields

Recall that in Section 2.4 we defined and briefly discussed ordered fields. *In this subsection K is an ordered field, so $\mathbb{Q} \subseteq K$.*

Note that $K^>$ with the induced ordering is an ordered multiplicative group. Obviously, the ordered additive group of K is archimedean iff $\operatorname{conv}(\mathbb{Q}) = K$; in this case we also call the ordered field K **archimedean**. Lemma 2.4.3 has an analogue for ordered fields:

LEMMA 3.5.1 (Hölder). *If K is archimedean, then the unique embedding $K \to \mathbb{R}$ of ordered additive groups sending $1 \in K$ to $1 \in \mathbb{R}$ is a ring morphism.*

We note the following easy bound on zeros of polynomials over ordered fields:

LEMMA 3.5.2. *Let $P = X^d + a_{d-1}X^{d-1} + \cdots + a_0$ with all $a_i \in K$. Set $M := 1 + |a_{d-1}| + \cdots + |a_0|$. Then for all $x \in K$ with $|x| \geqslant M$ we have $P(x) = x^d(1+\varepsilon)$ with $\varepsilon \in K$, $|\varepsilon| < 1$, so $P(x) \neq 0$.*

COROLLARY 3.5.3. *Let $K(x)$ be a field extension of K with x transcendental over K. Then there is a unique ordering of $K(x)$ that makes $K(x)$ an ordered field extension of K with $x > K$.*

PROOF. For any such ordering on $K(x)$ and monic polynomial $P \in K[X]$ we have $P(x) > 0$ by Lemma 3.5.2, and so there can be at most one such ordering. This also shows how to define such an ordering; alternatively, the existence of such an ordering follows by a routine model-theoretic compactness argument. □

Real closed fields

Call a field K **orderable** if some ordering of K makes K an ordered field; note that then $\operatorname{char}(K) = 0$. No algebraically closed field is orderable.

Call K **euclidean** if $x^2 + y^2 \neq -1$ for all $x, y \in K$, and

$$K = \{x^2 : x \in K\} \cup \{-x^2 : x \in K\}.$$

If K is euclidean, then K is an ordered field for a unique ordering, namely

$$a \geqslant 0 \iff a = x^2 \text{ for some } x \in K.$$

THEOREM 3.5.4 (Artin & Schreier). *The following conditions on K are equivalent:*

(i) *K is orderable, and K has no orderable proper algebraic field extension;*

(ii) *K is euclidean, and every $P \in K[X]$ of odd degree has a zero in K;*

(iii) *K is not algebraically closed, and $K(\mathrm{i})$, where $\mathrm{i}^2 = -1$, is algebraically closed;*

(iv) *K is not algebraically closed and has an algebraically closed field extension $L \supseteq K$ with $[L : K] < \infty$.*

Call K **real closed** if it satisfies the (equivalent) conditions of Theorem 3.5.4.

COROLLARY 3.5.5. *Let K' be a subfield of the real closed field K. Then:*

$$K' \text{ is real closed} \iff K' \text{ is algebraically closed in } K.$$

PROOF. If K' is real closed, then K' is algebraically closed in every orderable field extension, in particular, in K. Conversely, suppose K' is algebraically closed in K. Then K' is euclidean and every $P \in K'[X]$ of odd degree has a zero in K', since K' inherits these properties from K. □

Here is an obvious consequence of Corollary 3.5.5:

COROLLARY 3.5.6. *Let $(K_i)_{i \in I}$ with $I \neq \emptyset$ be a family of real closed subfields of a real closed field K. Then $\bigcap_i K_i$ is a real closed subfield of K.*

The archetypical example of a real closed field is the field $\mathbb{R}$ of real numbers. By the previous corollary, the algebraic closure of $\mathbb{Q}$ in $\mathbb{R}$ (known as the field of real algebraic numbers) is also real closed. Below we always consider a real closed field as equipped with the unique ordering making it an ordered field. Here are some further properties of real closed fields.

PROPOSITION 3.5.7. *Suppose K is a real closed field and $P \in K[X]$. Then:*

(i) *P is monic and irreducible in $K[X]$ iff $P = X - a$ for some $a \in K$, or $P = (X - a)^2 + b^2$ for some $a, b \in K$ with $b \neq 0$;*

(ii) *the map $x \mapsto P(x)\colon K \to K$ has the intermediate value property.*

PROOF. The quadratic polynomials in (i) take only values in $K^{>}$, and are thus irreducible. Conversely, suppose P is monic, irreducible, and of degree > 1. Then P is the minimum polynomial over K of $a + bi$ for some $a, b \in K$ with $b \neq 0$, so

$$P = \big(X - (a + bi)\big) \cdot \big(X - (a - bi)\big) = (X - a)^2 + b^2.$$

This proves (i). As to (ii), we can reduce to the case that $P(a) < 0 < P(b)$ with $a < b$ in K; it is enough to show that then $P(x) = 0$ for some $x \in K$ with $a < x < b$. The existence of such an x follows easily from (i) by factoring P into linear factors, and quadratic factors taking only values > 0. □

COROLLARY 3.5.8. *Let K be a real closed field and $K(x)$ a field extension of K with x transcendental over K. Let A be a cut in K. Then there is a unique ordering of $K(x)$ that makes it an ordered field extension of K for which x realizes the cut A.*

PROOF. Let $P \in K[X]$ be monic. Then $P(X) = Q(X)\prod_{i=1}^{n}(X - a_i)$ where $a_1, \dots, a_n \in K$ and $Q(X)$ is a product of monic irreducible quadratic polynomials in $K[X]$ as described in Proposition 3.5.7(i). Hence for any ordering of $K(x)$ with the indicated properties we have: $P(x) > 0$ iff the number of $i \in \{1, \dots, n\}$ such that $a_i \notin A$ is even; so there can be at most one such ordering. This also shows how to define such an ordering; alternatively, the existence of such an ordering follows by a routine model-theoretic compactness argument. □

Now let K be an ordered field. Then a **real closure** of K is a real closed algebraic field extension of K whose ordering extends the ordering of K. Here is the key result on this notion:

THEOREM 3.5.9 (Artin & Schreier). *K has a real closure. If K' is a real closure of K, then every ordered field embedding $K \to L$ into a real closed field L has a unique extension to an ordered field embedding $K' \to L$.*

Therefore, if K_1, K_2 are real closures of K, then there is a unique isomorphism $K_1 \xrightarrow{\cong} K_2$ over K. Thus we can speak of *the* real closure of K, denoted by K^{rc}.

Note that by Lemma 3.5.2 there is for each $a \in K^{\mathrm{rc}}$ an element $b \in K^{>0}$ such that $|a| \leqslant b$. It is tempting to jump to the conclusion that K is dense in K^{rc} (with respect to the order topology), and this jump is indeed a notorious source of error in the subject. A counterexample is the ordered field $K = \mathbb{R}(x)$ with $x > \mathbb{R}$: in its real closure the interval $(\sqrt{x} - 1, \sqrt{x} + 1)$ contains no element of K.

Suppose K is given as an ordered subfield of the real closed (ordered) field F. Then the algebraic closure $\{a \in F : a \text{ is algebraic over } K\}$ of K in F is a real closure of K, by Corollary 3.5.5, and is called the **real closure of K in F**; it is clearly the only field extension of K inside F that is a real closure of K.

Convex valuations

In this subsection K is an ordered field.

LEMMA 3.5.10. *Let A be a subring of K.*

(i) *The convex hull of A in K is a subring of K;*

(ii) *A is convex in K iff $[0,1] \subseteq A$; and*

(iii) *if A is convex, then A is a valuation ring of K.*

PROOF. Part (i) is clear. In (ii), $[0,1] \subseteq A$ is clearly necessary for convexity of A; conversely, if $[0,1] \subseteq A$ and $a \in A$, $x \in K$ with $0 < x < a$, then $xa^{-1} \in A$, hence $x = xa^{-1} \cdot a \in A$. For (iii), suppose A is convex. Let $x \in K^{\times}$; if $|x| \leqslant 1$, then $x \in A$, and if $|x| > 1$ then $|x^{-1}| < 1$, so $x^{-1} \in A$. Thus A is a valuation ring of K. □

LEMMA 3.5.11. *Let $\mathcal{O}$ be a valuation ring of K. The following are equivalent:*

(i) *$\mathcal{O}$ is convex;*

(ii) *$\smallo$ is convex;*

(iii) *$\smallo \subseteq (-1,1)$;*

(iv) *$|a| < 1/n$ for all $a \in \smallo$ and $n \geqslant 1$;*

(v) $\operatorname{conv}(\mathbb{Q}) \subseteq \mathcal{O}$.

PROOF. For (i) $\Rightarrow$ (ii), assume (i). Let $a, x \in K$, $0 < x < a$ and $a \in \smallo$. From $a^{-1} \notin \mathcal{O}$ and $0 < a^{-1} < x^{-1}$ we get $x^{-1} \notin \mathcal{O}$, so $x \in \smallo$. The implications (ii) $\Rightarrow$ (iii) and (iii) $\Rightarrow$ (iv) are obvious. For (iv) $\Rightarrow$ (v), assume (iv), let $a \in \operatorname{conv}(\mathbb{Q})$, $a > 0$, and take $n \geqslant 1$ with $a \leqslant n$. Then $a^{-1} \geqslant 1/n$, so $a^{-1} \notin \smallo$, and thus $a \in \mathcal{O}$. Finally, (v) $\Rightarrow$ (i) follows from part (ii) of the previous lemma. □

We say that a valuation v on K is **convex** if its valuation ring $\mathcal{O}$ satisfies the equivalent conditions in the previous lemma. Thus a valuation on the ordered field K is convex iff it is convex as a valuation on the ordered additive group of K as defined in Section 2.4. In terms of the dominance relation $\preccurlyeq$ associated to v:

$$v \text{ is convex} \iff \text{for all } x, y \in K \text{ with } |x| \leqslant |y| \text{ we have } x \preccurlyeq y.$$

Suppose v is a convex valuation on K. Then $\smallo$ is a convex subgroup of the ordered additive group $\mathcal{O}$, and the resulting ordering on the quotient group $\mathcal{O}/\smallo$ makes the residue field an ordered field. (Whenever a convex valuation on an ordered field is given we regard the residue field as an ordered field in this way.) If $\dot{v}$ is a coarsening of v by a convex subgroup of the value group of v, then $\dot{v}$ is also convex, and so is the valuation $\bar{v}$ on the ordered residue field $\dot{K}$ of $\dot{v}$. By Lemma 2.4.1, if v is nontrivial, then the v-topology on K coincides with the order topology on K.

Every ordered field carries a canonical convex valuation:

EXAMPLE. The standard valuation $v\colon K \to [K]$ with the reverse ordering on $[K]$ is given by $vx = [x]$. It is not just a valuation of the ordered additive group of K, but even a convex valuation $K^\times \to [K^\times]$ on the ordered field K, where $[K^\times]$ is made an ordered abelian (additive) group by $[x] + [y] = [x \cdot y]$ for $x, y \in K^\times$ and $0 = [1]$. The valuation ring of v is $\mathcal{O} = \operatorname{conv}(\mathbb{Q})$ of K.

The ordered field analogue of Lemma 2.4.7 follows from that lemma and its proof:

COROLLARY 3.5.12. *Let $\mathcal{O}$ be a convex subring of K, and $(L, \mathcal{O}_L)$ an immediate valued field extension of $(K, \mathcal{O})$. Then just one ordering on L makes L an ordered field extension of K such that $\mathcal{O}_L$ is convex. Moreover, $\mathcal{O}_L$ is the convex hull of $\mathcal{O}$ with respect to this ordering on L.*

Here is another useful extension result:

LEMMA 3.5.13. *Let $\mathcal{O}$ be a convex subring of K and $(K(y), \mathcal{O}_y)$ a valued field extension of $(K, \mathcal{O})$ such that $nvy \notin \Gamma$ for all $n \geqslant 1$. Then just one ordering on $K(y)$ makes it an ordered field extension of K such that $y > 0$ and $\mathcal{O}_y$ is convex.*

PROOF. By Corollary 3.1.9 and Lemma 3.1.30, y is transcendental over K, the value group of $(K(y), \mathcal{O}_y)$ is $\Gamma \oplus \mathbb{Z}vy$ (internal direct sum), and $\operatorname{res}(K) = \operatorname{res}(K(y))$. Let a field ordering on $K(y)$ be given with $y > 0$ that extends the ordering of K and with respect to which the valuation of $K(y)$ is convex. Let $f \in K(y)^\times$. Then $f = y^k gu$ with $k \in \mathbb{Z}$, $g \in K^>$, and $u \asymp 1$ in $K(y)$, and so $\operatorname{res}(u) \in \operatorname{res}(K)^\times$, hence

$$f > 0 \text{ in } K(y) \iff \operatorname{res}(u) > 0 \text{ in } \operatorname{res}(K).$$

This equivalence shows that there can only be one such ordering. It is routine to check that this equivalence also yields a definition of such an ordering. □

Of course, Lemma 3.5.13 goes through with $y < 0$ instead of $y > 0$.

LEMMA 3.5.14. *Every henselian valuation ring of K is convex.*

PROOF. Suppose $\mathcal{O}$ is a henselian valuation ring of K. Let $a \in \smallo$. Then for $x \in \mathcal{O}$ we have $x^2 + x + a \equiv x(x+1) \bmod \smallo$, so by Proposition 3.3.11 the polynomial $P := X^2 + X + a \in \mathcal{O}[X]$ has two distinct zeros in $\mathcal{O}$. Hence the discriminant $1 - 4a$ of P is positive, that is, $a < \frac{1}{4}$. This also holds for $-a$ in place of a, so $|a| < 1$. Thus $\mathcal{O}$ is convex, by (i) ⇔ (iii) in Lemma 3.5.11. □

LEMMA 3.5.15. *Let $\mathcal{O}$ be a convex subring of K. Then $\mathcal{O} + \mathcal{O}i$ is the unique valuation ring of $K[i]$ that lies over $\mathcal{O}$. The maximal ideal of $\mathcal{O} + \mathcal{O}i$ is $\smallo + \smallo i$.*

PROOF. Let $a, b \in K$ and $a + bi \notin \mathcal{O} + \mathcal{O}i$. Then $|a| > \mathcal{O}$ or $|b| > \mathcal{O}$, so

$$\frac{1}{a+bi} = \frac{a}{a^2+b^2} - \frac{b}{a^2+b^2}i \in \smallo + \smallo i \subseteq \mathcal{O} + \mathcal{O}i.$$

Thus $\mathcal{O} + \mathcal{O}i$ is a valuation ring of $K[i]$ lying over $\mathcal{O}$. Since i is integral over $\mathcal{O}$, any valuation ring of $K[i]$ lying over $\mathcal{O}$ includes $\mathcal{O} + \mathcal{O}i$ and thus equals $\mathcal{O} + \mathcal{O}i$. □

We have already seen that if a valued field is algebraically closed, then its value group is divisible and its residue field is also algebraically closed (Corollary 3.1.17). Here is an analogue for real closed valued fields:

THEOREM 3.5.16. *Suppose K is real closed, and K is equipped with a valuation ring $\mathcal{O}$ of K. Then the value group Γ of K is divisible, and $\operatorname{res}(K)$ is either real closed or algebraically closed. Moreover, the following are equivalent:*

(i) $\operatorname{res}(K)$ *is real closed;*

(ii) K *is henselian;*

(iii) $\mathcal{O}$ *is convex.*

PROOF. For every $a \in K^{>}$ and $n \geqslant 1$ the polynomial $X^n - a$ has a zero in K, so Γ is divisible. Equip the algebraic closure $K^{\mathrm{a}} = K(i)$ of K, $i^2 = -1$, with a valuation ring of K^{a} lying over $\mathcal{O}$. Then $\operatorname{res}(K^{\mathrm{a}})$ is an algebraic closure of $\operatorname{res}(K)$ and

$$\left[\operatorname{res}(K^{\mathrm{a}}) : \operatorname{res}(K)\right] \leqslant [K^{\mathrm{a}} : K] = 2.$$

Hence either $\operatorname{res}(K^{\mathrm{a}}) = \operatorname{res}(K)$, in which case $\operatorname{res}(K)$ is algebraically closed, or $\left[\operatorname{res}(K^{\mathrm{a}}) : \operatorname{res}(K)\right] = 2$, in which case $\operatorname{res}(K)$ is real closed (by Theorem 3.5.4).

If $\operatorname{res}(K)$ is real closed, then the valued field K is algebraically maximal, and hence henselian by Corollary 3.3.4. Thus (i) ⇒ (ii) holds, and (ii) ⇒ (iii) follows from Lemma 3.5.14. For (iii) ⇒ (i), note that if $\mathcal{O}$ is convex, then $\operatorname{res}(K)$ is orderable, so not algebraically closed, and hence real closed. □

Ordered fields with a convex valuation

In this subsection K is an ordered field with a convex valuation ring $\mathcal{O}$ of K, so K is an ordered and a valued field. This includes the case $\mathcal{O} = K$, where the corresponding valuation is trivial; we can then make it nontrivial by considering an ordered field extension $K(x)$ with $x > K$ as in Corollary 3.5.3, and taking the convex hull $\mathcal{O}_x$ of K in $K(x)$: then $(K(x), \mathcal{O}_x)$ is an ordered and valued field extension of $(K, \mathcal{O})$ for $\mathcal{O} = K$, with $\mathcal{O}_x \neq K(x)$.

COROLLARY 3.5.17. *Suppose K is real closed and C is a maximal subfield of $\mathcal{O}$. Then C is a lift of* $\operatorname{res}(K)$, *and* $\mathcal{O} = \operatorname{conv}(C)$.

PROOF. Proposition 3.3.8 and Theorem 3.5.16 imply that C is a lift of $\operatorname{res}(K)$. For $x \in \mathcal{O}$ we have $c \in C$ with $x - c \in \smallo$, so $c - 1 < x < c + 1$, thus $x \in \operatorname{conv}(C)$. □

COROLLARY 3.5.18. *There exists a unique convex valuation ring of the real closure K^{rc} of K lying over the valuation ring $\mathcal{O}$ of K. Equipping K^{rc} with this valuation ring we have $\Gamma_{K^{\text{rc}}} = \mathbb{Q}\Gamma$ and* $\operatorname{res}(K^{\text{rc}}) = \operatorname{res}(K)^{\text{rc}}$.

PROOF. Let $\mathcal{O}^{\text{rc}}$ be the convex hull of $\mathcal{O}$ in K^{rc}. Then $\mathcal{O}^{\text{rc}}$ is a convex valuation ring of K^{rc} lying over $\mathcal{O}$. By the remarks following the proof of Proposition 3.1.20, $\mathcal{O}^{\text{rc}}$ is the only convex valuation ring of K^{rc} lying over $\mathcal{O}$. Turn K^{rc} into a valued field with valuation ring $\mathcal{O}^{\text{rc}}$. Then $\Gamma_{K^{\text{rc}}}$ is divisible, so $\Gamma_{K^{\text{rc}}} = \mathbb{Q}\Gamma$. Since $\operatorname{res}(K^{\text{rc}})$ is an algebraic ordered field extension of $\operatorname{res}(K)$ and $\operatorname{res}(K^{\text{rc}})$ is real closed (by Theorem 3.5.16), we have $\operatorname{res}(K^{\text{rc}}) = \operatorname{res}(K)^{\text{rc}}$. □

COROLLARY 3.5.19. *For our valued field K we have:*

$$K \textit{ is real closed} \iff K \textit{ is henselian, } \operatorname{res}(K) \textit{ is real closed, and } \Gamma \textit{ is divisible.}$$

PROOF. The direction $\Rightarrow$ follows from Theorem 3.5.16. Conversely, suppose the conditions on the right side are satisfied. Corollary 3.5.18 gives a valuation ring of K^{rc} making $K^{\text{rc}} \supseteq K$ an immediate algebraic extension of valued fields. It now follows from Corollary 3.3.21 that $K = K^{\text{rc}}$ is real closed. □

EXAMPLE. Let C be an ordered field and $\mathfrak{M}$ a (multiplicative) ordered abelian group. Viewing $C[[\mathfrak{M}]]$ as a Hahn product, its Hahn ordering is given by

$$f > 0 \iff f_{\mathfrak{m}} > 0 \qquad (f \in C[[\mathfrak{M}]]^{\neq},\ \mathfrak{m} = \mathfrak{d}(f)).$$

Then $C[[\mathfrak{M}]]$ with its Hahn ordering is called an **ordered Hahn field**; it is indeed an ordered field containing C as an ordered subfield and with $\mathfrak{M}$ as an ordered subgroup of $C[[\mathfrak{M}]]^{>}$. The Hahn valuation on $C[[\mathfrak{M}]]$ has valuation ring $\operatorname{conv}(C)$ with respect to the Hahn ordering, and is thus a convex valuation. Therefore:

$$C[[\mathfrak{M}]] \text{ is real closed} \iff C \text{ is real closed and } \mathfrak{M} \text{ is divisible.}$$

Thus the Hahn field $\mathbb{R}((t^{\mathbb{Q}}))$ is real closed. The field $\mathrm{P}(C)$ of Puiseux series over C (Example 3.3.23) is an ordered subfield of the ordered Hahn field $C((t^{\mathbb{Q}}))$. If C is real closed, then $\mathrm{P}(C)$ is the real closure of its ordered subfield $C((t))$ in $C((t^{\mathbb{Q}}))$.

COROLLARY 3.5.20. *Suppose K is real closed. Then the completion K^{c} of the valued field K is real closed, and the valuation of K^{c} is convex.*

PROOF. Since $K \subseteq K^{\mathrm{c}}$ is an immediate extension, the residue field of K^{c} is real closed, by Theorem 3.5.16, hence K^{c} is not algebraically closed, by Corollary 3.1.17. Extend the valuation of K to the algebraic closure $K(i)$, $i^2 = -1$, of K. By Corollary 3.2.22, $K(i)^{\mathrm{c}}$ is algebraically closed, and by Corollary 3.2.29, $K(i)^{\mathrm{c}} = K^{\mathrm{c}}(i)$. Hence K^{c} is real closed, and by Theorem 3.5.16, its valuation is convex. □

Completion of ordered fields

In this subsection K is an ordered field. Call an ordered field extension $L \supseteq K$ **dense** if K is dense in L, in the order topology on L. We have an analogue of Theorem 3.2.13 for *ordered* fields:

THEOREM 3.5.21. *There is a dense ordered field extension $K^{\mathrm{d}} \supseteq K$ such that any dense ordered field extension $L \supseteq K$ embeds uniquely over K into K^{d}.*

PROOF. Let K^{d} be the completion of the ordered additive group of K; see Section 2.4. It is easy to check that Lemma 3.2.14 goes through for c-sequences in the ordered field K, and from this we obtain that the multiplication $(x, y) \mapsto x \cdot y$ and inversion $x \mapsto 1/x$ $(x \neq 0)$ on K have unique extensions to continuous maps

$$K^{\mathrm{d}} \times K^{\mathrm{d}} \to K^{\mathrm{d}}, \qquad (K^{\mathrm{d}})^{\neq} \to K^{\mathrm{d}},$$

with the product topology on $K^{\mathrm{d}} \times K^{\mathrm{d}}$, and that K^{d} is an ordered field extension of K with the first map as multiplication and the second map as inversion. It is now routine to check that this ordered field extension has the desired properties. □

The properties of the ordered field extension K^{d} of K postulated in Theorem 3.5.21 determine K^{d} up to a unique (ordered field) isomorphism over K. We call K^{d} the **completion of K.** Note that by construction, K^{d} is indeed complete: every c-sequence in K^{d} converges in K^{d}. When is K^{d} real closed?

PROPOSITION 3.5.22. *K^{d} is real closed if and only if K is dense in its real closure.*

Consider first the case that K is archimedean. Then K is isomorphic to a unique ordered subfield of $\mathbb{R}$, and identifying K with this subfield we can take $K^{\mathrm{d}} = \mathbb{R}$, and K^{rc} as a subfield of $\mathbb{R}$. Thus K^{d} is real closed and K is dense in K^{rc}. In general, if K^{d} is real closed, then clearly K is dense in K^{rc}. For the converse, we can assume K is not archimedean, and so K has a convex subring $\mathcal{O} \neq K$.

Accordingly, consider K below as equipped with a convex valuation ring $\mathcal{O} \neq K$ of K. Let K^{c} be the completion of the valued field K. Then $K^{\mathrm{c}} \supseteq K$ is an immediate valued field extension, so by Lemma 3.5.12 we can take a unique ordering on K^{c} making K^{c} an ordered field extension of K such that the valuation ring of K^{c} is convex with respect to this ordering. The valuation topology and the order topology on K^{c} coincide, by Lemma 2.4.1, so with this ordering K^{c} is also complete as an ordered field, and K is dense in K^{c} in the order topology. Thus:

LEMMA 3.5.23. *There is a unique isomorphism $K^{\mathrm{c}} \to K^{\mathrm{d}}$ of ordered fields which is the identity on K.*

In view of Corollary 3.5.20, this yields:

COROLLARY 3.5.24. *If K is real closed, then so is K^{d}.*

To conclude the proof of Proposition 3.5.22 it remains to use this corollary and to note that if K is dense in an ordered field extension L, then we can take $K^{\mathrm{d}} = L^{\mathrm{d}}$.

EXAMPLE. Let C be an ordered field. By Example 3.2.19 the completion in the Hahn field $C((t^{\mathbb{Q}})) = C[[x^{\mathbb{Q}}]]$ $(t = x^{-1})$ of its subfield $C(t^{\mathbb{Q}})$ is

$$L := \{f \in C[[x^{\mathbb{Q}}]] : \operatorname{supp} f|_{x^q} \text{ is finite for all } q \in \mathbb{Q}\}.$$

Now consider L as an ordered subfield of the ordered Hahn field $C((t^{\mathbb{Q}}))$. Then L is an ordered field extension of the ordered subfield $C(t^{\mathbb{Q}})$ of $C((t^{\mathbb{Q}}))$, and is as such also a completion of this ordered field $C(t^{\mathbb{Q}})$. If C is real closed, then so is L, by Corollaries 3.3.5 and 3.5.19.

Notes and comments

The notion of a real closed field and Theorems 3.5.4 and 3.5.9 are due to Artin and Schreier [12, 13]. For proofs of these theorems, see for example [216] or [322]; for the equivalence of (iii) and (iv) in Theorem 3.5.4, see also Leicht [254]. Lemma 3.5.1 is from [188]. Theorem 3.5.16 is in Knebusch and Wright [217], and Corollary 3.5.19 in Prestel [322]. Corollary 3.5.24 has been noticed by various authors; see [33, 170, 394].

3.6 SOME MODEL THEORY OF VALUED FIELDS

In this section we assume familiarity with Appendix B. We establish here quantifier elimination for *algebraically closed fields with a nontrivial valuation* and *real closed fields with a nontrivial convex valuation.* These well-known results foreshadow the deeper elimination theorem about the valued differential field $\mathbb{T}$ in Chapter 16. We also need Proposition 3.6.13 below in Section 16.6.

Algebraically closed valued fields

We augment the language $\{0, 1, -, +, \cdot\}$ of rings by a binary relation symbol $\preccurlyeq$ to obtain the language $\mathcal{L}_{\preccurlyeq}$. Let ACVF be the $\mathcal{L}_{\preccurlyeq}$-theory whose models are the structures $(K, \preccurlyeq)$ where K is an algebraically closed field and $\preccurlyeq$ is a nontrivial dominance relation on K. Here a dominance relation on a field K is said to be **trivial** if its corresponding valuation ring is K.

THEOREM 3.6.1. ACVF *has* QE.

First some remarks about dominance relations. Let R be an integral domain. We define a **dominance relation** on R to be a binary relation $\preccurlyeq$ on R such that conditions (D1)–(D6) in Section 3.1 hold for all $f, g, h \in R$. The **trivial** dominance relation $\preccurlyeq_{\mathrm{t}}$ on R

is the one with $r \preccurlyeq_{\rm t} s$ for all $r, s \in R$ with $s \neq 0$. For any dominance relation $\preccurlyeq$ on R there is a unique dominance relation $\preccurlyeq_F$ on $F = \mathrm{Frac}(R)$ such that $(R, \preccurlyeq) \subseteq (F, \preccurlyeq_F)$; it is given by

$$\frac{r_1}{s} \preccurlyeq_F \frac{r_2}{s} \iff r_1 \preccurlyeq r_2 \qquad (r_1, r_2, s \in R,\ s \neq 0).$$

EXAMPLE 3.6.2. The nontrivial dominance relations on $\mathbb{Z}$ are exactly the dominance relations $\preccurlyeq_p$, where p is a prime number:

$$a \preccurlyeq_p b \iff \text{for all } n\text{: } b \in p^n\mathbb{Z} \Rightarrow a \in p^n\mathbb{Z}.$$

Every substructure of a model of ACVF is a pair $(R, \preccurlyeq)$ with R an integral domain and $\preccurlyeq$ a dominance relation on R. Conversely, for any integral domain R with a dominance relation $\preccurlyeq$ on it, $(R, \preccurlyeq)$ is a substructure of a model of ACVF: first extend $(R, \preccurlyeq)$ to $(F, \preccurlyeq_F)$ as above; then extend $\preccurlyeq_F$ to a dominance relation on the algebraic closure of F; in case a valuation (in the form of a dominance relation) is trivial, adjoin a transcendental to make it nontrivial.

Let $(R, \preccurlyeq)$ be an integral domain with a dominance relation on it. Let

$$i : (R, \preccurlyeq) \to (K, \preccurlyeq_K)$$

be an embedding into an algebraically closed field K with dominance relation $\preccurlyeq_K$ on K. Let $\preccurlyeq^{\rm a}$ be a dominance relation on the algebraic closure $F^{\rm a}$ of $F = \mathrm{Frac}(R)$ such that $(R, \preccurlyeq) \subseteq (F^{\rm a}, \preccurlyeq^{\rm a})$. Then by Corollary 3.1.23 we can extend i to an embedding $(F^{\rm a}, \preccurlyeq_{F^{\rm a}}) \to (K, \preccurlyeq_K)$.

PROOF OF THEOREM 3.6.1. By the remarks preceding this proof and B.11.11 it suffices to show the following: Let E, F be nontrivially valued algebraically closed fields such that F is $|K|^+$-saturated, where K is a proper algebraically closed subfield of E; view subfields of E as valued subfields of E, and let $i\colon K \to F$ be a valued field embedding. Then there exists $x \in E \setminus K$ and an extension of i to a valued field embedding $j\colon K(x) \to F$.

To find such x and j we distinguish three cases. To simplify notation we identify the valued field K with the valued field iK via i.

CASE 1: $\mathrm{res}(K) \neq \mathrm{res}(E)$. Take $x \in \mathcal{O}_E$ such that $\overline{x} \notin \mathrm{res}(K)$. Since $\mathrm{res}(K)$ is algebraically closed, $\overline{x}$ is transcendental over $\mathrm{res}(K)$. Also $x \notin K$, so x is transcendental over K. By the saturation assumption on F we can find $y \in \mathcal{O}_F$ with $\overline{y} \notin \mathrm{res}(K)$. So $\overline{y}$ is transcendental over $\mathrm{res}(K)$ and y is transcendental over K. Then the field embedding $j\colon K(x) \to F$ over K with $j(x) = y$ is a valued field embedding by the uniqueness part of Lemma 3.1.31.

CASE 2: $\Gamma \neq \Gamma_E$. Note that Γ is divisible. Take any $\alpha \in \Gamma_E \setminus \Gamma$. Since F is $|K|^+$-saturated, so is Γ_F as an ordered set. Also, $\Gamma_F \neq \{0\}$. Hence we can take $\beta \in \Gamma_F$ realizing the same cut in Γ as α. So we have an isomorphism $\Gamma + \mathbb{Z}\alpha \to \Gamma + \mathbb{Z}\beta$ of ordered abelian groups over Γ which sends α to β. Take $x \in E^\times, y \in F^\times$ with $vx = \alpha$,

$vy = \beta$. Then $x \notin K$, so x is transcendental over K; likewise, y is transcendental over K. By the uniqueness part of Lemma 3.1.30, the field embedding $j\colon K(x) \to F$ over K with $j(x) = y$ is even a valued field embedding.

CASE 3: $\operatorname{res}(K) = \operatorname{res}(E)$ *and* $\Gamma = \Gamma_E$. Take any $x \in E \setminus K$. Then the valuation on $K(x)$ is uniquely determined by the valuation on K and by the map

$$a \mapsto v(x-a) : K \to \Gamma,$$

since each monic $f \in K[x]$ factors as $f = \prod_{i=1}^{n}(x-a_i)$ with all $a_i \in K$, so $vf = \sum_i v(x-a_i)$. It follows that for $y \in F \setminus K$ with $v(x-a) = v(y-a)$ for all $a \in K$, the field embedding $K(x) \to F$ over K that sends x to y is a valued field embedding. Such an element y exists by saturation and the next general lemma. □

LEMMA 3.6.3. *Let $K \subseteq L$ be a valued field extension such that* $\operatorname{res} K = \operatorname{res} L$. *Let $a_1, \dots, a_n \in K$, $n \geqslant 1$, and let $x \in L \setminus K$ be such that $v(x-a_i) \in \Gamma$ for $i = 1, \dots, n$. Then there exists $a \in K$ such that $v(x-a_i) = v(a-a_i)$ for $i = 1, \dots, n$.*

PROOF. Any $a \in K$ such that $v(a-x) > v(a-a_i)$ for $i = 1, \dots, n$ has the desired property. We may assume $v(x-a_1) \geqslant v(x-a_i)$ for $i = 2, \dots, n$. Since $v(x-a_1) \in \Gamma$ we can take $b \in K$ such that $v(x-a_1) = vb$. So $v\left(\frac{x-a_1}{b}\right) = 0$, and since $\operatorname{res} K = \operatorname{res} L$, we have $\frac{x-a_1}{b} = c + \varepsilon$ with $c \in K$, $c \asymp 1$, $\varepsilon \prec 1$. Then $a = a_1 + bc$ works because $x - a = b\varepsilon$ and $v(b\varepsilon) > v(x-a_i)$. □

COROLLARY 3.6.4. ACVF *is the model completion of the $\mathcal{L}_{\preccurlyeq}$-theory of pairs $(R, \preccurlyeq)$ where R is an integral domain and $\preccurlyeq$ is a dominance relation on R.*

For a valued field K with residue field $\boldsymbol{k}$, the pair $(\operatorname{char} K, \operatorname{char} \boldsymbol{k})$ is among the following, where p is a prime number:

$(0,0)$: equicharacteristic 0;

$(0,p)$: mixed characteristic p;

(p,p): equicharacteristic p.

Each of these actually occurs: if $\boldsymbol{k}$ is a field of characteristic p, with $p = 0$ or p a prime number, then the Hahn fields $\boldsymbol{k}((t^\Gamma))$ have equicharacteristic p; if p is a prime number, then the unique dominance relation $\preccurlyeq$ on $\mathbb{Q}$ such that $(\mathbb{Z}, \preccurlyeq_p) \subseteq (\mathbb{Q}, \preccurlyeq)$ yields a valued field of mixed characteristic p.

The **characteristic** of a valued field K is $(\operatorname{char} K, \operatorname{char} \boldsymbol{k})$; it equals the characteristic of any valued field extension of K. Let (m,n) be the characteristic of some valued field; define $\mathrm{ACVF}_{(m,n)}$ as the $\mathcal{L}_{\preccurlyeq}$-theory whose models are the structures $(K, \preccurlyeq) \models$ ACVF that are of characteristic (m,n) as a valued field.

COROLLARY 3.6.5. $\mathrm{ACVF}_{(m,n)}$ *is complete.*

PROOF. Let p range over prime numbers. Construing valued fields as fields with a dominance relation, the classification of dominance relations on $\mathbb{Z}$ from 3.6.2 gives: $(\mathbb{Z}, \preccurlyeq_{\mathrm{t}})$ embeds into every valued field of characteristic $(0,0)$, and $(\mathbb{Z}, \preccurlyeq_p)$ into every valued field of characteristic $(0,p)$. Clearly, $(\mathbb{F}_p, \preccurlyeq_{\mathrm{t}})$ embeds into every valued field of characteristic (p,p). Now use Corollary B.11.7. □

Real closed valued fields

An **ordered integral domain** is an integral domain R with a (total) ordering $\leqslant$ of R such that for all $x, y, z \in R$,

$$x \leqslant y \Rightarrow x + z \leqslant y + z, \quad x \leqslant y \,\&\, z \geqslant 0 \Rightarrow xz \leqslant yz.$$

Given an ordered integral domain $(R, \leqslant)$ there is a unique field ordering $\leqslant_F$ of its fraction field F such that $(R, \leqslant) \subseteq (F, \leqslant_F)$; we call $(F, \leqslant_F)$ the **ordered fraction field** of $(R, \leqslant)$. We augment the language $\{0, 1, -, +, \cdot\}$ of rings by a binary relation symbol $\leqslant$ to obtain the language $\mathcal{L}_{\text{OR}}$ of ordered rings, and we construe ordered integral domains as $\mathcal{L}_{\text{OR}}$-structures in the natural way.

We augment $\mathcal{L}_{\text{OR}}$ by a binary relation symbol $\preccurlyeq$ to obtain the language $\mathcal{L}_{\text{OR},\preccurlyeq}$, and construe each valued ordered field as an $\mathcal{L}_{\text{OR},\preccurlyeq}$-structure in the obvious way. A dominance relation $\preccurlyeq$ on an ordered integral domain R is said to be **convex** if for all $r, s \in R$ we have: $0 \leqslant r \leqslant s \Rightarrow r \preccurlyeq s$. So a dominance relation on an ordered field is convex iff its corresponding valuation ring is convex. If $\preccurlyeq$ is a convex dominance relation on the ordered integral domain R, then $\preccurlyeq_F$ as in the previous subsection is a convex dominance relation on its ordered fraction field F.

Let RCVF be the $\mathcal{L}_{\text{OR},\preccurlyeq}$-theory whose models are the $\mathcal{L}_{\text{OR},\preccurlyeq}$-structures $(K, \preccurlyeq)$ where K is a real closed ordered field and $\preccurlyeq$ is a nontrivial convex dominance relation on K. The substructures of the models of RCVF are exactly the pairs $(R, \preccurlyeq)$ where R is an ordered integral domain and $\preccurlyeq$ is a convex dominance relation on R; this follows easily from the remarks above and Corollary 3.5.18.

Let R be an ordered integral domain, $\preccurlyeq$ a convex dominance relation on R, and $i\colon (R, \preccurlyeq) \to (K, \preccurlyeq_K)$ an embedding into a real closed ordered field K with a convex dominance relation $\preccurlyeq_K$ on K. By Corollary 3.5.18 there is a unique convex dominance relation $\preccurlyeq_{F^{\text{rc}}}$ on the real closure F^{rc} of the ordered fraction field F of R with $(R, \preccurlyeq) \subseteq (F^{\text{rc}}, \preccurlyeq_{F^{\text{rc}}})$; it follows that the unique extension of i to an ordered field embedding $F^{\text{rc}} \to K$ is also an embedding $(F^{\text{rc}}, \preccurlyeq_{F^{\text{rc}}}) \to (K, \preccurlyeq_K)$.

THEOREM 3.6.6. RCVF *has* QE.

PROOF. By the above remarks and B.11.11 it suffices to show the following: Let $E, F \models$ RCVF be such that F is $|K|^+$-saturated where K is a proper real closed subfield of E; view subfields of E as valued ordered subfields of E, and let $i\colon K \to F$ be an embedding of ordered valued fields. Then there is an $x \in E \setminus K$ and an extension of i to an embedding $j\colon K(x) \to F$ of ordered valued fields.

To find such x and j we distinguish the same three cases as in the proof of Theorem 3.6.1. To simplify notation we identify the valued ordered field K with the valued ordered field iK via i.

CASE 1: $\operatorname{res}(K) \neq \operatorname{res}(E)$. Take $x \in \mathcal{O}_E$ such that $\overline{x} \notin \operatorname{res}(K)$. Since $\operatorname{res}(K)$ is real closed, $\overline{x}$ is transcendental over $\operatorname{res}(K)$. Also $x \notin K$, so x is transcendental over K. By the saturation assumption on F we can find $y \in \mathcal{O}_F$ such that $\overline{y}$ realizes the same cut in $\operatorname{res}(K)$ as $\overline{x}$. In particular $\overline{y} \notin \operatorname{res}(K)$, so $\overline{y}$ is transcendental over $\operatorname{res}(K)$ and y is transcendental over K. The field embedding $j\colon K(x) \to F$ over K with $j(x) = y$

is a valued field embedding by the uniqueness part of Lemma 3.1.31. It is easy to check that $x \in \mathcal{O}_E$ and $y \in \mathcal{O}_F$ realize the same cut in K, so j preserves order by Corollary 3.5.8.

CASE 2: $\Gamma \neq \Gamma_E$. Take any $x \in E^>$ with $\alpha := vx \notin \Gamma$. As in the proof of Case 2 of Theorem 3.6.1 we get $y \in F^>$ such that $\beta := vy$ realizes the same cut in Γ as α. As in that proof, x and y are transcendental over K, and the field embedding $j\colon K(x) \to F$ over K with $j(x) = y$ is a valued field embedding. It is easy to check that x and y realize the same cut in K, so j preserves order as in Case 1.

CASE 3: $\operatorname{res}(K) = \operatorname{res}(E)$ *and* $\Gamma = \Gamma_E$. Take any $x \in E \setminus K$. With v extended to a valuation on $E[i]$ we have $v\big(x - (a + bi)\big) = \min\big(v(x-a), vb\big)$ for $a, b \in K$ by Lemma 3.5.15. Thus we can proceed as in the proof of Case 3 of Theorem 3.6.1 to obtain a valued field embedding $j\colon K(x) \to F$ over K. By Corollary 3.5.12, j preserves order. □

The separation in three cases in the proofs of Theorem 3.6.1 and 3.6.6, according to whether the residue field extends, the value group extends, or neither, is a common feature of many proofs for QE or model completeness of (expansions of) valued fields; we see this again in proving Theorem 3.6.11 below, and in establishing QE for $\mathbb{T}$ in Chapter 16. The case of immediate extensions is usually the hardest.

In view of the remarks preceding Theorem 3.6.6 we have:

COROLLARY 3.6.7. RCVF *is the model completion of the* $\mathcal{L}_{\mathrm{OR},\preccurlyeq}$*-theory of ordered integral domains* R *equipped with a convex dominance relation on* R.

COROLLARY 3.6.8. RCVF *is complete.*

PROOF. Use that the ordered ring of integers with its trivial dominance relation embeds into every model of RCVF. □

COROLLARY 3.6.9. RCVF *has* NIP.

PROOF. Let $(K, \preccurlyeq) \models$ RCVF and suppose that the relation $R \subseteq K^m \times K^n$ is 0-definable in $(K, \preccurlyeq)$; we need to show that R is dependent. Let K^* be a $|K|^+$-saturated elementary extension of the ordered field K, and take $a^* > 0$ in K^* such that $\mathcal{O} = (-a^*, a^*)_{K^*} \cap K$. Let $\varphi(x, y)$ be a quantifier-free $\mathcal{L}_{\mathrm{OR},\preccurlyeq}$-formula that defines R in $(K, \preccurlyeq)$, with $x = (x_1, \dots, x_m)$, $y = (y_1, \dots, y_n)$. Boolean combinations of dependent relations are dependent (Lemma B.13.6). So we can assume that $\varphi(x, y)$ is of the form "$P(x,y) \leqslant Q(x,y)$" or "$P(x,y) \preccurlyeq Q(x,y)$" where $P, Q \in \mathbb{Z}[x, y]$. We associate to φ a quantifier-free formula φ^* in the language of ordered rings: in the first case we take $\varphi^* := \varphi$, and in the second case φ^* expresses

$$P(x,y) = Q(x,y) = 0 \vee \big(Q(x,y) \neq 0 \;\&\; |P(x,y)| < a^*|Q(x,y)|\big).$$

Let R^* be the subset of $(K^*)^{m+n}$ defined by φ^*. Then $R^* \cap K^{m+n} = R$. Since R^* is dependent by Corollary B.13.8, so is R. □

Tame pairs

A **tame pair** (tacitly: of real closed fields) is a pair (K, C) where $K \models \mathrm{RCVF}$ and C is a real closed subfield of K such that $\mathcal{O} = C + \smallo$, where $\mathcal{O}$ is the valuation ring of K (corresponding to the distinguished dominance relation $\preccurlyeq$ on K); note that then $\mathcal{O}$ is the convex hull of C in K and C is a lift of $\operatorname{res}(K)$. Conversely, if $K \models \mathrm{RCVF}$, then by Proposition 3.3.8 and Zorn, $\mathcal{O}$ contains a lift of $\operatorname{res}(K)$, and for any such lift C we have a tame pair (K, C).

It is worth noting that a tame pair (K, C) has a *standard part map* $\operatorname{st}\colon \mathcal{O} \to C$: it assigns to $a \in \mathcal{O}$ the unique $c \in C$ such that $a - c \in \smallo$; thus $\operatorname{st}\colon \mathcal{O} \to C$ is a ring morphism, and if $a, b \in \mathcal{O}$, $a \leqslant b$, then $\operatorname{st}(a) \leqslant \operatorname{st}(b)$.

Let $\mathcal{L}_{\text{tame}}$ be the language $\mathcal{L}_{\mathrm{OR},\preccurlyeq}$ augmented by a unary relation symbol U. We view each tame pair (K, C) as an $\mathcal{L}_{\text{tame}}$-structure in the natural way, interpreting U as the underlying set of C. Note that for tame pairs (K, C) and (L, C_L) with $(K, C) \subseteq (L, C_L)$, the standard part map of (L, C_L) extends the standard part map of (K, C). We let $\mathrm{RCF}_{\text{tame}}$ be the $\mathcal{L}_{\text{tame}}$-theory whose models are exactly the tame pairs (K, C).

LEMMA 3.6.10. *Assume $\boldsymbol{K} = (K, C)$ and $\boldsymbol{E} = (E, C_E)$ are models of $\mathrm{RCF}_{\text{tame}}$ such that $\boldsymbol{K} \subseteq \boldsymbol{E}$. Let $e \in C_E$, and let $K(e)^{\mathrm{rc}}$ and $C(e)^{\mathrm{rc}}$ be the real closures of $K(e)$ and $C(e)$ in E and C_E, respectively. Then $C_E \cap K(e)^{\mathrm{rc}} = C(e)^{\mathrm{rc}}$, and*

$$\boldsymbol{K}_1 := \big(K(e)^{\mathrm{rc}}, C(e)^{\mathrm{rc}}\big) \models \mathrm{RCF}_{\text{tame}}, \qquad \boldsymbol{K} \subseteq \boldsymbol{K}_1 \subseteq \boldsymbol{E},$$

where $K(e)^{\mathrm{rc}}$ is construed as a valued ordered subfield of E.

PROOF. This is trivial if $e \in C$. Assume $e \notin C$, so $e \notin K$. Also,

$$K_1 := K(e)^{\mathrm{rc}} \models \mathrm{RCVF}, \qquad C_1 := C(e)^{\mathrm{rc}} \subseteq C_E \cap K_1 \subseteq \mathcal{O}_1 := \mathcal{O}_E \cap K_1.$$

By Lemma 3.1.31 for $x := e$, and Corollary 3.5.18, C_1 is a lift of the residue field of K_1, so $C_1 = C_E \cap K_1$ and $\boldsymbol{K}_1 \models \mathrm{RCF}_{\text{tame}}$. Thus $\boldsymbol{K} \subseteq \boldsymbol{K}_1 \subseteq \boldsymbol{E}$. □

THEOREM 3.6.11. $\mathrm{RCF}_{\text{tame}}$ *is model complete.*

PROOF. Let $\boldsymbol{K} = (K, C)$, $\boldsymbol{E} = (E, C_E)$, $\boldsymbol{F} = (F, C_F)$ be models of $\mathrm{RCF}_{\text{tame}}$, where $\boldsymbol{K} \subseteq \boldsymbol{E}$ and $\boldsymbol{K} \preccurlyeq \boldsymbol{F}$, and $\boldsymbol{F}$ is $|E|^+$-saturated. By Corollary B.10.4 and Zorn it is enough to show that there is a substructure $\boldsymbol{K}_1$ of $\boldsymbol{E}$ that properly contains $\boldsymbol{K}$, is a model of $\mathrm{RCF}_{\text{tame}}$, and embeds over K into $\boldsymbol{F}$. To simplify notation we view subfields of E, respectively F, as valued ordered subfields of E, respectively F. As before we distinguish three cases:

CASE 1: $C \neq C_E$. Take $e \in C_E \setminus C$; then $e \notin K$. The saturation assumption on $\boldsymbol{F}$ gives $f \in C_F$ realizing the same cut in K as e. Arguing as in Case 1 in the proof of Theorem 3.6.6 this yields an ordered and valued field embedding $j\colon K(e) \to F$ over K sending e to f. With notations as in Lemma 3.6.10 and its proof we can extend j to an ordered field embedding $j_1\colon K_1 \to F$, which by Corollary 3.5.18 is also a valued field embedding. On the $\boldsymbol{F}$-side, $j_1(K_1)$ is the real closure of $K(f)$ in F and $j_1(C_1)$ is the real closure of $C(f)$ in C_F, so j_1 embeds $\boldsymbol{K}_1$ into $\boldsymbol{F}$.

CASE 2: $\Gamma \neq \Gamma_E$. We first argue as in Case 2 in the proof of Theorem 3.6.6, then take real closures as in Case 1 above, and follow the argument there, using instead of Lemma 3.6.10 the statement about residue fields in Lemma 3.1.30.

CASE 3: $C = C_E$ *and* $\Gamma = \Gamma_E$. Argue as in Case 3 in the proof of Theorem 3.6.6, and then extend $j\colon K(x) \to F$ to the real closure of $K(x)$ inside E. □

COROLLARY 3.6.12. $\mathrm{RCF}_{\mathrm{tame}}$ *is complete.*

PROOF. Take x in an ordered field extension of $\mathbb{Q}$ with $x > \mathbb{Q}$, and let $\mathbb{Q}^{\mathrm{rc}}$ be the real closure of $\mathbb{Q}$ in $\mathbb{Q}(x)^{\mathrm{rc}}$. Consider $\mathbb{Q}(x)^{\mathrm{rc}}$ as a valued field whose valuation ring is the convex hull of $\mathbb{Q}$ in $\mathbb{Q}(x)^{\mathrm{rc}}$. Then $\mathbb{Q}(x)^{\mathrm{rc}} \models \mathrm{RCVF}$, and $\mathbb{Q}^{\mathrm{rc}}$ is a maximal subfield of the valuation ring of $\mathbb{Q}(x)^{\mathrm{rc}}$, so $\big(\mathbb{Q}(x)^{\mathrm{rc}}, \mathbb{Q}^{\mathrm{rc}}\big) \models \mathrm{RCF}_{\mathrm{tame}}$. We claim that $\big(\mathbb{Q}(x)^{\mathrm{rc}}, \mathbb{Q}^{\mathrm{rc}}\big)$ embeds into every model of $\mathrm{RCF}_{\mathrm{tame}}$. To prove this claim, let $(K, C) \models \mathrm{RCF}_{\mathrm{tame}}$ and take any $y \in K^{>}$ with $y \succ 1$. Then $y > \mathbb{Q} \subseteq K$, so we have an ordered field embedding $j\colon \mathbb{Q}(x)^{\mathrm{rc}} \to K$ with $j(x) = y$. Consider any subfield of K as an ordered subfield of K, and let $\mathbb{Q}(y)^{\mathrm{rc}}$ be the real closure of $\mathbb{Q}(y)$ in K. Then $j\big(\mathbb{Q}(x)^{\mathrm{rc}}\big) = \mathbb{Q}(y)^{\mathrm{rc}}$. The j-image of the valuation ring of $\mathbb{Q}(x)^{\mathrm{rc}}$ is the convex hull of $\mathbb{Q}$ in $\mathbb{Q}(y)^{\mathrm{rc}}$. This convex hull is a convex valuation ring of $\mathbb{Q}(y)^{\mathrm{rc}}$; it is contained in $\mathcal{O} \cap \mathbb{Q}(y)^{\mathrm{rc}}$, but does not contain y, and thus it equals $\mathcal{O} \cap \mathbb{Q}(y)^{\mathrm{rc}}$ by Corollary 3.1.12, so j is a valued field embedding.

Finally, $j(\mathbb{Q}^{\mathrm{rc}})$ is the real closure of $\mathbb{Q}$ in K, so $j(\mathbb{Q}^{\mathrm{rc}}) \subseteq C \cap \mathbb{Q}(y)^{\mathrm{rc}}$. Now $C \cap \mathbb{Q}(y)^{\mathrm{rc}}$ is a real closed subfield of $\mathbb{Q}(y)^{\mathrm{rc}}$ by Corollary 3.5.6, and $y \notin C \cap \mathbb{Q}(y)^{\mathrm{rc}}$, so $j(\mathbb{Q}^{\mathrm{rc}}) = C \cap \mathbb{Q}(y)^{\mathrm{rc}}$. Thus j embeds $\big(\mathbb{Q}(x)^{\mathrm{rc}}, \mathbb{Q}^{\mathrm{rc}}\big)$ into (K, C). □

PROPOSITION 3.6.13. *Let* $\boldsymbol{K} = (K, C) \models \mathrm{RCF}_{\mathrm{tame}}$. *If* $X \subseteq K^n$ *is definable (with parameters) in* $\boldsymbol{K}$, *then* $X \cap C^n$ *is semialgebraic in the sense of* C. *Thus if* $X \subseteq K^n$ *is semialgebraic in the sense of* K, *then* $X \cap C^n$ *is semialgebraic in the sense of* C.

PROOF. Let $x = (x_1, \dots, x_n)$ be a tuple of distinct variables. Note that $C \preccurlyeq K$ as $\mathcal{L}_{\mathrm{OR}}$-structures, by B.12.13. Thus we have to show for every $\mathcal{L}_{\mathrm{tame},K}$-formula $\varphi'(x)$ that $U(x) \wedge \varphi'(x)$ is equivalent in $\boldsymbol{K}$ to $U(x) \wedge \varphi(x)$ for some $\mathcal{L}_{\mathrm{OR},C}$-formula $\varphi(x)$, where $U(x) := U(x_1) \wedge \cdots \wedge U(x_n)$. By Lemma B.9.2, this reduces to showing:

(∗) *Let* $\boldsymbol{E} = (E, C_E)$ *and* $\boldsymbol{F} = (F, C_F)$ *be elementary extensions of* $\boldsymbol{K}$ *and suppose* $e \in C_E^n$ *and* $f \in C_F^n$ *realize the same type over* C *in the real closed fields* C_E *and* C_F, *respectively. Then* e *and* f *realize the same type over* K *in* $\boldsymbol{E}$ *and* $\boldsymbol{F}$, *respectively.*

Taking real closures in C_E and C_F, the hypothesis of (∗) yields a unique ordered field isomorphism $C(e)^{\mathrm{rc}} \to C(f)^{\mathrm{rc}}$ over C sending e to f. We claim that this isomorphism extends to an ordered field isomorphism $K(e)^{\mathrm{rc}} \to K(f)^{\mathrm{rc}}$ over K (taking real closures in E and F), and that

$$C_E \cap K(e)^{\mathrm{rc}} = C(e)^{\mathrm{rc}}, \qquad \big(K(e)^{\mathrm{rc}}, C(e)^{\mathrm{rc}}\big) \models \mathrm{RCF}_{\mathrm{tame}},$$
$$C_F \cap K(f)^{\mathrm{rc}} = C(f)^{\mathrm{rc}}, \qquad \big(K(f)^{\mathrm{rc}}, C(f)^{\mathrm{rc}}\big) \models \mathrm{RCF}_{\mathrm{tame}}.$$

In view of Theorem 3.6.11, the conclusion of $(*)$ follows from this claim.

By induction on n we reduce the claim to the case $n = 1$. If $e \in C$, the claim holds trivially, so assume $e \notin C$. Then $f \notin C$, and e and f realize the same cut in C, and therefore the same cut in K. Thus we have an ordered field isomorphism $K(e)^{\mathrm{rc}} \to K(f)^{\mathrm{rc}}$ over K sending e to f, and the rest holds by Lemma 3.6.10. □

Notes and comments

The first explicitly model-theoretic result on valued fields is due to A. Robinson [350]: model completeness of ACVF (close to Theorem 3.6.1). Theorems 3.6.6 and 3.6.11 are from Cherlin-Dickmann [75] and Macintyre [274]. Model completeness of ACVF and RCVF also follow from the more general results of Ax & Kochen [28, 29] and Eršov [131], in view of B.11.12, B.12.1, B.12.13. (See [75, Section 1.3].) Corollary 3.6.7 was noticed by Becker [39]. Proposition 3.6.13 is a special case of [108, Proposition 8.1]; see also [286]. The proof that $\mathbb{T}$ has NIP in Section 16.6 is modeled on that of Corollary 3.6.9 above. For proofs that ACVF and $\mathrm{RCF}_{\mathrm{tame}}$ have NIP, see [98] and [159], respectively.

In this section we construed valued fields as one-sorted structures by encoding the valuation as a dominance relation. It is often more informative to view valued fields as three-sorted structures, with sorts for the underlying field, for the value group, and for the residue field; for example, see [315]. In Chapter 8 we use this setting in dealing with certain valued *differential* fields.

3.7 THE NEWTON TREE OF A POLYNOMIAL OVER A VALUED FIELD

In this section we use Newton diagrams to construct a *Newton tree* for any given non-constant polynomial over a henselian valued field of equicharacteristic zero. Such a Newton tree is a finite tree of *approximate zeros* of the polynomial, and induces a partition of the field into finitely many simple pieces on each of which the polynomial behaves in a simple way. In Chapters 13 and 14 we use more delicate Newton diagrams for differential polynomials over suitable valued differential fields, and in preparation for this, the reader may find the exposition of the Newton diagram method for ordinary polynomials below helpful. The present section also has an application in the proof of Proposition 9.7.1 below, and will be useful in the next volume. This section does not depend on Sections 3.4–3.6.

NOTATION. Let F be a field and $P \in F[Y]^{\neq}$. Then $\deg P \in \mathbb{N}$ denotes the **degree** of P, and $\operatorname{mul} P \in \mathbb{N}$ denotes the **multiplicity** of P at 0: the largest m such that $P \in Y^m F[Y]$. For any $f \in F$ we let $P_{+f} := P(f+Y)$ be the **additive conjugate** of P by f, and $P_{\times f} := P(fY)$ the **multiplicative conjugate** of P by f.

Throughout this section we fix a valued field K with valuation ring $\mathcal{O} \neq K$, residue field $\boldsymbol{k}$ of characteristic 0, residue map $a \mapsto \overline{a}\colon \mathcal{O} = K^{\preccurlyeq 1} \to \boldsymbol{k}$, value group Γ and valuation $v\colon K^\times \to \Gamma$. We also choose a subset $\mathfrak{M}$ of $K^\times$ which is mapped bijectively onto Γ by v. (We do not assume that $\mathfrak{M}$ is a subgroup of $K^\times$.) For $\gamma \in \Gamma$ we let $s\gamma$ be the unique element of $\mathfrak{M}$ with $v(s\gamma) = \gamma$. "Equivalence" in this section refers to *the equivalence relation $\sim$ on $K^\times$ induced by the valuation. We let f, g, y, z range over K, and $\mathfrak{m}, \mathfrak{n}$ over $\mathfrak{M}$. Finally, we fix throughout a polynomial $P(Y) \in K[Y]^{\neq}$: $P = a_0 + a_1 Y + \cdots + a_n Y^n$, $a_0, a_1, \dots, a_n \in K$, $a_n \neq 0$.*

Dominant part

The **dominant monomial** of P is the unique element $\mathfrak{d}_P$ of $\mathfrak{M}$ with $\mathfrak{d}_P \asymp P$. Then $\mathfrak{d}_P^{-1} P \in \mathcal{O}[Y]$, and we call the polynomial

$$D_P := \sum_i \overline{(a_i/\mathfrak{d}_P)}\, Y^i \in \boldsymbol{k}[Y]$$

the **dominant part** of P. Clearly D_P is nonzero with

$$\operatorname{mul} P \leqslant \operatorname{mul} D_P \leqslant \deg D_P \leqslant \deg P.$$

We call $\operatorname{dmul} P := \operatorname{mul} D_P$ the **dominant multiplicity** of P at 0 and $\operatorname{ddeg} P := \deg D_P$ the **dominant degree** of P. Note that

$$\begin{aligned} P_{\times\mathfrak{m}} &= \sum_i a_i \mathfrak{m}^i Y^i, \quad \text{so} \\ D_{P_{\times\mathfrak{m}}} &= \sum_i \overline{(a_i \mathfrak{m}^i/\mathfrak{d})}\, Y^i \qquad \text{where } \mathfrak{d} = \mathfrak{d}_{P_{\times\mathfrak{m}}}. \end{aligned}$$

LEMMA 3.7.1. $\mathfrak{m} \prec \mathfrak{n} \Rightarrow \operatorname{dmul} P_{\times\mathfrak{m}} \leqslant \operatorname{ddeg} P_{\times\mathfrak{m}} \leqslant \operatorname{dmul} P_{\times\mathfrak{n}} \leqslant \operatorname{ddeg} P_{\times\mathfrak{n}}$.

PROOF. Suppose $\mathfrak{m} \prec \mathfrak{n}$; it suffices to show that then $\operatorname{ddeg} P_{\times\mathfrak{m}} \leqslant \operatorname{dmul} P_{\times\mathfrak{n}}$. Let $d = \operatorname{dmul} P_{\times\mathfrak{n}}$. Then for $i > d$ we have $a_i \mathfrak{n}^i \preccurlyeq a_d \mathfrak{n}^d$ and so

$$a_i \mathfrak{m}^i = (a_i \mathfrak{n}^i) \cdot (\mathfrak{m}/\mathfrak{n})^i \prec (a_d \mathfrak{n}^d) \cdot (\mathfrak{m}/\mathfrak{n})^d = a_d \mathfrak{m}^d,$$

and thus $\operatorname{ddeg} P_{\times\mathfrak{m}} \leqslant d$ as required. □

Approximate zeros

An **approximate zero** of P is an element y such that

$$v\big(P(y)\big) > \min_i v(a_i y^i) \qquad \text{(in particular } n \geqslant 1 \text{ and } y \neq 0\text{);}$$

equivalently, $y \neq 0$ and $D_{P_{\times\mathfrak{m}}}(c) = 0$, where $\mathfrak{m} \asymp y$ and $c = \overline{(y/\mathfrak{m})}$. In this case the polynomial $D_{P_{\times\mathfrak{m}}} \in \boldsymbol{k}[Y]$ is not homogeneous, since $c \neq 0$, and so with $\delta := \min_i v(a_i y^i)$, there at least two elements $i \in \{0, \dots, n\}$ such that $v(a_i y^i) = \delta$. We say that $\mathfrak{m}$ is a **starting monomial** for P if $D_{P_{\times\mathfrak{m}}}$ is not homogeneous. Note that if y is an approximate zero of P and $y \sim z$, then z is an approximate zero of P. If $P(y) = 0$ and $y \neq 0$, then y is an approximate zero of P.

REMARK. We have $P_{+f}(Y) = P(f+Y) = b_0 + b_1 Y + \cdots + b_n Y^n$ with

$$b_i = \sum_{j=0}^{n-i} \binom{i+j}{i} a_{i+j} f^j, \qquad \text{in particular, } b_0 = P(f).$$

Suppose that $f \preccurlyeq 1$; then $P_{+f} \asymp P$: the identities above give $P_{+f} \preccurlyeq P$, and hence $P \preccurlyeq P_{+f}$ by $P = (P_{+f})_{+g}$ for $g = -f$. Next, suppose that $vf = \beta \leqslant 0$ and f is *not* an approximate zero of P. Then $v(P_{+f})$ depends only on β, not on f:

$$v(P_{+f}) = v\big(P(f)\big) = \min_i v(a_i) + i\beta.$$

To see this, put $\gamma := v(b_0) = \min_i v(a_i) + i\beta$, so for $0 \leqslant i \leqslant n$ and $0 \leqslant j \leqslant n-i$,

$$v(a_{i+j} f^j) = v(a_{i+j}) + j\beta = v(a_{i+j}) + (i+j)\beta - i\beta \geqslant \gamma - i\beta.$$

Hence $v(b_i) \geqslant \gamma - i\beta \geqslant \gamma$ for $i > 0$. The assertion follows.

Geometric interpretation

Recall that $\mathbb{Q}\Gamma$ is the divisible hull of Γ. We refer to $\mathbb{Z} \times \mathbb{Q}\Gamma$ as the *plane*, and to its elements as *points*. The **abscissa** of a point (i, α) is the integer $\operatorname{abscis}(i, \alpha) := i$. We view $\mathbb{Z} \times \{0\}$ as the horizontal axis of the plane, and $\{0\} \times \mathbb{Q}\Gamma$ as its vertical axis. For $\beta \in \mathbb{Q}\Gamma$, define the additive function

$$L_\beta \colon \mathbb{Z} \times \mathbb{Q}\Gamma \to \mathbb{Q}\Gamma, \qquad L_\beta(i, \alpha) := \alpha + i\beta.$$

Given any $\beta, \delta \in \mathbb{Q}\Gamma$ we refer to the set

$$\big\{(i, \alpha) \in \mathbb{Z} \times \mathbb{Q}\Gamma : L_\beta(i, \alpha) = \delta\big\}$$

as *the line* $L_\beta = \delta$. A point (i, α) is said to lie *above* (respectively *on*, respectively *below*) this line if $L_\beta(i, \alpha) > \delta$ (respectively $L_\beta(i, \alpha) = \delta$, respectively $L_\beta(i, \alpha) < \delta$). Since these are the only kind of lines we need to consider, by "line" we shall always mean a line of the form $L_\beta = \delta$ as above. Note that if p, q are points on the line ℓ and $p \neq q$, then $\operatorname{abscis}(p) \neq \operatorname{abscis}(q)$. Each line contains infinitely many points, and is of the form $L_\beta = \delta$ for exactly one pair $(\beta, \delta) \in \mathbb{Q}\Gamma \times \mathbb{Q}\Gamma$. For any two points (i_1, α_1) and (i_2, α_2) with $i_1 \neq i_2$ there is exactly one line containing both points, namely $L_\beta = \delta$ with $\beta = -\big(\frac{\alpha_2 - \alpha_1}{i_2 - i_1}\big)$, $\delta = \alpha_1 + i_1\beta$. We call β the **antislope** of the line $L_\beta = \delta$. (Its slope is $-\beta$.)

We define the **Newton diagram** of P to be the finite nonempty set of points

$$\mathcal{N}(P) := \big\{\big(i, v(a_i)\big) : i = 0, \dots, n,\ a_i \neq 0\big\} \subseteq \mathbb{Z} \times \Gamma \subseteq \mathbb{Z} \times \mathbb{Q}\Gamma.$$

An **edge** of $\mathcal{N}(P)$ is a line ℓ that contains at least two points of $\mathcal{N}(P)$ and such that all points of $\mathcal{N}(P)$ lie on or above ℓ. In this case, the point of $\ell \cap \mathcal{N}(P)$ with least abscissa is called the **left vertex** of ℓ in $\mathcal{N}(P)$, and the point of $\ell \cap \mathcal{N}(P)$ with largest

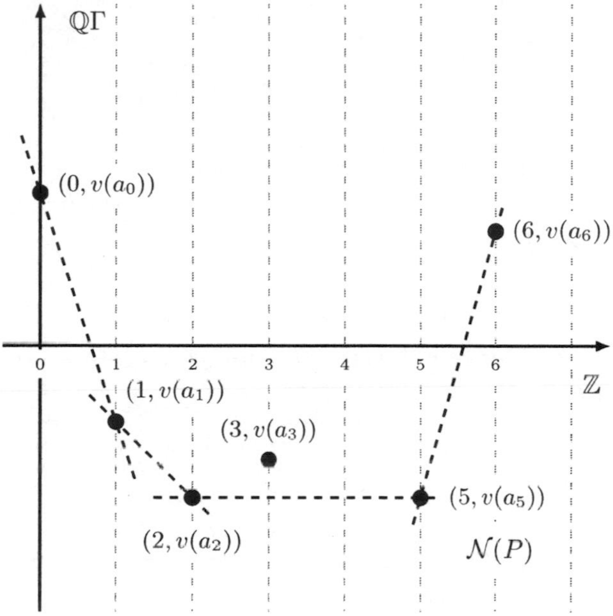

Figure 3.2: Picture of a Newton diagram.

abscissa is called the **right vertex** of ℓ in $\mathcal{N}(P)$. Figure 3.2 shows a Newton diagram with 6 points, 4 edges, and $n = 6$; this diagram is missing a point with abscissa 4, since $a_4 = 0$.

Note that if ℓ and ℓ' are edges of $\mathcal{N}(P)$ and $\ell \neq \ell'$, then antislope(ℓ) $\neq$ antislope(ℓ'). The **antislopes** of $\mathcal{N}(P)$ are by definition the antislopes of the edges of $\mathcal{N}(P)$. If ℓ and ℓ' are edges of $\mathcal{N}(P)$ with antislopes $\beta > \beta'$, then

$$\text{abscis(right vertex of } \ell) \leqslant \text{abscis(left vertex of } \ell').$$

In more graphic terms: the antislope increases when moving from right to left in the diagram along the edges. Now let ℓ be a line given by $L_\beta = \delta$, where $\beta, \delta \in \mathbb{Q}\Gamma$. Then for $i = 0, \dots, n$,

$$\begin{aligned} \big(i, v(a_i)\big) \text{ lies above } \ell \quad &\Longleftrightarrow \quad v(a_i) + i\beta > \delta, \\ \big(i, v(a_i)\big) \text{ lies on } \ell \quad &\Longleftrightarrow \quad v(a_i) + i\beta = \delta, \\ \big(i, v(a_i)\big) \text{ lies below } \ell \quad &\Longleftrightarrow \quad v(a_i) + i\beta < \delta. \end{aligned}$$

Hence

$$\left.\begin{array}{l} \text{the line } \ell \text{ contains at least one point of } \mathcal{N}(P) \text{ and} \\ \text{all points of } \mathcal{N}(P) \text{ lie on or above } \ell \end{array}\right\} \quad \Longleftrightarrow \quad \delta = \min_i v(a_i) + i\beta.$$

The antislope β of ℓ does not necessarily lie in Γ, but suppose it does, and suppose also that $\delta = \min_i v(a_i) + i\beta$. Then, with $\mathfrak{m} = s\beta$,

$$D_{P_{\times \mathfrak{m}}} = \sum_i \overline{(a_i \mathfrak{m}^i/\mathfrak{d})}\, Y^i \qquad \text{where } \mathfrak{d} = \mathfrak{d}_{P_{\times \mathfrak{m}}} = s\delta,$$

and so ℓ is an edge of $\mathcal{N}(P)$ if and only if $\mathfrak{m}$ is a starting monomial for P. If ℓ is an edge of $\mathcal{N}(P)$, then we set $P_\beta := D_{P_{\times \mathfrak{m}}}$, and we have

$$\text{abscis(left vertex of } \ell) = \operatorname{mul} P_\beta, \qquad \text{abscis(right vertex of } \ell) = \deg P_\beta.$$

The relation to the notion of approximate zero is as follows:

(1) If $y \asymp \mathfrak{m}$ is an approximate zero of P, then $\mathcal{N}(P)$ has an antislope $\beta := v\mathfrak{m}$, and $\overline{(y/\mathfrak{m})}$ is a zero of P_β.

(2) If $\beta = v\mathfrak{m} \in \Gamma$ is an antislope of $\mathcal{N}(P)$ and $c \in \boldsymbol{k}^\times$ is a zero of P_β, then P has an approximate zero $y \asymp \mathfrak{m}$ with $\overline{(y/\mathfrak{m})} = c$.

Note that y in (2) is determined up to equivalence by β and c. Clearly there are at most n antislopes of $\mathcal{N}(P)$. In this way the geometric interpretation of approximate zeros shows that P has, up to equivalence, at most n approximate zeros.

Asymptotic equations

A set $\mathcal{E} \subseteq K^\times$ is called **$\preccurlyeq$-closed** if $\mathcal{E} \neq \emptyset$, and $f \in \mathcal{E}$ whenever $0 \neq f \preccurlyeq g \in \mathcal{E}$. Let us consider an **asymptotic equation**

$$\text{(E)} \qquad P(Y) = 0, \qquad Y \in \mathcal{E}$$

where $\mathcal{E} \subseteq K^\times$ is $\preccurlyeq$-closed. A **solution** of (E) is an element y of $\mathcal{E}$ such that $P(y) = 0$. An **approximate solution** of (E) is an approximate zero $y \in \mathcal{E}$ of P. If $\mathcal{N}(P)$ has an antislope $\geqslant v\mathfrak{m}$ for some $\mathfrak{m} \in \mathcal{E}$, let β(E) be the least among these antislopes, and define the **dominant degree** of (E) to be the abscissa of the right vertex of the edge of $\mathcal{N}(P)$ with antislope β(E), denoted by ddeg (E); thus $\operatorname{mul} P < \operatorname{ddeg}(\mathrm{E}) \leqslant \deg P$. If in addition $\beta(\mathrm{E}) \in \Gamma$, then $\operatorname{ddeg}(\mathrm{E}) = \deg P_{\beta(\mathrm{E})}$, and we call $P_{\beta(\mathrm{E})}$ the **primary dominant part** of (E). If all antislopes of $\mathcal{N}(P)$ are $< v(\mathcal{E})$, then we define the dominant degree ddeg (E) of (E) to be 0. (In that case (E) has no approximate solution.) Related to ddeg (E) is the **dominant degree** of P on $\mathcal{E}$, defined to be the natural number

$$\operatorname{ddeg}_{\mathcal{E}} P := \max\big\{ \operatorname{ddeg} P_{\times \mathfrak{m}} : \mathfrak{m} \in \mathcal{E} \big\}.$$

Note that for all $\mathfrak{m}$,

$$\operatorname{mul} P = \operatorname{mul} P_{\times \mathfrak{m}} \leqslant \operatorname{mul} D_{P_{\times \mathfrak{m}}} \leqslant \deg D_{P_{\times \mathfrak{m}}} \leqslant \deg P_{\times \mathfrak{m}} = \deg P,$$

and so $\operatorname{mul} P \leqslant \operatorname{ddeg}_{\mathcal{E}} P \leqslant \deg P$. Moreover:

LEMMA 3.7.2. *If* $\operatorname{ddeg}(\mathrm{E}) = 0$, *then* $\operatorname{mul} P = \operatorname{ddeg}_{\mathcal{E}} P$. *If on the other hand* $\operatorname{ddeg}(\mathrm{E}) > 0$, *then* $\operatorname{mul} P < \operatorname{ddeg}(\mathrm{E}) = \operatorname{ddeg}_{\mathcal{E}} P$.

PROOF. Let K^{a} be an algebraic closure of K, equipped with a valuation extending that of K, let $\mathfrak{M}^{\mathrm{a}} \supseteq \mathfrak{M}$ be a subset of $(K^{\mathrm{a}})^{\times}$ which maps bijectively onto the value group $\mathbb{Q}\Gamma$ of K^{a} under this valuation, and let

$$\mathcal{E}^{\mathrm{a}} := \{a \in (K^{\mathrm{a}})^{\times} : a \preccurlyeq \mathfrak{m} \text{ for some } \mathfrak{m} \in \mathcal{E}\}$$

be the smallest $\preccurlyeq$-closed subset of $(K^{\mathrm{a}})^{\times}$ containing $\mathcal{E}$. Then ddeg (E) does not change if we replace K, $\mathfrak{M}$, $\mathcal{E}$ by K^{a}, $\mathfrak{M}^{\mathrm{a}}$, $\mathcal{E}^{\mathrm{a}}$, and $\operatorname{ddeg}_{\mathcal{E}} P = \operatorname{ddeg}_{\mathcal{E}^{\mathrm{a}}} P$ by Lemma 3.7.1. Thus we may assume that Γ is divisible, and do so below.

We establish the first implication by proving the contrapositive. So assume that $m := \operatorname{mul} P < d := \operatorname{ddeg}_{\mathcal{E}} P$, and take $\mathfrak{m} \in \mathcal{E}$ with $\operatorname{ddeg} P_{\times\mathfrak{m}} = d$. Then $\big(m, v(a_m)\big)$ lies on or above the line through $\big(d, v(a_d)\big)$ with antislope $v\mathfrak{m}$. Hence $\mathcal{N}(P)$ has an edge with left vertex $\big(m, v(a_m)\big)$ and antislope $\geqslant v\mathfrak{m}$, and thus ddeg (E) > 0.

For the second implication, assume ddeg (E) > 0. It is clear that then $\operatorname{mul} P <$ ddeg (E) $\leqslant \operatorname{ddeg}_{\mathcal{E}} P$. Assume towards a contradiction that $\mathfrak{m} \in \mathcal{E}$ is such that $d :=$ ddeg (E) $< i := \operatorname{ddeg} P_{\times\mathfrak{m}}$. Then $v(a_d)+dv\mathfrak{m} \geqslant v(a_i)+iv\mathfrak{m}$. Also, the edge of $\mathcal{N}(P)$ with antislope $\beta = \beta(\mathrm{E})$ passes through the point $\big(d, v(a_d)\big)$ but not through $\big(i, v(a_i)\big)$, so $v(a_i)+i\beta > v(a_d)+d\beta$. Let ℓ_1 be the line passing through $\big(d, v(a_d)\big)$ and $\big(i, v(a_i)\big)$; then the antislope $\beta_1 = -\left(\frac{v(a_i)-v(a_d)}{i-d}\right)$ of ℓ_1 satisfies $v\mathfrak{m} \leqslant \beta_1 < \beta$. Thus $\mathcal{N}(P)$ has an edge with left vertex $\big(d, v(a_d)\big)$ and antislope β^* such that $v\mathfrak{m} \leqslant \beta_1 \leqslant \beta^* < \beta$, contradicting the minimality of β. □

REMARK 3.7.3. Suppose K is henselian and ddeg (E) $= 1$. Then (E) has a unique solution. To see this, let $\beta = \beta(\mathrm{E})$, and note that $\beta \in \Gamma$. Let $\mathfrak{m} = s\beta$, $\mathfrak{d} = \mathfrak{d}_{P_{\times\mathfrak{m}}}$. The image of $\mathfrak{d}^{-1}P_{\times\mathfrak{m}} \in \mathcal{O}[Y]$ in $\boldsymbol{k}[Y]$ is $P_\beta = D_{P_{\times\mathfrak{m}}}$, which has degree 1 and multiplicity 0, and thus has a (unique) zero in $\boldsymbol{k}^{\times}$. Now K being henselian, it follows that $\mathfrak{d}^{-1}P_{\times\mathfrak{m}}$ has a unique zero y in $\mathcal{O}$. For this y we have $y \asymp 1$, and so $\mathfrak{m}y$ is a solution of (E). It is the only solution of (E) because $\mathcal{N}(P)$ has only one antislope that is $\geqslant v\mathfrak{n}$ for some $\mathfrak{n} \in \mathcal{E}$.

Refinements

A **refinement** of (E) is an asymptotic equation of the form

$$(\mathrm{E}') \qquad P_{+f}(Y) = 0, \qquad Y \in \mathcal{E}'$$

where $f \in \mathcal{E} \cup \{0\}$ and $\mathcal{E}' \subseteq \mathcal{E}$ is $\preccurlyeq$-closed. If y is a solution of (E′) and $f + y \neq 0$, then $f + y$ is a solution of (E). Moreover, if $y \not\sim -f$ is an approximate solution of (E′), then $f + y$ is an approximate solution of (E). To see this, let $y \not\sim -f$ be an approximate solution of (E′). We have

$$P_{+f}(Y) = b_0 + b_1 Y + \cdots + b_n Y^n, \qquad b_i = \frac{P^{(i)}(f)}{i!} = \sum_{j=i}^{n} \binom{j}{i} a_j f^{j-i}.$$

From $y \not\sim -f$ we get $vf, vy \geqslant v(f+y)$, and so

$$v\big(P(f+y)\big) = v\big(P_{+f}(y)\big) > \min_i v(b_i y^i) = \min_i v\left(\sum_{j=i}^{n}\binom{j}{i} a_j f^{j-i} y^i\right)$$
$$\geqslant \min_i \min_{j\geqslant i} v\left(a_j f^{j-i} y^i\right) \geqslant \min_j v\big(a_j (f+y)^j\big).$$

For $g \neq 0$ and $\mathcal{E} = \{y : 0 \neq y \prec g\}$ we also indicate (E) by

$$P(Y) = 0, \qquad Y \prec g,$$

and we set $\operatorname{ddeg}_{\prec g} P := \operatorname{ddeg}_{\mathcal{E}} P$. This notation is used in the next lemmas.

LEMMA 3.7.4. *Let f be an approximate solution of* (E). *Put $\beta := vf$, $\mathfrak{m} := s\beta$, $\mu :=$ multiplicity of the zero $\overline{f/\mathfrak{m}}$ of P_β, and $P_{+f}(Y) = b_0 + b_1 Y + \cdots + b_n Y^n$ with $b_0, \ldots, b_n \in K$. Consider the refinement*

$$(\mathrm{E}_{+f}) \qquad P_{+f}(Y) = 0, \qquad Y \prec f$$

of (E). *Then*

(i) *$b_\mu \neq 0$, and all points of $\mathcal{N}(P_{+f})$ with abscissa $< \mu$ lie above the line through $\big(\mu, v(b_\mu)\big)$ with antislope β, and all points of $\mathcal{N}(P_{+f})$ with abscissa $> \mu$ lie on or above that line;*

(ii) *if $b_i = 0$ for all $i < \mu$, then* (E_{+f}) *has dominant degree* 0*;*

(iii) *if $b_i \neq 0$ for some $i < \mu$, then* (E_{+f}) *has dominant degree μ.*

PROOF. Note that $b_i = \frac{P^{(i)}(f)}{i!}$. Set $\delta := \min_i v(a_i f^i)$, $\mathfrak{d} := s\delta$, and

$$Q(Y) := \mathfrak{d}^{-1} P_{\times\mathfrak{m}}(Y) = \sum_i (a_i \mathfrak{m}^i/\mathfrak{d})\, Y^i \in \mathcal{O}[Y].$$

Hence $Q^{(i)}(f/\mathfrak{m}) = (\mathfrak{m}^i/\mathfrak{d}) P^{(i)}(f)$, so $b_i = \frac{\mathfrak{d}}{i!\mathfrak{m}^i} Q^{(i)}(f/\mathfrak{m})$, $i = 0, \ldots, n$. Note that

$$\begin{aligned} v\big(Q^{(i)}(f/\mathfrak{m})\big) &> 0 \quad \text{for } 0 \leqslant i < \mu, \\ v\big(Q^{(\mu)}(f/\mathfrak{m})\big) &= 0, \\ v\big(Q^{(j)}(f/\mathfrak{m})\big) &\geqslant 0 \quad \text{for } \mu < j \leqslant n. \end{aligned}$$

Thus

$$\begin{aligned} v(b_i) &> \delta - i\beta \quad \text{for } 0 \leqslant i < \mu, \\ v(b_\mu) &= \delta - \mu\beta, \\ v(b_j) &\geqslant \delta - j\beta \quad \text{for } \mu < j \leqslant n. \end{aligned}$$

These inequalities give (i). Also, if $0 \neq y \prec f$ and $\mu < j \leqslant n$, then

$$v(b_\mu y^\mu) = (\delta - \mu\beta) + \mu vy = \delta + \mu(vy - \beta) < \delta + j(vy - \beta) \leqslant v(b_j y^j).$$

Hence the dominant degree of (E_{+f}) is at most μ. If $b_i = 0$ for all $i < \mu$ (equivalently, f is a zero of multiplicity μ of P), then (E_{+f}) has dominant degree 0. Suppose $b_i \neq 0$ for some $i < \mu$. The inequalities above show that each line through the point $\big(\mu, v(b_\mu)\big)$ and some point $\big(i, v(b_i)\big)$ with $i < \mu$ and $b_i \neq 0$ has antislope $> \beta$. Among these lines, let ℓ' be the one with minimal antislope $\beta' > \beta$. One checks easily that all points $\big(j, v(b_j)\big)$ with $\mu < j \leqslant n$ and $b_j \neq 0$ lie above ℓ', so ℓ' is an edge of $\mathcal{N}(P_{+f})$. Thus (E_{+f}) has dominant degree μ and $\beta(\mathrm{E}_{+f}) = \beta'$. □

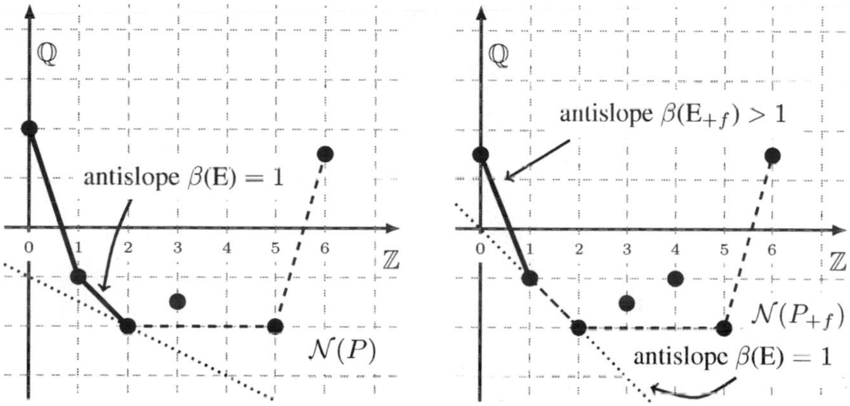

Figure 3.3: Behavior of Newton diagrams under refinement.

Figure 3.3 illustrates Lemma 3.7.4 above in the case where our value group Γ is the ordered abelian group $\mathbb{Q}$, $\mathcal{E} = \{y \in K^\times : vy > 1/2\}$, and

$$P = t^2 + 2t^{-1}Y + t^{-2}Y^2 + t^{-3/2}Y^3 + t^{-2}Y^5 + t^{3/2}Y^6 \quad \text{where } t \in \mathfrak{m},\ vt = 1.$$

In the Newton diagram on the left, the dotted line has antislope $1/2$, and the edges of antislope $> 1/2$ have been highlighted. On the right is the Newton diagram of the additive conjugate P_{+f} of P by the approximate zero $f = -2t$ of P.

In the previous lemma the dominant degree of (E_{+f}) is less than the dominant degree of (E), except when $\beta = \beta(\mathrm{E})$ and $\mu = \deg(P_\beta)$, in which case (E) and (E_{+f}) have the same dominant degree. The next lemma handles this exceptional case. In the proof we tacitly use that if $n \geqslant 1$ (so $P' \neq 0$), then the Newton diagram $\mathcal{N}(P')$ is obtained by shifting each point of $\mathcal{N}(P)$ not on the vertical axis $\{0\} \times \mathbb{Q}\Gamma$ by one unit to the left:

$$\mathcal{N}(P') = \big\{\big(i-1, v(a_i)\big) : i \geqslant 1,\ a_i \neq 0\big\}.$$

LEMMA 3.7.5. *Let K be henselian. Suppose that* (E) *has dominant degree $d \geqslant 1$,* $\beta(\mathrm{E}) \in \Gamma$, *and $P_{\beta(\mathrm{E})}$ has a zero in $\boldsymbol{k}$ of multiplicity d. Then:*

(i) $P^{(d-1)}$ *has a unique zero* $f \in \mathcal{E}$;

(ii) $vf = \beta(\mathrm{E})$ *and* f *is an approximate solution of* (E)*; moreover, if* y *is any approximate solution of* (E)*, then* $y \sim f$;

(iii) *if the refinement* (E_{+f}) *of* (E) *still has dominant degree* d *with* $\beta(\mathrm{E}_{+f}) \in \Gamma$*, then* $P_{\beta(\mathrm{E}_{+f})}$ *has no zero in* $\boldsymbol{k}$ *of multiplicity* d.

PROOF. To keep notations simple, put $\beta = \beta(\mathrm{E})$. Then $P_\beta = c_1(Y - c_2)^d$ with $c_1, c_2 \in \boldsymbol{k}$, $c_1 \neq 0$. Also $c_2 \neq 0$, since $\operatorname{mul} P_\beta < \deg P_\beta$. Set

$$\delta := \min_i v(a_i) + i\beta, \quad \mathfrak{m} := s\beta, \quad \mathfrak{d} := s\delta, \quad Q(Y) := \mathfrak{d}^{-1} P_{\times\mathfrak{m}}(Y) \in \mathcal{O}[Y],$$

so Q has image P_β in $\boldsymbol{k}[Y]$. Note that

$$Q^{(d-1)}(Y) = \frac{\mathfrak{m}^{d-1}}{\mathfrak{d}} P^{(d-1)}(\mathfrak{m}Y),$$

with image $d!c_1(Y - c_2)$ in $\boldsymbol{k}[Y]$. It follows that the asymptotic equation

$$P^{(d-1)}(Y) = 0, \qquad Y \in \mathcal{E}$$

has dominant degree 1 with primary dominant part $c(Y - c_2)$ for some $c \in \boldsymbol{k}^\times$. Hence (i) holds by Remark 3.7.3. It also follows that $f = \mathfrak{m}g$ where g is the unique zero of $Q^{(d-1)}$ in $\mathcal{O}$. From $\overline{g} = c_2$ we obtain that $vf = \beta$ and that f is an approximate zero of (E). Since $\beta = \beta(\mathrm{E})$ and $P_\beta(0) \neq 0$, β is the only antislope of $\mathcal{N}(P)$ that is $\geqslant v\mathfrak{n}$ for some $\mathfrak{n} \in \mathcal{E}$. Hence, if y is an approximate solution of (E), then $vy = \beta$, so $\overline{y/\mathfrak{m}} = c_2$ and $y \sim f$. As to (iii), this follows from $P^{(d-1)}(f) = 0$, that is, the coefficient of Y^{d-1} in $P_{+f}(Y)$ is 0. □

The exceptional case treated in this lemma does actually occur:

EXAMPLE. Let $K = \mathbb{R}((t^{\mathbb{Q}}))$ with its usual valuation $v\colon K^\times \to \Gamma = \mathbb{Q}$, so $vt = 1$; see Section 3.1. We take $\mathfrak{M} = t^{\mathbb{Q}}$, and identify the residue field $\boldsymbol{k}$ of K with $\mathbb{R}$ in the usual way. We set

$$P(Y) := (g^2 - t^2) - 2gY + Y^2 = (Y - g)^2 - t^2, \text{ where}$$

$$g := \sum_{k=1}^{\infty} t^{1-(1/k)} = 1 + t^{\frac{1}{2}} + t^{\frac{2}{3}} + t^{\frac{3}{4}} + \cdots \in K.$$

Then $\mathcal{N}(P) = \{(0,0),\ (1,0),\ (2,0)\}$. Let $\mathcal{E} := \{y \in K^\times : y \prec t^{-1}\}$; then the asymptotic equation (E) has $\beta(\mathrm{E}) = 0$, $\operatorname{ddeg}(\mathrm{E}) = 2$, $P_{\beta(\mathrm{E})}(Y) = (Y-1)^2$ in $\mathbb{R}[Y]$, and 1 is an approximate solution of (E). The polynomial

$$P_{+1}(Y) = ((g-1)^2 - t^2) - 2(g-1)Y + Y^2 = \big(Y - (g-1)\big)^2 - t^2$$

has Newton diagram $\mathcal{N}(P_{+1}) = \{(0,1),\ (1,\tfrac{1}{2}),\ (2,0)\}$, and the refinement

$$P_{+1}(Y) = 0, \qquad Y \prec 1$$

of (E) has dominant degree 2 with primary dominant part $(Y-1)^2$. In fact, for every $m \geqslant 1$ and $g_m := \sum_{k=1}^{m} t^{1-(1/k)}$, the refinement

$$P_{+g_m}(Y) = 0, \qquad Y \prec t^{1-(1/m)}$$

of (E) still has primary dominant part $(Y-1)^2$. On the other hand, the polynomial $P'(Y) = 2(Y-g)$ has the unique zero $f := g$ in K. We obtain $P_{+f}(Y) = Y^2 - t^2$, $\beta(\mathrm{E}_{+f}) = 1$, so the primary dominant part of (E_{+f}) is $Y^2 - 1$, with zeros 1 and -1 in $\mathbb{R}$, each of multiplicity 1. See Figure 3.4.

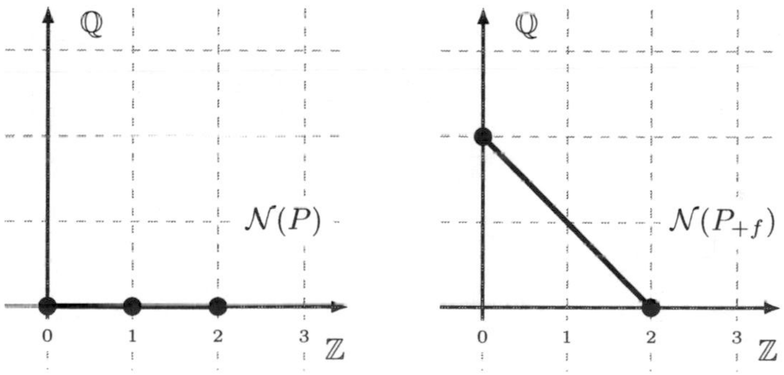

Figure 3.4: Unraveling an asymptotic equation.

The **complexity** of (E) is the pair $c(\mathrm{E}) = (d, \ell) \in \mathbb{N} \times \{0,1\}$ defined as follows:

$$d = \text{dominant degree of (E)},$$

$$\ell = \begin{cases} 1 & \text{if } d \geqslant 1,\ \beta(\mathrm{E}) \in \Gamma, \text{ and } P_{\beta(\mathrm{E})} \text{ has a zero in } \boldsymbol{k} \text{ of multiplicity } d, \\ 0 & \text{otherwise.} \end{cases}$$

In particular, if $d = 0$, then $c(\mathrm{E}) = (0,0)$. We order the cartesian product $\mathbb{N} \times \{0,1\}$ lexicographically. In Lemma 3.7.4, if $\beta \neq \beta(\mathrm{E})$ or $c(\mathrm{E}) = (d,0)$ with $d \geqslant 1$, then $c(\mathrm{E}) > c(\mathrm{E}_{+f})$. In Lemma 3.7.5 we have $c(\mathrm{E}) = (d,1)$, and either $c(\mathrm{E}_{+f}) = (d,0)$ or $c(\mathrm{E}_{+f}) = (0,0)$; thus $c(\mathrm{E}) > c(\mathrm{E}_{+f})$ in Lemma 3.7.5 as well.

Passing to refinements (E_{+f}) of (E) for suitably chosen f in order to lower the complexity of (E) in some sense *unravels* the asymptotic equation. In Sections 13.8, 14.3, and 14.4 we likewise try to unravel asymptotic *differential* equations.

Newton trees

In this subsection K is henselian. The lemmas above and the decrease in complexity upon refinement yield a family $(f_\sigma)_{\sigma\in\Sigma}$ of elements $f_\sigma \in K^\times$ indexed by a finite (possibly empty) set Σ of finite sequences $\sigma = (\sigma_1, \dots, \sigma_k)$, where $k \geqslant 1$ and each σ_j is a positive integer, such that:

(1) Σ is closed under taking initial segments of positive length, that is, whenever $\sigma = (\sigma_1, \ldots, \sigma_k) \in \Sigma$ and $1 \leqslant j \leqslant k$, then $\sigma|j := (\sigma_1, \ldots, \sigma_j) \in \Sigma$;

(2) $\{i \in \mathbb{N}^{\geqslant 1} : (i) \in \Sigma\} = \{1, \ldots, q\}$ with $q \in \mathbb{N}$; the elements $f_{(1)}, \ldots, f_{(q)}$ are approximate zeros of P, and any approximate zero of P is equivalent to $f_{(i)}$ for exactly one $i \in \{1, \ldots, q\}$;

(3) for each $\sigma = (\sigma_1, \ldots, \sigma_k) \in \Sigma$ we have

$$\{i \in \mathbb{N}^{\geqslant 1} : \sigma * i = (\sigma_1, \ldots, \sigma_k, i) \in \Sigma\} = \{1, \ldots, q_\sigma\}, \quad q_\sigma \in \mathbb{N};$$

the elements $f_{\sigma*1}, \ldots, f_{\sigma*q_\sigma}$ are approximate solutions of

$$(\mathrm{E}_\sigma) \qquad P_{+f(\sigma)}(Y) = 0, \qquad Y \prec f_\sigma$$

where $f(\sigma) := \sum_{j=1}^{k} f_{\sigma|j}$; any approximate solution of (E_σ) is equivalent to $f_{\sigma*i}$ for exactly one $i \in \{1, \ldots, q_\sigma\}$; and if $c(\mathrm{E}_\sigma) = (d, 1)$, then $q_\sigma = 1$ and $P^{(d-1)}_{+f(\sigma)}(f_{\sigma*1}) = 0$.

We call such a family $(f_\sigma)_{\sigma\in\Sigma}$ a **Newton tree for P in K**. (Strictly speaking, it is a forest rather than a tree, but below we shall give it a root.) Note that items (2) and (3) contain the instructions for growing a Newton tree for P in K, which is in general not unique. Besides K and the polynomial P this notion of Newton tree also involves the "monomial" set $\mathfrak{M}$, but changing $\mathfrak{M}$ does not affect whether the primary dominant part of (E) has a zero in $\boldsymbol{k}$ of multiplicity equal to its degree. Thus a Newton tree for P in K with respect to $\mathfrak{M}$ remains a Newton tree for P in K with another choice of $\mathfrak{M}$.

Let $(f_\sigma)_{\sigma\in\Sigma}$ be a Newton tree for P in K. If $\sigma = (\sigma_1, \ldots, \sigma_k) \in \Sigma$, then

$$f_{\sigma|1} \succ f_{\sigma|2} \succ \cdots \succ f_{\sigma|k} = f_\sigma,$$

so $f(\sigma) \sim f_{\sigma|1}$, $f(\sigma)$ is an approximate zero of P, and whenever $y - f(\sigma) \prec f_\sigma$ and $1 \leqslant j < k$, then $y - f(\sigma|j)$ is an approximate solution of $(\mathrm{E}_{\sigma|j})$. It follows that if y is an approximate zero of P, then there is a unique $\sigma \in \Sigma$ such that $y - f(\sigma) \prec f_\sigma$ and $y - f(\sigma)$ is not an approximate zero of $P_{+f(\sigma)}$: take $\sigma \in \Sigma$ such that $y - f(\sigma) \prec f_\sigma$ and σ is not a proper initial segment of any $\sigma' \in \Sigma$ with $y - f(\sigma') \prec f_{\sigma'}$.

The case that y is *not* an approximate zero of P is put under this roof as follows: put $\Sigma_0 := \Sigma \cup \{\emptyset\}$, set $f_\emptyset := \infty$, $f(\emptyset) := 0$, and let $y \prec \infty$ for each y, by convention. (Thus the forest Σ becomes a single tree Σ_0 with root $\emptyset$.) Then there is for each y a unique $\sigma \in \Sigma_0$ such that $y - f(\sigma) \prec f_\sigma$ and $y - f(\sigma)$ is not an approximate zero of $P_{+f(\sigma)}$. See Figure 3.5.

This leads to a piecewise uniform description of $v\big(P(y)\big)$ in terms of functions of the form $v\big(y - f(\sigma)\big)$, namely:

LEMMA 3.7.6. *Let $\sigma \in \Sigma_0$ and $P_{+f(\sigma)} = \sum_i b_i Y^i$. If $y - f(\sigma) \prec f_\sigma$ and $y - f(\sigma)$ is not an approximate zero of $P_{+f(\sigma)}$, then*

$$v\big(P(y)\big) = \min\big\{v(b_i) + i \cdot v\big(y - f(\sigma)\big) : i = 0, \ldots, n\big\} \qquad (0 \cdot \infty := 0 \text{ in } \Gamma_\infty).$$

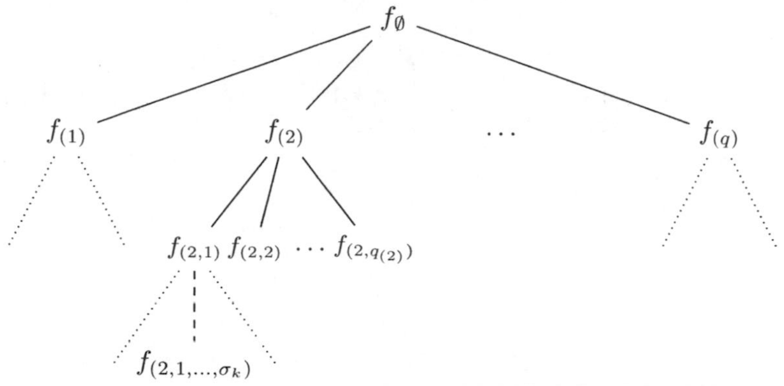

Figure 3.5: Newton tree.

For $\sigma \in \Sigma_0$, let G_σ be the set of all y satisfying the condition in this lemma:

$$G_\sigma := \{y : y - f(\sigma) \prec f_\sigma \text{ and } y - f(\sigma) \text{ is not an approximate zero of } P_{+f(\sigma)}\}.$$

For a more geometric description of G_σ, define a **special disk in** K to be a subset of K that is either K itself or an open ball of the form $\{y : v(y - f) > vf\}$ where $f \in K^\times$. A **special disk in** K **with holes** is a subset of K of the form $D \setminus (D_1 \cup \cdots \cup D_k)$ where D is a special disk in K and $D_1, \dots, D_k$ are disjoint special disks in K properly contained in D. We can now summarize some of the above as follows:

COROLLARY 3.7.7. *For each $\sigma \in \Sigma_0$ the set G_σ is a special disk in K with holes, and each y belongs to G_σ for exactly one $\sigma \in \Sigma_0$. The zeros of P in $K^\times$ are among the $f(\sigma)$ with $\sigma \in \Sigma$.*

Behavior under extension

Suppose K is henselian, and let L be a henselian valued field extension of K such that the residue field of K is algebraically closed in the residue field of L, and $\Gamma_L \cap \mathbb{Q}\Gamma = \Gamma$ (inside $\mathbb{Q}\Gamma_L$). Then any approximate zero of P in L is equivalent to some element in K. It follows that a Newton tree for P in K remains a Newton tree for P in L.

More generally, let $(f_\sigma)_{\sigma \in \Sigma}$ be a Newton tree for P in K and let L be any henselian valued field extension of K. Then we can extend Σ to a finite set $\Sigma_L \supseteq \Sigma$ of finite sequences of positive integers and extend $(f_\sigma)_{\sigma \in \Sigma}$ to a Newton tree $(f_\sigma)_{\sigma \in \Sigma_L}$ of P in L. This is because if a polynomial in $\boldsymbol{k}[Y]$ has degree $d \geqslant 1$ and has a zero of multiplicity d in an extension field of $\boldsymbol{k}$, then this zero lies in $\boldsymbol{k}$.

Notes and comments

Newton diagrams were introduced by Newton [303] (1676) in constructing fractional power series solutions $y = y(x)$ to polynomial equations $P(x, y) = 0$. Puiseux [326] rediscovered this in a complex-analytic setting: the field of convergent Puiseux series with complex coefficients is algebraically closed; see [4, §17], [63, Section 8.3], [80, p. 396], and [176, §38]. The use of derivatives as in the proof of Lemma 3.7.5 goes back to Smith [428], but Tschirnhaus [451] already observed that for a monic polynomial $P \in F[Y]$ of degree $n \geqslant 2$ over a field F and $f \in F$ with $P^{(n-1)}(f) = 0$ one has $P_{+f} = Y^n +$ (terms of degree $\leqslant n-2$). Dumas [116] applied Newton diagrams to algebraic equations over the field of p-adic numbers, and Ostrowski [313, 314] and Rella [332] extended this to more general henselian valued fields.

Early uses of Newton diagrams in the study of algebraic differential equations over the differential field $\mathrm{P}(\mathbb{C})$ of Puiseux series over $\mathbb{C}$ are by Briot-Bouquet [64] and by Fine [137, 138]. For recent uses of Newton diagrams in connection with differential equations over $\mathbb{C}[[x^{\mathbb{R}}]]$ and various differential subfields of it we refer to [67, 99, 135, 155, 283, 328].

The Newton diagrams for differential polynomials in our Chapters 13 and 14 do not seem to be related to the Newton polyhedrons for polynomials in several indeterminates over a discrete valued field in [426, 447].

Chapter Four

Differential Polynomials

Our differential fields are of characteristic zero with one distinguished derivation. We prove here some basic facts about these differential fields and their differential field extensions. In our work we often decompose differential polynomials into parts of a special form and operate formally on differential polynomials in various ways: additive and multiplicative conjugation, Ritt division, composition. Here we study these decompositions and operations in their natural setting.

We also consider valued differential fields, the property of continuity of the derivation with respect to the valuation topology, and the gaussian extension of the valuation to the ring of differential polynomials. Valued differential fields will be studied further in Chapter 6. We finish this chapter with some basic results on simple differential rings, and on differentially closed fields.

4.1 DIFFERENTIAL FIELDS AND DIFFERENTIAL POLYNOMIALS

Recall that derivations on commutative rings were introduced in Section 1.7. When we say below that a commutative ring K *contains* $\mathbb{Q}$ we are abusing language: the meaning of this phrase is that there exists a (necessarily unique) ring embedding of the field $\mathbb{Q}$ into K; as usual we identify $\mathbb{Q}$ in that case with its image in K under this embedding.

Differential rings

A **differential ring** is by definition a commutative ring K containing $\mathbb{Q}$, equipped with a derivation ∂ on K; when ∂ is clear from the context, we set $a' := \partial(a)$, and similarly, $a^{(n)} := \partial^n(a)$, with ∂^n the nth iterate of ∂. The Leibniz identity $(ab)' = a'b + ab'$ then extends as follows: for elements $a_1, \dots, a_m$ $(m \geqslant 1)$ of a differential ring,

$$(a_1 \cdots a_m)^{(n)} = \sum \frac{n!}{i_1! \cdots i_m!} a_1^{(i_1)} \cdots a_m^{(i_m)},$$

where the sum on the right is over all $(i_1, \dots, i_m) \in \mathbb{N}^m$ with $i_1 + \dots + i_m = n$. Given a differential ring K and an element a of a differential ring extension L of K we let $K\{a\} := K[a, a', a'', \dots]$ denote the smallest subring of L containing $K \cup \{a\}$ and closed under ∂: *the differential ring generated by a over K.*

Let K be a differential ring. The subring $\{a \in K : a' = 0\}$ of K contains $\mathbb{Q}$. It is called the **ring of constants** of K, and denoted by C_K (or just C if K is clear from the context). If $c \in C$, then $(ca)' = ca'$ for all $a \in K$. For $a \in K^\times$ we put $a^\dagger := a'/a$, the **logarithmic derivative** of a. Let $a, b \in K^\times$; then $(ab)^\dagger = a^\dagger + b^\dagger$, in particular, $(1/a)^\dagger = -a^\dagger$, and $a^\dagger = b^\dagger$ iff $a = bc$ for some $c \in C$.

Localization

Let R be a differential ring and S a multiplicative subset of R with $0 \notin S$. Then there is a unique derivation on $S^{-1}R$ making $S^{-1}R$ into a differential ring such that the natural map $r \mapsto r/1 : R \to S^{-1}R$ is a morphism of differential rings; it is given by

$$(r/s)' = (r's - rs')/s^2 \quad \text{for } r \in R,\, s \in S,$$

and we always consider $S^{-1}R$ as a differential ring in this way. In particular, if R is a differential integral domain (that is, a differential ring whose underlying ring is an integral domain), then the derivation ∂ of R extends uniquely to a derivation on the fraction field of R.

Differential fields

A **differential field** is a differential ring whose underlying ring is a field (of characteristic 0 since it contains $\mathbb{Q}$). If K is a differential field, then $C = C_K = \{c \in K : c' = 0\}$ is a subfield of K, called the **field of constants** of K. If K is a differential field and L is a field extension of K that is algebraic over K, then we always consider L as the differential field extension of K obtained as in Lemma 1.9.2 by extending the derivation of K to a derivation on L.

LEMMA 4.1.1. *Let K be a differential field. Any element of K that is algebraic over C lies in C. Thus if K is algebraically closed as a field, then so is C, and if K is real closed, then so is C.*

PROOF. Suppose $a \in K$ is algebraic over C, with minimum polynomial $P \in C[X]$ over C. Then $P'(a) \neq 0$ since $\text{char}(C) = 0$, and by the case $n = 1$ of Lemma 1.9.1 we obtain $0 = P'(a)a'$, so $a' = 0$, hence $a \in C$. □

LEMMA 4.1.2. *Let K be a differential field, L a differential field extension of K, and suppose $a \in C_L$ is algebraic over K. Then a is algebraic over C.*

PROOF. Let $P(X)$ be the minimum polynomial of a over K. Then

$$P(a)' = P^\partial(a) + P'(a) \cdot a' = P^\partial(a) = 0,$$

with $\deg P^\partial < \deg P$, so $P^\partial = 0$, and thus $P \in C[X]$. □

Let K be a field of characteristic zero. Then $P \mapsto P' = \partial P/\partial X$ is a derivation on the ring $K[X]$ of polynomials in the indeterminate X over K. This derivation extends to a derivation $R \mapsto \partial R/\partial X$ on the fraction field $K(X)$, with K as its field of constants. Taking logarithmic derivatives gives the following:

LEMMA 4.1.3. *Let* $R = a\cdot(X-b_1)^{k_1}\cdots(X-b_n)^{k_n}$ *where* $a \in K^{\times}$, $b_1,\dots,b_n \in K$, $k_1,\dots,k_n \in \mathbb{Z}$. *Then*

$$\frac{\partial R/\partial X}{R} = \sum_{i=1}^{n} \frac{k_i}{X-b_i}.$$

Assume next that K is a differential field. Then its derivation extends to the derivation $P \mapsto P^{\partial}$ on $K[X]$ with $X^{\partial} = 0$, and we extend this further to a derivation $R \mapsto R^{\partial}$ on $K(X)$. With these notations we get by a routine computation:

COROLLARY 4.1.4. *Let* $P, Q \in K[X]$ *with* $Q \neq 0$, *and set* $R := P/Q \in K(X)$. *Then*

$$R(a)' = R^{\partial}(a) + (\partial R/\partial X)(a)\cdot a' \text{ for } a \in K \text{ with } Q(a) \neq 0.$$

Differential automorphisms

Let K be a differential ring. An **automorphism** of K is an automorphism σ of the ring K such that $\sigma \circ \partial = \partial \circ \sigma$, with ∂ the derivation of K. Let L be a differential ring extension of K. The set of ring automorphisms σ of L with $\sigma(a) = a$ for all $a \in K$ is a group under composition, denoted by $\operatorname{Aut}(L|K)$. The set of automorphisms σ of the differential ring L with $\sigma(a) = a$ for all $a \in K$ is a subgroup of $\operatorname{Aut}(L|K)$, denoted by $\operatorname{Aut}_{\partial}(L|K)$. If both K and L are differential fields and the field extension $L|K$ is algebraic, then $\operatorname{Aut}_{\partial}(L|K) = \operatorname{Aut}(L|K)$. If L is an integral domain and F is the differential fraction field of L, then every $\sigma \in \operatorname{Aut}_{\partial}(L|K)$ extends uniquely to an automorphism $\sigma_F \in \operatorname{Aut}_{\partial}(F|K)$, and the map $\sigma \mapsto \sigma_F\colon \operatorname{Aut}_{\partial}(L|K) \to \operatorname{Aut}_{\partial}(F|K)$ is an embedding of groups.

Differential polynomials

Let K be a differential ring with derivation ∂ and Y a differential indeterminate over K. Then $K\{Y\}$ denotes the ring of differential polynomials in Y over K. As a ring, $K\{Y\}$ is just the polynomial ring $K[Y, Y', Y'', \dots]$ in the distinct indeterminates $Y^{(n)}$ ($n \in \mathbb{N}$) over K, where as usual we write Y, Y', Y'' instead of $Y^{(0)}$, $Y^{(1)}$, $Y^{(2)}$. We consider $K\{Y\}$ as the differential ring whose derivation, extending the derivation of K and also denoted by ∂, is given by $\partial(Y^{(n)}) = Y^{(n+1)}$ for every n. Given $P \in K\{Y\}$, the smallest $r \in \mathbb{N}$ such that $P \in K\big[Y, Y', \dots, Y^{(r)}\big]$ is called the **order** of the differential polynomial P, and the **degree** $\deg P$ of P is its (total) degree as an element of the polynomial ring $K[Y, Y', \dots]$ (with $\deg 0 = -\infty$). It is easy to check that if $P \in K\{Y\}$ with $P \notin K$ has order r, then P' has order $r+1$ and degree 1 in $Y^{(r+1)}$. For $P \in K\{Y\}$ and y an element of a differential ring extension of K, we let $P(y)$ be the element of that extension obtained by substituting $y, y', \dots$ for $Y, Y', \dots$ in P, respectively. In particular, we have $P = P(Y)$ for $P \in K\{Y\}$.

Let L be a differential ring extension of K and $y \in L$. Then the map

$$P \mapsto P(y)\colon\ K\{Y\} \to L$$

is a differential ring morphism, and is in fact the unique differential ring morphism $K\{Y\} \to L$ that is the identity on K and sends Y to y. With $L = K\{Y\}$ this gives a composition operation $(P,Q) \mapsto P \circ Q := P(Q)\colon K\{Y\}^2 \to K\{Y\}$. From the uniqueness we easily get $(P \circ Q)(y) = P\big(Q(y)\big)$ for all $P,Q \in K\{Y\}$ and all y in all differential ring extensions of K, and thus the associativity of composition:

$$(P \circ Q) \circ R \ = \ P \circ (Q \circ R) \quad \text{for } P,Q,R \in K\{Y\}.$$

Let $Y_1,\dots,Y_m$ be distinct differential indeterminates. We define the differential ring $K\{Y_1,\dots,Y_m\}$ inductively by letting it be K for $m=0$ and setting

$$K\{Y_1,\dots,Y_m\} \ := \ K\{Y_1,\dots,Y_{m-1}\}\{Y_m\} \qquad (m>0).$$

This is a domain if K is a domain. *In the rest of this subsection K is a differential field.* The fraction field of $K\{Y_1,\dots,Y_m\}$ is denoted by $K\langle Y_1,\dots,Y_m\rangle$, is given the unique derivation that extends the derivation of $K\{Y_1,\dots,Y_m\}$, and is called the field of **differential rational functions** in $Y_1,\dots,Y_m$ with coefficients in K.

Let L be a differential field extension of K and $y \in L$. Then

$$K\langle y\rangle \ := \ K(y,y',y'',\dots)$$

denotes the differential subfield of L generated over K by y. Likewise, $K\langle y_1,\dots,y_m\rangle$ is the differential subfield of L generated over K by elements $y_1,\dots,y_m \in L$. We say that y is **differentially algebraic** over K if $P(y)=0$ for some $P \in K\{Y\}^{\neq}$, equivalently, $K\langle y\rangle$ has finite transcendence degree over K. We also say that L is **differentially algebraic** over K if each element of L is differentially algebraic over K. If $L = K\langle y_1,\dots,y_m\rangle$ and each $y_i \in L$ is differentially algebraic over K, then L is differentially algebraic over K. If L is differentially algebraic over K and M is a differential field extension of L that is differentially algebraic over L, then M is differentially algebraic over K. If y is not differentially algebraic over K, then y is said to be **differentially transcendental** over K. Thus the differential indeterminate Y, as an element of $K\langle Y\rangle$, is differentially transcendental over K. The equality $Y' = Y\cdot Y^\dagger$ shows that Y is differentially algebraic over $K\langle Y^\dagger\rangle$, so $Y^\dagger$ must be differentially transcendental over K as well. More generally:

LEMMA 4.1.5. *Suppose $y \in L$ is* d*-transcendental over K, and $z \in K\langle y\rangle$, $z \notin K$. Then y is* d*-algebraic over $K\langle z\rangle$, and so z is* d*-transcendental over K.*

Here and below we use the prefix d to abbreviate *differential* or *differentially.* Thus d-*algebraic* stands for *differentially algebraic.*

PROOF. We have $z = P(y)/Q(y)$ where $P,Q \in K\{Y\}^{\neq}$ and $P \neq aQ$ for all $a \in K$. Set $F(Y) := P(Y) - zQ(Y) \in K\langle z\rangle\{Y\}$. From $z \notin K$ we get $F \neq 0$. Then $F(y)=0$ yields that y is d-algebraic over $K\langle z\rangle$. □

Minimal annihilators

In this subsection, K is a differential field and y is an element of a differential field extension of K. A **minimal annihilator** of y over K is an irreducible $P \in K\{Y\}^{\neq}$, say of order r, such that $P(y)=0$ and $Q(y)\neq 0$ for every $Q \in K\{Y\}^{\neq}$ of order at most r with $\deg_{Y^{(r)}} Q < \deg_{Y^{(r)}} P$.

LEMMA 4.1.6. *Let $P \in K\{Y\}^{\neq}$ of order r be irreducible with $P(y) = 0$ and $Q(y) \neq 0$ for all $Q \in K\{Y\}^{\neq}$ of order $< r$. Then P is a minimal annihilator of y over K.*

PROOF. Towards a contradiction, assume that P is not a minimal annihilator of y over K. Take $Q \in K\{Y\}^{\neq}$ of order r with $Q(y) = 0$ and $\deg_{Y^{(r)}} Q < \deg_{Y^{(r)}} P$. So $d := \deg_{Y^{(r)}} Q \geqslant 1$, and we choose Q with these properties such that d is minimal. By polynomial division in $K[Y, Y', \dots, Y^{(r)}]$ we obtain $A \in K[Y, Y', \dots, Y^{(r-1)}]^{\neq}$ and $B \in K[Y, Y', \dots, Y^{(r)}]$ with $AP = BQ + R$ where $\deg_{Y^{(r)}} R < d$. Then $R(y) = 0$, so $R = 0$ by minimality of d, hence P divides BQ. So P divides B, since P is irreducible and $\deg_{Y^{(r)}} Q < \deg_{Y^{(r)}} P$. Hence $B \in P\,K\{Y\}$, so $AB \in AP\,K\{Y\} = BQ\,K\{Y\}$ and thus $0 \neq A \in Q\,K\{Y\}$, contradicting $\operatorname{order}(A) < r = \operatorname{order}(Q)$. □

COROLLARY 4.1.7. *Suppose that y is* d*-algebraic over K. Then y has a minimal annihilator over K. Such a minimal annihilator of y is unique up to multiplication by a factor from $K^{\times}$.*

PROOF. Take $P \in K\{Y\}^{\neq}$ of minimal order r such that $P(y) = 0$. Replacing P by some factor we arrange that P is irreducible. Then by the preceding lemma, P is a minimal annihilator of y over K. Let $P^* \in K\{Y\}^{\neq}$ also be a minimal annihilator of y over K. Then $r = \operatorname{order}(P) = \operatorname{order}(P^*)$ and $d := \deg_{Y^{(r)}} P = \deg_{Y^{(r)}} P^*$, so we have $A, A^* \in K\{Y\}^{\neq}$ of order $< r$ such that $Q := AP + A^*P^*$ has degree $< d$ in $Y^{(r)}$. Then $Q(y) = 0$, so $Q = 0$, and thus by irreducibility of P and P^* we get $P^* = aP$ with $a \in K^{\times}$. □

COROLLARY 4.1.8. *Let y be* d*-algebraic over K with minimal annihilator $P \in K\{Y\}$ over K. Suppose $Q \in K\{Y\}^{\neq}$ is such that $Q(y) = 0$ and* $\operatorname{order}(Q) \leqslant \operatorname{order}(P)$. *Then $Q \in P\,K\{Y\}$, and hence* $\operatorname{order}(Q) = \operatorname{order}(P)$.

PROOF. Replacing Q by a suitable factor we arrange that Q is irreducible, and thus a minimal annihilator of y over K by Lemma 4.1.6. Now apply Corollary 4.1.7. □

Let $P \in K\{Y\}^{\neq}$ be irreducible. Then there is an element a in a differential field extension of K with P as a minimal annihilator over K. For any such a, if b in a differential field extension of K also has P as a minimal annihilator over K, then there is a differential field isomorphism $K\langle a\rangle \to K\langle b\rangle$ over K that sends a to b.

The following two lemmas contain the proofs of these facts, which we include here since they will serve as templates for similar more involved proofs later, for differential fields equipped with additional structure.

LEMMA 4.1.9. *Let $P \in K\{Y\}^{\neq}$ be irreducible. Then there is an element a in a differential field extension of K such that a is* d*-algebraic over K with minimal annihilator P over K.*

PROOF. Let $r = \operatorname{order}(P)$, and take a polynomial $p \in K[Y_0, \dots, Y_r]$ with $P = p(Y, Y', \dots, Y^{(r)})$, so p is irreducible. Consider the integral domain

$$K[y_0, \dots, y_r] := K[Y_0, \dots, Y_r]/(p), \qquad y_i := Y_i + (p) \text{ for } i = 0, \dots, r,$$

and let $F = K(y)$, where $y = (y_0, \dots, y_r)$, be its fraction field. Note that $p(y) = 0$ and $q(y) \neq 0$ for $q \in K[Y_0, \dots, Y_r]$ with $\deg_{Y_r} q < \deg_{Y_r} p$; in particular, we have $(\partial p/\partial Y_r)(y) \neq 0$. We are going to extend ∂ to a derivation on F such that $y_i' = y_{i+1}$ for $i = 0, \dots, r-1$. We first set

$$y_{r+1} := -\frac{p^{\partial}(y) + \sum_{i=0}^{r-1}(\partial p/\partial Y_i)(y) \cdot y_{i+1}}{(\partial p/\partial Y_r)(y)} \quad \text{in } K(y),$$

which by Lemma 1.9.1 is the value that y_r' will necessarily have, for any derivation on F extending ∂ with $y_i' = y_{i+1}$ for $i = 0, \dots, r-1$. Next we define the additive map $d \colon K[Y_0, \dots, Y_r] \to F$ by

$$d(f) = f^{\partial}(y) + \sum_{i=0}^{r} \frac{\partial f}{\partial Y_i}(y) \cdot y_{i+1}.$$

As in the proof of Lemma 1.9.2 we check that the kernel of d contains (p) and that the induced additive map $K[y] \to F$ is a derivation into F, which by Corollary 1.8.6 extends uniquely to a derivation on F. This derivation extends ∂, and setting $a := y_0$ we have $a^{(i)} = y_i$ for $i = 0, \dots, r$. This a has the desired property. □

LEMMA 4.1.10. *Let a and b in differential field extensions of K be* d*-algebraic over K with common minimal annihilator $P \in K\{Y\}$ over K. Then there is a differential field isomorphism $K\langle a\rangle \to K\langle b\rangle$ over K sending a to b.*

PROOF. Take $p \in K[Y_0, \dots, Y_r]$ be such that $P = p(Y, Y', \dots, Y^{(r)})$, where $r := \operatorname{order}(P)$. Put $\vec{a} := \big(a, a', \dots, a^{(r)}\big)$ and $\vec{b} := \big(b, b', \dots, b^{(r)}\big)$. The ring morphism

$$f \mapsto f(\vec{a}) : K[Y_0, \dots, Y_r] \to K[\vec{a}\,]$$

has kernel (p), and likewise with b instead of a, so we have a ring isomorphism $K[\vec{a}\,] \to K[\vec{b}\,]$ that sends $a^{(i)}$ to $b^{(i)}$ for $i = 0, \dots, r$. Since P is a minimal annihilator of a over K, we have $(\partial p/\partial Y_r)(\vec{a}\,) \neq 0$, so Lemma 1.9.1 gives

$$a^{(r+1)} = -\frac{p^{\partial}(\vec{a}\,) + \sum_{i=0}^{r-1}(\partial p/\partial Y_i)(\vec{a}\,) \cdot a^{(i+1)}}{(\partial p/\partial Y_r)(\vec{a}\,)} \in K(\vec{a}\,),$$

so $K[\vec{a}\,]' \subseteq K(\vec{a}\,)$ (as sets) and hence $K[\vec{a}] \subseteq K\{a\} \subseteq K(\vec{a}\,)$ (as rings). Likewise with b instead of a, so the ring isomorphism $K[\vec{a}\,] \to K[\vec{b}\,]$ from above extends to a differential field isomorphism $K\langle a\rangle \to K\langle b\rangle$ as desired. □

Let $P \in K\{Y\}^{\neq}$ be irreducible of order $r \in \mathbb{N}$, and set

$$S := \partial P/\partial Y^{(r)} \in K\big[Y, \dots, Y^{(r)}\big].$$

(In Section 4.3 we call S the *separant* of P.) Let a be an element of a differential field extension of K with minimal annihilator P over K. The proofs above show that then

$S(a) \neq 0$ and that the subring $K\left[a, a', \dots, a^{(r)}, \frac{1}{S(a)}\right]$ of $K\langle a \rangle$ is closed under the derivation of $K\langle a \rangle$. Therefore,

$$K\{a\} \subseteq K\left[a, a', \dots, a^{(r)}, \frac{1}{S(a)}\right], \qquad K\langle a \rangle = K\big(a, a', \dots, a^{(r)}\big)$$

with $a, a', \dots, a^{(r-1)}$ as a transcendence basis for the field extension $K\langle a \rangle \supseteq K$. Thus if b is an element of a differential field extension of K with $P(b) = 0$ and $S(b) \neq 0$, then we have a differential ring morphism $K\left[a, a', \dots, a^{(r)}, \frac{1}{S(a)}\right] \to K\langle b \rangle$ over K sending a to b, which restricts to a differential ring morphism $K\{a\} \to K\{b\}$.

COROLLARY 4.1.11. *Let y be* d*-algebraic over K. Then*

$$\operatorname{trdeg}\big(K\langle y \rangle | K\big) = \textit{order of a minimal annihilator of } y \textit{ over } K.$$

Differential transcendence bases

Let $L \supseteq K$ be an extension of differential fields. A set $E \subseteq L$ generates over K the differential field $K\langle E \rangle \subseteq L$, and we put

$$\operatorname{cl} E := \big\{ f \in L : f \text{ is d-algebraic over } K\langle E \rangle \big\}.$$

Then $\operatorname{cl} E$ is a differential subfield of L.

LEMMA 4.1.12. *Let $E \subseteq L$ and $a, b \in L$. Then:*

(i) $E \subseteq \operatorname{cl} E$;

(ii) $\operatorname{cl} E = \bigcup \{\operatorname{cl} E_0 : E_0 \subseteq E \textit{ is finite}\}$;

(iii) $\operatorname{cl}(\operatorname{cl} E) = \operatorname{cl} E$;

(iv) $a \notin \operatorname{cl} E,\ b \notin \operatorname{cl}\big(E \cup \{a\}\big) \Rightarrow a \notin \operatorname{cl}\big(E \cup \{b\}\big)$.

PROOF. Parts (i)–(iii) are clear. For (iv), put $F := \operatorname{cl} E$, and assume $a \notin F$ and b is d-transcendental over $F\langle a \rangle$. Then the family $\big(a^{(m)}\big)$ is algebraically independent over F and the family $\big(b^{(n)}\big)$ is algebraically independent over $F\langle a \rangle$. It follows that their "union" $\big(a^{(m)}, b^{(n)}\big)$ is algebraically independent over F. Hence $\big(a^{(m)}\big)$ is algebraically independent over $F\langle b \rangle$, that is, a is d-transcendental over $F\langle b \rangle$. □

Lemma 4.1.12 says that the operation $E \mapsto \operatorname{cl} E$ (for $E \subseteq L$) is a *pregeometry* on L. This yields a notion of independence, analogous to linear independence in vector spaces and algebraic independence for field extensions. Below we just formulate this independence notion in our situation and state the relevant facts about it. (These facts are special cases of generalities about pregeometries that can be found in many places, for example in [61, V, §5, Exercice 14], in [85, VII, §2], and in [284, Section 8.1].)

Let E range over subsets of L in the rest of this subsection. We say that E is d**-algebraically independent** over K if $x \notin \operatorname{cl}(E \setminus \{x\})$ for all $x \in E$. So E

is d-algebraically independent over K iff the family $\left(x^{(n)}\right)_{x\in E,\ n=0,1,2,\dots}$ is algebraically independent over K. Call E a **d-transcendence basis** of L over K if E is d-algebraically independent over K and L is d-algebraic over $K\langle E\rangle$. If E is d-algebraically independent over K, then E is contained in a d-transcendence basis of L over K. If $L = \operatorname{cl} E$, then E contains a d-transcendence basis of L over K. All d-transcendence bases of L over K have the same cardinality, which is called the **differential transcendence degree** of L over K, denoted by $\operatorname{trdeg}_\partial L|K$. So $\operatorname{trdeg}_\partial L|K = 0$ iff L is d-algebraic over K. Hence $\operatorname{trdeg}_\partial L|K = 0$ if the derivation of L is trivial. If M is a differential field extension of L, then

$$\operatorname{trdeg}_\partial M|K = \operatorname{trdeg}_\partial M|L + \operatorname{trdeg}_\partial L|K.$$

The zeros of a linear differential polynomial

Let K be a differential field and suppose the differential polynomial $A(Y) \in K\{Y\}$ is homogeneous of degree 1 and order $r \in \mathbb{N}$. Then the map $y \mapsto A(y)\colon K \to K$ is C-linear, so the set of zeros

$$Z(A) := \{y \in K : A(y) = 0\}$$

of A in K is a C-linear subspace of K. Towards proving the well-known fact that the dimension of this C-linear space is at most r, we define the **Wronskian matrix** of a $(1+n)$-tuple $(y_0,\dots,y_n)$ of elements of a differential ring by

$$\operatorname{Wr}(y_0,\dots,y_n) := \begin{pmatrix} y_0 & y_1 & \cdots & y_n \\ y_0' & y_1' & \cdots & y_n' \\ \vdots & \vdots & \ddots & \vdots \\ y_0^{(n)} & y_1^{(n)} & \cdots & y_n^{(n)} \end{pmatrix},$$

with determinant

$$\operatorname{wr}(y_0,\dots,y_n) := \det \operatorname{Wr}(y_0,\dots,y_n),$$

the **Wronskian** of $(y_0,\dots,y_n)$. We first show:

LEMMA 4.1.13. *Let $y_0,\dots,y_n \in K$. Then*

$$\operatorname{wr}(y_0,\dots,y_n) = 0 \iff y_0,\dots,y_n \text{ are } C\text{-linearly dependent.}$$

PROOF. Suppose $c_0,\dots,c_n \in C$ are not all zero and $\sum_{i=0}^n c_i y_i = 0$. Taking successive derivatives yields $\sum_{i=0}^n c_i y_i^{(j)} = 0$ for each j, showing that the columns of $\operatorname{Wr}(y_0,\dots,y_n)$ are linearly dependent (over C), so $\operatorname{wr}(y_0,\dots,y_n) = 0$. This yields "$\Leftarrow$," and we prove "$\Rightarrow$" by induction on n. The case $n = 0$ being trivial, suppose $\operatorname{wr}(y_0,\dots,y_n) = 0$ and $n > 0$. Thus there are $a_0,\dots,a_n \in K$, not all zero, such that $\sum_{i=0}^n a_i y_i^{(j)} = 0$ for $j = 0,\dots,n$. After reindexing and normalization, we may assume that $a_0 = 1$, so

$$y_0^{(j)} + \sum_{i=1}^n a_i y_i^{(j)} = 0 \qquad \text{for } j = 0,\dots,n. \tag{4.1.1}$$

Taking derivatives in (4.1.1) for $j = 0, \dots, n-1$ yields

$$y_0^{(j+1)} + \sum_{i=1}^{n} a_i y_i^{(j+1)} + \sum_{i=1}^{n} a_i' y_i^{(j)} = 0 = \sum_{i=1}^{n} a_i' y_i^{(j)} \qquad \text{for } j = 0, \dots, n-1.$$

Hence, if $a_i' \neq 0$ for some $i \in \{1, \dots, n\}$, then $\operatorname{wr}(y_1, \dots, y_n) = 0$, so $y_1, \dots, y_n$ are C-linearly dependent by inductive hypothesis. If $a_i' = 0$ for all $i \in \{1, \dots, n\}$, then $y_0, \dots, y_n$ are C-linearly dependent by (4.1.1) for $j = 0$. $\square$

Thus if $y_0, \dots, y_n \in K$ are linearly independent over C and L is a differential field extension of K, then $y_0, \dots, y_n$ remain linearly independent over C_L.

COROLLARY 4.1.14. $\dim_C Z(A) \leqslant r$.

PROOF. We may assume that $A(Y) = Y^{(r)} + \sum_{i=0}^{r-1} a_i Y^{(i)}$ where $a_0, \dots, a_{r-1} \in K$. Let $y_0, \dots, y_r \in Z(A)$. Then the last row of

$$\operatorname{Wr}(y_0, \dots, y_r) = \begin{pmatrix} y_0 & y_1 & \cdots & y_r \\ y_0' & y_1' & \cdots & y_r' \\ \vdots & \vdots & \ddots & \vdots \\ -\sum a_i y_0^{(i)} & -\sum a_i y_1^{(i)} & \cdots & -\sum a_i y_r^{(i)} \end{pmatrix}$$

is a K-linear combination of the preceding rows, so $\operatorname{wr}(y_0, \dots, y_r) = 0$. Now Lemma 4.1.13 tells us that $y_0, \dots, y_r$ are C-linearly dependent. $\square$

COROLLARY 4.1.15. *Let $y_0, \dots, y_r \in K$ be linearly independent over C and let $z_0, \dots, z_r \in K$. Then there is a unique homogeneous $A \in K\{Y\}$ of degree 1 and order at most r such that $A(y_i) = z_i$ for $i = 0, \dots, r$.*

PROOF. By Lemma 4.1.13, the matrix $W := \operatorname{Wr}(y_0, \dots, y_r)$ is invertible. Take $a_0, \dots, a_r \in K$ such that $(a_0, \dots, a_r)W = (z_0, \dots, z_r)$, and set

$$A := a_r Y^{(r)} + a_{r-1} Y^{(r-1)} + \cdots + a_0 Y.$$

Then $A(y_i) = z_i$ for $i = 0, \dots, r$. If $B \in K\{Y\}$ is also homogeneous of degree 1 and order at most r with $B(y_i) = z_i$ for $i = 0, \dots, r$, then $\dim_C Z(A - B) > r$ and hence $A = B$ by Corollary 4.1.14. $\square$

We record a few more useful and well-known facts about Wronskians. Let

$$A = Y^{(r)} + a_{r-1} Y^{(r-1)} + \cdots + a_0 Y \qquad \text{where } a_0, \dots, a_{r-1} \in K,\ r \geqslant 1.$$

COROLLARY 4.1.16. *Let $y_1, \dots, y_r$ be a basis of $Z(A)$ (so $\dim_C Z(A) = r$). Then*

$$A(Y) = \frac{\operatorname{wr}(y_1, \dots, y_r, Y)}{\operatorname{wr}(y_1, \dots, y_r)} \quad \text{in } K\langle Y\rangle.$$

PROOF. The right-hand quotient equals $B(Y) = Y^{(r)} + b_{r-1}Y^{(r-1)} + \cdots + b_0 Y$ with $b_0, \ldots, b_{r-1} \in K$, so $A - B$ is homogeneous of degree 1 and of order $< r$. Since $y_i \in Z(A - B)$ for $i = 1, \ldots, r$, this gives $A = B$ by Corollary 4.1.14. □

The following fact is known as Abel's identity:

LEMMA 4.1.17. *Let $y_1, \ldots, y_r \in Z(A)$ and $w := \mathrm{wr}(y_1, \ldots, y_r)$. Then*

$$w' = -a_{r-1}w.$$

PROOF. Expressing the determinant w in the usual way as a sum of $r!$ products, and differentiating gives $w' = w_1 + \cdots + w_r$ where w_i is the determinant of the matrix W_i obtained from $\mathrm{Wr}(y_1, \ldots, y_r)$ by differentiating its ith row. For $i = 1, \ldots, r-1$, each matrix W_i contains two identical rows, so $w_i = 0$. Thus

$$w' = w_r = \det W_r = \det \begin{pmatrix} y_1 & \cdots & y_r \\ y_1' & \cdots & y_r' \\ \vdots & \ddots & \vdots \\ y_1^{(r-2)} & \cdots & y_r^{(r-2)} \\ y_1^{(r)} & \cdots & y_r^{(r)} \end{pmatrix}.$$

Using row operations and $A(y_i) = 0$ for $i = 1, \ldots, r$, this equals

$$\det \begin{pmatrix} y_1 & \cdots & y_r \\ y_1' & \cdots & y_r' \\ \vdots & \ddots & \vdots \\ y_1^{(r-2)} & \cdots & y_r^{(r-2)} \\ -a_{r-1}y_1^{(r-1)} & \cdots & -a_{r-1}y_r^{(r-1)} \end{pmatrix} = -a_{r-1}w,$$

as required. □

LEMMA 4.1.18. *Let M be an $n \times n$ matrix over C, $n \geqslant 1$, and let $y = (y_1, \ldots, y_n)$ and $z = (z_1, \ldots, z_n) \in K^n$ be row vectors with $z = yM$. Then*

$$\mathrm{Wr}(z_1, \ldots, z_n) = \mathrm{Wr}(y_1, \ldots, y_n) \cdot M, \quad \mathrm{wr}(z_1, \ldots, z_n) = \mathrm{wr}(y_1, \ldots, y_n) \cdot \det M.$$

PROOF. We have $(z_1^{(i)}, \ldots, z_n^{(i)}) = (y_1^{(i)}, \ldots, y_n^{(i)}) \cdot M$ for each $i \in \mathbb{N}$. □

Notes and comments

Abstract differential fields were introduced by Baer [32]. In the older literature one also finds *hypertranscendental* [280] and *transcendentally transcendental* [294] instead of the terminology *differentially transcendental* introduced by Kolchin [218]. The notion of d-algebraic independence and the basic facts about it are due to Raudenbush [329]. Many of the foundational results in differential algebra are due to Ritt and Kolchin. A comprehensive reference on this subject is Kolchin's book [221]. In some places, like [209], a *differential ring* is a commutative ring K equipped with a derivation on K, without requiring that K contains $\mathbb{Q}$ as a subring. (What we call a differential ring is called a *Ritt algebra* in [209].) A history of differential algebra can be found in [66]. Lemma 4.1.17 dates back to [1].

4.2 DECOMPOSITIONS OF DIFFERENTIAL POLYNOMIALS

In this section K is a differential ring and Y a differential indeterminate. We let ∂ denote the derivation of K as well as of $K\{Y\}$. Let $P = P(Y) \in K\{Y\}$ be a differential polynomial of order at most $r \in \mathbb{N}$. We indicate some useful ways of expressing P as a sum of differential polynomials of a special form.

Natural decomposition

For $\boldsymbol{i} = (i_0, \dots, i_r) \in \mathbb{N}^{1+r}$ and $Q \in K\{Y\}$, put

$$Q^{\boldsymbol{i}} := Q^{i_0}(Q')^{i_1} \cdots (Q^{(r)})^{i_r}.$$

In particular, $Y^{\boldsymbol{i}} = Y^{i_0}(Y')^{i_1} \cdots (Y^{(r)})^{i_r}$, and $y^{\boldsymbol{i}} = y^{i_0}(y')^{i_1} \cdots (y^{(r)})^{i_r}$ for $y \in K$. Let $P_{\boldsymbol{i}} \in K$ be the coefficient of $Y^{\boldsymbol{i}}$ in P; then

$$P(Y) = \sum_{\boldsymbol{i}} P_{\boldsymbol{i}} Y^{\boldsymbol{i}} \quad \textbf{(natural decomposition)}.$$

Decomposition into homogeneous parts

For each $\boldsymbol{i}$ as above we put $|\boldsymbol{i}| := i_0 + \cdots + i_r$, and for $d \in \mathbb{N}$ we let

$$P_d(Y) := \sum_{|\boldsymbol{i}|=d} P_{\boldsymbol{i}} Y^{\boldsymbol{i}},$$

the homogeneous part of degree d of P. Note that then

$$P(Y) = \sum_d P_d(Y) \quad \textbf{(decomposition into homogeneous parts)}.$$

For $d = 0$ and $d = 1$ this gives $P_0 = P(0) \in K$, and

$$P_1 = \sum_{n=0}^{r} \frac{\partial P}{\partial Y^{(n)}}(0) Y^{(n)}, \quad P_1 \in \sum_{n=0}^{r} K \cdot Y^{(n)},$$

where $\frac{\partial P}{\partial Y^{(n)}}$ is the formal partial derivative of P with respect to the variable $Y^{(n)}$, which has nothing to do with the derivation ∂ of $K\{Y\}$. Given also $Q \in K\{Y\}$ we have $(PQ)_1 = P(0)Q_1 + Q(0)P_1$. For nonzero P we have

$$\deg(P) = \max\{d : P_d \neq 0\},$$

and we define the **multiplicity of P at** 0 as

$$\operatorname{mul}(P) := \min\{d : P_d \neq 0\}.$$

We also set $\operatorname{mul}(0) := \infty > \mathbb{N}$. Then for $Q \in K\{Y\}$, with the usual addition in $\mathbb{N} \cup \{\infty\}$:

$$\operatorname{mul}(PQ) \geqslant \operatorname{mul}(P) + \operatorname{mul}(Q), \qquad \operatorname{mul}(P+Q) \geqslant \min\big(\operatorname{mul}(P), \operatorname{mul}(Q)\big),$$

and if K is an integral domain, then $\operatorname{mul}(PQ) = \operatorname{mul}(P) + \operatorname{mul}(Q)$. We say that P is **homogeneous of degree** d if $P = P_d$.

Decomposition along orders

Let S^* be the set of words on a set S. For any word $\boldsymbol{\sigma} = \sigma_1 \cdots \sigma_n \in \{0, \dots, r\}^*$ of length $|\boldsymbol{\sigma}| = n$ and $Q \in K\{Y\}$ we put

$$Q^{[\boldsymbol{\sigma}]} := Q^{(\sigma_1)} \cdots Q^{(\sigma_n)},$$

so for each permutation π of $\{1, \dots, n\}$ we have $Q^{[\boldsymbol{\sigma}]} = Q^{[\pi(\boldsymbol{\sigma})]}$, where $\pi(\boldsymbol{\sigma}) := \sigma_{\pi(1)} \cdots \sigma_{\pi(n)}$. Thus

$$P(Y) = \sum_{\boldsymbol{\sigma}} P_{[\boldsymbol{\sigma}]} Y^{[\boldsymbol{\sigma}]} \quad \textbf{(decomposition along orders)}$$

where the coefficients $P_{[\boldsymbol{\sigma}]} \in K$ are uniquely determined by the requirements that $P_{[\boldsymbol{\sigma}]} = P_{[\pi(\boldsymbol{\sigma})]}$ for each permutation π of $\{1, \dots, |\boldsymbol{\sigma}|\}$ (and of course $P_{[\boldsymbol{\sigma}]} = 0$ for all but finitely many $\boldsymbol{\sigma}$). For example, with $P = YY'$ and $r = 1$, the words $\boldsymbol{\sigma}$ on $\{0, 1\}^*$ with $|\boldsymbol{\sigma}| = \deg P = 2$ are 00, 01, 10, and 11, with $Y^{[00]} = Y^2$, $Y^{[01]} = Y^{[10]} = YY'$, and $Y^{[11]} = (Y')^2$, so $P_{[01]} = P_{[10]} = 1/2$, and $P_{[00]} = P_{[11]} = 0$.

To find expressions relating the $P_{\boldsymbol{i}}$'s and $P_{[\boldsymbol{\sigma}]}$'s, we first note that $Y^{[\boldsymbol{\sigma}]} = Y^{\boldsymbol{i}(\boldsymbol{\sigma})}$, where $\boldsymbol{i}(\boldsymbol{\sigma}) = (i_0, \dots, i_r)$ with

$$i_k = \big|\{j \in \{1, \dots, |\boldsymbol{\sigma}|\} : \sigma_j = k\}\big|.$$

Thus $|\boldsymbol{i}(\boldsymbol{\sigma})| = |\boldsymbol{\sigma}|$. Using this notation we have

$$P_{\boldsymbol{i}} = \sum_{\boldsymbol{i}(\boldsymbol{\sigma}) = \boldsymbol{i}} P_{[\boldsymbol{\sigma}]}.$$

Putting $\boldsymbol{i}! := i_0! \cdots i_r!$ and $\binom{n}{\boldsymbol{i}} := \frac{n!}{\boldsymbol{i}!}$ we have

$$P_{[\boldsymbol{\sigma}]} = \frac{P_{\boldsymbol{i}(\boldsymbol{\sigma})}}{\big|\{\boldsymbol{\tau} : |\boldsymbol{\tau}| = |\boldsymbol{\sigma}|,\ \boldsymbol{i}(\boldsymbol{\tau}) = \boldsymbol{i}(\boldsymbol{\sigma})\}\big|} = \binom{|\boldsymbol{\sigma}|}{\boldsymbol{i}(\boldsymbol{\sigma})}^{-1} P_{\boldsymbol{i}(\boldsymbol{\sigma})}.$$

Taylor expansion

For $\boldsymbol{i} = (i_0, \dots, i_r)$ and $\boldsymbol{j} = (j_0, \dots, j_r)$ in $\mathbb{N}^{1+r}$ we define

$$\boldsymbol{i} \leqslant \boldsymbol{j} \quad :\Longleftrightarrow \quad i_0 \leqslant j_0, \dots, i_r \leqslant j_r,$$

and if $\boldsymbol{i} \leqslant \boldsymbol{j}$ we put $\binom{\boldsymbol{j}}{\boldsymbol{i}} := \binom{j_0}{i_0} \cdots \binom{j_r}{i_r}$. It is also convenient to define, for $\boldsymbol{i} \in \mathbb{N}^{1+r}$:

$$P_{(\boldsymbol{i})} := \frac{P^{(\boldsymbol{i})}}{\boldsymbol{i}!} \qquad \text{where } P^{(\boldsymbol{i})} := \frac{\partial^{|\boldsymbol{i}|} P}{\partial^{i_0} Y \cdots \partial^{i_r} Y^{(r)}}.$$

(Here the right-hand side is an iterated partial derivative of P with respect to the variables $Y, \dots, Y^{(r)}$ and has nothing to do with the derivation ∂ of $K\{Y\}$.) Thus, if $|\boldsymbol{i}| = 0$, then $P_{(\boldsymbol{i})} = P$, and if $|\boldsymbol{i}| = 1$, then $P_{(\boldsymbol{i})}$ is one of the $\frac{\partial P}{\partial Y^{(k)}}$. We have

$$\deg P_{(\boldsymbol{i})} \leqslant \deg P - |\boldsymbol{i}|, \qquad \text{so } P_{(\boldsymbol{i})} = 0 \text{ if } |\boldsymbol{i}| > \deg P.$$

Repeated application of Lemma 3.2.2 gives $(P_{(\boldsymbol{i})})_{(\boldsymbol{j})} = \binom{\boldsymbol{i}+\boldsymbol{j}}{\boldsymbol{j}} P_{(\boldsymbol{i}+\boldsymbol{j})}$ for $\boldsymbol{i}, \boldsymbol{j} \in \mathbb{N}^{1+r}$, where $\mathbb{N}^{1+r}$ is construed as a monoid under pointwise addition. Let Z be a new differential indeterminate. Then

$$P(Y+Z) \;=\; \sum_{\boldsymbol{i}} P_{(\boldsymbol{i})}(Y) Z^{\boldsymbol{i}} \quad \textbf{(Taylor expansion)}.$$

Here is an application:

LEMMA 4.2.1. *Suppose K is a differential field and $y' \neq 0$ for some $y \in K$. If $P \neq 0$, then $P(y) \neq 0$ for some $y \in K$.*

PROOF. If $y \in K$ is algebraic over C, then $y \in C$. Thus K has infinite dimension as a vector space over C. It follows that if $a_0, \dots, a_r \in K$ are not all zero, then $a_0 b + \cdots + a_r b^{(r)} \neq 0$ for some $b \in K$, by Corollary 4.1.14. Assume $P \notin K$ has order r and (total) degree $d \geqslant 1$. Then we have for all $b \in K$,

$$P(Y+b) - P(Y) \;=\; \sum_{k=0}^{r} \frac{\partial P}{\partial Y^{(k)}} \cdot b^{(k)} + \sum_{|\boldsymbol{i}| \geqslant 2} P_{(\boldsymbol{i})}(Y) \cdot b^{\boldsymbol{i}} \;=:\; Q_b(Y),$$

with $\deg(Q_b) < d$. Take $\boldsymbol{i} = (i_0, \dots, i_r) \in \mathbb{N}^{1+r}$ such that $|\boldsymbol{i}| = d$ and $Y^{\boldsymbol{i}}$ occurs in P. Take $k \in \{0, \dots, r\}$ with $i_k \neq 0$. Then $\frac{\partial P}{\partial Y^{(k)}}$ contains the monomial $Y^{\boldsymbol{j}}$ with $\boldsymbol{j} = (i_0, \dots, i_{k-1}, i_k - 1, i_{k+1}, \dots, i_r)$ of degree $d-1$. This monomial does not occur in $\sum_{|\boldsymbol{i}| \geqslant 2} P_{(\boldsymbol{i})}(Y) \cdot b^{\boldsymbol{i}}$, so the coefficient of this monomial in $Q_b(Y)$ is $a_0 b + \cdots + a_r b^{(r)}$ where $a_0, \dots, a_r \in K$ are independent of b, with $a_k \neq 0$. Now take $b \in K$ such that $a_0 b + \cdots + a_r b^{(r)} \neq 0$, so $Q_b \neq 0$. We can assume inductively that $Q_b(y) \neq 0$ for some $y \in K$. Then $P(y+b) \neq 0$ or $P(y) \neq 0$ for such y, by the identity above. □

Suppose K is a differential field with nontrivial derivation, that is, $y' \neq 0$ for some $y \in K$. Call a set $S \subseteq K$ **thin** (in K) if there is an $F \in K\{Y\}^{\neq}$ such that $F(y) = 0$ for all $y \in S$. Note that a finite union of thin subsets of K is thin, and that K is not thin by Lemma 4.2.1. Given any $F \in K\{Y\}^{\neq}$ the set

$$\{y \in K^{\times} : F(y^{\dagger}) = 0\}$$

is thin, since in $K\langle Y \rangle$ we have $0 \neq Y^n F(Y^{\dagger}) \in K\{Y\}$ for some n.

LEMMA 4.2.2. *Let $L|K$ be a differential field extension of the differential field K. Assume $y' \neq 0$ for some $y \in K$ and $S \subseteq L$ is thin in L. Then $S \cap K$ is thin in K.*

PROOF. Take $P \in L\{Y\}^{\neq}$ such that $S \subseteq \{y \in L : P(y) = 0\}$. We have

$$P \;=\; b_1 P_1 + \cdots + b_n P_n$$

where $b_1, \dots, b_n \in L$ are linearly independent over K, $P_1, \dots, P_n \in K\{Y\}$ are nonzero, and $n \geqslant 1$. Then $S \cap K \subseteq \{y \in K : P_1(y) = 0\}$. □

Decomposition into isobaric parts

In this subsection $P, Q \in K\{Y\}$ are of order at most $r \in \mathbb{N}$. Also, $\boldsymbol{i} = (i_0, \dots, i_r) \in \mathbb{N}^{1+r}$, $\|\boldsymbol{i}\| := i_1 + 2i_2 + \cdots + ri_r$, we let $\boldsymbol{\sigma}$ range over words in $\{0, \dots, r\}^*$, and for $\boldsymbol{\sigma} = \sigma_1 \cdots \sigma_d$ we put $\|\boldsymbol{\sigma}\| := \sigma_1 + \cdots + \sigma_d$, so $\|\boldsymbol{\sigma}\| = \|\boldsymbol{i}(\boldsymbol{\sigma})\|$. Define the **weight** $\operatorname{wt}(P) \in \mathbb{N} \cup \{-\infty\}$ of P by

$$\operatorname{wt}(P) := \max\left\{\|\boldsymbol{\sigma}\| : P_{[\boldsymbol{\sigma}]} \neq 0\right\} \text{ if } P \neq 0, \qquad \operatorname{wt}(0) := -\infty < \mathbb{N}.$$

In particular,

$$\operatorname{wt}(Y^{\boldsymbol{i}}) = \|\boldsymbol{i}\|, \quad \operatorname{wt}(P_{(\boldsymbol{i})}) \leqslant \operatorname{wt}(P) - \|\boldsymbol{i}\|,$$

and

$$\operatorname{wt}(P+Q) \leqslant \max\bigl(\operatorname{wt}(P), \operatorname{wt}(Q)\bigr).$$

Define the **weighted multiplicity** $\operatorname{wm}(P) \in \mathbb{N} \cup \{\infty\}$ of P at 0 by

$$\operatorname{wm}(P) := \min\left\{\|\boldsymbol{\sigma}\| : P_{[\boldsymbol{\sigma}]} \neq 0\right\} \text{ if } P \neq 0, \qquad \operatorname{wm}(0) := \infty > \mathbb{N}.$$

In particular, $\operatorname{wm}(Y^{\boldsymbol{i}}) = \operatorname{wt}(Y^{\boldsymbol{i}}) = \|\boldsymbol{i}\|$, and $\operatorname{wm}(P+Q) \geqslant \min\bigl(\operatorname{wm}(P), \operatorname{wm}(Q)\bigr)$. We say that P is **isobaric** if $P_{[\boldsymbol{\sigma}]} = 0$ for all $\boldsymbol{\sigma}$ with $\|\boldsymbol{\sigma}\| \neq \operatorname{wt}(P)$. Note that 0 is isobaric, and that if $P \neq 0$, then P is isobaric iff $\operatorname{wm}(P) = \operatorname{wt}(P)$. (Although $\deg 0 = \operatorname{wt} 0 = -\infty$, we do consider 0 to be homogeneous of any degree $d \in \mathbb{N}$ and to be isobaric of any weight $w \in \mathbb{N}$.) Degree and weight behave as follows under differentiation: $(Y^{\boldsymbol{i}})'$ is homogeneous of the same degree $|\boldsymbol{i}|$ as $Y^{\boldsymbol{i}}$, and is isobaric of weight $\operatorname{wt}(Y^{\boldsymbol{i}})+1 = \|\boldsymbol{i}\|+1$. Thus if $P \in C\{Y\}$ is isobaric of weight $w \in \mathbb{N}$, then P' is isobaric of weight $w+1$. For $w \in \mathbb{N}$, let

$$P_{[w]} := \sum_{\|\boldsymbol{\sigma}\|=w} P_{[\boldsymbol{\sigma}]} Y^{[\boldsymbol{\sigma}]}$$

be the **isobaric part of P of weight w.** Then $P_{[w]} = 0$ for $w > \operatorname{wt}(P)$, and

$$P = \sum_w P_{[w]} \qquad \textbf{(decomposition into isobaric parts)}.$$

Note that $(PQ)_{[w]} = \sum_{i+j=w} P_{[i]} Q_{[j]}$. In particular, if P and Q are isobaric, then so is PQ. Moreover:

LEMMA 4.2.3. *If K is an integral domain, then*

$$\operatorname{wm}(PQ) = \operatorname{wm}(P) + \operatorname{wm}(Q), \qquad \operatorname{wt}(PQ) = \operatorname{wt}(P) + \operatorname{wt}(Q).$$

Decomposition into subhomogeneous parts

In Chapter 13 we use yet another decomposition of P. For $\boldsymbol{i} = (i_0, \dots, i_r) \in \mathbb{N}^{1+r}$ define the **subdegree of $\boldsymbol{i}$** as $|\boldsymbol{i}|' := i_1 + \cdots + i_r$, and for $d \in \mathbb{N}$, set $P_{|d|'} := \sum_{|\boldsymbol{i}|'=d} P_{\boldsymbol{i}} Y^{\boldsymbol{i}}$, so

$$P = \sum_d P_{|d|'} \quad \textbf{(decomposition into subhomogeneous parts)}.$$

The **subdegree of** P, $\operatorname{sdeg}(P)$, is the largest d with $P_{|d|'} \neq 0$ if $P \neq 0$, and $\operatorname{sdeg}(P) := -\infty$ if $P = 0$. We say that P is **subhomogeneous** of subdegree d if $P = P_{|d|'}$. Thus P is subhomogeneous of subdegree d iff there are homogeneous $P_j \in K\{Y'\} \subseteq K\{Y\}$ of degree d such that $P = \sum_j P_j(Y)Y^j$. Suppose $P \neq 0$ is subhomogeneous of subdegree d; then $\operatorname{wt}(P) \geqslant d$, with equality iff $P \in K[Y] \cdot (Y')^d$.

Logarithmic decomposition

In this subsection we consider iterated logarithmic derivatives. Let K be a differential field and $y \in K$. We set $y^{\langle 0 \rangle} := y$, and inductively, if $y^{\langle n \rangle} \in K$ is defined and nonzero, $y^{\langle n+1 \rangle} := (y^{\langle n \rangle})^\dagger$ (and otherwise $y^{\langle n+1 \rangle}$ is not defined). The results we state below follow by easy inductions on n using Lemma 1.9.1. First, if $y^{\langle n \rangle}$ is defined, then

$$y^{(n)} = y \cdot L_n(y^{\langle 1 \rangle}, \dots, y^{\langle n \rangle}) = y^{\langle 0 \rangle} \cdot L_n(y^{\langle 1 \rangle}, \dots, y^{\langle n \rangle})$$

where the polynomial $L_n \in \mathbb{Z}[X_1, \dots, X_n]$ depends only on n (not on y or K) and is homogeneous of degree n with nonnegative coefficients:

$$\begin{aligned} L_0 &= 1 \\ L_1 &= X_1 \\ L_2 &= X_1^2 + X_1X_2 \\ L_3 &= X_1^3 + 3X_1^2X_2 + X_1X_2^2 + X_1X_2X_3 \\ &\vdots \end{aligned}$$

The L_n are given by the recursion

$$L_0 = 1, \qquad L_{n+1} = X_1L_n + \sum_{j=1}^{n} X_jX_{j+1}\frac{\partial L_n}{\partial X_j}.$$

So if $y^{\langle n \rangle}$ is defined, then $\mathbb{Q}[y, y', \dots, y^{(n)}] \subseteq \mathbb{Q}[y^{\langle 0 \rangle}, y^{\langle 1 \rangle}, \dots, y^{\langle n \rangle}]$. Similarly:

(1) if $y^{\langle n \rangle}$ is defined, then $\mathbb{Q}(y, y', \dots, y^{(n)}) = \mathbb{Q}(y^{\langle 0 \rangle}, y^{\langle 1 \rangle}, \dots, y^{\langle n \rangle})$;

(2) if $y, y', \dots, y^{(n)}$ are algebraically independent over $\mathbb{Q}$, then $y^{\langle n \rangle}$ is defined, and so $y^{\langle 0 \rangle}, y^{\langle 1 \rangle}, \dots, y^{\langle n \rangle}$ are algebraically independent over $\mathbb{Q}$ by (1).

Thus in $K\langle Y \rangle$ each $Y^{\langle n \rangle}$ is defined, $Y^{\langle 0 \rangle}, Y^{\langle 1 \rangle}, \dots, Y^{\langle n \rangle}$ are algebraically independent over K, and $K\langle Y \rangle = K(Y^{\langle n \rangle} : n = 0, 1, 2, \dots)$. If $y^{\langle n \rangle}$ is defined and $\boldsymbol{i} = (i_0, \dots, i_n) \in \mathbb{N}^{1+n}$, we set

$$y^{\langle \boldsymbol{i} \rangle} := (y^{\langle 0 \rangle})^{i_0}(y^{\langle 1 \rangle})^{i_1} \cdots (y^{\langle n \rangle})^{i_n}.$$

Using the L_n it follows that P has a unique decomposition

$$P = \textstyle\sum_{\boldsymbol{i}} P_{\langle \boldsymbol{i} \rangle} Y^{\langle \boldsymbol{i} \rangle} \quad \textbf{(logarithmic decomposition)},$$

with $\boldsymbol{i}$ ranging over $\mathbb{N}^{1+r}$, all $P_{\langle \boldsymbol{i} \rangle} \in K$, and $P_{\langle \boldsymbol{i} \rangle} = 0$ for all but finitely many $\boldsymbol{i}$.

Notes and comments

Lemma 4.2.1 appears in [342, p. 35].

4.3 OPERATIONS ON DIFFERENTIAL POLYNOMIALS

In this section K is a differential ring with derivation ∂ and Y is a differential indeterminate. We discuss here additive and multiplicative conjugation of differential polynomials, Ritt reduction, composition, and substituting powers. The operation of compositional conjugation is more involved and is treated in Section 5.7.

Additive and multiplicative conjugation

Let $P = P(Y) \in K\{Y\}$ be of order at most r, and let $h, h_1, h_2 \in K$. We define the differential polynomials

$$\begin{aligned} P_{+h} &= P_{+h}(Y) := P(h+Y) \in K\{Y\} && \textbf{(additive conjugate of } P\textbf{)}, \\ P_{\times h} &= P_{\times h}(Y) := P(hY) \in K\{Y\} && \textbf{(multiplicative conjugate of } P\textbf{)}. \end{aligned}$$

Thus

$$P_{+(h_1+h_2)} = (P_{+h_1})_{+h_2}, \quad P_{\times h_1 h_2} = (P_{\times h_1})_{\times h_2}, \quad (P_{\times h_1})_{+h_2} = (P_{+h_1 h_2})_{\times h_1}.$$

These conjugations commute with ∂, that is,

$$(P_{+h})' = (P')_{+h}, \qquad (P_{\times h})' = (P')_{\times h}.$$

Additive conjugation with h is an automorphism $Q \mapsto Q_{+h}$ of the differential ring $K\{Y\}$, with inverse $Q \mapsto Q_{-h} := Q_{+(-h)}$, and if h is a unit of K, then also multiplicative conjugation with h is an automorphism $Q \mapsto Q_{\times h}$ of this differential ring, with inverse $Q \mapsto Q_{/h} := Q_{\times h^{-1}}$. These automorphisms are the identity on K. If P is homogeneous of degree d, then so is $P_{\times h}$. Thus (without restricting P):

$$(P_{\times h})_d = (P_d)_{\times h}, \qquad (d \in \mathbb{N}).$$

Note also that $(P_{+h})_1 = \sum_{n=0}^{r} \frac{\partial P}{\partial Y^{(n)}}(h) Y^{(n)}$.

EXAMPLE. Consider a homogeneous linear $A \in K\{Y\}$:

$$A(Y) = a_0 Y + a_1 Y' + \cdots + a_n Y^{(n)} \qquad (a_0, \ldots, a_n \in K).$$

Then

$$A_{\times h}(Y) = b_0 Y + b_1 Y' + \cdots + b_n Y^{(n)}$$

where

$$b_i = \sum_{j=i}^{n} \binom{j}{i} a_j h^{(j-i)},$$

in particular $b_0 = a_0 h + a_1 h' + \cdots + a_n h^{(n)} = A(h)$ and $b_n = a_n h$.

We now derive expressions for the coefficients of the additive and multiplicative conjugates of P. Given words $\boldsymbol{\sigma} = \sigma_1 \cdots \sigma_m$ and $\boldsymbol{\tau} = \tau_1 \cdots \tau_n$ in $\{0, \dots, r\}^*$, define

$$\boldsymbol{\sigma} \leqslant \boldsymbol{\tau} \quad :\Longleftrightarrow \quad m = n \text{ and } \sigma_1 \leqslant \tau_1, \dots, \sigma_n \leqslant \tau_n,$$

and if $\boldsymbol{\sigma} \leqslant \boldsymbol{\tau}$, put

$$\binom{\boldsymbol{\tau}}{\boldsymbol{\sigma}} := \binom{\tau_1}{\sigma_1} \cdots \binom{\tau_n}{\sigma_n}, \qquad \boldsymbol{\tau} - \boldsymbol{\sigma} := (\tau_1 - \sigma_1) \cdots (\tau_n - \sigma_n) \in \{0, \dots, r\}^*.$$

With these notations we have:

LEMMA 4.3.1. *Let* $\boldsymbol{i} = (i_0, \dots, i_r) \in \mathbb{N}^{1+r}$ *and* $\boldsymbol{\sigma} \in \{0, \dots, r\}^*$. *Then*

$$(P_{+h})_{\boldsymbol{i}} = \sum_{\boldsymbol{j} \geqslant \boldsymbol{i}} \binom{\boldsymbol{j}}{\boldsymbol{i}} P_{\boldsymbol{j}} h^{\boldsymbol{j}-\boldsymbol{i}}, \tag{4.3.1}$$

$$(P_{\times h})_{[\boldsymbol{\sigma}]} = \sum_{\boldsymbol{\tau} \geqslant \boldsymbol{\sigma}} \binom{\boldsymbol{\tau}}{\boldsymbol{\sigma}} P_{[\boldsymbol{\tau}]} h^{[\boldsymbol{\tau}-\boldsymbol{\sigma}]}. \tag{4.3.2}$$

PROOF. By Taylor expansion,

$$P_{+h}(Y) = \sum_{\boldsymbol{j}} P_{(\boldsymbol{j})}(h) Y^{\boldsymbol{j}},$$

hence $(P_{+h})_{\boldsymbol{i}} = P_{(\boldsymbol{i})}(h)$. Now $P(Y) = \sum_{\boldsymbol{j}} P_{\boldsymbol{j}} Y^{\boldsymbol{j}}$ gives

$$P_{(\boldsymbol{i})} = \sum_{\boldsymbol{j}} P_{\boldsymbol{j}} \frac{(Y^{\boldsymbol{j}})^{(\boldsymbol{i})}}{\boldsymbol{i}!} = \sum_{\boldsymbol{j} \geqslant \boldsymbol{i}} \binom{\boldsymbol{j}}{\boldsymbol{i}} P_{\boldsymbol{j}} Y^{\boldsymbol{j}-\boldsymbol{i}},$$

since $(Y^{\boldsymbol{j}})^{(\boldsymbol{i})} = \frac{\boldsymbol{j}!}{(\boldsymbol{j}-\boldsymbol{i})!} Y^{\boldsymbol{j}-\boldsymbol{i}}$ if $\boldsymbol{j} \geqslant \boldsymbol{i}$ and $(Y^{\boldsymbol{j}})^{(\boldsymbol{i})} = 0$ otherwise, as an easy induction shows. This yields (4.3.1). A simple induction on $n = |\boldsymbol{\tau}|$ shows that for $\boldsymbol{\tau} = \tau_1 \cdots \tau_n \in \{0, \dots, r\}^*$ we have

$$(ab)^{[\boldsymbol{\tau}]} = \sum_{\boldsymbol{\sigma} \leqslant \boldsymbol{\tau}} \binom{\boldsymbol{\tau}}{\boldsymbol{\sigma}} a^{[\boldsymbol{\tau}-\boldsymbol{\sigma}]} b^{[\boldsymbol{\sigma}]} \qquad \text{for } a, b \in K\{Y\}.$$

Hence

$$\begin{aligned} P(hY) = \sum_{\boldsymbol{\tau}} P_{[\boldsymbol{\tau}]} (hY)^{[\boldsymbol{\tau}]} &= \sum_{\boldsymbol{\tau}} \left(P_{[\boldsymbol{\tau}]} \sum_{\boldsymbol{\sigma} \leqslant \boldsymbol{\tau}} \binom{\boldsymbol{\tau}}{\boldsymbol{\sigma}} h^{[\boldsymbol{\tau}-\boldsymbol{\sigma}]} Y^{[\boldsymbol{\sigma}]} \right) \\ &= \sum_{\boldsymbol{\sigma}} \left(\sum_{\boldsymbol{\tau} \geqslant \boldsymbol{\sigma}} \binom{\boldsymbol{\tau}}{\boldsymbol{\sigma}} P_{[\boldsymbol{\tau}]} h^{[\boldsymbol{\tau}-\boldsymbol{\sigma}]} \right) Y^{[\boldsymbol{\sigma}]}, \end{aligned}$$

and this yields (4.3.2). □

COROLLARY 4.3.2. *We have* $\deg P_{+h} = \deg P$ *and* $\operatorname{wt} P_{+h} = \operatorname{wt} P$. *If* h *is a unit of* K, *then* $\deg P_{\times h} = \deg P$ *and* $\operatorname{wt} P_{\times h} = \operatorname{wt} P$.

The following identities have routine proofs, using the Chain Rule:

$$\frac{\partial P_{+h}}{\partial Y^{(i)}} = \left(\frac{\partial P}{\partial Y^{(i)}}\right)_{+h}, \tag{4.3.3}$$

$$\frac{\partial P_{\times h}}{\partial Y^{(i)}} = \sum_{j\geqslant i} \binom{j}{i} h^{(j-i)} \left(\frac{\partial P}{\partial Y^{(j)}}\right)_{\times h}. \tag{4.3.4}$$

Suppose now that $P \neq 0$. We have

$$0 \leqslant \operatorname{mul} P_{+h} \leqslant \deg P, \qquad \operatorname{mul} P_{+h} > 0 \Longleftrightarrow P(h) = 0.$$

We call $\operatorname{mul} P_{+h}$ the **multiplicity** of P at h. Note that if $P \in K[Y]$, then

$$P(Y) = Q(Y) \cdot (Y - h)^{\mu}, \text{ where } Q(Y) \in K[Y],\, Q(h) \neq 0,\, \mu = \operatorname{mul} P_{+h}.$$

LEMMA 4.3.3. *Suppose* $0 \neq P \in K[Y](Y')^{\mathbb{N}}$ *and* $h' = 0$, *and let* $\mu \in \mathbb{N}$. *Then*

$$\operatorname{mul} P_{+h} = \mu \quad \Longleftrightarrow \quad \text{there are } i, j \in \mathbb{N} \text{ and } Q \in K[Y] \text{ such that } i + j = \mu,$$
$$P(Y) = Q(Y) \cdot (Y - h)^i \cdot (Y')^j, \text{ and } Q(h) \neq 0.$$

PROOF. The implication $\Leftarrow$ is obvious. For the direction $\Rightarrow$, note that we have $P(Y) = R(Y) \cdot (Y')^j$ with $j \in \mathbb{N}$ and $R \in K[Y]$. Set $i := \operatorname{mul} R_{+h}$; then $\operatorname{mul} P_{+h} = i + j$ and $R(Y) = Q(Y) \cdot (Y - h)^i$ where $Q \in K[Y]$, $Q(h) \neq 0$. □

COROLLARY 4.3.4. *For* $0 \neq P \in K[Y](Y')^{\mathbb{N}}$ *and* $h' = 0$ *we have:*

$$\operatorname{mul} P_{+h} = \deg P \Longleftrightarrow P = a \cdot (Y - h)^i \cdot (Y')^j \text{ where } a \in K^{\neq} \text{ and } i + j = \deg P.$$

Complexity and Ritt division

Let K *be a differential field,* $P, Q \in K\{Y\}$, *and* $a \in K$. For $P \notin K$, denote the order of P by r_P, the degree of P in $Y^{(r_P)}$ by s_P, and the total degree of P by t_P (so $s_P, t_P \geqslant 1$), and define the **complexity** of P to be the triple $c(P) = (r_P, s_P, t_P) \in \mathbb{N}^3$.

For $P \in K$ we set $c(P) = (0, 0, 0)$. We order $\mathbb{N}^3$ lexicographically. Note that if $P, Q \notin K$, then $c(P), c(Q) < c(PQ)$. For $P \notin K$ and $r = r_P$, $s = s_P$ we have

$$P = F_0 + F_1 \cdot Y^{(r)} + \cdots + F_s \cdot (Y^{(r)})^s, \qquad F_0, \dots, F_s \in K\big[Y, \dots, Y^{(r-1)}\big],$$

and we define the **initial** of P to be $I_P := F_s$, and the **separant** of P to be

$$S_P := \frac{\partial P}{\partial Y^{(r)}} = \sum_{i=1}^{s} iF_i \cdot (Y^{(r)})^{i-1},$$

so $I_P \neq 0$, $S_P \neq 0$, and $c(I_P) < c(P)$, $c(S_P) < c(P)$. The quantities above transform as follows under various conjugations:

$$c(P_{+a}) = c(P), \quad I_{P_{+a}} = I_{P,+a}, \quad S_{P_{+a}} = S_{P,+a},$$
$$\text{and for } a \neq 0, \quad c(P_{\times a}) = c(P), \quad I_{P_{\times a}} = a^{s_P} \cdot I_{P,\times a}, \quad S_{P_{\times a}} = a \cdot S_{P,\times a}.$$

LEMMA 4.3.5. *Suppose $P \notin K$, and set $r = r_P$. Then for $n \geqslant 1$,*

$$P^{(n)} = G_n + S_P Y^{(r+n)} \qquad \text{with } G_n \in K\big[Y, Y', \dots, Y^{(r+n-1)}\big].$$

In particular, $P^{(n)} \notin K$ and $P^{(n)}$ has order $r+n$, for every n.

PROOF. Let $s = s_P$ be the degree of P in $Y^{(r)}$, so $s \geqslant 1$, and

$$P = F_0 + F_1 \cdot Y^{(r)} + \cdots + F_s \cdot (Y^{(r)})^s, \quad F_0, \dots, F_s \in K\big[Y, Y', \dots, Y^{(r-1)}\big].$$

Then $P' = G_1 + S_P Y^{(r+1)}$ with

$$G_1 := F_0' + F_1' \cdot Y^{(r)} + \cdots + F_s' \cdot (Y^{(r)})^s \in K\big[Y, Y', \dots, Y^{(r)}\big].$$

This gives the case $n = 1$, and the rest goes by induction on n. □

COROLLARY 4.3.6. *Assume $P \notin K$. Then $P' \notin PK\{Y\}$.*

PROOF. With the notations above and $r = r_P$ we have $P' = S_P Y^{(r+1)} + G_1$. Assume $P' = PQ$. Then $Q = AY^{(r+1)} + B$ with $A, B \in K[Y, \dots, Y^{(r)}]$, so $S_P = PA$, contradicting $\mathrm{c}(S_P) < \mathrm{c}(P)$. □

COROLLARY 4.3.7. *Assume $F \in K\langle Y \rangle$, $F \notin K$. Then $F^\dagger \notin K\{Y\}$.*

PROOF. We have $F = P/Q$ with relatively prime $P, Q \in K\{Y\}^{\neq}$. Towards a contradiction, assume $F^\dagger = R \in K\{Y\}$. Then $P'Q - PQ' = PQR$, so P divides $P'Q$ in $K\{Y\}$, hence P divides P' in $K\{Y\}$, and so $P \in K$ by Corollary 4.3.6. Likewise we get $Q \in K$, so $F \in K$, a contradiction. □

Next an analogue for differential polynomials of division with remainder:

THEOREM 4.3.8 (Ritt division). *Let $P \notin K$. Then*

$$I_P^k S_P^l Q = A_0 P + A_1 P' + \cdots + A_n P^{(n)} + R$$

for some $k, l \in \mathbb{N}$, $A_0, \dots, A_n \in K\{Y\}$, and $R \in K\{Y\}$ with $\mathrm{c}(R) < \mathrm{c}(P)$.

We break up the proof into two steps, and first show:

LEMMA 4.3.9. *Let $P \notin K$. There are $l \in \mathbb{N}$ and $A_0, \dots, A_n \in K\{Y\}$ such that*

$$S_P^l Q = A_0 P + A_1 P' + \cdots + A_n P^{(n)} + R, \qquad \operatorname{order}(R) \leqslant \operatorname{order}(P).$$

PROOF. Let $r := r_P$ and $Q \in K[Y, \dots, Y^{(r+i)}]$. By induction on i we show:

$$S_P^l Q \equiv R \mod (P, P', \dots, P^{(i)}) \quad \text{in the ring } K\big[Y, Y', \dots, Y^{(r+i)}\big]$$

for suitable $l \in \mathbb{N}$ and $R \in K\{Y\}$ with $\operatorname{order}(R) \leqslant r$. If $i = 0$, then we can take $l = 0$, $R = Q$. So suppose $i > 0$. From Lemma 4.3.5 we obtain

$$P^{(i)} = S_P Y^{(r+i)} + G_i, \quad G_i \in K\big[Y, Y', \dots, Y^{(r+i-1)}\big].$$

As $Q = \sum_{k=0}^{s} F_k \cdot (Y^{(r+i)})^k$ with $s = s_Q$, $F_k \in K\big[Y, Y', \dots, Y^{(r+i-1)}\big]$, we get

$$S_P^s Q = \sum_{k=0}^{s} F_k \cdot (P^{(i)} - G_i)^k S_P^{s-k} \equiv Q_i \mod (P^{(i)}) \quad \text{in } K\big[Y, \dots, Y^{(r+i)}\big]$$

$$\text{with } Q_i = \sum_{k=0}^{s} F_k \cdot (-G_i)^k S_P^{s-k} \in K\big[Y, Y', \dots, Y^{(r+i-1)}\big].$$

The inductive assumption gives $l \in \mathbb{N}$ and $R \in K\{Y\}$ with $\operatorname{order}(R) \leqslant r$ and

$$S_P^l Q_i \equiv R \mod \big(P, P', \dots, P^{(i-1)}\big) \quad \text{in the ring } K\big[Y, Y', \dots, Y^{(r+i-1)}\big].$$

In combination with the previous congruence, this gives

$$S_P^{l+s} Q \equiv R \mod \big(P, P', \dots, P^{(i)}\big) \quad \text{in the ring } K\big[Y, Y', \dots, Y^{(r+i)}\big],$$

which finishes the induction. □

PROOF OF THEOREM 4.3.8. Take $l, A_0, \dots, A_n, R$ as in Lemma 4.3.9. We are done if $\mathrm{c}(R) < \mathrm{c}(P)$, so assume $\mathrm{c}(R) \geqslant \mathrm{c}(P)$. Then $r := r_P = r_R$ and $s_R \geqslant s_P$. Ordinary division by P for polynomials in $Y^{(r)}$ over $K\big[Y, Y', \dots, Y^{(r-1)}\big]$ yields

$$I_P^k R = A^* P + R^*, \quad k = 1 + s_R - s_P, \ A^*, R^* \in K\big[Y, Y', \dots, Y^{(r)}\big], \ s_{R^*} < s_P.$$

Hence

$$I_P^k S_P^l Q = A_0^* P + A_1^* P' + \dots + A_n^* P^{(n)} + R^*$$

where $A_0^* := I_P^k A_0 + A^*$, $A_i^* := I_P^k A_i$ for $i = 1, \dots, n$, and $\mathrm{c}(R^*) < \mathrm{c}(P)$. □

Additive conjugation in the Ritt division of the theorem above gives:

$$I_{P_{+a}}^k S_{P_{+a}}^l Q_{+a} = (A_0)_{+a} P_{+a} + (A_1)_{+a} P'_{+a} + \dots + (A_n)_{+a} P_{+a}^{(n)} + R_{+a}.$$

Corollary 4.1.8 and Lemma 4.3.9 yield:

COROLLARY 4.3.10. *Let y be an element of a differential field extension of K, d-algebraic over K, and let $P \in K\{Y\}^{\neq}$ be a minimal annihilator of y over K. Then for $Q \in K\{Y\}$ we have: $Q(y) = 0$ if and only if there exist $l \in \mathbb{N}$ and $A_0, \dots, A_n \in K\{Y\}$ such that $S_P^l Q = A_0 P + A_1 P' + \dots + A_n P^{(n)}$.*

Composition of differential polynomials

Let K be a differential field, and $P, Q \in K\{Y\}$. We prove here some elementary facts about $P(Q)$.

LEMMA 4.3.11. *For each $d \in \mathbb{N}$ we have $(P^{(n)})_d = (P_d)^{(n)}$. If P is homogeneous, then so is $P^{(n)}$. If $P \notin K$, then* $\deg P^{(n)} = \deg P$ *and* $\operatorname{wt} P^{(n)} = (\operatorname{wt} P) + n$, *and*

$$\operatorname{mul} P^{(n)} \geqslant \operatorname{mul} P, \qquad \operatorname{wm} P^{(n)} \geqslant \operatorname{wm} P.$$

PROOF. For any word $\boldsymbol{\sigma} = \sigma_1 \cdots \sigma_d \in \{0, \dots, r\}^*$ of length $|\boldsymbol{\sigma}| = d$ we have

$$(Y^{[\boldsymbol{\sigma}]})' = \sum_{i=1}^{d} Y^{(\sigma_i+1)}\big(Y^{[\boldsymbol{\sigma}]}/Y^{(\sigma_i)}\big),$$

so $(P')_d = (P_d)'$, which gives the claim about homogeneous P. Let $P \notin K$. Then $P' \neq 0$ by Lemma 4.3.5, so $\deg P' = \deg P$. With $w := \operatorname{wt} P$ we have $\operatorname{wt} P' \leqslant w+1$. It remains to show that $(P')_{[w+1]} \neq 0$. Let K_c be the field K with the trivial derivation. Then $K_c\{Y\}$ and $K\{Y\}$ have the same underlying ring, and $(P')_{[w+1]}$ equals in this underlying ring the derivative of $P_{[w]}$ in $K_c\{Y\}$. As $P_{[w]} \notin K$, this gives $(P')_{[w+1]} \neq 0$. Verifying the claims about $\operatorname{mul} P^{(n)}$ and $\operatorname{wm} P^{(n)}$ is routine. □

Thus we let $P_d^{(n)}$ denote both $(P^{(n)})_d$ and $(P_d)^{(n)}$.

LEMMA 4.3.12. *If $P \neq 0$ and $Q \notin K$, then $P(Q) \neq 0$.*

PROOF. If $Q \notin K$, then Q is d-transcendental over K by Lemma 4.1.5. □

Thus if $Q \notin K$, then $P \mapsto P \circ Q = P(Q)$ is an injective differential ring endomorphism of $K\{Y\}$, and so extends uniquely to a differential field embedding

$$F \mapsto F \circ Q = F(Q) : K\langle Y\rangle \to K\langle Y\rangle.$$

COROLLARY 4.3.13. *Suppose $P \neq 0$, $Q \notin K$, $\deg(P) = d$, $\deg(Q) = e$. Then*

$$\deg(P \circ Q) = de, \qquad (P \circ Q)_{de} = P_d \circ Q_e.$$

If P and Q are homogeneous, then so is $P \circ Q$.

PROOF. Let $\boldsymbol{i} = (i_0, \dots, i_r) \in \mathbb{N}^{1+r}$. By Lemma 4.3.11 we have $\deg Q^{(n)} = e$ for all n, so

$$\begin{aligned} Q^{\boldsymbol{i}} &= Q^{i_0}(Q')^{i_1} \cdots (Q^{(r)})^{i_r} \\ &= Q_e^{i_0}(Q_e')^{i_1} \cdots (Q_e^{(r)})^{i_r} + R \quad \text{where } \deg(R) < |\boldsymbol{i}|e. \end{aligned}$$

Thus $\deg(P \circ Q) \leqslant de$, and $(P \circ Q)_{de} = P_d \circ Q_e$, and so by Lemma 4.3.12 we get $\deg(P \circ Q) = de$. □

COROLLARY 4.3.14. *Suppose $P \neq 0$ has order $\leqslant r$ and $Q \notin K$. Then*

$$\operatorname{wm}(P \circ Q) \geqslant (\operatorname{mul} P)(\operatorname{wm} Q).$$

Moreover, with $\boldsymbol{i}$ ranging over $\mathbb{N}^{1+r}$,

$$\operatorname{wt}(P \circ Q) = \max\big\{|\boldsymbol{i}|(\operatorname{wt} Q) + \|\boldsymbol{i}\| : P_{\boldsymbol{i}} \neq 0\big\},$$

and so if P is homogeneous or isobaric, then

$$\operatorname{wt}(P \circ Q) = (\deg P)(\operatorname{wt} Q) + (\operatorname{wt} P).$$

PROOF. Let $\boldsymbol{i}$ range over $\mathbb{N}^{1+r}$. Lemma 4.3.11 gives $\operatorname{wm} Q^{(\boldsymbol{i})} \geqslant |\boldsymbol{i}|(\operatorname{wm} Q)$ for all $\boldsymbol{i}$, so $\operatorname{wm}(P \circ Q) \geqslant (\operatorname{mul} P)(\operatorname{wm} Q)$. Set

$$w := \operatorname{wt} Q, \qquad \mu := \max\big\{|\boldsymbol{i}|w + \|\boldsymbol{i}\| : P_{\boldsymbol{i}} \neq 0\big\}.$$

By Lemma 4.3.11, $\operatorname{wt} Q^{(n)} = w + n$, so with $Q^{(n)}_{[w+n]} := (Q^{(n)})_{[w+n]}$ we have

$$\begin{aligned} Q^{(\boldsymbol{i})} &= Q^{i_0}(Q')^{i_1}\cdots(Q^{(r)})^{i_r} \\ &= Q^{i_0}_{[w]}(Q'_{[w+1]})^{i_1}\cdots(Q^{(r)}_{[w+r]})^{i_r} + S \quad \text{where } \operatorname{wt}(S) < |\boldsymbol{i}|w + \|\boldsymbol{i}\|. \end{aligned}$$

Thus $\operatorname{wt}(P \circ Q) \leqslant \mu$. Let K_c be the field K with the trivial derivation. Then $K\{Y\}$ and $K_c\{Y\}$ have the same underlying ring, and $Q^{(n)}_{[w+n]} \in K\{Y\}$ equals in this underlying ring the nth derivative of $Q_{[w]}$ in $K_c\{Y\}$. So $(P \circ Q)_{[\mu]} \in K\{Y\}$ equals in this underlying ring the composition $\widetilde{P} \circ (Q_{[w]})$ computed in $K_c\{Y\}$, where

$$\widetilde{P} := \sum_{|\boldsymbol{i}|w + \|\boldsymbol{i}\| = \mu} P_{\boldsymbol{i}} Y^{\boldsymbol{i}} \neq 0.$$

Thus $\operatorname{wt}(P \circ Q) = \mu$ by Lemma 4.3.12. □

COROLLARY 4.3.15. *Let $F \in K\langle Y\rangle$, and suppose $P, Q \notin K$. Then*

$$P \circ Q \notin K, \qquad F \circ (P \circ Q) = (F \circ P) \circ Q.$$

PROOF. We have $P \circ Q \notin K$ by Corollary 4.3.13. The associativity of the composition operator on $K\{Y\}$ then leads to $F \circ (P \circ Q) = (F \circ P) \circ Q$. □

Substituting powers

Let K be a differential field and $c \in C^\times$. We consider here the effect of substituting a power y^c in a differential polynomial, where y and y^c are nonzero elements of a differential field extension L of K such that $(y^c)^\dagger = cy^\dagger$.

LEMMA 4.3.16. *For each n there is a homogeneous and isobaric $E_n \in C\{Y\}$ of order n, degree n, and weight n, such that for any differential field extension L of K and any elements $y, z \in L^\times$,*

$$z^\dagger = cy^\dagger \implies z^{(n)} = z \cdot \frac{E_n(y)}{y^n}.$$

PROOF. By induction on n. We can take $E_0 = 1$, and $E_1 = cY'$. Suppose $E_n \in C\{Y\}$ has the required property for a certain $n \geqslant 1$, and let L be a differential field extension of K and let $y, z \in L^\times$ satisfy $z^\dagger = cy^\dagger$. Then

$$\begin{aligned} z^{(n+1)} &= z' \cdot \frac{E_n(y)}{y^n} + z \cdot \left(\frac{E_n(y)}{y^n}\right)' \\ &= z \cdot \frac{cy'E_n(y)}{y^{n+1}} + z \cdot \frac{y^n E_n(y)' - ny^{n-1}y'E_n(y)}{y^{2n}} \\ &= z \cdot \frac{(c-n)y'E_n(y) + yE_n(y)'}{y^{n+1}}. \end{aligned}$$

Thus $E_{n+1}(Y) := (c-n)Y'E_n(Y)+YE_n(Y)' \in C\{Y\}$ has the desired property for $n+1$ in place of n, in view of Lemmas 4.3.5 and 4.3.11. □

COROLLARY 4.3.17. *Let $P \in K\{Y\}^{\neq}$ be homogeneous and isobaric, and $d = \deg(P)$, $w = \operatorname{wt}(P)$. Then there is a homogeneous and isobaric $E \in K\{Y\}^{\neq}$ of degree w and weight w such that for any differential field extension L of K and $y, z \in L^{\times}$,*

$$z^{\dagger} = cy^{\dagger} \implies P(z) = z^d \cdot \frac{E(y)}{y^w}.$$

This follows easily from Lemma 4.3.16.

COROLLARY 4.3.18. *Let $P \in K\{Y\}^{\neq}$ be homogeneous, $d = \deg(P)$, $w = \operatorname{wt}(P)$. Then there is a homogeneous $E \in K\{Y\}^{\neq}$ of degree w and weight w such that for any differential field extension L of K and $y, z \in L^{\times}$,*

$$z^{\dagger} = cy^{\dagger} \implies P(z) = z^d \cdot \frac{E(y)}{y^w}.$$

PROOF. Apply the previous corollary to the isobaric parts of P. □

Corollary 4.3.18 in the case $c = -1$ yields:

COROLLARY 4.3.19. *If $P \in K\{Y\}^{\neq}$ is homogeneous, then*

$$P(1/Y) = \frac{E(Y)}{Y^{d+w}} \qquad \text{in } K\langle Y\rangle,\ d := \deg P,\ w := \operatorname{wt} P,$$

where $E \in K\{Y\}^{\neq}$ is homogeneous of degree w, and $\operatorname{wt}(E) = w$.

Notes and comments

Ritt's division theorem (Theorem 4.3.8 above) appears in [342, p. 5]; see also [210, Lemma 7.3].

4.4 VALUED DIFFERENTIAL FIELDS AND CONTINUITY

We define a **valued differential field** to be a differential field K equipped with a valuation ring $\mathcal{O} \supseteq \mathbb{Q}$ of K. Here are some basic examples:

(1) Let $\boldsymbol{k}$ be a differential field and Γ an ordered abelian group. Then we make the Hahn field $\boldsymbol{k}((t^{\Gamma}))$ into a differential field extension of $\boldsymbol{k}$ by

$$\partial\Big(\sum a_{\gamma}t^{\gamma}\Big) = \sum a_{\gamma}'\, t^{\gamma} \qquad (\text{so } t' = 0).$$

We refer to this valued differential field as the *Hahn differential field* $\boldsymbol{k}((t^{\Gamma}))$. It has constant field $C_{\boldsymbol{k}}((t^{\Gamma}))$.

(2) Let $\boldsymbol{k}$ be a differential field, and let $g \in \boldsymbol{k}((t^{\mathbb{Q}}))$. We make the Hahn field $\boldsymbol{k}((t^{\mathbb{Q}}))$ into a valued differential field and a differential field extension of $\boldsymbol{k}$ by

$$\partial\Big(\sum a_q t^q\Big) = \sum a_q' t^q + \Big(\sum q a_q t^{q-1}\Big)\, g \qquad (\text{so } t' = g).$$

(3) Let C be a field of characteristic zero (so $C \supseteq \mathbb{Q}$, with the usual identification). We make the field $C(x)$ of rational functions in x over C into a differential field with constant field C and $x' = 1$. Let $v\colon C(x)^\times \to \mathbb{Z}$ be the discrete valuation on the field $C(x)$ given by $v(C^\times) = \{0\}$ and $v(x) = -1$. Then $C(x)$ equipped with the valuation ring $\mathcal{O}_v$ of v is a valued differential field.

(4) $\mathbb{T}$ and $\mathbb{T}_{\log}$: see Appendix A.

In the rest of this section we fix a valued differential field K. We begin with some easy observations:

LEMMA 4.4.1. *Let $a, b \in K$, $b \neq 0$. Then*

$$a' \preccurlyeq a,\ b' \preccurlyeq b \implies (ab)' \preccurlyeq ab, \quad (1/b)' \preccurlyeq 1/b, \quad (a/b)' \preccurlyeq a/b.$$

In particular, if $b' \preccurlyeq b$, then $(b^k)' \preccurlyeq b^k$ for every $k \in \mathbb{Z}$.

Our valued differential fields often have the property that the maximal ideal of its valuation ring is closed under the derivation: then the valuation ring is also closed under the derivation and various other useful inclusions hold:

LEMMA 4.4.2. *Suppose that $\partial\smallo \subseteq \smallo$. Then also $\partial\mathcal{O} \subseteq \mathcal{O}$. Moreover, let $a \in K$ be such that $a' \preccurlyeq a$. Then $\partial(a^m\smallo) \subseteq a^m\smallo$ and $\partial(a^m\mathcal{O}) \subseteq a^m\mathcal{O}$ for all m, hence*

$$\partial^n(a^m\smallo) \subseteq a^m\smallo, \quad \partial^n(a^m\mathcal{O}) \subseteq a^m\mathcal{O} \qquad \text{for all } m, n,$$

and if $a' \prec a$, then $\partial^n(a) \prec a$ for all $n \geqslant 1$.

PROOF. Let $x \in \mathcal{O}$, and suppose $x' \notin \mathcal{O}$. Set $y := 1/x'$, $z := x/x'$. Then $y, z \in \smallo$, so $y', z' \in \smallo$. But $z = xy$, so $z' = xy' + 1$, a contradiction. This proves the first assertion. For the second, let $a \in K$ with $a' \preccurlyeq a$. Then $\partial(a\smallo) \subseteq a'\smallo + a\partial\smallo \subseteq a\smallo$ and $\partial(a\mathcal{O}) \subseteq a'\mathcal{O} + a\partial\mathcal{O} \subseteq a\mathcal{O}$, so we are done for $m = 1$. Now use that $a' \preccurlyeq a$ gives $(a^m)' \preccurlyeq a^m$ for all m by the previous lemma. If $a' \prec a$, then $\partial(a) \in a\smallo$ and hence $\partial^n(a) \in \partial^{n-1}(a\smallo) \subseteq a\smallo$ for $n \geqslant 1$. □

If $\partial\smallo \subseteq \smallo$, then by Lemma 4.4.2 the residue field $\boldsymbol{k} = \mathcal{O}/\smallo$ is a differential field with derivation $a + \smallo \mapsto a' + \smallo$ $(a \in \mathcal{O})$. (From $\mathbb{Q} \subseteq \mathcal{O}$ we get that $\boldsymbol{k}$ is of characteristic zero, and so with the obvious identification $\mathbb{Q}$ is a subfield of $\boldsymbol{k}$.) The derivation ∂ of K is said to be **small** if $\partial\smallo \subseteq \smallo$; in that case we refer to $\boldsymbol{k}$ with the induced derivation as the **differential residue field** of K. Note that if K has small derivation, then so does every valued differential subfield of K.

The derivation of the Hahn differential field $\boldsymbol{k}((t^{\Gamma}))$ of Example (1) above is small, and its differential residue field is isomorphic to $\boldsymbol{k}$. In Example (2) above, the derivation of $\boldsymbol{k}((t^{\mathbb{Q}}))$ is small if $vg \geqslant 1$, in which case its differential residue field is again isomorphic to $\boldsymbol{k}$. The derivation of $C(x)$ in Example (3) is small, and its differential residue field is isomorphic to C with the trivial derivation. The derivations of $\mathbb{T}$ and $\mathbb{T}_{\log}$ of Example (4) are also small, and in both cases the differential residue field is isomorphic to $\mathbb{R}$ with the trivial derivation.

LEMMA 4.4.3. *Suppose $\partial o \subseteq o$, and $y \in K$. Then*

$$y \preccurlyeq 1 \Rightarrow (y')^2 \preccurlyeq y, \qquad y \succcurlyeq 1 \Rightarrow (y')^2 \preccurlyeq y^3.$$

If moreover $\partial \mathcal{O} \subseteq o$, then

$$0 \neq y \preccurlyeq 1 \Rightarrow (y')^2 \prec y, \qquad y \succcurlyeq 1 \Rightarrow (y')^2 \prec y^3.$$

PROOF. Let $y \preccurlyeq 1$, and suppose for a contradiction that $(y')^2 \succ y$. Then $y = \varepsilon(y')^2$ with $\varepsilon \prec 1$; differentiating yields $y' = \varepsilon'(y')^2 + 2\varepsilon y' y''$, contradicting $y' \succ \varepsilon'(y')^2$ and $y' \succ \varepsilon y' y''$. Next, suppose $y \succcurlyeq 1$; then $y^{-1} \preccurlyeq 1$, hence $(y'/y^2)^2 = \left((y^{-1})'\right)^2 \preccurlyeq y^{-1}$, so $(y')^2 \preccurlyeq y^3$. The proof of the part assuming $\partial \mathcal{O} \subseteq o$ is similar. □

COROLLARY 4.4.4. *Suppose $\partial o \subseteq o$. Let Δ be a convex subgroup of Γ, and $\dot{o}$ the maximal ideal of the valuation ring $\dot{\mathcal{O}}$ of the Δ-coarsening of K. Then $\partial \dot{o} \subseteq \dot{o}$.*

PROOF. If $y \in \dot{o}$, then $(y')^2 \in \dot{o}$ by Lemma 4.4.3, so $y' \in \dot{o}$. □

Continuity

We now give the valued differential field K its valuation topology. While not part of our definition of *valued differential field*, we are only interested in the case that $\partial\colon K \to K$ is continuous. Note that if ∂ is continuous, then for every differential polynomial $P \in K\{Y\}$ the function $y \mapsto P(y)\colon K \to K$ is continuous. Here is a slightly stronger version of this fact:

LEMMA 4.4.5. *Suppose $\partial\colon K \to K$ is continuous. Then for each $P(Y) \in K\{Y\}$ the function $y \mapsto P(y)\colon K \to K$ is c-continuous.*

PROOF. Since ∂ is continuous at 0 and additive, ∂ is even uniformly continuous. Hence for every n the map $y \mapsto y^{(n)}\colon K \to K$ is uniformly continuous and thus c-continuous. The c-continuity of differential polynomial functions on K now follows easily from Lemmas 2.2.23 and 3.2.14. □

LEMMA 4.4.6. *If $\partial o \subseteq o$, then $\partial\colon K \to K$ is continuous.*

PROOF. Assume $\partial o \subseteq o$. If $\gamma \in \Gamma^{>}$, and $y \in K$, $vy > 2\gamma$, then $vy' > \gamma$, by Lemma 4.4.3. So ∂ is continuous at 0, and since ∂ is additive, it is continuous. □

Up to a factor from $K^{\times}$ the condition $\partial o \subseteq o$ captures exactly continuity of ∂:

LEMMA 4.4.7. *The following conditions on K are equivalent:*

(i) *$\partial\colon K \to K$ is continuous;*

(ii) *for some* $a \in K^\times$ *we have* $\partial o \subseteq ao$;

(iii) *for some* $a \in K^\times$ *we have* $\delta o \subseteq o$, *for* $\delta := a^{-1}\partial$.

PROOF. As to (i) ⇒ (ii), assume ∂ is continuous. Take $\beta \in \Gamma$ such that $f' \in o$ for all $f \in K$ with $vf > \beta$, and take $b \in K$ with $vb = \beta$. Then for $g \in o$ we have $v(bg) > \beta$, so $(bg)' = b'g + bg' \in o$, hence $g' \in b^{-1}o + b^\dagger o$. Taking $a \in K^\times$ with $va = \min(-vb, vb^\dagger)$ we obtain $\partial o \subseteq ao$. The implication (ii) ⇒ (iii) is obvious, and (iii) ⇒ (i) follows from Lemma 4.4.6 applied to δ. □

A favorable situation is when ∂ is small *and* the derivation of the differential residue field $\boldsymbol{k}$ is nontrivial: this often allows properties of the differential residue field $\boldsymbol{k}$ to be lifted to useful information about the valued differential field K.

In this connection, consider the set A of all $a \in K^\times$ such that $\partial o \subseteq ao$ (that is, $a^{-1}\partial$ is small), and the derivation of the differential residue field of $(K, a^{-1}\partial)$ is nontrivial. Here $(K, a^{-1}\partial)$ denotes the valued differential field K with derivation $a^{-1}\partial$ instead of ∂. If $a \in A$ and $a \asymp b$ in K, then clearly $b \in A$. We wish to record the following observation:

LEMMA 4.4.8. *If* $A \neq \emptyset$, *then* $v(A)$ *consists of just one element.*

PROOF. Suppose ∂ is small and the derivation of the differential residue field $\boldsymbol{k}$ is nontrivial, and let $a \in K^\times$. It suffices to note that if $a \prec 1$, then $a^{-1}\partial$ is no longer small, and if $a \succ 1$, then $a^{-1}\partial$ is small, but $a^{-1}\partial\mathcal{O} \subseteq a^{-1}\mathcal{O} \subseteq o$, so the derivation of the differential residue field of $(K, a^{-1}\partial)$ is trivial. □

LEMMA 4.4.9. *Assume that* $C \subseteq \mathcal{O}$. *Let* $P(Y) = a_0 Y + \cdots + a_n Y^{(n)}$, *with all* $a_i \in K$, $a_n \neq 0$. *Then each level set* $P^{-1}(s)$ $(s \in K)$ *is a discrete subset of* K.

PROOF. Such a level set is empty or a translate of the C-linear subspace $V := P^{-1}(0)$ of K. Any $y_0, \dots, y_n \in K$ with $y_0 \succ y_1 \succ \cdots \succ y_n \succ 0$ are linearly independent over C. Since $\dim_C V \leqslant n$, there are no such $y_0, \dots, y_n \in V$. □

LEMMA 4.4.10. *Suppose* ∂ *is nontrivial, and the valuation of* K *is nontrivial. Then no differential polynomial* $P(Y) \in K\{Y\}^{\neq}$ *vanishes identically on any nonempty open subset of* K.

PROOF. Assume towards a contradiction that $P(Y) \in K\{Y\}^{\neq}$ vanishes identically on $\{y \in K : v(y - a) \geqslant \gamma\}$, $a \in K$, $\gamma \in \Gamma$. Take $g \in K$ with $vg = \gamma$ and set $Q := P(a + gY) \in K\{Y\}^{\neq}$. Then Q vanishes identically on $\mathcal{O}$, and so $Q(Y) \cdot Q(Y^{-1}) \cdot Y^n$ is for sufficiently large n a nonzero differential polynomial that vanishes identically on K. As the derivation of K is nontrivial, this is impossible by Lemma 4.2.1. □

Completion

Recall from Section 3.2 that any valued field E can be *completed*: it is dense in a valued field extension E^c such that for each valued field extension $E \subseteq F$ with E dense in F there is a unique valued field embedding $F \to E^c$ that is the identity on E. (Here "dense" is with respect to the relevant valuation topology.) These properties determine E^c up to unique valued field isomorphism over E, and E^c is called the **completion of** E. Recall that $E^c|E$ is an immediate extension.

LEMMA 4.4.11. *Suppose $\partial\colon K \to K$ is continuous. Then there is a unique continuous derivation on K^c extending the derivation of K.*

PROOF. The derivation of K being additive, it is even uniformly continuous, that is, for each $\gamma \in \Gamma$ there is $\delta \in \Gamma$ such that whenever $x, y \in K$ and $v(x-y) > \delta$, then $v(x'-y') > \gamma$. It follows that ∂ extends uniquely to a continuous map $K^c \to K^c$ (Lemma 2.2.31); this map is a derivation. □

Suppose K is as in the lemma. Then we consider K^c as the valued differential field whose derivation is the unique continuous derivation on K^c that extends the one of K. If $K \subseteq L$ is a valued differential field extension such that K is dense in L and the derivation of L is continuous with respect to the valuation topology, then the unique valued field embedding $L \to K^c$ that is the identity on K is a differential field embedding.

COROLLARY 4.4.12. *If K has small derivation, then so does K^c.*

Traces and norms

Let $L|K$ be an extension of valued differential fields with $[L:K] < \infty$ and put $n := [L:K]$. We let K^a be an algebraic closure of K equipped with the unique derivation extending the derivation of K. For $f \in L$, let

$$\operatorname{tr}_{L|K}(f) := \sum_{i=1}^{n} \sigma_i(f), \qquad \operatorname{N}_{L|K}(f) := \prod_{i=1}^{n} \sigma_i(f)$$

denote the trace of f in $L|K$ and the norm of f in $L|K$, respectively. Here $\sigma_1, \dots, \sigma_n$ are the distinct field embeddings $L \to K^a$ which are the identity on K. Note that each σ_i is an embedding $L \to K^a$ of differential fields. Hence for all $f \in L$:

$$\operatorname{tr}_{L|K}(f') = \operatorname{tr}_{L|K}(f)', \qquad \operatorname{tr}_{L|K}(f^\dagger) = \operatorname{N}_{L|K}(f)^\dagger \text{ if } f \neq 0.$$

We pick a valuation ring of K^a lying over the valuation ring of K to make K^a a valued field extension of K. If K is henselian, then each σ_i is a valued field embedding, and after identifying L with a valued subfield of K^a via σ_1, say, we have $\sigma_i f \asymp f$ for $f \in L$ and $i = 1, \dots, n$, hence (for $f \in L$):

$$\operatorname{tr}_{L|K}(f) \preccurlyeq f, \qquad \operatorname{N}_{L|K}(f) \asymp f^n.$$

It follows that for henselian K and $f \in L^\times$ we have

$$\operatorname{N}_{L|K}(f)^\dagger = \operatorname{tr}_{L|K}(f^\dagger) \preccurlyeq f^\dagger.$$

Monotonicity

We say that the valued differential field K is **monotone** if $a' \preccurlyeq a$ for all $a \prec 1$ in K; in that case the derivation of K is small, and $a' \preccurlyeq a$ for all $a \in K$, by Lemma 4.4.1. The Hahn differential field $\boldsymbol{k}((t^{\Gamma}))$ of Example (1) above is monotone. The valued differential field $\boldsymbol{k}((t^{\mathbb{Q}}))$ of Example (2) above is monotone if $vg \geqslant 1$. Also $C(x)$ of Example (3) is monotone. In Example (4), $\mathbb{T}_{\log}$ is monotone, but $\mathbb{T}$ is not. Note that if K is monotone, then so is every valued differential subfield of K, and every coarsening of K by a convex subgroup of $\Gamma = v(K^{\times})$.

Many constants and few constants

For later use we introduce the following conditions, where as usual $C = C_K$ and $\Gamma = v(K^{\times})$:

(1) K has **many constants** if $v(C^{\times}) = \Gamma$;

(2) K has **few constants** if $v(C^{\times}) = \{0\}$, equivalently, $C \subseteq \mathcal{O}$.

The Hahn differential field $\boldsymbol{k}((t^{\Gamma}))$ of Example (1) above has many constants. The valued differential fields of Examples (3) and (4) have few constants. If K has small derivation and many constants, then K is monotone.

If K has many constants, and L is a valued differential field extension of K with $\Gamma_L = \Gamma$, then L also has many constants. Although we don't really need this, here are two easy results on extensions with possibly bigger value group.

LEMMA 4.4.13. *Suppose K is henselian, K has many constants, $\boldsymbol{k}$ is real closed or algebraically closed, and L is a valued differential field extension of K and algebraic over K. Then L has many constants.*

PROOF. Let $a \in L^{\times}$, and take $n \geqslant 1$ such that $nv(a) \in \Gamma$. Then $a^n \asymp c$ with $c \in C^{\times}$, so $a^n = bc$ with $b \in L^{\times}$, $b \asymp 1$. Now L is also henselian with real closed or algebraically closed residue field, so $b = d^n$ or $-b = d^n$ with $d \in L^{\times}$. Then $c = (a/d)^n$ or $-c = (a/d)^n$, and thus $a/d \in C_L$ and $v(a) = v(a/d)$. □

LEMMA 4.4.14. *Suppose K is monotone with many constants, and let $\Gamma + \mathbb{Z}\beta$ be an ordered group extension of $\Gamma = v(K^{\times})$ such that $n\beta \notin \Gamma$ for all $n \geqslant 1$. Let $K(b)$ be a field extension of K with b transcendental over K and make $K(b)$ into a valued differential field extension of K by requiring $b' = 0$ and by extending the valuation of K to a valuation $v\colon K(b)^{\times} \to \Gamma + \mathbb{Z}\beta$ with $vb = \beta$. Then $K(b)$ is monotone, $C_{K(b)} = C(b)$, and $K(b)$ has many constants.*

PROOF. For $a_0, \dots, a_n \in K$, not all zero, and $P = \sum_i a_i Y^i$ we have

$$vP(b) = v(a_0 + a_1 b + \cdots + a_n b^n) = \min_i \big(v(a_i) + i\beta\big),$$

$$P(b)' = P^{\partial}(b), \text{ so } vP(b)' = \min_i \big(v(a_i') + i\beta\big),$$

from which we get $P(b)' \preccurlyeq P(b)$. Thus $K(b)$ is monotone.

Let $P, Q \in K[Y]^{\neq}$ be coprime, with Q monic, such that $P(b)/Q(b) \in C_{K(b)}$. Then $P^{\partial}(b)Q(b) = P(b)Q^{\partial}(b)$, so $Q|Q^{\partial}$ in $K[Y]$. Since $\deg Q^{\partial} < \deg Q$, this gives $Q^{\partial} = 0$, and so $P^{\partial} = 0$ as well. Then $P, Q \in C[Y]$, so $P(b)/Q(b) \in C(b)$. This gives $C_{K(b)} = C(b)$, and also shows that $K(b)$ has many constants. □

Classifying pc-sequences

Let (a_ρ) be a pc-sequence in K.

We say that (a_ρ) is of **differential-algebraic type over** K (for short: d**-algebraic type over** K) if $G(b_\lambda) \rightsquigarrow 0$ for some $G(Y) \in K\{Y\}$ and some pc-sequence (b_λ) in K equivalent to (a_ρ). A **minimal differential polynomial of** (a_ρ) over K is a differential polynomial $G(Y) \in K\{Y\}$ with the following properties:

(1) $G(b_\lambda) \rightsquigarrow 0$ for some pc-sequence (b_λ) in K equivalent to (a_ρ) (so $G \notin K$);

(2) $H(b_\lambda) \not\rightsquigarrow 0$ whenever $H \in K\{Y\}$ has lower complexity than G and the pc-sequence (b_λ) in K is equivalent to (a_ρ).

Thus (a_ρ) is of d-algebraic type over K if and only if (a_ρ) has a minimal differential polynomial over K. If G is a minimal differential polynomial of (a_ρ) over K and $a \in K$, then G_{+a} is a minimal differential polynomial of $(a_\rho - a)$.

We say that (a_ρ) is of **differential-transcendental type over** K (for short: d**-transcendental type over** K) if it is not of d-algebraic type over K, that is, $G(b_\lambda) \not\rightsquigarrow 0$ for each $G \in K\{Y\}$ and each pc-sequence (b_λ) in K equivalent to (a_ρ).

Notes and comments

Lemmas 4.4.1, 4.4.2, and 4.4.3 are partly borrowed from Section 2 of [87], but the terminology there is different from ours. The property of monotonicity appears in [87]; see also [299, Proposition 2.2]. Monotonicity together with having many constants is a key assumption in Scanlon's [382].

4.5 THE GAUSSIAN VALUATION

Let K be a valued differential field. Then we extend its valuation v to the valuation $v\colon K\langle Y\rangle \to \Gamma_\infty$ on $K\langle Y\rangle$ by

$$v(P) := \min_i v(P_i) = \min_\omega v(P_{[\omega]}) \in \Gamma_\infty,$$

for $P(Y) = \sum_i P_i Y^i \in K\{Y\}$. Because of its familiar connection to Gauss's lemma on unique factorization, this extended valuation is called the **gaussian extension** of v. (See also Section 3.1.) Note that for $P, Q \in K\{Y\}$ we have $v(P+Q) = \min\big(v(P), v(Q)\big)$ whenever $v(P) \neq v(Q)$, and also whenever P, Q have no common monomials, that is, for all $\boldsymbol{i}$ either $P_{\boldsymbol{i}} = 0$ or $Q_{\boldsymbol{i}} = 0$. Hence $v(P) = \min_d v(P_d)$.

Recall that $\mathcal{O}$ is the valuation ring of K with maximal ideal $\smallo$. In the rest of this section we assume $\partial\smallo \subseteq \smallo$ (and hence $\partial\mathcal{O} \subseteq \mathcal{O}$). Also, ϕ, f, g range over K.

The function v_P

LEMMA 4.5.1. *Let $P \in K\{Y\}$. Then we have:*

(i) *if $f \preccurlyeq 1$, then $P_{+f} \asymp P$; if $f \prec 1$ and $P \neq 0$, then $P_{+f} \sim P$;*

(ii) *the element $v(P_{\times f})$ of Γ_∞ depends only on vf;*

(iii) *if $P \neq 0$ and $P(0) = 0$, then the map*

$$vf \mapsto v(P_{\times f}) : \Gamma_\infty \to \Gamma_\infty$$

is strictly increasing.

PROOF. Assume $f \preccurlyeq 1$. From (4.3.1) we obtain $P_{+f} \preccurlyeq P$. Since $P = (P_{+f})_{+g}$ with $g := -f$, this also gives $P \preccurlyeq P_{+f}$. Next, assume $f \prec 1$ and $P \neq 0$. Then (4.3.1) yields $P_{+f} = P + Q$ with $Q \prec P$, and so $P_{+f} \sim P$. This proves (i).

Suppose that $v\phi = 0$. By (4.3.2), this gives $v(P_{\times\phi}) \geqslant v(P)$. Using $P = (P_{\times\phi})_{\times\phi^{-1}}$ we can reverse this inequality to get $v(P) \geqslant v(P_{\times\phi})$. Hence $v(P) = v(P_{\times\phi})$. Suppose now that $vf = vg$, and write $f = g\phi$ with $v\phi = 0$. By what we just proved and using $P_{\times f} = (P_{\times g})_{\times\phi}$ we obtain $v(P_{\times f}) = v(P_{\times g})$. This shows (ii). Next, suppose $P \neq 0$, $P(0) = 0$ and $vf > vg$. Write $f = g\phi$ where $v\phi > 0$. Then the identity (4.3.2) restricted to $\boldsymbol{\sigma}$ of length > 0, together with the assumption that $\partial\mathcal{o} \subseteq \mathcal{o}$, yields $v(P_{\times\phi}) > v(P)$. Using $P_{\times f} = (P_{\times g})_{\times\phi}$, we conclude $v(P_{\times f}) > v(P_{\times g})$. □

For $P \in K\{Y\}$ we define $v_P \colon \Gamma_\infty \to \Gamma_\infty$ by

$$v_P(\gamma) := v(P_{\times f}) \quad \text{whenever } vf = \gamma.$$

(Thus for $P = 0$ the function v_P takes the constant value ∞.) For $P, Q \in K\{Y\}$ we have

$$(PQ)_{\times f} = P_{\times f} Q_{\times f} \quad \text{and} \quad (P+Q)_{\times f} = P_{\times f} + Q_{\times f},$$

so

$$v_{PQ}(\gamma) = v_P(\gamma) + v_Q(\gamma) \quad \text{and} \quad v_{P+Q}(\gamma) \geqslant \min\bigl(v_P(\gamma), v_Q(\gamma)\bigr).$$

Note also that

$$v_P(\gamma) = \min_d v_{P_d}(\gamma).$$

Thus in some sense the properties of the functions v_P reduce to the case that P is homogeneous. In Chapter 6 we consider these functions v_P in more detail, and then the following results will be very useful. The first one says that if the derivation of the differential residue field $\boldsymbol{k}$ is nontrivial and $v_P(\alpha) = \beta$, then $v(P(f)) = \beta$ for "almost all" f with $vf = \alpha$. More precisely:

LEMMA 4.5.2. *Assume the derivation of $\boldsymbol{k}$ is nontrivial. Let $P \in K\{Y\}^{\neq}$ and $\alpha, \beta \in \Gamma$ be such that $v_P(\alpha) = \beta$, and let $a \in K$ be such that $va = \alpha$. Then there is a thin set $S \subseteq \boldsymbol{k}$ such that $vP(ay) = \beta$ for all $y \asymp 1$ in K with $\overline{y} \notin S$.*

PROOF. Take $b \in K$ with $vb = \beta$, so $vF = 0$ for $F := b^{-1}P_{\times a}$. Then the thin set $S := \{\overline{y} \in \boldsymbol{k} : \overline{F}(\overline{y}) = 0\} \subseteq \boldsymbol{k}$ has the property that $F(y) \asymp 1$ for all $y \asymp 1$ in K with $\overline{y} \notin S$, and so $vP(ay) = \beta$ for all such y. □

LEMMA 4.5.3. *Suppose $g \in K^\times$ and $g' \preccurlyeq g$. Then $g^{(n)} \preccurlyeq g$ for all n, and with $\gamma := vg$ we have $v_P(\gamma) = v(P) + d\gamma$ for homogeneous $P \in K\{Y\}^{\neq}$ of degree d.*

PROOF. From $g' \preccurlyeq g$ and Lemma 4.4.2 we obtain $g^{(n)} \preccurlyeq g$ for all n. Next, consider the case $d = 1$ and $P = Y^{(n)}$. Then

$$P_{\times g} = (gY)^{(n)} = gY^{(n)} + ng'Y^{(n-1)} + \cdots + g^{(n)}Y,$$

so $v_P(\gamma) = \gamma$. Thus for $\boldsymbol{i} = (i_0, \dots, i_r) \in \mathbb{N}^{1+r}$, $d = |\boldsymbol{i}|$, and $P = Y^{\boldsymbol{i}}$,

$$P_{\times g} \;=\; g^d Y^{\boldsymbol{i}} + R, \quad R \in K\{Y\},$$

where $vR \geqslant d\gamma$ and all monomials in R are lower than $Y^{\boldsymbol{i}}$ in the antilexicographic ordering on the set of monomials. For general homogeneous $P \in K\{Y\}^{\neq}$ of degree d we take among the terms $aY^{\boldsymbol{i}}$ in P with $va = vP$ the one for which $Y^{\boldsymbol{i}}$ is largest in the antilexicographic ordering on the set of monomials. For this term $aY^{\boldsymbol{i}}$ in P we have $|\boldsymbol{i}| = d$ and

$$P_{\times g} \;=\; ag^d Y^{\boldsymbol{i}} + R + S, \quad R, S \in K\{Y\},$$

where $vR \geqslant vP + d\gamma$, all monomials in R are antilexicographically lower than $Y^{\boldsymbol{i}}$, and $vS > vP + d\gamma$. □

COROLLARY 4.5.4. *If K is monotone and $P \in K\{Y\}^{\neq}$ is homogeneous of degree d, then $v_P(\gamma) \;=\; v(P) + d\gamma$ for all $\gamma \in \Gamma$.*

Let Δ be a convex subgroup of Γ, and let $\dot{v}\colon K^\times \to \dot{\Gamma} = \Gamma/\Delta$ be the Δ-coarsening of K with valuation ring $\dot{\mathcal{O}}$ and maximal ideal $\dot{o}$ of $\dot{\mathcal{O}}$. Then $\partial\dot{o} \subseteq \dot{o}$ by Corollary 4.4.4. The following is obvious.

LEMMA 4.5.5. *For $P \in K\{Y\}^{\neq}$ and $\gamma \in \Gamma$ we have*

$$\dot{v}_P(\dot{\gamma}) \;=\; v_P(\gamma) + \Delta.$$

Dominant weight

In this subsection we assume $P, Q \in K\{Y\}^{\neq}$, and set

$$\begin{aligned} \operatorname{dwt}(P) &:= \max\{\|\boldsymbol{\sigma}\| : v(P_{[\boldsymbol{\sigma}]}) = v(P)\}, \\ \operatorname{dwm}(P) &:= \min\{\|\boldsymbol{\sigma}\| : v(P_{[\boldsymbol{\sigma}]}) = v(P)\}, \end{aligned}$$

so

$$\operatorname{wm}(P) \leqslant \operatorname{dwm}(P) \leqslant \operatorname{dwt}(P) \leqslant \operatorname{wt}(P).$$

We call $\operatorname{dwt}(P)$ the **dominant weight** of P and $\operatorname{dwm}(P)$ the **dominant weighted multiplicity** of P. These quantities will be needed in Section 11.1.

LEMMA 4.5.6.

$$\mathrm{dwt}(PQ) = \mathrm{dwt}(P) + \mathrm{dwt}(Q), \quad \mathrm{dwm}(PQ) = \mathrm{dwm}(P) + \mathrm{dwm}(Q).$$

PROOF. Take f with $vf = v(P)$, so $P_0 := f^{-1}P \in \mathcal{O}\{Y\}$; then $\mathrm{dwt}(P) = \mathrm{wt}(\overline{P_0})$ and $\mathrm{dwm}(P) = \mathrm{wm}(\overline{P_0})$ where $\overline{P_0}$ is the image of P_0 under the ring morphism

$$\mathcal{O}\{Y\} \to \boldsymbol{k}\{Y\}$$

that extends the residue map $a \mapsto \overline{a}\colon \mathcal{O} \to \boldsymbol{k} := \mathrm{res}(K)$ and sends $Y^{(n)}$ to $Y^{(n)}$ for all n. The lemma now follows from Lemma 4.2.3. □

LEMMA 4.5.7. *The quantity* $\mathrm{dwt}(P_{\times g})$ *depends only on* vg, *for* $g \neq 0$.

PROOF. Take $\boldsymbol{\sigma}$ such that $\mathrm{dwt}(P) = \|\boldsymbol{\sigma}\|$ and $v(P_{[\boldsymbol{\sigma}]}) = v(P)$, and let $vg = 0$. Then by (4.3.2) we have $v\big((P_{\times g})_{[\boldsymbol{\sigma}]}\big) = v(P_{[\boldsymbol{\sigma}]}) = v(P) = v(P_{\times g})$, so $\mathrm{dwt}(P_{\times g}) \geqslant \mathrm{dwt}(P)$. To reverse this inequality, use $P = (P_{\times g})_{\times g^{-1}}$. This proves the lemma for $vg = 0$, and the general case follows easily from this special case. □

Thus for $\gamma \in \Gamma$ we can define $\mathrm{dwt}_P(\gamma) := \mathrm{dwt}(P_{\times g})$ where $\gamma = vg$.

LEMMA 4.5.8. *Suppose P is homogeneous and $g \in K^\times$ satisfies $g' \preccurlyeq g$. Then we have* $\mathrm{dwt}(P_{\times g}) = \mathrm{dwt}(P)$.

PROOF. By Lemma 4.5.3 we have $g^{(n)} \preccurlyeq g$ for all n and $v(P_{\times g}) = v(P) + d\gamma$, where $\gamma := vg$ and $d := \deg P$. Therefore, if $|\boldsymbol{\sigma}| = d$ and $\|\boldsymbol{\sigma}\| > \mathrm{dwt}(P)$, then by formula (4.3.2) we have $v\big((P_{\times g})_{[\boldsymbol{\sigma}]}\big) > v(P_{\times g})$. Now take $\boldsymbol{\sigma}$ such that $|\boldsymbol{\sigma}| = d$, $\|\boldsymbol{\sigma}\| = \mathrm{dwt}(P)$, and $v(P_{[\boldsymbol{\sigma}]}) = v(P)$. Then by (4.3.2),

$$(P_{\times g})_{[\boldsymbol{\sigma}]} = P_{[\boldsymbol{\sigma}]}g^d + f, \qquad vf > v(P_{[\boldsymbol{\sigma}]}g^d) = v(P) + d\gamma = v(P_{\times g}),$$

so $v\big((P_{\times g})_{[\boldsymbol{\sigma}]}\big) = v(P_{\times g})$ and thus $\mathrm{dwt}(P_{\times g}) = \mathrm{dwt}(P)$. □

The following is proved just like Lemma 4.5.7:

LEMMA 4.5.9. *If $\partial\mathcal{O} \subseteq o$, then* $\mathrm{dwm}(P_{\times g})$ *depends only on* vg, *for* $g \neq 0$.

If $\partial\mathcal{O} \subseteq o$ and $\gamma \in \Gamma$, then we define $\mathrm{dwm}_P(\gamma) := \mathrm{dwm}(P_{\times g})$ where $\gamma = vg$.

LEMMA 4.5.10. *Suppose $\partial\mathcal{O} \subseteq o$ and P is homogeneous. Let $g \in K^\times$, $g' \prec g$. Then $g^{(n)} \prec g$ for all $n \geqslant 1$, and* $\mathrm{dwm}(P_{\times g}) = \mathrm{dwm}(P)$.

PROOF. By Lemma 4.4.2 we have $g^{(n)} \prec g$ for all $n \geqslant 1$. By Lemma 4.5.3 we have $v(P_{\times g}) = v(P) + d\gamma$, where $\gamma := vg$ and $d := \deg P$. Take $r \in \mathbb{N}$ such that P has order $\leqslant r$. Below, $\boldsymbol{\sigma}, \boldsymbol{\tau} \in \{0, \dots, r\}^*$ are subject to $|\boldsymbol{\sigma}| = |\boldsymbol{\tau}| = d$. Then

$$\text{(4.5.1)} \qquad \boldsymbol{\tau} \geqslant \boldsymbol{\sigma} \text{ and } \boldsymbol{\tau} \neq \boldsymbol{\sigma} \quad \Rightarrow \quad v\big(P_{[\boldsymbol{\tau}]}g^{[\boldsymbol{\tau}-\boldsymbol{\sigma}]}\big) > v(P) + d\gamma.$$

So by (4.3.2) and (4.5.1), if $\|\boldsymbol{\sigma}\| < \mathrm{dwm}(P)$, then $v\big((P_{\times g})_{[\boldsymbol{\sigma}]}\big) > v(P_{\times g})$. Take $\boldsymbol{\sigma}$ such that $\|\boldsymbol{\sigma}\| = \mathrm{dwm}(P)$, and $v(P_{[\boldsymbol{\sigma}]}) = v(P)$. Then by (4.3.2) and (4.5.1),

$$(P_{\times g})_{[\boldsymbol{\sigma}]} = P_{[\boldsymbol{\sigma}]}g^d + f, \qquad vf > v\big(P_{[\boldsymbol{\sigma}]}g^d\big) = v(P) + d\gamma = v(P_{\times g}),$$

so $v\big((P_{\times g})_{[\boldsymbol{\sigma}]}\big) = v(P_{\times g})$ and thus $\mathrm{dwm}(P_{\times g}) = \mathrm{dwm}(P)$. □

4.6 DIFFERENTIAL RINGS

We gather here some basic facts about differential rings to be used at various places. This concerns differential ideals, simple differential rings, and linear disjointness over constant fields. *Throughout this section R is a differential ring.* Recall from Section 4.1 that this includes $\mathbb{Q}$ being a subring of R.

Radical differential ideals

A **differential ideal** of R is an ideal I of R with $f' \in I$ for all $f \in I$. If I and J are differential ideals of R, then so are $I \cap J$ and IJ. Note that if I is a differential ideal of R with $I \neq R$, then $a + I \mapsto a' + I$ is a derivation on R/I, making R/I a differential ring and the natural surjection $R \to R/I$ a morphism of differential rings. Conversely, if S is a differential ring and $h\colon R \to S$ is a morphism of differential rings, then $\ker h$ is a differential ideal of R. Given a subset S of R, we denote by $[S]$ the smallest differential ideal of R containing S; so $[S]$ is the ideal generated by the derivatives $s^{(n)}$ of the elements $s \in S$. A **radical differential ideal of** R is a differential ideal of R that is radical as an ideal of R.

In the rest of this subsection I is a differential ideal of R. Suppose that S is a multiplicative subset of R with $0 \notin S$. Recall from Section 4.1 the differential ring morphism $r \mapsto \iota(r) := \frac{r}{1}\colon R \to S^{-1}R$. The ideal $S^{-1}I$ of $S^{-1}R$ generated by $\iota(I)$ is a differential ideal, hence

$$\iota^{-1}(S^{-1}I) \;=\; \{r \in R : \ rs \in I \text{ for some } s \in S\}$$

is a differential ideal of R containing I.

LEMMA 4.6.1. *Let $a \in R$ with $a^n \in I$, where $n \geqslant 1$. Then $(a')^{2n-1} \in I$.*

PROOF. By induction on $k = 1, \dots, n$ we show that $a^{n-k}(a')^{2k-1} \in I$. The case $k = 1$ holds since $a^{n-1}a' = \frac{1}{n}(na^{n-1}a') = \frac{1}{n}(a^n)' \in I$. Suppose $1 \leqslant k < n$ and $a^{n-k}(a')^{2k-1} \in I$. Differentiating yields

$$(n-k)a^{n-k-1}(a')^{2k} + (2k-1)a^{n-k}(a')^{2k-2}a'' \in I,$$

and then multiplying by $\frac{1}{n-k}a'$ gives $a^{n-k-1}(a')^{2k+1} \in I$, as required. □

Thus the radical of a differential ideal of R is a differential ideal of R.

LEMMA 4.6.2. *Suppose I is radical and $a, b \in R$, $ab \in I$. Then $a'b \in I$.*

PROOF. Use $a'b + ab' = (ab)' \in I$ and multiply by $a'b$. □

COROLLARY 4.6.3. *If I is radical and $S \subseteq R$, then $(I : S)$ is a radical differential ideal of R.*

Given $S \subseteq R$, the smallest radical differential ideal of R containing S is $\sqrt{[S]}$.

COROLLARY 4.6.4. *Let $S, T \subseteq R$. Then $\sqrt{[S]} \cdot \sqrt{[T]} \subseteq \sqrt{[ST]}$.*

PROOF. Let $r \in R$. By the previous corollary, $\left(\sqrt{[rT]} : r\right)$ is a radical differential ideal of R; it contains T and thus also $\sqrt{[T]}$. Thus $r\sqrt{[T]} \subseteq \sqrt{[rT]}$ for all $r \in R$. Hence the radical differential ideal $\left(\sqrt{[ST]} : \sqrt{[T]}\right)$ of R contains S and therefore $\sqrt{[S]}$; it follows that $\sqrt{[S]} \cdot \sqrt{[T]} \subseteq \sqrt{[ST]}$. □

A differential ideal I of R is said to be **prime** if it is prime as an ideal of R, that is, R/I is an integral domain. A differential ideal of R is said to be **maximal** if it is proper (not equal to R) and maximal among proper differential ideals of R (which does not imply that it is a maximal ideal of R; see the next subsection). Here is a differential analogue of a well-known fact from commutative algebra:

PROPOSITION 4.6.5. *Every radical differential ideal of R is the intersection in R of a collection of prime differential ideals of R.*

In the proof we use:

LEMMA 4.6.6. *Let S be a multiplicative subset of R. Let I be a radical differential ideal of R disjoint from S and maximal with these properties. Then I is prime.*

PROOF. Let $a, b \in R$ with $ab \in I$, and suppose for a contradiction that $a, b \notin I$. Then $\sqrt{[I,a]}$ and $\sqrt{[I,b]}$ are radical differential ideals which properly contain I; thus we have $s, t \in S$ with $s \in \sqrt{[I,a]}$ and $t \in \sqrt{[I,b]}$. Then by Corollary 4.6.4,

$$st \in S \cap \sqrt{[I,a]}\sqrt{[I,b]} \subseteq S \cap I = \emptyset,$$

a contradiction. □

PROOF OF PROPOSITION 4.6.5. Let I be a radical differential ideal of R and $r \in R \setminus I$. Take a radical differential ideal J of R with $J \supseteq I$, disjoint from $S := \{1, r, r^2, \dots\}$, and maximal with these properties; then J is prime. □

COROLLARY 4.6.7. *Every maximal differential ideal of R is prime.*

COROLLARY 4.6.8. *Let K be a differential field, and let E and F be differential field extensions of K. Then there is a differential field L with differential field embeddings $E \to L$ and $F \to L$ that agree on K.*

PROOF. Corollary 1.7.5 makes the commutative K-algebra $E \otimes_K F$ into a differential ring R such that the maps $a \mapsto a \otimes 1 \colon E \to R$ and $b \mapsto 1 \otimes b \colon F \to R$ are (injective) differential ring morphisms. Take a prime differential ideal $\mathfrak{p}$ of R. Composing these maps $E \to R$ and $F \to R$ with the natural map $R \to R/\mathfrak{p}$ and extending $R/\mathfrak{p}$ to its differential fraction field L yields the desired result. □

Let K be a differential field and $R = K\{Y_1, \dots, Y_n\}$, where $Y_1, \dots, Y_n$ are distinct differential indeterminates over K. We have a Differential Nullstellensatz:

COROLLARY 4.6.9. *Let $P_1, \dots, P_m, Q \in R$. Then the following are equivalent:*

(i) $Q \notin \sqrt{[P_1, \dots, P_m]}$;

(ii) *there exists a differential field extension L of K and $\vec{y} \in L^n$ such that $P_i(\vec{y}) = 0$ for $i = 1, \dots, m$ and $Q(\vec{y}) \neq 0$.*

PROOF. Suppose (i) holds. Then we have a prime differential ideal I of R with $I \supseteq \sqrt{[P_1, \dots, P_m]}$ and $Q \notin I$. Identifying K in the obvious way with a differential subfield of the differential integral domain R/I, we get (ii) with L the differential fraction field of R/I and $\vec{y} = (y_1, \dots, y_n) \in L^n$ where $y_j = Y_j + I \in R/I$ for $j = 1, \dots, n$. The direction (ii) $\Rightarrow$ (i) is obvious. □

REMARK. If $P_1, \dots, P_m$ in the previous corollary are homogeneous of degree 1, then $\sqrt{[P_1, \dots, P_m]} = [P_1, \dots, P_m]$, since as an ideal of R, the latter is generated by homogeneous differential polynomials of degree 1, and thus prime by Lemma 1.1.16.

Simple differential rings

We say that R is **simple** if the only differential ideals of R are $\{0\}$ and R, that is, $\{0\}$ is a maximal differential ideal of R. Simple differential rings are integral domains, by Corollary 4.6.7. If c is a constant of R, then cR is a differential ideal of R. It follows that the ring of constants C_R of a simple differential ring R is a field.

LEMMA 4.6.10. *Suppose R is simple and $r \in R \setminus \partial(R)$. Let x be an element of a differential ring extension of R with $x' = r$. Then the differential ring $R[x]$ is simple and x is transcendental over R.*

PROOF. We equip the polynomial ring $R[X]$ over R with the derivation extending the derivation of R such that $X' = r$. We claim that $R[X]$ is simple. Let I be a nonzero differential ideal of $R[X]$, and let n be the smallest degree of a nonzero element of I. Let J be the set of all $a \in R$ such that there is some $P \in I$ with $P = aX^n +$ terms of degree $< n$. It is easy to see that J is a nonzero differential ideal of R, hence $1 \in J$. Let $P = X^n + a_{n-1}X^{n-1} + \cdots + a_0 \in I$ with $a_0, \dots, a_{n-1} \in R$. Suppose $n \geqslant 1$. We have $P' \in I$ and

$$P' = (nr + a'_{n-1})X^{n-1} + \text{terms of degree} < n - 1,$$

hence $P' = 0$ and thus $nr + a'_{n-1} = 0$, so $y := -\frac{1}{n}a_{n-1} \in R$ satisfies $y' = r$, a contradiction. Hence $n = 0$, so $P = 1 \in I$. Thus $R[X]$ is simple. The ring morphism $\phi\colon R[X] \to R[x]$ over R sending X to x is a surjective morphism of differential rings, so ϕ is an isomorphism. □

LEMMA 4.6.11. *Suppose R is simple, $r \in R$, and $y' \neq mry$ for all $y \in R^{\neq}$ and $m \geqslant 1$. Let x be a unit of a differential ring extension A of R with $x' = rx$. Then the differential subring $R[x, x^{-1}]$ of A is simple, and x is transcendental over R.*

PROOF. Let X be an indeterminate over R and equip $R[X, X^{-1}]$ with the derivation extending the derivation of R such that $X' = rX$. We claim that $R[X, X^{-1}]$ is simple. (As in the proof of Lemma 4.6.10, it follows that then x is transcendental over R and $R[x, x^{-1}]$ is simple.) For this, let $I \neq \{0\}$ be a differential ideal of $R[X, X^{-1}]$; then $I \cap R[X] \neq \{0\}$. Let n be the smallest degree of a nonzero element of $I \cap R[X]$.

Let J be the set of all $a \in R$ such that there is some $P \in I \cap R[X]$ with $P = aX^n +$ terms of degree $< n$. Then J is a nonzero differential ideal of R, hence $1 \in J$. Let $P = X^n + a_{n-1}X^{n-1} + \cdots + a_0 \in I$ with $a_0, \dots, a_{n-1} \in R$. If $n = 0$, then $P = 1 \in I$. Suppose $n \geqslant 1$. Then

$$P' = nrX^n + \big(a'_{n-1} + (n-1)a_{n-1}r\big)X^{n-1} + \cdots + (a'_1 + a_1 r)X + a'_0 \in I,$$

hence $nrP - P' \in I$ has degree less than n, so $nrP = P'$, hence $a'_i = (n-i)ra_i$ for $i = 0, \dots, n-1$. By hypothesis we obtain $a_0 = \cdots = a_{n-1} = 0$, that is, $P = X^n$. Hence $1 = X^{-n}P \in I$. □

For R, r, x as in Lemma 4.6.11 the differential subring $R[x]$ of $R[x, x^{-1}]$ is not simple, since it has $xR[x]$ as a differential ideal.

In the rest of this subsection K is a differential field with constant field C.

COROLLARY 4.6.12. *Suppose R is a simple differential ring extension of K. Then every constant element of the differential fraction field of R lies in R. If R is finitely generated as a K-algebra, then every such element is algebraic over C.*

PROOF. Let a be a constant element of the differential fraction field of R. Then $I := \{b \in R : ab \in R\}$ is a nonzero differential ideal of R, so $I = R$ and hence $a \in R$. Thus for every $c \in C$ with $a \neq c$, $(a-c)R$ is a nonzero differential ideal of R, so $a - c \in R^\times$. Assume also that R is finitely generated as a K-algebra. Then by Lemma 1.3.10, a is algebraic over K. Let $P(X) \in K[X]$ be the minimum polynomial of a over K. Differentiating the equality $P(a) = 0$ shows $P(X) \in C[X]$. □

COROLLARY 4.6.13. *Let $a \in K \setminus \partial(K)$. Let y be an element of a differential field extension of K with $y' = a$. Then every $r \in K(y)^\times$ with $r^\dagger \in K$ lies in K.*

PROOF. By Lemma 4.6.10, y is transcendental over K and $K[y]$ is simple. Let $r \in K(y)^\times$ be such that $s := r^\dagger \in K$. Take coprime $f, g \in K[y]^{\neq}$ and $k \in \mathbb{Z}$ such that $r = (f/g)y^k$ and y does not divide fg in $K[y]$. To show $r \in K$ we can multiply r with an element of $K^\times$ and arrange that f, g are monic in y. Then $f' \in K[y]$ is of lower degree in y than f, and $g' \in K[y]$ is of lower degree in y than g, and

$$r^\dagger = (f'/f) - (g'/g) + (ka/y) = s.$$

Multiplying both sides by fgy yields

$$(f'g - fg')y + fg \cdot ka = fg \cdot ys.$$

Hence $fg | (f'g - fg')y$, so $f | f'gy$ and $g | fg'y$, and thus $f | f'$ and $g | g'$ in $K[y]$. Hence $f' = g' = 0$, which by Corollary 4.6.12 yields $f, g \in C \subseteq K$, so $f = g = 1$. Then $k(a/y) = s \in K$, which in view of $a \neq 0$ gives $k = 0$. Therefore $r = 1 \in K$. □

COROLLARY 4.6.14. *Let $P \in K\{Y\}^{\neq}$ be irreducible of order r, and let $Q \in K\{Y\}^{\neq}$ have order $< r$. Then there is an element y of a differential field extension of K such that $P(y) = 0$, $S_P(y) \neq 0$, $Q(y) \neq 0$, and $C_{K\langle y\rangle}$ is algebraic over C.*

PROOF. Lemma 4.1.9 gives an element a in a differential field extension of K with minimal annihilator P over K. The differential subring $R := K\{a, S(a)^{-1}, Q(a)^{-1}\}$ of $K\langle a\rangle$ with $S := S_P$ satisfies

$$R = K[a, a', \dots, a^{(r)}, S(a)^{-1}, Q(a)^{-1}].$$

Let M be a maximal differential ideal of R. Then $\overline{R} := R/M$ is a simple differential ring, finitely generated as a K-algebra. For $y := a + M \in \overline{R}$ we get $P(y) = 0$, $S(y) \neq 0$, $Q(y) \neq 0$ in $K\langle y\rangle$, and $C_{K\langle y\rangle}$ is algebraic over C by Corollary 4.6.12. □

We use this to derive a technical fact needed in the next volume:

LEMMA 4.6.15. *Let $P \in K\{Y\}^{\neq}$ be irreducible of order r, and let $Q \in K\{Y\}^{\neq}$ have order $< r$. Suppose P is a minimal annihilator over K of every element y of every differential field extension L of K such that $P(y) = 0$, $Q(y) \neq 0$, and C_L is algebraic over C. Let a in a differential field extension of K satisfy $P(a) = 0$, $Q(a) \neq 0$, and $S_P(a) \neq 0$. Then P is a minimal annihilator of a over K.*

PROOF. Put $S := S_P$. Corollary 4.6.14 and its proof give an element y in a differential field extension of K such that $P(y) = 0$, $S(y) \neq 0$, $Q(y) \neq 0$, $C_{K\langle y\rangle}$ is algebraic over C, and the differential subring $R := K\{y, S(y)^{-1}, Q(y)^{-1}\}$ of $K\langle y\rangle$ is simple. Then P is a minimal annihilator of y by the hypothesis of the lemma. The remarks after Lemma 4.1.10 give a surjective differential ring morphism $R \to K\{a, S(a)^{-1}, Q(a)^{-1}\}$ over K sending y to a. Since R is simple, this morphism is an isomorphism. □

Linear disjointness of constant fields

Let K, L be subfields of a field F, and let C be a subfield of both K and L. We say that K **is linearly disjoint from** L **over** C if every finite set of elements of K which is C-linearly independent is also L-linearly independent. Equivalently, K is linearly disjoint from L over C iff the ring morphism $K \otimes_C L \to F$ with $a \otimes b \mapsto ab$ is injective. In particular, if K is linearly disjoint from L over C, then L is linearly disjoint from K over C. For proofs of this and other facts on linear disjointness, see [249, VIII, §3].

In the rest of this subsection, K is a differential field, and L is a differential field extension of K. The following lemma shows that K is linearly disjoint from the constant field C_L of L over the constant field C of K.

LEMMA 4.6.16. *Let R be a differential ring extension of K. The ring morphism*

$$K \otimes_C C_R \to R \qquad \textit{with } a \otimes c \mapsto ac \textit{ for } a \in K,\ c \in C_R,$$

is injective. (Its image is the differential subring $K[C_R]$ of R.)

PROOF. Suppose not. Then we have $c_1, \dots, c_n \in C_R$ $(n \geqslant 1)$, linearly independent over C, and $a_1, \dots, a_n \in K$, not all zero, such that $\sum_{i=1}^n a_i c_i = 0$. Let n be minimal. Then $a_1 \in K^\times$, so $a_1 c_1 \neq 0$, hence $n \geqslant 2$. Arranging $a_1 = 1$ and differentiating gives $\sum_{i=2}^n a_i' c_i = 0$, so $a_i \in C$ for $i = 1, \dots, n$ by the minimality of n, but this contradicts the C-linear independence of $c_1, \dots, c_n$. □

COROLLARY 4.6.17. *Let $A \in K\{Y\}$ be homogeneous of degree 1 and let $g \in K$. If $A(f) = g$ for some $f \in K[C_L]$, then $A(f) = g$ for some $f \in K$.*

PROOF. Let B be a C-vector space basis of C_L with $1 \in B$, let b range over B, and assume $f \in K[C_L]$. By Lemma 4.6.16 we have a unique admissible family (f_b) in K such that $f = \sum_b f_b b$. If $f \in K[C_L]$ and $A(f) = g$, then $\sum_b A(f_b)b = g$ and hence $A(f_1) = g$. □

In the same way we obtain:

COROLLARY 4.6.18. *Let D be a subring of C_L containing C. Then $K[D]$ is a differential subring of L with D as its ring of constants.*

COROLLARY 4.6.19. *Let R be a differential ring extension of K and D a subring of C_R containing C such that $R = K[D]$. Let J be a K-linear subspace of R such that $\partial J \subseteq J$, and set $I := J \cap D$. Then $J = IK$. If D is a field, then R is simple.*

PROOF. Let B be a C-vector space basis of D, and let b range over B. For any $f \in R$ there is by Lemma 4.6.16 a unique admissible family (f_b) in K such that $f = \sum_b f_b b$, and define the length $\ell(f)$ of f by $\ell(f) := |\{b : f_b \neq 0\}|$. By induction on $\ell(f)$ we show that if $f \in J$, then $f \in IK$. This is clear if $\ell(f) = 0$, so suppose $f \in J$ satisfies $\ell(f) \geqslant 1$. Take $b_1 \in B$ with $f_{b_1} \neq 0$; we may assume $f_{b_1} = 1$. We may also assume $f \notin D$ and thus take $b_2 \in B$ with $f_{b_2} \notin C$. Now $f' = \sum_b f_b' b$ and so $\ell(f') < \ell(f)$, hence $f' \in IK$. Setting $a := f_{b_2}$, we similarly have $(a^{-1}f)' \in IK$ and thus $(a^{-1})'f = (a^{-1}f)' - a^{-1}f' \in IK$. Since $a^{-1} \notin C$, we obtain $f \in IK$. This proves $J = IK$. If D is a field, it follows that R is simple as a differential ring. □

Thus with R and D as in Corollary 4.6.19 the map $I \mapsto IR$ is a bijection from the set of ideals of D onto the set of differential ideals of R, with inverse $J \mapsto J \cap D$.

From Corollaries 4.6.18, 4.6.19, and 4.6.12 we obtain:

COROLLARY 4.6.20. *Let D be a subfield of C_L containing C. Then $K(D)$ is a differential subfield of L with D as its field of constants.*

Adjoining constants

Let K be a differential field and L an extension field of K with subfield $D \supseteq C$ such that $L = K(D)$. If the field L has a derivation extending that of K with constant field D, then by the previous subsection K and D are linearly disjoint over C. In Section 10.5 we need a converse of this fact:

LEMMA 4.6.21. *Suppose K and D are linearly disjoint over C. Then there is a unique derivation on the field L that extends that of K and is trivial on D; this derivation has D as its constant field.*

PROOF. There is clearly at most one such derivation. For existence, let B be a transcendence basis of D over C. Linear disjointness gives algebraic independence of B over K; see [249, VIII, Proposition 3.2]. So the derivation of K extends to a derivation of the field $K(B)$ that annuls each element of B (by Corollary 1.9.4), hence to a derivation of the algebraic extension L of $K(B)$, and this derivation will annul each element of D, since D is algebraic over $C(B)$ (by Corollary 1.9.6). By Corollary 4.6.20 the constant field of this derivation on L equals D. □

Notes and comments

Proposition 4.6.5 is due to Ritt; the proof as given is from [209]. The equivalence of (i) and (ii) in Corollary 4.6.9 is in Raudenbush [330], with a constructive proof in Seidenberg [396]. Corollary 4.6.12 is in Levelt [255, Lemma A3]. Lemma 4.6.16 is from [363, p. 292]. For Corollary 4.6.18 see [220, Corollary 5 on p. 768], and for Corollary 4.6.19, see [255, Proposition 1].

4.7 DIFFERENTIALLY CLOSED FIELDS

This section presents L. Blum's proof of A. Robinson's theorem that the theory of differential fields has a model completion, namely the theory of *differentially closed fields*; see B.10 and B.11 for the concept of model completion. We shall meet differentially closed fields briefly in Chapter 5 and Section 10.7. We also construct *differential closures* of differential fields, but do not use these later.

The language of differential rings is the language $\mathcal{L}_\partial = \{0, 1, -, +, \cdot, \partial\}$ obtained by augmenting the language $\{0, 1, -, +, \cdot\}$ of rings by an extra unary function symbol ∂. We view differential rings as structures for this language in the obvious way. A differential ring morphism is a morphism with respect to this language. Likewise, a differential subring of a differential ring R is a substructure of R with respect to the language $\mathcal{L}_\partial$.

Let K be a differential field. Note that $\mathbb{Z}$ with its trivial derivation is a substructure of K that is not a differential subring of K, in view of our convention that differential rings contain $\mathbb{Q}$. In fact, the substructures of K are exactly the subrings of K closed under the derivation of K and equipped with the restriction of this derivation to the subring.

DEFINITION 4.7.1. A differential field K is said to be **differentially closed** if for all $P \in K[Y, \dots, Y^{(r)}]^{\neq}$ and $Q \in K[Y, \dots, Y^{(r-1)}]^{\neq}$ such that $Y^{(r)}$ occurs in P there is $y \in K$ with $P(y) = 0$ and $Q(y) \neq 0$. (In this definition, we may restrict to irreducible P.) Taking $P \in K[Y]$ we see that if K is differentially closed, then K is algebraically closed.

Applying Corollary 4.6.14 iteratively, we see that every differential field K has a differential field extension L such that L is differentially closed and d-algebraic over K, and C_L is algebraic over C.

Let DCF be the theory of differentially closed fields, formulated in the language $\mathcal{L}_\partial$.

THEOREM 4.7.2. DCF *admits quantifier elimination.*

PROOF. Let E and F be differentially closed fields such that F is $|E|^+$-saturated, and let R be a proper substructure of E and $\phi\colon R \to F$ be an embedding; we shall extend ϕ to an embedding of a differential subring of E, properly containing R, into F. If R is not a field, then we can extend ϕ to the fraction field of R inside E (which is also a differential subfield of E). So assume R is a differential field K. Take any $a \in E \setminus K$. Consider first the case that a is d-transcendental over K. Using that F is $|E|^+$-saturated with nontrivial derivation we get a $b \in F$ that is d-transcendental over the differential subfield $\phi(K)$ of F. Then ϕ extends to an embedding $K\{a\} \to F$ sending a to b. Next assume that a is d-algebraic over K, and let $P \in K\{Y\}$, of order $r \in \mathbb{N}$, be a minimal annihilator of a over K, so

$$P(a) = 0, \qquad Q(a) \neq 0 \text{ for all } Q \in K\{Y\}^{\neq} \text{ of order} < r.$$

Using that F is $|E|^+$-saturated and differentially closed we can take $b \in F$ such that this equation and these inequations hold with b instead of a and with P and the Q's replaced by their ϕ-images. Then by Lemma 4.1.10 we can extend ϕ to an embedding of $K\{a\}$ into F that sends a to b. □

COROLLARY 4.7.3. DCF *is complete, and* DCF *is the model completion of the theory of differential fields.*

PROOF. The field $\mathbb{Q}$ with the trivial derivation embeds into every differentially closed field, hence the first part follows from Theorem 4.7.2. The second statement follows from that theorem in view of the fact that every differential field extends to a differentially closed field. □

As a consequence, we can now add to the Differential Nullstellensatz 4.6.9 two conditions, each equivalent to condition (i) in that theorem:

(iii) for some differentially closed differential field extension L of K and some $\vec{y} \in L^n$, we have $P_i(\vec{y}) = 0$ for $i = 1, \dots, m$ and $Q(\vec{y}) \neq 0$;

(iv) for every differentially closed differential field extension L of K, there is $\vec{y} \in L^n$ such that $P_i(\vec{y}) = 0$ for $i = 1, \dots, m$ and $Q(\vec{y}) \neq 0$.

COROLLARY 4.7.4. *Suppose K is a differentially closed field. If $X \subseteq K^n$ is definable in K, then $X \cap C^n$ is definable in the algebraically closed field C.*

PROOF. Let $P = P(Y_1, \dots, Y_n) \in K\{Y_1, \dots, Y_n\}$ be a differential polynomial in the distinct indeterminates $Y_1, \dots, Y_n$ over K. Upon removing from P the monomials involving any $Y_i^{(r)}$ with $r \geqslant 1$ we obtain an ordinary polynomial $p \in K[Y_1, \dots, Y_n]$ such that for all $y_1, \dots, y_n \in C$,

$$P(y_1, \dots, y_n) = p(y_1, \dots, y_n).$$

By QE for DCF (Theorem 4.7.2), this fact reduces our job to showing that for any polynomial $p \in K[Y_1, \dots, Y_n]$ the subset $\{c \in C^n : p(c) = 0\}$ of C^n is definable in the algebraically closed field C; for this, use the following Lemma 4.7.5. □

LEMMA 4.7.5. *Let $E \subseteq F$ be a field extension, $Y_1, \dots, Y_n$ distinct indeterminates, and $p \in F[Y_1, \dots, Y_n]$. Then there are polynomials $p_1, \dots, p_m \in E[Y_1, \dots, Y_n]$ such that for all $a \in E^n$,*

$$p(a) = 0 \iff p_1(a) = \cdots = p_m(a) = 0.$$

PROOF. Take a basis $b_1, \dots, b_m$ of the E-linear subspace of F generated by the coefficients of P. Then $p = b_1p_1 + \cdots + b_mp_m$ with $p_1, \dots, p_m \in E[Y_1, \dots, Y_n]$, and then $p_1, \dots, p_m$ have the desired property. □

Differential closures

This subsection is not used later.

LEMMA 4.7.6. *Let K be a differential field, let $P \in K\{Y\}^{\neq}$ be irreducible of order r, and let $Q \in K\{Y\}^{\neq}$ have order $< r$. Then there is an element y of a differential field extension of K with $P(y) = 0$, $Q(y) \neq 0$, such that $K\langle y\rangle$ embeds over K into any differentially closed differential field extension of K.*

PROOF. Let $S = S_P$, and let a and $R = K\{a, S(a)^{-1}, Q(a)^{-1}\}$ be as in the proof of Corollary 4.6.14. Let M be a maximal differential ideal of R. Then $\overline{R} := R/M$ is a simple differential ring, hence an integral domain; we identify K with its image in $\overline{R}$ via the natural embedding $K \to \overline{R}$. With $y := a + M \in \overline{R}$ we have $P(y) = 0$, $Q(y) \neq 0$. Since the ring R is noetherian, we can take $P_1, \dots, P_m \in K\{Y\}$ of order $\leqslant r$ such that $P_1(a), \dots, P_m(a)$ generate the ideal M in R. Let L be a differentially closed differential field extension of K. By the equivalence of (iii) and (iv) above, we can take $z \in L$ with $P(z) = 0$, $P_i(z) = 0$ for $i = 1, \dots, m$, $S(z) \neq 0$, $Q(z) \neq 0$. The remarks after Lemma 4.1.10 give a differential ring morphism $R \to L$ over K sending a to z. Now M is part of the kernel of this differential ring morphism, so we get a differential ring morphism $\overline{R} \to L$ over K, which is an embedding since $\overline{R}$ is simple. □

Let K be a differential field. A **differential closure** of K is a differential field extension K^{dc} of K which is differentially closed and such that every embedding of K into a differentially closed field L extends to an embedding of K^{dc} into L. Every differential field has a differential closure: this follows by a straightforward transfinite construction using Lemma 4.7.6. For the sake of completeness we also mention the following results without proof.

THEOREM 4.7.7. *Any two differential closures of K are isomorphic over K.*

Thanks to this theorem, we may refer to *the* differential closure K^{dc} of K. Note that K^{dc} is d-algebraic over K, and that the constant field of K^{dc} is the algebraic closure of the constant field C of K inside K^{dc}.

In contrast to the corresponding notions for fields, differential fields always have proper d-algebraic extensions, and the differential closure of a differential field K is not always minimal over K:

PROPOSITION 4.7.8. *If the derivation of K is trivial, then K^{dc} properly contains a differentially closed differential subfield containing K.*

Notes and comments

Most of the material above stems from Robinson [353] and Blum [46]. Theorem 4.7.7 is in Shelah [406]. Proposition 4.7.8 is due, independently, to Kolchin [222], Rosenlicht [362], and Shelah [408]. See [285] for an exposition of 4.7.7 and 4.7.8.

Singer [418] shows that the theory of *ordered* differential fields has a model completion. (No relation between ordering and derivation is imposed, and the relevant language is that of differential rings with a binary relation symbol $\leqslant$ for the ordering.) Given a model K of this model completion, Singer [419] also showed that the differential field $K[i]$, where $i^2 = -1$, is differentially closed (and hence $K[i]$ is the differential closure of K). This fact together with Corollary 4.7.4 shows that such K, as a subset of $K[i]$, is not definable in the differential field $K[i]$. By Proposition 10.7.10 below this is in contrast to what happens for $K = \mathbb{T}$.

By Michaux [291], the theory of valued differential fields (no relation between derivation and valuation being imposed) has a model completion. Similar results for certain theories of topological differential fields are in Guzy and Point [161].

Chapter Five

Linear Differential Polynomials

Linear differential polynomials and their zero sets play a special role in our work. Their theory is of course much better understood than for differential polynomials in general. For us this fact is particularly relevant since the property of a valued differential field being differential-henselian (studied in Chapter 7) involves the linear part of an arbitrary differential polynomial in an essential way. Also, the key operation of compositional conjugation on arbitrary differential polynomials is defined by transformations of linear differential polynomials; see Section 5.7.

Accordingly we consider in this chapter homogeneous linear differential polynomials and the corresponding linear operators, and prove various basic results on them as needed later. In particular, we study the property of a linear differential operator over a differential field K of defining a surjective map $K \to K$, and the transformation of a system of linear differential equations in several unknowns to an equivalent system of several linear differential equations in a single unknown.

In the final section of this chapter we apply this material on linear differential polynomials (plus some commutative algebra) to prove a result on zeros of systems of non-linear algebraic differential equations.

5.1 LINEAR DIFFERENTIAL OPERATORS

This section contains definitions and basic results about linear differential operators. *Throughout K is a differential ring with derivation ∂ and ring of constants C. We let a, b, sometimes with subscripts, range over K.*

The ring $K[\partial]$

A homogeneous linear differential polynomial

$$A(Y) \;=\; a_0Y + a_1Y' + \cdots + a_nY^{(n)} \in K\{Y\}$$

defines a C-linear operator $y \mapsto A(y) = a_0y + a_1y' + \cdots + a_ny^{(n)}$ on K and the composition of two such operators is again an operator of this form. In its role as an operator the above A is more conveniently written as $a_0 + a_1\partial + \cdots + a_n\partial^n$. Formally, $K[\partial]$ is a ring that contains K as a subring and with a distinguished element, denoted ∂, such that $K[\partial]$, as a left-module over K, is free with basis

$$\partial^0, \partial^1, \partial^2, \partial^3, \ldots, \qquad \text{with } \partial^0 = 1,\ \partial^1 = \partial,\ \partial^m \neq \partial^n \text{ whenever } m \neq n,$$

and such that $\partial a = a\partial + a'$. (So $K[\partial]$ is commutative iff $a' = 0$ for all a.) This description determines $K[\partial]$ up to isomorphism over K as a ring extension of K with a distinguished element ∂. Note the following identity:

$$\partial^n a = \sum_{i=0}^{n} \binom{n}{i} a^{(n-i)}\partial^i = a^{(n)} + na^{(n-1)}\partial + \cdots + a\partial^n. \tag{5.1.1}$$

The element ∂ of $K[\partial]$ should not be confused with the derivation ∂ of K: $\partial \neq 0$ in $K[\partial]$ even if the derivation ∂ is trivial on K. Every $A \in K[\partial]$ has the form

$$A = a_0 + a_1\partial + \cdots + a_n\partial^n,$$

and for such A we put

$$A(y) := a_0 y + a_1 y' + \cdots + a_n y^{(n)}$$

for y in a differential ring extension of K. Multiplication of elements of $K[\partial]$ corresponds to composition: $(AB)(y) = A\big(B(y)\big)$ for $A, B \in K[\partial]$ and y in a differential ring extension of K. The map $K[\partial] \to \operatorname{Hom}_C(K, K)$ that associates to each $A \in K[\partial]$ the C-linear operator $y \mapsto A(y)$ on K is itself C-linear. If K is a differential field and $C \neq K$, then this map $K[\partial] \to \operatorname{Hom}_C(K, K)$ is injective, by Lemma 4.2.1. Multiplication on the right by an element of K corresponds to multiplicative conjugation of the corresponding linear differential polynomial:

$$\begin{aligned} \text{if} \quad (a_0 + a_1\partial + \cdots + a_n\partial^n)b &= b_0 + b_1\partial + \cdots + b_n\partial^n, \\ \text{then} \quad \big(a_0 Y + a_1 Y' + \cdots + a_n Y^{(n)}\big)_{\times b} &= b_0 Y + b_1 Y' + \cdots + b_n Y^{(n)}. \end{aligned}$$

If L is a differential ring, then any differential ring morphism $K \to L$ extends uniquely to a ring morphism $K[\partial] \to L[\partial]$ sending $\partial \in K[\partial]$ to $\partial \in L[\partial]$. Thus every automorphism σ of the differential ring K extends to an automorphism, also denoted by σ, of the ring $K[\partial]$ with $\sigma\partial = \partial$ (so $\sigma\big(A(y)\big) = (\sigma A)(\sigma y)$ for $A \in K[\partial]$ and $y \in K$).

Call $A \in K[\partial]$ **monic of order** n if it has the form $\partial^n + a_{n-1}\partial^{n-1} + \cdots + a_0$. If $A, B \in K[\partial]$ are monic of order m, n, then AB is monic of order $m+n$. To $P \in K\{Y\}$ we associate its **linear part** $L_P \in K[\partial]$,

$$L_P := \sum_n \frac{\partial P}{\partial Y^{(n)}}(0)\partial^n \qquad \Big(\text{so } L_{P_{+a}} = \sum_n \frac{\partial P}{\partial Y^{(n)}}(a)\partial^n\Big).$$

Hence $L_P(y) = P_1(y)$ for all y in all differential ring extensions of K, and then for $P, Q \in K\{Y\}$ we have $L_{PQ} = P(0)L_Q + Q(0)L_P$, $L_{P_{\times a}} = L_P \cdot a$ and $\partial L_P = L_{P'}$.

Let $A \in K[\partial]$. Then its **kernel**

$$\ker A := \{y \in K : A(y) = 0\}$$

is a C-submodule of K. If we need to stress the dependence of $\ker A$ on K, we write $\ker_K A$. If L is a differential ring extension of K, then $\ker_K A$ is a C-submodule of $\ker_L A$, and $\ker_K A = K \cap \ker_L A$.

Let a be a unit of K. We define the **twist** of A by a to be $A_{\ltimes a} := a^{-1}Aa \in K[\partial]$. In particular, $\partial_{\ltimes a} = \partial + a^\dagger$. We have $\ker A = a \ker A_{\ltimes a}$. Note that if A is monic of order n, then so is $A_{\ltimes a}$. The map $B \mapsto B_{\ltimes a}$ is an automorphism of the ring $K[\partial]$; it is the identity on K, with inverse $B \mapsto B_{\ltimes a^{-1}}$.

Right-inverses of linear differential operators

Suppose now that ∂^{-1} is a *right-inverse* to the derivation ∂ of K, that is, $\partial^{-1}\colon K \to K$ satisfies $\partial \circ \partial^{-1} = \mathrm{id}_K$. (Think of ∂^{-1} as integration.) Then, if $b \in K$ and $b = a^\dagger$ with a a unit of K, we obtain a right-inverse $(\partial + b)^{-1} := a^{-1}\partial^{-1}a$ to $\partial + b$, since

$$(\partial + b) \circ a^{-1}\partial^{-1}a \;=\; (a^{-1}\partial a) \circ a^{-1}\partial^{-1}a \;-\; \mathrm{id}_K\,.$$

Here $a^{-1}\partial^{-1}a\colon K \to K$ sends each $x \in K$ to $a^{-1}\partial^{-1}(ax)$. Next, if $A \in K[\partial]$ is monic of order $n \geqslant 1$ and factors as $A = (\partial + b_1)\cdots(\partial + b_n)$ with each $b_i = a_i^\dagger$ and a_i a unit of K, then the operator A on K has right-inverse

$$A^{-1} \;:=\; (a_n^{-1}\partial^{-1}a_n) \circ \cdots \circ (a_1^{-1}\partial^{-1}a_1)$$

in the sense that $A \circ A^{-1} = \mathrm{id}_K$.

The derivative of a linear differential operator

For an operator

$$A \;=\; a_0 + a_1\partial + \cdots + a_n\partial^n \in K[\partial]$$

we define its **derivative** $A' \in K[\partial]$ by

$$A' \;:=\; a_1 + 2a_2\partial + \cdots + na_n\partial^{n-1} \qquad (\text{so } A' = 0 \text{ if } n = 0).$$

The map $A \mapsto A'$ is a derivation on the ring $K[\partial]$:

LEMMA 5.1.1. *Let $A, B \in K[\partial]$. Then*

$$(A+B)' \;=\; A' + B', \qquad (AB)' \;=\; A'B + AB'.$$

PROOF. The first identity is clear. The second follows from its special cases

$$(A\partial^j)' \;=\; A'\partial^j + jA\partial^{j-1}, \qquad (aB)' \;=\; aB', \qquad (\partial^j b)' \;=\; j\partial^{j-1}b,$$

where $j \in \mathbb{N}^{\geqslant 1}$. (For the last equality, use (5.1.1).) □

For $A \in K[\partial]$ and $i \in \mathbb{N}$ we define $A^{(i)} \in K[\partial]$ by

$$A^{(0)} \;:=\; A, \qquad A^{(i+1)} \;:=\; (A^{(i)})'.$$

Induction on i yields that if $A = \sum_{j=0}^n a_j\partial^j$, then

$$\frac{A^{(i)}}{i!} \;=\; \sum_{j=i}^{n} \binom{j}{i} a_j \partial^{j-i}, \tag{5.1.2}$$

and thus for all $f \in K$:

$$(5.1.3) \qquad Af = \sum_{i=0}^{n} \frac{A^{(i)}(f)}{i!}\partial^i.$$

In the next lemma, k is an integer, and

$$\binom{k}{n} := \frac{1}{n!} \cdot k(k-1)\cdots(k-n+1) \qquad \text{(a rational number).}$$

So $\binom{k}{n} = \frac{k!}{(k-n)!\cdot n!} \in \mathbb{N}^{\geqslant 1}$ if $k \geqslant n$ and $\binom{k}{n} = 0$ if $0 \leqslant k < n$.

LEMMA 5.1.2. *Let $A \in K[\partial]$, $f \in K$, and suppose $x \in K^\times$ satisfies $x' = 1$. Then*

$$A(fx^k) = \sum_{i \geqslant 0} \binom{k}{i} A^{(i)}(f) \cdot x^{k-i}.$$

PROOF. By the identity (5.1.3) we have

$$A(fx^k) = (Af)(x^k) = \sum_{i \geqslant 0} \frac{A^{(i)}(f)}{i!}(x^k)^{(i)}.$$

It remains to note that for every $i \in \mathbb{N}$,

$$(x^k)^{(i)} = k(k-1)\cdots(k-i+1)\cdot x^{k-i} = i!\cdot \binom{k}{i} x^{k-i}. \qquad \square$$

The order of a linear differential operator

In this subsection we assume that K is a differential field. Let $A, B \in K[\partial]$. Define the **order** of A by:

$$\operatorname{order} 0 = -\infty, \qquad \operatorname{order} A = n \text{ for } A = a_0 + a_1\partial + \cdots + a_n\partial^n,\ a_n \neq 0.$$

It is easy to check that $\operatorname{order} AB = \operatorname{order} A + \operatorname{order} B$; in particular, if $A \neq 0$ and $B \neq 0$, then $AB \neq 0$. With $[A, B] := AB - BA$ it follows easily from (5.1.1) that $\operatorname{order}[A, B] < \operatorname{order} A + \operatorname{order} B$ for nonzero A, B.

In the rest of this subsection we assume $K \neq C$.

LEMMA 5.1.3. *Let $A \in K[\partial]$ and $A \notin K$. Then $[A, b] \neq 0$ for some $b \in K$.*

PROOF. We have $A = a_0 + a_1\partial + \cdots + a_n\partial^n$, $n \geqslant 1$, $a_n \neq 0$. Take $b \in K$ with $b' \neq 0$. Then $bA = a_n b\partial^n + a_{n-1}b\partial^{n-1}$ + terms of order $< n-1$, and by (5.1.1), $Ab = a_n b\partial^n + (a_n n b' + a_{n-1}b)\partial^{n-1}$ + terms of order $< n-1$, so $bA \neq Ab$. $\square$

In view of $\partial a = a' + a\partial$ we obtain:

COROLLARY 5.1.4. *Let $A \in K[\partial]$. Then $[A, B] = 0$ for all $B \in K[\partial]$ iff $A \in C$.*

An **ideal** of $K[\partial]$ is an additive subgroup I of $K[\partial]$ such that $FA \in I$ and $AF \in I$ whenever $F \in K[\partial]$ and $A \in I$.

COROLLARY 5.1.5. *The ring $K[\partial]$ has no ideals except $\{0\}$ and $K[\partial]$.*

PROOF. Let I be an ideal of $K[\partial]$, and suppose $I \neq \{0\}$. Take a nonzero $A \in I$ of minimal order. Then for all $b \in K$ we have $[A, b] \in I$ and order $[A, b] <$ order A, hence $[A, b] = 0$. Thus $A \in K$ by Lemma 5.1.3 and hence $I = K[\partial]$. □

Euclidean division in $K[\partial]$

In this subsection K is a differential field. The ring $K[\partial]$ admits *division with remainder on the left*: for $A, B \in K[\partial]$, $B \neq 0$, there exist unique $Q, R \in K[\partial]$ with $A = QB + R$ and order $R <$ order B. It follows that for every left ideal I of $K[\partial]$ there exists an $A \in K[\partial]$ such that $I = K[\partial]A$. (A *left ideal* of $K[\partial]$ is an additive subgroup I of $K[\partial]$ such that $FA \in I$ whenever $F \in K[\partial]$ and $A \in I$.) Using division with remainder on the left and induction on min(order A, order B) it follows easily that $K[\partial]A \cap K[\partial]B \neq \{0\}$ for $A, B \in K[\partial]^{\neq}$. Thus for $A_1, \dots, A_r \in K[\partial]^{\neq}$, $r \geqslant 1$, we can define the *least common left multiple* of $A_1, \dots, A_r$ to be the unique monic $A \in K[\partial]$ such that

$$K[\partial]A = K[\partial]A_1 \cap \dots \cap K[\partial]A_r.$$

Note that if $A, B \in K[\partial]$, then $1 \notin K[\partial]A + K[\partial]B$ iff A and B have a nontrivial common right divisor, that is, a $D \in K[\partial]$ of positive order such that there are $A_1, B_1 \in K[\partial]$ with $A = A_1 D$ and $B = B_1 D$.

LEMMA 5.1.6. *Let $A, B \in K[\partial]$ and a satisfy $(\partial - a)A \in K[\partial]B$ and $\operatorname{order}(B) \geqslant 2$. Then $1 \notin K[\partial]A + K[\partial]B$.*

PROOF. Suppose $1 = \alpha A + \beta B$ with $\alpha, \beta \in K[\partial]$. Then $\alpha = q \cdot (\partial - a) + b$ with $q \in K[\partial]$. Also $(\partial - a)A = \gamma B$ with $\gamma \in K[\partial]$, hence

$$1 = bA + (q\gamma + \beta)B.$$

Then $b \neq 0$ since $\operatorname{order}(B) > 0$, and so $b^{-1}(\partial - c)b = \partial - a$ for $c = a + b^{\dagger}$. Hence $b^{-1}(\partial - c) \in K[\partial]B$, contradicting $\operatorname{order}(B) \geqslant 2$. □

LEMMA 5.1.7. *Let $A, B, D \in K[\partial]$ and a be such that*

$$K[\partial]A + K[\partial]B = K[\partial]D \neq K[\partial](\partial - a)A + K[\partial]B.$$

Then $K[\partial](\partial - a)A + K[\partial]B = K[\partial](\partial - c)D$ for some $c \in K$.

PROOF. Since $A, B \in K[\partial]D$ we can divide by D on the right and reduce to the case that $D = 1$, and so $K[\partial](\partial - a)A + K[\partial]B = K[\partial]E$ with $\operatorname{order}(E) \geqslant 1$. But also $1 \in K[\partial]A + K[\partial]B \subseteq K[\partial]A + K[\partial]E$, so $\operatorname{order}(E) \geqslant 2$ would contradict the previous lemma (with E instead of B). Thus $\operatorname{order}(E) = 1$. □

For $A = \sum_i a_i \partial^i \in K[\partial]$ we define its **adjoint** $A^* \in K[\partial]$ by

$$A^* := \sum_i (-1)^i \partial^i a_i,$$

so $a^* = a$ and $\partial^* = -\partial$.

LEMMA 5.1.8. *The map $A \mapsto A^*$ is an involution: for all $A, B \in K[\partial]$,*

$$(A+B)^* = A^* + B^*, \qquad (AB)^* = B^* A^*, \qquad A^{**} = A.$$

PROOF. The first identity is obvious. Let $K[\partial]^{\text{opp}}$ be the opposite ring of $K[\partial]$; it has the same underlying additive abelian group as $K[\partial]$, and its multiplication $*$ is given by $A * B := BA$. Then $K[\partial]^{\text{opp}}$ contains K as a subring and as a left K-module is free with basis $(-\partial)^0, (-\partial)^1, (-\partial)^2, \dots$, by (5.1.1). Moreover $(-\partial) * a = a * (-\partial) + a'$ for all $a \in K$. So the K-linear map $K[\partial] \to K[\partial]^{\text{opp}}$ of left K-modules sending ∂^n to $(-\partial)^n$ for each n, is a ring isomorphism. This map is nothing but $A \mapsto A^*$, so $(AB)^* = (A^*) * (B^*) = B^* A^*$ for all $A, B \in K[\partial]$. □

Using this involution "left" results imply similar "right" results. For example, we obtain in this way *division with remainder on the right*: for $A, B \in K[\partial]$, $B \neq 0$, there exist unique $Q, R \in K[\partial]$ such that $A = BQ + R$ and $\operatorname{order}(R) < \operatorname{order}(B)$. Hence for every right ideal I of $K[\partial]$ there is $B \in K[\partial]$ such that $I = BK[\partial]$. The right analogues of the lemmas above now follow easily:

LEMMA 5.1.9. *Let $A, B \in K[\partial]$ and a satisfy $A(\partial - a) \in BK[\partial]$ and* $\operatorname{order}(B) \geqslant 2$. *Then $1 \notin AK[\partial] + BK[\partial]$.*

LEMMA 5.1.10. *Let $A, B, D \in K[\partial]$ and a be such that*

$$AK[\partial] + BK[\partial] = DK[\partial] \neq A(\partial - a)K[\partial] + BK[\partial].$$

Then $A(\partial - a)K[\partial] + BK[\partial] = D(\partial - c)K[\partial]$ for some $c \in K$.

We also note:

LEMMA 5.1.11. *Let L be a differential field extension of K, and $A \in K[\partial]$. Then $L[\partial]A \cap K[\partial] = K[\partial]A$ and $AL[\partial] \cap K[\partial] = AK[\partial]$.*

PROOF. We may assume that $A \neq 0$. Let $B \in L[\partial]A \cap K[\partial]$. Then $B = QA$ where $Q \in L[\partial]$. Uniqueness of division with remainder on the left gives $Q \in K[\partial]$. Thus $L[\partial]A \cap K[\partial] = K[\partial]A$, and $AL[\partial] \cap K[\partial] = AK[\partial]$ follows likewise. □

The kernel of a linear differential operator

Let K be a differential field in this subsection. Each $A \in K[\partial]$ acts as a C-linear operator on K, and its kernel

$$\ker A = \{y \in K : A(y) = 0\}$$

is a C-linear subspace of K, which is of dimension $\leqslant n$ if order $A \leqslant n$, $A \neq 0$ (Corollary 4.1.14). If $A \in K[\partial]^{\neq}$, $\dim_C \ker A = \operatorname{order} A$, and L is a differential field extension of K with $C_L = C$, then $\ker_L A = \ker A$. For $A, B \in K[\partial]^{\neq}$ we have an exact sequence

$$0 \to \ker B \to \ker AB \to \ker A$$

of C-linear spaces, where the map on the right is given by $y \mapsto B(y)$, so

$$\dim_C \ker AB \leqslant \dim_C \ker A + \dim_C \ker B,$$

with equality if and only if $B(K) \supseteq \ker A$. In particular:

LEMMA 5.1.12. *Let $A, B \in K[\partial]^{\neq}$, and $m = \operatorname{order}(A)$, $n = \operatorname{order}(B)$. Then*

$$\dim_C \ker AB = m + n \quad \Rightarrow \quad \dim_C \ker A = m \text{ and } \dim_C \ker B = n.$$

We leave the proof of the next lemma to the reader:

LEMMA 5.1.13. *Let $A = a_0 + a_1\partial + \cdots + a_n\partial^n \in K[\partial]$ be of order $n \geqslant 1$. Suppose $g \in K^\times$ satisfies $g^\dagger = -(na_n)^{-1}a_{n-1}$, and put $\widetilde{A} = a_n^{-1}g^{-1}Ag \in K[\partial]$. Then*

$$\widetilde{A} = \widetilde{a}_0 + \widetilde{a}_1\partial + \cdots + \widetilde{a}_{n-2}\partial^{n-2} + \partial^n$$

for suitable $\widetilde{a}_0, \dots, \widetilde{a}_{n-2} \in K$. Moreover, we have an isomorphism

$$y \mapsto gy : \ker \widetilde{A} \to \ker A$$

of C-linear spaces.

EXAMPLE. If $A = a_0 + a_1\partial + \partial^2$, then $\widetilde{A} = \left(a_0 - \frac{1}{2}a_1' - \frac{1}{4}a_1^2\right) + \partial^2$.

LEMMA 5.1.14. *Let $A \in K[\partial]$, $b \in K^\times$, and $B := (\partial - b^\dagger)A$. Let also $y \in K$.*

(i) *If $A(y) = b$, then $B(y) = 0$ and* $\ker B = Cy \oplus \ker A$.

(ii) *If $A(y) \neq 0$ and $B(y) = 0$, then $A(cy) = b$ for some $c \in C^\times$.*

PROOF. Part (i) is easy to see, and for (ii) note that if $A(y) \neq 0$ and $B(y) = 0$, then $A(y)^\dagger = b^\dagger$ and hence $A(cy) = cA(y) = b$ for some $c \in C^\times$. □

The following lemma uses kernels to formulate a criterion for a differential operator to be contained in a given left ideal of $K[\partial]$:

LEMMA 5.1.15. *Let $A, B \in K[\partial]^{\neq}$, $m = \operatorname{order} A$, $n = \operatorname{order} B$.*

(i) *If* $\dim_C \ker A = m$ *and* $\ker A \subseteq \ker B$, *then* $B \in K[\partial]A$;

(ii) *if* $\dim_C \ker B = n$ *and* $B \in K[\partial]A$, *then* $\ker A \subseteq \ker B$ *and* $\dim_C \ker A = m$.

Thus if $\dim_C \ker A = m$ *and* $\dim_C \ker B = n$, *then*

$$B \in K[\partial]A \iff \ker A \subseteq \ker B.$$

PROOF. For (i), suppose $\dim_C \ker A = m$ and $\ker A \subseteq \ker B$. Take $Q, R \in K[\partial]$ with $B = QA + R$, order $R < m$; then $\ker A \subseteq \ker R$, so order $R < m \leqslant \dim_C \ker R$ and hence $R = 0$. Part (ii) is immediate from Lemma 5.1.12. □

Linear differential equations with constant coefficients

In this subsection K *is a differential field.* The ring $C[\partial]$ is commutative, and so we have a C-algebra isomorphism $P(Y) \mapsto P(\partial)\colon C[Y] \to C[\partial]$.

Let $A \in C[\partial]^{\neq}$, *and take* $P = P(Y) \in C[Y]^{\neq}$ *with* $A = P(\partial)$, *so* order $A = \deg P$. *We also assume there is given an element* $x \in K$ *with* $x' = 1$, *and that for each* $c \in C$ *there exists* $a \in K^{\times}$ *with* $a' = ca$. *For each* $c \in C$ *we pick such an* a *and denote it suggestively by* e^{cx}. Note that x is transcendental over C, by Lemma 4.1.1. For $f \in C[x]$ we let $\deg_x f \in \mathbb{N} \cup \{-\infty\}$ be the degree of f viewed as a polynomial in x over C. Note also that for $c \in C$ we have $(\partial - c)\, e^{cx} = e^{cx}\,\partial$ in $K[\partial]$, and thus $(\partial - c)^i\, e^{cx} = e^{cx}\,\partial^i$ in $K[\partial]$ for all $i \in \mathbb{N}$.

LEMMA 5.1.16. *Let* $f \in C[x]$ *and let* $c \in C$ *be such that* $P(c) \neq 0$. *Then*

$$A(e^{cx} f) = e^{cx} g \qquad \text{for some } g \in C[x] \text{ with } \deg_x f = \deg_x g.$$

PROOF. Take $a_0, \dots, a_n \in C$ such that $P(Y) = \sum_{i=0}^n a_i (Y - c)^i$. Then $A = \sum_{i=0}^n a_i(\partial - c)^i$. Using the above identity for $(\partial - c)^i\, e^{cx}$ we get

$$A(e^{cx} f) = e^{cx}\left(\sum_{i=0}^n a_i f^{(i)}\right) = e^{cx} g \text{ for } g = \sum_{i=0}^n a_i f^{(i)} \in C[x],$$

and $\deg_x g = \deg_x f$, since $a_0 = P(c) \neq 0$. □

COROLLARY 5.1.17. *Let* $c_1, \dots, c_n \in C$ *be distinct, where* $n \geqslant 1$. *Then the elements* $e^{c_1 x}, \dots, e^{c_n x}$ *of* K *are linearly independent over* $C(x)$.

PROOF. By induction on n. The case $n = 1$ being obvious, let $n \geqslant 2$, and let $f_1, \dots, f_n \in C[x]$ satisfy $\sum_{i=1}^n e^{c_i x} f_i = 0$. Take $d \in \mathbb{N}$ with $d > \deg_x f_n$. Then by Lemma 5.1.16 applied to $(Y - c_n)^d$ instead of P we get

$$0 = (\partial - c_n)^d \left(\sum_{i=1}^n e^{c_i x} f_i\right) = \sum_{i=1}^{n-1} e^{c_i x} g_i + e^{c_n x} f_n^{(d)} = \sum_{i=1}^{n-1} e^{c_i x} g_i$$

where $g_i \in C[x]$ with $\deg_x f_i = \deg_x g_i$ for $i = 1, \dots, n-1$. By inductive hypothesis we have $g_i = 0$ and hence $f_i = 0$, for $i = 1, \dots, n-1$, and thus $f_n = 0$. □

Zeros of $P(Y)$ in C yield elements of the kernel of A:

PROPOSITION 5.1.18. *Let $c_1, \dots, c_n \in C$ be distinct zeros of P, of respective multiplicities $m_1, \dots, m_n \in \mathbb{N}^{\geqslant 1}$, $n \geqslant 1$. Then $A(\mathrm{e}^{c_i x} x^j) = 0$ for $1 \leqslant i \leqslant n$, $0 \leqslant j < m_i$, and the family $\left(\mathrm{e}^{c_i x} x^j\right)_{1 \leqslant i \leqslant n,\ 0 \leqslant j < m_i}$ is linearly independent over C.*

PROOF. Let $1 \leqslant i \leqslant n, 0 \leqslant j < m_i$. Then $P = Q \cdot (Y - c_i)^{m_i}$ where $Q \in C[Y]$. Set $B := Q(\partial) \in C[\partial]$. Then $A = B \cdot (\partial - c_i)^{m_i}$, so

$$A(\mathrm{e}^{c_i x} x^j) \;=\; B\big(\mathrm{e}^{c_i x}(x^j)^{(m_i)}\big) \;=\; B(0) \;=\; 0.$$

The linear independence statement is immediate from Corollary 5.1.17. □

NOTATION. Let $\mathrm{m}(P)$ denote the number of zeros of P in C, counted with multiplicity; thus $\mathrm{m}(P) = \sum_{c \in C} \mathrm{mul}(P_{+c})$, where only finitely many terms in this sum are nonzero. Note that $\mathrm{m}(P) = \mathrm{m}(P_{+a})$ for $a \in C$, and $\mathrm{m}(P) = \mathrm{m}(P_{\times b})$ for $b \in C^{\times}$. Also, $\mathrm{m}(P) = \mathrm{m}(P_1) + \cdots + \mathrm{m}(P_n)$ if $P = P_1 \cdots P_n$, $P_1, \dots, P_n \in C[Y]^{\neq}$ $(n \geqslant 1)$. We also set $\mathrm{m}(A) := \mathrm{m}(P)$; so $\mathrm{m}(A) = \mathrm{m}(A_1) + \cdots + \mathrm{m}(A_n)$ if $A = A_1 \cdots A_n$, $A_1, \dots, A_n \in C[\partial]^{\neq}$ $(n \geqslant 1)$.

From Proposition 5.1.18 we obtain:

$$\mathrm{m}(A) \;\leqslant\; \dim_C \ker A. \tag{5.1.4}$$

If C is algebraically closed, then $\mathrm{order}(A) = \mathrm{m}(A) = \dim_C \ker A$. Lemma 5.2.11 below gives another situation where $\mathrm{m}(A) = \dim_C \ker A$.

The type of a linear differential operator

In this subsection K is a differential field. We say that $A, B \in K[\partial]$ have **the same type** if the $K[\partial]$-modules $K[\partial]/K[\partial]A$ and $K[\partial]/K[\partial]B$ are isomorphic. For $a \in K^{\times}$ and $A \in K[\partial]$ we have $K[\partial]A = K[\partial]aA$, and the automorphism $B \mapsto Ba$ of the (left) $K[\partial]$-module $K[\partial]$ maps $K[\partial]A$ onto $K[\partial]Aa$, so A, aA, and Aa have the same type. For $A \in K[\partial]^{\neq}$, the order of A equals the K-vector space dimension of $K[\partial]/K[\partial]A$, so if $A, B \in K[\partial]^{\neq}$ have the same type, then $\mathrm{order}(A) = \mathrm{order}(B)$.

LEMMA 5.1.19. *Let $A, B \in K[\partial]^{\neq}$. Then A and B have the same type if and only if $\mathrm{order}(A) = \mathrm{order}(B)$ and there is $R \in K[\partial]$ of order less than $\mathrm{order}(A)$ with $1 \in K[\partial]R + K[\partial]A$ and $BR \in K[\partial]A$.*

PROOF. Let $\phi\colon K[\partial]/K[\partial]B \to K[\partial]/K[\partial]A$ be an isomorphism of $K[\partial]$-modules. Set $n := \mathrm{order}(A) = \mathrm{order}(B)$. Then $\phi(1 + K[\partial]B) = R + K[\partial]A$ with $R \in K[\partial]$ of order $< n$. One checks easily that then $1 \in K[\partial]R + K[\partial]A$ and $BR \in K[\partial]A$. Conversely, assume $\mathrm{order}(A) = \mathrm{order}(B) = n$, and $1 \in K[\partial]R + K[\partial]A$ and $BR \in K[\partial]A$ with $R \in K[\partial]$ of order $< n$. Then the $K[\partial]$-linear map $K[\partial] \to K[\partial]/K[\partial]A$ sending 1 to $R + K[\partial]A$ is surjective and its kernel contains $K[\partial]B$, hence induces a $K[\partial]$-linear map $K[\partial]/K[\partial]B \to K[\partial]/K[\partial]A$, which is bijective since $\dim_K K[\partial]/K[\partial]A = n = \dim_K K[\partial]/K[\partial]B$. □

EXAMPLE. Let $A, B \in K[\partial]$ be monic of order 1. Then A and B have the same type iff $B = A_{\ltimes a}$ for some $a \in K^\times$.

COROLLARY 5.1.20. *Let $A, B \in K[\partial]^{\neq}$ have the same type, and let $R \in K[\partial]$ be as in Lemma 5.1.19. Then $R(\ker A) \subseteq \ker B$, and the map $x \mapsto R(x)\colon \ker A \to \ker B$ is an isomorphism of C-linear spaces.*

PROOF. Take $L_1, L_2, L \in K[\partial]$ with $1 = L_1 R + L_2 A$ and $BR = LA$. Then for each $x \in \ker A$ we have $B\big(R(x)\big) = L\big(A(x)\big) = L(0) = 0$, so $R(x) \in \ker B$, and $x = L_1\big(R(x)\big)$; hence we have an injective C-linear map $x \mapsto R(x)\colon \ker A \to \ker B$. By symmetry, we also have an injective C-linear map $\ker B \to \ker A$, showing that $\dim_C \ker A = \dim_C \ker B$, so $x \mapsto R(x)\colon \ker A \to \ker B$ is surjective. □

Factorization in $K[\partial]$

In this subsection K is a differential field. We call $A \in K[\partial]$ of positive order **irreducible** if there are no $A_1, A_2 \in K[\partial]$ of positive order with $A = A_1 A_2$. If $A \in K[\partial]$ has positive order, then $A = A_1 \cdots A_r$ with $r \in \mathbb{N}^{\geqslant 1}$ and each $A_i \in K[\partial]$ irreducible. We say that $A \in K[\partial]$ **splits** over K if $A \neq 0$ and $A = c(\partial - a_1)\cdots(\partial - a_n)$ for some $c \in K^\times$ and some $a_1, \dots, a_n$. If $A, B \in K[\partial]$ split over K, so does AB (use twisting to eliminate factors $c \in K^\times$). In Section 14.5 we shall prove under natural assumptions on K that each nonzero $A \in K[\partial]$ splits over K. In this connection the following facts are relevant.

LEMMA 5.1.21. *Let $A \in K[\partial]$ be of order $n \geqslant 1$, and $u^\dagger = a \in K$ with nonzero u from some differential field extension of K. Then*

$$A \in K[\partial](\partial - a) \iff A(u) = 0.$$

PROOF. If $A = B \cdot (\partial - a) + b$ with $B \in K[\partial]$, $b \in K$, then $A(u) = bu$. □

In particular, if $A \in K[\partial]$ and $A(u) = 0$, $u \in K^\times$, then $A \in K[\partial](\partial - u^\dagger)$.

PROPOSITION 5.1.22. *Suppose $A, B, D \in K[\partial]$, $A = BD$, and A splits over K. Then B and D split over K.*

PROOF. We shall use Lemma 5.1.10 to show that B splits over K. (A similar use of Lemma 5.1.7 shows that D splits over K.) Let $A = (\partial - a_1)\cdots(\partial - a_n)$, $n \geqslant 1$, and for $i = 0, \dots, n$, take the unique monic $B_i \in K[\partial]$ such that

$$B_i K[\partial] \;=\; (\partial - a_1)\cdots(\partial - a_i)K[\partial] + BK[\partial],$$

so $B_0 = 1$ and $B_n = bB$, $b \neq 0$. By Lemma 5.1.10 we have, for $0 \leqslant i < n$, either $B_{i+1} = B_i$, or $B_{i+1} = B_i(\partial - c)$ for some $c \in K$. Thus each B_i splits over K, and so does B. □

The next lemma shows how this proposition applies to linear parts of differential polynomials. If y in a differential field extension of K is d-algebraic over K and P is a minimal annihilator of y over K (so $P \in K\{Y\}^{\neq}$ has minimal complexity subject to $P(y) = 0$), then $S_P(y) \neq 0$ and so $L_{P_{+y}}$ has order r_P.

LEMMA 5.1.23. *Let E and F be differential subfields of K with $E \subseteq F \subseteq K$, and let $f \in K$ be* d*-algebraic over E with minimal annihilator P over E and Q over F. Then $L_{P_{+f}} \in K[\partial]L_{Q_{+f}}$, so if $L_{P_{+f}}$ splits over K, then so does $L_{Q_{+f}}$.*

PROOF. From $P(f) = 0$ and Corollary 4.3.10 we obtain

$$HP = A_0 Q + A_1 Q' + \cdots + A_n Q^{(n)}, \qquad H = S_Q^l,$$

where $l \in \mathbb{N}$, $A_0, \dots, A_n \in F\{Y\}$. By additive conjugation,

$$H_{+f} P_{+f} = (A_0)_{+f} Q_{+f} + \cdots + (A_n)_{+f} Q_{+f}^{(n)}.$$

Taking linear parts and using $P(f) = Q(f) = 0$ gives

$$\begin{aligned} H(f) L_{P_{+f}} &= A_0(f) L_{Q_{+f}} + \cdots + A_n(f) L_{Q_{+f}^{(n)}} \\ &= \big(A_0(f) + \cdots + A_n(f)\partial^n\big) L_{Q_{+f}}. \end{aligned}$$

It only remains to note that $H(f) \neq 0$. □

Factorization into irreducibles and composition series

In this subsection K is a differential field, $R = K[\partial]$, and $A \in R^{\neq}$ is monic. We discuss in what sense a factorization of A into irreducible elements of R is unique. Given $A_1, A_2 \in R$ with $A = A_1 A_2$, we have $RA \subseteq RA_2$, which gives an exact sequence

$$0 \longrightarrow R/RA_1 \longrightarrow R/RA \longrightarrow R/RA_2 \longrightarrow 0$$

of R-linear maps, with $R/RA_1 \to R/RA$ induced by the R-linear endomorphism $B \mapsto BA_2$ of R. Conversely, let M_1 be a submodule of the R-module $M = R/RA$. Take monic $A_2 \in R$ such that RA_2 is the kernel of the composition

$$R \to R/RA = M \to M/M_1$$

of the natural surjections. Then $A \in RA_2$, we have monic $A_1 \in R$ with $A = A_1 A_2$, and so we obtain a commutative diagram of R-linear maps

$$\begin{array}{ccccccccc} 0 & \longrightarrow & M_1 & \hookrightarrow & M & \longrightarrow & M/M_1 & \longrightarrow & 0 \\ & & \uparrow & & \uparrow{=} & & \uparrow & & \\ 0 & \longrightarrow & R/RA_1 & \longrightarrow & R/RA & \longrightarrow & R/RA_2 & \longrightarrow & 0 \end{array}$$

where the map $R/RA_1 \to R/RA$ is induced by the endomorphism $B \mapsto BA_2$ of the R-module R, and the vertical arrows are isomorphisms, the rightmost one being $B + RA_2 \mapsto B + M_1$ for $B \in R$. Thus A is irreducible (in R) iff the R-module R/RA is irreducible. Using this recursively, a factorization $A = A_1 \cdots A_m$ of A into irreducible $A_1, \dots, A_m \in R$ gives rise to a composition series

$$\{0\} = M_0 \subset M_1 \subset \cdots \subset M_m = M$$

of length m of the R-module $M = R/RA$ and for $i = 1, \dots, m$ isomorphisms $M_i \cong R/RA_1 \cdots A_i$ and $M_i/M_{i-1} \cong R/RA_i$ of R-modules.

Usually, A has many factorizations into irreducible monic operators; e.g., if $K = C(x)$ where $x \notin C$ and $x' = 1$, then $\partial^2 = (\partial + f^\dagger)(\partial - f^\dagger)$ for all $f = ax + b$ with $a, b \in C$, not both zero. However, the *number* and the *types* of the irreducible factors in such a factorization do not depend on the particular factorization:

COROLLARY 5.1.24. *If $A_1, \dots, A_m, B_1, \dots, B_n$ are irreducible elements of R such that $A_1 \cdots A_m = B_1 \cdots B_n$, then $m = n$ and there is a permutation $i \mapsto i'$ of $\{1, \dots, m\}$ such that A_i and $B_{i'}$ are of the same type for $i = 1, \dots, m$.*

This corollary follows from the remarks preceding it in combination with the Jordan-Hölder Theorem (Corollary 1.2.2). Note that Corollary 5.1.24 gives rise to another proof of Proposition 5.1.22. The remarks above also yield:

COROLLARY 5.1.25. *The linear differential operator A splits over K if and only if the R-module $M = R/RA$ has a composition series*

$$\{0\} = M_0 \subset M_1 \subset \cdots \subset M_m = M$$

with $\dim_K M_i/M_{i-1} = 1$ *for* $i = 1, \dots, m$.

Linear closedness and linear surjectivity

In this subsection we let K be a differential field, and we let r range over $\mathbb{N}^{\geqslant 1}$. We define K to be **r-linearly closed** if each nonzero $A \in K[\partial]$ of order $\leqslant r$ splits over K. We define K to be **linearly closed** if it is r-linearly closed for each r. If the derivation of K is trivial, that is, $K = C$, then $K[\partial] = C[\partial]$ is the usual (commutative) polynomial ring with ∂ as an indeterminate over C, so in that case K is linearly closed iff C is algebraically closed. The property of being linearly closed has a nice first-order axiomatization:

LEMMA 5.1.26. *Let $n \geqslant 1$ be given. Then there exists a differential polynomial*

$$d_n(Y_1, \dots, Y_n, Y) \in \mathbb{Q}\{Y_1, \dots, Y_n, Y\} \qquad \text{(with all its coefficients in } \mathbb{Z}\text{)}$$

such that for each differential ring R and all $a_1, \dots, a_n, a \in R$,

$$d_n(a_1, \dots, a_n, a) = 0 \iff \partial^n + a_1\partial^{n-1} + \cdots + a_n = (\partial - a)B \text{ for some } B \in R[\partial].$$

PROOF. Just note that in the ring $\mathbb{Q}\{Y_1, \dots, Y_n, Y\}[\partial]$ we have

$$\partial^n + Y_1\partial^{n-1} + \cdots + Y_n = (\partial - Y)B + d_n$$

with $B \in \mathbb{Q}\{Y_1, \dots, Y_n, Y\}[\partial]$ and $d_n \in \mathbb{Q}\{Y_1, \dots, Y_n, Y\}$. □

Thus K is linearly closed iff for each $n \geqslant 2$,

$$K \models \forall y_1 \cdots \forall y_n \exists y\, d_n(y_1, \dots, y_n, y) = 0.$$

We say that K is r-**linearly surjective** if $A(K) = K$ for each nonzero $A \in K[\partial]$ of order at most r, equivalently, for all $a_0, \dots, a_r \in K$ such that $a_i \neq 0$ for some i, the inhomogeneous linear differential equation $1 + a_0 y + \cdots + a_r y^{(r)} = 0$ has a solution in K. We say that K is **linearly surjective** if it is r-linearly surjective for each r. One shows easily:

LEMMA 5.1.27. *Linear surjectivity has the following properties:*

(i) *if $A(K) = K$ for every monic irreducible $A \in K[\partial]$ of order $\leqslant r$, then K is r-linearly surjective;*

(ii) *if K is linearly closed and 1-linearly surjective, then K is linearly surjective;*

(iii) *if K is a directed union of r-linearly surjective differential subfields, then K is r-linearly surjective;*

(iv) *if K is r-linearly surjective and $A_1, \dots, A_n \in K[\partial]^{\neq}$ $(n \geqslant 1)$ are of order at most r, then* $\dim_C \ker A_1 \cdots A_n = \dim_C \ker A_1 + \cdots + \dim_C \ker A_n$.

Note that (ii) follows from (i). Next we establish a descent property of linear surjectivity. Let L be a differential field extension of K with $d := [L : K] < \infty$. Let K^{a} be an algebraic closure of K with the unique derivation extending the derivation of K. For $y \in L$, let

$$\operatorname{tr}_{L|K}(y) := \sum_{i=1}^{d} \sigma_i(y)$$

be the trace of y in $L|K$. Here $\sigma_1, \dots, \sigma_d$ are the distinct field embeddings $L \to K^{\mathrm{a}}$ which are the identity on K. The map

$$y \mapsto \tau_{L|K}(y) := \frac{1}{d} \operatorname{tr}_{L|K}(y) : L \to K$$

is K-linear and the identity on K. Moreover, $\tau_{L|K}(f') = \tau_{L|K}(f)'$ for each $f \in L$. It follows that if $A \in K[\partial]$, $g \in K$, and the equation $A(y) = g$ has a solution $f \in L$, then it has the solution $\tau_{L|K}(f)$ in K. Since every algebraic extension field of K is a directed union of algebraic extension fields of finite degree, this gives the following:

COROLLARY 5.1.28. *If K has an r-linearly surjective algebraic differential field extension, then K is r-linearly surjective.*

In Section 5.4 below we show that algebraic differential field extensions of linearly surjective differential fields are linearly surjective. This requires some generalities on systems of linear differential equations. In Section 5.5 we prove the same with *linearly closed* instead of *linearly surjective*.

By convention we consider any K as being 0-linearly surjective.

Picard-Vessiot closed differential fields

In this subsection K is a differential field, and r ranges over $\mathbb{N}^{\geqslant 1}$. We assume here familiarity with the facts stated about zeros of linear differential polynomials in Section 4.1. We say that K is r-**Picard-Vessiot closed** (or r-**pv-closed**) if $\dim_C \ker A = \operatorname{order} A$ for all $A \in K[\partial]^{\neq}$ of order at most r. If K is a directed union of r-pv-closed differential subfields, then K is r-pv-closed. Also, K is 1-pv-closed iff $(K^\times)^\dagger = K$. We say that K is **Picard-Vessiot closed** (or **pv-closed**) if K is r-pv-closed for all r. This subsection, with some additional material in Section 5.5, will mainly get used in the next volume.

LEMMA 5.1.29. *The following two conditions on K are equivalent:*

(i) *K is r-linearly closed and 1-pv-closed;*

(ii) $\ker A \neq \{0\}$ *for all $A \in K[\partial] \setminus K$ of order at most r.*

PROOF. Use Lemma 5.1.21. □

LEMMA 5.1.30. *Suppose $r \geqslant 2$. Then the following are equivalent:*

(i) *K is r-pv-closed;*

(ii) *K is r-linearly closed, r-linearly surjective, and 1-pv-closed;*

(iii) *K is r-linearly closed, 1-linearly surjective, and 1-pv-closed.*

In particular,

$$K \text{ is pv-closed} \iff K \text{ is linearly closed, linearly surjective, and } (K^\times)^\dagger = K.$$

PROOF. To show (i) ⇒ (ii), suppose K is r-pv-closed. Then K is r-linearly closed and 1-pv-closed, by the previous lemma. To show that K is r-linearly surjective, it suffices to show that K is 1-linearly surjective. Let $A \in K[\partial]^{\neq}$ with $\operatorname{order} A = 1$, and $b \in K^\times$; we need to show $b \in A(K)$. Put $B := (\partial - b^\dagger) \cdot A$, so $\operatorname{order} B = 2 \leqslant r$; then $\dim_C \ker A = 1 < 2 = \dim_C \ker B$, so there exists $y \in K$ with $A(y) \neq 0$, $B(y) = 0$, and then $A(cy) = b$ for some $c \in C$, by Lemma 5.1.14(ii).

The implication (ii) ⇒ (iii) is trivial, and (iii) ⇒ (i) follows from the fact that if $A, B \in K[\partial]^{\neq}$ with $B(K) = K$, then $\dim_C \ker AB = \dim_C \ker A + \dim_C \ker B$. □

By Lemmas 5.1.29 and 5.1.30, every differentially closed field is linearly closed, linearly surjective, and pv-closed.

LEMMA 5.1.31. *Suppose K is pv-closed and C is algebraically closed. Then K is algebraically closed.*

PROOF. Let $P \in K[Y]$ have degree $\geqslant 2$. Take an algebraic closure K^{a} of K and give it the unique derivation extending the derivation of K. Then $C_{K^{\mathrm{a}}} = C$ by

Lemma 4.1.2. Let V be the C-linear subspace of K^{a} generated by the zeros of P in K^{a}. Let $y_1, \dots, y_n$ be a basis for V (so $n \geqslant 1$), and let

$$A(Y) := \frac{\operatorname{wr}(y_1, \dots, y_n, Y)}{\operatorname{wr}(y_1, \dots, y_n)} = Y^{(n)} + a_1 Y^{(n-1)} + \dots + a_n Y, \quad a_1, \dots, a_n \in K^{\mathrm{a}}.$$

Then $V = Z(A)$. If also $V = Z(B)$, $B = Y^{(n)} + b_1 Y^{(n-1)} + \dots + b_n Y$ where $b_1, \dots, b_n \in K^{\mathrm{a}}$, then $A = B$ by Corollary 4.1.16. Each $\sigma \in \operatorname{Aut}(K^{\mathrm{a}}|K) = \operatorname{Aut}_\partial(K^{\mathrm{a}}|K)$ satisfies $\sigma(V) = V$ and so fixes the coefficients of A. Thus $A \in K\{Y\}$. Since K is pv-closed, this yields $V \subseteq K$; in particular $P(y) = 0$ for some $y \in K$. □

Relating $K[i]$ and K

In this subsection K is a differential field in which -1 is not a square in K. It follows that K has a (differential) field extension $K[i]$, $i^2 = -1$. The results below indicate how factorization in $K[\partial]$ relates to factorization in $K[i][\partial]$. This will be used in later chapters where K is real closed, and thus $K[i]$ is algebraically closed.

LEMMA 5.1.32. *If every $A \in K[\partial]^{\neq}$ splits over $K[i]$, then $K[i]$ is linearly closed.*

PROOF. The "complex conjugation" automorphism $a + bi \mapsto \overline{a + bi} := a - bi$ $(a, b \in K)$ of the differential field $K[i]$ extends uniquely to an automorphism $A \mapsto \overline{A}$ of the ring $K[i][\partial]$ with $\overline{\partial} = \partial$. Then $\overline{A} = A \iff A \in K[\partial]$, for all $A \in K[i][\partial]$. Let $A \in K[i][\partial]^{\neq}$ be monic. Let B be the least common left multiple of A and $\overline{A}$. Then $\overline{B} = B$, so $B \in K[\partial]$. If B splits over $K[i]$, then so does A, by Proposition 5.1.22. □

LEMMA 5.1.33. *Suppose u is a nonzero element in a differential field extension of $K[i]$ such that $u^\dagger \in K[i]$. Then there is $B \in K[\partial]$ of order 2 such that $B(u) = 0$.*

PROOF. Write $u^\dagger = a + ib$, so $u' = au + ibu$. Differentiating this relation yields

$$u'' = (a' + a^2 - b^2)u + i(2ab + b')u.$$

If $b = 0$, then we can take $B = \partial^2 - (a' + a^2)$. Suppose that $b \neq 0$. Then we eliminate i by forming a suitable linear combination of u'' and u':

$$bu'' - (2ab + b')u' = (-ab' + a'b - a^2 b - b^3)u.$$

It follows that we can take $B := b\partial^2 - (2ab + b')\partial + (ab' - a'b + a^2 b + b^3)$. □

LEMMA 5.1.34. *Suppose $A \in K[\partial]$ is monic and irreducible over K of order 2 and splits over $K[i]$. Then*

$$A = \big(\partial - (a - bi + b^\dagger)\big) \cdot \big(\partial - (a + bi)\big) \quad \textit{for some } a \in K,\ b \in K^\times.$$

PROOF. We have $A = (\partial - c)\big(\partial - (a + bi)\big)$ with $c \in K[i]$ and $a, b \in K$. Then $b \neq 0$ and $c = a - bi + b^\dagger$, and so A has the desired form. Note also that

$$A = \partial^2 - (2a + b^\dagger)\partial + (-a' + a^2 + ab^\dagger + b^2).$$

□

LEMMA 5.1.35. *Suppose $A \in K[\partial]$ is monic and splits over $K[i]$. Then $A = A_1 \cdots A_m$ with all $A_i \in K[\partial]$ monic and irreducible of order 1 or order 2.*

PROOF. We proceed by induction on $\operatorname{order}(A)$ and can assume A has order $n > 1$. By Lemma 5.1.21 we have a nonzero u in some differential field extension of $K[i]$ such that $u^\dagger \in K[i]$ and $A(u) = 0$. By Lemma 5.1.33 this gives a monic $D \in K[\partial]$ of order 1 or 2 such that $D(u) = 0$. Take such D of least order. Then $A = BD + R$ with $B, R \in K[\partial]$ and $\operatorname{order}(R) < \operatorname{order}(D)$, hence $R(u) = 0$, so $R = 0$ and thus $A = BD$. Now B and D split over $K[i]$ by Lemma 5.1.10, and so we can apply the inductive assumption to B. □

For $f = a + bi \in K[i]$ $(a, b \in K)$ we set $\overline{f} := a - bi$, $\Re(f) := a$, $\Im(f) := b$, so $f \mapsto \overline{f}$ is a differential field automorphism of $K[i]$. Now let F be a differential subfield of K, and suppose $f \in K[i]$ is d-algebraic over $F[i]$ with minimal annihilator $P(Y)$ over $F[i]$. Then f, $\overline{f}$, and $\Re(f)$ are d-algebraic over F. Set $r := r_P$; then $L_{P_{+f}}$ has order r, with coefficient $S_P(f)$ of ∂^r.

LEMMA 5.1.36. *Suppose $L_{P_{+f}}$ splits over $K[i]$, and let R be a minimal annihilator of $\Re f$ over $F[i]\langle f\rangle$. Then $L_{R_{+\Re f}}$ splits over $K[i]$.*

PROOF. Set $Q := R_{\times 1/2, +f}$, so $R = Q_{-f, \times 2}$. From $\Re(f) = \frac{1}{2}(f + \overline{f})$ we get $Q(\overline{f}) = 0$. Thus Q is a minimal annihilator of $\overline{f}$ over $F[i]\langle f\rangle$. Also, $\overline{P}$ is a minimal annihilator of $\overline{f}$ over $F[i]$, hence $L_{Q_{+\overline{f}}}$ splits over $K[i]$ by Lemma 5.1.23. Now

$$Q_{+\overline{f}} = R_{+\Re f, \times 1/2}, \qquad L_{Q_{+\overline{f}}} = L_{R_{+\Re f, \times 1/2}},$$

so $L_{R_{+\Re f}} = L_{R_{+\Re f, \times 1/2, \times 2}} = 2L_{R_{+\Re f, \times 1/2}}$ splits over $K[i]$. □

LEMMA 5.1.37. *Suppose $K[i]$ is r-linearly closed, and $S \in K\{Y\}$ is a minimal annihilator of $\Re f$ over F. Then $L_{S_{+\Re f}}$ splits over $K[i]$.*

PROOF. Since $K[i]$ is r-linearly closed, $L_{P_{+f}}$ splits over $K[i]$. Let R be a minimal annihilator of $\Re f$ over $F[i]\langle f\rangle$. Then $L_{R_{+\Re f}}$ splits over $K[i]$ by Lemma 5.1.36. We now apply Lemma 5.1.23 with $\Re f$, S, R in the role of f, P, Q, and obtain

$$L_{S_{+\Re f}} = A \cdot L_{R_{+\Re f}} \quad \text{with } A \in K[i][\partial], \quad \operatorname{order} A = r_S - r_R.$$

So it is enough to show that A splits over $K[i]$. Let $Q := R_{\times 1/2, +f}$ be as in the proof of Lemma 5.1.36, so $r_Q = r_R$. By considering the inclusions among the fields

$$F[i], \quad E := F[i]\langle f\rangle, \quad E\langle \overline{f}\rangle = E\langle \Re f\rangle, \quad F[i]\langle \Re f\rangle$$

and the corresponding transcendence degrees, we obtain

$$r + r_R = r + r_Q = r_S + \operatorname{trdeg}\big(E\langle \Re f\rangle | F[i]\langle \Re f\rangle\big),$$

so $r_S - r_R = r - \operatorname{trdeg}\big(E\langle \Re f\rangle | F[i]\langle \Re f\rangle\big) \leqslant r$. Thus A splits over $K[i]$. □

Linear elements

In this subsection K is a differential field, y lies in a differential field extension of K, and $r \in \mathbb{N}$. We say that y is r-**linear** over K if there is an $A \in K[\partial]^{\neq}$ of order $\leqslant r$ such that $A(y) = 0$. Note that $0 \in K$ is 0-linear over K, and each $a \in K^{\times}$ is 1-linear over K (take $A = \partial - a^{\dagger}$). If y is r-linear over K, then

$$y^{(n)} \in Ky + \cdots + Ky^{(r-1)} \text{ for all } n,$$

and thus $K\langle y \rangle = K(y, y', \ldots, y^{(r-1)})$ as fields. Lemma 5.1.14(i) shows that y is $(r+1)$-linear over K if $A(y) \in K$ for some $A \in K[\partial]^{\neq}$ of order $\leqslant r$. We say that y is **linear** over K if it is r-linear over K for some r. Thus y is linear over K iff $A(y) \in K$ for some $A \in K[\partial]^{\neq}$. "Algebraic over K" implies "linear over K":

LEMMA 5.1.38. *Suppose $r = [K(y) : K] < \infty$. Then y is r-linear over K.*

PROOF. Let $P \in K[Y]$ be the minimum polynomial of y over K. From $P'(y) \neq 0$ we get $y' = -P^{\partial}(y)/P'(y) \in K(y)$, by Lemma 1.9.1. Hence $y^{(n)} \in K(y)$ for all n; so a nontrivial K-linear combination of $y, y', \ldots, y^{(r)}$ is zero, that is, $A(y) = 0$ for some $A \in K[\partial]^{\neq}$ of order $\leqslant r$. □

Sums, products, and derivatives of linear elements remain linear:

LEMMA 5.1.39. *Let L be a differential field extension of K and let $y, z \in L$ be m-linear and n-linear over K, respectively. Then y' is m-linear over K, $y+z$ is $(m+n)$-linear over K, and yz is mn-linear over K.*

PROOF. All $y^{(i)}$ $(i \geqslant 0)$ lie in the K-linear space $Ky + \cdots + Ky^{(m-1)}$ of dimension at most m, so $y', \ldots, y^{(m+1)}$ are K-linearly dependent, and thus y' is m-linear over K. Also $z^{(i)} \in Kz + \cdots + Kz^{(n-1)}$ for all i, and so all $(y+z)^{(i)}$ lie in the K-linear space $Ky + \cdots + Ky^{(m-1)} + Kz + \cdots + Kz^{(n-1)}$ of dimension at most $m+n$. Hence $y+z, (y+z)', \ldots, (y+z)^{(m+n)}$ are K-linearly dependent, and thus $y+z$ is $(m+n)$-linear over K. Finally, $(yz)^{(i)} \in \sum_{j<m, k<n} Ky^{(j)}z^{(k)}$ for all i, and so by the same reasoning yz is mn-linear over K. □

COROLLARY 5.1.40. *If L is a differential field extension of K and $y_1, \ldots, y_n \in L$ are linear over K, then all elements of $K\{y_1, \ldots, y_n\}$ are linear over K.*

There is no such result for reciprocals:

LEMMA 5.1.41. *Let $a \in K \setminus \partial(K)$, and $y' = a$. Then y is transcendental over K and 2-linear over K, but $z = 1/y$ is not linear over K.*

PROOF. Suppose towards a contradiction that y is algebraic over K, and let $P = \sum_{i=0}^{r} a_i Y^i$ be the minimum polynomial of y over K ($r \geqslant 1$, $a_i \in K$ for $i = 0, \ldots, r$, $a_r = 1$). By Lemma 1.9.1, we then have $P^{\partial}(y) + P'(y)a = 0$ and hence

$$P^{\partial}(Y) + P'(Y)a = 0 \text{ in } K[Y].$$

In particular, $a'_{r-1} + ra = 0$, so $a = (-a_{r-1}/r)' \in \partial(K)$, a contradiction. Thus y, and hence also $z = 1/y$, are transcendental over K. To see that z is not linear over K, one first shows by an easy induction on $n \geqslant 1$ that

$$z^{(n)} = n!(-a)^n z^{n+1} + f_n \qquad \text{where } f_n \in Kz^2 + \cdots + Kz^n.$$

Thus from $A(z) = 0$ with monic $A \in K[\partial]$ of order r we obtain

$$r!(-a)^r z^{r+1} + g(z) = 0 \qquad \text{where } g(z) \in Kz + \cdots + Kz^r,$$

contradicting the transcendence of z over K. □

Notes and comments

Early algebraic studies of the ring $K[\partial]$ are Ore [309, 310], but the *type* of a linear differential operator occurs already in Poincaré [319]. Lemma 5.1.15 is [421, Lemma 2.1]. Proposition 5.1.18 goes back to Euler [133]. Corollary 5.1.24 is due to Landau [248] and Loewy [267]. Lemma 5.1.39 and its Corollary 5.1.40 are from [420], but have a classical origin [387, Sec. 167]. In connection with Lemma 5.1.41 we note that by [166], if y is a nonzero element of a differential field extension of K such that both y and $1/y$ are linear over K, then $y^\dagger$ is algebraic over K. See [421, Section 2.4] for some history of the algebraic study of linear differential operators.

5.2 SECOND-ORDER LINEAR DIFFERENTIAL OPERATORS

Lemma 5.1.35 above is a first indication that linear differential operators of order 1 and 2 are going to play a special role. In preparation for this, we now make a few basic observations concerning linear differential operators of order 2.

Let K be a differential field, and let $A \in K[\partial]$ be of order 2. Lemma 5.1.13 and the example following it show that if $(K^\times)^\dagger = K$, then in order to describe the kernel $\ker A$ of A, one may reduce to the case that

$$A = 4\partial^2 + f \qquad (f \in K)$$

which we assume in the rest of this section. Note that $\dim_C \ker A \leqslant 2$, and A is irreducible in $K[\partial]$ iff A does not split over K. We introduce the function

$$z \mapsto \omega(z) := -(2z' + z^2) : K \to K.$$

Then for $y \in K^\times$ we have $\omega(2y^\dagger) = -4y''/y$, hence $A(y) = y(f - \omega(2y^\dagger))$, and thus

$$A(y) = 0 \iff 4y'' + fy = 0 \iff \omega(2y^\dagger) = f.$$

Hence

$$\dim_C \ker A \geqslant 1 \iff f \in \omega\big(2(K^\times)^\dagger\big).$$

Another way to formulate this involves the set

$$\Omega(K) := \{f \in K : 4y'' + fy = 0 \text{ for some } y \in K^\times\}.$$

Thus $\Omega(K)$ is existentially definable in K, and if $(K^\times)^\dagger = K$, then $\Omega(K) = \omega(K)$.

If A splits over K, then $A = 4(\partial + a)(\partial - a)$ and $f = \omega(2a)$ for some $a \in K$. Conversely, if $z \in K$ and $\omega(z) = f$, then $A = 4\left(\partial + \frac{z}{2}\right)\left(\partial - \frac{z}{2}\right)$. Together with Lemma 5.1.21, the above yields

$$\dim_C \ker A \geqslant 1 \Rightarrow A \text{ splits over } K \iff f \in \omega(K).$$

If $(K^\times)^\dagger = K$ (equivalently, $\dim_C \ker B = 1$ for each $B \in K[\partial]$ of order 1), then

$$\dim_C \ker A \geqslant 1 \iff A \text{ splits over } K \iff f \in \omega(K).$$

The following computations involving ω will be used several times later on:

LEMMA 5.2.1. *Let $w, y, z \in K$ be such that $y = z - w \neq 0$. Then*

$$\omega(w) - \omega(z) = y \cdot \big(2(y^\dagger + w) + y\big).$$

PROOF.
$$\begin{aligned}\omega(w) - \omega(z) &= 2(z-w)' + 2(z-w)w + (z-w)^2 \\ &= (z-w)\cdot\big(2(z-w)^\dagger + 2w + (z-w)\big) \\ &= y \cdot \big(2y^\dagger + 2w + y\big) \\ &= y \cdot \big(2(y^\dagger + w) + y\big).\end{aligned}$$
□

COROLLARY 5.2.2. *Let $y \in K^\times$, $z = -y^\dagger$. Then $\omega(z+y) = \omega(z) - y^2$.*

Membership in $\omega(K)$ governs solvability of certain differential equations:

COROLLARY 5.2.3. *Let $f, g, h \in K$, $f \neq 0$. Then*

$$\exists z \in K\,\big[z' = fz^2 + gz + h\big] \iff 4fh - (g + f^\dagger)^2 + 2(g + f^\dagger)' \in \omega(K).$$

PROOF. Let z range over K, and set $f_* := -2f$, $g_* := -2g$, $h_* := -2h$. Then

$$\begin{aligned} z' = fz^2 + gz + h &\iff -2z' = f_* z^2 + g_* z + h_* \\ &\iff -2f_* z' = (f_* z)^2 + g_*(f_* z) + f_* h_* \\ &\iff -2\big((f_* z)' - f_*' z\big) = (f_* z)^2 + g_*(f_* z) + f_* h_* \\ &\iff -2(f_* z)' = (f_* z)^2 + (g_* - 2f_*^\dagger)(f_* z) + f_* h_*. \end{aligned}$$

So with $g_{**} := g_* - 2f_*^\dagger = -2(g + f^\dagger)$ and $h_{**} := f_* h_* = 4fh$,

$$\exists z\,\big[z' = fz^2 + gz + h\big] \iff \exists z\,\big[-2z' = z^2 + g_{**}z + h_{**}\big].$$

Now completing the square yields

$$-2z' = z^2 + g_{**}z + h_{**} \iff -2\left(z + \tfrac{g_{**}}{2}\right)' = \left(z + \tfrac{g_{**}}{2}\right)^2 + \left(h_{**} - \tfrac{g_{**}^2}{4} - g_{**}'\right),$$

and $h_{**} - \frac{g_{**}^2}{4} - g_{**}' = 4fh - (g + f^\dagger)^2 + 2(g + f^\dagger)'$. □

Relation to the Schwarzian

We now consider the function

$$y \mapsto \mathrm{s}(y) := \omega(z+y) = \omega(z) - y^2 : K^\times \to K \qquad \text{where } z = -y^\dagger.$$

For a nonconstant element u of K,

$$\mathrm{S}(u) := (u'^\dagger)' - \frac{1}{2}(u'^\dagger)^2 = \frac{u'''}{u'} - \frac{3}{2}\left(\frac{u''}{u'}\right)^2$$

is known as the **Schwarzian derivative** of u. It is related to $\mathrm{s}(u^\dagger)$ as follows:

LEMMA 5.2.4. *Let $u \in K \setminus C$. Then $2\,\mathrm{S}(u) = \mathrm{s}(u^\dagger)$.*

PROOF. To see this note that $u'^\dagger = (uu^\dagger)^\dagger = u^\dagger + u^{\dagger\dagger}$ and hence

$$\begin{aligned}(u'^\dagger)' &= (u^\dagger)' + (u^{\dagger\dagger})',\\ (u'^\dagger)^2 &= (u^\dagger)^2 + (u^{\dagger\dagger})^2 + 2u^\dagger u^{\dagger\dagger} = (u^\dagger)^2 + (u^{\dagger\dagger})^2 + 2(u^\dagger)',\end{aligned}$$

so

$$\begin{aligned}2\,\mathrm{S}(u) &= 2(u'^\dagger)' - (u'^\dagger)^2\\ &= 2(u^\dagger)' + 2(u^{\dagger\dagger})' - (u^\dagger)^2 - (u^{\dagger\dagger})^2 - 2(u^\dagger)'\\ &= 2(u^{\dagger\dagger})' - (u^{\dagger\dagger})^2 - (u^\dagger)^2 = \mathrm{s}(u^\dagger).\end{aligned}$$

□

Note that for each $y \in K^\times$ we have $\mathrm{s}(-y) = \mathrm{s}(y)$; in particular, for $u \in K \setminus C$ we have $\mathrm{S}(u) = \mathrm{S}(1/u)$. More generally, consider the (left) action of the group $\mathrm{GL}_2(C)$ on the set $K \setminus C$ given by

$$(T,u) \mapsto Tu := \frac{t_{11}u + t_{12}}{t_{21}u + t_{22}} \qquad \text{for } T = \left(\begin{smallmatrix} t_{11} & t_{12} \\ t_{21} & t_{22}\end{smallmatrix}\right) \in \mathrm{GL}_2(C).$$

Then S is invariant under this action:

LEMMA 5.2.5. *Let $T \in \mathrm{GL}_2(C)$ and $u \in K \setminus C$. Then $\mathrm{S}(Tu) = \mathrm{S}(u)$.*

PROOF. Let $T = \left(\begin{smallmatrix} t_{11} & t_{12} \\ t_{21} & t_{22}\end{smallmatrix}\right)$. Set $u_* := Tu$, $y := t_{21}u + t_{22}$, $c := t_{21}$ (so $y' = cu'$), and $d := t_{11}t_{22} - t_{12}t_{21}$ (so $d \in C^\times$). Then

$$u_*' = \frac{d}{y^2}u', \qquad u_*'^\dagger = u'^\dagger - 2c\frac{u'}{y},$$

hence

$$\begin{aligned}(u_*'^\dagger)' &= (u'^\dagger)' + 2c^2\left(\frac{u'}{y}\right)^2 - 2c\frac{u''}{y},\\ (u_*'^\dagger)^2 &= (u'^\dagger)^2 + 4c^2\left(\frac{u'}{y}\right)^2 - 4c\frac{u''}{y},\end{aligned}$$

and thus $\mathrm{S}(u_*) = \mathrm{S}(u)$. □

LEMMA 5.2.6. *Let $y_1, y_2 \in K^\times$ be such that* $\operatorname{wr}(y_1, y_2) \in C^\times$. *Then*

$$u := y_1/y_2 \notin C, \qquad 2\,\mathrm{s}(u^\dagger) \;=\; \omega(2y_1^\dagger) + \omega(2y_2^\dagger).$$

PROOF. Set $c := \operatorname{wr}(y_1, y_2)$, so $c = y_1y_2' - y_1'y_2 = -y_1y_2u^\dagger \in C^\times$ and hence $u \notin C$ and $u^\dagger = -c/(y_1y_2)$. Therefore $u^\dagger = y_1^\dagger - y_2^\dagger$ and $-u^{\dagger\dagger} = y_1^\dagger + y_2^\dagger$, so

$$\begin{aligned} 2\,\mathrm{s}(u^\dagger) &= 2\omega(-u^{\dagger\dagger}) - 2(u^\dagger)^2 \\ &= -4\big((y_1^\dagger)' + (y_2^\dagger)'\big) - 2(y_1^\dagger + y_2^\dagger)^2 - 2(y_1^\dagger - y_2^\dagger)^2 \\ &= -2\big(2(y_1^\dagger)' + 2(y_2^\dagger)'\big) - (2y_1^\dagger)^2 - (2y_2^\dagger)^2, \end{aligned}$$

and the lemma follows. □

We can use the function s to detect whether $\dim_C \ker A = 2$:

COROLLARY 5.2.7. *If $y_1, y_2 \in \ker A$ are C-linearly independent, then $\mathrm{s}(u^\dagger) = f$ for $u = y_1/y_2 \in K \setminus C$. Conversely, if $u \in K \setminus C$, $\mathrm{s}(u^\dagger) = f$, $r \in K^\times$, $r^\dagger = \frac{1}{2}u'^\dagger$, then $y_1 = \frac{u}{r}$ and $y_2 = \frac{1}{r}$ form a basis of the C-linear space $\ker A$ with $y_1/y_2 = u$.*

Note that for $u \in K \setminus C$ and $r \in K^\times$, we have: $r^\dagger = \frac{1}{2}u'^\dagger \iff r^2 \in C^\times u'$.

PROOF. For the first claim, let $y_1, y_2 \in \ker A$ be C-linearly independent, and set $w := \operatorname{wr}(y_1, y_2)$, $u := y_1/y_2$. Then $w' = 0$ by Lemma 4.1.17, so $w \in C^\times$, and thus $\mathrm{s}(u^\dagger) = f$ using Lemma 5.2.6 and $\omega(2y_1^\dagger) = \omega(2y_2^\dagger) = f$. For the second claim, let $u \in K \setminus C$, $\mathrm{s}(u^\dagger) = f$, $r \in K^\times$, $r^\dagger = \frac{1}{2}u'^\dagger$, and set $y_1 = \frac{u}{r}$ and $y_2 = \frac{1}{r}$. Then

$$2y_1^\dagger \;=\; 2u^\dagger - u'^\dagger \;=\; -u^{\dagger\dagger} + u^\dagger$$

and hence $\omega(2y_1^\dagger) = \mathrm{s}(u^\dagger) = f$; also

$$2y_2^\dagger \;=\; -u'^\dagger \;=\; -u^{\dagger\dagger} - u^\dagger \;=\; -(1/u)^{\dagger\dagger} + (1/u)^\dagger,$$

so $\omega(2y_2^\dagger) = \mathrm{s}\left((1/u)^\dagger\right) = \mathrm{s}(-u^\dagger) = \mathrm{s}(u^\dagger) = f$. Thus $A(y_1) = A(y_2) = 0$. □

COROLLARY 5.2.8. *Suppose $(K^\times)^\dagger = K$. Then*

$$\dim_C \ker A = 2 \iff f \in \mathrm{s}(K^\times).$$

If in addition K is 1-linearly surjective, then $\omega(K) = \mathrm{s}(K^\times)$.

PROOF. The first part follows from Corollary 5.2.7. Suppose now that K is also 1-linearly surjective. By the remark preceding Lemma 5.1.12, if A splits over K, then $\dim_C \ker A = 2$; this gives $\omega(K) = \mathrm{s}(K^\times)$ by the first part. □

COROLLARY 5.2.9. *Suppose $(K^\times)^\dagger = K$, and let $u, u_* \in K \setminus C$. Then*

$$\mathrm{s}(u^\dagger) = \mathrm{s}(u_*^\dagger) \iff u_* = Tu \text{ for some } T \in \mathrm{GL}_2(C).$$

PROOF. The forward direction is a consequence of Corollary 5.2.7, and the backward direction follows from Lemma 5.2.5. □

The function σ

We now assume that -1 is not a square in K, and work in the differential field extension $K[i]$ ($i^2 = -1$) of K. Note that for $a, b \in K$ we have

$$A(a + bi) = A(a) + A(b)i,$$

so if the equation $A(y) = 0$ has a nonzero solution in $K[i]$, then it has a nonzero solution in K. Thus

$$\omega\big(2(K[i]^\times)^\dagger\big) \cap K = \omega\big(2(K^\times)^\dagger\big),$$

and by Corollary 5.2.8:

COROLLARY 5.2.10. *Suppose K is 1-linearly surjective and $(K^\times)^\dagger = K$, and let $d := \dim_C \ker A$. Then either $d = 0$ or $d = 2$, and d also equals the $C[i]$-vector space dimension of the kernel of A viewed as an element of $K[i][\partial]$.*

Consider now also the function $\sigma\colon K^\times \to K$ defined by

$$\sigma(y) := \mathrm{s}(yi) = \omega(z + yi) = \omega(z) + y^2 \qquad \text{where } z = -y^\dagger.$$

If $f = \sigma(y)$ with $y \in K^\times$, and $z = -y^\dagger$, then A factors over $K[i]$ as

$$A = 4\left(\partial + \frac{z + yi}{2}\right)\left(\partial - \frac{z + yi}{2}\right).$$

Note also that $\sigma(y) = \sigma(-y)$ for $y \in K^\times$. For $a, b \in K$ we have

$$\omega(a + bi) = \omega(a) + b^2 - 2(ab + b')i,$$

so if $b \neq 0$, then: $\omega(a + bi) \in K \Leftrightarrow a = -b^\dagger$. Hence

$$\omega\big(K[i]\big) \cap K = \omega(K) \cup \sigma(K^\times),$$

and thus

(5.2.1) $$A \text{ splits over } K[i] \iff f \in \omega(K) \cup \sigma(K^\times).$$

Suppose that A splits over $K[i]$. Then $f \in \omega(K)$ or $f \in \sigma(K^\times)$. We now indicate a differential field extension L of K such that $\ker_L A$ has dimension 2 as a vector space over C_L, and we specify a basis for this vector space.

CASE 1: $f \in \omega(K)$. Let $z \in K$ be such that $f = \omega(z)$. Take a nonzero element y_1 in a differential field extension of K with $2y_1^\dagger = z$, and next an element y_2 of a differential field extension of $K\langle y_1\rangle = K(y_1)$ with $y_2' - (z/2)y_2 = 1/y_1$. Then $L := K\langle y_1, y_2\rangle = K(y_1, y_2)$ is a differential field extension of K such that $y_1, y_2 \in \ker_L A$, and $\mathrm{wr}(y_1, y_2) = y_1y_2' - y_1'y_2 = 1$. So L is as promised, with y_1, y_2 as a basis of the vector space $\ker_L A$ over C_L. Note that if $(K^\times)^\dagger = K$ and K is 1-linearly surjective, then we can take $y_1, y_2 \in K$, so that $L = K$.

CASE 2: $f \in \sigma(K^\times)$. Let $y \in K^\times$ satisfy $\sigma(y) = f$. First take an element r in a field extension of $K[i]$ with $r^2 = y$, and next an element $e(y) \neq 0$ in a differential field extension of $K[i, r]$ such that $e(y)^\dagger = \frac{1}{2}yi$. Thinking of $e(y)$ as $\exp(\frac{1}{2}i \int y)$, set

$$e(-y) := e(y)^{-1}, \qquad y_1 := \frac{e(y)}{r}, \qquad y_2 := \frac{e(-y)}{r}, \qquad L := K[i]\langle y_1, y_2\rangle.$$

Then $2y_1^\dagger = -y^\dagger + yi$ and $2y_2^\dagger = -y^\dagger - yi$, so $L = K[i](y_1, y_2)$ and

$$\omega(2y_1^\dagger) = \sigma(y) = \sigma(-y) = \omega(2y_2^\dagger) = f.$$

Also $(y_1/y_2)^\dagger = y_1^\dagger - y_2^\dagger = yi \neq 0$, so y_1, y_2 are linearly independent over C_L. Thus L is as we promised, with y_1, y_2 as a basis of the vector space $\ker_L A$ over C_L.

Application to linear differential operators with constant coefficients

In this subsection we assume that C is real closed, and that i is an element of a differential field extension of K with $i^2 = -1$. We have

$$\omega(C) = \{-c^2 : c \in C\} = C^{\leqslant}, \qquad \sigma(C^\times) = \{c^2 : c \in C^\times\} = C^{>},$$

so $C = \omega(C) \cup \sigma(C^\times)$. Hence if $f \in C$, then A splits over $K[i]$, by (5.2.1). For the next lemma, recall the definition of $\mathrm{m}(B)$ for $B \in C[\partial]^{\neq}$ from Section 5.1.

LEMMA 5.2.11. *Suppose K is 1-linearly surjective, $(K^\times)^\dagger = K$, and $\omega(K) \cap C \subseteq C^{\leqslant}$. Let $B \in C[\partial]^{\neq}$. Then $\dim_C \ker B = \mathrm{m}(B)$.*

PROOF. We have $\mathrm{m}(B) \leqslant \dim_C \ker B$ by (5.1.4). To show $\dim_C \ker B \leqslant \mathrm{m}(B)$, we can assume B is monic of order $\geqslant 1$. Then Proposition 3.5.7(i) yields $B = B_1 \cdots B_n$ ($n \geqslant 1$), with $B_i \in C[\partial]$ monic of order 1 or 2 for $i = 1, \dots, n$. Hence

$$\mathrm{m}(B) = \mathrm{m}(B_1) + \cdots + \mathrm{m}(B_n), \qquad \dim_C \ker B \leqslant \dim_C \ker B_1 + \cdots + \dim_C \ker B_n.$$

This gives a reduction to the case that B has order 1 or 2. If $\mathrm{order}(B) = 1$ then $\ker B \neq \{0\}$ (since $(K^\times)^\dagger = K$), so $\dim_C \ker B = 1 = \mathrm{m}(B)$. Suppose $\mathrm{order}(B) = 2$, say $B = \partial^2 + b\partial + c$ with $b, c \in C$. Put $\widetilde{B} := \partial^2 + (c - \frac{b^2}{4})$; then $\dim_C \ker B = \dim_C \ker \widetilde{B}$; see Lemma 5.1.13 and the example following it. Also $\mathrm{m}(B) = \mathrm{m}(\widetilde{B})$. So replacing B by $\widetilde{B}$ we arrange $b = 0$. Then $\mathrm{m}(B) = 0$ if $c > 0$ and $\mathrm{m}(B) = 2$ if $c \leqslant 0$. By Corollary 5.2.10 applied to $A = 4B$ we either have $\dim_C \ker B = 0$ or $\dim_C \ker B = 2$. From $(K^\times)^\dagger = K$ and $\omega(K) \cap C = C^{\leqslant}$, we get

$$\dim_C \ker B \geqslant 1 \iff 4c \in \omega(K) \iff c \leqslant 0,$$

and this yields $\dim_C \ker B = \mathrm{m}(B)$. $\square$

Notes and comments

The Schwarzian derivative plays a role in the analytic theory of linear differential equations, where versions of Corollaries 5.2.7 and 5.2.9 are well-known; see [185, Chapter 10].

5.3 DIAGONALIZATION OF MATRICES

Given a differential field K, the ring $K[\partial]$ is euclidean as defined below. Here we establish a basic result about matrices over any euclidean ring. In the next section we use this to reduce a system of linear differential equations in several unknowns to several linear differential equations in one unknown. Throughout this section R is a (possibly non-commutative) ring with $1 \neq 0$ and we let a, b, c range over R. We say that R is a **domain** if for all a, b, if $ab = 0$, then $a = 0$ or $b = 0$.

Total divisibility

We say that a **totally divides** b (notation: $a \| b$) if $Rb \subseteq aR$ and $bR \subseteq Ra$ (and thus $RbR \subseteq aR \cap Ra$). Note:

(1) $a \| 0$,

(2) $0 \| b \iff b = 0$,

(3) $a \| a \iff aR = Ra$,

(4) $a \| b \ \&\ b \| c \Rightarrow a \| c$,

(5) if a is a unit of R, then $a \| b$ for all b, and

(6) given a the set $\{b : a \| b\}$ is a (two-sided) ideal of R.

The ring R is said to be **simple** if the only (two-sided) ideals of R are $\{0\}$ and R. If R is simple and $a \| b$, then $b = 0$ or $a \in R^{\times}$.

Degree functions

Addition on $\mathbb{N}$ and the usual ordering on $\mathbb{N}$ are extended as usual to $\mathbb{N} \cup \{-\infty\}$. A **degree function** on R is a map $\mathrm{d}\colon R \to \mathbb{N} \cup \{-\infty\}$ such that for all a, b,

(D1) $\mathrm{d}(a) = -\infty \iff a = 0$;

(D2) $\mathrm{d}(-a) = \mathrm{d}(a)$;

(D3) $\mathrm{d}(a + b) \leqslant \max\{\mathrm{d}(a), \mathrm{d}(b)\}$; and

(D4) $\mathrm{d}(ab) = \mathrm{d}(a) + \mathrm{d}(b)$.

Let d be a degree function on R. Then $\mathrm{d}(1) = 0$ by (D1) and (D4). It follows from (D1), (D2), (D3) that $\mathrm{d}(a + b) = \max\{\mathrm{d}(a), \mathrm{d}(b)\}$ if $\mathrm{d}(a) \neq \mathrm{d}(b)$. By (D1) and (D4) the set $R^{\neq}$ is closed under multiplication, so R is a domain.

Euclidean rings

Let d be a degree function on R. Note that then for all a, b with $a \neq 0$, there is at most one pair $(q, r) \in R^2$ with $b = qa + r$ and $\mathrm{d}(r) < \mathrm{d}(a)$, and also at most one pair $(q^*, r^*) \in R^2$ with $b = aq^* + r^*$ and $\mathrm{d}(r^*) < \mathrm{d}(a)$. We say that R is **left euclidean with respect to** d if for all a, b, $a \neq 0$, there is $q \in R$ with $\mathrm{d}(b - qa) < \mathrm{d}(a)$; similarly, R is **right euclidean with respect to** d if for all a, b, $a \neq 0$, there is $q^* \in R$ with $\mathrm{d}(b - aq^*) < \mathrm{d}(a)$. We say that R is **euclidean with respect to** d if it is both left and right euclidean with respect to d. Note that if R is euclidean with respect to d, then $R^\times = \{a : \mathrm{d}(a) = 0\}$.

If R is a domain, we call a **irreducible** (in R) if $a \notin R^\times$, and there are no $a_1, a_2 \in R \setminus R^\times$ with $a = a_1 a_2$. Note that if a is irreducible in the domain R, then $a \neq 0$, and au, ua are also irreducible for $u \in R^\times$. Call R **euclidean** if it is euclidean with respect to some degree function on R. If R is euclidean, then R is a domain, every nonzero element of $R \setminus R^\times$ equals $a_1 \cdots a_n$ for some $n \geqslant 1$ and irreducible $a_1, \dots, a_n$ in R, every left ideal of R is principal (of the form Ra), and every right ideal of R is principal (of the form aR).

LEMMA 5.3.1. *Suppose every left ideal of R is principal. Then every submodule M of the left R-module R^n is generated by n elements.*

PROOF. The case $n = 0$ holds trivially, so suppose $n \geqslant 1$, and identify R^{n-1} with $R^{n-1} \times \{0\} \subseteq R^n$ in the natural way. Inductively, the module $M \cap R^{n-1}$ is generated by elements $b_1, \dots, b_{n-1}$. Let $e := (0, \dots, 0, 1) \in R^n$, and consider the left ideal $I := \{a \in R : ae \in M + R^{n-1}\}$ of R. Let $a \in R$ with $I = Ra$, and pick $b_n \in M$ with $ae \in b_n + R^{n-1}$. Then $b_1, \dots, b_n$ generate M. □

Ore domains

A **right Ore domain** is a domain R such that for all $a, b \in R^{\neq}$ we have $aR \cap bR \neq \{0\}$. A **left Ore domain** is defined similarly. If R is both a left and right Ore domain, then R is called an **Ore domain**. (For example, every integral domain is an Ore domain.) Call R **right noetherian** if every right ideal of R is finitely generated (as a right R-module); likewise with *left* instead of *right*.

LEMMA 5.3.2. *Let R be a right noetherian domain. Then R is a right Ore domain.*

PROOF. Let $a, b \in R^{\neq}$. Since the right ideal of R generated by the $a^m b$ is finitely generated, we have n such that $a^{n+1} b = \sum_{i=0}^{n} a^i b c_i$ with $c_0, \dots, c_n \in R$. By canceling some power a^m with $m \leqslant n$ on both sides we arrange $c_0 \neq 0$. Then

$$0 \neq bc_0 = a^{n+1} b - \sum_{i=1}^{n} a^i b c_i \in aR \cap bR,$$

as desired. □

Similarly, every left noetherian domain is left Ore. In particular, every euclidean ring is an Ore domain.

Now let R be a left Ore domain and M an R-module. Then

$$M_{\text{tor}} := \{x \in M : ax = 0 \text{ for some } a \in R^{\neq}\}$$

is a submodule of M: for $x, y \in M_{\text{tor}}$, take $a, b \in R^{\neq}$ with $ax = by = 0$; take $r, s, t \in R$ with $ra = sb = t \neq 0$; then $t(x + y) = rax + sby = 0$, so $x + y \in M_{\text{tor}}$; it follows likewise that if $x \in M_{\text{tor}}$ and $a \in R$, then $ax \in M_{\text{tor}}$. We call M_{tor} the **torsion submodule** of M. Call M a **torsion module** if $M_{\text{tor}} = M$ and **torsion-free** if $M_{\text{tor}} = \{0\}$. So M_{tor} is a torsion module and M/M_{tor} is torsion-free.

Diagonalization

In this subsection R is euclidean and $m, n \geqslant 1$. When m, n are clear from the context, we let 0 denote the zero element of the additive group $R^{m \times n}$ of $m \times n$ matrices over R. For $m = n$ we denote the multiplicative group of units of the ring $R^{n \times n}$ by $\mathrm{GL}_n(R)$. An $m \times n$ matrix $A = (a_{ij})$ over R is said to be **diagonal** if $a_{ij} = 0$ for all $i \in \{1, \ldots, m\}$ and $j \in \{1, \ldots, n\}$ with $i \neq j$. Given $m \times n$ matrices A and B over R, we say that A and B are **equivalent** (over R) if there are $P \in \mathrm{GL}_m(R)$ and $Q \in \mathrm{GL}_n(R)$ with $A = PBQ$; in symbols: $A \sim B$. Clearly $\sim$ is an equivalence relation on the set of $m \times n$ matrices over R, with the equivalence class of 0 being $\{0\}$. We are going to show that every matrix over a euclidean ring is equivalent to a diagonal matrix; more precisely:

THEOREM 5.3.3. *Every $m \times n$ matrix over R is equivalent to a diagonal $m \times n$ matrix $D = (d_{ij})$ over R such that $d_{ii} \| d_{jj}$ for $1 \leqslant i < j \leqslant \min\{m, n\}$.*

The proof involves row operations on an $m \times n$ matrix A over R:

(R1) interchange two rows;

(R2) add a left multiple of the ith row to the jth row ($i \neq j$);

(R3) multiply a row on the left by a unit of R.

If B arises from A by applying one of the operations (R1)–(R3), then $B = PA$ for some $P \in \mathrm{GL}_m(R)$, so $A \sim B$. We also have column operations on A:

(C1) interchange two columns;

(C2) add a right multiple of the ith column to the jth column ($i \neq j$);

(C3) multiply a column on the right by a unit of R.

If B arises from A by applying one of the operations (C1)–(C3), then $B = AQ$ for some $Q \in \mathrm{GL}_n(R)$, hence $A \sim B$.

We fix a degree function d on R with respect to which R is euclidean. Given an $m \times n$ matrix $A = (a_{ij})$ over R, let $\mathrm{d}(A)$ be the minimum of the $\mathrm{d}(a_{ij})$ with $a_{ij} \neq 0$ if $A \neq 0$, and $d(A) := -\infty$ if $A = 0$.

LEMMA 5.3.4. *Let $A = \left(\begin{smallmatrix} a & 0 \\ 0 & b \end{smallmatrix}\right)$ where $a, b \in R$, $a \neq 0$, and suppose $\mathrm{d}(a) \leqslant \mathrm{d}(B)$ for all 2×2 matrices B over R with $A \sim B$. Then $a \| b$.*

PROOF. Let $c \in R$, and take $q, r \in R$ with $bc = qa + r$ and $\mathrm{d}(r) < \mathrm{d}(a)$. First applying the operation (R2) and then (C2) we see that $A \sim \left(\begin{smallmatrix} a & 0 \\ r & b \end{smallmatrix}\right)$, hence $r = 0$. This shows $bR \subseteq Ra$. Similarly one obtains $Rb \subseteq aR$. □

Towards the proof of Theorem 5.3.3, consider an $m \times n$ matrix A over R.

CLAIM: *A is equivalent to a diagonal matrix D over R with $\mathrm{d}(D) \leqslant \mathrm{d}(A)$.*

PROOF OF CLAIM. Set $k := \min\{m, n\} \in \mathbb{N}^{\geqslant 1}$ and $d := \mathrm{d}(A) \in \mathbb{N} \cup \{-\infty\}$. We order the set $\mathbb{N}^{\geqslant 1} \times (\mathbb{N} \cup \{-\infty\})$ lexicographically and prove the claim by induction on (k, d). If $A = 0$, then the claim holds trivially, so assume $A \neq 0$. Applying the operations (R1) and (C1) we first replace A by an equivalent $m \times n$ matrix, without changing (k, d), to reduce to the case that $\mathrm{d}(a_{11}) = d$. Now using (R2) and (C2) and euclidean division by a_{11}, we see that $A \sim B$ where $B = (b_{ij})$ is an $m \times n$ matrix over R with $b_{11} = a_{11}$ (hence $\mathrm{d}(B) \leqslant d$) and $\mathrm{d}(b_{i1}) < \mathrm{d}(a_{11})$ for $i = 2, \dots, m$ and $\mathrm{d}(b_{1j}) < \mathrm{d}(a_{11})$ for $j = 2, \dots, n$. If $m > 1$ and $b_{i1} \neq 0$ for some $i \in \{2, \dots, m\}$, then $\mathrm{d}(B) < \mathrm{d}(a_{11}) = d$, and we can apply the inductive hypothesis to B. Similarly, if $n > 1$ and $b_{1j} \neq 0$ for some $j \in \{2, \dots, n\}$, then $\mathrm{d}(B) < \mathrm{d}(A)$, and the inductive hypothesis applies to B. Thus we may assume $b_{i1} = 0$ for all $i \in \{2, \dots, m\}$ and $b_{1j} = 0$ for all $j \in \{2, \dots, n\}$. If $k = 1$, then B is already diagonal, so suppose $k > 1$. Then we have an $(m-1) \times (n-1)$ matrix B' over R such that

$$B = \begin{pmatrix} a_{11} & \\ & \boxed{B'} \end{pmatrix}.$$

The claim now follows by applying the inductive hypothesis to B'. □

PROOF OF THEOREM 5.3.3. Let A be an $m \times n$ matrix over R. To show that A is equivalent over R to a diagonal matrix as in Theorem 5.3.3, we can assume $A \neq 0$. The claim gives a diagonal matrix $D = (d_{ij})$ over R with $A \sim D$ and $\mathrm{d}(D) \leqslant \mathrm{d}(B)$ for all $m \times n$ matrices B over R with $A \sim B$. By applying the operations (R1) and (C1) to D we arrange $\mathrm{d}(D) = \mathrm{d}(d_{11})$. Set $k := \min\{m, n\}$. We are done if $k = 1$, so assume $k > 1$. By Lemma 5.3.4 and the minimality of $\mathrm{d}(D)$ we have $d_{11} \| d_{ii}$ for $i = 2, \dots, k$. We can assume inductively that the $(m-1) \times (n-1)$-matrix $(d_{ij})_{i,j \geqslant 2}$ is equivalent to a diagonal matrix as in Theorem 5.3.3, the entries of which will then be in the (two-sided) ideal of R generated by the d_{ii} with $i = 2, \dots, k$, and so d_{11} totally divides each of those entries. □

COROLLARY 5.3.5. *Suppose R is simple. Let $A \neq 0$ be an $m \times n$ matrix over R, and $k := \min\{m, n\}$. Then A is equivalent to a diagonal $m \times n$ matrix $D = (d_{ij})$ over R such that for some $r \in \{1, \dots, k\}$ we have*

$$d_{11} = \cdots = d_{r-1,r-1} = 1, \qquad d_{rr} \neq 0, \quad d_{r+1,r+1} = \cdots = d_{kk} = 0.$$

PROOF. Theorem 5.3.3 gives a diagonal $m \times n$ matrix $B = (b_{ij})$ over R with $A \sim B$ and $b_{ii} \| b_{jj}$ for $1 \leqslant i < j \leqslant k$. Since $A \neq 0$, we have $b_{rr} \neq 0$ for some $r \in \{1, \ldots, k\}$; take r maximal with this property. For $i \in \{1, \ldots, r-1\}$ we have $b_{ii} \| b_{rr}$, and so b_{ii} is a unit of R (as R is simple). Now use (R3). □

The next result is an application of Theorem 5.3.3 to finitely generated modules over euclidean rings. Here and below, *module* means *left module.*

COROLLARY 5.3.6. *Let M be a finitely generated R-module. Then*

$$M \cong (R/Rd_1) \oplus \cdots \oplus (R/Rd_r) \oplus R^s$$

where $d_1, \ldots, d_r \in R^{\neq}$, $r, s \in \mathbb{N}$, and $d_1 \| d_2 \| \cdots \| d_r$. If R is simple, then

$$M \cong (R/Rd) \oplus R^s \qquad (d \in R^{\neq},\ s \in \mathbb{N}).$$

PROOF. Take $n \geqslant 1$ and a surjective R-linear map $R^n \to M$. If the kernel is trivial, then the above holds with $r = 0$, $s = n$ (and $d = 1$ for simple R). Assume the kernel is not trivial. Then Lemma 5.3.1 yields an $m \geqslant 1$ and an $m \times n$ matrix A over R such that $M \cong R^n/N$ where $N = \{yA : y \in R^m\}$ and the elements of R^m and R^n are viewed as row vectors. Applying Theorem 5.3.3 and Corollary 5.3.5 to A yields the desired result with $r + s = n$. □

Hence a finitely generated R-module M is torsion-free iff $M \cong R^n$ for some n.

Independence and rank

Let R be a euclidean domain, and let M range over R-modules. We refer to the Conventions and Notations in the beginning of this volume for the terminology in dealing with linear (in)dependence over R.

Let $m, n \geqslant 1$ and let $A = (a_{ij})$ be an $m \times n$ matrix over R. We say that the rows of A are R-independent if $A_1, \ldots, A_m$ are R-independent, where $A_i = (a_{i1}, \ldots, a_{in})$ is the ith row, considered as a vector of the (left) R-module R^n. For $r_1, \ldots, r_m \in R$ we may view $(r_1, \ldots, r_m)$ as a $1 \times m$ matrix over R, and so

$$r_1 A_1 + \cdots + r_m A_m = (r_1, \ldots, r_m)A.$$

Therefore, if the rows of A are R-independent and $A \sim B$ for the $m \times n$ matrix B over R, then the rows of B are R-independent. Suppose $A \sim D$, where $D = (d_{ij})$ is a diagonal $m \times n$ matrix over R. Then the rows of A are R-independent iff $m \leqslant n$ and $d_{ii} \neq 0$ for $i = 1, \ldots, m$. For R-linear $f \colon R^m \to R^n$ this yields:

COROLLARY 5.3.7. *If f is injective, then $m \leqslant n$.*

PROOF. Let $e_1, \ldots, e_m$ be the usual basis vectors of R^m. Assume f is injective. Let A be the $m \times n$ matrix with ith row $A_i := f(e_i)$. Then the rows of A are R-independent, so $m \leqslant n$ by Theorem 5.3.3 and the remarks above. □

COROLLARY 5.3.8. *If f is surjective, then $m \geqslant n$.*

PROOF. Let $e_1, \dots, e_n$ be the usual basis vectors of R^n. Assume f is surjective. Take $b_j \in R^m$ with $f(b_j) = e_j$ for $j = 1, \dots, n$. Then $R^n \cong Rb_1 + \cdots + Rb_n \subseteq R^m$, so $n \leqslant m$ by Corollary 5.3.7. □

Of course these two corollaries also hold for $m = 0$ or $n = 0$, which we allow below. By these two corollaries, if $R^m \cong R^n$ as R-modules, then $m = n$. So for each finitely generated free R-module M we may define the **rank** of M to be the unique n such that $M \cong R^n$; in this case,

$$\begin{aligned} \operatorname{rank} M &= \text{largest } m \text{ such that there are } R\text{-independent } x_1, \dots, x_m \in M \\ &= \text{least } m \text{ such that } M \text{ is generated by some } x_1, \dots, x_m \in M. \end{aligned}$$

Now let M be any finitely generated R-module. Then the finitely generated torsion-free R-module M/M_{tor} is free, and we set $\operatorname{rank}(M) := \operatorname{rank}(M/M_{\text{tor}})$; thus M is a torsion module iff $\operatorname{rank}(M) = 0$. Also,

$$\operatorname{rank} M = \text{largest } m \text{ such that there are } R\text{-independent } x_1, \dots, x_m \in M.$$

Clearly if M, N are finitely generated R-modules, then

$$\operatorname{rank}(M \oplus N) = \operatorname{rank}(M) + \operatorname{rank}(N).$$

In fact, the next lemma shows that $M \mapsto \operatorname{rank}(M)$, for finitely generated M, is a $\mathbb{Z}$-valued Euler-Poincaré map on R-modules in the sense of Section 1.2.

LEMMA 5.3.9. *Let an exact sequence of R-modules and R-linear maps be given:*

$$0 \longrightarrow K \xrightarrow{\iota} M \xrightarrow{\pi} N \longrightarrow 0.$$

If M is finitely generated, then so are K and N, and $\operatorname{rank}(M) = \operatorname{rank}(K) + \operatorname{rank}(N)$.

PROOF. Suppose M is finitely generated. Then N is finitely generated, and by Lemma 5.3.1, so is K. To prove the rank formula, assume until further notice that N is a torsion module. Suppose $x_1, \dots, x_m \in M$ are R-independent. We have $r_1, \dots, r_m \in R^{\neq}$ with $r_1x_1, \dots, r_mx_m \in \ker \pi = \iota(K)$, and as $r_1x_1, \dots, r_mx_m$ are also R-independent, we get $m \leqslant \operatorname{rank}(K)$. This shows $\operatorname{rank}(M) \leqslant \operatorname{rank}(K)$. It is obvious that $\operatorname{rank}(K) \leqslant \operatorname{rank}(M)$, and so the two ranks are equal.

We now drop the assumption that N is a torsion module. The canonical map $\nu\colon N \to N/N_{\text{tor}}$ yields the R-linear surjection $\nu \circ \pi\colon M \to N/N_{\text{tor}}$ with kernel $M_1 := \pi^{-1}(N_{\text{tor}})$. Since the R-module N/N_{tor} is free, we have $M \cong M_1 \oplus (N/N_{\text{tor}})$, so $\operatorname{rank}(M) = \operatorname{rank}(M_1) + \operatorname{rank}(N)$. We have an exact sequence

$$0 \longrightarrow K \xrightarrow{\iota} M_1 \xrightarrow{\pi|M_1} N_{\text{tor}} \longrightarrow 0$$

and so $\operatorname{rank}(K) = \operatorname{rank}(M_1)$ by what we showed before. □

COROLLARY 5.3.10. *Suppose M is generated by elements $x_1, \dots, x_n$, and $m \leqslant n$ is such that $x_1, \dots, x_m$ are R-independent and $x_1, \dots, x_m, x_j$ are R-dependent for all j with $m < j \leqslant n$. Then* $\operatorname{rank}(M) = m$.

PROOF. Let $K := Rx_1 + \cdots + Rx_m$. Then M/K is a torsion module and $K \cong R^m$, so $\operatorname{rank}(K) = m$. Now use Lemma 5.3.9. □

Notes and comments

Lemma 5.3.2 is due to Goldie [149]. Theorem 5.3.3 for $R = \mathbb{Z}$ was proved by Smith [427]. The version here, for euclidean R, is due to Wedderburn [460] and Jacobson [202]; our presentation follows [86, Section 1.4].

5.4 SYSTEMS OF LINEAR DIFFERENTIAL EQUATIONS

In this section K is a differential field. We apply the previous section to $R = K[\partial]$. The ring $K[\partial]$ is euclidean with respect to the degree function d on $K[\partial]$ given by $\mathrm{d}(A) := \operatorname{order}(A)$. By Corollary 5.1.5, if $C \neq K$, then $K[\partial]$ is simple.

Inhomogeneous equations

In this subsection, $R = K[\partial]$. Let $m, n \geqslant 1$, and let A be an $m \times n$ matrix over R. Given a column vector $f = (f_1, \dots, f_n)^{\mathrm{t}} \in K^n$ we let $A(f)$ be the column vector in K^m with ith entry

$$\sum_{j=1}^{n} A_{ij}(f_j) \quad (i = 1, \dots, m),$$

not to be confused with the matrix product $Af \in R^m$ that has ith entry $\sum_{j=1}^{n} A_{ij} f_j$. The map $f \mapsto A(f) \colon K^n \to K^m$ is C-linear. It is easy to check that if B is an $n \times p$ matrix over R with $p \in \mathbb{N}^{\geqslant 1}$, and $g = (g_1, \dots, g_p)^{\mathrm{t}} \in K^p$, then $(AB)(g) = A\big(B(g)\big)$. For the $n \times n$ identity matrix I over R we have $I(f) = f$ for all $f \in K^n$.

Let $a = (a_1, \dots, a_m)^{\mathrm{t}} \in K^m$ be a column vector. The pair (A, a) gives rise to a system $A(y) = a$ of linear differential equations over K. A **solution** in K to this system is a column vector $f = (f_1, \dots, f_n)^{\mathrm{t}} \in K^n$ such that $A(f) = a$. We say that the system $A(y) = a$ is R-**consistent** if for every row vector $r = (r_1, \dots, r_m) \in R^m$ such that $rA = 0$ (a matrix product) we have $r(a) = 0$, where

$$r(a) := r_1(a_1) + \cdots + r_m(a_m).$$

If $A(y) = a$ has a solution in K, then $A(y) = a$ is clearly R-consistent.

We can increase m while keeping n fixed by adding extra zero rows to A and extra zero entries to a; in this way we can arrange that $m \geqslant n$. Such a change in A, a does not change the solutions to the system in K nor its R-consistency status. So we assume below that $m \geqslant n$.

Using Theorem 5.3.3, take $P \in \mathrm{GL}_m(R)$ and $Q \in \mathrm{GL}_n(R)$ such that PAQ is diagonal, and set $b = P(a) \in K^m$. Then each column $z \in K^n$ with $(PAQ)(z) = b$ yields a column $y = Q(z) \in K^n$ with $A(y) = a$; this gives a bijective correspondence $z \mapsto Q(z)$ between the solutions of $(PAQ)(z) = b$ in K and the solutions of $A(y) = a$ in K. It is also easy to see that $A(y) = a$ is R-consistent iff $(PAQ)(z) = b$ is R-consistent. Put $B_i = (PAQ)_{ii} \in R$ for $i = 1, \dots, n$. Then the system $(PAQ)(z) = b$ with $z = (z_1, \dots, z_n)^{\mathrm{t}} \in K^n$ takes the form

$$B_1(z_1) = b_1, \dots, B_n(z_n) = b_n, \qquad 0 = b_i \text{ for } n < i \leqslant m,$$

that is, $(PAQ)(z) = b$ has no solution in K if $b_i \neq 0$ for some i with $n < i \leqslant m$, and otherwise its solutions are the columns $z \in K^n$ such that

$$B_1(z_1) = b_1, \dots, B_n(z_n) = b_n.$$

The system $(PAQ)(z) = b$ is R-consistent if and only if for all $r_1, \dots, r_m \in R$ with $r_1 B_1 = \cdots = r_n B_n = 0$ we have $r_1(b_1) + \cdots + r_m(b_m) = 0$, which in turn is equivalent to $b_i = 0$ for all $i \in \{1, \dots, n\}$ with $B_i = 0$, and $b_i = 0$ for all $i \in \{n+1, \dots, m\}$. Thus:

LEMMA 5.4.1. *If K is linearly surjective and a system $A(y) = a$ as above is R-consistent, then it has a solution in K.*

Next we consider a more common way of presenting a system of linear differential equations, namely as a matrix differential equation $y' = Ay + b$ where A is an $n \times n$ matrix over K (not over $K[\partial]$ as above), with $n \geqslant 1$, and $b = (b_1, \dots, b_n)^{\mathrm{t}} \in K^n$ is a given column vector. A solution of this equation in K is a column $f = (f_1, \dots, f_n)^{\mathrm{t}} \in K^n$ such that $f' = Af + b$ where $f' := (f_1', \dots, f_n')^{\mathrm{t}} \in K^n$. Let such an equation $y' = Ay + b$ be given. Define the $n \times n$-matrix B over $R = K[\partial]$ by $B_{ij} := -A_{ij}$ for $i \neq j$ and $B_{ii} := \partial - A_{ii}$. Then the equation $y' = Ay + b$ has clearly the same solutions in K as the system $B(y) = b$. We claim that the system $B(y) = b$ is automatically R-consistent. To see why, let $r = (r_1, \dots, r_n) \in R^n$ be a row vector and $rB = 0$ in R^n. Then $r_j \partial = rA_j$ for $j = 1, \dots, n$ where A_j is the jth column of A. Hence $r_1 = \cdots = r_n = 0$, since otherwise an equality $r_j \partial = rA_j$ with nonzero r_j of highest order gives a contradiction. This proves our claim.

COROLLARY 5.4.2. *If K is linearly surjective, then each matrix differential equation $y' = Ay + b$ over K as above has a solution in K.*

Let $n \geqslant 1$, $a_1, \dots, a_n \in K$, $L = \partial^n + a_1 \partial^{n-1} + \cdots + a_n \in K[\partial]$, and $b \in K$. The solutions of the equation $L(z) = b$ in K are the $f \in K$ such that $L(f) = b$. Setting $y_0 := z$, this equation is equivalent to the system

$$(*) \qquad y_0' = y_1, \dots, y_{n-2}' = y_{n-1}, \quad y_{n-1}' = -(a_1 y_{n-1} + \cdots + a_n y_0) + b$$

in the unknowns $y_0, \dots, y_{n-1}$, in the sense that $z \mapsto (z, z', \dots, z^{(n-1)})$ maps the set of solutions in K of $L(z) = b$ bijectively onto the set of solutions $(y_0, \dots, y_{n-1}) \in K^n$

of (∗). The system (∗) can be written as a matrix equation $y' = A_L y + (0,\dots,0,b)^{\mathrm{t}}$ where A_L is the $n \times n$ matrix

$$A_L := \begin{pmatrix} 0 & 1 & 0 & \cdots & 0 \\ 0 & 0 & 1 & \cdots & 0 \\ \vdots & \vdots & \vdots & \ddots & \vdots \\ 0 & 0 & 0 & \cdots & 1 \\ -a_n & -a_{n-1} & -a_{n-2} & \cdots & -a_1 \end{pmatrix}$$

over K, called the **companion matrix** of L. Thus the converse of Corollary 5.4.2 is also valid. We can now derive:

COROLLARY 5.4.3. *Suppose K is linearly surjective and E is a differential field extension of K and algebraic over K. Then E is linearly surjective.*

PROOF. Let $n \geqslant 1$, let A be an $n \times n$ matrix over E, and $b \in E^n$ a column vector. We have to show that the matrix equation $y' = Ay + b$ has a solution in E. The entries of A and b lie in a finite degree extension of K inside E, so we can arrange that E is of finite degree over K, say with basis $e_1, \dots, e_m$ over K. Writing the e_i', the $e_i e_j$, and the entries of A and b as K-linear combinations of $e_1, \dots, e_m$ and making the substitution

$$y_j = z_{j1}e_1 + \cdots + z_{jm}e_m \quad (j = 1, \dots, n),$$

one obtains an $mn \times mn$ matrix $A^{\diamond}$ over K and a column vector $b^{\diamond} \in K^{mn}$ such that any solution of the matrix equation $z' = A^{\diamond}z + b^{\diamond}$ in K yields a solution of $y' = Ay + b$ in E. □

Independence and finite-dimensionality

Let $n \geqslant 1$ and let $Y = (Y_1, \dots, Y_n)$ be a tuple of distinct differential indeterminates. Throughout, r ranges over $\mathbb{N}$.

Let $K[\partial]^n$ be the (left) $K[\partial]$-module of row vectors $(L_1, \dots, L_n)$ with components $L_j \in K[\partial]$, and $L(L_1, \dots, L_n) = (LL_1, \dots, LL_n)$ for $L \in K[\partial]$. In order to relate homogeneous differential polynomials in $K\{Y\}$ of degree 1 to vectors in $K[\partial]^n$ we consider the K-linear space

$$K\{Y\}_1 := \sum_{j=1}^{n} \sum_{r} KY_j^{(r)}$$

of homogeneous differential polynomials in $K\{Y\}$ of degree 1, and the K-linear bijection

$$A \mapsto A^{\partial} := (A_1^{\partial}, \dots, A_n^{\partial}) : K\{Y\}_1 \to K[\partial]^n$$

of (left) K-linear spaces, such that for $A = \sum_{j=1}^{n} \sum_r a_{jr} Y_j^{(r)}$, all $a_{jr} \in K$, we have $A_j^{\partial} := \sum_r a_{jr}\partial^r \in K[\partial]$ for $j = 1, \dots, n$. Let $A \in K\{Y\}_1$. Then clearly

$$A(y) = \sum_{j=1}^{n} A_j^{\partial}(y_j) \quad \text{for } y = (y_1, \dots, y_n) \in K^n.$$

Also $A' \in K\{Y\}_1$ and $(A')^\partial = \partial A^\partial$ in the (left) $K[\partial]$-module $K[\partial]^n$, and so $(A^{(r)})^\partial = \partial^r A^\partial$ by induction on r. Thus for $A_1, \dots, A_m \in K\{Y\}_1$ the following are equivalent:

(1) the family $\left(A_i^{(r)}\right)$ $(i = 1, \dots, m,\ r = 0, 1, \dots)$ is linearly dependent over K;

(2) there exist $L_1, \dots, L_m \in K[\partial]$, not all equal to 0, such that

$$L_1 A_1^\partial + \cdots + L_m A_m^\partial = 0,$$

that is, $A_1^\partial, \dots, A_m^\partial$ are $K[\partial]$-dependent as defined in the previous section.

If these conditions are satisfied we say that $A_1, \dots, A_m$ are d-**dependent**; if not, we say that $A_1, \dots, A_m$ are d-**independent**. (We might add *over* K, but if $A_1, \dots, A_m$ are d-independent over K, then $A_1, \dots, A_m$ are d-independent over any differential field extension.) If $A_1, \dots, A_m$ are d-independent, then $m \leqslant n$ by Section 5.3.

Let $A_1, \dots, A_m \in K\{Y\}_1$ be given, where $m \geqslant 1$. The above K-linear bijection $K\{Y\}_1 \to K[\partial]^n$ maps A_i for $i = 1, \dots, m$ to the vector

$$(A_{i,1}^\partial, \dots, A_{i,n}^\partial) \in K[\partial]^n,$$

which we take as the ith row of the $m \times n$ matrix $A^\partial = (A_{i,j}^\partial)$ over $K[\partial]$. For this matrix and $y = (y_1, \dots, y_n)^{\mathrm{t}} \in K^n$ we have

$$\big(A_1(y), \dots, A_m(y)\big)^{\mathrm{t}} = A^\partial(y).$$

Consider the case $m = n$, so we are given $A_1, \dots, A_n \in K\{Y\}_1$. Let A^∂ be the corresponding $n \times n$ matrix over $K[\partial]$. The zero set

$$\begin{aligned} Z(A_1, \dots, A_n) &:= \big\{y \in K^n : A_1(y) = \cdots = A_n(y) = 0\big\} \\ &= \big\{y \in K^n : A^\partial(y) = 0\big\} \end{aligned}$$

is a C-linear subspace of K^n. When is it finite-dimensional?

LEMMA 5.4.4. *Assume $K \neq C$. Then the vector space $Z(A_1, \dots, A_n)$ over C is finite-dimensional iff $A_1, \dots, A_n$ are* d*-independent. If $A_1, \dots, A_n$ are* d*-independent and K is linearly surjective, then for every $a \in K^n$ there is a $y \in K^n$ with $A(y) = a$.*

PROOF. Take $P, Q \in \mathrm{GL}_n\big(K[\partial]\big)$ such that $PA^\partial Q = D$ is diagonal. The C-linear bijection $z \mapsto Q(z)\colon K^n \to K^n$ maps the C-linear space $\big\{z \in K^n : D(z) = 0\big\}$ onto the C-linear space $\big\{y \in K^n : A^\partial(y) = 0\big\}$. Thus the latter vector space over C is finite-dimensional iff $D_{ii} \neq 0$ for $i = 1, \dots, n$, which in turn is equivalent to $A_1, \dots, A_n$ being d-independent, using the equivalence of (1) and (2) above. The last claim of the lemma now follows because if K is linearly surjective and $D_{ii} \neq 0$ for $i = 1, \dots, n$, then there is for every $b \in K^n$ a $z \in K^n$ with $D(z) = b$. □

COROLLARY 5.4.5. *Let $A_1, \dots, A_n \in K\{Y\}_1$ be* d*-independent and $b_1, \dots, b_n \in K$. Let L be a differential field extension of K and let $y = (y_1, \dots, y_n)^{\mathrm{t}} \in L^n$ be such that $A_i(y) = b_i$ for $i = 1, \dots, n$. Then $y_1, \dots, y_n$ are linear over K.*

PROOF. Take $P, Q \in \mathrm{GL}_n(K[\partial])$ such that $PA^{\partial}Q = D$ is diagonal. Then $D_{ii} \neq 0$ for $i = 1, \dots, n$ and for $z = (z_1, \dots, z_n)^{\mathrm{t}} := Q^{-1}(y)$ we have $D(z) = P(b)$. Hence each z_i is linear over K, and thus the components of $y = Q(z)$ are linear over K. □

Note that we have an isomorphism

$$A = (A_1, \dots, A_m) \mapsto A^{\partial} \; : \; K\{Y\}_1^m \to K[\partial]^{m \times n}$$

of left K-modules. Let $A = (A_1, \dots, A_m) \in K\{Y\}_1^m$, and recall that then

$$A^{\partial}(y) \; = \; \big(A_1(y), \dots, A_m(y)\big)^{\mathrm{t}}$$

for all $y = (y_1, \dots, y_n)^{\mathrm{t}} \in K^n$. Also let a tuple $B = (B_1, \dots, B_n) \in K\{Z\}_1^n$ be given, with $Z = (Z_1, \dots, Z_p)$ a tuple of p distinct indeterminates, $p \geqslant 1$, and let B^{∂} be the corresponding $n \times p$ matrix over $K[\partial]$. Substituting B_j for Y_j in A yields

$$\Big(A_1\big(B_1(Z), \dots, B_n(Z)\big), \dots, A_m\big(B_1(Z), \dots, B_n(Z)\big)\Big) \in K\{Z\}_1^m$$

whose corresponding $m \times p$ matrix over $K[\partial]$ is $A^{\partial}B^{\partial}$, as is easily verified.

Now let L be a differential field extension of K such that $[L : K] = n$. Let $b_1, \dots, b_n$ be a basis of L as a vector space over K. Let X be a new differential indeterminate, and $Y = (Y_1, \dots, Y_n)$ be as before. Let $A = A(X) \in L\{X\}_1$. Then

$$A(b_1Y_1 + \cdots + b_nY_n) \; = \; A_1(Y)b_1 + \cdots + A_n(Y)b_n \in L\{Y\}_1$$

with uniquely determined $A_1, \dots, A_n \in K\{Y\}_1$. We now have:

LEMMA 5.4.6. *If $K \neq C$ and $A \neq 0$, then $A_1, \dots, A_n$ are* d*-independent.*

PROOF. The K-linear map

$$y = (y_1, \dots, y_n) \mapsto b_1y_1 + \cdots + b_ny_n \colon K^n \to L$$

maps $Z(A_1, \dots, A_n) \subseteq K^n$ bijectively onto $Z(A) \subseteq L$. If $A \neq 0$, then $Z(A)$ has finite dimension as a vector space over C_L, and so $Z(A_1, \dots, A_n)$ has finite dimension as a vector space over C in view of $[C_L : C] \leqslant n$. Thus the desired result follows, in view of Lemma 5.4.4. □

Let $P \in K\{Y\}$. The homogeneous part of P of degree 1 is

$$P_1 \; := \; \sum_{j=1}^{n} \left(\sum_r \frac{\partial P}{\partial Y_j^{(r)}}(0) Y_j^{(r)} \right) \in K\{Y\}_1.$$

Note that $(P_1)' \; = \; (P')_1$. Let $y = (y_1, \dots, y_n) \in K^n$. We set

$$P_{+y} \; := \; P(y + Y) \; = \; P(y_1 + Y_1, \dots, y_n + Y_n) \in K\{Y\}$$

and we note that $(P_{+y})' = (P')_{+y}$, and

$$(P_{+y})_1 = \sum_{j=1}^{n}\left(\sum_r \frac{\partial P}{\partial Y_j^{(r)}}(y)Y_j^{(r)}\right).$$

Let $P_1,\dots,P_m \in K\{Y\}$ and $y \in K^n$. Then we say that $P_1,\dots,P_m$ are **d-dependent at** y if $(P_{1,+y})_1,\dots,(P_{m,+y})_1$ are d-dependent, and **d-independent at** y otherwise. Note that if L is a differential field extension of K, then $P_1,\dots,P_m$ are d-dependent at y with respect to K iff they are d-dependent at y with respect to L. If $P_1,\dots,P_m$ are d-independent at some point of K^n, then $m \leqslant n$.

The following lemma gives a simple sufficient condition for d-independence. First, for $P \in K\{Y\}$ and $\vec{r} = (r_1,\dots,r_n) \in \mathbb{N}^n$ we say that P has order at most $\vec{r}$ if $P \in K\big[Y_j^{(r)} : 1 \leqslant j \leqslant n,\ 0 \leqslant r \leqslant r_j\big]$. Thus if $A \in K\{Y\}_1$ has order at most $\vec{r} \in \mathbb{N}^n$, then $A = \sum_{j=1}^n \big(\sum_{r=0}^{r_j} a_{jr} Y_j^{(r)}\big)$ with all $a_{jr} = \partial A/\partial Y_j^{(r)} \in K$.

LEMMA 5.4.7. *Let $\vec{r} = (r_1,\dots,r_n) \in \mathbb{N}^n$, and assume $A_1,\dots,A_m \in K\{Y\}_1$ have order at most $\vec{r}$ and the $m \times n$ matrix $\big(\partial A_i/\partial Y_j^{(r_j)}\big)$ over K has rank m. Then $A_1,\dots,A_m$ are* d*-independent.*

PROOF. Suppose not. Take $s \in \mathbb{N}$ minimal such that $\big(A_i^{(r)}\big)_{1\leqslant i\leqslant m,\ 0\leqslant r\leqslant s}$ is linearly dependent over K. Take $f_{ir} \in K$ for $1 \leqslant i \leqslant m$ and $0 \leqslant r \leqslant s$ such that $f_{is} \neq 0$ for some i and $\sum_{i,r} f_{ir}A_i^{(r)} = 0$. Now $A_i^{(r)}$ has order at most $(r_1+r,\dots,r_n+r)$ and $\partial A_i^{(r)}/\partial Y_j^{(r_j+r)} = \partial A_i/\partial Y_j^{(r_j)}$. Comparing coefficients of $Y_j^{(r_j+s)}$ yields

$$f_{1s}\big(\partial A_1/\partial Y_j^{(r_j)}\big) + \cdots + f_{ms}\big(\partial A_m/\partial Y_j^{(r_j)}\big) = 0 \qquad (j = 1,\dots,n),$$

so $\big(\partial A_i/\partial Y_j^{(r_j)}\big)$ has rank $< m$. □

We cannot reverse Lemma 5.4.7: take $m = n = 2$, $\vec{r} = (1,1)$ and $A_1 := Y_1' + Y_2'$, $A_2 := Y_1$; then A_1, A_2 are d-independent, but $\big(\partial A_i/\partial Y_j'\big)$ has only rank 1.

Homogeneous equations

In this subsection we let $n \geqslant 1$.

NOTATION. Let R be a differential ring. For an $n \times n$ matrix $A = (a_{ij})$ over R we set $A' := (a_{ij}')$. Then $A \mapsto A'$ is a derivation on the ring of $n \times n$ matrices over R:

$$(A+B)' = A' + B', \quad (AB)' = A'B + AB' \qquad \text{for } n \times n \text{ matrices } A, B \text{ over } R,$$

and also $(A^{\mathrm{t}})' = (A')^{\mathrm{t}}$ for such A. For $A \in \mathrm{GL}_n(R)$ we have $(A^{-1})' = -A^{-1}A'A^{-1}$. For a column vector $f = (f_1,\dots,f_n)^{\mathrm{t}} \in R^n$ we set $f' = (f_1',\dots,f_n')^{\mathrm{t}}$. Then

$$(Af)' = A'f + Af' \quad \text{for column vectors } f \in R^n \text{ and } n \times n \text{ matrices } A \text{ over } R.$$

Let A be an $n \times n$ matrix over K. Given a differential ring extension R of K, the solutions to the matrix differential equation $y' = Ay$ form a C_R-submodule of R^n, which we denote by $\mathrm{sol}_R(A)$; we also set $\mathrm{sol}(A) := \mathrm{sol}_K(A)$.

LEMMA 5.4.8. *If $f_1, \ldots, f_r \in \mathrm{sol}(A)$ are K-linearly dependent, then $f_1, \ldots, f_r$ are C-linearly dependent. (In particular, $\dim_C \mathrm{sol}(A) \leqslant n$.)*

PROOF. Let $f_1, \ldots, f_r \in \mathrm{sol}(A)$ with $r \geqslant 1$ be such that $f_r = \sum_{i=1}^{r-1} a_i f_i$ where $a_1, \ldots, a_{r-1} \in K$, and $f_1, \ldots, f_{r-1}$ are K-linearly independent. It is enough to show that then $f_1, \ldots, f_r$ are C-linearly dependent. Now

$$0 = f_r' - Af_r = \sum_{i=1}^{r-1} a_i' f_i + \sum_{i=1}^{r-1} a_i(f_i' - Af_i) = \sum_{i=1}^{r-1} a_i' f_i.$$

Then $a_i' = 0$ for $i = 1, \ldots, r-1$, hence $a_i \in C$ for $i = 1, \ldots, r-1$. □

Some other aspects of homogeneous linear equations are better understood in the setting of differential modules; see the next section.

5.5 DIFFERENTIAL MODULES

Throughout this section K is a differential field. We define here differential modules over K, and use these to show, among other things, that if K is linearly closed (respectively, pv-closed), then so is any algebraic differential field extension of K.

Let R be a differential ring. If M is a (left) $R[\partial]$-module, then the additive map $\partial_M \colon M \to M$ given by $\partial_M(x) = \partial x$ for $x \in M$ (with $\partial \in K[\partial]$ in the module product ∂x) satisfies $\partial_M(ax) = a'x + a\partial_M(x)$ for $a \in R \subseteq R[\partial]$, $x \in M$, so ∂_M is a ∂-compatible derivation on the R-module M as defined in Section 1.7. Conversely, let M be an R-module and ∂_M a ∂-compatible derivation on the R-module M. Then there is a unique (left) $R[\partial]$-module with M as its underlying R-module such that $\partial_M \colon M \to M$ equals the multiplication $x \mapsto \partial x$ by the scalar $\partial \in R[\partial]$.

In particular, the derivation ∂ of R is a ∂-compatible derivation on the R-module R, so this makes R an $R[\partial]$-module with $\partial a = a'$ for $a \in R$. The differential ideals of R are exactly the submodules of this $R[\partial]$-module R.

Note also that for $R[\partial]$-modules M and N, a map $f : M \to N$ is $R[\partial]$-linear iff it is R-linear and $f(\partial x) = \partial f(x)$ for all $x \in M$. For $R = K$ this suggests the following notion which turns out to be very useful:

DEFINITION 5.5.1. A **differential module** over K is a finite-dimensional vector space M over K together with a ∂-compatible derivation on M; we construe such M as a (left) $K[\partial]$-module as indicated in the remarks preceding this definition. The **dimension** of a differential module M over K is the dimension $\dim_K M$ of M as a vector space over K.

Let $A = (a_{ij})$ be an $n \times n$ matrix over K. We make the K-linear space K^n with standard basis $e_1, \dots, e_n$ into a differential module M_A over K by requiring

$$\partial(e_j) = -\sum_{i=1}^{n} a_{ij} e_i \qquad (j = 1, \dots, n).$$

This determines M_A, and we call M_A the differential module associated to A. Note that for $e = \sum_j f_j e_j$ ($f_1, \dots, f_n \in K$) we have

$$\partial(e) = \sum_j f_j' e_j - \sum_i \left(\sum_j a_{ij} f_j \right) e_i = e' - Ae,$$

from which it follows that $\operatorname{sol}(A) = \{e \in M_A : \partial(e) = 0\}$. Conversely, if M is a differential module over K of dimension n with basis $b_1, \dots, b_n$ as a K-linear space, then there is a unique $n \times n$ matrix $A = (a_{ij})$ over K such that the K-linear map $M \to M_A$ with $b_i \mapsto e_i$ for $i = 1, \dots, n$ is an isomorphism of $K[\partial]$-modules, and we call this A the matrix associated to M with respect to the basis $b_1, \dots, b_n$.

EXAMPLE. For $L \in K[\partial]^{\neq}$ the submodule $K[\partial]L$ of the (left) $K[\partial]$-module $K[\partial]$ yields the quotient module $K[\partial]/K[\partial]L$, which has dimension $\operatorname{order}(L)$ as a K-linear space, so $K[\partial]/K[\partial]L$ is a differential module over K.

Let A, B be $n \times n$ matrices over K. A matrix $P \in \mathrm{GL}_n(K)$ defines an isomorphism $e \mapsto Pe \colon M_A \to M_B$ of $K[\partial]$-modules if and only if $\partial(Pe_j) = P\partial(e_j)$ for $j = 1, \dots, n$, if and only if $BP - PA = P'$; in this case, any differential ring extension R of K yields an isomorphism $f \mapsto Pf \colon \operatorname{sol}_R(A) \to \operatorname{sol}_R(B)$ of C_R-modules. We call the matrix differential equations $y' = Ay$ and $y' = By$ **equivalent** if $M_A \cong M_B$. Figure 5.1 illustrates the various incarnations of homogeneous linear differential equations.

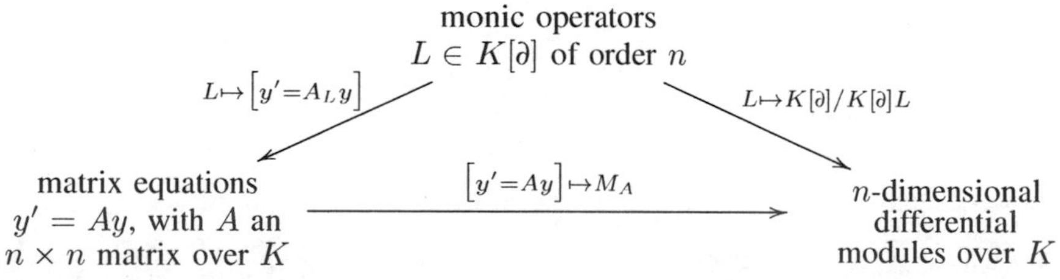

Figure 5.1: The correspondence between linear differential operators, matrix differential equations, and differential modules.

Given a differential module M over K, the elements $f \in M$ with $\partial(f) = 0$ are said to be **horizontal**. The set of horizontal elements of M is a finite-dimensional C-linear subspace of M. Indeed, if A is an $n \times n$ matrix over K, then $\operatorname{sol}(A)$ is the C-linear subspace of the underlying K-vector space K^n of M_A consisting of the horizontal elements of M_A, and $\dim_C \operatorname{sol}(A) \leqslant n$, with equality iff the K-linear space M_A has a basis consisting of horizontal elements (by Lemma 5.4.8).

EXAMPLE 5.5.2. Turn K into a $K[\partial]$-module with scalar multiplication

$$(L, f) \mapsto L(f) : K[\partial] \times K \to K.$$

Then K becomes a differential module over K, and $1 \in K$ is horizontal.

Let M be a differential module over K. If $f \in M^{\neq}$ is horizontal, then Kf is a submodule of the $K[\partial]$-module M, and $a \mapsto af \colon K \to Kf$ is an isomorphism with the above differential module K. We call M **horizontal** if M is isomorphic as a $K[\partial]$-module to a direct sum of copies of the above differential module K; equivalently, M has a basis consisting of horizontal elements.

Corollary 5.3.6 yields the following for any (left) $K[\partial]$-module M:

COROLLARY 5.5.3. *Suppose $C \neq K$. Then M is a differential module over K if and only if $M \cong K[\partial]/K[\partial]L$ for some monic $L \in K[\partial]^{\neq}$.*

A **cyclic vector** of a differential module M over K of dimension n is a vector $e \in M$ such that $e, \partial e, \dots, \partial^{n-1}e$ is a basis of M as a K-linear space. For example, if $L = \partial^n + a_1\partial^{n-1} + \cdots + a_n \in K[\partial]$ $(a_1, \dots, a_n \in K)$, then the differential module $M := K[\partial]/K[\partial]L$ of dimension n has cyclic vector $e := 1 + K[\partial]L$ with $Le = 0$, and the matrix of M with respect to the basis $e, \partial e, \dots, \partial^{n-1}e$ is $-A_L^{\mathrm{t}}$. Conversely, a cyclic vector e of a differential module M over K of dimension n with $Le = 0$ and $L \in K[\partial]$ of order n yields an isomorphism $K[\partial]/K[\partial]L \to M$ of differential modules over K sending $1 + K[\partial]L$ to e. If $C \neq K$, then by Corollary 5.5.3 every differential module $M \neq \{0\}$ over K has a cyclic vector. In the next subsection we explain the role of cyclic vectors in connection with the correspondences in Figure 5.1 above.

If $C \neq K$, then by Corollary 5.1.25 and the existence of cyclic vectors the differential field K is linearly closed if and only if every differential module $M \neq \{0\}$ over K has a 1-dimensional differential submodule; it is easy to see that this equivalence also holds if $C = K$.

COROLLARY 5.5.4. *Suppose K is linearly closed. Let L be a differential field extension of K that is algebraic over K. Then L is also linearly closed.*

PROOF. It suffices to consider the case $[L : K] < \infty$. For this case, let $M \neq \{0\}$ be a differential module over L and let M_K denote M viewed as a $K[\partial]$-module. Then M_K is a differential module over K. Hence we can take a 1-dimensional differential submodule N of M_K. Then the L-linear subspace LN of M generated by N is an $L[\partial]$-submodule of M with $\dim_L LN = 1$. □

Duality

Let M and N be $K[\partial]$-modules. The K-linear space $\operatorname{Hom}_K(M,N)$ of all K-linear maps $M \to N$ is made into a $K[\partial]$-module by defining

$$(\partial\phi)(f) = \partial(\phi f) - \phi(\partial f) \qquad \text{for } \phi \in \operatorname{Hom}_K(M,N) \text{ and } f \in M.$$

If M and N are differential modules over K, then so is $\operatorname{Hom}_K(M,N)$, with

$$\dim_K \operatorname{Hom}_K(M,N) = \dim_K M \cdot \dim_K N,$$

and if in addition M and N are horizontal, then so is $\operatorname{Hom}_K(M,N)$. A special case of this construction is the **dual** $M^* := \operatorname{Hom}_K(M,K)$ of M. Here we view K as a horizontal differential module over K as explained in Example 5.5.2. Writing $\langle\phi, f\rangle := \phi(f) \in K$ for $\phi \in M^*$ and $f \in M$, we have the identity

$$\text{(5.5.1)} \qquad \partial\langle\phi, f\rangle = \langle\partial\phi, f\rangle + \langle\phi, \partial f\rangle.$$

The natural K-linear map $f \mapsto \langle -, f\rangle : M \to M^{**}$ is a morphism of $K[\partial]$-modules, and if M is a differential module over K, then this morphism is an isomorphism. In the rest of this subsection, we let $M \neq \{0\}$ be a differential module over K and $e_0, \dots, e_{n-1}$ be a basis of M, and we let A be the matrix associated to M with respect to $e_0, \dots, e_{n-1}$, so $A = (a_{ij})_{0 \leqslant i \leqslant n-1,\, 0 \leqslant j \leqslant n-1}$ with $\partial(e_j) = -\sum_i a_{ij} e_i$ for all j. We also let A^* be the matrix associated to M^* with respect to the basis $e_0^*, \dots, e_{n-1}^*$ of M^* dual to $e_0, \dots, e_{n-1}$.

LEMMA 5.5.5. *$e_0, \dots, e_{n-1}$ are horizontal iff $e_0^*, \dots, e_{n-1}^*$ are horizontal.*

PROOF. Let i and j range over $\{0, \dots, n-1\}$. From (5.5.1) we obtain

$$\text{(5.5.2)} \qquad \langle\partial e_i^*, e_j\rangle + \langle e_i^*, \partial e_j\rangle = \partial\langle e_i^*, e_j\rangle = \partial\delta_{ij} = 0,$$

where δ_{ij} is the Kronecker delta. If each e_j is horizontal, then for each i,

$$\langle\partial e_i^*, e_j\rangle = -\langle e_i^*, \partial e_j\rangle = -\langle e_i^*, 0\rangle = 0 \quad \text{for each } j,$$

so $\partial e_i^* = 0$. This shows the forward direction, and the converse goes likewise. □

By the previous lemma, if M is horizontal, then so is M^*.

LEMMA 5.5.6. $A^* = -A^{\mathrm{t}}$.

PROOF. By (5.5.2) we obtain for $j, k = 0, \dots, n-1$,

$$\langle\partial e_j^*, e_k\rangle = -\langle e_j^*, \partial e_k\rangle = \sum_i a_{ik}\langle e_j^*, e_i\rangle = a_{jk}$$

and thus $\partial e_j^* = \sum_i a_{ji} e_i^*$ as required by the lemma. □

The matrix differential equation $y' = A^*y$ is called the **adjoint equation** of the matrix differential equation $y' = Ay$. In the next lemma, given $L \in K[\partial]$, we let $L^* \in K[\partial]$ denote the adjoint of L as defined in Section 5.1.

LEMMA 5.5.7. *Suppose e_0 is a cyclic vector of M and $e_i = \partial^i e$ for $i = 0, \dots, n-1$. Let $L \in K[\partial]$ be monic of order n with $Le = 0$, $e := e_0$. Then $e^* := e^*_{n-1}$ is a cyclic vector of M^* and $L^*e^* = 0$.*

PROOF. We have $L = \partial^n + \sum_{i=1}^n a_{n-i}\partial^{n-i}$ with $a_0, \dots, a_{n-1} \in K$. Recall that $\partial^* = -\partial$, so $L^* = (\partial^*)^n + \sum_{i=1}^n (\partial^*)^{n-i} a_{n-i}$. By the previous lemma, $A^* = A_L$, so

$$\partial e_i^* + e_{i-1}^* = a_i e_{n-1}^* \qquad \text{for } i = 0, \dots, n-1,$$

where $e_{-1}^* := 0$. These identities give $M^* = Ke^* + K\partial e^* + \cdots + K\partial^{n-1}e^*$, so e^* is a cyclic vector of M^*. By induction on i they also yield

$$\Big((\partial^*)^i + \sum_{j=1}^{i} (\partial^*)^{i-j} a_{n-j}\Big)e^* = e^*_{n-i-1} \qquad \text{for } i = 0, \dots, n.$$

Taking $i = n$ we obtain $L^*e^* = 0$ as claimed. □

COROLLARY 5.5.8. *Let $L \in K[\partial]$ be monic of order $n \geqslant 1$. Then*

$$(K[\partial]/K[\partial]L)^* \cong K[\partial]/K[\partial]L^* \cong M_{A_L},$$

as differential modules over K.

PROOF. Set $M := K[\partial]/K[\partial]L$ and $e := 1 + K[\partial]L \in M$. Then M has matrix $-A_L^{\mathrm{t}}$ with respect to the basis $e, \partial e, \dots, \partial^{n-1}e$, so M^* has matrix A_L with respect to the dual basis, and thus $M^* \cong M_{A_L}$. By the lemma above, $M^* \cong K[\partial]/K[\partial]L^*$. □

COROLLARY 5.5.9. *If $C \neq K$ and $n \geqslant 1$, then any matrix differential equation $y' = Ay$, where A is an $n \times n$ matrix over K is equivalent to a matrix differential equation $y' = A_L y$ where $L \in K[\partial]$ is monic of order n.*

PROOF. Assume $C \neq K$ and A is an $n \times n$ matrix over K, $n \geqslant 1$. By Corollary 5.5.3 we have $M_A \cong K[\partial]/K[\partial]L^*$ where $L \in K[\partial]$ is monic of order n; then $M_A \cong M_{A_L}$ by the preceding corollary. □

In the next subsection we use the following *lemma on integrating factors*. This concerns the situation where $e_0, \dots, e_{n-1}$ are horizontal, $b \in M$ is given, and we wish to integrate b, that is, determine $f \in M$ such that $\partial f = b$. This can be done if we can integrate the "Fourier coefficients" $\langle e_i^*, b\rangle$ of b:

LEMMA 5.5.10. *Let $e_0, \dots, e_{n-1}$ be horizontal, and $b \in M$. Suppose $a_i \in K$ are such that $a_i' = \langle e_i^*, b\rangle$, for $i = 0, \dots, n-1$. Then $f := \sum_i a_i e_i$ satisfies $\partial f = b$.*

PROOF. From $\partial e_i = 0$ we get

$$\partial f = \sum_i \partial a_i e_i = \sum_i (a_i' e_i + a_i \partial e_i) = \sum_i \langle e_i^*, b\rangle e_i = b. \qquad \square$$

Fundamental matrices

In this subsection n ranges over $\mathbb{N}^{\geqslant 1}$. Let A be an $n \times n$ matrix over K. Note that if F is an $n \times n$ matrix over K, then $F' = AF$ iff the columns of F lie in $\operatorname{sol}(A)$, and in this case, $F \in \operatorname{GL}_n(K)$ iff the columns of F form a basis of the C-linear space $\operatorname{sol}(A)$. Let R be a differential ring extension of K. A matrix $F \in \operatorname{GL}_n(R)$ is called a **fundamental matrix** for the differential equation $y' = Ay$ if $F' = AF$. We say that R **contains a fundamental matrix for** $y' = Ay$ if $\operatorname{GL}_n(R)$ contains one.

LEMMA 5.5.11. *Let $F, G \in \operatorname{GL}_n(R)$ be fundamental matrices for $y' = Ay$. Then $F^{-1}G$ lies in the subgroup $\operatorname{GL}_n(C_R)$ of $\operatorname{GL}_n(R)$. Thus the set of all fundamental matrices for $y' = Ay$ in $\operatorname{GL}_n(R)$ is $F\operatorname{GL}_n(C_R)$.*

PROOF. Set $P := F^{-1}G$; then $FP = G$, so

$$AG = G' = F'P + FP' = AFP + FP' = AG + FP',$$

hence $P' = 0$, and thus P has all its entries in C_R. Likewise, the inverse $G^{-1}F$ of P has all its entries in C_R, so $P \in \operatorname{GL}_n(C_R)$. □

LEMMA 5.5.12. *Let $F \in \operatorname{GL}_n(R)$. Then F is a fundamental matrix for $y' = Ay$ iff $G := (F^{\mathrm{t}})^{-1}$ is a fundamental matrix for the adjoint equation $y' = A^*y$.*

PROOF. First note that $G \in \operatorname{GL}_n(R)$. We have $G' = -G(F')^{\mathrm{t}}G$, and a straightforward computation now shows that $G' = A^*G \iff F' = AF$. □

LEMMA 5.5.13. *Let $F \in \operatorname{GL}_n(R)$ be a fundamental matrix for $y' = Ay$. Then*

$$(\det F)' = \operatorname{tr} A \cdot \det F.$$

PROOF. Let i, j range over $\{1, \dots, n\}$, and let $f_j \in R^n$ be the jth row of F. By expanding the determinant and the product formula:

$$(\det F)' = \sum_i \det F_i \qquad \text{where } F_i = \begin{pmatrix} f_1 \\ \vdots \\ f_i' \\ \vdots \\ f_n \end{pmatrix}.$$

We have $f_i' = \sum_j a_{ij} f_j$ since $F' = AF$. Subtracting from the ith row of F_i the linear combination $\sum_{j \neq i} a_{ij} f_j$ replaces therefore the ith row of F_i by $a_{ii} f_i$, and so $\det F_i = a_{ii} \det F$. Hence $(\det F)' = \sum_i a_{ii} \det F = \operatorname{tr} A \cdot \det F$. □

Consider a monic operator

$$L = \partial^n + a_{n-1}\partial^{n-1} + \cdots + a_0 \in K[\partial] \qquad (a_0, \dots, a_{n-1} \in K),$$

with companion matrix A_L. The map

$$f \mapsto \left(f, f', \dots, f^{(n-1)}\right)^{\mathrm{t}} : \ \ker_R L \to \operatorname{sol}_R(A_L) \tag{5.5.3}$$

is an isomorphism of C_R-modules. Thus, given an $n \times n$ matrix F over R, we have: F is a fundamental matrix for $y' = A_L y$ iff there are $f_1, \dots, f_n \in \ker_R L$ such that $F = \mathrm{Wr}(f_1, \dots, f_n)$ and $\mathrm{wr}(f_1, \dots, f_n) \in R^\times$.

LEMMA 5.5.14. *Suppose $C \neq K$ and R contains a fundamental matrix for each equation $y' = A_L y$ where $L \in K[\partial]$ is monic of order n. Then R contains a fundamental matrix for each equation $y' = Ay$, where A is an $n \times n$ matrix over K.*

PROOF. Let A be an $n \times n$ matrix over K. Corollary 5.5.9 provides a monic $L \in K[\partial]$ of order n such that $y' = Ay$ is equivalent to $y' = A_L y$. Take $P \in \mathrm{GL}_n(K)$ such that $P \operatorname{sol}_R(A_L) = \operatorname{sol}_R(A)$, and let $F \in \mathrm{GL}_n(K)$ be a fundamental matrix for $y' = A_L y$. Then $PF \in \mathrm{GL}_n(R)$ is a fundamental matrix for $y' = Ay$. □

In the rest of this subsection we assume that our differential ring extension R of K is an integral domain whose ring of constants C_R is a field that equals the field of constants of the differential fraction field of R. Let $L \in K[\partial]$ be monic of order n as displayed above. Then $\ker_R L$ is a C_R-linear subspace of R and $\dim_{C_R} \ker_R L \leqslant n$. Moreover, by Lemma 4.1.13, $y_1, \dots, y_n \in R$ are C_R-linearly independent iff $\mathrm{wr}(y_1, \dots, y_n) \neq 0$. In particular, if $\mathrm{Wr}(y_1, \dots, y_n)$ is a fundamental matrix for $y' = A_L y$, then $y_1, \dots, y_n$ is a basis of the C_R-linear space $\ker_R L$. The following lemma contains a partial converse.

LEMMA 5.5.15. *Suppose that* $\ker_R(\partial - a_{n-1}) \neq \{0\}$. *Let* $y_1, \dots, y_n \in \ker_R L$ *be such that* $w := \mathrm{wr}(y_1, \dots, y_n) \neq 0$. *Then* $w \in R^\times$.

PROOF. By Lemma 4.1.17 we have $w' + a_{n-1} w = 0$; so $w^{-1} \in \mathrm{Frac}(R)$ satisfies $(w^{-1})' - a_{n-1} w^{-1} = 0$. By assumption, we also have $v \in R^{\neq}$ with $v' - a_{n-1} v = 0$. Hence $w^{-1} = cv$ for some $c \in C_R^\times$; in particular, $w \in R^\times$. □

COROLLARY 5.5.16. *Suppose $C \neq K$, and let $r \in \mathbb{N}^{\geqslant 1}$. The following are equivalent:*

(i) *R contains a fundamental matrix for $y' = Ay$, for every $n \leqslant r$ and $n \times n$ matrix A over K;*

(ii) $\dim_{C_R} \operatorname{sol}_R A = n$ *for every $n \leqslant r$ and $n \times n$ matrix A over K;*

(iii) $\dim_{C_R} \ker_R L = \operatorname{order} L$ *for every $L \in K[\partial]^{\neq}$ of order at most r;*

(iv) *R contains a fundamental matrix for $y' = A_L y$, for every monic $L \in K[\partial]$ of positive order at most r.*

PROOF. We clearly have (i) ⇒ (ii). The implication (ii) ⇒ (iii) is immediate from the isomorphism (5.5.3), (iii) ⇒ (iv) holds by Lemma 5.5.15, and (iv) ⇒ (i) by Lemma 5.5.14. (Only the last implication used the hypothesis $C \neq K$.) □

Next an analogue of Corollary 5.4.2 for homogeneous linear differential equations:

COROLLARY 5.5.17. *Let $r \in \mathbb{N}^{\geqslant 1}$. The following are equivalent:*

(i) *K contains a fundamental matrix for $y' = Ay$, for every $n \leqslant r$ and $n \times n$ matrix A over K;*

(ii) $\dim_C \operatorname{sol} A = n$ *for every $n \leqslant r$ and $n \times n$ matrix A over K;*

(iii) *K is r-pv-closed;*

(iv) *K contains a fundamental matrix for $y' = A_L y$, for every monic $L \in K[\partial]$ of positive order at most r;*

(v) *every differential module M over K with* $\dim_K M \leqslant r$ *is horizontal.*

PROOF. The equivalence of (i)–(iv) follows from Corollary 5.5.16. The equivalence of (iv) and (v) involves Lemma 5.4.8. (Each of (i)–(v) implies $C \neq K$.) □

Corollary 5.5.19 below is an analogue of Corollary 5.4.3, and follows easily from Corollary 5.5.17 and the next lemma:

LEMMA 5.5.18. *Suppose $C \neq K$. Let E be a differential subring of R containing K such that E is an algebraic field extension of K. Suppose R contains a fundamental matrix for $y' = Ay$, for every n and $n \times n$ matrix A over K. Then R also contains a fundamental matrix for $y' = Ay$, for every n and $n \times n$ matrix A over E.*

PROOF. Let $n \geqslant 1$ and let A be an $n \times n$ matrix over E; by Corollary 5.5.16 it suffices to show that $\dim_{C_R} \operatorname{sol}_R A = n$. We may assume that E is of finite degree over K, say with basis $e_1, \dots, e_m$ over K. Take the $mn \times mn$ matrix $A^\diamond$ over K from the proof of Corollary 5.4.3. As in that proof we see that for all $z \in \operatorname{sol}_R A^\diamond$ we have $Bz \in \operatorname{sol}_R A$, where B is the $n \times mn$ matrix

$$B = \begin{pmatrix} e_1 & e_2 & \cdots & e_m & & & & \\ & & & & \ddots & & & \\ & & & & & e_1 & e_2 & \cdots & e_m \end{pmatrix}$$

over E. Let $Z \in \mathrm{GL}_{mn}(R)$ be a fundamental matrix for $z' = A^\diamond z$. Then the $n \times mn$ matrix BZ over R has rank n over $\operatorname{Frac}(R)$, and its mn columns lie in $\operatorname{sol}_R A$, so $\operatorname{sol}_R A$ has dimension at least n as a vector space over C_R. By Lemma 5.4.8 this dimension is also at most n, so equal to n. □

The situations considered in the next two corollaries, with $E \neq K$ in the first one, actually occur, by a fact mentioned in the *Notes and comments* to this section.

COROLLARY 5.5.19. *If K is pv-closed and E is a differential field extension of K and algebraic over K, then E is pv-closed.*

From this corollary we now obtain a variant of Lemma 5.1.31:

COROLLARY 5.5.20. *Suppose K is pv-closed and C is real closed. Then K is real closed.*

PROOF. The differential field extension $E = K[i]$, where $i^2 = -1$, is algebraic over K and has constant field $C_E = C[i]$. By the previous corollary, E is pv-closed, and since C is real closed, C_E is algebraically closed. Hence E is algebraically closed by Lemma 5.1.31 and so K is real closed by Theorem 3.5.4. □

Let A be an $n \times n$ matrix over K, and identify the K-linear space $M = M_A$ with its dual M^* via the isomorphism $f \mapsto \langle f, - \rangle : M \to M^*$, where $\langle\ ,\ \rangle$ denotes the usual inner product $M \times M \to K$ given by

$$\langle f, g \rangle := \sum_i f_i g_i \qquad (f = (f_1, \dots, f_n),\ g = (g_1, \dots, g_n) \in K^n)$$

on $M = K^n$. Let $F \in \mathrm{GL}_n(K)$ be a fundamental matrix of $y' = Ay$, with columns $f_1, \dots, f_n$, and let $G = (F^{\mathrm{t}})^{-1}$, with columns $g_1, \dots, g_n$. Then $\langle g_i, f_j \rangle = \delta_{ij}$ for $i, j = 1, \dots, n$, so $g_1, \dots, g_n$ is the basis of M^* dual to the horizontal basis $f_1, \dots, f_n$ of M. The next result goes under the name of *variation of constants.*

LEMMA 5.5.21. *Let $b \in K^n$ be a column vector, and suppose $a_i \in K$ satisfies $a_i' = \langle g_i, b \rangle$, for $i = 1, \dots, n$. Then $f := \sum_i a_i f_i \in K^n$ satisfies $f' = Af + b$.*

This is immediate from Lemma 5.5.10. By Lemma 5.1.30, if K is pv-closed, then K is linearly surjective. The following corollary complements this fact: if one is already given a basis of the kernel of a monic differential operator $L \in K[\partial]^{\neq}$, then solving inhomogeneous linear differential equations $L(y) = b$ reduces to integration.

COROLLARY 5.5.22. *Let $L \in K[\partial]^{\neq}$ be monic of order n and suppose $y_1, \dots, y_n$ is a basis of* $\ker L$. *Let $(z_1, \dots, z_n)$ be the bottom row of the matrix $(W^{\mathrm{t}})^{-1}$, where $W = \mathrm{Wr}(y_1, \dots, y_n)$. Let $b \in K$, and suppose $a_i \in K$ satisfies $a_i' = bz_i$, for $i = 1, \dots, n$. Then $y := \sum_i a_i y_i \in K$ satisfies $L(y) = b$.*

PROOF. Apply the previous lemma with the companion matrix A_L of L and the column vector $(0, \dots, 0, b)^{\mathrm{t}}$ in place of A and b. □

EXAMPLE. Let $L = \partial - f$ where $f \in K$, let $y_1 \in K^\times$ be such that $y_1' = fy_1$, and let $b \in K$. Then any $a \in K$ with $a' = b/y_1$ yields $L(y) = b$ for $y := ay_1$.

Notes and comments

Lemma 5.4.1 is from [94, pp. 91–94]. Corollary 5.5.3 was apparently first shown by Loewy [268] for the case where K is a field of meromorphic functions, and the argument used here (via Corollary 5.3.5) was suggested in [6]; see [81] for a historical discussion. Lemma 5.5.7 is credited to O. Gabber in [211, Lemma (1.5.3)]. The notion of adjoint equation and Lemmas 5.5.12, 5.5.21, and Corollary 5.5.22 stem from the classical literature on linear differential equations, see [320, Chapter III, §10].

By [95, 208], any real closed K has a pv-closed differential field extension L with $C_L = C$; note that any such L is real closed by Corollary 5.5.20 applied to L.

5.6 LINEAR DIFFERENTIAL OPERATORS IN THE PRESENCE OF A VALUATION

In this section K is a valued differential field with valuation v and derivation ∂, such that $\partial o \subseteq o$. We let a, b (possibly with subscripts) range over K.

Note that ∂ induces a derivation on the residue field $\boldsymbol{k} = \mathcal{O}/o$ which we also denote by ∂, so $\partial(a+o) = \partial(a) + o$ for $a \in \mathcal{O}$. We have $\mathcal{O}[\partial]$ as a subring of $K[\partial]$, with a ring homomorphism

$$\sum_i a_i\partial^i \mapsto \sum_i (a_i + o)\partial^i \,:\, \mathcal{O}[\partial] \to \boldsymbol{k}[\partial].$$

We extend v to a valuation on the additive group of $K[\partial]$ by setting

$$v(A) := \min_i v(a_i) \in \Gamma_\infty \text{ for } A = \sum_i a_i\partial^i \in K[\partial].$$

Note that then $v(aA) = va + v(A)$. Also $v(Aa)$ depends, for fixed $A \in K[\partial]$, only on $\gamma = va$, by Lemma 4.5.1(ii), but generally in a more complicated way; thus we can define $v_A(\gamma) := v(Aa)$ for $\gamma = va$.

LEMMA 5.6.1. *Let $A, B \in K[\partial]^{\neq}$ and $\gamma \in \Gamma$. Then:*

(i) *if $v(B) = 0$, then $v(AB) = v(A)$;*

(ii) $v_{AB}(\gamma) = v_A\big(v_B(\gamma)\big)$.

PROOF. For (i) we use $v(aA) = va + v(A)$ to reduce to the case $v(A) = v(B) = 0$. Then $v(AB) = 0$ follows by applying the homomorphism $\mathcal{O}[\partial] \to \boldsymbol{k}[\partial]$. For (ii), take $y \in K^\times$ with $vy = \gamma$, so $v_B(\gamma) = v(By)$ and $v_{AB}(\gamma) = v(ABy)$. Now $By = b\widehat{B}$ with $vb = v(By)$ and $v(\widehat{B}) = 0$. Hence $ABy = Ab\widehat{B}$, so by (i):

$$v_{AB}(\gamma) = v(ABy) = v(Ab\widehat{B}) = v(Ab) = v_A(vb) = v_A\big(v(By)\big) = v_A\big(v_B(\gamma)\big)$$

as claimed. □

COROLLARY 5.6.2. *If K is r-linearly closed, then so is $\boldsymbol{k}$.*

PROOF. Let $A \in \mathcal{O}[\partial]$ be monic, and $A = BD$ with $B, D \in K[\partial]$ and B, D of order $\geqslant 1$. Take $a \in K^\times$ with $va = -vD$. Then $A = (Ba^{-1})\cdot(aD)$ with $v(aD) = 0$, and thus $v(Ba^{-1}) = vA = 0$. Therefore, if $A \in \mathcal{O}[\partial]$ is monic of order $\geqslant 1$ and its image in $\boldsymbol{k}[\partial]$ is irreducible, then A is irreducible in $K[\partial]$. The lemma is an easy consequence of this fact. □

COROLLARY 5.6.3. *Let $A, B \in K[\partial]$ be monic and $AB \in \mathcal{O}[\partial]$. Then $A, B \in \mathcal{O}[\partial]$.*

PROOF. By Lemma 5.6.1(ii) and Lemma 4.5.1(iii) we have

$$0 = v(AB) = v_{AB}(0) = v_A\big(v_B(0)\big) \leqslant v_A(0) \leqslant 0,$$

so $v(A) = v_A(0) = 0$, and hence $v(B) = v_B(0) = 0$ by Lemma 4.5.1(iii). □

For $A = \sum_i a_i \partial^i \in K[\partial]$ we have $a_0 = A(1)$ and we put

$$\operatorname{dwm}(A) := \min\{i : v(a_i) = v(A)\} \in \mathbb{N},$$

so $\operatorname{dwm}(A) = 0 \iff v(a_0) = v(A)$. Here is a special case of Lemma 4.5.9:

LEMMA 5.6.4. *Suppose that $\partial\mathcal{O} \subseteq \mathcal{o}$. Then, given any $A \in K[\partial]^{\neq}$, the quantity $\operatorname{dwm}(Ag)$ for $g \in K^\times$ depends only on vg.*

If $\partial\mathcal{O} \subseteq \mathcal{o}$, then we define for $A \in K[\partial]$, $A \neq 0$:

$$\operatorname{dwm}_A(\gamma) := \operatorname{dwm}(Ag), \quad \text{where } g \in K^\times,\ vg = \gamma.$$

LEMMA 5.6.5. *Suppose that $\partial\mathcal{O} \subseteq \mathcal{o}$. Let $A, B \in K[\partial]^{\neq}$. Then*

$$\operatorname{dwm}_{AB}(\gamma) = \operatorname{dwm}_A\big(v_B(\gamma)\big) + \operatorname{dwm}_B(\gamma) \qquad (\gamma \in \Gamma).$$

PROOF. The induced derivation on $\boldsymbol{k}$ is trivial, so $\boldsymbol{k}[\partial]$ is commutative. Therefore, if $v(A) = v(B) = 0$, then $\operatorname{dwm}(AB) = \operatorname{dwm}(A) + \operatorname{dwm}(B)$. With b, y, $\gamma = vy$, and $\widehat{B}$ as in the proof of part (ii) of Lemma 5.6.1, write $Ab = a\widehat{A}$ where $va = v_A(vb) = v_A\big(v_B(\gamma)\big)$ and $v(\widehat{A}) = 0$. Then $ABy = a\widehat{A}\widehat{B}$, so

$$\operatorname{dwm}_{AB}(\gamma) = \operatorname{dwm}(\widehat{A}\widehat{B}) = \operatorname{dwm}(\widehat{A}) + \operatorname{dwm}(\widehat{B}).$$

The claim now follows from $\operatorname{dwm}(\widehat{B}) = \operatorname{dwm}(b\widehat{B}) = \operatorname{dwm}(By) = \operatorname{dwm}_B(\gamma)$ and

$$\operatorname{dwm}(\widehat{A}) = \operatorname{dwm}(a\widehat{A}) = \operatorname{dwm}(Ab) = \operatorname{dwm}_A(vb) = \operatorname{dwm}_A\big(v_B(\gamma)\big). \qquad \square$$

Suppose that $C \subseteq \mathcal{O}$ and $A \in K[\partial]^{\neq}$. If $y_1, \dots, y_r \in \ker^{\neq} A$ and $y_1 \succ \cdots \succ y_r$, then $y_1, \dots, y_r$ are C-linearly independent. So $|v(\ker^{\neq} A)| \leqslant \dim_C \ker A$. Moreover:

LEMMA 5.6.6. *Suppose that $\mathcal{O} = C + \mathcal{o}$, and let $A \in K[\partial]^{\neq}$. Then*

(i) *$v(\ker^{\neq} A)$ is finite, of size $\dim_C \ker A$;*

(ii) *if $v(\ker^{\neq} A) = \{vy_1, \dots, vy_m\}$ where $y_1, \dots, y_m \in \ker^{\neq} A$, $vy_1 < \cdots < vy_m$, then $y_1, \dots, y_m$ is a basis of $\ker A$.*

PROOF. This follows from Lemma 2.3.13, in view of the fact that K with the valuation v is a Hahn space over C. $\square$

The assumptions $\partial\mathcal{O} \subseteq \mathcal{o}$ and $\mathcal{O} = C + \mathcal{o}$ of the lemmas above are satisfied if K is *differential-valued* as defined in Section 9.1.

For $A \in K[\partial]^{\neq}$, $A = \sum_i a_i \partial^i$, we define

$$\operatorname{dwt}(A) := \max\{i : v(a_i) = v(A)\}.$$

Then the analogues of Lemmas 5.6.4 and 5.6.5 hold without the assumption $\partial\mathcal{O} \subseteq \mathcal{o}$: $\operatorname{dwt}(Ag)$ for $g \in K^\times$ depends only on vg, and setting $\operatorname{dwt}_A(\gamma) := \operatorname{dwt}(Ag)$ when $\gamma = vg$, $g \in K^\times$, we have $\operatorname{dwt}_{AB}(\gamma) = \operatorname{dwt}_A\big(v_B(\gamma)\big) + \operatorname{dwt}_B(\gamma)$ for $A, B \in K[\partial]^{\neq}$ and $\gamma \in \Gamma$. (Same proofs.)

Exceptional values

In this subsection we let γ range over Γ and y range over $K^\times$. Let $A \in K[\partial]^{\neq}$, $A = a_0 + a_1\partial + \cdots + a_n\partial^n$, $a_0, \ldots, a_n \in K$, so $A(1) = a_0$. Recall that $\operatorname{dwm}(A) = \min\{i : v(a_i) = v(A)\}$, so $\operatorname{dwm}(A) > 0$ is equivalent to $A(1) \prec A$. We call γ an **exceptional value** for A if there is a y such that $vy = \gamma$ and $A(y) \prec Ay$, and we let $\mathscr{E}(A)$ be the set of exceptional values for A, in other words

$$\mathscr{E}(A) := \{vy : A(y) \prec Ay\} = \{vy : \operatorname{dwm}(Ay) > 0\}.$$

For $a \in K^\times$ we have

$$\mathscr{E}(aA) = \mathscr{E}(A), \qquad \mathscr{E}(Aa) = \mathscr{E}(A) - va.$$

LEMMA 5.6.7. $v(\ker^{\neq} A) \subseteq \mathscr{E}(A)$.

PROOF. If $g \in \ker^{\neq} A$, then $A(g) = 0 \prec Ag$, so $vg \in \mathscr{E}(A)$. □

Lemma 5.6.7 is useful in solving a differential equation $A(y) = 0$: knowing $\mathscr{E}(A)$ helps to pin down the valuation of the potential solutions.

LEMMA 5.6.8. *Let $f_1, f_2 \in K^\times$ with $v(f_2) \notin \mathscr{E}(A)$. Then:*

$$f_1 \asymp f_2 \Rightarrow A(f_1) \asymp A(f_2), \qquad f_1 \prec f_2 \Rightarrow A(f_1) \prec A(f_2),$$

and thus $f_1 \sim f_2 \Rightarrow A(f_1) \sim A(f_2)$.

PROOF. If $f_1 \asymp f_2$, then $A(f_1) \asymp Af_1 \asymp Af_2 \asymp A(f_2)$, and if $f_1 \prec f_2$, then $A(f_1) \preccurlyeq Af_1 \prec Af_2 \asymp A(f_2)$. □

If $\partial\mathcal{O} \subseteq \scriptstyle\mathcal{O}$, then $\mathscr{E}(A) = \{\gamma : \operatorname{dwm}_A(\gamma) > 0\}$, so for $B \in K[\partial]^{\neq}$, by Lemma 5.6.5:

$$\mathscr{E}(AB) = v_B^{-1}(\mathscr{E}(A)) \cup \mathscr{E}(B).$$

Note that $\Gamma^\flat := \{va : a' \prec a\}$ is a subgroup of Γ. By Lemma 4.5.10 we have:

LEMMA 5.6.9. *If $\partial\mathcal{O} \subseteq \scriptstyle\mathcal{O}$, then $\mathscr{E}(A)$ is a union of cosets of $\Gamma^\flat$.*

EXAMPLE. Suppose $\partial\mathcal{O} \subseteq \scriptstyle\mathcal{O}$ and $\operatorname{order}(A) = 1$. Then

$$vy \in \mathscr{E}(A) \iff (a_0/a_1) + y^\dagger \prec 1.$$

Hence either $\mathscr{E}(A) = \emptyset$ or $\mathscr{E}(A)$ is a coset of $\Gamma^\flat$.

As a special case, the valued differential field $K = C(x)$ from Example (3) at the beginning of Section 4.4 satisfies $y^\dagger \prec 1$ for all y, so $\partial\mathcal{O} \subseteq \scriptstyle\mathcal{O}$. For this K and $a \in K$, if $a \prec 1$, then $\mathscr{E}(\partial + a) = \Gamma^\flat$, and if $a \succcurlyeq 1$, then $\mathscr{E}(\partial + a) = \emptyset$.

To indicate the dependence of $\mathscr{E}(A)$ on K we write it as $\mathscr{E}_K(A)$. If L is a valued differential field extension of K with small derivation, then $\mathscr{E}_K(A) \subseteq \mathscr{E}_L(A)$, and if in addition $\partial\mathcal{O}_L \subseteq \scriptstyle\mathcal{O}_L$, then $\mathscr{E}_K(A) = \mathscr{E}_L(A) \cap \Gamma$ by Lemma 5.6.4.

Linear surjectivity in the presence of a valuation

Let $A \in K[\partial]^{\neq}$. Recall that if $a \in K^{\times}$ and $\operatorname{dwm}(Aa) = 0$, then

$$v\big(A(a)\big) = v(Aa) = v_A(va).$$

We call A **neatly surjective** if for all $b \in K^{\times}$ there is $a \in K^{\times}$ such that $A(a) = b$ and $v_A(va) = vb$; note that then $A\colon K \to K$ and $v_A\colon \Gamma \to \Gamma$ are surjective. In Chapter 6 we prove that $v_A\colon \Gamma \to \Gamma$ *is* indeed surjective, but this is a rather difficult result, so we prefer to add it explicitly as an assumption in some results below.

LEMMA 5.6.10. *Neat surjectivity has the following basic properties:*

(i) *each $\phi \in K^{\times} \subseteq K[\partial]^{\neq}$ is neatly surjective;*

(ii) *if $A, B \in K[\partial]^{\neq}$ are neatly surjective, then so is AB;*

(iii) *if K is linearly closed and each $A \in K[\partial]^{\neq}$ of order 1 is neatly surjective, then every $A \in K[\partial]^{\neq}$ is neatly surjective;*

(iv) *if $A \in K[\partial]^{\neq}$, $A(K) = K$, and* $\operatorname{dwm}(Aa) = 0$ *for all $a \in K^{\times}$, then A is neatly surjective.*

LEMMA 5.6.11. *If K is 1-linearly surjective, ∂ is neatly surjective, and $\mathcal{O} \subseteq (K^{\times})^{\dagger}$, then every $A \in K[\partial]$ of order 1 is neatly surjective (so $\boldsymbol{k}$ is 1-linearly surjective).*

PROOF. Let $A = \partial + \phi \in K[\partial]$, and suppose $A(K) = K$. If $\operatorname{dwm}(Aa) = 0$ for all $a \in K^{\times}$, then A is neatly surjective by Lemma 5.6.10(iv). So let $\operatorname{dwm}(Aa) > 0$, with $a \in K^{\times}$. Then $a^{-1}Aa = \partial + (a^{\dagger} + \phi)$ with $a^{\dagger} + \phi \in \mathcal{O}$, and so, assuming $\mathcal{O} \subseteq (K^{\times})^{\dagger}$, we have $a^{\dagger} + \phi = b^{\dagger}$ with $b \in K^{\times}$, hence $a^{-1}Aa = b^{-1}\partial b$. Assuming next that ∂ is neatly surjective, A is neatly surjective by Lemma 5.6.10(i)(ii). □

If every $A \in K[\partial]^{\neq}$ of order $\leqslant n$ is neatly surjective, then $\boldsymbol{k}$ is n-linearly surjective. We have the following approximate converse to this fact:

LEMMA 5.6.12. *Suppose $\boldsymbol{k}$ is n-linearly surjective, $A \in K[\partial]^{\neq}$ has order $\leqslant n$ and $v_A\colon \Gamma \to \Gamma$ is surjective. Then there is for each $b \in K^{\times}$ an $a \in K^{\times}$ such that $A(a) \sim b$ and $v_A(va) = vb$.*

PROOF. Let $b \in K^{\times}$ and take $\alpha \in \Gamma$ with $vb = v_A(\alpha)$. Take $\phi \in K^{\times}$ with $v\phi = \alpha$, so $B := b^{-1}A\phi$ satisfies $v(B) = 0$. Take $u \in K$ with $u \asymp 1$ such that $B(u) \sim 1$ (possible since $\boldsymbol{k}$ is n-linearly surjective). Then $A(u\phi) \sim b$, so $a = u\phi$ works. □

This suggests a way to "neatly" solve equations $A(y) = g$:

COROLLARY 5.6.13. *Assume K is spherically complete, $\boldsymbol{k}$ is n-linearly surjective, $A \in K[\partial]^{\neq}$ has order $\leqslant n$, and $v_A\colon \Gamma \to \Gamma$ is surjective. Then A is neatly surjective.*

PROOF. Let $g \in K^\times$; we wish to find $f \in K^\times$ such that $A(f) = g$ and $v_A(vf) = vg$. Take $f_0 \in K^\times$ with $A(f_0) \sim g$ and $v_A(vf_0) = vg$. If $A(f_0) = g$, we are done. Suppose $A(f_0) \neq g$. Then we take $y \in K^\times$ such that $A(y) \sim g - A(f_0) \prec g$ and $v_A(vy) = v(g - A(f_0))$ and set $f_1 := f_0 + y$. Then $g - A(f_1) \prec g - A(f_0)$, and $f_0 \sim f_1$. If $A(f_1) = g$, we are done, and otherwise we continue as before, with f_1 instead of f_0. In general, we have a sequence $(f_\lambda)_{\lambda<\rho}$ in $K^\times$, indexed by an ordinal $\rho > 0$, with f_0 as chosen initially, such that the following conditions hold:

(1) $v\big(g - A(f_\lambda)\big)$ is strictly increasing as a function of λ,

(2) $v_A\big(v(f_\mu - f_\lambda)\big) = v\big(g - A(f_\lambda)\big)$ for $\lambda < \mu < \rho$.

In particular, $f_\lambda \sim f_0$ for all λ. Consider first the case that $\rho = \nu + 1$ is a successor ordinal. If $A(f_\nu) = g$, we are done, so assume $A(f_\nu) \neq g$. The same way we got f_1 from f_0 we take $f_{\nu+1} \in K^\times$ with $g - A(f_{\nu+1}) \prec g - A(f_\nu)$ and $v_A\big(v(f_{\nu+1} - f_\nu)\big) = v\big(g - A(f_\nu)\big)$. Then the extended sequence $(f_\lambda)_{\lambda<\rho+1}$ has the above properties with $\rho + 1$ instead of ρ.

Suppose ρ is a limit ordinal. Then (f_λ) is a pc-sequence, so $f_\lambda \rightsquigarrow f_\rho \in K$. Then the extended sequence $(f_\lambda)_{\lambda<\rho+1}$ has the above properties with $\rho + 1$ instead of ρ. Eventually, this building process must result in an f as desired. □

Neat surjectivity is preserved under adjoining i, as we explain now. Assume -1 is not a square in the residue field $\boldsymbol{k}$. Then $K[i]$ is a valued differential field with valuation given by $v(a + bi) = \min(va, vb)$ for $a, b \in K$, so its valuation ring is $\mathcal{O} + \mathcal{O}i$, with maximal ideal $\sigma + \sigma i$ satisfying $\partial(\sigma + \sigma i) \subseteq \sigma + \sigma i$. The value group of $K[i]$ is again $v(K^\times) = \Gamma$, and:

LEMMA 5.6.14. *Let $A \in K[\partial]^{\neq}$. Then $A\colon K \to K$ is neatly surjective if and only if $A\colon K[i] \to K[i]$ is neatly surjective.*

PROOF. Assume first that $A\colon K \to K$ is neatly surjective. Let $f \in K[i]^\times$ and take $\gamma \in \Gamma$ such that $v_A(\gamma) = vf$. Then $f = g + hi$, with $g, h \in K$, and $vf = \min(vg, vh)$. Take $a, b \in K$ such that $A(a) = g$, $v_A(va) = vg$, $A(b) = h$, $v_A(vb) = vh$. Then $\gamma = \min(va, vb) = v(a + bi)$, and $A(a + bi) = f$. Thus $A\colon K[i] \to K[i]$ is neatly surjective.

Next, assume $A\colon K[i] \to K[i]$ is neatly surjective. Let $f \in K^\times$, and take $a, b \in K$ such that $A(a + bi) = f$ and $v_A(\gamma) = vf$ with $\gamma = \min(va, vb)$. Then $A(a) = f$ and $A(b) = 0$. If $va \leqslant vb$, then $va = \gamma$, and if $va > vb$, then $\gamma = v(a + b)$ and $A(a + b) = f$. □

Independence and diagonalization in the presence of a valuation

In the next two lemmas $Y = (Y_1, \dots, Y_n)$ is a tuple of distinct differential indeterminates, $n \geqslant 1$. These lemmas involve independence notions defined earlier in this chapter.

LEMMA 5.6.15. *Suppose $A_1, \dots, A_m$ are in $\mathcal{O}\{Y\}_1$ and their images $\overline{A}_1, \dots, \overline{A}_m$ in $\boldsymbol{k}\{Y\}_1$ are* d*-independent (over $\boldsymbol{k}$). Then $A_1, \dots, A_m$ are* d*-independent.*

PROOF. This follows easily from the definition of d-dependence in Section 5.4. □

LEMMA 5.6.16. *Let $n \geqslant 1$ and let A be an $n \times n$ matrix over $\mathcal{O}[\partial]$ such that its image $\overline{A}$ as an $n \times n$ matrix over $\boldsymbol{k}[\partial]$ has $\boldsymbol{k}[\partial]$-independent rows. Then there are $S, T \in \mathrm{GL}_n\big(\mathcal{O}[\partial]\big)$ such that $SAT = (D_{ij})$ with $D_{ij} \prec 1$ for all $i \neq j$ in $\{1, \dots, n\}$, and $D_{ii} \asymp 1$ for $i = 1, \dots, n$.*

PROOF. Diagonalize $\overline{A}$ by row and column operations, and lift each step to a row or column operation over $\mathcal{O}[\partial]$. □

Notes and comments

For archimedean Γ one can find Lemma 5.6.1(ii) and Lemma 5.6.5 in Section 1.6 of [343].

5.7 COMPOSITIONAL CONJUGATION

In this section K is a differential ring with derivation ∂ and $\phi \in K^\times$. We define K^ϕ to be the differential ring with the same underlying ring as K but with the derivation δ given by $\delta(a) = \phi^{-1} \cdot \partial(a)$ for $a \in K$, so $\delta = \phi^{-1}\partial$. This gives rise to the ring $K^\phi\{Y\}$ of differential polynomials over K^ϕ. Thus $K^\phi\{Y\}$ has the same underlying ring as $K\{Y\}$ but its derivation extends $\delta = \phi^{-1}\partial$. We denote this extended derivation also by δ, so $\delta(Y^{(n)}) = Y^{(n+1)}$ for all n. For a differential polynomial $P \in K^\phi\{Y\}$ written as an ordinary polynomial $P = p(Y, Y', \dots, Y^{(n)}) \in K\big[Y, Y', \dots, Y^{(n)}\big]$, we have $P(y) = p\big(y, \delta(y), \dots, \delta^n(y)\big)$ for $y \in K$.

Transformation formulas

In order to relate the differential rings $K\{Y\}$ and $K^\phi\{Y\}$ we take δ as the element $\phi^{-1}\partial$ of $K[\partial]$ and express the powers of ∂ as linear combinations of powers of δ in the ring $K[\partial]$, with $\phi^{(k)} = \partial^k(\phi)$:

$$\begin{aligned}
\partial^1 &= \phi\delta \\
\partial^2 &= \phi^2\delta^2 + \phi'\delta \\
\partial^3 &= \phi^3\delta^3 + 3\phi\phi'\delta^2 + \phi''\delta \\
\partial^4 &= \phi^4\delta^4 + 6\phi^2\phi'\delta^3 + \big(4\phi\phi'' + 3(\phi')^2\big)\delta^2 + \phi^{(3)}\delta \\
&\ \ \vdots \\
\partial^n &= F^n_n(\phi)\delta^n + F^n_{n-1}(\phi)\delta^{n-1} + \dots + F^n_1(\phi)\delta
\end{aligned}$$

where the differential polynomials $F^n_k(X) \in \mathbb{Q}\{X\} \subseteq K\{X\}$ have nonnegative integer coefficients and are independent of K and ϕ: $F^n_n(X) = X^n$ and $F^n_1(X) = X^{(n-1)}$ for $n \geqslant 1$, and we have the recursion formula

$$F^{n+1}_k(X) = F^n_k(X)' + XF^n_{k-1}(X) \qquad (k = 1, \dots, n) \tag{5.7.1}$$

where $F_0^n := 0$ for $n \geqslant 1$, to make the last identity true for $k = 1$. For later use we also set $F_0^0 := 1$. So F_k^n is homogeneous of degree k and isobaric of weight $n - k$ for $0 \leqslant k \leqslant n$, and of order $n - k$ for $1 \leqslant k \leqslant n$. The identities relating the powers of ∂ and δ suggest that we consider the ring morphism

$$P(Y) \mapsto P^\phi(Y) \colon K\{Y\} \to K^\phi\{Y\}$$

that is the identity on $K[Y]$ and sends $Y^{(n)}$, for each $n \geqslant 1$, to

$$(Y^{(n)})^\phi := F_n^n(\phi)Y^{(n)} + F_{n-1}^n(\phi)Y^{(n-1)} + \cdots + F_1^n(\phi)Y'.$$

In particular, $Y^\phi = Y$, $(Y')^\phi = \phi Y'$, $(Y'')^\phi = \phi^2 Y'' + \phi' Y'$. Note that $P \mapsto P^\phi$ is a ring isomorphism: it maps each subring $K\big[Y, \dots, Y^{(n)}\big]$ of $K\{Y\}$ isomorphically onto the same subring of $K^\phi\{Y\}$. We call K^ϕ the **compositional conjugate of K by ϕ** and P^ϕ the **compositional conjugate of P by ϕ.**

LEMMA 5.7.1. *Let $P \in K\{Y\}$ and $y \in K$. Then*

$$P(y) = P^\phi(y), \qquad (\partial P)^\phi = \phi \cdot \delta(P^\phi),$$
$$(P^\phi)_{+y} = (P_{+y})^\phi, \qquad (P^\phi)_{\times y} = (P_{\times y})^\phi.$$

PROOF. For $P = Y^{(n)}$ these identities follow from the definitions, using induction on n if necessary. The general case reduces easily to this special case. □

In view of the last two identities we let P_{+y}^ϕ denote both $(P^\phi)_{+y}$ and $(P_{+y})^\phi$, and let $P_{\times y}^\phi$ denote both $(P^\phi)_{\times y}$ and $(P_{\times y})^\phi$.

If θ is also a unit of K, then $K^{\phi\theta} = (K^\phi)^\theta$, and we have ring isomorphisms

$$\begin{array}{rcccc} P \mapsto P^\phi & : & K\{Y\} & \longrightarrow & K^\phi\{Y\} \\ Q \mapsto Q^\theta & : & K^\phi\{Y\} & \longrightarrow & K^{\phi\theta}\{Y\} \\ P \mapsto P^{\phi\theta} & : & K\{Y\} & \longrightarrow & K^{\phi\theta}\{Y\} \end{array}$$

and the last map is the composition of the preceding two:

$$P^{\phi\theta} = (P^\phi)^\theta, \quad P \in K\{Y\}.$$

For $P = Y^{(n)}$ this is easily checked by an induction on n, using the second identity of Lemma 5.7.1; the identity for all P is then an easy consequence. Note also that for $\phi = 1$ we have $K^\phi = K$ and $P^\phi = P$.

Another induction on n using the second identity of Lemma 5.7.1 yields

$$\left(\partial^n Q\right)^\phi = P^\phi\left(Q^\phi\right) \text{ for } P = Y^{(n)} \in K\{Y\} \text{ and all } Q \in K\{Y\}.$$

It follows that $P(Q)^\phi = P^\phi(Q^\phi)$ for all $P, Q \in K\{Y\}$. Thus we have a ring isomorphism

$$A \mapsto A^\phi : K[\partial] \to K^\phi[\delta], \quad a^\phi = a \text{ for } a \in K, \quad \partial^\phi = \phi\delta.$$

It sends ∂^n for $n \geqslant 1$ to $F_n^n(\phi)\delta^n + F_{n-1}^n(\phi)\delta^{n-1} + \cdots + F_1^n(\phi)\delta$, and $A^\phi(y) = A(y)$ for all $A \in K[\partial]$, $y \in K$. We identify the rings $K[\partial]$ and $K^\phi[\delta]$ via this isomorphism. Thus for $A = a_0 + a_1\partial + \cdots + a_r\partial^r \in K[\partial]$ with $a_0, \dots, a_r \in K$ we have

$$\begin{aligned} A = A^\phi &= b_0 + b_1\delta + \cdots + b_r\delta^r \in K^\phi[\delta], \text{ where} \\ b_i &= \sum_{j=i}^{r} F_i^j(\phi)a_j \quad \text{for } i = 0, \dots, r. \end{aligned}$$

Here are some easy consequences:

LEMMA 5.7.2. *Let K be a differential field and $r \in \mathbb{N}^{\geqslant 1}$. If K is r-linearly closed (respectively r-linearly surjective, respectively r-pv-closed), then so is K^ϕ.*

The following identities for the powers ∂^n in the ring $K^\phi[\delta]$ will also be useful:

$$\begin{aligned} \partial^1 &= \phi\cdot\delta \\ \partial^2 &= \phi^2\cdot\delta^2 + \phi\cdot\delta(\phi)\cdot\delta \\ \partial^3 &= \phi^3\cdot\delta^3 + 3\phi^2\cdot\delta(\phi)\cdot\delta^2 + \big(\phi^2\cdot\delta^2(\phi) + \phi\cdot\delta(\phi)^2\big)\cdot\delta \\ &\ \vdots \\ \partial^n &= G_n^n(\phi)\cdot\delta^n + G_{n-1}^n(\phi)\cdot\delta^{n-1} + \cdots + G_1^n(\phi)\cdot\delta \end{aligned}$$

where the differential polynomials $G_k^n(X) \in \mathbb{Q}\{X\} \subseteq K^\phi\{X\}$ have nonnegative integer coefficients and are independent of K and ϕ: $G_n^n(X) = X^n$ for $n \geqslant 1$, and

$$(5.7.2) \qquad G_k^{n+1} = X\cdot\big(\delta(G_k^n) + G_{k-1}^n\big) \qquad (k = 1, \dots, n)$$

where $G_0^n := 0$ for $n \geqslant 1$, to make the last identity true for $k = 1$. For later use we also set $G_0^0 := 1$. The differential polynomial $G_k^n \in K^\phi\{X\}$ is homogeneous of degree n and isobaric of weight $n - k$, and $G_k^n(\phi) = F_k^n(\phi)$, for $0 \leqslant k \leqslant n$.

LEMMA 5.7.3. *As rings, $\mathbb{Q}[\phi, \dots, \partial^n(\phi)] = \mathbb{Q}[\phi, \dots, \delta^n(\phi)]$. In particular, if K is a differential field, then the differential subfields $\mathbb{Q}\langle\phi\rangle = \mathbb{Q}(\partial^n(\phi) : n = 0, 1, 2, \dots)$ of K and $\mathbb{Q}\langle\phi\rangle = \mathbb{Q}(\delta^n(\phi) : n = 0, 1, 2, \dots)$ of K^ϕ have the same underlying field.*

In the rest of this subsection $P \in K\{Y\}$ has order at most $r \in \mathbb{N}$, and $\boldsymbol{\sigma}, \boldsymbol{\tau} \in \{0, \dots, r\}^*$ have equal length. For $\boldsymbol{\sigma} \leqslant \boldsymbol{\tau}$, $\boldsymbol{\sigma} = \sigma_1 \cdots \sigma_d$, $\boldsymbol{\tau} = \tau_1 \cdots \tau_d$ we define

$$F_{\boldsymbol{\sigma}}^{\boldsymbol{\tau}} := F_{\sigma_1}^{\tau_1} \cdots F_{\sigma_d}^{\tau_d} \in \mathbb{Q}\{X\},$$

so $F_{\boldsymbol{\sigma}}^{\boldsymbol{\tau}}$ is homogeneous of degree $\|\boldsymbol{\sigma}\|$, and isobaric of weight $\|\boldsymbol{\tau}\| - \|\boldsymbol{\sigma}\|$; in particular, $F_\varepsilon^\varepsilon := 1$ for the the empty word ε in $\{0, \dots, r\}^*$, and $F_{\boldsymbol{\sigma}}^{\boldsymbol{\tau}} = 0$ if $\tau_i > \sigma_i = 0$ for some $i \in \{1, \dots, d\}$.

LEMMA 5.7.4. $\quad (P^\phi)_{[\boldsymbol{\sigma}]} = \sum_{\boldsymbol{\tau} \geqslant \boldsymbol{\sigma}} F_{\boldsymbol{\sigma}}^{\boldsymbol{\tau}}(\phi) P_{[\boldsymbol{\tau}]}$.

PROOF. For $\boldsymbol{\tau} = \tau_1 \cdots \tau_d$ we have

$$(Y^{[\boldsymbol{\tau}]})^{\phi} = (Y^{(\tau_1)})^{\phi} \cdots (Y^{(\tau_d)})^{\phi} = \sum_{\boldsymbol{\sigma} \leqslant \boldsymbol{\tau}} F^{\boldsymbol{\tau}}_{\boldsymbol{\sigma}}(\phi) Y^{[\boldsymbol{\sigma}]}.$$

With $P = \sum_{\boldsymbol{\tau}} P_{[\boldsymbol{\tau}]} Y^{[\boldsymbol{\tau}]}$, this gives

$$\begin{aligned} P^{\phi} &= \sum_{\boldsymbol{\tau}} P_{[\boldsymbol{\tau}]} (Y^{[\boldsymbol{\tau}]})^{\phi} = \sum_{\boldsymbol{\tau}} P_{[\boldsymbol{\tau}]} \sum_{\boldsymbol{\sigma} \leqslant \boldsymbol{\tau}} F^{\boldsymbol{\tau}}_{\boldsymbol{\sigma}}(\phi) Y^{[\boldsymbol{\sigma}]} \\ &= \sum_{\boldsymbol{\sigma}} \left(\sum_{\boldsymbol{\tau} \geqslant \boldsymbol{\sigma}} F^{\boldsymbol{\tau}}_{\boldsymbol{\sigma}}(\phi) P_{[\boldsymbol{\tau}]} \right) Y^{[\boldsymbol{\sigma}]}, \end{aligned}$$

from which the lemma follows. □

COROLLARY 5.7.5. *P and P^{ϕ} have the same order, degree, weight, and complexity. If P is homogeneous, then so is P^{ϕ}. If P is subhomogeneous, then so is P^{ϕ}.*

PROOF. The differential polynomials $(Y^{(n)})^{\phi}$ are homogeneous of degree 1, so if P is homogeneous of degree d, then P^{ϕ} is too. Considering the homogeneous parts of P, this yields $\deg(P^{\phi}) = \deg(P)$. Similarly one shows that if P is subhomogeneous of subdegree d, then P^{ϕ} is too. Lemma 5.7.4 shows that if $\|\boldsymbol{\sigma}\| > \operatorname{wt}(P)$, then $(P^{\phi})_{[\boldsymbol{\sigma}]} = 0$. Hence $\operatorname{wt}(P^{\phi}) \leqslant \operatorname{wt}(P)$, and since $P = (P^{\phi})^{\phi^{-1}}$, we obtain $\operatorname{wt}(P^{\phi}) = \operatorname{wt}(P)$. In the same way we get $\operatorname{order}(P^{\phi}) = \operatorname{order}(P)$ and $\mathrm{c}(P^{\phi}) = \mathrm{c}(P)$. □

Compositional conjugation and upward shift

In this subsection we assume familiarity with Appendix A. We study here K^{ϕ} for $K = \mathbb{T}$ and special ϕ.

EXAMPLE. Consider $\mathbb{T}$ with its usual derivation $\partial = \frac{d}{dx}$, and set $t = \frac{1}{x}$. The *upward shift* of $f = f(x) \in \mathbb{T}$ is defined by $f{\uparrow} := f(\mathrm{e}^x)$. Then $f \mapsto f{\uparrow}$ is an isomorphism $\mathbb{T}^t = (\mathbb{T}, x\partial) \to (\mathbb{T}, \partial)$ of differential fields, with inverse $f \mapsto f{\downarrow} = f(\log x)$. Let $P(Y) \in \mathbb{T}\{Y\}$, and let $P{\uparrow}$ be the differential polynomial in $\mathbb{T}\{Y\}$ obtained by applying $f \mapsto f{\uparrow}$ to the coefficients of $P^t \in \mathbb{T}^t\{Y\}$. Thus $(Y'){\uparrow} = \mathrm{e}^{-x} Y'$. We have $P(y){\uparrow} = P{\uparrow}(y{\uparrow})$ for $y \in \mathbb{T}$. (It suffices to check this for $P = Y^{(n)}$.) Note:

$$(P_{+a}){\uparrow} = (P{\uparrow})_{+a{\uparrow}}, \quad (P_{\times a}){\uparrow} = (P{\uparrow})_{\times a{\uparrow}} \qquad (P \in \mathbb{T}\{Y\},\ a \in \mathbb{T}),$$

and $P \mapsto P{\uparrow}$ is an $\mathbb{R}[Y]$-algebra automorphism of $\mathbb{T}\{Y\}$ agreeing with $f \mapsto f{\uparrow}$ on $\mathbb{T}$. It is shown in [194] that upward shifting is a very useful tool for analyzing the zero sets of differential polynomials over $\mathbb{T}$. Although in an arbitrary differential field the operation $f \mapsto f{\uparrow}$ is not available, in cases of interest we have a nonzero element x with $x' = 1$, and then compositional conjugation by $t = \frac{1}{x}$ makes sense and is a substitute for upward shifting, especially for differential polynomials with constant coefficients. This is worked out in Section 12.8, where we relate compositional conjugation by such t to more general compositional conjugations.

In the rest of this subsection x is a unit of K with $x' = 1$, we set $t = \frac{1}{x}$, and work in K^t whose derivation is $\delta := x\partial$, so $\delta(x) = x$ and $\delta(t) = -t$.

Let $\left[{n \atop k}\right]$ $(k = 0, \dots, n)$ be the (unsigned) Stirling numbers of the first kind; see [90, §§5.5, 6.3], [151, §6.1]. They are defined recursively by $\left[{n \atop 0}\right] := 0$ for $n \geqslant 1$, $\left[{n \atop n}\right] := 1$,

$$\begin{bmatrix} n+1 \\ k \end{bmatrix} = \begin{bmatrix} n \\ k-1 \end{bmatrix} + n \cdot \begin{bmatrix} n \\ k \end{bmatrix} \qquad \text{for } k = 1, \dots, n.$$

Thus $\left[{n \atop k}\right] \in \mathbb{N}^{\geqslant 1}$ for $k = 1, \dots, n$. We also set $\left[{n \atop k}\right] := 0$ for $k > n$. See Table 5.1. As the table suggests, $\left[{n+1 \atop n}\right] = \binom{n+1}{2}$ and $\left[{n+1 \atop 1}\right] = n!$, which is easily verified.

$k \backslash n$	0	1	2	3	4	5	
0	1	0	0	0	0	0	$\dots$
1		1	1	2	6	24	$\dots$
2			1	3	11	50	$\dots$
3				1	6	35	$\dots$
4					1	10	$\dots$
5						1	$\dots$
$\vdots$							$\ddots$

Table 5.1: Unsigned Stirling numbers of the first kind.

For $k, n \geqslant 0$ we let $s(n,k) := (-1)^{n-k}\left[{n \atop k}\right]$, so $s(n,0) = 0$ for $n \geqslant 1$, $s(n,n) = 1$, and

$$s(n+1,k) = s(n,k-1) - n \cdot s(n,k) \qquad \text{for } k = 1, \dots, n. \tag{5.7.3}$$

The $s(n,k)$ are known as the signed Stirling numbers of the first kind.

LEMMA 5.7.6. *For $k = 0, \dots, n$ we have*

$$F^n_k(t) = s(n,k) \cdot t^n, \tag{5.7.4}$$

and the sum of the coefficients of G^n_k equals $\left[{n \atop k}\right]$.

PROOF. An easy induction on n using (5.7.1) shows (5.7.4). Since $\delta(t) = -t$ and each differential polynomial G^n_k is isobaric of weight $n-k$ and homogeneous of degree n, we have $G^n_k(t) = (-1)^{n-k} c_{n,k} t^n$ where $c_{n,k}$ = sum of the coefficients of G^n_k. Since $G^n_k(t) = F^n_k(t)$ we get $c_{n,k} = \left[{n \atop k}\right]$. □

EXAMPLE 5.7.7. For $n \geqslant 1$ we have $G^n_{n-1} = \left[{n \atop n-1}\right] X^{n-1} X'$, and thus

$$F^n_{n-1}(\phi) = \begin{bmatrix} n \\ n-1 \end{bmatrix} \phi^{n-1}\, \phi^{\dagger}.$$

To see this, use (5.7.2) to show that G^n_{n-1} is an integer multiple of $X^{n-1}X'$, and then apply the previous lemma.

For $\boldsymbol{\tau} = \tau_1 \cdots \tau_d \geqslant \boldsymbol{\sigma} = \sigma_1 \cdots \sigma_d$ we have

$$F^{\boldsymbol{\tau}}_{\boldsymbol{\sigma}}(t) = s(\boldsymbol{\tau}, \boldsymbol{\sigma}) \cdot t^{\|\boldsymbol{\tau}\|} \qquad \text{where } s(\boldsymbol{\tau}, \boldsymbol{\sigma}) := s(\tau_1, \sigma_1) \cdots s(\tau_d, \sigma_d),$$

hence

$$(P^t)_{[\boldsymbol{\sigma}]} = \sum_{\boldsymbol{\tau} \geqslant \boldsymbol{\sigma}} s(\boldsymbol{\tau}, \boldsymbol{\sigma}) t^{\|\boldsymbol{\tau}\|} P_{[\boldsymbol{\tau}]}$$

by Lemma 5.7.4.

Compositional conjugation in the algebraic closure of $K\langle Y \rangle$

The material in this subsection and the next one will only be used in Section 13.5.

Let K be a differential field. Since $P \mapsto P^\phi \colon K\{Y\} \to K^\phi\{Y\}$ is a ring isomorphism, it extends uniquely to a field isomorphism $P \mapsto P^\phi \colon K\langle Y \rangle \to K^\phi\langle Y \rangle$. Fix an algebraic closure $F = K\langle Y \rangle^{\mathrm{a}}$ of the field $K\langle Y \rangle$. Since $K\langle Y \rangle = K^\phi\langle Y \rangle$ as *fields,* this isomorphism $K\langle Y \rangle \to K^\phi\langle Y \rangle$ may be viewed as an automorphism of the field $K\langle Y \rangle$, and we extend it to an automorphism $P \mapsto P^\phi$ of F. The derivation ∂ of $K\langle Y \rangle$ and the derivation δ of $K^\phi\langle Y \rangle$ both extend uniquely to derivations of F, also denoted by ∂ and δ, respectively, with the same constant field $C_F = C^{\mathrm{a}}$. Note that $P \mapsto P^\phi \colon F \to F$ is the identity on K, and maps C_F onto itself.

The second identity in Lemma 5.7.1 continues to hold if P is an element of F:

LEMMA 5.7.8. *For each $P \in F$, we have $(\partial P)^\phi = \phi \cdot \delta(P^\phi)$.*

PROOF. Let σ be the automorphism $P \mapsto P^\phi$ of F. The derivation $\phi\delta$ of F yields a derivation $\sigma^{-1} \circ (\phi\delta) \circ \sigma$ on F that agrees with ∂ on $K\{Y\}$ by the second identity in Lemma 5.7.1. Hence $\partial = \sigma^{-1} \circ (\phi\delta) \circ \sigma$, and thus $\sigma \circ \partial = (\phi\delta) \circ \sigma$. □

COROLLARY 5.7.9. *Let $P \in K\{Y\}$ and $Q \in F$. Then $P(Q)^\phi = P^\phi(Q^\phi)$.*

PROOF. An induction on n using the previous lemma yields the identity $(\partial^n Q)^\phi = (Y^{(n)})^\phi(Q^\phi)$ and this gives what we want. □

LEMMA 5.7.10. *Suppose $Q \in F$ is transcendental over K, that is, $Q \notin K^{\mathrm{a}} \subseteq F$. Then Q is even* d*-transcendental over K.*

PROOF. Since Y is d-transcendental over K, it is enough to show that $Y \in F$ is d-algebraic over $K\langle Q \rangle$. As Q is algebraic over $K\langle Y \rangle$, we can take $R_0, \dots, R_n \in K\{Y\}$ with $R_n \neq 0$, such that

$$R_n Q^n + R_{n-1} Q^{n-1} + \cdots + R_0 = 0.$$

Put

$$R := R_n(Z) Q^n + R_{n-1}(Z) Q^{n-1} + \cdots + R_0(Z) \in K\langle Q \rangle\{Z\},$$

where Z is a differential indeterminate different from Y. Take r with $R_0, \dots, R_n$ of order $\leqslant r$ and take $\boldsymbol{i} \in \mathbb{N}^{1+r}$ with $(R_n)_{\boldsymbol{i}} \neq 0$. Then in $K\langle Q \rangle$ we have the equality

$$R_{\boldsymbol{i}} = (R_n)_{\boldsymbol{i}} Q^n + (R_{n-1})_{\boldsymbol{i}} Q^{n-1} + \cdots + (R_0)_{\boldsymbol{i}},$$

so $R_{\boldsymbol{i}} \neq 0$ since Q is transcendental over K. Hence $R \neq 0$ in $K\langle Q \rangle\{Z\}$, and $R(Y) = 0$. Thus Y is d-algebraic over $K\langle Q \rangle$. $\square$

Compositional conjugation and rational powers

Let K be a differential field. As before, we fix an algebraic closure F of $K\langle Y \rangle$, taking it as a differential field extension of $K\langle Y \rangle$, and extend the ring isomorphism $P \mapsto P^\phi \colon K\{Y\} \to K^\phi\{Y\}$ to an automorphism $P \mapsto P^\phi$ of the field F.

NOTATION. For elements f and g of a differential field E we define

$$f =_{\mathrm{c}} g \ :\Longleftrightarrow\ f = c \cdot g \text{ for some } c \in C_E^\times,$$

so if $f, g \neq 0$, then $f =_{\mathrm{c}} g \Longleftrightarrow f^\dagger = g^\dagger$.

We use this for $E := F$. For $P, Q \in F$ we have $P =_{\mathrm{c}} Q \Rightarrow P^\phi =_{\mathrm{c}} Q^\phi$.

Next, we extend the usual power map

$$(P, k) \mapsto P^k \colon F^\times \times \mathbb{Z} \to F^\times$$

of the multiplicative group $F^\times$ to a map

$$(P, q) \mapsto P^q \colon F^\times \times \mathbb{Q} \to F^\times$$

such that

$$(P^q)^\dagger = qP^\dagger \qquad \text{for all } q \in \mathbb{Q} \text{ and } P \in F^\times.$$

Note that

$$(P^{k/l})^l =_{\mathrm{c}} P^k \qquad \text{for } k, l \in \mathbb{Z},\ l \neq 0.$$

The following rules are also easy to verify, for $P, Q \in F^\times$ and $q, q_1, q_2 \in \mathbb{Q}$:

(1) $P =_{\mathrm{c}} Q \Rightarrow P^q =_{\mathrm{c}} Q^q$;

(2) $P^{q_1} P^{q_2} =_{\mathrm{c}} P^{q_1 + q_2}$;

(3) $(P^{q_1})^{q_2} =_{\mathrm{c}} P^{q_1 q_2}$;

(4) $(PQ)^q =_{\mathrm{c}} P^q Q^q$.

We allow of course $\phi \in \mathbb{Q}^\times \subseteq K^\times$, but a "power" P^q should not be confused with the "compositional conjugate" P^ϕ of $P \in F^\times$ by ϕ. Powers behave as follows under compositional conjugation:

LEMMA 5.7.11. *Let $P \in F^\times$, $q \in \mathbb{Q}$. Then $(P^q)^\phi =_{\mathrm{c}} (P^\phi)^q$.*

PROOF. For $k \in \mathbb{Z}$ we have $(P^k)^\phi = (P^\phi)^k$ since P^k has its usual meaning in the multiplicative group $F^\times$. Take $n \geqslant 1$ such that $nq = k \in \mathbb{Z}$, and set $Q := (P^q)^\phi$, $R := (P^\phi)^q$. Then $(P^q)^n =_c P^k$ gives

$$Q^n = \big((P^q)^n\big)^\phi =_c (P^k)^\phi = (P^\phi)^k =_c R^n,$$

so $Q =_c R$. □

Recall that by convention $G_0^0 = 1$ and $G_0^n = 0$ for $n \geqslant 1$. We have

$$(5.7.5) \qquad (Y^{(n)})^\phi = \sum_{m=0}^{n} G_m^n(\phi) Y^{(m)} \in K^\phi\{Y\}.$$

Let $q \in \mathbb{Q}$ and suppose $\phi^q \in K$. Then in $K^\phi\{Y\}$,

$$(Y^{(m)})_{\times \phi^q} = \sum_{i=0}^{m} \binom{m}{i} \delta^{m-i}(\phi^q) Y^{(i)},$$

and thus by identity (5.7.5) we have in $K^\phi\{Y\}$,

$$(5.7.6) \qquad (Y^{(n)})^\phi_{\times \phi^q} = \sum_{i=0}^{n} \left(\sum_{m=i}^{n} \binom{m}{i} G_m^n(\phi) \delta^{m-i}(\phi^q) \right) Y^{(i)}.$$

We consider in more detail the case $q = 1$, with $\phi^1 = \phi \in K^\times$, since Section 13.4 requires us to deal with $P^\phi_{\times\phi}$ where compositional and multiplicative conjugation are combined. The key to this is the fact that $P^\phi_{\times\phi}(Y') = P(Y')^\phi$. More formally, for $P \in K\{Y\}^{\neq}$, set $P^{\times\phi} := P^\phi_{\times\phi} \in K^\phi\{Y\}$, and $P^\times := P(Y') \in K\{Y'\}$. Then

$$(5.7.7) \qquad (P^\times)^\phi = P^\phi\big((Y')^\phi\big) = P^\phi(\phi Y') = P^{\times\phi}(Y')$$

in $K^\phi\langle Y\rangle$. Setting

$$R_i^n(Y) := \sum_{m=i}^{n} \binom{m}{i} G_m^n(Y)\, Y^{(m-i)} \in \mathbb{Q}\{Y\} \subseteq K^\phi\{Y\} \qquad (i = 0, \dots, n),$$

the identity (5.7.6) gives the following identity in $K^\phi\{Y\}$:

$$(Y^{(n)})^{\times\phi} = R_0^n(\phi) Y + R_1^n(\phi) Y' + \cdots + R_n^n(\phi) Y^{(n)}.$$

We have $Y^{(n+1)} = (Y^{(n)})^\times$ and hence by (5.7.7) applied to $P = Y^{(n)}$,

$$(Y^{(n+1)})^\phi = (Y^{(n)})^{\times\phi}(Y') = R_0^n(\phi) Y' + R_1^n(\phi) Y'' + \cdots + R_n^n(\phi) Y^{(n+1)}.$$

In view of (5.7.5) this gives $R_i^n(\phi) = G_{i+1}^{n+1}(\phi)$ for $i = 0, \dots, n$. Lemma 5.7.3 shows that $\phi \in K^\phi$ is d-transcendental over $\mathbb{Q}$ for suitable K, ϕ, and thus

$$R_i^n = G_{i+1}^{n+1} \text{ in } \mathbb{Q}\{Y\} \qquad \text{for } i = 0, \dots, n.$$

Next, let $q \in \mathbb{Q}$ be such that $\phi^q \in K$. For $P \in K\{Y\}^{\neq}$ we set

$$P^{\times q,\phi} := P^{\phi}_{\times \phi^q} \in K^{\phi}\{Y\}, \qquad P^{\times q} := P\big((Y')^q\big) \in F.$$

Now suppose that $P \in K\{Y\}^{\neq}$ is homogeneous; let $d = \deg(P)$ and $w = \operatorname{wt}(P)$. Then Corollary 4.3.18 gives

$$P^{\times q} =_{\mathrm{c}} (Y')^{dq-w} \cdot E(Y')$$

with homogeneous $E \in K\{Y\}^{\neq}$, $\deg(E) = \operatorname{wt}(E) = w$. Computing in F gives

$$\begin{aligned}(P^{\times q})^{\phi} &= P^{\phi}\big(((Y')^q)^{\phi}\big) \\ &=_{\mathrm{c}} P^{\phi}\big(((Y')^{\phi})^q\big) \\ &= P^{\phi}\big((\phi Y')^q\big) \\ &=_{\mathrm{c}} P^{\phi}\big(\phi^q (Y')^q\big) = P^{\times q,\phi}\big((Y')^q\big),\end{aligned}$$

using Corollary 5.7.9 for the first equality; for the second, use the homogeneity of P^{ϕ} and Lemma 5.7.11. For future reference we summarize:

LEMMA 5.7.12. *Let $P \in K\{Y\}^{\neq}$ be homogeneous, $d = \deg(P)$, $w = \operatorname{wt}(P)$, $q \in \mathbb{Q}$. Then there is a homogeneous $E \in K\{Y\}^{\neq}$ with $\deg(E) = \operatorname{wt}(E) = w$ such that for all $\phi \in K^{\times}$ with $\phi^q \in K$ we have (in F):*

$$P^{\times q,\phi}\big((Y')^q\big) =_{\mathrm{c}} (\phi Y')^{dq-w} \cdot E^{\times\phi}(Y').$$

Suppose $q \in \mathbb{N}^{\geqslant 1}$. If $P \in K\{Y\}^{\neq}$ is homogeneous, then $P^{\times q} = P\big((Y')^q\big) \in K\{Y\}$ is homogeneous with $\deg(P^{\times q}) = q\deg(P)$, by Corollary 4.3.13. Hence for arbitrary $P \in K\{Y\}^{\neq}$ and $d \in \mathbb{N}$ we have $(P^{\times q})_{qd} = (P_d)^{\times q}$.

Notes and comments

An explicit formula for the transformation coefficients F^n_k appears in [448, Theorem 1] and a formula for the G^n_k in [89, (8)]. (Combining Propositions 12.5.1 and 12.8.3 below also gives a closed form expression for the F^n_k.) The formula for F^n_k in the special case of Lemma 5.7.6 dates back to at least the 1823 dissertation of Scherk [385]; see [45, Appendix]. The Stirling number $\left[{n \atop k}\right]$ counts the number of permutations of n objects with k disjoint cycles; these numbers were introduced by Stirling [433, p. 11] in 1730.

The K-algebra underlying both $K\{Y\}$ and $K^{\phi}\{Y\}$ is $A = K[Y, Y', \dots]$, and $P \mapsto P^{\phi}$ is an automorphism of A; in Chapter 12 we study this automorphism.

5.8 THE RICCATI TRANSFORM

In this section K is a differential *field*, y ranges over $K^{\times}$, and $z := y^{\dagger}$. Then

$$\frac{y^{(n)}}{y} = R_n(z)$$

where the differential polynomial $R_n(Z) \in K\{Z\}$ has nonnegative integer coefficients and is independent of K and y:

$$\begin{aligned} R_0(Z) &= 1 \\ R_1(Z) &= Z \\ R_2(Z) &= Z^2 + Z' \\ R_3(Z) &= Z^3 + 3ZZ' + Z'' \\ &\vdots \end{aligned}$$

These R_n are defined by the recursion

$$(5.8.1) \qquad R_0(Z) := 1, \qquad R_{n+1}(Z) := ZR_n(Z) + R_n(Z)'.$$

An easy induction yields $R_n(Z) = Z^n + A_n(Z)$ with $\deg(A_n) \leqslant n-1$ and where every monomial in A_n has the form $Z^{i_0}(Z')^{i_1}\cdots(Z^{(k)})^{i_k}$ with $1 \leqslant k \leqslant n-1$, $i_k \geqslant 1$. Note that $R_2(Z) = -\frac{1}{4}\omega(2Z)$ where $\omega(Z) = -(Z^2+2Z')$ is as in Section 5.2. The R_n are related to the F_k^n from Section 5.7 as follows:

LEMMA 5.8.1. *$R_n(Z) = F_n^n(Z) + F_{n-1}^n(Z) + \cdots + F_0^n(Z)$, so the homogeneous and isobaric parts of R_n are given by $R_n(Z)_k = R_n(Z)_{[n-k]} = F_k^n(Z)$ for $k = 0, \dots, n$.*

PROOF. Let Y be a differential indeterminate over $\mathbb{Q}$, let ∂ be the usual derivation of $K\langle Y\rangle$, set $\phi := Y^\dagger \in \mathbb{Q}\langle Y\rangle$, and take $\delta = \phi^{-1}\partial$. Then $\delta^k(Y) = Y$ for all $k \in \mathbb{N}$, and thus

$$R_n(\phi) = \frac{Y^{(n)}}{Y} = \frac{\sum_{k=0}^n F_k^n(\phi)\,\delta^k(Y)}{Y} = \sum_{k=0}^n F_k^n(\phi).$$

It remains to note that ϕ is differentially transcendental over $\mathbb{Q}$. □

Hence $R_n(Z)_{[0]} = F_n^n(Z) = Z^n$, so the $R_n(Z)$ are linearly independent over K. If $n \geqslant 1$, then $R_n(Z)_1 = R_n(Z)_{[n-1]} = F_1^n(Z) = Z^{(n-1)}$, so R_n has order $n-1$.

For each n we have the following binomial identity in the differential polynomial ring $\mathbb{Q}\{Z_1, Z_2\}$:

$$(5.8.2) \qquad R_n(Z_1 + Z_2) = \sum_{i+j=n} \binom{n}{i} R_i(Z_1)R_j(Z_2).$$

To see this, note that in the differential field $\mathbb{Q}\langle Y_1, Y_2\rangle$,

$$\begin{aligned} R_n(Y_1^\dagger + Y_2^\dagger) = R_n\big((Y_1Y_2)^\dagger\big) &= \frac{(Y_1Y_2)^{(n)}}{Y_1Y_2} = \frac{\sum_{i=0}^n \binom{n}{i} Y_1^{(i)} Y_2^{(n-i)}}{Y_1Y_2} \\ &= \sum_{i+j=n} \binom{n}{i} \frac{Y_1^{(i)}}{Y_1}\frac{Y_2^{(j)}}{Y_2} = \sum_{i+j=n} \binom{n}{i} R_i(Y_1^\dagger)R_j(Y_2^\dagger). \end{aligned}$$

By induction on m this gives:

COROLLARY 5.8.2. *For all* $z_1, \dots, z_m \in K$ *we have*

$$R_n(z_1 + \cdots + z_m) = \sum \frac{n!}{i_1! \cdots i_m!} R_{i_1}(z_1) \cdots R_{i_m}(z_m)$$

where the sum on the right is over all tuples $(i_1, \dots, i_m) \in \mathbb{N}^m$ *with* $i_1 + \cdots + i_m = n$.

DEFINITION 5.8.3. Let $\mathrm{Ri}\colon K\{Y\} \to K\{Z\}$ be the K-algebra homomorphism such that $\mathrm{Ri}\left(Y^{(n)}\right) = R_n(Z)$ for all n. We call $\mathrm{Ri}(P)$ the **Riccati transform** of P. If $P \in K\{Y\}$ is homogeneous of degree d, then

$$\mathrm{Ri}(P)(z) = \frac{P(y)}{y^d}.$$

REMARK. If $P \in K\{Y\}^{\neq}$ is homogeneous of order $r \geqslant 1$, then

$$\mathrm{Ri}(P) \neq 0, \quad \operatorname{order} \mathrm{Ri}(P) = r - 1, \quad \deg \mathrm{Ri}(P) \leqslant \operatorname{wt} P \leqslant r \cdot \operatorname{sdeg} P.$$

A straightforward computation yields:

LEMMA 5.8.4. *If* $P \in K\{Y\}$ *is homogeneous of degree d and* $g \in K$, *then*

$$\mathrm{Ri}(P)_{+g}(Z) = \sum_{\boldsymbol{\sigma}} \left(\sum_{\boldsymbol{\tau} \geqslant \boldsymbol{\sigma}} P_{[\boldsymbol{\tau}]} \binom{\boldsymbol{\tau}}{\boldsymbol{\sigma}} R_{\boldsymbol{\tau}-\boldsymbol{\sigma}}(g) \right) R_{\boldsymbol{\sigma}}(Z).$$

Here $\boldsymbol{\sigma} = \sigma_1 \cdots \sigma_d$ *and* $R_{\boldsymbol{\sigma}} := R_{\sigma_1} \cdots R_{\sigma_d} = \mathrm{Ri}(Y^{[\boldsymbol{\sigma}]})$.

COROLLARY 5.8.5. *Let* $P \in K\{Y\}$ *be homogeneous of degree d and let* $h \in K^{\times}$. *Then*

$$\mathrm{Ri}(P_{\times h}) = h^d \cdot \mathrm{Ri}(P)_{+h^\dagger}.$$

PROOF. With $g := h^\dagger$ and $\boldsymbol{\sigma}$ ranging over words $\sigma_1 \cdots \sigma_d$ of length d we have

$$\begin{aligned} h^d \cdot \mathrm{Ri}(P)_{+g}(Z) &= \sum_{\boldsymbol{\sigma}} \left(\sum_{\boldsymbol{\tau} \geqslant \boldsymbol{\sigma}} P_{[\boldsymbol{\tau}]} \binom{\boldsymbol{\tau}}{\boldsymbol{\sigma}} h^d R_{\boldsymbol{\tau}-\boldsymbol{\sigma}}(g) \right) R_{\boldsymbol{\sigma}}(Z) \\ &= \sum_{\boldsymbol{\sigma}} \left(\sum_{\boldsymbol{\tau} \geqslant \boldsymbol{\sigma}} P_{[\boldsymbol{\tau}]} \binom{\boldsymbol{\tau}}{\boldsymbol{\sigma}} h^{[\boldsymbol{\tau}-\boldsymbol{\sigma}]} \right) R_{\boldsymbol{\sigma}}(Z) = \mathrm{Ri}(P_{\times h})(Z) \end{aligned}$$

by the previous lemma and (4.3.2). □

Here is how the Riccati transform interacts with compositional conjugation:

LEMMA 5.8.6. *Let* $\phi \in K^{\times}$ *and* $P \in K\{Y\}$. *Then*

$$\mathrm{Ri}(P^{\phi}) = \mathrm{Ri}(P)^{\phi}_{\times\phi} \text{ in } K^{\phi}\{Z\}.$$

PROOF. We can reduce to the case that P is homogeneous of degree d. Then for $y \in K^\times$ and $z = y^\dagger$, and setting $\delta = \phi^{-1}\partial$, we have

$$\begin{aligned}\operatorname{Ri}(P)^{\phi}_{\times\phi}(\delta(y)/y) &= \operatorname{Ri}(P)^{\phi}(z) = \operatorname{Ri}(P)(z)\\ &= \frac{P(y)}{y^d} = \frac{P^{\phi}(y)}{y^d} = \operatorname{Ri}(P^{\phi})(\delta(y)/y).\end{aligned}$$

This remains true for y in differential field extensions of K^ϕ, and so for all a in all such extensions we have $\operatorname{Ri}(P^\phi)(a) = \operatorname{Ri}(P)^{\phi}_{\times\phi}(a)$. □

The Riccati transform of a linear differential operator

Let

$$A = a_0 + a_1\partial + \cdots + a_n\partial^n \in K[\partial] \qquad (a_0, \dots, a_n \in K).$$

The **Riccati transform** $\operatorname{Ri}(A)$ of A is defined to be the Riccati transform of the corresponding differential polynomial:

$$\operatorname{Ri}(A)(Z) := a_0R_0(Z) + a_1R_1(Z) + \cdots + a_nR_n(Z) \in K\{Z\}.$$

Using (5.8.2) and (5.1.2) we have for $g \in K$:

$$\operatorname{Ri}(A)_{+g}(Z) = b_0R_0(Z) + b_1R_1(Z) + \cdots + b_nR_n(Z),$$

where

$$b_i = \sum_{j=i}^{n} \binom{j}{i} a_jR_{j-i}(g) = \frac{1}{i!}\operatorname{Ri}(A^{(i)})(g).$$

In particular, $b_0 = \operatorname{Ri}(A)(g)$ and $b_n = a_n$.

LEMMA 5.8.7. *Let $A \in K[\partial]$, set $R := \operatorname{Ri}(A)$, and let $a \in K$. Then*

$$A \in K[\partial](\partial - a) \iff R(a) = 0.$$

PROOF. Immediate from Lemma 5.1.21. □

We define a **Riccati polynomial** over K to be a differential polynomial

$$R(Z) = a_0R_0(Z) + a_1R_1(Z) + \cdots + a_nR_n(Z) \qquad (a_0, \dots, a_n \in K),$$

that is, a Riccati polynomial over K is the Riccati transform of some $A \in K[\partial]$. It follows that if R is a Riccati polynomial over K and $g \in K$, then R_{+g} is a Riccati polynomial over K of the same order.

LEMMA 5.8.8. *Let $A \in K[\partial]$, and let h be an element of a differential field extension of K with $h \neq 0$ and $h^\dagger = g \in K$. Then $A_{\ltimes h} \in K[\partial]$.*

PROOF. By Corollary 5.8.5 we have $\operatorname{Ri}(A_{\ltimes h}) = \operatorname{Ri}(A)_{+g} \in K\{Z\}$. Next one shows by an easy induction on n and using $R_n = Z^n +$ lower degree terms: if L is a differential field extension of K, and $R = a_0R_0 + \cdots + a_nR_n$ $(a_0, \dots, a_n \in L)$ is a Riccati polynomial over L with $R \in K\{Z\}$, then $a_0, \dots, a_n \in K$. □

The decrease in order under Riccati transformation is the basis of several inductive proofs. The next lemma is an example. For $r \in \mathbb{N}$ we say that K is **weakly r-differentially closed** if every $P \in K\{Y\} \setminus K$ of order $\leqslant r$ has a zero in K. (Thus K is weakly 0-differentially closed iff K is algebraically closed.) We also say that K is **weakly differentially closed** if K is weakly r-differentially closed for each $r \in \mathbb{N}$. Clearly differentially closed $\Rightarrow$ weakly differentially closed.

LEMMA 5.8.9. *Suppose K is weakly r-differentially closed, $r \in \mathbb{N}$. Then K is $(r+1)$-linearly closed.*

PROOF. By induction on r. The case $r = 0$ is obvious, so let $r > 0$, and let $A \in K[\partial]$ have order $r+1$. Then its Riccati transform $\operatorname{Ri}(A) \in K\{Z\}$ has order r, so has a zero $a \in K$, which gives $A = B \cdot (\partial - a)$ with $B \in K[\partial]$ of order r. Now apply the inductive assumption to B. □

Riccati transforms in the presence of a valuation

In this subsection K is a valued differential field such that $\partial\smallo \subseteq \smallo$, and thus $\partial\mathcal{O} \subseteq \mathcal{O}$.

LEMMA 5.8.10. *Let $P \in K\{Y\}$ be homogeneous. Then $v(P) = v(\operatorname{Ri}(P))$.*

PROOF. By K-linearity of the Riccati transform we can reduce to the case $vP = 0$. Then $\operatorname{Ri}(P) \in \mathcal{O}\{Z\}$. Let $\overline{P}$ be the image of $P \in \mathcal{O}\{Y\}$ in $\boldsymbol{k}\{Y\}$. Then $\overline{P}$ is homogeneous and nonzero, so $\operatorname{Ri}(\overline{P}) \neq 0$. Since $\operatorname{Ri}(\overline{P})$ is the image of $\operatorname{Ri}(P)$ under the natural map $\mathcal{O}\{Z\} \to \boldsymbol{k}\{Z\}$, we get $v\big(\operatorname{Ri}(P)\big) = 0$. □

In view of Corollary 5.8.5, this lemma yields:

COROLLARY 5.8.11. *If $P \in K\{Y\}^{\neq}$ is homogeneous of degree d, and $a \in K^\times$, then*

$$v_P(\alpha) \;=\; d\alpha + v(R_{+a^\dagger}), \qquad \textit{with } \alpha := va,\ R := \operatorname{Ri}(P).$$

For a nonzero Riccati polynomial $R = a_0R_0 + \cdots + a_nR_n$ $(a_0, \ldots, a_n \in K)$, put

$$\mu(R) \;:=\; \min\{i : v(a_i) = v(R)\}, \qquad \nu(R) \;:=\; \max\{i : v(a_i) = v(R)\}.$$

Note that for nonzero $A \in K[\partial]$ and $R = \operatorname{Ri}(A)$, we have

$$v(A) \;=\; v(R), \quad \operatorname{dwm}(A) \;=\; \mu(R), \quad \operatorname{dwt}(A) \;=\; \nu(R),$$

and by Corollary 5.8.5:

$$v(Ay) = vy + v(R_{+z}), \quad \operatorname{dwm}(Ay) = \mu(R_{+z}), \quad \operatorname{dwt}(Ay) = \nu(R_{+z}).$$

The quotients of K by its $\mathbb{Q}$-linear subspaces $\smallo$ and $\mathcal{O}$ are $K/\smallo$ and $K/\mathcal{O}$, with canonical maps

$$g \mapsto g + \smallo \colon K \to K/\smallo, \qquad g \mapsto g + \mathcal{O} \colon K \to K/\mathcal{O}.$$

If R is a nonzero Riccati polynomial over K, then by Lemma 4.5.1(i), the value $v(R_{+g})$ for $g \in K$ depends only on the coset $g + \mathcal{O}$.

LEMMA 5.8.12. *Let R be a nonzero Riccati polynomial over K. Then $\mu(R_{+g})$ and $\nu(R_{+g})$ depend only on $g + \mathcal{o}$ and $g + \mathcal{O}$, respectively, for $g \in K$.*

PROOF. Let $R = a_0R_0 + \cdots + a_nR_n$ with $a_0, \ldots, a_n \in K$, and let $g \in K^{\prec 1}$. Then $R_j(g) \in \mathcal{o}$ for $j > 0$, since $\mathcal{o}$ is closed under ∂. Now

$$R_{+g}(Z) \;=\; b_0R_0(Z) + b_1R_1(Z) + \cdots + b_nR_n(Z)$$

where

$$b_i \;=\; \sum_{j-i}^{n} \binom{j}{i} a_jR_{j-i}(g) \;=\; a_i + (i+1)a_{i+1}R_1(g) + \cdots + \binom{n}{i} a_nR_{n-i}(g).$$

Hence $v(b_i) \geqslant v(R)$ for all i, and $v(b_i) = v(a_i) = v(R)$ for $i = \mu(R)$; so $\mu(R_{+g}) \leqslant \mu(R)$. Using $R = (R_{+g})_{-g}$ we get $\mu(R_{+g}) = \mu(R)$. The general case follows easily from this special case $vg > 0$. The argument for $\nu(R_{+g})$ is similar. □

For a nonzero Riccati polynomial R over K we define

$$\mu_R\colon K/\mathcal{o} \to \mathbb{N}, \qquad \mu_R(g+\mathcal{o}) := \mu(R_{+g}) \qquad (g \in K),$$
$$\nu_R\colon K/\mathcal{O} \to \mathbb{N}, \qquad \nu_R(g+\mathcal{O}) := \nu(R_{+g}) \qquad (g \in K).$$

Notes and comments

Riccati polynomials are named after J. Riccati (1676–1754), who studied the differential equation $Z' + aZ^2 = bx^r$ for $a, b, r \in \mathbb{R}$ in [337].

The results in Section 14.5 and Chapter 15 below yield that $\mathbb{T}[i]$ and $\mathbb{T}_{\log}[i]$ ($i^2 = -1$) are weakly differentially closed.

5.9 JOHNSON'S THEOREM

Below $Y = (Y_1, \ldots, Y_n)$ is a tuple of distinct d-indeterminates, $n \geqslant 1$. The goal of this section is to prove the following result due to Joseph Johnson [205].

THEOREM 5.9.1. *Let $K \subseteq L$ be a differential field extension, let $P_1, \ldots, P_n \in K\{Y\}$, and let $y = (y_1, \ldots, y_n) \in L^n$ be such that $P_1(y) = \cdots = P_n(y) = 0$ and $P_1, \ldots, P_n$ are* d*-independent at y. Then $y_1, \ldots, y_n$ are* d*-algebraic over K.*

Theorem 5.9.1 is an analogue of Corollary 1.9.7, with differential field extensions replacing field extensions. Accordingly, we begin by establishing systematically analogues of results in Sections 1.7, 1.8, and 1.9 for differential K-algebras A instead of K-algebras, with $\Omega_{A|K}$ construed as an $A[\partial]$-module instead of just an A-module, and differential transcendence degree instead of transcendence degree.

Note also that Theorem 5.9.1 partially extends Corollary 5.4.5 to non-linear systems of algebraic differential equations.

Johnson considers more generally fields of arbitrary characteristic with finitely many commuting derivations, as is usual in the Kolchin tradition. In our setting where differential fields are of characteristic 0 with a single derivation, things are a bit simpler, but nevertheless we need some nontrivial facts on regular local rings from Chapter 1. We give the proof of Theorem 5.9.1 at the end of this section; we precede it with some generalities on the tensor product of $K[\partial]$-modules and the module of differentials of a given differential field extension.

We shall apply Theorem 5.9.1 in Section 7.6 to differential-henselian fields.

Tensor products of $K[\partial]$-modules

Let K be a differential ring, let M, N be $K[\partial]$-modules, and consider the K-module $T := M \otimes_K N$. Corollary 1.7.3 gives a ∂-compatible derivation on T making T a $K[\partial]$-module such that

$$\partial(x \otimes y) \;=\; (\partial x) \otimes y + x \otimes (\partial y) \qquad \text{for all } x \in M, y \in N.$$

If K is a differential field and M and N are differential modules over K, then T is a differential module over K with $\dim_K T = \dim_K M \cdot \dim_K N$, and if in addition M and N are horizontal, then so is T.

Let A be a **differential K-algebra,** that is, A is a differential ring equipped with a differential ring morphism $K \to A$; the latter also makes A a K-algebra. We view A as a $K[\partial]$-module with $\partial a = a'$ for $a \in A$. By Corollary 1.7.4 we obtain the $A[\partial]$-module $A \otimes_K M$, called the **base change** of M to A. Note that

$$\partial(a \otimes x) \;=\; a' \otimes x + a \otimes (\partial x) \qquad \text{for } a \in A, x \in M.$$

Let (x_i) be a family in M. Then the $K[\partial]$-module M is free on (x_i) iff the K-module M is free on $(\partial^n x_i)_{i,n}$. If the $K[\partial]$-module M is free on (x_i), then the $A[\partial]$-module $A \otimes_K M$ is free on $(1 \otimes x_i)$: use that $\partial^n(1 \otimes x) = 1 \otimes \partial^n x$ for $x \in M$. The differential ring morphism $K \to A$ extends to a ring morphism $K[\partial] \to A[\partial]$ sending ∂ to ∂, and any $A[\partial]$-module is construed as a $K[\partial]$-module via this ring morphism $K[\partial] \to A[\partial]$.

Let I be a differential ideal of K, and $A = K/I$, viewed as a differential K-algebra via the canonical map $K \to A$. Then IM is a $K[\partial]$-submodule of M, which makes M/IM a $K[\partial]$-module as well as an A-module. We give M/IM its unique structure of $A[\partial]$-module that induces its given A-module structure as well as its $K[\partial]$-module structure coming from the natural ring morphism $K[\partial] \to A[\partial]$. Then the isomorphism $A \otimes_K M \xrightarrow{\cong} M/IM$ of (1.7.3) is $A[\partial]$-linear.

The next lemma is for use in the second volume.

LEMMA 5.9.2. *Let K be a differential field, M a differential module over K, and F a differential field extension of K. Then the base change $F \otimes_K M$ of M to F is a differential module over F with $\dim_F F \otimes_K M = \dim_K M$. Moreover, if $M = K[\partial]/K[\partial]L$ where $L \in K[\partial]^{\neq}$, then $F \otimes_K M \cong F[\partial]/F[\partial]L$ as $F[\partial]$-modules.*

PROOF. The first part holds by the remarks preceding the lemma. Suppose $M = K[\partial]/K[\partial]L$ where $L \in K[\partial]^{\neq}$ has order $r \geqslant 1$. Then $e = 1 + K[\partial]L$ is a cyclic vector of M, since $M = Ke \oplus K\partial e \oplus \cdots \oplus K\partial^{r-1}e$ (internal direct sum of K-linear subspaces) and $Le = 0$. Setting $e_F := 1 \otimes e$ we have $\partial^n e_F = 1 \otimes \partial^n e$ for each n and so $F \otimes_K M = Fe_F \oplus F\partial e_F \oplus \cdots \oplus F\partial^{r-1}e_F$ with $Le_F = 1 \otimes Le = 0$. □

If K is a differential field, and A, B are differential K-algebras, then the derivation on the ring $A \otimes_K B$ provided by Corollary 1.7.5 is the unique derivation on this ring for which the (injective) maps $a \mapsto a \otimes 1 \colon A \to A \otimes_K B$ and $b \mapsto 1 \otimes b \colon B \to A \otimes_K B$ are morphisms of differential rings.

The module of differentials as a $K[\partial]$-module

Let K be a differential ring and A a differential K-algebra. The map ∂^* of Corollary 1.8.14 makes the A-module $\Omega_{A|K}$ of Kähler differentials an $A[\partial]$-module (and thus a $K[\partial]$-module), and the universal K-derivation $\mathrm{d}_{A|K} \colon A \to \Omega_{A|K}$ is then $K[\partial]$-linear. Below we view $\Omega_{A|K}$ as an $A[\partial]$-module in this way; so $\partial\omega = \partial^*(\omega)$ for $\omega \in \Omega_{A|K}$.

Let B be a second differential K-algebra and let $h \colon A \to B$ be a morphism of differential K-algebras. Then B is via h also a differential A-algebra, and so we have the $B[\partial]$-module $\Omega_{B|A}$, and the base change $B \otimes_A \Omega_{A|K}$ of the $A[\partial]$-module $\Omega_{A|K}$ to B is a $B[\partial]$-module. By Proposition 1.8.16 we have an exact sequence

$$B \otimes_A \Omega_{A|K} \xrightarrow{\alpha} \Omega_{B|K} \xrightarrow{\beta} \Omega_{B|A} \longrightarrow 0$$

of B-modules and B-linear maps, where $\alpha(1 \otimes \mathrm{d}\,a) = \mathrm{d}\,h(a)$ for all $a \in A$ and $\beta(\mathrm{d}_{B|K}\, b) = \mathrm{d}_{B|A}\, b$ for all $b \in B$. One verifies easily that α and β are $B[\partial]$-linear.

Next, let $I \neq A$ be a differential ideal of A, let $B := A/I$ be the quotient differential K-algebra with derivation ∂_B, with the canonical morphism $h \colon A \to B$ of differential K-algebras. Then the ∂_B-compatible derivation $a + I^2 \mapsto a' + I^2$ (for $a \in I$) on the B-module I/I^2 makes the latter a $B[\partial]$-module. Proposition 1.8.17 yields the exact sequence of B-modules and B-linear maps

$$I/I^2 \xrightarrow{\gamma} B \otimes_A \Omega_{A|K} \xrightarrow{\alpha} \Omega_{B|K} \longrightarrow 0,$$

with $\gamma(a + I^2) = 1 \otimes \mathrm{d}\,a$ for $a \in I$. It is easy to check that γ is even $B[\partial]$-linear.

Finally, let R be a differential K-algebra, let $\mathfrak{p}$ be a differential prime ideal of R, and set $A := R_{\mathfrak{p}}$. Then A is a differential K-algebra and a local ring whose maximal ideal $\mathfrak{m} = \mathfrak{p}A$ is a differential ideal of A, with differential residue field $F := A/\mathfrak{m}$. It is easy to check that the maps α_0 and γ_0 in the exact sequence

$$\mathfrak{m}/\mathfrak{m}^2 \xrightarrow{\gamma_0} F \otimes_R \Omega_{R|K} \xrightarrow{\alpha_0} \Omega_{F|K} \longrightarrow 0$$

of F-linear spaces and F-linear maps from (1.8.6) are $F[\partial]$-linear.

Modules of differentials for differential field extensions

In the rest of this section K is a differential field and L is a differential field extension of K. For any family $(a_i)_{i \in I}$ in L, the following conditions are equivalent:

(1) the family (a_i) in L is d-algebraically independent over K;

(2) the family $\big(a_i^{(n)}\big)_{i,n}$ in L is algebraically independent over K;

(3) the family $\big(\mathrm{d}\, a_i^{(n)}\big)_{i,n}$ in the vector space $\Omega_{L|K}$ is linearly independent over L;

(4) the family $(\mathrm{d}\, a_i)$ in the $L[\partial]$-module $\Omega_{L|K}$ is $L[\partial]$-independent.

For (1) $\Leftrightarrow$ (2) we refer to the subsection on differential transcendence bases in Section 4.1, (2) $\Leftrightarrow$ (3) follows from Lemma 1.9.8, and (3) $\Leftrightarrow$ (4) holds because $\mathrm{d}\, a^{(n)} = \partial^n \mathrm{d}\, a$ for $a \in L$. In particular, (1) $\Leftrightarrow$ (4) yields for $a \in L$:

$$a \text{ is d-algebraic over } K \iff \mathrm{d}\, a \in (\Omega_{L|K})_{\mathrm{tor}}.$$

Thus:

$$L \text{ is d-algebraic over } K \iff \Omega_{L|K} \text{ is a torsion } L[\partial]\text{-module.}$$

Suppose now that $L = K\langle x_1, \dots, x_n\rangle$. Then $L = K\big(x_j^{(m)} : m \in \mathbb{N},\ j = 1, \dots, n\big)$, so the vector space $\Omega_{L|K}$ over L is generated by the $\mathrm{d}\big(x_j^{(m)}\big) = \partial^m \mathrm{d}\, x_j$ with $m \in \mathbb{N}$ and $j = 1, \dots, n$. Hence as an $L[\partial]$-module, $\Omega_{L|K}$ is generated by $\mathrm{d}\, x_1, \dots, \mathrm{d}\, x_n$. Moreover:

COROLLARY 5.9.3. $\operatorname{rank} \Omega_{L|K} = \operatorname{trdeg}_\partial L|K$.

PROOF. By permuting the indices we arrange that $m \leqslant n$ is such that $x_1, \dots, x_m$ is a d-transcendence base of L over K; in particular, $\operatorname{trdeg}_\partial L|K = m$. Also, the elements $\mathrm{d}\, x_1, \dots, \mathrm{d}\, x_m$ of $\Omega_{L|K}$ are $L[\partial]$-independent by (1) $\Leftrightarrow$ (4), and likewise, if $m < j \leqslant n$, then $x_1, \dots, x_m, x_j$ are d-algebraically dependent, so $\mathrm{d}\, x_1, \dots, \mathrm{d}\, x_m, \mathrm{d}\, x_j$ are $L[\partial]$-dependent. Thus $\operatorname{rank} \Omega_{L|K} = m$ by Corollary 5.3.10. □

Independence at a prime

Let R be the differential K-algebra $K\{Y_1, \dots, Y_n\}$. By Lemma 1.8.11 the R-module $\Omega_{R|K}$ is free on its generating family $\big(\mathrm{d}(Y_j^{(r)})\big)_{j,r}$. As an $R[\partial]$-module with $\partial\, \mathrm{d}\, P = \mathrm{d}(P')$ for $P \in R$, it is free on $\mathrm{d}\, Y_1, \dots, \mathrm{d}\, Y_n$.

Let $P_1, \dots, P_m \in R$ and $\mathfrak{p} \in \operatorname{Spec}(R)$. Then we say that $P_1, \dots, P_m$ are **d-independent at** $\mathfrak{p}$ if the family $\big(\mathrm{d}(P_i^{(r)})\big)_{i,r}$ in the R-module $\Omega_{R|K}$ is independent at $\mathfrak{p}$ as defined in Section 1.7. If $P_1, \dots, P_m$ are d-independent at $\mathfrak{p}$ and $\mathfrak{p} \supseteq \mathfrak{q} \in \operatorname{Spec}(R)$, then $P_1, \dots, P_m$ are d-independent at $\mathfrak{q}$, by Lemma 1.7.6. In view of the isomorphism (1.7.3), the following are equivalent:

(1) $P_1, \dots, P_m$ are d-independent at $\mathfrak{p}$;

(2) the family $\big(1 \otimes \mathrm{d}(P_i^{(r)})\big)_{i,r}$ in $(R/\mathfrak{p}) \otimes_R \Omega_{R|K}$ is $(R/\mathfrak{p})$-linearly independent.

The R-module $\Omega_{R|K}$ is free on $\big(\mathrm{d}(Y_j^{(r)})\big)_{j,r}$, so the $(R/\mathfrak{p})$-module $(R/\mathfrak{p}) \otimes_R \Omega_{R|K}$ is free on $\big(1 \otimes \mathrm{d}(Y_j^{(r)})\big)_{j,r}$, and likewise the vector space $F \otimes_R \Omega_{R|K}$ over $F := \operatorname{Frac}(R/\mathfrak{p})$ is free on $\big(1 \otimes \mathrm{d}(Y_j^{(r)})\big)_{j,r}$. Thus (1) and (2) are also equivalent to

(3) the family $\big(1 \otimes \mathrm{d}(P_i^{(r)})\big)_{i,r}$ in $F \otimes_R \Omega_{R|K}$ is F-linearly independent.

If $\mathfrak{p}$ is also a differential ideal of R, then (2) is equivalent to (4), and (3) to (5):

(4) $1 \otimes \mathrm{d}\, P_1, \dots, 1 \otimes \mathrm{d}\, P_m \in (R/\mathfrak{p}) \otimes_R \Omega_{R|K}$ are $(R/\mathfrak{p})[\partial]$-independent;

(5) $1 \otimes \mathrm{d}\, P_1, \dots, 1 \otimes \mathrm{d}\, P_m \in F \otimes_R \Omega_{R|K}$ are $F[\partial]$-independent.

Suppose $\mathfrak{p} \in \mathrm{Spec}(R)$ is a differential ideal of R. The inclusion $K \to R$ composed with the canonical map $R \to R/\mathfrak{p}$ yields an injective differential ring morphism $K \to R/\mathfrak{p}$ via which we identify K below with a differential subring of $R/\mathfrak{p}$.

LEMMA 5.9.4. *Let $y \in L^n$ and let $\mathfrak{p}$ be the kernel of the differential K-algebra morphism $R \to L$ that sends Y_j to y_j for $j = 1, \dots, n$. Then $\mathfrak{p}$ is a differential prime ideal of R, and this morphism $R \to L$ induces a differential ring isomorphism $R/\mathfrak{p} \to K\{y\}$, which extends to a differential field isomorphism $F \to K\langle y \rangle$, where* $F := \mathrm{Frac}(R/\mathfrak{p})$. *Moreover, we have the equivalence*

$$P_1, \dots, P_m \text{ are d-independent at } y \iff P_1, \dots, P_m \text{ are d-independent at } \mathfrak{p}.$$

PROOF. Obvious, except for the final equivalence. For that we use the relevant definitions, the equality (1.8.5), the isomorphism of free $F[\partial]$-modules

$$(A_1, \dots, A_n) \mapsto A_1 \cdot (1 \otimes \mathrm{d}\, Y_1) + \dots + A_n \cdot (1 \otimes \mathrm{d}\, Y_n) : \; F[\partial]^n \to F \otimes_R \Omega_{R|K},$$

and the above equivalence (1) $\Leftrightarrow$ (5). □

An abstract version of Johnson's theorem

We keep the notations introduced in the previous subsection. Theorem 5.9.1 will be derived from the next result.

PROPOSITION 5.9.5. *Let $P_1, \dots, P_n \in R$, $I := [P_1, \dots, P_n]$, and let $\mathfrak{p} \in \mathrm{Spec}(R)$ be such that $I \subseteq \mathfrak{p}$ and $P_1, \dots, P_n$ are d-independent at $\mathfrak{p}$. Set*

$$\mathfrak{q} := \{a \in R : \; ab \in I \text{ for some } b \in R \setminus \mathfrak{p}\}.$$

Then $\mathfrak{q}$ is a differential ideal of R with the following properties:

(i) *$\mathfrak{q}$ is the smallest prime ideal of R containing I and contained in $\mathfrak{p}$;*

(ii) *the differential fraction field of $R/\mathfrak{q}$ is d-algebraic over K.*

PROOF. We consider R and its localization $A := R_{\mathfrak{p}}$ as differential subrings of the differential field $\mathrm{Frac}(R) = K\langle Y \rangle$. Clearly $\mathfrak{q}$ is a differential ideal of R with $I \subseteq \mathfrak{q} \subseteq \mathfrak{p}$, and any prime ideal of R containing I and contained in $\mathfrak{p}$ contains $\mathfrak{q}$; so to finish the proof of (i) it remains to show that $\mathfrak{q}$ is prime. Note that A is a local domain with maximal ideal $\mathfrak{m} := \mathfrak{p}A$, and $(IA) \cap R = \mathfrak{q}$. We shall prove that IA is a prime ideal of A (so $\mathfrak{q}$ is prime as a consequence). To reduce to a more finitary situation, set for $r \in \mathbb{N}$:

$$R_r := K\big[Y_j^{(k)} : j = 1,\dots,n,\ k = 0,\dots,r\big], \quad \mathfrak{p}_r := \mathfrak{p} \cap R_r, \quad A_r := (R_r)_{\mathfrak{p}_r},$$

so A_r is a local subring of A with maximal ideal $\mathfrak{m}_r := \mathfrak{p}_r A_r = \mathfrak{m} \cap A_r$. For IA to be a prime ideal of A, it is clearly enough to show:

CLAIM: *Let $r, N \in \mathbb{N}$ and let distinct*

$$G_1,\dots,G_N \in R_r \cap \big\{P_i^{(k)} : i = 1,\dots,n,\ k \in \mathbb{N}\big\}$$

be given. Then $G_1,\dots,G_N$ generate a prime ideal of A_r.

To prove this claim, set $F := A/\mathfrak{m}$ and $F_r := A_r/\mathfrak{m}_r$, so F_r is a subfield of the field F after the usual identification. We have an exact sequence

$$\text{(5.9.1)} \qquad \mathfrak{m}/\mathfrak{m}^2 \xrightarrow{\gamma_0} F \otimes_R \Omega_{R|K} \longrightarrow \Omega_{F|K} \longrightarrow 0$$

of F-linear spaces, where $\gamma_0(a + \mathfrak{m}^2) = 1 \otimes \mathrm{d}\,a$ for $a \in \mathfrak{p}$. Composing γ_0 with the natural F_r-linear map $\mathfrak{m}_r/\mathfrak{m}_r^2 \to \mathfrak{m}/\mathfrak{m}^2$ yields an F_r-linear map

$$\mathfrak{m}_r/\mathfrak{m}_r^2 \to F \otimes_R \Omega_{R|K}, \qquad a + \mathfrak{m}_r^2 \mapsto 1 \otimes \mathrm{d}\,a \text{ for } a \in \mathfrak{p}_r.$$

Since $P_1,\dots,P_n$ are d-independent at $\mathfrak{p}$, the natural isomorphism $\mathrm{Frac}(R/\mathfrak{p}) \cong F$ and the equivalence (1) $\Leftrightarrow$ (3) preceding Lemma 5.9.4 yield that in $F \otimes_R \Omega_{R|K}$ the elements $1 \otimes \mathrm{d}\,G_1,\dots,1 \otimes \mathrm{d}\,G_N$ are linearly independent over F. Therefore the elements $G_1 + \mathfrak{m}_r^2,\dots,G_N + \mathfrak{m}_r^2$ of $\mathfrak{m}_r/\mathfrak{m}_r^2$ are linearly independent over F_r. Now A_r is a regular local ring by Corollary 1.6.7, so $G_1,\dots,G_N$ generate a prime ideal in A_r by Corollaries 1.4.4(ii) and 1.6.3. This concludes the proof of (i).

Towards proving (ii), first note that by Lemma 1.4.15 we have $\mathfrak{q}R_{\mathfrak{q}} = IR_{\mathfrak{q}}$ and

$$\mathfrak{q} = \{a \in R : ab \in I \text{ for some } b \in R \setminus \mathfrak{q}\}.$$

To show (ii) we replace $\mathfrak{p}$ by $\mathfrak{q}$ to arrange that $\mathfrak{p}$ is a differential ideal of R and $\mathfrak{m} = \mathfrak{p}A = IA$ is the differential ideal of A generated by $P_1,\dots,P_n$. Then (5.9.1) is an exact sequence of $F[\partial]$-modules and $F[\partial]$-linear maps. As $P_1,\dots,P_n$ are d-independent at $\mathfrak{p}$, the elements $1 \otimes \mathrm{d}\,P_1,\dots,1 \otimes \mathrm{d}\,P_n$ of $F \otimes_R \Omega_{R|K}$ are $F[\partial]$-independent. Since the $F[\partial]$-module $\mathfrak{m}/\mathfrak{m}^2$ is generated by $P_1 + \mathfrak{m}^2,\dots,P_n + \mathfrak{m}^2$, the $F[\partial]$-linear map γ_0 yields that the $F[\partial]$-module $\mathfrak{m}/\mathfrak{m}^2$ is free on $P_1 + \mathfrak{m}^2,\dots,P_n + \mathfrak{m}^2$, and γ_0 is injective. The $F[\partial]$-module $F \otimes_R \Omega_{R|K}$ being free on $1 \otimes \mathrm{d}\,Y_1,\dots,1 \otimes \mathrm{d}\,Y_n$, taking ranks of $F[\partial]$-modules, we get $\mathrm{rank}(\Omega_{F|K}) = 0$ from (5.9.1) and Lemma 5.3.9. As the canonical field isomorphism $\mathrm{Frac}(R/\mathfrak{p}) \cong F$ respects the natural derivations on these fields, this "rank 0" fact gives (ii) in view of Corollary 5.9.3. □

PROOF OF THEOREM 5.9.1. We are given $P_1,\dots,P_n \in K\{Y\}$ and $y \in L^n$ such that $P_1(y) = \cdots = P_n(y) = 0$ and $P_1,\dots,P_n$ are d-independent at y; we have to show that then $K\langle y\rangle$ is d-algebraic over K. Let $\mathfrak{p}$ be the kernel of the differential K-algebra morphism $R \to L$ that sends Y_j to y_j for $j = 1, \dots, n$. Then $P_1, \dots, P_n$ are d-independent at $\mathfrak{p}$ by Lemma 5.9.4. Let $\mathfrak{q}$ be as in Proposition 5.9.5. Then (ii) of that proposition yields an element $Q_j \in \mathfrak{q} \cap K\{Y_j\}^{\neq}$ for $j = 1, \dots, n$. Then $Q_j \in \mathfrak{p}$ for $j = 1, \dots, n$, so $K\langle y\rangle$ is d-algebraic over K. □

We finish with an application of Theorem 5.9.1 and Lemma 5.4.7:

COROLLARY 5.9.6. *Suppose $K \subseteq L$ is a differential field extension, $P_1, \dots, P_n$ in $K\{Y\}$ have order at most $(r_1, \dots, r_n) \in \mathbb{N}^n$, and $y = (y_1, \dots, y_n) \in L^n$ satisfies*

$$P_1(y) = \cdots = P_n(y) = 0, \qquad \det\Big(\big(\partial P_i/\partial Y_j^{(r_j)}\big)(y)\Big) \neq 0.$$

Then $y_1, \dots, y_n$ are d*-algebraic over K.*

PROOF. By the hypothesis $P_1, \dots, P_n$ are d-independent at y. □

EXAMPLE. Let $Q_1, \dots, Q_n \in K\{Y\}$ have order at most $(r_1, \dots, r_n) \in \mathbb{N}^n$. If L is a differential field extension of K and $y = (y_1, \dots, y_n) \in L^n$ satisfies

$$y_i^{(r_i+1)} = Q_i(y) \qquad (i = 1, \dots, n),$$

then $y_1, \dots, y_n$ are d-algebraic over K. (Take $P_i := Y_i^{(r_i+1)} - Q_i$.)

Notes and comments

Proposition 5.9.5 is from [205]. See also [204] for more information on Kähler differentials in the context of (partial) differential fields. An analytic version of Corollary 5.9.6 is due to Rubel [372], whose proof is elementary and independent of Johnson's theorem.

Chapter Six

Valued Differential Fields

Throughout this chapter, K is a valued differential field whose derivation is small in the sense that $\partial\smallo \subseteq \smallo$. In Chapter 4 we already derived some consequences of this assumption, and here we go much further with it, sometimes under mild additional conditions. As noted before, one thing we get from $\partial\smallo \subseteq \smallo$ is that $\partial\mathcal{O} \subseteq \mathcal{O}$, and so the residue field $\boldsymbol{k} = \mathcal{O}/\smallo$ is naturally a differential field. Sometimes we impose as extra condition that the derivation of $\boldsymbol{k}$ is nontrivial. While this extra condition is not satisfied for $K = \mathbb{T}$, it does hold in suitable coarsenings of compositional conjugates of $\mathbb{T}$, and this is how the results of this chapter apply to $\mathbb{T}$, and other valued differential fields of interest.

Section 6.1 considers the asymptotic behavior of the function $v_P\colon \Gamma \to \Gamma$ for homogeneous $P \in K\{Y\}^{\neq}$. In Section 6.2 we show among other things that the derivation of any valued differential field extension of K that is algebraic over K is also small. In Section 6.3 we show how differential field extensions of the residue differential field $\boldsymbol{k}$ give rise to valued differential field extensions of K with small derivation and the same value group. We also study there monotone extensions.

In Section 6.4 we show how the derivation and the valuation on K jointly give rise to a valuation on Γ. This helps in improving the estimates on v_P from Section 6.1. The ordered abelian group Γ with this valuation is an *asymptotic couple* in the sense of Rosenlicht, and for frequent later use we prove in Section 6.5 some basic facts about asymptotic couples in general. In Section 6.6 we define the *dominant part* of a differential polynomial, and establish its basic properties. The key facts about the functions v_P, the valuation on Γ, and dominant parts are then used to prove the important Equalizer Theorem in Section 6.7:

THEOREM 6.0.1. *Let $P, Q \in K\{Y\}^{\neq}$ be homogeneous of degree d and e, respectively, with $d > e$, and suppose $(d-e)\Gamma = \Gamma$. Then there is a unique $\alpha \in \Gamma$ such that $v_P(\alpha) = v_Q(\alpha)$.*

We shall use this mainly for $d = 1$, $e = 0$, or when Γ is divisible, and in either case the condition $(d-e)\Gamma = \Gamma$ is trivially satisfied. (For $d = 1$, $e = 0$ the theorem says that $v_A\colon \Gamma \to \Gamma$ is bijective for each $A \in K[\partial]^{\neq}$.)

In Section 6.8 we consider a pc-sequence (a_ρ) in K and a differential polynomial $G \in K\{Y\} \setminus K$. If the derivation of $\boldsymbol{k}$ is nontrivial, then we show how to replace (a_ρ) by an equivalent pc-sequence to arrange that $G(a_\rho)$ is also a pc-sequence. We also prove some variants of this important fact, and use this in Section 6.9 to construct immediate extensions of such K in analogy with Kaplansky's lemmas 3.2.6 and 3.2.7. This has the effect that any such K has a spherically complete immediate valued differential field extension with small derivation.

Notes and comments

The key result in this chapter that is needed later in connection with the model theory of $\mathbb{T}$ is the Equalizer Theorem of Section 6.7. Other sections are clearly part of any reasonably broad theory of valued differential fields, and play a role in the next chapter on differential-henselian fields. Sections 6.8 and 6.9 are inspired by analogous results on certain kinds of valued difference fields due to Bélair, Macintyre, and Scanlon [41].

6.1 ASYMPTOTIC BEHAVIOR OF v_P

Here is a key consequence of our assumption $\partial \mathcal{O} \subseteq \mathcal{O}$:

LEMMA 6.1.1. *If $y \in \mathcal{O}$, then $(y')^{n+1} \preccurlyeq y^n$ for all n.*

PROOF. Suppose there is a counterexample. Take n least with $(y')^{n+1} \succ y^n$ for some $y \in \mathcal{O}$, and fix such y. Then $n \geqslant 1$ and $y^n = \varepsilon(y')^{n+1}$, $\varepsilon \prec 1$, so

$$ny^{n-1}y' = \varepsilon'(y')^{n+1} + (n+1)\varepsilon(y')^n y''.$$

After dividing by y' this gives

$$ny^{n-1} = \varepsilon'(y')^n + (n+1)\varepsilon(y')^{n-1}y''.$$

By the minimality of n we have $\varepsilon'(y')^n \prec y^{n-1}$, so $y^{n-1} \asymp \varepsilon(y')^{n-1}y''$, and hence $y^{n-1}y' \asymp \varepsilon(y')^n y''$, so by taking nth powers,

$$y^{n(n-1)}(y')^n \asymp \varepsilon^n (y')^{n^2}(y'')^n.$$

Thus, using $y^n = \varepsilon(y')^{n+1}$,

$$\varepsilon^{n-1}(y')^{(n+1)(n-1)}(y')^n \asymp \varepsilon^n (y')^{n^2}(y'')^n,$$

so $(y')^{n-1} \asymp \varepsilon(y'')^n$, contradicting the minimality of n. □

COROLLARY 6.1.2. *If Γ has rank 1, then K is monotone.*

In the remainder of this section α, β, γ range over Γ.

COROLLARY 6.1.3. *Let $P \in K\{Y\}^{\neq}$ be homogeneous of degree d and $\gamma \neq 0$. Then $v_P(\gamma) = v(P) + d\gamma + o(\gamma)$. More generally, if $\alpha \neq \beta$, then*

$$v_P(\alpha) - v_P(\beta) = d \cdot (\alpha - \beta) + o(\alpha - \beta).$$

PROOF. Let $\gamma > 0$, take $g \in K^\times$ with $vg = \gamma$, and set $\Delta := \{\delta \in \Gamma : \delta = o(\gamma)\}$. In the Δ-coarsening of K we have $\dot{v}g' \geqslant \dot{v}g$, so $\dot{v}_P(\dot{\gamma}) = \dot{v}(P) + d\dot{\gamma}$ by Lemma 4.5.3, and thus $v_P(\gamma) = v(P) + d\gamma + o(\gamma)$. Next, for $Q := P_{\times g^{-1}}$,

$$v(P) = v_Q(\gamma) = v(Q) + d\gamma + o(\gamma), \qquad v(Q) = v_P(-\gamma),$$

so $v_P(-\gamma) = v(P) - d\gamma + o(-\gamma)$. Thus the first part of the lemma also holds if $\gamma < 0$. For the second part, let $\alpha \neq \beta$ and take $a, b \in K^\times$ with $va = \alpha$, $vb = \beta$. Then

$$v_P(\alpha) = v(P_{\times a}) = v(P_{\times b, \times (a/b)}) = v_{P_{\times b}}(\alpha - \beta),$$

so by the first part,

$$v_P(\alpha) = v(P_{\times b}) + d(\alpha - \beta) + o(\alpha - \beta) = v_P(\beta) + d(\alpha - \beta) + o(\alpha - \beta),$$

which gives the desired result. □

Here is some convenient notation. Let $\gamma > 0$. Then $\alpha \geqslant \beta + o(\gamma)$ is defined to mean that $\alpha \geqslant \beta + \delta$ for some $\delta = o(\gamma)$ in Γ; equivalently, it means that $\alpha \geqslant \beta - \frac{1}{n}\gamma$ for all $n \geqslant 1$. We also declare that $\infty \geqslant \beta + o(\gamma)$ for $\infty \in \Gamma_\infty$.

COROLLARY 6.1.4. *Let $P \in K\{Y\}^{\neq}$, $P(0) = 0$, and $\gamma > 0$. Then*

$$v_P(\gamma) \geqslant v(P) + \gamma + o(\gamma).$$

COROLLARY 6.1.5. *Let $P, Q \in K\{Y\}^{\neq}$ be homogeneous of degrees d, e. Then*

$$\gamma \neq 0 \implies v_P(\gamma) - v_Q(\gamma) = v(P) - v(Q) + (d - e)\gamma + o(\gamma).$$

Also, if $d > e$, then $v_P - v_Q \colon \Gamma \to \Gamma$ is strictly increasing.

PROOF. The second part here follows from the second part of 6.1.3. □

LEMMA 6.1.6. *Let $P, Q \in K\{Y\}^{\neq}$ be homogeneous of the same degree and let $\gamma > 0$ be such that $v(P) \geqslant v(Q) + \gamma$. Then for $\alpha = O(\gamma)$ we have $v_P(\alpha) > v_Q(\alpha)$.*

PROOF. This is clear if $\alpha = 0$, so let $0 \neq \alpha = O(\gamma)$. Then by Corollary 6.1.5,

$$v_P(\alpha) - v_Q(\alpha) = v(P) - v(Q) + o(\gamma) \geqslant \gamma + o(\gamma) > 0. \qquad \square$$

The next consequence of Lemma 6.1.6 is needed in Section 7.2.

COROLLARY 6.1.7. *Let $A, B \in K[\partial]^{\neq}$, $B = A + E$, $\gamma > 0$, $v(E) \geqslant v(A) + \gamma$, and $\alpha = O(\gamma)$. Then $v_A(\alpha) = v_B(\alpha)$. Moreover, for any $a \in K^\times$ with $va = \alpha$ we have $A(a) \prec Aa \iff B(a) \prec Ba$. Therefore, $\alpha \in \mathscr{E}(A) \iff \alpha \in \mathscr{E}(B)$.*

PROOF. By Corollary 6.1.6 we have $v_E(\alpha) > v_A(\alpha)$ and thus $v_A(\alpha) = v_B(\alpha)$. Let $a \in K^\times$ be such that $va = \alpha$ and $A(a) \prec Aa$. Then $E(a) \preccurlyeq Ea \prec Aa \asymp Ba$, so $B(a) = A(a) + E(a) \prec Ba$. We can interchange A and B in this argument. □

To get useful information from Corollary 6.1.5 concerning $v_F\colon \Gamma \to \Gamma$ when the differential polynomial $F \in K\{Y\}^{\neq}$ is not homogeneous, we need two order-theoretic facts. To formulate those facts, let X and Y be nonempty ordered sets.

LEMMA 6.1.8. *Let $f_0, \dots, f_n\colon X \to Y$ be strictly increasing bijections. Then the map $f\colon X \to Y$ given by $f(x) := \min_i f_i(x)$ is a strictly increasing bijection.*

PROOF. Define $g\colon Y \to X$ by $g(y) := \max_i f_i^{-1}(y)$. One verifies easily that then $g \circ f = \mathrm{id}_X$ and $f \circ g = \mathrm{id}_Y$. □

Given maps $f, g\colon X \to Y$ we define $f <_c g$ to mean that there are convex subsets $A < B < C$ in X such that $X = A \cup B \cup C$, $f > g$ on A, $f = g$ on B, and $f < g$ on C. (Some of A, B, C can be empty.)

LEMMA 6.1.9. *Let $f_0, \dots, f_n\colon X \to Y$ be such that $f_i <_c f_j$ for all $i < j$. Define $f\colon X \to Y$ by $f(x) := \min_i f_i(x)$. Then there are $i_0 < \cdots < i_m$ in $\{0, \dots, n\}$ and convex subsets $D_m < \cdots < D_0$ of X such that*

(i) *$X = D_m \cup \cdots \cup D_0$, and $D_m, \dots, D_0$ are nonempty;*

(ii) *$f = f_{i_k}$ on D_k, for $k = 0, \dots, m$.*

PROOF. For $n = 0$ the lemma holds with $m = 0$ and $D_0 = X$. Let $n \geqslant 1$, and for $i = 1, \dots, n$, take convex subsets $A_i < B_i < C_i$ in X such that $X = A_i \cup B_i \cup C_i$, $f_0 > f_i$ on A_i, $f_0 = f_i$ on B_i, and $f_0 < f_i$ on C_i. Take $i^* \in \{1, \dots, n\}$ such that $C_{i^*} \subseteq C_i$ for $i = 1, \dots, n$. Then $f_0 < f_i$ on $D_0 := C_{i^*}$ for $i = 1, \dots, n$, so if $D_0 = X$, then the lemma holds with $m = 0$, $i_0 = 0$. Suppose that $D_0 \neq X$. Then for $X' := X \setminus C$ we have $f_0 \geqslant f_{i^*}$ on X', and then we can use as an inductive assumption that the lemma holds for the restrictions of $f_1, \dots, f_n$ to X'. □

COROLLARY 6.1.10. *Let $F \in K\{Y\}^{\neq}$ have $F_{d_0}, \dots, F_{d_n}$ with $d_0 < \cdots < d_n$ as its nonzero homogeneous parts. Set $f_i := v_{F_{d_i}}\colon \Gamma \to \Gamma$ for $i = 0, \dots, n$. Then there are $i_0 < \cdots < i_m$ in $\{0, \dots, n\}$ and convex subsets $D_m < \cdots < D_0$ of Γ such that*

(i) *$\Gamma = D_m \cup \cdots \cup D_0$, and $D_m, \dots, D_0$ are nonempty;*

(ii) *$v_F(\gamma) = f_{i_k}(\gamma)$ for all $\gamma \in D_k$ and $k = 0, \dots, m$.*

If $F(0) = 0$ and each f_i is bijective, then so is v_F.

PROOF. By Corollary 6.1.5 we have $f_i <_c f_j$ for $0 \leqslant i < j \leqslant n$, in the sense defined earlier. We can now apply Lemma 6.1.9, since $v_F(\gamma) = \min_i f_i(\gamma)$ for all γ, and if $F(0) = 0$, then each function f_i is strictly increasing. □

Notes and comments

Lemma 6.1.1 and Corollary 6.1.2 are in [87], Section 2.

6.2 ALGEBRAIC EXTENSIONS

Recall the standing assumption of this chapter that the derivation of K is small, that is, $\partial\smallo \subseteq \smallo$. At the end of this section we prove:

PROPOSITION 6.2.1. *If L is a valued differential field extension of K and L is algebraic over K, then the derivation of L is also small.*

We actually work in the more general setting of an ambient valued differential field L with subfields $E \subseteq F$ such that F is algebraic over E. In the lemmas below we consider E and F as *valued* subfields of L, and first deal with the case that $F|E$ is immediate, next the case that E is henselian and $F|E$ is unramified, and finally the purely ramified case. It is important for later use in Section 6.9 not to assume in the first two cases that E or F is closed under the derivation ∂ of L.

Immediate algebraic extensions

LEMMA 6.2.2. *Suppose $F|E$ is immediate, and $\partial\smallo_E \subseteq \smallo_L$ and $\partial\mathcal{O}_E \subseteq \mathcal{O}_L$. Then $\partial\smallo_F \subseteq \smallo_L$ and $\partial\mathcal{O}_F \subseteq \mathcal{O}_L$.*

PROOF. Since we have $\mathcal{O}_F = \mathcal{O}_E + \smallo_F$, it suffices to show that $\partial\smallo_F \subseteq \smallo_L$. If $\mathcal{O}_E = E$, then $\mathcal{O}_F = F$ and we are done. Assume $\mathcal{O}_E \neq E$. We can also arrange that $1 < [F : E] < \infty$. Then by Lemma 3.3.30 we have $F = E(y)$ where $y \in L^\times$, $y \prec 1$ and y has minimum polynomial

$$P(Y) = Y^n + a_{n-1}Y^{n-1} + \cdots + a_1 Y + a_0$$

over E with coefficients $a_i \in \mathcal{O}_E$, $a_1 \asymp 1$, $a_0 \prec 1$. (Note that $a_0 \neq 0$, since $n = [F : E] > 1$.)

CLAIM 1: *We have $y \asymp a_0$, and $y' \preccurlyeq (aa_0)'$ for some $a \in \mathcal{O}_E$.*

We get $y \asymp a_0$ from $a_1 y \sim -a_0$. Next, $P(y) = 0$ gives

$$0 = P(y)' = P^{\partial}(y) + P'(y)y',$$

and since $P'(y) \sim a_1 \asymp 1$, this gives

$$y' \asymp P^{\partial}(y) = \sum_{i=0}^{n-1} a_i' y^i,$$

so we get $i \in \{0, \dots, n-1\}$ with $y' \preccurlyeq a_i' y^i$. If $i = 0$, this gives $y' \preccurlyeq a_0'$. If $i \geqslant 1$, then we get $y' \preccurlyeq a_i' a_0 = (a_i a_0)' - a_i a_0'$, so $y' \preccurlyeq (a_i a_0)'$, or $y' \preccurlyeq a_i a_0' \preccurlyeq a_0'$.

By Lemma 3.3.31,

$$\smallo(y) := \smallo_E \mathcal{O}_E[y] + y\mathcal{O}_E[y]$$

is a maximal ideal of $\mathcal{O}_E[y]$, and

$$\begin{aligned} \mathcal{O}_F &= \mathcal{O}_E[y]_{o(y)} = S^{-1}\mathcal{O}_E[y] \text{ with } S := 1 + o(y), \\ o_F &= o(y)\mathcal{O}_F = S^{-1}o(y). \end{aligned}$$

CLAIM 2: *For each $b \in \mathcal{O}_E$ there exists $u \in \mathcal{O}_E$ such that $(by)' \preccurlyeq (ua_0)'$.*

To see why, let $b \in \mathcal{O}_E$, so $(by)' = b'y + by'$. Now $b'y \asymp b'a_0 = (ba_0)' - ba_0'$, so $b'y \preccurlyeq (ba_0)'$ or $b'y \preccurlyeq ba_0' \preccurlyeq a_0'$. Also, with a as in Claim 1 we have $by' \preccurlyeq (aa_0)'$. Thus $(by)' \preccurlyeq (ua_0)'$ for $u = b$ or $u = 1$ or $u = a$.

CLAIM 3: *For each $\phi \in o(y)$ there exists $\varepsilon \in o_E$ such that $\phi' \preccurlyeq \varepsilon'$.*

This property holds for $\phi \in o_E$, and for $\phi = y$ by Claim 1, and it is inherited under taking sums and products of elements in $o(y)$. This yields Claim 3 using also Claim 2.

Let $f \in o_F$. Then $f = \phi/(1+e)$ with $\phi, e \in o(y)$; differentiating this quotient, we obtain from Claim 3 that there is $\varepsilon \in o_E$ such that $f' \preccurlyeq \varepsilon'$. □

Here is a further reduction of the problem of showing $\partial o_F \subseteq o_L$ and $\partial \mathcal{O}_F \subseteq \mathcal{O}_L$:

LEMMA 6.2.3. *Suppose $F|E$ is immediate and $S \subseteq \mathcal{O}_E$ satisfies*

$$\mathcal{O}_E = \{f/g : f, g \in S, g \asymp 1\}, \quad \partial S \subseteq \mathcal{O}_L, \quad \partial(S \cap o_E) \subseteq o_L.$$

Then $\partial o_F \subseteq o_L$ and $\partial \mathcal{O}_F \subseteq \mathcal{O}_L$.

PROOF. Let $a \in \mathcal{O}_E$, so $a = f/g$ with $f, g \in S$, $g \asymp 1$. Then $f', g' \preccurlyeq 1$, and thus $a' = (f'g - fg')/g^2 \preccurlyeq 1$. Also, if $a \prec 1$, then $f \prec 1$, so $f' \prec 1$, and thus $a' \prec 1$. Now apply Lemma 6.2.2. □

Unramified and purely ramified algebraic extensions

LEMMA 6.2.4. *Suppose E is henselian, $\Gamma_F = \Gamma_E$, and $\partial o_E \subseteq o_L$ and $\partial \mathcal{O}_E \subseteq \mathcal{O}_L$. Then $\partial o_F \subseteq o_L$ and $\partial \mathcal{O}_F \subseteq \mathcal{O}_L$.*

PROOF. We can arrange $[F : E] = [\text{res}(F) : \text{res}(E)] = n > 1$. Take $y \asymp 1$ in F such that $\text{res}(F) = \text{res}(E)[\overline{y}]$, where $\overline{y}$ is the residue class of y in $\text{res}(F)$. Corollary 3.3.12 gives $a_0, \dots, a_{n-1} \preccurlyeq 1$ in E such that

$$P(Y) = Y^n + a_{n-1}Y^{n-1} + \cdots + a_1 Y + a_0 \in E[Y]$$

is the minimum polynomial of y over E, so its reduction $\overline{P}(Y)$ in $\text{res}(E)[Y]$ is the minimum polynomial of $\overline{y}$ over $\text{res}(E)$; in particular $P'(y) \asymp 1$. Moreover,

$$P'(y)y' = -\sum_{i=0}^{n-1} a_i' y^i,$$

hence $y' \preccurlyeq 1$. Let $f \in F$, and take $f_0, \dots, f_{n-1} \in E$ such that

$$f = f_0 + f_1 y + \cdots + f_{n-1} y^{n-1}, \text{ so}$$
$$f' = \sum_{i=0}^{n-1} f_i' y^i + \left(\sum_{j=1}^{n-1} j f_j y^{j-1} \right) y'.$$

If $f \prec 1$, then $f_0, \dots, f_{n-1} \prec 1$, and thus $f' \prec 1$. Likewise, if $f \preccurlyeq 1$, then $f_0, \dots, f_{n-1} \preccurlyeq 1$, and thus $f' \preccurlyeq 1$. □

LEMMA 6.2.5. *Let $F = E(u^{1/p})$ where p is a prime number, $u \in E^\times$ and $vu \notin p\Gamma_E$, and assume that $\partial E \subseteq E$ and $\partial o_E \subseteq o_E$. Then $\partial o_F \subseteq o_F$.*

PROOF. Let $u^{i/p} := (u^{1/p})^i$ for $i \in \mathbb{Z}$. By Lemma 3.1.28, $\operatorname{res}(E) = \operatorname{res}(F)$. Let $a \in F^\times$, $a \prec 1$; our job is to show that $a' \prec 1$. We have

$$a = a_0 + a_1 u^{1/p} + \cdots + a_{p-1} u^{(p-1)/p}$$

with $a_0, \dots, a_{p-1} \in E$. Then $a_i u^{i/p} \prec 1$ for $i = 0, \dots, p-1$, so we may reduce to the case $a = a_i u^{i/p}$, and so $a^p = b \in E^\times$. Then $pa^{p-1}a' = b'$, and so by Lemma 6.1.1,

$$va' = vb' - (p-1)va = vb' - \big((p-1)/p\big)vb > 0,$$

as desired. □

PROOF OF PROPOSITION 6.2.1. Let the valued differential field extension L of K be algebraic over K. Note that any field between K and L is closed under the derivation of L. By extending L we arrange that L is an algebraic closure of K. Next, by Lemma 6.2.2, we arrange that K is henselian. We then reach L in two steps. In the first step we pass from K to its maximal unramified extension K^{unr} in L. Then the derivation on K^{unr} is small by Lemma 6.2.4. In the second step we apply Proposition 3.3.48 and Lemma 6.2.5 to L as an extension of K^{unr}. □

6.3 RESIDUE EXTENSIONS

The differential field $K\langle Y \rangle$ with the gaussian extension of the valuation of K is a valued differential field extension of K, with differential subring $\mathcal{O}\{Y\}$ of $K\{Y\}$, and we claim that

(1) $\mathcal{O}\{Y\} \subseteq \mathcal{O}_{K\langle Y \rangle}$ and $\partial o_{K\langle Y \rangle} \subseteq o_{K\langle Y \rangle}$;

(2) $Y \preccurlyeq 1$, and the image y of Y in the differential residue field of $K\langle Y \rangle$ is d-transcendental over the differential residue field $\boldsymbol{k}$ of K;

(3) $\boldsymbol{k}\langle y \rangle$ is the differential residue field of $K\langle Y \rangle$.

For (1), consider first the case of $P \in K\{Y\}$ with $vP > 0$. Then $vP' > 0$, because $P = gF$ with $g \in \mathcal{o}$ and $F \in \mathcal{O}\{Y\}$, so $P' = g'F + gF'$. Next, let $f(Y) \in \mathcal{o}_{K\langle Y\rangle}$. Then $f = P/Q$ with $P, Q \in K\{Y\}$ and $vQ = 0$ and $vP > 0$, so $f' = (P'Q - PQ')/Q^2$, and $v(P'Q) = vP' > 0$ and $v(PQ') \geqslant vP > 0$, so $vf' > 0$. For (2), let $g(Y) \in \boldsymbol{k}\{Y\}^{\neq}$ and take $P \in \mathcal{O}\{Y\}$ with $\overline{P} = g$. Then $vP = 0$ and so $g(y)$, the image of P under the residue map, is $\neq 0$. For (3), let $f(Y) \in K\langle Y\rangle$ and $vf = 0$. Then $f = P/Q$ with $P, Q \in K\{Y\}$ and $vP = vQ = 0$, so the image of f in the residue field of $K\langle Y\rangle$ is $\overline{P}(y)/\overline{Q}(y) \in \boldsymbol{k}\langle y\rangle$.

This valued differential field extension $K\langle Y\rangle$ has the following universal property:

LEMMA 6.3.1. *Let L be any valued differential field extension of K with $\partial\mathcal{o}_L \subseteq \mathcal{o}_L$ and with an element $a \preccurlyeq 1$ whose image in the differential residue field of L is* d-*transcendental over $\boldsymbol{k}$. Then there is a unique valued differential field embedding $K\langle Y\rangle \to L$ over K sending Y to a.*

Here is an analogue of the above for d-*algebraic* residue field extensions:

THEOREM 6.3.2. *Let $K\langle a\rangle$ be a differential field extension of K such that a is* d-*algebraic over K with minimal annihilator F over K of order r. Assume that $vF = 0$ and $vI = 0$ where I is the initial of F, and that the image $\overline{F}$ of F in $\boldsymbol{k}\{Y\}$ is irreducible. Then v extends uniquely to a valuation $v\colon K\langle a\rangle^{\times} \to \Gamma$ such that $\partial\mathcal{o}_{K\langle a\rangle} \subseteq \mathcal{o}_{K\langle a\rangle}$, $a \preccurlyeq 1$, and* $\operatorname{res} a$ *is* d-*algebraic over $\boldsymbol{k}$ with minimal annihilator $\overline{F}$ over $\boldsymbol{k}$. For this extended valuation the differential residue field of $K\langle a\rangle$ is $\boldsymbol{k}\langle\operatorname{res} a\rangle$.*

PROOF. Let F have order r, and set $d := \deg_{Y^{(r)}} F$, so $d \geqslant 1$. Then $\overline{F}$ has order r and $d = \deg_{Y^{(r)}} \overline{F}$. Note that if $b \in K\langle a\rangle^{\times}$, then $b = P(a)/Q(a)$ where $Q \in K\big[Y, \dots, Y^{(r-1)}\big]^{\neq}$ and $P \in K\big[Y, \dots, Y^{(r)}\big]^{\neq}$ with $\deg_{Y^{(r)}} P < d$.

Suppose an extension of the valuation of K to a valuation v of $K\langle a\rangle^{\times}$ as in the theorem is given. Let $P \in K\big[Y, \dots, Y^{(r)}\big]^{\neq}$ and $\deg_{Y^{(r)}} P < d$; we claim that then $vP(a) = vP$. To prove this claim we can multiply P by an element of $K^{\times}$ and arrange that $vP = 0$. Then $\overline{P} \neq 0$ and $\deg_{Y^{(r)}} \overline{P} < d$, so $\operatorname{res} P(a) = \overline{P}(\operatorname{res} a) \neq 0$, and thus $vP(a) = 0$, as claimed. Thus there can be at most one such extension.

To construct such an extension, let $b = P(a)/Q(a)$ be as above; we claim that then $vP - vQ$ depends only on b and not on the choice of P and Q. To prove this claim, let also $b = G(a)/H(a)$ where $H \in K\big[Y, \dots, Y^{(r-1)}\big]^{\neq}$ and $G \in K\big[Y, \dots, Y^{(r)}\big]^{\neq}$ with $\deg_{Y^{(r)}} G < d$. Then

$$\deg_{Y^{(r)}} HP - QG \; < \; d, \qquad (HP - QG)(a) \; = \; 0,$$

so $HP = QG$, and thus $vP - vQ = vG - vH$ as claimed. This allows us to extend the valuation v on K to a function $v\colon K\langle a\rangle^{\times} \to \Gamma$ by setting $vb := vP - vQ$ when $b = P(a)/Q(a)$ for b, P, Q as above. Next we claim that $v\colon K\langle a\rangle^{\times} \to \Gamma$ is a valuation on the field $K\langle a\rangle$. To prove this, let $b, c \in K\langle a\rangle^{\neq}$, and $b = P(a)/Q(a)$ as above, and also $c = G(a)/H(a)$ with $H \in K\big[Y, \dots, Y^{(r-1)}\big]^{\neq}$ and $G \in K\big[Y, \dots, Y^{(r)}\big]^{\neq}$ with $\deg_{Y^{(r)}} G < d$. We can arrange $Q = H$. Then $b + c = (P + G)(a)/Q(a)$, and so, if $b + c \neq 0$,

$$v(b + c) \; = \; v(P + G) - vQ \; \geqslant \; \min(vP - vQ, vG - vQ) \; = \; \min(vb, vc).$$

As to $v(bc) = vb + vc$, this holds if $G \in K[Y, \dots, Y^{(r-1)}]$, because then $bc = (PG)(a)/Q^2(a)$. In particular, it holds for $c \in K(a, \dots, a^{(r-1)})$. This allows us to reduce the general case to the case that $b = P(a)$ and $c = G(a)$ with P, G as above satisfying the additional condition $vP = vG = 0$ (so $vb = vc = 0$). Since $vI = 0$, division with remainder in $\mathcal{O}[Y, \dots, Y^{(r)}]$ gives $I^m PG = BF + R$ with $B, R \in \mathcal{O}[Y, \dots, Y^{(r)}]$ and $\deg_{Y^{(r)}} R < d$. Then $vR = 0$, since $vR > 0$ would give $\overline{I}^m \cdot \overline{P} \cdot \overline{G} = \overline{B} \cdot \overline{F}$ in $\boldsymbol{k}\{Y\}$, contradicting the irreducibility of $\overline{F}$. Now $bc = R(a)/I^m(a)$, so $v(bc) = vR - mvI = 0$, as required. We have now shown that v is a valuation on $K\langle a\rangle$.

Towards proving that $\partial \mathcal{o}_{K\langle a\rangle} \subseteq \mathcal{o}_{K\langle a\rangle}$ we first note that $a^{(i)} \preccurlyeq 1$ for $i < r$. We have $F_0, \dots, F_d \in \mathcal{O}[Y, \dots, Y^{(r-1)}]$ with $F_d = I$ such that

$$F = F_d \cdot \left(Y^{(r)}\right)^d + F_{d-1} \cdot \left(Y^{(r)}\right)^{d-1} + \cdots + F_0.$$

Thus $a^{(r)}$ is integral over the subring $\mathcal{O}[a, \dots, a^{(r-1)}]_{I(a)}$ of the valuation ring $\mathcal{O}_{K\langle a\rangle}$ of $K\langle a\rangle$, so $a^{(r)} \preccurlyeq 1$. In view of $f_i := F_i(a)$ and $v(F_i) \geqslant 0$, this gives $f_i' \preccurlyeq 1$ for $i = 0, \dots, d$. Now with $g := a^{(r)}$ we have

$$0 = F(a)' = f_d' g^d + f_{d-1}' g^{d-1} + \cdots + f_0' + \frac{\partial F}{\partial Y^{(r)}}(a) \cdot g',$$

with $\deg_{Y^{(r)}} \frac{\partial F}{\partial Y^{(r)}} < d$ and $v \frac{\partial F}{\partial Y^{(r)}} = 0$ (using $vI = 0$), so $g' = a^{(r+1)} \preccurlyeq 1$. Now let $b = P(a)/Q(a)$ be as above with $b \in \mathcal{o}_{K\langle a\rangle}$. We can arrange $vP > 0$ and $vQ = 0$, and then

$$b' = \frac{P(a)'Q(a) - P(a)Q(a)'}{Q(a)^2},$$

which in view of $a^{(i)} \preccurlyeq 1$ for $i = 0, \dots, r+1$ yields $P(a)' \prec 1$, $Q(a)' \preccurlyeq 1$, in addition to $P(a) \prec 1$ and $Q(a) \asymp 1$. Thus $b' \in \mathcal{o}_{K\langle a\rangle}$, as promised. From $F(a) = 0$ we get $\overline{F}(\operatorname{res} a) = 0$. Suppose $A \in \boldsymbol{k}\{Y\}^{\neq}$ has order at most r and $\deg_{Y^{(r)}} A < d$. Take $P \in \mathcal{O}\{Y\}^{\neq}$ of order at most r with $\deg_{Y^{(r)}} P < d$ such that $\overline{P} = A$. Then $vP = 0$, so $vP(a) = 0$, and thus $A(\operatorname{res} a) = \operatorname{res} P(a) \neq 0$. Thus $\overline{F}$ is a minimal annihilator of $\operatorname{res} a$ over $\boldsymbol{k}$.

Now let $b = P(a)/Q(a)$ as before be such that $vb = 0$. Then we can assume $vP = vQ = 0$, so $\operatorname{res} b = \overline{P}(\operatorname{res} a)/\overline{Q}(\operatorname{res} a) \in \boldsymbol{k}\langle \operatorname{res} a\rangle$. Thus the differential residue field of $K\langle a\rangle$ is $\boldsymbol{k}\langle \operatorname{res} a\rangle$. □

This gives a partial analogue of Proposition 3.1.34 for valued differential fields:

COROLLARY 6.3.3. *Let $\boldsymbol{k}_L$ be a differential field extension of $\boldsymbol{k}$. Then K has a valued differential field extension L with $\partial \mathcal{o}_L \subseteq \mathcal{o}_L$, with the same value group as K and with differential residue field isomorphic to $\boldsymbol{k}_L$ over $\boldsymbol{k}$.*

PROOF. We can reduce to the case $\boldsymbol{k}_L = \boldsymbol{k}\langle y\rangle$. If y is d-transcendental over $\boldsymbol{k}$, the corollary holds with $L = K\langle Y\rangle$ equipped with the gaussian extension of the valuation of K. Next, suppose y is d-algebraic over $\boldsymbol{k}$, with minimal annihilator $\overline{F}(Y) \in \boldsymbol{k}\{Y\}$

over $\boldsymbol{k}$. Take $F \in \mathcal{O}\{Y\}$ to have image $\overline{F}$ in $\boldsymbol{k}\{Y\}$ and to have the same complexity as $\overline{F}$. Then $vF = vI = 0$, where I is the initial of F, and F is irreducible in $K\{Y\}$. Take a differential field extension $L = K\langle a\rangle$ of K such that a has minimal annihilator F over K. Then L with the valuation defined in Theorem 6.3.2 has the desired properties. □

Preserving monotonicity

Let L be an ambient valued differential field with derivation ∂, and consider subfields of L as valued subfields. Let E be a subfield of L such that $a' \preccurlyeq a$ for all $a \in E$. In particular $\partial o_E \subseteq o_L$ and $\partial \mathcal{O}_E \subseteq \mathcal{O}_L$. We do not assume $\partial E \subseteq E$. After proving some lemmas we shall derive:

PROPOSITION 6.3.4. *If $b \in L$ is algebraic over E, then $b' \preccurlyeq b$.*

LEMMA 6.3.5. *Let $F \supseteq E$ be a subfield of L with $\partial\mathcal{O}_F \subseteq \mathcal{O}_L$ and $\Gamma_E = \Gamma_F$. Then $b' \preccurlyeq b$ for all $b \in F$.*

PROOF. Let $b \in F^\times$, and take $a \in E^\times$, $u \in F$, with $b = au$ and $u \asymp 1$. Now use $a' \preccurlyeq a \asymp b$ and $u' \preccurlyeq 1$ to get $b' = a'u + au' \preccurlyeq b$. □

COROLLARY 6.3.6. *If L has small derivation and a monotone valued differential subfield K with $\Gamma_L = \Gamma$, then L is monotone.*

PROOF. Apply Lemma 6.3.5 with $E = K$, $F = L$. □

COROLLARY 6.3.7. *If K is monotone, then so is the differential field $K\langle Y\rangle$ with the gaussian valuation, as well as any extension $K\langle a\rangle$ as in Theorem 6.3.2, and any extension L as in Corollary 6.3.3.*

LEMMA 6.3.8. *Let $\mathcal{B}$ be a valuation-independent subset of the valued vector space L over the valued field E, such that $b' \preccurlyeq b$ for all $b \in \mathcal{B}$. Then $f' \preccurlyeq f$ for all f in the E-linear span of $\mathcal{B}$ in L.*

PROOF. Let $f = \sum_{i=1}^n a_i b_i \neq 0$ with $a_1, \dots, a_n \in E$ and distinct $b_1, \dots, b_n \in \mathcal{B}$. Then $v(f) = \min_i v(a_i b_i)$ and $f' = \sum a_i' b_i + \sum a_i b_i'$, and

$$v\Big(\sum_i a_i' b_i\Big) \geqslant \min_i v(a_i' b_i) \geqslant \min_i v(a_i b_i) = v(f),$$

$$v\Big(\sum_i a_i b_i'\Big) \geqslant \min_i v(a_i b_i') \geqslant \min_i v(a_i b_i) = v(f),$$

so $v(f') \geqslant v(f)$. □

LEMMA 6.3.9. *Let $y \in L$ be such that $y^p = a \in E^\times$ with p a prime number and $vy \notin \Gamma_E$. Then $b' \preccurlyeq b$ for all $b \in E(y)$.*

PROOF. We have $pvy = va$ and $py^{p-1}y' = a'$, so

$$y' \asymp a'/y^{p-1} \preccurlyeq a/y^{p-1} = y.$$

The desired result now follows from Lemmas 3.1.28 and 6.3.8, the latter applied to $\mathcal{B} = \{1, y, \dots, y^{p-1}\}$. □

PROOF OF PROPOSITION 6.3.4. By passing to an algebraic closure we arrange that L is algebraically closed. Let E^{a} be the algebraic closure of E in L, let E^{h} be the henselization of E in E^{a}, and let F be the maximal unramified extension of E^{h} in E^{a}. Then $\partial\mathcal{O}_F \subseteq \mathcal{O}_L$ by Lemmas 6.2.2 and 6.2.4, and so $b' \preccurlyeq b$ for all $b \in F$ by Lemma 6.3.5. Now use Lemma 6.3.9 to conclude that $b' \preccurlyeq b$ for all $b \in E^{\mathrm{a}}$. □

COROLLARY 6.3.10. *If K is monotone, then every valued differential field extension of K that is algebraic over K is also monotone.*

COROLLARY 6.3.11. *Suppose $E \subseteq \mathcal{O}_L$ and L has transcendence degree 1 over E. Then L^ϕ is monotone for some $\phi \in L^\times$.*

PROOF. The case $\Gamma_L = \{0\}$ is trivial, so assume $\Gamma_L \neq \{0\}$. Take $y \in L^\times$ with $y \prec 1$. Then y is transcendental over E. Let $\mathcal{B} = \{1, y, y^2, \dots\}$ and note that $\mathcal{B}$ is valuation-independent over E and L is algebraic over $E(y)$. If $y' \preccurlyeq y$, then $b' \preccurlyeq b$ for all $b \in E(y)$ by Lemmas 6.3.8 and 4.4.1, and so L is monotone by Proposition 6.3.4. Suppose that $y' \succ y$. Then we set $\phi = y^\dagger$, and $\delta = \phi^{-1}\partial$, so $\delta(a) \preccurlyeq a$ for all $a \in E$. Moreover $\delta(y) = y$. Thus the previous argument applies to L^ϕ instead of L, so L^ϕ is monotone. □

Notes and comments

Proposition 6.3.4 and Corollary 6.3.11 are slight extensions of [299, Theorem 4.1], [117, Theorem 3] and [299, Proposition 5.1].

6.4 THE VALUATION INDUCED ON THE VALUE GROUP

Let a, b range over K and α, β, γ over Γ. Note: if $a' \preccurlyeq a$ and $a \asymp b$, then $b' \preccurlyeq b$. Also, if $a' \succ a$ and $a \asymp b$, then $b' \succ b$ and $a^\dagger \sim b^\dagger$, and thus $a^\dagger \asymp b^\dagger$. This allows us to define a function $\nabla : \Gamma^{\neq} \to \Gamma$ as follows: for $\alpha \neq 0$, $\alpha = va$,

$$a' \preccurlyeq a \Rightarrow \nabla(\alpha) := 0, \qquad a' \succ a \Rightarrow \nabla(\alpha) := v(a^\dagger) = v(a') - v(a) < 0.$$

Thus $\nabla(\Gamma^{\neq}) \subseteq \Gamma^{\leqslant}$. We extend ∇ to all of Γ by $\nabla(0) := \infty$.

LEMMA 6.4.1. *The function $\nabla : \Gamma \to \Gamma_\infty$ has the following properties:*

(i) *∇ is a (non-surjective) valuation on the abelian group Γ;*

(ii) *$\nabla(k\alpha) = \nabla(\alpha)$ for nonzero $k \in \mathbb{Z}$;*

(iii) *$\nabla(\alpha) = o(\alpha)$ for $\alpha \neq 0$;*

(iv) *if* $a' \succ a$ *and* $va = \alpha$, *then* $v(a^{(n)}) = \alpha + n\triangledown(\alpha)$ *for all* n.

PROOF. It is clear that $\triangledown(-\alpha) = \triangledown(\alpha)$. Let $\alpha, \beta \neq 0$, and take a, b such that $va = \alpha$, $vb = \beta$. Then $v(ab) = \alpha + \beta$, and by considering separately the cases $(ab)' \preccurlyeq ab$ and $(ab)' \succ ab$ we obtain $\triangledown(\alpha + \beta) \geqslant \min\big(\triangledown(\alpha), \triangledown(\beta)\big)$. This proves (i), and (ii) follows from $(a^k)^\dagger = ka^\dagger$ for $a \neq 0$ and $k \in \mathbb{Z}$. As to (iii), replacing α by $-\alpha$ if necessary, we arrange $\alpha > 0$, and then taking a with $va = \alpha$, we have $v(a^\dagger) \geqslant -\alpha/n$ for all $n \geqslant 1$, by Lemma 6.1.1, which gives the desired estimate.

Let $a' \succ a$ and set $\alpha := va$, so $\triangledown(\alpha) < 0$. It is obvious that for $n = 0, 1$ we have $v\big(a^{(n)}\big) = \alpha + n\triangledown(\alpha)$. Assume this equality holds for a certain $n \geqslant 1$. Applying (iii) to $\triangledown(\alpha)$ in the role of α gives $\triangledown\big(\triangledown(\alpha)\big) = o\big(\triangledown(\alpha)\big)$, so $\triangledown(\alpha) < \triangledown\big(\triangledown(\alpha)\big) = \triangledown\big(n\triangledown(\alpha)\big)$, and thus $\triangledown\big(\alpha + n\triangledown(\alpha)\big) = \triangledown(\alpha) < 0$. Therefore,

$$v\big(a^{(n+1)}\big) \;=\; \alpha + n\triangledown(\alpha) + \triangledown\big(\alpha + n\triangledown(\alpha)\big) \;=\; \alpha + (n+1)\triangledown(\alpha),$$

which proves (iv). □

In the next section we show that $(\Gamma, \triangledown)$ is a so-called asymptotic couple, and there we establish some useful general facts about asymptotic couples. The above lemma is good enough for the application that follows now.

A more precise estimate on v_P

We first consider the case $P = Y^{(n)}$.

LEMMA 6.4.2. *Suppose* $\triangledown(\gamma) < 0$. *Then*

$$\begin{aligned} v_{Y^{(n)}}(\gamma) &= \gamma + n\triangledown(\gamma), \\ v_{Y^{[\sigma]}}(\gamma) &= d\gamma + \|\boldsymbol{\sigma}\|\triangledown(\gamma) \text{ for } \boldsymbol{\sigma} = \sigma_1 \dots \sigma_d. \end{aligned}$$

PROOF. To obtain the first identity, take $g \in K^\times$ with $vg = \gamma$. Then

$$\begin{aligned} \big(Y^{(n)}\big)_{\times g} &= (gY)^{(n)} = \sum_{i=0}^{n} \binom{n}{i} g^{(i)} Y^{(n-i)}, \text{ so} \\ v_{Y^{(n)}}(\gamma) &= \min_{0 \leqslant i \leqslant n} vg^{(i)} \;=\; \min_{0 \leqslant i \leqslant n} \gamma + i\triangledown(\gamma) \;=\; \gamma + n\triangledown(\gamma). \end{aligned}$$

The second identity follows from the first using $Y^{[\sigma]} = Y^{(\sigma_1)} \cdots Y^{(\sigma_d)}$. □

Let $P \in K\{Y\}^{\neq}$ be homogeneous of degree d. Lemma 4.5.3 says that if $\triangledown(\gamma) \geqslant 0$, then $v_P(\gamma) = v(P) + d\gamma$. Here we deal with the case $\triangledown(\gamma) < 0$:

PROPOSITION 6.4.3. *Suppose* $\triangledown(\gamma) < 0$. *Then*

$$vP + d\gamma + \mathrm{wt}(P)\triangledown(\gamma) \;\leqslant\; v_P(\gamma) \;\leqslant\; vP + d\gamma + \big(\mathrm{wt}(P) - \mathrm{dwt}(P)\big) \cdot |\triangledown(\gamma)|.$$

In particular, $v_P(\gamma) = vP + d\gamma + O\big(\triangledown(\gamma)\big)$.

PROOF. If $P = aY^{\boldsymbol{\sigma}}$, $a \in K^\times$, then by Lemma 6.4.2,

$$v_P(\gamma) \;=\; va + d\gamma + \|\boldsymbol{\sigma}\|\nabla(\gamma).$$

This gives the lower bound for all P. For the upper bound we first prove:

CLAIM: *Let $\boldsymbol{\sigma}_0$ be such that $P_{[\boldsymbol{\sigma}_0]} \neq 0$. Let $\boldsymbol{\sigma} \geqslant \boldsymbol{\sigma}_0$ be maximal with the property $vP_{[\boldsymbol{\sigma}]} \leqslant vP_{[\boldsymbol{\sigma}_0]} + \|\boldsymbol{\sigma}-\boldsymbol{\sigma}_0\| \cdot |\nabla(\gamma)|$, and let $vg = \gamma$. Then $v\big((P_{\times g})_{[\boldsymbol{\sigma}]}\big) \;=\; vP_{[\boldsymbol{\sigma}]} + d\gamma$.*

Towards proving the claim, recall that

$$(P_{\times g})_{[\boldsymbol{\sigma}]} \;=\; \sum_{\boldsymbol{\tau} \geqslant \boldsymbol{\sigma}} \binom{\boldsymbol{\tau}}{\boldsymbol{\sigma}} P_{[\boldsymbol{\tau}]} g^{[\boldsymbol{\tau}-\boldsymbol{\sigma}]} \;=\; P_{[\boldsymbol{\sigma}]} g^d + \sum_{\boldsymbol{\tau} > \boldsymbol{\sigma}} \binom{\boldsymbol{\tau}}{\boldsymbol{\sigma}} P_{[\boldsymbol{\tau}]} g^{[\boldsymbol{\tau}-\boldsymbol{\sigma}]}.$$

Let $\boldsymbol{\tau} > \boldsymbol{\sigma}$, and note that maximality of $\boldsymbol{\sigma}$ gives

$$vP_{[\boldsymbol{\tau}]} \;>\; vP_{[\boldsymbol{\sigma}_0]} + \|\boldsymbol{\tau}-\boldsymbol{\sigma}_0\| \cdot |\nabla(\gamma)|,$$

which in view of $\|\boldsymbol{\tau}-\boldsymbol{\sigma}_0\| = \|\boldsymbol{\tau}-\boldsymbol{\sigma}\| + \|\boldsymbol{\sigma}-\boldsymbol{\sigma}_0\|$ gives $vP_{[\boldsymbol{\tau}]} > vP_{[\boldsymbol{\sigma}]} + \|\boldsymbol{\tau}-\boldsymbol{\sigma}\| \cdot |\nabla(\gamma)|$. Thus, using $|\nabla(\gamma)| = -\nabla(\gamma)$, we get

$$\begin{aligned} v\big(P_{[\boldsymbol{\tau}]} g^{[\boldsymbol{\tau}-\boldsymbol{\sigma}]}\big) \;&=\; vP_{[\boldsymbol{\tau}]} + d\gamma + \|\boldsymbol{\tau}-\boldsymbol{\sigma}\| \cdot \nabla(\gamma) \\ &>\; vP_{[\boldsymbol{\sigma}]} + \|\boldsymbol{\tau}-\boldsymbol{\sigma}\| \cdot |\nabla(\gamma)| + d\gamma + \|\boldsymbol{\tau}-\boldsymbol{\sigma}\| \cdot \nabla(\gamma) \\ &=\; vP_{[\boldsymbol{\sigma}]} + d\gamma \;=\; v\big(P_{[\boldsymbol{\sigma}]} g^d\big), \end{aligned}$$

so $v(P_{\times g})_{[\boldsymbol{\sigma}]} \;=\; v(P_{[\boldsymbol{\sigma}]} g^d) \;=\; vP_{[\boldsymbol{\sigma}]} + d\gamma$ as claimed. Now, take $\boldsymbol{\sigma}_0$ such that $vP_{[\boldsymbol{\sigma}_0]} = vP$ and $\|\boldsymbol{\sigma}_0\| = \operatorname{dwt}(P)$, and take $\boldsymbol{\sigma} \geqslant \boldsymbol{\sigma}_0$ as in the claim. Then

$$v_P(\gamma) \;\leqslant\; vP_{[\boldsymbol{\sigma}]} + d\gamma \;\leqslant\; v(P) + d\gamma + \|\boldsymbol{\sigma}-\boldsymbol{\sigma}_0\| \cdot |\nabla(\gamma)|,$$

by the claim, which gives the desired upper bound. □

Notes and comments

To our knowledge the valuation ∇ induced on the value group has not been used before in this generality.

6.5 ASYMPTOTIC COUPLES

In this section Γ is an arbitrary ordered abelian group, unless we specify it as the value group of our valued differential field K. We let α, β, γ range over Γ.

An **asymptotic couple** is a pair (Γ, ψ) with $\psi\colon \Gamma^{\neq} \to \Gamma$, such that for all $\alpha, \beta \neq 0$,

(AC1) $\alpha + \beta \neq 0 \implies \psi(\alpha+\beta) \geqslant \min\big(\psi(\alpha), \psi(\beta)\big)$;

(AC2) $\psi(k\alpha) = \psi(\alpha)$ for all $k \in \mathbb{Z}^{\neq}$, in particular, $\psi(-\alpha) = \psi(\alpha)$;

(AC3) $\alpha > 0 \implies \alpha + \psi(\alpha) > \psi(\beta)$.

If in addition for all α, β,

(HC) $0 < \alpha \leqslant \beta \Rightarrow \psi(\alpha) \geqslant \psi(\beta)$,

then (Γ, ψ) is said to be of H**-type**, or to be an H**-asymptotic couple**. (We chose the prefix H because H-asymptotic couples originate from Hardy fields; for more on this, see Chapter 9.) Trivial examples of H-asymptotic couples are obtained by taking any Γ and any constant function $\psi\colon \Gamma^{\neq} \to \Gamma$. In the next lemma Γ is the value group of our valued differential field K.

LEMMA 6.5.1. *Let the function* $\nabla\colon \Gamma^{\neq} \to \Gamma$ *be as defined in the previous section. Then* (Γ, ∇) *is an asymptotic couple.*

PROOF. Conditions (AC1) and (AC2) correspond to (i) and (ii) of Lemma 6.4.1. As to (AC3), if $\alpha > 0$, $\beta \neq 0$, then $\alpha + \nabla(\alpha) > 0 \geqslant \nabla(\beta)$ by (iii) of Lemma 6.4.1. □

Let (Γ, ψ) be an asymptotic couple. When convenient we consider ψ as extended to a function $\Gamma_\infty \to \Gamma_\infty$ by $\psi(0) = \psi(\infty) := \infty$. Then $\psi(\alpha + \beta) \geqslant \min\big(\psi(\alpha), \psi(\beta)\big)$ holds for all α, β, and $\psi\colon \Gamma \to \Gamma_\infty$ is a (non-surjective) valuation on the abelian group Γ, as defined in Section 2.2. In particular,

$$\psi(\alpha) < \psi(\beta) \implies \psi(\alpha + \beta) = \psi(\alpha).$$

Each α yields subgroups

$$\{\gamma : \psi(\gamma) \geqslant \alpha\}, \quad \{\gamma : \psi(\gamma) > \alpha\}$$

of Γ. If (Γ, ψ) is of H-type, then these two subgroups are convex in Γ, and for all α, β, if $\psi(\alpha) > \psi(\beta)$, then $[\alpha] < [\beta]$, by (AC2) and (HC). In particular, if (Γ, ψ) is of H-type, then $\psi\colon \Gamma \to \psi(\Gamma)$ is a convex valuation on the ordered abelian group Γ, as defined in Section 2.4.

We define the **shift** $\psi + \alpha\colon \Gamma^{\neq} \to \Gamma$ by

$$(\psi + \alpha)(\gamma) := \psi(\gamma) + \alpha.$$

Then $(\Gamma, \psi + \alpha)$ is also an asymptotic couple (a **shift** of (Γ, ψ)), and if (Γ, ψ) is of H-type, so is $(\Gamma, \psi + \alpha)$. The next lemma gives two other useful constructions of asymptotic couples.

LEMMA 6.5.2. *If* (Γ, ψ_1) *and* (Γ, ψ_2) *are asymptotic couples, then so is* (Γ, ψ) *with* $\psi\colon \Gamma^{\neq} \to \Gamma$ *given by* $\psi(\gamma) := \min\big(\psi_1(\gamma), \psi_2(\gamma)\big)$. *If* (Γ, ψ) *is an asymptotic couple,* $\gamma_0 \in \Gamma$, $0 \leqslant \gamma_0 < \{\gamma + \psi(\gamma) : \gamma > 0\}$, *then so is* (Γ, ψ_0) *with* $\psi_0\colon \Gamma^{\neq} \to \Gamma$ *given by*

$$\psi_0(\gamma) = \psi(\gamma) \text{ if } \psi(\gamma) \leqslant 0, \qquad \psi_0(\gamma) = \gamma_0 \text{ if } \psi(\gamma) > 0.$$

The lemma goes through with *H-asymptotic* in place of *asymptotic*. The proofs are routine verifications. Less routine is the following.

LEMMA 6.5.3. *Let (Γ, ψ) be an asymptotic couple and $\alpha, \beta \neq 0$. Then*

$$n\big(\psi(\beta) - \psi(\alpha)\big) \; < \; |\alpha| \quad \textit{for all } n.$$

PROOF. Replacing α if necessary by $-\alpha$ we can assume $\alpha > 0$ and likewise we arrange $\beta > 0$. We can also assume $\psi(\beta) > \psi(\alpha)$. Next, replacing ψ by its shift $\psi - \psi(\alpha)$ we arrange $\psi(\alpha) = 0$. Then $\psi(\beta) > 0$, and our job is to show that $n\psi(\beta) < \alpha$ for all n. Note that $\psi(\beta) < \alpha + \psi(\alpha) = \alpha$. Also

$$\psi(\beta) < \psi(\beta) + \psi\big(\psi(\beta)\big),$$

hence $\psi\big(\psi(\beta)\big) > 0$, so $\psi\big(n\psi(\beta)\big) > 0$ for all $n \geqslant 1$, and thus $n\psi(\beta) \neq \alpha$ for all $n \geqslant 1$. Assume towards a contradiction that $n\psi(\beta) > \alpha$ for some $n \geqslant 1$. Then for the least such n we have $n\psi(\beta) = \alpha + \gamma$ with $0 < \gamma < \psi(\beta)$, so $\psi(\gamma) = \psi\big(n\psi(\beta) - \alpha\big) = 0$ since $\psi(\alpha) = 0 < \psi\big(n\psi(\beta)\big)$. Therefore, $\psi(\beta) < \gamma + \psi(\gamma) = \gamma$, a contradiction. □

Let (Γ, ψ) be an asymptotic couple. Using Lemma 6.5.3, ψ extends uniquely to a map $(\mathbb{Q}\Gamma)^{\neq} \to \mathbb{Q}\Gamma$, also denoted by ψ, such that $(\mathbb{Q}\Gamma, \psi)$ is an asymptotic couple. Then $\psi\big((\mathbb{Q}\Gamma)^{\neq}\big) = \psi(\Gamma^{\neq})$ and if (Γ, ψ) is of H-type, so is $(\mathbb{Q}\Gamma, \psi)$.

Suppose Γ is divisible, and thus a vector space over $\mathbb{Q}$. Now ψ is a valuation on Γ, and so if (α_i) is a family of elements of $\Gamma^{\neq}$ with $\psi(\alpha_i) \neq \psi(\alpha_j)$ for all distinct indices i, j, then (α_i) is a $\mathbb{Q}$-linearly independent family. In particular, if $\dim_{\mathbb{Q}} \Gamma$ is finite, then $\psi(\Gamma^{\neq})$ is finite of size at most $\dim_{\mathbb{Q}} \Gamma$.

The next two lemmas state other basic facts about asymptotic couples.

LEMMA 6.5.4. *Let (Γ, ψ) be an asymptotic couple.*

(i) *If $\alpha, \beta < \gamma + \psi(\gamma)$ for all $\gamma > 0$, then $\psi(\alpha - \beta) > \min(\alpha, \beta)$. In particular, if $\alpha, \beta \neq 0$, then $\psi\big(\psi(\alpha) - \psi(\beta)\big) > \min\big(\psi(\alpha), \psi(\beta)\big)$.*

(ii) *If $\alpha, \beta \neq 0$ and $\alpha \neq \beta$, then $\psi(\alpha) - \psi(\beta) = o(\alpha - \beta)$.*

(iii) *The map $\gamma \mapsto \gamma + \psi(\gamma)\colon \Gamma^{\neq} \to \Gamma$ is strictly increasing.*

PROOF. For (i), let $\alpha < \beta < \gamma + \psi(\gamma)$ for all $\gamma > 0$. Then $\beta < (\beta - \alpha) + \psi(\beta - \alpha)$, so $\psi(\beta - \alpha) > \alpha$, as required. For (ii), let $\alpha, \beta \neq 0$ with $\gamma := \alpha - \beta \neq 0$. We have to show that then $n\big|\psi(\alpha) - \psi(\beta)\big| < |\gamma|$ for all n. If $\psi(\gamma) > \psi(\beta)$, then $\psi(\alpha) = \psi(\beta)$ by axioms (AC1) and (AC2). Suppose $\psi(\gamma) \leqslant \psi(\beta)$. Then by axiom (AC1) again we have $\psi(\gamma) \leqslant \psi(\alpha)$, hence by Lemma 6.5.3:

$$n\psi(\gamma) \leqslant n\psi(\beta) < n\psi(\gamma) + |\gamma|, \qquad n\psi(\gamma) \leqslant n\psi(\alpha) < n\psi(\gamma) + |\gamma|.$$

Thus $n\big|\psi(\alpha) - \psi(\beta)\big| < |\gamma|$ in all cases. Property (iii) follows easily from (ii). □

Item (ii) justifies thinking of $\psi(\gamma)$ as a *slowly varying* function of $\gamma \in \Gamma^{\neq}$. From (i) we obtain something needed in Section 13.3:

COROLLARY 6.5.5. *Suppose* (Γ, ψ) *is of H-type and* $\alpha > 0$ *and* β *are such that* $\alpha = \psi(\alpha) \leqslant \beta < \gamma + \psi(\gamma)$ *for all* $\gamma > 0$. *Then* $\beta = \alpha + o(\alpha)$.

LEMMA 6.5.6. *Let* Γ *be an ordered abelian group and* $\psi \colon \Gamma^{\neq} \to \Gamma$. *The pair* (Γ, ψ) *is an asymptotic couple if and only if*

(i) $\psi(k\gamma) = \psi(\gamma)$ *for all* $k \in \mathbb{Z}^{\neq}$, $\gamma \in \Gamma^{\neq}$,

(ii) $\psi(\alpha - \beta) \geqslant \min\big(\psi(\alpha), \psi(\beta)\big)$ *for all* $\alpha, \beta \in \Gamma^{\neq}$, $\alpha \neq \beta$, *and*

(iii) *the map* $\gamma \mapsto \gamma + \psi(\gamma) \colon \Gamma^{>} \to \Gamma$ *is strictly increasing.*

PROOF. If (Γ, ψ) is an asymptotic couple, then clearly (i)–(iii) hold. Conversely, suppose (Γ, ψ) satisfies (i)–(iii). To verify (AC3), let $\alpha, \beta \in \Gamma^{>}$; we have to derive that $\psi(\beta) < \alpha + \psi(\alpha)$. From $\alpha + \beta > \beta > 0$ and (iii) we get $\alpha + \beta + \psi(\alpha + \beta) > \beta + \psi(\beta)$. Also $\alpha + \beta + \psi(\beta) > \beta + \psi(\beta)$, hence by (ii)

$$\alpha + \beta + \psi\big((\alpha + \beta) - \beta\big) > \beta + \psi(\beta),$$

that is, $\alpha + \beta + \psi(\alpha) > \beta + \psi(\beta)$, so $\alpha + \psi(\alpha) > \psi(\beta)$ as required. □

Notes and comments

Rosenlicht introduced and studied asymptotic couples in [363, 364, 365]. Lemma 6.5.3 is Theorem 5 in [364]; the proof here follows [15].

6.6 DOMINANT PART

In this section we choose for every $P \in K\{Y\}^{\neq}$ *an element* $\mathfrak{d}_P \in K^{\times}$ *with* $\mathfrak{d}_P \asymp P$, *such that* $\mathfrak{d}_P = \mathfrak{d}_Q$ *whenever* $P \sim Q$, $P, Q \in K\{Y\}^{\neq}$. *Let* $P \in K\{Y\}^{\neq}$.

DEFINITION 6.6.1. We have $\mathfrak{d}_P^{-1} P \in \mathcal{O}\{Y\}$; we call the differential polynomial

$$D_P := \overline{\mathfrak{d}_P^{-1} P} = \sum_{\boldsymbol{i}} \big(\overline{P_{\boldsymbol{i}}/\mathfrak{d}_P}\big) Y^{\boldsymbol{i}} = \sum_{\boldsymbol{\omega}} \big(\overline{P_{[\boldsymbol{\omega}]}/\mathfrak{d}_P}\big) Y^{[\boldsymbol{\omega}]} \in \boldsymbol{k}\{Y\}$$

the **dominant part** of P. Clearly D_P is nonzero with

$$\begin{aligned} \deg D_P &\leqslant \deg P, & \operatorname{order} D_P &\leqslant \operatorname{order} P, \\ \operatorname{wm}(D_P) &= \operatorname{dwm}(P), & \operatorname{wt}(D_P) &= \operatorname{dwt}(P). \end{aligned}$$

If P is homogeneous of degree d, respectively isobaric of weight w, so is D_P. It will also be convenient to set $\mathfrak{d}_Q := 0 \in K$ and $D_Q := 0 \in \boldsymbol{k}\{Y\}$ for $Q = 0 \in K\{Y\}$.

Another choice of $\mathfrak{d}_P$ multiplies D_P by a factor from $\boldsymbol{k}^{\times}$. In fact, only quantities like $\deg D_P$ and $\operatorname{mul} D_P$ that are independent of the choice of $\mathfrak{d}_P$ will matter in this chapter. Here are a few simple rules:

LEMMA 6.6.2. *Let $Q \in K\{Y\}$. Then*

(i) *if $P \succ Q$, then $\mathfrak{d}_{P+Q} = \mathfrak{d}_P$ and $D_{P+Q} = D_P$;*

(ii) *given any $a \in K^\times$ we have $D_{aP} = e\,D_P$ for some $e \in \boldsymbol{k}^\times$;*

(iii) *$\mathfrak{d}_{PQ} = u^{-1}\,\mathfrak{d}_P\mathfrak{d}_Q$ and $D_{PQ} = \overline{u}\,D_P D_Q$, for some $u \asymp 1$ in K.*

PROOF. Item (i) is clear; (ii) follows from (iii). For (iii) we may assume $Q \neq 0$. We have $v(PQ) = v(P) + v(Q)$, so we have $u \asymp 1$ in K with $\mathfrak{d}_{PQ} = u^{-1}\mathfrak{d}_P\mathfrak{d}_Q$, hence with $\boldsymbol{i}, \boldsymbol{j}, \boldsymbol{l}$ ranging over $\mathbb{N}^{1+r}$ with $r = \max\{\text{order } P, \text{order } Q\}$:

$$\begin{aligned} D_{PQ} &= \sum_{\boldsymbol{i}} \overline{(P\cdot Q)_{\boldsymbol{i}}/\mathfrak{d}_{PQ}}\, Y^{\boldsymbol{i}} = \overline{u} \sum_{\boldsymbol{i}} \overline{(P\cdot Q)_{\boldsymbol{i}}/(\mathfrak{d}_P\mathfrak{d}_Q)}\, Y^{\boldsymbol{i}} \\ &= \overline{u} \sum_{\boldsymbol{i}} \left(\sum_{\boldsymbol{j}+\boldsymbol{l}=\boldsymbol{i}} \overline{P_{\boldsymbol{j}}/\mathfrak{d}_P} \cdot \overline{Q_{\boldsymbol{l}}/\mathfrak{d}_Q} \right) Y^{\boldsymbol{i}} = \overline{u} D_P D_Q. \end{aligned}$$

□

LEMMA 6.6.3. *Let $Q \in \mathcal{O}\{Y\}$ be such that $\overline{Q} \notin \boldsymbol{k}$. Then*

$$P(Q) \asymp P, \qquad D_{P(Q)} \in \boldsymbol{k}^\times \cdot D_P(\overline{Q}).$$

PROOF. We have $\mathfrak{d}_P^{-1} P(Q) = \sum_{\boldsymbol{i}} (P_{\boldsymbol{i}}/\mathfrak{d}_P) Q^{\boldsymbol{i}} \in \mathcal{O}\{Y\}$, and

$$\overline{\mathfrak{d}_P^{-1} P(Q)} = \sum_{\boldsymbol{i}} \overline{(P_{\boldsymbol{i}}/\mathfrak{d}_P)}\, \overline{Q}^{\boldsymbol{i}} = D_P(\overline{Q}) \neq 0$$

by Lemma 4.3.12, from which the claims follow. □

We have $P = \mathfrak{d}_P D + R$ with $D \in \mathcal{O}\{Y\}$, $\overline{D} = D_P$, and $R \in K\{Y\}$, $R \prec P$. We also observe the following:

LEMMA 6.6.4. *If $P = Q + R$, $Q, R \in K\{Y\}$, $R \prec Q$, then $P \sim Q$, $D_P = D_Q$. If $P = \mathfrak{d}_P D + R$, $D, R \in K\{Y\}$, $R \prec P$, then $D \in \mathcal{O}\{Y\}$ and $\overline{D} = D_P$.*

Recall that $\operatorname{mul}(P)$ is the least $d \in \mathbb{N}$ such that $P_d \neq 0$. Thus

$$\operatorname{mul} D_P > 0 \iff D_P(0) = 0 \iff P(0) \prec P.$$

Note that the quantity $\operatorname{mul} D_P$ does not depend on the choice of $\mathfrak{d}_P$. The main part of Lemma 6.6.5 is the implication $a \prec 1 \Rightarrow \deg D_{P_{\times a}} \leqslant \operatorname{mul} D_P$.

LEMMA 6.6.5. *Let $a \in K$, $a \preccurlyeq 1$. Then:*

(i) *$D_{P_{+a}} \in \boldsymbol{k}^\times \cdot (D_P)_{+\overline{a}}$, and thus $\deg D_{P_{+a}} = \deg D_P$;*

(ii) *if $a \asymp 1$, then $D_{P_{\times a}} \in \boldsymbol{k}^\times \cdot (D_P)_{\times\overline{a}}$, $\operatorname{mul} D_{P_{\times a}} = \operatorname{mul} D_P$, $\deg D_{P_{\times a}} = \deg D_P$;*

(iii) *if $a \prec 1$, then $D_{P_{+a}} = D_P$, $\deg D_{P_{\times a}} \leqslant \operatorname{mul} D_P$.*

PROOF. Taking $Q = Y + a$ in Lemma 6.6.3 gives (i). Taking $Q = aY$ in that lemma gives (ii). If $a \prec 1$, then $P_{+a} \sim P$, which gives the equality in (iii). For the inequality in (iii) we can assume $P \asymp 1$. Then $d := \operatorname{mul}(D_P)$ gives

$$\begin{aligned} P &= \sum_{i<d} P_i + P_d + \sum_{i>d} P_i, \quad P_i \prec 1 \text{ for } i < d, \quad P_d \asymp 1, \quad P_i \preccurlyeq 1 \text{ for } i > d, \\ P_{\times a} &= \sum_{i<d} (P_i)_{\times a} + (P_d)_{\times a} + \sum_{i>d} (P_i)_{\times a}. \end{aligned}$$

Assume $0 \neq a \prec 1$. Then we obtain from Corollary 6.1.3 that for $i > d$,

$$v\big((P_d)_{\times a}\big) \;=\; dva + o(va) \;<\; v\big((P_i)_{\times a}\big) \;=\; v(P_i) + iva + o(va).$$

The inequality in (ii) now follows, since $(P_{\times a})_i = (P_i)_{\times a}$ for all i. □

Set $\operatorname{dmul}(P) := \operatorname{mul} D_P$ and $\operatorname{ddeg}(P) := \deg D_P$. We call $\operatorname{dmul}(P)$ the **dominant multiplicity** of P at 0, and $\operatorname{ddeg}(P)$ the **dominant degree** of P.

COROLLARY 6.6.6. *Let $a, b \in K$, $g \in K^\times$ be such that $a - b \preccurlyeq g$. Then*

$$\operatorname{ddeg} P_{+a,\times g} \;=\; \operatorname{ddeg} P_{+b,\times g}.$$

PROOF. Just note that $d := (b-a)/g \in \mathcal{O}$ and $P_{+b,\times g} = P_{+a,\times g,+d}$. □

COROLLARY 6.6.7. *Let $\mathfrak{m}, \mathfrak{n} \in K^\times$. Then* $\operatorname{mul} P = \operatorname{mul}(P_{\times\mathfrak{m}}) \leqslant \operatorname{ddeg} P_{\times\mathfrak{m}}$ *and*

$$\mathfrak{m} \prec \mathfrak{n} \implies \operatorname{dmul} P_{\times\mathfrak{m}} \;\leqslant\; \operatorname{ddeg} P_{\times\mathfrak{m}} \;\leqslant\; \operatorname{dmul} P_{\times\mathfrak{n}} \;\leqslant\; \operatorname{ddeg} P_{\times\mathfrak{n}}.$$

LEMMA 6.6.8. *Let $a \in K^\times$, and suppose* $\operatorname{ddeg} P_{\times a} = k$. *Then*

$$0 \;\leqslant\; v(P_k) - v(P) \;\leqslant\; (1 + \deg P)\,|va|.$$

PROOF. This is clear when $vP = vP_k$, so assume $vP < vP_k$, and take l with $vP = vP_l$. Then $l \neq k$. Recall that $P_{\times a,k} = P_{k,\times a}$, and $P_{\times a,l} = P_{l,\times a}$. Set $\alpha := va$, and consider first the case that $\nabla(\alpha) \geqslant 0$. Then by Lemma 4.5.3,

$$v(P_{\times a,k}) = vP_k + k\alpha, \quad v(P_{\times a,l}) = vP_l + l\alpha = vP + l\alpha,$$

so $v(P_k) + k\alpha \leqslant vP + l\alpha$, and thus $0 < vP_k - vP \leqslant |(l-k)\alpha|$, which gives what we want. Next, assume $\nabla(\alpha) < 0$. Then by Proposition 6.4.3,

$$v(P_{\times a,k}) = vP_k + k\alpha + o(\alpha), \quad v(P_{\times a,l}) = vP_l + l\alpha + o(\alpha) = vP + l\alpha + o(\alpha),$$

and we can argue as before. □

The dominant degree

As before, $P \in K\{Y\}^{\neq}$. Let $\mathfrak{m}$ and $\mathfrak{n}$ range over $K^{\times}$, and let $\mathcal{E} \subseteq K^{\times}$ be nonempty such that $\mathfrak{m} \in \mathcal{E}$ whenever $\mathfrak{m} \preccurlyeq \mathfrak{n} \in \mathcal{E}$. The **dominant degree** of P on $\mathcal{E}$, $\operatorname{ddeg}_{\mathcal{E}} P$, is the natural number given by

$$\operatorname{ddeg}_{\mathcal{E}} P \;:=\; \max\left\{\operatorname{ddeg}(P_{\times\mathfrak{m}}) :\; \mathfrak{m} \in \mathcal{E}\right\}.$$

By Corollary 6.6.7 we have $\operatorname{mul} P \leqslant \operatorname{ddeg}_{\mathcal{E}} P$. If $Q \in K\{Y\}$, $Q \neq 0$, then clearly

$$\operatorname{ddeg}_{\mathcal{E}} PQ \;=\; \operatorname{ddeg}_{\mathcal{E}} P + \operatorname{ddeg}_{\mathcal{E}} Q.$$

For $\phi \in K^{\times}$ we have $\operatorname{ddeg}_{\mathcal{E}} P_{\times\phi} = \operatorname{ddeg}_{\phi\mathcal{E}} P$.

LEMMA 6.6.9. *Suppose $v(\mathcal{E})$ does not have a smallest element. Then*

$$\operatorname{ddeg}_{\mathcal{E}} P = \max\left\{\operatorname{dmul}(P_{\times\mathfrak{m}}) :\; \mathfrak{m} \in \mathcal{E}\right\}.$$

In particular, there exists $\mathfrak{m} \in \mathcal{E}$ such that for all $\mathfrak{n} \in \mathcal{E}$ with $\mathfrak{m} \prec \mathfrak{n}$, the differential polynomial $D_{P_{\times\mathfrak{n}}}$ is homogeneous of degree $\operatorname{ddeg}_{\mathcal{E}} P$.

PROOF. We have $\operatorname{dmul} P_{\times\mathfrak{m}} \leqslant \operatorname{ddeg} P_{\times\mathfrak{m}} \leqslant \operatorname{ddeg}_{\mathcal{E}} P$ for $\mathfrak{m} \in \mathcal{E}$. Take $\mathfrak{m} \in \mathcal{E}$ with $\operatorname{ddeg} P_{\times\mathfrak{m}} = \operatorname{ddeg}_{\mathcal{E}} P$, and let $\mathfrak{n} \in \mathcal{E}$ be such that $\mathfrak{m} \prec \mathfrak{n}$. Then by Corollary 6.6.7 we have $\operatorname{ddeg}_{\mathcal{E}} P = \operatorname{ddeg} P_{\times\mathfrak{m}} = \operatorname{dmul} P_{\times\mathfrak{n}} = \operatorname{ddeg} P_{\times\mathfrak{n}}$. This yields the claim. □

LEMMA 6.6.10. *Suppose that $f \in \mathcal{E}$. Then* $\operatorname{ddeg}_{\mathcal{E}} P_{+f} = \operatorname{ddeg}_{\mathcal{E}} P$.

PROOF. It is enough that for $\mathfrak{m} \in \mathcal{E}$ with $\mathfrak{m} \succcurlyeq f$ we have $\operatorname{ddeg} P_{+f,\times\mathfrak{m}} = \operatorname{ddeg} P_{\times\mathfrak{m}}$, and this is a special case of Corollary 6.6.6. □

COROLLARY 6.6.11. *Suppose* $\operatorname{ddeg}_{\mathcal{E}} P = 1$. *Let $y \in \mathcal{E} \cup \{0\}$ be a zero of P and $\mathfrak{m} \in \mathcal{E}$. Then* $\operatorname{mul} P_{+y,\times\mathfrak{m}} = \operatorname{dmul} P_{+y,\times\mathfrak{m}} = \operatorname{ddeg} P_{+y,\times\mathfrak{m}} = 1$.

PROOF. Since $(P_{+y})_0 = P(y) = 0$, we have

$$1 \leqslant \operatorname{mul} P_{+y} = \operatorname{mul} P_{+y,\times\mathfrak{m}} \leqslant \operatorname{dmul} P_{+y,\times\mathfrak{m}} \leqslant \operatorname{ddeg} P_{+y,\times\mathfrak{m}} \leqslant \operatorname{ddeg}_{\mathcal{E}} P_{+y} = 1,$$

using Lemma 6.6.10 for the last step. □

Let $\mathcal{E}' \subseteq \mathcal{E}$ be nonempty such that $\mathfrak{m} \in \mathcal{E}'$ whenever $\mathfrak{m} \preccurlyeq \mathfrak{n} \in \mathcal{E}'$. Then for $f \in \mathcal{E}$ we have $\operatorname{ddeg}_{\mathcal{E}'} P_{+f} \leqslant \operatorname{ddeg}_{\mathcal{E}} P$ by Lemma 6.6.10. For $\gamma \in \Gamma$ and $\mathcal{E} = \{\mathfrak{n} : v\mathfrak{n} \geqslant \gamma\}$, we set $\operatorname{ddeg}_{\geqslant\gamma} P := \operatorname{ddeg}_{\mathcal{E}} P$, so if $v\mathfrak{m} = \gamma$, then $\operatorname{ddeg}_{\geqslant\gamma} P = \operatorname{ddeg} P_{\times\mathfrak{m}}$.

COROLLARY 6.6.12. *Let $a, b \in K$ and $\alpha, \beta \in \Gamma$ be such that $v(b-a) \geqslant \alpha$ as well as $\beta \geqslant \alpha$. Then* $\operatorname{ddeg}_{\geqslant\beta} P_{+b} \leqslant \operatorname{ddeg}_{\geqslant\alpha} P_{+a}$.

PROOF. Since $P_{+b} = P_{+a,+(b-a)}$, we have $\operatorname{ddeg}_{\geqslant\alpha} P_{+b} = \operatorname{ddeg}_{\geqslant\alpha} P_{+a}$ by Lemma 6.6.10. It remains to note that $\operatorname{ddeg}_{\geqslant\beta} P_{+b} \leqslant \operatorname{ddeg}_{\geqslant\alpha} P_{+b}$. □

Suppose $\Gamma \neq \{0\}$ and $\Gamma^{>}$ has no least element. Then $\mathcal{E} = \{\mathfrak{n} : \mathfrak{n} \prec \mathfrak{m}\}$ is nonempty, and we set $\operatorname{ddeg}_{\prec\mathfrak{m}} P := \operatorname{ddeg}_{\mathcal{E}} P$. Also $v(\{\mathfrak{n} : \mathfrak{n} \prec \mathfrak{m}\})$ has no least element, hence

$$\operatorname{ddeg}_{\prec\mathfrak{m}} P \;=\; \max\left\{\operatorname{ddeg} P_{\times\mathfrak{n}} :\; \mathfrak{n} \prec \mathfrak{m}\right\} \;=\; \max\left\{\operatorname{dmul} P_{\times\mathfrak{n}} :\; \mathfrak{n} \prec \mathfrak{m}\right\}.$$

6.7 THE EQUALIZER THEOREM

We stated this theorem in the introduction to this chapter. *In this section a, b, f, g, y range over K, and α, β, γ over Γ. We fix $P \in K\{Y\}^{\neq}$.*

Lemmas on equalizing

These lemmas are technical facts to be used in the proof of the Equalizer Theorem.

LEMMA 6.7.1. *Suppose $va = \alpha < 0$, $\operatorname{ddeg} P_{\times a} = m$, $\operatorname{wt}(P) = w$, $m < n$. Then*

$$\nabla(\alpha) = 0 \implies v(P_n) - v(P_m) > (n-m)|\alpha|,$$
$$\nabla(\alpha) < 0 \implies v(P_n) - v(P_m) > (n-m)|\alpha| + 2w \cdot \nabla(\alpha).$$

PROOF. Since $0 \neq P_{\times a,m} = P_{m,\times a}$, we have $P_m \neq 0$, so $v(P_n) - v(P_m) \in \Gamma_\infty$ is certainly defined. From $m < n$ we get $v(P_{m,\times a}) < v(P_{n,\times a})$, so

$$\nabla(\alpha) = 0 \implies v(P_m) + m\alpha < v(P_n) + n\alpha,$$
$$\nabla(\alpha) < 0 \implies v(P_m) + m\alpha + w \cdot \nabla(\alpha) < v(P_n) + n\alpha - w \cdot \nabla(\alpha),$$

by Lemma 4.5.3 and Proposition 6.4.3, respectively. □

LEMMA 6.7.2. *Suppose $va = \alpha < 0$, $vb = \beta \geqslant \alpha - (1/d)\alpha$ for some $d \geqslant 1$, and $\operatorname{ddeg}_{\geqslant \alpha} P = m$. Then $P_{+a,m} \sim P_{+(a+b),m}$.*

PROOF. Set $w := \operatorname{wt} P$ and $Q := P_{+a}$, so $P_{+(a+b),m} = Q_{+b,m}$. We have to show $Q_m \sim Q_{+b,m}$ (which includes $Q_m \neq 0$, $Q_{+b,m} \neq 0$). By Lemma 6.6.10 the assumption $\operatorname{ddeg}_{\geqslant\alpha} P = m$ gives $\operatorname{ddeg}_{\geqslant\alpha} Q = m$, so $\operatorname{ddeg} Q_{\times a} = m$. Define $\delta \in \Gamma$ by $\delta := 0$ if $\nabla(\alpha) \geqslant 0$, and $\delta := 2w \cdot \nabla(\alpha)$ if $\nabla(\alpha) < 0$; in any case, $\delta = o(\alpha)$ by Lemma 6.4.1. By Lemma 6.7.1 applied to Q we have $Q_m \neq 0$ and

(6.7.1) $$v(Q_n) - v(Q_m) > (n-m)|\alpha| + \delta \text{ for all } n > m.$$

We can assume $b \neq 0$. We have $Q_{+b} = \sum_n Q_{n,+b}$ and by Taylor expansion,

$$Q_{n,+b} = \sum_{|\boldsymbol{i}| \leqslant n} Q_{n,(\boldsymbol{i})}(Y) \cdot b^{\boldsymbol{i}},$$

where each $Q_{n,(\boldsymbol{i})}$ is homogeneous of degree $n - |\boldsymbol{i}|$. It follows that

$$Q_{n,+b,m} = 0 \text{ for } n < m, \quad \text{and } Q_{m,+b,m} = Q_m, \text{ which gives}$$
$$Q_{+b,m} = Q_m + \sum_{n>m} Q_{n,+b,m}, \quad Q_{n,+b,m} = \sum_{|\boldsymbol{i}|=n-m} Q_{n,(\boldsymbol{i})} \cdot b^{\boldsymbol{i}} \text{ for } n > m.$$

Hence it is enough to get $v(Q_{n,(\boldsymbol{i})} b^{\boldsymbol{i}}) > v(Q_m)$ for all $n > m$ and $|\boldsymbol{i}| = n - m$. For such n and $\boldsymbol{i}$ with $Q_{n,(\boldsymbol{i})} \neq 0$ we have $\|\boldsymbol{i}\| \leqslant w$, so $v(b^{\boldsymbol{i}}) \geqslant (n-m)\beta + \varepsilon$, where

$\varepsilon := 0 \in \Gamma$ if $\nabla(\beta) \geqslant 0$, and $\varepsilon := w \cdot \nabla(\beta)$ if $\nabla(\beta) < 0$; in any case, $\varepsilon = o(\beta)$ if $\beta \neq 0$. Thus it suffices to get $v(Q_n) + (n-m)\beta + \varepsilon > v(Q_m)$ for all $n > m$, that is,

$$v(Q_n) - v(Q_m) \;>\; (n-m)(-\beta) - \varepsilon \quad \text{for all } n > m.$$

It remains to use (6.7.1) and $(n-m)|\alpha| + \delta > (n-m)(-\beta) - \varepsilon$ for $n > m$. □

LEMMA 6.7.3. *Let $P, Q \in K\{Y\}^{\neq}$ be homogeneous of degrees d, e, respectively, with $d > e$, and assume $(d-e)\Gamma = \Gamma$. Set $\alpha := (vQ - vP)/(d-e)$ and $\beta := \big(v_Q(\alpha) - v_P(\alpha)\big)/(d-e)$. Then*

(i) $\nabla(\alpha) \geqslant 0 \implies v_P(\alpha) = v_Q(\alpha)$;

(ii) $\alpha \neq 0 \implies \nabla(\beta) > \nabla(\alpha)$;

(iii) $\nabla(\alpha) < 0,\ \nabla(\beta) < 0 \implies \nabla(\beta) = o\big(\nabla(\alpha)\big)$.

PROOF. If $\nabla(\alpha) \geqslant 0$, then $v_P(\alpha) = vP + d\alpha$ and $v_Q(\alpha) = vQ + e\alpha$, by Lemma 4.5.3, so $v_P(\alpha) = v_Q(\alpha)$. This proves (i). For (ii) and (iii), assume $\alpha \neq 0$. Then $\nabla(\alpha) = o(\alpha)$ by Lemma 6.4.1(iii). If $\beta = 0$, we are done, so assume $\beta \neq 0$. Then $\nabla(\alpha) < 0$ by (i). If $\nabla(\beta) \geqslant 0$, then $\nabla(\beta) > \nabla(\alpha)$, and we are done. So assume $\nabla(\beta) < 0$. Set $w := \max(\operatorname{wt} P, \operatorname{wt} Q)$. By Proposition 6.4.3,

$$\begin{aligned} v_Q(\alpha) &= vQ + e\alpha + \varepsilon, &\qquad |\varepsilon| \leqslant w|\nabla(\alpha)|,\\ v_P(\alpha) &= vP + d\alpha + \delta, &\qquad |\delta| \leqslant w|\nabla(\alpha)|, \end{aligned}$$

so $|\beta| = |\varepsilon - \delta|/(d-e) \leqslant 2w|\nabla(\alpha)|/(d-e)$. From $\nabla(\beta) < 0$ we get $\nabla(\beta) = o(\beta)$, so $\nabla(\beta) = o\big(\nabla(\alpha)\big)$. In view of $\nabla(\alpha) < 0$, this also gives $\nabla(\beta) > \nabla(\alpha)$. □

In the next lemma we consider Γ as equipped with its valuation ∇.

LEMMA 6.7.4. *Let $P, Q \in K\{Y\}^{\neq}$ be homogeneous of degrees d, e, respectively, with $d > e$. Let $(\alpha_\rho)_{\rho<\nu}$ be a pc-sequence in Γ such that $\alpha_\rho \rightsquigarrow \alpha$, with $\gamma_\rho := \nabla(\alpha_{\rho+1} - \alpha_\rho) < 0$ for all ρ, and $\gamma_\sigma = o(\gamma_\rho)$ whenever $\rho < \sigma < \nu$. Then we have $v_P(\alpha_\rho) - v_Q(\alpha_\rho) \rightsquigarrow v_P(\alpha) - v_Q(\alpha)$.*

PROOF. Set $i := v_P - v_Q$. We have to show that $i(\alpha_\rho) \rightsquigarrow i(\alpha)$. We can arrange that $\nabla(\alpha - \alpha_\rho) = \gamma_\rho$ for all ρ. Take $a_\rho \in K$ with $va_\rho = \alpha_\rho$. By Proposition 6.4.3,

$$\begin{aligned} v_P(\alpha) &= v_P\big(\alpha_\rho + (\alpha - \alpha_\rho)\big) \;=\; v_{P_{\times a_\rho}}(\alpha - \alpha_\rho)\\ &= v(P_{\times a_\rho}) + d(\alpha - \alpha_\rho) + O(\gamma_\rho) \;=\; v_P(\alpha_\rho) + d(\alpha - \alpha_\rho) + O(\gamma_\rho), \end{aligned}$$

and likewise with $v_Q(\alpha)$. Hence $i(\alpha) = i(\alpha_\rho) + (d-e)(\alpha - \alpha_\rho) + O(\gamma_\rho)$. Then by Lemma 6.5.4(ii) and (AC2) we have

$$\nabla\big(i(\alpha) - i(\alpha_\rho)\big) - \nabla(\alpha - \alpha_\rho) \;=\; o(\gamma_\rho),$$

that is, $\nabla\big(i(\alpha) - i(\alpha_\rho)\big) = \gamma_\rho + o(\gamma_\rho)$. It follows that $\nabla\big(i(\alpha) - i(\alpha_\rho)\big)$ as a function of ρ is strictly increasing, so $i(\alpha_\rho) \rightsquigarrow i(\alpha)$. □

Proof of the Equalizer Theorem

Let $P, Q \in K\{Y\}^{\neq}$ be homogeneous of degrees d and e with $d > e$, and assume $(d-e)\Gamma = \Gamma$. We wish to find an equalizer for P, Q, that is, an α such that $v_P(\alpha) = v_Q(\alpha)$. (Since $v_P - v_Q$ is strictly increasing, there is at most one equalizer for P, Q.) If $vP = vQ$, then $\alpha = 0$ works. Assume $vP \neq vQ$. Since for $\alpha \neq 0$ we have $v_P(\alpha) = vP + d\alpha + o(\alpha)$ and $v_Q(\alpha) = vQ + e\alpha + o(\alpha)$, we expect the α such that $vP + d\alpha = vQ + e\alpha$, that is, $\alpha = (vQ - vP)/(d-e)$, to be a good approximation to an equalizer. This leads to the following approximation scheme: Set $\alpha_0 = 0$ and take this as the initial term of a sequence $(\alpha_\rho)_{\rho<\nu}$ in Γ indexed by the ordinals ρ less than a certain ordinal $\nu > 0$, such that the following conditions hold:

(1) $v_P(\alpha_\rho) \neq v_Q(\alpha_\rho)$ for $\rho < \nu$;

(2) $\alpha_{\rho+1} = \alpha_\rho + \big(v_Q(\alpha_\rho) - v_P(\alpha_\rho)\big)/(d-e)$ when $\rho + 1 < \nu$;

(3) $\triangledown\big(v_Q(\alpha_\sigma) - v_P(\alpha_\sigma)\big) > \triangledown\big(v_Q(\alpha_\rho) - v_P(\alpha_\rho)\big)$ for $\rho < \sigma < \nu$;

(4) $\triangledown(\alpha_\sigma - \alpha_\rho) = \triangledown\big(v_Q(\alpha_\rho) - v_P(\alpha_\rho)\big)$ for $\rho < \sigma < \nu$.

Note that by (3) we have $\alpha_\rho \neq \alpha_\sigma$ whenever $\rho < \sigma < \nu$. We also pick for each $\rho < \nu$ an element $a_\rho \in K$ such that $va_\rho = \alpha_\rho$. Consider first the case that ν is a successor ordinal, $\nu = \mu + 1$. Then we set

$$\alpha_\nu := \alpha_\mu + \big(v_Q(\alpha_\mu) - v_P(\alpha_\mu)\big)/(d-e).$$

If $v_P(\alpha_\nu) = v_Q(\alpha_\nu)$, we are done. Assume $v_P(\alpha_\nu) \neq v_Q(\alpha_\nu)$. Then (1)–(4) continue to hold with ν replaced by $\nu+1$. This is clear for (1) and (2); for (3), apply Lemma 6.7.3 to $P_{\times a_\mu}$ and $Q_{\times a_\mu}$ in the role of P and Q, so that $\alpha = \alpha_\nu - \alpha_\mu$. Now (4) with $\nu + 1$ instead of ν follows easily.

Next, consider the case that $\nu > 0$ is a limit ordinal. Then by (3) and (4) we have a pc-sequence (α_ρ) in Γ (with respect to the valuation $\triangledown$ on Γ). Set

$$\gamma_\rho := \triangledown(\alpha_{\rho+1} - \alpha_\rho) = \triangledown\big(v_Q(\alpha_\rho) - v_P(\alpha_\rho)\big).$$

Then $\gamma_\rho < 0$ for all ρ: if this would fail for a certain $\rho < \nu$, then Lemma 6.7.3 applied to $P_{\times a_\rho}$ and $Q_{\times a_\rho}$ in the role of P and Q (which corresponds to $\alpha = \alpha_{\rho+1} - \alpha_\rho$) gives $v_P(\alpha_{\rho+1}) = v_Q(\alpha_{\rho+1})$, contradicting (1). Likewise, (3) and Lemma 6.7.3 yield:

$$\rho < \sigma < \nu \implies \gamma_\rho < \gamma_\sigma < 0, \quad \gamma_\sigma = o(\gamma_\rho).$$

We shall construct a pseudolimit of (α_ρ) in Γ, but before we do this, let us assume that $\alpha_\nu \in \Gamma$ is such a pseudolimit. If $v_P(\alpha_\nu) = v_Q(\alpha_\nu)$, we are done, so assume $v_P(\alpha_\nu) \neq v_Q(\alpha_\nu)$. We claim that then (1)–(4) holds with $\nu + 1$ in place of ν. This claim is clearly valid for (1) and (2). As to (4), let $\rho < \nu$ be given, and take σ with $\rho < \sigma < \nu$ so large that $\triangledown(\alpha_\nu - \alpha_\sigma) = \gamma_\sigma$. Then $\gamma_\sigma > \gamma_\rho$ gives

$$\triangledown(\alpha_\nu - \alpha_\rho) = \triangledown\big((\alpha_\nu - \alpha_\sigma) + (\alpha_\sigma - \alpha_\rho)\big) = \gamma_\rho.$$

A similar argument using Lemma 6.7.4 gives (3) with $\nu+1$ in place of ν.

We now turn to the construction of a pseudolimit of (α_ρ). Note:

$$\rho<\sigma<\nu \implies v(a^\dagger_\sigma - a^\dagger_\rho) \;=\; \nabla(\alpha_\sigma-\alpha_\rho) \;=\; \gamma_\rho.$$

Let $R := \mathrm{Ri}(P)$ and $S := \mathrm{Ri}(Q)$. By Corollary 6.6.12,

$$\rho<\sigma<\nu \implies \begin{cases} \operatorname{ddeg}_{\geqslant\gamma_\rho} R_{+a^\dagger_\rho} \geqslant \operatorname{ddeg}_{\geqslant\gamma_\sigma} R_{+a^\dagger_\sigma}, \\ \operatorname{ddeg}_{\geqslant\gamma_\rho} S_{+a^\dagger_\rho} \geqslant \operatorname{ddeg}_{\geqslant\gamma_\sigma} S_{+a^\dagger_\sigma}, \end{cases}$$

which gives m, n and an index ρ_0 such that for all $\rho \geqslant \rho_0$,

$$\operatorname{ddeg}_{\geqslant\gamma_\rho} R_{+a^\dagger_\rho} = m, \qquad \operatorname{ddeg}_{\geqslant\gamma_\rho} S_{+a^\dagger_\rho} = n.$$

Applying Lemma 6.7.2 to $R_{+a^\dagger_{\rho_0}}$ in the role of P and $a := a^\dagger_\rho - a^\dagger_{\rho_0}$ and $\alpha := \gamma_{\rho_0}$, we obtain

$$\begin{array}{lcll} \rho>\rho_0,\ vb\geqslant\gamma_\rho & \implies & R_{+a^\dagger_\rho,m} \sim R_{+a^\dagger_\rho+b,m}, & \text{and likewise,} \\ \rho>\rho_0,\ vb\geqslant\gamma_\rho & \implies & S_{+a^\dagger_\rho,n} \sim S_{+a^\dagger_\rho+b,n}, & \text{and therefore} \\ \rho_0<\rho<\sigma<\nu & \implies & R_{+a^\dagger_\rho,m} \sim R_{+a^\dagger_\sigma,m}, & S_{+a^\dagger_\rho,n} \sim S_{+a^\dagger_\sigma,n}. \end{array}$$

Thus the following element of Γ does not depend on the choice of $\rho>\rho_0$:

$$\alpha \;:=\; \frac{1}{d-e}\big(v(S_{+a^\dagger_\rho,n}) - v(R_{+a^\dagger_\rho,m})\big) \qquad (\rho>\rho_0). \tag{6.7.2}$$

We claim that $\alpha_\rho \rightsquigarrow \alpha$. First, for $\rho>\rho_0$,

$$\begin{aligned} \alpha-\alpha_\rho \;&=\; \frac{1}{d-e}\Big[\big(e\alpha_\rho + v(S_{+a^\dagger_\rho,n})\big) - \big(d\alpha_\rho + v(R_{+a^\dagger_\rho,m})\big)\Big], \\ v_P(\alpha_\rho) \;&=\; vP_{\times a_\rho} \;=\; d\alpha_\rho + v(R_{+a^\dagger_\rho}). \end{aligned}$$

Next, by Lemma 6.6.8 applied to $R_{+a^\dagger_\rho}$ instead of P we have for $\rho>\rho_0$,

$$\begin{aligned} 0 \;&\leqslant\; v(R_{+a^\dagger_\rho,m}) - v(R_{+a^\dagger_\rho}) \;\leqslant\; \big(1+\operatorname{wt}(P)\big)|\gamma_\rho|, \ \text{ so} \\ 0 \;&\leqslant\; d\alpha_\rho + v(R_{+a^\dagger_\rho,m}) - v_P(\alpha_\rho) \;\leqslant\; \big(1+\operatorname{wt}(P)\big)|\gamma_\rho|, \ \text{ and likewise,} \\ 0 \;&\leqslant\; e\alpha_\rho + v(S_{+a^\dagger_\rho,n}) - v_Q(\alpha_\rho) \;\leqslant\; \big(1+\operatorname{wt}(Q)\big)|\gamma_\rho|, \ \text{ and thus} \\ \alpha-\alpha_\rho \;&=\; \frac{1}{d-e}\big(v_Q(\alpha_\rho)-v_P(\alpha_\rho)\big) + O(\gamma_\rho) \;=\; \alpha_{\rho+1}-\alpha_\rho + O(\gamma_\rho). \end{aligned}$$

The above together with Lemma 6.5.4(ii) gives

$$\nabla(\alpha-\alpha_\rho) - \nabla(\alpha_{\rho+1}-\alpha_\rho) \;=\; o(\gamma_\rho),$$

that is, $\nabla(\alpha - \alpha_\rho) = \gamma_\rho + o(\gamma_\rho)$, for $\rho > \rho_0$. Now, for $\rho_0 < \rho < \sigma < \nu$ we have $\gamma_\sigma = o(\gamma_\rho)$, so $\nabla(\alpha - \alpha_\rho) < \nabla(\alpha - \alpha_\sigma)$, which proves our claim that $\alpha_\rho \rightsquigarrow \alpha$.

The arguments above show that if there were no equalizer for P, Q, we could extend our sequence $(\alpha_\rho)_{\rho<\nu}$ indefinitely, that is, make the ordinal ν as large as we want, which contradicts $|\nu| \leqslant |\Gamma|$. This concludes the proof of the Equalizer Theorem. □

In the rest of this section we keep assuming that $P, Q \in K\{Y\}^{\neq}$ are homogeneous of degrees d, e with $d > e$ (so $d \geqslant 1$), and that $(d-e)\Gamma = \Gamma$.

COROLLARY 6.7.5. *The function $v_P - v_Q \colon \Gamma \to \Gamma$ is bijective. If $d\Gamma = \Gamma$, then the function $v_P \colon \Gamma \to \Gamma$ is bijective.*

PROOF. $(v_P - v_Q)(\alpha) = \beta$ is equivalent to $v_P(\alpha) = v_{bQ}(\alpha)$ where $vb = \beta$. Also, $v_P = v_P - v_Q$ for $Q = 1$. □

The transfinite part of the proof of the Equalizer Theorem is rather bizarre and can probably be eliminated. (Allen Gehret noticed that this transfiniteness is very mild: any sequence $(\alpha_\rho)_{\rho<\nu}$ in the proof above must have length $\nu \leqslant \omega p$ for some $p \in \omega$; this is because $\nu = \omega^2$ leads to the equalities (6.7.2) for $\rho_0 < \rho < \omega^2$, with $\rho_0 = \omega q$, $q \in \omega$, and thus $\alpha_{\omega(q+1)} = \alpha_{\omega(q+2)}$, a contradiction.) In any case, since the theorem is true, logical considerations suggest that an equalizer is obtainable with a finite bound on the number of steps used, where the bound depends only on the orders and degrees of P and Q.

Acting on this suggestion we indicate an algorithm to compute the equalizer of P, Q. As before we set $R := \mathrm{Ri}(P)$ and $S := \mathrm{Ri}(Q)$, so $\deg R \leqslant \mathrm{wt}\, P$ and $\deg S \leqslant \mathrm{wt}\, Q$. We also assume that for each $\alpha \in \Gamma$ there is given a "monomial" $\mathfrak{m}_\alpha \in K^\times$ with $v\mathfrak{m}_\alpha = \alpha$. We define the function

$$v_{P,Q} : \Gamma \to \Gamma, \qquad v_{P,Q}(\alpha) := \alpha + \frac{v_Q(\alpha) - v_P(\alpha)}{d-e}.$$

Define an **equalizer sequence** for P, Q to be a sequence $\alpha_0, \dots, \alpha_N$ in Γ with $N \in \mathbb{N}$ such that $\alpha_0 = 0$, and for each $i < N$, either $\alpha_{i+1} = v_{P,Q}(\alpha_i)$, or

$$\alpha_{i+1} = \frac{v(S_{+\mathfrak{m}_\alpha^\dagger, n}) - v(R_{+\mathfrak{m}_\alpha^\dagger, m})}{d-e}$$

for $\alpha := \alpha_i$ and some $m \leqslant \mathrm{wt}\, P$ and $n \leqslant \mathrm{wt}\, Q$ with $R_{+\mathfrak{m}_\alpha^\dagger, m} \neq 0$, $S_{+\mathfrak{m}_\alpha^\dagger, n} \neq 0$.

COROLLARY 6.7.6. *Assume P and Q have order $\leqslant r$. Then there exists an equalizer sequence $\alpha_0, \dots, \alpha_N$ for P, Q such that α_N is an equalizer for P, Q and such that $N \leqslant N(r,d)$ with $N(r,d) \in \mathbb{N}$ depending only on r and d.*

PROOF. Recall that there is no strictly descending sequence (β_n) in a well-ordered set. Thus, starting with the equalizer obtained from the proof of the Equalizer Theorem, and going back appropriately in the (possibly transfinite) sequence of that proof, we obtain the reversal of an equalizer sequence for P, Q. A uniform bound as claimed must exist by model-theoretic compactness. □

Note that the validity of the algorithm implicit in Corollary 6.7.6 depends on the Equalizer Theorem, whose proof is hardly constructive.

Notes and comments

Suppose $P \in K\{Y\}^{\neq}$ is homogeneous of degree 1. Then the Equalizer Theorem says that the function $v_P \colon \Gamma \to \Gamma$ is a bijection. We do have a very different and more constructive (but longer) proof of this fact if K is H-asymptotic as defined in Section 9.1 below. This unpublished proof predates the material above, and also yields the definability of the function v_P in the asymptotic couple (Γ, ∇) when K is a Liouville closed H-field as defined in Section 10.6.

We also have a more constructive proof of the Equalizer Theorem for P, Q of order $\leqslant 1$ and H-asymptotic K.

6.8 EVALUATION AT PSEUDOCAUCHY SEQUENCES

In this section we assume that the induced derivation on the residue field $\boldsymbol{k}$ *is nontrivial.* In addition L denotes a valued differential field extension of K with $\partial\mathcal{O}_L \subseteq \mathcal{O}_L$, and (a_ρ) is a pc-sequence in K with $a_\rho \rightsquigarrow a \in L$, and $G(Y) \in L\{Y\} \setminus L$. We set $\gamma_\rho := v\big(a_{s(\rho)} - a_\rho\big) \in \Gamma_\infty$, where $s(\rho) :=$ immediate successor of ρ.

LEMMA 6.8.1. *There is a pc-sequence* (b_ρ) *in* K *equivalent to* (a_ρ) *such that* $\big(G(b_\rho)\big)$ *is a pc-sequence and* $G(b_\rho) \rightsquigarrow G(a)$.

PROOF. After removing some initial ρ's we can assume $\gamma_\rho = v(a - a_\rho) \in \Gamma$ for all ρ and $\gamma_{\rho'} > \gamma_\rho$ whenever $\rho' > \rho$. Take $g_\rho \in K$ with $v(g_\rho) = \gamma_\rho$ and define $u_\rho \in L$ by $a_\rho - a = g_\rho u_\rho$, so $u_\rho \asymp 1$. Let $x_\rho \in K$ be such that $x_\rho \asymp 1$ and $u_\rho + x_\rho \asymp 1$ in L. Put $y_\rho := u_\rho + x_\rho$ and $b_\rho := a_\rho + g_\rho x_\rho \in K$, so $b_\rho - a = g_\rho y_\rho$. It follows that (b_ρ) pseudoconverges to a and has the same width as (a_ρ), so by Lemma 2.2.17 it is a pc-sequence in K equivalent to (a_ρ). We have

$$G(b_\rho) - G(a) = \sum_{|\boldsymbol{i}| \geqslant 1} G_{(\boldsymbol{i})}(a)(g_\rho y_\rho)^{\boldsymbol{i}}$$

where $G_{(\boldsymbol{i})} = \frac{G^{(\boldsymbol{i})}}{\boldsymbol{i}!}$. Put $g_{\boldsymbol{i}} := G_{(\boldsymbol{i})}(a) \in L$ for $|\boldsymbol{i}| \geqslant 1$. Then

$$G(b_\rho) - G(a) \;=\; \sum_{|\boldsymbol{i}| \geqslant 1} g_{\boldsymbol{i}}(g_\rho y_\rho)^{\boldsymbol{i}} \;=\; P(g_\rho y_\rho) \;=\; P_{\times g_\rho}(y_\rho), \text{ where}$$

$$P(Y) \;:=\; G(a+Y) - G(a) \;=\; \sum_{|\boldsymbol{i}| \geqslant 1} g_{\boldsymbol{i}} Y^{\boldsymbol{i}} \in L\{Y\}, \text{ so } \deg P \geqslant 1,\ P(0) = 0.$$

By Lemma 4.5.2 with L instead of K, we have for each ρ a thin set $T_\rho \subseteq \boldsymbol{k}_L$, independent of the choice of x_ρ, such that $0 \in T_\rho$ and for all $y \in L$,

$$y \asymp 1,\ \overline{y} \notin T_\rho \implies v\big(P(g_\rho y)\big) \;=\; v_P(\gamma_\rho).$$

Note that $v_P(\gamma_\rho)$ is strictly increasing as a function of ρ. By Lemma 4.2.2 we can take for each ρ a thin set S_ρ in $\boldsymbol{k}$ such that for all $e \in \boldsymbol{k}$, if $\overline{u}_\rho + e \in T_\rho$, then $e \in S_\rho$. Therefore, by choosing the x_ρ such that $\overline{x}_\rho \notin S_\rho$ we get $\overline{y}_\rho \notin T_\rho$, and then $\big(G(b_\rho)\big)$ is a pc-sequence and $G(b_\rho) \rightsquigarrow G(a)$. □

Note that $v(b_\rho - a) = \gamma_\rho$ and $v\big(P(b_\rho - a)\big) = v_P(\gamma_\rho)$, eventually, for (b_ρ) as in the above proof. In addition we can arrange that for $e = 1, \dots, \deg P$, if $P_e \neq 0$, then $v\big(P_e(b_\rho - a)\big) = v_{P_e}(\gamma_\rho)$, eventually. Indeed, all this works just as well for finitely many differential polynomials:

LEMMA 6.8.2. *Let $\mathcal{H}$ be a finite subset of $L\{Y\}$. Then there is a pc-sequence (b_ρ) in K equivalent to (a_ρ) such that for each $H \in \mathcal{H}$, if $H \notin L$, then $\big(H(b_\rho)\big)$ is a pc-sequence with $H(b_\rho) \rightsquigarrow H(a)$.*

When $G(a_\rho) \rightsquigarrow 0$ we can improve these lemmas as follows:

LEMMA 6.8.3. *Suppose that $G(a_\rho) \rightsquigarrow 0$, and let $\mathcal{H}$ be a finite subset of $L\{Y\}$. Then there is a pc-sequence (b_ρ) in K that is equivalent to (a_ρ), such that $G(b_\rho) \rightsquigarrow 0$, and for each $H \in \mathcal{H}$, if $H \notin L$, then $\big(H(b_\rho)\big)$ is a pc-sequence with $H(b_\rho) \rightsquigarrow H(a)$.*

PROOF. We can assume $G \in \mathcal{H}$ and $v\big(G(a_\rho)\big)$ strictly increases with ρ. We now make the same reductions as in the proof of Lemma 6.8.1, introducing g_ρ, u_ρ, x_ρ, y_ρ, b_ρ, $P = \sum_{|\boldsymbol{i}|\geqslant 1} g_{\boldsymbol{i}} Y^{\boldsymbol{i}} \in L\{Y\}$, accordingly. The proof of that lemma shows how to arrange that (b_ρ) is a pc-sequence in K such that $H(b_\rho) \rightsquigarrow H(a)$, for each $H \in \mathcal{H}$ with $H \notin L$, and

$$v\big(P(g_\rho y_\rho)\big) = v_P(\gamma_\rho), \text{ eventually.}$$

We claim that then $G(b_\rho) \rightsquigarrow 0$. To prove this claim, note that for all ρ,

$$G(a_\rho) - G(a) = \sum_{|\boldsymbol{i}|\geqslant 1} g_{\boldsymbol{i}}(a_\rho - a)^{\boldsymbol{i}} = P_{\times g_\rho}(u_\rho) \preccurlyeq P_{\times g_\rho}(y_\rho) = G(b_\rho) - G(a).$$

There can only be one ρ with $G(a_\rho) - G(a) \prec G(a)$, because for such ρ we have $v\big(G(a_\rho)\big) = v\big(G(a)\big)$. So $G(a_\rho) - G(a) \succcurlyeq G(a)$, eventually, hence $G(b_\rho) - G(a) \succcurlyeq G(a)$, eventually, and thus $G(b_\rho) - G(a) \succ G(a)$, eventually. It now follows that $v\big(G(b_\rho)\big) = v\big(G(b_\rho) - G(a)\big)$, eventually, so $G(b_\rho) \rightsquigarrow 0$. □

6.9 CONSTRUCTING CANONICAL IMMEDIATE EXTENSIONS

In this section we assume the derivation on $\boldsymbol{k}$ is nontrivial. Thus Lemmas 6.8.1, 6.8.2, and 6.8.3 are available for our K.

In Lemma 6.9.1, Corollary 6.9.2, and Lemma 6.9.3 we fix a pc-sequence (a_ρ) in K. Recall from the end of Section 4.4 the notions of (a_ρ) being of d-algebraic type over K, that of $G \in K\{Y\}$ being a minimal differential polynomial of (a_ρ) over K, and of (a_ρ) being of d-transcendental type over K. We are going to associate to K and the pc-sequence (a_ρ) an immediate valued differential field extension $K\langle a\rangle$. If (a_ρ) is of d-transcendental type, this extension is canonical, and if it is of d-algebraic type, it is canonical modulo a choice of minimal differential polynomial.

LEMMA 6.9.1. *Suppose (a_ρ) is of* d*-transcendental type over K. Then K has an immediate valued differential field extension $K\langle a\rangle$ such that:*

(i) $\partial\mathcal{O}_{K\langle a\rangle} \subseteq \mathcal{O}_{K\langle a\rangle}$, $a_\rho \rightsquigarrow a$, *and a is* d*-transcendental over K;*

(ii) *for any valued differential field extension L of K with $\partial\mathcal{O}_L \subseteq \mathcal{O}_L$ and any $b \in L$ with $a_\rho \rightsquigarrow b$ there is a unique valued differential field embedding $K\langle a\rangle \longrightarrow L$ over K that sends a to b.*

PROOF. Let F be an elementary extension of K containing a pseudolimit a of (a_ρ). Let $K\langle a\rangle$ be the valued differential subfield of F generated by a over K. Let $G(Y) \in K\{Y\}$, $G \notin K$. Lemma 6.8.1 gives a pc-sequence (b_ρ) in K equivalent to (a_ρ) such that $G(b_\rho) \rightsquigarrow G(a)$. Now, $G(b_\rho) \not\rightsquigarrow 0$, since (a_ρ) is of d-transcendental type. So $G(a) \neq 0$ and $G(b_\rho) \sim G(a)$, eventually. Since G was arbitrary, we see that a is d-transcendental over K and $K\langle a\rangle$ is an immediate extension of K. From $\partial\mathcal{O}_F \subseteq \mathcal{O}_F$ we get $\partial\mathcal{O}_{K\langle a\rangle} \subseteq \mathcal{O}_{K\langle a\rangle}$.

Let L and b be as in (ii). By the proof of Lemma 6.8.1 we can arrange in the argument above, in addition to $G(b_\rho) \rightsquigarrow G(a)$, that $G(b_\rho) \rightsquigarrow G(b)$; hence $v_{K\langle a\rangle}\big(G(a)\big) = v_L\big(G(b)\big) \in \Gamma$. □

The following is immediate from Lemma 6.9.1:

COROLLARY 6.9.2. *Let L be a valued differential field extension of K satisfying $\partial\mathcal{O}_L \subseteq \mathcal{O}_L$, and assume $a_\rho \rightsquigarrow b$, where $b \in L$ is* d*-algebraic over K. Then (a_ρ) is of* d*-algebraic type over K.*

LEMMA 6.9.3. *Suppose P is a minimal differential polynomial of (a_ρ) over K. Then K has an immediate valued differential field extension $K\langle a\rangle$ such that:*

(i) $\partial\mathcal{O}_{K\langle a\rangle} \subseteq \mathcal{O}_{K\langle a\rangle}$, $a_\rho \rightsquigarrow a$ *and $P(a) = 0$;*

(ii) *for any valued differential field extension L of K with $\partial\mathcal{O}_L \subseteq \mathcal{O}_L$ and any $b \in L$ with $a_\rho \rightsquigarrow b$ and $P(b) = 0$ there is a unique valued differential field embedding $K\langle a\rangle \to L$ over K that sends a to b.*

PROOF. Let P have order r and take $p \in K[Y_0, \dots, Y_r]$ such that

$$P = p\big(Y, Y', \dots, Y^{(r)}\big).$$

Then p is irreducible by Corollary 3.2.5. Consider the domain

$$K[y_0, \dots, y_r] := K[Y_0, \dots, Y_r]/(p), \qquad y_i := Y_i + (p) \text{ for } i = 0, \dots, r,$$

and let $F = K(y_0, \dots, y_r)$ be its fraction field.

We extend the valuation v on K to a valuation $v\colon F^\times \to \Gamma$ as follows. Pick a pseudolimit e of (a_ρ) in some valued differential field extension E of K with $\partial\mathcal{O}_E \subseteq \mathcal{O}_E$. We let v also denote the valuation of E. Let $\phi \in F^\times$, so

$$\phi = f(y_0, \dots, y_r)/g(y_0, \dots, y_{r-1})$$

with $f \in K[Y_0, \dots, Y_r]$ of lower degree in Y_r than p and $g \in K[Y_0, \dots, Y_{r-1}]^{\neq}$. Set $\vec{e} := (e, e', \dots, e^{(r)})$, and also $\vec{b} = (b, b', \dots, b^{(r)})$ for $b \in K$.

By Lemma 6.8.2 we can take a pc-sequence (b_ρ) in K equivalent to (a_ρ) such that if $f \notin K$, then $f(\vec{b}_\rho) \rightsquigarrow f(\vec{e})$, and if $g \notin K$, then $g(\vec{b}_\rho) \rightsquigarrow g(\vec{e})$. Also $f(\vec{b}_\rho) \not\rightsquigarrow 0$ and $g(\vec{b}_\rho) \not\rightsquigarrow 0$ by the minimality of P, so eventually, $f(\vec{b}_\rho) \sim f(\vec{e})$ and $g(\vec{b}_\rho) \sim g(\vec{e})$, in particular, $f(\vec{e}) \neq 0$ and $g(\vec{e}) \neq 0$, and $v(f(\vec{e})) \in \Gamma$ and $v(g(\vec{e})) \in \Gamma$.

CLAIM: *$f(\vec{e})/g(\vec{e})$ depends only on ϕ and not on the choice of (f, g).*

To see why this claim is true, suppose that also

$$\phi = f_1(y_0, \dots, y_r)/g_1(y_0, \dots, y_{r-1})$$

with $f_1 \in K[Y_0, \dots, Y_r]$ of lower degree in Y_r than p and $g_1 \in K[Y_0, \dots, Y_{r-1}]^{\neq}$. Then $fg_1 \equiv f_1 g \bmod p$ in $K[Y_0, \dots, Y_r]$, and thus $fg_1 = f_1 g$ since fg_1 and $f_1 g$ have lower degree in Y_r than p. Thus $f(\vec{e})/g(\vec{e}) = f_1(\vec{e})/g_1(\vec{e})$, as promised. This proves the claim and allows us to define $v\colon F^\times \to \Gamma$ by

$$v(\phi) := v\big(f(\vec{e})/g(\vec{e})\big) = v\big(f(\vec{e})\big) - v\big(g(\vec{e})\big).$$

Clearly, this v extends the valuation of K. Let $\phi_1, \phi_2 \in F^\times$. It is easy to check that if $\phi_1 + \phi_2 \neq 0$, then $v(\phi_1 + \phi_2) \geqslant \min(v\phi_1, v\phi_2)$. We have

$$\phi_1 = \frac{f_1(y_0, \dots, y_r)}{g_1(y_0, \dots, y_{r-1})}, \quad \phi_2 = \frac{f_2(y_0, \dots, y_r)}{g_2(y_0, \dots, y_{r-1})}, \quad \phi_1\phi_2 = \frac{f_3(y_0, \dots, y_r)}{g_3(y_0, \dots, y_{r-1})}$$

where $f_1, f_2, f_3 \in K[Y_0, \dots, Y_r]$ have lower degree in Y_r than p and where g_1, g_2, g_3 are nonzero polynomials in $K[Y_0, \dots, Y_{r-1}]$. Then

$$\frac{f_1}{g_1}\frac{f_2}{g_2} = \frac{pq}{g_1 g_2 g_3} + \frac{f_3}{g_3}$$

with $q \in K[Y_0, \dots, Y_r]$. Lemma 6.8.3 gives a pc-sequence (b_λ) in K equivalent to (a_ρ), such that $p(\vec{b}_\lambda) \rightsquigarrow 0$, and if $q \notin K$, then $q(\vec{b}_\lambda) \rightsquigarrow q(\vec{e})$, and such that for $i = 1, 2, 3$ we have $f_i(\vec{b}_\lambda) \sim f_i(\vec{e})$ and $g_i(\vec{b}_\lambda) \sim g_i(\vec{e})$, eventually. This gives $f_1(\vec{b}_\lambda)f_2(\vec{b}_\lambda)/g_1(\vec{b}_\lambda)g_2(\vec{b}_\lambda) \sim f_3(\vec{b}_\lambda)/g_3(\vec{b}_\lambda)$, eventually, and thus

$$v(\phi_1\phi_2) = v(\phi_1) + v(\phi_2).$$

Thus $v\colon F^\times \to \Gamma$ is indeed a valuation on F, and below we consider F as a valued field accordingly. Now let $f \in K[Y_0, \dots, Y_r]$ be of lower degree in Y_r than p, $f \notin K$. Take a pc-sequence (b_ρ) in K equivalent to (a_ρ) such that $f(\vec{b}_\rho) \rightsquigarrow f(\vec{e})$. Then

$$v\big(f(y_0, \dots, y_r) - f(\vec{b}_\rho)\big) = v\big(f(\vec{e}) - f(\vec{b}_\rho)\big), \quad \text{for all } \rho.$$

As $f(\vec{e}) \sim f(\vec{b}_\rho)$, eventually, this gives $f(y_0, \dots, y_r) \sim f(\vec{b}_\rho)$, eventually. Thus F is an immediate extension of K.

We now equip F with the derivation extending the derivation of K such that $y_i' = y_{i+1}$ for $0 \leqslant i < r$. Setting $a := y_0$ we have $a^{(i)} = y_i$ for $i = 0, \ldots, r$, $K\langle a\rangle = F$, and $P(a) = 0$. We claim that $a_\rho \rightsquigarrow a$. Consider first the case that $r = 0$ and p has degree 1 in Y_0. Then $p = fY_0 + g$ with $f, g \in K$, $f \neq 0$, so $a = y_0 = -g/f \in K$, and we have a pc-sequence (b_λ) in K equivalent to (a_ρ) such that $p(b_\lambda) \rightsquigarrow 0$, that is, $b_\lambda \rightsquigarrow a$, so $a_\rho \rightsquigarrow a$. If $r = 0$ and p has degree > 1 in Y_0, or $r > 0$, then $a \notin K$ and $v(a - a_\rho) = v(y_0 - a_\rho) = v(e - a_\rho)$ for all ρ, which again gives $a_\rho \rightsquigarrow a$.

To get $\partial o_F \subseteq o_F$, we set

$$S := \big\{g(a) : g \in K\big[Y, \ldots, Y^{(r-1)}\big],\ g(a) \preccurlyeq 1\big\}.$$

(If $r = 0$, then we have $K\big[Y, \ldots, Y^{(r-1)}\big] = K$, so $S = \mathcal{O}$.) By Lemma 6.2.3 applied to $K\big(a, \ldots, a^{(r-1)}\big)$ in the role of E, it is enough to show that $\partial S \subseteq \mathcal{O}_F$ and $\partial(S \cap o_F) \subseteq o_F$. We prove the first of these inclusions. The second follows in the same way.

Let $g \in K\big[Y, \ldots, Y^{(r-1)}\big] \setminus K$ with $g(a) \preccurlyeq 1$; we have to show $g(a)' \preccurlyeq 1$. We can assume $g(a)' \neq 0$. Take $g_1(Y), g_2(Y) \in K\big[Y, \ldots, Y^{(r-1)}\big]$ such that

$$g(Y)' = g_1(Y) + g_2(Y)Y^{(r)} \quad \text{in } K\{Y\}.$$

Then

$$g(a)' = g_1(a) + g_2(a)a^{(r)},$$

and for all $y \in K$,

$$g(y)' = g_1(y) + g_2(y)y^{(r)}.$$

Take a pc-sequence (b_λ) in K equivalent to (a_ρ) such that $g(b_\lambda) \rightsquigarrow g(e)$. Hence $g(b_\lambda) \sim g(e)$, eventually, so $g(b_\lambda) \sim g(a)$, eventually, and thus $g(b_\lambda)' \preccurlyeq 1$, eventually. We now distinguish two cases:

CASE 1: *P has degree > 1 in $Y^{(r)}$, or $g_2 = 0$.* Then we can assume that (b_λ) has been chosen such that in addition we have, eventually,

$$g(b_\lambda)' = g_1(b_\lambda) + g_2(b_\lambda)b_\lambda^{(r)} \sim g_1(a) + g_2(a)a^{(r)} = g(a)'.$$

Therefore $g(a)' \preccurlyeq 1$, as desired.

CASE 2: *P has degree 1 in $Y^{(r)}$, and $g_2 \neq 0$.* Then

$$g_1 + g_2Y^{(r)} = \frac{h_1P + h_2}{h}, \qquad h, h_1, h_2 \in K[Y, \ldots, Y^{(r-1)}],\ h, h_1 \neq 0,$$

so $0 \neq g(a)' = h_2(a)/h(a)$, so $h_2 \neq 0$. As in the previous case we can assume (b_λ) to have been chosen such that in addition we have, eventually,

$$P(b_\lambda) \rightsquigarrow 0, \quad h(b_\lambda) \sim h(a), \quad h_1(b_\lambda) \sim h_1(a), \quad h_2(b_\lambda) \sim h_2(a).$$

Now, eventually $g(b_\lambda)' \preccurlyeq 1$, so eventually $h_1(b_\lambda)P(b_\lambda) + h_2(b_\lambda) \preccurlyeq h(b_\lambda)$. From this it follows easily that $h_2(b_\lambda) \preccurlyeq h(b_\lambda)$, eventually, so $h_2(a) \preccurlyeq h(a)$, that is, $g(a)' \preccurlyeq 1$. This finishes the proof of (i). We now turn to (ii).

Suppose L is a valued differential field extension of K with $\partial\mathcal{O}_L \subseteq \mathcal{O}_L$, and $b \in L$ satisfies $P(b) = 0$ and $a_\rho \rightsquigarrow b$. Let $Q \in K[Y, \dots, Y^{(r)}]$, $Q \notin K$, and suppose Q has lower degree in $Y^{(r)}$ than P. Then $Q(b) \neq 0$ by the argument in the beginning of the proof with b in the role of e. Thus P is a minimal annihilator of b over K. By Lemma 6.8.1 (or rather its proof) we have a pc-sequence (b_ρ) in K equivalent to (a_ρ) such that $Q(b_\rho) \rightsquigarrow Q(e)$ as well as $Q(b_\rho) \rightsquigarrow Q(b)$. As in the beginning of the proof this gives

$$v\big(Q(b_\rho)\big) = v\big(Q(e)\big) = v\big(Q(a)\big), \quad \text{eventually,}$$

and also $v\big(Q(b_\rho)\big) = v_L\big(Q(b)\big)$, eventually. In particular, $v\big(Q(a)\big) = v_L\big(Q(b)\big)$. Thus the differential field embedding $K\langle a\rangle \to L$ over K sending a to b is also a valued field embedding. □

Here are two immediate consequences of Lemmas 6.9.1 and 6.9.3:

COROLLARY 6.9.4. *If K has no proper immediate valued differential field extension with small derivation, then K is spherically complete.*

COROLLARY 6.9.5. *K has an immediate valued differential field extension with small derivation that is spherically complete.*

By Corollary 3.3.41, any two extensions of K as in the last corollary are isomorphic over K as valued fields; we would like to improve this to being isomorphic over K as valued differential fields. In Section 7.4 we establish this improvement under some extra assumptions on K. This involves an extension of the notion of "henselian" to the setting of valued differential fields, to be developed in the next chapter.

Chapter Seven

Differential-Henselian Fields

In this chapter K is a valued differential field with small derivation. As usual, $\Gamma := v(K^{\times})$ and $\boldsymbol{k} := \operatorname{res}(K)$, the latter a *differential* field. By an *extension of* K we mean a valued differential field extension of K whose derivation is small.

In this setting we study *differential-henselianity* and establish useful results about this notion in analogy with various facts about henselian valued fields in Section 3.3. We say that K is **differential-henselian** (for short: **d-henselian**) if the two conditions below are satisfied:

(DH1) $\boldsymbol{k}$ is linearly surjective;

(DH2) for every $P \in \mathcal{O}\{Y\}$ with $P_0 \prec 1$ and $P_1 \asymp 1$, there is $y \prec 1$ in K such that $P(y) = 0$.

If $\Gamma = \{0\}$ this is the same as K being linearly surjective. Note that for $P \in \mathcal{O}\{Y\}$ the condition $P_0 \prec 1$ and $P_1 \asymp 1$ in (DH2) implies $\operatorname{dmul} P = 1$.

Define K to be **maximal** if K has no proper immediate extension. If the derivation of $\boldsymbol{k}$ is nontrivial, then by Corollary 6.9.4,

$$K \text{ is maximal} \iff K \text{ is spherically complete.}$$

By Zorn's Lemma, K does have a maximal immediate extension. Define K to be **differential-algebraically maximal** (for short: **d-algebraically maximal**) if K has no proper immediate d-algebraic extension. If the derivation of $\boldsymbol{k}$ is nontrivial, then by Lemma 6.9.3, K is d-algebraically maximal iff there is no divergent pc-sequence in K of d-algebraic type over K. By Zorn, K does have an immediate d-algebraic extension that is d-algebraically maximal.

It is obvious that d-*henselianity* can be formulated in the language of valued differential fields as a first-order axiom scheme. For d-*algebraic maximality* this is not obvious, and perhaps false. We study below how these two notions are related.

After preliminaries about d-henselianity in Section 7.1 we prove the following in Section 7.2, in analogy with one direction of Corollary 3.3.21:

THEOREM 7.0.1. *If $\boldsymbol{k}$ is linearly surjective and K is* d-*algebraically maximal, then K is* d-*henselian.*

This depends critically on the $d = 1$ case of the Equalizer Theorem. An immediate consequence is a differential analogue of Hensel's Lemma:

COROLLARY 7.0.2. *If $\boldsymbol{k}$ is linearly surjective and K is spherically complete, then K is* d-*henselian.*

In particular, if $\boldsymbol{k}$ is a linearly surjective differential field, then the Hahn differential field $\boldsymbol{k}((t^{\Gamma}))$ defined in Section 4.4 is d-henselian.

For monotone K with linearly surjective $\boldsymbol{k}$ we prove in Section 7.4 the uniqueness-up-to-isomorphism-over-K of maximal immediate extensions. This will be crucial in the next chapter. In Section 7.5 we assume $C \subseteq \mathcal{O}$ and show that in the presence of monotonicity (perhaps unnecessary) we have a converse to Theorem 7.0.1:

THEOREM 7.0.3. *If $C \subseteq \mathcal{O}$ and K is monotone and* d*-henselian, then K is* d*-algebraically maximal.*

Finally, we consider differential-henselianity in several variables and obtain a partial analogue of Proposition 3.3.16. In detail, let $Y = (Y_1, \dots, Y_n)$ be a tuple of distinct differential indeterminates, $n \geqslant 1$. Let $P_1(Y), \dots, P_n(Y) \in \mathcal{O}\{Y\}$. We consider the system of equations

$$P_1(Y) = \cdots = P_n(Y) = 0.$$

Let $A_i \in \mathcal{O}\{Y\}$ be the homogeneous part of P_i of degree 1, and let $\overline{A}_i$ be its image in $\boldsymbol{k}\{Y\}_1$. Recall the notion of d-independence defined in Section 5.4.

THEOREM 7.0.4. *Suppose $\boldsymbol{k}$ is linearly surjective, K is* d*-algebraically maximal, $\overline{A}_1, \dots, \overline{A}_n \in \boldsymbol{k}\{Y\}_1$ are* d*-independent, and $P_1(0) \prec 1, \dots, P_n(0) \prec 1$, with $0 := (0, \dots, 0) \in K^n$. Then there exists a tuple $y = (y_1, \dots, y_n) \in K^n$ such that*

$$P_1(y) = \cdots = P_n(y) = 0, \qquad y_1 \prec 1, \dots, y_n \prec 1.$$

In Section 7.6 we prove this first for spherically complete K by approximation arguments, using also heavily the material in Section 5.4. The theorem then follows by appealing to the result from Section 5.9 due to Johnson [205].

7.1 PRELIMINARIES ON DIFFERENTIAL-HENSELIANITY

Throughout this chapter we assume $r \in \mathbb{N}$. To allow certain kinds of inductive arguments, we define K to be **r-differential-henselian** (for short: **r-d-henselian**) if the two conditions below are satisfied:

(DHr1) $\boldsymbol{k}$ is r-linearly surjective;

(DHr2) for every $P \in \mathcal{O}\{Y\}$ of order $\leqslant r$ with $P_0 \prec 1$ and $P_1 \asymp 1$, there is $y \prec 1$ in K such that $P(y) = 0$.

For $r = 0$ this is the same as K being henselian as a valued field. If $\Gamma = \{0\}$ it is the same as K being r-linearly surjective. Note that K is d-henselian iff K is r-d-henselian for each r.

LEMMA 7.1.1. *Suppose K is r-*d*-henselian, $P \in \mathcal{O}\{Y\}$ has order $\leqslant r$, $P_1 \asymp 1$, and $P_i \prec 1$ for all $i \geqslant 2$. Then $P(y) = 0$ for some $y \in \mathcal{O}$.*

PROOF. The assumption on P gives $\operatorname{ddeg} P = 1$. Use (DH$r$1) to get $u \in \mathcal{O}$ with $D_P(\overline{u}) = 0$, and thus $\operatorname{mul}(D_P)_{+\overline{u}} = \deg(D_P)_{+\overline{u}} = 1$. Therefore $\operatorname{dmul} P_{+u} = \operatorname{ddeg} P_{+u} = 1$ by Lemma 6.6.5(i). Also $P_{+u} \asymp 1$ by Lemma 4.5.1(i). Now apply (DHr2) to P_{+u} in the role of P to get $y \in \smallo$ with $P(u+y) = 0$. □

COROLLARY 7.1.2. *Suppose K is r-d-henselian, and $A \in K[\partial]$ with $v(A) = 0$ has order at most r. Then $A(\smallo) = \smallo$ and $A(\mathcal{O}) = \mathcal{O}$.*

PROOF. The first statement follows from (DHr2), and the second statement from Lemma 7.1.1. □

For $P \in \mathcal{O}\{Y\}$ and $a \in \mathcal{O}$ we say that P is in **differential-hensel position at** a (abbreviated as **dh-position at** a) if $P(a) \prec 1$ and $P_{+a,1} \asymp 1$. Note that then P is in dh-position at b for each $b \in \mathcal{O}$ with $a - b \in \smallo$. Note also that if K is r-d-henselian, $P \in \mathcal{O}\{Y\}$ is of order $\leqslant r$ and P is in dh-position at $a \in \mathcal{O}$, then there is $b \in \mathcal{O}$ with $a - b \in \smallo$ such that $P(b) = 0$. This gives the following analogue of an important result (Proposition 3.3.8) about henselian valued fields:

PROPOSITION 7.1.3. *Suppose K is d-henselian. Then $\boldsymbol{k}$ can be lifted to a differential subfield of K, that is, there is a differential subfield F of K such that $F \subseteq \mathcal{O}$ and F maps (isomorphically) onto $\boldsymbol{k}$ under the residue map $\mathcal{O} \to \boldsymbol{k}$.*

PROOF. This is the case $E = \mathbb{Q}$ of the following more general result: Let $E \subseteq \mathcal{O}$ be a differential subfield of K; then there is a differential subfield F of K such that $E \subseteq F \subseteq \mathcal{O}$ and F maps onto $\boldsymbol{k}$ under the residue map $\mathcal{O} \to \boldsymbol{k}$. Suppose $\operatorname{res}(E) \neq \boldsymbol{k}$. Take $a \in \mathcal{O}$ such that $\overline{a} \notin \operatorname{res}(E)$. If $P(a) \asymp 1$ for all nonzero $P(Y) \in E\{Y\}$, then $E\langle a \rangle$ is a proper differential field extension of E contained in $\mathcal{O}$. Next, consider the case that $P(a) \prec 1$ for some nonzero $P(Y) \in E\{Y\}$. Pick such P of minimal complexity, say P has order r. Then $Q(a) \asymp 1$ for all nonzero $Q(Y) \in E\{Y\}$ of lower complexity, hence $\frac{\partial P}{\partial Y^{(r)}}(a) \asymp 1$, so P is in dh-position at a. This gives $b \in \mathcal{O}$ with $P(b) = 0$ and $\overline{a} = \overline{b}$. Since $\overline{a} \notin \operatorname{res}(E)$, we have $b \notin E$. It follows that $E\langle b \rangle$ is a proper differential field extension of E contained in $\mathcal{O}$. We finish the proof by invoking Zorn. □

An embedding result

We now relate d-henselianity to the material on residue extensions in Section 6.3. Let L and F be extensions of K. Then $\boldsymbol{k}$ is a *differential* subfield of $\boldsymbol{k}_L$ and of $\boldsymbol{k}_F$. Assume also that L and F are r-d-henselian, and let $i\colon \boldsymbol{k}_L \to \boldsymbol{k}_F$ be a differential field embedding over $\boldsymbol{k}$.

LEMMA 7.1.4. *Suppose $\overline{a} \in \boldsymbol{k}_L$ is d-algebraic over $\boldsymbol{k}$, with minimal annihilator $\overline{P}(Y) \in \boldsymbol{k}\{Y\}$ of order r over $\boldsymbol{k}$. Then there exist $b \in \mathcal{O}_L$ with $\operatorname{res}(b) = \overline{a}$, $\boldsymbol{k}_{K\langle b \rangle} = \boldsymbol{k}\langle \overline{a} \rangle$, and $v(K\langle b \rangle^\times) = \Gamma$, and a valued differential field embedding $j\colon K\langle b \rangle \to F$, such that $\operatorname{res}(jy) = i(\operatorname{res}(y))$ for all $y \in \mathcal{O}_{K\langle b \rangle}$.*

PROOF. Take $P \in \mathcal{O}\{Y\}$ such that P has image $\overline{P}$ in $\boldsymbol{k}\{Y\}$ and P has the same complexity as $\overline{P}$. Then $v(I) = 0$ where I is the initial of P. Take also $a \in \mathcal{O}_L$ with $\operatorname{res}(a) = \overline{a}$. Then $\frac{\partial P}{\partial Y^{(r)}}(a) \asymp 1$, so P is in dh-position at a. This gives $b \in \mathcal{O}_L$ with $P(b) = 0$ and $\operatorname{res}(b) = \overline{a}$. If $Q \in K\{Y\}^{\neq}$ has lower complexity than P, then $Q(b) \neq 0$ by an easy reduction to the case $Q \asymp 1$. Thus P is a minimal annihilator of b over K. Likewise, we get $f \in \mathcal{O}_F$ with $P(f) = 0$ and $\operatorname{res}(f) = i(\overline{a})$. Then P is also a minimal annihilator of f over K, and so Theorem 6.3.2 gives a valued differential field embedding $j\colon K\langle b\rangle \to F$ with $jb = f$. This j has the desired property. □

In the above set-up, assume that L and F are even d-henselian (rather than r-d-henselian). Then by Lemma 7.1.4 and the result on d-transcendental residue extensions preceding Theorem 6.3.2:

COROLLARY 7.1.5. *There exist a valued differential subfield $E \supseteq K$ of L such that $\boldsymbol{k}_E = \boldsymbol{k}_L$ and $v(E^{\times}) = \Gamma$, and a valued differential field embedding $j\colon E \to F$ such that* $\operatorname{res}(jb) = i(\operatorname{res}(b))$ *for all $b \in \mathcal{O}_E$.*

Differential-henselianity and specialization

We now consider the behavior of d-henselianity with respect to coarsening and specialization. Let Δ be a convex subgroup of Γ, and $\dot{v} = v_\Delta$ the Δ-coarsening of v, with valuation ring $\dot{\mathcal{O}}$ and maximal ideal $\dot{o}$ of $\dot{\mathcal{O}}$. Then $\partial\dot{o} \subseteq \dot{o}$ by Corollary 4.4.4, and thus $\partial\dot{\mathcal{O}} \subseteq \dot{\mathcal{O}}$, which gives a differential residue field $\dot{K}$. Moreover $\partial o_{\dot{K}} \subseteq o_{\dot{K}}$, where for convenience ∂ denotes also the derivation of $\dot{K}$. Thus the residue field $\operatorname{res}(\dot{K})$ of $\dot{K}$ is naturally a differential field, and the canonical ring isomorphism $\operatorname{res}(K) \cong \operatorname{res}(\dot{K})$ is a differential ring isomorphism.

LEMMA 7.1.6. *If K is r-d-henselian, then so is $\dot{K}$.*

PROOF. Let $P \in \dot{\mathcal{O}}\{Y\}$ of order $\leqslant r$ have image Q in $\dot{K}\{Y\}$. Then Q has coefficients in the valuation ring of $\dot{K}$. It now remains to note that the differential residue field of K is isomorphic to the differential residue field of $\dot{K}$, and that if $Q_0 \prec 1$, $Q_1 \asymp 1$, then $P_0 \prec 1$, $P_1 \asymp 1$. □

LEMMA 7.1.7. *Suppose $(K, \dot{\mathcal{O}})$ and $\dot{K}$ are r-d-henselian. Then so is K.*

PROOF. Since K and $\dot{K}$ have isomorphic differential residue fields, K satisfies condition (DHr1). Next, let $P \in \mathcal{O}\{Y\}$ of order $\leqslant r$ satisfy $P_0 \prec 1$ and $P_1 \asymp 1$; it is enough to find $b \prec 1$ in K such that $P(b) = 0$. Let Q be the image of P in $\dot{K}\{Y\}$; then $Q \in \mathcal{O}_{\dot{K}}\{Y\}$. Then $Q_0 \prec 1$ and $Q_1 \asymp 1$, which gives $a \in o$ with $Q(\dot{a}) = 0$. Then $P(a) \mathrel{\dot{\prec}} 1$. Now $P_{+a} \sim P$ by Lemma 4.5.1, and so $P_{+a,0} = P(a) \mathrel{\dot{\prec}} 1$ and $vP_{+a,1} = 0$, from which we get $\dot{v}P_{+a,1} = 0$. As $(K, \dot{\mathcal{O}})$ is r-d-henselian, this gives $y \in \dot{o}$ such that $P_{+a}(y) = 0$, and then $b := a + y$ satisfies $b \prec 1$ and $P(b) = 0$. □

So far we have not used the Equalizer Theorem in studying d-henselianity, but to progress further we need the $d = 1$ case of this theorem, which says that for $A \in K[\partial]^{\neq}$ the function $v_A\colon \Gamma \to \Gamma$ is bijective. Recall also that this function is strictly increasing and satisfies $v_A(\gamma) = v(A) + \gamma + o(\gamma)$ for $\gamma \in \Gamma^{\neq}$.

Relation to neat surjectivity

If K is d-henselian, then not only $\boldsymbol{k}$ but also K itself is linearly surjective. More precisely:

LEMMA 7.1.8. *Suppose K is r-d-henselian. Then every $A \in K[\partial]^{\neq}$ of order $\leqslant r$ is neatly surjective.*

PROOF. Let $A \in K[\partial]^{\neq}$ have order $\leqslant r$, let $b \in K^{\times}$, and take $\alpha \in \Gamma$ such that $v_A(\alpha) = \beta := vb$. We have to find $a \in K^{\times}$ such that $va = \alpha$ and $A(a) = b$. First, take any $\phi \in K^{\times}$ with $v\phi = \alpha$. Then $v(A\phi) = vb$, so $v(B) = 0$ with $B := b^{-1}A\phi$. Then Lemma 7.1.1 gives $y \in \mathcal{O}$ with $B(y) = 1$, so $y \asymp 1$ and $A(\phi y) = b$. Thus $a := \phi y$ has the desired property. □

COROLLARY 7.1.9. *If K is 1-d-henselian, then $\smallo = (1 + \smallo)^{\dagger}$.*

PROOF. Let $a \in \smallo$. We look for $y \in \smallo$ such that $y'/(1+y) = a$, that is, $ay - y' = -a$. Now use that if $a - \partial$ is neatly surjective, the equation $ay - y' = -a$ does indeed have a solution y in $\smallo$. □

Application to having many constants

For use in the next chapter we show that under certain conditions d-henselianity yields many constants.

LEMMA 7.1.10. *Suppose K is 1-d-henselian and monotone, and let $b \in K$, $b' \prec b$. Then $b \asymp c$ for some $c \in C$.*

PROOF. Assume that $a \in K^{\times}$.

CLAIM: *$y \asymp a$ and $y' = a$ for some $y \in K$.*

To see this, note that $(aZ)' - a = a(a^{\dagger}Z + Z' - 1)$ and $a^{\dagger} \preccurlyeq 1$, so by Lemma 7.1.1 we get $z \in \mathcal{O}$ with $a^{\dagger}z + z' = 1$. Then $z \asymp 1$, so $y := az \asymp a$ satisfies $y' = a$.

Returning to b, note first that if $b' = 0$, then $c := b$ gives $c \in C$ with $c \asymp b$. Assume $b' \neq 0$. Then with $a := b'$ the claim above gives $y \asymp b'$ such that $y' = b'$, so $c := y - b \in C$ and $c \sim -b$. □

COROLLARY 7.1.11. *Suppose K is 1-d-henselian, monotone, and $(\boldsymbol{k}^{\times})^{\dagger} = \boldsymbol{k}$. Then K has many constants and $(K^{\times})^{\dagger} = (\mathcal{O}^{\times})^{\dagger} = \mathcal{O}$.*

PROOF. Let $b \in K^{\times}$, and suppose $b' \asymp b$; to show K has many constants, it is enough by Lemma 7.1.10 to get $y \asymp 1$ in K such that $(by)' \prec b$, that is, $(by)^{\dagger} \prec 1$, that is, $b^{\dagger} + y^{\dagger} \prec 1$. Since $b^{\dagger} \asymp 1$, the assumption of the corollary provides such a $y \asymp 1$. As to $(\mathcal{O}^{\times})^{\dagger} = \mathcal{O}$, let $a \in \mathcal{O}$. If $a \prec 1$, then use $\smallo = (1 + \smallo)^{\dagger}$. Assume $a \asymp 1$. Take $u \asymp 1$ with $a \sim u^{\dagger}$. Then $a - u^{\dagger} = (1 + \varepsilon)^{\dagger}$, $\varepsilon \in \smallo$, so $a = (u(1 + \varepsilon))^{\dagger} \in (\mathcal{O}^{\times})^{\dagger}$. This proves the claimed equalities. □

7.2 MAXIMALITY AND DIFFERENTIAL-HENSELIANITY

We begin with proving a converse to Lemma 7.1.1.

LEMMA 7.2.1. *Suppose for all $P \in \mathcal{O}\{Y\}$ of order $\leqslant r$ with $P_1 \asymp 1$ and $P_i \prec 1$ for all $i \geqslant 2$, there is $z \in \mathcal{O}$ with $P(z) = 0$. Then K is r-d-henselian.*

PROOF. It is clear that (DHr1) holds. For (DHr2), let $P \in \mathcal{O}\{Y\}$ be of order $\leqslant r$ with $P_0 \prec 1$ and $P_1 \asymp 1$; we have to get $y \prec 1$ in K such that $P(y) = 0$. If $P_0 = 0$, then $y = 0$ works. Assume $P_0 \neq 0$. Take $\gamma \in \Gamma$ such that $v_{P_1}(\gamma) = v(P_0)$. Then $\gamma > 0$, so

$$v_{P_1}(\gamma) = \gamma + o(\gamma) \ < \ v_{P_i}(\gamma) \quad \text{for all } i \geqslant 2.$$

Take $g \prec 1$ in K with $vg = \gamma$. Then $P_0 \asymp P_{1,\times g}$ and $P_{i,\times g} \prec P_{1,\times g}$ for all $i \geqslant 2$. Now take $h \in K^\times$ with $h \asymp P_{1,\times g}$ and apply our hypothesis to $h^{-1}P_{\times g}$ in the role of P to get $z \in \mathcal{O}$ with $P(gz) = 0$. Then $y := gz \prec 1$ and $P(y) = 0$. $\square$

In the next four lemmas we consider more closely a differential polynomial $P \in \mathcal{O}\{Y\}$ of order $\leqslant r$ and $a \in \mathcal{O}$ such that P is in dh-position at a, that is, with $Q := P_{+a}$ we have $Q(0) = P(a) \prec 1$ and $Q_1 \asymp 1$. Let $Q_1 = a_0 Y + a_1 Y' + \cdots + a_r Y^{(r)}$ ($a_0, \dots, a_r \in \mathcal{O}$) and set

$$A := L_{P_{+a}} = a_0 + a_1\partial + \cdots + a_r\partial^r \in K[\partial], \ \text{ so } vA = 0.$$

We define $v(P,a)$ to be the unique $\gamma \in \Gamma_\infty$ such that $v_A(\gamma) = v\big(P(a)\big)$, with $v(P,a) = \infty$ by convention if $P(a) = 0$. Thus $v(P,a) > 0$, and by Corollary 6.1.3,

$$P(a) \neq 0 \implies v\big(P(a)\big) = v(P,a) + o\big(v(P,a)\big).$$

Note: if K is monotone, then $v(P,a) = v\big(P(a)\big)$ by Corollary 4.5.4.

LEMMA 7.2.2. *Suppose K is r-d-henselian. Then there is $b \in \mathcal{O}$ with $P(b) = 0$ and $v(a-b) \geqslant v(P,a)$; any such b satisfies $v(a-b) = v(P,a)$.*

PROOF. If $P(a) = 0$, then we must take $b = a$. Assume $P(a) \neq 0$, set $\gamma := v(P,a) \in \Gamma^>$, take $g \in K$ with $vg = \gamma$. Then $P(a+gY) = Q_{\times g}$, and

$$Q_{\times g} = P(a) + Q_{1,\times g} + \sum_{j \geqslant 2} Q_{j,\times g}, \qquad v\big(Q_{1,\times g}\big) = v\big(P(a)\big),$$

$$v\big(Q_{j,\times g}\big) \geqslant j\gamma + o(\gamma) \ \text{ for } j \geqslant 2, \text{ so}$$

$$P(a+gY) = P(a) \cdot \left(1 + \sum_{i=0}^{r} b_i Y^{(i)} + R(Y)\right), \quad \operatorname{mul} R \geqslant 2,$$

$$b_0, \dots, b_r \in \mathcal{O},\ b_i \asymp 1 \text{ for some } i, \qquad v(R) \geqslant \gamma + o(\gamma) > 0.$$

Now K is r-d-henselian, so Lemma 7.1.1 provides $y \in \mathcal{O}$ such that

$$1 + \sum_{i=0}^{r} b_i y^{(i)} + R(y) = 0.$$

Any such y satisfies $y \asymp 1$, and so for $b := a + gy$ we have $P(b) = 0$ and $v(a - b) = v(P, a)$. $\square$

Under a weaker assumption on K we can still draw a useful conclusion:

LEMMA 7.2.3. *Suppose $\boldsymbol{k}$ is r-linearly surjective and $P(a) \neq 0$. Then there is $b \in \mathcal{O}$ such that P is in dh-position at b, $v(a - b) \geqslant v(P, a)$ and $P(b) \prec P(a)$; any such b satisfies $v(a - b) = v(P, a)$ and $v(P, b) > v(P, a)$.*

PROOF. At the point in the proof of Lemma 7.2.2 where we invoke that K is r-d-henselian, we choose instead $y \in \mathcal{O}$ such that $1 + \sum_{i=0}^{r} b_i y^{(i)} \prec 1$. Then $y \asymp 1$, so $v(b - a) = v(P, a)$ and $P(b) \prec P(a)$ with $b := a + gy$. To show that P is in dh-position at b, we use Taylor expansion. Let $\boldsymbol{i} = (i_0, \dots, i_r)$ and $\boldsymbol{j}$ range over $\mathbb{N}^{1+r}$, recall that $P_{(\boldsymbol{i})} = \frac{P^{(\boldsymbol{i})}}{\boldsymbol{i}!} \in K\{Y\}$, and

$$P(a+Y) \;=\; P(a) + \sum_{|\boldsymbol{i}|\geqslant 1} P_{(\boldsymbol{i})}(a) Y^{\boldsymbol{i}}, \qquad P_{+a,1} = \sum_{|\boldsymbol{i}|=1} P_{(\boldsymbol{i})}(a) Y^{(\boldsymbol{i})}.$$

Taylor expanding $P_{(\boldsymbol{i})}$ at a gives

$$P_{(\boldsymbol{i})}(b) \;=\; P_{(\boldsymbol{i})}(a) + \sum_{|\boldsymbol{j}|\geqslant 1} P_{(\boldsymbol{i})(\boldsymbol{j})}(a) \cdot (gy)^{\boldsymbol{j}}.$$

Since $P_{(\boldsymbol{i})(\boldsymbol{j})}(a) = \binom{\boldsymbol{i}+\boldsymbol{j}}{\boldsymbol{j}} P_{(\boldsymbol{i}+\boldsymbol{j})}(a)$, this gives for $|\boldsymbol{i}| \geqslant 1$:

$$\begin{aligned} P_{(\boldsymbol{i})}(b) &\sim P_{(\boldsymbol{i})}(a) && \text{if } P_{(\boldsymbol{i})}(a) \asymp 1, \\ P_{(\boldsymbol{i})}(b) &\prec 1 && \text{if } P_{(\boldsymbol{i})}(a) \prec 1. \end{aligned}$$

Thus P is in dh-position at b. It only remains to show $v(P, b) > v(P, a)$. The same way (P, a) gives rise to $A = L_{P_{+a}} \in K[\partial]$, the pair (P, b) yields $B = L_{P_{+b}} \in K[\partial]$, and the arguments above show that $B = A + E$ with $v(E) \geqslant \gamma + o(\gamma)$ where $\gamma := v(P, a)$. In combination with $v_A(\gamma) = \gamma + o(\gamma)$, this gives

$$v_E(\gamma) \;=\; v(E) + \gamma + o(\gamma) \;\geqslant\; 2\gamma + o(\gamma) \;>\; v_A(\gamma),$$

Hence $v_B(\gamma) = v_A(\gamma)$, and so $P(b) \prec P(a)$ forces $v(P, b) > v(P, a)$. $\square$

Without even assuming that $\boldsymbol{k}$ is r-linearly surjective, the arguments above and Corollary 6.1.7 give something that will be useful later:

LEMMA 7.2.4. *Suppose $P(a) \neq 0$, $b \in \mathcal{O}$, $v(a-b) \geqslant v(P, a)$ and $P(b) \prec P(a)$. Then $v(a - b) = v(P, a)$, and for all $b^* \in \mathcal{O}$ with $v(b - b^*) > v(P, a)$ and $B^* := L_{P_{+b^*}}$,*

(i) *P is in dh-position at b^*;*

(ii) *$P(b^*) \prec P(a)$ and $v(P, b^*) > v(P, a)$;*

(iii) *for all $y \in K^\times$, if $vy = O\big(v(P, a)\big)$ and $A(y) \prec Ay$, then $B^*(y) \prec B^* y$;*

(iv) $\big\{\alpha \in \mathscr{E}(A) : \alpha = O\big(v(P, a)\big)\big\} \subseteq \mathscr{E}(B^*)$.

LEMMA 7.2.5. *Let $\boldsymbol{k}$ be r-linearly surjective, and suppose there is no $b \in K$ with $P(b) = 0$ and $v(a-b) = v(P, a)$. Then there exists a divergent pc-sequence (a_ρ) in K such that $P(a_\rho) \rightsquigarrow 0$.*

PROOF. Let $(a_\rho)_{\rho<\lambda}$ be a sequence in $\mathcal{O}$ with λ an ordinal > 0, $a_0 = a$, and

(1) P is in dh-position at a_ρ, for all $\rho < \lambda$;

(2) $v(a_{\rho'} - a_\rho) = v(P, a_\rho)$ whenever $\rho < \rho' < \lambda$; and

(3) $P(a_{\rho'}) \prec P(a_\rho)$ and $v(P, a_{\rho'}) > v(P, a_\rho)$ whenever $\rho < \rho' < \lambda$.

Note that there is such a sequence if $\lambda = 1$. Suppose $\lambda = \mu + 1$ is a successor ordinal. Then Lemma 7.2.3 yields $a_\lambda \in K$ such that $v(a_\lambda - a_\mu) = v(P, a_\mu)$, $P(a_\lambda) \prec P(a_\mu)$ and $v(P, a_\lambda) > v(P, a_\mu)$. Then the extended sequence $(a_\rho)_{\rho<\lambda+1}$ has the above properties with $\lambda + 1$ instead of λ.

Suppose λ is a limit ordinal. Then (a_ρ) is a pc-sequence and $P(a_\rho) \rightsquigarrow 0$. If (a_ρ) has no pseudolimit in K we are done. Assume otherwise, and take a pseudolimit $a_\lambda \in K$ of (a_ρ). The extended sequence $(a_\rho)_{\rho<\lambda+1}$ clearly satisfies condition (2) with $\lambda + 1$ instead of λ. Applying Lemma 7.2.4 to a_ρ, $a_{\rho+1}$ and a_λ in the place of a, b and b^*, where $\rho < \lambda$, we see that conditions (1) and (3) are also satisfied with $\lambda + 1$ instead of λ. This building process must come to an end. □

COROLLARY 7.2.6. *If $\boldsymbol{k}$ is r-linearly surjective and K is spherically complete, then K is r-*d*-henselian.*

COROLLARY 7.2.7. *If $\boldsymbol{k}$ is r-linearly surjective, then K has an immediate r-*d*-henselian extension. If $\boldsymbol{k}$ is linearly surjective, then K has an immediate* d*-henselian extension.*

PROOF. For $r = 0$, take the henselization. For $r \geqslant 1$, use Corollary 6.9.5. □

PROOF OF THEOREM 7.0.1. Assume $\boldsymbol{k}$ is linearly surjective, K is d-algebraically maximal, and suppose towards a contradiction that $P \in \mathcal{O}\{Y\}$, $P_0 \prec 1$, $P_1 \asymp 1$, and there is no $b \in \mathcal{o}$ with $P(b) = 0$. Then Lemma 7.2.5 provides a divergent pc-sequence (a_ρ) in K with $P(a_\rho) \rightsquigarrow 0$. Thus (a_ρ) is of d-algebraic type over K, and so Lemma 6.9.3 yields a proper immediate d-algebraic extension of K. □

Step-completeness and differential-henselianity

The spherical completeness in Corollary 7.2.6 can be replaced by step-completeness, at the cost of a stronger linear hypothesis: Call $\mathcal{O}$ **r-linearly surjective** if for all $a_0, \dots, a_r \in \mathcal{O}$ such that $a_i \asymp 1$ for some i, the inhomogeneous linear differential equation

$$1 + a_0 y + \cdots + a_r y^{(r)} = 0$$

has a solution in $\mathcal{O}$. It is easy to check that $\mathcal{O}$ is r-linearly surjective iff each $A \in K[\partial]^{\neq}$ of order $\leqslant r$ is neatly surjective. Thus by Lemma 7.1.8, if K is r-d-henselian, then $\mathcal{O}$ is r-linearly surjective. Call $\mathcal{O}$ **linearly surjective** if it is r-linearly surjective for each r.

LEMMA 7.2.8. *Suppose $\mathcal{O}$ is r-linearly surjective. Let $P \in K\{Y\}$ have order $\leqslant r$ with $P \asymp 1$ and $\deg P = \operatorname{ddeg} P = 1$, and let $a \in \mathcal{O}$ be such that $P(a) \prec 1$. Then P is in dh-position at a, there is $b \in \mathcal{O}$ with $P(b) = 0$ and $v(a-b) \geqslant v(P,a)$, and any such b satisfies $v(a-b) = v(P,a)$.*

PROOF. By Lemma 4.5.1(i) and Lemma 6.6.5(i) we see that P is in dh-position at a. Now argue as in the proof of Lemma 7.2.2, using the fact that $R = 0$ and the hypothesis that $\mathcal{O}$ is r-linearly surjective in place of Lemma 7.1.1. □

LEMMA 7.2.9. *Suppose $\mathcal{O}$ is r-linearly surjective, $P \in \mathcal{O}\{Y\}$ of order $\leqslant r$ is in dh-position at $a \in \mathcal{O}$, and $P(a) \neq 0$. Then there is $b \in \mathcal{O}$ such that P is in dh-position at b, $v(a-b) = v(P,a)$, and $v\big(P(b)\big) \geqslant 2v\big(P(a)\big) + o\big(v(P(a))\big)$.*

PROOF. Put $\gamma := v(P,a)$. We follow the proof of Lemma 7.2.2. There we took $y \in \mathcal{O}$ such that $1 + \sum_{i=0}^{r} b_i y^{(i)} + R(y) = 0$, and here we take $y \in \mathcal{O}$ such that $1 + \sum_{i=0}^{r} b_i y^{(i)} = 0$. Then $b := a + gy$ gives $P(b) = P(a)R(y)$, so

$$\begin{aligned} v\big(P(b)\big) &= v\big(P(a)\big) + v\big(R(y)\big) \geqslant v\big(P(a)\big) + \gamma + o(\gamma) \\ &= 2v\big(P(a)\big) + o\big(v(P(a))\big). \end{aligned}$$ □

Suppose $\mathcal{O}$, P, a, b are as in Lemma 7.2.9. Then

$$v(P,b) \geqslant 2v(P,a) + o(v(P,a)).$$

This is because $v(P,b) = v\big(P(b)\big) + o\big(v(P(b))\big)$ if $P(b) \neq 0$.

LEMMA 7.2.10. *Suppose $\mathcal{O}$ is r-linearly surjective, and $P(Y) \in \mathcal{O}\{Y\}$ of order $\leqslant r$ is in dh-position at $a \in \mathcal{O}$. Suppose that there is no $b \in K$ with $P(b) = 0$ and $v(a-b) = v(P,a)$. Then there exists a divergent special pc-sequence (a_ρ) in K such that $P(a_\rho) \rightsquigarrow 0$.*

PROOF. We follow the proof of Lemma 7.2.5 except that in condition (3) on $(a_\rho)_{\rho<\lambda}$ the second inequality is replaced by the stronger

$$v(P,a_{\rho'}) > (3/2)v(P,a_\rho) \text{ whenever } \rho < \rho' < \lambda,$$

the appeal to Lemma 7.2.3 is replaced by an appeal to Lemma 7.2.9, which yields $v(P,a_\lambda) > (3/2)v(P,a_\mu)$ instead of $v(P,a_\lambda) > v(P,a_\mu)$. Also, when λ is a limit ordinal, (a_ρ) is a *special* pc-sequence. □

COROLLARY 7.2.11. *If $\mathcal{O}$ is r-linearly surjective and each special pc-sequence in K with a minimal differential polynomial over K of order $\leqslant r$ pseudoconverges in K, then K is r-*d*-henselian.*

COROLLARY 7.2.12. *If $\mathcal{O}$ is linearly surjective and K is step-complete, then K is* d*-henselian.*

Lifting zeros of linear differential operators

Let $A \in K[\partial]$ have order $\leqslant r$, with $vA = 0$. Let $\overline{A}$ be the image of A under the natural map $\mathcal{O}[\partial] \to \boldsymbol{k}[\partial]$.

LEMMA 7.2.13. *Suppose $\mathcal{O}$ is r-linearly surjective. Then the additive map*

$$a \mapsto \overline{a}: \ \mathcal{O} \cap \ker A \ \to \ \ker \overline{A}$$

is surjective.

PROOF. Suppose $a \in \mathcal{O}$ and $\overline{A}(\overline{a}) = 0$. Then $A(a) \prec 1$, so we have $y \prec 1$ in K with $A(y) = A(a)$. Then $A(a-y) = 0$ and $\overline{a-y} = \overline{a}$. □

If $\mathcal{O}$ is r-linearly surjective, then $\dim_{C_{\boldsymbol{k}}} \ker \overline{A} \leqslant \dim_C \ker A$ by Lemma 7.2.13. Under an additional condition on K we have equality:

PROPOSITION 7.2.14. *Suppose $\mathcal{O}$ is r-linearly surjective and K has many constants (that is, $v(C^\times) = \Gamma$). Then* $\dim_{C_{\boldsymbol{k}}} \ker \overline{A} = \dim_C \ker A$.

PROOF. By Lemma 7.2.13 we can take $f_1, \dots, f_m \in \mathcal{O} \cap \ker A$ whose residues $\overline{f_1}, \dots, \overline{f_m}$ form a basis of the $C_{\boldsymbol{k}}$-linear space $\ker \overline{A}$. Let V be the C linear subspace of $\ker A$ generated by $f_1, \dots, f_m$ and consider the homogeneous differential polynomial

$$P(Y) \ := \ \operatorname{wr}(Y, f_1, \dots, f_m) \in \mathcal{O}\{Y\}$$

of degree 1 and order $\leqslant m$. Then by Lemma 4.1.13 we have

$$V \ = \ \{y \in K : \ P(y) = 0\}.$$

Moreover

$$P(Y) \ = \ \sum_{i=0}^{m} (-1)^i (\det W_i)\, Y^{(i)},$$

where W_i is the $m \times m$ matrix whose jth column is

$$\big(f_j, \dots, f_j^{(i-1)}, f_j^{(i+1)}, \dots, f_j^{(m)}\big)^{\text{t}};$$

in particular, $W_m = \operatorname{Wr}(f_1, \dots, f_m)$. Since $\operatorname{wr}(\overline{f_1}, \dots, \overline{f_m}) \neq 0$, this gives $\deg D_P = 1$ and order $D_P = \operatorname{order} P = m \leqslant r$. For all $y \in \mathcal{O}$,

$$P \text{ is in dh-position at } y \Longleftrightarrow P(y) \prec 1 \Longleftrightarrow \operatorname{wr}(\overline{y}, \overline{f_1}, \dots, \overline{f_m}) = 0 \Longleftrightarrow \overline{A}(\overline{y}) = 0.$$

It is enough to show $V = \ker A$. Suppose towards a contradiction that $V \neq \ker A$. Take $a \in \ker A$, $a \notin V$. Since K has many constants we can multiply a by a nonzero constant to arrange $a \asymp 1$. From $\overline{A}(\overline{a}) = 0$ we get $0 \neq P(a) \prec 1$, so $\gamma := v(P, a) \in \Gamma^{>}$. By Lemma 7.2.8 we can take $b \in V$ with $v(a-b) = \gamma$. Next, take $c \in C$ such that $vc = -\gamma$ and put $g := c(a-b)$; then $g \asymp 1$, $\overline{A}(\overline{g}) = 0$, so Lemma 7.2.8 gives $h \in V$ with $v(g-h) = v(P, g) > 0$. Put $b^* := b + c^{-1}h \in V$. Then $a - b^* = (a-b) - c^{-1}h = c^{-1}(g-h)$, so

$$v(a-b^*) \ = \ \gamma + v(g-h) \ > \ \gamma \ = \ v(P, a),$$

in contradiction to Lemma 7.2.8. □

Differential-henselianity and completion

By Corollary 4.4.12 the completion K^c of K is an immediate extension of K. Our aim in this subsection is:

PROPOSITION 7.2.15. *If K is* d-*henselian and* $\mathrm{cf}(\Gamma)=\omega$, *then K^c is* d-*henselian.*

Let $\mathcal{O}^c$ be the valuation ring of K^c. The proof uses a variant of Lemma 7.2.3:

LEMMA 7.2.16. *Assume K is* d-*henselian and $P \in \mathcal{O}^c\{Y\}$ is in dh-position at the point $a \in \mathcal{O}^c$, with $P(a) \neq 0$. Let $\alpha \in \Gamma^>$ be given. Then there is $b \in \mathcal{O}^c$ such that P is in dh-position at b, $v(a-b) = v(P,a)$ and $P(b) \prec P(a)$, $vP(b) \geqslant \alpha$, $v(P,b) > v(P,a)$.*

PROOF. Take $g \in K^c$ with $vg = \gamma := v(P,a)$ as in the proof of Lemma 7.2.2. As in that proof we get

$$P(a+gY) \;=\; P(a)\cdot\left(1+\sum_{i=0}^{r} b_i Y^{(i)} + R(Y)\right), \quad R \in \mathcal{O}^c\{Y\},\ \mathrm{mul}\, R \geqslant 2,$$

$$b_0,\dots,b_r \in \mathcal{O}^c,\ b_i \asymp 1 \text{ for some } i, \qquad v(R) \geqslant \gamma + o(\gamma) > 0.$$

As K is d-henselian and dense in K^c we have $y \in \mathcal{O}$ such that

$$v\left(1+\sum_{i=0}^{r} b_i y^{(i)} + R(y)\right) \;\geqslant\; \alpha.$$

Any such y satisfies $y \asymp 1$ and so for $b := a+gy$ we have $P(b) \prec P(a)$ and $vP(b) \geqslant \alpha$ and $v(a-b) = v(P,a) > 0$. Thus P is in dh-position at b, and so $v(P,b) > v(P,a)$ by Lemma 7.2.3. □

PROOF OF PROPOSITION 7.2.15. Assume K is d-henselian and $\mathrm{cf}(\Gamma)=\omega$. Let $P \in \mathcal{O}^c\{Y\}$ be given such that $0 \neq P(0) \prec 1$ and $P_1 \asymp 1$; it suffices to show that then P has a zero in the maximal ideal $\mathcal{o}^c$ of $\mathcal{O}^c$. Fix a strictly increasing cofinal sequence (α_n) in $\Gamma^>$ with $\alpha_0 = vP(0)$. We proceed as in the proof of Lemma 7.2.10, with some differences: Let $(a_m)_{m<n}$ be a finite sequence in $\mathcal{o}^c$ with $n \geqslant 1$ and

(1) $a_0 = 0$, and P is in dh-position at a_m for all $m < n$,

(2) $v(a_{m'} - a_m) = v(P, a_m)$ whenever $m < m' < n$,

(3) $P(a_{m'}) \prec P(a_m)$, $v(P, a_{m'}) > v(P, a_m)$ whenever $m < m' < n$,

(4) $vP(a_m) \geqslant \alpha_m$ whenever $m < n$.

Note that we have such a sequence for $n = 1$. Let $n = \mu + 1$. If $P(a_\mu) = 0$, we are done, so assume $P(a_\mu) \neq 0$. Then Lemma 7.2.16 yields $a_n \in K$ such that $vP(a_n) \geqslant \alpha_n$, $v(a_n - a_\mu) = v(P, a_\mu)$, $P(a_n) \prec P(a_\mu)$ and $v(P, a_n) > v(P, a_\mu)$. Then the extended sequence $(a_m)_{m<n+1}$ has the above properties with $n+1$ instead of n. Iterating this extension procedure yields an infinite sequence (a_n) in $\mathcal{o}^c$. This is a c-sequence, and so $a_n \to a \in \mathcal{o}^c$. Then $P(a) = 0$. □

Notes and comments

Proposition 7.2.14 is a variant of [9, Theorem 4.1.6].

7.3 DIFFERENTIAL-HENSEL CONFIGURATIONS

In this section we assume that $\Gamma \neq \{0\}$. The notion of *dh-position* is too closely tied to differential polynomials of a special form, so we relax it as follows. Let $P \in K\{Y\}$ have order $\leqslant r$, and $a \in K$. Then

$$P(a+Y) = P(a) + A(Y) + R(Y), \qquad A, R \in K\{Y\},$$

where $A = \sum_{i=0}^{r} P_{(i)}(a)Y^{(i)}$ is homogeneous of degree 1, and all terms in R have degree $\geqslant 2$. Let us say that P is in **differential-hensel configuration** (abbreviated as **dh-configuration**) at a if $A \neq 0$, and there is $\gamma \in \Gamma$ such that $v(P(a)) \geqslant v_A(\gamma) < v_R(\gamma)$; equivalently, $A \neq 0$, and either $P(a) = 0$ or there is $\gamma \in \Gamma$ such that $v(P(a)) = v_A(\gamma) < v_R(\gamma)$. To prove this equivalence, assume $v(P(a)) > v_A(\gamma) < v_R(\gamma)$, $\gamma \in \Gamma$. Increasing γ to $\gamma + \delta$ ($\delta \in \Gamma^{>}$) such that $v(P(a)) = v_A(\gamma + \delta)$ we note that $v_A(\gamma+\delta) = v_A(\gamma)+\delta+o(\delta)$ and $v_R(\gamma+\delta) \geqslant v_R(\gamma)+2\delta+o(\delta)$, by Corollary 6.1.3, so $v_A(\gamma + \delta) < v_R(\gamma + \delta)$.

For any extension L of K, the Equalizer Theorem yields: P is in dh-configuration at a with respect to K iff P is in dh-configuration at a with respect to L.

For differential polynomials of order 0, "differential-hensel configuration" agrees with "hensel configuration" as defined in Section 3.3.

For P in dh-configuration at a we define $v(P, a) \in \Gamma_\infty$ as follows: if $P(a) \neq 0$, then $v(P, a)$ is the unique $\gamma \in \Gamma$ with $v\big(P(a)\big) = v_A(\gamma)$, and if $P(a) = 0$, then $v(P, a) := \infty$. Note: if $P \in \mathcal{O}\{Y\}$ and $a \in \mathcal{O}$, then

$$P \text{ is in dh-position at } a \implies P \text{ is in dh-configuration at } a.$$

Suppose P is in dh-configuration at a with $P(a) \neq 0$. Take $g \in K^\times$ with $vg = v(P, a)$, and put $G(Y) := P(a+gY)/P(a)$, so $G(Y) = 1+B(Y)+S(Y)$ where $B \in K\{Y\}$ is homogeneous of degree 1 with $v(B) = 0$ and all terms in $S \in K\{Y\}$ have degree $\geqslant 2$ with $v(S) > 0$. Assuming now that $\boldsymbol{k}$ is r-linearly surjective, we can take $y \in K$ with $y \asymp 1$ and $1 + B(y) \prec 1$. Then G is in dh-position at y. If K is r-d-henselian we can take y as above such that $G(y) = 0$, and then $b := a + gy$ satisfies $P(b) = 0$ and $v(a - b) = v(P, a)$. So we have shown:

LEMMA 7.3.1. *If K is r-d-henselian and $P \in K\{Y\}$ of order $\leqslant r$ is in dh-configuration at $a \in K$, then $P(b) = 0$ and $v(a - b) = v(P, a)$ for some $b \in K$.*

LEMMA 7.3.2. *Let $a \in K$ and $P \in K\{Y\}$. Then*

$$P \textit{ is in dh-configuration at } a \iff \operatorname{ddeg} P_{+a,\times g} = 1 \textit{ for some } g \in K^\times.$$

The proof is straightforward and left to the reader.

LEMMA 7.3.3. *Let $P \in K\{Y\}$ be in dh-configuration at $a \in K$ and $P(a) \neq 0$. Then P is in dh-configuration at b for every $b \in K$ with $v(a-b) \geqslant v(P,a)$.*

PROOF. By passing to the algebraic closure of K we arrange that Γ has no least positive element. Take $g \in K^\times$ with $vg = v(P,a)$. Then

$$Q := P_{+a,\times g} = P(a) + A + R, \quad P(a) \asymp A \succ R$$

where $A \in K\{Y\}$ is homogeneous of degree 1 and $\operatorname{mul} R \geqslant 2$. Since Γ has no least positive element, we can take $h \in K$ such that $h \succ 1$ and $A_{\times h} \succ R_{\times h}$. Let $b \in K$ be such that $v(a-b) > v(gh)$. Then $b = a + ghy$ with $y \prec 1$, so

$$P_{+b,\times gh} = P_{+a,\times gh,+y} \sim P_{+a,\times gh} = Q_{\times h},$$

and thus $\operatorname{ddeg} P_{+b,\times gh} = 1$. □

If P is in dh-configuration at a (with $P \in K\{Y\}$, $a \in K$), and $\phi \in K^\times$, then $P_{\times\phi}$ is in dh-configuration at a/ϕ, with $v(P_{\times\phi}, a/\phi) = v(P,a) - v\phi$.

Compositional conjugation

Let $\phi \in K$ and $\phi \asymp 1$. Then the derivation of K^ϕ is also small, and the residue differential field of K^ϕ is $\boldsymbol{k}^{\overline{\phi}}$, where $\overline{\phi}$ is the residue class $\phi + o$ in $\boldsymbol{k} = \mathcal{O}/o$. In particular, $\boldsymbol{k}$ is r-linearly surjective iff $\boldsymbol{k}^{\overline{\phi}}$ is r-linearly surjective. For $P \in K\{Y\}$ we have $(P^\phi)_d = (P_d)^\phi$ for all $d \in \mathbb{N}$, and $v(P^\phi) = v(P)$. Thus $\mathcal{O}$ is r-linearly surjective iff $\mathcal{O}^\phi$ is r-linearly surjective, and if K is r-d-henselian, then so is K^ϕ.

Coarsening and specialization

In this subsection Δ is a convex subgroup of Γ. Let $\dot{\mathcal{O}}$ be the valuation ring of the corresponding coarsening $\dot{v}\colon K^\times \to \dot{\Gamma} = \Gamma/\Delta$.

LEMMA 7.3.4. *If K is r-d-henselian, then so is $(K, \dot{\mathcal{O}})$.*

PROOF. Recall that $\Gamma \neq \{0\}$. Let $P \in \dot{\mathcal{O}}\{Y\}$ and $a \in \dot{\mathcal{O}}$, and suppose P is in dh-position at a with respect to the valuation $\dot{v}$.

CLAIM: *P is in dh-configuration at a with respect to the original valuation v.*

To see why, let $P(a+Y) = P(a) + L(Y) + R(Y)$ where $\dot{v}(P(a)) > 0$, $L \in \dot{\mathcal{O}}\{Y\}$ is homogeneous of degree 1 with $\dot{v}(L) = 0$, and all terms of $R \in \dot{\mathcal{O}}\{Y\}$ have degree $\geqslant 2$. We can assume $P(a) \neq 0$. Take $\gamma \in \Gamma$ such that $v\big(P(a)\big) = v_L(\gamma)$. Then $\dot{v}\big(P(a)\big) = \dot{v}_L(\dot{\gamma})$ by Lemma 4.5.5, so $\gamma > \Delta$. Then $v_L(\gamma) = v(L) + \gamma + o(\gamma)$, and $v_R(\gamma) \geqslant v(R) + 2\gamma + o(\gamma)$ by Corollary 6.1.3. Since $v(L) \in \Delta$ and $v(R) \geqslant \delta$ for some $\delta \in \Delta$, it follows that $v_L(\gamma) < v_R(\gamma)$. This proves the claim, and also gives $v(P,a) > \Delta$. Assume now that K is r-d-henselian. Then the claim and Lemma 7.3.1 yield $b \in K$ with $P(b) = 0$ and $v(a-b) > \Delta$, so $\dot{v}(a-b) > 0$. It remains to note that $\dot{K}$ is r-linearly surjective by Lemma 7.1.6. □

In combination with Lemmas 7.1.6 and 7.1.7, we obtain:

COROLLARY 7.3.5. *K is r-d-henselian if and only if both $(K, \dot{\mathcal{O}})$ and $\dot{K}$ are r-d-henselian.*

7.4 MAXIMAL IMMEDIATE EXTENSIONS IN THE MONOTONE CASE

In this section K is a monotone valued differential field, the induced derivation on $\boldsymbol{k}$ is nontrivial, and $\Gamma \neq \{0\}$. Note that any extension L of K with $\Gamma_L = \Gamma$ is monotone, by Corollary 6.3.6.

We begin with proving the key result that will give uniqueness of maximal immediate extensions when the differential residue field is linearly surjective.

The differential-henselian configuration theorem

Proposition 7.4.1 below is analogous to Proposition 3.3.19 and plays a similar critical role.

Let L be a monotone extension of K, and (a_ρ) a pc-sequence in K such that $a_\rho \rightsquigarrow a \in L$. We set $\gamma_\rho := v(a_{s(\rho)} - a_\rho)$, and let α, β, γ range over Γ.

PROPOSITION 7.4.1. *Let $G(Y) \in L\{Y\} \setminus L$ have order $\leqslant r$ and suppose*

(i) *$G(a_\rho) \rightsquigarrow 0$;*

(ii) *for every $\boldsymbol{i} \in \mathbb{N}^{1+r}$ with $|\boldsymbol{i}| = 1$ and every pc-sequence (e_λ) in K equivalent to (a_ρ), we have $G_{(\boldsymbol{i})}(e_\lambda) \not\rightsquigarrow 0$.*

Then G is in dh-configuration at a, and $v(G, a) > v(a - a_\rho)$, eventually. There is also an index ρ_0 such that for all $\rho > \rho_0$ and all $g \in L$,

$$g \asymp a - a_\rho \implies \operatorname{dmul} G_{+a,\times g} = \operatorname{ddeg} G_{+a,\times g} = 1.$$

PROOF. We let $\boldsymbol{i}$ range over the elements of $\mathbb{N}^{1+r}$ with $|\boldsymbol{i}| = 1$, and $\boldsymbol{j}$ over all elements in $\mathbb{N}^{1+r}$. By removing some initial ρ's we can assume $\gamma_\rho = v(a - a_\rho) \in \Gamma^{\neq}$ for all ρ, and $\gamma_{\rho'} > \gamma_\rho$ whenever $\rho' > \rho$. Next, set

$$g_{\boldsymbol{j}} := G_{(\boldsymbol{j})}(a), \quad P(Y) := G(a+Y) - G(a) = \sum_{|\boldsymbol{j}| \geqslant 1} g_{\boldsymbol{j}} Y^{\boldsymbol{j}} = \sum_{e=1}^{N} P_e(Y) \in L\{Y\},$$

where $N := \deg P \geqslant 1$. Lemma 6.8.3 and the remarks after Lemma 6.8.1 provide a pc-sequence (b_ρ) in K that is equivalent to (a_ρ) such that:

(1) $G(b_\rho) \rightsquigarrow G(a), \quad G(b_\rho) \rightsquigarrow 0$,

(2) $G_{(\boldsymbol{i})}(b_\rho) \rightsquigarrow G_{(\boldsymbol{i})}(a)$ whenever $G_{(\boldsymbol{i})} \notin L$,

(3) $v(b_\rho - a) = \gamma_\rho$, eventually,

(4) $v\big(P_e(b_\rho - a)\big) = v_{P_e}(\gamma_\rho)$, eventually, whenever $1 \leqslant e \leqslant N$ and $P_e \neq 0$.

If $e, e' \in \{1, \dots, N\}$, $e \neq e'$, $P_e \neq 0$, $P_{e'} \neq 0$, then either $v_{P_e}(\gamma_\rho) < v_{P_{e'}}(\gamma_\rho)$, eventually, or $v_{P_{e'}}(\gamma_\rho) < v_{P_e}(\gamma_\rho)$, eventually, by Corollary 6.1.5. This yields a $d \in \{1, \dots, N\}$ such that, after removing some initial ρ's, we have

$$e \in \{1, \dots, N\}, d \neq e \implies v_{P_d}(\gamma_\rho) < v_{P_e}(\gamma_\rho) \text{ for all } \rho.$$

Now $G(b_\rho) - G(a) = P(b_\rho - a)$ for all ρ, and $G(b_\rho) \succ G(a)$, eventually, and

$$G(a + Y) = G(a) + A(Y) + R(Y), \quad A := P_1, \quad R := P_2 + \cdots + P_N.$$

Suppose that $d = 1$. Then $G(b_\rho) \sim A(b_\rho - a)$, eventually, and so, eventually,

$$v\big(G(a)\big) > v\big(A(b_\rho - a)\big) = v_A(\gamma_\rho) < v_R(\gamma_\rho),$$

so G is in dh-configuration at a, and $v(G, a) > \gamma_\rho = v(a - a_\rho)$, eventually. Also the second part of the desired conclusion follows easily.

It only remains to show that $d = 1$. For each $\boldsymbol{i}$ we have

$$G_{(\boldsymbol{i})}(b_\rho) - G_{(\boldsymbol{i})}(a) = \sum_{\boldsymbol{j} > \boldsymbol{i}} \binom{\boldsymbol{j}}{\boldsymbol{i}} G_{(\boldsymbol{j})}(a)(b_\rho - a)^{\boldsymbol{j} - \boldsymbol{i}} = \sum_{\boldsymbol{j} > \boldsymbol{i}} \binom{\boldsymbol{j}}{\boldsymbol{i}} g_{\boldsymbol{j}}(b_\rho - a)^{\boldsymbol{j} - \boldsymbol{i}}.$$

If $G_{(\boldsymbol{i})} \neq 0$, then $G_{(\boldsymbol{i})}(b_\rho) \sim G_{(\boldsymbol{i})}(a) = g_{\boldsymbol{i}}$, eventually, so

$$g_{\boldsymbol{i}} \succ \sum_{\boldsymbol{j} > \boldsymbol{i}} \binom{\boldsymbol{j}}{\boldsymbol{i}} g_{\boldsymbol{j}}(b_\rho - a)^{\boldsymbol{j} - \boldsymbol{i}}, \text{ eventually, and thus}$$

$$g_{\boldsymbol{i}}(b_\rho - a)^{\boldsymbol{i}} \succ \sum_{\boldsymbol{j} > \boldsymbol{i}} \binom{\boldsymbol{j}}{\boldsymbol{i}} g_{\boldsymbol{j}}(b_\rho - a)^{\boldsymbol{j}}, \text{ eventually.}$$

From $\sum_{\boldsymbol{i}} \binom{\boldsymbol{j}}{\boldsymbol{i}} = j_0 + j_1 + \cdots + j_r$ we get

$$\sum_{\boldsymbol{i}} \left(\sum_{\boldsymbol{j} > \boldsymbol{i}} \binom{\boldsymbol{j}}{\boldsymbol{i}} g_{\boldsymbol{j}}(b_\rho - a)^{\boldsymbol{j}} \right) = \sum_{e=2}^{N} e \left(\sum_{|\boldsymbol{j}| = e} g_{\boldsymbol{j}}(b_\rho - a)^{\boldsymbol{j}} \right) = \sum_{e=2}^{N} e P_e(b_\rho - a).$$

Now $G_{(\boldsymbol{i})} \neq 0$ for at least one $\boldsymbol{i}$, and if $G_{(\boldsymbol{i})} = 0$ and $\boldsymbol{j} > \boldsymbol{i}$, then $G_{(\boldsymbol{j})} = 0$, so

$$\min_{\boldsymbol{i}} v\big(g_{\boldsymbol{i}}(b_\rho - a)^{\boldsymbol{i}}\big) < v\left(\sum_{e=2}^{N} e P_e(b_\rho - a)\right) = \min\big\{v_{P_e}(\gamma_\rho) : e = 2, \dots, N\big\},$$

eventually. Since $A = P_1$ we have $v\big(P_1(b_\rho - a)\big) = v_A(\gamma_\rho)$ for all ρ, so to obtain $d = 1$ it is enough to get, cofinally in ρ,

$$v_A(\gamma_\rho) \leqslant \min_{\boldsymbol{i}} v\big(g_{\boldsymbol{i}}(b_\rho - a)^{\boldsymbol{i}}\big).$$

To derive $d = 1$ from the above, we now use that L is monotone, which gives

$$v\big(g_{\boldsymbol{i}}(b_\rho - a)^{\boldsymbol{i}}\big) \geqslant v(g_{\boldsymbol{i}}) + \gamma_\rho \text{ for all } \boldsymbol{i} \text{ and } \rho, \text{ and so}$$

$$v_A(\gamma_\rho) = v(A) + \gamma_\rho = \big(\min_{\boldsymbol{i}} v(g_{\boldsymbol{i}})\big) + \gamma_\rho \leqslant \min_{\boldsymbol{i}} v\big(g_{\boldsymbol{i}}(b_\rho - a)^{\boldsymbol{i}}\big)$$

for all ρ, so $d = 1$. □

In this proof the assumption that K and L are monotone is used only at the end when it is explicitly invoked. The earlier part of the proof goes through if this assumption is dropped, and so with the weaker assumption the proposition holds for $\deg G = 1$.

Uniqueness of maximal immediate extensions

LEMMA 7.4.2. *Let L be a monotone* d*-algebraically maximal extension of K such that $\boldsymbol{k}_L$ is linearly surjective. Let (a_ρ) be a divergent pc-sequence in K with minimal differential polynomial $G(Y)$ over K. Then $a_\rho \rightsquigarrow b$ and $G(b) = 0$ for some $b \in L$.*

PROOF. Take a pseudolimit $a \in L$ of (a_ρ), and take a pc-sequence (b_λ) in K equivalent to (a_ρ) such that $G(b_\lambda) \rightsquigarrow 0$. Then G is in dh-configuration at a, and $v(G,a) > v(a - b_\lambda)$, eventually, by Proposition 7.4.1. Now L is d-henselian by Theorem 7.0.1, so by Lemma 7.3.1 we have $b \in L$ such that

$$v_L(a-b) = v(G,a) \text{ and } G(b) = 0.$$

This gives $v_L(a-b) > v(a-b_\lambda)$, eventually, so $b_\lambda \rightsquigarrow b$, and thus $a_\rho \rightsquigarrow b$. □

Together with Lemmas 6.3.5, 6.9.1, and 6.9.3 this yields:

THEOREM 7.4.3. *Suppose $\boldsymbol{k}$ is linearly surjective. Then any two maximal immediate extensions of K are isomorphic over K and* d*-henselian. Also, any two* d*-algebraic immediate extensions of K that are* d*-algebraically maximal are isomorphic over K and* d*-henselian.*

We now state minor variants of these results using the notion of saturation from model theory (see B.9), as needed in the proof of the Equivalence Theorem 8.2.5. Let $|X|$ denote the cardinality of a set X, and let κ be a cardinal.

LEMMA 7.4.4. *Let K be* d*-henselian. Let E be a valued differential subfield of K such that the derivation of $\boldsymbol{k}_E$ is nontrivial. Assume K is κ-saturated with $\kappa > |v(E^\times)|$. If (a_ρ) is a divergent pc-sequence in E with minimal differential polynomial $G(Y)$ over E, then $a_\rho \rightsquigarrow b$ and $G(b) = 0$ for some $b \in K$.*

PROOF. Let (a_ρ) be a divergent pc-sequence in E with minimal differential polynomial $G(Y)$ over E. By saturation we have a pseudolimit $a \in K$ of (a_ρ). Take a pc-sequence (b_λ) in E equivalent to (a_ρ) such that $G(b_\lambda) \rightsquigarrow 0$. By Proposition 7.4.1, G is in dh-configuration at a with $v(G,a) > v(a-b_\lambda)$, eventually. Take $b \in K$ with $v(a-b) = v(G,a)$ and $G(b) = 0$. Then $b_\lambda \rightsquigarrow b$, so $a_\rho \rightsquigarrow b$. □

In combination with Lemmas 6.9.1, 6.9.3, and 7.4.2, this yields:

COROLLARY 7.4.5. *Let K and E be as in the previous lemma, and assume also that $\boldsymbol{k}_E$ is linearly surjective and $v(E^\times) \neq \{0\}$. Then any maximal immediate extension of E can be embedded in K over E.*

Recall that a valued field of equicharacteristic 0 is henselian if and only if it is algebraically maximal. We already established an analogue for valued differential fields in one direction: d-algebraically maximal valued differential fields with small derivation and linearly surjective differential residue field are d-henselian. The converse fails, even in the monotone case:

EXAMPLE. Let $\boldsymbol{k}$ be a countable linearly surjective differential field and let Γ be a countable ordered abelian group, $\Gamma \neq \{0\}$. Then $\boldsymbol{k}(t^\Gamma)$ is a countable valued differential subfield of the Hahn differential field $\boldsymbol{k}(\!(t^\Gamma)\!)$. The latter is d-henselian, so we can take a countable d-henselian K such that

$$\boldsymbol{k}(t^\Gamma) \subseteq K \subseteq \boldsymbol{k}(\!(t^\Gamma)\!) \quad \text{(as valued differential fields).}$$

The constant field of $\boldsymbol{k}(\!(t^\Gamma)\!)$ is $C_{\boldsymbol{k}}(\!(t^\Gamma)\!)$, which is uncountable. Take $c \in C_{\boldsymbol{k}}(\!(t^\Gamma)\!)$ with $c \notin K$. Then $K\langle c\rangle = K(c)$ is a proper immediate d-algebraic extension of K, so K is not d-algebraically maximal.

In this example K has many constants. With few constants, we do have a converse in the monotone case, as we shall see in the next section.

7.5 THE CASE OF FEW CONSTANTS

In this section we consider in more detail the situation where $C \subseteq \mathcal{O}$, that is, the valuation is trivial on C. Key facts are Lemma 7.5.5 and Proposition 7.5.6 below.

Valuation properties of linear differential operators

In this subsection we fix $r \geqslant 1$, and assume: K is r-d-henselian, $A \in K[\partial]^{\neq}$ has order $\leqslant r$.

LEMMA 7.5.1. *Suppose $A(1) \prec A$. Then $A(y) = 0$ for some $y \sim 1$ in K.*

PROOF. We have $A = a_0 + a_1\partial + \cdots + a_r\partial^r$ with $a_0, a_1, \dots, a_r \in K$, so $A(1) = a_0$. We can arrange that $a_0 \in \smallo$ and $a_1, \dots, a_r \in \mathcal{O}$, $a_i \asymp 1$ for some $i \in \{1, \dots, r\}$. With $R(Z) := \operatorname{Ri}(A)$ we have $R(Z) \in \mathcal{O}\{Z\}$, $R(0) = a_0 \prec 1$ and

$$R(Z)_1 = a_1 Z + \cdots + a_r Z^{(r-1)} \asymp 1,$$

so we get $z \in \smallo$ with $R(z) = 0$. By Corollary 7.1.9 we can take $y \in 1 + \smallo$ such that $y^\dagger = z$, and then $A(y) = 0$. □

The proof shows that in Lemma 7.5.1 we can relax the assumption that K is r-d-henselian to K being $(r-1)$-d-henselian, with $(1+\smallo)^\dagger = \smallo$ in case $r = 1$. The same holds for Lemma 7.5.2, Corollaries 7.5.3 and 7.5.4, and Lemma 7.5.5.

LEMMA 7.5.2. *Suppose $C \subseteq \mathcal{O}$. Then there are no $b_0 \succ b_1 \succ \cdots \succ b_r$ in $K^\times$ with $A(b_i) \prec Ab_i$ for $i = 0, \dots, r$.*

PROOF. Let $b \in K^\times$ and $A(b) \prec Ab$. Then $B(1) \prec B$ for $B = Ab$, so we have $y \sim 1$ in K with $B(y) = 0$, and thus $A(by) = 0$ with $by \sim b$. It remains to note that if $b_0 \succ b_1 \succ \cdots \succ b_r$ in $K^\times$, then $b_0, \dots, b_r$ are C-linearly independent. □

Here is a reformulation of Lemma 7.5.2 and its proof in terms of the set

$$\mathscr{E}(A) = \{vb : b \in K^\times,\ A(b) \prec Ab\}$$

of exceptional values for A that was introduced in Section 5.6:

COROLLARY 7.5.3. *If $C \subseteq \mathcal{O}$, then $\mathscr{E}(A) = v(\ker^{\neq} A)$, so $|\mathscr{E}(A)| \leqslant \dim_C \ker A$.*

COROLLARY 7.5.4. *Suppose $C \subseteq \mathcal{O}$. Let (a_ρ) be a well-indexed sequence in K such that $a_\rho \rightsquigarrow a$, with $a \in K$. Then $A(a_\rho) \rightsquigarrow A(a)$.*

PROOF. Replacing a_ρ by $a_\rho - a$ we reduce to the case $a = 0$. By omitting some initial terms from (a_ρ) we can further assume that va_ρ is strictly increasing as a function of ρ. So $v(Aa_\rho) = v_A(va_\rho)$ is strictly increasing as a function of ρ. It remains to note that $A(a_\rho) \prec Aa_\rho$ for only finitely many ρ, by Lemma 7.5.2, and $A(a_\rho) \asymp Aa_\rho$ for all other ρ. □

Extension to dominant degree 1

Lemma 7.5.2 leads to:

LEMMA 7.5.5. *Suppose $r \geqslant 1$, K is r-d-henselian, $C \subseteq \mathcal{O}$ and $G \in K\{Y\} \setminus K$ has order $\leqslant r$. Then there do not exist $y_0, \dots, y_{r+1} \in K$ such that:*

(i) $y_{i-1} - y_i \succ y_i - y_{i+1}$ *for all* $i \in \{1, \dots, r\}$, *and* $y_r \neq y_{r+1}$;

(ii) $G(y_0) = \cdots = G(y_{r+1}) = 0$;

(iii) $\operatorname{ddeg} G_{+y_{r+1}, \times g} = 1$ *and* $y_0 - y_{r+1} \preccurlyeq g$ *for some* $g \in K^\times$.

PROOF. Towards a contradiction, suppose $y_0, \dots, y_{r+1} \in K$ satisfy (i), (ii), (iii). Set $a_i := y_i - y_{r+1}$ for $i = 0, \dots, r$ and $P := G_{+y_{r+1}}$. Then $a_i \sim y_i - y_{i+1}$ for $i = 0, \dots, r$ and so by (i) and (ii),

$$a_0 \succ a_1 \succ \cdots \succ a_r \neq 0, \qquad P(a_0) = \cdots = P(a_r) = P(0) = 0.$$

Now $P = A + R$ with $A = P_1$ and $\operatorname{mul} R \geqslant 2$. Taking g as in (iii) we have $g \succcurlyeq a_0$, so $a_i = gb_i$ with $b_i \preccurlyeq 1$ for $i = 0, \dots, r$, and $b_0 \succ b_1 \succ \cdots \succ b_r$. Also $P_{\times g} = A_{\times g} + R_{\times g}$ with $R_{\times g} \prec A_{\times g}$ by (iii), and for $i = 0, \dots, r$,

$$\begin{aligned} P(a_i) &= A_{\times g}(b_i) + R_{\times g}(b_i) = 0, \text{ so} \\ A_{\times g}(b_i) &= -R_{\times g}(b_i) \preccurlyeq R_{\times g b_i} \prec A_{\times g b_i} \end{aligned}$$

which contradicts Lemma 7.5.2. □

PROPOSITION 7.5.6. *Let $r \geqslant 1$, and assume K is r-d-henselian and $C \subseteq \mathcal{O}$. Let $G \in K\{Y\}$ with* $\operatorname{order} G \leqslant r$ *and* $\operatorname{ddeg} G = 1$. *Let E be an immediate extension of K. Then G has the same zeros in $\mathcal{O}$ as in $\mathcal{O}_E$.*

PROOF. Note first that $\operatorname{ddeg} G_{+y} = 1$ for all $y \in \mathcal{O}_E$. Towards a contradiction, suppose $G(\ell) = 0$ with $\ell \in \mathcal{O}_E \setminus \mathcal{O}$. Now $\ell \preccurlyeq 1$ gives $\operatorname{ddeg} G_{+\ell} = 1$, and from $G(\ell) = 0$ it follows easily that $\operatorname{ddeg} G_{+\ell,\times g} = 1$ for all $g \preccurlyeq 1$ in E.

CLAIM: *Let $\gamma \in v(\ell - K)^{\geqslant 0}$. Then there is a $y \in \mathcal{O}$ with $G(y) = 0$ and $v(\ell - y) \geqslant \gamma$.*

To prove this claim, take $a \in K$ and $g \in K^\times$ such that $v(\ell - a) = vg = \gamma$. Then by Corollary 6.6.6 and the observation preceding the claim, $\operatorname{ddeg} G_{+a,\times g} = \operatorname{ddeg} G_{+\ell,\times g} = 1$, so we get $b \in \mathcal{O}$ such that $G(a + gb) = 0$, so $y := a + gb$ satisfies the claim. Having proved the claim, we get $y_0, \dots, y_r, y_{r+1} \in \mathcal{O}$ such that

$$\ell - y_0 \succ \ell - y_1 \succ \cdots \succ \ell - y_{r+1}, \qquad G(y_0) = G(y_1) = \cdots = G(y_{r+1}) = 0,$$

contradicting Lemma 7.5.5: take $g = 1$ in (iii). □

We also have the following variant:

LEMMA 7.5.7. *Assume that $C \subseteq \mathcal{O}$. Let $A \in K[\partial]^{\neq}$ be neatly surjective. Let E be an immediate extension of K. Then* $\ker_E A = \ker A$.

PROOF. Take $r \geqslant 1$ and $a_0, \dots, a_r \in K$ such that $A = a_0 + a_1\partial + \cdots + a_r\partial^r$. Set $G := a_0 Y + \cdots + a_r Y^{(r)} \in K\{Y\}$. We claim that G has the same zeros in $\mathcal{O}$ as in $\mathcal{O}_E$, and for this we follow the proof by contradiction of Proposition 7.5.6, deriving $\operatorname{ddeg} G_{+a,\times g} = 1$ as in that proof. Now $G_{+a,\times g}(Y) = G(a) + G_{\times g}(Y)$, and $G(a) = A(a)$, while $G_{\times g}$ corresponds to Ag. Thus $v(Ag) \leqslant v\big(A(a)\big)$, and as A is neatly surjective, we get $b \in \mathcal{O}$ with $A(gb) = -A(a)$, so $y = a + gb$ gives $A(y) = 0$ and $v(\ell - y) \geqslant v(\ell - a)$. As in the proof of Proposition 7.5.6, this argument yields $y_0, \dots, y_r, y_{r+1} \in K$ such that

$$\ell - y_0 \succ \ell - y_1 \succ \cdots \succ \ell - y_{r+1}, \qquad A(y_0) = A(y_1) = \cdots = A(y_{r+1}) = 0,$$

so $y_0 - y_{r+1} \succ y_1 - y_{r+1} \succ \cdots \succ y_r - y_{r+1}$ are C-linearly independent elements of $\ker A$, which is impossible. This proves our claim. For $y \in \ker_E A$ with $y \succ 1$, take $f \in K^\times$ with $f \asymp y$, and note that then $f^{-1}y \in (\ker_E Af) \cap \mathcal{O}_E$. Applying the above to Af instead of A we conclude that $f^{-1}y \in \mathcal{O}$, so $y \in K$. □

The following special case is worth recording:

COROLLARY 7.5.8. *If $C \subseteq \mathcal{O}$ and $\partial \in K[\partial]^{\neq}$ is neatly surjective, then $C_E = C$ for any immediate extension E of K.*

It turns out that any K satisfying the assumptions of Corollary 7.5.8 (and thus any d-henselian K with few constants) is *asymptotic* in the sense of Chapter 9, by Lemma 9.1.1. For more on such K, see Section 9.4, in particular, Corollary 9.4.11. When does K extend to a d-henselian valued differential field with few constants? (The conclusion of Lemma 7.5.2 holds for such K, any $r \geqslant 1$, and any $A \in K[\partial]^{\neq}$ of order $\leqslant r$.) Corollary 10.1.14 and subsequent remarks address this question.

Few constants and monotonicity

Having few constants leads to the promised converse of Theorem 7.0.1 in the monotone case:

PROOF OF THEOREM 7.0.3. Assume $C \subseteq \mathcal{O}$ and K is monotone and d-henselian; our job is to show that K is d-algebraically maximal. Towards a contradiction, let $K\langle a\rangle$ be an immediate d-algebraic extension of K with $a \notin K$. Take a divergent pc-sequence (a_ρ) in K such that $a_\rho \rightsquigarrow a$. Then (a_ρ) is of d-algebraic type over K by Corollary 6.9.2, so we have a minimal differential polynomial $G(Y)$ of (a_ρ) over K. Replacing (a_ρ) by an equivalent pc-sequence in K we arrange that $G(a_\rho) \rightsquigarrow 0$. Then the assumptions of Proposition 7.4.1 are satisfied, so taking $g_\rho \in K$ with $g_\rho \asymp a - a_\rho$ we have $\operatorname{ddeg} G_{+a,\times g_\rho} = 1$ eventually, and we can arrange this holds for all ρ. Then $\operatorname{ddeg} G_{+a_\rho,\times g_\rho} = 1$ for all ρ by Corollary 6.6.6. This gives for each ρ an element $z_\rho \in K$ with $G(z_\rho) = 0$ and $a_\rho - z_\rho \preccurlyeq g_\rho$, so $a - z_\rho \preccurlyeq g_\rho$.

Take $r \geqslant 1$ such that G has order $\leqslant r$. Pick some index ρ_0 and set $g := g_{\rho_0}$. In view of Lemma 2.2.19 the above yields indices $\rho_0 < \rho_1 < \cdots < \rho_{r+1}$ such that $a - z_{\rho_i} \succ a - z_{\rho_j}$ whenever $0 \leqslant i < j \leqslant r+1$. Set $y_i := z_{\rho_i}$ for $i = 0,\dots,r+1$. Then conditions (i) and (ii) of Lemma 7.5.5 are satisfied. Also $a - y_{r+1} \prec a - y_0 \preccurlyeq g$, hence $y_0 - y_{r+1} \preccurlyeq g$ and $a_{\rho_0} - y_{r+1} \preccurlyeq g$. In view of Corollary 6.6.6 the latter gives $\operatorname{ddeg} G_{+y_{r+1},\times g} = \operatorname{ddeg} G_{+a_{\rho_0},\times g} = 1$, so condition (iii) in Lemma 7.5.5 also holds, which contradicts that lemma. □

7.6 DIFFERENTIAL-HENSELIANITY IN SEVERAL VARIABLES

In this section we prove Theorem 7.0.4 in several stages. After some preliminaries we first handle the case of spherically complete K with archimedean Γ; this goes by diagonalization and successive approximation. Next we treat arbitrary spherically complete K using a reduction to the previous case by iterated coarsening. The last stage is an appeal to a result in commutative differential algebra due to J. Johnson.

In this section n ranges over $\mathbb{N}^{\geqslant 1}$, and for $y = (y_1,\dots,y_n) \in K^n$ we set $vy := \min(vy_1,\dots,vy_n)$. This makes K^n into a valued abelian group, and in particular, $vy \geqslant 0$ means $y \in \mathcal{O}^n$, and $vy > 0$ means $y \in o^n$. Recall that if K is spherically complete, then K^n is spherically complete as a valued abelian group.

Notations and some easy equivalences

Let $Y = (Y_1,\dots,Y_n)$ be a tuple of distinct differential indeterminates, and equip $K\{Y\}$ with the gaussian extension of the valuation of K. Let $P = (P_1,\dots,P_n) \in K\{Y\}^n$. Then we set

$$\begin{aligned} vP &:= \min(vP_1,\dots,vP_n),\\ P(y) &:= \big(P_1(y),\dots,P_n(y)\big) \in K^n \quad \text{for } y \in K^n. \end{aligned}$$

A **solution of the system** $P(Y) = 0$ **in** K is a point $y \in K^n$ such that $P(y) = 0$. Also, a **solution of** $P(Y) = 0$ **in** $\mathcal{O}$ is a point $y \in \mathcal{O}^n$ with $P(y) = 0$, and likewise, with o

instead of $\mathcal{O}$. Recall that for $F \in K\{Y\}$ and $y = (y_1, \dots, y_n) \in K^n$,

$$F_{+y} = F(y+Y) = F(y_1 + Y_1, \dots, y_n + Y_n) \in K\{Y\}.$$

We also set $P_{+y} := (P_{1,+y}, \dots, P_{n,+y}) \in K\{Y\}^n$ for $y \in K^n$.

In the rest of this section, let $P \in \mathcal{O}\{Y\}^n$ and consider the associated system

$$(*) \qquad P(Y) = 0.$$

Let $A_i = P_{i,1}$ be the homogeneous part of P_i of degree 1, with image $\overline{A}_i$ in $\boldsymbol{k}\{Y\}_1$.

We say that the system $(*)$ is in **differential-hensel position** (dh-position) if the differential polynomials $\overline{A}_1, \dots, \overline{A}_n$ are d-independent and $P_1(0) \prec 1, \dots, P_n(0) \prec 1$. (For $n = 1$ this means "dh-position at 0" as defined in Section 7.1.) Note that if $(*)$ is in dh-position and $y \in \smallo^n$, then $P_{+y} = 0$ is in dh-position.

We say that the system $(*)$ is in **diagonal differential-hensel position** (ddh-position) if for $i = 1, \dots, n$,

$$0 \neq \overline{A}_i \in \boldsymbol{k}\{Y_i\}, \qquad P_i(0) \prec 1.$$

Note that if $(*)$ is in ddh-position, then it is in dh-position. Suppose now that $(*)$ is in ddh-position, so for $i = 1, \dots, n$,

$$P_i(Y) = a_i + A_i(Y) + R_i(Y), \qquad a_i \in \smallo,$$

where $R_i \in \mathcal{O}\{Y\}$ has only terms of degree $\geqslant 2$. Set $a := (a_1, \dots, a_n)$. Then a solution of $(*)$ cannot be much smaller than a:

LEMMA 7.6.1. *Suppose $a \neq 0$ and y is a solution of $(*)$ in K. Then*

$$vy \leqslant va + o(va).$$

PROOF. Suppose $vy \geqslant (1 + \varepsilon)va$ where $\varepsilon \in \mathbb{Q}^>$. Then for $i = 1, \dots, n$ we have $A_i(y), R_i(y) \prec a$, which for i with $va_i = va$ is impossible. □

LEMMA 7.6.2. *Assume $\Gamma \neq \{0\}$. The following are equivalent:*

(i) *every system $(*)$ such that $\overline{A}_1, \dots, \overline{A}_n$ are* d*-independent and the coefficients of the monomials in P of degree $\geqslant 2$ are in $\smallo$ has a solution in $\mathcal{O}$;*

(ii) *$\boldsymbol{k}$ is linearly surjective, and every $(*)$ in ddh-position has a solution in $\smallo$;*

(iii) *$\boldsymbol{k}$ is linearly surjective, and every $(*)$ in dh-position has a solution in $\smallo$.*

PROOF. Suppose (i) holds. For $n = 1$ this yields by Lemma 7.2.1 that K is d-henselian, so $\boldsymbol{k}$ is linearly surjective. Let $(*)$ be in ddh-position. Using the notations above, $P_i = a_i + A_i + R_i$, $a_i \prec 1$ and $A_i(Y) = D_i(Y_i) + E_i(Y)$ with $D_i \in K\{Y_i\}_1$, $D_i \asymp 1$, and $E_i \in K\{Y\}_1$, $E_i \prec 1$, for $i = 1, \dots, n$. To get (ii) we need to find a solution of $(*)$ in $\smallo$. We can take nonzero $g \in \smallo$ with $vg > 0$ so small that

$$vD_i(gY_i) \leqslant va_i, \qquad vD_i(gY_i) < vE_i(gY), vR_i(gY) \qquad (i = 1, \dots, n).$$

Taking nonzero $h_i \in \mathcal{o}$ with $vh_i = vD_i(gY_i)$, we obtain a system

$$h_1^{-1}P_1(gY) = \cdots = h_n^{-1}P_n(gY) = 0.$$

Considering the homogeneous parts of degree 1 of the $h_i^{-1}P_i(gY)$, this system has a solution y in $\mathcal{O}$ by (i), and so gy is a solution of $(*)$ in $\mathcal{o}$.

Next, assume (ii). Let $(*)$ be in dh-position. By Lemma 5.6.16 we can transform this into a system $Q(Y) = 0$ in ddh-position such that any solution of $Q(Y) = 0$ in $\mathcal{o}$ gives rise to a solution of $(*)$ in $\mathcal{o}$. Now $Q(Y) = 0$ has a solution in $\mathcal{o}$ by (ii), and so $(*)$ has a solution in $\mathcal{o}$. This gives (iii).

Finally, assume (iii). Let $(*)$ be such that $\overline{A}_1, \dots, \overline{A}_n$ are d-independent and the coefficients of the monomials in P of degree $\geqslant 2$ are in $\mathcal{o}$. To get (i) we need to find a solution of $(*)$ in $\mathcal{O}$. Since $\boldsymbol{k}$ is linearly surjective we can use Lemma 5.4.4 to get $y \in \mathcal{O}^n$ such that $vP(y) > 0$. Then $P_{+y}(Y) = 0$ is a system in dh-position, and so has a solution b in $\mathcal{o}$ by (iii), and then $y + b$ is a solution of $(*)$ in $\mathcal{O}$. □

The case of an archimedean value group

We define $\mathcal{O}$ to be **strongly linearly surjective** if for all $n \geqslant 1$ and $D_1(Y_1) \in \mathcal{O}\{Y_1\}_1, \dots, D_n(Y_n) \subset \mathcal{O}\{Y_n\}_1$ with $D_1, \dots, D_n \asymp 1$, all $E_1(Y), \dots, E_n(Y) \prec 1$ in $\mathcal{O}\{Y\}_1$, and all $a = (a_1, \dots, a_n) \in \mathcal{O}^n$ there exists a $y \in \mathcal{O}^n$ such that $y \asymp a$ and

$$D_1(y_1) + E_1(y) = a_1, \ \dots, \ D_n(y_n) + E_n(y) = a_n.$$

LEMMA 7.6.3. *Suppose $\Gamma := v(K^\times)$ is archimedean, $\boldsymbol{k}$ is linearly surjective, and K is spherically complete. Then $\mathcal{O}$ is strongly linearly surjective.*

PROOF. Note that K is monotone by Corollary 6.1.2. In addition, K is d-henselian. Therefore, if $A \asymp 1$ in $K[\partial]$, then by Corollary 4.5.4 and Lemma 7.1.8 there is for each $a \in K$ an element $y \in K$ such that $A(y) = a$ and $y \asymp a$.

Let $D_1, \dots, D_n, E_1, \dots, E_n$ be as above. Define $\mathbb{Q}$-linear maps

$$D, E \colon \mathcal{O}^n \to \mathcal{O}^n, \quad D(y) := \big(D_1(y_1), \dots, D_n(y_n)\big), \ E(y) := \big(E_1(y), \dots, E_n(y)\big).$$

From the fact stated in the beginning of the proof and by Lemma 2.3.11 we obtain a valuation preserving right-inverse D^* to D, that is, $D^* \colon \mathcal{O}^n \to \mathcal{O}^n$ is $\mathbb{Q}$-linear, $D \circ D^* = \mathrm{id}_{\mathcal{O}^n}$, and $D^*(a) \asymp a$ for all $a \in \mathcal{O}^n$. Consider the $\mathbb{Q}$-linear map $F := (D + E) \circ D^* = \mathrm{id}_{\mathcal{O}^n} + G \colon \mathcal{O}^n \to \mathcal{O}^n$, with $G := E \circ D^*$. Then $G(a) \prec a$ for all nonzero $a \in \mathcal{O}^n$, and so by Corollary 2.2.13, $F \colon \mathcal{O}^n \to \mathcal{O}^n$ is a valuation-preserving $\mathbb{Q}$-linear bijection. Given $a \in \mathcal{O}^n$, we get $b \in \mathcal{O}^n$ with $F(b) = a$ and $b \asymp a$, so $y := D^*(b)$ yields $(D + E)(y) = a$ and $y \asymp a$. □

Assume Γ is archimedean. Let $(*)$ be in ddh-position, so for $i = 1, \dots, n$ we have $P_i = a_i + D_i + E_i + R_i$ where $a_i \in \mathcal{o}$, $D_i \in \mathcal{O}\{Y_i\}_1$, $D_i \asymp 1$, $E_i \in \mathcal{O}\{Y\}_1$, $E_i \prec 1$, and all terms in $R_i \in \mathcal{O}\{Y\}$ have degree $\geqslant 2$. We try to construct a solution $y \prec 1$

in K to $(*)$. Set $a := (a_1, \dots, a_n)$. If $a = 0$, then $y = 0$ is such a solution, so assume $a \neq 0$. We associate to $(*)$ the *linear* system

$$a_1 + D_1(Y_1) + E_1(Y) = \cdots = a_n + D_n(Y_n) + E_n(Y) = 0.$$

Suppose it has a solution y in K with $vy = va$. Substituting $y + Y$ for Y gives

$$P_{i,+y}(Y) = P_i(y+Y) = a_i^* + D_i(Y_i) + E_i^*(Y) + R_i^*(Y), \quad \text{where}$$

$$a_i^* = R_i(y),\ E_i^* = E_i + dR_i(y,Y),\ dR_i(y,Y) := \sum_{j=1}^{n}\sum_{k\in\mathbb{N}} \frac{\partial R_i}{\partial Y_j^{(k)}}(y)Y_j^{(k)},$$

and $R_i^* \in \mathcal{O}\{Y\}$ has only terms of degree $\geqslant 2$, and thus

$$va_i^* \geqslant 2vy = 2va, \qquad v\big(dR_i(y,Y)\big) \geqslant vy = va.$$

In particular, $va^* \geqslant 2va$ for $a^* := (a_1^*, \dots, a_n^*)$, and $E_i^* \prec 1$. Thus the modified system $P_{+y}(Y) = 0$ with $P_{+y} := (P_{1,+y}, \dots, P_{n,+y})$ is still in ddh-position.

LEMMA 7.6.4. *Suppose Γ is archimedean, $\boldsymbol{k}$ is linearly surjective, K is spherically complete, and $(*)$ is in ddh-position. Then there is $f \in \mathcal{O}^n$ such that $P(f) = 0$.*

PROOF. Let $(f_i)_{i<m}$ be a sequence in $\mathcal{O}^n$ with $m \geqslant 1$, such that

(1) $P_{+f_i}(Y) = 0$ is in ddh-position, for all $i < m$,

(2) $v(f_j - f_i) = vP(f_i)$ whenever $i < j < m$,

(3) $vP(f_j) \geqslant 2vP(f_i)$ whenever $i < j < m$.

Taking $f_0 = 0$ shows that such a sequence exists for $m = 1$. Let $m = \mu + 1$. By Lemma 7.6.3 we can take $y \in \mathcal{O}^n$ such that $vy = vP(f_\mu)$ and y is a solution of the linear system associated to $P_{+f_\mu}(Y) = 0$. Since $P(f_\mu) \neq 0$, the arguments preceding the lemma applied to $P_{+f_\mu}(Y)$ instead of $P(Y)$ show that for $f_m = f_\mu + y$ we have $vP(f_m) \geqslant 2vP(f_\mu)$ and $P_{+f_m}(Y) = 0$ is in ddh-position. Then the extended sequence $(f_i)_{i<m+1}$ has the above properties with $m+1$ instead of m.

This construction yields a c-sequence $(f_i)_{i\in\mathbb{N}}$ in $\mathcal{O}^n$ with a limit $f \in \mathcal{O}^n$, and then $P(f) = 0$. □

Combining Lemmas 7.6.2 and 7.6.4 yields:

COROLLARY 7.6.5. *If Γ is archimedean, $\boldsymbol{k}$ is linearly surjective, and K is spherically complete, then conditions* (i), (ii), (iii) *of Lemma 7.6.2 are satisfied.*

Reduction to the case of an archimedean value group

This reduction rests on iterated coarsening in combination with the following lemma.

LEMMA 7.6.6. *Assume $\boldsymbol{k}$ is linearly surjective, K is spherically complete, and Γ has a smallest nontrivial convex subgroup Δ. Let $(*)$ be in dh-position, with $vP(0) \in \Delta$. Then $vy \in \Delta$ and $vP(y) > \Delta$ for some $y \in \mathcal{O}^n$.*

PROOF. The valued differential residue field $\dot{K}$ of the Δ-coarsening of K has linearly surjective residue field, has archimedean value group Δ, is maximal as a valued field, and $(*)$ yields a system $\dot{P}(Y) = 0$ in dh-position over $\dot{K}$. Then Corollary 7.6.5 yields $y = (y_1, \dots, y_n) \in \mathcal{O}^n$ such that for $\dot{y} := (\dot{y}_1, \dots, \dot{y}_n)$ we have $\dot{P}(\dot{y}) = 0$. Since $\dot{y} \neq 0$ we have $vy \in \Delta$, and so y has the desired property. □

PROPOSITION 7.6.7. *Suppose* $\boldsymbol{k}$ *is linearly surjective, and* K *is spherically complete. Then every system* $(*)$ *in dh-position has a solution in* $\mathcal{O}$.

PROOF. Since K is d-henselian by Corollary 7.0.2, the assumptions on K are inherited by any coarsening of K by a convex subgroup of Γ. This will be tacitly used in what follows. Let $(*)$ be in dh-position; our job is to show that $(*)$ has a solution in $\mathcal{O}$. If $P(0) = 0$ we are done, so assume $P(0) \neq 0$. Then $\Gamma \neq \{0\}$. Let $\lambda > 0$ be an ordinal, $(f_\rho)_{\rho<\lambda}$ a sequence in $\mathcal{O}^n$, and $(\Delta_\rho)_{\rho<\lambda}$ an increasing sequence of convex subgroups of Γ, such that for all $\rho < \lambda$,

(1) $vP(f_\rho) > \Delta_\rho$;

(2) Δ_ρ is the largest convex subgroup of Γ not containing $vP(f_\rho)$;

(3) $vP(f_\rho) \in \Delta_{\rho+1}$ whenever $\rho + 1 < \lambda$;

(4) $v(f_{\rho'} - f_\rho) > \Delta_\rho$ whenever $\rho < \rho' < \lambda$;

(5) $v(f_{\rho+1} - f_\rho) \in \Delta_{\rho+1}$ whenever $\rho + 1 < \lambda$.

Such sequences exist for $\lambda = 1$: take $f_0 = 0$ and let Δ_0 be the largest convex subgroup of Γ not containing $vP(0)$.

Suppose $\lambda = \mu + 1$ is a successor ordinal. If $P(f_\mu) = 0$ we are done, so assume $P(f_\mu) \neq 0$. Using (1) for $\rho := \mu$ and Lemma 5.6.15, the system $P_{+f_\mu}(Y) = 0$ is in dh-position with respect to the Δ_μ-coarsening of K. By (2) for $\rho = \mu$ we have a smallest convex subgroup Δ of Γ that properly contains Δ_μ. Then $vP(f_\mu) \in \Delta$, so Lemma 7.6.6 (applied to the Δ_μ-coarsening of K instead of K) yields a $y \in \mathcal{O}^n$ such that $vy > \Delta_\mu$, $vy \in \Delta$, and $vP_{+f_\mu}(y) > \Delta$. If $P(f_\mu + y) = 0$, we are done. Suppose $P(f_\mu + y) \neq 0$. Then we set $f_\lambda := f_\mu + y$ and let Δ_λ be the largest convex subgroup of Γ not containing $vP(f_\lambda)$, in particular, $\Delta \subseteq \Delta_\lambda$. It is clear that then the conditions (1)–(5) are satisfied with $\lambda + 1$ instead of λ.

Next assume that λ is a limit ordinal. Then $(f_\rho)_{\rho<\lambda}$ is a pc-sequence by (4) and (5), and thus has a pseudolimit f_λ in $\mathcal{O}^n$. Then $v(f_\lambda - f_\rho) > \Delta_\rho$ for all $\rho < \lambda$, so $vP(f_\lambda) > \Delta_\rho$ for all $\rho < \lambda$, by (1). If $P(f_\lambda) = 0$ we are done, so assume $P(f_\lambda) \neq 0$. Let Δ_λ be the largest convex subgroup of Γ not containing $vP(f_\lambda)$. Then $\Delta_\lambda \supseteq \Delta_\rho$ for all $\rho < \lambda$. Thus conditions (1)–(5) hold for $\lambda + 1$ instead of λ.

The construction above cannot continue indefinitely, so must end in producing a solution of $(*)$ in $\mathcal{O}$. □

The final step and an application

Let $Y = (Y_1, \ldots, Y_n)$ be as before.

PROOF OF THEOREM 7.0.4. We have d-algebraically maximal K with linearly surjective $\boldsymbol{k}$, and a system $(*)$ in dh-position; our job is to show that $(*)$ has a solution in $\mathcal{O}$. Take a maximal immediate extension L of K. Then Proposition 7.6.7 yields $y \in \mathcal{O}_L^n$ such that $P(y) = 0$. By Lemma 5.6.15 and Theorem 5.9.1 the immediate extension $K\langle y\rangle$ of K is d-algebraic, so $y \in \mathcal{O}^n$. □

Using also Lemma 5.4.6, this has the following consequence:

COROLLARY 7.6.8. *Suppose $\boldsymbol{k}$ is linearly surjective, K is* d*-algebraically maximal, and L is an algebraic extension of K with $\Gamma_L = \Gamma$. Then L is* d*-henselian.*

PROOF. We can reduce to the case that $[L : K] = n < \infty$, and as K is henselian, this yields a basis $b_1, \ldots, b_n$ of the vector space L over K such that $b_i \asymp 1$ for all i and $\overline{b}_1, \ldots, \overline{b}_n$ is a basis of $\boldsymbol{k}_L$ over $\boldsymbol{k}$. Note that $\boldsymbol{k}_L$ is linearly surjective, by Corollary 5.4.3. Let X be a single differential indeterminate and $F \in \mathcal{O}_L\{X\}$ such that $A := F_1 \asymp 1$ and $F(0) \prec 1$; our job is to show that F has a zero in $\mathcal{O}_L$. Making the substitution $X = b_1Y_1 + \cdots + b_nY_n$ we have

$$\begin{aligned}
F(b_1Y_1 + \cdots + b_nY_n) &= P_1(Y)b_1 + \cdots + P_n(Y)b_n, && P_1, \ldots, P_n \in \mathcal{O}\{Y\},\\
A(b_1Y_1 + \cdots + b_nY_n) &= A_1(Y)b_1 + \cdots + A_n(Y)b_n, && A_1, \ldots, A_n \in \mathcal{O}\{Y\},\\
\overline{A}(\overline{b}_1Y_1 + \cdots + \overline{b}_nY_n) &= \overline{A}_1(Y)\overline{b}_1 + \cdots + \overline{A}_n(Y)\overline{b}_n, && \overline{A}_1, \ldots, \overline{A}_n \in \boldsymbol{k}\{Y\}.
\end{aligned}$$

Also A_i is the homogeneous part of degree 1 of P_i, and $\overline{A}_1, \ldots, \overline{A}_n \in \boldsymbol{k}\{Y\}_1$ are d-independent by Lemma 5.4.6. From $F(0) \prec 1$ we get $P_1(0) \prec 1, \ldots, P_n(0) \prec 1$, and so by Theorem 7.0.4 the system $P_1(Y) = \cdots = P_n(Y) = 0$ has a solution $y = (y_1, \ldots, y_n) \in \mathcal{O}^n$, which gives a zero $y_1b_1 + \cdots + y_nb_n \in \mathcal{O}_L$ of F. □

This corollary might not be optimal: we would like to drop the hypothesis $\Gamma_L = \Gamma$, and it would also be nice to weaken the assumption on K to d-henselianity, or strengthen the conclusion to L being d-algebraically maximal.

Chapter Eight

Differential-Henselian Fields with Many Constants

The results in this brief chapter will not be used later, but complement the earlier generalities. The valued differential fields considered here are easier to analyze as to their model-theoretic properties than $\mathbb{T}$, and so this may provide an illuminating contrast with our later focus on objects like $\mathbb{T}$. Note that d-*henselian* includes having small derivation, so d-henselian valued differential fields with many constants are *monotone*: this is a strong restriction, opposite in spirit to the *asymptotic* condition imposed in later chapters.

Our goal here is to derive Scanlon's extension in [382, 383] of the Ax-Kochen-Eršov theorems to d-henselian valued differential fields with many constants. This is largely an application of Chapter 7, in particular Section 7.4.

Given structures $\boldsymbol{M}$ and $\boldsymbol{N}$ for the same (first-order) language, $\boldsymbol{M} \equiv \boldsymbol{N}$ means that $\boldsymbol{M}$ and $\boldsymbol{N}$ are elementarily equivalent: they satisfy the same sentences in that language; see Appendix B. Among the results to be established is the following:

THEOREM 8.0.1. *Suppose K and L are* d-*henselian valued differential fields with many constants. Then $K \equiv L$ as valued differential fields if and only if* $\operatorname{res} K \equiv \operatorname{res} L$ *as differential fields and $\Gamma_K \equiv \Gamma_L$ as ordered abelian groups.*

In particular, if K is a d-henselian valued differential field with many constants, and with differential residue field $\boldsymbol{k}$ and value group Γ, then $K \equiv \boldsymbol{k}((t^\Gamma))$, where $\boldsymbol{k}((t^\Gamma))$ is the Hahn differential field of Example (1) in Section 4.4.

Theorem 8.0.1 refers to the logical notion of elementary equivalence. We derive from it a purely algebraic result for which we have no other proof:

COROLLARY 8.0.2. *Let K be a* d-*henselian valued differential field with many constants. Then every valued differential field extension of K that is algebraic over K is also* d-*henselian.*

By Corollary 6.3.10, if L is a valued differential field extension of a monotone valued differential field K and L is algebraic over K, then L is monotone, so has small derivation. Here is an easy proof of Corollary 8.0.2, using Theorem 8.0.1:

PROOF. First some remarks on an arbitrary valued field extension $E \subseteq F = E(a)$ where E is henselian, and $[F : E] = n$. We can express properties of the valued field F in terms of the valued field E, as follows. Let

$$P(X) = X^n + a_{n-1}X^{n-1} + \cdots + a_1 X + a_0$$

be the minimum polynomial of a over E. Let $f \in F = E(a)$, and let $(b_0, \ldots, b_{n-1})$ be the unique tuple in E^n such that $f = b_0 + b_1 a + \cdots + b_{n-1}a^{n-1}$. For $1 \leqslant m \leqslant n$ and $f_0, \ldots, f_{m-1} \in E$ we can then express "$X^m + f_{m-1}X^{m-1} + \cdots + f_1 X + f_0$ is the minimum polynomial of f over E" as a first-order condition

$$E \models C_{m,n}(a_0, \ldots, a_{n-1}, b_0, \ldots, b_{n-1}, f_0, \ldots, f_{m-1})$$

on the quantities indicated. By Corollary 3.3.15 we can then express the valuation on F in terms of the valuation on E by $v(f) = v(f_0)/m$ with $f, f_0, \ldots, f_{m-1}$ as above. If in addition ∂ is a derivation on $E \supseteq \mathbb{Q}$ and ∂ also stands for its unique extension to a derivation on F, then, with $P' := \partial P/\partial X \in E[X]$, we have

$$\partial(a) = -P^{\partial}(a)/P'(a), \qquad \partial(f) = \sum_{i=0}^{n-1} \partial(b_i)a^i + \sum_{i=1}^{n-1} ib_i\partial(a)^{i-1}.$$

Turning now to our K, let $n \geqslant 1$ be given. By the remarks above there is a set Σ_n of sentences in the language of valued differential fields, independent of K, such that $K \models \Sigma_n$ if and only if every valued differential field extension L of K with $[L : K] = n$ is d-henselian. Now with $\boldsymbol{k}$ the differential residue field of K and $\Gamma = v(K^\times)$ we have $K \equiv \boldsymbol{k}((t^\Gamma))$. But for the Hahn differential field $\boldsymbol{k}((t^\Gamma))$ we have $\boldsymbol{k}((t^\Gamma)) \models \Sigma_n$, since every valued differential field extension L of $\boldsymbol{k}((t^\Gamma))$ with $[L : \boldsymbol{k}((t^\Gamma))] = n$ is maximal as a valued field by Corollary 3.2.28, and hence d-henselian by Corollaries 5.4.3 and 7.2.6. Thus $K \models \Sigma_n$. As this holds for all $n \geqslant 1$, we obtain the desired result. □

In model-theoretic work on valued fields and their expansions we often prefer a many-sorted set-up. This can be useful even in stating results properly. Thus we shall consider here three-sorted structures

$$\boldsymbol{K} = (K, \boldsymbol{k}, \Gamma; \pi, v)$$

where K and $\boldsymbol{k}$ are fields, Γ is an ordered abelian group, $v \colon K^\times \to \Gamma$ is a (surjective) valuation making K into a valued field, and $\pi \colon \mathcal{O} \to \boldsymbol{k}$ with $\mathcal{O} := \mathcal{O}_v$ is a surjective ring morphism. Note that then π has kernel o_v, and thus induces a field isomorphism $\boldsymbol{k}_v \cong \boldsymbol{k}$ between the residue field $\boldsymbol{k}_v$ and $\boldsymbol{k}$ such that the diagram

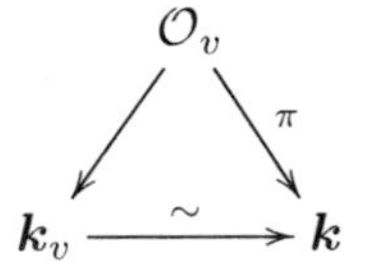

commutes. Let us call Γ the value group of $\boldsymbol{K}$, and $\boldsymbol{k}$ its residue field (even though the latter is only isomorphic to the residue field $\boldsymbol{k}_v$ of $\mathcal{O}_v$). We shall refer to $\boldsymbol{K}$ as a valued field, since it represents a way to construe the (one-sorted) valued field $(K, \mathcal{O}_v)$ as a three-sorted model-theoretic structure where the residue field and the value group are more explicitly present.

To adapt this setting to valued *differential* fields we consider three-sorted structures $\boldsymbol{K}$ as above, but with some additional features: K and $\boldsymbol{k}$ are differential fields (rather than just fields), and v makes K into a valued differential field with small derivation such that π is a differential ring morphism (rather than just a ring morphism). Note that then the above field isomorphism $\boldsymbol{k}_v \cong \boldsymbol{k}$ is actually a differential field isomorphism. When referring to $\boldsymbol{K}$ as a valued differential field, we assume these additional features are present.

In this three-sorted set-up we have field variables ranging over K, residue variables ranging over $\boldsymbol{k}$, and value group variables ranging over Γ. Given a possibly many-sorted language $\mathcal{L}$ and $\mathcal{L}$-structures $\boldsymbol{M}$ and $\boldsymbol{N}$, we have the usual notions of $\boldsymbol{M}$ being a substructure of $\boldsymbol{N}$ (notation: $\boldsymbol{M} \subseteq \boldsymbol{N}$), of $\boldsymbol{M}$ and $\boldsymbol{N}$ being elementarily equivalent (notation: $\boldsymbol{M} \equiv \boldsymbol{N}$), and of $\boldsymbol{M}$ being an elementary substructure of $\boldsymbol{N}$ (notation: $\boldsymbol{M} \preccurlyeq \boldsymbol{N}$); see Appendix B. The main result of this chapter is the Equivalence Theorem 8.2.5, among whose consequences are the following:

THEOREM 8.0.3. *Suppose $\boldsymbol{K}$ and $\boldsymbol{K}^*$ are* d*-henselian valued differential fields with many constants such that $\boldsymbol{K} \subseteq \boldsymbol{K}^*$. Then $\boldsymbol{K} \preccurlyeq \boldsymbol{K}^*$ if and only if $\boldsymbol{k} \preccurlyeq \boldsymbol{k}^*$ as differential fields, and $\Gamma \preccurlyeq \Gamma^*$ as ordered abelian groups.*

By *definable* we mean *definable with parameters from the ambient structure.*

THEOREM 8.0.4. *Suppose $\boldsymbol{K}$ is a* d*-henselian valued differential field with many constants. Then each subset of $\boldsymbol{k}^m \times \Gamma^n$ definable in $\boldsymbol{K}$ is a finite union of rectangles $Y \times Z$ with $Y \subseteq \boldsymbol{k}^m$ definable in the differential field $\boldsymbol{k}$ and $Z \subseteq \Gamma^n$ definable in the ordered abelian group Γ.*

The next sections lead up to the statement and proof of the Equivalence Theorem in Section 8.2, and to conclude this chapter we derive the above consequences.

Notes and comments

We refer to [382] for Scanlon's original treatment and to [383] for an update. The latter also deals with the mixed characteristic case.

8.1 ANGULAR COMPONENTS

Let $\boldsymbol{K} = (K, \boldsymbol{k}, \Gamma; \pi, v)$ be a valued field as explained in the introduction to this chapter, with valuation ring $\mathcal{O} := \{a \in K : va \geqslant 0\}$. For a subfield E of K we set $\mathcal{O}_E := \mathcal{O} \cap E$, a valuation ring of E, and $\boldsymbol{k}_E := \pi(\mathcal{O}_E)$, a subfield of $\boldsymbol{k}$. For such E we also let E stand for the valued subfield $(E, \mathcal{O}_E)$ of $(K, \mathcal{O})$, as well as for the substructure $(E, \boldsymbol{k}_E, v(E^\times); \dots)$ of $\boldsymbol{K}$.

An **angular component map** on $\boldsymbol{K}$ is a multiplicative group morphism

$$\mathrm{ac}\colon K^\times \to \boldsymbol{k}^\times$$

such that $\mathrm{ac}(a) = \pi(a)$ whenever $a \asymp 1$ in K; we extend it to $\mathrm{ac}\colon K \to \boldsymbol{k}$ by setting $\mathrm{ac}(0) = 0$ (so $\mathrm{ac}(ab) = \mathrm{ac}(a)\,\mathrm{ac}(b)$ for all $a, b \in K$), and also refer to this extension as an angular component map on $\boldsymbol{K}$.

EXAMPLE. Construing a Hahn field $\boldsymbol{k}((t^\Gamma))$ as a valued field

$$\boldsymbol{K} = (\boldsymbol{k}((t^\Gamma)), \boldsymbol{k}, \Gamma; \pi, v)$$

in the natural way, we have the angular component map $\mathrm{ac}\colon \boldsymbol{k}((t^\Gamma)) \to \boldsymbol{k}$ on $\boldsymbol{K}$ given by $\mathrm{ac}(ct^\gamma + g) = c$ for $c \in \boldsymbol{k}^\times$, $\gamma \in \Gamma$, and $g \in \boldsymbol{k}((t^\Gamma))$ with $v(g) > \gamma$.

A cross-section s on the valued field $\boldsymbol{K}$ yields an angular component map ac on $\boldsymbol{K}$ by setting $\mathrm{ac}(x) = \pi\big(x/s(vx)\big)$ for $x \in K^\times$. Thus by Lemma 3.3.39:

COROLLARY 8.1.1. *If the valued field $\boldsymbol{K}$ is $\aleph_1$-saturated, then there exists an angular component map on $\boldsymbol{K}$.*

Monotone valued differential fields with angular component

The presence of an angular component map simplifies the proof of the Equivalence Theorem 8.2.5, but in the aftermath we can often discard these maps again, by Corollary 8.1.3.

Let $\boldsymbol{K} = (K, \boldsymbol{k}, \Gamma; \pi, v)$ be a monotone valued differential field. By an **angular component map** on $\boldsymbol{K}$ we mean an angular component map ac on $\boldsymbol{K}$ as a valued field such that the following conditions are satisfied:

(1) $\mathrm{ac}(c) \in C_{\boldsymbol{k}} \subseteq \boldsymbol{k}$ for all $c \in C$;

(2) $\mathrm{ac}(a)' = \mathrm{ac}(a')$ for all $a \in K^\times$ with $v(a) = v(a')$.

Examples are the Hahn differential fields $\boldsymbol{k}((t^\Gamma))$ with angular component map given by $\mathrm{ac}(a) = a_{\gamma_0}$ for nonzero $a = \sum a_\gamma t^\gamma \in \boldsymbol{k}((t^\Gamma))$ and $\gamma_0 := va$.

LEMMA 8.1.2. *Suppose $\boldsymbol{K}$ has many constants. Then each angular component map on its valued subfield C extends uniquely to an angular component map on $\boldsymbol{K}$.*

PROOF. Given an angular component map ac on C the claimed extension to $\boldsymbol{K}$, also denoted by ac, is obtained as follows: for $x \in K^\times$ we have $x = uc$ with $u \in K^\times$, $u \asymp 1$, $c \in C^\times$; then $\mathrm{ac}(x) = \pi(u)\,\mathrm{ac}(c)$. □

Here is an immediate consequence of Corollary 8.1.1 and Lemma 8.1.2:

COROLLARY 8.1.3. *If $\boldsymbol{K}$ has many constants and is $\aleph_1$-saturated, then there is an angular component map on $\boldsymbol{K}$.*

Notes and comments

Angular component maps were introduced in [100]. In [316] there is an example of a valued field of residue characteristic 0 that has no angular component map (and thus also no cross-section). Angular components often facilitate access to the definable sets, and this is also their role in this chapter.

8.2 EQUIVALENCE OVER SUBSTRUCTURES

In this section we consider three-sorted structures

$$\boldsymbol{K} \;=\; (K, \boldsymbol{k}, \Gamma; \pi, v, \mathrm{ac})$$

where $(K, \boldsymbol{k}, \Gamma; \pi, v)$ is a monotone valued differential field with an angular component map $\mathrm{ac}\colon K \to \boldsymbol{k}$ on it. Such a structure will be called a **monotone** ac**-valued differential field**. The main result of this chapter is Theorem 8.2.5. It tells us when two d-henselian monotone ac-valued differential fields with many constants are elementarily equivalent over a common substructure. In Section 8.3 we derive from it in the usual way some attractive consequences on the elementary theories of such valued differential fields and on the induced structure on the value group and differential residue field.

If $\boldsymbol{K}$ is d-henselian, then by Theorem 7.1.3 there is a differential ring morphism $i\colon \boldsymbol{k} \to \mathcal{O}$ such that $\pi\big(i(a)\big) = a$ for all $a \in K$; we call such i a **lifting** of $\boldsymbol{k}$ to $\boldsymbol{K}$. This will play a minor role in the proof of the Equivalence Theorem.

A **good substructure** of $\boldsymbol{K} = (K, \boldsymbol{k}, \Gamma; \pi, v, \mathrm{ac})$ is a triple $\boldsymbol{E} = (E, \boldsymbol{k}_{\boldsymbol{E}}, \Gamma_{\boldsymbol{E}})$ such that

(GS1) E is a differential subfield of K,

(GS2) $\boldsymbol{k}_{\boldsymbol{E}}$ is a differential subfield of $\boldsymbol{k}$ with $\mathrm{ac}(E) \subseteq \boldsymbol{k}_{\boldsymbol{E}}$ (hence $\pi(\mathcal{O}_E) \subseteq \boldsymbol{k}_{\boldsymbol{E}}$),

(GS3) $\Gamma_{\boldsymbol{E}}$ is an ordered abelian subgroup of Γ with $v(E^{\times}) \subseteq \Gamma_{\boldsymbol{E}}$.

For good substructures $\boldsymbol{E} = (E, \boldsymbol{k}_{\boldsymbol{E}}, \Gamma_{\boldsymbol{E}})$ and $\boldsymbol{F} = (F, \boldsymbol{k}_{\boldsymbol{F}}, \Gamma_{\boldsymbol{F}})$, we define $\boldsymbol{E} \subseteq \boldsymbol{F}$ to mean that $E \subseteq F$, $\boldsymbol{k}_{\boldsymbol{E}} \subseteq \boldsymbol{k}_{\boldsymbol{F}}$, and $\Gamma_{\boldsymbol{E}} \subseteq \Gamma_{\boldsymbol{F}}$. If E is a differential subfield of K with $\mathrm{ac}(E) = \pi(\mathcal{O}_E)$, then $\big(E, \pi(\mathcal{O}_E), v(E^{\times})\big)$ is a good substructure of $\boldsymbol{K}$, and if in addition $F \supseteq E$ is a differential subfield of K such that $v(F^{\times}) = v(E^{\times})$, then $\mathrm{ac}(F) = \pi(\mathcal{O}_F)$. In the remainder of this section

$$\boldsymbol{K} \;=\; (K, \boldsymbol{k}, \Gamma; \pi, v, \mathrm{ac}), \qquad \boldsymbol{K}^* \;=\; (K^*, \boldsymbol{k}^*, \Gamma^*; \pi^*, v^*, \mathrm{ac}^*)$$

are monotone ac-valued differential fields, with valuation rings $\mathcal{O}$ and $\mathcal{O}^*$, and

$$\boldsymbol{E} \;=\; (E, \boldsymbol{k}_{\boldsymbol{E}}, \Gamma_{\boldsymbol{E}}), \qquad \boldsymbol{E}^* \;=\; (E^*, \boldsymbol{k}_{\boldsymbol{E}^*}, \Gamma_{\boldsymbol{E}^*})$$

are good substructures of $\boldsymbol{K}$, $\boldsymbol{K}^*$, respectively. We put $\mathcal{O}_{E^*} := \mathcal{O}^* \cap E^*$.

A **good map** $\boldsymbol{f}\colon \boldsymbol{E} \to \boldsymbol{E}^*$ is a triple $\boldsymbol{f} = (f, f_{\mathrm{r}}, f_{\mathrm{v}})$ consisting of a differential field isomorphism $f\colon E \to E^*$, a differential field isomorphism $f_{\mathrm{r}}\colon \boldsymbol{k}_{\boldsymbol{E}} \to \boldsymbol{k}_{\boldsymbol{E}^*}$, and an ordered group isomorphism $f_{\mathrm{v}}\colon \Gamma_{\boldsymbol{E}} \to \Gamma_{\boldsymbol{E}^*}$, such that

(GM1) $f_{\mathrm{r}}\big(\mathrm{ac}(a)\big) = \mathrm{ac}^*\big(f(a)\big)$ for all $a \in E$, and f_{r} is elementary as a partial map between the differential fields $\boldsymbol{k}$ and $\boldsymbol{k}^*$;

(GM2) $f_{\mathrm{v}}\big(v(a)\big) = v^*\big(f(a)\big)$ for all $a \in E^{\times}$, and f_{v} is elementary as a partial map between the ordered abelian groups Γ and Γ^*.

Let $\boldsymbol{f}\colon \boldsymbol{E} \to \boldsymbol{E}^*$ be a good map as above. Then the field part $f\colon E \to E^*$ of $\boldsymbol{f}$ is a valued differential field isomorphism, and $f_{\rm r}$ and $f_{\rm v}$ agree on $\pi(\mathcal{O}_E)$ and $v(E^\times)$ with the maps $\pi(\mathcal{O}_E) \to \pi^*(\mathcal{O}_{E^*})$ and $v(E^\times) \to v^*(E^{*\times})$ induced by f. We say that a good map $\boldsymbol{g} = (g, g_{\rm r}, g_{\rm v})\colon \boldsymbol{F} \to \boldsymbol{F}^*$ **extends** $\boldsymbol{f}$ if $\boldsymbol{E} \subseteq \boldsymbol{F}$, $\boldsymbol{E}^* \subseteq \boldsymbol{F}^*$, and g, $g_{\rm r}$, $g_{\rm v}$ extend f, $f_{\rm r}$, $f_{\rm v}$, respectively. The **domain** of $\boldsymbol{f}$ is $\boldsymbol{E}$. Note that if a good map $\boldsymbol{E} \to \boldsymbol{E}^*$ exists, then $\boldsymbol{k} \equiv \boldsymbol{k}^*$ as differential fields and $\Gamma \equiv \Gamma^*$ as ordered abelian groups.

The next two lemmas show that various parts of the conditions (GM1) and (GM2) are automatically satisfied by certain extensions of good maps.

LEMMA 8.2.1. *Let $\boldsymbol{f}\colon \boldsymbol{E} \to \boldsymbol{E}^*$ be a good map, and suppose $F \supseteq E$ and $F^* \supseteq E^*$ are differential subfields of K and K^*, respectively, such that $\pi(\mathcal{O}_F) \subseteq \boldsymbol{k}_{\boldsymbol{E}}$ and $v(F^\times) = v(E^\times)$. Let $g\colon F \to F^*$ be a valued differential field isomorphism such that g extends f and $f_{\rm r}\big(\pi(u)\big) = \pi^*\big(g(u)\big)$ for all $u \in \mathcal{O}_F$. Then $\operatorname{ac}(F) \subseteq \boldsymbol{k}_{\boldsymbol{E}}$ and $f_{\rm r}\big(\operatorname{ac}(a)\big) = \operatorname{ac}^*\big(g(a)\big)$ for all $a \in F$.*

PROOF. Let $a \in F$. Then $a = a_1 u$ where $a_1 \in E$ and $u \in \mathcal{O}_F$, $v(u) = 0$, so $\operatorname{ac}(a) = \operatorname{ac}(a_1)\pi(u) \in \boldsymbol{k}_{\boldsymbol{E}}$. It follows easily that $f_{\rm r}\big(\operatorname{ac}(a)\big) = \operatorname{ac}^*\big(g(a)\big)$. □

In the same way we obtain:

LEMMA 8.2.2. *Suppose $\pi(\mathcal{O}_E) = \boldsymbol{k}_{\boldsymbol{E}}$, let $\boldsymbol{f}\colon \boldsymbol{E} \to \boldsymbol{E}^*$ be a good map, and let $F \supseteq E$ and $F^* \supseteq E^*$ be differential subfields of K and K^*, respectively, such that $v(F^\times) = v(E^\times)$. Let $g\colon F \to F^*$ be a valued differential field isomorphism extending f. Then $\operatorname{ac}(F) = \pi(\mathcal{O}_F)$ and $g_{\rm r}\big(\operatorname{ac}(a)\big) = \operatorname{ac}^*\big(g(a)\big)$ for all $a \in F$, where the map $g_{\rm r}\colon \pi(\mathcal{O}_F) \to \pi^*(\mathcal{O}_{F^*})$ is induced by g (and thus extends $f_{\rm r}$).*

Lemma 8.2.2 is also useful in getting the following:

LEMMA 8.2.3. *Suppose $\pi(\mathcal{O}_E) = \boldsymbol{k}_{\boldsymbol{E}}$, let $\boldsymbol{f}\colon \boldsymbol{E} \to \boldsymbol{E}^*$ be a good map. Suppose that $F \supseteq E$ and $F^* \supseteq E^*$ are differential subfields of K and K^*, respectively, and are* immediate *extensions of E and E^*, respectively. Suppose that $g\colon F \to F^*$ is a valued differential field isomorphism that extends f. Then $\boldsymbol{g} = (g, f_{\rm r}, f_{\rm v})$ is a good map that extends $\boldsymbol{f}$.*

The following is useful in connection with having many constants:

LEMMA 8.2.4. *Let $b \in K^\times$. Then the following are equivalent:*

(i) *There is $c \in C^\times$ such that $b \asymp c$.*

(ii) *There is $a \in K^\times$ such that $a \asymp 1$ and $a^\dagger = b^\dagger$.*

PROOF. If $c \in C^\times$ and $b \asymp c$, then $a := bc^{-1}$ satisfies $a \asymp 1$ and $a^\dagger = b^\dagger$. Conversely, if $a \in K^\times$ and $a \asymp 1$, $a^\dagger = b^\dagger$, then $b \asymp c := a^{-1}b \in C^\times$. □

THEOREM 8.2.5. *Suppose $\boldsymbol{K}$, $\boldsymbol{K}^*$ are* d-*henselian with many constants. Then any good map $\boldsymbol{E} \to \boldsymbol{E}^*$ is a partial elementary map between $\boldsymbol{K}$ and $\boldsymbol{K}^*$.*

PROOF. The theorem holds trivially for $\Gamma = \{0\}$, so assume that $\Gamma \neq \{0\}$. Let $\boldsymbol{f} = (f, f_r, f_v)\colon \boldsymbol{E} \to \boldsymbol{E}^*$ be a good map. By passing to suitable elementary extensions of $\boldsymbol{K}$ and $\boldsymbol{K}^*$ we arrange that $\boldsymbol{K}$ and $\boldsymbol{K}^*$ are κ-saturated, where κ is an uncountable cardinal such that $|\boldsymbol{k}_{\boldsymbol{E}}|$, $|\Gamma_{\boldsymbol{E}}| < \kappa$. Call a good substructure $\boldsymbol{E}_1 = (E_1, \boldsymbol{k}_1, \Gamma_1)$ of $\boldsymbol{K}$ *small* if $|\boldsymbol{k}_1|$, $|\Gamma_1| < \kappa$. We shall prove that the good maps with small domain form a back-and-forth system from $\boldsymbol{K}$ to $\boldsymbol{K}^*$. (This clearly suffices to obtain the theorem: Proposition B.5.4.) In other words, we shall prove that under the present assumptions on $\boldsymbol{E}$, $\boldsymbol{E}^*$ and $\boldsymbol{f}$, there is for each $a \in K$ a good map $\boldsymbol{g}$ extending $\boldsymbol{f}$ such that $\boldsymbol{g}$ has small domain $\boldsymbol{F} = (F, \dots)$ with $a \in F$. The most delicate of the extension procedures we need comes from Corollary 7.4.5 (to extend domains) and Theorem 7.4.3 (to extend good maps). In addition we have several other basic extension procedures:

(1) *Given $d \in \boldsymbol{k}$, arranging that $d \in \boldsymbol{k}_{\boldsymbol{E}}$.*

By saturation and the definition of "good map" this can be achieved without changing $f, f_v, E, \Gamma_{\boldsymbol{E}}$ by extending f_r to a partial elementary map between $\boldsymbol{k}$ and $\boldsymbol{k}^*$ with d in its domain.

(2) *Given $\gamma \in \Gamma$, arranging that $\gamma \in \Gamma_{\boldsymbol{E}}$.*

This follows in the same way.

(3) *Arranging $\boldsymbol{k}_{\boldsymbol{E}} = \pi(\mathcal{O}_E)$.*

Suppose $d \in \boldsymbol{k}_{\boldsymbol{E}}$, $d \notin \pi(\mathcal{O}_E)$; set $e := f_r(d)$.

Assume first that d is d-transcendental over $\pi(\mathcal{O}_E)$. Pick $a \in \mathcal{O}$ and $b \in \mathcal{O}^*$ such that $\overline{a} = d$ and $\overline{b} = e$. Then $v(E\langle a\rangle^\times) = v(E^\times)$, and Lemmas 6.3.1 and 8.2.1 yield a good map $\boldsymbol{g} = (g, f_r, f_v)$ with small domain $(E\langle a\rangle, \boldsymbol{k}_{\boldsymbol{E}}, \Gamma_{\boldsymbol{E}})$ such that $\boldsymbol{g}$ extends $\boldsymbol{f}$ and $g(a) = b$.

Next, assume that d is d-algebraic over $\pi(\mathcal{O}_E)$. By introducing a minimal annihilator of d over $\pi(\mathcal{O}_E)$, Lemmas 7.1.4 and 8.2.1 provide an element $a \in \mathcal{O}$ with $\operatorname{res}(a) = d$, $\pi(\mathcal{O}_{E\langle a\rangle}) = \pi(\mathcal{O}_E)\langle d\rangle$, and $v(E\langle a\rangle^\times) = v(E^\times)$, and a good map (g, f_r, f_v) extending $\boldsymbol{f}$ with small domain $(E\langle a\rangle, \boldsymbol{k}_{\boldsymbol{E}}, \Gamma_{\boldsymbol{E}})$.

By iterating (3) we can arrange $\boldsymbol{k}_{\boldsymbol{E}} = \pi(\mathcal{O}_E)$; this condition is actually preserved in each of the extension procedures (4)–(8) below, as the reader may easily verify. We do assume in the rest of the proof that $\boldsymbol{k}_{\boldsymbol{E}} = \pi(\mathcal{O}_E)$. Let us say that $\boldsymbol{E}$ *has many constants* if $v(C_E^\times) = v(E^\times)$.

(4) *Extending $\boldsymbol{f}$ to a good map whose domain has many constants.*

Let $\beta \in v(E^\times)$. Pick $b \in E^\times$ such that $v(b) = \beta$. Since K has many constants, we can use Lemma 8.2.4 to get $a \in K$ such that $a \asymp 1$ and $P(a) = 0$ where

$$P(Y) := Y' - b^\dagger Y \in \mathcal{O}_E\{Y\}.$$

Note that $v(qa) = 0$ and $P(qa) = 0$ for all $q \in \mathbb{Q}^\times \subseteq E^\times$. Hence by saturation we can arrange that $\overline{a}$ is transcendental over $\boldsymbol{k}_{\boldsymbol{E}}$. Then $\overline{P}(Y)$ is a minimal annihilator of $\overline{a}$ over $\boldsymbol{k}_{\boldsymbol{E}}$. By Lemma 3.1.31,

$$E\langle a\rangle = E(a), \qquad v\big(E(a)^\times\big) = v(E^\times), \qquad \pi\big(\mathcal{O}_{E(a)}\big) = \boldsymbol{k}_{\boldsymbol{E}}(\overline{a}).$$

We shall find a good map extending $\boldsymbol{f}$ with domain $\big(E(a), \boldsymbol{k}_{\boldsymbol{E}}(\overline{a}), \Gamma_{\boldsymbol{E}}\big)$. Consider the differential polynomial $Q := f(P)$, that is,

$$Q(Y) \;=\; Y' - f(b)^\dagger Y.$$

By saturation we can find $e \in \boldsymbol{k}^*$ with $\overline{Q}(e) = 0$ and a differential field isomorphism $g_{\mathrm{r}}\colon \boldsymbol{k}_{\boldsymbol{E}}(\overline{a}) \to \boldsymbol{k}_{\boldsymbol{E}^*}(e)$ that extends f_{r}, sends $\overline{a}$ to e and is elementary as a partial map between the differential fields $\boldsymbol{k}$ and $\boldsymbol{k}^*$. Using again Lemma 8.2.4 we find $a^* \in K^*$ such that $a^* \asymp 1$ and $Q(a^*) = 0$. Since $\overline{Q}(\operatorname{res}(a^*)) = \overline{Q}(e) = 0$, we can multiply a^* by an element in C^* of valuation zero and arrange $\operatorname{res}(a^*) = e$. Then Theorem 6.3.2 and Lemma 8.2.2 yield a good map $\boldsymbol{g} = (g, g_{\mathrm{r}}, f_{\mathrm{v}})$ where $g\colon E(a) \to E^*(a^*)$ extends f and sends a to a^*. The domain $\big(E(a), \boldsymbol{k}_{\boldsymbol{E}}(\overline{a}), \Gamma_{\boldsymbol{E}}\big)$ of $\boldsymbol{g}$ is small, and $vc = \beta$ for $c := a^{-1}b \in C_{E(a)}$.

In the extension procedures (1)–(4) the value group $v(E^\times)$ does not change, so if the domain $\boldsymbol{E}$ of $\boldsymbol{f}$ has many constants, then so does the domain of the extension of $\boldsymbol{f}$ constructed in each of (1)–(4). Also $\Gamma_{\boldsymbol{E}}$ does not change in (1), (3), and (4), but at this stage we allow $\Gamma_{\boldsymbol{E}} \neq v(E^\times)$.

(5) *Arranging that* $\boldsymbol{k}_{\boldsymbol{E}} = \pi(\mathcal{O}_E)$ *is linearly surjective and* $\boldsymbol{E}$ *has many constants.*

This can be done by repeated applications of (1), (3), and (4).

(6) *Arranging that* $\boldsymbol{k}_{\boldsymbol{E}} = \pi(\mathcal{O}_E)$ *and* $\boldsymbol{E}$ *is* d-*henselian* (*by which we mean that* E *as a valued differential subfield of* $\boldsymbol{K}$ *is* d-*henselian*).

After arranging (5) above, we can additionally arrange that $\boldsymbol{E}$ is d-henselian by Corollary 7.4.5, Theorem 7.4.3, and Lemma 8.2.3.

(7) *Towards arranging* $\Gamma_{\boldsymbol{E}} = v(E^\times)$*; the case of no torsion modulo* $v(E^\times)$.

Suppose $\gamma \in \Gamma_{\boldsymbol{E}}$ has no torsion modulo $v(E^\times)$, that is, $n\gamma \notin v(E^\times)$ for all $n \geqslant 1$. Take $a \in C$ such that $v(a) = \gamma$. Let i be a lifting of the differential residue field $\boldsymbol{k}$ to $\boldsymbol{K}$. Since $\operatorname{ac}(a) \in C_{\boldsymbol{k}}^\times$, we have $a/i\big(\operatorname{ac}(a)\big) \in C$ and $v\big(a/i\big(\operatorname{ac}(a)\big)\big) = \gamma$. So replacing a by $a/i\big(\operatorname{ac}(a)\big)$ we arrange that $v(a) = \gamma$ and $\operatorname{ac}(a) = 1$. In the same way we obtain $a^* \in C_{K^*}$ such that $v^*(a^*) = \gamma^* := f_{\mathrm{v}}(\gamma)$ and $\operatorname{ac}^*(a^*) = 1$. Then Lemma 3.1.30 gives an isomorphism $g\colon E(a) \to E^*(a^*)$ of valued fields extending f with $g(a) = a^*$. Then $(g, f_{\mathrm{r}}, f_{\mathrm{v}})$ is a good map extending $\boldsymbol{f}$ with small domain $\big(E(a), \boldsymbol{k}_{\boldsymbol{E}}, \Gamma_{\boldsymbol{E}}\big)$; this domain has many constants if $\boldsymbol{E}$ does.

(8) *Towards arranging* $\Gamma_{\boldsymbol{E}} = v(E^\times)$*; the case of prime torsion modulo* $v(E^\times)$.

Here we assume that $\boldsymbol{E}$ has many constants and is d-henselian.

Let $\gamma \in \Gamma_{\boldsymbol{E}} \setminus v(E^\times)$ with $\ell\gamma \in v(E^\times)$, where ℓ is a prime number. As E has many constants we can pick $b \in C_E$ such that $v(b) = \ell\gamma$. Since E is d-henselian we have a lifting of its differential residue field $\boldsymbol{k}_{\boldsymbol{E}}$ to $\boldsymbol{E}$ and we can use this as in (7) to arrange that $\mathrm{ac}(b) = 1$. We shall find $c \in C$ such that $c^\ell = b$ and $\mathrm{ac}(c) = 1$. As in (7) we have $a \in C$ such that $v(a) = \gamma$ and $\mathrm{ac}(a) = 1$. Then the polynomial $P(Y) := Y^\ell - b/a^\ell \in \mathcal{O}[Y]$ satisfies $P(1) \prec 1$ and $P'(1) \asymp 1$. This gives $u \in K$ such that $P(u) = 0$ and $u \sim 1$. Now let $c = au$. Clearly $c^\ell = b$ (so $c \in C$) and $\mathrm{ac}(c) = 1$. Likewise we find $c^* \in C_{K^*}$ such that ${c^*}^\ell = f(b)$ and $\mathrm{ac}^*(c^*) = 1$. Then Lemma 3.1.28 gives an isomorphism $g\colon E(c) \to E^*(c^*)$ of valued fields extending f with $g(c) = c^*$, and so $(g, f_{\mathrm{r}}, f_{\mathrm{v}})$ is a good map extending $\boldsymbol{f}$ with small domain $\big(E(c), \boldsymbol{k}_{\boldsymbol{E}}, \Gamma_{\boldsymbol{E}}\big)$; this domain has many constants.

By iterating (7) and (8) we can assume in the rest of the proof that $\Gamma_{\boldsymbol{E}} = v(E^\times)$, and we shall do so. This condition is actually preserved in the earlier extension procedures (3) and (4), as the reader may easily verify. Anyway, we can refer from now on to $\Gamma_{\boldsymbol{E}}$ as the *value group* of E. Note that in (7) and (8) the differential residue field does not change.

Let $a \in K$ be given. We need to extend $\boldsymbol{f}$ to a good map whose domain is small and contains a. At this stage we can assume $\boldsymbol{k}_{\boldsymbol{E}} = \pi(\mathcal{O}_E)$, $\Gamma_{\boldsymbol{E}} = v(E^\times)$, and $\boldsymbol{E}$ has many constants. By Lemma 3.1.10, $\big|\pi\big(\mathcal{O}_{E\langle a\rangle}\big)\big| < \kappa$ and $\big|v\big(E\langle a\rangle^\times\big)\big| < \kappa$. Then (1)–(8) and Corollary 7.4.5 plus Theorem 7.4.3 allow us to extend $\boldsymbol{f}$ to a good map $\boldsymbol{f}_1 = (f_1, f_{1,\mathrm{r}}, f_{1,\mathrm{v}})$ with small domain $\boldsymbol{E}_1 \supseteq \boldsymbol{E}$ such that $\boldsymbol{E}_1 = (E_1, \boldsymbol{k}_1, \Gamma_1)$ has many constants, $\boldsymbol{k}_1$ is linearly surjective, and

$$\pi\big(\mathcal{O}_{E\langle a\rangle}\big) \subseteq \boldsymbol{k}_1 = \pi\big(\mathcal{O}_{E_1}\big), \qquad v\big(E\langle a\rangle^\times\big) \subseteq \Gamma_1 = v(E_1^\times).$$

As we extended $\boldsymbol{f}$ to $\boldsymbol{f}_1$, we extend $\boldsymbol{f}_1$ to a good map $\boldsymbol{f}_2$ with small domain $\boldsymbol{E}_2 \supseteq \boldsymbol{E}_1$ such that $\boldsymbol{E}_2 = (E_2, \boldsymbol{k}_2, \Gamma_2)$ has many constants, $\boldsymbol{k}_2$ linearly surjective, and

$$\pi\big(\mathcal{O}_{E_1\langle a\rangle}\big) \subseteq \boldsymbol{k}_2 = \pi\big(\mathcal{O}_{E_2}\big), \qquad v\big(E_1\langle a\rangle^\times\big) \subseteq \Gamma_2 = v(E_2^\times).$$

Continuing this way and taking a union of the resulting domains and good maps gives a small good substructure $\boldsymbol{E}_\infty \supseteq \boldsymbol{E}$, such that $\boldsymbol{E}_\infty = (E_\infty, \boldsymbol{k}_\infty, \Gamma_\infty)$ has many constants, $\boldsymbol{k}_\infty$ is linearly surjective, and

$$\boldsymbol{k}_\infty = \pi\big(\mathcal{O}_{E_\infty\langle a\rangle}\big) = \pi\big(\mathcal{O}_{E_\infty}\big), \qquad \Gamma_\infty = v\big(E_\infty\langle a\rangle^\times\big) = v(E_\infty^\times),$$

together with an extension of $\boldsymbol{f}$ to a good map $\boldsymbol{f}_\infty = (f_\infty, \dots)$ with domain $\boldsymbol{E}_\infty$. Thus the valued differential subfield $E_\infty\langle a\rangle$ of $\boldsymbol{K}$ is an immediate extension of its valued differential subfield E_∞. By Corollary 7.4.5 the valued differential subfield $E_\infty\langle a\rangle$ of $\boldsymbol{K}$ has a maximal immediate valued differential field extension F inside $\boldsymbol{K}$. Then F is a maximal immediate extension of E_∞ as well. This gives a good substructure $\boldsymbol{F} = (F, \Gamma_\infty, \boldsymbol{k}_\infty)$ of $\boldsymbol{K}$. Likewise, the valued differential subfield $f_\infty(E_\infty)$ of $\boldsymbol{K}^*$

has a maximal immediate valued differential field extension F^* in $\boldsymbol{K}^*$. Use Theorem 7.4.3 and Lemma 8.2.2 to extend $\boldsymbol{f}_\infty$ to a good map $\boldsymbol{F} \to \boldsymbol{F}^* = (F^*, \dots)$, and note that $a \in F$. □

8.3 RELATIVE QUANTIFIER ELIMINATION

Here we derive various consequences of the Equivalence Theorem of Section 8.2. Let $\mathcal{L}$ be the three-sorted language of valued fields, with sorts f (the field sort), r (the residue sort), and v (the value group sort). This language consists of two copies of the one-sorted language $\mathcal{L}_{\mathrm{R}} = \{0, 1, -, +, \cdot\}$ of rings, one in the sort f and one in the sort r, a copy of the language $\mathcal{L}_{\mathrm{OA}} = \{\leqslant, 0, -, +\}$ of ordered abelian groups in the sort v, and function symbols v of sort f v (for the valuation) and π (for the residue morphism) of sort f r. We view a valued field $(K, \boldsymbol{k}, \Gamma; \dots)$ as an $\mathcal{L}$-structure in the natural way, with f-variables ranging over K, r-variables over $\boldsymbol{k}$, and v-variables over Γ. We augment $\mathcal{L}$ with a function symbol ∂ of sort f f, a function symbol $\bar{\partial}$ of sort r r, and a function symbol ac of sort f r to get the language $\mathcal{L}(\partial, \bar{\partial}, \mathrm{ac})$ of ac-valued differential fields. If we do not indicate otherwise, then in this section

$$\boldsymbol{K} = (K, \boldsymbol{k}, \Gamma; \dots), \qquad \boldsymbol{K}^* = (K^*, \boldsymbol{k}^*, \Gamma^*; \dots)$$

are d-henselian monotone ac-valued differential fields with many constants; they are considered as $\mathcal{L}(\partial, \bar{\partial}, \mathrm{ac})$-structures in the obvious way.

COROLLARY 8.3.1. *$\boldsymbol{K} \equiv \boldsymbol{K}^*$ if and only if $\boldsymbol{k} \equiv \boldsymbol{k}^*$ as differential fields and $\Gamma \equiv \Gamma^*$ as ordered abelian groups.*

PROOF. The "only if" direction is obvious. Suppose $\boldsymbol{k} \equiv \boldsymbol{k}^*$ as differential fields, and $\Gamma \equiv \Gamma^*$ as ordered groups. This gives good substructures $\boldsymbol{E} := (\mathbb{Q}, \mathbb{Q}, \{0\})$ of $\boldsymbol{K}$, and $\boldsymbol{E}^* := (\mathbb{Q}, \mathbb{Q}, \{0\})$ of $\boldsymbol{K}^*$, and an obvious good map $\boldsymbol{E} \to \boldsymbol{E}^*$. Now apply Theorem 8.2.5. □

Thus $\boldsymbol{K}$ is elementarily equivalent to the Hahn differential field $\boldsymbol{k}((t^\Gamma))$ with angular component map as in Section 8.1.

COROLLARY 8.3.2. *Suppose $\boldsymbol{E} = (E, \boldsymbol{k}_E, \Gamma_E; \dots) \subseteq \boldsymbol{K}$ is a* d*-henselian* ac*-valued differential subfield of $\boldsymbol{K}$ with many constants, such that $\boldsymbol{k}_E \preccurlyeq \boldsymbol{k}$ as differential fields, and $\Gamma_E \preccurlyeq \Gamma$ as ordered abelian groups. Then $\boldsymbol{E} \preccurlyeq \boldsymbol{K}$.*

PROOF. Take an elementary extension $\boldsymbol{K}^*$ of $\boldsymbol{E}$. Then $\boldsymbol{K}^*$ has many constants, $(E, \boldsymbol{k}_E, \Gamma_E)$ is a good substructure of both $\boldsymbol{K}$ and $\boldsymbol{K}^*$, and the identity on $(E, \Gamma, \boldsymbol{k}_E)$ is a good map. Hence by Theorem 8.2.5 we have $\boldsymbol{K} \equiv_{\boldsymbol{E}} \boldsymbol{K}^*$. Since $\boldsymbol{E} \preccurlyeq \boldsymbol{K}^*$, this gives $\boldsymbol{E} \preccurlyeq \boldsymbol{K}$. □

The proofs of these corollaries use only a weak form of the Equivalence Theorem, but now we turn to a result that uses its full strength: a relative elimination of quantifiers for the $\mathcal{L}(\partial, \bar{\partial}, \mathrm{ac})$-theory T of d-henselian monotone ac-valued differential fields with many constants. We specify that the function symbols π and v of $\mathcal{L}(\partial, \bar{\partial}, \mathrm{ac})$ are to be interpreted as *total* functions in any $\boldsymbol{K}$ as follows: extend $\pi\colon \mathcal{O} \to \boldsymbol{k}$ to $\pi\colon K \to \boldsymbol{k}$ by $\pi(a) = 0$ for $a \notin \mathcal{O}$, and extend $v\colon K^\times \to \Gamma$ to $v\colon K \to \Gamma$ by $v(0) = 0$. Let $\mathcal{L}_{\mathrm{r}}$ be the sublanguage of $\mathcal{L}(\partial, \overline{\partial}, \mathrm{ac})$ involving only the sort r, that is, $\mathcal{L}_{\mathrm{r}}$ is a copy of the language of differential fields, with $\overline{\partial}$ as its symbol for the derivation operator. Let $\mathcal{L}_{\mathrm{v}}$ be the sublanguage of $\mathcal{L}(\partial, \overline{\partial}, \mathrm{ac})$ involving only the sort v, that is, $\mathcal{L}_{\mathrm{v}} = \mathcal{L}_{\mathrm{OA}}$ is the language of ordered abelian groups.

Let $x = (x_1, \dots, x_l)$ be a tuple of distinct f-variables, $y = (y_1, \dots, y_m)$ a tuple of distinct r-variables, and $z = (z_1, \dots, z_n)$ a tuple of distinct v-variables. Set

$$\mathbb{Z}\{x\} := \big\{P \in \mathbb{Q}\{x_1, \dots, x_l\} : \text{ all coefficients of } P \text{ are in } \mathbb{Z}\big\}.$$

Define a *special* r*-formula in* (x, y) to be an $\mathcal{L}(\partial, \overline{\partial}, \mathrm{ac})$-formula

$$\psi(x, y) := \psi_{\mathrm{r}}\big(\mathrm{ac}(P_1(x)), \dots, \mathrm{ac}(P_k(x)), y\big)$$

where $k \in \mathbb{N}$, $\psi_{\mathrm{r}}(u_1, \dots, u_k, y)$ is an $\mathcal{L}_{\mathrm{r}}$-formula, and $P_1(x), \dots, P_k(x) \in \mathbb{Z}\{x\}$. Also, a *special* v*-formula in* (x, z) is an $\mathcal{L}(\partial, \overline{\partial}, \mathrm{ac})$-formula

$$\theta(x, z) := \theta_{\mathrm{v}}\big(v(P_1(x)), \dots, v(P_k(x)), z\big)$$

where $k \in \mathbb{N}$, $\theta_{\mathrm{v}}(v_1, \dots, v_k, y)$ is an $\mathcal{L}_{\mathrm{v}}$-formula, and $P_1(x), \dots, P_k(x) \in \mathbb{Z}\{x\}$. Note that these special formulas do not have quantified f-variables. We can now state our relative quantifier elimination:

COROLLARY 8.3.3. *Every $\mathcal{L}(\partial, \overline{\partial}, \mathrm{ac})$-formula $\phi(x, y, z)$ is T-equivalent to*

$$\big(\psi_1(x, y) \wedge \theta_1(x, z)\big) \vee \dots \vee \big(\psi_N(x, y) \wedge \theta_N(x, z)\big)$$

for some $N \in \mathbb{N}$ and some special r*-formulas $\psi_1(x, y), \dots, \psi_N(x, y)$ in (x, y), and some special* v*-formulas $\theta_1(x, z), \dots, \theta_N(x, z)$ in (x, z).*

PROOF. Let $\Theta(x, y, z)$ be the set of $\mathcal{L}(\partial, \overline{\partial}, \mathrm{ac})$-formulas

$$\big(\psi_1(x, y) \wedge \theta_1(x, z)\big) \vee \dots \vee \big(\psi_N(x, y) \wedge \theta_N(x, z)\big)$$

displayed in the statement of the corollary. It is clear that Θ is closed under taking disjunctions, and easy to check that Θ is closed under taking negations, modulo logical equivalence. Thus by Corollary B.9.3 it is enough to show that every T-realizable (x, y, z)-type is completely determined by its intersection with Θ. This guides the argument that follows.

Let $\psi(x, y)$ and $\theta(x, z)$ range over special formulas as described above. For $\boldsymbol{K} = (K, \Gamma, \boldsymbol{k}; \dots) \models T$ and $a \in K^l$, $r \in \boldsymbol{k}^m$, $\gamma \in \Gamma^n$, let

$$\begin{aligned}
\mathrm{tp}_{\mathrm{r}}^{\boldsymbol{K}}(a, r) &:= \big\{\psi(x, y) : \boldsymbol{K} \models \psi(a, r)\big\}, \\
\mathrm{tp}_{\mathrm{v}}^{\boldsymbol{K}}(a, \gamma) &:= \big\{\theta(x, z) : \boldsymbol{K} \models \theta(a, \gamma)\big\}.
\end{aligned}$$

Let $\boldsymbol{K}$ and $\boldsymbol{K}^*$ be any models of T, and let

$$(a, r, \gamma) \in K^l \times \boldsymbol{k}^m \times \Gamma^n, \qquad (a^*, r^*, \gamma^*) \in (K^*)^l \times (\boldsymbol{k}^*)^m \times (\Gamma^*)^n$$

be such that $\mathrm{tp}_{\mathrm{r}}^{\boldsymbol{K}}(a,r) = \mathrm{tp}_{\mathrm{r}}^{\boldsymbol{K}^*}(a^*,r^*)$ and $\mathrm{tp}_{\mathrm{v}}^{\boldsymbol{K}}(a,\gamma) = \mathrm{tp}_{\mathrm{v}}^{\boldsymbol{K}^*}(a^*,\gamma^*)$. By the above, it suffices to show that under these assumptions we have

$$\mathrm{tp}^{\boldsymbol{K}}(a,r,\gamma) \;=\; \mathrm{tp}^{\boldsymbol{K}^*}(a^*,r^*,\gamma^*).$$

Let $\boldsymbol{E} := (E, \Gamma_{\boldsymbol{E}}, \boldsymbol{k}_{\boldsymbol{E}})$ where $E := \mathbb{Q}\langle a\rangle$, $\boldsymbol{k}_{\boldsymbol{E}}$ is the differential subfield of $\boldsymbol{k}$ generated by $\mathrm{ac}(E)$ and r, and $\Gamma_{\boldsymbol{E}}$ is the ordered subgroup of Γ generated by γ over $v(E^\times)$, so $\boldsymbol{E}$ is a good substructure of $\boldsymbol{K}$. Likewise we define the good substructure $\boldsymbol{E}^*$ of $\boldsymbol{K}^*$. For each $P(x) \in \mathbb{Z}\{x\}$ we have $P(a) = 0$ iff $\mathrm{ac}\big(P(a)\big) = 0$, and also $P(a^*) = 0$ iff $\mathrm{ac}^*\big(P(a^*)\big) = 0$. In view of this fact, the assumptions give us a good map $\boldsymbol{E} \to \boldsymbol{E}^*$ sending a to a^*, γ to γ^* and r to r^*. It remains to apply Theorem 8.2.5. □

In the proof of the corollary above it is important that our notion of a good substructure $\boldsymbol{E} = (E, \Gamma_{\boldsymbol{E}}, \boldsymbol{k}_{\boldsymbol{E}})$ did not require $\Gamma_{\boldsymbol{E}} = v(E^\times)$ or $\boldsymbol{k}_{\boldsymbol{E}} = \pi(\mathcal{O}_E)$. Related to it is that in Corollary 8.3.3 we have a separation of r- and v-variables; this makes the next result almost obvious.

COROLLARY 8.3.4. *Each subset of $\boldsymbol{k}^m \times \Gamma^n$ definable in $\boldsymbol{K}$ is a finite union of rectangles $Y \times Z$ with $Y \subseteq \boldsymbol{k}^m$ definable in the differential field $\boldsymbol{k}$ and $Z \subseteq \Gamma^n$ definable in the ordered abelian group Γ.*

PROOF. By Corollary 8.3.3 and using its notations it is enough to observe that for $a \in K^l$, a special r-formula $\psi(x,y)$ in (x,y), and a special v-formula $\theta(x,z)$ in (x,z), the set $\big\{r \in \boldsymbol{k}^m : \boldsymbol{K} \models \psi(a,r)\big\}$ is definable in the differential field $\boldsymbol{k}$, and the set $\big\{\gamma \in \Gamma^n : \boldsymbol{K} \models \theta(a,\gamma)\big\}$ is definable in the ordered abelian group Γ. □

Corollary 8.3.4 says in particular that the relations on $\boldsymbol{k}$ definable in $\boldsymbol{K}$ are definable in the differential field $\boldsymbol{k}$, and likewise, the relations on Γ definable in $\boldsymbol{K}$ are definable in the ordered abelian group Γ.

Theorems 8.0.1 and 8.0.4 from the introduction to this chapter do not mention angular component maps. To get these results from Corollaries 8.3.1 and 8.3.4 we first pass to suitable $\aleph_1$-saturated elementary extensions and then use Corollary 8.1.3 to get the necessary angular component maps.

To get Theorem 8.0.3 from Corollary 8.3.2, we arrange likewise that $\boldsymbol{K}$ and $\boldsymbol{K}^*$ from that theorem are $\aleph_1$-saturated (but not yet equipped with angular component maps). Then we have a cross-section $s\colon \Gamma \to C^\times$ of the valued subfield C of $\boldsymbol{K}$. Use Lemma 3.3.40 to extend s to a cross-section $s^*\colon \Gamma^* \to C_{K^*}^\times$ of the valued subfield C_{K^*} of $\boldsymbol{K}^*$. These cross-sections yield angular component maps on the valued fields C and C^*, which by Lemma 8.1.2 extend uniquely to angular component maps on $\boldsymbol{K}$ and $\boldsymbol{K}^*$. This allows us to use Corollary 8.3.2 to get $\boldsymbol{K} \preccurlyeq \boldsymbol{K}^*$.

Notes and comments

Corollary 8.3.3 is analogous to a result by Pas [315] for henselian valued fields of equicharacteristic zero. Readers familiar with the model-theoretic properties of *stable embeddedness* and *orthogonality* will observe that by Corollary 8.3.4, $\boldsymbol{k}$ and Γ are stably embedded in $\boldsymbol{K}$, and $\boldsymbol{k}$ and Γ are orthogonal in $\boldsymbol{K}$. We refer to Appendix B for the definition of the model-theoretic property NIP, the *Non-Independence Property*. Using Corollary 8.3.3 one can show (along the lines of [417, proof of Theorem A.15]) that if the differential residue field $\boldsymbol{k}$ has NIP (that is, its theory has NIP), then so does $\boldsymbol{K}$; this also uses the fact that by [160] every ordered abelian group has NIP.

8.4 A MODEL COMPANION

Let $\mathcal{L}_{\partial,\preccurlyeq} := \{0, 1, -, +, \cdot, \partial, \preccurlyeq\}$ be the one-sorted language of valued differential fields, and construe valued differential fields as $\mathcal{L}_{\partial,\preccurlyeq}$-structures in the obvious way.

PROPOSITION 8.4.1. *The theory of* d*-henselian valued differential fields with many constants, differentially closed differential residue field, and nontrivial divisible value group is complete and model complete. It is the model companion of the theory of monotone valued differential fields.*

PROOF. For completeness and model completeness, use Theorems 8.0.1 and 8.0.3 in combination with Corollary 4.7.3 and Example B.11.12. Let K be an arbitrary monotone valued differential field; to prove the model companion claim, it is enough to embed K into a d-henselian valued differential field with many constants, differentially closed differential residue field, and nontrivial divisible value group. By Example (1) at the beginning of Section 4.4, K has a monotone valued differential field extension K_1 with a nontrivial value group. By Corollary 6.3.10, K_1 has a monotone valued differential field extension K_2 with divisible value group. By Corollary 6.3.7, K_2 has a monotone valued differential field extension K_3 with differentially closed differential residue field and $\Gamma_{K_3} = \Gamma_{K_3}$. Theorem 7.0.1 then yields an immediate d-henselian valued differential field extension K_4 of K_3. Then K_4 is still monotone by Corollary 6.3.6, and thus has many constants by Corollary 7.1.11. □

Chapter Nine

Asymptotic Fields and Asymptotic Couples

The key restriction on valued differential fields in Chapter 6 was the *continuity* of the derivation. (Strictly speaking, we assumed the derivation ∂ to be *small*, but continuity of ∂ reduces by compositional conjugation to smallness of ∂.)

In this chapter we introduce asymptotic differential fields: valued differential fields with a much stronger interaction of the valuation and derivation. For brevity we just call them *asymptotic fields.* They include Rosenlicht's differential-valued fields [364] and share many of their basic properties. The advantage of the class of asymptotic fields over its subclass of differential-valued fields is that the former is closed under taking valued differential subfields, under coarsening, and even under specialization (subject to a mild restriction).

A key feature of an asymptotic field is its asymptotic couple, which is just its value group with some extra structure induced by the derivation. In Section 9.1 we define asymptotic fields, their asymptotic couples, and discuss Hardy fields. In Section 9.2 we consider asymptotic couples independent of their connection to asymptotic fields. This is used in Section 9.3 to describe the behavior of differential polynomials as functions on asymptotic fields. In Section 9.4 we consider asymptotic fields with small derivation and the operations of coarsening and specialization. In Section 9.5 we show that algebraic extensions of asymptotic fields are asymptotic. In Section 9.6 we adapt the results on immediate extensions from Section 6.9 to asymptotic fields. Section 9.7 treats differential polynomials of order one over H-asymptotic fields. In Section 9.8 we return to asymptotic couples and prove some useful extension results about them, and in Section 9.9 we establish a property of *closed* H-asymptotic couples as needed in Chapter 16. The present chapter and the next include some new material but are mainly based on [364, 18, 19, 20].

Some terminology: When a valued differential field K is given, then an *extension of* K is a valued differential field extension of K. Likewise, an *extension* of an ordered valued differential field K is an ordered valued differential field extension of K. The term "embedding" is used in a similar way: when K and L are given as valued differential fields, then an *embedding* of K into L is an embedding of valued differential fields, and when K and L are given as ordered valued differential fields, then an *embedding* of K into L is an embedding of ordered valued differential fields.

9.1 ASYMPTOTIC FIELDS AND THEIR ASYMPTOTIC COUPLES

In the first subsection we define asymptotic fields and differential-valued fields, in the second subsection we show how to visualize an asymptotic couple, in the third subsection we introduce the asymptotic couple of an asymptotic field, and in the fourth subsection we define comparability classes and the property of being grounded. In the last subsection we discuss Hardy fields as examples.

Asymptotic fields

An **asymptotic differential field**, or just **asymptotic field**, is a valued differential field K such that for all $f,g \in K^\times$ with $f,g \prec 1$,

(A) $f \prec g \Longleftrightarrow f' \prec g'$.

If in addition we have for all $f,g \in K^\times$ with $f,g \prec 1$,

(H) $f \prec g \Longrightarrow f^\dagger \succcurlyeq g^\dagger$,

then we say that K is an H**-asymptotic field** or an asymptotic field of H**-type**. Our main interest is in H-asymptotic fields, but many things go through without the H-type assumption.

LEMMA 9.1.1. *Let K be a valued differential field such that $C \subseteq \mathcal{O}$, $\partial\scriptstyle{\mathcal{O}}$ $\subseteq \mathcal{o}$, and ∂ is neatly surjective. Then K is asymptotic.*

PROOF. From $C \subseteq \mathcal{O}$ we obtain $C \cap \mathcal{o} = \{0\}$, so the restriction of ∂ to $\mathcal{o}$ is injective. From $\partial\mathcal{o} \subseteq \mathcal{o}$ we obtain a strictly increasing map $v_{Y'} = v_\partial : \Gamma \to \Gamma$ with $v_\partial(0) = 0$. Let $f,g \in \mathcal{o}^{\neq}$. Then $f',g' \in \mathcal{o}^{\neq}$, so $v_\partial(vf) = v(f')$ and $v_\partial(vg) = v(g')$ by the neat surjectivity of ∂. Thus $f \prec g \Longleftrightarrow f' \prec g'$. □

Suppose K is an asymptotic field. Then $C \cap \mathcal{o} = \{0\}$: if $0 \neq c \in C \cap \mathcal{o}$, then $c^2 \prec c \prec 1$, so $0 = (c^2)' \prec c' = 0$ by (A), which is impossible. Thus the valuation v is trivial on C, and the map $\varepsilon \mapsto \varepsilon' : \mathcal{o} \to K$ is injective. In particular, $C \subseteq \mathcal{O}$ and the residue map $a \mapsto \overline{a} : \mathcal{O} \to \boldsymbol{k}$ is injective on C. The following three conditions on K are clearly equivalent:

(1) $\mathcal{O} = C + \mathcal{o}$;

(2) $\{\overline{a} : a \in C\} = \boldsymbol{k}$;

(3) for all $f \asymp 1$ in K there exists $c \in C$ with $f \sim c$.

We say that K is **differential-valued** (or d**-valued**, for short) if it satisfies these three (equivalent) conditions. So the constant field of a d-valued field is also a lift of its residue field. The next lemma is clear from (3) above.

LEMMA 9.1.2. *If L is an asymptotic extension of a* d*-valued field K with* $\operatorname{res}(K) = \operatorname{res}(L)$, *then L is* d*-valued, with $C_L = C$.*

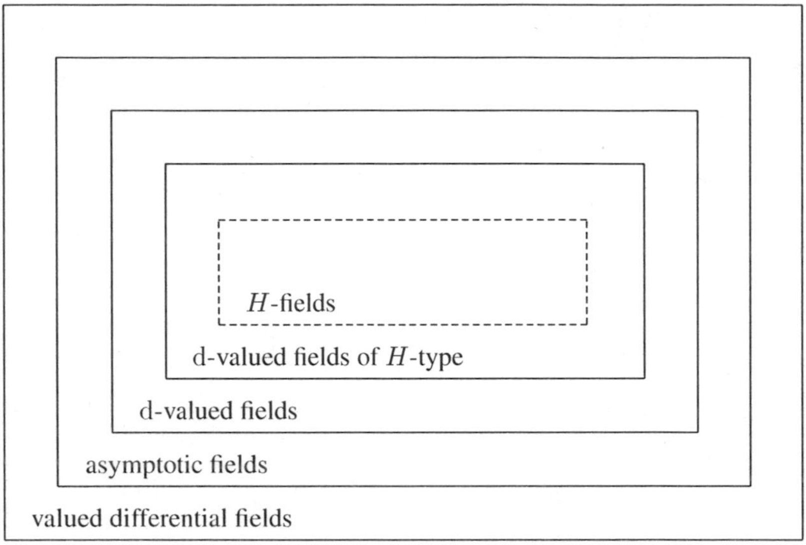

Figure 9.1: Classes of valued differential fields.

Our final results in Chapter 16 concern just d-valued fields of H-type such as $\mathbb{T}$, but towards this goal we need the wider setting of H-asymptotic fields where we can coarsen and pass to suitable differential subfields at our convenience.

Figure 9.1 indicates the inclusion relations among some classes of valued differential fields, with the class of H-fields from the next chapter as the smallest class. Strictly speaking, H-fields are more than just valued differential fields, since they also carry an ordering; that's why we use a dashed rectangle for this class.

Any differential subfield of an asymptotic field K with the restricted dominance relation is itself an asymptotic field. If K is an asymptotic field and $\phi \in K^{\times}$, then its compositional conjugate K^{ϕ} with the same dominance relation remains an asymptotic field with the same constant field. (These two statements remain true with *H-asymptotic* in place of *asymptotic*.) If K is a d-valued field and $\phi \in K^{\times}$, then K^{ϕ} (with same dominance relation) remains d-valued.

An asymptotic field is said to be **asymptotically maximal** if it has no proper immediate asymptotic extension. Likewise, an asymptotic field is **asymptotically d-algebraically maximal** if it has no proper immediate d-algebraic asymptotic extension. Any asymptotic field has, by Zorn, an immediate asymptotic extension that is asymptotically maximal, and also an immediate d-algebraic asymptotic extension that is asymptotically d-algebraically maximal. These notions will become important in Chapter 14.

Asymptotic couples

We defined asymptotic couples in Section 6.5. Consider an asymptotic couple (Γ, ψ). For reasons that will become clear in the next subsection we also use the notation

$$\alpha^\dagger := \psi(\alpha), \qquad \alpha' := \alpha + \psi(\alpha) \qquad (\alpha \in \Gamma^{\neq}).$$

Of course this is only used when ψ is understood from the context. Thus $\alpha' > \beta^\dagger$ for $\alpha \in \Gamma^{>}$, $\beta \in \Gamma^{\neq}$. The following subsets of Γ play special roles:

$$(\Gamma^{\neq})' := \{\gamma' : \gamma \in \Gamma^{\neq}\}, \qquad (\Gamma^{>})' := \{\gamma' : \gamma \in \Gamma^{>}\},$$
$$\Psi := \psi(\Gamma^{\neq}) = \{\gamma^\dagger : \gamma \in \Gamma^{\neq}\} = \{\gamma^\dagger : \gamma \in \Gamma^{>}\}.$$

We call Ψ the Ψ-set of (Γ, ψ), and set $\alpha^\dagger := \infty \in \Gamma_\infty$ for $\alpha = 0 \in \Gamma$.

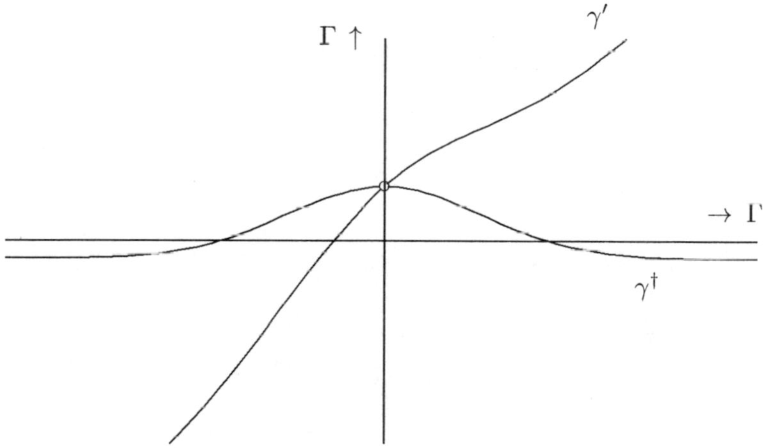

Figure 9.2: The graphs of $\gamma \mapsto \gamma'$ and $\gamma \mapsto \gamma^\dagger$.

Figure 9.2 gives a rough idea of the graphs of γ' and $\gamma^\dagger$ as functions of $\gamma \in \Gamma^{\neq}$. We have found this picture very helpful in getting a "feel" for asymptotic couples. It suggests for example that γ' is strictly increasing on $\Gamma^{\neq}$, and this is part (iii) of Lemma 6.5.4. It also suggests that γ' has the intermediate value property on $\Gamma^{>}$ as well as on $\Gamma^{<}$. This is the case for H-asymptotic couples by Lemma 9.2.14 below, but not for all asymptotic couples; see Example 2.8 in [19]. The picture is really meant just for H-asymptotic couples, where $\gamma^\dagger$ is increasing for $\gamma < 0$ and decreasing for $\gamma > 0$. Of course, we are unable to make the picture show, for H-asymptotic (Γ, ψ), that ψ is constant on each archimedean class $[\alpha]$, $\alpha \in \Gamma^{\neq}$.

The asymptotic couple of an asymptotic field

Let K be an asymptotic field. It is clear that then vg' is uniquely determined by vg for $g \in K^\times$ and $vg \neq 0$, that is, the derivation of K induces a function

$$\gamma \mapsto \gamma' : \Gamma^{\neq} \to \Gamma \qquad (\gamma = vg, \gamma' = vg', g \text{ as above})$$

on its value group Γ. We also consider the logarithmic-derivative analogue:

$$\gamma \mapsto \gamma^\dagger := \gamma' - \gamma : \Gamma^{\neq} \to \Gamma,$$

that is, $\gamma^\dagger = v(g^\dagger)$ for g as above. The asymptotic couple of K is just the value group of K equipped with this induced operation $\gamma \mapsto \gamma^\dagger$. To justify this terminology, we characterize asymptotic fields as follows:

PROPOSITION 9.1.3. *Let K be a valued differential field. Then the conditions below are equivalent:*

(i) *K is an asymptotic field;*

(ii) *there is an asymptotic couple (Γ, ψ) with $\Gamma := v(K^\times)$ such that for all $g \in K^\times$ with $g \not\asymp 1$ we have $\psi(vg) = v(g^\dagger)$;*

(iii) *for all $f, g \in K^\times$ with $f, g \not\asymp 1$ we have: $f \preccurlyeq g \Longleftrightarrow f' \preccurlyeq g'$;*

(iv) *for all $f, g \in K^\times$ we have:*

$$\begin{cases} f \prec 1,\ g \not\asymp 1 & \Rightarrow \quad f' \prec g^\dagger, \\ f \asymp 1,\ g \not\asymp 1 & \Rightarrow \quad f' \preccurlyeq g^\dagger. \end{cases}$$

PROOF. We first show the equivalence of (ii) and (iv). Suppose that (ii) holds. The first implication of (iv) is clear. For the second implication, let $f, g \in K^\times$, $f \asymp 1$, $g \not\asymp 1$. We have $g^\dagger \asymp (fg)^\dagger = f^\dagger + g^\dagger$. Hence $f' \asymp f^\dagger \preccurlyeq g^\dagger$. Conversely, suppose that (iv) holds. Then $g^\dagger \neq 0$ for $g \not\asymp 1$ in $K^\times$ by taking $f = g$ or $f = g^{-1}$ in the first implication of (iv). Let $f, g \in K^\times$ with $f \asymp g \prec 1$. Then $f^\dagger - g^\dagger = (f/g)^\dagger \asymp (f/g)'$ with $f/g \asymp 1$, so $f^\dagger - g^\dagger \preccurlyeq f^\dagger$ and $f^\dagger - g^\dagger \preccurlyeq g^\dagger$ by the second implication of (iv). It follows that $f^\dagger \asymp g^\dagger$. Thus $v(f^\dagger)$ only depends on vf, for $f \in K^\times$ with $f \not\asymp 1$. Now (ii) follows, with (AC3) a consequence of the first implication in (iv).

Note that (ii) $\Rightarrow$ (iii) is a consequence of Lemma 6.5.4(iii), and (iii) $\Rightarrow$ (i) is trivial. Lemma 6.5.6 gives (i) $\Rightarrow$ (ii). □

If K is an asymptotic field, we call (Γ, ψ) as defined in (ii) of Proposition 9.1.3 the **asymptotic couple of** K. An asymptotic field is of H-type iff its asymptotic couple is of H-type. If (Γ, ψ) is the asymptotic couple of the asymptotic field K and $a \in K^\times$, then (Γ, ψ^a) with $\psi^a := \psi - va$ is the asymptotic couple of K^a.

COROLLARY 9.1.4. *Let K be an asymptotic field and $f, g \in K$. Then*

(i) *if $f \prec g \not\asymp 1$, then $f' \prec g'$;*

(ii) *if* $f \preccurlyeq g \not\asymp 1$, *then* $f \sim g \Longleftrightarrow f' \sim g'$;

(iii) *if* $f \preccurlyeq g \not\asymp 1$ *and* $g' \preccurlyeq g$, *then* $f^{(n)} \preccurlyeq g'$ *for all* $n \geqslant 1$;

(iv) *if* $f \asymp 1$ *and* $0 \neq g \not\asymp 1$, *then* $f^\dagger \not\sim g^\dagger$.

PROOF. For (i), if $f \prec g$ and $f, g \not\asymp 1$, then $f' \prec g'$ by Proposition 9.1.3(iii), and if $1 \asymp f \prec g$, then $f' \preccurlyeq g^\dagger = g'/g \prec g'$ by Proposition 9.1.3(iv). For (ii), suppose that $g \not\asymp 1$. If $f \sim g$, then $1 \not\asymp g \succ f - g$, so $g' \succ f' - g'$ by (i), i.e., $f' \sim g'$. Conversely, suppose that $f \preccurlyeq g \not\asymp 1$ and $f' \sim g'$. If $f - g \not\asymp 1$, then Proposition 9.1.3(iii) yields $f \sim g$. If $f - g \asymp 1$, then $g \succ 1$, and thus $f - g \asymp 1 \prec g$. Part (iii) follows by induction on n from (i) and Proposition 9.1.3(iii). As to (iv), if $f \asymp 1$ and $0 \neq g \not\asymp 1$, then $g/f \asymp g \not\asymp 1$, so $(g/f)^\dagger = g^\dagger \quad f^\dagger \asymp g^\dagger$, hence $f^\dagger \not\sim g^\dagger$. □

In the next two corollaries we assume that K *is an asymptotic field.*

COROLLARY 9.1.5. *The derivation of* K *is continuous.*

PROOF. If $\Gamma = \{0\}$, then the valuation topology on K is discrete and so ∂ is continuous. Assume $\Gamma \neq \{0\}$, and take $g \in K^\times$ with $g \succ 1$. Then $a := g' \neq 0$, and so $\partial\smallo \subseteq a\smallo$ by Corollary 9.1.4(i). Thus ∂ is continuous by Lemma 4.4.7. □

As a consequence of Corollary 9.1.5 and Lemma 4.4.11, the completion K^{c} of our asymptotic field K is naturally a valued differential field, but more is true:

COROLLARY 9.1.6. *The valued differential field* K^{c} *is an asymptotic field. If* K *is* d-*valued, then so is* K^{c}.

PROOF. We claim that K^{c} is asymptotic with the same asymptotic couple (Γ, ψ) as K, and to show this we use (ii) of Proposition 9.1.3. Let $f \in K^{\mathrm{c}}$ and $0 \neq f \not\asymp 1$. Density of K in K^{c} gives $g \in K$ with $f \sim g$ and $f' \sim g'$. Then $\psi(vf) = \psi(vg) = v(g^\dagger) = v(f^\dagger)$. This proves our claim. If K is even d-valued, then so is K^{c} by Lemma 9.1.2, since $\operatorname{res} K = \operatorname{res} K^{\mathrm{c}}$. □

We say that an asymptotic couple (Γ, ψ) has **asymptotic integration** if

$$\Gamma = (\Gamma^{\neq})'.$$

An asymptotic field is said to have **asymptotic integration** if its asymptotic couple has asymptotic integration. The reason for this terminology is that if K is a d-valued field, then K has asymptotic integration iff for all $a \in K^\times$ there is $b \in K$ such that $a \sim b'$.

Comparability and groundedness

Let K be an asymptotic field. On the set $K^\times$ we define the binary relations $\asymp\!\!\!\asymp$, $\prec\!\!\!\prec$, $\preccurlyeq\!\!\!\preccurlyeq$ as follows:

$$\begin{aligned} f \asymp\!\!\!\asymp g \quad &:\Longleftrightarrow \quad f^\dagger \asymp g^\dagger, \\ f \prec\!\!\!\prec g \quad &:\Longleftrightarrow \quad f^\dagger \prec g^\dagger, \\ f \preccurlyeq\!\!\!\preccurlyeq g \quad &:\Longleftrightarrow \quad f^\dagger \preccurlyeq g^\dagger. \end{aligned}$$

For the meaning of $\preceq\!\!\!\prec$ in Hardy fields, see Corollary 9.1.11. Note: these relations on $K^\times$ do not change when passing from K to a compositional conjugate K^ϕ with $\phi \in K^\times$. For $f, g \in K^\times$ with $f, g \not\asymp 1$ we say that f and g are **comparable** if $f \asymp\!\!\!\asymp g$, and we say that f is **flatter** than g if $f \prec\!\!\!\prec g$. Comparability is an equivalence relation on $\{f \in K^\times : f \not\asymp 1\}$. The equivalence class of an element f from this set is called its **comparability class**, written as $\mathrm{Cl}(f)$. The relation $\prec\!\!\!\prec$ induces a linear ordering on the set of comparability classes by

$$\mathrm{Cl}(f) < \mathrm{Cl}(g) \ :\Longleftrightarrow\ f \prec\!\!\!\prec g.$$

Let (Γ, ψ) be the asymptotic couple of K. Then for $f \in K^\times$ with $f \not\asymp 1$ we have: $\mathrm{Cl}(f)$ is the smallest comparability class of K iff $\psi(vf)$ is the largest element of Ψ. Thus K has a smallest comparability class iff Ψ has a largest element. Note also that if K is of H-type and $f, g \in K$, then $1 \prec f \preccurlyeq g \Rightarrow f \preceq\!\!\!\prec g$. We define the asymptotic field K to be **grounded** if K has a smallest comparability class, equivalently, Ψ has a largest element; an asymptotic field that is not grounded is called **ungrounded**.

Hardy fields

Hardy fields, defined below, are H-asymptotic fields of a classical origin, and will often serve as examples and counterexamples. They are not just asymptotic fields but also carry a natural ordering.

Let $\mathcal{G}$ be the ring of germs at $+\infty$ of real-valued functions whose domain is a subset of $\mathbb{R}$ containing an interval $(a, +\infty)$, $a \in \mathbb{R}$; the domain may vary and the ring operations are defined as usual. We call a germ $g \in \mathcal{G}$ *continuous*, respectively *differentiable*, if it is the germ of a continuous, respectively differentiable, function $(a, +\infty) \to \mathbb{R}$ for some $a \in \mathbb{R}$; for differentiable $g \in \mathcal{G}$ we let $g' \in \mathcal{G}$ denote the germ of the derivative of that function. If $g \in \mathcal{G}$ is the germ of a real-valued function on some interval $(a, +\infty)$, $a \in \mathbb{R}$, then we simplify notation by letting g also denote this function if the resulting ambiguity is harmless. With this convention, given a property P of real numbers and $g \in \mathcal{G}$ we say that $P\big(g(t)\big)$ holds eventually if $P\big(g(t)\big)$ holds for all sufficiently large real t. We identify each real number r with the germ at $+\infty$ of the function $\mathbb{R} \to \mathbb{R}$ that takes the constant value r. This makes the field $\mathbb{R}$ into a subring of $\mathcal{G}$.

DEFINITION 9.1.7 (N. Bourbaki). A **Hardy field** is a subring K of $\mathcal{G}$ such that K is a field, all $g \in K$ are differentiable, and $g' \in K$ for all $g \in K$.

EXAMPLE. Every subfield of $\mathbb{R}$ is a Hardy field. Given polynomials $p(x), q(x) \in \mathbb{R}[x]$ with $q \neq 0$, identify the rational function $p(x)/q(x) \in \mathbb{R}(x)$ with the germ at $+\infty$ of $t \mapsto p(t)/q(t)$, for $t \in \mathbb{R}$ with $q(t) \neq 0$. This makes the rational function field $\mathbb{R}(x)$ into a Hardy field, with x the germ of the identity function on $\mathbb{R}$.

In the rest of this subsection K is a Hardy field, and f, g range over K. We consider K as a differential field with derivation $f \mapsto f'$. It has constant field $C = K \cap \mathbb{R}$, with intersection taken inside $\mathcal{G}$. Every nonzero f has a multiplicative inverse in K, so eventually $f(t) \neq 0$, hence either eventually $f(t) < 0$ or eventually $f(t) > 0$ (by eventual continuity of f). We make K an ordered field by declaring

$$f > 0 :\Longleftrightarrow f(t) > 0, \text{ eventually.}$$

Thus C is a common ordered subfield of K and of $\mathbb{R}$. Since $f' \in K$, either $f' < 0$, or $f' = 0$, or $f' > 0$, and accordingly, f is either eventually strictly decreasing, or eventually constant, or eventually strictly increasing, hence the limit $\lim_{t\to\infty} f(t)$ always exists, as an element of the extended real line $\mathbb{R} \cup \{\pm\infty\}$. Our Hardy field K is a valued field with convex valuation ring

$$\mathcal{O} = \{f : |f| \leqslant n \text{ for some } n\} = \{f : |f| \leqslant c \text{ for some } c \in C\}.$$

The natural dominance relation $\preccurlyeq$ of a Hardy field and the derived asymptotic relations $\prec, \asymp, \sim$ from Section 3.1 have the following meaning in terms of limits:

LEMMA 9.1.8. *For $g \neq 0$, we have:*

$$f \preccurlyeq g \Longleftrightarrow \lim_{t\to+\infty} \frac{f(t)}{g(t)} \in \mathbb{R}, \qquad f \prec g \Longleftrightarrow \lim_{t\to+\infty} \frac{f(t)}{g(t)} = 0,$$

$$f \asymp g \Longleftrightarrow \lim_{t\to+\infty} \frac{f(t)}{g(t)} \in \mathbb{R}^{\times}, \qquad f \sim g \Longleftrightarrow \lim_{t\to+\infty} \frac{f(t)}{g(t)} = 1.$$

Thus $f \succ 1$ iff $\lim_{t\to+\infty} |f(t)| = +\infty$.

EXAMPLE. Let $K = \mathbb{R}(x)$. Then for $a_0, a_1, \dots, a_n \in \mathbb{R}$, $a_n \neq 0$ we have in K: $a_0 + a_1x + \cdots + a_nx^n \sim a_nx^n$. Thus $\Gamma = \mathbb{Z}v(x)$ with $v(x) < 0 = v(1)$, and the valuation is given by $v(p/q) = (\deg p - \deg q)v(x)$, for $p, q \in \mathbb{R}[x]^{\neq}$.

Recall that $f^\dagger := f'/f = \big(\log|f|\big)'$ for $f \neq 0$.

PROPOSITION 9.1.9. *Let $f, g \neq 0$.*

(i) *If $f \prec g$, then $f^\dagger < g^\dagger$.*

(ii) *If $f \prec g \prec 1$, then $f^\dagger \succcurlyeq g^\dagger$.*

(iii) *If $f \asymp 1$ and $\mathbb{R} \subseteq K$, then $f \sim c$ for some $c \in \mathbb{R}^{\times}$.*

(iv) *If $f \preccurlyeq 1$, $g \not\asymp 1$, then $f' \prec g^\dagger$.*

PROOF. For $f = 1$, item (i) is clear: if $g > \mathcal{O}$, say, then g is ultimately strictly increasing, hence $g^\dagger = g'/g > 0$. In general, if $f \prec g$, then $1 \prec g/f$ and hence $0 < (g/f)^\dagger = g^\dagger - f^\dagger$. As to (ii): if $f \prec g \prec 1$, then $f^\dagger < g^\dagger < 1^\dagger = 0$ by (i), so $f^\dagger \succcurlyeq g^\dagger$. If $f \asymp 1$, then $c := \lim_{t\to+\infty} f(t)$ works in (iii). Next, let $f \preccurlyeq 1$, $g \not\asymp 1$. To get $f' \prec g^\dagger$, replace f by $f+1$ and g by $1/g$, if necessary, to arrange $f \asymp 1 \prec g$. Then $f^k \prec g$ for all $k \in \mathbb{Z}$, so $kf^\dagger < g^\dagger$ for all $k \in \mathbb{Z}$, by (i), hence $f^\dagger \prec g^\dagger$, and thus $f' \prec fg^\dagger \asymp g^\dagger$. This proves (iv). □

By the equivalence of (i) and (iv) in Proposition 9.1.3 and items (ii) and (iv) in Proposition 9.1.9, every Hardy field is H-asymptotic. Moreover, every Hardy field containing $\mathbb{R}$ is d-valued, by Proposition 9.1.9(iii).

Next we indicate without proof three ways of extending our Hardy field K:

(1) let K^{rc} consist of the continuous germs $y \in \mathcal{G}$ such that $P(y) = 0$ for some $P(Y) \in K[Y]^{\neq}$; then K^{rc} is the unique real closed Hardy field extension of K that is algebraic over K, and is thus a real closure of the ordered field K;

(2) any differentiable $h \in \mathcal{G}$ with $h' \in K$ yields a Hardy field $K(h) \supseteq K$;

(3) e^f generates a Hardy field extension $K(\mathrm{e}^f)$ of K.

All three are proved in [366], but (1) earlier in [355], and (2) and (3) in [62, Appendice]. An easy consequence of (2) is that there is a Hardy field $K(\mathbb{R}) \supseteq K$ generated as a field over K by $\mathbb{R}$. It has $\mathbb{R}$ as its constant field. A special case of (3) is that for $f \in K^{>}$ we have a Hardy field $K(\log f)$, since $(\log f)' = f^\dagger \in K$.

COROLLARY 9.1.10. *The derivation of K is small.*

PROOF. We have $x' = 1$ for the germ $x \in \mathcal{G}$ of the identity function on $\mathbb{R}$, so by extending K we can arrange $x \in K$. Then $f \preccurlyeq 1$ gives $f \prec x$, and hence $f' \prec x' = 1$, since K is asymptotic. □

COROLLARY 9.1.11. *Suppose that $f, g \succ 1$. Then*

$$f \preccurlyeq\!\!\!\prec g \iff |f| \leqslant |g|^n \text{ for some } n.$$

PROOF. Assume $f \preccurlyeq\!\!\!\prec g$. Then $f^\dagger \preccurlyeq g^\dagger$, that is, $(\log|f|)' \preccurlyeq (\log|g|)'$. Working in a Hardy field extension of K containing $\log|f|$ and $\log|g|$ we have $\log|f| \succ 1$ and $\log|g| \succ 1$, so $\log|f| \preccurlyeq \log|g|$, that is, $\log|f| \leqslant n\log|g|$ for some n, and thus $|f| \leqslant |g|^n$ for some n. The converse follows by reversing this reasoning. □

EXAMPLE (of a Hardy field that is not d-valued). The Hardy subfield $\mathbb{Q}(x)$ of $\mathbb{R}(x)$ is d-valued with constant field $\mathbb{Q}$, but its algebraic extension

$$\mathbb{Q}\left(\sqrt{2 + x^{-1}}\right) \qquad \text{(a subfield of the Hardy field } \mathbb{R}(x)^{\mathrm{rc}}\text{)}$$

is a Hardy field with the same constant field $\mathbb{Q}$ and with $\sqrt{2}$ in the residue field. Thus $\mathbb{Q}\left(\sqrt{2 + x^{-1}}\right)$ is not d-valued, even though it is a valued differential subfield of the d-valued Hardy field $\mathbb{R}(x)^{\mathrm{rc}}$.

Notes and comments

Rosenlicht introduced differential-valued fields and their asymptotic couples in [364]; besides Hardy fields that paper has nice examples of a complex-analytic nature. To our knowledge the larger class of asymptotic fields has not been singled out previously for special attention.

Basic references on Hardy fields are [62, Appendice] and [366]. Corollary 9.1.11 is from [367]. The next chapter defines H-fields and pre-H-fields as ordered valued differential fields that share certain key elementary properties with Hardy fields.

9.2 H-ASYMPTOTIC COUPLES

We begin this section with proving some basic facts about arbitrary asymptotic couples, but after that we focus on H-asymptotic couples. Using an intermediate value property for suitable functions on ordered abelian groups, we derive a key trichotomy for H-asymptotic couples: Corollary 9.2.16. We also introduce a contraction map and study ψ-maps on suitable H-asymptotic couples. We finish with a look at the device of coarsening asymptotic couples and asymptotic fields.

Throughout this section (Γ, ψ) *is an asymptotic couple.*

Further basic facts on asymptotic couples

In this subsection α, β, γ range over Γ. By axiom (AC3) for asymptotic couples (see Section 6.5) we have $\Psi < (\Gamma^{>})'$. Recall from Section 2.1 that

$$\Psi^{\downarrow} = \{\alpha : \alpha \leqslant \gamma \text{ for some } \gamma \in \Psi\}.$$

If $\alpha < 0$, then $\alpha' = \alpha^{\dagger} + \alpha < \alpha^{\dagger}$. In particular, $(\Gamma^{<})' \subseteq \Psi^{\downarrow}$.

THEOREM 9.2.1. *The set* $\Gamma \setminus (\Gamma^{\neq})'$ *has at most one element. If* Ψ *has a largest element* $\max \Psi$, *then* $\Gamma \setminus (\Gamma^{\neq})' = \{\max \Psi\}$.

For the proof we need the following lemmas:

LEMMA 9.2.2. *Suppose* $\beta \neq 0$, $\alpha = \beta'$, $\gamma \neq \alpha$, *and* $\gamma \geqslant \beta^{\dagger}$. *Then* $(\alpha - \gamma)^{\dagger} \leqslant \gamma$.

PROOF. We have $\alpha - \gamma = \beta + \beta^{\dagger} - \gamma \leqslant \beta$ with $\alpha - \gamma \neq 0$ and $\beta \neq 0$ and hence $(\alpha - \gamma)' \leqslant \beta'$, that is, $\alpha - \gamma + (\alpha - \gamma)^{\dagger} \leqslant \alpha$, and thus $(\alpha - \gamma)^{\dagger} \leqslant \gamma$, as claimed. □

LEMMA 9.2.3. *The following conditions on* α *are equivalent:*

(i) $\alpha \in (\Gamma^{\neq})'$;

(ii) $(\alpha - \gamma)^{\dagger} = \gamma$ *for some* $\gamma \in \Psi$;

(iii) $(\alpha - \gamma)^{\dagger} \leqslant \gamma$ *for some* $\gamma \in \Psi$.

PROOF. Assume $\alpha = \beta'$, $\beta \neq 0$, and put $\gamma := \beta^{\dagger} \in \Psi$. Then $\gamma \neq \alpha$, so $(\alpha - \gamma)^{\dagger} \leqslant \gamma$ by Lemma 9.2.2. This shows (i) $\Rightarrow$ (iii). For (iii) $\Rightarrow$ (ii), we first reduce to the case $\alpha = 0$ by passing from ψ and γ to $\psi - \alpha$ and $\gamma - \alpha$. So assume towards a contradiction that $\gamma \in \Psi$, $\gamma^{\dagger} \leqslant \gamma$, but $\delta^{\dagger} \neq \delta$ for all $\delta \in \Psi$. Note that then $\gamma \neq 0$ and $\gamma^{\dagger} < \gamma < (\Gamma^{>})'$ and $\gamma^{\dagger\dagger} \neq \gamma^{\dagger}$ (take $\delta := \gamma^{\dagger}$). Hence by Lemma 6.5.4(i),

$$\min(\gamma^{\dagger}, \gamma^{\dagger\dagger}) = (\gamma - \gamma^{\dagger})^{\dagger} > \min(\gamma, \gamma^{\dagger}) = \gamma^{\dagger},$$

a contradiction. For (ii) $\Rightarrow$ (i), assume $\gamma \in \Psi$ and $(\alpha - \gamma)^{\dagger} = \gamma$. Then $\alpha - \gamma \neq 0$ and $\alpha = (\alpha - \gamma) + \gamma = (\alpha - \gamma) + (\alpha - \gamma)^{\dagger} = (\alpha - \gamma)'$, so $\alpha \in (\Gamma^{\neq})'$. □

PROOF OF THEOREM 9.2.1. Suppose $\alpha \neq \beta$ and $\alpha, \beta \notin (\Gamma^{\neq})'$. Then Lemma 9.2.3 gives for $\gamma \in \Psi$ that $(\alpha - \gamma)^{\dagger} > \gamma$ and $(\beta - \gamma)^{\dagger} > \gamma$. Thus for $\gamma := (\alpha - \beta)^{\dagger}$,

$$(\alpha - \beta)^{\dagger} = \big((\alpha - \gamma) - (\beta - \gamma)\big)^{\dagger} \geqslant \min\big((\alpha - \gamma)^{\dagger}, (\beta - \gamma)^{\dagger}\big) > \gamma = (\alpha - \beta)^{\dagger},$$

a contradiction. Suppose that Ψ has a largest element $\max \Psi$. If $\max \Psi = \alpha'$, $\alpha \neq 0$, then $\alpha < 0$, hence $\alpha^{\dagger} = \max \Psi - \alpha > \max \Psi \geqslant \alpha^{\dagger}$, a contradiction. □

COROLLARY 9.2.4. *There is at most one β such that*

$$\Psi < \beta < (\Gamma^{>})'.$$

If Ψ has a largest element, there is no such β.

An element β as in Corollary 9.2.4 is called a **gap** in (Γ, ψ). If K is an asymptotic field with asymptotic couple (Γ, ψ), then such an element is also called a gap in K. Call (Γ, ψ) **grounded** if Ψ has a largest element, and **ungrounded** otherwise. So an asymptotic field is grounded iff its asymptotic couple is grounded.

LEMMA 9.2.5. *Suppose K is an asymptotic field. Then:*

(i) *K has at most one gap;*

(ii) *if K is grounded, then K has no gap;*

(iii) *if L is an H-asymptotic field extension of K such that $\Gamma^{<}$ is cofinal in $\Gamma_L^{<}$, then a gap in K remains a gap in L;*

(iv) *if K is the union of a directed family (K_i) of asymptotic subfields such that no K_i has a gap, then K has no gap.*

Theorem 9.2.1 and Lemma 6.5.4(iii) yield:

LEMMA 9.2.6. *If $\Gamma \neq \{0\}$, then $(\Gamma^{>})'$ is cofinal in Γ, and $(\Gamma^{<})'$ is coinitial in Γ.*

LEMMA 9.2.7. *Suppose Γ is divisible and (Γ, ψ) has asymptotic integration. Then, given $q \in \mathbb{Q}^{\times}$ and α, the map $\gamma \mapsto \gamma + q\psi(\gamma - \alpha) \colon \Gamma^{\neq \alpha} \to \Gamma$ is bijective.*

PROOF. First, reduce to the case $\alpha = 0$. Let θ be the map in the lemma for $\alpha = 0$. Now use that $\theta(q\gamma) = q\big(\gamma + \psi(q\gamma)\big) = q\big(\gamma + \psi(\gamma)\big) = q\gamma'$ for $\gamma \neq 0$. □

Recall from Section 6.5 that ψ extends uniquely to a map $(\mathbb{Q}\Gamma)^{\neq} \to \Gamma$, also denoted by ψ, such that $(\mathbb{Q}\Gamma, \psi)$ is an asymptotic couple. Here $\mathbb{Q}\Gamma$ denotes the ordered divisible hull of Γ. Note that $\Psi = \psi(\Gamma^{\neq}) = \psi\big((\mathbb{Q}\Gamma)^{\neq}\big)$. Thus (Γ, ψ) is grounded iff $(\mathbb{Q}\Gamma, \psi)$ is grounded. Using Lemma 9.2.3, we also get:

COROLLARY 9.2.8. $(\Gamma^{\neq})' = ((\mathbb{Q}\Gamma)^{\neq})' \cap \Gamma.$

We say that (Γ, ψ) **has small derivation** if $\alpha' > 0$ for all $\alpha > 0$. So an asymptotic field has small derivation iff its asymptotic couple has small derivation.

LEMMA 9.2.9. *(Γ, ψ) has small derivation iff there is no $\gamma < 0$ with $\Psi \leqslant \gamma$.*

PROOF. Suppose $\Psi \leqslant \gamma < 0$. For $\alpha < 0$ we have $\alpha' < \alpha^\dagger \leqslant \gamma$. Thus $\gamma, 0 \notin (\Gamma^<)'$, and hence γ or 0 is in $(\Gamma^>)'$ by Theorem 9.2.1. In each case (Γ, ψ) does not have small derivation. Conversely, suppose (Γ, ψ) does not have small derivation. If $\Gamma^>$ has a least element ε, then $\min(\Gamma^>)' = \varepsilon'$ and so $\Psi < \varepsilon' \leqslant 0$, hence $\Psi \leqslant -\varepsilon$. If $\Gamma^>$ does not have a least element, we take any $\alpha > 0$ with $\alpha' \leqslant 0$, and then β with $0 < \beta < \alpha$ satisfies $\Psi \leqslant \gamma := \beta' < 0$. □

LEMMA 9.2.10. *Suppose that (Γ, ψ) has small derivation. Then:*

(i) $(\mathbb{Q}\Gamma, \psi)$ *has small derivation;*

(ii) $\gamma^\dagger \leqslant \gamma \Rightarrow \gamma^\dagger > -\gamma/n$ *for all* $n \geqslant 1$*;*

(iii) $\gamma^\dagger > \gamma \Rightarrow \gamma^\dagger > \gamma/n$ *for all* $n \geqslant 1$*;*

(iv) $\gamma^\dagger \leqslant 0 \Rightarrow \gamma^\dagger = o(\gamma)$.

PROOF. If $\Gamma^>$ has no least element, then $\Gamma^>$ is coinitial in $(\mathbb{Q}\Gamma)^>$, therefore $(\Gamma^>)'$ is coinitial in $((\mathbb{Q}\Gamma)^>)'$, which yields (i). If $\Gamma^>$ has a least element ε, then $\varepsilon' \geqslant \varepsilon > 0$, and thus $((\mathbb{Q}\Gamma)^>)' > \varepsilon^\dagger \geqslant 0$.

Let $n \geqslant 1$ in the rest of the proof. Suppose that $\gamma^\dagger \leqslant \gamma$. Then

$$(-\gamma)' = -\gamma + \gamma^\dagger \leqslant 0,$$

so $\gamma > 0$. Applying (i) yields $(\gamma/n) + \gamma^\dagger = (\gamma/n) + (\gamma/n)^\dagger > 0$, which yields (ii). For (iii), assume $\gamma^\dagger > \gamma$. If $\gamma < 0$, then $(-\gamma/n) + \gamma^\dagger = (-\gamma/n) + (-\gamma/n)^\dagger > 0$ by (i), so $\gamma^\dagger > \gamma/n$. If $\gamma \geqslant 0$, then $\gamma^\dagger > \gamma \geqslant \gamma/n$. This proves (iii). Part (iv) follows from (ii) by taking $\gamma > 0$. □

LEMMA 9.2.11. *Let (Γ, ψ) be an ungrounded H-asymptotic couple, and let $\alpha \in \Psi^\downarrow$. Then there are $\gamma_0 \in \Psi^{>\alpha}$ and $\delta_0 \in (\Gamma^>)'$ such that the map*

$$\gamma \mapsto \psi(\gamma - \alpha) : \Gamma \to \Gamma_\infty$$

is constant on the set $[\gamma_0, \delta_0] := \{\gamma : \gamma_0 \leqslant \gamma \leqslant \delta_0\}$, with value $> \alpha$.

PROOF. Take $\beta_0 \in \Gamma^>$ so small that $\psi(\beta_0) > \alpha$ and $[\beta_0] < [\psi(\beta_0) - \alpha]$, and set $\gamma_0 := \psi(\beta_0)$ and $\delta_0 := \beta_0'$. Then

$$[\gamma_0 - \alpha] = [\psi(\beta_0) - \alpha] = [\beta_0 + \psi(\beta_0) - \alpha] = [\delta_0 - \alpha],$$

so $\psi(\gamma_0 - \alpha) = \psi(\delta_0 - \alpha)$. Since the map $\gamma \mapsto \psi(\gamma - \alpha) \colon \Gamma^{>\alpha} \to \Gamma$ is decreasing, γ_0 and δ_0 have the desired property. The map takes value $> \alpha$ by Lemma 6.5.4(i). □

REMARK. With (Γ, ψ) and α, γ_0, δ_0 as in Lemma 9.2.11 and any H-asymptotic couple (Γ_1, ψ_1) extending (Γ, ψ), the map $\gamma_1 \mapsto \psi_1(\gamma_1 - \alpha)$ is constant on $[\gamma_0, \delta_0]_{\Gamma_1}$.

We let $\psi(* - \alpha)$ denote the constant value of $\psi(\gamma - \alpha)$ for $\gamma \in [\gamma_0, \delta_0]$ in Lemma 9.2.11; it does not depend on the choice of γ_0, δ_0 in that lemma.

COROLLARY 9.2.12. *Let* (Γ, ψ) *be as in Lemma 9.2.11. If* $\alpha \in \Psi^{\downarrow}$, *then* $\alpha < \psi(*-\alpha)$. *If* $\alpha, \beta \in \Psi^{\downarrow}$ *and* $\alpha \leqslant \beta$, *then* $\psi(* - \alpha) \leqslant \psi(* - \beta)$.

At one point in Section 9.9 we also need a variant of the above for $\alpha \in (\Gamma^{>})'$:

LEMMA 9.2.13. *Let* (Γ, ψ) *be an ungrounded* H*-asymptotic couple and* $\alpha \in (\Gamma^{>})'$. *Then there are* $\gamma_0 \in \Psi$ *and* $\delta_0 \in (\Gamma^{>})'$ *with* $\delta_0 < \alpha$ *such that the map*

$$\gamma \mapsto \psi(\gamma - \alpha) : \Gamma \to \Gamma_{\infty}$$

is constant on the set $[\gamma_0, \delta_0]$.

PROOF. Take the unique $\beta > 0$ with $\beta' = \alpha$ and set $\gamma_0 := \psi(\beta)$. Then $\gamma_0 - \alpha = -\beta$, so $\psi(\gamma_0 - \alpha) = \psi(\beta)$. Next, in $(\mathbb{Q}\Gamma, \psi)$ we have $(\beta/2)' < \beta' = \alpha$ and $(\beta/2)' - \alpha = (\beta/2) + \psi(\beta) - \alpha = -\beta/2$, so $\psi\big((\beta/2)' - \alpha\big) = \psi(\beta)$ as well. So γ_0 and any $\delta_0 = \beta_0'$ with $0 < \beta_0 \leqslant \beta/2$ and $\beta_0 \in \Gamma$ have the required property. $\square$

With (Γ, ψ) and α as in Lemma 9.2.13, we set $\psi(* - \alpha) := \psi(\beta)$ for the unique $\beta > 0$ with $\beta' = \alpha$, so $\psi(* - \alpha)$ is the constant value of the map in that lemma.

Steady functions and slow functions on H-asymptotic couples

In this subsection (Γ, ψ) *is of* H*-type,* and α, β, γ range over Γ. Our first goal is to show:

LEMMA 9.2.14. *The functions*

$$\gamma \mapsto \gamma' : \Gamma^{>} \to \Gamma, \qquad \gamma \mapsto \gamma' : \Gamma^{<} \to \Gamma$$

have the intermediate value property. (Recall from Lemma 6.5.4 that these functions are strictly increasing.)

The proof is based on the results about steady and slow functions from Section 2.4. From this section we recall some terminology. Let $v\colon \Gamma \to S_{\infty}$ be a convex valuation on the ordered abelian group Γ. Then $v\alpha > v\beta \;\Rightarrow\; \alpha = o(\beta)$. We can replace here "$\Rightarrow$" by "$\Longleftrightarrow$" by taking for v the standard valuation $\gamma \mapsto [\gamma] \colon \Gamma \to [\Gamma]$, which assigns to each γ its archimedean class $[\gamma]$, with the reversed natural ordering on $[\Gamma]$ so as to make $[0]$ its largest element. Also important for us is the convex valuation $\psi\colon \Gamma \to \Gamma_{\infty}$. We let $o_v(\beta)$ stand for any element α with $v\alpha > v\beta$.

Let U be a nonempty convex subset of Γ. In Section 2.4 we defined a function $i\colon U \to \Gamma$ to be v-steady if i has the intermediate value property and $i(x) - i(y) = x - y + o_v(x - y)$ for all distinct $x, y \in U$. We also defined a function $\eta\colon U \to \Gamma$ to be v-slow on the right if for all $x, y, z \in U$,

(s1) $\eta(x) - \eta(y) = o_v(x - y)$ if $x \neq y$;

(s2) $\eta(y) = \eta(z)$ if $x < y < z$ and $z - y = o_v(z - x)$.

In the same way we defined $\eta\colon U \to G$ to be v-slow on the left, except that in clause (s2) we replace "$x < y < z$" by "$x > y > z$." By Lemma 2.4.9, the sum of a v-steady function $U \to \Gamma$ and a v-slow (on the right or on the left) function $U \to \Gamma$ is v-steady.

PROOF OF LEMMA 9.2.14. With v the standard valuation, the identity function on $\Gamma^{>}$ is clearly v-steady. Also, the restriction of ψ to $\Gamma^{>}$ is v-slow on the right: part (ii) of Lemma 6.5.4 shows that (s1) is satisfied, and the H-type assumption implies easily that (s2) is satisfied. Thus by Lemma 2.4.9 the map

$$\gamma \mapsto \gamma' = \gamma + \psi(\gamma) : \Gamma^{>} \to \Gamma$$

is v-steady; in particular, it has the intermediate value property. The other part of Lemma 9.2.14 follows in the same way, with $\Gamma^{<}$ and "v-slow on the left" in place of $\Gamma^{>}$ and "v-slow on the right." □

LEMMA 9.2.15. *For all γ we have: $\gamma \in (\Gamma^{<})' \iff \gamma < \psi(\alpha)$ for some $\alpha > 0$. If $\Psi^{>0} \neq \emptyset$, then there is a unique element $1 \in \Gamma^{>}$ with $\psi(1) = 1$.*

PROOF. If $\gamma = (-\alpha)'$, $\alpha > 0$, then $\gamma = \psi(\alpha) - \alpha < \psi(\alpha)$. Thus the forward direction of the equivalence holds. The other direction holds trivially if $\Gamma = (\Gamma^{\neq})'$, so assume $\Gamma \neq (\Gamma^{\neq})'$. Then Theorem 9.2.1 gives $\Gamma \setminus (\Gamma^{\neq})' = \{\beta\}$. By Corollary 9.2.6 and Lemma 9.2.14, $(\Gamma^{<})'$ is downward closed in Γ and $(\Gamma^{>})'$ upward closed in Γ, so $(\Gamma^{<})' = \Gamma^{<\beta}$ and $(\Gamma^{>})' = \Gamma^{>\beta}$. Since $(\Gamma^{<})' \subseteq \Psi^{\downarrow} < (\Gamma^{>})'$, this yields $\Psi \leqslant \beta$. So if $\gamma < \psi(\alpha)$ for some $\alpha > 0$, then $\gamma < \beta$ and so $\gamma \in (\Gamma^{<})'$.

Suppose now that $\Psi^{>0} \neq \emptyset$. Then $0 \in (\Gamma^{<})'$, so there is some $1 \in \Gamma^{>}$ such that $0 = (-1)'$, that is, $\psi(1) = 1$; uniqueness of 1 follows from Lemma 6.5.4(iii). □

If $\Psi^{>0} \neq \emptyset$, then the element 1 as in Lemma 9.2.15 serves as a unit of length, and we identify $\mathbb{Z}$ with the subgroup $\mathbb{Z} \cdot 1$ of Γ via $k \mapsto k \cdot 1$.

COROLLARY 9.2.16. *(Γ, ψ) has exactly one of the following three properties:*

(i) *there is β such that $\Psi < \beta < (\Gamma^{>})'$, that is, (Γ, ψ) has a gap;*

(ii) *Ψ has a largest element, that is, (Γ, ψ) is grounded;*

(iii) *$\Gamma = (\Gamma^{\neq})'$, that is, (Γ, ψ) has asymptotic integration.*

PROOF. We know from Lemma 9.2.4 that (i) and (ii) exclude each other. Also, β as in (i) cannot be in $(\Gamma^{<})'$, since $(\Gamma^{<})' \subseteq \Psi^{\downarrow}$. If Ψ has a largest element, this element is not in $(\Gamma^{\neq})'$, by Theorem 9.2.1. Thus (i) as well as (ii) excludes (iii). If we are not in case (i) or (ii), then we are in case (iii) by Lemma 9.2.15. □

The order in which the three possibilities are listed is natural: The trivial H-asymptotic couple with $\Gamma = \{0\}$ falls under (i), with $\beta = 0$. When Γ is not trivial, but divisible and of finite dimension as a vector space over $\mathbb{Q}$, then Ψ is nonempty and finite, and so we are in case (ii). Case (iii) typically requires a more infinitary construction: for example, the asymptotic couple of $\mathbb{T}$ falls under (iii). In all cases the set Ψ has an upper bound in Γ, with $\sup\Psi = \beta$ in case (i), but the set Ψ has no supremum in Γ in case (iii).

In view of Corollary 9.2.8 it follows that a gap in (Γ,ψ) remains a gap in $(\mathbb{Q}\Gamma,\psi)$, and that if $(\mathbb{Q}\Gamma,\psi)$ has asymptotic integration, then so does (Γ,ψ). But there are (Γ,ψ) with asymptotic integration such that $(\mathbb{Q}\Gamma,\psi)$ does not have asymptotic integration; see [20, Example 12.9]. This situation needs to be dealt with in parts of Section 16.3 where the next lemma will be used. This lemma concerns the special role of the set $2\Psi = \{2\psi(\gamma) : \gamma \neq 0\}$ in that section.

LEMMA 9.2.17. *Suppose (Γ,ψ) has asymptotic integration, and let γ be given. Then the following conditions are equivalent:*

(i) *γ is a supremum of 2Ψ in the ordered set Γ;*

(ii) *$\gamma > 2\Psi$, and there are no $\alpha,\beta > \Psi$ with $\alpha+\beta=\gamma$;*

(iii) *$\frac{1}{2}\gamma$ is a gap in $(\mathbb{Q}\Gamma,\psi)$;*

(iv) *$\frac{1}{2}\gamma$ is a supremum of Ψ in the ordered set $\mathbb{Q}\Gamma$.*

PROOF. Assume $\gamma = \sup 2\Psi$, and $\alpha,\beta > \Psi$ satisfy $\alpha+\beta=\gamma$. Take $\alpha_1 \in \Gamma$ with $\Psi < \alpha_1 < \alpha$. Then $2\Psi < \alpha_1+\beta < \gamma$, contradicting (i). This gives (i) $\Rightarrow$ (ii).

To get (ii) $\Rightarrow$ (iii), we prove the contrapositive. So assume $\frac{1}{2}\gamma$ is not a gap in $(\mathbb{Q}\Gamma,\psi)$. Then either $\Psi \not< \frac{1}{2}\gamma$ or $\frac{1}{2}\gamma \not< \big((\mathbb{Q}\Gamma)^{>}\big)'$. If $\Psi \not< \frac{1}{2}\gamma$, then $\gamma \not> 2\Psi$. Suppose $\frac{1}{2}\gamma \not< \big((\mathbb{Q}\Gamma)^{>}\big)'$. Since $\Gamma^{>}$ is coinitial in $(\mathbb{Q}\Gamma)^{>}$, we have $\delta \in \Gamma^{>}$ with $\frac{1}{2}\gamma \geqslant \delta'$. Then $\gamma - \delta' \geqslant \frac{1}{2}\gamma > \Psi$, so $\alpha := \delta'$ and $\beta := \gamma - \alpha$ satisfy $\alpha,\beta > \Psi$ and $\alpha+\beta=\gamma$.

Lemma 9.2.4 gives (iii) $\Rightarrow$ (iv), and (iv) $\Rightarrow$ (i) is clear. □

The H-asymptotic couple (Γ,ψ) is said to have **rational asymptotic integration** if $(\mathbb{Q}\Gamma,\psi)$ has asymptotic integration, and an H-asymptotic field is said to have rational asymptotic integration if its asymptotic couple has rational asymptotic integration.

Contraction

In this subsection (Γ,ψ) is H-asymptotic and ungrounded, and we let α, β, γ range over Γ. So we are in case (i) or case (iii) of Corollary 9.2.16. By Lemma 9.2.15 we can define the so-called contraction map $\chi\colon \Gamma^{<} \to \Gamma^{<}$ by $\chi(\alpha)' = \alpha^{\dagger}$. It has the following basic properties:

LEMMA 9.2.18. *Let $\alpha,\beta < 0$. Then:*

(i) *the map χ is increasing: $\alpha < \beta \implies \chi(\alpha) \leqslant \chi(\beta)$;*

(ii) $\alpha^{\dagger} = \beta^{\dagger} \implies \chi(\alpha) = \chi(\beta)$;

(iii) *$\chi(\alpha) = o_{\psi}(\alpha)$, and thus $\chi(\alpha) = o(\alpha)$;*

(iv) $\alpha \neq \beta \implies \chi(\alpha) - \chi(\beta) = o(\alpha - \beta)$.

Moreover, χ, while defined in terms of ψ, does not change upon replacing ψ by a shift $\psi+\gamma$. If (Γ,ψ) has small derivation, then

$$\alpha^\dagger<0,\ \psi(\alpha^\dagger)<0\ \Rightarrow\ \alpha^\dagger=o(\alpha),\ \psi(\alpha^\dagger)=o(\alpha^\dagger),\ \chi(\alpha)=\alpha^\dagger-\psi(\alpha^\dagger).$$

PROOF. Properties (i) and (ii) hold because the map $\alpha\mapsto\alpha'$ is strictly increasing and the map $\alpha\mapsto\alpha^\dagger$ is increasing (where $\alpha<0$). To get (iii), we note that

$$\chi(\alpha)<0,\qquad \chi(\alpha)'\ =\ \chi(\alpha)+\chi(\alpha)^\dagger\ =\ \alpha^\dagger,$$

so $\chi(\alpha)^\dagger>\alpha^\dagger$. As to (iv), this follows from (ii) if $\alpha^\dagger=\beta^\dagger$. If $\alpha^\dagger<\beta^\dagger$, then $\beta=o(\alpha)$, so $\chi(\alpha)=o(\alpha-\beta)$ and $\chi(\beta)=o(\alpha-\beta)$ by (iii), so $\chi(\alpha)-\chi(\beta)=o(\alpha-\beta)$.

The invariance of χ under shifts follows easily from the definition of χ. The last statement is easily deduced from Lemma 9.2.10(iv). □

LEMMA 9.2.19. *Let Δ be a subgroup of Γ with $\psi(\Delta^{\neq})\subseteq\Delta$, and suppose $\alpha\in\Delta^{<}$ is such that $\alpha^\dagger$ is not maximal in $\psi(\Delta^{\neq})$. Then $\chi(\alpha)\in\Delta$.*

PROOF. By the assumptions on Δ and α we can take $\beta\in\Delta^{<}$ such that $\beta'=\alpha^\dagger$. Then $\beta=\chi(\alpha)$. □

The next result will soon be needed in Section 9.3.

LEMMA 9.2.20. *Assume $\Gamma\neq\{0\}$ and $d,e_0,e_1,\dots,e_n\in\mathbb{Z}$ are not all 0, and let $\alpha\in\Gamma$. Then there exists a $\beta<0$ such that:*

$$\begin{aligned}&\textit{either for all } \gamma\in(\beta,0),\ \alpha+d\psi(\gamma)+e_0\gamma+e_1\chi(\gamma)+\cdots+e_n\chi^n(\gamma)\ <\ 0,\\ &\textit{or for all } \gamma\in(\beta,0),\ \alpha+d\psi(\gamma)+e_0\gamma+e_1\chi(\gamma)+\cdots+e_n\chi^n(\gamma)\ >\ 0.\end{aligned}$$

PROOF. Since $\Gamma^{<}$ is cofinal in $(\mathbb{Q}\Gamma)^{<}$ we can pass to the divisible hull and arrange that Γ is divisible. Consider first the case that (Γ,ψ) has asymptotic integration.

SUBCASE 1: $d\geqslant 1$. Then $\alpha+d\psi(\gamma)$ is increasing as a function of $\gamma<0$, and $\{\alpha+d\psi(\gamma):\gamma<0\}$ has no supremum in Γ (using divisibility of Γ). Thus we have $\beta\in\Gamma^{<}$ and $\varepsilon\in\Gamma^{>}$ such that either $\alpha+d\psi(\gamma)<-\varepsilon$ for all $\gamma\in(\beta,0)$, or $\alpha+d\psi(\gamma)>\varepsilon$ for all $\gamma\in(\beta,0)$. By increasing β if necessary, we have also $|e_0\gamma+e_1\chi(\gamma)+\cdots+e_n\chi^n(\gamma)|<\varepsilon/2$ for all $\gamma\in(\beta,0)$, and so this β does the job.

SUBCASE 2: $d\leqslant -1$. This follows by symmetry from Subcase 1.

SUBCASE 3: $d=0$. If $\alpha\neq 0$, then the desired result follows since for all $\gamma<0$,

$$|e_0\gamma+e_1\chi(\gamma)+\cdots+e_n\chi^n(\gamma)|\ <\ \big(|e_0|+1\big)|\gamma|.$$

If $\alpha=0$, then we take i least with $e_i\neq 0$ and use that $e_i\chi^i(\gamma)$ is the dominant term in the sum $e_0\gamma+e_1\chi(\gamma)+\cdots+e_n\chi^n(\gamma)$ for $\gamma<0$.

This concludes the asymptotic integration case. Before we continue, we make a general observation: for $\delta\in\Gamma$ the shift $(\Gamma,\psi-\delta)$ has the same χ-map as (Γ,ψ), and $\alpha+d\psi(\gamma)=(\alpha+d\delta)+d(\psi-\delta)(\gamma)$ for $\gamma<0$, so we may replace (Γ,ψ) by its shift $(\Gamma, \psi - \delta)$, keeping the same $d, e_0, \dots, e_n$ and replacing α by $\alpha + d\delta$. The remaining case is that Ψ has a supremum in Γ. Then we arrange by shifting that this supremum is 0, so $\sup \Psi = 0 \notin \Psi$. If $\alpha \neq 0$, the desired result follows from

$$|d\psi(\gamma) + e_0\gamma + e_1\chi(\gamma) + \dots + e_n\chi^n(\gamma)| \ < \ \big(|e_0| + 1\big)|\gamma| \qquad (\gamma < 0),$$

which in turn depends on Lemma 9.2.10(iv). Suppose $\alpha = 0$. If also $d = 0$, then we reach the desired result by arguing as in Subcase 3 above, so assume $d \neq 0$. If $e_0 \neq 0$, then we use that $e_0\gamma$ is the dominant term in the sum

$$d\psi(\gamma) + e_0\gamma + e_1\chi(\gamma) + \dots + e_n\chi^n(\gamma).$$

So we can assume $\alpha = 0$, $d \neq 0$, $e_0 = 0$. In this case we proceed by induction on n. If $n = 0$, the desired result holds, so let $n \geqslant 1$. Using $\psi(\gamma) = \chi(\gamma) + \psi\big(\chi(\gamma)\big)$ and setting $\gamma^* := \chi(\gamma)$ for $\gamma < 0$, we have

$$d\psi(\gamma) + e_1\chi(\gamma) + \dots + e_n\chi^n(\gamma) \ = \ d\psi(\gamma^*) + (d + e_1)\gamma^* + \sum_{i=1}^{n-1} e_{i+1}\chi^i(\gamma^*),$$

and so a suitable inductive hypothesis takes care of this. □

By Lemma 9.2.15 the image of the strictly increasing map $\gamma \mapsto \gamma + \psi(\gamma)\colon \Gamma^< \to \Gamma$ is a cofinal subset of $\Psi^\downarrow$. In contrast to this, we have:

LEMMA 9.2.21. *Suppose (Γ, ψ) has asymptotic integration and Γ is divisible. Let $\alpha \in \Gamma$, $d_0 \in \mathbb{N}^{\geqslant 1}$, $d \in \mathbb{N}^{\geqslant 2}$ and the map $i\colon \Gamma^< \to \Gamma$ be such that for all $\gamma < 0$,*

$$i(\gamma) \ = \ \alpha + d_0\,\gamma + d\,\psi(\gamma) + \varepsilon(\gamma) \quad \textit{with } \varepsilon(\gamma) = o(\gamma),$$

and i is increasing. Then there are $\beta < 0$ and $\gamma_0 \in \Psi$ and $\delta_0 \in (\Gamma^>)'$ such that the i-image of the interval $(\beta, 0)$ is disjoint from the interval (γ_0, δ_0).

PROOF. If $i(\beta) > \Psi$ for some $\beta < 0$, we take such β such that $i(\beta) \in (\Gamma^>)'$, and then the lemma holds with $\delta_0 := i(\beta)$ and any $\gamma_0 \in \Psi$. So we can assume that $i(\Gamma^<) \subseteq \Psi^\downarrow$, and it remains to show that $i(\Gamma^<)$ is not cofinal in $\Psi^\downarrow$. Towards a contradiction, suppose $i(\Gamma^<)$ is cofinal in $\Psi^\downarrow$. Pick $\alpha_0 \in \Gamma$ such that for all $\gamma < 0$,

$$i(\gamma) - \alpha_0 \ = \ d_0\gamma + d(\psi(\gamma) - \alpha_0) + \varepsilon(\gamma),$$

in other words, $(1 - d)\alpha_0 = \alpha$. Replacing (Γ, ψ) by its shift $(\Gamma, \psi - \alpha_0)$ and i by the map $i - \alpha_0$, we arrange that $\alpha = 0$. We now distinguish two cases.

CASE 1: *(Γ, ψ) has small derivation.* Then $\Psi \cap \Gamma^> \neq \emptyset$, so we have $1 \in \Gamma^>$ with $\psi(1) = 1$. Identifying $\mathbb{Q}$ with its image under $r \mapsto r \cdot 1\colon \mathbb{Q} \to \Gamma$, we get $\frac{3}{2} = \frac{1}{2} + \psi\left(\frac{1}{2}\right) > \Psi$. Let $-1 \leqslant \gamma < 0$. Then $\psi(\gamma) \geqslant \psi(-1) = 1$ and $\varepsilon(\gamma) > \gamma$, hence

$$i(\gamma) \ = \ d_0\,\gamma + d\,\psi(\gamma) + \varepsilon(\gamma) \ > \ (d_0 + 1)\gamma + d.$$

Taking in addition $\gamma \geqslant \frac{1}{d_0+1}\left(\frac{3}{2}-d\right)$, we get $i(\gamma) > \left(\frac{3}{2}-d\right)+d = \frac{3}{2}$, so $i(\gamma) \notin \Psi^{\downarrow}$, and we have a contradiction.

CASE 2: (Γ,ψ) *does not have small derivation.* Then by Lemma 9.2.9 we can take $\delta > 0$ such that $\Psi \leqslant -\delta$. For $\gamma < 0$ we have

$$\begin{aligned} i(\gamma) &= \psi(\gamma) + \big(d_0\gamma + (d-1)\psi(\gamma) + \varepsilon(\gamma)\big), \text{ and} \\ d_0\gamma + (d-1)\psi(\gamma) + \varepsilon(\gamma) &\leqslant \psi(\gamma) \leqslant -\delta, \end{aligned}$$

so the i-image of $\Gamma^{<}$ is contained in $(\Psi-\delta)^{\downarrow}$, which is contained $\Psi^{\downarrow}$ but not cofinal in it, since $\psi(\delta) > \Psi - \delta$. □

COROLLARY 9.2.22. *Suppose (Γ,ψ) has rational asymptotic integration. Let $\alpha \in \Gamma$ and $d_0 \in \mathbb{N}^{\geqslant 1}$, $d \in \mathbb{N}^{\geqslant 2}$, $e_1,\dots,e_n \in \mathbb{Z}$. Let $i\colon \Gamma^{<} \to \Gamma$ be given by*

$$i(\gamma) = \alpha + d_0\,\gamma + d\psi(\gamma) + \sum_{i=1}^{n} e_i\,\chi^i(\gamma). \tag{9.2.1}$$

Then i is strictly increasing, and there are $\beta < 0$ and $\gamma_0 \in \Psi$ and $\delta_0 \in (\Gamma^{>})'$ such that the i-image of the interval $(\beta,0)$ is disjoint from the interval (γ_0,δ_0).

PROOF. By passing to the divisible hull of Γ and extending i according to (9.2.1), we arrange that Γ is divisible. Clearly i is strictly increasing by Lemma 9.2.18(iv). Thus Lemma 9.2.21 applies. □

Some facts about ψ-functions

In this subsection (Γ,ψ) is of H-type with small derivation, $\Gamma \neq \{0\}$ and Γ is divisible; α, β, γ range over Γ. We set

$$\Gamma_\psi := \{\gamma : \psi^n(\gamma) < 0 \text{ for all } n \geqslant 1\},$$

so $\psi(\Gamma_\psi) \subseteq \Gamma_\psi^{<0} = -\Gamma_\psi^{>0}$, and $\Gamma_\psi^{<0}$ is downward closed: if $\alpha \leqslant \beta \in \Gamma_\psi^{<0}$, then $\psi(\alpha) \leqslant \psi(\beta)$, and so by induction $\alpha \in \Gamma_\psi^{<0}$. In particular, $\Gamma_\psi^{<0}$ is a convex (possibly empty) subset of $\Gamma^{<}$. Note also that for $\gamma \in \Gamma_\psi$ we have

$$\psi^{n+1}(\gamma) = o\big(\psi^n(\gamma)\big).$$

Let α be given as well as natural numbers r and $d, d_1 \dots, d_r$. Then we have a function $i = i_{\alpha,d,d_1,\dots,d_r} : \Gamma_\psi \to \Gamma$,

$$i(\gamma) := \alpha + d\gamma + d_1\psi(\gamma) + \cdots + d_r\psi^r(\gamma).$$

We call i a **ψ-function of slope d and order r**, and also a **ψ-function of type (α,d,r)** if we want to specify α. By part (i) of Lemma 6.5.4, if α, β are distinct and nonzero, then $\psi\big(\psi(\alpha)-\psi(\beta)\big) > \psi(\alpha-\beta)$, that is,

$$\psi(\alpha) - \psi(\beta) = o_\psi(\alpha-\beta).$$

It follows that the function

$$\gamma \mapsto d_1\psi(\gamma) + \cdots + d_r\psi^r(\gamma) : \Gamma_\psi \to \Gamma$$

is ψ-slow on the left when restricted to $\Gamma_\psi^{<0}$ and ψ-slow on the right when restricted to $\Gamma_\psi^{>0}$. Also, if $d \geqslant 1$, then for all distinct $x, y \in \Gamma_\psi$,

$$i(x) - i(y) \;=\; d(x-y) + o_\psi(x-y)$$

and so i is strictly increasing, and i has the intermediate value property on $\Gamma_\psi^{<0}$ as well as on $\Gamma_\psi^{>0}$. The next result is needed in Section 9.3.

LEMMA 9.2.23. *Let i, j be ψ-functions of slopes $d > e$. Then either $i < j$ on $\Gamma_\psi^{<0}$, or there is a unique $\beta \in \Gamma_\psi^{<0}$ such that $i(\beta) = j(\beta)$; in the latter case, $i(\gamma) > j(\gamma)$ if $\beta < \gamma \in \Gamma_\psi^{<0}$. Likewise, either $i > j$ on $\Gamma_\psi^{>0}$, or there is a unique $\beta \in \Gamma_\psi^{>0}$ such that $i(\beta) = j(\beta)$; in the latter case, $i(\gamma) < j(\gamma)$ if $\beta > \gamma \in \Gamma_\psi^{>0}$.*

PROOF. Use that $i - j$ is strictly increasing, and that it has the intermediate value property on $\Gamma_\psi^{<0}$ as well as on $\Gamma_\psi^{>0}$. □

Coarsening

Given also an asymptotic couple (Γ_1, ψ_1), a **morphism**

$$h\colon (\Gamma, \psi) \to (\Gamma_1, \psi_1)$$

is a $\leqslant$-preserving morphism $h\colon \Gamma \to \Gamma_1$ of abelian groups such that

$$h\big(\psi(\gamma)\big) = \psi_1\big(h(\gamma)\big) \text{ for all } \gamma \in \Gamma \setminus \ker h.$$

Let Δ be a convex subgroup of Γ. Then we have the ordered quotient group $\dot\Gamma := \Gamma/\Delta$: if $\gamma \geqslant 0$ in Γ, then $\dot\gamma := \gamma + \Delta \geqslant 0$ in $\dot\Gamma$. Lemma 9.2.24 below yields an asymptotic couple $(\dot\Gamma, \dot\psi)$.

LEMMA 9.2.24. *There is a unique map $\dot\psi = \psi_\Delta\colon \dot\Gamma^{\neq} \to \dot\Gamma$ such that $(\dot\Gamma, \dot\psi)$ is an asymptotic couple and $\gamma \mapsto \dot\gamma\colon \Gamma \to \dot\Gamma$ is a morphism $(\Gamma, \psi) \to (\dot\Gamma, \dot\psi)$. It is given by*

$$\dot\psi(\dot\gamma) \;=\; \psi(\gamma) + \Delta \qquad \textit{for } \gamma \in \Gamma \setminus \Delta.$$

If $\Delta \neq \{0\}$, then $\Psi_\Delta := \dot\psi(\dot\Gamma^{\neq})$ has supremum $\psi(\delta) + \Delta$, where $\delta \in \Delta^{\neq}$ is arbitrary. If (Γ, ψ) has small derivation, then $(\dot\Gamma, \dot\psi)$ has small derivation. If (Γ, ψ) is H-asymptotic, then $(\dot\Gamma, \dot\psi)$ is H-asymptotic.

PROOF. It suffices to prove existence. In what follows, let $\alpha, \beta \in \Gamma^{\neq}$ and $\alpha \neq \beta$; then by part (ii) of Lemma 6.5.4, we have $\big[\psi(\alpha) - \psi(\beta)\big] < [\alpha - \beta]$. Hence, if $\alpha - \beta \in \Delta$, then $\psi(\alpha) - \psi(\beta) \in \Delta$. Therefore we have a map

$$\dot\psi\colon \dot\Gamma^{\neq} \to \dot\Gamma, \quad \dot\gamma \mapsto \psi(\gamma) + \Delta \quad (\gamma \in \Gamma \setminus \Delta).$$

If $\beta-\alpha > \Delta$, then $\beta'-\alpha' = (\beta+\psi(\beta))-(\alpha+\psi(\alpha)) > \Delta$, and thus by Lemma 6.5.6, $(\dot{\Gamma}, \dot{\psi})$ is an asymptotic couple. To prove the claim about Ψ_Δ, assume in addition that $\alpha \in \Delta$, $\alpha > 0$, and $\beta > \Delta$. Then $(\beta + \psi(\beta)) - \psi(\alpha) > \Delta$, so in $\dot{\Gamma}$,

$$\psi(\alpha) + \Delta \; < \; \beta' + \Delta \; = \; (\beta + \Delta)'.$$

Moreover, $\psi(\beta) < \alpha + \psi(\alpha) \in \psi(\alpha) + \Delta$, so $\Psi_\Delta \leqslant \psi(\alpha) + \Delta$.

If (Γ, ψ) has small derivation, then $\alpha > \Delta$ implies $\alpha' = \alpha + \alpha^\dagger > \Delta$ by Lemma 9.2.10(iv), and so $(\dot{\Gamma}, \dot{\psi})$ has small derivation. □

In connection with coarsening we sometimes use the following lemma.

LEMMA 9.2.25. *Assume $\Delta \neq \{0\}$. The following conditions are equivalent:*

(i) $\psi(\Delta^{\neq}) \cap \Delta \neq \emptyset$;

(ii) $\psi(\Delta^{\neq}) \subseteq \Delta$;

(iii) $(\Delta^{\neq})' \cap \Delta \neq \emptyset$;

(iv) $(\Delta^{\neq})' \subseteq \Delta$.

If (Γ, ψ) is of H-type and (i) *holds, then $\Psi^{>0} \subseteq \Delta$.*

PROOF. Let $\delta \in \Delta^{\neq}$ be such that $\delta^\dagger \in \Delta$. Then we have for $\delta_1 \in \Delta^{\neq}$:

$$|\delta^\dagger - \delta_1^\dagger| \leqslant |\delta - \delta_1| \in \Delta,$$

so $\delta_1^\dagger \in \Delta$. The equivalences easily follow. Now assume (Γ, ψ) is of H-type, (i) holds, and $\Psi^{>0} \neq \emptyset$. Then $(\Delta, \psi|\Delta^{\neq})$ is an H-asymptotic couple with $\psi(\delta) > 0$ for some $\delta \in \Delta^{\neq}$, so by Lemma 9.2.15 the unique element $1 \in \Gamma^{>}$ with $\psi(1) = 1$ lies in Δ. Then $\Psi^{>0} < 1 + 1 \in \Delta$, and thus $\Psi^{>0} \subseteq \Delta$. □

Now let K be an asymptotic field with valuation v and asymptotic couple (Γ, ψ). The convex subgroup Δ of Γ then leads to the coarsening of v by Δ:

$$\dot{v} = v_\Delta\colon\; K^\times \to \dot{\Gamma} = \dot{v}(K^\times), \qquad \dot{v}(f) := v(f) + \Delta.$$

The dominance relation on K corresponding to the coarsened valuation $\dot{v}$ is denoted by $\dot{\preccurlyeq}$ or $\preccurlyeq_\Delta$, so $f \preccurlyeq g \Rightarrow f \mathrel{\dot{\preccurlyeq}} g$ for $f, g \in K$. (See Section 3.4.) Let $(K, \dot{\preccurlyeq})$ be the valued differential field K with v replaced by $\dot{v}$.

COROLLARY 9.2.26. *The valued differential field $(K, \dot{\preccurlyeq})$ has the following properties:*

(i) *$(K, \dot{\preccurlyeq})$ is an asymptotic field with asymptotic couple $(\dot{\Gamma}, \dot{\psi})$;*

(ii) *if $a, b \in K$, $a, b \not\asymp 1$, then $a \mathrel{\dot{\prec}} b \iff a' \mathrel{\dot{\prec}} b'$.*

PROOF. For (i), use Lemma 9.2.24 and Proposition 9.1.3. For (ii), let $a, b \in K$ and $a, b \not\asymp 1$. Then $a \prec b \iff a' \prec b'$. Set $\alpha = va$, $\beta = vb$, so $va' = \alpha + \alpha^\dagger$ and $vb' = \beta + \beta^\dagger$. Now Lemma 6.5.4(ii) gives $a \mathrel{\dot{\prec}} b \iff a' \mathrel{\dot{\prec}} b'$. □

Notes and comments

Lemmas 9.2.20, 9.2.21, Corollary 9.2.22, and the results on ψ-functions are new, but the rest of this section is essentially taken from [18, 19, 20, 363]. (For example, Theorem 9.2.1 is from [363], but its first part only for well-ordered Ψ. In the above generality, it is [19, Theorem 2.6].)

A more explicit version of Lemma 9.2.11 is in [146]. The H-type assumption in Lemma 9.2.14 cannot be dropped; see [19, Examples 2.8, 2.9]. Ordered abelian groups with a contraction map are studied in [232, 233] and [239, Appendix A]; see also [15, Section 5]. Constructions of d-valued fields with prescribed asymptotic couple can be found in [240, 363, 365] and [20, Section 11].

9.3 APPLICATION TO DIFFERENTIAL POLYNOMIALS

In this section K is an H-asymptotic field with divisible value group $\Gamma \neq \{0\}$. We let β and γ range over Γ. We also fix a differential polynomial $P \in K\{Y\}^{\neq}$ of order r. We shall use here the ψ-functions from Section 9.2 to describe for some $\beta < 0$ the behavior of $vP(y)$ for vy in the interval $(\beta, 0) \subseteq \Gamma$.

Special case

Here is a key step in our work:

LEMMA 9.3.1. *Assume $\partial\mathcal{o} \subseteq \mathcal{o}$. Let B be a nonempty convex subset of $\Gamma^{<}$ such that $\psi(B) \subseteq B$ (so $B \subseteq \Gamma_{\psi}^{<0}$). Then there are $\beta \in B$ and a ψ-function $i = i_{\alpha,d,d_1,\dots,d_r}$ such that $P_d \neq 0$, and for all $y \in K$ with $vy \in B^{>\beta}$:*

(i) *$P(y) \sim P_d(y)$; and $P_d(y) \succ P_e(y)$ whenever $e \in \mathbb{N}$, $e \neq d$;*

(ii) *$vP(y) = i(vy)$;*

(iii) *$d_j \leqslant d \cdot r!$ for $j = 1, \dots, r$.*

The inductive proof goes as follows: if P is inhomogeneous, consider the homogeneous parts of P; reduce the homogeneous case of positive order to that of lower order by taking the Riccati transform.

PROOF OF LEMMA 9.3.1. We proceed by induction on r. For $r = 0$ and homogeneous $P = aY^d$, $a \in K^{\times}$, the desired result holds with any $\beta \in B$ and $i = i_{\alpha,d}$, $\alpha = va$. Assume it holds whenever P (of order r) is homogeneous. Suppose now that P of order r is not necessarily homogeneous, and let $P_{d_0}, \dots, P_{d_k}$ with $d_0 < \dots < d_k$ be the nonzero homogeneous parts of P. Take $\beta \in B$ and ψ-functions $i_0, \dots, i_k$ with $i_p = i_{\alpha_p, d_p, d_{p,1}, \dots, d_{p,r}}$ and $d_{p,1}, \dots, d_{p,r} \leqslant d_p r!$ for $p = 0, \dots, k$, and

$$vP_{d_0}(y) = i_0(vy), \dots, vP_{d_k}(y) = i_k(vy) \text{ whenever } y \in K,\ vy \in B^{>\beta},$$

and such that for all distinct $p, q \in \{0, \dots, k\}$, either $i_p > i_q$ on $B^{>\beta}$, or $i_p < i_q$ on $B^{>\beta}$. This gives $p \in \{0, \dots, k\}$ such that for all $y \in K$ with $vy \in B^{>\beta}$,

$$P(y) \sim P_{d_p}(y), \qquad P_{d_p}(y) \succ P_e(y) \text{ whenever } e \in \mathbb{N},\, e \neq d_p.$$

Suppose $H \in K\{Y\}$ has order $r+1$, and is homogeneous of degree d. Put $G := \mathrm{Ri}(H)$, so G is of order r, $v(G) = v(H)$, and $\deg G \leqslant \mathrm{wt}(H) \leqslant (r+1)d$. Take β and i_p as above for G in the role of P, and note that $vH(y) = dvy + i_p\big(\psi(vy)\big)$ for all $y \in K$ with $vy \in B^{>\beta}$. Thus the lemma holds for H instead of P. □

The same induction on r shows that in Lemma 9.3.1, if $P \in C\{Y\}^{\neq}$ and P is homogeneous of degree d, then we can take i to be of type $(0, d, r)$. For later use it is important to find out more about the values of α and d in Lemma 9.3.1 for general P. The next lemma provides such information for special K and B.

LEMMA 9.3.2. *Suppose* $\sup \Psi = 0$ *and* $0 \notin \Psi$. *Set* $B := \Gamma^{<}$; *so* $\psi(B) \subseteq B$. *Then there are* $\beta < 0$ *and a* ψ*-function* $i = i_{\alpha,d,d_1,\dots,d_r}$ *with the properties described in Lemma 9.3.1 such that in addition* $\alpha = v(P)$ *and* $d = \max\{e \in \mathbb{N} : v(P) = v(P_e)\}$.

PROOF. Follow the proof of Lemma 9.3.1, taking into account that $\Gamma_\psi = \Gamma^{\neq}$, and that for any ψ-functions $i_1 = i_{\alpha_1,d,d_1,\dots,d_m}$ and $i_2 = i_{\alpha_2,e,e_1,\dots,e_n}$, if either $\alpha_1 < \alpha_2$, or $\alpha_1 = \alpha_2$ and $d > e$, then $i_1 < i_2$ on some interval $(\beta, 0)$ with $\beta < 0$. □

COROLLARY 9.3.3. *Suppose* $\sup \Psi = 0$ *and* $0 \notin \Psi$. *Assume also that* $P \succ P_0$. *Then there is* $\beta < 0$ *such that* $P(y_1) \succ P(y_2) \succ P(z)$ *whenever* $y_1, y_2, z \in K$ *and* $\beta < vy_1 < vy_2 < vz = 0$.

PROOF. Take β and $i = i_{\alpha,d,d_1,\dots,d_r}$ as in Lemma 9.3.2 above. Since $P \neq 0$ and $P \succ P_0$ we have $d \geqslant 1$, and so the desired result holds for this β. □

A useful coarsening

In order to better pin down the value of α in Lemma 9.3.1 we introduce a coarsening that gives a reduction to the case $\sup \Psi = 0 \notin \Psi$. This will be used in the proof of Proposition 11.3.4. *In this subsection we assume* $\partial\mathcal{o} \subseteq \mathcal{o}$.

Let B be a nonempty convex subset of $\Gamma^{<}$ such that $\psi(B) \subseteq B$, just as in Lemma 9.3.1 (so $B \subseteq \Gamma_\psi^{<}$). Then we have the convex subgroup

$$\Delta = \Delta(B) := \{\gamma : \psi(\gamma) > \psi(B)\}$$

of Γ. Note that also $\Delta = \{\gamma : \psi(\gamma) > B\}$. We are now going to use the coarsening $\dot{v} = v_\Delta$ of v by Δ. Recall from the end of Section 9.2 that K with $\dot{v}$ is an H-asymptotic field with asymptotic couple $(\dot{\Gamma}, \dot{\psi})$ and $\Psi_\Delta := \dot{\psi}(\dot{\Gamma}^{\neq})$.

LEMMA 9.3.4. *Let* α *be as in Lemma 9.3.1. Then* $\dot{\alpha} = \dot{v}(P)$. *Moreover, if* $P \mathrel{\dot{\succ}} P_0$, *then there is* $\beta \in B$ *such that* $P(y_1) \mathrel{\dot{\succ}} P(y_2) \mathrel{\dot{\succ}} P(z)$ *whenever* $y_1, y_2, z \in K$ *and* $vy_1, vy_2 \in B^{>\beta}$, $v(z) \in \Delta$, $y_1 \mathrel{\dot{\succ}} y_2$.

PROOF. We first show that $\Delta = \{\gamma : B < -|\gamma|\}$. If $\beta \in B$ and $\gamma \in \Delta^<$, then $\psi(\beta) < \psi(\gamma)$, so $\beta < \gamma$. This gives $B < \Delta$. Next, assume $\gamma \in \Gamma^<$ and $B < \gamma$. Then $\beta < \psi(\beta) \leqslant \psi(\gamma)$ for all $\beta \in B$, so $\gamma \in \Delta$. This proves our claim about Δ.

Note that $\dot{\Gamma} = \Gamma/\Delta$ is divisible, and $\dot{B}$ is a convex subset of $\dot{\Gamma}^<$ with $[\dot{\beta}, \dot{0}) \subseteq \dot{B}$ for all $\beta \in B$. So for $\beta \in B$ we have $\beta < \psi(\beta) \in B$, so $\dot{\beta} \leqslant \dot{\psi}(\dot{\beta}) = \psi(\beta) + \Delta < \dot{0}$. Thus $\dot{0} \notin \Psi_\Delta$ and $\sup \Psi_\Delta = \dot{0}$. We can now apply Lemma 9.3.2 and Corollary 9.3.3 to K equipped with $\dot{v}$, and this gives the desired result. □

Reduction to the special case

In this subsection K is ungrounded. So we have the contraction map $\chi\colon \Gamma^< \to \Gamma^<$. We now consider a convex nonempty set $B \subseteq \Gamma^<$ that satisfies a *weaker* condition than in Lemma 9.3.1, namely, $\chi(B) \subseteq B$. For example, this holds for $B = \Gamma^<$. (In this subsection we do not assume $\partial o \subseteq o$.)

PROPOSITION 9.3.5. *There are $\alpha \in \Gamma$ and $\beta \in B$, and $d_0, d_1, \dots, d_r \in \mathbb{N}$ such that $d_0 \leqslant \deg P$, and for all $y \in K$,*

$$vy \in B^{>\beta} \implies vP(y) = \alpha + d_0 vy + \sum_{i=1}^{r} d_i \psi\big(\chi^{i-1}(vy)\big).$$

If $P(0) = 0$ we can take $d_0 \geqslant 1$. If P is homogeneous, we can take $d_0 = \deg P$.

PROOF. Consider first the special case that $\partial o \subseteq o$ and $\psi(B) < 0$. Then we have $\psi(B) \subseteq B$: given $\gamma \in B$ we have $\psi(\gamma) < 0$ and $\chi(\gamma) \in B$ and so $\psi\big(\chi(\gamma)\big) < 0$, hence

$$\gamma < \psi(\gamma) = \chi(\gamma) + \psi\big(\chi(\gamma)\big) < \chi(\gamma) \in B,$$

and thus $\psi(\gamma) \in B$ since B is convex. Also $\chi(\gamma) = \psi(\gamma) - \psi^2(\gamma)$ for $\gamma \in B$, by Lemma 9.2.18. This gives $\psi(\gamma) < \chi(\gamma) < \psi(\gamma)/2$ for $\gamma \in B$, so $\psi^2(\gamma) = \psi\big(\chi(\gamma)\big)$ for such γ, and thus by induction on m,

$$\psi^m(\gamma) = \psi\big(\chi^{m-1}(\gamma)\big) \qquad (\gamma \in B,\ m \geqslant 1).$$

This yields the desired result in view of Lemma 9.3.1.

The general case is reduced to this special case by compositional conjugation. In this connection recall that, unlike ψ, the contraction map χ is invariant under compositional conjugation. Take an elementary extension L of K with an element $\theta \in L^\times$ such that $\psi(B) < v\theta < (\Gamma_L^>)'$ (so L^θ has small derivation). Let B_L be the convex hull of B in Γ_L. Then $\chi(B_L) \subseteq B_L$ and $\psi^\theta(B_L) < 0$, so we are in the special case considered before, with P^θ, B_L, and L^θ in place of P, B, and K. This gives $\alpha_0 \in \Gamma_L$, $\beta \in B$, and $d_0, \dots, d_r \in \mathbb{N}$, $d_0 \leqslant \deg P^\theta$, such that for all $y \in L$,

$$vy \in B_L^{>\beta} \implies vP^\theta(y) = \alpha_0 + d_0 vy + \sum_{i=1}^{r} d_i \psi^\theta\big(\chi^{i-1}(vy)\big),$$

and thus with $d := d_1 + \cdots + d_r$ we have for all $y \in L$ with $vy \in B_L^{>\beta}$,

$$vP(y) = vP^\theta(y) = (\alpha_0 - dv\theta) + d_0 vy + \sum_{i=1}^{r} d_i \psi(\chi^{i-1}(vy)).$$

This holds in particular for $y \in K$ with $vy \in B^{>\beta}$, and so $\alpha := \alpha_0 - dv\theta \in \Gamma$. □

By taking $B = \Gamma^<$ in Proposition 9.3.5, we obtain the existence of some $\beta < 0$ such that $P(y) \neq 0$ for all $y \in K$ with $\beta < vy < 0$.

COROLLARY 9.3.6. *Suppose $P(0) = 0$. Then there are $\beta < 0$ and a strictly increasing function $i\colon (\beta, 0) \to \Gamma$ with the intermediate value property such that $vP(y) = i(vy)$ for all $y \in K$ with $\beta < vy < 0$.*

PROOF. With $B = \Gamma^<$ and notations from the proof of Proposition 9.3.5, the function

$$i\colon B_L \to \Gamma_L, \quad i(\beta) := \alpha + d_0\beta + \sum_{i=1}^{r} d_i \psi(\chi^{i-1}(\beta))$$

is strictly increasing and has the intermediate value property. Now use that (Γ_L, ψ_L) is an elementary extension of (Γ, ψ). □

We have $\psi(\chi^m(\gamma)) = \psi(\gamma) - \sum_{i=1}^{m} \chi^i(\gamma)$ for $\gamma < 0$. This yields a formulation of Proposition 9.3.5 for $B = \Gamma^<$ that is more illuminating for vy tending to 0:

COROLLARY 9.3.7. *There are $\alpha \in \Gamma$ and $d, d_0, e_1, \dots, e_{r-1} \in \mathbb{N}$ with $d_0 \leqslant \deg P$, $d \geqslant e_1 \geqslant \dots \geqslant e_{r-1}$, such that for some $\beta < 0$ we have, for all $y \in K$,*

$$\beta < vy < 0 \implies vP(y) = \alpha + d\psi(vy) + d_0 vy - \sum_{i=1}^{r-1} e_i \chi^i(vy).$$

If $P(0) = 0$ we can take here $d_0 \geqslant 1$. If P is homogeneous, we can take $d_0 = \deg P$.

Here we adopt the convention that $\sum_{i=1}^{r-1} e_i \chi^i(\gamma) = 0$ for $r \leqslant 1$ and $\gamma < 0$. With α and $d_0, \dots, d_r$ as in Proposition 9.3.5 for $B = \Gamma^<$ we can take in Corollary 9.3.7 the same values for α and d_0, and $d = d_1 + \cdots + d_r$, $e_i = d_{i+1} + \cdots + d_r$. Note:

$$y \in K,\ vy < 0 \implies \sum_{i=1}^{r-1} e_i \chi^i(vy) = o_\psi(vy).$$

By Lemma 9.2.20 there is a unique tuple $(\alpha, d, d_0, e_1, \dots, e_{r-1})$ as in Corollary 9.3.7. Here is a slight extension of Corollary 9.3.6:

COROLLARY 9.3.8. *Suppose $P(a) = 0$, $a \in K$. Then for any γ there are $\beta < \gamma$ and a strictly increasing $i\colon (\beta, \gamma) \to \Gamma$ with the intermediate value property such that $vP(a + y) = i(vy)$ for all $y \in K$ with $\beta < vy < \gamma$.*

PROOF. Take $g \in K$ with $vg = \gamma$, put $Q := P_{+a, \times g} = P(a + gY)$, and apply Corollary 9.3.6 to Q in the role of P. □

Notes and comments

The material in this section is new and plays a key role in Chapter 11.

9.4 BASIC FACTS ABOUT ASYMPTOTIC FIELDS

Next we consider asymptotic fields in connection with having a small derivation, d-henselianity, coarsening, and specialization. *Throughout this section K is an asymptotic field with asymptotic couple (Γ, ψ), valuation ring $\mathcal{O}$, maximal ideal $\smallo$ of $\mathcal{O}$, and residue field $\boldsymbol{k} = \mathcal{O}/\smallo$.*

The case of a small derivation

Recall that the derivation of a valued differential field is said to be small if the derivative of every infinitesimal is infinitesimal. Any asymptotic field has a compositional conjugate with small derivation, so this enables us to reduce problems to the case where the derivation is small.

Suppose K has small derivation ∂. By Lemma 4.4.2, not only do we have $\partial\smallo \subseteq \smallo$, but even $\partial\mathcal{O} \subseteq \mathcal{O}$, and so ∂ induces a derivation on $\boldsymbol{k}$. As we saw in Sections 4.5 and 5.6 this allows us to associate to $P \in K\{Y\}^{\neq}$ and $A \in K[\partial]^{\neq}$ functions $v_P \colon \Gamma \to \Gamma$ and $v_A \colon \Gamma \to \Gamma$ such that

$$v_P(\gamma) = v(P_{\times g}), \qquad v_A(\gamma) = v(Ag) \qquad (g \in K^\times,\ vg = \gamma).$$

For example, for $A = \partial$ we have

$$v_\partial(\gamma) \;=\; \min(\gamma, \gamma') \text{ if } \gamma \in \Gamma^{\neq}, \qquad v_\partial(0) \;=\; 0.$$

An asymptotic field has small derivation iff its asymptotic couple has. Thus:

LEMMA 9.4.1. *If K has small derivation and L is an asymptotic field extension of K with $\Gamma_L = \Gamma$, then L has small derivation.*

When referring to $\sup \Psi$ we mean the supremum of Ψ in the ordered set Γ_∞, and the condition $\sup \Psi = \alpha$ (for $\alpha \in \Gamma_\infty$) includes the requirement that $\sup \Psi$ exists. Note that if $\sup \Psi = 0$, then K has small derivation. While $\sup \Psi = 0$ is a strong restriction, it is often possible and useful to reduce to this case. (If $\Gamma = \{0\}$, then $\Psi = \emptyset$, so $\sup \Psi = 0$.) In the next two lemmas we describe situations where $\sup \Psi = 0$ holds. The first lemma is in the spirit of Lemma 4.4.8.

LEMMA 9.4.2. *Suppose K has small derivation and the differential residue field $\boldsymbol{k}$ has nontrivial derivation. Then* $\sup \Psi = 0$.

PROOF. Take $f \asymp 1$ in K such that $f' \asymp 1$. Then $g^\dagger \succcurlyeq 1$ for all $g \in K^\times$ with $g \not\asymp 1$, by Proposition 9.1.3, and thus $\Psi \leqslant 0$. It remains to use Lemma 9.2.9. □

LEMMA 9.4.3. *Suppose $\partial\smallo = \smallo$. Then $(\Gamma^>)' = \Gamma^>$, and thus* $\sup \Psi = 0$. *Also, C maps onto the constant field $C_{\boldsymbol{k}}$ of $\boldsymbol{k}$ under the residue map $\mathcal{O} \to \boldsymbol{k}$.*

This is an easy consequence of Corollary 9.2.4. The condition $\partial\sigma = \sigma$ is satisfied when K has small derivation and ∂ is neatly surjective. The following partial converse to Lemma 9.4.3 is useful in verifying the hypothesis of Lemma 5.6.11.

LEMMA 9.4.4. *Suppose* $\sup\Psi = 0$, *C maps onto the constant field $C_{\boldsymbol{k}}$ of $\boldsymbol{k}$ under the residue map $\mathcal{O} \to \boldsymbol{k}$, and $\partial K = K$. Then ∂ is neatly surjective.*

PROOF. The condition $\sup\Psi = 0$ gives $v_\partial(\gamma) = \gamma'$ for all $\gamma \in \Gamma$, where $0' = 0 \in \Gamma$ by convention. Hence the only nonobvious case is when $a, b \in K^\times$, $a' = b$ and $a \asymp 1$, $b \prec 1$. Then the image $\overline{a}$ of a in $\boldsymbol{k}$ satisfies $\overline{a}' = 0$, so $\overline{a} = \overline{c}$ with $c \in C$, by our assumption on the constant field of $\boldsymbol{k}$. Hence $(a-c)' = b$ and $a - c \prec 1$, so $v_\partial\big(v(a-c)\big) = vb$, as desired. □

In cases like $K = \mathbb{T}$ where $\sup\Psi = 0$ fails, it is often useful to arrange it by compositional conjugation and coarsening.

LEMMA 9.4.5. *Suppose* $\sup\Psi = 0$ *and L is an immediate valued differential field extension of K with small derivation. Then L is asymptotic.*

PROOF. We use the equivalence of (i) and (ii) in Proposition 9.1.3. Let $b \in L^\times$ be such that $b \not\asymp 1$. Take $a \in K^\times$ with $b = a(1+\varepsilon)$, $\varepsilon \prec 1$. Then $b^\dagger = a^\dagger + \frac{\varepsilon'}{1+\varepsilon}$ with $a^\dagger \succcurlyeq 1$ and $\frac{\varepsilon'}{1+\varepsilon} \prec 1$, so $b^\dagger \sim a^\dagger$, and thus $\psi(vb) = \psi(va) = v(a^\dagger) = v(b^\dagger)$. □

The condition $\sup\Psi = 0$ arises naturally in some situations:

LEMMA 9.4.6. *Suppose that $\partial\sigma \subseteq \sigma$ and that $K\langle Y\rangle$ equipped with its gaussian valuation is asymptotic. Then* $\sup\Psi = 0$.

PROOF. We have $Y \asymp 1$ in $K\langle Y\rangle$, so $f' \prec 1 \asymp Y' \preccurlyeq g^\dagger$ for all $f, g \in K^\times$ with $f \prec 1$ and $g \not\asymp 1$, by (i) $\Leftrightarrow$ (iv) in Proposition 9.1.3. Thus $\sup\Psi = 0$. □

LEMMA 9.4.7. *Assume $\partial\sigma \subseteq \sigma$. Let $K\langle a\rangle$ be as in Theorem 6.3.2 with minimal annihilator F of a over K of order $r \geqslant 1$, and $\overline{F} \notin \boldsymbol{k}^\times Y'$. If $K\langle a\rangle$ is asymptotic, then* $\sup\Psi = 0$.

PROOF. We have $a \asymp 1$. We claim that $a' \asymp 1$. This claim clearly holds for $r \geqslant 2$, and also when $r = 1$ and F has degree $\geqslant 2$ in Y'. Suppose $r = 1$ and F has degree 1 in Y'. Then $F = IY' + G$ with $I, G \in \mathcal{O}[Y]$, $vI = 0$. The assumption on $\overline{F}$ and its irreducibility in $\boldsymbol{k}\{Y\}$ yields $vG = 0$, and so again $a' \asymp 1$. With the claim established, argue as in the proof of Lemma 9.4.6 with a replacing Y. □

We now turn to the case $\max\Psi = 0$:

LEMMA 9.4.8. *Assume* $\max\Psi = 0$. *Let L be a valued differential field extension of K with small derivation, $\Gamma_L = \Gamma$, and $(\boldsymbol{k}_L^\times)^\dagger \cap \boldsymbol{k} \subseteq (\boldsymbol{k}^\times)^\dagger$. Then L is asymptotic.*

PROOF. Let $g \in L^\times$, $g \prec 1$. By Proposition 9.1.3 it is enough to show that $v(g^\dagger) = \psi(vg)$. We have $u \in L$, $f \in K$ such that $g = uf$, $u \asymp 1$, and $f \asymp g$. Then $g^\dagger = u^\dagger + f^\dagger$, and $u^\dagger \preccurlyeq 1$, $f^\dagger \succcurlyeq 1$. If $u^\dagger \prec 1$ or $f^\dagger \succ 1$, then $v(g^\dagger) = v(f^\dagger) = \psi(vf) = \psi(vg)$. It remains to consider the case that $u^\dagger \asymp 1$ and $f^\dagger \asymp 1$. If $u^\dagger + f^\dagger \asymp 1$, we have again $v(g^\dagger) = v(f^\dagger) = \psi(vf) = \psi(vg)$, so assume $u^\dagger + f^\dagger \prec 1$. Then $\overline{u}^\dagger \in (\boldsymbol{k}_L^\times)^\dagger \cap \boldsymbol{k}$, and so we have $\phi \in K$ such that $\phi \asymp 1$ and $u^\dagger \sim \phi^\dagger$. Then $0 \neq \phi f \prec 1$ and $(\phi f)^\dagger = \phi^\dagger + f^\dagger \prec 1$, contradicting $\max \Psi = 0$. □

COROLLARY 9.4.9. *Suppose that* $\max \Psi = 0$. *Then* $K\langle Y\rangle$ *equipped with its gaussian valuation is asymptotic.*

PROOF. Apply Lemma 9.4.8 to $L = K\langle Y\rangle$, and Corollary 4.3.7 with K as $\boldsymbol{k}$. □

Differential-henselian asymptotic fields

In this subsection we assume that the asymptotic field K has small derivation. Let $\boldsymbol{k}$ be its differential residue field. Recall that if K is 1-differential-henselian, then $\mathcal{O}$ is 1-linearly surjective. Thus by the next lemma, if K is 1-differential-henselian, then $\sup \Psi = 0$.

LEMMA 9.4.10. *Suppose $\mathcal{O}$ is 1-linearly surjective. Then* $\partial o = o$, $\sup \Psi = 0$, *and the field embedding* $a \mapsto \overline{a}\colon C \to C_{\boldsymbol{k}}$ *is bijective.*

PROOF. From $v(\partial) = 0$ we get $v_\partial(\Gamma^>) = \Gamma^>$ by the Equalizer Theorem. Since $\mathcal{O}$ is 1-linearly surjective, ∂ is neatly surjective, so $\partial o = o$. Now use Lemma 9.4.3. □

Since $\sup \Psi = 0$ fails for $\mathbb{T}$, this asymptotic field is not d-henselian. Chapter 15 and Lemma 14.1.2, however, show that $\mathbb{T}$ becomes d-henselian if we coarsen its valuation by any convex subgroup of $\Gamma_{\mathbb{T}}$ that contains $v(x)$.

COROLLARY 9.4.11. *Assume $\boldsymbol{k}$ is linearly surjective. Then K has an immediate asymptotic field extension L such that:*

(i) *L is* d-*algebraic over K;*

(ii) *L has small derivation and is* d-*henselian;*

(iii) *no proper differential subfield of L containing K is* d-*henselian.*

PROOF. Corollary 7.2.7 yields an immediate valued differential field extension F of K with small derivation such that F is d-henselian. Then F is asymptotic by Lemma 9.4.5. By keeping only the elements of F that are d-algebraic over K we arrange that F is d-algebraic over K.

Let L be the intersection inside F of the collection of all differential subfields E of F that contain K and are d-henselian. Proposition 7.5.6 applied to the extensions $E \subseteq F$ shows that L has the desired property. □

We conjecture that there is only one L as in Corollary 9.4.11, up to isomorphism over K. For any $r \in \mathbb{N}$, Corollary 9.4.11 goes through with "linearly surjective" and "d-henselian" replaced by "r-linearly surjective" and "r-d-henselian."

Specialization of asymptotic fields

Let K have small derivation, and let Δ be a convex subgroup of $\Gamma = v(K^\times)$. It follows from Corollary 4.4.4 that the valuation ring $\dot{\mathcal{O}}$ of the coarsening $\dot{v} = v_\Delta \colon K^\times \to \dot{\Gamma}$ of v and the maximal ideal $\dot{o}$ of $\dot{\mathcal{O}}$ are closed under ∂. Note that if $\sup \Psi = 0$, then $\psi(\Delta^{\neq}) \subseteq \Delta$ and $\sup \psi(\Delta^{\neq}) = 0$ in Δ, by Lemma 9.2.10(iv).

In the rest of this subsection we assume that $\psi(\Delta^{\neq}) \subseteq \Delta$. The residue field $\dot{K} = \dot{\mathcal{O}}/\dot{o}$ of $\dot{\mathcal{O}}$ with its induced valuation $v \colon \dot{K}^\times \to \Delta$ and the induced derivation is then an asymptotic field with asymptotic couple $(\Delta, \psi|\Delta^{\neq})$. (This follows from the equivalence of (i) and (ii) in Proposition 9.1.3.) Moreover, $\dot{K}$ has small derivation, the field isomorphism $\operatorname{res}(K) \to \operatorname{res}(\dot{K})$ from Section 3.4 is a differential field isomorphism, and if K is of H-type, then so is $\dot{K}$.

LEMMA 9.4.12. *Suppose K has asymptotic integration and $\Delta \neq \{0\}$. Then $\dot{K}$ has asymptotic integration.*

PROOF. Take $\alpha \in \Delta^{\neq}$, and note that $\alpha + \psi(\alpha) \in \Delta$. Let $\delta \in \Delta$, $\delta \neq \alpha + \psi(\alpha)$, and take $\beta \in \Gamma^{\neq}$ with $\beta + \psi(\beta) = \delta$. Then by Lemma 6.5.4(ii),

$$\big(\alpha + \psi(\alpha)\big) - \big(\beta + \psi(\beta)\big) = \alpha - \beta + o(\alpha - \beta) \in \Delta,$$

so $\alpha - \beta \in \Delta$, and thus $\beta \in \Delta$. □

We say that the differential field F is **closed under integration** if for every $f \in F$ there is a $y \in F$ with $y' = f$, and we say that F is **closed under logarithms** if for every $f \in F^\times$ there is a $y \in F$ with $y' = f^\dagger$. So if F is closed under integration, then F is closed under logarithms. If the differential field F is closed under integration (closed under logarithms) and $\phi \in F^\times$, then F^ϕ is closed under integration (closed under logarithms, respectively).

LEMMA 9.4.13. *Suppose $\Delta \neq \{0\}$. If K is closed under integration, then so is $\dot{K}$. If K is closed under logarithms, then so is $\dot{K}$.*

PROOF. Take $\delta \in \Delta^{\neq}$; then $\psi(\delta) \in \Delta$. Suppose K is closed under integration, and let $f \in \dot{\mathcal{O}}$. We need to find $y \in \dot{\mathcal{O}}$ with $y' - f \in \dot{o}$. If $vf > \Delta$, then we may take $y := 0$, so we may assume $vf \in \Delta$. Take $y \in K$ with $y' = f$; we claim that $y \in \dot{\mathcal{O}}$. Suppose $y \notin \dot{\mathcal{O}}$. Then $\gamma := vy < \Delta$, in particular $[\gamma] > [\delta]$ and hence $[\psi(\gamma) - \psi(\delta)] < [\gamma - \delta] = [\gamma]$ by Lemma 6.5.4(ii). Since $\gamma + \psi(\gamma) = vf \in \Delta$, we therefore have $[\gamma] = [\gamma + \psi(\gamma) - \psi(\delta)] \in [\Delta]$ and hence $\gamma \in \Delta$, a contradiction. Thus $\dot{K}$ is closed under integration. Similarly one shows that if K is closed under logarithms, then so is $\dot{K}$. □

Let L be an asymptotic field extension of K with small derivation. Let Δ_L be the convex hull of Δ in Γ_L, and let

$$\dot{\mathcal{O}}_L := \{y \in L : vy \geqslant \delta \text{ for some } \delta \in \Delta_L\}$$

be the valuation ring of the corresponding coarsening $\dot{v}\colon L^\times \to \dot{\Gamma}_L = \Gamma_L/\Delta_L$. Then $\dot{\mathcal{O}}_L$ and its maximal ideal $\dot{o}_L$ are closed under the derivation ∂ of L. Note that $\dot{\mathcal{O}}_L$ lies over $\dot{\mathcal{O}}$. Let $\dot{L} := \dot{\mathcal{O}}_L/\dot{o}_L$ be the residue field of $\dot{\mathcal{O}}_L$, equipped with the induced valuation $v\colon \dot{L}^\times \to \Delta_L$. We have $\psi_L(\Delta_L^{\neq}) \subseteq \Delta_L$, so the field $\dot{L}$ with its valuation v and the induced derivation is an asymptotic field with asymptotic couple $(\Delta_L, \psi_L|\Delta_L^{\neq})$. With the usual identifications we have $\dot{K} \subseteq \dot{L}$, not only as (residue) fields, but also as asymptotic fields.

Flattening

For certain *definable* ways of coarsening we use special notation and terminology, borrowed from [194]. These particular coarsenings are called flattenings and will be very useful in Chapters 13 and 14. (Some readers might prefer to skip this subsection until it gets used in those chapters.) *Assume K is of H-type, and let γ range over* Γ. We have the convex subgroup

$$\Gamma^\flat := \{\gamma : \psi(\gamma) > 0\}$$

of Γ. For example, if $K = \mathbb{T}$, then

$$\Gamma^\flat = \{vh : h \in K^\times,\ h \prec\!\!\prec \mathrm{e}^x\}.$$

Let

$$v^\flat : K^\times \to \Gamma^\sharp \qquad \text{where } \Gamma^\sharp := \Gamma/\Gamma^\flat$$

be the coarsening of the valuation v by $\Gamma^\flat$, with associated dominance relation $\preccurlyeq^\flat$ on K, and valuation ring $\mathcal{O}^\flat$. We call the valuation $v^\flat$ the **flattening** of v. Note that the definition of $\Gamma^\flat$ depends on the derivation of K.

Let $\phi \in K^\times$, and use a subscript ϕ to indicate the flattened objects $\Gamma^\flat_\phi$, $v^\flat_\phi$, $\preccurlyeq^\flat_\phi$, $\prec^\flat_\phi$, $\asymp^\flat_\phi$, and $\sim^\flat_\phi$ associated to the asymptotic field K^ϕ. In particular

$$\Gamma^\flat_\phi = \{\gamma : \psi(\gamma) > v\phi\}.$$

Clearly $\Gamma^\flat_\phi = \{0\}$ iff $v\phi \notin (\Gamma^<)'$, and $\Gamma^\flat_\phi = \Gamma$ iff $\Psi > v\phi$. The flattening $v^\flat_\phi$ of the valuation v on K^ϕ will be useful for ungrounded K and $v\phi \in \Psi^\downarrow$. In that case, with $v\phi$ increasing cofinally in $\Psi^\downarrow$, the convex subgroup $\Gamma^\flat_\phi$ of Γ becomes arbitrarily small, and so $v^\flat_\phi$ (with value group $\Gamma/\Gamma^\flat_\phi$) approximates v in some sense.

We denote the gaussian extension of $v^\flat_\phi$ to $K^\phi\langle Y\rangle$ also by $v^\flat_\phi$. The binary relations $\preccurlyeq^\flat_\phi$, $\prec^\flat_\phi$, $\asymp^\flat_\phi$, and $\sim^\flat_\phi$, on K extend likewise, with the same notations, to binary relations on $K^\phi\langle Y\rangle$. As $K\langle Y\rangle$ and $K^\phi\langle Y\rangle$ have the same underlying field, we also use the same notations for elements of $K\langle Y\rangle$; for example, given $f \in K$ and $P \in K\langle Y\rangle$, $f \preccurlyeq^\flat_\phi P$ means that $v^\flat_\phi(f) \geqslant v^\flat_\phi(P)$ in $\Gamma/\Gamma^\flat_\phi$, where in the latter inequality f and P are taken as elements of $K^\phi \subseteq K^\phi\langle Y\rangle$ and $K^\phi\langle Y\rangle$, respectively.

LEMMA 9.4.14. *Suppose $\Phi \in K^\times$ with $\Phi \not\asymp 1$ and $\Phi^\dagger \asymp \phi$, and $f,g \in K$. Then*

$$\begin{aligned} f \preccurlyeq^\flat_\phi g &\iff v^\flat_\phi(f) \geqslant v^\flat_\phi(g) \iff f \preccurlyeq hg \text{ for some } h \in K^\times \text{ with } h \prec\!\!\prec \Phi,\\ f \prec^\flat_\phi g &\iff v^\flat_\phi(f) > v^\flat_\phi(g) \iff f \prec hg \text{ for all } h \in K^\times \text{ with } h \prec\!\!\prec \Phi,\\ f \asymp^\flat_\phi g &\iff v^\flat_\phi(f) = v^\flat_\phi(g) \iff f \asymp hg \text{ for some } h \in K^\times \text{ with } h \prec\!\!\prec \Phi.\end{aligned}$$

PROOF. These equivalences follow from their validity for $g = 1$, and in that case we argue as follows:

$$\begin{aligned} f \preccurlyeq^\flat_\phi 1 &\iff vf \geqslant vh \text{ for some } h \in K^\times \text{ with } vh \in \Gamma^\flat_\phi\\ &\iff f \preccurlyeq h \text{ for some } h \in K^\times \text{ with } h \prec\!\!\prec \Phi,\end{aligned}$$

where we use that $1 \prec\!\!\prec \Phi$ to get the last equivalence when $vh = 0$. We proceed likewise with $f \prec^\flat_\phi 1$ and $f \asymp^\flat_\phi 1$, treating the case $vh = 0$ separately. □

The valuation ring of $v^\flat_\phi$ is

$$\mathcal{O}^\flat_\phi = \{f \in K : vf \geqslant 0 \text{ or } v(f') > vf + v\phi\},$$

with maximal ideal

$$\mathcal{o}^\flat_\phi = \{f \in K : vf > 0 \text{ and } v(f') \leqslant vf + v\phi\}.$$

Note that if $\Gamma^\flat = \{0\}$, then the canonical map $\Gamma \to \Gamma^\sharp$ is an isomorphism of ordered groups via which Γ is identified with $\Gamma^\sharp$. If $\Gamma^\flat \neq \{0\}$, then K has small derivation and there exists a unique positive element 1 in $\Gamma^\flat$ with $\psi(1) = 1$, so $\psi\big((\Gamma^\flat)^{\neq}\big) \subseteq \Gamma^\flat$ by Lemma 9.2.25(i). Note also that if $\Gamma^\flat \neq \{0\}$, then by Lemma 9.2.24 we have $\sup \Psi^\flat = 0$ in $\Gamma^\sharp$, where

$$\Psi^\flat := \{v^\flat(f^\dagger) : f \in K^\times,\ f \not\asymp^\flat 1\}.$$

If $\Gamma^\flat \neq \{0\}$, then we denote the asymptotic residue field of $v^\flat$ by $K^\flat := \mathcal{O}^\flat/\mathcal{o}^\flat$, with asymptotic couple $\big(\Gamma^\flat, \psi|\Gamma^\flat\big)$.

Let $\Phi \in K^\times$ with $\Phi \not\asymp 1$. The binary relations $\preccurlyeq^\flat_{\Phi^\dagger}$, $\prec^\flat_{\Phi^\dagger}$, $\asymp^\flat_{\Phi^\dagger}$ on K, and on $K\langle Y\rangle$, are often denoted by $\preccurlyeq_\Phi$, $\prec_\Phi$, $\asymp_\Phi$, respectively. Thus for $f,g \in K$:

$$\begin{aligned} f \preccurlyeq_\Phi g &\iff f \preccurlyeq hg \text{ for some } h \in K^\times \text{ with } h \prec\!\!\prec \Phi,\\ f \prec_\Phi g &\iff f \prec hg \text{ for all } h \in K^\times \text{ with } h \prec\!\!\prec \Phi,\\ f \asymp_\Phi g &\iff f \asymp hg \text{ for some } h \in K^\times \text{ with } h \prec\!\!\prec \Phi.\end{aligned}$$

An advantage of this notation is that the relations $\preccurlyeq_\Phi$, $\prec_\Phi$, $\asymp_\Phi$ do not change in passing from K to a compositional conjugate K^ϕ ($\phi \in K^\times$). Note that $\Phi \not\asymp_\Phi 1$ by Corollary 9.1.4(iv), and if $\Phi \prec 1$, then $\Phi \prec_\Phi 1$. Also, for $f \not\asymp 1$ in $K^\times$,

$$f \asymp_\Phi 1 \iff f \prec\!\!\prec \Phi.$$

Here are some further rules:

LEMMA 9.4.15. *Let $f, g \in K^\times$, $f, g \not\asymp 1$, $f \prec_g g$. Then $f \prec_f g$.*

PROOF. We have $f/g \prec^{\flat}_{g^\dagger} 1$, so $\psi(vf - vg) \leqslant \psi(vg)$, hence $\psi(vf - vg) \leqslant \psi(vf)$, which in view of $vf > vg$ yields $f/g \prec^{\flat}_{f^\dagger} 1$, so $f \prec_f g$. □

LEMMA 9.4.16. *Let $\Phi_1, \Phi_2 \in K^\times$ with $\Phi_1, \Phi_2 \not\asymp 1$ and $\Phi_1 \preccurlyeq\!\!\!\prec \Phi_2$. Then for $f, g \in K$:*

$$f \preccurlyeq_{\Phi_1} g \Rightarrow f \preccurlyeq_{\Phi_2} g \qquad \textit{and} \qquad f \prec_{\Phi_2} g \Rightarrow f \prec_{\Phi_1} g.$$

LEMMA 9.4.17. *Suppose $\Phi \in K^\times$, $\Phi \not\asymp^\flat 1$. Then for all $f, g \in K$:*

$$f \preccurlyeq^\flat g \Rightarrow f \preccurlyeq_\Phi g \qquad \textit{and} \qquad f \prec_\Phi g \Rightarrow f \prec^\flat g.$$

PROOF. The second part follows from the first. To prove the first part, arrange $f \neq 0$ and $g = 1$. Assume $f \preccurlyeq^\flat 1$. Then $f \preccurlyeq 1$ or $f^\dagger \prec 1$. From $\Phi \not\asymp^\flat 1$ we get $\Phi^\dagger \succcurlyeq 1$, so $f \preccurlyeq 1$ or $f^\dagger \prec \Phi^\dagger$, and thus $f \preccurlyeq_\Phi 1$. □

In the next lemma and its corollaries below we assume that K is ungrounded and has small derivation. Also $P \in K\{Y\}^{\neq}$ and $g \in K^\times$, $g \not\asymp 1$.

LEMMA 9.4.18. *Suppose P is homogeneous of degree d. Then $P_{\times g} \asymp_g P g^d$.*

PROOF. If $g^\dagger \preccurlyeq 1$, then $P_{\times g} \asymp Pg^d$ by Lemma 4.5.3. If $g^\dagger \succ 1$, then $g^{\dagger\dagger} \prec g^\dagger$ by Lemma 9.2.10(iv), and thus $P_{\times g} \asymp_g Pg^d$ by Proposition 6.4.3. □

COROLLARY 9.4.19. *Suppose $g \prec 1$ and $d := \operatorname{dmul}(P) = \operatorname{mul}(P)$. Then*

$$P_{\times g} \asymp_g P g^d.$$

PROOF. We have $P_i \preccurlyeq P_d$ for $i \geqslant d$, and $g \prec_g 1$, hence $P_i g^i \prec_g P_d g^d$ for $i > d$, and the claim follows from Lemma 9.4.18. □

For $F \in K\{Y\}$ and $d \in \mathbb{N}$ we put

$$F_{\leqslant d} := F_0 + F_1 + \cdots + F_d, \qquad F_{>d} := F - F_{\leqslant d}.$$

With this notation, we have:

COROLLARY 9.4.20. *Suppose $g \succ 1$ and $d := \operatorname{ddeg}(P) = \operatorname{ddeg}(P_{\times g})$. Then*

$$gP_{>d} \preccurlyeq_g P.$$

PROOF. The case $P_{>d} = 0$ being trivial, assume $P_{>d} \neq 0$. Take $i > d$ such that $P_i \asymp P_{>d}$. Then by Lemma 9.4.18 and $g \succ 1$ we have

$$(P_{\times g})_i = (P_i)_{\times g} \asymp_g P_i g^i \succcurlyeq P_i g^{d+1} \asymp g^{d+1} P_{>d}$$

and so $(P_{\times g})_{>d} \succcurlyeq_g g^{d+1} P_{>d}$. Since $\operatorname{ddeg}(P_{\times g}) = d$, we have $(P_{\times g})_d \succcurlyeq (P_{\times g})_{>d}$, and by Lemma 9.4.18 again, $(P_{\times g})_d = (P_d)_{\times g} \asymp_g g^d P_d$. Since $\operatorname{ddeg}(P) = d$, we have $P \asymp P_d$, so

$$g^d P \asymp g^d P_d \asymp_g (P_{\times g})_d \succcurlyeq (P_{\times g})_{>d} \succcurlyeq_g g^{d+1} P_{>d},$$

and thus $P \succcurlyeq_g gP_{>d}$ as claimed. □

COROLLARY 9.4.21. *Let $f \in K^\times$ with $f \not\asymp 1$ and $f \prec\!\!\prec g$. Then $P_{\times f} \asymp_g P$.*

PROOF. Take $d \in \mathbb{N}$ with $P_{\times f} \asymp (P_d)_{\times f}$. Lemma 9.4.18 gives $(P_d)_{\times f} \asymp_f P_d f^d$, so $P_{\times f} \asymp_f P_d f^d$, which by Lemma 9.4.16 yields $P_{\times f} \asymp_g P_d f^d$. From $f \prec\!\!\prec g$ we also get $f \asymp_g 1$, so $P_d f^d \asymp_g P_d \preccurlyeq P$, hence $P_{\times f} \preccurlyeq_g P$. Applying the same argument to $P_{\times f}, f^{-1}$ in place of P, f yields $P = (P_{\times f})_{\times f^{-1}} \preccurlyeq_g P_{\times f}$. □

9.5 ALGEBRAIC EXTENSIONS OF ASYMPTOTIC FIELDS

In this section we prove analogues for asymptotic fields of results in Section 6.2 on algebraic extensions of valued differential fields with small derivation.

Immediate algebraic extensions

In this subsection we fix a valued differential field L with a subfield K. We consider K as a valued subfield of L with valuation ring $\mathcal{O}$, maximal ideal o of $\mathcal{O}$, and value group Γ. In order for L to be asymptotic it is of course necessary that for all $f, g \in K^\times$ with $f, g \prec 1$ we have

$$f \prec g \iff f' \prec g'.$$

Call K **asymptotic in** L if this condition is satisfied. Below we prove:

PROPOSITION 9.5.1. *Suppose K is asymptotic in L and L is an immediate algebraic extension of K. Then L is asymptotic.*

For later use it is important that in this proposition we do not require K to be closed under the derivation of L. First some remarks and a lemma. Suppose K is asymptotic in L, and set $C := \{a \in K : a' = 0\}$, a subfield of K. Then $C \cap o = \{0\}$ by the same argument as for asymptotic fields. Thus the valuation v is trivial on C, and the map $\varepsilon \mapsto \varepsilon' : o \to L$ is injective. A variant of Proposition 9.1.3 goes through with the same proof:

LEMMA 9.5.2. *Suppose $v(L^\times) = v(K^\times)$ $(= \Gamma)$. The following are equivalent:*

(i) *K is asymptotic in L;*

(ii) *there is an asymptotic couple (Γ, ψ) such that for all $f \in K^\times$ with $f \not\asymp 1$ we have $\psi(vf) = v(f^\dagger)$;*

(iii) *for all $f, g \in K^\times$ with $f, g \not\asymp 1$ we have: $f \preccurlyeq g \iff f' \preccurlyeq g'$;*

(iv) *for all $f, g \in K^\times$ we have:*

$$\begin{cases} f \prec 1,\ g \not\asymp 1 & \Rightarrow \quad f' \prec g^\dagger, \\ f \asymp 1,\ g \not\asymp 1 & \Rightarrow \quad f' \preccurlyeq g^\dagger. \end{cases}$$

If K is asymptotic in L and $v(L^\times) = v(K^\times)$, then we call (Γ, ψ) as defined in (ii) of the lemma above the **asymptotic couple of** K **in** L, so

$$\Psi := \psi(\Gamma^{\neq}) = \{v(g^\dagger) : 1 \not\asymp g \in K^\times\}.$$

PROOF OF PROPOSITION 9.5.1. Let (Γ, ψ) be the asymptotic couple of K in L. We shall prove that L is asymptotic with asymptotic couple (Γ, ψ). Let $g \in L^\times$, $g \not\asymp 1$; by Proposition 9.1.3, (i) $\Leftrightarrow$ (ii), it suffices to show that then $v(g^\dagger) = \psi(vg)$.

Before doing so, note that we may replace L by a compositional conjugate: if $\phi \in L^\times$, then K is also asymptotic in L^ϕ with $(\Gamma, \psi - v\phi)$ as the asymptotic couple of K in L^ϕ, and with $\phi^{-1}g^\dagger$ as the logarithmic derivative of g in L^ϕ.

Take $f \in K^\times$ with $f \sim g$, so $g = f(1+\varepsilon)$ with $\varepsilon \in \mathcal{o}_L$. Then $g^\dagger = f^\dagger + \frac{\varepsilon'}{1+\varepsilon}$. Now $v(f^\dagger) = \psi(vf) = \psi(vg) \in \Psi$. Compositional conjugation by $\phi := f^\dagger$ (and renaming) arranges that $v(f^\dagger) = 0 \in \Psi$, so (Γ, ψ) has small derivation. Then $\partial\mathcal{o} \subseteq \mathcal{o}_L$ and $\partial\mathcal{O} \subseteq \mathcal{O}_L$ by Lemma 9.5.2(iv), so $\partial\mathcal{o}_L \subseteq \mathcal{o}_L$ by Lemma 6.2.2. This gives $f^\dagger \asymp 1 \succ \frac{\varepsilon'}{1+\varepsilon}$, so $g^\dagger \sim f^\dagger$, and thus $v(g^\dagger) = v(f^\dagger) = \psi(vg)$. □

Algebraic extensions

In this subsection we assume that K is an asymptotic field and $L|K$ is an extension of valued differential fields. We shall prove:

PROPOSITION 9.5.3. *If $L|K$ is algebraic, then L is an asymptotic field.*

LEMMA 9.5.4. *Suppose that K is henselian and $L|K$ is of finite degree. Assume also that for every $f \preccurlyeq 1$ in L there exists $a \preccurlyeq 1$ in K such that*

(i) *$f' \preccurlyeq a'$, and*

(ii) *if $f' \asymp a'$ and $f \prec 1$, then $a \prec 1$.*

Then L is an asymptotic field.

PROOF. We verify that L satisfies condition (iv) of Proposition 9.1.3. For this, let $f, g \in L^\times$ with $f \preccurlyeq 1$ and $g \not\asymp 1$. Let $a \preccurlyeq 1$ in K satisfy (i) and (ii). Put $b := \mathrm{N}_{L|K}(g) \in K^\times$; then $b \not\asymp 1$ and $b^\dagger \preccurlyeq g^\dagger$ by a result at the end of Section 4.4. Hence if $f \asymp 1$, then by (i): $f' \preccurlyeq a' \preccurlyeq b^\dagger \preccurlyeq g^\dagger$. If $f \prec 1$, then we similarly get, using (ii), that $f' \prec a' \preccurlyeq b^\dagger \preccurlyeq g^\dagger$, or $a \prec 1$ and $f' \asymp a' \prec b^\dagger \preccurlyeq g^\dagger$. □

LEMMA 9.5.5. *Suppose that K is henselian and $L|K$ is of finite degree, such that $[L : K] = \big[\mathrm{res}(L) : \mathrm{res}(K)\big] = n > 1$. Then L is an asymptotic field.*

PROOF. Take $y \asymp 1$ in L such that $\mathrm{res}(L) = \mathrm{res}(K)[\overline{y}]$, where $\overline{y}$ is the residue class of y in $\mathrm{res}(L)$. Let $a_0, \dots, a_{n-1} \preccurlyeq 1$ in K be such that

$$P(Y) = Y^n + a_{n-1}Y^{n-1} + \cdots + a_1 Y + a_0 \in K[Y]$$

is the minimum polynomial of y over K, so its reduction $\overline{P}(Y)$ in $\mathrm{res}(K)[Y]$ is the minimum polynomial of $\overline{y}$ over $\mathrm{res}(K)$; in particular $P'(y) \asymp 1$. Moreover, $P'(y)y' = -\sum_{i=0}^{n-1} a_i' y^i$, hence $y' \asymp a_{i_0}'$ with $i_0 \in \{0, \dots, n-1\}$. Let $f \preccurlyeq 1$ in L. Then we have $f_0, \dots, f_{n-1} \preccurlyeq 1$ in K such that

$$f = f_0 + f_1 y + \cdots + f_{n-1} y^{n-1}.$$

Then

$$f' = \sum_{i=0}^{n-1} f_i' y^i + \left(\sum_{j=1}^{n-1} j f_j y^{j-1} \right) y'$$

and hence

$$v(f') \geqslant \gamma := \min \left\{ \min_{0 \leqslant i < n} v(f_i'), \min_{0 < j < n} v(f_j) + v(a_{i_0}') \right\}.$$

We now define an element $a \in \mathcal{O}$ as follows: If $\gamma = \min_{0 \leqslant i < n} v(f_i')$, then $a := f_{i_1}$ where $0 \leqslant i_1 < n$ and $v(f_{i_1}') = \gamma$; if $\gamma < \min_{0 \leqslant i < n} v(f_i')$, then $a := a_{i_0}$. Then a satisfies conditions (i) and (ii) of Lemma 9.5.4, so L is an asymptotic field. □

The next lemma is a variant of Lemma 3.3 in [19]:

LEMMA 9.5.6. *Suppose* $\operatorname{res}(K) = \operatorname{res}(L)$, $T \supseteq K^\times$ *is a subgroup of* $L^\times$ *such that* $L = K(T)$ *(as fields), each element of* $K[T] \setminus \{0\}$ *has the form* $t_1 + \cdots + t_k$ *with* $k \geqslant 1$, $t_1, \dots, t_k \in T$ *and* $t_1 \succ t_i$ *for* $2 \leqslant i \leqslant k$, *and for all* $a, b \in T$,

$$a, b \prec 1 \implies a' \prec b^\dagger \tag{9.5.1}$$

$$a \asymp 1,\ b \prec 1 \implies a' \preccurlyeq b^\dagger. \tag{9.5.2}$$

Then L *is an asymptotic field.*

PROOF. Let $g \in L^\times$, $g \prec 1$. Then by the assumptions on T we have

$$g = b \cdot \frac{1 + \sum_{i=1}^m a_i}{1 + \sum_{j=1}^n b_j}$$

where $b \in T$, and $a_i, b_j \in T$ with $a_i, b_j \prec 1$, for $1 \leqslant i \leqslant m$, $1 \leqslant j \leqslant n$. So $g \asymp b \prec 1$; we claim that $g' \asymp b'$. For this, note that $b^\dagger \neq 0$ by (9.5.1), so

$$g^\dagger = b^\dagger \left(1 + \frac{\sum_{i=1}^m a_i'/b^\dagger}{1 + \sum_{i=1}^m a_i} - \frac{\sum_{j=1}^n b_j'/b^\dagger}{1 + \sum_{j=1}^n b_j} \right).$$

By (9.5.1) we have $a_i'/b^\dagger, b_j'/b^\dagger \prec 1$ for $1 \leqslant i \leqslant m$, $1 \leqslant j \leqslant n$. Therefore $g^\dagger \asymp b^\dagger$ and thus $g' \asymp b'$ as claimed.

Now if $f, g \in L^\times$ satisfy $f, g \prec 1$, then this claim and (9.5.1) yield $f' \prec g^\dagger$. Suppose $f, g \in L^\times$ with $f \asymp 1$ and $g \prec 1$. Take $b \in T$ with $b \asymp g$ and $b' \asymp g'$ as above. Since $\operatorname{res}(K) = \operatorname{res}(L)$ we have $f = a + h$ with $a \in K$, $a \asymp 1$ and $h \in L$, $h \prec 1$. By the claim above we find $t \in T \cup \{0\}$ with $h \asymp t$ and $h' \asymp t'$. Now $a' \preccurlyeq b^\dagger$ by (9.5.2) and $t' \prec b^\dagger$ by (9.5.1), so $f' = a' + h' \preccurlyeq b^\dagger \asymp g^\dagger$. The implication (iv) ⇒ (i) of Proposition 9.1.3 now yields that L is an asymptotic field. □

LEMMA 9.5.7. *Let* p *be a prime number, and suppose that* $L = K(u^{1/p})$ *where* $u \in K^\times$ *with* $vu \notin p\Gamma$. *Then* L *is an asymptotic field.*

PROOF. Let $u^{i/p} := (u^{1/p})^i$ for $i \in \mathbb{Z}$, and put $T := \bigcup_{i=0}^{p-1} K^\times u^{i/p}$. Then T is a multiplicative subgroup of $L^\times$. By Lemma 3.1.28, $\operatorname{res}(K) = \operatorname{res}(L)$, and each element of L has the form $t_1 + \cdots + t_k$ with $t_1, \dots, t_k \in T$, $t_1 \succ \cdots \succ t_k$. Now let $a, b \in T$; then $a^p \in K^\times$ and $(a^p)^\dagger = pa^\dagger \asymp a^\dagger$, and similarly $b^p \in K^\times$, $(b^p)^\dagger \asymp b^\dagger$. Hence if $a, b \prec 1$, then $v(a') = va + v\big((a^p)^\dagger\big) = \frac{1}{p}\alpha + \psi(\alpha)$ and $v(b^\dagger) = v\big((b^p)^\dagger\big) = \psi(\beta)$ where $\alpha := pva$, $\beta := pvb \in \Gamma$, and in the asymptotic couple $(\mathbb{Q}\Gamma, \psi)$ we have $\frac{1}{p}\alpha + \psi(\alpha) = (\operatorname{id} + \psi)\big(\frac{1}{p}\alpha\big) > \psi(\beta)$, so $a' \prec b^\dagger$. If $a \asymp 1 \succ b$, then $a' \asymp (a^p)' \preccurlyeq (b^p)^\dagger \asymp b^\dagger$. Hence L is an asymptotic field by Lemma 9.5.6. □

PROOF OF PROPOSITION 9.5.3. We assume that $L|K$ is algebraic. We need to show that then L is an asymptotic field. The property of being an asymptotic field is inherited by valued differential subfields, so we can assume that L is an algebraic closure of K. Next, by Lemma 9.5.1, we can arrange that K is henselian. We then reach L in two steps. In the first step we pass from K to its maximal unramified extension K^{unr} inside L. Then K^{unr} is asymptotic by Lemma 9.5.5. In the second step we obtain L as a purely ramified extension of K^{unr} (Proposition 3.3.48), and now Lemma 9.5.7 applies. □

If K has small derivation and $L|K$ is algebraic, then L has small derivation, by Proposition 6.2.1. If K is of H-type and $L|K$ is algebraic, then L is of H-type.

An application

In this subsection K is an ungrounded H-asymptotic field, and $\Gamma = v(K^\times) \neq \{0\}$. For use in Chapter 11 we extend slightly some results from Section 9.3 by dropping the assumption that Γ is divisible:

COROLLARY 9.5.8. *Let $P \in K\{Y\}^{\neq}$ have order r. Then Corollary 9.3.6 goes through if we drop "with the intermediate value property." Also Corollary 9.3.7 goes through, for a unique tuple $(\alpha, d, d_0, e_1, \dots, e_{r-1})$. In particular, there exists $\beta \in \Gamma^<$ such that $P(y) \neq 0$ for all $y \in K$ with $\beta < vy < 0$.*

PROOF. This follows by applying Corollaries 9.3.6 and 9.3.7 to the algebraic closure K^{a} of K, taking into account that the value group of K^{a} is $\mathbb{Q}\Gamma$, that $\Gamma^<$ is cofinal in $(\mathbb{Q}\Gamma)^<$, that $\Psi_{K^{\mathrm{a}}} = \Psi$, and that the χ-map of the H-asymptotic couple $(\mathbb{Q}\Gamma, \psi)$ extends the χ-map of (Γ, ψ). The "uniqueness" uses Lemma 9.2.20. □

Here is a striking consequence that will not be needed, since in Chapter 13 we prove more precise results for so-called ω-free K (using heavier machinery):

COROLLARY 9.5.9. *Suppose K is existentially closed in some grounded H-asymptotic field extension of K. Then Corollary 9.3.7 holds with all $e_i = 0$, for every P in $K\{Y\}^{\neq}$ of order r.* (See end of B.7 for *existentially closed.*)

PROOF. The assumption on K yields an elementary extension of K with an element $y \succ 1$ such that $K\langle y\rangle$ has $\operatorname{Cl}(y)$ as its smallest comparability class. Let $P \in K\{Y\}^{\neq}$ have order r, let β and $(\alpha, d, d_0, e_1, \dots, e_{r-1})$ be as in Corollary 9.3.7, and suppose towards a contradiction that some $e_i \neq 0$. We have $\beta < vy < 0$, and taking $a \in K^\times$ with $va = \alpha$ we get an element $f = P(y)/ay^{d_0}(y^\dagger)^d$ of $K\langle y\rangle^\times$ such that $vf = -\sum_{i=1}^{r-1} e_i\chi^i(vy)$, so $f \not\asymp 1$ and $\mathrm{Cl}(f) < \mathrm{Cl}(y)$, a contradiction. □

If K is a directed union of grounded asymptotic subfields, then K is existentially closed in some grounded H-asymptotic field extension of K; see B.7.9. Note that $K = \mathbb{T}$ and $K = \mathbb{T}_{\log}$ are such directed unions.

Notes and comments

The above material on algebraic extensions is a mild generalization of Section 3 in [19].

9.6 IMMEDIATE EXTENSIONS OF ASYMPTOTIC FIELDS

We begin by recording analogues for asymptotic fields of the results on maximal immediate extensions from Chapter 6. Next we prove some easy technical facts that are often needed. We also show how to construct immediate extensions by (un)coarsening, and we finish by constructing fluent completions.

Asymptotically maximal immediate extensions

In this subsection K is an asympotic field with small derivation and differential residue field $\boldsymbol{k}$ such that the derivation on $\boldsymbol{k}$ is nontrivial. This is the same condition as in Section 6.9. Recall that the derivation of any immediate asymptotic extension of K is small. Here are two easy consequences of Section 6.9 and Lemmas 9.4.2 and 9.4.5:

COROLLARY 9.6.1. *If K is asymptotically maximal, then K is spherically complete.*

COROLLARY 9.6.2. *K has an immediate asymptotic extension that is spherically complete.*

Easy technical lemmas

In this subsection F is a valued differential field with a subfield K that is asymptotic in F, and with L as an intermediate field, so that $K \subseteq L \subseteq F$. Note that then L is asymptotic in F if for each $f \in L^\times$ with $f \prec 1$ there is $a \in K^\times$ such that $f \asymp a$ and $f' \asymp a'$.

LEMMA 9.6.3. *Suppose $L|K$ is immediate and for all $a \in K^\times$ and $f \in L$,*

$$(a \prec 1, f \prec 1) \implies f' \prec a^\dagger. \tag{9.6.1}$$

Then L is asymptotic in F.

PROOF. Let $a \in K^\times$, $h \in L^\times$ with $h \prec a \prec 1$; we claim that $h' \prec a'$. To see this, set $g := h/a \in L^\times$, so $g \prec 1$. Now apply (9.6.1) with $f = g$ to get $\frac{h}{a} - \frac{h'}{a'} = -\frac{g'a}{a'} \prec 1$; hence $h'/a' \prec 1$ and thus $h' \prec a'$.

Let now $f \in L^\times$, $f \prec 1$. Take $a \in K$ with $f \sim a$, so $h = f - a \prec a \prec 1$. Then $h' \prec a'$, so $f' = a' + h' \asymp a'$ as required. □

LEMMA 9.6.4. *Suppose $L|K$ is immediate, $U \supseteq K$ is a K-linear subspace of L with $L = \{u/w : u, w \in U, w \neq 0\}$, and for all $a \in K^\times$ and $u \in U$,*

$$(a \prec 1, u \prec 1) \implies u' \prec a^\dagger, \tag{9.6.2}$$

$$(a \prec 1, u \asymp 1) \implies u' \preccurlyeq a^\dagger. \tag{9.6.3}$$

Then L is asymptotic in F.

PROOF. We shall verify (9.6.1). Let $f \in L$, $f \prec 1$. Then $f = u/w$ with $u, w \in U$ and $w \neq 0$. After dividing u and w by an element of K asymptotic to w, we may assume $u \asymp f$ and $w \asymp 1$. We have $f' = u'/w - fw'/w$ with $u'/w \asymp u' \prec a^\dagger$ and $fw'/w \asymp fw' \prec a^\dagger$ for all $a \in K^\times$ with $a \prec 1$, by (9.6.2) and (9.6.3), respectively; hence $f' \prec a^\dagger$ for all such a. □

Immediate extensions by coarsening

Let K be an asymptotic field and Δ a convex subgroup of $\Gamma = v(K^\times)$, giving rise to the coarsened valuation

$$\dot{v}\colon K^\times \to \dot{\Gamma} = \Gamma/\Delta$$

with residue field $\dot{K} = \dot{\mathcal{O}}/\dot{o}$. Let $(L, \dot{v})$ be an immediate valued field extension of $(K, \dot{v})$, and denote its residue field by $\dot{L}$, so $\dot{L} = \dot{K}$ after the usual identification. In Section 3.4 we extended the valuation v on K to a valuation $v\colon L^\times \to \Gamma$ such that if $f \in L^\times$ and $f = gu$ with $g \in K^\times$ and $u \in L^\times$, $\dot{v}(u) = 0$, then $v(f) = v(g) + v(\dot{u})$. By Lemma 3.4.5, L with v is an immediate valued field extension of the valued field K, and its coarsening by Δ is the $(L, \dot{v})$ we started with. Let L also be equipped with a derivation ∂ making $(L, \dot{v})$ an asymptotic extension of $(K, \dot{v})$. Then:

LEMMA 9.6.5. *With the above valuation v on L, the valued differential field L is a Δ-immediate asymptotic extension of K.*

PROOF. Let $f \in L$ and $\dot{v}f > 0$; we claim that then $v(f') > \Psi$. To see this, note that $f = g(1+h)$ with $g \in K$, $h \in L$, and $\dot{v}h > 0$, since $(L, \dot{v})$ is an immediate extension of $(K, \dot{v})$. Then $vh > 0$, so $vg = vf > \Delta$. Now

$$f' = g'(1+h) + gh', \quad \text{with } v(g'(1+h)) = vg' > \Psi.$$

Also, $0 < \dot{v}f = \dot{v}g$ gives $\dot{v}(f') = \dot{v}(g') = v(g') + \Delta$, so $v(f') = v(g') + \delta$ with $\delta \in \Delta$, hence $vf' > \Psi + \delta$. With h instead of f this gives $vh' > \Psi + \delta$ for some $\delta \in \Delta$, so $v(gh') > \Psi$ in view of $vg > \Delta$. Thus $v(f') > \Psi$, as claimed.

Now (L, v) is an immediate valued field extension of K by Lemma 3.4.5. Using this fact, the claim above, and Lemma 9.6.3 we show that (L, v) is an asymptotic field. Let $f \in L$ with $vf > 0$; it is enough to show that then $v(f') > \Psi$. As before, $f = g(1+h)$ with $g \in K$, $h \in L$, and $\dot{v}h > 0$. Then $vg > 0$ and $f' = g'(1+h) + gh'$ with $v(g'(1+h)) = vg' > \Psi$. By the claim above applied to h instead of f we have $v(h') > \Psi$, so $v(gh') \geqslant v(h') > \Psi$. Thus $v(f') > \Psi$. □

Fluency and Δ-immediate extensions

Let K be an asymptotic field, Δ a convex subgroup of Γ, and $\dot{v} = v_\Delta$ the coarsening of v by Δ. By Lemma 9.6.5 any proper immediate asymptotic extension of the asymptotic field $(K, \dot{v})$ yields by "uncoarsening" a proper Δ-immediate asymptotic extension of K.

LEMMA 9.6.6. *Assume $\Delta \neq \{0\}$. Then the following are equivalent:*

(i) *K has no proper Δ-immediate asymptotic extension;*

(ii) *every Δ-fluent pc-sequence in K pseudoconverges in K.*

PROOF. The direction (ii) $\Rightarrow$ (i) follows from Corollary 3.4.18. As to (i) $\Rightarrow$ (ii), we prove its contrapositive. Assume there exists a divergent Δ-fluent pc-sequence in K. Take $\delta \in \Delta^{\neq}$ and $\phi \in K^\times$ with $v\phi = \psi(\delta)$. Then $\psi^\phi(\delta) = 0$, so $0 \in \psi^\phi(\Delta^{\neq}) \cap \Delta$. Replacing K by its compositional conjugate K^ϕ we arrange $0 \in \psi(\Delta^{\neq}) \cap \Delta$, so K has small derivation and $\psi(\Delta^{\neq}) \subseteq \Delta$. Then $(K, \dot{v})$ has small derivation, and the derivation of the differential residue field of $(K, \dot{v})$ is nontrivial. Hence $(K, \dot{v})$ has a proper immediate asymptotic extension by Corollary 9.6.1, and this yields a proper Δ-immediate asymptotic extension of K by Lemma 9.6.5. $\square$

By Zorn there exists a Δ-immediate asymptotic extension of K that has no proper Δ-immediate asymptotic extension. Such an extension of K is called a *maximal Δ-immediate asymptotic extension* of K. Let $K(\Delta)$ be a maximal Δ-immediate asymptotic extension of K. By Lemma 9.6.6, if $\Delta \neq \{0\}$, then $K(\Delta)$ is also a maximal Δ-immediate extension of K as a valued field, and so its Δ-coarsening $(K, \dot{v})$ is henselian. Here is a weak "asymptotic" version of Proposition 3.4.22:

PROPOSITION 9.6.7. *Let K be an asymptotic field such that $\Gamma \neq \{0\}$ and $[\Gamma^{\neq}]$ has no least element. Then K has an immediate asymptotic extension which, as a valued field extension of K, is a fluent completion of K.*

PROOF. We follow the construction in the proof of Proposition 3.4.22. Fix a decreasing coinitial sequence (Δ_α) of nontrivial convex subgroups of Γ indexed by the ordinals $\alpha < \lambda$ for some infinite limit ordinal λ. For each α we pick a maximal Δ_α-immediate asymptotic extension $K(\Delta_\alpha)$ of K. We arrange this so that $K(\Delta_\alpha)$ is a valued differential subfield of $K(\Delta_\beta)$ whenever $\alpha < \beta < \lambda$. Put

$$K^{\mathrm{f}} := \bigcup_{\alpha<\lambda} K(\Delta_\alpha).$$

Then K^{f} is an asymptotic extension of K, and K^{f} as a valued field extension of K is a semifluent completion of K as defined in the proof of Proposition 3.4.22. Iterating as in that proof the construction $K \rightsquigarrow K^{\mathrm{f}}$ we eventually arrive at an asymptotic extension of K that is also a fluent completion of K. $\square$

9.7 DIFFERENTIAL POLYNOMIALS OF ORDER ONE

Throughout this section K is an H-asymptotic field with asymptotic couple (Γ, ψ), $\Gamma \neq \{0\}$. We let γ range over Γ and y over $K^\times$, and assume $P \in K\{Y\}^{\neq}$ has order $\leqslant 1$. For such P we improve on what Sections 6.9 and 9.3 yield for differential polynomials of arbitrary order.

Behavior of $v(P(y))$

The goal of this subsection is to show the following:

PROPOSITION 9.7.1. *Suppose $P(0) = 0$. Then there is a finite set $\Delta(P) \subseteq \Gamma$ and a finite partition of $\Gamma \setminus \Delta(P)$ into convex subsets of Γ such that for each set U in this partition we have a strictly increasing function $i_U : U \to \Gamma$ for which*

$$v\big(P(y)\big) = i_U(vy) \text{ whenever } vy \in U.$$

We derive this from a more precise result for homogeneous P. We start with some generalities about functions on ordered abelian groups. Let G be an ordered abelian group and $U \subseteq G$. A function $\eta\colon U \to G$ is said to be **slowly varying** if $\eta(\alpha) - \eta(\beta) = o(\alpha - \beta)$ for all distinct $\alpha, \beta \in U$. Note that then the function $\alpha \mapsto \alpha + \eta(\alpha)\colon U \to G$ is strictly increasing. A key example of a slowly varying function is of course $\psi\colon \Gamma^{\neq} \to \Gamma$. Note that each constant function $U \to G$ is slowly varying, and that if $\eta_1, \eta_2\colon U \to G$ are slowly varying, so are $\eta_1 + \eta_2, \eta_1 - \eta_2$, and

$$\alpha \mapsto \min\big(\eta_1(\alpha), \eta_2(\alpha)\big)\colon\ U \to G.$$

LEMMA 9.7.2. *Let $s \in K$ be given. Then there is a $\gamma_0 \in \Gamma$ and a slowly varying function $\eta\colon \Gamma \setminus \{\gamma_0\} \to \Gamma$ with the following properties:*

(i) *$v(y^\dagger - s) = \eta(vy)$ for all y with $vy \neq \gamma_0$;*

(ii) *for each $\alpha \in \Gamma$ the set $\{\gamma : \gamma \neq \gamma_0,\ \eta(\gamma) \leqslant \alpha\}$ is a union of finitely many disjoint convex subsets of Γ.*

PROOF. Suppose first that $s = a^\dagger$ with $a \in K^\times$. Then $y^\dagger - s = (y/a)^\dagger$, so $v(y^\dagger - s) = \psi(vy - va)$ if $vy \neq va$. In this case we can take $\gamma_0 = va$ and $\eta(\gamma) = \psi(\gamma - va)$.

Next assume that $s \neq y^\dagger$ for all y. Then we take a nonzero ϕ in an elementary extension L of K such that $v(y^\dagger - s) \leqslant v(\phi^\dagger - s)$ for all y. (This ϕ could be an element of K.) Let $\gamma_1 := v(\phi^\dagger - s) \in \Gamma_L$. Then we claim that for all y with $vy \neq v\phi$:

$$v(y^\dagger - s) = \begin{cases} \psi_L(vy - v\phi) & \text{if } \psi_L(vy - v\phi) \leqslant \gamma_1 \\ \gamma_1 & \text{if } \psi_L(vy - v\phi) \geqslant \gamma_1. \end{cases}$$

To see this, let $vy \neq v\phi$. From $v(y^\dagger - s) \leqslant v(\phi^\dagger - s)$ we get $y^\dagger - \phi^\dagger \not\prec s - \phi^\dagger$. Since

$$y^\dagger - s = (y^\dagger - \phi^\dagger) - (s - \phi^\dagger) \text{ and } y^\dagger - \phi^\dagger = (y/\phi)^\dagger,$$

this gives the claim. Note that the claim can also be expressed as:

$$v(y^\dagger - s) = \min\{\psi_L(vy - v\phi), \gamma_1\} \text{ whenever } vy \neq v\phi.$$

If $v\phi \in \Gamma$, then the lemma clearly holds with $\gamma_0 := v\phi$, and if $v\phi \notin \Gamma$, then the lemma holds for any $\gamma_0 \in \Gamma$. □

LEMMA 9.7.3. *Suppose P is homogeneous of degree d. There is a finite set $\Delta(P) \subseteq \Gamma$ and a finite partition of $\Gamma \setminus \Delta(P)$ into convex subsets of Γ such that for each set U in this partition we have a slowly varying function $\eta_U : U \to \Gamma$ for which*

$$v\big(P(y)\big) \;=\; d\,vy + \eta_U(vy) \;\; \textit{whenever } vy \in U.$$

PROOF. We have $P = a_0 Y^d + a_1 Y^{d-1} Y' + \cdots + a_d (Y')^d$ with $a_0, \dots, a_d \in K$, so $P(y) = y^d(a_0 + a_1 z + \cdots + a_d z^d)$, where $z = y^\dagger$. The henselization of K is still H-asymptotic with the same asymptotic couple, so we can assume K is henselian. Now apply Lemma 3.7.6 and the surrounding remarks to the polynomial $a_0 + a_1 Z + \cdots + a_d Z^d$ to partition K into sets $G_1, \dots, G_k$ such that for $i = 1, \dots, k$: G_i is a special disk in K with holes, and we have $b_{i0}, \dots, b_{id}, s_i \in K$ with

$$v\big(P(y)\big) \;=\; d\,vy + \min\{v(b_{ij}) + j\,v(z - s_i) : 0 \leqslant j \leqslant d\} \qquad \text{if } z \in G_i.$$

In addition, either $s_i = 0$ and the condition $z \in G_i$ is of the form

$$v(z - s_{i,1}) \leqslant v(s_{i,1}), \; \dots, \; v(z - s_{i,n(i)}) \leqslant v(s_{i,n(i)}),$$

or $s_i \neq 0$ and the condition $z \in G_i$ is of the form

$$v(z - s_i) > v(s_i),\; v(z - s_{i,1}) \leqslant v(s_{i,1}), \; \dots, \; v(z - s_{i,n(i)}) \leqslant v(s_{i,n(i)})$$

where in both cases $s_{i,1}, \dots, s_{i,n(i)} \in K^\times$. Now use Lemma 9.7.2. □

PROOF OF PROPOSITION 9.7.1. We have

$$P(Y) \;=\; P_1(Y) + \cdots + P_n(Y) \;\; \text{(decomposition into homogeneous parts).}$$

Let d below range over the numbers $1, \dots, n$ for which $P_d \neq 0$; likewise with e. Applying Lemma 9.7.3 to all P_d simultaneously we obtain a finite $\Delta(P) \subseteq \Gamma$ and a finite partition of $\Gamma \setminus \Delta(P)$ into convex subsets of Γ such that for each set U in the partition and each d we have a slowly varying function $\eta_{U,d} : U \to \Gamma$ for which

$$v\big(P_d(y)\big) \;=\; d\,vy + \eta_{U,d}(vy) \;\; \text{whenever } vy \in U.$$

If $d < e$, then the function

$$\gamma \mapsto \big(e\,\gamma + \eta_{U,e}(\gamma)\big) - \big(d\,\gamma + \eta_{U,d}(\gamma)\big) : \; U \to \Gamma$$

is strictly increasing. Hence, after increasing $\Delta(P)$ and refining our partition if necessary, we can arrange that for each U in the partition there is a $d = d_U$ such that for all $e \neq d$:

$$v\big(P_d(y)\big) < v\big(P_e(y)\big) \text{ whenever } vy \in U.$$

Thus $v\big(P(y)\big) = v\big(P_d(y)\big)$ for $d = d_U$ and $vy \in U$, and the proposition follows. □

Evaluation at pc-sequences

In this subsection we fix a pc-sequence (a_ρ) *in* K. Proposition 9.7.6 below is an analogue of Kaplansky's theorem (Lemma 3.2.7) about pc-sequences of algebraic type. It is stronger than Lemma 6.9.3 in not requiring the derivation of K to be small with nontrivial differential residue field, but weaker in assuming that K is H-asymptotic and P has order $\leqslant 1$. Moreover, we never need to replace in our situation (a_ρ) by an equivalent pc-sequence.

LEMMA 9.7.4. *Assume* $P \notin K$ *and* $a_\rho \rightsquigarrow a \in K$. *Then* $P(a_\rho) \rightsquigarrow P(a)$.

PROOF. Replacing a_ρ by $a_\rho - a$ and $P(Y)$ by $P_{+a}(Y) - P(a)$ we can assume that $a_\rho \rightsquigarrow 0$, $P(0) = 0$, and have to show that then $P(a_\rho) \rightsquigarrow 0$. Proposition 9.7.1 gives a finite subset $\Delta(P)$ of Γ and a finite partition of $\Gamma \setminus \Delta(P)$ into finitely many convex subsets of Γ such that for each set U in this partition, $v\big(P(y)\big)$ is strictly increasing as a function of $vy \in U$. Taking U in the partition such that $v(a_\rho) \in U$ eventually, we conclude that $P(a_\rho) \rightsquigarrow 0$. □

COROLLARY 9.7.5. *If* $P \notin K$, *then* $\big(P(a_\rho)\big)$ *is a pc-sequence.*

PROOF. Use that (a_ρ) has a pseudolimit in some elementary extension of K. □

PROPOSITION 9.7.6. *Assume* $P \notin K[Y]$, $P(a_\rho) \rightsquigarrow 0$, *and* $Q(a_\rho) \not\rightsquigarrow 0$ *for every* $Q \in K\{Y\}$ *with* $\mathrm{c}(Q) < \mathrm{c}(P)$. *Then* K *has an immediate asymptotic extension field* $K\langle a\rangle$ *with the following properties:*

(i) $P(a) = 0$ *and* $a_\rho \rightsquigarrow a$;

(ii) *for any* H*-asymptotic extension field* L *of* K *and any* $b \in L$ *with* $P(b) = 0$ *and* $a_\rho \rightsquigarrow b$ *there is a unique embedding* $K\langle a\rangle \to L$ *over* K *that sends* a *to* b.

PROOF. Note first that $P \in K[Y, Y']$ is irreducible. Consider the domain

$$K[y_0, y_1] \;=\; K[Y, Y']/(P), \quad y_0 := Y + (P), \quad y_1 := Y' + (P),$$

and let $K(y_0, y_1)$ be its field of fractions. We extend the valuation v on K to a valuation $v\colon K(y_0, y_1)^\times \to \Gamma$ as follows. Pick a pseudolimit e of (a_ρ) in some H-asymptotic field extension of K. Let $\phi \in K(y_0, y_1)$, $\phi \neq 0$, so $\phi = f(y_0, y_1)/g(y_0)$ with $f \in K[Y, Y']$ of lower degree in Y' than P and $g \in K[Y]^{\neq}$.

CLAIM: $v\big(f(e, e')\big), v\big(g(e)\big) \in \Gamma$, *and* $v\big(f(e, e')\big) - v\big(g(e)\big)$ *depends only on* ϕ *and not on the choice of* (f, g).

To see why this claim is true, note that $f(a_\rho, a_\rho') \not\rightsquigarrow 0$ by the minimality of P, and that $f(a_\rho, a_\rho') \rightsquigarrow f(e, e')$ if $f \notin K$, by Lemma 9.7.4. Hence

$$f(a_\rho, a_\rho') \sim f(e, e') \text{ and } v\big(f(a_\rho, a_\rho')\big) = v\big(f(e, e')\big), \quad \text{eventually.}$$

In particular, $v\big(f(e, e')\big) \in \Gamma$, and likewise, $v\big(g(e)\big) \in \Gamma$. Suppose that also $\phi = f_1(y_0, y_1)/g_1(y_0)$ with $f_1 \in K[Y, Y']$ of lower degree in Y' than P and $g_1 \in K[Y]^{\neq}$.

Then $fg_1 \equiv f_1 g \bmod P$ in $K[Y, Y']$, and thus $fg_1 = f_1 g$ since fg_1 and $f_1 g$ have lower degree in Y' than P. Hence $v\big(f(e,e')\big) - v\big(g(e)\big) = v\big(f_1(e,e')\big) - v\big(g_1(e)\big)$, thus establishing the claim.

This allows us to define $v\colon K(y_0, y_1)^\times \to \Gamma$ by

$$v\phi := v\big(f(e,e')\big) - v\big(g(e)\big).$$

It is routine to check that this map v is a valuation on the field $K(y_0, y_1)$, except maybe for the multiplicative law. For this, for $i = 1, 2$ let $\phi_i \in K(y_0, y_1)^\times$, so $\phi_i = f_i(y_0, y_1)/g_i(y_0)$ with $f_i \in K[Y, Y']^{\neq}$ of lower degree in Y' than P and $g_i \in K[Y]^{\neq}$. Then

$$v(\phi_i) = v\big(f_i(a_\rho, a'_\rho)\big) - v\big(g_i(a_\rho)\big) \quad \text{eventually } (i = 1, 2)$$

and hence

$$v(\phi_1) + v(\phi_2) = v\big((f_1 f_2)(a_\rho, a'_\rho)\big) - v\big((g_1 g_2)(a_\rho)\big) \quad \text{eventually.}$$

We have $f_1 f_2 = (qP)/h + r/h$ where $q, r \in K[Y, Y']$, $h \in K[Y]^{\neq}$ and $\deg_{Y'} r < \deg_{Y'} P$. Then $\phi_1\phi_2 = r(y_0, y_1)/s(y_0)$ where $s := g_1 g_2 h \in K[Y]^{\neq}$, and hence

$$v(\phi_1\phi_2) = v\big(r(a_\rho, a'_\rho)\big) - v\big(s(a_\rho)\big) \qquad \text{eventually.}$$

Now $(f_1 f_2)(a_\rho, a'_\rho) \not\rightsquigarrow 0$ since $f_1(a_\rho, a'_\rho) \not\rightsquigarrow 0$ and $f_2(a_\rho, a'_\rho) \not\rightsquigarrow 0$. Moreover, $P(a_\rho, a'_\rho) \rightsquigarrow 0$ and $r(a_\rho, a'_\rho) \not\rightsquigarrow 0$; hence

$$v\big((f_1 f_2)(a_\rho, a'_\rho)\big) = v\big(r(a_\rho, a'_\rho)\big) - v\big(h(a_\rho)\big) \quad \text{eventually.}$$

Therefore $v(\phi_1\phi_2) = v(\phi_1) + v(\phi_2)$ as desired. Obviously, the value group of the valuation v on $K(y_0, y_1)$ is Γ. Its residue field is that of K. To see this, let $\phi \in K(y_0, y_1)^\times$ and $v\phi = 0$; we shall find $s \in K$ with $v(\phi - s) > 0$. We have $\phi = f(y_0, y_1)/g(y_0)$ with $f \in K[Y, Y']$ of lower degree in Y' than P and $g \in K[Y]^{\neq}$. Multiplying f and g by a suitable element of $K^\times$ we may assume that $v\big(f(e,e')\big) = v\big(g(e)\big) = 0$. By the above we have $f(e,e') \sim f(a_\rho, a'_\rho)$ and $g(e) \sim g(a_\rho)$, eventually, hence $v\left(\frac{f(e,e')}{g(e)} - s\right) > 0$ where $s := \frac{f(a_\rho, a'_\rho)}{g(a_\rho)}$ with large enough ρ, so $v(\phi - s) > 0$ for such s.

We now equip $K(y_0, y_1)$ with the derivation extending the derivation of K such that $y_0' = y_1$. We also set $a := y_0$, so $a' = y_1$ and $K(y_0, y_1) = K\langle a\rangle$. Then $P(a) = 0$, trivially, and $a_\rho \rightsquigarrow a$, as is easily checked.

To show that $K\langle a\rangle$ is an asymptotic field, we use Proposition 9.5.1 with $K(a)$ in the role of K and $L = K\langle a\rangle$. By that proposition it is enough to show that $K(a)$ is asymptotic in $K\langle a\rangle$. To do that we apply Lemma 9.6.4 with $L = K(a)$, $F = K\langle a\rangle$ and $U = K[a]$. Consider elements $u = g(a)$ with $g \in K[Y] \setminus K$, and $b \in K^\times$ such that $b \prec 1$; it suffices to show that then $u' \prec b^\dagger$ if $u \prec 1$, and $u' \preccurlyeq b^\dagger$ if $u \asymp 1$. If $u \asymp 1$, then we take $s \in K$ with $u \sim s$ and use $s' \preccurlyeq b^\dagger$ to reduce to the case $u \prec 1$. So we assume $u \prec 1$. Note that $v(u)$ = eventual value of $v\big(g(a_\rho)\big)$, so $v\big(g(a_\rho)\big) > 0$ eventually, hence $v\big(g(a_\rho)'\big)$ is eventually constant as a function of ρ, and $v\big(g(a_\rho)'\big) > v(b^\dagger)$ eventually. Therefore it is enough to show:

CLAIM: $v(u') =$ *eventual value of* $v\big(g(a_\rho)'\big)$.

Let $g = c_0 + c_1Y + \cdots + c_nY^n$ ($c_i \in K$), and put $g^\partial := c_0' + c_1'Y + \cdots + c_n'Y^n$, so

$$u' = g(a)' = g^\partial(a) + \frac{\partial g}{\partial Y}(a)a', \qquad g(a_\rho)' = g^\partial(a_\rho) + \frac{\partial g}{\partial Y}(a_\rho)a_\rho'.$$

Therefore the claim holds if $P(Y, Y')$ is of degree > 1 in Y', so we can assume

$$P(Y, Y') = P_0(Y) + P_1(Y)Y' \qquad \text{where } P_0, P_1 \in K[Y],\ P_1 \neq 0.$$

Put $R := -P_0/P_1$. Then $P_1(a) \neq 0$ and $a' = R(a)$. Also $P_1(a_\rho) \neq 0$ eventually. We may assume that $P_1(a_\rho) \neq 0$ for all ρ. Then for each ρ we have

$$(9.7.1) \qquad g(a_\rho)' = g^\partial(a_\rho) + \frac{\partial g}{\partial Y}(a_\rho)R(a_\rho) + \frac{\partial g}{\partial Y}(a_\rho)\big(a_\rho' - R(a_\rho)\big).$$

Now

$$v\big(g^\partial(a_\rho) + \frac{\partial g}{\partial Y}(a_\rho)R(a_\rho)\big) = v\big(g^\partial(a) + \frac{\partial g}{\partial Y}(a)R(a)\big)$$

eventually. Also $a_\rho' - R(a_\rho) = P(a_\rho, a_\rho')/P_1(a_\rho)$, hence $v\big(\frac{\partial g}{\partial Y}(a_\rho)\big(a_\rho' - R(a_\rho)\big)\big)$ is eventually strictly increasing. We have

$$v\big(\frac{\partial g}{\partial Y}(a_\rho)\big(a_\rho' - R(a_\rho)\big)\big) > v\big(g^\partial(a) + \frac{\partial g}{\partial Y}(a)R(a)\big)$$

eventually: otherwise,

$$v\big(\frac{\partial g}{\partial Y}(a_\rho)\big(a_\rho' - R(a_\rho)\big)\big) < v\big(g^\partial(a) + \frac{\partial g}{\partial Y}(a)R(a)\big)$$

eventually, and then, by (9.7.1), $v\big(g(a_\rho)'\big)$ would be both eventually constant and eventually strictly increasing. Hence

$$v\big(g(a_\rho)'\big) = v\big(g^\partial(a) + \frac{\partial g}{\partial Y}(a)R(a)\big) = v(u')$$

eventually, proving the claim. Thus $K\langle a\rangle$ is an asymptotic field.

It remains to prove item (ii). But this follows easily from the above, since any b as in (ii) is transcendental over K, and can serve as e in the arguments above. $\square$

Notes and comments

Suppose K has small derivation and P (of order $\leqslant 1$) is homogeneous of degree $d \geqslant 1$, with $d\Gamma = \Gamma$. Then the function $v_P\colon \Gamma \to \Gamma$ is a strictly increasing bijection, by the Equalizer Theorem. For this case, however, we have a more constructive (unpublished) argument leading to the stronger result that the function $\gamma \mapsto \frac{1}{d}v_P(\gamma)\colon \Gamma \to \Gamma$ is ψ-steady.

9.8 EXTENDING H-ASYMPTOTIC COUPLES

Let (Γ, ψ) and (Γ_1, ψ_1) be asymptotic couples. An **embedding**

$$h\colon (\Gamma, \psi) \to (\Gamma_1, \psi_1)$$

is an embedding $h\colon \Gamma \to \Gamma_1$ of ordered abelian groups such that

$$h\big(\psi(\gamma)\big) \;=\; \psi_1\big(h(\gamma)\big) \text{ for } \gamma \in \Gamma^{\neq}.$$

If $\Gamma \subseteq \Gamma_1$ and the inclusion $\Gamma \hookrightarrow \Gamma_1$ is an embedding $(\Gamma, \psi) \to (\Gamma_1, \psi_1)$, then we call (Γ_1, ψ_1) an **extension** of (Γ, ψ).

In the rest of this section we assume (Γ, ψ) to be an H-asymptotic couple, and (Γ_1, ψ_1) to be an asymptotic couple, not necessarily of H-type. Thus ψ is constant on every archimedean class of Γ: for $\alpha, \beta \in \Gamma^{\neq}$ with $[\alpha] = [\beta]$ we have $\psi(\alpha) = \psi(\beta)$.

Most of the extension results in this section come from [19, Section 2], but the restriction to H-asymptotic (Γ, ψ) leads to fewer case distinctions.

LEMMA 9.8.1. *Let $i\colon \Gamma \to G$ be an embedding of ordered abelian groups inducing a bijection $[\Gamma] \to [G]$. Then there is a unique function $\psi_G\colon G^{\neq} \to G$ such that (G, ψ_G) is an H-asymptotic couple and $i\colon (\Gamma, \psi) \to (G, \psi_G)$ is an embedding.*

PROOF. Define $\psi_G(g) := i\big(\psi(\gamma)\big)$ for $g \in G^{\neq}$ and $\gamma \in \Gamma^{\neq}$ with $[g] = \big[i(\gamma)\big]$. Then $\psi_G\colon G^{\neq} \to G$ has the required properties: to check (AC3), pass to the divisible hulls of Γ and G. □

The next lemma and its proof show how to remove a gap.

LEMMA 9.8.2. *Let β be a gap in (Γ, ψ). Then there is an H-asymptotic couple $(\Gamma + \mathbb{Z}\alpha, \psi^{\alpha})$ extending (Γ, ψ) such that*

(i) $\alpha > 0$ *and* $\alpha' = \beta$*;*

(ii) *if* $i\colon (\Gamma, \psi) \to (\Gamma_1, \psi_1)$ *is an embedding and* $\alpha_1 \in \Gamma_1$, $\alpha_1 > 0$, $\alpha_1' = i(\beta)$, *then* i *extends uniquely to an embedding* $j\colon \big(\Gamma + \mathbb{Z}\alpha, \psi^{\alpha}\big) \to (\Gamma_1, \psi_1)$ *with* $j(\alpha) = \alpha_1$.

PROOF. Suppose $(\Gamma + \mathbb{Z}\alpha, \psi^{\alpha})$ is an asymptotic couple that extends (Γ, ψ) and satisfies (i). At this point we do not assume $(\Gamma + \mathbb{Z}\alpha, \psi^{\alpha})$ to be of H-type. Then $\alpha' < (\Gamma^{>})'$ gives $0 < \alpha < \Gamma^{>}$. Since Ψ has no largest element, $[\Gamma^{\neq}]$ has no least element, so $0 < n\alpha < \Gamma^{>}$ for all $n \geqslant 1$, in particular, $\Gamma + \mathbb{Z}\alpha = \Gamma \oplus \mathbb{Z}\alpha$. Hence

$$\psi^{\alpha}(\alpha) \;=\; \alpha' - \alpha \;=\; \beta - \alpha \;>\; \Psi,$$

and thus for all $\gamma \in \Gamma$ and $k \in \mathbb{Z}$ with $\gamma + k\alpha \neq 0$,

$$\psi^{\alpha}(\gamma + k\alpha) \;=\; \begin{cases} \psi(\gamma), & \text{if } \gamma \neq 0, \\ \beta - \alpha, & \text{otherwise.} \end{cases} \tag{9.8.1}$$

It easily follows that $(\Gamma+\mathbb{Z}\alpha,\psi^\alpha)$ is of H-type, and has the universal property (ii). All this assumes the existence of an asymptotic couple $(\Gamma+\mathbb{Z}\alpha,\psi^\alpha)$ that extends (Γ,ψ) and satisfies (i).

To get such a couple, take an ordered abelian group extension $\Gamma^\alpha=\Gamma+\mathbb{Z}\alpha$ of Γ such that $0<n\alpha<\Gamma^>$ for all $n\geqslant 1$ and extend ψ to $\psi^\alpha\colon(\Gamma^\alpha)^{\neq}\to\Gamma^\alpha$ according to (9.8.1); in particular $\alpha+\psi^\alpha(\alpha)=\beta$. It remains to show that $(\Gamma+\mathbb{Z}\alpha,\psi^\alpha)$ is an asymptotic couple. It is tedious but routine to check that (AC1) is satisfied, and (AC2) holds trivially. As to (AC3), let $\gamma+r\alpha,\delta+s\alpha\in\Gamma^{\neq}$ ($\gamma,\delta\in\Gamma$, $r,s\in\mathbb{Z}$) with $\delta+s\alpha>0$ (hence $\delta\geqslant 0$); we have to show $\psi^\alpha(\gamma+r\alpha)<\psi^\alpha(\delta+s\alpha)+(\delta+s\alpha)$. The case $\psi^\alpha(\gamma+r\alpha)\leqslant\psi^\alpha(\delta+s\alpha)$ is obvious, so assume $\psi^\alpha(\gamma+r\alpha)>\psi^\alpha(\delta+s\alpha)$. Then $\delta\neq 0$ by (9.8.1), so $\delta>0$. If moreover $\gamma=0$, then $[0]<[\alpha]<[\Gamma^{\neq}]$ and $\beta<(\Gamma^>)'$ give $\beta-(s+1)\alpha<\delta+\psi(\delta)$, that is,

$$\psi^\alpha(\gamma+r\alpha)-\psi^\alpha(\delta+s\alpha)\ =\ \beta-\alpha-\psi(\delta)\ <\ \delta+s\alpha,$$

as required. Similarly, if $\gamma\neq 0$, we get

$$\psi^\alpha(\gamma+r\alpha)\ =\ \psi(\gamma)\ <\ \psi(\delta)+(\delta+s\alpha),$$

by (AC3) for (Γ,ψ), and $[0]<[\alpha]<[\Gamma^{\neq}]$. □

The universal property (ii) determines $(\Gamma+\mathbb{Z}\alpha,\psi^\alpha)$ up to isomorphism over (Γ,ψ). Note also that $[\Gamma+\mathbb{Z}\alpha]=[\Gamma]\cup\{[\alpha]\}$, so for $\Psi^\alpha:=\psi^\alpha\big((\Gamma+\mathbb{Z}\alpha)^{\neq}\big)$ we have:

(9.8.2) $$\Psi^\alpha\ =\ \Psi\cup\{\beta-\alpha\},\qquad \beta-\alpha\ =\ \max\Psi^\alpha.$$

Lemma 9.8.2 goes through with $\alpha<0$ and $\alpha_1<0$ in place of $\alpha>0$ and $\alpha_1>0$, respectively. In the setting of this modified lemma we have $\Gamma^<<n\alpha<0$ for all $n\geqslant 1$, and (9.8.1) goes through for $\gamma\in\Gamma$ and $k\in\mathbb{Z}$ with $\gamma+k\alpha\neq 0$, and (9.8.2) goes through. So we have really two ways to remove a gap, and this is a pervasive *fork in the road.* In any case, removal of a gap as above leads by (9.8.2) to a grounded H-asymptotic couple, and this is the situation we consider next.

LEMMA 9.8.3. *Assume Ψ has a largest element β. Then there is an H-asymptotic couple $(\Gamma+\mathbb{Z}\alpha,\psi^\alpha)$ extending (Γ,ψ) with $\alpha\neq 0$, $\alpha'=\beta$, such that for any embedding $i\colon(\Gamma,\psi)\to(\Gamma_1,\psi_1)$ and any $\alpha_1\in\Gamma_1^{\neq}$ with $\alpha_1'=i(\beta)$ there is a unique extension of i to an embedding $j\colon(\Gamma+\mathbb{Z}\alpha,\psi^\alpha)\to(\Gamma_1,\psi_1)$ with $j(\alpha)=\alpha_1$.*

PROOF. Suppose $(\Gamma+\mathbb{Z}\alpha,\psi^\alpha)$ is an asymptotic couple extending (Γ,ψ) with $\alpha\neq 0$, $\alpha'=\beta$. In the divisible hull of $(\Gamma+\mathbb{Z}\alpha,\psi^\alpha)$ we have for $\gamma\in(\mathbb{Q}\Gamma)^<$,

$$\gamma'\ =\ \gamma+\psi(\gamma)\ <\ \psi(\gamma)\ \leqslant\ \beta\ =\ \alpha',$$

so $\gamma<\alpha<0$ by Lemma 6.5.4(iii). Hence $\Gamma^<<n\alpha<0$ for all $n\geqslant 1$, in particular, $\Gamma+\mathbb{Z}\alpha\ =\ \Gamma\oplus\mathbb{Z}\alpha$. Also $\psi^\alpha(\alpha)\ =\ \beta-\alpha\ >\ \Psi$, and thus (9.8.1) holds for all $\gamma\in\Gamma$ and $k\in\mathbb{Z}$ with $\gamma+k\alpha\neq 0$. It easily follows that $(\Gamma+\mathbb{Z}\alpha,\psi^\alpha)$ is of H-type, and has the required universal property. All this assumes the existence of an asymptotic couple $(\Gamma+\mathbb{Z}\alpha,\psi^\alpha)$ extending (Γ,ψ) with $\alpha\neq 0$, $\alpha'=\beta$, but the above also suggests how to construct it. The details are similar to those in the proof of Lemma 9.8.2 and are left to the reader. □

Let $(\Gamma + \mathbb{Z}\alpha, \psi^\alpha)$ be as in Lemma 9.8.3. Then $[\Gamma + \mathbb{Z}\alpha] = [\Gamma] \cup \{[\alpha]\}$, so (9.8.2) holds for $\Psi^\alpha := \psi^\alpha\big((\Gamma + \mathbb{Z}\alpha)^{\neq}\big)$. Thus our new Ψ-set Ψ^α still has a maximum, but this maximum is larger than the maximum β of the original Ψ-set Ψ. By iterating this construction indefinitely, taking a union, and passing to the divisible hull, we obtain a divisible H-asymptotic couple with asymptotic integration. Once we have a divisible H-asymptotic couple with asymptotic integration, we can create an extension with a gap as follows:

LEMMA 9.8.4. *Suppose (Γ, ψ) has asymptotic integration and Γ is divisible. Then there is an H-asymptotic couple $(\Gamma + \mathbb{Q}\beta, \psi_\beta)$ extending (Γ, ψ) such that:*

(i) $\Psi < \beta < (\Gamma^>)'$;

(ii) *for any divisible H-asymptotic (Γ_1, ψ_1) extending (Γ, ψ) and $\beta_1 \in \Gamma_1$ with $\Psi < \beta_1 < (\Gamma^>)'$ there is a unique embedding $(\Gamma + \mathbb{Q}\beta, \psi_\beta) \to (\Gamma_1, \psi_1)$ of asymptotic couples that is the identity on Γ and sends β to β_1.*

PROOF. Since Ψ has no largest element, we can take an elementary extension (Γ^*, ψ^*) of (Γ, ψ) with an element $\beta \in \Gamma^*$ such that $\Psi < \beta < (\Gamma^>)'$. Moreover, for each $\alpha \in \Gamma^>$ we have $\alpha^\dagger < \beta < \alpha'$ and $\alpha' - \alpha^\dagger = \alpha$, so Γ is dense in $\Gamma + \mathbb{Q}\beta$ by Lemma 2.4.17. Hence $[\Gamma + \mathbb{Q}\beta] = [\Gamma]$, and thus ψ^* maps $(\Gamma + \mathbb{Q}\beta)^{\neq}$ into Ψ, and with ψ_β the restriction of ψ^* to $(\Gamma + \mathbb{Q}\beta)^{\neq}$ we have an H-asymptotic couple $(\Gamma + \mathbb{Q}\beta, \psi_\beta)$ extending (Γ, ψ) satisfying (i). Let (Γ_1, ψ_1) be a divisible H-asymptotic couple extending (Γ, ψ) and $\beta_1 \in \Gamma_1$, $\Psi < \beta_1 < (\Gamma^>)'$. By the universal property of Lemma 2.4.16 we have a unique embedding $\Gamma + \mathbb{Q}\beta \to \Gamma_1$ of ordered vector spaces over $\mathbb{Q}$ that is the identity on Γ and sends β to β_1. It is routine to check that this embedding is also an embedding $(\Gamma + \mathbb{Q}\beta, \psi_\beta) \to (\Gamma_1, \psi_1)$ of asymptotic couples. □

Let (Γ, ψ) be divisible with asymptotic integration and let $(\Gamma + \mathbb{Q}\beta, \psi_\beta)$ be an H-asymptotic couple as in Lemma 9.8.4. If $(\Gamma + \mathbb{Q}\alpha, \psi_\alpha)$ is also a divisible H-asymptotic couple extending (Γ, ψ) with $\Psi < \alpha < (\Gamma^>)'$, then by (ii) we have an isomorphism $(\Gamma + \mathbb{Q}\beta, \psi_\beta) \to (\Gamma + \mathbb{Q}\alpha, \psi_\alpha)$ of asymptotic couples that is the identity on Γ and sends β to α. In this sense, $(\Gamma + \mathbb{Q}\beta, \psi_\beta)$ is unique up to isomorphism over (Γ, ψ). Thus the construction of $(\Gamma + \mathbb{Q}\beta, \psi_\beta)$ in the proof of Lemma 9.8.4 gives the following extra information, with Ψ_β the set of values of ψ_β on $(\Gamma + \mathbb{Q}\beta)^{\neq}$:

COROLLARY 9.8.5. *The set Γ is dense in the ordered abelian group $\Gamma + \mathbb{Q}\beta$, so $[\Gamma] = [\Gamma + \mathbb{Q}\beta]$, $\Psi_\beta = \Psi$ and β is a gap in $(\Gamma + \mathbb{Q}\beta, \psi_\beta)$.*

Here is a version of Lemma 9.8.4 for possibly non-divisible Γ:

COROLLARY 9.8.6. *Suppose (Γ, ψ) has rational asymptotic integration. Then there is an H-asymptotic couple $(\Gamma + \mathbb{Z}\beta, \psi_\beta)$ extending (Γ, ψ) such that:*

(i) $\Psi < \beta < (\Gamma^>)'$, $k\beta \notin \Gamma$ *for all* $k \in \mathbb{Z}^{\neq}$, *and* $[\Gamma + \mathbb{Z}\beta] = [\Gamma]$;

(ii) *for any H-asymptotic couple (Γ_1, ψ_1) extending (Γ, ψ) and any $\beta_1 \in \Gamma_1$ with $\Psi < \beta_1 < (\Gamma^>)'$ there is a unique embedding $(\Gamma \oplus \mathbb{Z}\beta, \psi_\beta) \to (\Gamma_1, \psi_1)$ of asymptotic couples that is the identity on Γ and sends β to β_1.*

Moreover, β is a gap in $(\Gamma + \mathbb{Z}\beta, \psi_\beta)$ for any such extension of (Γ, ψ).

PROOF. By assumption, $(\mathbb{Q}\Gamma, \psi)$ has asymptotic integration, so we can take an H-asymptotic couple $(\mathbb{Q}\Gamma + \mathbb{Q}\beta, \psi_\beta)$ extending $(\mathbb{Q}\Gamma, \psi)$ and satisfying (i) and (ii) in Lemma 9.8.4, with $(\mathbb{Q}\Gamma, \psi)$ replacing (Γ, ψ). Note that $k\beta \notin \Gamma$ for all $k \in \mathbb{Z}^{\neq}$. By Corollary 9.8.5 we have $\psi_\beta\big((\Gamma + \mathbb{Z}\beta)^{\neq}\big) = \Psi$. Denoting the restriction of ψ_β to $(\Gamma + \mathbb{Z}\beta)^{\neq}$ also by ψ_β we obtain therefore an H-asymptotic couple $(\Gamma \oplus \mathbb{Z}\beta, \psi_\beta)$ extending (Γ, ψ) satisfying (i) and (ii) in the present corollary. □

Recall from Section 2.1 that a cut in an ordered set S is just a downward closed subset of S, and that an element a of an ordered set extending S is said to realize a cut C in S if $C < a < S \setminus C$ (so $a \notin S$).

LEMMA 9.8.7. *Let C be a cut in $[\Gamma^{\neq}]$ and let $\beta \in \Gamma$ be such that $\beta < (\Gamma^{>})'$, $\gamma^\dagger \leqslant \beta$ for all $\gamma \in \Gamma^{\neq}$ with $[\gamma] > C$, and $\beta \leqslant \delta^\dagger$ for all $\delta \in \Gamma^{\neq}$ with $[\delta] \in C$. Then there exists an H-asymptotic couple $(\Gamma \oplus \mathbb{Z}\alpha, \psi^\alpha)$ extending (Γ, ψ), with $\alpha > 0$, such that:*

(i) *$[\alpha]$ realizes the cut C in $[\Gamma^{\neq}]$, and $\psi^\alpha(\alpha) = \beta$;*

(ii) *given any embedding i of (Γ, ψ) into an H-asymptotic couple (Γ_1, ψ_1) and any element $\alpha_1 \in \Gamma_1^{>}$ such that $[\alpha_1]$ realizes the cut $\big\{[i(\delta)] : [\delta] \in C\big\}$ in $\big[i(\Gamma^{\neq})\big]$ and $\psi_1(\alpha_1) = i(\beta)$, there is a unique extension of i to an embedding $j\colon (\Gamma \oplus \mathbb{Z}\alpha, \psi^\alpha) \to (\Gamma_1, \psi_1)$ with $j(\alpha) = \alpha_1$.*

If (Γ, ψ) has asymptotic integration, then so does $(\Gamma \oplus \mathbb{Z}\alpha, \psi^\alpha)$. If (Γ, ψ) has rational asymptotic integration, then so does $(\Gamma \oplus \mathbb{Z}\alpha, \psi^\alpha)$.

PROOF. By Lemma 2.4.5 we can extend Γ to an ordered abelian group $\Gamma^\alpha := \Gamma \oplus \mathbb{Z}\alpha$ with $\alpha > 0$ such that $[\alpha]$ realizes the cut C in $[\Gamma^{\neq}]$. Then $[\Gamma^\alpha] = [\Gamma] \cup \big\{[\alpha]\big\}$. We extend $\psi\colon \Gamma^{\neq} \to \Gamma$ to $\psi^\alpha\colon (\Gamma^\alpha)^{\neq} \to \Gamma$ by

$$\psi^\alpha(\gamma + k\alpha) := \min\{\psi(\gamma), \beta\} \quad \text{for } \gamma \in \Gamma, k \in \mathbb{Z}^{\neq}.$$

(So $\psi^\alpha\big((\Gamma^\alpha)^{\neq}\big) = \Psi \cup \{\beta\}$.) A tedious but routine checking of cases shows that ψ^α decreases on $(\Gamma^\alpha)^{>}$, and that axioms (AC1) and (AC2) for asymptotic couples hold for $(\Gamma^\alpha, \psi^\alpha)$. To verify (AC3), let $\delta = \gamma + k\alpha$ and $\delta^* = \gamma^* + k^*\alpha$ be elements of $(\Gamma^\alpha)^{>}$ ($\gamma, \gamma^* \in \Gamma$, $k, k^* \in \mathbb{Z}$); we have to show that $\psi^\alpha(\delta^*) < \delta + \psi^\alpha(\delta)$. We can assume $[\delta^*] < [\delta]$, since otherwise $\psi^\alpha(\delta^*) \leqslant \psi^\alpha(\delta) < \delta + \psi^\alpha(\delta)$. We distinguish the following cases, using Lemma 6.5.4(i) in Cases 2 and 3:

CASE 1: $[\delta^*] = [\gamma^*]$, $[\delta] = [\gamma]$. Then $\big[\psi(\gamma^*) - \psi(\gamma)\big] < [\gamma^* - \gamma] = [\gamma] = [\delta]$, hence $\psi^\alpha(\delta^*) = \psi(\gamma^*) < \psi(\gamma) + \delta = \psi^\alpha(\delta) + \delta$.

CASE 2: $[\delta^*] = [\alpha]$, $[\delta] = [\gamma]$. Then $[\gamma] > C$, so $\psi(\gamma) \leqslant \beta$. Hence $\psi\big(\beta - \psi(\gamma)\big) > \psi(\gamma)$, so $\big[\beta - \psi(\gamma)\big] < [\gamma] = [\delta]$, and thus $\psi^\alpha(\delta^*) = \beta < \psi(\gamma) + \delta = \psi^\alpha(\delta) + \delta$.

CASE 3: $[\delta^*] = [\gamma^*]$, $[\delta] = [\alpha]$. Then $[\gamma^*] \in C$, so

$$\psi\big(\psi(\gamma^*) - \beta\big) > \min\big\{\psi(\gamma^*), \beta\big\} = \beta,$$

so $[\psi(\gamma^*) - \beta] \in C$ or $\psi(\gamma^*) = \beta$. Hence $[\psi(\gamma^*) - \beta] < [\alpha] = [\delta]$, and thus $\psi^\alpha(\delta^*) = \psi(\gamma^*) < \beta + \delta = \psi^\alpha(\delta) + \delta$.

So $(\Gamma^\alpha, \psi^\alpha)$ is indeed an H-asymptotic couple satisfying (i). That it satisfies (ii) follows easily from the universal property of Lemma 2.4.5.

Assume (Γ, ψ) has asymptotic integration; given $\gamma \in \Gamma$ and $k \in \mathbb{Z}^{\neq}$ we shall find an antiderivative of $\gamma + k\alpha$ in $(\Gamma^\alpha, \psi^\alpha)$. If $\gamma = \beta$, then $k\alpha$ is such an antiderivative, so assume $\gamma \neq \beta$. Lemma 6.5.2 gives the asymptotic couple $(\Gamma, \min(\psi, \beta))$ whose Ψ-set has maximum element β. Since $\gamma \neq \beta$, Theorem 9.2.1 gives $\gamma^* \in \Gamma$ such that $\gamma^* + \min(\psi(\gamma^*), \beta) = \gamma$ and so $\gamma^* + k\alpha + \psi^\alpha(\gamma^* + k\alpha) = \gamma + k\alpha$, that is, $\gamma^* + k\alpha$ is an antiderivative as required.

Preserving rational asymptotic integration is done likewise. □

For $C = \emptyset$ and β a gap in (Γ, ψ), this gives:

COROLLARY 9.8.8. *Let $\beta \in \Gamma$ be a gap in (Γ, ψ). Then there exists an H-asymptotic couple $(\Gamma + \mathbb{Z}\alpha, \psi^\alpha)$ extending (Γ, ψ), such that:*

(i) $0 < n\alpha < \Gamma^>$ *for all* $n \geqslant 1$, *and* $\psi^\alpha(\alpha) = \beta$;

(ii) *for any embedding i of (Γ, ψ) into an H-asymptotic couple (Γ_1, ψ_1) and any $\alpha_1 \in \Gamma_1^>$ with $\psi_1(\alpha_1) = i(\beta)$, there is a unique extension of i to an embedding $j \colon (\Gamma + \mathbb{Z}\alpha, \psi^\alpha) \to (\Gamma_1, \psi_1)$ with $j(\alpha) = \alpha_1$.*

Notes and comments

As already mentioned, much of this section comes from [19, Section 2]. For example, Lemma 9.8.3 is a special case of [19, Lemma 2.12], and Lemma 9.8.7 combines Lemma 2.15 of [19] and a remark that follows the proof of that lemma.

9.9 CLOSED H-ASYMPTOTIC COUPLES

An H-asymptotic couple (Γ, ψ) is said to be **closed** if it is divisible with asymptotic integration and $\Psi := \psi(\Gamma^{\neq})$ is downward closed. At the beginning of Section 10.6 we indicate why the H-asymptotic couple of $\mathbb{T}$ is closed.

By [18], closed H-asymptotic couples admit quantifier elimination. A step in that direction is Proposition 9.9.2 below, which is needed in Section 16.1. In this book we do not need quantifier elimination for closed H-asymptotic couples, but to give some orientation for what follows we mention a consequence of it:

closed = existentially closed (for H-asymptotic couples).

In the first subsection we prove an easy part of this fact, namely that every H-asymptotic couple extends to a closed H-asymptotic couple.

Embedding H-asymptotic couples into closed H-asymptotic couples

Let us construe an asymptotic couple (Γ, ψ) as an ordered group Γ equipped with a binary relation on Γ, namely the graph of $\psi\colon \Gamma^{\neq} \to \Gamma$. In this way the H-asymptotic couples are exactly the models of a set of $\forall\exists$-sentences in the language of ordered abelian groups augmented by a binary relation symbol. By Section B.10 and Lemma B.10.8 this gives the notion of an H-asymptotic couple being *existentially closed*, and the fact that every H-asymptotic couple extends to an existentially closed one.

LEMMA 9.9.1. *Existentially closed H-asymptotic couples are closed.*

PROOF. Let (Γ, ψ) be an existentially closed H-asymptotic couple. Remarks after the proof of Lemma 6.5.3 show that Γ is divisible. Next, it follows from Lemma 9.8.2 that (Γ, ψ) has no gap, from Lemma 9.8.3 that (Γ, ψ) is not grounded (so it has asymptotic integration) and from Lemma 9.8.7 that Ψ is downward closed. □

A closure property of closed H-asymptotic couples

Let (Γ, ψ) be an asymptotic couple. Recall from 6.5 that we extended $\psi\colon \Gamma^{\neq} \to \Gamma$ to a function $\psi\colon \Gamma_\infty \to \Gamma_\infty$ by $\psi(0) = \psi(\infty) := \infty$. For $\alpha_1, \dots, \alpha_n \in \Gamma$, $n \geqslant 1$, we define the function $\psi_{\alpha_1,\dots,\alpha_n}\colon \Gamma_\infty \to \Gamma_\infty$ by recursion on n:

$$\psi_{\alpha_1}(\gamma) := \psi(\gamma - \alpha_1), \qquad \psi_{\alpha_1,\dots,\alpha_n}(\gamma) := \psi\big(\psi_{\alpha_1,\dots,\alpha_{n-1}}(\gamma) - \alpha_n\big) \text{ for } n \geqslant 2.$$

PROPOSITION 9.9.2. *Let (Γ, ψ) be a closed H-asymptotic couple and let (Γ^*, ψ^*) be an H-asymptotic couple extending (Γ, ψ). Suppose $n \geqslant 1$, $\alpha_1, \dots, \alpha_n \in \Gamma$, $q_1, \dots, q_n \in \mathbb{Q}$ and $\gamma \in \Gamma^*$ are such that*

$$\psi^*_{\alpha_1,\dots,\alpha_n}(\gamma) \neq \infty \quad (\textit{so } \psi^*_{\alpha_1,\dots,\alpha_i}(\gamma) \neq \infty \textit{ for } i = 1, \dots, n), \ \textit{and}$$
$$\gamma + q_1\psi^*_{\alpha_1}(\gamma) + \cdots + q_n\psi^*_{\alpha_1,\dots,\alpha_n}(\gamma) \in \Gamma \qquad (\textit{in } \mathbb{Q}\Gamma^*).$$

Then $\gamma \in \Gamma$.

In the rest of this section we establish Proposition 9.9.2.

Some lemmas

Let D be a subset of an ordered abelian group Γ. We say that D is **bounded** if $D \subseteq [p, q]$ for some $p \leqslant q$ in Γ, and otherwise we call D **unbounded**. (These notions and the next one are with respect to the ambient Γ.) A **(convex) component of** D is by definition a nonempty convex subset C of Γ such that $C \subseteq D$ and C is maximal with these properties. The components of D partition the set D: for $d \in D$ the unique component of D containing d is

$$\big\{\gamma \in D^{\leqslant d} : [\gamma, d] \subseteq D\big\} \cup \big\{\gamma \in D^{\geqslant d} : [d, \gamma] \subseteq D\big\}.$$

Now let (Γ, ψ) be an H-asymptotic couple, $n \geqslant 1$, and let α be a sequence $\alpha_1, \dots, \alpha_n$ from Γ. We set

$$D_\alpha := \big\{\gamma \in \Gamma : \psi_\alpha(\gamma) \neq \infty\big\}.$$

Thus

$$D_\alpha = \Gamma \setminus \{\alpha_1\} \qquad \text{for } n = 1, \text{ and}$$
$$D_\alpha = \{\gamma \in D_{\alpha'} : \psi_{\alpha'}(\gamma) \neq \alpha_n\} \qquad \text{for } n > 1 \text{ and } \alpha' = \alpha_1, \dots, \alpha_{n-1}.$$

One checks easily by induction on n that for distinct $\gamma, \gamma' \in D_\alpha$,

$$\psi_\alpha(\gamma) - \psi_\alpha(\gamma') = o(\gamma - \gamma').$$

LEMMA 9.9.3. *Assume Γ is divisible, and let $q_1, \dots, q_n \in \mathbb{Q}$. Then the function*

$$\gamma \mapsto \gamma + q_1\psi_{\alpha_1}(\gamma) + q_2\psi_{\alpha_1,\alpha_2}(\gamma) + \cdots + q_n\psi_\alpha(\gamma) : D_\alpha \to \Gamma$$

is strictly increasing. Moreover, this function has the intermediate value property on every component of D_α.

PROOF. Let $\eta\colon D_\alpha \to \Gamma$ be the function given by

$$\eta(\gamma) := q_1\psi_{\alpha_1}(\gamma) + q_2\psi_{\alpha_1,\alpha_2}(\gamma) + \cdots + q_n\psi_\alpha(\gamma).$$

The function $\gamma \mapsto \gamma + \eta(\gamma)\colon D_\alpha \to \Gamma$ is strictly increasing since $\eta(\gamma) - \eta(\gamma') = o(\gamma - \gamma')$ for all distinct $\gamma, \gamma' \in D_\alpha$. Let C be a component of D_α with $\alpha_1 < C$, and let $\beta < \gamma_1 < \gamma_2$ in C, with $\gamma_2 - \gamma_1 \leqslant \gamma_1 - \beta$. Then

$$0 < \gamma_1 - \alpha_1 < \gamma_2 - \alpha_1 = (\gamma_2 - \gamma_1) + (\gamma_1 - \alpha_1) \leqslant 2(\gamma_1 - \alpha_1),$$

so $\psi(\gamma_1 - \alpha_1) = \psi(\gamma_2 - \alpha_1)$. Hence $\eta(\gamma_1) = \eta(\gamma_2)$, since $\eta(\gamma)$ depends only on $\psi(\gamma - \alpha_1)$. Using terminology of Section 2.4 we have shown that $\eta|C$ is v-slow on the right. Thus by Lemma 2.4.9 the function $\gamma \mapsto \gamma + \eta(\gamma)$ on C has the intermediate value property. For the components $< \alpha_1$ of D_α we can argue likewise. □

In the rest of this subsection the H-asymptotic couple (Γ, ψ) is closed.

LEMMA 9.9.4. *The set D_α has at most 2^n components, and on each of these the function ψ_α is monotone and has the intermediate value property.*

PROOF. For $n = 1$ we have $\alpha = \alpha_1$, and the two components of D_α are $\Gamma^{<\alpha}$, on which ψ_α is increasing, and $\Gamma^{>\alpha}$, on which ψ_α is decreasing. On each of these ψ_α has the intermediate value property, since Ψ is downward closed. Suppose the lemma holds for a certain $\alpha = \alpha_1, \dots, \alpha_n$, let $\alpha_{n+1} \in \Gamma$, and set $\alpha+ := \alpha_1, \dots, \alpha_n, \alpha_{n+1}$. Consider a component C of D_α. Then ψ_α is monotone on C, say increasing on C, and has the intermediate value property on C. Put

$$C_1 := \{\gamma \in C : \psi_\alpha(\gamma) < \alpha_{n+1}\},$$

and similarly define C_2 and C_3, with $=$ and $>$, respectively, replacing $<$. Thus C is the disjoint union of its convex subsets C_1, C_2 and C_3, and $C_1 < C_2 < C_3$. Also $C \cap D_{\alpha+} = C_1 \cup C_3$, $\psi_{\alpha+}$ is clearly increasing on C_1 and decreasing on C_3, and has the intermediate value property on C_1 and on C_3. If both C_1 and C_3 are nonempty, then C_2 is nonempty (because of the intermediate value property of ψ_α on C), and thus C_1 and C_3 are the components of $D_{\alpha+}$ that are contained in C. Otherwise C only contributes one component to $D_{\alpha+}$, or none at all, depending on whether just one or both of C_1 and C_3 are empty. □

At this point we need more terminology. Let $f\colon \Gamma_\infty \to \Gamma_\infty$ be a function, and let C be a nonempty convex subset of Γ on which f does not take the value ∞. Let $p, q \in \Gamma$, and let $S \subseteq \Gamma$ be downward closed. (We only use this for $f = \psi_\alpha$ with C a component of D_α, and $S = \Psi$.)

(1) f **increases on** C **from** p **to** q if $f|C$ is increasing, $p \leqslant q$, and $f(C) = [p, q]$;

(2) f **decreases on** C **from** p **to** q if $f|C$ is decreasing, $p \geqslant q$, and $f(C) = [q, p]$;

(3) f **increases on** C **from** p **to** S if $f|C$ is increasing, and $f(C) = S^{\geqslant p}$;

(4) f **decreases on** C **from** S **to** q if $f|C$ is decreasing, and $f(C) = S^{\geqslant q}$;

(5) f **decreases on** C **from** S **to** $-\infty$ if $f|C$ is decreasing, and $f(C) = S$;

(6) f **decreases on** C **from** p **to** $-\infty$ if $f|C$ is decreasing, and $f(C) = (-\infty, p]$.

Next, let (Γ^*, ψ^*) be a closed H-asymptotic couple that extends our closed H-asymptotic couple (Γ, ψ). Besides $D_\alpha \subseteq \Gamma$ we now also have the set $D^*_\alpha \subseteq \Gamma^*$ with its components, taken relative to the ambient (Γ^*, ψ^*); note that $D^*_\alpha \cap \Gamma = D_\alpha$.

LEMMA 9.9.5. *The components C of D_α have the following properties:*

(i) *C is contained in a (necessarily unique) component C^* of D^*_α, and the map $C \mapsto C^*$ is a bijection from the set of components of D_α onto the set of components of D^*_α, with $C^* \cap \Gamma = C$ for each C;*

(ii) *D_α has a (necessarily unique) unbounded component $C_\infty > \alpha_1$; the corresponding component C^*_∞ of D^*_α is $> \alpha_1$ and unbounded in Γ^*;*

(iii) *for bounded C there are $p, q \in \Gamma$ such that one of the following holds:*

  (a) *ψ_α increases on C from p to q and ψ^*_α increases on C^* from p to q,*

  (b) *ψ_α decreases on C from p to q and ψ^*_α decreases on C^* from p to q,*

  (c) *ψ_α increases on C from p to Ψ and ψ^*_α increases on C^* from p to Ψ^*,*

  (d) *ψ_α decreases on C from Ψ to q and ψ^*_α decreases on C^* from Ψ^* to q;*

(iv) *for the unbounded component $C_\infty > \alpha_1$ of D_α and the corresponding component C^*_∞ of D^*_α, one of the following holds:*

  (a) *ψ_α decreases on C_∞ from Ψ to $-\infty$ and ψ^*_α decreases on C^*_∞ from Ψ^* to $-\infty$,*

  (b) *there is $p \in \Gamma$ such that ψ_α decreases on C_∞ from p to $-\infty$ and ψ^*_α decreases on C^*_∞ from p to $-\infty$.*

Before we start the proof we note that $\alpha_1 \notin D_\alpha$ and $\psi_\alpha(\alpha_1+\gamma) = \psi_\alpha(\alpha_1-\gamma)$ for all $\gamma \in \Gamma$. Thus $\alpha_1+\gamma \mapsto \alpha_1-\gamma \colon \Gamma \to \Gamma$ maps each component $> \alpha_1$ of D_α onto a component $< \alpha_1$ of D_α. The lemma therefore also gives a unique unbounded component $< \alpha_1$ of D_α, with properties symmetric to those for C_∞.

PROOF. We proceed by induction on n. The case $n = 1$ is easy to verify. Suppose the lemma holds for a certain $\alpha = \alpha_1, \dots, \alpha_n$, let $\alpha_{n+1} \in \Gamma$, and set $\alpha+ := \alpha_1, \dots, \alpha_n, \alpha_{n+1}$. By the remark preceding this proof we only need to consider components $> \alpha_1$. Let $C > \alpha_1$ be a component of D_α, and C^* the corresponding component of D^*_α. Set

$$C_1 := \{\gamma \in C : \psi_\alpha(\gamma) < \alpha_{n+1}\}.$$

Similarly define C_2 and C_3, with $=$ and $>$, respectively, replacing $<$, and define the sets C^*_i for $i = 1,2,3$ likewise, replacing C by C^* and ψ_α by ψ^*_α. Hence $C^*_i \cap \Gamma = C_i$, for $i = 1,2,3$. The components of $D_{\alpha+}$ contained in C are the nonempty sets among C_1 and C_3, and the components of $D^*_{\alpha+}$ contained in C^* are the nonempty sets among C^*_1 and C^*_3. The inductive assumption easily gives that for $i = 1,2,3$ we have: $C_i \neq \emptyset \Leftrightarrow C^*_i \neq \emptyset$. This proves (i) and (ii).

Assume that C is bounded in Γ (and hence C^* is bounded in Γ^*). We also assume ψ_α increases on C and ψ^*_α increases on C^*. (The case that ψ_α decreases on C and ψ^*_α decreases on C^* is similar and left to the reader.) We distinguish cases:

CASE 1: *There exist $p, q \in \Gamma$ such that ψ_α increases on C from p to q and ψ^*_α increases on C^* from p to q.* Fix such p, q. Then we have several subcases:

(a) $q \leqslant \alpha_{n+1}$. Then $C_3, C^*_3 = \emptyset$. If $q < \alpha_{n+1}$, then $C_1, C^*_1 \neq \emptyset$, $C_2, C^*_2 = \emptyset$, $\psi_{\alpha+}$ increases on C_1 from $\psi(p-\alpha_{n+1})$ to $\psi(q-\alpha_{n+1})$, and $\psi^*_{\alpha+}$ increases on C^*_1 from $\psi(p-\alpha_{n+1})$ to $\psi(q-\alpha_{n+1})$. If $\alpha_{n+1} = q > p$, then $C_1, C^*_1 \neq \emptyset$, $C_2, C^*_2 \neq \emptyset$, and $\psi_{\alpha+}$ increases on C_1 from $\psi(p-\alpha_{n+1})$ to Ψ, and $\psi^*_{\alpha+}$ increases on C^*_1 from $\psi(p-\alpha_{n+1})$ to Ψ^*. If $\alpha_{n+1} = p = q$, then $C_1 = \emptyset$ and $C^*_1 = \emptyset$.

(b) $\alpha_{n+1} \leqslant p$ *and* $q \neq \alpha_{n+1}$. Then $C_1, C^*_1 = \emptyset$ and $C_3, C^*_3 \neq \emptyset$. If $\alpha_{n+1} < p$, then $C_2, C^*_2 = \emptyset$, and $\psi_{\alpha+}$ decreases on C_3 from $\psi(p-\alpha_{n+1})$ to $\psi(q-\alpha_{n+1})$, and $\psi^*_{\alpha+}$ decreases on C^*_3 from $\psi(p-\alpha_{n+1})$ to $\psi(q-\alpha_{n+1})$. If $\alpha_{n+1} = p$, then $C_2, C^*_2 \neq \emptyset$, $\psi_{\alpha+}$ decreases on C_3 from Ψ to $\psi(q-\alpha_{n+1})$, and $\psi^*_{\alpha+}$ decreases on C^*_3 from Ψ^* to $\psi(q-\alpha_{n+1})$.

(c) $p < \alpha_{n+1} < q$. Then $C_i, C^*_i \neq \emptyset$ $(i = 1,2,3)$. Here, $\psi_{\alpha+}$ increases on C_1 from $\psi(p-\alpha_{n+1})$ to Ψ, and $\psi^*_{\alpha+}$ increases on C^*_1 from $\psi(p-\alpha_{n+1})$ to Ψ^*. Similarly, $\psi_{\alpha+}$ decreases on C_3 from Ψ to $\psi(q-\alpha_{n+1})$, and $\psi^*_{\alpha+}$ decreases on C^*_3 from Ψ^* to $\psi(q-\alpha_{n+1})$.

CASE 2: *There exists $p \in \Gamma$ such that ψ_α increases on C from p to Ψ, and ψ^*_α increases on C^* from p to Ψ^*.* Fix such p, and note that then $p \in \Psi$. This case is essentially treated as Case 1, using Lemmas 9.2.11 and 9.2.13 and some notation introduced after their proofs. If, for example, $\alpha_{n+1} < p$, so that $C_1, C^*_1 = \emptyset$, $C_2, C^*_2 = \emptyset$ and $C_3, C^*_3 \neq \emptyset$, then $\psi_{\alpha+}$ decreases on C_3 from $\psi(p-\alpha_{n+1})$ to $\psi(*-\alpha_{n+1})$, and $\psi^*_{\alpha+}$ decreases on C_3^* from $\psi(p-\alpha_{n+1})$ to $\psi^*(*-\alpha_{n+1})$, which equals $\psi(*-\alpha_{n+1})$. We leave the details to the reader.

Now suppose $C = C_\infty$ is the unbounded component $> \alpha_1$ of D_α. Then $C^* = C_\infty^*$ is the unbounded component $> \alpha_1$ of D_α^*. We have two cases again:

CASE 3: *ψ_α decreases on C from Ψ to $-\infty$, and ψ_α^* decreases on C^* from Ψ^* to $-\infty$.* If $\alpha_{n+1} > \Psi$, then $C_1, C_1^* \neq \emptyset$ and $C_2, C_2^*, C_3, C_3^* = \emptyset$, so $\psi_{\alpha+}$ decreases on C_1 from $\psi(*-\alpha_{n+1})$ to $-\infty$, $\psi_{\alpha+}^*$ decreases on C_1^* from $\psi(*-\alpha_{n+1})$ to $-\infty$, C_1 is the unbounded component $> \alpha_1$ of $D_{\alpha+}$, and C_1^* is the unbounded component $> \alpha_1$ of $D_{\alpha+}^*$. If, on the other hand, $\alpha_{n+1} \in \Psi$, then $C_1, C_1^* \neq \emptyset$, $C_3, C_3^* \neq \emptyset$, $C_1 > C_2 > C_3$, $C_1^* > C_2^* > C_3^*$, $\psi_{\alpha+}$ decreases on C_1 from Ψ to $-\infty$, $\psi_{\alpha+}^*$ decreases on C_1^* from Ψ^* to $-\infty$, $\psi_{\alpha+}$ increases on C_3 from $\psi(*-\alpha_{n+1})$ to Ψ, $\psi_{\alpha+}^*$ increases on C_3^* from $\psi(*-\alpha_{n+1})$ to Ψ^*, the unbounded component $> \alpha_1$ of $D_{\alpha+}$ is C_1, and the unbounded component $> \alpha_1$ of $D_{\alpha+}^*$ is C_1^*.

CASE 4: *There is $p \in \Gamma$ such that ψ_α decreases on C from p to $-\infty$ and ψ_α^* decreases on C^* from p to $-\infty$.* This case is treated like Case 3, except that we now have three subcases, according to whether $\alpha_{n+1} > p$, $\alpha_{n+1} = p$, or $\alpha_{n+1} < p$.

This finishes the inductive step, hence the proof of the lemma. □

Let $q_1, \dots, q_n \in \mathbb{Q}$ and let $\theta\colon D_\alpha \to \Gamma$ be given by

$$\theta(\gamma) := \gamma + q_1\psi_{\alpha_1}(\gamma) + q_2\psi_{\alpha_1,\alpha_2}(\gamma) + \cdots + q_n\psi_\alpha(\gamma).$$

LEMMA 9.9.6. *Let C_∞ be the unbounded component $> \alpha_1$ of D_α. Then θ is not bounded from above on C_∞: for any $\beta \in \Gamma$ there exists $\gamma \in C_\infty$ with $\theta(\gamma) > \beta$. Similarly, θ is not bounded from below on the unbounded component $< \alpha_1$ of D_α.*

PROOF. Note that $[\theta(\gamma) - \theta(\delta)] = [\gamma - \delta]$ for all $\gamma, \delta \in D_\alpha$, and that $[\Gamma]$ has no maximum, by closedness of (Γ, ψ). Given $\beta \in \Gamma$ we pick $\delta \in C_\infty$, and $\gamma > \delta$ such that $[\gamma] > [\delta], [\beta - \theta(\delta)]$. Then $[\theta(\gamma) - \theta(\delta)] > [\beta - \theta(\delta)]$, in particular $\theta(\gamma) > \beta$. □

Proof of Proposition 9.9.2

Recall the setting: (Γ, ψ) is a closed H-asymptotic couple, (Γ^*, ψ^*) is an H-asymptotic couple that extends (Γ, ψ); also, α is a sequence $\alpha_1, \dots, \alpha_n$ with $n \geqslant 1$ and $\alpha_1, \dots, \alpha_n \in \Gamma$, and $q_1, \dots, q_n \in \mathbb{Q}$. This yields the function $\theta\colon D_\alpha \to \Gamma$ as defined at the end of the previous subsection, and likewise we have the function $\theta^*: D_\alpha^* \to \Gamma^*$ given by

$$\theta^*(\gamma) := \gamma + q_1\psi_{\alpha_1}^*(\gamma) + q_2\psi_{\alpha_1,\alpha_2}^*(\gamma) + \cdots + q_n\psi_\alpha^*(\gamma) \qquad (\gamma \in D_\alpha^*).$$

It is clear that $D_\alpha^* \cap \Gamma = D_\alpha$ and that θ^* extends θ. The claim we have to establish is that every $\gamma \in D_\alpha^*$ with $\theta^*(\gamma) \in \Gamma$ lies in Γ. We first note that by extending (Γ^*, ψ^*) further, if necessary, we can arrange that the H-asymptotic couple (Γ^*, ψ^*) is also closed; this uses Lemmas 9.9.1 and B.10.8. In view of the results in the previous subsection it suffices to prove under these conditions:

LEMMA 9.9.7. *Let C be a component of D_α, with corresponding component C^* of D^*_α. If $\delta \in C^* \setminus C$, then $\theta^*(\delta) \in \Gamma^* \setminus \Gamma$.*

PROOF. By induction on n. The case $n = 1$ is easily checked using Lemmas 9.9.3 and 9.2.7. Assume the lemma holds for a certain sequence $\alpha = \alpha_1, \dots, \alpha_n$, and certain $q_1, \dots, q_n \in \mathbb{Q}$. Let $\alpha_{n+1} \in \Gamma$, put $\alpha+ = \alpha_1, \dots, \alpha_n, \alpha_{n+1}$ and let $q_{n+1} \in \mathbb{Q}$. Then we have corresponding functions $\theta_+ : D_{\alpha+} \to \Gamma$ and $\theta^*_+ : D^*_{\alpha+} \to \Gamma^*$ given by

$$\theta_+(\gamma) := \theta(\gamma) + q_{n+1}\psi_{\alpha+}(\gamma),$$
$$\theta^*_+(\gamma) := \theta^*(\gamma) + q_{n+1}\psi^*_{\alpha+}(\gamma).$$

Let C be a component of D_α with corresponding component C^* of D^*_α. Define C_i, C^*_i (for $i = 1, 2, 3$) as in the proof of Lemma 9.9.5. Then the components of $D_{\alpha+}$ contained in C are the nonempty sets among C_1, C_3, and the components of $D^*_{\alpha+}$ contained in C^* are the nonempty sets among C^*_1, C^*_3. We assume $\delta \in C^*_i \setminus C_i$ for $i = 1$ or $i = 3$, and have to show that $\theta^*_+(\delta) \notin \Gamma$. If δ lies in the convex hull of C_i in C^*_i, that is, if there are $p, q \in C_i$ such that $p < \delta < q$, then the injectivity of θ^*_+ and intermediate value property of $\theta_+|[p, q]$ already guarantee that $\theta^*_+(\delta) \in \Gamma^* \setminus \Gamma$, without use of the induction hypothesis. So from now on, we assume that δ does not lie in the convex hull of C_i in C^*_i.

Suppose there exists an element $\beta \in \Gamma$ lying strictly between δ and α_1, and set $\varepsilon := \frac{1}{2}|\beta - \alpha_1| > 0$. Then $\psi^*_{\alpha_1}$ is constant on the segment

$$I = I_\beta := \{\gamma \in \Gamma^* : \delta - \varepsilon \leqslant \gamma \leqslant \delta + \varepsilon\},$$

since $[\gamma - \alpha_1] = [\delta - \alpha_1]$ for all $\gamma \in I$. An easy induction on k gives $I \subseteq D^*_{\alpha_1,\dots,\alpha_k}$ and $\psi^*_{\alpha_1,\dots,\alpha_k}$ is constant on I, for $k = 1, \dots, n+1$. Hence $I \subseteq C^*_i$, $\psi^*_{\alpha+}$ is constant on I, and $\theta^*_+(\gamma) = \theta^*_+(\delta) + \gamma - \delta$ for all $\gamma \in I$. If $I \cap C_i \neq \emptyset$, say $\xi \in I \cap C_i$, then

$$\theta^*_+(\delta) = \theta^*(\delta) + q_{n+1}\psi^*_{\alpha+}(\delta) = \theta^*(\delta) + q_{n+1}\psi_{\alpha+}(\xi) \notin \Gamma,$$

since $\theta^*(\delta) \notin \Gamma$, by the induction hypothesis. Thus for the rest of the proof we assume $I_\beta \cap C_i = \emptyset$ for all $\beta \in \Gamma$ strictly between δ and α_1. Next, by the remark preceding the proof of Lemma 9.9.5 we arrange that δ, C and C^* are all $> \alpha_1$.

We now first consider the case that C is bounded, ψ_α increases on C, ψ^*_α increases on C^*, and $i = 1$. Then $C_1 < C_2 < C_3$ and $C^*_1 < C^*_2 < C^*_3$. The following possibilities arise (see proof of Lemma 9.9.5):

CASE 1: *We have $p \in \Gamma$ such that $\psi_{\alpha+}$ increases on C_1 from p to Ψ and $\psi^*_{\alpha+}$ increases on C^*_1 from p to Ψ^*.* By the proof of Lemma 9.9.5 this gives $C_2 \neq \emptyset$. Since δ is not in the convex hull of C_1 in C^*_1, either $\delta > C_1$ or $\delta < C_1$.

(a) $\delta > C_1$. Taking any $\beta \in C_1$ we have $\alpha_1 < \beta < \delta$, and thus $C_1 < I < C_2$, with $I = I_\beta$ as defined above. Let $\varepsilon = \frac{1}{2}|\beta - \alpha_1| \in \Gamma^>$ be as before, and choose $\xi \in C_1$ so large that $|\alpha_{n+1} - \psi_\alpha(\xi)| \leqslant \varepsilon$. Hence, in $[\Gamma^*]$,

$$[\psi^*_{\alpha+}(\delta) - \psi_{\alpha+}(\xi)] < [\psi^*_\alpha(\delta) - \psi_\alpha(\xi)] \leqslant [\alpha_{n+1} - \psi_\alpha(\xi)] \leqslant [\varepsilon].$$

Let $f(\gamma) := \theta^*(\gamma) + q_{n+1}\psi_{\alpha+}(\xi)$, for $\gamma \in I$. Then $\theta^*_+(\delta) > f(\delta - \varepsilon)$ since

$$\begin{aligned} \theta^*_+(\delta) - f(\delta-\varepsilon) &= \varepsilon + \theta^*_+(\delta-\varepsilon) - f(\delta-\varepsilon) \\ &= \varepsilon + q_{n+1}\big(\psi^*_{\alpha+}(\delta) - \psi_{\alpha+}(\xi)\big). \end{aligned}$$

Likewise $\theta^*_+(\delta) < f(\delta+\varepsilon)$. Hence, by the intermediate value property for f on I (Lemma 9.9.3) we get $\gamma \in I$ with $f(\gamma) = \theta^*_+(\delta)$. Since $I \cap C_1 = \emptyset$, we have $\gamma \notin \Gamma$, so $f(\gamma) \notin \Gamma$ by inductive hypothesis. Thus $\theta^*_+(\delta) \notin \Gamma$.

(b) $\delta < C_1$. Then $\psi^*_{\alpha+}(\gamma) = p$ for all γ such that $\delta \leqslant \gamma < C_1$. In particular $\theta^*_+(\delta) = \theta^*(\delta) + q_{n+1}p \notin \Gamma$, by the induction hypothesis.

CASE 2: *We have $p, q \in \Gamma$ such that $\psi_{\alpha+}$ increases on C_1 from p to q and $\psi^*_{\alpha+}$ increases on C^*_1 from p to q.* Again, either $\delta < C_1$ or $\delta > C_1$. Both subcases are treated as in Case 1(b).

Next we consider the case that C is bounded, ψ_α increases on C and ψ^*_α increases on C^*, and $i = 3$. Either $\psi_{\alpha+}$ decreases on C_3 from Ψ to q and $\psi^*_{\alpha+}$ decreases on C^*_3 from Ψ^* to q, for some $q \in \Gamma$, or $\psi_{\alpha+}$ decreases on C_3 from p to q and $\psi^*_{\alpha+}$ decreases on C^*_3 from p to q, for some $p, q \in \Gamma$. The latter subcase is treated as in Case 2 above. In the first subcase, suppose that $\delta < C_3$. Then $C_2 \neq \emptyset$, hence there exists $\beta \in \Gamma$ with $\alpha_1 < \beta < \delta$, and thus $C_2 < I_\beta < C_3$, with I_β as defined previously. Now for any $\varepsilon \in \Gamma^>$, in particular for $\varepsilon = \frac{1}{2}(\beta - \alpha_1)$, we can choose $\xi \in C_3$ such that $|\alpha_{n+1} - \psi_\alpha(\xi)| \leqslant \varepsilon$. Now continue as in Case 1(a) above. If $\delta > C_3$, argue as in Case 1(b). The case that C is bounded and ψ_α is decreasing on C can be handled in a similar way, and is left to the reader.

Now assume C is unbounded and $i = 1$. Then ψ_α decreases on C by Lemma 9.9.5(iv), so $C_3 < C_2 < C_1$, and C_1 is necessarily the unbounded component $> \alpha_1$ of $D_{\alpha+}$. We have the following cases:

CASE 3: *$\psi_{\alpha+}$ decreases on C_1 from p to $-\infty$ and $\psi^*_{\alpha+}$ decreases on C^*_1 from p to $-\infty$, for some $p \in \Gamma$.* Again, either $\delta < C_1$ or $\delta > C_1$. If $\delta < C_1$, proceed as in Case 1(b) above; if $\delta > C_1$, then $\theta^*_+(\delta) \notin \Gamma$ follows from Lemmas 9.9.6 and 9.9.3.

CASE 4: *$\psi_{\alpha+}$ decreases on C_1 from Ψ to $-\infty$ and $\psi^*_{\alpha+}$ decreases on C^*_1 from Ψ^* to $-\infty$.* If $\delta < C_1$, then inspection of the proof of Lemma 9.9.5 gives $C_2 \neq \emptyset$. Hence there exists $\beta \in \Gamma$ with $\alpha_1 < \beta < \delta$. Now adopt the argument in Case (1)(a) above. If $\delta > C_1$, we again apply Lemma 9.9.6.

Finally, consider the case that C is unbounded and $i = 3$. Then $\psi_{\alpha+}$ increases on C_3 from p to Ψ and $\psi^*_{\alpha+}$ increases on C_3 from p to Ψ^*, for some $p \in \Gamma$. If $\delta > C_3$, note that any $\beta \in C_3$ will satisfy $\alpha_1 < \beta < \delta$, and continue as in Case 1(a). If $\delta < C_3$, argue as in Case 1(b). This finishes the induction. □

Notes and comments

Proposition 9.9.2 is essentially Property (B) on p. 333 of [18], proved there on pp. 336–342. The H-asymptotic couples considered there are equipped with extra structure, but this can be dropped, as pointed out in Section 6 of that paper.

Chapter Ten

H-Fields

Valued differential subfields of differential-valued fields are not always differential-valued, as shown by an example at the end of Section 9.1. They do satisfy an axiom that defines the notion of a *pre-differential-valued* (*pre*-d-*valued*) *field.* In Section 10.1 we upgrade some basic facts on asymptotic fields to pre-d-valued fields; for example, algebraic extensions of pre-d-valued fields are pre-d-valued, not just asymptotic. In Section 10.2 we adjoin integrals (solutions $y = \int f$ of equations $y' = f$) to pre-d-valued fields of H-type; the expression $\int f$ here is purely suggestive and we attach no formal meaning to it. This is used in Section 10.3 to show that every pre-d-valued field of H-type has a canonical d-valued extension. In Section 10.4 we adjoin exponential integrals (solutions $y = \exp(\int g)$ of equations $y^\dagger = g$) to pre-d-valued fields of H-type; again, the use of exp here is only suggestive.

Hardy fields are pre-d-valued fields, but also have a field ordering that interacts with the valuation and derivation. Axiomatizing this interaction yields the notion of a *pre-H-field*; *H-fields* are d-valued pre-H-fields. Our main goal (only reached in Chapter 16) is the model theory of the particular H-field $\mathbb{T}$, but this requires considering H-fields in general and their ordered valued differential subfields, the pre-H-fields. We begin their study in Section 10.5, showing among other things that each pre-H-field has a canonical H-field extension. Applying and adapting earlier adjunction results we show in Section 10.6 that every H-field can be extended in a minimal way to one that is *Liouville closed,* that is, real closed and closed under integration and exponential integration; there are at most two such minimal Liouville closed extensions, up-to-isomorphism. We finish this chapter with some miscellaneous facts about asymptotic fields in Section 10.7.

10.1 PRE-DIFFERENTIAL-VALUED FIELDS

Throughout this section K is a valued differential field. We say that K is **pre-differential-valued** (for short: **pre-d-valued**) if the following holds:

(PDV) for all $f, g \in K^\times$, if $f \preccurlyeq 1$, $g \prec 1$, then $f' \prec g^\dagger$.

Any differential field with the trivial valuation is pre-d-valued. If K is pre-d-valued, then so is K^ϕ for $\phi \in K^\times$. If K is d-valued, then K is pre-d-valued. Any valued differential subfield of a pre-d-valued field is pre-d-valued.

LEMMA 10.1.1. *The following conditions on K are equivalent:*

(i) *K is pre-*d*-valued;*

(ii) *for all $f,h \in K^\times$, if $f \preccurlyeq 1 \not\asymp h$, then $f' \prec h^\dagger$;*

(iii) *K is asymptotic and for all $f,g \in K^\times$, if $f \asymp 1 \not\asymp g$, then $f' \prec g^\dagger$.*

PROOF. The equivalence (i) $\Leftrightarrow$ (ii) is clear. The equivalence (ii) $\Leftrightarrow$ (iii) follows from the equivalence of (i) and (iv) in Proposition 9.1.3. □

From this lemma and $u^\dagger \asymp u'$ for $u \asymp 1$ in K we obtain:

COROLLARY 10.1.2. *If K is pre-*d*-valued and $f,g \in K^\times$, $f \asymp g \not\asymp 1$, then $f^\dagger \sim g^\dagger$.*

From Proposition 9.1.3 and Lemma 10.1.1 we conclude:

COROLLARY 10.1.3. *Ungrounded asymptotic fields are pre-*d*-valued.*

See Section 10.7 for examples of asymptotic fields that are not pre-d-valued. The next lemma characterizes pre-d-valued fields as exactly the valued differential fields whose valuation is trivial on the constant field and that obey a valuation-theoretic version of l'Hospital's Rule:

LEMMA 10.1.4. *The following conditions on K are equivalent:*

(i) *K is pre-*d*-valued;*

(ii) *$C \subseteq \mathcal{O}$, and for all $f,g \in K^\times$, if $f \preccurlyeq g \prec 1$, then $\frac{f}{g} - \frac{f'}{g'} \prec 1$;*

(iii) *$C \subseteq \mathcal{O}$, and for all $f,g \in K^\times$, if $1 \prec f \preccurlyeq g$, then $\frac{f}{g} - \frac{f'}{g'} \prec 1$.*

PROOF. Suppose K is pre-d-valued. Then K is asymptotic by Lemma 10.1.1. Hence $C \subseteq \mathcal{O}$. Let $f,g \in K^\times$ and $f \preccurlyeq g \prec 1$. Then $g' \neq 0$, and

$$\frac{f}{g} - \frac{f'}{g'} = \frac{fg' - f'g}{gg'} = -\frac{(f/g)'}{g^\dagger}.$$

As $f/g \preccurlyeq 1$ and $g \prec 1$, this gives $(f/g)' \prec g^\dagger$, and we have shown (i) $\Rightarrow$ (ii). To prove (ii) $\Rightarrow$ (iii), suppose (ii) holds, and let $f,g \in K$ with $1 \prec f \preccurlyeq g$ and set $a := 1/g$, $b := 1/f$. Then $a \preccurlyeq b \prec 1$, hence $a^2/b \preccurlyeq a \prec 1$. Also,

$$\frac{f}{g} - \frac{f'}{g'} = \frac{1/b}{1/a} - \frac{(1/b)'}{(1/a)'} = \frac{a}{b} - \frac{a^2 b'}{b^2 a'} = \frac{(a^2/b)'}{a'} - \frac{a^2/b}{a},$$

which is infinitesimal by (ii). Finally, we show (iii) $\Rightarrow$ (i). Suppose (iii) holds, and let $f,h \in K^\times$ with $f \preccurlyeq 1 \not\asymp h$. To get $f' \prec h^\dagger$, we may replace f by $f+1$ and h by $1/h$ to arrange $f \asymp 1 \prec h$. Then $1 \prec fh \asymp h$, so

$$f'\frac{h}{h'} = \frac{(fh)'}{h'} - \frac{fh}{h} \prec 1$$

by (iii), and therefore $f' \prec h^\dagger$. Hence K is pre-d-valued by Lemma 10.1.1. □

EXAMPLE. Let $\boldsymbol{k}$ be a differential field, and consider the Hahn field $\boldsymbol{k}((t^{\mathbb{Q}}))$. Let $g \in \boldsymbol{k}((t^{\mathbb{Q}}))$ and make $\boldsymbol{k}((t^{\mathbb{Q}}))$ into a differential field extension of $\boldsymbol{k}$ by

$$\Big(\sum a_q t^q\Big)' = \sum a_q' t^q + \Big(\sum q a_q t^{q-1}\Big) g \qquad (\text{so } t' = g.)$$

Suppose $vg < 1$. Then $\boldsymbol{k}((t^{\mathbb{Q}}))$ is pre-d-valued: just note that for $f \in \boldsymbol{k}((t^{\mathbb{Q}}))^\times$ we have $v(f') = vf + vg - 1$ if $vf \neq 0$, and $v(f') > vg - 1$ if $vf = 0$.

Pre-d-valued fields by coarsening

In this subsection K is an asymptotic field, and Δ is a convex subgroup of $\Gamma = v(K^\times)$. With notation as at the end of Section 9.2, we have the Δ-coarsening $(K, \dot{\preccurlyeq})$ of K, which is again an asymptotic field. The asymptotic fields that arise naturally and are not pre-d-valued are obtained by coarsening the valuation of a pre-d-valued field. This raises the question whether every asymptotic field arises by coarsening the valuation of a pre-d-valued field. In Section 10.7 we show that this is not the case. The next result answers another question: when is a coarsening of an asymptotic field pre-d-valued?

PROPOSITION 10.1.5. *Suppose that $\{0\} \neq \Delta \neq \Gamma$. The following are equivalent:*

(i) $(K, \dot{\preccurlyeq})$ *is a pre-*d*-valued field;*

(ii) *the subset Ψ_Δ of $\dot{\Gamma}$ does not have a largest element;*

(iii) $\psi(\gamma) - \psi(\delta) \notin \Delta$ *for all* $\gamma \in \Gamma \setminus \Delta$, $\delta \in \Delta^{\neq}$;

(iv) $\psi(\gamma) - \psi(\delta) < \Delta$ *for all* $\gamma \in \Gamma \setminus \Delta$, $\delta \in \Delta^{\neq}$.

PROOF. The implication (iv) $\Rightarrow$ (iii) is trivial, (iii) $\Rightarrow$ (ii) follows by Lemma 9.2.24, and (ii) $\Rightarrow$ (i) by Corollary 10.1.3. To show (i) $\Rightarrow$ (iv), let $(K, \dot{\preccurlyeq})$ be pre-d-valued, and let $\gamma \in \Gamma$, $\gamma > \Delta$, $\delta \in \Delta^{\neq}$. Then $\gamma = vg$, $\delta = vf$ with $f, g \in K^\times$, so $\dot{v}(f') > \dot{v}(g^\dagger)$, hence $\Delta > v(g^\dagger) - v(f') = \psi(\gamma) - \big(\delta + \psi(\delta)\big)$. Thus $\psi(\gamma) - \psi(\delta) < \Delta$. □

In view of Lemma 9.2.25 this yields:

COROLLARY 10.1.6. *Suppose $\psi(\Delta^{\neq}) \cap \Delta \neq \emptyset$. Then*

$$(K, \dot{\preccurlyeq}) \text{ is pre-d-valued} \iff \psi(\Gamma \setminus \Delta) \subseteq \Gamma \setminus \Delta.$$

EXAMPLE 10.1.7. Assume K is of H-type and $1 \in \Gamma^>$ satisfies $\psi(1) = 1$. Set

$$\Delta := \big\{\gamma \in \Gamma : \psi^n(\gamma) \geqslant 0 \text{ for some } n \geqslant 1\big\}.$$

Then Δ is a convex subgroup of Γ with $1 \in \psi(\Delta^{\neq}) \cap \Delta$, and $\psi(\Gamma \setminus \Delta) \subseteq \Gamma \setminus \Delta$. Therefore $(K, \dot{\preccurlyeq})$ is pre-d-valued by Corollary 10.1.6. Moreover, any $b \in K^\times$ with $1 \not\asymp b \mathrel{\dot{\asymp}} 1$ yields a gap $\dot{v}(b') = \dot{v}(b^\dagger) = \psi\big(v(b)\big) + \Delta$ of $(K, \dot{\preccurlyeq})$, by Lemma 9.2.24 and the equivalence of (i) and (ii) in Proposition 10.1.5. Since $\Delta \neq \{0\}$, such b exist.

If K is an $\aleph_0$-saturated elementary extension of $\mathbb{T}$, then $\Delta \neq \Gamma$ for this Δ, since for each $n \geqslant 1$ and $\gamma := v\big(\exp^{n+1}(x)\big)$ we have $\psi^m(\gamma) < 0$ for $m = 1, \dots, n$.

Specialization

In this subsection K is an asymptotic field with small derivation and Δ is a convex subgroup of $\Gamma = v(K^\times)$ with $\psi(\Delta^{\neq}) \subseteq \Delta$. We have the residue field $\dot{K} = \dot{\mathcal{O}}/\dot{o}$ of the valuation ring $\dot{\mathcal{O}}$ of $(K, \preccurlyeq)$. Recall from Section 9.4 that $\dot{K}$ with its induced valuation and derivation is an asymptotic field with small derivation having asymptotic couple $(\Delta, \psi|\Delta^{\neq})$. The residue map $a \mapsto \dot{a} \colon \dot{\mathcal{O}} \to \dot{K}$ restricts to a surjective differential ring morphism $\mathcal{O} \to \mathcal{O}_{\dot{K}}$, which in turn restricts to a field embedding $C \to C_{\dot{K}}$, and we identify C with a subfield of $C_{\dot{K}}$ via this embedding. In addition we have a differential field isomorphism $\operatorname{res}(K) \to \operatorname{res}(\dot{K})$, making the diagram

$$\begin{array}{ccccc} C & \hookrightarrow & \mathcal{O} & \longrightarrow & \operatorname{res}(K) \\ \downarrow & & \downarrow & & \downarrow \cong \\ C_{\dot{K}} & \hookrightarrow & \mathcal{O}_{\dot{K}} & \longrightarrow & \operatorname{res}(\dot{K}) \end{array}$$

of differential ring morphisms commutative.

LEMMA 10.1.8. *Suppose K is pre-*d*-valued. Then $\dot{K}$ is pre-*d*-valued. If K is* d*-valued, then $\dot{K}$ is* d*-valued with $C_{\dot{K}} = C$.*

PROOF. Let $f \in \mathcal{O}$ and $g \in o \setminus \dot{o}$, so $\dot{f} \in \mathcal{O}_{\dot{K}}$ and $\dot{g} \in o_{\dot{K}} \setminus \{0\}$. Then $vg \in \Delta^{\neq}$, so $vg' \in \Delta$, hence $v(\dot{f}') \geqslant v(f') > v(g^\dagger) = v(\dot{g}^\dagger)$, so $\dot{K}$ is pre-d-valued. Now assume that K is d-valued. Then the composition $C \hookrightarrow \mathcal{O} \to \operatorname{res}(K)$ is an isomorphism, hence the composition $C_{\dot{K}} \hookrightarrow \mathcal{O}_{\dot{K}} \to \operatorname{res}(\dot{K})$ is also an isomorphism. Therefore $\dot{K}$ is d-valued with $C_{\dot{K}} = C$. □

The case $\sup \Psi = 0$

In this subsection K is an asymptotic field and (Γ, ψ), $\mathcal{O}$, o, and $\boldsymbol{k} = \mathcal{O}/o$ have the usual meaning. We complement here the remarks on small derivation in Section 9.4, focusing on the case $\sup \Psi = 0 \notin \Psi$, where necessarily K has small derivation, is ungrounded, and thus pre-d-valued.

LEMMA 10.1.9. *Suppose $\sup \Psi = 0 \notin \Psi$, and L is a valued differential field extension of K with small derivation and $\Gamma_L = \Gamma$. Then L is pre-*d*-valued.*

PROOF. Since L has small derivation, we have $f' \preccurlyeq 1$ for all $f \preccurlyeq 1$ in L. So it is enough to show that $g^\dagger \succ 1$ whenever $g \in L^\times$ and $g \not\asymp 1$. For such g, take $a \in K^\times$ and $u \in L^\times$ with $g = au$ and $u \asymp 1$. Then $g^\dagger = a^\dagger + u^\dagger$ with $u^\dagger \asymp u' \preccurlyeq 1 \prec a^\dagger$ and hence $g^\dagger \sim a^\dagger \succ 1$. □

Note: $\sup \Psi = 0 \notin \Psi \iff 0 \in \Gamma$ is a gap in K. Thus to get a K with $\Gamma \neq \{0\}$ and $\sup \Psi = 0 \notin \Psi$ we can start with any asymptotic field with a gap and nontrivial valuation—Example 10.1.7 yields pre-d-valued fields of H-type with this property—and then arrange by compositional conjugation that this gap is 0.

EXAMPLE 10.1.10. Suppose $\sup \Psi = 0 \notin \Psi$. Let a in a differential field extension of K be transcendental over K with $a' = 1$; equip $K(a)$ with the valuation in Theorem 6.3.2 for $F := Y' - 1$. Then by Lemma 10.1.9 the valued differential field $K(a)$ is pre-d-valued with value group Γ and $a \asymp 1$.

In view of Corollary 6.3.3 we obtain from Lemma 10.1.9:

COROLLARY 10.1.11. *If* $\sup \Psi = 0 \notin \Psi$ *and* $\boldsymbol{k}_L$ *is a differential field extension of* $\boldsymbol{k}$, *then* K *has a pre*-d-*valued field extension* L *with small derivation, the same value group as* K, *and differential residue field isomorphic to* $\boldsymbol{k}_L$ *over* $\boldsymbol{k}$.

The assumption $\sup \Psi = 0 \notin \Psi$ in Lemma 10.1.9 may seem overly restrictive, but is in fact appropriate in view of the next two lemmas.

LEMMA 10.1.12. *Suppose* $\partial \smallo \subseteq \smallo$ *and* $K\langle Y \rangle$ *with its gaussian valuation is pre*-d-*valued. Then* $\sup \Psi = 0 \notin \Psi$.

PROOF. By Lemma 9.4.6 we have $\sup \Psi = 0$. Also, $f' \prec 1 \asymp Y' \prec g^\dagger$ for all $f, g \in K^\times$ with $f \prec 1$ and $g \not\asymp 1$, and thus $0 \notin \Psi$. □

LEMMA 10.1.13. *Assume* $\partial \smallo \subseteq \smallo$. *Let* $K\langle a \rangle$ *be as in Theorem 6.3.2 with minimal annihilator* F *of* a *over* K *of order* $r \geqslant 1$, *and* $\overline{F} \notin \boldsymbol{k}^\times Y'$. *If* $K\langle a \rangle$ *is pre*-d-*valued, then* $\sup \Psi = 0 \notin \Psi$.

PROOF. By Lemma 9.4.7 and its proof we have $\sup \Psi = 0$ and $a' \asymp 1$. Now argue as in the proof of Lemma 10.1.12 with a replacing Y. □

In Section 9.4 we already mentioned the significant constraint $\sup \Psi = 0$ on d-henselian K. This condition is necessary for K to be merely *embeddable* into a d-henselian asymptotic field; is it also sufficient? Here is a partial answer.

COROLLARY 10.1.14. *Suppose* $\sup \Psi = 0 \notin \Psi$. *Then* K *has a* d-*henselian pre*-d-*valued field extension* L *with* $\Gamma_L = \Gamma$.

PROOF. Since $\boldsymbol{k}$ can be extended to a linearly surjective differential field, Corollary 10.1.11 yields a pre-d-valued field extension E of K with small derivation, the same value group as K, and linearly surjective differential residue field $\boldsymbol{k}_E \supseteq \boldsymbol{k}$. Then Corollary 6.9.5 and Lemma 10.1.9 yield an immediate spherically complete pre-d-valued field extension L of E; any such L is d-henselian. □

If $\max \Psi = 0$, then K also has a d-henselian asymptotic field extension with the same value group, by Lemma 9.4.8 together with facts about the *linear surjective closure of a differential field*, a notion to be developed in the next volume.

Extensions of pre-differential-valued fields I

*In this subsection we assume that L is an asymptotic field extension of the pre-*d*-valued field K.*

LEMMA 10.1.15. *Suppose Ψ is cofinal in Ψ_L, and for all $f \in L^{\preccurlyeq 1}$, either $f' \preccurlyeq a'$ for some $a \in K^{\preccurlyeq 1}$, or $f' \preccurlyeq h'$ for some $h \in L^{\prec 1}$. Then L is pre-*d*-valued.*

PROOF. Let $f, g \in L^\times$, $f \preccurlyeq 1$, $g \not\asymp 1$. Take $b \in K^\times$ with $b \not\asymp 1$ and $b^\dagger \preccurlyeq g^\dagger$. If $a \in K^{\preccurlyeq 1}$ and $f' \preccurlyeq a'$, then $a' \prec b^\dagger$, so $f' \prec g^\dagger$. If $h \in L^{\prec 1}$ and $f' \preccurlyeq h'$, then $h' \prec g^\dagger$, so again $f' \prec g^\dagger$. Thus L satisfies (PDV). □

COROLLARY 10.1.16. *Suppose Ψ is cofinal in Ψ_L, and* $\operatorname{res}(L) = \operatorname{res}(K)$. *Then L is pre-*d*-valued.*

PROOF. Let $f \in L^{\preccurlyeq 1}$. Take $a \in K$ such that $h := f - a \prec 1$. Then $a \preccurlyeq 1$ and $f' = a' + h'$, so $f' \preccurlyeq a'$ or $f' \preccurlyeq h'$. Thus L is pre-d-valued by Lemma 10.1.15. □

COROLLARY 10.1.17. *If $L|K$ is immediate, then L is pre-*d*-valued.*

Corollary 9.1.6 and the previous corollary yield:

COROLLARY 10.1.18. *The completion K^{c} of K is pre-*d*-valued.*

Extensions of pre-differential-valued fields II

*In this subsection K is a pre-*d*-valued field and $L|K$ is a valued differential field extension.* We first record a pre-d-valued version of Lemma 9.5.6 with the same proof:

LEMMA 10.1.19. *Suppose* $\operatorname{res}(K) = \operatorname{res}(L)$, *$T \supseteq K^\times$ is a subgroup of $L^\times$ with $L = K(T)$, each element of $K[T]^{\neq}$ has the form $t_1 + \cdots + t_k$ with $k \geqslant 1$, $t_1, \dots, t_k \in T$ and $t_1 \succ t_i$ for $2 \leqslant i \leqslant k$, and for all $a, b \in T$,*

$$a \preccurlyeq 1,\ b \prec 1 \quad \Longrightarrow \quad a' \prec b^\dagger.$$

*Then L is pre-*d*-valued.*

If $L|K$ is algebraic, then L is asymptotic by Proposition 9.5.3, and so Ψ_L is defined and equals Ψ. Thus by Corollary 10.1.16:

LEMMA 10.1.20. *If $L|K$ is algebraic and* $\operatorname{res}(K) = \operatorname{res}(L)$, *then L is pre-*d*-valued.*

LEMMA 10.1.21. *Suppose that K is henselian and $L|K$ is of finite degree such that* $[L:K] = [\operatorname{res}(L) : \operatorname{res}(K)]$. *Then L is pre-*d*-valued.*

PROOF. The proof of Lemma 9.5.5 shows that for all $f \in L^{\preccurlyeq 1}$ we have $f' \preccurlyeq a'$ for some $a \in K^{\preccurlyeq 1}$. Hence L is pre-d-valued by Lemma 10.1.15. □

PROPOSITION 10.1.22. *Suppose that $L|K$ is algebraic. Then L is pre-*d*-valued.*

PROOF. We first arrange that L is an algebraic closure of K. By Lemma 10.1.20 we next arrange that K is henselian. Using 10.1.21 and 10.1.20 we now obtain that L is pre-d-valued as in the proof of Proposition 9.5.3. □

COROLLARY 10.1.23. *If K is* d*-valued, then its algebraic closure K^{a}, equipped with the unique extension of the derivation of K to a derivation on K^{a} and any valuation on K^{a} extending that of K, is also* d*-valued.*

PROOF. The constant field of K^{a} is an algebraic closure of C (Lemmas 4.1.1, 4.1.2), and the residue field of K^{a} is an algebraic closure of $\operatorname{res}(K)$ (Corollary 3.1.9). □

REMARK. The end of Section 9.1 has an example of a d-valued Hardy field $\mathbb{Q}(x)$ with an algebraic Hardy field extension that is not d-valued.

Notes and comments

Pre-d-valued fields are implicit in [364, Theorem 1] and explicit in [19]. The statement of [364, Theorem 6] is not quite correct; its proof does give Corollary 10.1.23. Proposition 10.1.22 is from [19].

10.2 ADJOINING INTEGRALS

We perform here mild variants of adjunctions done in Sections 4 and 5 of [19]. In contrast to that paper we consider only the case of H-type; this avoids some case distinctions. *In this section we assume that K is an H-asymptotic field.*

LEMMA 10.2.1. *Suppose K is pre-*d*-valued and vs is a gap in K, with $s \in K$. Then K has a pre-*d*-valued extension $K(y)$ of H-type with $y' = s$, $y \prec 1$, and $\operatorname{res} K(y) = \operatorname{res} K$, such that for any asymptotic extension L of K and $z \in L$ with $z' = s$ and $z \prec 1$ there is a unique K-embedding $K(y) \to L$ sending y to z.*

PROOF. Suppose $K(y)$ is an asymptotic extension of K with $y' = s$ and $y \prec 1$. Then $\alpha := vy > 0$ and $\alpha' = vs$, so $0 < n\alpha < \Gamma^{>}$ for all $n \geqslant 1$, by Lemma 9.8.2 and its proof. Hence y is transcendental over K and the asymptotic couple of $K(y)$ is just the H-asymptotic couple $(\Gamma + \mathbb{Z}\alpha, \psi^{\alpha})$ with $\alpha' = \beta$ from Lemma 9.8.2, with $\beta := vs$. Thus $K(y)$ is H-asymptotic and has the desired universal property.

To construct such an extension, take a valued differential field extension $K(y)$ of K with y transcendental over K, $y' = s$, and $0 < n\alpha < \Gamma^{>}$ for all $n \geqslant 1$, where $\alpha := vy$. Note that then $\operatorname{res} K = \operatorname{res} K(y)$ by Lemma 3.1.30. It remains to show that $K(y)$ is pre-d-valued. We have $\Gamma_{K(y)} := v\big(K(y)^{\times}\big) = \Gamma + \mathbb{Z}\alpha = \Gamma \oplus \mathbb{Z}\alpha$. We verify the conditions of Lemma 10.1.19 for $L = K(y)$ and $T := K^{\times} y^{\mathbb{Z}}$. It is clear that every element of $K[T]^{\neq}$ has the form $t_1 + \cdots + t_n$ with $n \geqslant 1$, $t_1, \dots, t_n \in T$, and $t_1 \succ \cdots \succ t_n$. Let $a \in T$; it clearly suffices to show for $\beta := vs$: $a \preccurlyeq 1 \Rightarrow v(a') \geqslant \beta$, and $a \prec 1 \Rightarrow v(a^{\dagger}) < \beta$. If $a \asymp 1$, then $a \in K^{\times}$, so $v(a') \geqslant \beta$; accordingly we assume $a \prec 1$ below. With $a = fy^k$, $f \in K^{\times}$, we have

$$a' = f'y^k + kfy^{k-1}y' = f'y^k + kas/y, \qquad a^{\dagger} = f^{\dagger} + ky^{\dagger}.$$

Either $f \prec 1$, or $f \asymp 1$ and $k > 0$. Consider first the case $f \prec 1$. Then $v(f'y^k) = v(f') + k\alpha > \beta$ and $v(kas/y) \geqslant v(a) + \beta - \alpha \geqslant \beta$, so $v(a') \geqslant \beta$; also $v(f^\dagger) < \beta - \alpha$ and $v(y^\dagger) = \beta - \alpha$, so $v(a^\dagger) < \beta$. Next assume $f \asymp 1$ and $k > 0$. Then we get $v(a') \geqslant \beta$ as before, and $v(f^\dagger) = v(f') \geqslant \beta$ and so $v(a^\dagger) = \beta - \alpha < \beta$. □

REMARKS. Let K and $K(y)$ be as in Lemma 10.2.1. Then $K(y)$ is grounded by (9.8.2). If K is d-valued, then so is $K(y)$ with $C_{K(y)} = C$, by Lemma 9.1.2.

What if we replace $y \prec 1$ in Lemma 10.2.1 by $y \succ 1$? To get a concise answer we restrict ourselves to d-valued K:

LEMMA 10.2.2. *Let K be* d*-valued, and let vs be a gap in K, with $s \in K$. Then K has a* d*-valued extension $K(y)$ of H-type with $y' = s$ and $y \succ 1$ such that for any asymptotic extension L of K and $z \in L$ with $z' = s$ and $z \succ 1$ there is a unique K-embedding $K(y) \to L$ sending y to z.*

PROOF. Similar to that of Lemma 10.2.1, using the remarks following (9.8.2) on the variant of Lemma 9.8.2 when $\alpha < 0$. With α, β, T as in the proof of Lemma 10.2.1 and $a \in T$, one shows: $a \preccurlyeq 1 \Rightarrow v(a') > \beta - \alpha$, and $a \prec 1 \Rightarrow v(a^\dagger) \leqslant \beta - \alpha$; this uses the d-valued assumption on K. □

REMARK. In Lemma 10.2.2 we have $C_{K(y)} = C$ and $K(y)$ is grounded, by (9.8.2).

LEMMA 10.2.3. *Let K be pre-*d*-valued, and let $s \in K$ be such that $vs = \max \Psi$. Then K has a pre-*d*-valued extension $K(y)$ of H-type with $y' = s$ such that for any pre-*d*-valued extension L of K and $z \in L$ with $z' = s$ there is a unique K-embedding $K(y) \to L$ sending y to z.*

PROOF. Suppose $K(y)$ is a pre-d-valued extension of K such that $y' = s$. The assumption that $K(y)$ is pre-d-valued gives $\alpha := vy < 0$, and as at the beginning of the proof of Lemma 9.8.3 (working in the divisible hull of the asymptotic couple of $K(y)$) we see that $\Gamma^{<} < n\alpha < 0$ for all $n \geqslant 1$. Thus y is transcendental over K, and the asymptotic couple of $K(y)$ is the H-asymptotic couple $(\Gamma + \mathbb{Z}\alpha, \psi^\alpha)$ with $\alpha' = \beta := vs$ from Lemma 9.8.3. This makes it clear why $K(y)$ has the universal property stated in the lemma, and it also indicates a way to construct such an extension $K(y)$ along the lines of the proof of Lemma 10.2.1. □

Let K, s, and $K(y)$ be as in Lemma 10.2.3 and set $\alpha := vy$. In view of the remarks following the proof of Lemma 9.8.3 we have $\Psi_{K(y)} = \Psi \cup \{\alpha^\dagger\}$ with $\Psi < \alpha^\dagger$. By Lemma 3.1.30 we have $\operatorname{res} K(y) = \operatorname{res} K$, and by Corollary 4.6.13 and $s \notin \partial K$ we have $C_{K(y)} = C$. If K is d-valued, then so is $K(y)$, by Lemma 9.1.2.

LEMMA 10.2.4. *Suppose K is henselian, $s \in K$, $vs \in (\Gamma^{>})'$, $s \notin \partial\mathcal{O}$, and*

$$S := \{v(s - a') : a \in \mathcal{O}\}$$

has no largest element. Let $L = K(y)$ be a field extension of K with y transcendental over K, and let L be equipped with the unique derivation extending the derivation of K such that $y' = s$. Then there is a unique valuation of L that makes it an H-asymptotic *extension of K with $y \not\asymp 1$. With this valuation L is an immediate extension of K with $y \prec 1$, and so L is pre-*d*-valued if K is.*

PROOF. Let $\kappa = \mathrm{cf}(S)$, so κ is an infinite cardinal, and let ρ, σ, τ range over ordinals $< \kappa$. Let (a_ρ) be a sequence in $\mathcal{O}$ such that $\big(v(s - a'_\rho)\big)$ is a strictly increasing sequence in S, and cofinal in S. Then $s - a'_\rho \asymp (a_\sigma - a_\rho)'$ for $\sigma > \rho$, hence $(a_\tau - a_\sigma)' \prec (a_\sigma - a_\rho)'$ and so $a_\tau - a_\sigma \prec a_\sigma - a_\rho$ for $\tau > \sigma > \rho$; thus (a_ρ) is a pc-sequence in K. Also, (a_ρ) has no pseudolimit in K: suppose $a_\rho \rightsquigarrow a \in K$; then $a - a_\rho \asymp a_\sigma - a_\rho \prec 1$ for $\sigma > \rho$, so $a \in \mathcal{O}$; moreover, $a' - a'_\rho \asymp a'_\sigma - a'_\rho$ for $\sigma > \rho$, hence $v(s - a') > v(s - a'_\rho)$ for all ρ, contradicting the cofinality property of the $v(s - a'_\rho)$. With $P(Y) := Y' - s$ we have $P(a_\rho) \rightsquigarrow 0$, and since K is henselian, we have $Q(a_\rho) \not\rightsquigarrow 0$ for all $Q(Y) \in K[Y]^{\neq}$. Hence the hypotheses of Proposition 9.7.6 are satisfied. The first part of that proposition then yields a valuation of L that makes it an immediate H-asymptotic extension of K with $y \prec 1$. Let any valuation of L be given that makes it an H-asymptotic extension of K with $y \not\asymp 1$. Then $y \prec 1$, since $y \succ 1$ implies $vs = v(y') < (\Gamma^{>})'$. Also $y' = s$ gives that $\big(v(y' - a'_\rho)\big)$ is strictly increasing, and so is $\big(v(y - a_\rho)\big)$, and thus $a_\rho \rightsquigarrow y$. It remains to use Proposition 9.7.6, in particular part (ii). □

REMARKS. Suppose K and s are as in Lemma 10.2.4 and E is an asymptotic field extension of K with an element $z \prec 1$ such that $z' = s$. Then z is transcendental over K. This is because $a'_\rho \rightsquigarrow z'$ for the pc-sequence (a_ρ) in the proof of that lemma, so $a_\rho \rightsquigarrow z$; it remains to recall that (a_ρ) is of transcendental type over K.

The assumption in Lemma 10.2.4 that $\big\{v(s - a') : a \in \mathcal{O}\big\}$ has no maximum is always satisfied for pre-d-valued K, by part (iii) of the next lemma.

LEMMA 10.2.5. *Suppose K is pre-*d*-valued and $s \in K$. Then:*

(i) *if $\big\{v(s - a') :\ a \in K\big\}$ has a maximum $\beta \in \Gamma_\infty$, then $\beta \notin (\Gamma^{\neq})'$;*

(ii) *if K has asymptotic integration and $s \notin \partial K$, then $\big\{v(s - a') :\ a \in K\big\}$ has no maximum;*

(iii) *if $vs \in (\Gamma^{>})'$, and $s \notin \partial\mathcal{O}$, then $\big\{v(s - a') : a \in \mathcal{O}\big\}$ has no maximum.*

PROOF. Suppose $\max\{v(s - a') : a \in K\} = \beta \in (\Gamma^{\neq})'$. Take $a \in K$ and $b \in K^\times$, $b \not\asymp 1$, such that $\beta = v(s - a') = v(b')$. Next, take $u \in K$ with $u \asymp 1$ and $s - a' = ub'$. Then $u' \prec b^\dagger$ as K is pre-d-valued, hence $s - (a + ub)' = -u'b \prec b'$, contradicting the maximality of β. This proves (i), and (ii) is an immediate consequence. The proof of (iii) is like that of (i), with $(\Gamma^{>})'$ and $a, b \in \mathcal{O}$ instead of $(\Gamma^{\neq})'$ and $a, b \in K$. □

LEMMA 10.2.6. *Suppose K is henselian, $s \in K$, $v(s - a') < (\Gamma^{>})'$ for all $a \in K$, and $S := \big\{v(s - a') :\ a \in K\big\}$ has no largest element. Let $L = K(y)$ be a field extension of K with y transcendental over K, and let L be equipped with the unique derivation extending the derivation of K such that $y' = s$. Then there is a unique valuation of L making it an H-asymptotic extension of K. With this valuation L is an immediate extension of K with $y \succ 1$, and so L is pre-*d*-valued if K is.*

PROOF. Note that $vs \in S$. Since S has no largest element, each $\alpha \in S$ satisfies $\alpha < \gamma$ for some $\gamma \in \Psi$. Set $\kappa = \mathrm{cf}(S)$, and let ρ, σ, τ range over ordinals $< \kappa$. Take a sequence (a_ρ) in K such that $v(s - a'_\rho)$ is strictly increasing and cofinal in S as a function of ρ, and $s - a'_\rho \prec s$ for all ρ. Then $s - a'_\rho \sim a'_\sigma - a'_\rho$ for $\sigma > \rho$, and hence $(a_\tau - a_\sigma)' \prec (a_\sigma - a_\rho)'$ for $\tau > \sigma > \rho$. Note that $a_\sigma - a_\rho \succ 1$ for $\sigma > \rho$, since otherwise $s - a'_\sigma \sim (a_{\sigma+1} - a_\sigma)' \prec (a_\sigma - a_\rho)'$, so $v(s - a'_\sigma) \in (\Gamma^>)'$, contradicting the hypothesis. Hence $a_\tau - a_\sigma \prec a_\sigma - a_\rho$ for $\tau > \sigma > \rho$, so (a_ρ) is a pc-sequence in K. From $a_\sigma - a_\rho \succ 1$ for $\sigma < \rho$ we get $a_\rho \preccurlyeq 1$ for at most one ρ, and so we can arrange that $a_\rho \succ 1$ for all ρ. If $a_\rho \rightsquigarrow a \in K$, then for some index ρ_0 we have $a - a_\rho \asymp a_\sigma - a_\rho \succ 1$ for $\sigma > \rho > \rho_0$, and so $a' - a'_\rho \asymp a'_\sigma - a'_\rho$ for $\sigma > \rho > \rho_0$, hence $v(s - a') > v(s - a'_\rho)$ for all ρ, contradicting the cofinality property of the $v(s - a'_\rho)$. Thus the pc-sequence (a_ρ) in K is divergent.

Now apply Proposition 9.7.6 to $P(Y) := Y' - s$ as in the proof of Lemma 10.2.4 to get a valuation of L making it an H-asymptotic extension of K. Assume any such valuation is given. Then $y \succ 1$, since $y \preccurlyeq 1$ implies $vs = v(y') \geqslant \Psi$, so $v(s - a') \in (\Gamma^>)'$ for some $a \in K$, a contradiction. Also, $y' = s$ gives that $\big(v(y' - a'_\rho)\big)$ is strictly increasing, and so is $\big(v(y - a_\rho)\big)$, and thus $a_\rho \rightsquigarrow y$. It remains to use Proposition 9.7.6, in particular part (ii). □

By alternating the passage to a henselization with the extension procedures of Lemmas 10.2.4 and 10.2.6 we obtain the following:

PROPOSITION 10.2.7. *Let K be* d*-valued with asymptotic integration. Then K has an immediate asymptotic extension $K(\int)$ such that:*

(i) *$K(\int)$ is henselian and closed under integration;*

(ii) *$K(\int)$ embeds over K into any henselian* d*-valued H-asymptotic extension of K that is closed under integration.*

PROOF. Define an *integration tower* on K to be a strictly increasing chain $(K_\lambda)_{\lambda \leqslant \mu}$ of immediate asymptotic extensions of K, indexed by the ordinals less than or equal to some ordinal μ, such that: $K_0 = K$; if λ is a limit ordinal, $0 < \lambda \leqslant \mu$, then $K_\lambda = \bigcup_{\iota < \lambda} K_\iota$; for $\lambda < \lambda + 1 \leqslant \mu$, *either* $K_{\lambda+1}$ is a henselization of K_λ, *or* K_λ is already henselian, $K_{\lambda+1} = K_\lambda(y_\lambda)$ with $y_\lambda \notin K_\lambda$ (hence y_λ is transcendental over K_λ), $y'_\lambda = s_\lambda \in K_\lambda$, and (1) or (2) below holds, where $(\Gamma_\lambda, \psi_\lambda)$ denotes the asymptotic couple of K_λ:

(1) $v(s_\lambda) \in (\Gamma_\lambda^>)'$, and $s_\lambda \neq \varepsilon'$ for all $\varepsilon \in K_\lambda^{\prec 1}$;

(2) $S_\lambda := \big\{v(s_\lambda - a') : a \in K_\lambda\big\} < (\Gamma_\lambda^>)'$.

Note that in (1) we can arrange $y_\lambda \prec 1$ by subtracting a constant from y_λ, and that in (2) the set S_λ has no largest element and $y_\lambda \succ 1$. This is relevant in connection with the uniqueness of the valuations in Lemmas 10.2.4 and 10.2.6.

Take a maximal integration tower $(K_\lambda)_{\lambda \leqslant \mu}$ on K, where "maximal" means that it cannot be extended to an integration tower $(K_\lambda)_{\lambda \leqslant \mu+1}$ on K. Then the top K_μ of the tower has the properties stated about $K(\int)$. □

COROLLARY 10.2.8. *With K and $K(\int)$ as in Proposition 10.2.7, the only henselian asymptotic subfield of $K(\int)$ containing K and closed under integration is $K(\int)$.*

PROOF. Let F be the intersection of all henselian asymptotic subfields $L \supseteq K$ of $K(\int)$ that are closed under integration. Then F itself is among these L, and so F is the smallest such L. Since F is a d-valued field extension of K closed under integration, $K(\int)$ embeds into F over K, and the image of any such embedding is also among these L, and thus equals F. In particular, $K(\int)$ is K-isomorphic to F, and so inherits the minimality property of F. □

Suppose K is d-valued and has asymptotic integration. The minimality property of Corollary 10.2.8 yields that an asymptotic extension $K(\int)$ as in Proposition 10.2.7 is unique up to isomorphism over K: if L is also an asymptotic extension of K with the properties of $K(\int)$ as in that proposition, then there is a K-embedding of L into $K(\int)$, and the image of this embedding is then $K(\int)$.

Here is an obvious consequence of Proposition 10.2.7:

COROLLARY 10.2.9. *Spherically complete* d*-valued fields of H-type with asymptotic integration are closed under integration.*

Notes and comments

Lemmas 10.2.1–10.2.5 are variants of [19, 4.1, 4.2, 4.3, 5.1]. That paper also has 10.2.7–10.2.9 without H-type assumption. Another proof of 10.2.9, also without assuming H-type, is in [235]. If C is algebraically closed, then the differential field $K(\int)$ from Proposition 10.2.7 contains the Picard-Vessiot antiderivative closure of K constructed in [279].

10.3 THE DIFFERENTIAL-VALUED HULL

Valued differential subfields of d-valued fields are pre-d-valued. We prove here a strong converse in the H-type case. *Throughout this section K is a pre-*d*-valued field of H-type.*

THEOREM 10.3.1. *The pre-*d*-valued field K of H-type has a* d*-valued extension* $\operatorname{dv}(K)$ *of H-type such that any embedding of K into any* d*-valued field L of H-type extends uniquely to an embedding of* $\operatorname{dv}(K)$ *into L.*

The universal property in the theorem determines $\operatorname{dv}(K)$ up to unique isomorphism over K of valued differential field extensions of K. We call $\operatorname{dv}(K)$ the **differential-valued hull of K.**

PROOF OF THEOREM 10.3.1. If K is not d-valued, then for some $b \asymp 1$ in K we have $b' \notin \partial\smallo$, and as K is of H-type, either $v(b') \in (\Gamma^{>})'$ or $v(b') < (\Gamma^{>})'$ for such b, and in the latter case K has a gap $v(b')$. For the purpose of this proof, call K *nice* if there is no $b \asymp 1$ in K such that $v(b')$ is a gap in K. Thus if K has no gap (in particular, if K is grounded), then K is nice. Also, if K is d-valued, then K is nice. If K is nice, so is any immediate pre-d-valued extension. If K is nice, put $K_0 := K$; otherwise, take $b \asymp 1$ in K such that $v(b')$ is a gap in K, put $s := b'$, and take $K_0 := K(y)$ with $y \prec 1$ as in Lemma 10.2.1. Thus $\operatorname{res}(K_0) = \operatorname{res}(K)$ and K_0 is nice. Starting with K_0 and iterating and alternating applications of Proposition 10.1.22 and Lemma 10.2.4 we obtain an immediate henselian d-valued extension K_1 of K_0 such that any embedding of K into any henselian d-valued field L of H-type extends to an embedding of K_1 into L; this also uses the remarks following Lemma 10.2.4.

Let D be the constant field of K_1; so D maps isomorphically onto $\operatorname{res}(K_1) = \operatorname{res}(K)$ under the residue map $\mathcal{O}_{K_1} \to \operatorname{res}(K_1)$. Put $\operatorname{dv}(K) := K(D)$, a d-valued subfield of K_1. To show that $\operatorname{dv}(K)$ has the desired universal property, let $i\colon K \to L$ be any embedding of K into a d-valued field L of H-type. Extend i to an embedding $i_1\colon K_1 \to L^{\mathrm{h}}$. Then $i_1(D) \subseteq C_{L^{\mathrm{h}}} = C_L \subseteq L$, so $i_1\big(\operatorname{dv}(K)\big) \subseteq L$. This gives an embedding $i_1|\operatorname{dv}(K)\colon \operatorname{dv}(K) \to L$ that extends i. Given $d \in D$, any such embedding must send d to the element of C_L whose residue class in $\operatorname{res}(L)$ equals the natural i-image of $\operatorname{res}(d) \in \operatorname{res}(K)$ in $\operatorname{res}(L)$. This gives the required uniqueness. □

The proof of the theorem above provides extra information:

(a) $\operatorname{res} K = \operatorname{res} \operatorname{dv}(K)$;

(b) $\operatorname{dv}(K) = K(D)$ where D is the constant field of $\operatorname{dv}(K)$;

(c) $\operatorname{dv}(K)$ is d-algebraic over K, as a consequence of (b).

COROLLARY 10.3.2. *The value group of* $\operatorname{dv}(K)$ *is as follows:*

(i) *Assume K has no gap; then* $\Gamma_{\operatorname{dv}(K)} = \Gamma$.

(ii) *Assume K has a gap β and no $b \asymp 1$ in K satisfies $v(b') = \beta$; then* $\Gamma_{\operatorname{dv}(K)} = \Gamma$.

(iii) *Assume K has a gap β and $b \asymp 1$ in K satisfies $v(b') = \beta$. Let a be the unique element of* $\operatorname{dv}(K)$ *with $a' = b'$ and $a \prec 1$. Then the asymptotic couple of* $\operatorname{dv}(K)$ *is $(\Gamma + \mathbb{Z}\alpha, \psi^{\alpha})$ as in Lemma 9.8.2, with $\alpha := va$.*

PROOF. We use the notation and terminology from the proof of Theorem 10.3.1. If K has no gap, or K has a gap β and there is no $b \asymp 1$ in K with $v(b') = \beta$, then K is nice, hence K_1 is an immediate extension of $K_0 = K$, and so $\operatorname{dv}(K)$ is an immediate extension of K. This shows (i) and (ii). Suppose K has a gap β and $b \asymp 1$ in K satisfies $v(b') = \beta$. Then $K_0 = K(y)$ where $y' = b'$ and $y \prec 1$, and K_1 is an immediate extension of K_0. Thus $\operatorname{dv}(K)$ is an immediate extension of K_0, and (iii) follows from Lemma 10.2.1, taking $a := y \in K(D) = \operatorname{dv}(K)$. □

If K is d-valued, then of course $\operatorname{dv}(K) = K$, and the assumption in (i) or (ii) of Corollary 10.3.2 holds. The assumption in Corollary 10.3.2(iii) holds for $(K, \dot{\preccurlyeq})$ as in Example 10.1.7. If K, a are as in Example 10.1.10, then $K(a)$, 0, a also satisfy the assumption on K, β, b in Corollary 10.3.2(iii).

In Section 16.3 we shall need the following:

LEMMA 10.3.3. *Suppose K has asymptotic integration. Let L be a pre-*d*-valued field extension of K of H-type and suppose L is algebraic over K and has a gap β. Then there is no $b \asymp 1$ in L with $v(b') = \beta$.*

PROOF. Identify $\mathrm{dv}(K)$ with a valued differential subfield of $\mathrm{dv}(L)$ via the embedding $\mathrm{dv}(K) \to \mathrm{dv}(L)$ that extends the inclusion $K \to L$. Now $\mathrm{dv}(K)$ is an immediate extension of K by Corollary 10.3.2 and item (a) preceding that corollary, and $\mathrm{dv}(L)$ is algebraic over $\mathrm{dv}(K)$ by items (a) and (b) preceding the corollary. It follows that $\Gamma_{\mathrm{dv}(L)}/\Gamma$ is a torsion group, and so $\Gamma_{\mathrm{dv}(L)}/\Gamma_L$ is a torsion group. Now apply part (iii) of Corollary 10.3.2 to L in the role of K. □

Notes and comments

Theorem 10.3.1 without H-type restriction is in [19].

10.4 ADJOINING EXPONENTIAL INTEGRALS

In this section K is an H-asymptotic field with asymptotic couple (Γ, ψ), $\Gamma \neq \{0\}$, and a, b range over $K^\times$ and j, k over $\mathbb{Z}$. Given $s \in K$ we wish to adjoin an exponential integral $\exp(\int s)$ to K. *In Proposition 10.4.1 and Lemmas 10.4.2–10.4.6 below, we assume $s \in K^\times$ is such that $s \neq a^\dagger$ for all a, and $K(f)$ is a field extension of K with f transcendental over K, equipped with the unique derivation extending the derivation of K such that $f^\dagger = s$.* If K is algebraically closed, then the constant field of $K(f)$ is C, by Lemmas 4.1.1, 4.6.11, and Corollary 4.6.12.

PROPOSITION 10.4.1. *Suppose K is algebraically closed and* d*-valued. Then there is a valuation of $K(f)$ that makes it a* d*-valued extension of H-type of K, and for any such valuation with asymptotic couple (Γ_f, ψ_f) of $K(f)$,*

$$\Gamma_f = \Gamma \iff v(s - a^\dagger) \in (\Gamma^>)' \text{ for some } a.$$

The proof consists of several lemmas, which give more precise information.

LEMMA 10.4.2. *Let $K(f)$ carry a valuation making it a* d*-valued extension of K with value group $\Gamma_f = \Gamma$. Then $v(s - a^\dagger) \in (\Gamma^>)'$ for some a.*

PROOF. Since $vf \in \Gamma$, we have a and g such that $f = ag$, $g \in K(f)$, and $g \asymp 1$. Then $s - a^\dagger = g^\dagger \asymp g'$. Thus $v(s - a^\dagger) = v(g') \in (\Gamma^>)'$. □

LEMMA 10.4.3. *Assume K is henselian and pre-*d*-valued, and $vs \in (\Gamma^>)'$. Then there is a unique valuation of $K(f)$ that makes it an H-asymptotic extension of K with $f - 1 \not\prec 1$. With this valuation $K(f)$ is pre-*d*-valued, and an immediate extension of K with $f \sim 1$.*

PROOF. Put $S := \{v(s - (1+\varepsilon)^\dagger) : \varepsilon \in o\} \subseteq (\Gamma^>)'$.

CLAIM 1: *The set S has no largest element.*

To see this, note first that $vs \in S$. Let $\gamma \in S$ with $\gamma \geqslant vs$, and take $\varepsilon \in \mathcal{O}$ with $\gamma = v\big(s - (1+\varepsilon)^\dagger\big)$. Since $(\Gamma^>)'$ is upward closed, we have $b \in \mathcal{O}$ with $v(b') = \gamma$. Take $u \in K$ with $u \asymp 1$ and $s - (1+\varepsilon)^\dagger = ub'$. Now $v(u'b) > v(b') = \gamma$, so with $\delta \in \mathcal{O}$ such that $(1+\varepsilon)(1+ub) = 1+\delta$ we get

$$s - (1+\delta)^\dagger \;=\; s - (1+\varepsilon)^\dagger - (1+ub)^\dagger \;=\; ub' - \frac{(ub)'}{1+ub} \;=\; \frac{u^2bb' - u'b}{1+ub},$$

hence $v\big(s - (1+\delta)^\dagger\big) > \gamma$. This proves Claim 1.

Let $\kappa = \operatorname{cf}(S)$, so κ is an infinite cardinal; let ρ, σ, τ range over the ordinals $< \kappa$. Take a sequence (ε_ρ) in $\mathcal{O}$ such that $v\big(s - (1+\varepsilon_\rho)^\dagger\big)$ is strictly increasing as a function of ρ, and cofinal in S. In particular, $(1+\varepsilon_\rho)^\dagger \rightsquigarrow s$.

CLAIM 2: *(ε_ρ) is a pc-sequence in K.*

For $\rho < \sigma$ we have $s - (1+\varepsilon_\rho)^\dagger \sim (1+\varepsilon_\sigma)^\dagger - (1+\varepsilon_\rho)^\dagger$. Also, for $\rho \neq \sigma$, $(\varepsilon_\sigma - \varepsilon_\rho)' \succ \varepsilon_\rho'(\varepsilon_\sigma - \varepsilon_\rho)$, hence

$$(1+\varepsilon_\sigma)^\dagger - (1+\varepsilon_\rho)^\dagger \;=\; \frac{(1+\varepsilon_\rho)(\varepsilon_\sigma - \varepsilon_\rho)' - \varepsilon_\rho'(\varepsilon_\sigma - \varepsilon_\rho)}{(1+\varepsilon_\rho)(1+\varepsilon_\sigma)} \;\sim\; (\varepsilon_\sigma - \varepsilon_\rho)'.$$

It follows that $(\varepsilon_\tau - \varepsilon_\sigma)' \prec (\varepsilon_\sigma - \varepsilon_\rho)'$ for $\tau > \sigma > \rho$. Hence $\varepsilon_\tau - \varepsilon_\sigma \prec \varepsilon_\sigma - \varepsilon_\rho$ for $\tau > \sigma > \rho$. Thus (ε_ρ) is a pc-sequence.

CLAIM 3: *The pc-sequence (ε_ρ) has no pseudolimit in K.*

To see this, suppose $\varepsilon_\rho \rightsquigarrow \varepsilon \in K$. Then $\varepsilon \in \mathcal{O}$ and $\varepsilon - \varepsilon_\rho \asymp \varepsilon_\sigma - \varepsilon_\rho$ for $\sigma > \rho$, so $(\varepsilon - \varepsilon_\rho)' \asymp (\varepsilon_\sigma - \varepsilon_\rho)'$ for $\sigma > \rho$. Computations as in the proof of Claim 2 then give $(1+\varepsilon)^\dagger - (1+\varepsilon_\rho)^\dagger \asymp (1+\varepsilon_\sigma)^\dagger - (1+\varepsilon_\rho)^\dagger$ for $\sigma > \rho$. Hence $(1+\varepsilon_\rho)^\dagger \rightsquigarrow (1+\varepsilon)^\dagger$, which in view of $(1+\varepsilon_\rho)^\dagger \rightsquigarrow s$ gives $v\big(s - (1+\varepsilon)^\dagger\big) > v\big(s - (1+\varepsilon_\rho)^\dagger\big)$ for all ρ, contradicting the choice of (ε_ρ) . This proves Claim 3.

With $P(Y) := Y' - (1+Y)s$ we have $P(\varepsilon_\rho) \rightsquigarrow 0$ and $P(y) = 0$ for $y := f - 1$. By the claims we can apply Proposition 9.7.6 to $P(Y)$ to get a valuation of $K(f)$ that makes it an immediate H-asymptotic extension of K with $\varepsilon_\rho \rightsquigarrow y$.

As to uniqueness, assume $K(f)$ is given a valuation making it an H-asymptotic extension of K with $y \not\asymp 1$. Then $y \prec 1$, since $y \succ 1$ gives $y \sim y+1 = f$, and so $v(y^\dagger) = v(f^\dagger) = vs \in (\Gamma^>)'$, which is impossible. From $(1+\varepsilon_\rho)^\dagger \rightsquigarrow s = (1+y)^\dagger$ we obtain as in the proof of Claim 2 that $\varepsilon_\rho \rightsquigarrow y$, and so part (ii) of Proposition 9.7.6 yields the desired uniqueness. □

LEMMA 10.4.4. *Let K be algebraically closed and* d*-valued, and suppose a satisfies $v(s - a^\dagger) \in (\Gamma^>)'$. Then some valuation of $K(f)$ makes it a* d*-valued extension of H-type of K; any such valuation makes $K(f)$ an immediate extension of K.*

PROOF. By Lemma 10.4.3 with $s - a^\dagger$ and f/a instead of s and f we get a valuation of $K(f)$ making it a d-valued extension of H-type of K. Let $K(f)$ be equipped with any such valuation. Then $(f/a)^\dagger = s - a^\dagger$, hence $f \asymp a$, so $f/a = c(1+z)$ with $c \in C^\times$ and $z \prec 1$, and then $(1+z)^\dagger = s - a^\dagger$. Thus by Lemma 10.4.3, the valued field $K(f) = K(1+z)$ is an immediate extension of K. □

This takes care of Proposition 10.4.1 in the case that $v(s - a^\dagger) \in (\Gamma^>)'$ for some a. It remains to consider the case that $v(s - a^\dagger) < (\Gamma^>)'$ for all a. This condition is for d-valued K equivalent to:

$$s - a^\dagger \succ u' \quad \text{for all } a \text{ and for all } u \in K^{\preccurlyeq 1}. \tag{10.4.1}$$

In the next two lemmas we presuppose that (10.4.1) *holds,* but do *not* assume that K is algebraically closed or d-valued, since the extra generality will be of some use. (Of course we keep the assumption that K is an H-asymptotic field with asymptotic couple (Γ, ψ), $\Gamma \neq \{0\}$.) Put

$$S := \{v(s - a^\dagger) : a \in K^\times\} \subseteq \Gamma,$$

so $S < (\Gamma^>)'$. Taking $a = 1$ shows that $vs \in S$. The next lemma is the most substantial result of this section.

LEMMA 10.4.5. *Suppose S has no maximum and Γ is divisible. Then there is a unique valuation on $K(f)$ that makes $K(f)$ an H-asymptotic extension of K with asymptotic couple (Γ_f, ψ_f). Moreover, for this valuation we have:*

(i) $vf \notin \Gamma$, $\Gamma_f = \Gamma \oplus \mathbb{Z}vf$, $[\Gamma_f] = [\Gamma]$, *and* $\operatorname{res} K(f) = \operatorname{res} K$*;*

(ii) *if K has small derivation, then so does $K(f)$;*

(iii) *if K is pre-d-valued, then so is $K(f)$;*

(iv) *if K is d-valued, then so is $K(f)$, with the same constant field.*

PROOF. Since S has no largest element, Lemma 9.2.15 and Corollary 9.2.16 yield $S \subseteq (\Gamma^<)'$. Let the infinite cardinal κ be the cofinality of S, let ρ, σ, τ range over the ordinals $< \kappa$, and let (a_ρ) be a sequence in $K^\times$ such that the sequence $\big(v(s - a_\rho^\dagger)\big)$ is strictly increasing and cofinal in S with $v(s - a_\rho^\dagger) > vs$ for all ρ. Then $s \sim a_\rho^\dagger$ for all ρ. Since $vs \in (\Gamma^<)'$, and $b^\dagger \asymp b'$ if $b \asymp 1$, this gives $a_\rho \not\asymp 1$ for all ρ by (i) of Corollary 9.1.4. Also,

$$s - a_\rho^\dagger \asymp a_\sigma^\dagger - a_\rho^\dagger \asymp a_\rho^\dagger - a_\sigma^\dagger \qquad \text{for } \rho < \sigma,$$

since s is a pseudolimit of the pc-sequence $(a_\rho^\dagger)$. Hence

$$(a_\rho/a_\sigma)^\dagger = a_\rho^\dagger - a_\sigma^\dagger \succ a_\sigma^\dagger - a_\tau^\dagger = (a_\sigma/a_\tau)^\dagger \qquad \text{for } \rho < \sigma < \tau.$$

We have $v(a_\rho/a_\sigma) \neq 0$ for $\rho < \sigma$: otherwise $s - a_\rho^\dagger \asymp (a_\rho/a_\sigma)^\dagger \asymp (a_\rho/a_\sigma)'$ and $v\big((a_\rho/a_\sigma)'\big) > (\Gamma^<)'$ by (i) of Corollary 9.1.4, a contradiction. Put $\alpha_\rho := v(a_\rho)$, so $\alpha_\rho - \alpha_\sigma = v(a_\rho/a_\sigma)$ and $\psi(\alpha_\rho - \alpha_\sigma) = v(s - a_\rho^\dagger)$ for $\rho < \sigma$. Thus (α_ρ) is a pc-sequence with respect to the valuation ψ on the ordered abelian group Γ.

CLAIM 1: *For $\alpha = va$ we have $\psi(\alpha - \alpha_\rho) = v(a^\dagger - a_\rho^\dagger) = v(a^\dagger - s)$, eventually.*

This is because $a^\dagger - a_\rho^\dagger = (a^\dagger - s) + (s - a_\rho^\dagger)$, and $a^\dagger - s \succ s - a_\rho^\dagger$, eventually. Since $\alpha \in \Gamma$ in Claim 1 is arbitrary, it follows that (α_ρ) has no pseudolimit in (Γ, ψ).

Suppose now that $K(f)$ is equipped with a valuation that makes it an H-asymptotic field extension of K with asymptotic couple (Γ_f, ψ_f). Then $\eta := vf \notin \Gamma$: otherwise $f = ua$ for some a and $u \in K(f)$ with $u \asymp 1$; for such a and u we have $u^\dagger \asymp u' \prec b'$ for all $b \succ 1$ by (i) of Corollary 9.1.4, in particular $u^\dagger \prec s - a_\rho^\dagger$ and therefore

$$a^\dagger - a_\rho^\dagger \;=\; f^\dagger - u^\dagger - a_\rho^\dagger \;=\; (s - a_\rho^\dagger) - u^\dagger \;\asymp\; s - a_\rho^\dagger$$

for all ρ, contradicting Claim 1. This yields $\Gamma_f = \Gamma \oplus \mathbb{Z}\eta$. Now $\psi_f(\eta - \alpha_\rho) = v(s - a_\rho^\dagger)$ for all ρ, so $\alpha_\rho \rightsquigarrow \eta$ in (Γ_f, ψ_f), and for $\alpha = va$, Claim 1 gives

$$\psi_f(\alpha - \eta) \;=\; v(a^\dagger - s) \;=\; v(a^\dagger - a_\rho^\dagger) \;=\; \psi(\alpha - \alpha_\rho), \quad \text{eventually.}$$

Hence the sequence $\big([\eta - \alpha_\rho]\big)$ is strictly decreasing in $[\Gamma_f]$, and $[\alpha - \eta] = [\alpha - \alpha_\rho]$ eventually. Hence $[\Gamma_f] = [\Gamma]$. It also follows that for all $\alpha \in \Gamma$,

$$\alpha < \eta \Longleftrightarrow \alpha < \alpha_\rho \text{ eventually}, \qquad \alpha > \eta \Longleftrightarrow \alpha > \alpha_\rho \text{ eventually}.$$

This determines the ordering on Γ_f. Hence there is at most one valuation on $K(f)$ making $K(f)$ an H-asymptotic field extension of K.

We now construct such a valuation. The valuation $\psi : \Gamma^{\neq} \to \Psi$ is coarser than the standard valuation of Γ, so (α_ρ) has no pseudolimit in Γ with respect to the standard valuation by Lemma 2.2.21. Hence Lemma 2.4.21 gives an element η in an ordered abelian group extending Γ such that $\eta \notin \Gamma$, the sequence $\big([\eta - \alpha_\rho]\big)$ is eventually strictly decreasing, and $[\Gamma_f] = [\Gamma]$, where $\Gamma_f := \Gamma \oplus \mathbb{Z}\eta$. By Lemma 9.8.1 we have a unique extension of $\psi\colon \Gamma^{\neq} \to \Gamma$ to a map $\psi_f\colon \Gamma_f^{\neq} \to \Gamma$ such that (Γ_f, ψ_f) is an asymptotic couple of H-type. Now Γ is divisible, so $\Gamma^{>}$ is coinitial in $\Gamma_f^{>}$. Therefore, if (Γ, ψ) has small derivation, then so does (Γ_f, ψ_f).

CLAIM 2: *$\psi_f(va + j\eta) = v(a^\dagger + js) \in S$ for all a and all $j \neq 0$.*

To prove Claim 2, let a and $j \neq 0$ be given, and take $u, d \in K^\times$ such that $a = ud^{-j}$ and $u \asymp 1$ (which is possible since Γ is divisible). Then $a^\dagger = -jd^\dagger + u^\dagger$ and $u^\dagger \asymp u' \prec s - d^\dagger$, so

$$\begin{aligned} v(a^\dagger + js) \;&=\; v(js - jd^\dagger + u^\dagger) \;=\; v(s - d^\dagger) \in S, \\ \psi_f(va + j\eta) \;&=\; \psi_f\big(j(-vd + \eta)\big) \;=\; \psi_f(\eta - vd), \end{aligned}$$

so it only remains to show that $\psi_f(\eta - vd) = v(s - d^\dagger)$. Since $\big([\eta - \alpha_\rho]\big)$ is strictly decreasing, there are arbitrarily large ρ with $[\eta - vd] = [\alpha_\rho - vd]$. For those ρ we have $\alpha_\rho \neq vd$ (since $\eta \neq vd$), and thus

$$\psi_f(\eta - vd) \;=\; \psi(\alpha_\rho - vd) \;=\; v(a_\rho^\dagger - d^\dagger).$$

By cofinality of $\big(v(s - a_\rho^\dagger)\big)$ in S we get $v(a_\rho^\dagger - d^\dagger) = v(s - d^\dagger)$ for all sufficiently large ρ. Hence $\psi_f(\eta - vd) = v(s - d^\dagger)$ as required.

We now equip $K(f)$ with the valuation $v\colon K(f)^\times \to \Gamma_f$ that extends the valuation of K and such that $vf = \eta$.

CLAIM 3: *$K(f)$ is an asymptotic field with asymptotic couple (Γ_f, ψ_f).*

By Claim 2 we have:

$$v\big((af^j)^\dagger\big) \;=\; \psi_f(va+j\eta) \qquad \text{if } va+j\eta \neq 0. \tag{10.4.2}$$

Therefore, if $va+j\eta, vb+k\eta > 0$, then

$$v\big((af^j)'\big) \;=\; (va+j\eta)' \;>\; \psi_f(vb+k\eta) \;=\; v\big((bf^k)^\dagger\big).$$

Now suppose $va+j\eta = 0$; then $j=0$, so $v\big((af^j)'\big) = v(a') > S$. In particular, if $k \neq 0$, then $v\big((af^j)'\big) > v\big((bf^k)^\dagger\big)$ by (10.4.2) and Claim 2. If $vb \neq 0$, then $v\big((af^j)'\big) = v(a') \geqslant v(b^\dagger)$. Since $\Gamma_f \neq \Gamma$ we have $\operatorname{res} K(f) = \operatorname{res} K$. Hence K, $L := K(f)$ and $T := K^\times f^{\mathbb{Z}}$ satisfy the hypotheses of Lemma 9.5.6, so L is an asymptotic field. By (10.4.2), its asymptotic couple is (Γ_f, ψ_f). If K is pre-d-valued, then so is $K(f)$ by Corollary 10.1.16. If K is d-valued, then so is $K(f)$ with $C_{K(f)} = C$, by Lemma 9.1.2. □

The uniqueness part of Lemma 10.4.5 strengthens Proposition 10.4.1 significantly in the case considered.

LEMMA 10.4.6. *Suppose S has a maximum and Γ is divisible. Then there is a valuation on $K(f)$ making it an H-asymptotic extension of K with $[v(af)] \notin [\Gamma]$ for some a. For any such valuation, with asymptotic couple (Γ_f, ψ_f) of $K(f)$,*

(i) *$vf \notin \Gamma$, $\Gamma_f = \Gamma \oplus \mathbb{Z}vf$, $\operatorname{res} K(f) = \operatorname{res} K$, and $\Psi_f = \Psi \cup \{\max S\} \subseteq \Gamma$;*

(ii) *if K has small derivation, then so does $K(f)$; and*

(iii) *if K is d-valued, then so is $K(f)$, with the same constant field.*

PROOF. Let vt with $t = s - b^\dagger$ be the largest element of S. Then $v(t-a^\dagger) = v\big(s-(ab)^\dagger\big) \leqslant vt$ for all a. By renaming f/b as f and t as s the hypotheses of Lemma 10.4.6 remain valid; then $v(s-a^\dagger) \leqslant vs$ for all a. Also

$$v(a^\dagger + js) \;=\; \min\big(vs, v(a^\dagger)\big) \in S \qquad \text{for all } a \text{ and all } j \neq 0. \tag{10.4.3}$$

To see this, let a and $j \neq 0$ be given. Since Γ is divisible we can take $u, d \in K^\times$ such that $u \asymp 1$ and $a = ud^{-j}$. Then $a^\dagger = -jd^\dagger + u^\dagger$ and $u^\dagger \asymp u' \prec s - d^\dagger$, so

$$v(a^\dagger + js) \;=\; v(js - jd^\dagger + u^\dagger) \;=\; v(s-d^\dagger) \;=\; \min\big(vs, v(d^\dagger)\big) \;=\; \min\big(vs, v(a^\dagger)\big).$$

(To get the last two equalities, consider the cases $vs = v(d^\dagger)$ and $vs = v(a^\dagger)$ separately.) Since (Γ, ψ) is of H-type,

$$M \;:=\; \big\{\gamma \in \Gamma^< :\; \psi(\gamma) \leqslant vs\big\}$$

is downward closed in Γ. Let η be an element of an ordered abelian group extending Γ such that $M < \eta < \Gamma \setminus M$, so $\eta < 0$. Put $\Gamma_f := \Gamma \oplus \mathbb{Z}\eta$. Then $[\eta] \notin [\Gamma]$: otherwise

$n\gamma < \eta < \gamma$ where $\gamma \in \Gamma^<$ and $n \geqslant 2$, so $\psi(\gamma) = \psi(n\gamma) \leqslant vs$, a contradiction. Hence $[\Gamma_f] = [\Gamma] \cup \{[\eta]\}$. We can now apply Lemma 9.8.7 with

$$C := \{[\gamma] : \gamma \in \Gamma,\ M < \gamma < 0\}, \quad \beta := vs.$$

This gives an H-asymptotic couple (Γ_f, ψ_f) that extends (Γ, ψ) with

$$\psi_f : \Gamma_f^{\neq} \to \Gamma, \qquad \psi_f(\gamma + j\eta) := \min\big(\psi(\gamma), vs\big) \quad \text{for } \gamma \in \Gamma \text{ and } j \neq 0.$$

Equip $K(f)$ with the valuation $v\colon K(f)^\times \to \Gamma_f$ extending the valuation of K such that $vf = \eta$. Since $\Gamma_f \neq \Gamma$ we have $\operatorname{res} K(f) = \operatorname{res} K$. If $va + j\eta \neq 0$, then

$$v\big((af^j)^\dagger\big) = v(a^\dagger + js) = \psi_f(va + j\eta)$$

by (10.4.3). Therefore, if $va + j\eta, vb + k\eta > 0$, then

$$v\big((af^j)'\big) = (va + j\eta)' > \psi_f(vb + k\eta) = v\big((bf^k)^\dagger\big).$$

Next suppose $va + j\eta = 0$; then $va = 0$, $j = 0$, so $v\big((af^j)'\big) = v(a')$. Now $v(a') > vs$, and if $vb \neq 0$, then $v(a') \geqslant v(b^\dagger)$; hence $v\big((af^j)'\big) \geqslant v\big((bf^k)^\dagger\big)$ if $vb + k\eta \neq 0$. We have now shown that K, $L = K(f)$ and $T = K^\times f^{\mathbb{Z}}$ satisfy the hypotheses of Lemma 9.5.6, so $K(f)$ is an asymptotic field.

Now let $K(f)$ be equipped with any valuation that makes it an H-asymptotic extension of K with asymptotic couple (Γ_f, ψ_f) such that $[v(af)] \notin [\Gamma]$ for some a. Fix such a. Then $vf \notin \Gamma$, so $\Gamma_f = \Gamma \oplus \mathbb{Z}vf$ and $\operatorname{res} K(f) = \operatorname{res} K$. Also $[\Gamma_f] = [\Gamma] \cup \{[v(af)]\}$ by Lemma 2.4.4, and $\psi_f\big(v(af)\big) = v(a^\dagger + s)$, hence

$$\Psi_f = \Psi \cup \{v(a^\dagger + s)\}.$$

To finish (i) we show that $v(a^\dagger + s) = \max S$: clearly, $v(a^\dagger + s) \in S$, and

$$\begin{aligned} v(s - b^\dagger) &= v\big((af/ab)^\dagger\big) = \psi_f\big(v(af) - v(ab)\big) \\ &= \min\{\psi_f\big(v(af)\big), \psi\big(v(ab)\big)\} \leqslant \psi_f\big(v(af)\big) = v(a^\dagger + s), \end{aligned}$$

where the third equality holds because $[v(af)] \neq [v(ab)]$ and (Γ_f, ψ_f) is of H-type. To prove (ii), assume that K has small derivation. We wish to show that $K(f)$ has small derivation. With a as above we have $\psi_f\big(v(af)\big) = v(a^\dagger + s) \in \Psi_f$, so we are done if $v(a^\dagger + s) \geqslant 0$. Suppose $v(a^\dagger + s) < 0$. Then Lemma 9.2.9 gives $\gamma \in \Gamma^<$ such that $v(a^\dagger + s) < \psi(\gamma)$. Then $v(af) < \gamma < 0$ or $-v(af) < \gamma < 0$. Because $[\Gamma_f] = [\Gamma] \cup \{[v(af)]\}$, it follows that $\Gamma^<$ is cofinal in $\Gamma_f^<$, and we are done by Lemma 9.2.9. Lemma 9.1.2 gives (iii). □

Lemmas 10.4.2, 10.4.4, 10.4.5, 10.4.6 above cover all claims of Proposition 10.4.1.

EXAMPLE. Let $K = \mathbb{C}[[x^{\mathbb{Q}}]]$ with its usual derivation ($x' = 1$ and $C = \mathbb{C}$) and valuation $v\colon K^\times \to \Gamma = \mathbb{Q}$, $vx = -1$, so K is an algebraically closed d-valued field of H-type with $\Psi = \{1\}$. Let $s = i \in \mathbb{C}$, $i^2 = -1$, so $s \neq a^\dagger$ for all a. Let f be as in the beginning of this subsection: transcendental over K with the derivation on $K(f)$ extending that of K with $f^\dagger = s$. (One may think of f as e^{ix}.) We have $S = \{0\} = \{vs\}$, and $K(f)$ with the valuation constructed in the proof of Lemma 10.4.6 is d-valued and satisfies $vf < \Gamma$ and $\Psi_f = \{0, 1\}$.

COROLLARY 10.4.7. *Suppose that K is algebraically closed,* d*-valued, and has small derivation. Then K has a* d*-valued extension L of H-type with small derivation and the following properties:*

(i) *L is algebraically closed with constant field C;*

(ii) *for each $s \in K$ there exists $f \in L^\times$ with $f^\dagger = s$;*

(iii) *L is algebraic over its subfield generated over K by the $f \in L^\times$ with $f^\dagger \in K$;*

(iv) *there exists a family $(f_i)_{i\in I}$ of elements of $L^\times$ such that $f_i^\dagger \in K$ for all $i \in I$, and $\Gamma_L = \Gamma \oplus \bigoplus_{i\in I} \mathbb{Q}v(f_i)$ (internal direct sum of $\mathbb{Q}$-linear subspaces).*

Moreover, any such L has the following additional properties:

(v) *$\Psi_L \subseteq \Gamma$, and if Ψ has a maximum α, then $\Psi_L = \Gamma^{\leqslant\alpha}$;*

(vi) *L has no proper differential field extension M satisfying* (i), (ii), (iii) *with M instead of L.*

PROOF. Iterating the extension steps of Lemmas 10.4.4, 10.4.5, and 10.4.6, and alternating these steps by taking algebraic closures, we obtain a d-valued field extension L of K, with small derivation, of H-type, and satisfying (i)–(iv).

Let any such L be given. To prove (v), consider any element of Ψ_L. It equals $v(f^\dagger)$ with $f \in L^\times$, $f \not\asymp 1$. By (iv) we have a nonzero $k \in \mathbb{Z}$ such that

$$f^k \asymp af_1^{k_1}\cdots f_m^{k_m}, \qquad f_1,\dots,f_m \in L^\times,\ k_1,\dots,k_m \in \mathbb{Z}.$$

Then $kf^\dagger \asymp a^\dagger + k_1f_1^\dagger + \cdots + k_mf_m^\dagger \in K$, so $v(f^\dagger) \in \Gamma$, as claimed. Now assume also that Ψ has largest element α. Then by $\Psi_L \subseteq \Gamma$ and Lemma 9.2.4 we have $\Psi_L \subseteq \Gamma^{\leqslant\alpha}$. The reverse inclusion is clear from (ii).

If M is a differential field extension of L satisfying (i), (ii), (iii), then for any element $g \neq 0$ of M with $g^\dagger \in K$ we have $f \in L^\times$ such that $f^\dagger = g^\dagger$, hence $g = cf$ for some $c \in C$, and thus $g \in L$. This proves (vi). □

Notes and comments

Lemma 10.4.3 is a variant of [19, Lemma 5.2]. The other results in this section are new.

10.5 H-FIELDS AND PRE-H-FIELDS

An H**-field** is by definition an ordered differential field K such that:

(H1) for all $f \in K$, if $f > C$, then $f' > 0$;

(H2) $\mathcal{O} = C + o$ where $\mathcal{O} = \{g \in K : |g| \leqslant c \text{ for some } c \in C\}$ and o is the maximal ideal of the convex subring $\mathcal{O}$ of K.

Hardy fields containing $\mathbb{R}$ as a subfield are H-fields, by Proposition 9.1.9(i), (iii). Also $\mathbb{T}$ with its usual ordering and derivation is an H-field. Any compositional conjugate K^ϕ of an H-field K with $\phi \in K^>$ is an H-field (with the same ordering).

We regard any H-field K as an ordered valued differential field by taking the valuation given by the valuation ring $\mathcal{O}$ defined in (H2). Note that in an H-field the ordering determines the valuation, but in our experience the valuation is the more robust and useful feature. Nevertheless, the ordering deserves attention.

LEMMA 10.5.1. *Let K be an H-field, $f, g \in K$, $f \preccurlyeq 1$, $0 \neq g \not\asymp 1$. Then $f' \prec g^\dagger$.*

PROOF. Subtracting a constant from f and using (H2), we may assume $f \prec 1$. Replacing g by $-g$ if necessary we may assume that $g > 0$; and replacing g by $1/g$ if necessary we may assume that $g > C$. Let $c \in C^>$. Then $c+f, c-f$ are > 0 and $\asymp 1$, so $g(c+f), g(c-f) > C$ and hence $g'(c+f) + gf' > 0$ and $g'(c-f) - gf' > 0$ by (H1). Dividing by $g' > 0$ gives $-c-f < f'/g^\dagger < c-f$. This holds for all $c \in C^>$, so $f'/g^\dagger \prec 1$. □

Note that H-fields are d-valued by the previous lemma and axiom (H2). The example at the end of Section 9.1 shows that a differential subfield of an H-field with the induced ordering and valuation is not always an H-field; it is, however, always a pre-H-field in the following sense: A **pre-H-field** is an ordered pre-d-valued field K whose ordering, valuation, and derivation interact as follows:

(PH1) the valuation ring $\mathcal{O}$ is convex with respect to the ordering;

(PH2) for all $f \in K$, if $f > \mathcal{O}$, then $f' > 0$.

If K is a pre-H-field, then so is any ordered valued differential subfield of K, and any compositional conjugate K^ϕ with $\phi \in K^>$. Hardy fields are pre-H-fields. Any ordered differential field with the trivial valuation is a pre-H-field. Since we construe H-fields as ordered valued differential fields, they are in particular pre-H-fields. By part (ii) of the next lemma, pre-H-fields are H-asymptotic fields:

LEMMA 10.5.2. *Let K be a pre-H-field and $f, g \in K^\times$.*

(i) *If $f \prec g$, then $f^\dagger < g^\dagger$.*

(ii) *If $f \prec g \prec 1$, then $f^\dagger \succcurlyeq g^\dagger$.*

PROOF. Suppose $f \prec g$. Then $g/f \succ 1$ and hence $g^\dagger = f^\dagger + (g/f)^\dagger > f^\dagger$ by (PH2). This shows (i), and (ii) follows from (i) by taking inverses. □

LEMMA 10.5.3. *Let K be a* d*-valued field and let $(K_i)_{i \in I}$ be a family of* d*-valued subfields of K with $I \neq \emptyset$. Then $\bigcap_i K_i$ is a* d*-valued subfield of K.*

PROOF. With $\mathcal{O}_i$ the valuation ring of K_i, the valuation ring of $\bigcap_i K_i$ is $\bigcap_i \mathcal{O}_i$. Now, given any $a \in \left(\bigcap_i \mathcal{O}_i\right)^{\neq}$, there are unique $c \in C$ and $c_i \in C_{K_i}$ such that $a \sim c$ and $a \sim c_i$. Hence all c_i are equal to c, and thus $c \in \bigcap_i K_i$. It follows that $\bigcap_i K_i$ is d-valued. □

It is easy to check that for any pre-H-field K,

$$K \text{ is an } H\text{-field} \iff K \text{ is d-valued.}$$

Thus Lemma 10.5.3 goes through if d-*valued field* and d-*valued subfield* are replaced by *H-field* and *H-subfield*, respectively.

Algebraic extensions of pre-H-fields

In this subsection K is a pre-H-field.

PROPOSITION 10.5.4. *Let $L|K$ be an algebraic extension of ordered valued differential fields such that $\mathcal{O}_L$ is the convex hull of $\mathcal{O}$ in L. Then L is a pre-H-field.*

This follows from Proposition 10.1.22 and the case $T = K^\times$ of the next lemma:

LEMMA 10.5.5. *Let L be an ordered pre-*d*-valued extension of K with convex valuation ring $\mathcal{O}_L$. Let T be a subgroup of $L^\times$ such that $\Gamma_L \subseteq \mathbb{Q}v(T)$ (in $\mathbb{Q}\Gamma_L$) and for each $t \in T$ with $t \succ 1$ we have $t^\dagger > 0$. Then L is a pre-H-field.*

PROOF. Since L satisfies (PH1), it remains to verify (PH2). For this, let $f \in L$ and $f > \mathcal{O}_L$. Take $n \geqslant 1$ and $t \in T$ with $f^n \asymp t$. Then $nf^\dagger = (f^n)^\dagger \sim t^\dagger$ by Corollary 10.1.2, and so $t^\dagger > 0$ gives $f' > 0$. □

By Corollary 3.5.18 and its proof, the convex hull $\mathcal{O}^{\mathrm{rc}}$ of $\mathcal{O}$ in the real closure K^{rc} of K is the only convex valuation ring of K^{rc} that lies over $\mathcal{O}$. Consider K^{rc} as equipped with the valuation corresponding to $\mathcal{O}^{\mathrm{rc}}$. Then K^{rc} is a pre-H-field extension of K, by Proposition 10.5.4. The constant field of K^{rc} is the real closure of the ordered subfield C of K in K^{rc}, by Lemmas 4.1.1 and 4.1.2. Consequently:

COROLLARY 10.5.6. *If K is an H-field, then so is K^{rc}.*

We equip the differential field extension $K[i]$ ($i^2 = -1$) of K with the valuation ring $\mathcal{O} + \mathcal{O}i$ of $K[i]$; see Lemma 3.5.15. This makes $K[i]$ a valued differential field extension of K with the same value group as K, and we have $v(a + bi) = \min(va, vb)$ for all $a, b \in K$. Moreover:

LEMMA 10.5.7. *The valued differential field $K[i]$ is pre-*d*-valued and of H-type. If K is an H-field, then $K[i]$ is* d*-valued.*

PROOF. $K[i]$ is pre-d-valued by Proposition 10.1.22. Since K and $K[i]$ have the same asymptotic couple, $K[i]$ is of H-type. For the last claim of the lemma, use that the constant field of $K[i]$ is $C + Ci$. □

Our main interest is in the model theory of the H-field $\mathbb{T}$, but it can be useful to work in algebraically closed extensions like $\mathbb{T}[i]$ that are not H-fields but still d-valued of H-type. (In Section 10.7 we show that the subset $\mathbb{T}$ of $\mathbb{T}[i]$ is definable in the differential field $\mathbb{T}[i]$.)

Immediate extensions of pre-H-fields

In this subsection K is a pre-H-field. The next lemma is often used and elaborates on Corollary 3.5.12.

LEMMA 10.5.8. *Let L be an immediate asymptotic extension of K. Then L has a unique field ordering extending that of K in which $\mathcal{O}_L$ is convex. With this ordering L is a pre-H-field and $\mathcal{O}_L$ is the convex hull of $\mathcal{O}$ in L. If K is an H-field, then so is L with this ordering.*

PROOF. L is pre-d-valued by Corollary 10.1.17. If K is an H-field, then L is d-valued by Lemma 9.1.2. Corollary 3.5.12 yields a unique ordering on L as claimed. With this ordering L is a pre-H-field, by Lemma 10.5.5 with $T = K^\times$. □

We always consider any immediate asymptotic extension of K as a pre-H-field according to Lemma 10.5.8. In view of Corollary 9.1.6 we get:

COROLLARY 10.5.9. *The completion K^{c} of K is a pre-H-field. If K is an H-field, then so is K^{c}.*

Adjoining integrals

In this subsection K is a pre-H-field. The next three results are pre-H-field versions of Lemmas 10.2.1, 10.2.2 and 10.2.3.

COROLLARY 10.5.10. *Let s and $K(y)$ be as in Lemma 10.2.1. Then there is a unique ordering on $K(y)$ making it a pre-H-field extension of K. If K is an H-field, then so is $K(y)$ with that unique ordering, and $C_{K(y)} = C$.*

PROOF. By passing from s, y to $-s$, $-y$ if necessary, we arrange that $s < 0$. Then $o < y < K^{>o} = \{f \in K : f > o\}$ for any ordering of $K(y)$ as in the corollary.

Equip $K(y)$ with the unique ordering that according to Lemma 3.5.13 makes it an ordered field extension of K with $y > 0$ and convex $\mathcal{O}_{K(y)}$. To show that then $K(y)$ is a pre-H-field, we verify the hypothesis of Lemma 10.5.5 for $L = K(y)$ and $T = K^\times y^{\mathbb{Z}}$. So let $t = gy^j \succ 1$ with $g \in K^\times$ and $j \in \mathbb{Z}$. Then $t^\dagger = g^\dagger + js/y$, and $\gamma + j\alpha < 0$, where $\gamma := vg$ and $\alpha := vy$. So either $\gamma < 0$, or $\gamma = 0$ and $j < 0$. If $\gamma < 0$, then $v(g^\dagger) = \psi(\gamma) < v(s/y) = vs - \alpha$, that is, $g^\dagger \succ s/y$, which in view of $g^\dagger > 0$ gives $t^\dagger > 0$. If $\gamma = 0$ and $j < 0$, then $v(g^\dagger) \geqslant vs > v(s/y)$, so $s/y \succ g^\dagger$, which by $j < 0$ and $s/y < 0$ yields again $t^\dagger > 0$.

For the H-field case, see the remark following Lemma 10.2.1. □

COROLLARY 10.5.11. *Let the H-field K and s and $K(y)$ be as in Lemma 10.2.2. Then there is a unique ordering on $K(y)$ making it a pre-H-field extension of K. In fact, $K(y)$ with that unique ordering is an H-field and $C_{K(y)} = C$.*

PROOF. Similar to that of Corollary 10.5.10. (If $s > 0$ and $K(y)$ is equipped with an ordering making $K(y)$ a pre-H-field extension of K, then $\mathcal{O} < y < K^{>\mathcal{O}}$.) □

COROLLARY 10.5.12. *With K, s, and $K(y)$ as in Lemma 10.2.3, there is a unique ordering on $K(y)$ making it a pre-H-field extension of K.*

PROOF. Like that of Corollary 10.5.11: here also $s > 0$ forces $\mathcal{O} < y < K^{>\mathcal{O}}$ for any ordering on $K(y)$ making it a pre-H-field extension of K. □

As a pre-H-field, K is pre-d-valued and so has a d-valued hull $\operatorname{dv}(K)$. In view of 10.5.10 and 10.5.8, the construction of $\operatorname{dv}(K)$ in the proof of Theorem 10.3.1 yields an analogue for pre-H-fields:

COROLLARY 10.5.13. *A unique field ordering on* $\operatorname{dv}(K)$ *makes* $\operatorname{dv}(K)$ *a pre-H-field extension of K. Let $H(K)$ be* $\operatorname{dv}(K)$ *equipped with this ordering. Then $H(K)$ is an H-field and embeds uniquely over K into any H-field extension of K.*

Of course this universal property determines $H(K)$ up to unique isomorphism (of ordered valued differential fields) over K. We call $H(K)$ the **H-field hull** of K.

Extending the constant field

Here we show that d-valued fields have d-valued extensions by constants; likewise for H-fields. First a variant of Lemma 9.6.4:

LEMMA 10.5.14. *Let K be a pre-*d*-valued field and L a valued differential field extension of K such that $\Gamma_L = \Gamma$. Let $U \supseteq K$ be a K-linear subspace of L such that $L = \{u/w : u, w \in U,\ w \neq 0\}$, and for all $u \in U^{\neq}$ and $a \in K^{\times}$,*

$$u \preccurlyeq 1,\ a \prec 1 \quad \Longrightarrow \quad u' \prec a^{\dagger}.$$

*Then L is pre-*d*-valued.*

PROOF. Let $f, g \in L^{\times}$, $f \preccurlyeq 1$, $g \prec 1$. Then $f = u/w$ with $u, w \in U^{\neq}$ and $w \neq 0$. Using $U \supseteq K$ and $\Gamma_L = \Gamma$ we arrange $u \asymp f$ and $w \asymp 1$. Also $g = ah$ with $a \in K^{\times}$, $a \prec 1$, and $h \asymp 1$. Then $f' = (u' - fw')/w \asymp u' - fw' \prec a^{\dagger}$. Thus $f' \prec a^{\dagger}$, and likewise $h^{\dagger} \asymp h' \prec a^{\dagger}$. Hence $g^{\dagger} = a^{\dagger} + h^{\dagger} \sim a^{\dagger}$, so $f' \prec g^{\dagger}$. □

Let K be a d-valued field. Note that then K, as a valued vector space over C, is a Hahn space as defined in Section 2.3. Next, let L be an extension field of K with a subfield $D \supseteq C$ such that K and D are linearly disjoint over C and $L = K(D)$. Then Lemma 4.6.21 provides a unique derivation on L that extends the one of K and is trivial on D; this derivation has D as its constant field; below we consider L as a differential field in this way.

PROPOSITION 10.5.15. *There exists a unique valuation on the field L extending that of K and trivial on D. This valuation has the same value group as K. Equipped with this valuation, L is* d*-valued.*

PROOF. The linear disjointness gives a D-linear isomorphism $D \otimes_C K \to K[D]$ given by $d \otimes a \mapsto da$. By Corollary 2.3.15 this yields a valuation $v\colon K[D]^{\neq} \to \Gamma$ on the abelian group $K[D]$ that extends the valuation of K and makes $K[D]$ a Hahn space over D. An easy consequence of Lemma 2.3.14 is that $v(fg) = v(f) + v(g)$ for $f, g \in K[D]^{\neq}$, so v extends to a (field) valuation $v\colon L^\times \to \Gamma$. Corollary 2.3.15 also shows that v is the only field valuation on L that extends the valuation of K and is trivial on D. Lemma 10.5.14 for $U = K[D]$ and Lemma 2.3.14 show that L with v is pre-d-valued. Lemma 2.3.14 also gives for $f \in K[D]^{\neq}$ that $f \sim da$ (with respect to v) for suitable $d \in D^\times$ and $a \in K^\times$. Hence L with v is d-valued. □

In the next proposition we consider the differential field L to be equipped with the valuation of Proposition 10.5.15. Thus L extends the d-valued field K.

PROPOSITION 10.5.16. *Let K and D be given orderings that make K an H-field and D an ordered field extension of C. Then there is a unique field ordering of L extending the orderings of K and D in which the valuation ring of L is convex. With this ordering L is an H-field.*

PROOF. Let $f \in K[D]^{\neq}$. Lemma 2.3.14 gives $f \sim da$ with $d \in D^{\neq}$ and $a \in K^>$ (and thus $f \asymp a$). We claim that the sign of d in the ordered field D depends only on f, not on the choice of d and a. To see this, suppose also $f \sim eb$ with $e \in D^{\neq}$ and $b \in K^>$. Then $d/e \sim b/a \asymp 1$ in the d-valued field L, and $b/a > 0$ in K. Hence $d/e \in C^>$, and so d and e have the same sign. Thus we can make $K[D]$ uniquely into an ordered vector space over D such that for any f as above we have $f > 0$ iff $d > 0$. This ordering is clearly compatible with multiplication: if $0 < f, g \in K[D]$, then $0 < fg$. We extend this ordering to the fraction field L of $K[D]$ to make L an ordered field. Clearly this ordering on L is the only candidate for meeting the requirements. It does extend the orderings of K and D, and it is an easy exercise to check, first, that $\mathcal{O}_L$ is convex in L for this ordering, and next, that it is the convex hull of D in L for this ordering. With this ordering L is a pre-H-field by Lemma 10.5.5 applied to $T = K^\times$, and thus an H-field, since L is d-valued. □

COROLLARY 10.5.17. *Let F be a pre-H-field, E a differential subfield of F, and $F = E(C_F)$. Then $v(F^\times) = v(E^\times) + \mathbb{Z}\alpha$ for some $\alpha \in v(F^\times)$.*

PROOF. Extending F to its H-field hull we arrange that F is an H-field. Next, consider E as a pre-H-subfield of F and identify its H-field hull $H(E)$ with an H-subfield of F via its embedding over E into F. Then $H(E)$ and C_F are linearly disjoint over $C_{H(E)}$ by a remark preceding Lemma 4.6.16. It remains to note that $v(F^\times) = v\big(H(E)^\times\big)$ by Proposition 10.5.15, and that $v\big(H(E)^\times\big) = v(E^\times) + \mathbb{Z}\alpha$ for some $\alpha \in v\big(H(E)^\times\big)$, by Corollary 10.3.2. □

Adjoining exponential integrals

In this subsection K is a pre-H-field with asymptotic couple (Γ, ψ), $\Gamma \neq \{0\}$, and a, b range over $K^\times$ and j, k over $\mathbb{Z}$. We assume $s \in K^\times$ is such that $s \neq a^\dagger$ for each a, and we take a field extension $K(f)$ of K with f transcendental over K, equipped with the unique derivation extending the derivation of K such that $f^\dagger = s$. We prove here some pre-H-field versions of extension lemmas in Section 10.4.

LEMMA 10.5.18. *Suppose K is henselian and $v(s) \in (\Gamma^{>})'$. Then there is a unique valuation on $K(f)$ making it an H-asymptotic extension of K with $f \sim 1$. Equipped with this valuation, $K(f)$ is an immediate extension of K, and thus, by Lemma 10.5.8, a pre-H-field extension of K for a unique ordering of $K(f)$.*

PROOF. Immediate from Lemma 10.4.3. □

Next we establish a pre-H-version of Lemma 10.4.5, and accordingly we now assume that $s - a^{\dagger} \succ u'$ for all a and all $u \in K^{\preccurlyeq 1}$, and put

$$S := \{v(s - a^{\dagger}) : a \in K^{\times}\} \subseteq \Gamma,$$

so $S < (\Gamma^{>})'$.

LEMMA 10.5.19. *Suppose S has no maximum and Γ is divisible. With the valuation on $K(f)$ of Lemma 10.4.5, there is a unique field ordering on $K(f)$ making it a pre-H-field extension of K with $f > 0$.*

PROOF. We can assume $s < 0$. (If $s > 0$, replace s and f by $-s$ and f^{-1}.)

CLAIM: *For all $a \prec 1$ we have: $f \succ a \Longleftrightarrow s > a^{\dagger}$.*

To prove this claim, let $a \prec 1$ and set $\alpha := va$. Then

$$\begin{aligned} f \succ a \iff \eta < \alpha &\iff \alpha_{\lambda} < \alpha \text{ eventually} \\ &\iff a_{\lambda} \succ a \text{ eventually} \\ &\iff a_{\lambda}^{\dagger} > a^{\dagger} \text{ eventually} \iff s > a^{\dagger}. \end{aligned}$$

We order $K(f)$ as indicated in Lemma 3.5.13 with f in the role of y, and show that this makes $K(f)$ a pre-H-field extension of K. Now $K(f)$ is pre-d-valued by item (iii) of Lemma 10.4.5, and so with T as in the proof of that lemma, it remains to show by Lemma 10.5.5 that for all $t \in T$ with $t \succ 1$ we have $t^{\dagger} > 0$.

Let $t \in T$, $t \succ 1$, $t = af^j$. Consider first the case $j > 0$. Take $d \in K^{\times}$ with $ad^j \asymp 1$. Then $t \asymp (f/d)^j \succ 1$ gives $f \succ d$, so $s > d^{\dagger}$ by the Claim, hence $js > jd^{\dagger} = -a^{\dagger} + g$ with $g \in K$, $vg > \Psi$, and thus $a^{\dagger} + js > g$. Also $v(a^{\dagger} + js) \in S < vg$ by Claim 2 in the proof of Lemma 10.4.5, and thus $t^{\dagger} = a^{\dagger} + js > 0$. In the same way we treat the case that $j < 0$. The case $j = 0$ is trivial. □

Since $(-f)^{\dagger} = f^{\dagger} = s$, Lemma 10.5.19 goes through with $f < 0$ instead of $f > 0$.

The next result, Lemma 10.5.20, is essential for dealing with "Liouville closure" in Section 10.6. The case where the set $\{v(s - a^{\dagger}) : a \in K^{\times}\}$ in Lemma 10.5.20 has no maximum is basically contained in Lemma 10.5.19, but the rather lengthy proof we give here (we have no other) makes no use of that lemma.

LEMMA 10.5.20. *Suppose K is a real closed H-field, $s < 0$, and $v(s - a^{\dagger}) \in \Psi^{\downarrow}$ for all a. Then there is a unique pair consisting of a valuation of $L = K(f)$ and a field ordering on L making it a pre-H-field extension of K with $f > 0$. With this valuation and ordering L is an H-field, and we have:*

(i) $vf \notin \Gamma$, $\Gamma_L = \Gamma \oplus \mathbb{Z}vf$, $f \prec 1$;

(ii) $C = C_L$ *and* Ψ *is cofinal in* Ψ_L;

(iii) *a gap in* K *remains a gap in* L;

(iv) *if* L *has a gap which is not in* Γ, *then* $[\Gamma] = [\Gamma_L]$.

PROOF. Suppose $L = K(f)$ is equipped with a valuation and an ordering making L a pre-H-field extension of K with $f > 0$.

CLAIM 1: $vf \notin \Gamma$. Otherwise $f = au$ for some a and some $u \in L$ with $u \asymp 1$. For such a, u we have $s - a^\dagger = u^\dagger$, so $v(s - a^\dagger) > \Psi$, contradicting an assumption.

CLAIM 2: $f \prec 1$. This is because $f \succ 1$ would give $s = f^\dagger > 0$.

CLAIM 3: $f \succ b \iff s > b^\dagger$. This holds by Lemma 10.5.2(i) and Claim 1.

Claim 3 shows how vf determines a cut in Γ. Thus in constructing a valuation of L and ordering of L with the desired properties, the three claims above leave no choice: we equip L with the unique valuation extending the valuation of K such that $0 < vf \notin \Gamma$ realizes the cut in Γ described in Claim 3 above (Lemmas 2.4.16 and 3.1.30), and with the unique ordering extending the ordering of K in which the valuation ring of L is convex, and with $f > 0$ (Lemma 3.5.13). We are going to show that with this valuation and ordering L is an H-field.

Put $\eta := vf$, so $\Gamma_L = \Gamma \oplus \mathbb{Z}\eta$. We claim that $va + j\eta > 0$ if and only if

$$\textit{either} \quad (1)\ va = 0 \text{ and } j > 0, \quad \textit{or} \quad (2)\ va \neq 0 \text{ and } a^\dagger + js < 0.$$

Since $\eta > 0$, this is clear if $va = 0$, or $va \neq 0$ and $j = 0$. Assume $va \neq 0$ and $j < 0$. Let $d \in K$ be a solution to the equation $d^j = 1/|a|$. We have $va + j\eta > 0$ if and only if $\eta < vd$, which is equivalent to $s > (-1/j) \cdot a^\dagger$, by definition of the cut in Γ realized by $\eta = vf$, that is, to $a^\dagger + js < 0$. If $va \neq 0$, $j > 0$, one argues similarly.

We observe that for $va + j\eta \neq 0$ we have $v(a^\dagger + js) \in \Psi^\downarrow$. To see this, we may assume $j \neq 0$; take $d \in K^\times$ such that $d^j = 1/|a|$; then $v(a^\dagger + js) = v(s - d^\dagger) \in \Psi^\downarrow$. Suppose $va + j\eta \neq 0$ and $va = vb$; then $v(a^\dagger + js) = v(b^\dagger + js)$: $a = ub$ with $u \in K$, $u \asymp 1$ gives $v(u^\dagger) = v(u') > \Psi$, hence $v(u^\dagger) > v(b^\dagger + js)$, and $a^\dagger + js = (b^\dagger + js) + u^\dagger$, and thus $v(a^\dagger + js) = v(b^\dagger + js)$, as promised. We can therefore extend $\psi\colon \Gamma^{\neq} \to \Gamma$ to a map $\psi_L\colon \Gamma_L^{\neq} \to \Gamma$ by

$$\psi_L(va + j\eta) := v(a^\dagger + js) \quad \text{for } va + j\eta \neq 0.$$

CLAIM 4: *The function* ψ_L *is decreasing on* $\Gamma_L^>$. Let $va + j\eta > vb + k\eta > 0$; our job is to show that $v(a^\dagger + js) \leqslant v(b^\dagger + ks)$. The above criterion "either (1) or (2)" for an element of Γ_L to be positive yields:

(3) either $va = vb$ and $j > k$, or $va \neq vb$ and $a^\dagger + js < b^\dagger + ks$;

(4) either $vb = 0$ and $k > 0$, or $vb \neq 0$ and $b^\dagger + ks < 0$.

This leads to four cases:

CASE 1: *$va = vb = 0$ and $j > k > 0$.* Then $v(a^\dagger + js) = vs = v(b^\dagger + ks)$.

CASE 2: *$va = vb \neq 0$, $j > k$, and $b^\dagger + ks < 0$.* Then $v(b^\dagger + js) \in \Psi^\downarrow$ gives $(a/b)^\dagger \prec b^\dagger + js$, hence $a^\dagger + js \sim b^\dagger + js < b^\dagger + ks < 0$, so $v(a^\dagger + js) \leqslant v(b^\dagger + ks)$.

CASE 3: *$va \neq vb$, $a^\dagger + js < b^\dagger + ks$, $vb = 0$, and $k > 0$.* Then $b^\dagger + ks \sim ks < 0$ and thus $v(a^\dagger + js) \leqslant v(b^\dagger + ks)$.

CASE 4: *$va \neq vb$, $a^\dagger + js < b^\dagger + ks$, $vb \neq 0$, and $b^\dagger + ks < 0$.* Then clearly $a^\dagger + js < b^\dagger + ks < 0$, and thus $v(a^\dagger + js) \leqslant v(b^\dagger + ks)$.

This proves Claim 4. Since $\psi_L(k\gamma) = \psi_L(\gamma)$ for $\gamma \in \Gamma_L^{\neq}$, $k \neq 0$, the function ψ_L is constant on every archimedean class of Γ_L.

CLAIM 5: *(Γ_L, ψ_L) is an H-asymptotic couple and Ψ is cofinal in $\Psi_L := \psi_L(\Gamma_L^{\neq})$.* First assume $[\Gamma] = [\Gamma_L]$. For $va + j\eta \neq 0$, let b be such that $[vb] = [va + j\eta]$; then $\psi_L(va + j\eta) = \psi_L(vb) = v(b^\dagger) = \psi(vb)$. Thus (Γ_L, ψ_L) is an H-asymptotic couple by Lemma 9.8.1 and its proof. Next suppose $[\Gamma] \neq [\Gamma_L]$, and take any $\gamma_0 \in \Gamma_L^{>}$ such that $[\gamma_0] \notin [\Gamma]$. Then we have for $j \neq 0$,

$$[va] < [\gamma_0] \Rightarrow \psi_L(va + j\gamma_0) = \psi_L(\gamma_0), \quad [\gamma_0] < [va] \Rightarrow \psi_L(va + j\gamma_0) = v(a^\dagger),$$
$$\text{so } \psi_L(va + j\gamma_0) = \min\{v(a^\dagger), \psi_L(\gamma_0)\}.$$

Set $\Gamma_0 := \Gamma \oplus \mathbb{Z}\gamma_0 \subseteq \Gamma_L$. Then $(\Gamma_0, \psi_L|\Gamma_0^{\neq})$ is an H-asymptotic couple: apply Lemmas 2.4.5 and 9.8.7 to the cut $C := [\Gamma^{\neq}]^{<[\gamma_0]}$ and $\beta := \psi_L(\gamma_0)$, using also a fact from the proof of the latter. Since $\Gamma_L \subseteq \mathbb{Q}\Gamma_0$, (Γ_L, ψ_L) is an H-asymptotic couple as well. The cofinality is because $v(a^\dagger + js) \in \Psi^\downarrow$ whenever $va + j\eta \neq 0$.

CLAIM 6: *A gap in (Γ, ψ) remains a gap in (Γ_L, ψ_L). If (Γ_L, ψ_L) has a gap that is not in Γ, then $[\Gamma] = [\Gamma_L]$.* Suppose $\beta \in \Gamma$ is a gap in (Γ, ψ), but not in (Γ_L, ψ_L). This gives $\alpha \in \Gamma_L^{>}$ with $\beta \leqslant \psi_L(\alpha)$ or $\alpha' \leqslant \beta$. In either case, $0 < \alpha < \Gamma^{>}$, so $\psi_L(\alpha) \geqslant \Psi$, contradicting that $\Psi_L \subseteq \Psi^\downarrow$ and that in the presence of a gap in (Γ, ψ) the set Ψ cannot have a maximum.

For the second part of Claim 6, suppose towards a contradiction that (Γ_L, ψ_L) has a gap that is not in Γ, but $[\Gamma] \neq [\Gamma_L]$. Then Ψ_L has no maximum, so Ψ has no maximum. By the first part of Claim 6, (Γ, ψ) has no gap. Thus (Γ, ψ) has rational asymptotic integration. With Γ_0 as in the proof of Claim 5, we obtain from the last statement in Lemma 9.8.7 that $(\mathbb{Q}\Gamma_L, \psi_L) = (\mathbb{Q}\Gamma_0, \psi_L)$ has asymptotic integration, contradicting that $(\mathbb{Q}\Gamma_0, \psi_L)$ has a gap.

By Claim 1 and Lemma 3.1.30 we have $\operatorname{res} K = \operatorname{res} L$. Using this fact and Claim 5 above we can prove very quickly:

CLAIM 7: *L is* d*-valued.* To see this, put $T := K^\times f^{\mathbb{Z}}$, and let $t = af^j \in T$. Then $vt = va + j\eta$, and if $vt \neq 0$, then $\psi_L(vt) = v(a^\dagger + js) = v(t^\dagger)$ and so $v(t') = vt + \psi_L(vt)$. Now use Lemma 10.1.19 and Claim 5.

To complete the proof of the lemma, it remains to show:

CLAIM 8: *L is an H-field.* Let $t \in T$, $t \succ 1$; by Lemma 10.5.5 it is enough to derive $t^\dagger > 0$. We can assume $t \notin K$ and $t = af^j$, so $j \neq 0$. Take $d \in K^\times$ with $d^j = 1/|a|$. First assume $j > 0$. Then $va + j\eta = vt < 0$ gives $\eta < vd$, so $s > d^\dagger = -a^\dagger/j$, by the definition of the cut in Γ realized by $\eta = vf$. Thus $t^\dagger = a^\dagger + js > 0$, as required. Suppose $j < 0$. Then $vd < vf$, and we distinguish the cases $vd > 0$ (similar to the case $j > 0$), $vd = 0$ (where we use $s < 0$ and $vs < v(d^\dagger) = v(a^\dagger)$), and $vd < 0$ (where we use $v(a^\dagger) = v(d^\dagger) < vs$ and $a^\dagger > 0$). □

Notes and comments

A. Robinson [357] derived some Hardy field asymptotics from first-order axioms about ordered differential fields, and he showed where his axioms fall short. The concepts of H-field and pre-H-field, introduced in [19], do not suffer from that defect. A variant of the notion of H-field with an analogue of Corollary 10.5.9 is in [241, 242]. Hahn fields $\mathbb{R}((t^\Gamma))$ with H-field derivations respecting infinite sums are studied in [20, 240]; see also the survey [289].

Corollary 10.5.13 is [19, Corollary 4.6], but its proof there incorrectly claims a unique field ordering on $\mathrm{dv}(K)$ extending that of K in which the valuation ring of $\mathrm{dv}(K)$ is convex. Propositions 10.5.15 and 10.5.16 are [364, Theorem 3] and [20, Proposition 9.1], respectively. Lemma 10.5.20 is a combination of Lemma 5.3 and the remarks following it in [19].

10.6 LIOUVILLE CLOSED H-FIELDS

A pre-H-field K is said to be **Liouville closed** if it is a real closed H-field and for all $a \in K$ there exist $y, z \in K$ such that $y' = a$ and $z \neq 0$, $z^\dagger = a$; note that then any equation $y' + ay = b$ with $a, b \in K$ has a solution in $K^\times$.

EXAMPLES. The H-field $\mathbb{T}$ is Liouville closed: see Appendix A. A Hardy field containing $\mathbb{R}$ as a subfield is Liouville closed if and only if it is real closed, closed under integration and closed under exponentiation. For a Hardy field $K \supseteq \mathbb{R}$ we define $\mathrm{Li}(K)$ as the smallest Hardy field extension of K that is real closed and closed under integration and exponentiation; thus $\mathrm{Li}(K)$ is the smallest Liouville closed Hardy field containing K. If K is a Liouville closed H-field and $\phi \in K^>$, then K^ϕ is a Liouville closed H-field.

Let K be a Liouville closed H-field. Then K is closed under integration (and thus under logarithms) as defined in Section 9.4. Also, K is *closed under powers* in the following sense: for all $c \in C$ and $f \in K^\times$ there exists $y \in K^\times$ such that $y^\dagger = cf^\dagger$; such y behaves like f^c. The H-asymptotic couple (Γ, ψ) of K is closed in the sense of Section 9.9: it has asymptotic integration, Γ is divisible, and Ψ is downward closed. Thus (Γ, ψ) has a contraction map $\chi\colon \Gamma^< \to \Gamma^<$ as defined in Section 9.2. This contraction map is induced by a logarithm map on $K^>$ as follows: choosing for each $a \in K^>$ a "logarithm" $\mathrm{L}(a) \in K$, that is, $\mathrm{L}(a)' = a^\dagger$, we have

$$\chi(\alpha) = v\big(\mathrm{L}(a)\big) \text{ whenever } \alpha \in \Gamma^<, \alpha = va, a \in K^>.$$

EXAMPLE. For $K = \mathbb{T}$ and $a \in \mathbb{T}^>$ we may take $\mathrm{L}(a) = \log(a)$; if $a > \mathbb{R}$, then the sequence $a, \log a, \log\log a, \ldots$ of iterated logarithms of a is coinitial in $\mathbb{T}^{>\mathbb{R}}$, by results in Appendix A. Thus for any $\alpha \in \Gamma_{\mathbb{T}}^{<}$ the sequence $(\chi^n(\alpha))$ is cofinal in $\Gamma_{\mathbb{T}}^{<}$.

For use in Section 11.8 we mention two easy results:

LEMMA 10.6.1. *Let K and L be Liouville closed H-subfields of an H-field M such that $C_L = C_M$. Then $K \cap L$ is a Liouville closed H-subfield of M.*

PROOF. It is easy to check that $K \cap L$ is a real closed H-subfield of M with constant field $C = C_K$. Let $a \in K \cap L$. First, take $f \in K$ and $g \in L$ such that $f' = a = g'$. Then $f - g \in C_M$, so $f - g \in L$, and thus $f \in K \cap L$. Next, take $f \in K^\times$ and $g \in L^\times$ such that $f^\dagger = a = g^\dagger$. Then $f/g \in C_M$, so $f/g \in L$, and thus $f \in K \cap L$. □

Using Lemma 10.5.3 and Corollary 3.5.6 we obtain likewise:

LEMMA 10.6.2. *If K is an H-field and $(K_i)_{i\in I}$ ($I \neq \emptyset$) is a family of Liouville closed H-subfields of K, all with the same constants as K, then $\bigcap_i K_i$ is a Liouville closed H-subfield of K.*

Completion

Recall that by Corollary 9.1.6 the completion of a d-valued field is d-valued. In this subsection we show that taking the completion preserves some further properties, like being a Liouville closed H-field.

COROLLARY 10.6.3. *Suppose the* d*-valued field K is closed under integration. Then the completion K^{c} of K is also closed under integration.*

PROOF. Note that (Γ, ψ) has asymptotic integration, in particular $\Gamma \neq \{0\}$. Let $b \in K^{\mathrm{c}}$. To get $a \in K^{\mathrm{c}}$ with $a' = b$ we may subtract from b an element of K and arrange in this way that $vb > \Psi$. Take a c-sequence (b_ρ) in K with $b_\rho \to b$ and $vb_\rho > \Psi$ for all ρ. Take for each ρ the unique $a_\rho \in K$ with $a_\rho' = b_\rho$ and $a_\rho \prec 1$. Then (a_ρ) is clearly a c-sequence in K, which gives $a \in K^{\mathrm{c}}$ with $a_\rho \to a$, and thus, taking derivatives, $a' = b$. □

COROLLARY 10.6.4. *Suppose the* d*-valued field K satisfies $(K^\times)^\dagger = K$. Then the completion K^{c} of K also satisfies $((K^{\mathrm{c}})^\times)^\dagger = K^{\mathrm{c}}$.*

PROOF. Let $b \in K^{\mathrm{c}}$. To get nonzero $a \in K^{\mathrm{c}}$ with $a^\dagger = b$ we may subtract from b an element of K and arrange in this way that $vb > \Psi$. Take a c-sequence (b_ρ) in K with $b_\rho \to b$ and $vb_\rho > \Psi$ for all ρ. Take for each ρ the unique $a_\rho \in \smallo$ with $(1 + a_\rho)^\dagger = b_\rho$. Now $(1+\delta)^\dagger - (1+\varepsilon)^\dagger \sim (\delta - \varepsilon)'$ for distinct $\delta, \varepsilon \in \smallo$, so (a_ρ) is a c-sequence in K. This gives $a \in K^{\mathrm{c}}$ with $a_\rho \to a$, and thus $a \prec 1$ and $(1 + a)^\dagger = b$. □

Corollaries 3.5.20, 10.5.9, 10.6.3, and 10.6.4 yield a result used in Section 14.1:

LEMMA 10.6.5. *If K is a Liouville closed H-field, then so is its completion K^{c}.*

Liouville extensions

Let K be a differential field. A **Liouville extension** of K is a differential field extension L of K such that C_L is algebraic over C and for each $a \in L$ there are $t_1, \dots, t_n \in L$ with $a \in K(t_1, \dots, t_n)$ and for $i = 1, \dots, n$,

(1) t_i is algebraic over $K(t_1, \dots, t_{i-1})$, or

(2) $t_i' \in K(t_1, \dots, t_{i-1})$, or

(3) $t_i \neq 0$ and $t_i^\dagger \in K(t_1, \dots, t_{i-1})$.

We leave the routine proofs of the next two lemmas to the reader.

LEMMA 10.6.6. *Let $M|L$ and $L|K$ be differential field extensions. If $M|K$ is a Liouville extension, then so is $M|L$. If $M|L$ and $L|K$ are Liouville extensions, then so is $M|K$.*

LEMMA 10.6.7. *Let $M|K$ be a differential field extension such that C_M is algebraic over C. Then the subfield of M generated by any nonempty set of intermediate Liouville extensions of K is also a Liouville extension of K. Thus there exists a largest differential subfield of M that contains K and is a Liouville extension of K.*

The assumption in Lemma 10.6.7 that C_M is algebraic over K cannot be dropped: for a transcendental real number r, the Hardy subfields $\mathbb{Q}(x)$ and $\mathbb{Q}(x+r)$ of $\mathbb{R}(x)$ are both Liouville extensions of $\mathbb{Q}$ with $\mathbb{Q}$ as common field of constants, but the Hardy subfield $\mathbb{Q}(x, x+r) = \mathbb{Q}(x, r)$ of $\mathbb{R}(x)$ is not a Liouville extension of $\mathbb{Q}$.

Let K be a differential field, and $|K| :=$ cardinality of K. We observe:

LEMMA 10.6.8. *Suppose L is a Liouville extension of K. Then $|L| = |K|$.*

PROOF. Define a chain of differential subfields $K = K_0 \subseteq K_1 \subseteq K_2 \subseteq \cdots$ of L:

$$K_{n+1} = \begin{cases} \text{algebraic closure of } K_n \text{ in } L & \text{for } n \equiv 0 \pmod 3 \\ K_n(\{a \in L : a' \in K_n\}) & \text{for } n \equiv 1 \pmod 3 \\ K_n(\{a \in L^\times : a^\dagger \in K_n\}) & \text{for } n \equiv 2 \pmod 3 \end{cases}$$

Clearly $|K_n| = |K|$ for all n (by induction), and $L = \bigcup_n K_n$, so $|L| = |K|$. □

LEMMA 10.6.9. *Let K be a Liouville closed H-field. Then K has no proper Liouville extension with the same constants as K.*

PROOF. Suppose L is a proper Liouville extension of the differential field K with the same constants as K. Up to K-isomorphism the only proper algebraic extension field of K is $K(i)$ with $i^2 = -1$, and as a differential field extension of K it contains the constant $i \notin C$. Hence L must contain a solution $y \notin K$ to an equation $y' = a$ with $a \in K$, or a solution $z \notin K$ with $z \neq 0$ to an equation $z^\dagger = b$ with $b \in K$. But given y as above, take $y_0 \in K$ with $y_0' = a$, and then $y - y_0 \in C_L \setminus C$, a contradiction. Similarly, given z as above, take $z_0 \in K^\times$ with $z_0^\dagger = b$, and note that then $z/z_0 \in C_L \setminus C$, a contradiction. □

Liouville closure

Let K be an H-field. A **Liouville closure** of K is a Liouville closed H-field extension L of K such that L is also a Liouville extension of K. Note that if L is a Liouville closure of K, then by Lemma 10.6.9 there is no proper H-subfield of L that contains K and is Liouville closed.

LEMMA 10.6.10. *Let $K \subseteq M$ be an extension of H-fields such that M is Liouville closed and C_M is algebraic over C. Then M has a unique H-subfield $L \supseteq K$ that is a Liouville closure of K.*

PROOF. Lemma 10.6.7 gives a largest Liouville extension L of K in M. Then $C_L = C_M$, so L is an H-subfield of M. It is also clear that L is real closed, and that for any $a \in L$ and $y, z \in M$ with $y' = a$ and $z \neq 0$, $z^\dagger = a$ we have $y, z \in L$. Thus L is a Liouville closure of K; uniqueness follows from Lemma 10.6.9. □

COROLLARY 10.6.11. *Let K be a Hardy field containing $\mathbb{R}$ as a subfield. Then* $\mathrm{Li}(K)$ *is a Liouville closure of K.*

PROOF. Use $C_{\mathrm{Li}(K)} = \mathbb{R}$ and Lemma 10.6.10. □

The main result about Liouville closures

THEOREM 10.6.12. *Let K be an H-field. Then one of the following occurs:*

(I) *K has exactly one Liouville closure up to isomorphism over K,*

(II) *K has exactly two Liouville closures up to isomorphism over K.*

Moreover, for any H-field K we have:

(1) If no Liouville H-field extension of K has a gap, then K falls under Case (I). Special case: H-subfields of $\mathbb{T}$ properly containing $\mathbb{R}$ fall under Case (I).

(2) If K is grounded, then K falls under Case (I).

(3) Suppose K has a gap γ. Then K falls under Case (II): in one Liouville closure L_1 of K, all $s \in K$ with $vs = \gamma$ have the form b' with $b \succ 1$, while in another Liouville closure L_2 of K all $s \in K$ with $vs = \gamma$ have the form b' with $b \prec 1$.

We prove Theorem 10.6.12 and statements (1), (2), (3) later in this section.

Liouville towers

In this subsection K is an H-field. A **Liouville tower on** K is a strictly increasing chain $(K_\lambda)_{\lambda \leqslant \mu}$ of H-fields, indexed by the ordinals less than or equal to some ordinal μ, such that

(1) $K_0 = K$;

(2) if λ is a limit ordinal, $0 < \lambda \leqslant \mu$, then $K_\lambda = \bigcup_{\iota < \lambda} K_\iota$;

(3) for $\lambda < \lambda + 1 \leqslant \mu$, *either*

(a) K_λ is not real closed and $K_{\lambda+1}$ is a real closure of K_λ,

or K_λ is real closed, $K_{\lambda+1} = K_\lambda(y_\lambda)$ with $y_\lambda \notin K_\lambda$ (so y_λ is transcendental over K_λ), and one of the following holds, with $(\Gamma_\lambda, \psi_\lambda)$ the asymptotic couple of K_λ and $\Psi_\lambda := \psi_\lambda(\Gamma_\lambda^{\neq})$:

(b) $y'_\lambda = s_\lambda \in K_\lambda$ with $y_\lambda \prec 1$ and $v(s_\lambda)$ is a gap in K_λ,

(c) $y'_\lambda = s_\lambda \in K_\lambda$ with $y_\lambda \succ 1$ and $v(s_\lambda)$ is a gap in K_λ,

(d) $y'_\lambda = s_\lambda \in K_\lambda$ with $v(s_\lambda) = \max \Psi_\lambda$,

(e) $y'_\lambda = s_\lambda \in K_\lambda$ with $y_\lambda \prec 1$, $v(s_\lambda) \in (\Gamma_\lambda^{>})'$, and $s_\lambda \neq \varepsilon'$ for all $\varepsilon \in K_\lambda^{\prec 1}$,

(f) $y'_\lambda = s_\lambda \in K_\lambda$ such that $S_\lambda := \{v(s_\lambda - a') : a \in K_\lambda\} < (\Gamma_\lambda^{>})'$, and S_λ has no largest element,

(g) $y_\lambda^\dagger = s_\lambda \in K_\lambda$ with $y_\lambda \sim 1$, $v(s_\lambda) \in (\Gamma_\lambda^{>})'$, and $s_\lambda \neq a^\dagger$ for all $a \in K_\lambda^\times$,

(h) $y_\lambda^\dagger = s_\lambda \in K_\lambda^<$ with $y_\lambda > 0$, and $v(s_\lambda - a^\dagger) \in \Psi_\lambda^\downarrow$ for all $a \in K_\lambda^\times$.

The H-field K_μ is called the **top** of the tower $(K_\lambda)_{\lambda \leqslant \mu}$. Note that (a), (b), (c), (d) correspond to Corollaries 10.5.6, 10.5.10, 10.5.11, 10.5.12, respectively, and (e), (f), (g), (h) to Lemmas 10.2.4, 10.2.6, 10.5.18, 10.5.20, respectively.

LEMMA 10.6.13. *Let a Liouville tower on K as above be given. Then:*

(i) *K_μ is a Liouville extension of K;*

(ii) *the constant field C_μ of K_μ is a real closure of C if $\mu > 0$;*

(iii) $|K_\mu| = |K|$, *hence $\mu < |K|^+$, where $|K|^+$ is the least cardinal $> |K|$.*

PROOF. For (i) and (ii), use results of Sections 10.2 and 10.4 to show by induction on $\lambda \leqslant \mu$ that K_λ is a Liouville extension of K, and that the constant field of K_λ is a real closure of C for $\lambda > 0$. Item (iii) follows from (i) by Lemma 10.6.8. □

By Lemma 10.6.13(iii) there is a *maximal* Liouville tower $(K_\lambda)_{\lambda \leqslant \mu}$ on K, "maximal" meaning that it cannot be extended to a Liouville tower $(K_\lambda)_{\lambda \leqslant \mu+1}$ on K.

LEMMA 10.6.14. *Let L be the top of a maximal Liouville tower on K. Then L is Liouville closed, and hence a Liouville closure of K.*

PROOF. Using (a) and 10.5.6 we see that L is real closed. Likewise, L has no gap by (b) and 10.5.10, and L is not grounded by (d) and 10.5.12. So L has asymptotic integration. Then (e) with 10.2.4, and (f) with 10.2.6 show that L is closed under integration. In the same way, (g) with 10.5.18 and (h) with 10.5.20 show that $(L^\times)^\dagger = L$. Thus L is Liouville closed. □

By the last two lemmas, each H-field has a Liouville closure. The following result (where C_M is not necessarily algebraic over C) has a straightforward proof:

LEMMA 10.6.15. *Let M be a Liouville closed H-field extension of K and $(K_\lambda)_{\lambda\leqslant\mu}$ a Liouville tower on K. Suppose this tower is in M (consists of H-subfields of M), and maximal in M, that is, it cannot be cannot be extended to a Liouville tower $(K_\lambda)_{\lambda\leqslant\mu+1}$ on K in M. Then $(K_\lambda)_{\lambda\leqslant\mu}$ is a maximal Liouville tower on K.*

From Lemmas 10.6.9, 10.6.14, and 10.6.15 we obtain:

COROLLARY 10.6.16. *A Liouville closed H-field extension L of K is a Liouville closure of K iff no proper H-subfield of L containing K is Liouville closed.*

Uniqueness of Liouville closure

The uniqueness properties in Sections 10.2, 10.4, and 10.5, such as Lemma 10.5.8, together with Lemma 10.6.10, yield:

LEMMA 10.6.17. *Let K be an H field and $(K_\lambda)_{\lambda\leqslant\mu}$ a Liouville tower on K such that no K_λ with $\lambda<\mu$ has a gap. Then every embedding of K into a Liouville closed H-field L extends to an embedding of K_μ into L. If K_μ is also Liouville closed, then K_μ is up to isomorphism over K the unique Liouville closure of K.*

Combining Lemmas 10.6.14 and 10.6.17 yields item (1) after Theorem 10.6.12:

COROLLARY 10.6.18. *If no Liouville H-field extension of the H-field K has a gap, then K has up to isomorphism over K a unique Liouville closure.*

To apply this to H-subfields of $\mathbb{T}$ we first note:

LEMMA 10.6.19. *No H-subfield of $\mathbb{T}$ properly containing $\mathbb{R}$ has a gap.*

PROOF. More generally, let $\Delta\neq\{0\}$ be a subgroup of $\Gamma_{\mathbb{T}}$ with $\psi(\Delta^{\neq})\subseteq\Delta$, where $\psi=\psi_{\mathbb{T}}$; we claim that then the asymptotic couple $(\Delta,\psi|\Delta^{\neq})$ does not have a gap. This is clear if this asymptotic couple is grounded. Suppose it is ungrounded, and take $\alpha\in\Delta^{<}$. Then $\chi^n(\alpha)\in\Delta^{<}$ for all n, by Lemma 9.2.19, so $\Delta^{<}$ is cofinal in $\Gamma_{\mathbb{T}}^{<}$, and thus $(\Delta,\psi|\Delta^{\neq})$ has asymptotic integration. □

Combining 10.6.14, 10.6.15, 10.6.17, and 10.6.19 gives:

COROLLARY 10.6.20. *If K is an H-subfield of $\mathbb{T}$ properly containing $\mathbb{R}$, then any two Liouville closures of K are isomorphic over K.*

COROLLARY 10.6.21. *Let $K\supseteq\mathbb{R}$ be a Hardy field and $e\colon K\to\mathbb{T}$ an H-field embedding with $e|\mathbb{R}=\mathrm{id}_{\mathbb{R}}$. Then e extends to an H-field embedding $\mathrm{Li}(K)\to\mathbb{T}$.*

PROOF. By Corollary 10.6.11, $\mathrm{Li}(K)$ is a Liouville closure of K. If $K=\mathbb{R}$, then $\mathbb{R}(x)\subseteq\mathrm{Li}(K)$ and we can extend e to $\mathbb{R}(x)$ to reduce to the case that K properly contains $\mathbb{R}$. In that case Corollary 10.6.20 applies. □

Let K be an H-field. To construct useful Liouville towers on K with the property stated in Lemma 10.6.17, let $\Lambda\subseteq\{(a),(b),\dots,(h)\}$ with $(a)\in\Lambda$. Then the definition of Λ-*tower on K* is identical to that of *Liouville tower on K*, except that in clause (3) of that definition only the items from Λ occur.

LEMMA 10.6.22. *Let K be a real closed H-field without a gap, and let $(K_\lambda)_{\lambda \leqslant \mu}$ be a Λ-tower on K with* (h) $\notin \Lambda$. *Then no K_λ with $\lambda \leqslant \mu$ has a gap.*

PROOF. By induction on $\lambda \leqslant \mu$ one shows easily that the asymptotic couple of K_λ is divisible without a gap, or grounded. □

Lemma 10.6.22 indicates that the Liouville extension of type (h) considered in Lemma 10.5.20 is special: while none of the extensions of type (b)–(g) can produce a gap that wasn't already there, (h) can *create* a gap; Section 13.9 below contains a concrete example. In Section 11.5 we show that *every* real closed H-field with asymptotic integration has an immediate H-field extension K with an element s satisfying the hypotheses of Lemma 10.5.20, and such that for any nonzero y in any H-field extension of K with $y^\dagger = s$, the pre-H-field $K(y)$ has a gap. (Thus no spherically complete H-field can be Liouville closed.) This explains the perhaps curious arrangement of the results leading to the proof of the main theorem below.

Next we turn to statement (2) after Theorem 10.6.12:

PROPOSITION 10.6.23. *If K is a grounded H-field, then all Liouville closures of K are isomorphic over K.*

Towards its proof we first show:

LEMMA 10.6.24. *Let K be a grounded H-field. There exists a Liouville tower on K with top L such that:*

(i) *every H-field in the tower, in particular L, is grounded;*

(ii) *for every $a \in K$ there exist $y, z \in L$ with $y' = a$ and $z \neq 0$, $z^\dagger = a$.*

PROOF. Let $\Lambda := \{(a), (e), (f), (g), (h)\}$. Take a maximal Λ-tower $(K_\lambda)_{\lambda \leqslant \mu}$ on K. Induction on λ using Lemmas 10.2.4, 10.2.6, 10.5.18 and 10.5.20 shows that each Ψ_λ has maximum $\max \Psi$. Maximality with respect to (a), (g), (h) and Lemmas 10.5.18 and 10.5.20 yield: K_μ is real closed and for all $a \in K$ there is $z \in K_\mu^\times$ with $z^\dagger = a$. Take $s \in K_\mu$ with $vs = \max \Psi$. Lemma 10.2.3 and the remarks following it give an H-field extension $L := K_\mu(y)$ of K_μ such that y is transcendental over K_μ and $y' = s$. Then Ψ_L again has a maximum, namely $\psi_L(vy) > \max \Psi$. It only remains to show that $K \subseteq \partial L$. Suppose $t \in K$ and $t \notin \partial K_\mu$. Then the maximality property of the tower with respect to (a), (e), (f) together with Lemma 10.2.5(i) gives $\max \Psi = v(t - a')$ for some $a \in K_\mu$. For such a we have $t - a' = cs + d$ with $c \in C_\mu$ and $d \in K_\mu$, $vd > \max \Psi$. Then $d = e'$ with $e \in K_\mu$, and so $t = (a + cy + e)'$ with $a + cy + e \in L$. □

Let K be a grounded H-field, and let $\ell(K)$ be the real closure of an H-field extension L of K as in Lemma 10.6.24. Then $\Psi_{\ell(K)} = \Psi_L$, so $\ell(K)$ is grounded as well. Thus we can iterate this operation, and form $\ell^2(K) := \ell(\ell(K))$, and so on. Taking the union of the increasing sequence of H-fields $\ell^n(K)$ built in this way, and applying Lemma 10.6.17, we obtain Proposition 10.6.23.

We now prove statement (3) following Theorem 10.6.12:

PROPOSITION 10.6.25. *Let K be an H-field with a gap $\gamma \in \Gamma$. Then K has Liouville closures L_1 and L_2, such that any embedding of K into a Liouville closed H-field M extends to an embedding of L_1 or of L_2 into M, depending on whether the image of γ in Γ_M lies in $(\Gamma_M^<)'$ or in $(\Gamma_M^>)'$. Each Liouville closure of K is K-isomorphic to L_1 or to L_2, but L_1 and L_2 are not K-isomorphic.*

PROOF. Take $s \in K$ such that $vs = \gamma$. Let $K_1 := K(y_1)$ and $K_2 := K(y_2)$ be H-field extensions of K with y_i transcendental over K and $y_i' = s$, for $i = 1, 2$, such that $y_1 \succ 1$ and $y_2 \prec 1$. (Such K_i exist by Corollaries 10.5.10 and 10.5.11.) Then both K_1 and K_2 are grounded. For $i = 1, 2$ let L_i be a Liouville closure of K_i. Let an embedding of K into a Liouville closed H-field M be given. If the image of γ in Γ_M lies in $(\Gamma_M^<)'$, then we can extend that embedding to an embedding of K_1 into M, and hence by Proposition 10.6.23, to an embedding of L_1 into M. If the image of γ in Γ_M lies in $(\Gamma_M^>)'$, then we can similarly extend that embedding to an embedding of L_2 into M. It is now routine to show that L_1 and L_2 as defined here have all the properties claimed in the proposition. □

EXAMPLE. Let K be an ordered field, and equip K with the trivial derivation and trivial valuation. Then K is an H-field with $\Gamma = \{0\}$, and has gap $0 = v(1)$. The two Liouville closures L_1 and L_2 of K in Proposition 10.6.25 satisfy $0 \in \Psi_{L_1}$ and $\Psi_{L_2} < 0$. Replacing the derivation ∂ of L_1 by a suitable multiple $a\partial$, $a \in L_1^>$, we obtain a K-isomorphic copy of L_2. (For a more interesting H-field with a gap, see Section 13.9.)

Propositions 10.6.23 and 10.6.25 above concern two special cases, and we now turn to the general situation in the course of proving Theorem 10.6.12. Let K be an H-field. Take a maximal Liouville tower $(K_\lambda)_{\lambda \leqslant \mu}$ on K. Then K_μ is a Liouville closure of K. We have two cases:

(A) No K_λ has a gap. Then K falls under Case (I) by Lemma 10.6.17.

(B) Some K_λ in the tower has a gap. Take λ minimal with this property. Let L_1 and L_2 be the two Liouville closures of K_λ as in Proposition 10.6.25. Then L_1 and L_2 are also Liouville closures of K. Given any embedding of K into a Liouville closed H-field M, we can first extend it to an embedding of K_λ into M, and then by Proposition 10.6.25 to an embedding of L_1 or L_2 into M. Applying this to the inclusion of K into any Liouville closure M of K, it follows that L_1 or L_2 is K-isomorphic to M. Thus if L_1 and L_2 are K-isomorphic, then K falls under Case (I), and otherwise K falls under Case (II).

This yields the following more precise version of Theorem 10.6.12:

THEOREM 10.6.26. *Let K be an H-field. Then K has at least one and at most two Liouville closures, up to isomorphism over K. Any embedding of K into a Liouville closed H-field M extends to an embedding of some Liouville closure of K into M. Moreover, if K has two Liouville closures, not isomorphic over K, then K has a Liouville H-field extension L with a gap such that L embeds over K into any Liouville closed H-field extension of K.*

Notes and comments

The differential-algebraic notion of *Liouville extension* was motivated by Liouville's work [262, 263] on explicit solutions of second-order linear differential equations; see Kolchin [219, p. 5] and Rosenlicht-Singer [371].

Rosenlicht [369, Theorem 3] considers the Liouville closure $\mathrm{Li}(K)$ of the Hardy field $K = \mathbb{R}$; it contains Hardy's field of logarithmico-exponential functions [163]. Liouville closed H-fields and the results of this section are from [19] (with errata at the end of [20]), except for Lemma 10.6.5, which is [20, Lemma 10.2]. We point out that the material of the present section demands the H-field setting: Hardy fields and H-subfields of $\mathbb{T}$ tend to obscure the fork in the road caused by gaps.

If an H-field K has a Liouville H-field extension with a gap, then K has two Liouville closures that are not isomorphic over K, according to [19]. In reviewing that paper we realized that it doesn't contain a proof of that claim, but Allen Gehret has since provided us with one; we do not not use the claim in this book.

In the next chapter we introduce a first-order condition (ω-freeness) on an H-field that makes it fall under Case (I) of Theorem 10.6.12; see Corollary 13.6.2.

A miniature version of "Liouville closure" is *closure under powers*, studied in the setting of Hardy fields in [368], and for H-fields in [20, Sections 7, 8]: every H-field has a closure under powers, and up to isomorphism it has at most two.

10.7 MISCELLANEOUS FACTS ABOUT ASYMPTOTIC FIELDS

This section indicates some possible and impossible features of asymptotic fields. Only Lemma 10.7.8 will be used later. Can a differentially closed field have a nontrivial valuation making it an asymptotic field? This was answered negatively by Scanlon [384]. The solution given here, inspired by Rosenlicht [364], is a little different. Scanlon used the logarithmic derivative map of an elliptic curve. The differential-algebraic properties of this map are used here for another purpose: to construct asymptotic fields that cannot be obtained by coarsening pre-d-valued fields. Unrelated to this, we also indicate an algebraically closed d-valued field of H-type that is not an algebraic closure of a real closed d-valued field, and we finish with the result that $\mathbb{T}$ as a subset of $\mathbb{T}[i]$ is definable in the differential field $\mathbb{T}[i]$.

Differentially closed fields cannot be asymptotic

In this subsection K is a differential field, with constant field C, and y ranges over K.

PROPOSITION 10.7.1. *If K is differentially closed, then there is no valuation ring $\mathcal{O} \neq K$ of K with $\partial\mathcal{O} \subseteq \mathcal{O}$.*

PROOF. Let K be differentially closed and suppose for a contradiction that $\mathcal{O} \neq K$ is a valuation ring of K with $\partial\mathcal{O} \subseteq \mathcal{O}$. Since K remains differentially closed upon replacing ∂ by $a\partial$ with $a \in \sigma^{\neq}$, we can assume $\partial\mathcal{O} \subseteq \sigma$. Take y such that $y+(y')^2 = y^3$ and $y \neq y^3$. The argument below uses the second part of Lemma 4.4.3 several times. If $y \succ 1$, then $y + (y')^2 \prec y^3$, a contradiction. Thus $y \in \mathcal{O}$, and so $y \equiv y^3 \bmod \sigma$, hence $y \equiv -1$, 0, or 1 mod o. The case $y \equiv 0 \bmod o$ is impossible, since $y \neq 0$ gives $(y')^2 \prec y$ and $y^3 \prec y$. If $y \equiv 1 \bmod o$, set $y = z + 1$, and we get a similar contradiction from $-2z+(z')^2 = z^3+3z^2$ and $0 \neq z \prec 1$, and the case $y \equiv -1 \bmod o$ is likewise impossible. □

The logarithmic derivative map on an elliptic curve

In this subsection K is a differential field with constant field C. Let $c \in C$, $c \neq 0, 1$, and put

$$P(X) := X(X-1)(X-c) = X^3 - (c+1)X^2 + cX \in C[X].$$

Consider the projective plane curve E defined over C by the equation

$$Y^2 Z = X^3 - (c+1)X^2 Z + cXZ^2.$$

The affine part of E (in standard coordinates) is given by $Y^2 = P(X)$, and E is an elliptic curve whose group law has $(0:1:0)$, the unique point at infinity on E, as its zero element.

We define the logarithmic derivative map $\ell_E \colon E(K) \to K$ as follows:

$$\ell_E(0:1:0) := 0, \quad \ell_E(x:y:1) := x'/y \text{ if } y \neq 0, \quad \ell_E(x:0:1) := 0.$$

By Lemma 2 on p. 805 of [220], the map ℓ_E is a group homomorphism from $E(K)$ to the additive group of K; its kernel is $E(C)$.

LEMMA 10.7.2. *Let $\mathcal{O}$ be a valuation ring of K such that $\partial o \subseteq o$ and $c, c-1 \in \mathcal{O} \setminus o$. Then $\ell_E\big(E(K)\big) \subseteq \mathcal{O}$.*

PROOF. Let $x, y \in K$ with $(x:y:1) \in E(K)$. If $y = 0$, then $\ell_E(x:y:1) = 0 \in \mathcal{O}$. Assume $y \neq 0$; we need to show $x'/y \preccurlyeq 1$. We distinguish two cases. Suppose first that $x \preccurlyeq 1$; then $y^2 = P(x) \preccurlyeq 1$, hence $y \preccurlyeq 1$. Differentiating both sides of the equality $y^2 = P(x)$ we obtain $2yy' = P'(x) \cdot x'$. Hence if $P'(x) \asymp 1$ then $x'/y \asymp 2y' \preccurlyeq 1$. Otherwise $P'(x) \prec 1$ and therefore $y^2 = P(x) \asymp 1$, since the reduced polynomial $\overline{P}(X) \in (\operatorname{res} K)[X]$ has no multiple zeros; hence $x'/y \asymp x' \preccurlyeq 1$. Now suppose that $x \succ 1$. Then $y^2 = P(x) \asymp x^3$ and $(x')^2 \preccurlyeq x^3$ by Lemma 4.4.3, therefore $x'/y \preccurlyeq 1$. □

COROLLARY 10.7.3. *If $\ell_E \colon E(K) \to K$ is surjective, then K has no nontrivial valuation making it an asymptotic field.*

For use in the next subsection we prove:

PROPOSITION 10.7.4. *Suppose C is algebraically closed. There exists a differential field extension L of K, d-algebraic over K with $C_L = C$, such that*

(i) *$\ell_E \colon E(L) \to L$ is surjective (so L has no nontrivial valuation making it an asymptotic field);*

(ii) *for each $y \in L^\times$ with $y^\dagger \in C^\times$ there exists $n \geqslant 1$ with $y^n \in K$.*

The main work goes into proving the next two lemmas, in both of which C is assumed to be algebraically closed.

LEMMA 10.7.5. *There is no group homomorphism $E(C) \to C^\times$ with finite kernel.*

PROOF. Suppose $\varphi\colon E(C) \to C^\times$ is a group homomorphism with $k := |\ker \varphi| < \infty$. For an additive abelian group G and $n \geqslant 1$, put $G[n] := \{g \in G : ng = 0\}$, a subgroup of G. Then, for $n \geqslant 1$, $E(C)[n]$ is finite with $|E(C)[n]| = n^2$, see [414, Theorem VI.6.1], and $\varphi\big(E(C)[n]\big) \subseteq \{\zeta \in C^\times : \zeta^n = 1\}$. Take a prime number p not dividing k. Then $\varphi|E(C)[p]$ is injective, contradicting $|\{\zeta \in C^\times : \zeta^p = 1\}| \leqslant p$. □

LEMMA 10.7.6. *Let $a \in K^\times$, and let f be an element in a differential field extension of K with the same constant field C as K, satisfying $(f')^2 = a^2 P(f) \neq 0$. Let y be a nonzero element of the differential field $K\langle f\rangle = K(f, f')$ with $y^\dagger \in C^\times$. Then $y^n \in K$ for some $n \geqslant 1$.*

PROOF. From Section 4.1 recall that given a differential field extension L of K, we denote by $\operatorname{Aut}_\partial(L|K)$ the group of differential automorphisms of L which are the identity on K. Suppose for a contradiction that $y^n \notin K$ for all $n > 0$. Then y is transcendental over K, so f is transcendental over K. Also $\sigma(y)/y \in C^\times$ for every $\sigma \in \operatorname{Aut}_\partial\big(K(y)|K\big)$, and the map

$$\sigma \mapsto \sigma(y)/y\colon \operatorname{Aut}_\partial\big(K(y)|K\big) \to C^\times \tag{10.7.1}$$

is a group isomorphism [220, p. 803]. Writing the group operation on $E(K)$ additively, we have for $\sigma \in \operatorname{Aut}_\partial\big(K\langle f\rangle|K\big)$,

$$p_\sigma := \big(\sigma(f) : \sigma(f'/a) : 1\big) - (f : f'/a : 1) \in E(C),$$

and the map

$$\sigma \mapsto p_\sigma\colon \operatorname{Aut}_\partial\big(K\langle f\rangle|K\big) \to E(C) \tag{10.7.2}$$

is an injective group homomorphism [220, p. 807]. The homomorphism

$$\sigma \mapsto \sigma \restriction K(y)\colon \operatorname{Aut}_\partial\big(K\langle f\rangle|K\big) \to \operatorname{Aut}_\partial\big(K(y)|K\big) \tag{10.7.3}$$

is surjective by [220, Theorem 3 on p. 797]; in particular, $\operatorname{Aut}_\partial\big(K\langle f\rangle|K\big)$ is infinite. Hence the map (10.7.2) is an isomorphism, by [220, p. 807]. Moreover, the kernel of (10.7.3) is finite by [220, p. 796, Theorem 2]. Composing the inverse of the map (10.7.2) with (10.7.3) and (10.7.1) yields a homomorphism $E(C) \to C^\times$ with finite kernel; this is impossible by the previous lemma. □

To prove Proposition 10.7.4, fix a differential closure K^{dc} of K; then K^{dc} is d-algebraic over K and $C_{K^{\mathrm{dc}}} = C$. (See Section 4.7.) By Zorn we can take a differential subfield L of K^{dc} containing K which is maximal with respect to the property that for each $y \in L^\times$ with $y^\dagger \in C^\times$ there exists $n \geqslant 1$ with $y^n \in K$. Then $\ell_E\big(E(L)\big) = L$. To see this, let $a \in L^\times$ and $f \in (K^{\mathrm{dc}})^\times$ be such that $(f')^2 = a^2 P(f) \neq 0$. Suppose that y is a nonzero element of $L\langle f\rangle$ such that $y^\dagger \in C^\times$. Lemma 10.7.6 (applied to L in place of K) yields an $n \geqslant 1$ with $y^n \in L$; since $(y^n)^\dagger = ny^\dagger \in C^\times$, we have $y^{nm} \in K$ for some $m \geqslant 1$. So $L\langle f\rangle = L$ by maximality of L; in particular $f \in L$ and thus $(f : f'/a : 1) \in E(L)$ with $\ell_E(f : f'/a : 1) = a$. $\square$

An asymptotic field that is not a coarsening of a pre-d-valued field

Let $\boldsymbol{k}$ be a differential field such that $a'+na \neq 0$ for all $a \in \boldsymbol{k}^\times$ and $n \geqslant 1$. Equip $\boldsymbol{k}(\!(t)\!)$ with the valuation v that has $\boldsymbol{k}[[t]]$ as its valuation ring, and with the unique derivation extending that of $\boldsymbol{k}$ such that $t' = t$ and $\boldsymbol{k}[[t]]' \subseteq \boldsymbol{k}[[t]]$. Then $\boldsymbol{k}(\!(t)\!)$ is an asymptotic field. (To see this, note that $(at^n)' = (a' + na)t^n$ for $a \in \boldsymbol{k}^\times$ and $n \geqslant 1$.) It has small derivation, with $\Psi = \{0\}$ (so its asymptotic couple is of H-type); its differential residue field is $\boldsymbol{k}$, and its constant field is $C_{\boldsymbol{k}}$. The valued differential subfield $K := \boldsymbol{k}(t)$ of $\boldsymbol{k}(\!(t)\!)$ is an asymptotic field with valuation ring $\mathcal{O} = \boldsymbol{k}[t]_{(t)}$. If $\boldsymbol{k}$ has an element b with $b' \neq 0$, then $b \asymp b' \asymp 1$, and so K is not pre-d-valued.

LEMMA 10.7.7. *Suppose $\boldsymbol{k}$ has an element b with $b' \neq 0$, and $\boldsymbol{k}$ has no nontrivial valuation making it a pre-d-valued field. Then $\boldsymbol{k}(t)$ has no valuation v_1 with $v_1(t) > 0$ that makes $\boldsymbol{k}(t)$ a pre-d-valued field.*

PROOF. Suppose v_1 is a valuation on $\boldsymbol{k}(t)$ with $v_1(t) > 0$ that makes $\boldsymbol{k}(t)$ a pre-d-valued field. Let $\mathcal{O}_1$ be the valuation ring of v_1. Then $\mathcal{O}_1 \cap \boldsymbol{k}$ is a valuation ring of $\boldsymbol{k}$ making $\boldsymbol{k}$ a pre-d-valued field, so $\mathcal{O}_1 \cap \boldsymbol{k} = \boldsymbol{k}$. Then $\mathcal{O}_1 = \boldsymbol{k}[t]_{(t)} = \mathcal{O}$, which contradicts the observation preceding the lemma. $\square$

If $\boldsymbol{k}$ satisfies the hypothesis of the lemma, then K cannot be obtained by coarsening a pre-differential valuation of $\boldsymbol{k}(t)$: there is no valuation $v_1\colon \boldsymbol{k}(t)^\times \to \Gamma_1$ making $\boldsymbol{k}(t)$ a pre-d-valued field such that for some convex subgroup Δ of Γ_1 the coarsened valuation $\dot{v}_1\colon \boldsymbol{k}(t)^\times \to \Gamma_1/\Delta$ has valuation ring $\mathcal{O}$.

To obtain a differential field $\boldsymbol{k}$ satisfying the requirements above we first take any algebraically closed field C of characteristic zero. Let $C(x)$ be a field extension of C with x transcendental over C, equipped with the unique derivation with constant field C and $x' = 1$. Proposition 10.7.4 yields a differential field extension $\boldsymbol{k}$ of $C(x)$ such that for every $y \in \boldsymbol{k}^\times$ with $y^\dagger \in C^\times$ there exists $n > 0$ with $y^n \in C(x)$, and such that $\boldsymbol{k}$ has no nontrivial valuation making it an asymptotic field. Since there is no $y \in C(x)^\times$ with $y^\dagger \in C^\times$ (by Corollary 4.6.13), there is no $y \in \boldsymbol{k}^\times$ with $y^\dagger \in C^\times$. Hence, as required, $a' + na \neq 0$ for all $a \in \boldsymbol{k}^\times$ and $n \geqslant 1$, $C_{\boldsymbol{k}} \neq \boldsymbol{k}$, and $\boldsymbol{k}$ has no nontrivial valuation making it a pre-d-valued field. Thus the asymptotic field K defined just before Lemma 10.7.7 cannot be obtained by coarsening a pre-differential valuation.

An algebraically closed d-valued field that is not an algebraic closure of a real closed d-valued field

Every algebraically closed field L of characteristic zero has a real closed subfield K such that $L = K(i)$, where $i^2 = -1$. This suggests the following question:

Let L be an algebraically closed d*-valued field of H-type; is there a real closed* d*-valued subfield K of L such that $L = K(i)$?*

We show that the answer is negative for L as in Corollary 10.4.7. This result follows from the next lemma, also used in the next subsection. Let K be a real closed d-valued field and $K[i]$ its algebraic closure, $i^2 = -1$. The residue field of K is isomorphic to C, hence real closed. Thus the valuation ring $\mathcal{O}$ of K is convex by Theorem 3.5.16. For $y = a + bi \in K[i]$ $(a, b \in K)$ we let $|y| := \sqrt{a^2 + b^2} \in K^{\geqslant}$ be the absolute value of y. We have the subgroup

$$S := \{y \in K[i] : |y| = 1\}$$

of the multiplicative group $K[i]^\times$, with $S \subseteq \mathcal{O} + \mathcal{O}i$, $K[i]^\times = K^> \cdot S$, $K^> \cap S = \{1\}$.

LEMMA 10.7.8. *Let $a + bi \in S$ $(a, b \in K)$. Then $(a + bi)^\dagger = \mathrm{wr}(a, b)i$. Thus*

$$\big(K[i]^\times\big)^\dagger = (K^>)^\dagger \oplus S^\dagger, \text{ an internal direct sum of subgroups of } \big(K[i]^\times\big)^\dagger,$$
$$S^\dagger \subseteq \{f \in K : f \preccurlyeq g' \text{ for some } g \in \smallo\} \cdot i.$$

PROOF. Since $(a + bi)(a - bi) = 1$ we have

$$\begin{aligned}(a + bi)^\dagger &= (a' + b'i)(a - bi) = (aa' + bb') + (ab' - a'b)i \\ &= \tfrac{1}{2}(a^2 + b^2)' + (ab' - a'b)i = (ab' - a'b)i = \mathrm{wr}(a, b)i.\end{aligned}$$

From $a, b \in \mathcal{O} = C + \smallo$ we get $\mathrm{wr}(a, b) \preccurlyeq g'$ for some $g \in \smallo$. It remains to note that $K[i]^\times = K^> \cdot S$ gives $\big(K[i]^\times\big)^\dagger = (K^>)^\dagger + S^\dagger$. □

COROLLARY 10.7.9. *If L is a* d*-valued field with small derivation and an element $y \in L^\times$ such that $y^\dagger = i$, $i^2 = -1$, then L has no real closed* d*-valued subfield K with $L = K(i)$.*

Definability of $\mathbb{T}$ in $\mathbb{T}[i]$

It is well-known that $\mathbb{R}$ as a subset of $\mathbb{C} = \mathbb{R}[i]$ is not definable (even allowing parameters) in the field $\mathbb{C}$ of complex numbers; see B.12.3. In stark contrast, the *differential* field structure of $\mathbb{T}[i]$ is rich enough to define $\mathbb{T}$:

PROPOSITION 10.7.10. *The subset $\mathbb{T}$ of $\mathbb{T}[i]$ is definable in the differential field $\mathbb{T}[i]$.*

PROOF. Below, "definable" means definable without parameters in the differential field $\mathbb{T}[i]$. Let $\mathcal{O}$ be the valuation ring of $\mathbb{T}$ with maximal ideal $\smallo$. Then $\mathcal{O} + \mathcal{O}i$ is the valuation ring of $\mathbb{T}[i]$ with maximal ideal $\smallo + \smallo i$. We begin with noting the definability of $\mathbb{R}$, as a consequence of Lemma 10.7.8:

$$\mathbb{R} = \big\{y \in \mathbb{T}[i] : y' = 0 \text{ and } y = f^\dagger \text{ for some } f \in \mathbb{T}[i]^\times\big\}.$$

Indeed, this lemma gives the more precise result $\big(\mathbb{T}[i]^\times\big)^\dagger \subseteq \mathbb{T} + i\partial\smallo$. We claim that $\big(\mathbb{T}[i]^\times\big)^\dagger = \mathbb{T} + i\partial\smallo$. To see this, note that for $b \in \smallo$ we have

$$\sin b := b - \frac{b^3}{6} + \frac{b^5}{120} - \cdots \in \mathbb{T}, \qquad \cos b := 1 - \frac{b^2}{2} + \frac{b^4}{24} - \cdots \in \mathbb{T}^{\neq}$$

with $(\sin b)' = b' \cos b$ and $(\cos b)' = -b' \sin b$, and so

$$f := \exp(ib) := \cos b + i \sin b \in \mathbb{T}[i]^\times$$

satisfies $f^\dagger = ib'$. Hence $i\partial\mathcal{o} \subseteq (\mathbb{T}[i]^\times)^\dagger$, and thus $\mathbb{T} + i\partial\mathcal{o} \subseteq (\mathbb{T}[i]^\times)^\dagger$, and the claim follows. Think of $\mathbb{T} + i\partial\mathcal{o}$ as a thin strip around the "real axis" $\mathbb{T}$ in the "complex plane" $\mathbb{T}[i]$. Intersecting $\mathbb{T} + i\partial\mathcal{o}$ with its multiple $i(\mathbb{T} + i\partial\mathcal{o})$ yields the definable set $\partial\mathcal{o} + i\partial\mathcal{o}$, and taking integrals gives the definability of $\mathcal{O} + \mathcal{O}i$, and of its maximal ideal $\mathcal{o} + \mathcal{o}i$. Hence $\big(\mathbb{T}[i]^\times\big)^\dagger + (\mathcal{o} + \mathcal{o}i) = \mathbb{T} + \mathcal{o}i$ is definable. Thus $\mathcal{O} + \mathcal{o}i = (\mathcal{O} + \mathcal{O}i) \cap (\mathbb{T} + \mathcal{o}i)$ is definable. It is easy to check that for $f \in \mathcal{O} + \mathcal{o}i$,

$$f \cdot (\mathbb{T} + \mathcal{o}i) \subseteq \mathbb{T} + \mathcal{o}i \iff f \in \mathcal{O},$$

so $\mathcal{O}$ is definable, and therefore its fraction field $\mathbb{T}$ in $\mathbb{T}[i]$ is definable. □

Chapter Eleven

Eventual Quantities, Immediate Extensions, and Special Cuts

Our main interest is ultimately in H-fields with asymptotic integration and small derivation such as $\mathbb{T}$. The induced derivation on the residue field of such asymptotic fields is trivial, however, so these asymptotic fields are not covered by Corollary 9.6.2 on spherically complete immediate asymptotic extensions. One goal in the present chapter is to remedy this defect by establishing the following result in Section 11.4:

THEOREM 11.0.1. *Every asymptotically maximal H-asymptotic field with rational asymptotic integration is spherically complete.*

Proving Theorem 11.0.1 requires some tools that are also important later. These tools arise from the fact that for asymptotic fields K with asymptotic integration and $P \in K\{Y\}^{\neq}$, certain quantities associated to its compositional conjugates P^{ϕ}, such as its dominant degree $\operatorname{ddeg} P^{\phi}$, become constant for sufficiently high $v\phi \in \Psi^{\downarrow}$. These "eventual quantities" will be studied in Sections 11.1, 11.2, 11.3.

In Sections 11.5, 11.6, 11.7 we consider special (definable) cuts in H-asymptotic fields K with asymptotic integration, and introduce some key elementary properties of K, namely λ-freeness and ω-freeness, which express that these cuts are not realized in K. We show that $\mathbb{T}$ has these properties, but a full exploitation of these subtle but powerful elementary properties must be left to Chapter 13, where the machinery of triangular automorphisms of $K\{Y\}$ from Chapter 12 is available.

In Section 11.8 we consider certain special existentially definable subsets of Liouville closed H-fields K, and the behavior of the functions ω and σ (introduced in Section 5.2) on these sets. This will play a role in our Elimination of Quantifiers for $\mathbb{T}$ in Chapter 16.

11.1 EVENTUAL BEHAVIOR

In this section K is an asymptotic field with value group $\Gamma \neq \{0\}$. We let ϕ range over $K^{\times}$, and $\boldsymbol{\sigma}$, $\boldsymbol{\tau}$ over $\mathbb{N}^*$. We also fix a differential polynomial $P \in K\{Y\}^{\neq}$.

The function $y \mapsto P(y)$ defined by P on K does not give up its secrets easily, but we do have some tricks up our sleeve. First, things are more transparent if $\partial\smallo \subseteq \smallo$, and the "best case" is when $\sup \Psi = 0$. If $\sup \Psi$ exists we can reduce to that best case by compositional conjugation. In general we try to simulate this best case by working in compositional conjugates K^{ϕ} with small derivation, that is, $v\phi < (\Gamma^{>})'$, but $v\phi$ as high as possible. When K is ungrounded, it turns out that certain quantities associated to P^ϕ such as $\operatorname{dwt}(P^\phi)$ eventually stabilize for increasing $v\phi \in \Psi^\downarrow$. These "eventual" quantities associated to P turn out to be important invariants. This section is devoted to proving their existence.

Behavior of $vF^n_k(\phi)$

In order to better understand $v(P^\phi)$ as a function of ϕ we use from Lemma 5.7.4 the identity

$$(P^\phi)_{[\sigma]} = \sum_{\tau \geqslant \sigma} F^{\tau}_{\sigma}(\phi) P_{[\tau]}. \tag{11.1.1}$$

This leads to the study of $vF^{\tau}_{\sigma}(\phi)$. Recall that for $\boldsymbol{\tau} = \tau_1 \cdots \tau_d \geqslant \boldsymbol{\sigma} = \sigma_1 \cdots \sigma_d$,

$$F^{\boldsymbol{\tau}}_{\boldsymbol{\sigma}} := F^{\tau_1}_{\sigma_1} \cdots F^{\tau_d}_{\sigma_d}.$$

LEMMA 11.1.1. *If $\partial\mathcal{O} \subseteq \mathcal{O}$ and $\phi \preccurlyeq 1$, then $v(P^\phi) \geqslant v(P)$, with equality if $\phi \asymp 1$.*

PROOF. Assume $\partial\mathcal{O} \subseteq \mathcal{O}$ and $\phi \preccurlyeq 1$. For $0 \leqslant k \leqslant n$ we have $F^n_k \in \mathbb{Q}\{X\}$, so $F^n_k(\phi) \preccurlyeq 1$. Thus $v(P^\phi) \geqslant v(P)$ by (11.1.1). If $\phi \asymp 1$, use $P = (P^\phi)^{\phi^{-1}}$. □

In studying $vF^n_k(\phi)$ we consider the case $\phi^\dagger \preccurlyeq \phi$ in the next lemma, and take up the case $\phi^\dagger \succ \phi$ in the next subsection. We set $\delta = \phi^{-1}\partial$.

LEMMA 11.1.2. *Suppose that $\delta\smallo \subseteq \smallo$, and let $0 \leqslant k \leqslant n$.*

(i) *If $\phi^\dagger \preccurlyeq \phi$, then $v\big(F^n_k(\phi)\big) \geqslant nv\phi$, with equality if $k = n$.*

(ii) *If $\phi^\dagger \prec \phi$ and $k < n$, then $v\big(F^n_k(\phi)\big) > nv\phi$.*

PROOF. Note that $\phi^\dagger \preccurlyeq \phi$ means $v\big(\delta(\phi)\big) \geqslant v\phi$. Now use (5.7.1) and induction on n, and Lemma 4.4.2 applied to K^ϕ in place of K. □

COROLLARY 11.1.3. *Suppose that $\delta\smallo \subseteq \smallo$ and $\phi^\dagger \preccurlyeq \phi$, and $\boldsymbol{\tau} \geqslant \boldsymbol{\sigma}$. Then $v\big(F^{\boldsymbol{\tau}}_{\boldsymbol{\sigma}}(\phi)\big) \geqslant \|\boldsymbol{\tau}\| v\phi$, with equality if $\boldsymbol{\tau} = \boldsymbol{\sigma}$.*

Behavior of $v(P^\phi)$

Note that for $\delta := \phi^{-1}\partial$ we have

$$\delta\smallo \subseteq \smallo \iff v\phi < (\Gamma^>)'.$$

Accordingly we restrict ϕ in the rest of this subsection to satisfy $v\phi < (\Gamma^>)'$. (Thus $\phi^\dagger \prec 1$ if $\phi \prec 1$.) The main goal of this subsection is:

PROPOSITION 11.1.4. *Suppose that $\partial\smallo \subseteq \smallo$. Then for all $\phi \preccurlyeq 1$,*

$$v(P) + \operatorname{dwt}(P^\phi)v\phi \leqslant v(P^\phi) \leqslant v(P) + \operatorname{dwm}(P)v\phi.$$

Here $\phi \preccurlyeq 1$ is in addition to the standing assumption that $v\phi < (\Gamma^{>})'$. We first prove some lemmas on the behavior of $vF_k^n(\phi)$ for $\phi^\dagger \succ \phi$. Combining this with the case $\phi^\dagger \preccurlyeq \phi$ from the previous subsection, we shall derive Proposition 11.1.4.

Assume P has order at most r, let $\boldsymbol{\sigma}, \boldsymbol{\tau} \in \{0, \dots, r\}^*$ have equal length, and set

$$\begin{bmatrix}\boldsymbol{\tau}\\ \boldsymbol{\sigma}\end{bmatrix} := \begin{bmatrix}\tau_1\\ \sigma_1\end{bmatrix} \cdots \begin{bmatrix}\tau_d\\ \sigma_d\end{bmatrix} \qquad \text{for } \boldsymbol{\sigma} = \sigma_1 \cdots \sigma_d,\ \boldsymbol{\tau} = \tau_1 \cdots \tau_d,$$

where $\left[\begin{smallmatrix}n\\ m\end{smallmatrix}\right]$ is the signless Stirling number of the first kind from Section 5.7.

LEMMA 11.1.5. *Assume $\partial\mathcal{O} \subseteq \mathcal{O}$. Let $z \in K$, $z \succ 1$. Then*

$$R_n(z) = z^n(1+\varepsilon) \ \textit{with}\ \varepsilon \preccurlyeq z^\dagger/z \prec 1.$$

PROOF. This is clear for $n = 0$ and $n = 1$. Suppose $n \geqslant 1$ and $R_n(z) = z^n(1+\varepsilon)$ with ε as in the lemma. Then

$$\begin{aligned} R_{n+1}(z) = zR_n(z) + R_n(z)' &= z^{n+1}(1+\varepsilon) + nz^{n-1}z'(1+\varepsilon) + z^n\varepsilon' \\ &= z^{n+1}\left(1 + \varepsilon + n\frac{z^\dagger}{z}(1+\varepsilon) + \frac{\varepsilon'}{z}\right). \end{aligned}$$

Next, use that $\varepsilon' \prec z^\dagger$, hence $\varepsilon'/z \prec z^\dagger/z$. □

LEMMA 11.1.6. *Assume $\partial\mathcal{O} \subseteq \mathcal{O}$. Let $f \in K^\times$ be such that $f^\dagger \succ 1$. Then*

$$f^{[\boldsymbol{\sigma}]} \sim f^{|\boldsymbol{\sigma}|}(f^\dagger)^{\|\boldsymbol{\sigma}\|}.$$

Hence if $Q \in \mathcal{O}\{Y\}$ is homogeneous of degree d, isobaric of weight w, and $c \asymp 1$ where $c := \sum_{\boldsymbol{i}} Q_{\boldsymbol{i}}$ is the sum of its coefficients, then

$$Q(f) \sim cf^d(f^\dagger)^w.$$

PROOF. It suffices to do the case $|\boldsymbol{\sigma}| = 1$ of a single factor, that is, it suffices to show $f^{(n)}/f \sim (f^\dagger)^n$. Now $R_n(f^\dagger) = f^{(n)}/f$, so Lemma 11.1.5 applied to $z = f^\dagger$ yields the desired result. □

REMARK. With the assumptions of this lemma, the proof gives

$$f^{[\boldsymbol{\sigma}]} = f^{|\boldsymbol{\sigma}|}(f^\dagger)^{\|\boldsymbol{\sigma}\|}(1+\varepsilon) \qquad \text{where } \varepsilon \preccurlyeq f^{\dagger\dagger}/f^\dagger \prec 1.$$

LEMMA 11.1.7. *Suppose $\phi^\dagger \succ \phi$. Then for $0 < k \leqslant n$ we have*

$$F_k^n(\phi) \sim \begin{bmatrix}n\\ k\end{bmatrix}\phi^k(\phi^\dagger)^{n-k}.$$

PROOF. Let $0 < k \leqslant n$. Recall from Section 5.7 the definition of the differential polynomial $G_k^n \in K^\phi\{Y\}$ with nonnegative integer coefficients, homogeneous of degree n

and isobaric of weight $n-k$, satisfying $F_k^n(\phi) = G_k^n(\phi)$. By the previous lemma applied to $Q = G_k^n$ and with K^ϕ in place of K we obtain

$$F_k^n(\phi) = G_k^n(\phi) \sim c\phi^n \left(\frac{\delta(\phi)}{\phi}\right)^{n-k} = c\phi^k(\phi^\dagger)^{n-k},$$

where c is the sum of the coefficients of G_k^n, that is, $c = \left[\begin{smallmatrix} n \\ k \end{smallmatrix}\right]$ by Lemma 5.7.6. □

We set $\operatorname{supp}\boldsymbol{\sigma} := \{i : \sigma_i \neq 0\}$. So if $\boldsymbol{\tau} \geqslant \boldsymbol{\sigma}$, then $\operatorname{supp}\boldsymbol{\sigma} \subseteq \operatorname{supp}\boldsymbol{\tau}$. If $\boldsymbol{\tau} \geqslant \boldsymbol{\sigma}$ and $\phi^\dagger \succ \phi$, then we have the equivalences

$$F_{\boldsymbol{\sigma}}^{\boldsymbol{\tau}}(\phi) \neq 0 \iff \begin{bmatrix} \boldsymbol{\tau} \\ \boldsymbol{\sigma} \end{bmatrix} \neq 0 \iff \operatorname{supp}\boldsymbol{\tau} = \operatorname{supp}\boldsymbol{\sigma}.$$

LEMMA 11.1.8. *Suppose $\boldsymbol{\tau} \geqslant \boldsymbol{\sigma}$ and* $\operatorname{supp}\boldsymbol{\tau} = \operatorname{supp}\boldsymbol{\sigma}$, *and $\phi^\dagger \succ \phi$. Then*

$$F_{\boldsymbol{\sigma}}^{\boldsymbol{\tau}}(\phi) \sim \begin{bmatrix} \boldsymbol{\tau} \\ \boldsymbol{\sigma} \end{bmatrix} \phi^{\|\boldsymbol{\sigma}\|}(\phi^\dagger)^{\|\boldsymbol{\tau}\|-\|\boldsymbol{\sigma}\|}.$$

Also, if $\phi \not\asymp 1$, then $v\big(F_{\boldsymbol{\sigma}}^{\boldsymbol{\tau}}(\phi)\big) = \|\boldsymbol{\tau}\| v\phi + o(v\phi)$.

PROOF. The first statement follows from the previous lemma. Note that

$$\phi^{\|\boldsymbol{\sigma}\|}(\phi^\dagger)^{\|\boldsymbol{\tau}\|-\|\boldsymbol{\sigma}\|} = \phi^{\|\boldsymbol{\tau}\|}\big(\delta(\phi)/\phi\big)^{\|\boldsymbol{\tau}\|-\|\boldsymbol{\sigma}\|}.$$

If $\phi \not\asymp 1$, then $\psi^\phi(v\phi) = v(\phi^\dagger/\phi) < 0$, so $v\big(\delta(\phi)/\phi\big) = o(v\phi)$ by Lemma 9.2.10(iv). □

LEMMA 11.1.9. *Suppose $\phi \prec 1$ and $\boldsymbol{\tau} \geqslant \boldsymbol{\sigma}$. Then $v\big(F_{\boldsymbol{\sigma}}^{\boldsymbol{\tau}}(\phi)\big) \geqslant \|\boldsymbol{\sigma}\| v\phi$. Moreover,*

$$v\big(F_{\boldsymbol{\sigma}}^{\boldsymbol{\tau}}(\phi)\big) = \|\boldsymbol{\sigma}\| v\phi \iff \boldsymbol{\tau} = \boldsymbol{\sigma}.$$

PROOF. From $\phi \prec 1$ we obtain $\phi^\dagger \prec 1$. If $\phi^\dagger \succ \phi$, then we appeal to Lemma 11.1.8. If $\phi^\dagger \preccurlyeq \phi$, then we use Corollary 11.1.3. □

LEMMA 11.1.10. *Assume $\phi \prec 1$. Then $v\big((P^\phi)_{[\boldsymbol{\sigma}]}\big) \geqslant \|\boldsymbol{\sigma}\| v\phi + v(P)$, and*

$$\begin{cases} (P^\phi)_{[\boldsymbol{\sigma}]} \sim \phi^{\|\boldsymbol{\sigma}\|} P_{[\boldsymbol{\sigma}]} & \text{if } v(P_{[\boldsymbol{\sigma}]}) = v(P), \\ v\big((P^\phi)_{[\boldsymbol{\sigma}]}\big) > \|\boldsymbol{\sigma}\| v\phi + v(P) & \text{otherwise.} \end{cases}$$

PROOF. Suppose $\boldsymbol{\tau} \geqslant \boldsymbol{\sigma}$. Then $v\big(F_{\boldsymbol{\sigma}}^{\boldsymbol{\tau}}(\phi) P_{[\boldsymbol{\tau}]}\big) \geqslant \|\boldsymbol{\sigma}\| v\phi + v(P)$ by Lemma 11.1.9, hence

$$v\big((P^\phi)_{[\boldsymbol{\sigma}]}\big) \geqslant \|\boldsymbol{\sigma}\| v\phi + v(P)$$

by (11.1.1), showing the first statement. Lemma 11.1.9 yields that if $v(P_{[\boldsymbol{\sigma}]}) = v(P)$ and $\boldsymbol{\tau} \neq \boldsymbol{\sigma}$, then

$$v\big(F_{\boldsymbol{\sigma}}^{\boldsymbol{\tau}}(\phi) P_{[\boldsymbol{\tau}]}\big) > \|\boldsymbol{\sigma}\| v\phi + v(P) = v(\phi^{\|\boldsymbol{\sigma}\|} P_{[\boldsymbol{\sigma}]}).$$

Hence $(P^{\phi})_{[\sigma]} \sim \phi^{\|\sigma\|}P_{[\sigma]}$ if $v(P_{[\sigma]}) = v(P)$. Suppose that $v(P_{[\sigma]}) > v(P)$. Then $v(\phi^{\|\sigma\|}P_{[\sigma]}) > \|\sigma\|v\phi + v(P)$, and by the previous lemma again

$$v\big(F^{\tau}_{\sigma}(\phi)P_{[\tau]}\big) > \|\sigma\|v\phi + v(P) \quad \text{if } \tau \neq \sigma,$$

hence $v\big((P^{\phi})_{[\sigma]}\big) > \|\sigma\|v\phi + v(P)$. □

PROOF OF PROPOSITION 11.1.4. Assume $\partial o \subseteq o$ and $\phi \preccurlyeq 1$. We need to show

$$v(P) + \operatorname{dwt}(P^{\phi})v\phi \;\leqslant\; v(P^{\phi}) \;\leqslant\; v(P) + \operatorname{dwm}(P)v\phi.$$

If $\phi \asymp 1$, this holds by Lemma 11.1.1. Let $\phi \prec 1$ and take σ with $\|\sigma\| = \operatorname{dwt}(P^{\phi})$. Then by Lemma 11.1.10 we have

$$v(P) + \operatorname{dwt}(P^{\phi})v\phi \;=\; \|\sigma\|v\phi + v(P) \;\leqslant\; v\big((P^{\phi})_{[\sigma]}\big) \;=\; v(P^{\phi}).$$

For σ such that $\|\sigma\| = \operatorname{dwm}(P)$, this same lemma gives

$$v(P^{\phi}) \;\leqslant\; v\big((P^{\phi})_{[\sigma]}\big) \;=\; v(P) + \operatorname{dwm}(P)v\phi,$$

as required. □

We record a few consequences of Proposition 11.1.4 and Lemma 11.1.10.

COROLLARY 11.1.11. *Assume* $\partial o \subseteq o$ *and set* $w := \operatorname{dwm}(P)$. *Then*

(i) *If* $\phi \preccurlyeq 1$ *and* $\operatorname{dwt}(P^{\phi}) = \operatorname{dwm}(P)$, *then* $v(P^{\phi}) = v(P) + wv\phi$.

(ii) *If* $\phi \prec 1$, *then* $\operatorname{dwm}(P^{\phi}) \leqslant \operatorname{dwt}(P^{\phi}) \leqslant \operatorname{dwm}(P) \leqslant \operatorname{dwt}(P)$.

(iii) *If* $\phi \prec 1$ *and* $\operatorname{dwm}(P^{\phi}) = \operatorname{dwt}(P)$, *then for all* σ,

$$\begin{aligned} v(P_{[\sigma]}) = v(P) \qquad &\Longleftrightarrow \qquad v\big((P^{\phi})_{[\sigma]}\big) = v(P^{\phi}), \\ v(P_{[\sigma]}) = v(P) \qquad &\Longrightarrow \qquad (P^{\phi})_{[\sigma]} \sim \phi^{w}P_{[\sigma]}. \end{aligned}$$

PROOF. Item (i) follows from Proposition 11.1.4. Suppose $\phi \prec 1$. Then by the same Proposition we have $\operatorname{dwt}(P^{\phi}) \leqslant \operatorname{dwm}(P)$, which gives (ii). Assume also that $\operatorname{dwm}(P^{\phi}) = \operatorname{dwt}(P)$. Then the four numbers in (ii) are all equal to w. Therefore, if $v(P_{[\sigma]}) = v(P)$, then $\|\sigma\| = \operatorname{dwm}(P) = \operatorname{dwt}(P) = w$, so $(P^{\phi})_{[\sigma]} \sim \phi^{w}P_{[\sigma]}$ by Lemma 11.1.10, and thus $v\big((P^{\phi})_{[\sigma]}\big) = v(P^{\phi})$ by (i). Conversely, if $v\big((P^{\phi})_{[\sigma]}\big) = v(P^{\phi})$, then $\|\sigma\| = \operatorname{dwm}(P^{\phi}) = \operatorname{dwt}(P^{\phi}) = w$, so $v(P^{\phi}) = v(P) + \|\sigma\|v\phi$ by (i), and thus $v(P_{[\sigma]}) = v(P)$ by Lemma 11.1.10. □

Now let $\phi_1, \phi_2 \in K^{\times}$ and $v\phi_1, v\phi_2 < (\Gamma^{>})'$, $\phi_1 \preccurlyeq \phi_2$. Then $\phi_3 := \phi_1\phi_2^{-1}$ satisfies $K^{\phi_1} = (K^{\phi_2})^{\phi_3}$ and $P^{\phi_1} = (P^{\phi_2})^{\phi_3}$, in particular, the derivation of $(K^{\phi_2})^{\phi_3}$ is small. Thus if $\phi_1 \prec \phi_2$, then $\phi_3 \prec 1$, so we can apply Corollary 11.1.11 with K^{ϕ_2} and P^{ϕ_2} instead of K and P, with ϕ_3 in the role of ϕ:

COROLLARY 11.1.12. *Suppose* $\phi_1, \phi_2 \in K^\times$ *and* $v\phi_1, v\phi_2 < (\Gamma^>)'$, $\phi_1 \prec \phi_2$. *Then*

$$\operatorname{dwm}(P^{\phi_1}) \leqslant \operatorname{dwt}(P^{\phi_1}) \leqslant \operatorname{dwm}(P^{\phi_2}) \leqslant \operatorname{dwt}(P^{\phi_2}),$$

and if $\operatorname{dwt}(P^{\phi_1}) = \operatorname{dwm}(P^{\phi_2}) = w$, *then*

$$v(P^{\phi_1}) - wv(\phi_1) = v(P^{\phi_2}) - wv(\phi_2).$$

COROLLARY 11.1.13. *Assume* K *is of* H*-type and* $\partial\mathcal{o} \subseteq \mathcal{o}$. *Let* $\Phi \in K^\times$ *be such that* $\Phi \not\asymp^\flat 1$. *Then for all* $\phi \preccurlyeq 1$ *we have* $P^\phi \asymp_\Phi P$.

PROOF. This is clear for $\phi \asymp 1$. Assume $\phi \prec 1$. Then $\phi^\dagger \prec 1 \preccurlyeq \Phi^\dagger$, so $\phi \prec\!\!\prec \Phi$, and hence the claim follows from Proposition 11.1.4. □

Newton weight, Newton degree, and Newton multiplicity

We recall that $\Psi^\downarrow < (\Gamma^>)'$. Moreover, if K has no gap, then

$$\Psi^\downarrow = \{\gamma : \gamma < (\Gamma^>)'\}.$$

Call ϕ **active** (in K) if $v\phi \in \Psi^\downarrow$. If ϕ is active in K, then ϕ remains active in every asymptotic field extension of K.

In the rest of this subsection we assume that K *is ungrounded,* and we restrict ϕ and ϕ_0 to be active in K, in particular, $\phi, \phi_0 \in K^\times$. This restriction implies the restriction on ϕ in the previous subsection, and coincides with it if K has no gap.

LEMMA 11.1.14. *There exists* ϕ_0 *such that for all* $\phi \preccurlyeq \phi_0$,

$$\operatorname{dwm}(P^\phi) = \operatorname{dwt}(P^\phi) = \operatorname{dwm}(P^{\phi_0}) = \operatorname{dwt}(P^{\phi_0}).$$

PROOF. Clear from the first part of Corollary 11.1.12. □

We say that a property $S(\phi)$ of (active) elements ϕ holds **eventually** if there exists ϕ_0 such that $S(\phi)$ holds for all $\phi \preccurlyeq \phi_0$. Let ϕ_0 be as in Lemma 11.1.14 and note that $\operatorname{dwt}(P^{\phi_0})$ does not depend on the choice of such ϕ_0. We set

$$\operatorname{nwt}(P) := \operatorname{dwt}(P^{\phi_0}) = \text{eventual value of } \operatorname{dwt}(P^\phi) = \text{eventual value of } \operatorname{dwm}(P^\phi),$$

and call it the **Newton weight** of P. Thus $v(P^\phi) - \operatorname{nwt}(P)v(\phi)$ is independent of $\phi \preccurlyeq \phi_0$, by the second part of Corollary 11.1.12, and we set

$$v^{\mathrm{e}}(P) := v(P^\phi) - \operatorname{nwt}(P)v(\phi) \qquad (\phi \preccurlyeq \phi_0).$$

COROLLARY 11.1.15. $v(P^\phi) = v^{\mathrm{e}}(P) + \operatorname{nwt}(P)v(\phi)$, *eventually.*

If P is homogeneous, then $P^\phi \in K^\phi\{Y\}$ is homogeneous with $\deg P^\phi = \deg P$, by Corollary 5.7.5. Therefore $v(P^\phi) = \min_{i\in\mathbb{N}} v(P_i^\phi)$ for each ϕ. If $P_i \neq 0$, then $v(P_i^\phi) = v^{\mathrm{e}}(P_i) + \operatorname{nwt}(P_i)v\phi$ eventually, and thus:

COROLLARY 11.1.16. *There is an* $i \in \mathbb{N}$ *such that* $v(P^\phi) = v(P_i^\phi)$ *eventually, and for any such* i *we have* $v^{\mathrm{e}}(P) = v^{\mathrm{e}}(P_i)$ *and* $\mathrm{nwt}(P) = \mathrm{nwt}(P_i)$.

It follows from Lemmas 4.5.7 and 5.7.1 that for $g \in K^\times$ the natural number $\mathrm{nwt}(P_{\times g})$ depends only on vg, so we can define a function $\mathrm{nwt}_P \colon \Gamma \to \mathbb{N}$ by

$$\mathrm{nwt}_P(\gamma) := \mathrm{nwt}(P_{\times g}) \qquad \text{for } g \in K^\times \text{ with } vg = \gamma.$$

Let $g \in K^\times$. If $g \asymp 1$, then $v(P^\phi_{\times g}) = v(P^\phi)$ and $\mathrm{nwt}(P) = \mathrm{nwt}(P_{\times g})$, so $v^{\mathrm{e}}(P) = v^{\mathrm{e}}(P_{\times g})$. Thus $v^{\mathrm{e}}(P_{\times g})$ depends only on $\gamma = vg$, and so we can set $v^{\mathrm{e}}_P(\gamma) := v^{\mathrm{e}}(P_{\times g})$ when $vg = \gamma$. Note also that $0 \leqslant \mathrm{nwt}(P_{\times g}) \leqslant \mathrm{wt}(P)$ for all $g \in K^\times$.

Let $\boldsymbol{k}$ be the residue field of K. Since Ψ has no largest element, the Ψ-set $\Psi - v\phi$ of any compositional conjugate K^ϕ contains positive elements, so by the equivalence (i) $\Longleftrightarrow$ (iv) of Proposition 9.1.3 we have $\delta\mathcal{O} \subseteq \mathcal{o}$ for the derivation $\delta = \phi^{-1}\partial$ of K^ϕ. Thus for all ϕ the differential residue field of K^ϕ is $\boldsymbol{k}$ with the trivial derivation.

With ϕ_0 as above, Corollary 11.1.11(iii) yields that the dominant part $D_{P^{\phi_0}}$ of P^{ϕ_0} is isobaric of weight $\mathrm{nwt}(P)$, and that for all $\phi \prec \phi_0$,

$$D_{P^\phi} = u(\phi) D_{P^{\phi_0}} \text{ for some } u(\phi) \in \boldsymbol{k}^\times.$$

Thus the degree of $D_{P^{\phi_0}}$ is independent of the choice of ϕ_0; we set

$$\mathrm{ndeg}\, P := \deg D_{P^{\phi_0}} = \text{eventual value of } \mathrm{ddeg}\, P^\phi,$$

and call it the **Newton degree** of P. Likewise, we set

$$\mathrm{nmul}\, P := \mathrm{mul}\, D_{P^{\phi_0}} = \text{eventual value of } \mathrm{dmul}\, P^\phi,$$

and call it the **Newton multiplicity** of P at 0. Thus

$$\mathrm{mul}\, P \leqslant \mathrm{nmul}\, P \leqslant \mathrm{ndeg}\, P \leqslant \deg P.$$

Given $P \in K\{Y\}^{\neq}$, the definitions of $v^{\mathrm{e}}P \in \Gamma$ and $\mathrm{nwt}\, P, \mathrm{ndeg}\, P, \mathrm{nmul}\, P \in \mathbb{N}$ involve K, but these quantities do not change when K is replaced by an asymptotic extension L such that Ψ is cofinal in Ψ_L. In particular, these quantities remain the same when passing to the algebraic closure of K.

Behavior under compositional conjugation

We continue to assume that K is ungrounded, and ϕ continues to range over active elements in K. Let $f \in K^\times$. The assumptions on K remain valid for K^f, and the active elements in K^f are the quotients ϕ/f, hence

$$\mathrm{nwt}\, P^f = \mathrm{nwt}\, P, \quad \mathrm{nmul}\, P^f = \mathrm{nmul}\, P, \quad \mathrm{ndeg}\, P^f = \mathrm{ndeg}\, P.$$

LEMMA 11.1.17. $v^{\mathrm{e}}(P^f) = v^{\mathrm{e}}(P) + \mathrm{nwt}(P)vf$.

PROOF. By Corollary 11.1.15 we have, eventually,

$$v((P^f)^{\phi/f}) = v^{\mathrm{e}}(P^f) + \mathrm{nwt}(P)(v\phi - vf) = v^{\mathrm{e}}(P) + \mathrm{nwt}(P)v\phi,$$

which gives the desired result. □

Newton weight of linear differential operators

*In this subsection we assume that K is ungrounded (and so pre-*d*-valued). We let ϕ, ϕ_1, ϕ_2 range over active elements of K, g over $K^\times$, and γ over Γ.*

Let $A = a_0 + a_1\partial + \cdots + a_r\partial^r \in K[\partial]^{\neq}$ ($a_0, \dots, a_r \in K$, $a_r \neq 0$). Recall that in Section 5.6 we defined $\operatorname{dwm}(A), \operatorname{dwt}(A) \in \mathbb{N}$ if K has small derivation, and so $\operatorname{dwm}(A^\phi)$ and $\operatorname{dwt}(A^\phi)$ are defined. Set $P := a_0Y + a_1Y' + \cdots + a_rY^{(r)} \in K\{Y\}^{\neq}$. Then $\operatorname{dwm}(A^\phi) = \operatorname{dwm}(P^\phi)$ and $\operatorname{dwt}(A^\phi) = \operatorname{dwt}(P^\phi)$. We call $\operatorname{nwt}(A) := \operatorname{nwt}(P)$ the **Newton weight** of A; so

$$\operatorname{nwt}(A) \;=\; \text{eventual value of } \operatorname{dwt}(A^\phi) \;=\; \text{eventual value of } \operatorname{dwm}(A^\phi).$$

Thus eventually $D_{P^\phi} = u(\phi)Y^{(w)}$ where $u(\phi) \in \boldsymbol{k}^\times$ and $w := \operatorname{nwt}(A)$. We also define the function $\operatorname{nwt}_A \colon \Gamma \to \mathbb{N}$ by

$$\operatorname{nwt}_A(\gamma) \;:=\; \operatorname{nwt}(Ag) \;=\; \operatorname{nwt}_P(\gamma) \quad \text{for } \gamma = vg,$$

so for $a \in K^\times$ we have $\operatorname{nwt}_{aA}(\gamma) = \operatorname{nwt}_A(\gamma)$ and $\operatorname{nwt}_{Aa}(\gamma) = \operatorname{nwt}_A(va + \gamma)$. We set $v^{\mathrm{e}}(A) := v^{\mathrm{e}}(P)$ and define $v^{\mathrm{e}}_A := v^{\mathrm{e}}_P : \Gamma \to \Gamma$. So, given any γ,

$$v_{A^\phi}(\gamma) \;=\; v^{\mathrm{e}}_A(\gamma) + \operatorname{nwt}_A(\gamma)v\phi, \quad \text{eventually,} \tag{11.1.2}$$

by Corollary 11.1.15. It follows that for $a \in K^\times$,

$$v^{\mathrm{e}}_{aA}(\gamma) \;=\; va + v^{\mathrm{e}}_A(\gamma), \qquad v^{\mathrm{e}}_{Aa}(\gamma) \;=\; v^{\mathrm{e}}_A(va + \gamma).$$

EXAMPLE. We have $(\partial g)^\phi = g' + \phi g \delta$ in $K^\phi[\delta]$. If $g \asymp 1$, then $(\partial g)^\phi \sim \phi g\delta$ for all ϕ (since K is pre-d-valued), and if $g \not\asymp 1$, then eventually $(\partial g)^\phi \sim g'$. Hence $\operatorname{nwt}_\partial(0) = 1$ and $\operatorname{nwt}_\partial(\gamma) = 0$ for $\gamma \neq 0$. From this and (11.1.2) we get

$$v^{\mathrm{e}}_\partial(\gamma) \;=\; \begin{cases} 0 & \text{if } \gamma = 0, \\ \gamma' & \text{if } \gamma \neq 0. \end{cases}$$

Recall from Section 5.6 that if $\partial\mathcal{O} \subseteq o$, then $\mathscr{E}(A) = \{\gamma : \operatorname{dwm}_A(\gamma) \geqslant 1\}$. Thus for each ϕ we have $\mathscr{E}(A^\phi) = \{\gamma : \operatorname{dwm}_{A^\phi}(\gamma) \geqslant 1\}$. Define

$$\mathscr{E}^{\mathrm{e}}(A) \;:=\; \{\gamma : \operatorname{nwt}_A(\gamma) \geqslant 1\}.$$

If $\phi_1 \prec \phi_2$, then $\mathscr{E}(A^{\phi_1}) \subseteq \mathscr{E}(A^{\phi_2})$ by Corollary 11.1.12, so

$$\mathscr{E}^{\mathrm{e}}(A) \;=\; \bigcap_\phi \mathscr{E}(A^\phi) \;=\; \{vg : A(g) \prec A^\phi g \text{ for all } \phi\}.$$

Call $\mathscr{E}^{\mathrm{e}}(A)$ the set of **eventual exceptional values** for A. By Lemma 5.6.7 we have $v(\ker^{\neq} A) = v(\ker^{\neq} A^\phi) \subseteq \mathscr{E}(A^\phi)$ for each ϕ, so $v(\ker^{\neq} A) \subseteq \mathscr{E}^{\mathrm{e}}(A)$. For $a \in K^\times$ we have $\mathscr{E}^{\mathrm{e}}(aA) = \mathscr{E}^{\mathrm{e}}(A)$ and $\mathscr{E}^{\mathrm{e}}(Aa) = \mathscr{E}^{\mathrm{e}}(A) - va$.

EXAMPLE. If $r = 0$, then $\mathscr{E}^{\mathrm{e}}(A) = \emptyset$. Suppose $r = 1$, and set $a := a_0/a_1$. Then

$$vg \in \mathscr{E}^{\mathrm{e}}(A) \quad\Longleftrightarrow\quad \phi^{-1}a + \phi^{-1}g^\dagger \prec 1 \text{ for all } \phi \quad\Longleftrightarrow\quad v(a + g^\dagger) > \Psi,$$

where the first equivalence follows from a similar equivalence on exceptional values in Section 5.6. Thus $\mathscr{E}^{\mathrm{e}}(A)$ has at most one element. In particular, $\mathscr{E}^{\mathrm{e}}(\partial) = \{0\}$.

In Chapter 14 we show that for certain K the set $\mathscr{E}^{\mathrm{e}}(A)$ is finite of size at most r.

11.2 NEWTON DEGREE AND NEWTON MULTIPLICITY

In this section K is an ungrounded asymptotic field with value group $\Gamma \neq \{0\}$. We also restrict $\phi \in K^\times$ to be active, and we fix a differential polynomial $P \in K\{Y\}^{\neq}$. We let $\mathfrak{m}$ and $\mathfrak{n}$ range over $K^\times$.

LEMMA 11.2.1. *Let $\mathfrak{m}$ be given, and suppose $P(f) = 0$ for some $f \preccurlyeq \mathfrak{m}$ in some asymptotic field extension of K. Then* $\operatorname{ndeg} P_{\times\mathfrak{m}} \geqslant 1$.

PROOF. Let L be an asymptotic field extension of K and suppose $f \in L$, $f \preccurlyeq \mathfrak{m}$, and $P(f) = 0$. Then $f = a\mathfrak{m}$ with $a \preccurlyeq 1$. We have $Q(a) = 0$ for $Q := P_{\times\mathfrak{m}}$, so $Q^\phi(a) = 0$, and as L^ϕ has small derivation, this gives $D_{Q^\phi}(\overline{a}) = 0$, and thus $\deg D_{Q^\phi} \geqslant 1$. As this holds for all ϕ, we get $\operatorname{ndeg} Q \geqslant 1$. □

LEMMA 11.2.2. *There are $\mathfrak{m}_0, \mathfrak{m}_1 \in K^\times$ such that* $\operatorname{nmul} P_{\times\mathfrak{m}} = \operatorname{ndeg} P_{\times\mathfrak{m}} = \operatorname{mul} P$ *for all $\mathfrak{m} \preccurlyeq \mathfrak{m}_0$, and* $\operatorname{nmul} P_{\times\mathfrak{m}} = \operatorname{ndeg} P_{\times\mathfrak{m}} = \deg P$ *for all $\mathfrak{m} \succcurlyeq \mathfrak{m}_1$.*

PROOF. Replacing K and P by K^{ϕ_0} and P^{ϕ_0} for some active ϕ_0 in K we arrange that $\partial\mathcal{O} \subseteq \mathcal{O}$. Then by Proposition 11.1.4 we have for all $\phi \preccurlyeq 1$,

$$v_P(\gamma) \leqslant v_{P^\phi}(\gamma) \leqslant v_P(\gamma) + \operatorname{wt}(P)v\phi.$$

This holds in particular also for each nonzero homogeneous part of P in place of P. Let $d = \operatorname{mul}(P)$, set $F := P_d$. If $d < e$ with $G := P_e \neq 0$, then by the above inequalities and Corollary 6.1.5 there is $\gamma_0 \in \Gamma$ such that $v_{F^\phi}(\gamma) < v_{G^\phi}(\gamma)$, for all $\phi \preccurlyeq 1$ and all $\gamma \in \Gamma^{\geqslant\gamma_0}$. This gives the first part of the lemma, and the second part follows in the same way. □

Next we state some results on *Newton degree* that follow easily from corresponding facts on *dominant degree* in Section 6.6, using Lemma 5.7.1.

LEMMA 11.2.3. *Let $a \in K$, $a \preccurlyeq 1$. Then we have*

(i) $\operatorname{ndeg} P_{+a} = \operatorname{ndeg} P$;

(ii) *if $a \prec 1$, then* $\operatorname{ndeg} P_{\times a} \leqslant \operatorname{nmul} P$;

(iii) *if $a \asymp 1$, then* $\operatorname{nmul} P_{\times a} = \operatorname{nmul} P$, $\operatorname{ndeg} P_{\times a} = \operatorname{ndeg} P$.

By (iii) of this lemma we have maps $\operatorname{nmul}_P, \operatorname{ndeg}_P : \Gamma \to \mathbb{N}$ given by

$$\operatorname{nmul}_P(\gamma) = \operatorname{nmul}(P_{\times g}), \quad \operatorname{ndeg}_P(\gamma) = \operatorname{ndeg}(P_{\times g}) \quad \text{for } g \in K^\times \text{ with } vg = \gamma.$$

COROLLARY 11.2.4. *Let $a, b \in K$, $g \in K^\times$ be such that $a - b \preccurlyeq g$. Then*

$$\operatorname{ndeg} P_{+a,\times g} = \operatorname{ndeg} P_{+b,\times g}.$$

Here is an analogue of Corollary 6.6.7:

COROLLARY 11.2.5. $\operatorname{mul} P = \operatorname{mul} P_{\times\mathfrak{m}} \leqslant \operatorname{ndeg} P_{\times\mathfrak{m}} \leqslant \deg P_{\times\mathfrak{m}} = \deg P$, *and*

$$\mathfrak{m} \prec \mathfrak{n} \implies \operatorname{nmul} P_{\times\mathfrak{m}} \leqslant \operatorname{ndeg} P_{\times\mathfrak{m}} \leqslant \operatorname{nmul} P_{\times\mathfrak{n}} \leqslant \operatorname{ndeg} P_{\times\mathfrak{n}}.$$

Thus the functions $\operatorname{nmul}_P, \operatorname{ndeg}_P : \Gamma \to \mathbb{N}$ are decreasing, taking values in the finite set of $d \in \mathbb{N}$ for which $P_d \neq 0$.

The Newton degree on a set $\mathcal{E}$

In this subsection $\mathcal{E} \subseteq K^\times$ is $\preccurlyeq$-closed, that is, $\mathcal{E} \neq \emptyset$, and $\mathfrak{m} \in \mathcal{E}$ whenever $\mathfrak{m} \preccurlyeq \mathfrak{n} \in \mathcal{E}$. The **Newton degree** of P on $\mathcal{E}$, $\operatorname{ndeg}_{\mathcal{E}} P$, is the natural number

$$\operatorname{ndeg}_{\mathcal{E}} P \ :=\ \max\big\{\operatorname{ndeg}(P_{\times\mathfrak{m}}) : \ \mathfrak{m} \in \mathcal{E}\big\}.$$

By Corollary 11.2.5 we have $\operatorname{mul} P \leqslant \operatorname{ndeg}_{\mathcal{E}} P$. If $Q \in K\{Y\}$, $Q \neq 0$, then clearly

$$\operatorname{ndeg}_{\mathcal{E}} PQ \ =\ \operatorname{ndeg}_{\mathcal{E}} P + \operatorname{ndeg}_{\mathcal{E}} Q.$$

We have $\operatorname{ndeg}_{\mathcal{E}} P^{\phi} = \operatorname{ndeg}_{\mathcal{E}} P$. For $a \in K^\times$ we have $\operatorname{ndeg}_{\mathcal{E}} P_{\times a} = \operatorname{ndeg}_{a\mathcal{E}} P$.

LEMMA 11.2.6. *If $v(\mathcal{E})$ does not have a smallest element, then*

$$\operatorname{ndeg}_{\mathcal{E}} P \ =\ \max\big\{\operatorname{nmul}(P_{\times\mathfrak{m}}) : \ \mathfrak{m} \in \mathcal{E}\big\}.$$

LEMMA 11.2.7. *If $f \in \mathcal{E}$, then* $\operatorname{ndeg}_{\mathcal{E}} P_{+f} = \operatorname{ndeg}_{\mathcal{E}} P$.

Let $\mathcal{E}' \subseteq \mathcal{E}$ be $\preccurlyeq$-closed. Then for $f \in \mathcal{E}$ we have $\operatorname{ndeg}_{\mathcal{E}'} P_{+f} \leqslant \operatorname{ndeg}_{\mathcal{E}} P$ by Lemma 11.2.7. For $\gamma \in \Gamma$ and $\mathcal{E} = \{\mathfrak{n} : v\mathfrak{n} \geqslant \gamma\}$, we set $\operatorname{ndeg}_{\geqslant\gamma} P := \operatorname{ndeg}_{\mathcal{E}} P$, so if $v\mathfrak{m} = \gamma$, then $\operatorname{ndeg}_{\geqslant\gamma} P = \operatorname{ndeg} P_{\times\mathfrak{m}}$.

COROLLARY 11.2.8. *Let $a, b \in K$ and $\alpha, \beta \in \Gamma$ be such that $v(b-a) \geqslant \alpha$ as well as $\beta \geqslant \alpha$. Then* $\operatorname{ndeg}_{\geqslant\beta} P_{+b} \leqslant \operatorname{ndeg}_{\geqslant\alpha} P_{+a}$.

For $\mathcal{E} := \{\mathfrak{n} : \mathfrak{n} \prec \mathfrak{m}\}$ we set $\operatorname{ndeg}_{\prec\mathfrak{m}} P := \operatorname{ndeg}_{\mathcal{E}} P$. If $\Gamma^>$ has no least element, then $v\big(\{\mathfrak{n} : \mathfrak{n} \prec \mathfrak{m}\}\big)$ has no least element, and so

$$\operatorname{ndeg}_{\prec\mathfrak{m}} P \ =\ \max\big\{\operatorname{ndeg} P_{\times\mathfrak{n}} : \ \mathfrak{n} \prec \mathfrak{m}\big\} \ =\ \max\big\{\operatorname{nmul} P_{\times\mathfrak{n}} : \ \mathfrak{n} \prec \mathfrak{m}\big\}.$$

DEFINITION 11.2.9. An algebraic differential equation with asymptotic side condition (for short: an **asymptotic equation**) over K is of the form

(E) $$P(Y) = 0, \qquad Y \in \mathcal{E}.$$

For $g \in K^\times$ and $\mathcal{E} = \{y \in K^\times : y \prec g\}$ we also indicate (E) by

$$P(Y) = 0, \qquad Y \prec g.$$

Likewise with $\preccurlyeq$ in place of $\prec$.

A **solution** of (E) is a $y \in \mathcal{E}$ such that $P(y) = 0$. The **Newton degree** of (E) is defined to be $\operatorname{ndeg}_{\mathcal{E}} P$.

Let f be an element of a valued differential field extension of K. We say that a solution y of (E) **best approximates** f—tacitly: among solutions of (E)—if $y - f \preccurlyeq z - f$ for each solution z of (E). Of course, if $f \in K^\times$ is a solution of (E), then f is the unique solution of (E) that best approximates f. Also, if $f \succ \mathcal{E}$, then $y - f \asymp f$ for all $y \in \mathcal{E}$, so f is best approximated by each solution of (E) in K.

LEMMA 11.2.10. *Let f be an element of a valued differential field extension of K and $\mathfrak{m} \in \mathcal{E}$ with $f \preccurlyeq \mathfrak{m}$. Suppose y is a solution of the asymptotic equation*

$$(11.2.1) \qquad P_{\times\mathfrak{m}}(Y) = 0, \qquad Y \preccurlyeq 1$$

that best approximates $\mathfrak{m}^{-1}f$. Then the solution $\mathfrak{m}y$ of (E) *best approximates f.*

PROOF. Let z be a solution of (E). If $z \succ \mathfrak{m}$, then $z - f \sim z \succ \mathfrak{m} \succcurlyeq \mathfrak{m}y - f$, and if $z \preccurlyeq \mathfrak{m}$, then $\mathfrak{m}^{-1}z$ is a solution of (11.2.1) and so $y - \mathfrak{m}^{-1}f \preccurlyeq \mathfrak{m}^{-1}z - \mathfrak{m}^{-1}f$ and hence $\mathfrak{m}y - f \preccurlyeq z - f$. □

The Newton degree in a cut

In the next lemma, let (a_ρ) be a pc-sequence in K, and put $\gamma_\rho = v(a_{s(\rho)} - a_\rho)$, where $s(\rho)$ is the immediate successor of ρ.

LEMMA 11.2.11. *There is an index ρ_0 and $d \in \mathbb{N}$ such that for all $\rho > \rho_0$ we have $\gamma_\rho \in \Gamma$ and $\operatorname{ndeg}_{\geqslant\gamma_\rho} P_{+a_\rho} = d$. Denoting this number d by $d(P,(a_\rho))$ to indicate its dependence on P and (a_ρ), we have $d(P,(a_\rho)) = d(P,(b_\sigma))$ whenever (b_σ) is a pc-sequence in K equivalent to (a_ρ).*

PROOF. Take ρ_0 such that for all $\rho' > \rho \geqslant \rho_0$ we have

$$\gamma_{\rho'} > \gamma_\rho, \qquad v(a_{\rho'} - a_\rho) = \gamma_\rho \in \Gamma.$$

Then for such ρ' and ρ we have by Corollary 11.2.8,

$$\operatorname{ndeg}_{\geqslant\gamma_{\rho'}} P_{+a_{\rho'}} \leqslant \operatorname{ndeg}_{\geqslant\gamma_\rho} P_{+a_\rho}.$$

This yields the existence of $d(P,(a_\rho))$. For the second part, take ρ_0 as above such that also $\operatorname{ndeg}_{\geqslant\gamma_\rho} P_{+a_\rho} = d$ for all $\rho \geqslant \rho_0$. Let (b_σ) be a pc-sequence in K equivalent to (a_ρ), and take σ_0 such that for all $\sigma' > \sigma \geqslant \sigma_0$ we have

$$v(b_{\sigma'} - b_\sigma) = \delta_\sigma := v(b_{\sigma+1} - b_\sigma) \in \Gamma.$$

We can also assume that $e \in \mathbb{N}$ and ρ_0, σ_0 are such that $\operatorname{ndeg}_{\geqslant\delta_\sigma} P_{+b_\sigma} = e$ for all $\sigma \geqslant \sigma_0$, and $b_\sigma - a_\rho \prec a_\rho - a_{\rho_0}$ for all $\rho > \rho_0$ and $\sigma > \sigma_0$. Finally, we can assume that $\delta_\sigma \geqslant \gamma_{\rho_0}$ for all $\sigma > \sigma_0$. Then for $\sigma > \sigma_0$ we have $v(b_\sigma - a_{\rho_0}) = \gamma_{\rho_0}$ and so

$$e = \operatorname{ndeg}_{\geqslant\delta_\sigma} P_{+b_\sigma} \leqslant \operatorname{ndeg}_{\geqslant\gamma_{\rho_0}} P_{+a_{\rho_0}} = d$$

by Corollary 11.2.8. By symmetry we also have $d \leqslant e$, so $d = e$. □

We now associate to each pc-sequence (a_ρ) in K an object $c_K(a_\rho)$, the **cut** defined by (a_ρ) in K, such that if (b_σ) is also a pc-sequence in K, then

$$c_K(a_\rho) = c_K(b_\sigma) \iff (a_\rho) \text{ and } (b_\sigma) \text{ are equivalent.}$$

We do this in such a way that the cuts $c_K(a_\rho)$, with (a_ρ) a pc-sequence in K, are the elements of a set $c(K)$. For $\boldsymbol{a} \in c(K)$ we define

$$\operatorname{ndeg}_{\boldsymbol{a}} P := d(P,(a_\rho)) = \text{eventual value of } \operatorname{ndeg}_{\geqslant\gamma_\rho} P_{+a_\rho},$$

where (a_ρ) is any pc-sequence in K with $\boldsymbol{a} = c_K(a_\rho)$, using the notations of Lemma 11.2.11. We call $\operatorname{ndeg}_{\boldsymbol{a}} P$ the **Newton degree of P in the cut $\boldsymbol{a}$.** Let (a_ρ) be a pc-sequence in K and $\boldsymbol{a} = c_K(a_\rho)$. For $y \in K$ the cut $c_K(a_\rho + y)$ depends only on $(\boldsymbol{a}, y)$, and so we can set $\boldsymbol{a} + y := c_K(a_\rho + y)$. Likewise, for $y \in K^\times$ the cut $c_K(a_\rho y)$ depends only on $(\boldsymbol{a}, y)$, and so we can set $\boldsymbol{a} \cdot y := c_K(a_\rho y)$. We record some basic facts about $\operatorname{ndeg}_{\boldsymbol{a}} P$:

LEMMA 11.2.12. *Let (a_ρ) be a pc-sequence in K, $\boldsymbol{a} = c_K(a_\rho)$. Then*

(i) $\operatorname{ndeg}_{\boldsymbol{a}} P \leqslant \deg P$;

(ii) $\operatorname{ndeg}_{\boldsymbol{a}} P^\phi = \operatorname{ndeg}_{\boldsymbol{a}} P$;

(iii) $\operatorname{ndeg}_{\boldsymbol{a}} P_{+y} = \operatorname{ndeg}_{\boldsymbol{a}+y} P$ *for* $y \in K$;

(iv) *if $y \in K$ and vy is in the width of (a_ρ), then* $\operatorname{ndeg}_{\boldsymbol{a}} P_{+y} = \operatorname{ndeg}_{\boldsymbol{a}} P$;

(v) $\operatorname{ndeg}_{\boldsymbol{a}} P_{\times y} = \operatorname{ndeg}_{\boldsymbol{a}\cdot y} P$ *for* $y \in K^\times$;

(vi) *if $Q \in K\{Y\}^{\neq}$, then* $\operatorname{ndeg}_{\boldsymbol{a}} PQ = \operatorname{ndeg}_{\boldsymbol{a}} P + \operatorname{ndeg}_{\boldsymbol{a}} Q$;

(vii) *if $P(\ell) = 0$ for some pseudolimit ℓ of (a_ρ) in an asymptotic field extension of K, then* $\operatorname{ndeg}_{\boldsymbol{a}} P \geqslant 1$;

(viii) *if L is an asymptotic field extension of K and Ψ is cofinal in Ψ_L, then Ψ_L has no largest element and* $\operatorname{ndeg}_{\boldsymbol{a}} P = \operatorname{ndeg}_{\boldsymbol{a}_L} P$, *where $\boldsymbol{a}_L = c_L(a_\rho)$.*

PROOF. Most of these items are routine or follow easily from earlier lemmas; in particular, item (vii) from Lemma 11.2.1. Item (iv) follows from (iii). □

The case of order 1

In this case we derive properties of the Newton multiplicity and Newton degree that for higher order require stronger assumptions on K and a lot more work: see the introduction to Chapter 13. *In this subsection we assume that K is H-asymptotic with rational asymptotic integration.* Let $P \in K[Y, Y']^{\neq}$.

LEMMA 11.2.13. *There are $w \in \mathbb{N}$, $\alpha \in \Gamma^>$, and $A \in K[Y]^{\neq}$, such that eventually*

$$P^\phi = \phi^w A(Y)(Y')^w + R_\phi, \quad R_\phi \in K^\phi[Y, Y'], \quad v(R_\phi) > v(P^\phi) + \alpha.$$

PROOF. Let $P = \sum_{j\in J} A_j(Y)(Y')^j$ with finite nonempty $J \subseteq \mathbb{N}$ and $A_j \in K[Y]^{\neq}$ for all $j \in J$. Then

$$P^\phi = \sum_{j\in J} \phi^j A_j(Y)(Y')^j, \qquad v(\phi^j A_j) = jv(\phi) + v(A_j).$$

Since $\Psi^{\downarrow}$ has no largest element we have $w \in J$ such that eventually

$$wv(\phi) + v(A_w) \;<\; jv(\phi) + v(A_j) \;\text{ for all } j \in J \setminus \{w\}.$$

As K has rational asymptotic integration we also have $\alpha \in \Gamma^{>}$ such that eventually $wv(\phi) + v(A_w) + \alpha < jv(\phi) + v(A_j)$ for all $j \in J \setminus \{w\}$. Then the conclusion of the lemma holds with this w, α and $A := A_w$. □

Note that $\operatorname{nwt}(P) = w$ for w as in the lemma above.

PROPOSITION 11.2.14. *There exists $\gamma \in \Gamma^{>}$ such that for all $g \in K$,*

$$\begin{aligned} 0 < vg < \gamma \implies \operatorname{nmul}(P) \;&=\; \operatorname{nmul}(P_{\times g}) \;=\; \operatorname{ndeg}(P_{\times g}),\\ -\gamma < vg < 0 \implies \operatorname{ndeg}(P) \;&=\; \operatorname{ndeg}(P_{\times g}) \;=\; \operatorname{nmul}(P_{\times g}). \end{aligned}$$

PROOF. Take w, α, A as in Lemma 11.2.13. Subtracting from A a polynomial $D \in K[Y]$ with $D \prec A$ and decreasing α if necessary we arrange that $\operatorname{mul} A = \operatorname{dmul} A$ and $\deg A = \operatorname{ddeg} A$, so

$$\operatorname{nmul} P = d + w, \quad \operatorname{ndeg} P = e + w, \text{ where } d := \operatorname{mul} A,\ e := \deg A.$$

Set $B := A(Y)(Y')^w$ and take active ϕ_0 in K such that for all $\phi \preccurlyeq \phi_0$,

$$P^{\phi} \;=\; \phi^w B + R_{\phi}, \quad R_{\phi} \in K^{\phi}[Y, Y'], \quad v(R_{\phi}) \;>\; v(P^{\phi}) + \alpha.$$

So for $\phi \preccurlyeq \phi_0$ and $g \in K^{\times}$ with $g \prec 1$ we have, in $K^{\phi}[Y, Y']$,

$$\begin{aligned} P^{\phi}_{\times g} \;&=\; \phi^w B_{\times g} + (R_{\phi})_{\times g},\\ v(\phi^w B_{\times g}) \;&\leqslant\; wv(\phi) + v(B) + (d + w + 1)v(g) \;=\; v(P^{\phi}) + (d + w + 1)v(g),\\ v\big((R_{\phi})_{\times g}\big) \;&\geqslant\; v(R_{\phi}). \end{aligned}$$

Fix $g \in K^{\times}$ with $0 < (d + w + 1)v(g) < \alpha$. Then we have for $\phi \preccurlyeq \phi_0$,

$$v(P^{\phi}_{\times g}) \;=\; v(\phi^w B_{\times g}) \;<\; v\big((R_{\phi})_{\times g}\big).$$

Now in $K^{\phi}[Y, Y']$ we have $(Y')_{\times g} = \phi^{-1} g' Y + gY'$. From $\phi \prec g^{\dagger}$, eventually, we get $(Y')_{\times g} \sim \phi^{-1} g' Y$, eventually. Now $A = \sum_{i=d}^{e} a_i Y^i$ with all $a_i \in K$, $v(A) = v(a_d) = v(a_e)$. So eventually in $K^{\phi}[Y, Y']$,

$$B_{\times g} \;=\; \left(\sum_{i=d}^{e} a_i g^i Y^i\right) \cdot \big((Y')_{\times g}\big)^w \;\sim\; \phi^{-w} a_d g^d (g')^w Y^{d+w},$$

so $P^{\phi}_{\times g} \sim a_d g^d (g')^w Y^{d+w}$, eventually. Thus $\operatorname{nmul} P_{\times g} = \operatorname{ndeg} P_{\times g} = d + w$.

Next, let $g \in K^{\times}$, $g \succ 1$, $\phi \preccurlyeq \phi_0$. Then $v(\phi^w B_{\times g}) \leqslant wv(\phi) + v(B) = v(P^{\phi})$ and $v\big((R_{\phi})_{\times g}\big) \geqslant v(R_{\phi}) + (N + 1)vg$ where $N := \deg P$. A similar computation as for $g \prec 1$ also gives $B_{\times g} \sim \phi^{-w} a_e g^e (g')^w Y^{e+w}$, eventually. It follows that $\operatorname{ndeg}(P_{\times g}) \;=\; \operatorname{nmul}(P_{\times g}) = e + w$ for $-\alpha < (N + 1)vg < 0$. □

If the assumption of rational asymptotic integration is weakened to K having asymptotic integration, then Lemma 11.2.13, Proposition 11.2.14, and their proofs go through for $P \in K[Y, Y']^{\neq}$ of degree $\leqslant 1$ in Y'.

11.3 USING NEWTON MULTIPLICITY AND NEWTON WEIGHT

In this section K is an H-asymptotic field with rational asymptotic integration. Also $P \in K\{Y\}^{\neq}$ and ϕ ranges over the active elements of K.

Behavior of $vP(y)$

We establish here a key fact needed to construct immediate extensions in Section 11.4, namely Proposition 11.3.1. First we observe:

$$\operatorname{nmul} P \geqslant 1 \iff D_{P^\phi}(0) = 0, \text{ eventually} \iff P(0) \prec P^\phi, \text{ eventually.}$$

PROPOSITION 11.3.1. *Suppose* $\operatorname{nmul}(P) \geqslant 1$. *Then there are $\beta \in \Gamma^<$ and a strictly increasing function $i\colon (\beta, 0) \to \Gamma$ such that $vP(y) = i(vy)$ for all $y \in K$ with $\beta < vy < 0$. If Γ is divisible there is such i with the intermediate value property.*

Before establishing this proposition, we prove two lemmas.

LEMMA 11.3.2. *Let $Q \in K\{Y\}^{\neq}$ and suppose $P^\phi \succ Q^\phi$, eventually. Then there is $\beta \in \Gamma^<$ such that $P(y) \succ Q(y)$ for all $y \in K$ with $\beta < vy < 0$.*

PROOF. For $\mu = \operatorname{nwt}(P)$ and $\nu = \operatorname{nwt}(Q)$ we have eventually

$$vP^\phi = v^{\mathrm{e}}P + \mu v\phi, \qquad vQ^\phi = v^{\mathrm{e}}Q + \nu v\phi.$$

The algebraic closure of K has asymptotic integration, and the above eventual equalities remain true there, as does $vP^\phi < vQ^\phi$, eventually. It follows that we have $\gamma \in \Gamma^>$ such that $vP^\phi + \gamma \leqslant vQ^\phi$, eventually. Take ϕ such that $vP^\phi + \gamma \leqslant vQ^\phi$ and $v\phi \geqslant \psi(\gamma)$. Replace K, P, Q by K^ϕ, P^ϕ, Q^ϕ to arrange that K has small derivation and $v^\flat P < v^\flat Q$. Next, replace P, Q by P/a, Q/a for suitable $a \in K^\times$ to get $v^\flat P = 0$, in particular, $P, Q \in \mathcal{O}^\flat\{Y\}$. Let

$$y \mapsto y^\flat : \mathcal{O}^\flat \to K^\flat$$

be the residue map, and let $P^\flat$, $Q^\flat$ be the images of P, Q under the differential ring morphism $\mathcal{O}^\flat\{Y\} \to K^\flat\{Y\}$ that extends this residue map by sending each $Y^{(n)}$ to $Y^{(n)}$. Then $P^\flat \neq 0$ and $K^\flat$ is an H-asymptotic field with asymptotic couple $(\Gamma^\flat, \psi|(\Gamma^\flat)^{\neq})$. Applying Corollary 9.5.8 to $K^\flat$ and $P^\flat$ instead of K and P, we get $\beta \in (\Gamma^\flat)^<$ such that $P^\flat(y^\flat) \neq 0$ for all $y \in K$ with $\beta < vy < 0$. Also $Q^\flat = 0$, so $Q^\flat(y^\flat) = 0$ for those y. Thus $vP(y) \in \Gamma^\flat$ and $vQ(y) > \Gamma^\flat$ for those y. □

LEMMA 11.3.3. *Suppose* $\operatorname{nmul}(P) \geqslant 1$ *and set* $Q := P - P(0)$. *There is $\beta \in \Gamma^<$ such that $Q(y) \succ P(0)$ for all $y \in K$ with $\beta < vy < 0$, and so $P(y) \sim Q(y)$ for those y.*

PROOF. Note that $Q \neq 0$ and $Q(0) = 0$. From $\operatorname{nmul}(P) \geqslant 1$ we get $P(0) \prec P^\phi$, eventually, so $P^\phi \sim Q^\phi \succ P(0)$, eventually. It remains to apply Corollary 9.5.8 if $P(0) = 0$, and Lemma 11.3.2 if $P(0) \neq 0$. □

Proposition 11.3.1 now follows from Lemma 11.3.3 and Corollaries 9.3.6 and 9.5.8.

More on $vP(y)$

In this subsection we let y, z range over K and β, γ over Γ, and we assume that $P \in K\{Y\}^{\neq}$ has order r. Corollary 9.5.8 gives us an element α of Γ and natural numbers $d, d_0, e_1, \dots, e_{r-1}$ such that for some $\beta < 0$ we have, for all y,

$$\beta < vy < 0 \Longrightarrow vP(y) \ = \ \alpha + d\psi(vy) + d_0 vy - \sum_{i=1}^{r-1} e_i \chi^i(vy).$$

Moreover, this property determines $\alpha, d, d_0, e_1, \dots, e_{r-1}$ uniquely. Recall also that if P is homogeneous, then $d_0 = \deg P$, and that if $P(0) = 0$, then $d_0 \geqslant 1$. We consider $\alpha + d\psi(vy)$ as the main term in the sum above as vy tends to 0, and our goal is to show that α and d coincide with the "eventual" quantities from Section 11.1:

PROPOSITION 11.3.4. *We have $\alpha = v^{\mathrm{e}}(P)$ and $d = \operatorname{nwt}(P)$. If $P(0) = 0$, then for some $\beta < 0$ and all $y, z \in K$ with $\beta < vy < vz = 0$ we have $P(y) \succ P(z)$.*

PROOF. By the proof of Corollary 9.5.8 the statements in this subsection preceding Proposition 11.3.4 remain true when replacing K by its algebraic closure, with the same tuple $(\alpha, d, d_0, e_1, \dots, e_{r-1})$. So we assume in the rest of the proof that Γ is divisible. We now combine tricks from Section 9.3: passing to an elementary extension, compositional conjugation, and coarsening. Set $B := \Gamma^{<}$. Let L be an elementary extension of K and let $\theta \in L^{\times}$ be such that $\Psi < v\theta \in \Psi_L$. Let B_L be the convex hull of B in Γ_L. As in the proof of Proposition 9.3.5, our α and d are related by $\alpha = \alpha_0 - dv\theta$, with $\alpha_0 \in \Gamma_L$ obtained by applying Lemma 9.3.1 to L^{θ}, B_L, P^{θ} in the role of K, B, P. We now also apply Lemma 9.3.4 to L^{θ}, B_L, P^{θ} in the role of K, B, P. In our situation we have

$$\Delta = \Delta(B_L) := \big\{\delta \in \Gamma_L : \ |\delta| < \varepsilon \text{ for all } \varepsilon \in \Gamma^{>}\big\},$$

a convex subgroup of Γ_L, and so $\alpha_0 \equiv v(P^{\theta}) \bmod \Delta$ by Lemma 9.3.4. Therefore,

$$\alpha \equiv v(P^{\theta}) - dv\theta \mod \Delta.$$

In terms of the original structure K and using terminology from Section 11.1, it follows that for each $\varepsilon \in \Gamma^{>}$ we have

$$\big|\alpha - \big(v(P^{\phi}) - dv\phi\big)\big| < \varepsilon, \qquad \text{eventually.}$$

Also $v(P^{\phi}) = v^{\mathrm{e}}(P) + \operatorname{nwt}(P)v\phi$, eventually. Thus for each $\varepsilon \in \Gamma^{>}$,

$$\big|\big(\alpha - v^{\mathrm{e}}(P)\big) + \big(d - \operatorname{nwt}(P)\big)v\phi\big| < \varepsilon, \qquad \text{eventually.}$$

Since K has asymptotic integration, we get $d = \operatorname{nwt}(P)$, and then $\alpha = v^{\mathrm{e}}(P)$. The second part of the proposition follows from the second part of Lemma 9.3.4. □

Now we have the following variant of Corollary 9.3.7:

COROLLARY 11.3.5. *There is* $\beta < 0$ *such that for all* $y \in K$ *with* $\beta < vy < 0$,

$$vP(y) = v^{\mathrm{e}}(P) + \mathrm{nwt}(P)\psi(vy) + d_0 vy - \sum_{i=1}^{r-1} e_i \chi^i(vy), \quad \textit{and thus}$$

$$vP(y) = v^{\mathrm{e}}(P) + \mathrm{nwt}(P)\psi(vy) + \gamma(y), \quad |\gamma(y)| \leqslant (\deg P) \cdot |vy|.$$

Here is another interesting consequence of Proposition 11.3.4:

COROLLARY 11.3.6. *There exists* $\beta < 0$ *such that* $\mathrm{nwt}_P(\gamma) = 0$ *for all* $\gamma \in (\beta, 0)$.

PROOF. With $\beta < 0$ and $d, d_0, e_1, \dots, e_{r-1}$ as above, we have for all y,

$$\beta < vy < 0 \implies vP(y) = v^{\mathrm{e}}(P) + \mathrm{nwt}(P)\psi(vy) + d_0 vy - \sum_{i=1}^{r-1} e_i \chi^i(vy).$$

Now let $\beta < \gamma < 0$. We claim that then $\mathrm{nwt}_P(\gamma) = 0$. To see this, take $g \in K$ with $vg = \gamma$, and let $y \in K$ with $vy < 0$ be so small that $\beta < vy + \gamma < \gamma$ and $\psi(\gamma) < \psi(vy)$. Then $\psi(v(gy)) = \psi(\gamma)$, and thus $\chi(v(gy)) = \chi(\gamma)$, so

$$vP_{\times g}(y) = v^{\mathrm{e}}(P) + \mathrm{nwt}(P)\psi(\gamma) + d_0\gamma - \sum_{i=1}^{r-1} e_i \chi^i(\gamma) + d_0 vy.$$

Here only the term $d_0 vy$ depends on vy, so by Proposition 11.3.4 applied to $P_{\times g}$,

$$v^{\mathrm{e}}(P_{\times g}) = v^{\mathrm{e}}(P) + \mathrm{nwt}(P)\psi(\gamma) + d_0\gamma - \sum_{i=1}^{r-1} e_i \chi^i(\gamma), \qquad \mathrm{nwt}(P_{\times g}) = 0.$$

Note also that in replacing P by $P_{\times g}$, the quantity d_0 does not change and the quantities $e_1, \dots, e_{r-1}$ become 0. □

Evaluation at pc-sequences

In the discussion following Corollary 11.1.15 we defined for $P \in K\{Y\}^{\neq}$ a function $v^{\mathrm{e}}_P \colon \Gamma \to \Gamma$. This function behaves much like v_P, at least piecewise:

LEMMA 11.3.7. *The set* Γ *is a finite union of subsets* $\Gamma(\mu, d)$ *with* $0 \leqslant \mu \leqslant \mathrm{wt}(P)$ *and* $P_d \neq 0$, *such that for all distinct* $\alpha, \beta \in \Gamma(\mu, d)$ *we have*

$$v^{\mathrm{e}}_P(\alpha) - v^{\mathrm{e}}_P(\beta) = d \cdot (\alpha - \beta) + o(\alpha - \beta).$$

Thus if $P(0) = 0$, *then* v^{e}_P *is strictly increasing on each set* $\Gamma(\mu, d)$.

PROOF. Since Ψ has no largest element, we can take an elementary extension K^* of the asymptotic field K with an element $\theta \in K^*$ such that $v\theta \in \Psi_{K^*}$, and $v\theta > v\phi$ for all ϕ. For $\gamma \in \Gamma$ and $g \in K^\times$ such that $vg = \gamma$ we have

$$v^{\mathrm{e}}_P(\gamma) = v(P^{\theta}_{\times g}) - \mathrm{nwt}(P_{\times g})v\theta = v_{P^{\theta}}(\gamma) - \mathrm{nwt}(P_{\times g})v\theta.$$

The nonzero homogeneous parts of P^θ have the same degrees as the nonzero homogeneous parts of P. For $0 \leqslant \mu \leqslant \mathrm{wt}(P)$ and $P_d \neq 0$, define $\Gamma(\mu, d)$ to be the set of $\gamma \in \Gamma$ such that $\mathrm{nwt}(P_{\times g}) = \mu$ for $g \in K^\times$ with $vg = \gamma$, and $v_{P^\theta}(\gamma) = v_Q(\gamma)$ with $Q := P^\theta_d$. It remains to use Corollary 6.1.3. □

In the next lemma (a_ρ) is a pc-sequence in K with a pseudolimit e in an immediate asymptotic extension E of K. Let $G(Y) \in E\{Y\} \setminus E$ and set $\gamma_\rho := v\big(a_{s(\rho)} - a_\rho\big)$.

LEMMA 11.3.8. *There is a pc-sequence (b_λ) in K equivalent to (a_ρ) such that $\big(G(b_\lambda)\big)$ is a pc-sequence in E with $G(b_\lambda) \rightsquigarrow G(e)$.*

PROOF. By removing some initial ρ's we arrange that $\gamma_\rho = v(e - a_\rho) \in \Gamma$ for all ρ, and (γ_ρ) is strictly increasing. By removing also some indices ρ corresponding to limit ordinals we arrange in addition that for each ρ we have $\delta_\rho \in \Gamma^>$ such that $\gamma_{s(\rho)} - \gamma_\rho > \delta_\rho$ and $\gamma_\rho - \gamma_{\rho'} > \delta_\rho$ whenever $\rho > \rho'$. Take $g_\rho \in K$ with $v(g_\rho) = \gamma_\rho$ and define $u_\rho \in E$ by $a_\rho - e = g_\rho u_\rho$, so $u_\rho \asymp 1$. Take $x_\rho \in K$, subject for now only to $-\delta_\rho < vx_\rho < 0$, and put $b_\rho := a_\rho + g_\rho x_\rho \in K$ and $y_\rho := u_\rho + x_\rho \in E$. Then

$$y_\rho \sim x_\rho, \quad b_\rho - e = g_\rho y_\rho, \quad \gamma_\rho - \delta_\rho < v(b_\rho - e) < \gamma_\rho.$$

Thus (b_ρ) pseudoconverges to e and has the same width as (a_ρ), so by Lemma 2.2.17 it is a pc-sequence in K equivalent to (a_ρ). We have

$$G(b_\rho) - G(e) \;=\; \sum_{|\boldsymbol{i}|\geqslant 1} G_{(\boldsymbol{i})}(e)(g_\rho y_\rho)^{\boldsymbol{i}} \qquad \text{where } G_{(\boldsymbol{i})} = \frac{G^{(\boldsymbol{i})}}{\boldsymbol{i}!}.$$

Put $g_{\boldsymbol{i}} := G_{(\boldsymbol{i})}(e) \in E$ for $|\boldsymbol{i}| \geqslant 1$. Then

$$G(b_\rho) - G(e) \;=\; \sum_{|\boldsymbol{i}|\geqslant 1} g_{\boldsymbol{i}}(g_\rho y_\rho)^{\boldsymbol{i}} \;=\; P(g_\rho y_\rho) \;=\; P_{\times g_\rho}(y_\rho), \text{ where}$$

$$P(Y) \;:=\; \sum_{|\boldsymbol{i}|\geqslant 1} g_{\boldsymbol{i}} Y^{\boldsymbol{i}} \in E\{Y\}, \text{ so } \deg P \geqslant 1,\ P(0) = 0.$$

Corollary 11.3.5 with E instead of K gives for each ρ a $\mu_\rho \in \big\{0, \dots, \mathrm{wt}(P)\big\}$ and an $\varepsilon_\rho \in \Gamma^>$ such that for all $y \in E$ with $-\varepsilon_\rho < vy < 0$,

$$v\big(P(g_\rho y)\big) \;=\; v_P^{\mathrm{e}}(\gamma_\rho) + \mu_\rho \psi(vy) + \gamma(\rho, y), \quad |\gamma(\rho, y)| \leqslant (\deg P)|vy|.$$

Take these ε_ρ so small that $(\deg P)\varepsilon_\rho < \delta_\rho/4$. Lemma 11.3.7 with E in place of K gives sets $\Gamma(\mu, d)$ with $0 \leqslant \mu \leqslant \mathrm{wt}(P)$ and $1 \leqslant d \leqslant \deg P$. Passing to suitable cofinal subsequences we arrange that for a fixed such μ and d we have $\gamma_\rho \in \Gamma(\mu, d)$ for all ρ, and μ_ρ is constant as a function of ρ. Then

$$v_P^{\mathrm{e}}(\gamma_{\rho'}) - v_P^{\mathrm{e}}(\gamma_\rho) \;=\; d(\gamma_{\rho'} - \gamma_\rho) + o(\gamma_{\rho'} - \gamma_\rho) \text{ when } \rho' > \rho.$$

For $\rho' > \rho$ we have $\gamma_{\rho'} - \gamma_\rho > \delta_{\rho'}$ as well as $\gamma_{\rho'} - \gamma_\rho > \delta_\rho$, so $|vy_{\rho'} - vy_\rho| < \gamma_{\rho'} - \gamma_\rho$, and thus $\psi(vy_{\rho'}) - \psi(vy_\rho) = o(\gamma_{\rho'} - \gamma_\rho)$ by Lemma 6.5.4(ii). We now impose on x_ρ

the further restriction $-\varepsilon_\rho < vx_\rho < 0$. Since $vx_\rho = vy_\rho$, it follows that for $\rho' > \rho$ we have $|vy_\rho| + |vy_{\rho'}| < \varepsilon_\rho + \varepsilon_{\rho'}$, so in view of $(\deg P)\varepsilon_\rho < \delta_\rho/4$,

$$v\big(P(g_{\rho'}y_{\rho'})\big) - v\big(P(g_\rho y_\rho)\big) = d(\gamma_{\rho'} - \gamma_\rho) + o(\gamma_{\rho'} - \gamma_\rho) + \varepsilon, \quad |\varepsilon| < (\gamma_{\rho'} - \gamma_\rho)/2.$$

Then $v\big(G(b_\rho) - G(e)\big) = v\big(P(g_\rho y_\rho)\big)$ is strictly increasing as a function of ρ, so $G(b_\rho) \rightsquigarrow G(e)$, as promised. □

Substituting powers of Y'

In this subsection we assume $q \in \mathbb{N}^{\geqslant 1}$. For use in proving Lemma 11.7.5 we establish a lower bound on the Newton weight of the differential polynomial $P^{\times q} = P\big((Y')^q\big) \in K\{Y\}$ introduced in Section 5.7.

LEMMA 11.3.9. $\operatorname{nwt}(P^{\times q}) \geqslant dq$ *for some* $d \in \mathbb{N}$ *with* $P_d \neq 0$.

PROOF. We have $(P^{\times q})_{qi} = (P_i)^{\times q}$ for each $i \in \mathbb{N}$ and $(P^{\times q})_j = 0$ for $j \in \mathbb{N}\setminus q\mathbb{N}$, so $\operatorname{nwt}(P^{\times q}) = \operatorname{nwt}\big((P_i)^{\times q}\big)$ for some $i \in \mathbb{N}$ with $P_i \neq 0$, by Corollary 11.1.16. Hence we may assume that P is homogeneous. Setting $d = \deg P$ we then need to show that $\operatorname{nwt}(P^{\times q}) \geqslant dq$. By Corollary 4.3.18 we have a homogeneous $E \in K\{Y\}^{\neq}$ of degree $w := \operatorname{wt}(P)$ such that

$$P^{\times q} = (Y')^{dq-w} \cdot E^{\times}.$$

The differential polynomial $E^{\times\phi} = E^{\phi}_{\times\phi}$ is homogeneous of degree w, and so its dominant part $D_{E^{\times\phi}}$ is homogeneous of degree w, with

$$D_{(E^\times)^\phi} = D_{E^{\times\phi}(Y')} = u\, D_{E^{\times\phi}}(Y') \text{ for some } u \in \boldsymbol{k}^\times.$$

By Corollary 4.3.14 we have

$$\operatorname{wt}\big(D_{E^{\times\phi}}(Y')\big) = \deg D_{E^{\times\phi}} + \operatorname{wt} D_{E^{\times\phi}} = w + \operatorname{dwt}(E^{\times\phi}).$$

Hence $\operatorname{nwt}(E^\times) \geqslant w$ and so

$$\operatorname{nwt}(P^{\times q}) = \operatorname{nwt}\big((Y')^{dq-w}\big) + \operatorname{nwt}(E^\times) \geqslant (dq - w) + w = dq,$$

as desired. □

COROLLARY 11.3.10. *Suppose* $P(0) = 0$ *and* $q \geqslant 2$. *Then there is* $\beta \in \Gamma^<$ *such that*

$$\big\{vP\big((y')^q\big) : y \in K,\ \beta < vy < 0\big\}$$

is disjoint from some interval (γ_0, δ_0) *with* $\gamma_0 \in \Psi$, $\delta_0 \in (\Gamma^>)'$.

PROOF. Set $d := \operatorname{nwt}(P^{\times q})$. Then $d \geqslant 2$ by Lemma 11.3.9. Next, Corollary 11.3.5 gives $\alpha, \beta \in \Gamma$ with $\beta < 0$, $d_0, e_1, \dots, e_n \in \mathbb{N}$ such that for all $y \in K$,

$$\beta < vy < 0 \implies vP((y')^q) = \alpha + d_0 vy + d\psi(vy) - \sum_{i=1}^{n} e_i \chi^i(vy).$$

From the remarks preceding Proposition 11.3.4 it follows that $d_0 \geqslant 1$. It now remains to apply Corollary 9.2.22. □

11.4 CONSTRUCTING IMMEDIATE EXTENSIONS

In this section K is an H-asymptotic field with rational asymptotic integration. Here is the main result of this section:

THEOREM 11.4.1. *Every pc-sequence in K has a pseudolimit in some immediate asymptotic extension of K.*

This is of course just an alternative formulation of Theorem 11.0.1. We do not know a direct proof of Theorem 11.4.1 using evaluation at pc-sequences, along the lines of Sections 6.8 and 6.9. Instead we depend heavily on the preceding facts on Newton weight, Newton degree, and Newton multiplicity, in particular, on Proposition 11.3.1.

We let ϕ range over the active elements of K, and a, b, y over K. Also, $\mathfrak{m}$, $\mathfrak{n}$, $\mathfrak{d}$, $\mathfrak{v}$, and $\mathfrak{w}$ range over $K^\times$, and P, Q range over $K\{Y\}^{\neq}$.

Vanishing

Recall from Section 11.2 that

$$\operatorname{ndeg}_{\prec\mathfrak{v}} P = \max\{\operatorname{ndeg} P_{\times\mathfrak{m}} : \mathfrak{m} \prec \mathfrak{v}\} = \max\{\operatorname{nmul} P_{\times\mathfrak{m}} : \mathfrak{m} \prec \mathfrak{v}\}.$$

Let ℓ be an element in some asymptotic extension of K such that $\ell \notin K$ and $v(K-\ell) = \{v(a-\ell) : a \in K\}$ has no largest element (so ℓ has no best approximation in K). Note that then $v(K-\ell) \subseteq \Gamma$, and that there is a divergent pc-sequence in K with pseudolimit ℓ.

We say that P **vanishes at** (K,ℓ) if for all a and $\mathfrak{v}$ with $a-\ell \prec \mathfrak{v}$ we have $\operatorname{ndeg}_{\prec\mathfrak{v}} P_{+a} \geqslant 1$. (Intuitively, "$P$ vanishes at (K,ℓ)" means that K thinks P could have a zero near ℓ.) Let $Z(K,\ell)$ be the set of all P that vanish at (K,ℓ). Here are some frequently used basic facts:

(1) $P \in Z(K,\ell) \iff P_{+b} \in Z(K,\ell-b)$;

(2) $P \in Z(K,\ell) \iff P_{\times\mathfrak{m}} \in Z(K,\ell/\mathfrak{m})$;

(3) $P \in Z(K,\ell) \implies PQ \in Z(K,\ell)$;

(4) $P \in K \implies P \notin Z(K,\ell)$.

If $P \notin Z(K,\ell)$, we have a, $\mathfrak{v}$ with $a-\ell \prec \mathfrak{v}$ and $\operatorname{ndeg}_{\prec\mathfrak{v}} P_{+a} = 0$, and then also $\operatorname{ndeg}_{\prec\mathfrak{v}} P_{+b} = 0$ for any b with $b-\ell \prec \mathfrak{v}$, by Lemma 11.2.7.

LEMMA 11.4.2. $Y-b \notin Z(K,\ell)$.

PROOF. Take a and $\mathfrak{v}$ such that $a-\ell \prec \mathfrak{v} \asymp b-\ell$. Then for $P := Y-b$ and $\mathfrak{m} \prec \mathfrak{v}$ we have $P_{+a,\times\mathfrak{m}} = \mathfrak{m}Y + (a-b)$ and $\mathfrak{m} \prec a-b$, so $\operatorname{ndeg}_{\prec\mathfrak{v}} P_{+a} = 0$. □

LEMMA 11.4.3. *Suppose $P \notin Z(K,\ell)$, and let a, $\mathfrak{v}$ be such that $a-\ell \prec \mathfrak{v}$ and $\operatorname{ndeg}_{\prec\mathfrak{v}} P_{+a} = 0$. Then $P(f) \sim P(a)$ for all f in all asymptotic extensions of K with $f-a \asymp \mathfrak{m} \prec \mathfrak{v}$ for some $\mathfrak{m}$. (Recall: $\mathfrak{m} \in K^\times$ by convention.)*

PROOF. Let f in an asymptotic extension E of K satisfy $f - a \asymp \mathfrak{m} \prec \mathfrak{v}$, so $f = a + \mathfrak{m}u$ with $u \asymp 1$ in E. Now

$$P_{+a,\times\mathfrak{m}} = P(a) + R \qquad \text{with } R \in K\{Y\},\ R(0) = 0,$$

so

$$P^{\phi}_{+a,\times\mathfrak{m}} = P(a) + R^{\phi}.$$

From $\operatorname{ndeg} P_{+a,\times\mathfrak{m}} = 0$ we get $R^{\phi} \prec P(a)$, eventually. Thus

$$P(f) = P_{+a,\times\mathfrak{m}}(u) = P(a) + R^{\phi}(u) \quad \text{in } E^{\phi},$$

with $R^{\phi}(u) \preccurlyeq R^{\phi} \prec P(a)$, eventually, in E^{ϕ}, so $P(f) \sim P(a)$. □

Note that the conclusion applies to $f = \ell$, and so for P and a, $\mathfrak{v}$ as in the lemma we have $P(\ell) \sim P(a)$, hence $P(\ell) \neq 0$ and $vP(\ell) \in \Gamma$. In particular, if K is also an H-field and $P \notin Z(K,\ell)$, then $\operatorname{sign} P(a)$ is independent of $a \in K$ and $\mathfrak{v}$ subject to $a - \ell \prec \mathfrak{v}$ and $\operatorname{ndeg}_{\prec\mathfrak{v}} P_{+a} = 0$, and so $\operatorname{sign}(P,\ell) := \operatorname{sign} P(a)$ does not depend on the choice of such a, $\mathfrak{v}$.

LEMMA 11.4.4. *Suppose that $P, Q \notin Z(K,\ell)$. Then $PQ \notin Z(K,\ell)$.*

PROOF. Take $a, b, \mathfrak{v}, \mathfrak{w}$ such that $a - \ell \prec \mathfrak{v}$, $b - \ell \prec \mathfrak{w}$ and

$$\operatorname{ndeg}_{\prec\mathfrak{v}} P_{+a} = \operatorname{ndeg}_{\prec\mathfrak{w}} Q_{+b} = 0.$$

We can assume $a - \ell \preccurlyeq b - \ell$. Take $\mathfrak{n} \asymp a - \ell$ and $d \in K$ with $d - \ell \prec \mathfrak{n}$. Then $d - \ell \prec \mathfrak{v}$ and $d - \ell \prec \mathfrak{w}$, so $\operatorname{ndeg}_{\prec\mathfrak{v}} P_{+a} = \operatorname{ndeg}_{\prec\mathfrak{v}} P_{+d} = 0$, and so $\operatorname{ndeg}_{\prec\mathfrak{n}} P_{+d} = 0$. In the same way we obtain $\operatorname{ndeg}_{\prec\mathfrak{n}} Q_{+d} = 0$. Hence $\operatorname{ndeg}_{\prec\mathfrak{n}} (PQ)_{+d} = 0$. □

LEMMA 11.4.5. *Suppose $P \in Z(K,\ell)$. Then for each b there is a with $a - \ell \prec b - \ell$ and $P(a) \not\asymp P(b)$.*

PROOF. Let $\mathfrak{v} \asymp b - \ell$ and take $a_1 \in K$ with $a_1 - \ell \prec \mathfrak{v}$, so $\operatorname{ndeg}_{\prec\mathfrak{v}} P_{+a_1} \geqslant 1$, so we have $\mathfrak{m} \prec \mathfrak{v}$ with $\operatorname{nmul}(P_{+a_1,\times\mathfrak{m}}) \geqslant 1$. Then by Proposition 11.3.1, the set

$$\{vP(a_1 + \mathfrak{m}y) : \beta < vy < 0\}$$

is infinite for each $\beta \in \Gamma^{<}$, so we can take y with $vy < 0$ and $a_1 + \mathfrak{m}y - \ell \prec \mathfrak{v}$ and $P(a_1 + \mathfrak{m}y) \not\asymp P(b)$. Then $a := a_1 + \mathfrak{m}y$ has the desired property. □

LEMMA 11.4.6. *Suppose $P, Q \notin Z(K,\ell)$ and $P - Q \in Z(K,\ell)$. Then $P(\ell) \sim Q(\ell)$.*

PROOF. By Lemma 11.4.4 we have b and $\mathfrak{v}$ such that

$$\ell - b \prec \mathfrak{v}, \qquad \operatorname{ndeg}_{\prec\mathfrak{v}} P_{+b} = \operatorname{ndeg}_{\prec\mathfrak{v}} Q_{+b} = 0.$$

Replacing ℓ by $\ell - b$ and P, Q by P_{+b}, Q_{+b} we arrange $b = 0$, that is,

$$\ell \prec \mathfrak{v}, \qquad \operatorname{ndeg}_{\prec\mathfrak{v}} P = \operatorname{ndeg}_{\prec\mathfrak{v}} Q = 0,$$

in particular, $P(0) \neq 0$ and $Q(0) \neq 0$. By Lemma 11.4.3 we have for all $a \prec \mathfrak{v}$,

$$P(a) \sim P(0) \sim P(\ell), \qquad Q(a) \sim Q(0) \sim Q(\ell).$$

If $P(\ell) \not\sim Q(\ell)$, then $P(0) \not\sim Q(0)$, so $(P - Q)(a) \asymp (P - Q)(0)$ for all $a \prec \mathfrak{v}$, contradicting $P - Q \in Z(K,\ell)$ by Lemma 11.4.5. Thus $P(\ell) \sim Q(\ell)$. □

Constructing immediate extensions

As in the previous subsection, ℓ is an element in some asymptotic extension of K such that $\ell \notin K$ and $v(K-\ell)$ has no largest element.

LEMMA 11.4.7. *Suppose $Z(K,\ell)=\emptyset$. Then $P(\ell)\neq 0$ for all P, and $K\langle\ell\rangle$ is an immediate asymptotic extension of K. Suppose that g in an asymptotic extension L of K satisfies $v(a-g)=v(a-\ell)$ for all a. Then there is a unique valued differential field embedding $K\langle\ell\rangle\to L$ over K that sends ℓ to g.*

PROOF. Clearly $P(\ell)\neq 0$ for all P. Let any nonzero element $f=P(\ell)/Q(\ell)$ of the asymptotic field extension $K\langle\ell\rangle$ of K be given. Lemma 11.4.4 gives a and $\mathfrak{v}$ such that

$$a-\ell\prec\mathfrak{v},\qquad \operatorname{ndeg}_{\prec\mathfrak{v}}P_{+a}\ =\ \operatorname{ndeg}_{\prec\mathfrak{v}}Q_{+a}\ =\ 0,$$

and so $P(\ell)\sim P(a)$ and $Q(\ell)\sim Q(a)$ by Lemma 11.4.3, and thus $f\sim P(a)/Q(a)$. It follows that $K\langle\ell\rangle$ is an immediate extension of K.

It is clear that $Z(K,g)=Z(K,\ell)=\emptyset$, so g is d-transcendental over K and $K\langle g\rangle$ is an immediate extension of K, by the first part of the proof. Given any P we take a and $\mathfrak{v}$ such that $a-\ell\prec\mathfrak{v}$ and $\operatorname{ndeg}_{\prec\mathfrak{v}}P_{+a}=0$. Then $P(a)\sim P(g)$ and $P(a)\sim P(\ell)$, and thus $vP(g)=vP(\ell)$. Hence the unique differential field embedding $K\langle\ell\rangle\to L$ over K that sends ℓ to g is also a valued field embedding. □

LEMMA 11.4.8. *Suppose that $Z(K,\ell)\neq\emptyset$ and P is an element of $Z(K,\ell)$ of minimal complexity. Then K has an immediate asymptotic extension $K\langle f\rangle$ with $P(f)=0$ and $v(a-f)=v(a-\ell)$ for all a, and such that if E is an asymptotic extension of K and $e\in E$ satisfies $P(e)=0$ and $v(a-e)=v(a-\ell)$ for all a, then there is a unique valued differential field embedding $K\langle f\rangle\to E$ over K that sends f to e.*

PROOF. Let P have order r and take $p\in K[Y_0,\dots,Y_r]$ such that

$$P\ =\ p\big(Y,Y',\dots,Y^{(r)}\big).$$

It is clear from Lemma 11.4.4 that p is irreducible. Consider the domain

$$K[y_0,\dots,y_r]\ =\ K[Y_0,\dots,Y_r]/(p),\quad y_i=Y_i+(p)\text{ for }i=0,\dots,r,$$

and let $K(y_0,\dots,y_r)$ be its fraction field. We extend $v\colon K^\times\to\Gamma$ to

$$v\colon\ K(y_0,\dots,y_r)^\times\to\Gamma$$

as follows. Let $s\in K(y_0,\dots,y_r)^\times$ and take $g\in K[Y_0,\dots,Y_r]$, $h\in K[Y_0,\dots,Y_{r-1}]$ with $g\big(Y,Y',\dots,Y^{(r)}\big)\notin Z(K,\ell)$ (so $g\notin pK[Y_0,\dots,Y_r]$) and $h\neq 0$ such that $s=g(y_0,\dots,y_r)/h(y_0,\dots,y_{r-1})$. Then $vg\big(\ell,\dots,\ell^{(r)}\big)$ and $vh\big(\ell,\dots,\ell^{(r-1)}\big)$ lie in Γ by the comments following Lemma 11.4.3. We claim that

$$vg\big(\ell,\ell',\dots,\ell^{(r)}\big)-vh\big(\ell,\dots,\ell^{(r-1)}\big)\in\Gamma$$

depends only on s and not on the choice of g and h. To see this, let $g_1 \in K[Y_0,\dots,Y_r]$, $h_1 \in K[Y_0,\dots,Y_{r-1}]$ be such that $g_1(Y,\dots,Y^{(r)}) \notin Z(K,\ell)$, $h_1 \neq 0$, and $s = g_1(y_0,\dots,y_r)/h_1(y_0,\dots,y_{r-1})$. Then

$$gh_1 - g_1h \in pK[Y_0,\dots,Y_r], \quad (gh_1)(Y,\dots,Y^{(r)}),(g_1h)(Y,\dots,Y^{(r)}) \notin Z(K,\ell),$$

which yields the claim by Lemma 11.4.6. We now set, for g, h as above,

$$vs := vg(\ell,\ell',\dots,\ell^{(r)}) - vh(\ell,\dots,\ell^{(r-1)}),$$

or more suggestively,

$$vs = v\big(G(\ell)/H(\ell)\big) \in \Gamma, \text{ with } G = g(Y,\dots,Y^{(r)}), H = h(Y,\dots,Y^{(r-1)}).$$

Let $s_1,s_2 \in K(y_0,\dots,y_r)^\times$. Then $v(s_1s_2) = vs_1 + vs_2$ follows easily by means of Lemma 11.4.4. Next, assume also $s_1 + s_2 \neq 0$; to prove that $v\colon K(y_0,\dots,y_r)^\times \to \Gamma$ is a valuation it remains to show that then $v(s_1+s_2) \geqslant \min(vs_1, vs_2)$. For $i = 1,2$ we have $s_i = g_i(y_0,\dots,y_r)/h_i(y_0,\dots,y_{r-1})$ where

$$0 \neq g_i \in K[Y_0,\dots,Y_r], \quad 0 \neq h_i \in K[Y_0,\dots,Y_{r-1}],$$

and g_i has lower degree in Y_r than p. Then for $s := s_1 + s_2$ we have

$$s = g(y_0,\dots,y_r)/h(y_0,\dots,y_{r-1}), \quad g := g_1h_2 + g_2h_1, \quad h := h_1h_2,$$

and so $g \neq 0$ (because $s \neq 0$) and g also has lower degree in Y_r than p. In particular, $g(Y,\dots,Y^{(r)}) \notin Z(K,\ell)$, hence

$$vs = v\big(g(\ell,\dots,\ell^{(r)})/h(\ell,\dots,\ell^{(r-1)})\big),$$

and so by working in the valued field $K\langle\ell\rangle$ we see that $vs \geqslant \min(vs_1, vs_2)$, as promised. Thus we now have $K(y_0,\dots,y_r)$ as a valued field extension of K. To show that $K(y_0,\dots,y_r)$ has the same residue field as K, consider an element $s = g(y_0,\dots,y_r) \notin K$ with nonzero $g \in K[Y_0,\dots,Y_r]$ of lower degree in Y_r than p; it suffices to show that $s \sim b$ for some b. Set $G := g(Y,\dots,Y^{(r)})$ and take a and $\mathfrak{v}$ with $a - \ell \prec \mathfrak{v}$ and $\operatorname{ndeg}_{\prec\mathfrak{v}} G_{+a} = 0$. Then $G(\ell) \sim G(a)$ by Lemma 11.4.3, so for $b := G(a)$ we have

$$v(s-b) = v\big(g(y_0,\dots,y_r) - b\big) = v(G(\ell) - b) > vb,$$

that is, $s \sim b$. This finishes the proof that the valued field $K(y_0,\dots,y_r)$ is an immediate extension of K.

Next we equip $K(y_0,\dots,y_r)$ with the derivation extending the derivation of K such that $y_i' = y_{i+1}$ for $0 \leqslant i < r$. Setting $f := y_0$ we have $f^{(i)} = y_i$ for $i = 0,\dots,r$, $K\langle f\rangle = K(y_0,\dots,y_r)$, and $P(f) = 0$. Note that $v(f-a) = v(\ell - a)$ by Lemma 11.4.2.

In order for $K\langle f\rangle = K\big(f,\dots,f^{(r-1)},f^{(r)}\big)$ to be an asymptotic field, it suffices by Proposition 9.5.1 to show that $K\big(f,\dots,f^{(r-1)}\big)$ is asymptotic in $K\langle f\rangle$. To do that we are going to apply Lemma 9.6.4 with

$$L = K\big(f,\dots,f^{(r-1)}\big),\quad F = K\langle f\rangle,\ \text{ and } U = K\big[f,\dots,f^{(r-1)}\big].$$

Consider elements $s \in K\big[f,\dots,f^{(r-1)}\big]$ and $b \in K^\times$ such that $b \prec 1$; it suffices to show that then $s' \prec b^\dagger$ if $s \prec 1$, and $s' \preccurlyeq b^\dagger$ if $s \asymp 1$. If $s \asymp 1$, take $d \in K$ with $s \sim d$ and use $d' \preccurlyeq b^\dagger$ to reduce to the case $s \prec 1$. So we assume $s \prec 1$ and wish to show that $s' \prec b^\dagger$. This inequality certainly holds when $s' = 0$, so assume $s' \neq 0$. Take $g \in K[Y_0,\dots,Y_{r-1}]$, $g \neq 0$ with $s = g\big(f,\dots,f^{(r-1)}\big)$, and take $g_1, g_2 \in K[Y_0,\dots,Y_{r-1}]$ such that, in $K\{Y\}$,

$$g\big(Y,\dots,Y^{(r-1)}\big)' \;=\; g_1\big(Y,\dots,Y^{(r-1)}\big) + g_2\big(Y,\dots,Y^{(r-1)}\big)Y^{(r)}.$$

Then

$$s' \;=\; g\big(f,\dots,f^{(r-1)}\big)' \;=\; g_1\big(f,\dots,f^{(r-1)}\big) + g_2\big(f,\dots,f^{(r-1)}\big)f^{(r)},$$

and for all a,

$$g\big(a,\dots,a^{(r-1)}\big)' \;=\; g_1\big(a,\dots,a^{(r-1)}\big) + g_2\big(a,\dots,a^{(r-1)}\big)a^{(r)}.$$

There are two cases to consider:

CASE 1: *p has degree* >1 *in* Y_r, *or* $g_2 = 0$. Then we can take a such that

$$s \sim g\big(a,\dots,a^{(r-1)}\big),\qquad s' \sim g\big(a,\dots,a^{(r-1)}\big)'.$$

Since $s \prec 1$, the left-hand side gives $g\big(a,\dots,a^{(r-1)}\big)' \prec b^\dagger$, and then the right-hand side yields $s' \prec b^\dagger$.

CASE 2: *p has degree* 1 *in* Y_r, *and* $g_2 \neq 0$. Then

$$g_1 + g_2 Y_r = \frac{h_1 p + h_2}{h},\qquad h, h_1, h_2 \in K[Y_0,\dots,Y_{r-1}],\ h, h_1 \neq 0,$$

so $0 \neq s' = h_2\big(f,\dots,f^{(r-1)}\big)/h(f,\dots,f^{(r-1)})$, so $h_2 \neq 0$. Let

$$\begin{aligned} G &:= g(Y,\dots,Y^{(r-1)}),\quad H := h(Y,\dots,Y^{(r-1)}),\\ H_1 &:= h_1(Y,\dots,Y^{(r-1)}),\quad H_2 := h_2(Y,\dots,Y^{(r-1)}).\end{aligned}$$

By Lemma 11.4.4 there is $\mathfrak{v}$ such that for some a,

$$a - \ell \prec \mathfrak{v},\quad \operatorname{ndeg}_{\prec\mathfrak{v}} G_{+a} = \operatorname{ndeg}_{\prec\mathfrak{v}} H_{+a} = \operatorname{ndeg}_{\prec\mathfrak{v}} (H_1)_{+a} = \operatorname{ndeg}_{\prec\mathfrak{v}} (H_2)_{+a} = 0.$$

Fix such $\mathfrak{v}$, and let $A \subseteq K$ be the set of all a satisfying the above. Then for $a \in A$ we can apply Lemma 11.4.3 with ℓ in the role of f, so $h(a,\dots,a^{(r-1)}) \neq 0$ and

$$\begin{aligned} s &= g(f,\dots,f^{(r-1)}) \sim g(a,\dots,a^{(r-1)}), \\ &\frac{h_1(f,\dots,f^{(r-1)})}{h(f,\dots,f^{(r-1)})} \sim \frac{h_1(a,\dots,a^{(r-1)})}{h(a,\dots,a^{(r-1)})}, \\ s' &= \frac{h_2(f,\dots,f^{(r-1)})}{h(f,\dots,f^{(r-1)})} \sim \frac{h_2(a,\dots,a^{(r-1)})}{h(a,\dots,a^{(r-1)})}, \\ g(a,\dots,a^{(r-1)})' &= \frac{h_1(a,\dots,a^{(r-1)})}{h(a,\dots,a^{(r-1)})} P(a) + \frac{h_2(a,\dots,a^{(r-1)})}{h(a,\dots,a^{(r-1)})}. \end{aligned}$$

Now $g(a,\dots,a^{(r-1)}) \asymp s \prec 1$, and so $vg(a,\dots,a^{(r-1)})'$ as well as

$$v\frac{h_1(a,\dots,a^{(r-1)})}{h(a,\dots,a^{(r-1)})} \text{ and } v\frac{h_2(a,\dots,a^{(r-1)})}{h(a,\dots,a^{(r-1)})} = vs'$$

do not depend on $a \in A$. On the other hand $\operatorname{ndeg}_{\prec\mathfrak{v}} P_{+a} > 0$, so by Lemma 11.4.5, $vP(a)$ is not constant as a function of $a \in A$. Hence $s' \asymp g(a,\dots,a^{(r-1)})'$, and thus $s' \prec b^{\dagger}$. This finishes the proof that $K\langle f\rangle$ is an asymptotic field.

Suppose now that e in an asymptotic field extension E of K satisfies $P(e) = 0$ and $v(a-e) = v(a-\ell)$ for all a. By Lemma 11.4.3 we have $vQ(e) = vQ(f)$ for all $Q \notin Z(K,\ell)$, in particular, $Q(e) \neq 0$ for all Q of lower complexity than P. Thus we have a differential field embedding $K\langle f\rangle \to E$ over K sending f to e, and this is also a valued field embedding. □

Here are two immediate consequences of Lemmas 11.4.7 and 11.4.8.

COROLLARY 11.4.9. *If K is asymptotically maximal, then K is spherically complete.*

PROOF. Suppose K is not spherically complete. Then we have a divergent pc-sequence (a_ρ) in K, and thus a pseudolimit ℓ in an asymptotic extension of K with $\ell \notin K$. Now Lemmas 11.4.7 (if $Z(K,\ell) = \emptyset$) and 11.4.8 (if $Z(K,\ell) \neq \emptyset$) provide a proper immediate asymptotic extension of K. □

COROLLARY 11.4.10. *The asymptotic field K has a spherically complete immediate asymptotic extension.*

PROOF. This follows by Zorn from Corollary 11.4.9. □

Relation to the Newton degree in a cut

As before, ℓ is an element in some asymptotic extension of K such that $\ell \notin K$ and $v(K-\ell)$ has no largest element. Let (a_ρ) be a divergent pc-sequence in K with pseudolimit ℓ.

LEMMA 11.4.11. *If $P(a_\rho) \rightsquigarrow 0$, then $P \in Z(K,\ell)$.*

PROOF. Suppose $P \notin Z(K,\ell)$. Take a and $\mathfrak{v}$ such that $a-\ell \prec \mathfrak{v}$ and $\operatorname{ndeg}_{\prec\mathfrak{v}} P_{+a} = 0$. Now $v(a-a_\rho) = v(a-\ell)$, eventually, so by Lemma 11.4.3 we have $P(a_\rho) \sim P(a)$ eventually, so $v\big(P(a_\rho)\big) = v\big(P(a)\big) \neq \infty$ eventually. □

We now connect the notion of P vanishing at (K,ℓ) with the Newton degree $\operatorname{ndeg}_{\boldsymbol{a}} P$ of P in the cut $\boldsymbol{a} = c_K(a_\rho)$.

LEMMA 11.4.12. $\operatorname{ndeg}_{\boldsymbol{a}} P \geqslant 1 \iff P \in Z(K,\ell)$. *More precisely,*

$$\operatorname{ndeg}_{\boldsymbol{a}} P \;=\; \min\big\{\operatorname{ndeg}_{\prec\mathfrak{v}} P_{+a} : a-\ell \prec \mathfrak{v}\big\}.$$

PROOF. We may assume $v(\ell - a_\rho)$ is strictly increasing with ρ. Given any index ρ, take $\mathfrak{v} \asymp \ell - a_\rho$, take $\rho' > \rho$, and set $a := a_{\rho'}$. Then $a-\ell \prec \mathfrak{v}$. Now $\gamma_\rho := v(\ell-a_\rho) = v(a-a_\rho)$, and thus by Lemma 11.2.7,

$$\operatorname{ndeg}_{\prec\mathfrak{v}} P_{+a} \;\leqslant\; \operatorname{ndeg}_{\preccurlyeq\mathfrak{v}} P_{+a} \;=\; \operatorname{ndeg}_{\geqslant\gamma_\rho} P_{+a_\rho}.$$

It follows that $\min\big\{\operatorname{ndeg}_{\prec\mathfrak{v}} P_{+a} : a-\ell \prec \mathfrak{v}\big\} \leqslant \operatorname{ndeg}_{\boldsymbol{a}} P$. For the reverse inequality, let a and $\mathfrak{v}$ be such that $a-\ell \prec \mathfrak{v}$. Let ρ be such that $\ell - a_\rho \preccurlyeq \ell - a$. Then $a_\rho - a \prec \mathfrak{v}$ and $\gamma_\rho := v(\ell-a_\rho) > v(\mathfrak{v})$, so by Lemma 11.2.7:

$$\operatorname{ndeg}_{\geqslant\gamma_\rho} P_{+a_\rho} \;\leqslant\; \operatorname{ndeg}_{\prec\mathfrak{v}} P_{+a_\rho} \;=\; \operatorname{ndeg}_{\prec\mathfrak{v}} P_{+a}.$$

Therefore $\operatorname{ndeg}_{\boldsymbol{a}} P \leqslant \min\big\{\operatorname{ndeg}_{\prec\mathfrak{v}} P_{+a} : a-\ell \prec \mathfrak{v}\big\}$. □

Recall from the end of Section 4.4 the notion of a minimal differential polynomial over K of a pc-sequence in K.

COROLLARY 11.4.13. *The following conditions on P are equivalent:*

(i) *P is an element of $Z(K,\ell)$ of minimal complexity;*

(ii) *P is a minimal differential polynomial of (a_ρ) over K.*

PROOF. Assume (i). Then Lemma 11.4.8 provides an immediate extension $K\langle f\rangle$ of K with $P(f) = 0$ and $v(a-\ell) = v(a-f)$ for all a. Hence $a_\rho \rightsquigarrow f$, and so Lemma 11.3.8 gives a pc-sequence (b_λ) in K equivalent to (a_ρ) such that $P(b_\lambda) \rightsquigarrow 0$. Conversely, if (b_λ) is a pc-sequence in K equivalent to (a_ρ) and $Q(b_\lambda) \rightsquigarrow 0$, then $b_\lambda \rightsquigarrow \ell$, so $Q \in Z(K,\ell)$ by Lemma 11.4.11. Thus we have established (i) ⇒ (ii).

Now assume (ii). Again, from Lemma 11.4.11 we get $P \in Z(K,\ell)$. The direction (i) ⇒ (ii) shows that P is of minimal complexity in $Z(K,\ell)$. □

11.5 SPECIAL CUTS IN H-ASYMPTOTIC FIELDS

The cuts referred to in the title of this section are given by certain jammed pc-sequences in H-asymptotic fields K such as $\mathbb{T}$. Their divergence in K is equivalent to K having a rather subtle but important elementary property. To explain this for $K = \mathbb{T}$, consider the iterated logarithms $\ell_n \in \mathbb{T}$ defined by $\ell_0 := x$ and $\ell_{n+1} := \log \ell_n$. Then the corresponding sequence (λ_n) given by

$$\lambda_n := -(\ell_n{}^{\dagger\dagger}) = \frac{1}{\ell_0} + \frac{1}{\ell_0\ell_1} + \cdots + \frac{1}{\ell_0\ell_1\cdots\ell_n}$$

is a jammed pc-sequence in $\mathbb{T}$ by Example 11.5.1 and Lemma 11.5.2 below. It is easy to check that this pc-sequence has no pseudolimit in $\mathbb{T}$. By Corollary 11.6.1 below, the divergence of this pc-sequence in $\mathbb{T}$ implies that for all $s \in \mathbb{T}$ there is $g \succ 1$ in $\mathbb{T}$ such that $s - g^{\dagger\dagger} \succcurlyeq g^{\dagger}$. A related elementary property of $\mathbb{T}$ plays a key role in our work. It corresponds to the fact that the sequence (ω_n) with

$$\omega_n := -2\lambda_n' - \lambda_n^2 = \frac{1}{\ell_0^2} + \frac{1}{\ell_0^2\ell_1^2} + \cdots + \frac{1}{\ell_0^2\ell_1^2\cdots\ell_n^2}$$

is a divergent (jammed) pc-sequence in $\mathbb{T}$; see Corollary 11.7.8.

In this section and the next two we treat this material for ungrounded H-asymptotic fields. More precisely, we assume in this section:

> *K is an ungrounded H-asymptotic field with $\Gamma \neq \{0\}$. (Thus $\Psi := (\Gamma^{\neq})^{\dagger}$ is nonempty and has no largest element.)*

Hence K is a pre-differential-valued field by Corollary 10.1.3, with contraction map

$$\chi\colon \Gamma^{<} \to \Gamma^{<}.$$

Transfinitely iterated logarithms

For $f \in K^{\succ 1}$, take $\operatorname{L} f \in K^{\succ 1}$ such that $(\operatorname{L} f)' \asymp f^{\dagger}$, that is, $v(\operatorname{L} f) = \chi(vf)$. If K is closed under logarithms, then we can choose such $\operatorname{L} f$ with $(\operatorname{L} f)' = f^{\dagger}$. We now introduce "iterated logarithms" ℓ_ρ, for possibly transfinite ρ. More precisely, we construct a sequence (ℓ_ρ) in $K^{\succ 1}$, indexed by the ordinals ρ less than some infinite limit ordinal κ, such that

(1) $\ell_{\rho'} \prec \ell_\rho$ whenever $\rho' > \rho$;

(2) (ℓ_ρ) is coinitial in $K^{\succ 1}$: for each $f \in K^{\succ 1}$ there is an index ρ with $\ell_\rho \preccurlyeq f$.

We construct this sequence by transfinite recursion: take any element $\ell_0 \succ 1$ in K, and take $\ell_{\rho+1} := \operatorname{L}\ell_\rho$; if λ is an infinite limit ordinal such that all ℓ_ρ with $\rho < \lambda$ have already been chosen, then we pick ℓ_λ to be any element in $K^{\succ 1}$ such that $\ell_\lambda \prec \ell_\rho$ for all $\rho < \lambda$, if there is such an ℓ_λ, while if there is no such ℓ_λ, we put $\kappa := \lambda$. We call (ℓ_ρ) a **logarithmic sequence** for K. Note that if L is an H-asymptotic field extension of K

such that $\Gamma^<$ is cofinal in $\Gamma_L^<$, then Ψ is cofinal in Ψ_L (hence Ψ_L also does not have a largest element), and a logarithmic sequence for K remains a logarithmic sequence for L.

Set $e_\rho := v(\ell_\rho) \in \Gamma^<$, so $v(\ell_\rho^\dagger) = e_\rho^\dagger$, and $e_{\rho+1} = \chi(e_\rho)$, and thus

$$e_{\rho+1}^\dagger = e_\rho^\dagger - \chi(e_\rho) = e_\rho^\dagger - e_{\rho+1} > e_\rho^\dagger.$$

Therefore the sequence (e_ρ) is strictly increasing and cofinal in $\Gamma^<$, and the sequence $(e_\rho^\dagger)$ is strictly increasing and cofinal in Ψ. From (ℓ_ρ) we define the sequences (γ_ρ) in $K^\times$ and (λ_ρ) in K as follows:

$$\gamma_\rho := \ell_\rho^\dagger, \qquad \lambda_\rho := -\gamma_\rho^\dagger = -\ell_\rho^{\dagger\dagger} := -(\ell_\rho^{\dagger\dagger}).$$

Then $v(\gamma_\rho) = e_\rho^\dagger$, so $\gamma_{\rho'} \prec \gamma_\rho$ for $\rho < \rho' < \kappa$.

EXAMPLE 11.5.1. Suppose $K = \mathbb{T}$. Then the sequence (ℓ_n) given by $\ell_0 = x$ and $\ell_{n+1} = \log(\ell_n)$ is a logarithmic sequence for K, and for each n we have

$$\gamma_n = \frac{1}{\ell_0\ell_1\cdots\ell_n}, \qquad \lambda_n = \frac{1}{\ell_0} + \frac{1}{\ell_0\ell_1} + \cdots + \frac{1}{\ell_0\ell_1\cdots\ell_n}.$$

LEMMA 11.5.2. *For $\rho < \rho' < \kappa$ we have $\lambda_{\rho'} - \lambda_\rho \sim \gamma_{\rho+1}$. The sequence (λ_ρ) is a jammed pc-sequence of width $\{\gamma \in \Gamma_\infty : \gamma > \Psi\}$.*

PROOF. We have

$$\gamma_{\rho+1} = \ell_{\rho+1}'/\ell_{\rho+1} = (\mathrm{L}\,\ell_\rho)'/\ell_{\rho+1} \asymp \ell_\rho^\dagger/\ell_{\rho+1} = \gamma_\rho/\ell_{\rho+1},$$

so $\gamma_\rho/\gamma_{\rho+1} \asymp \ell_{\rho+1}$. Hence

$$\lambda_{\rho+1} - \lambda_\rho = \gamma_\rho^\dagger - \gamma_{\rho+1}^\dagger = (\gamma_\rho/\gamma_{\rho+1})^\dagger \sim \ell_{\rho+1}^\dagger = \gamma_{\rho+1}.$$

Let $\rho < \rho' < \kappa$; we want $\lambda_{\rho'} - \lambda_\rho \sim \gamma_{\rho+1}$, and have just shown this for $\rho' = \rho + 1$. When $\rho + 1 < \rho' < \kappa$, then by Lemma 6.5.4(i) and $\lambda_{\rho'} - \lambda_{\rho+1} = (\gamma_{\rho+1}/\gamma_{\rho'})^\dagger$,

$$v(\lambda_{\rho'} - \lambda_{\rho+1}) = \psi\big(v(\gamma_{\rho+1}/\gamma_{\rho'})\big) = \psi(e_{\rho+1}^\dagger - e_{\rho'}^\dagger) > e_{\rho+1}^\dagger = v(\gamma_{\rho+1}),$$

so $\lambda_{\rho'} - \lambda_\rho = (\lambda_{\rho'} - \lambda_{\rho+1}) + (\lambda_{\rho+1} - \lambda_\rho) \sim \gamma_{\rho+1}$, as claimed. Thus (λ_ρ) is a pc-sequence with $v(\lambda_{\rho'} - \lambda_\rho) = e_{\rho+1}^\dagger$ for $\rho < \rho' < \kappa$. By Lemma 6.5.4(ii),

$$e_{\rho'+1}^\dagger - e_{\rho+1}^\dagger = o(e_{\rho'+1} - e_{\rho+1}) < |e_\rho|.$$

Given any convex subgroup $\Delta \neq \{0\}$ of Γ, we can take an index $\rho(\Delta)$ such that $e_{\rho(\Delta)} \in \Delta$, and then by the last displayed inequality,

$$e_{\rho'+1}^\dagger - e_{\rho+1}^\dagger \in \Delta \text{ whenever } \kappa > \rho' > \rho > \rho(\Delta).$$

Thus (λ_ρ) is jammed. □

REMARK. By Lemmas 11.5.2 and 2.2.17 the sequence $(\lambda_\rho + \gamma_\rho)$ is also a pc-sequence in K, and is equivalent to (λ_ρ).

By Proposition 11.5.3 below, pc-sequences (λ_ρ) obtained from different choices of logarithmic sequence (ℓ_ρ) are equivalent. The proof uses the properties of $\psi(* - \alpha)$ for $\alpha \in \Psi^{\downarrow}$; see Corollary 9.2.12.

By part (i) of Lemma 6.5.4, if $f, g \in K$ are active and $f \succ g$, then $f^\dagger - g^\dagger \prec f$. Below we frequently use this fact without further mention. Each γ_ρ is active in K, and if $a \in K$ is active, then $\gamma_\rho \prec a$ for all big enough ρ.

PROPOSITION 11.5.3. *Let (u_σ) be a well-indexed sequence of active elements in K such that $\big(v(u_\sigma)\big)$ is strictly increasing and cofinal in $\Psi^{\downarrow}$. Then some cofinal subsequence of $(-u_\sigma^\dagger)$ is a pc-sequence equivalent to (λ_ρ).*

PROOF. Take a pseudolimit λ of (λ_ρ) in some H-asymptotic field extension of K. We claim that $v(\lambda + u_\sigma^\dagger) = \psi\big(* - v(u_\sigma)\big)$ for all σ. Let any σ be given. Then $\lambda + u_\sigma^\dagger = (\lambda + \gamma_\rho^\dagger) + (u_\sigma^\dagger - \gamma_\rho^\dagger)$ with $\lambda + \gamma_\rho^\dagger = \lambda - \lambda_\rho \sim \gamma_{\rho+1}$. Also, Lemma 9.2.11 shows that for all big enough ρ,

$$v(\gamma_\rho^\dagger - u_\sigma^\dagger) = \psi\big(v(\gamma_\rho) - v(u_\sigma)\big) = \psi\big(* - v(u_\sigma)\big) < v(\gamma_{\rho+1}),$$

so $\lambda + u_\sigma^\dagger \sim u_\sigma^\dagger - \gamma_\rho^\dagger$, which proves our claim. Next, given σ, take $\sigma' > \sigma$ such that $u_{\sigma'} \prec \lambda + u_\sigma^\dagger$. By the claim above (with σ' in place of σ) and Corollary 9.2.12 we have $\lambda + u_{\sigma'}^\dagger \prec u_{\sigma'}$, and thus $\lambda + u_{\sigma'}^\dagger \prec \lambda + u_\sigma^\dagger$.

It now follows easily that some cofinal subsequence of $(-u_\sigma^\dagger)$ pseudoconverges to λ. Replacing (u_σ) by a suitable cofinal subsequence we arrange $-u_\sigma^\dagger \rightsquigarrow \lambda$. We show that then $(-u_\sigma^\dagger)$ is equivalent to (λ_ρ). By the above, $(-u_\sigma^\dagger)$ and (λ_ρ) have a common pseudolimit λ. Given ρ we can take σ such that $u_\sigma \prec \gamma_\rho$. Then $u_\sigma^\dagger - \gamma_\rho^\dagger \prec \gamma_\rho$, which in view of $\lambda + \gamma_\rho^\dagger \prec \gamma_\rho$ yields $\lambda + u_\sigma^\dagger \prec \gamma_\rho$. Thus the width of the pc-sequence $(-u_\sigma^\dagger)$ equals the width of (λ_ρ), and it remains only to apply Lemma 2.2.17. □

COROLLARY 11.5.4. *Set $\lambda_\rho^* := -(\lambda_{\rho+1} - \lambda_\rho)^\dagger$ for $\rho < \kappa$. Then (λ_ρ^*) is a pc-sequence in K equivalent to (λ_ρ).*

PROOF. Let $u_\rho := \lambda_{\rho+1} - \lambda_\rho$. Then $\lambda_\rho^* = -u_\rho^\dagger$, and $u_\rho \sim \gamma_{\rho+1}$ by Lemma 11.5.2. For $\rho < \rho' < \kappa$ we obtain

$$\lambda_{\rho'}^* - \lambda_\rho^* = (u_\rho / u_{\rho'})^\dagger \sim (\gamma_{\rho+1}/\gamma_{\rho'+1})^\dagger = \lambda_{\rho'+1} - \lambda_{\rho+1} \sim \gamma_{\rho+2},$$

hence (λ_ρ^*) is a pc-sequence in K. Now apply Proposition 11.5.3 to (u_ρ). □

LEMMA 11.5.5. *Suppose K has asymptotic integration, and $\lambda_\rho \rightsquigarrow \lambda$ with λ in an immediate asymptotic extension E of K. Let $G \in E\{Y\} \setminus E$. Then $G(\lambda_\rho) \rightsquigarrow G(\lambda)$.*

PROOF. Use Corollary 10.3.2 to arrange that E is d-valued of H-type, and then Proposition 10.2.7 to get E also closed under integration. For each ρ, take $y_\rho \in E$ with

$y'_\rho = \lambda_\rho - \lambda$. Then $y'_\rho \asymp \gamma_{\rho+1}$ by Lemma 11.5.2. As $\gamma_{\rho+1} = \ell^\dagger_{\rho+1} \asymp (\mathrm{L}\,\ell_{\rho+1})' = \ell'_{\rho+2}$, and $\ell_{\rho+2} \succ 1$, this gives $y_\rho \asymp \ell_{\rho+2}$ for all ρ. Put

$$P := G(\lambda + Y') - G(\lambda) \in E\{Y\}.$$

Then $P \neq 0$, $P(0) = 0$, and $P(y_\rho) = G(\lambda_\rho) - G(\lambda)$ for each ρ. Corollary 9.5.8 gives $\beta \in \Gamma^<$ and a strictly increasing map $i\colon (\beta, 0) \to \Gamma$ such that $vP(y) = i(vy)$ for all $y \in E$ with $vy \in (\beta, 0)$. Thus $P(y_\rho) \rightsquigarrow 0$. □

We now characterize pseudolimits of (λ_ρ) in terms of active elements:

LEMMA 11.5.6. *Let λ be an element of a valued differential field extension of K. Then the following conditions on λ are equivalent:*

(i) $\lambda_\rho \rightsquigarrow \lambda$;

(ii) *for all active $a \in K$ there is an active $b \prec a$ in K with $\lambda + b^\dagger \prec a$;*

(iii) *for all active $a \in K$ we have $\lambda + a^\dagger \prec a$;*

(iv) *for all $g \succ 1$ in K we have $\lambda + g^{\dagger\dagger} \prec g^\dagger$.*

PROOF. Assume (i). Then for all ρ,

$$\lambda + \gamma^\dagger_\rho = \lambda - \lambda_\rho \sim \gamma_{\rho+1} \prec \gamma_\rho,$$

which gives (ii). To prove (ii) $\Rightarrow$ (iii), assume (ii), and suppose $a \in K$ is active. Take active $b \prec a$ in K with $\lambda + b^\dagger \prec a$. Then $a^\dagger - b^\dagger \prec a$ by the remark preceding Proposition 11.5.3, so $\lambda + a^\dagger = (\lambda + b^\dagger) + (a^\dagger - b^\dagger) \prec a$, which gives (iii). The direction (iii) $\Rightarrow$ (iv) is obvious. For (iv) $\Rightarrow$ (i), assume (iv). Let any ρ be given, and take $g \succ 1$ in K such that for $a = g^\dagger$ we have $a \prec \gamma_{\rho+1}$ in K. Then $\lambda + a^\dagger \prec a$ and $\gamma^\dagger_{\rho+1} - a^\dagger \prec \gamma_{\rho+1}$. In view of

$$\lambda + \gamma^\dagger_\rho = (\lambda + a^\dagger) + (\gamma^\dagger_{\rho+1} - a^\dagger) + (\gamma^\dagger_\rho - \gamma^\dagger_{\rho+1}),$$

we get $\lambda + \gamma^\dagger_\rho \sim \gamma_{\rho+1}$, so (i) holds. □

COROLLARY 11.5.7. *Let γ be an element in an H-asymptotic field extension of K with $\Psi < v\gamma < (\Gamma^>)'$, and set $\lambda := -\gamma^\dagger$. Then λ and $\lambda + \gamma$ are pseudolimits of (λ_ρ).*

PROOF. Let $a \in K$ be active. Then $\lambda + a^\dagger = (a/\gamma)^\dagger \prec a$ by Lemma 9.2.11 and the remark following its proof. Hence $(\lambda + \gamma) + a^\dagger = \gamma + (a/\gamma)^\dagger \prec a$. Thus $\lambda_\rho \rightsquigarrow \lambda$ and $\lambda_\rho \rightsquigarrow \lambda + \gamma$ by the equivalence of (i) and (iii) in Lemma 11.5.6. □

We now discuss the effect of compositional conjugation on the above. Let $\phi \in K^\times$, and consider the compositional conjugate K^ϕ; its derivation is $\delta = \phi^{-1}\partial$. The logarithmic sequence (ℓ_ρ) for K was obtained via a map $\mathrm{L}\colon K^{\succ 1} \to K^{\succ 1}$. The dependence of L on the contraction map χ (which is invariant under compositional conjugation) shows that we can use the same map L to obtain the same logarithmic sequence (ℓ_ρ) for the compositional conjugate K^ϕ. Let $(\gamma_\rho^\phi) := (\delta\ell_\rho/\ell_\rho)$ be the corresponding sequence of active elements in K^ϕ, so $\gamma_\rho^\phi = \gamma_\rho/\phi$ and

$$\lambda_\rho^\phi := -\frac{\delta\gamma_\rho^\phi}{\gamma_\rho^\phi} = -\left(\frac{\gamma_\rho^\dagger}{\phi} - \frac{\phi^\dagger}{\phi}\right) = \frac{\lambda_\rho}{\phi} + \frac{\phi^\dagger}{\phi}.$$

Thus, given $\lambda \in K$ we have

$$\lambda_\rho \rightsquigarrow \lambda \iff \lambda_\rho^\phi \rightsquigarrow (\lambda/\phi) + (\phi^\dagger/\phi).$$

COROLLARY 11.5.8. *The pc-sequence (λ_ρ) has a pseudolimit in K if and only if the corresponding pc-sequence (λ_ρ^ϕ) has a pseudolimit in K^ϕ.*

In the next subsection we relate the above to gap creation.

Gap creation

Recall from Section 9.2 that a gap in K is an element β of its value group Γ such that $\Psi < \beta < (\Gamma^>)'$. Thus by our standing assumption and Corollary 9.2.16, either K has a gap, or K has asymptotic integration.

LEMMA 11.5.9. *Suppose K has a gap vf where $f \in K^\times$, and let $s := f^\dagger$. Then for all active $a \in K$ we have $s - a^\dagger \prec a$.*

PROOF. Let $a \in K$ be active. Then $va < vf < (\Gamma^>)'$, so by Lemma 6.5.4(i),

$$v(s - a^\dagger) = v(f^\dagger - a^\dagger) = v\big((f/a)^\dagger\big) = \psi(vf - va) > va. \qquad \square$$

Thus by Lemmas 11.5.6 and 11.5.9, if K has a gap vf, $f \in K^\times$, then $\lambda_\rho \rightsquigarrow -f^\dagger$.

Let $s \in K$. We say that s **creates a gap over** K if some exponential integral of s introduces a gap, that is, vf is a gap in $K(f)$, for some element $f \neq 0$ in some H-asymptotic field extension of K with $f^\dagger = s$.

LEMMA 11.5.10. *Suppose that K has asymptotic integration and that $s \in K$ creates a gap over K. Then $s - a^\dagger$ is active, for all $a \in K^\times$.*

PROOF. Take f as in the definition above, and let $a \in K^\times$. Then $f \not\asymp a$, so $v(s - a^\dagger) = v\big((f/a)^\dagger\big) < (\Gamma^>)'$, and thus $v(s - a^\dagger) \in \Psi^\downarrow$. $\square$

LEMMA 11.5.11. *Suppose $s \in K$ creates a gap over K, and let $\phi \in K^\times$. Then $(s/\phi) - (\phi^\dagger/\phi) \in K^\phi$ creates a gap over K^ϕ.*

PROOF. Take some H-asymptotic field extension $L = K(f)$ of K with $f \neq 0$, $f^\dagger = s$, and $\Psi_L < vf < (\Gamma_L^>)'$. Then with $\delta = \phi^{-1}\partial$ we have, in L^ϕ,

$$\Psi_L^\phi < v(f/\phi) < (\mathrm{id} + \psi_L^\phi)(\Gamma_L^>), \qquad \delta(f/\phi)/(f/\phi) = (s/\phi) - (\phi^\dagger/\phi),$$

so $(s/\phi) - (\phi^\dagger/\phi) \in K^\phi$ creates a gap over K^ϕ. $\square$

LEMMA 11.5.12. *Suppose K has asymptotic integration and $s \in K$ creates a gap over K. Then $\lambda_\rho \rightsquigarrow -s$. Also, $y' + sy \not\asymp 1$ for all $y \in K$.*

PROOF. Take an H-asymptotic field extension $L = K(f)$ of K with gap vf, $f \neq 0$, $f^\dagger = s$. Then $s - a^\dagger \prec a$ for all active $a \in L$, by Lemma 11.5.9; so $s - a^\dagger \prec a$ for all active $a \in K$, and thus $\lambda_\rho \rightsquigarrow -s$ by Lemma 11.5.6.

Next, assume towards a contradiction that $y \in K$ satisfies $y' + sy \asymp 1$. Then $y \neq 0$ and $y' + sy = y(y^\dagger + s)$, so for $a = y^{-1}$ we have $s - a^\dagger \asymp a$. Then a is active by Lemma 11.5.10, so $s - a^\dagger \prec a$ by the first part of the proof, and we have a contradiction. □

Towards a partial converse to the first part of Lemma 11.5.12 we have:

LEMMA 11.5.13. *Suppose K has asymptotic integration, and $\lambda_\rho \rightsquigarrow \lambda \in K$. Set*

$$S := \{v(\lambda + y^\dagger) : y \in K^\times\} \subseteq \Gamma_\infty.$$

Then S is a cofinal subset of $\Psi^\downarrow$, and thus $\lambda \neq 0$. Moreover, if f is a nonzero element of some H-asymptotic field extension of K and $f^\dagger = -\lambda$, then $vf \notin \Gamma$.

PROOF. We first show that $\lambda + y^\dagger$ is active, for each $y \in K^\times$. To prove this, assume towards a contradiction that $y \in K^\times$ and $\lambda + y^\dagger$ is not active. Set $\alpha := vy$ and take $\beta \in \Gamma^{\neq}$ with $\alpha = \beta'$. Take an active $a \not\asymp y$ with $va \geqslant \beta^\dagger$. Then $\lambda + a^\dagger \prec a$, so

$$\psi(va - \alpha) = v(a^\dagger - y^\dagger) = v(a^\dagger - y^\dagger + \lambda + y^\dagger) = v(\lambda + a^\dagger) > va,$$

contradicting Lemma 9.2.2, for $\alpha = vy$ and $\gamma = va$. Thus $S \subseteq \Psi^\downarrow$, and S is cofinal in $\Psi^\downarrow$ by (iii) of Lemma 11.5.6. Now let $f \neq 0$ in some H-asymptotic field extension of K satisfy $f^\dagger = -\lambda$, and suppose towards a contradiction that $vf \in \Gamma$. Then $f = uy$ with $u \in K(f)$, $u \asymp 1$ and $y \in K^\times$, so

$$v(\lambda + y^\dagger) = v(f^\dagger - y^\dagger) = v(u^\dagger) = v(u') > \Psi$$

where the last inequality follows from Proposition 9.1.3. This is in contradiction to $S \subseteq \Psi^\downarrow$. □

Here is a partial converse to the first part of Lemma 11.5.12:

LEMMA 11.5.14. *Assume that K has asymptotic integration, that Γ is divisible, and that $\lambda_\rho \rightsquigarrow \lambda \in K$. Then $s = -\lambda$ creates a gap over K.*

PROOF. Lemmas 10.4.5 and 11.5.13 give an H-asymptotic field extension $L = K(f)$ with $vf \notin \Gamma$, $f^\dagger = s$, and $\Psi_L = \Psi$. We show that $\alpha := vf$ is a gap in L. Suppose it is not. Since $\Psi_L = \Psi$ has no maximum, Lemma 9.2.15 and its Corollary 9.2.16 then give a nonzero $\beta \in \Gamma_L$ such that $\alpha = \beta'$. Then $\beta^\dagger \in \Psi$. Take active a in K with $a \not\asymp f$ and $va \geqslant \beta^\dagger$. Then $s - a^\dagger \prec a$, hence

$$\psi_L(\alpha - va) = v(f^\dagger - a^\dagger) = v(s - a^\dagger) > va,$$

which contradicts Lemma 9.2.2, for $\gamma = va$. □

REMARK. Suppose that K has asymptotic integration, Γ is divisible, and $s \in K$ creates a gap over K. Then vf is a gap in $K(f)$ for *any* nonzero element f of any H-asymptotic field extension of K with $f^\dagger = s$. (To see this, note that by Lemmas 11.5.12 and 11.5.13, any such f satisfies $vf \notin \Gamma$ and hence is transcendental over K; it remains to use the uniqueness part of Lemma 10.4.5.)

COROLLARY 11.5.15. *Every pre-*d*-valued field of H-type has a grounded* d*-valued extension of H-type. Every pre-H-field has a grounded H-field extension.*

PROOF. Let E be a pre-d-valued field of H-type. Replacing E by $\operatorname{dv}(E)$ we assume below that E is d-valued. If E is grounded, we are done, so we assume that E has a gap or has asymptotic integration. If E has a gap, then Lemma 10.2.1 yields a grounded d-valued extension $E(y)$ of H-type of E. In general we can arrange (for example by passing to the algebraic closure using Corollary 10.1.23) that Γ_E is divisible. For such E it remains to consider the case that E has asymptotic integration. In that case Corollary 11.4.10 allows us to pass from E to an immediate extension and arrange that E is also spherically complete. Then some element of E creates a gap over E by Lemma 11.5.14, and so E has a d-valued extension $E(f)$ of H-type with a gap, by the remark preceding this corollary, and Lemma 10.4.5. The case of a gap has been treated earlier in the proof.

For a pre-H-field E, replace $\operatorname{dv}(E)$ in the proof above by $H(E)$, and appeal to 10.5.10, 10.5.6, 10.5.19 instead of 10.2.1, 10.1.23, 10.4.5. □

Notes and comments

The γ_n occur in classical logarithmic criteria for convergence of infinite series [2, 44, 295] and integrals [338, p. 229]; see [62, §§4.1, 3.2]. Attempts to sharpen these logarithmic criteria provided the initial impetus for du Bois-Reymond's "orders of infinity" [50]; see [139, Section 1]. In this connection, the role of gaps as an "ideal boundary between convergence and divergence" was anticipated in [52, §§10–17]. The ω_n appear in [167, 184] (investigating the nature of solutions y to $4y'' + fy = 0$), and implicitly already in [459, §25]; see [57, Section 17] and [370] for the setting of Hardy fields.

11.6 THE PROPERTY OF λ-FREENESS

The ultimate goal of our work is to analyze the elementary (= first-order) theory of $\mathbb{T}$. Accordingly we introduce and study in this section the elementary property of being λ-free and show that $\mathbb{T}$ is λ-free. *We keep the standing assumption from the previous section that K is an ungrounded H-asymptotic field with* $\Gamma \neq \{0\}$. We begin with a consequence of the material in the previous section.

COROLLARY 11.6.1. *The following four conditions on K are equivalent:*

(i) (λ_ρ) *has no pseudolimit in* K*;*

(ii) *for all* s *there is* $g \succ 1$ *such that* $s - g^{\dagger\dagger} \succcurlyeq g^\dagger$*;*

(iii) *for all* s *there is an active* a *such that* $s - a^\dagger \succcurlyeq a$*;*

(iv) *for all* s *there is an active* a *such that* $s - b^\dagger \succcurlyeq a$ *for all active* $b \prec a$.

Here a, b, g, s *range over* K. *Also, condition* (i) *implies condition* (v) *given by*

(v) K *has asymptotic integration, and no element of* K *creates a gap over* K.

Finally, if Γ *is divisible, then* (i) *is equivalent to* (v).

PROOF. The equivalence of (i), (ii), (iii), (iv) follows from Lemma 11.5.6. For the implication (i) $\Rightarrow$ (v), use the remark following Lemma 11.5.9, and Lemma 11.5.12. For the last claim, also use Lemma 11.5.14. $\square$

Condition (ii) in Corollary 11.6.1 is a first-order condition on the asymptotic field K. Conditions (iii) and (iv) are also first-order conditions on K, but are logically more complex. We say that K is **λ-free** if it satisfies condition (ii) of Corollary 11.6.1. Thus if K is λ-free, then K has asymptotic integration, and no element of K creates a gap over K. From Lemma 11.5.12 we conclude:

COROLLARY 11.6.2. *If* K *has asymptotic integration and is* 1*-linearly surjective, and* Γ *is divisible, then* K *is* λ*-free.*

Every Liouville closed H-field is λ-free, by Corollary 11.6.2. More generally, if K has asymptotic integration and $(K^\times)^\dagger = K$, then K is λ-free, by Lemma 11.5.13. By Proposition 11.6.17 below we can drop "Γ is divisible" in Corollary 11.6.2. The following also helps in identifying λ-free K.

LEMMA 11.6.3. *Let* $s \in K$ *lie in a differential subfield* F *of* K *for which* $\max \Psi_F$ *exists. Then there is an active* $a \in K$ *such that* $s - a^\dagger \succcurlyeq a$.

PROOF. Take $a \in F$ such that $va = \max \Psi_F$. Then $a \in K$ is active, so if $s - a^\dagger \succcurlyeq a$, we are done. Suppose $s - a^\dagger \prec a$. Then $v(s - a^\dagger) \in (\Gamma_F^>)'$, so $v(s - a^\dagger) > \Psi$. Take $b \succ 1$ in K such that $vb' = va$. Then $vb + vb^\dagger = va$, so $v(a/b) = vb^\dagger$, and thus $a/b \in K$ is active, and $s - (a/b)^\dagger = (s - a^\dagger) + b^\dagger \sim b^\dagger \asymp a/b$. $\square$

COROLLARY 11.6.4. *If* K *is a union of grounded* H*-asymptotic subfields, then* K *is* λ*-free.*

If K has a λ-free H-asymptotic field extension L such that $\Gamma^<$ is cofinal in $\Gamma_L^<$, then K is λ-free. In particular, if K has an immediate λ-free H-asymptotic field extension, then K is λ-free. In the other direction we have:

LEMMA 11.6.5. *If* K *is* λ*-free, then so is its completion* K^c.

PROOF. If $\lambda_\rho \rightsquigarrow \lambda \in K^c$, then $\lambda_\rho \rightsquigarrow a$ for any $a \in K$ with $v(a - \lambda) > \Psi$. $\square$

In connection with the next result, see Proposition 9.6.7.

LEMMA 11.6.6. *If K is λ-free, then so is any asymptotic extension which, as a valued field extension of K, is a fluent completion of K.*

PROOF. Immediate from Lemma 3.4.23. □

LEMMA 11.6.7. *If K is λ-free, then so is its henselization K^{h}.*

PROOF. If (λ_ρ) pseudoconverges in K^{h}, then it pseudoconverges in some fluent completion of K, and thus in K, by Lemma 3.4.23. □

COROLLARY 11.6.8. *If K is λ-free, then K has rational asymptotic integration. Also:*

$$K \text{ is } \lambda\text{-free} \iff \text{the algebraic closure } K^{\mathrm{a}} \text{ of } K \text{ is } \lambda\text{-free}.$$

PROOF. The first claim is a consequence of the subsequent equivalence, so we prove the equivalence. Suppose (λ_ρ) pseudoconverges in K^{a}. Then by Lemma 3.2.6, the pc-sequence (λ_ρ) is of algebraic type over K, and so pseudoconverges in K^{h}, by Corollaries 3.2.12 and 3.3.21, and thus in K by Lemma 11.6.7. This gives the forward direction of the equivalence, and the backward direction holds because $\Gamma^<$ is cofinal in $(\mathbb{Q}\Gamma)^<$. □

LEMMA 11.6.9. *Suppose K is λ-free, λ is a pseudolimit of (λ_ρ) in an H-asymptotic field extension of K, and $s \in K$. Then $s + \lambda \sim b$ for some active $b \in K$.*

PROOF. Take active $a \in K$ as in (iv) of Corollary 11.6.1, that is, $s - b^\dagger \succcurlyeq a$ for all active $b \prec a$ in K. For such b we have $s - a^\dagger = s - b^\dagger + (b/a)^\dagger \sim s - b^\dagger$, since $(b/a)^\dagger \prec a \preccurlyeq s - b^\dagger$. Take $\gamma_{\rho_0} \prec a$. Then $s + \lambda_\rho \sim s + \lambda_{\rho_0} \succ \gamma_{\rho_0}$ for $\rho \geqslant \rho_0$, so $s + \lambda_\rho$ is active for $\rho \geqslant \rho_0$. Now $\lambda - \lambda_\rho \sim \gamma_{\rho+1}$ for all ρ, and thus $\lambda - \lambda_\rho \prec \gamma_{\rho_0} \prec s + \lambda_\rho$ for $\rho \geqslant \rho_0$. Hence $s + \lambda = (s + \lambda_\rho) + (\lambda - \lambda_\rho) \sim s + \lambda_\rho$ for $\rho \geqslant \rho_0$. □

The following technical consequence of Corollary 11.6.8 and Lemma 11.6.9 will be used in Section 13.7. Recall the derivation $R \mapsto \partial R/\partial Y$ on $K(Y)$ from Section 4.1.

COROLLARY 11.6.10. *Suppose K is λ-free, and λ is a pseudolimit of (λ_ρ) in an H-asymptotic field extension of K. Then λ is transcendental over K, and for each $R \in K(Y)^\times$ there is an active $a \in K$ such that*

$$\left(\frac{\partial R/\partial Y}{R}\right)(\lambda) \preccurlyeq \frac{1}{a}.$$

PROOF. Take an algebraically closed H-asymptotic extension of K containing λ, and replace K by its algebraic closure in this extension. It remains to appeal to Corollary 11.6.8 and to use Lemmas 4.1.3 and 11.6.9. □

By Corollary 11.5.8, if K is λ-free and $\phi \in K^\times$, then the compositional conjugate K^ϕ is λ-free. Being λ-free is also preserved under suitable specializations:

LEMMA 11.6.11. *Suppose K is λ-free with small derivation. Let $\Delta \neq \{0\}$ be a convex subgroup of Γ with $\psi(\Delta^{\neq}) \subseteq \Delta$. Then the asymptotic residue field $\dot{K}$ of the coarsened valuation $\dot{v} = v_\Delta$ is of H-type, $\psi(\Delta^{\neq}) \neq \emptyset$, $\psi(\Delta^{\neq})$ is cofinal in Ψ (and so has no largest element) and $\dot{K}$ is λ-free.*

PROOF. It is clear that $\dot{K}$ is H-asymptotic and that the claims about $\psi(\Delta^{\neq})$ hold. As usual, let $\dot{\mathcal{O}} = \{a \in K : va \geqslant \delta$ for some $\delta \in \Delta\}$ denote the valuation ring of $\dot{v}$, with maximal ideal $\dot{o} = \{a \in K : va > \Delta\}$ and residue map $a \mapsto \dot{a} \colon \dot{\mathcal{O}} \to \dot{K} = \dot{\mathcal{O}}/\dot{o}$. Let $s \in \dot{\mathcal{O}}$; it is enough to get $a \in K^\times$ such that $va \in \Delta^{\neq}$, a is active in K, and $s - a^\dagger \succcurlyeq a$. By Corollary 11.6.1 and since $\Psi^{>0} \neq \emptyset$, we can take active $b \in K$ such that $vb > 0$ and $s - b^\dagger \succcurlyeq b$. Then $vb \in \Delta$ by Lemma 9.2.25, and so $a := b$ works. □

Being λ-free is preserved under adjunction of certain exponential integrals:

LEMMA 11.6.12. *Suppose K is λ-free and Γ is divisible. Let $s \in K$ be such that $s - a^\dagger$ is active for each $a \in K^\times$, and set $S := \{v(s - a^\dagger) : a \in K^\times\}$. Let $f^\dagger = s$, where $f \neq 0$ lies in an H-asymptotic field extension of K. Suppose that*

(i) *S does not have a largest element, or*

(ii) *S has a largest element and $[\gamma + vf] \notin [\Gamma]$ for some $\gamma \in \Gamma$.*

Then $K(f)$ is λ-free.

PROOF. Let L be the algebraic closure of K in an algebraic closure of $K(f)$. Then L is an H-asymptotic field extension of K with $\Gamma_L = \Gamma$. We claim that

$$S = \{v(s - b^\dagger) : b \in L^\times\}.$$

To see this, let $b \in L^\times$, and take $a \in K^\times$, $u \asymp 1$ with $b = au$; then

$$v(u^\dagger) = v(u') > v(s - a^\dagger) = v(s - a^\dagger - u^\dagger) = v(s - b^\dagger)$$

since $s - a^\dagger$ is active in L. This proves the claim. In particular, $s - b^\dagger$ is active in L, for each $b \in L^\times$. Since L is λ-free by Corollary 11.6.8, we have a reduction to the case that K is algebraically closed and f is transcendental over K.

If S satisfies condition (i), then $\Psi_{K(f)} = \Psi$ by Lemma 10.4.5. If S satisfies condition (ii), then $\Psi_{K(f)} = \Psi \cup \{v(s - a^\dagger)\}$ for some $a \in K^\times$, by Lemma 10.4.6(i) and its proof. In each case, $vf \notin \Gamma$ and Ψ is a cofinal subset of $\Psi_{K(f)}$, so our logarithmic sequence (ℓ_ρ) for K remains a logarithmic sequence for $K(f)$. Thus (λ_ρ) has no pseudolimit in $K(f)$ by Corollary 3.3.24. □

LEMMA 11.6.13. *Suppose K is λ-free and is equipped with an ordering making it a real closed H-field. Let $s \in K^<$ be such that $s - a^\dagger$ is active, for all $a \in K^\times$. Let f be a nonzero element in a pre-H-field extension of K such that $f^\dagger = s$. Then $K(f)$ is λ-free.*

PROOF. This is clear if $f \in K$, so suppose $f \notin K$; then f is transcendental over K. Lemma 10.5.20 gives $vf \notin \Gamma$ and Ψ is cofinal in $\Psi_{K(f)}$, so $\Psi_{K(f)}$ does not have a largest element, and $\Gamma^<$ is cofinal in $\Gamma^<_{K(f)}$. Hence the logarithmic sequence (ℓ_ρ) for K remains a logarithmic sequence for $K(f)$. Now use Corollary 3.3.24. □

Application to eventual equalizing

In this subsection we assume that K has asymptotic integration. For use in Section 14.2 we prove here Proposition 11.6.17.

We say that $A \in K[\partial]^{\neq}$ is **v-surjective** if for each $g \in K^\times$ there is an $f \in K^\times$ such that $A(f) \asymp g$. For example, ∂ is v-surjective since K has asymptotic integration. If $A \in K[\partial]^{\neq}$ is v-surjective and $a \in K^\times$, then so are aA and Aa for any $a \in K^\times$. If $A \in K[\partial]^{\neq}$ has order 1, then $aA = \partial - s$ for some $a \in K^\times$ and $s \in K$. Note that if $A = \partial - s$ with $s \in K$, then $a^{-1}Aa = \partial - (s - a^\dagger)$ for $a \in K^\times$.

LEMMA 11.6.14. *Let $s \in K$ and $A = \partial - s$. Then A is v-surjective if $v(s - a^\dagger) > \Psi$ for some $a \in K^\times$, or $\{v(s - a^\dagger) : a \in K^\times\}$ is a subset of $\Psi^\downarrow$ with a largest element.*

PROOF. Consider first the case that $v(s - a^\dagger) > \Psi$ with $a \in K^\times$. Renaming $a^{-1}Aa$ and $s - a^\dagger$ as A and s we have $vs > \Psi$. Let $g \in K^\times$ and take $f \in K^\times$ with $f \not\asymp 1$ and $f' \asymp g$. Then $A(f) = f' - sf$, and $f^\dagger \succ s$ gives $f' \succ sf$, so $A(f) \sim f' \asymp g$. Next, assume $\{v(s - a^\dagger) : a \in K^\times\}$ is a subset of $\Psi^\downarrow$ with a largest element. Take $a \in K^\times$ such that $v(s - a^\dagger)$ is maximal. Renaming $a^{-1}Aa$ and $s - a^\dagger$ as A and s we have $vs \in \Psi^\downarrow$ and $s - b^\dagger \succcurlyeq s$ for all $b \in K^\times$. Let $g \in K^\times$ and take $\beta \in \Gamma^{\neq}$ such that $\beta' = vg$. If $\beta^\dagger \leqslant v(s)$, then any $f \in K^\times$ with $vf = \beta$ satisfies $f^\dagger - s \asymp f^\dagger$, so $A(f) = f(f^\dagger - s) \asymp f' \asymp g$. Suppose $\beta^\dagger > v(s)$. Then we take $\alpha \in \Gamma$ with $\alpha + vs = vg$. Since $vg = \beta + \beta^\dagger > \beta + vs$, this gives $\alpha > \beta$, so $\alpha' > \beta' = vg = \alpha + vs$ (with $0' := \infty$ by convention). Take $f \in K^\times$ with $vf = \alpha$. Then $A(f) = f' - sf \sim sf \asymp g$, as desired. □

LEMMA 11.6.15. *Assume Γ is divisible, $s \in K$, and $\{v(s - a^\dagger) : a \in K^\times\}$ is a subset of $\Psi^\downarrow$ without largest element. Then the following are equivalent for $g \in K^\times$:*

(i) *$vg \notin v(A(K))$ for $A := \partial - s$;*

(ii) *$g^\dagger - s$ creates a gap over K.*

PROOF. Lemma 10.4.5 yields an H-asymptotic extension $L = K(b)$ with $b \neq 0$, $b^\dagger = s$, $\eta := vb \notin \Gamma$, and $\Gamma_L = \Gamma \oplus \mathbb{Z}\eta$, $[\Gamma_L] = [\Gamma]$ (and thus $\Psi_L = \Psi$). Set $A := \partial - s$ and let $f \in K^\times$. Then $A(f) = f \cdot (f/b)^\dagger$, so

$$v(A(f)) - \eta \;=\; (vf - \eta) + \psi_L(vf - \eta) \;=\; (vf - \eta)'.$$

Also, $\Psi_L \subseteq \Gamma$ gives $(\Gamma - \eta) \cap (\Gamma_L^{\neq})' \subseteq (\Gamma - \eta)'$, and thus for all $\gamma \in \Gamma$,

$$\gamma \in v(A(K)) \iff \gamma - \eta \in (\Gamma_L^{\neq})' \iff \gamma - \eta \text{ is not a gap in } L.$$

Now assume $\gamma := vg \notin v(A(K))$. Then $\gamma - \eta = v(g/b)$ is a gap in L, and so $(g/b)^\dagger = g^\dagger - s$ creates a gap over K. This proves (i) $\Rightarrow$ (ii). The converse is part of Lemma 11.5.12. □

COROLLARY 11.6.16. *Suppose Γ is divisible and $A \in K[\partial]$ has order* 1. *Then*

$$|\Gamma \setminus v(A(K))| \leqslant 1.$$

PROOF. We first reduce to the case that $A = \partial - s$, $s \in K$, and then set

$$S := \{v(s - a^\dagger) : a \in K^\times\}.$$

If $S \not\subseteq \Psi^\downarrow$ or S is a subset of $\Psi^\downarrow$ with a largest element, then A is v-surjective by Lemma 11.6.14. Suppose S is a subset of $\Psi^\downarrow$ without a largest element, and $g, h \in K^\times$, $vg, vh \notin v\big(A(K)\big)$. Then $g^\dagger - s$ and $h^\dagger - s$ create a gap over K by Lemma 11.6.15, and so $v\big((g/h)^\dagger\big) = v(g^\dagger - h^\dagger) > \Psi$ by Lemmas 11.5.2 and 11.5.12, thus $vg = vh$. □

EXAMPLE. If Γ is divisible and $\lambda_\rho \rightsquigarrow \lambda \in K$, then $v\big(A(K)\big) = \Gamma^{\neq}$ for $A = \partial - \lambda$.

We let ϕ range over the active elements in K. Let $P = aY + bY'$ with $a, b \in K$ not both zero. We say that P **has eventual equalizers** if for every $g \in K^\times$ there is $f \in K^\times$ such that eventually $P^\phi_{\times f} \asymp g$. Note that if P has eventual equalizers and $h \in K^\times$, then so do hP and $P_{\times h}$. The next result improves on Corollary 11.6.2.

PROPOSITION 11.6.17. *The following conditions on K are equivalent:*

(i) $\partial - s$ *is v-surjective for every* $s \in K$*;*

(ii) *every* $P = aY + bY'$ *with* $a, b \in K$ *not both zero has eventual equalizers;*

(iii) K *is* λ*-free.*

PROOF. Assume (i) and let $P = aY + bY'$ with $a, b \in K$ not both zero; our job is to show that P has eventual equalizers. The case $b = 0$ being trivial, assume $b \neq 0$. Replacing P by $b^{-1}P$ we get $P = Y' - sY$ with $s \in K$. Then for $f \in K^\times$,

$$P^\phi_{\times f} = f\phi Y' + (f' - sf)Y.$$

Consider first the case that $v(s - h^\dagger) > \Psi$ for some $h \in K^\times$. For such h we have $h^{-1}P_{\times h} = Y' - (s - h^\dagger)Y$, so by suitable renaming we arrange that $v(s) > \Psi$. Now, let $g \in K^\times$, and take $f \in K^\times$ with $f \not\asymp 1$ such that $f' \asymp g$. Then $f' - sf \sim f' \asymp g$, and eventually $f\phi \prec f'$, so by the identity above for $P^\phi_{\times f}$ we get $P^\phi_{\times f} \asymp g$, eventually.

Next, assume that $v(s - h^\dagger) \in \Psi^\downarrow$ for all $h \in K^\times$. Take $f \in K^\times$ such that $f' - sf \asymp g$. Then $f' - sf = f(f^\dagger - s)$ and $f^\dagger - s \succ \phi$, eventually, so $f' - sf \succ f\phi$ eventually, and thus $P^\phi_{\times f} \asymp g$, eventually. This finishes the proof of (i) $\Rightarrow$ (ii).

For (ii) $\Rightarrow$ (i), assume (ii), and let $s \in K$. Then $P := Y' - sY$ has eventual equalizers. Let $g \in K^\times$ and take $f \in K^\times$ such that $P^\phi_{\times f} \asymp g$, eventually. Then by the identity above we must have $f' - sf \asymp g$.

Next we prove the contrapositive of (i) $\Rightarrow$ (iii). Suppose K is not λ-free. Take $\lambda \in K$ such that $\lambda_\rho \rightsquigarrow \lambda$. If $y \in K^\times$ and $y' - \lambda y \asymp 1$, then $a := y^{-1}$ gives $a^\dagger + \lambda \asymp a$, so a is active by Lemma 11.5.13, but this contradicts the equivalence (i) $\Longleftrightarrow$ (iii) of Lemma 11.5.6. Thus $\partial - \lambda$ is not v-surjective.

Finally, we prove (iii) $\Rightarrow$ (ii), so assume (iii). Let $P = aY + bY'$ with $a, b \in K$ not both zero. Now the algebraic closure K^{a} of K is also λ-free. Suppose $g \in K^\times$ and

$f \in (K^{\mathrm{a}})^{\times}$ are such that eventually $P^{\phi}_{\times f} \asymp g$. By taking ϕ (in K) with high enough $v\phi$ and applying the Equalizer Theorem to $P^{\phi} \in K^{\phi}\{Y\}$ we see that $vf \in \Gamma$. Thus in trying to show that P has eventual equalizers, we can assume that Γ is divisible. But then the λ-freeness of K and Lemmas 11.6.14 and 11.6.15 show that (i) holds, and we already know that (i) $\Rightarrow$ (ii). □

Notes and comments

Section 12 of [20] contains some aspects of λ-freeness, but without naming this property. This occurs in connection with gap creation in real closed H-fields with asymptotic integration. Lemma 11.6.7 can also be deduced from Lemma 3.4.13 and Proposition 3.4.24, since (λ_ρ) is jammed. Allen Gehret has shown us a proof that if K is λ-free, then so is $\mathrm{dv}(K)$.

11.7 BEHAVIOR OF THE FUNCTION ω

In this section we keep the assumption that K is an ungrounded H-asymptotic field with $\Gamma \neq \{0\}$. Recall from Section 5.2 the function

$$\omega\colon K \to K, \qquad \omega(z) \;=\; -(2z' + z^2).$$

LEMMA 11.7.1. *For $\rho < \rho' < \kappa$ we have $\omega(\lambda_{\rho'}) - \omega(\lambda_\rho) \sim \gamma^2_{\rho+1}$.*

PROOF. Let $\rho < \rho' < \kappa$. Then $\lambda_{\rho'} - \lambda_\rho = \gamma_{\rho+1} + \varepsilon$ with $\varepsilon \prec \gamma_{\rho+1}$. By Lemma 5.2.1:

$$\begin{aligned}\omega(\lambda_\rho) - \omega(\lambda_{\rho'}) &= (\gamma_{\rho+1} + \varepsilon)\cdot\big(2(\gamma_{\rho+1} + \varepsilon)^\dagger + 2\lambda_\rho + (\gamma_{\rho+1} + \varepsilon)\big)\\ &= (\gamma_{\rho+1} + \varepsilon)\cdot\big(2\gamma^\dagger_{\rho+1} + 2(1+\tau)^\dagger + 2\lambda_\rho + (\gamma_{\rho+1} + \varepsilon)\big),\end{aligned}$$

where $\tau := \varepsilon/\gamma_{\rho+1}$. Thus $\tau \prec 1$ and so $(1+\tau)^\dagger \prec \gamma_{\rho+1}$, hence

$$\begin{aligned}\omega(\lambda_\rho) - \omega(\lambda_{\rho'}) &= (\gamma_{\rho+1} + \varepsilon)\cdot\big(-2\lambda_{\rho+1} + 2(1+\tau)^\dagger + 2\lambda_\rho + (\gamma_{\rho+1} + \varepsilon)\big)\\ &= (\gamma_{\rho+1} + \varepsilon)\cdot\big(2(\lambda_\rho - \lambda_{\rho+1}) + 2(1+\tau)^\dagger + (\gamma_{\rho+1} + \varepsilon)\big)\\ &= (\gamma_{\rho+1} + \varepsilon)\cdot\big(-\gamma_{\rho+1} + 2(1+\tau)^\dagger + \varepsilon_1\big) \text{ (with } \varepsilon_1 \prec \gamma_{\rho+1})\\ &\sim -\gamma^2_{\rho+1},\end{aligned}$$

which finishes the proof. □

In view of Lemma 11.5.2 and its proof, this gives:

COROLLARY 11.7.2. $(\omega_\rho) := \big(\omega(\lambda_\rho)\big)$ *is a jammed pc-sequence of width*

$$\{\gamma \in \Gamma_\infty : \gamma > 2\Psi\}.$$

The sequence $(\omega_\rho + \gamma^2_\rho)$ is also a pc-sequence in K, equivalent to (ω_ρ).

EXAMPLE. Suppose $K = \mathbb{T}$ and let (ℓ_n) be the logarithmic sequence for K from Example 11.5.1. Using Corollary 5.2.2 we see that for each n,

$$\omega_n = \omega(\lambda_n) = \frac{1}{\ell_0^2} + \frac{1}{(\ell_0\ell_1)^2} + \cdots + \frac{1}{(\ell_0\ell_1\cdots\ell_n)^2}.$$

COROLLARY 11.7.3. *Suppose $\lambda_\rho \rightsquigarrow \lambda$, with λ in an asymptotic field extension L of K. Then $\omega(\lambda_\rho) \rightsquigarrow \omega(\lambda)$ in L.*

PROOF. Eventually we have $\lambda - \lambda_\rho \sim \gamma_{\rho+1}$, so a computation as in the proof of Lemma 11.7.1 gives $\omega(\lambda) - \omega(\lambda_\rho) \sim \gamma_{\rho+1}^2$, eventually. □

It follows from Corollaries 11.7.2 and 11.7.3, Lemma 2.2.17, and Proposition 11.5.3 that the pc-sequences (ω_ρ) obtained from different choices of the logarithmic sequence (ℓ_ρ) for K are all equivalent.

LEMMA 11.7.4. *Suppose K has small derivation and asymptotic integration. Then $\lambda_\rho, \omega_\rho \prec 1$ eventually.*

PROOF. Eventually $0 < v(\gamma_\rho) < (\Gamma^>)'$, so $v(\lambda_\rho) = \psi\big(v(\gamma_\rho)\big) > 0$ eventually, by Lemma 6.5.4(i). Hence $\lambda_\rho \prec 1$ eventually. Now use $\omega_\rho = -(2\lambda_\rho' + \lambda_\rho^2)$. □

The next lemma will be used in proving Proposition 13.6.4.

LEMMA 11.7.5. *Suppose K has rational asymptotic integration, and $\omega_\rho \rightsquigarrow \omega$ with ω in an immediate asymptotic extension F of K. Let $H \in F\{Y\} \setminus F$. Then $H(\omega_\rho) \rightsquigarrow H(\omega)$, and there are $\gamma_0 \in \Psi$ and $\delta_0 \in (\Gamma^>)'$ and an index ρ_0 such that the subset $\big\{v\big(H(\omega_\rho) - H(\omega)\big) : \rho > \rho_0\big\}$ of Γ_∞ is disjoint from (γ_0, δ_0).*

PROOF. Replacing F by an algebraic closure F^{a} and K by the algebraic closure of K inside F^{a}, we arrange that F is algebraically closed. Passing from F to a suitable immediate asymptotic extension, we may further assume that F is closed under integration, using Corollaries 3.3.22 and 10.3.2 and Proposition 10.2.7. For each ρ, take $y_\rho \in F$ such that $(y_\rho')^2 = \omega_\rho - \omega$. Then $(y_\rho')^2 \asymp \gamma_{\rho+1}^2 \asymp (\ell_{\rho+2}')^2$, by Lemma 11.7.1, so $y_\rho \asymp \ell_{\rho+2}$. Now put $Q := H(Y + \omega) - H(\omega)$ and $P := Q\big((Y')^2\big)$. Then $P \neq 0$, $P(0) = 0$, and $P(y_\rho) = H(\omega_\rho) - H(\omega)$ for each ρ. By Corollary 9.5.8 we can take $\beta \in \Gamma^<$ and a strictly increasing map $i\colon (\beta, 0) \to \Gamma$ such that $vP(y) = i(vy)$ for all $y \in F$ with $vy \in (\beta, 0)$, hence $H(\omega_\rho) \rightsquigarrow H(\omega)$. It now remains to apply Corollary 11.3.10 to F, Q in the role of K, P, using $H(\omega_\rho) - H(\omega) = Q\big((y_\rho')^2\big)$. □

In connection with Proposition 11.5.3 we noted that if $a, b \in K$ are active with $a \succ b$, then $a^\dagger - b^\dagger \prec a$. For the function ω we have likewise:

LEMMA 11.7.6. *Let $a, b \in K$ be active with $a \succ b$. Then*

$$\omega(-a^\dagger) - \omega(-b^\dagger) \prec a^2.$$

PROOF. Apply Lemma 5.2.1 to $w := -a^\dagger$, $z := -b^\dagger$, and $y := z - w = a^\dagger - b^\dagger$, and note that $y^\dagger + w = y^\dagger - a^\dagger \prec a$ since $a \succ y$ and $y = (a/b)^\dagger$ is active. □

Here is an analogue of Lemma 11.5.6 for the sequence (ω_ρ):

LEMMA 11.7.7. *Let $\omega \in K$. Then the following are equivalent:*

(i) $\omega_\rho \rightsquigarrow \omega$;

(ii) *for all active $a \in K$ we have $\omega - \omega(-a^\dagger) \prec a^2$;*

(iii) *for all active $a \in K$ there is an active $b \prec a$ in K such that*

$$\omega - \omega(-b^\dagger) \prec a^2;$$

(iv) *for all $g \succ 1$ in K we have $\omega - \omega(-g^{\dagger\dagger}) \prec (g^\dagger)^2$.*

PROOF. Assume (i). To prove (ii), let $a \in K$ be active. Take ρ such that

$$\gamma_\rho \prec a, \qquad \omega - \omega(\lambda_\rho) \sim \omega(\lambda_{\rho'}) - \omega(\lambda_\rho) \text{ for } \rho < \rho' < \kappa.$$

Then $\omega - \omega(\lambda_\rho) \sim \gamma_{\rho+1}^2 \prec a^2$. Also $\omega(\lambda_\rho) - \omega(-a^\dagger) \prec a^2$ by Lemma 11.7.6 above for $b := \gamma_\rho$, hence $\omega - \omega(-a^\dagger) \prec a^2$.

From Lemma 11.7.6 we get (ii) $\Rightarrow$ (iii). To get (iii) $\Rightarrow$ (iv), assume (iii), and let $g \in K, g \succ 1$. Take active $b \prec a := g^\dagger$ in K with $\omega - \omega(-b^\dagger) \prec a^2$. By the above we also have $\omega(-b^\dagger) - \omega(-a^\dagger) \prec a^2$, so $\omega - \omega(-a^\dagger) \prec a^2$.

Finally, to prove (iv) $\Rightarrow$ (i), assume (iv), and let $\rho < \kappa$. Take $g \succ 1$ in K such that $a := g^\dagger \prec \gamma_{\rho+1}$ in K. Then $\omega - \omega(-a^\dagger) \prec a^2$, so

$$\omega - \omega(\lambda_\rho) = \big(\omega - \omega(-a^\dagger)\big) + \big(\omega(-a^\dagger) - \omega(\lambda_{\rho+1})\big) + \big(\omega(\lambda_{\rho+1}) - \omega(\lambda_\rho)\big),$$

and so $\omega - \omega(\lambda_\rho) \sim \gamma_{\rho+1}^2$, which gives (i). □

COROLLARY 11.7.8. *The following four conditions on K are equivalent:*

(i) *(ω_ρ) has no pseudolimit in K;*

(ii) *for every $f \in K$ there is $g \succ 1$ in K with $f - \omega(-g^{\dagger\dagger}) \succcurlyeq (g^\dagger)^2$;*

(iii) *for every $f \in K$ there is an active $a \in K$ with $f - \omega(-a^\dagger) \succcurlyeq a^2$;*

(iv) *for every $f \in K$ there is an active $a \in K$ such that $f - \omega(-b^\dagger) \succcurlyeq a^2$ for all active $b \prec a$ in K.*

Relating (λ_ρ) and (ω_ρ)

We introduce here the cuts

$$\lambda(K) := c_K(\lambda_\rho), \qquad \omega(K) := c_K(\omega_\rho),$$

in K defined by (λ_ρ) and (ω_ρ). By Lemmas 11.5.2 and 11.7.1,

$$\operatorname{ndeg}_{\lambda(K)} P = \text{eventual value of } \operatorname{ndeg} P_{+\lambda_\rho, \times\gamma_{\rho+1}}, \text{ and}$$
$$\operatorname{ndeg}_{\omega(K)} P = \text{eventual value of } \operatorname{ndeg} P_{+\omega_\rho, \times\gamma_{\rho+1}^2} \quad \text{for } P \in K\{Y\}^{\neq}.$$

LEMMA 11.7.9. *Suppose $\omega_\rho \rightsquigarrow \omega \in K$. Set $P(Z) := 2Z' + Z^2 + \omega \in K\{Z\}$. Let ϕ range over the elements of $K^\times$ with $v\phi \in \Psi^\downarrow$. Then for all ρ,*

$$P^\phi_{+\lambda_\rho, \times\gamma_{\rho+1}} \sim \gamma^2_{\rho+1}(Z-1)^2, \quad \textit{eventually.}$$

In particular, $\operatorname{ndeg}_{\lambda(K)} P = 2$.

PROOF. Recalling that $\omega_\rho = -(2\lambda'_\rho + \lambda^2_\rho)$, we get

$$\begin{aligned} P_{+\lambda_\rho, \times\gamma_{\rho+1}} &= 2(\gamma_{\rho+1}Z + \lambda_\rho)' + (\gamma_{\rho+1}Z + \lambda_\rho)^2 + \omega \\ &= (2\gamma'_{\rho+1}Z + 2\gamma_{\rho+1}Z' + 2\lambda'_\rho) + (\gamma^2_{\rho+1}Z^2 + 2\gamma_{\rho+1}\lambda_\rho Z + \lambda^2_\rho) + \omega \\ &= \gamma_{\rho+1}\big(2Z' + \gamma_{\rho+1}Z^2 + 2(\lambda_\rho - \lambda_{\rho+1})Z\big) + (\omega - \omega_\rho). \end{aligned}$$

Hence

$$P^\phi_{+\lambda_\rho, \times\gamma_{\rho+1}} = \gamma_{\rho+1}\big(2\phi Z' + \gamma_{\rho+1}Z^2 + 2(\lambda_\rho - \lambda_{\rho+1})Z\big) + (\omega - \omega_\rho),$$

and for fixed ρ, we have $\phi \prec \gamma_{\rho+1} \sim (\lambda_{\rho+1} - \lambda_\rho)$, eventually. Since $\omega - \omega_\rho \sim \gamma^2_{\rho+1}$ for all ρ, the lemma follows. □

COROLLARY 11.7.10. *If (ω_ρ) has a pseudolimit in K, then K has an immediate asymptotic extension L such that L is* d*-algebraic over K and not λ-free.*

PROOF. If K is not λ-free, take $L = K$. If K is λ-free, then K has rational asymptotic integration, and so we are in the setting of Section 11.4, and the desired result follows from Lemmas 11.7.9, 11.4.12, and 11.4.8. □

To sharpen the above we first prove a lemma:

LEMMA 11.7.11. *Suppose K is λ-free and λ is a pseudolimit of (λ_ρ) in an asymptotic field extension of K. Let $a, b \in K$ and set $Q(Z) := Z' + aZ + b \in K\{Z\}$. Then there is an $\alpha \in \Psi^\downarrow$ and $\rho_0 < \kappa$ such that*

$$v\big(Q(\lambda) - Q(\lambda_\rho)\big) = v(\gamma_{\rho+1}) + \alpha \quad \textit{for } \rho_0 < \rho < \kappa.$$

PROOF. Let $\lambda^*_\rho := -(\lambda_{\rho+1} - \lambda_\rho)^\dagger$; then $\lambda^*_\rho \not\rightsquigarrow a$ by Corollary 11.5.4, and so we have $\alpha \in \Psi^\downarrow$ and $\rho_0 < \kappa$ such that $v(a - \lambda^*_\rho) = \alpha$ for $\rho_0 < \rho < \kappa$. Also $\lambda - \lambda_\rho \sim \lambda_{\rho+1} - \lambda_\rho$ for all ρ. With $u_\rho := (\lambda - \lambda_\rho)/(\lambda_{\rho+1} - \lambda_\rho)$ we get $(\lambda - \lambda_\rho)^\dagger = u^\dagger_\rho + (\lambda_{\rho+1} - \lambda_\rho)^\dagger$ and $u_\rho \asymp 1$, so $v(u^\dagger_\rho) > \Psi$, for $\rho_0 < \rho < \kappa$. Therefore

$$\begin{aligned} Q(\lambda) - Q(\lambda_\rho) &= (\lambda - \lambda_\rho)\big((\lambda - \lambda_\rho)^\dagger + a\big) \\ &= (\lambda - \lambda_\rho)\big(u^\dagger_\rho + (a - \lambda^*_\rho)\big) \\ &\sim \gamma_{\rho+1}(a - \lambda^*_\rho) \end{aligned}$$

for $\rho_0 < \rho < \kappa$. □

PROPOSITION 11.7.12. *Suppose K is λ-free and $\omega_\rho \rightsquigarrow \omega \in K$. Then*

$$P(Z) := 2Z' + Z^2 + \omega \in K\{Z\}$$

is a minimal differential polynomial of the pc-sequence (λ_ρ) over K.

PROOF. We have $P(\lambda_\rho) = -\omega_\rho + \omega \rightsquigarrow 0$. Towards a contradiction, suppose that $Q \in K\{Z\} \setminus K$ is a differential polynomial of smaller complexity than P such that $Q(\widetilde{\lambda}_\sigma) \rightsquigarrow 0$ for some pc-sequence $(\widetilde{\lambda}_\sigma)$ in K, equivalent to (λ_ρ). Then there are two possibilities: either $\operatorname{order}(Q) = 0$ (so $Q \in K[Z]$), or $\operatorname{order}(Q) = \deg(Q) = 1$. In the first case, (λ_ρ) has a pseudolimit in the henselization of K, contradicting Lemma 11.6.7. So we are in the second case, and may assume that $Q = Z' + aZ + b$ where $a, b \in K$. Now Q is a minimal differential polynomial of (λ_ρ) over K, hence Lemma 11.4.8 and Corollary 11.4.13 give a pseudolimit λ of (λ_ρ) in an immediate asymptotic field extension of K with $Q(\lambda) = 0$. Lemma 11.7.11 yields $\alpha \in \Psi^{\downarrow}$ and $\rho_0 < \kappa$ such that $v\big(Q(\lambda_\rho)\big) = v(\gamma_{\rho+1}) + \alpha$ for $\rho_0 < \rho < \kappa$. Increasing ρ_0 if necessary, we may assume that also $\alpha < v(\gamma_{\rho+1})$ for $\rho_0 < \rho < \kappa$. Then by Lemma 11.7.1,

$$P(\lambda_\rho) = \omega - \omega_\rho \sim \gamma_{\rho+1}^2 \prec Q(\lambda_\rho) \qquad (\rho_0 < \rho < \kappa)$$

and thus for $R := P - 2Q = Z^2 - 2aZ + (\omega - 2b) \in K[Z]$ we get $R(\lambda_\rho) \asymp Q(\lambda_\rho)$, for $\rho_0 < \rho < \kappa$. So $R(\lambda_\rho) \rightsquigarrow 0$, contradicting what we showed above. □

Combining Proposition 11.7.12 with Lemma 11.4.8 and Corollary 11.4.13 yields:

COROLLARY 11.7.13. *Suppose K is λ-free and $\omega_\rho \rightsquigarrow \omega \in K$. Then K has an immediate asymptotic field extension $K(\lambda)$ with $\lambda_\rho \rightsquigarrow \lambda$ and $\omega(\lambda) = \omega$, such that if L is any asymptotic field extension of K and $\lambda^* \in L$ satisfies $\lambda_\rho \rightsquigarrow \lambda^*$ and $\omega(\lambda^*) = \omega$, then there is an embedding $K(\lambda) \to L$ over K sending λ to λ^*.*

Embedding here means of course *embedding of valued differential fields.*

ω-freeness

Let us say that our H-asymptotic field K is **ω-free** if (ω_ρ) has no pseudolimit in K (so K satisfies the conditions of Corollary 11.7.8). It follows from Corollary 11.7.3 that if K is ω-free, then K is λ-free. Thus for the H-asymptotic fields K considered in this section we have the implications

ω-free $\Rightarrow$ λ-free $\Rightarrow$ rational asymptotic integration $\Rightarrow$ asymptotic integration,

where the last two properties are determined by the asymptotic couple of K. The pc-sequence (ω_n) in the Liouville closed H-field $\mathbb{T}$ is divergent, so $\mathbb{T}$ is ω-free. See [22] for an example of a Liouville closed H-field that is not ω-free.

The λ-free K in Corollary 11.6.4 are actually ω-free:

LEMMA 11.7.14. *Let $f \in K$ lie in a differential subfield F of K for which $\max \Psi_F$ exists. Then there is an active $a \in K$ such that $f - \omega(-a^\dagger) \succcurlyeq a^2$.*

PROOF. Take $a \in F$ such that $va = \max \Psi_F$. Then $a \in K$ is active, so in case $f - \omega(-a^\dagger) \succcurlyeq a^2$, we are done. Suppose $f - \omega(-a^\dagger) \prec a^2$. In the algebraic closure K^{a} of K with its differential subfield F^{a} this gives

$$v\left(\sqrt{f-\omega(-a^\dagger)}\right) > \Psi_F = \Psi_{F^{\mathrm{a}}},$$

so $v\left(\sqrt{f-\omega(-a^\dagger)}\right) > \Psi$ by Corollary 9.2.4, and thus $v\big(f-\omega(-a^\dagger)\big) > 2\Psi$. Take $b \succ 1$ in K with $vb' = va$. Then $vb + vb^\dagger = va$, so $(a/b) \asymp b^\dagger$. Hence $(a/b) \in K$ is active, and $a \succ (a/b)$. Then Lemma 5.2.1 applied to $w = -a^\dagger$, $z = -(a/b)^\dagger$, $y = z - w = b^\dagger$ gives

$$\omega(-a^\dagger) - \omega(-(a/b)^\dagger) = b^\dagger \cdot \big(2(b^{\dagger\dagger} - a^\dagger) + b^\dagger\big).$$

We have $b^\dagger = (a/b)u$ with $u \asymp 1$, so $b^{\dagger\dagger} = a^\dagger - b^\dagger + u^\dagger$, hence $b^{\dagger\dagger} - a^\dagger = -b^\dagger + u^\dagger \sim -b^\dagger$ (since K is pre-differential-valued), and thus

$$\omega(-a^\dagger) - \omega(-(a/b)^\dagger) \sim -(b^\dagger)^2 \asymp (a/b)^2 \prec a^2.$$

Together with $f - \omega(-a^\dagger) \prec (a/b)^2$ (a consequence of $v(f - \omega(-a^\dagger)) > 2\Psi$), it follows that $f - \omega(-(a/b)^\dagger) \asymp (a/b)^2$. □

COROLLARY 11.7.15. *If K is a union of grounded H-asymptotic subfields, then K is ω-free.*

In view of how $\mathbb{T}$ is constructed, this shows again that $\mathbb{T}$ is ω-free.

Constructing ω-free H-asymptotic fields

Let F be a grounded pre-d-valued field of H-type, so we have $f \in F^\times$ with

$$f \succ 1, \qquad vf^\dagger = \max \Psi.$$

We shall associate canonically to the pair F, f a pre-d-valued extension F_ω of F with $\operatorname{res}(F_\omega) = \operatorname{res}(F)$, such that F_ω is of H-type, has asymptotic integration, and is ω-free. In fact, by Lemma 10.2.3 we can take an increasing sequence

$$F = F_0 \subseteq F_1 \subseteq F_2 \subseteq F_3 \subseteq \cdots$$

of pre-d-valued extensions F_n of F of H-type, with distinguished elements $f_n \in F_n^\times$ such that $f_0 = f$, and for each n we have

$$f_n \succ 1, \qquad f_n^\dagger = \max \Psi_{F_n}, \qquad F_{n+1} = F_n(f_{n+1}), \qquad f_{n+1}' = f_n^\dagger.$$

Then $F_\omega := \bigcup_n F_n$ is an extension of F with the properties we announced, its ω-freeness being a consequence of each F_n being grounded. From the universal property of Lemma 10.2.3 we obtain:

LEMMA 11.7.16. *Any embedding of F into a pre-*d*-valued field L of H-type that is closed under logarithms extends to an embedding $F_\omega \to L$.*

If F is d-valued, then each F_n in the construction above is d-valued, and so F_ω is d-valued. This applies in particular to $\mathrm{dv}(F)$, since $\mathrm{dv}(F)$ is a grounded d-valued field of H-type.

Suppose F is also equipped with an ordering making it a pre-H-field. Then by Corollary 10.5.12 there is a unique ordering on F_ω making it a pre-H-field extension of F. Let F_ω be equipped with this ordering. Note that if F is an H-field, then so is F_ω. Lemma 11.7.16 goes through (and "embedding" in the setting of pre-H-fields means "embedding of ordered valued differential fields"):

LEMMA 11.7.17. *Any embedding of F into a pre-H-field L closed under logarithms extends to an embedding $F_\omega \to L$.*

For any grounded pre-H-field F, we let F_ω be a pre-H-field extension of F as constructed above. In particular, if F is a grounded H-field, then F_ω is an H-field extension of F with $C_{F_\omega} = C_F$.

COROLLARY 11.7.18. *Every pre-*d*-valued field of H-type has an ω-free* d*-valued extension of H-type, and every pre-H-field has an ω-free H-field extension.*

PROOF. Let E be a pre-d-valued field of H-type. Then E has a grounded d-valued extension F of H-type, by Corollary 11.5.15, and thus F_ω is an ω-free d-valued extension of H-type of E. The argument for pre-H-fields is the same. □

COROLLARY 11.7.19. *Every pre-H-field has an H-field extension with a gap.*

PROOF. To extend a pre-H-field E to an H-field with a gap, we can arrange by Corollary 11.7.18 that E is an H-field with rational asymptotic integration. Taking the real closure we arrange further that E has divisible value group. The rest is done in the proof of Corollary 11.5.15. □

If K has an H-asymptotic field extension L with $\Gamma^<$ cofinal in $\Gamma_L^<$ and ω-free L, then K is ω-free. Analogues of Lemmas 11.6.5, 11.6.6, 11.6.7 have similar proofs:

LEMMA 11.7.20. *If K is ω-free, then so is its completion K^{c}.*

LEMMA 11.7.21. *If K is ω-free, then so is any immediate asymptotic extension of K that, as a valued field extension of K, is a fluent completion of K.*

LEMMA 11.7.22. *If K is ω-free, then so is its henselization K^{h}.*

As in Corollary 11.6.8 this yields:

COROLLARY 11.7.23. *K is ω-free if and only if K^{a} is ω-free.*

One direction of Corollary 11.7.23 will be vastly generalized in Section 13.6.

Specialization of K with small derivation preserves ω-freeness:

LEMMA 11.7.24. *Suppose K is ω-free with small derivation. Let $\Delta \neq \{0\}$ be a convex subgroup of Γ with $\psi(\Delta^{\neq}) \subseteq \Delta$. Then the asymptotic residue field $\dot{K}$ of $\dot{v} = v_\Delta$ is also ω-free.*

PROOF. We mimic the proof of Lemma 11.6.11 and use its notations. Let $f \in \dot{\mathcal{O}}$; it is enough to get $a \in K^\times$ such that $va \in \Delta^{\neq}$, a is active in K, and $f - \omega(-a^\dagger) \succcurlyeq a^2$. By Corollary 11.7.8 and since $\Psi^{>0} \neq \emptyset$, we can take active $b \in K$ such that $vb > 0$ and $f - \omega(-b^\dagger) \succcurlyeq b^2$. Then $vb \in \Delta$ by Lemma 9.2.25, and so $a := b$ works. □

Next, we consider the effect of compositional conjugation on the above. Let $\phi \in K^\times$ and let $\delta = \phi^{-1}\partial$ be the derivation of the compositional conjugate K^ϕ. As we saw before, the analogue of the sequence (γ_ρ) in K is the sequence $(\gamma_\rho^\phi) = (\gamma_\rho/\phi)$ in K^ϕ, and the analogue of the pc-sequence (λ_ρ) in K is the pc-sequence (λ_ρ^ϕ) in K^ϕ, where

$$\lambda_\rho^\phi = (\lambda_\rho/\phi) + (\phi^\dagger/\phi).$$

Also, the role of $\omega = \omega_K \colon K \to K$ is taken over in K^ϕ by

$$\omega^\phi := \omega_{K^\phi} \colon K^\phi \to K^\phi, \quad \omega^\phi(z) := -\big(2\delta(z) + z^2\big) = -(2/\phi)z' - z^2,$$

so

$$\begin{aligned}
\omega^\phi(\lambda_\rho^\phi) &= -\frac{2}{\phi}\cdot\left(\frac{\lambda_\rho}{\phi} + \frac{\phi^\dagger}{\phi}\right)' - \left(\frac{\lambda_\rho}{\phi} + \frac{\phi^\dagger}{\phi}\right)^2 \\
&= -\frac{2}{\phi}\cdot\left[\frac{\lambda_\rho'}{\phi} - \frac{\lambda_\rho\phi^\dagger}{\phi} + \left(\frac{\phi^\dagger}{\phi}\right)'\right] - \left(\frac{\lambda_\rho}{\phi}\right)^2 - \left(\frac{\phi^\dagger}{\phi}\right)^2 - 2\frac{\lambda_\rho\phi^\dagger}{\phi^2} \\
&= \frac{\omega(\lambda_\rho)}{\phi^2} + \omega^\phi\left(\frac{\phi^\dagger}{\phi}\right).
\end{aligned}$$

Hence, the sequence $(\omega_\rho) = \big(\omega(\lambda_\rho)\big)$ has a pseudolimit in K iff the analogous sequence $(\omega_\rho^\phi) := \big(\omega^\phi(\lambda_\rho^\phi)\big)$ has a pseudolimit in K^ϕ. In particular, K is ω-free iff K^ϕ is ω-free. For later use we note that

$$\begin{aligned}
\omega^\phi(\phi^\dagger/\phi) &= -(2/\phi)\cdot(\phi'/\phi^2)' - (\phi'/\phi^2)^2 \\
&= -(2/\phi)\cdot\big((\phi''/\phi^2) - 2(\phi')^2/\phi^3\big) - (\phi'/\phi^2)^2 \\
&= -2(\phi''/\phi^3) + 3(\phi'/\phi^2)^2 = -\phi^{-2}\omega(-\phi^\dagger),
\end{aligned}$$

so we can also express $\omega_\rho^\phi = \omega^\phi(\lambda_\rho^\phi)$ as

$$\omega_\rho^\phi = \phi^{-2}\big(\omega_\rho - \omega(-\phi^\dagger)\big),$$

in analogy to the equality

$$\lambda_\rho^\phi = \phi^{-1}\big(\lambda_\rho + \phi^\dagger\big).$$

More generally, let E be any differential field, $\phi \in E^\times$, and define $\omega\colon E \to E$ by $\omega(z) = -2z' - z^2$ and likewise $\omega^\phi\colon E^\phi \to E^\phi$ by $\omega^\phi(z) = -2(z'/\phi) - z^2$. With $z^\phi := \phi^{-1}(z + \phi^\dagger)$ for $z \in E$, computations as before give the identity

$$\omega^\phi(z^\phi) = \phi^{-2}\big(\omega(z) - \omega(-\phi^\dagger)\big).$$

Related to ω is the function $\sigma\colon E^\times \to E$ given by $\sigma(y) = \omega(-y^\dagger) + y^2$. The corresponding function for E^ϕ is $\sigma^\phi\colon (E^\phi)^\times \to E^\phi$ given by

$$\sigma^\phi(y) = \omega^\phi(-y^\dagger/\phi) + y^2.$$

It is routine to check that it satisfies a similar transformation formula:

$$\sigma^\phi(y/\phi) = \phi^{-2}\big(\sigma(y) - \omega(-\phi^\dagger)\big) \qquad (y \in E^\times).$$

11.8 SOME SPECIAL DEFINABLE SETS

Throughout this section K is a pre-H-field. If K has asymptotic integration, then we fix a logarithmic sequence (ℓ_ρ) for K as in Section 11.5, and corresponding sequences (γ_ρ), (λ_ρ), and (ω_ρ). We single out the $\mathcal{O}$-submodule

$$\mathrm{I}(K) := \{y \in K : y \preccurlyeq f' \text{ for some } f \in \mathcal{O}\}$$

of K. We have $\partial\mathcal{O} \subseteq \mathrm{I}(K)$ and $(\mathcal{O}^\times)^\dagger \subseteq \mathrm{I}(K)$. If the derivation ∂ of K is small and K is an H-field, then $\mathrm{I}(K) \subseteq o$. If K has no gap or K is an H-field, then

$$\mathrm{I}(K) = \{y \in K : y \preccurlyeq f' \text{ for some } f \in o\}.$$

LEMMA 11.8.1. *Suppose K has a gap β. Then:*

$$\begin{cases} v(b') \neq \beta \text{ for all } b \asymp 1 \text{ in } K & \Rightarrow\ \mathrm{I}(K) = \{y \in K : vy > \beta\}; \\ v(b') = \beta \text{ for some } b \asymp 1 \text{ in } K & \Rightarrow\ \mathrm{I}(K) = \{y \in K : vy \geqslant \beta\}. \end{cases}$$

For $\phi \in K^>$ we have $\phi\,\mathrm{I}(K^\phi) = \mathrm{I}(K)$. The following facts are also easy to verify:

LEMMA 11.8.2. *Suppose K has asymptotic integration. Then for $y \in K$,*

$$\begin{aligned} y \in \mathrm{I}(K) &\iff y \prec f^\dagger \text{ for all nonzero } f \in o \\ &\iff y \text{ is not active in } K \\ &\iff y \prec \gamma_\rho \text{ for all } \rho. \end{aligned}$$

If $\mathrm{I}(K) \subseteq \partial K$, then $\mathrm{I}(K) = \partial\mathcal{O}$. If $\mathrm{I}(K) \subseteq (K^\times)^\dagger$, then $\mathrm{I}(K) = (\mathcal{O}^\times)^\dagger$. Moreover, $(1/\ell_\rho)' \in \mathrm{I}(K)$ for all ρ, and for each $y \in \mathrm{I}(K)$ there is a ρ such that $y \prec (1/\ell_\rho)'$. Finally, if L is a pre-H-field extension of K, then $\mathrm{I}(K) = \mathrm{I}(L) \cap K$.

COROLLARY 11.8.3. *Suppose K has asymptotic integration and $\mathrm{I}(K) \subseteq \partial o$. Then K is an H-field.*

PROOF. Let $u \in \mathcal{O}$. Then $u' \in \mathrm{I}(K)$, so $u' = y'$ with $y \in o$, and thus $u - y \in C$. □

EXAMPLE. $\mathrm{I}(\mathbb{T}) = \left\{y \in \mathbb{T} : y \prec \frac{1}{\ell_0\ell_1\cdots\ell_n} \text{ for all } n\right\}$, where (ℓ_n) is the logarithmic sequence for $\mathbb{T}$ from Example 11.5.1.

We also define

$$\begin{aligned}\Gamma(K) &:= \{a^\dagger : a \in K^{\succ 1}\},\\ \Lambda(K) &:= \{-a^{\dagger\dagger} : a \in K^{\succ 1}\},\\ \Delta(K) &:= \{-a'^\dagger : 0 \neq a \in K^{\prec 1}\}.\end{aligned}$$

Note that

$$\Lambda(K) = -\Gamma(K)^\dagger = \{-a^{\dagger\dagger} : a \in K^\times,\ a \not\asymp 1\} = \{-a^{\dagger\dagger} : 0 \neq a \in K^{\prec 1}\}.$$

For $\phi \in K^>$ we have

$$\phi\Gamma(K^\phi) = \Gamma(K), \qquad \phi\Lambda(K^\phi) = \phi^\dagger + \Lambda(K), \qquad \phi\Delta(K^\phi) = \phi^\dagger + \Delta(K).$$

The set $\Gamma(K)$ is closed under addition. The elements of $\Gamma(K)$ are active in K. The sets $\mathrm{I}(K)$, $\Gamma(K)$, and $-\Gamma(K)$ are pairwise disjoint. If K has asymptotic integration, then there is for each $y \in \Gamma(K)$ an index ρ with $\gamma_\rho \prec y$. If K has asymptotic integration and $(K^\times)^\dagger = K$, then

$$K = \mathrm{I}(K) \cup \Gamma(K) \cup \big(-\Gamma(K)\big),$$

by Lemma 11.8.2. It is easy to see that if $(K^\times)^\dagger = K$ and L is a pre-H-field extension of K, then $\Gamma(L) \cap K = \Gamma(K)$.

Let $a \in K^{\succ 1}$, and set $y := a^\dagger \in \Gamma(K)$ and $z = -y^\dagger = -a^{\dagger\dagger} \in \Lambda(K)$. Then

$$z + y = -\big(a^{\dagger\dagger} + (1/a)^\dagger\big) = -\big((1/a)^\dagger\,(1/a)\big)^\dagger = -(1/a)'^\dagger \in \Delta(K).$$

In particular, if K has asymptotic integration, then $\lambda_\rho \in \Lambda(K)$ and $\lambda_\rho + \gamma_\rho \in \Delta(K)$ for all ρ, and $\Lambda(K) \cup \Delta(K)$ does not contain a pseudolimit of (λ_ρ), by Lemma 11.5.13. The sets $\Lambda(K)$ and $\Delta(K)$ are disjoint.

LEMMA 11.8.4. *Suppose K has asymptotic integration, $\mathrm{I}(K) = \partial\mathcal{o}$, and $(K^\times)^\dagger = K$. Then $K = \Lambda(K) \cup \Delta(K)$.*

PROOF. Let $f \in K$, and take $a \in K^\times$ with $a^\dagger \neq 0$ and $f = -a^{\dagger\dagger}$. If $a \not\asymp 1$, then $f \in \Lambda(K)$, so suppose $a \asymp 1$. Then $a^\dagger \in (\mathcal{O}^\times)^\dagger \subseteq \mathrm{I}(K) = \partial\mathcal{o}$. Take $b \in \mathcal{o}$ with $a^\dagger = b'$; then $b \neq 0$, and $f = -b'^\dagger \in \Delta(K)$. □

LEMMA 11.8.5. *Let $f \in K^\times$. Then*

$$f \in \mathrm{I}(K) \implies -f^\dagger \notin \Lambda(K),$$

and if K has asymptotic integration and $(K^\times)^\dagger = K$, then

$$f \in \mathrm{I}(K) \iff -f^\dagger \notin \Lambda(K).$$

PROOF. Suppose that $-f^\dagger \in \Lambda(K)$. Take $a \in K^\times$, $a \not\asymp 1$, such that $-f^\dagger = -a^{\dagger\dagger}$. Then $f \in C^\times a^\dagger$ and hence $vf = va^\dagger \in \Psi$, so $f \notin \mathrm{I}(K)$. Now suppose K has asymptotic integration, $(K^\times)^\dagger = K$, and $-f^\dagger \notin \Lambda(K)$. Take $a \in K^\times$ such that $f = a^\dagger$; then $-a^{\dagger\dagger} = -f^\dagger \notin \Lambda(K)$ and hence $a \asymp 1$, so $f = a^\dagger \in (\mathcal{O}^\times)^\dagger = \mathrm{I}(K)$. □

COROLLARY 11.8.6. *Suppose K has asymptotic integration and $(K^\times)^\dagger = K$, and let L be a pre-H-field extension of K. Then $\Lambda(K) = \Lambda(L) \cap K$.*

PROOF. Clearly $\Lambda(K) \subseteq \Lambda(L) \cap K$, so let $g \in \Lambda(L) \cap K$, and take $f \in K^\times$ with $g = -f^\dagger$. Then $f \notin \mathrm{I}(L)$ by the first part of Lemma 11.8.5, hence $f \notin \mathrm{I}(K)$ and so $g \in \Lambda(K)$, by the second part. □

LEMMA 11.8.7. *Let $f \in K^\times$. Then*

$$f \in \partial o \iff -f^\dagger \in \Delta(K).$$

PROOF. If $f \in \partial o$, say $f = a'$ with $a \in o$, then $-f^\dagger = -a'^\dagger \in \Delta(K)$. Conversely, if $-f^\dagger \in \Delta(K)$, then taking $a \in o^{\neq}$ with $f^\dagger = a'^\dagger$ gives $f \in Ca' \subseteq \partial o$. □

COROLLARY 11.8.8. *Suppose that K has asymptotic integration, $\mathrm{I}(K) = \partial o$, and $(K^\times)^\dagger = K$. Then for $f \in K^\times$ we have*

$$f \in \mathrm{I}(K) \iff -f^\dagger \in \Delta(K), \qquad f \notin \mathrm{I}(K) \iff -f^\dagger \in \Lambda(K).$$

LEMMA 11.8.9. *Suppose $(K^\times)^\dagger = K$. Then $\Lambda(K) + \mathrm{I}(K) \subseteq \Lambda(K)$.*

PROOF. Let $a \in K$, $a \succ 1$ and $y \in \mathrm{I}(K)$; we want to show $-a^{\dagger\dagger} + y \in \Lambda(K)$. Take $b \in K \setminus C$ with $b^{\dagger\dagger} = a^{\dagger\dagger} - y$. Then $(a^\dagger/b^\dagger)^\dagger = y \in \mathrm{I}(K)$, hence $a^\dagger/b^\dagger \asymp 1$ and therefore $b \not\asymp 1$. □

First-order linear differential equations and the predicate I

Throughout this subsection K is a Liouville closed H-field and y ranges over K.

LEMMA 11.8.10. *Suppose the derivation of K is small. Let $A = aY + bY'$ where $0 \neq a \succcurlyeq b$ in K. Then $A(\mathcal{O}) = a\mathcal{O}$.*

PROOF. Dividing by a we arrange $a = 1$. Let $g \in \mathcal{O}$; our job is to find $f \in \mathcal{O}$ such that $A(f) = g$. Take $f_0 \in K$ with $A(f_0) = g$, and $y_0 \in K^\times$ with $A(y_0) = 0$. Then

$$A^{-1}(g) := \{f \in K : A(f) = g\} = f_0 + Cy_0,$$

so $\left|v\big(A^{-1}(g)\big)\right| \leqslant 2$. Let $\phi \in K^>$ be active, $\phi \preccurlyeq 1$. Then the compositional conjugate K^ϕ is a Liouville closed H-field with small derivation, and $A^\phi = Y + b\phi Y'$ with $b\phi \preccurlyeq 1$. We equip K^ϕ with the coarsening $v^\flat_\phi$. Then $\max \Psi^\flat_\phi = 0$ in $\Gamma/\Gamma^\flat_\phi$, where $\Psi^\flat_\phi$ is the Ψ-set of the asymptotic couple of $(K^\phi, v^\flat_\phi)$. By Lemma 10.1.8 the residue map from $\mathcal{O}^\flat_\phi$ onto its residue field maps the constant field of K onto the constant field of that (differential) residue field. Thus we can apply Lemmas 9.4.4 and 5.6.11 to $(K^\phi, v^\flat_\phi)$ to get $f_\phi \in K$ with $A^\phi(f_\phi) = A(f_\phi) = g$ (so $f_\phi \in A^{-1}(g)$) and $v^\flat_\phi(f_\phi) \geqslant 0$. In view of $\left|v\big(A^{-1}(g)\big)\right| \leqslant 2$ we get $v(f_\phi) \geqslant 0$, eventually. □

LEMMA 11.8.11. *Let $f, g \in K$, $g \neq 0$. Then*

$$\exists y\,[y' = fy + g \,\&\, y \prec 1] \quad \Longleftrightarrow \quad f, g \in \mathrm{I}(K) \vee [f \notin \mathrm{I}(K) \wedge g \prec f].$$

PROOF. Put $P := Y' - fY - g \in K\{Y\}$. Take $b \neq 0$ with $b^\dagger = f$ and then $a \neq 0$ with $a \not\asymp b$ and $va + \psi(va - vb) = vg$. For active ϕ in K we have

$$P^\phi_{\times a} = a\phi Y' + a(a/b)^\dagger Y - g$$

with $a(a/b)^\dagger \asymp g$. Since eventually $\phi \preccurlyeq (a/b)^\dagger$, Lemma 11.8.10 yields a zero of $P_{\times a}$ in $\mathcal{O}$. But $P_{\times a}$ has no zero in o, so P has a zero $z \in K$ with $z \asymp a$, and for every zero y of P we have $y \succcurlyeq a$. Thus P has a zero $y \prec 1$ iff $a \prec 1$. Now suppose $f \in \mathrm{I}(K)$. Then $a \not\asymp b \asymp 1$ and $va + \psi(va) = vg$, that is, $a' \asymp g$; so $a \prec 1$ iff $g \in \mathrm{I}(K)$. Next assume $f \notin \mathrm{I}(K)$, so $b \not\asymp 1$. Then

$$\begin{aligned} a \prec 1 \quad &\Longleftrightarrow \quad va - vb > -vb \\ &\Longleftrightarrow \quad \underbrace{va - vb + \psi(va - vb)}_{= vg - vb} > \underbrace{-vb + \psi(-vb)}_{= vf - vb} \\ &\Longleftrightarrow \quad g \prec f. \end{aligned}$$

□

COROLLARY 11.8.12. *Let $f, g, h \in K$, $g, h \neq 0$. Then the following are equivalent:*

(i) *there exists y such that $y' = fy + g$ and $y \prec h$;*

(ii) $[f - h^\dagger \in \mathrm{I}(K)$ *and* $g/h \in \mathrm{I}(K)]$ *or* $[f - h^\dagger \notin \mathrm{I}(K)$ *and* $(g/h) \prec f - h^\dagger]$.

PROOF. Set $f_* := f - h^\dagger$, $g_* := g/h$. Then $(y/h)' = y'/h - h^\dagger(y/h)$, so $y' = fy + g$ and $y \prec h$ is equivalent to $(y/h)' = f_*(y/h) + g_*$ and $y/h \prec 1$. It remains to appeal to Lemma 11.8.11. □

Interaction with the ordering

So far the field ordering of our pre-H-field K has not played a role. As a first comment referring to this ordering we mention that $\mathrm{I}(K)$ is convex in the ordered set K. Below we characterize λ-freeness and ω-freeness in terms of certain cuts in the ordered set K, and we explore further the behavior of the differential polynomial function ω in this setting. We often tacitly use the fact that for $a, b \in K^\times$ we have $a \prec b \Rightarrow a^\dagger < b^\dagger$. One consequence of this fact is that $\Lambda(K) < \Delta(K)$, which in view of Lemma 11.8.4 yields:

COROLLARY 11.8.13. *If K is Liouville closed, then $\Lambda(K)$ is downward closed, $\Delta(K)$ is upward closed, and $K = \Lambda(K) \cup \Delta(K)$.*

PROOF. Immediate from Lemma 11.8.4 and Corollary 11.8.16. □

Here is another use of $a \prec b \Rightarrow a^\dagger < b^\dagger$, for $a, b \in K^\times$:

LEMMA 11.8.14. *Assume K is ungrounded and $L \supseteq K$ is a pre-H-field extension such that $\Gamma^{>}$ is coinitial in $\Gamma_L^{>}$. Then $\Delta(K)$ is coinitial in $\Delta(L)$, $\Gamma(K)$ is coinitial in $\Gamma(L)$, and $\Lambda(K)$ is cofinal in $\Lambda(L)$.*

PROOF. Let $a \in L^{\prec 1}$, $a \neq 0$. Take $b \in K$ with $a \prec b \prec 1$. Then $a' \prec b'$ and hence $-a'^{\dagger} > -b'^{\dagger}$. This shows $\Delta(K)$ is coinitial in $\Delta(L)$. Next, let $a \in L^{\succ 1}$. Take $b \in K^{\succ 1}$ with $a^{\dagger} \succ b^{\dagger}$. Then $a^{\dagger}, b^{\dagger} > 0$, hence $a^{\dagger} > b^{\dagger}$ and $-a^{\dagger\dagger} < -b^{\dagger\dagger}$. ☐

LEMMA 11.8.15. *Let K have asymptotic integration. Then (λ_ρ) is strictly increasing and cofinal in $\Lambda(K)$, and $(\lambda_\rho + \gamma_\rho)$ is strictly decreasing and coinitial in $\Delta(K)$. If K has small derivation, then $\lambda_\rho > 0$, eventually.*

PROOF. We have $\Lambda(K) = -\Gamma(K)^{\dagger}$ and $\lambda_\rho = -\gamma_\rho^{\dagger}$, so the first claim follows from the remark preceding the previous lemma and the properties of (γ_ρ). Let $a \in K$, $0 \neq a \prec 1$. Take ρ with $1/\ell_\rho \succ a$; then $(1/\ell_\rho)' \succ a'$, so $-(1/\ell_\rho)'^{\dagger} < -a'^{\dagger}$. Thus $(\lambda_\rho + \gamma_\rho)$ is coinitial in $\Delta(K)$. For $\rho < \rho'$ we have $1/\ell_\rho \prec 1/\ell_{\rho'} \prec 1$, hence $(1/\ell_\rho)' \prec (1/\ell_{\rho'})'$ and thus $-(1/\ell_\rho)'^{\dagger} > -(1/\ell_{\rho'})'^{\dagger}$; so $(\lambda_\rho+\gamma_\rho)$ is strictly decreasing. Suppose now that K has small derivation. Then 1 is active, so $\gamma_\rho \prec 1$ eventually, and thus $\lambda_\rho > 0$ eventually. ☐

COROLLARY 11.8.16. *Assume K has asymptotic integration. Let L be a pre-H-field extension of K and $\lambda \in L$. Then: $\lambda_\rho \rightsquigarrow \lambda \iff \Lambda(K) < \lambda < \Delta(K)$.*

PROOF. Use the previous lemma in combination with Lemma 11.5.2. ☐

COROLLARY 11.8.17. *Suppose K has asymptotic integration and L is a pre-H-field extension of K with $(L^{\times})^{\dagger} = L$. Then $\Lambda(L)^{\downarrow} \cap K$ equals $\Lambda(K)^{\downarrow}$ or $K \setminus \Delta(K)^{\uparrow}$.*

PROOF. Suppose $\Lambda(L)^{\downarrow} \cap K \neq \Lambda(K)^{\downarrow}$. Take $f \in \big(\Lambda(L)^{\downarrow} \cap K\big) \setminus \Lambda(K)^{\downarrow}$. Then $f < \Delta(L)$, so $\Lambda(K) < f < \Delta(K)$. To get $\Lambda(L)^{\downarrow} \cap K = K \setminus \Delta(K)^{\uparrow}$, let $g \in K \setminus \Delta(K)^{\uparrow}$; it is enough to show that then $g \in \Lambda(L)^{\downarrow}$. As the case $g \in \Lambda(K)^{\downarrow}$ is obvious, assume $g \notin \Lambda(K)^{\downarrow}$. Then $\Lambda(K) < g < \Delta(K)$. Then by Corollary 11.8.16 both f and g are pseudolimits of (λ_ρ), hence $v(g-f) > \Psi$ by Lemma 11.5.2 and thus $g - f \in \mathrm{I}(K) \subseteq \mathrm{I}(L)$. Therefore $g \in f + \mathrm{I}(L) \subseteq \Lambda(L)^{\downarrow}$ by Lemma 11.8.9. ☐

Our eventual quantifier elimination for $\mathbb{T}$ requires predicates for the sets $\Lambda(\mathbb{T})$ and $\omega(\mathbb{T})$, and for this reason we pay attention to properties of these sets expressible by universal sentences. In this connection it is convenient to include also $\mathrm{I}(\mathbb{T})$ and $\Delta(\mathbb{T})$ in an auxiliary role. Such universal properties are contained in Lemmas 11.8.5, 11.8.9, and Corollary 11.8.13. Here is another useful one:

LEMMA 11.8.18. *Assume K has asymptotic integration, and let $a, \phi \in K$. Then*

$$a, \phi > 0,\ a \succcurlyeq 1 \Rightarrow \phi a - \phi^{\dagger} \in \Delta(K)^{\uparrow}.$$

PROOF. We first establish the following:

(1) if the derivation of K is small, then $\sigma \cap \Delta(K) \neq \emptyset$;

(2) if the derivation of K is not small, then $K^< \cap \Delta(K) \neq \emptyset$.

For (1), assume the derivation of K is small. Take $1 \in \Gamma^>$ with $\psi(1) = 1$. Then $\Psi < 1+1$. Pick $a \in \sigma$ with $v(a') = 1+1$. Then $v(-a'^\dagger) = 1$ and so $-a'^\dagger \in \sigma \cap \Delta(K)$.

For (2), assume the derivation of K is not small. Then for each ρ we have $\gamma_\rho \succ 1$, so λ_ρ is active and $\lambda_\rho < 0$. By Lemma 11.5.13, $v(\lambda_\rho)$ is eventually constant, and in Ψ. Hence $\lambda_\rho \succ \gamma_\rho$, eventually, and thus $\lambda_\rho + \gamma_\rho \in K^< \cap \Delta(K)$, eventually.

Now let $a, \phi > 0$ and $a \succcurlyeq 1$. Then $a \in \Delta(K^\phi)^\uparrow$ by (1) and (2). Also $\Delta(K^\phi) = \phi^{-1}(\Delta(K) + \phi^\dagger)$, which gives $\phi a - \phi^\dagger \in \Delta(K)^\uparrow$. □

LEMMA 11.8.19. *Suppose K is Liouville closed. Then*

$$\Gamma(K) = (K^{>C})^\dagger = \{y \in K^> : vy \in \Psi\} = K^> \setminus \mathrm{I}(K),$$

and hence $\Gamma(K)$ is upward closed.

PROOF. The set equalities are easy consequences of the results above. To get $\Gamma(K)$ upward closed, use that $\mathrm{I}(K)$ is convex. □

EXAMPLE. Let (ℓ_n) be the logarithmic sequence for $\mathbb{T}$ from Example 11.5.1. Then for $y \in \mathbb{T}$ we have

$$y \in \Gamma(\mathbb{T}) \iff y \geqslant \frac{1}{\ell_0 \ell_1 \cdots \ell_n} \text{ for some } n.$$

Also, for $z \in \mathbb{T}$ we have

$$z \in \Lambda(\mathbb{T}) \iff z \leqslant \frac{1}{\ell_0} + \frac{1}{\ell_0\ell_1} + \cdots + \frac{1}{\ell_0\ell_1\cdots\ell_n} \text{ for some } n$$

and

$$\begin{aligned} z \in \Delta(\mathbb{T}) &\iff z \geqslant \left(\frac{1}{\ell_0} + \frac{1}{\ell_0\ell_1} + \cdots + \frac{1}{\ell_0\ell_1\cdots\ell_n}\right) + \frac{1}{\ell_0\ell_1\cdots\ell_n} \text{ for some } n \\ &\iff z > \frac{1}{\ell_0} + \frac{1}{\ell_0\ell_1} + \cdots + \frac{1}{\ell_0\ell_1\cdots\ell_n} \text{ for all } n. \end{aligned}$$

Figure 11.1 shows the sets $\mathrm{I}(\mathbb{T})$, $\Gamma(\mathbb{T})$, $\Lambda(\mathbb{T})$, and $\Delta(\mathbb{T})$.

Next we study the behavior of $\omega\colon K \to K$, $\omega(z) := -(2z' + z^2)$.

PROPOSITION 11.8.20. *Suppose K is Liouville closed. Then the restriction of ω to $\Lambda(K)$ is strictly increasing, and $\omega(\Lambda(K)) = \omega(\Delta(K))$.*

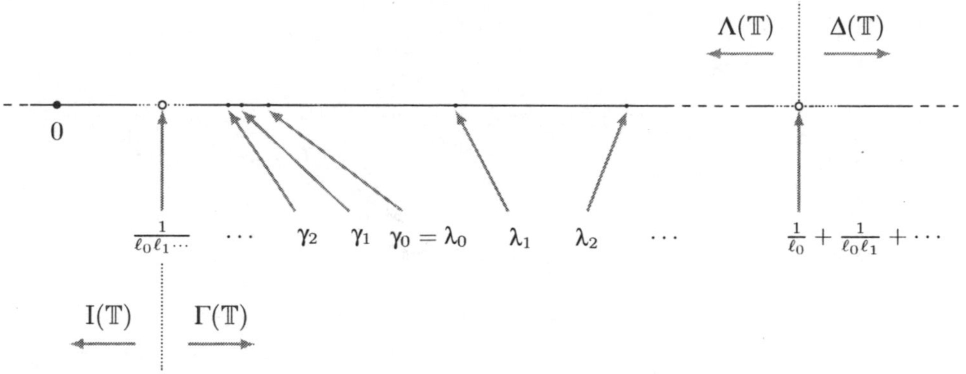

Figure 11.1: The sets $\mathrm{I}(\mathbb{T})$, $\Gamma(\mathbb{T})$, $\Lambda(\mathbb{T})$, and $\Delta(\mathbb{T})$.

PROOF. Pick for each $f \in K^{\succ 1}$ an $\mathrm{L}\, f \in K^{\succ 1}$ with $(\mathrm{L}\, f)' = f^\dagger$. Thus we have: $f \in K^{>C} \Rightarrow \mathrm{L}\, f > C$. We also pick for all $f \in K^>$ and $r \in \mathbb{Q}$ an element $f^r \in K^>$ such that $(f^r)^\dagger = rf^\dagger$. It follows easily that for $f \in K$, $r \in \mathbb{Q}$,

$$f > C,\ r > 0 \Rightarrow f^r > C, \qquad f > C,\ r < 0 \Rightarrow f^r \prec 1.$$

Now let $z \in \Lambda(K)$. We shall find an element $f \in \Delta(K)$ (depending on z) such that $\omega(z) < \omega(z^*)$ for all $z^* \in K$ with $z < z^* \leqslant f$. As z is arbitrary and $f > \Lambda(K)$, this will prove that ω is strictly increasing on $\Lambda(K)$. Take $a \in K^{>C}$ such that $z = -a^{\dagger\dagger}$, and let $t := (\mathrm{L}\, a)^{-r}$ where r is any rational number with $0 < r \leqslant 1$. (The value of r will be further specified later.) Then $0 \neq t \prec 1$, so $f := -t'^\dagger \in \Lambda(K)$. From $t' = -r(\mathrm{L}\, a)^\dagger t = -r\frac{a^\dagger}{\mathrm{L}\, a}t$ we get, with $y := a^\dagger = (\mathrm{L}\, a)'$:

$$f \ =\ -y^\dagger + (r+1)(\mathrm{L}\, a)^\dagger \ =\ z + (r+1)\frac{y}{\mathrm{L}\, a}.$$

Put

$$\varepsilon \ :=\ h(r+1)\frac{y}{\mathrm{L}\, a} \qquad \text{with } h \in K,\ 0 < h \leqslant 1,$$

so $\varepsilon > 0$ and $z + \varepsilon$ runs through all $w \in K$ with $z < w \leqslant f$, as h varies. Thus it suffices to show that $\omega(z+\varepsilon) > \omega(z)$. Since

$$\omega(z) - \omega(z+\varepsilon) \ =\ \varepsilon\big(2(\varepsilon^\dagger + z) + \varepsilon\big) \tag{11.8.1}$$

by Lemma 5.2.1, this reduces to showing that $2(\varepsilon^\dagger + z) + \varepsilon < 0$. Now

$$\varepsilon^\dagger \ =\ h^\dagger + y^\dagger - (\mathrm{L}\, a)^\dagger \ =\ h^\dagger - z - \frac{y}{\mathrm{L}\, a}$$

and thus

$$2(\varepsilon^\dagger + z) + \varepsilon \ =\ 2h^\dagger - \big(2 - h(r+1)\big)\frac{y}{\mathrm{L}\, a}, \tag{11.8.2}$$

so the inequality $2(\varepsilon^\dagger + z) + \varepsilon < 0$ becomes

$$2h^\dagger \ < \ \big(2 - h(r+1)\big)\frac{y}{\operatorname{L}a}.$$

This holds if $h \prec 1$, since then $h^\dagger < 0$, while $\frac{y}{\operatorname{L}a} > 0$. Suppose $h \asymp 1$. Then $v(h^\dagger) = v(h') > \Psi$, in particular $\frac{y}{\operatorname{L}a} = (-1/r)t^\dagger \succ h^\dagger$. Taking $0 < r < 1$ we get $0 < 2 - h(r+1)$ and $2 - h(r+1) \asymp 1$, so $2h^\dagger \prec \big(2 - h(r+1)\big)\frac{y}{\operatorname{L}a}$. The desired inequality follows, as its right-hand side is positive. Hence for any rational r with $0 < r < 1$, we obtain $f > \Lambda(K)$ such that $\omega(z) < \omega(z^*)$ for all $z^* \in K$ with $z < z^* \leqslant f$. Thus ω is indeed strictly increasing on $\Lambda(K)$.

For $r = 1$, we obtain an element $f = z + \frac{2y}{\operatorname{L}a} > \Lambda(K)$, and equations (11.8.1) and (11.8.2) with $h = 1$ then yield that $\omega(z) = \omega(f)$. Conversely, let $f \in \Delta(K)$. We shall construct $a \in K^{>C}$ such that $f = z + \frac{2y}{\operatorname{L}a}$ for $y = a^\dagger$ and $z = -a^{\dagger\dagger}$. This will finish the proof of the second claim. Take $g \in K^{<}$ with $f = -g^\dagger$. Then by Lemma 11.8.7 we can take $t \prec 1$ in K with $g = t'$. Then $t > 0$, hence $(1/t)' = a^\dagger$ where $a \in K^{>C}$. By readjusting our choice of $\operatorname{L}a$ we arrange $t = \frac{1}{\operatorname{L}a}$. Then, with $y = a^\dagger$ and $z = -a^{\dagger\dagger}$ we have $z \in \Lambda(K)$ and $f = z + \frac{2y}{\operatorname{L}a}$, as desired. □

Recall from Section 5.2 that

$$\Omega(K) := \{f \in K : \ 4y'' + fy = 0 \text{ for some } y \in K^\times\}.$$

In view of that section and Corollary 11.8.13 this gives for Liouville closed K:

$$\Omega(K) \ = \ \omega(K) \ = \ \omega\big(\Lambda(K)\big) \ = \ \omega\big(\Delta(K)\big).$$

Since every pre-H-field can be embedded into a Liouville closed H-field, the function ω is strictly increasing on $\Lambda(K)^\downarrow$ even without assuming that K is Liouville closed. Together with Lemma 11.8.15, we therefore obtain:

Corollary 11.8.21. *If K has asymptotic integration, then the sequence $(\omega_\rho) = \big(\omega(\lambda_\rho)\big)$ is strictly increasing and cofinal in $\omega\big(\Lambda(K)\big)$.*

The following is a consequence of Lemma 11.7.1:

Corollary 11.8.22. *Let $f \in K$ and suppose $f \geqslant \omega(z) + cy^2$ for some $y \in \Gamma(K)$ and some constant $c \in C^>$, with $z = -y^\dagger$. Then $f > \omega\big(\Lambda(K)\big)$.*

Proof. By passing to an extension if necessary we can arrange that K is Liouville closed. Let $y \in \Gamma(K)$ and $c \in C^>$ be such that $f \geqslant \omega(z) + cy^2$, with $z = -y^\dagger$. Take $a \in K^{>C}$ such that $y = a^\dagger$. We may choose our logarithmic sequence (ℓ_ρ) for K beginning with $\ell_0 = a$; then $\gamma_0 = \ell_0^\dagger = y$ and $\lambda_0 = -\gamma_0^\dagger = z$. For $\rho > 0$ we have $\gamma_1^2 \sim \omega(\lambda_\rho) - \omega(\lambda_0) > 0$ by Lemma 11.7.1. Hence $\omega(\lambda_\rho) - \omega(\lambda_0) < c\gamma_0^2$, so

$$f \ \geqslant \ \omega(z) + cy^2 \ = \ \big(\omega(\lambda_0) - \omega(\lambda_\rho) + c\gamma_0^2\big) + \omega(\lambda_\rho) \ > \ \omega(\lambda_\rho).$$

Since this holds for all $\rho > 0$, we conclude that $f > \omega\big(\Lambda(K)\big)$. □

COROLLARY 11.8.23. *Assume K is Liouville closed, and $f, g \in K^{>}$ are such that $-\frac{1}{2}f^{\dagger} \in \Lambda(K)$ and $f \preccurlyeq g$. Then $\omega(-\frac{1}{2}f^{\dagger}) + g > \omega\big(\Lambda(K)\big)$.*

PROOF. Take $y \in K^{>}$ with $y^2 = f$. Then $z := -y^{\dagger} = -\frac{1}{2}f^{\dagger} \in \Lambda(K)$, so $y > \mathrm{I}(K)$ by Lemma 11.8.5, hence $y \in \Gamma(K)$ by Lemma 11.8.19. Thus, with $c \in C^{>}$ such that $cf \leqslant g$, we have $\omega(-\frac{1}{2}f^{\dagger}) + g \geqslant \omega(z) + cy^2 > \omega\big(\Lambda(K)\big)$ by Corollary 11.8.22. □

COROLLARY 11.8.24. *Suppose K is Liouville closed and has small derivation. Then $\omega_\rho > 0$ eventually.*

PROOF. Take $x \in K$ with $x' = 1$. Then $x \succ 1$, so $1/x = -x^{\dagger\dagger} \in \Lambda(K)$, and $\omega(1/x) = 1/x^2 > 0$. Now use Corollary 11.8.21. □

COROLLARY 11.8.25. *Suppose K is Liouville closed with small derivation. Then for each $A \in C[\partial]^{\neq}$ we have $\mathrm{m}(A) = \dim_C \ker A$.*

PROOF. Lemma 11.7.4 and Corollary 11.8.24 give $\omega(K) \cap C \subseteq C^{\leqslant}$. It remains to use Lemma 5.2.11. □

COROLLARY 11.8.26. *Suppose K is Liouville closed with small derivation, and let $P(Y) \in C[Y]$. Set $A := P(\partial) \in C[\partial]$. Then*

$$\begin{aligned}
\exists y \in K^{\times}\big[y \prec 1 \;\&\; A(y) = 0\big] &\iff \exists c \in C\big[c < 0 \;\&\; P(c) = 0\big],\\
\exists y \in K^{\times}\big[y \preccurlyeq 1 \;\&\; A(y) = 0\big] &\iff \exists c \in C\big[c \leqslant 0 \;\&\; P(c) = 0\big],\\
\exists y \in K^{\times}\big[y \succcurlyeq 1 \;\&\; A(y) = 0\big] &\iff \exists c \in C\big[c \geqslant 0 \;\&\; P(c) = 0\big].
\end{aligned}$$

PROOF. Take $x \in K$ with $x' = 1$. As in the subsection on linear differential equations with constant coefficients in Section 5.1 we pick for each $c \in C$ an element of $K^{\times}$, denoted by e^{cx}, such that $(\mathrm{e}^{cx})' = c\mathrm{e}^{cx}$. Thus $\mathrm{e}^{0x} \in C^{\times}$, and it is routine to show that if $c \in C^{<}$, then $\mathrm{e}^{cx} \prec x^{-n}$ for all n, and if $c \in C^{>}$, then $\mathrm{e}^{cx} \succ x^n$ for all n. It follows easily that $\mathrm{e}^{c_1 x} x^{k_1} \not\asymp \mathrm{e}^{c_2 x} x^{k_2}$ for all distinct $(c_1, k_1), (c_2, k_2) \in C \times \mathbb{Z}$. It now remains to use Proposition 5.1.18 and Corollary 11.8.25. □

Here is a universal property similar to Lemma 11.8.9:

LEMMA 11.8.27. *Suppose K is Liouville closed. Then for all $g, h \in \mathrm{I}(K)$,*

$$\omega(K)^{\downarrow} + gh \subseteq \omega(K)^{\downarrow}.$$

PROOF. Let $a \in \omega(K)^{\downarrow}$ and $g, h \in \mathrm{I}(K)$. Take ρ so large that $a < \omega_\rho = \omega(\lambda_\rho)$. Then $g, h \prec \gamma_{\rho+2}$, so $a + gh < \omega_\rho + \gamma_{\rho+2}^2 < \omega_{\rho+1}$, and thus $a + gh \in \omega(K)^{\downarrow}$. □

COROLLARY 11.8.28. *Suppose K is Liouville closed, and $f \in K^{\times}$, $-\frac{1}{2}f^{\dagger} \notin \Lambda(K)$. Then $\omega(K)^{\downarrow} + f \subseteq \omega(K)^{\downarrow}$.*

PROOF. This is clear if $f < 0$, so assume $f > 0$, and take $y \in K^{>}$ with $y^2 = f$. Then $-y^{\dagger} = -\frac{1}{2}f^{\dagger} \notin \Lambda(K)$, so $y \in \mathrm{I}(K)$ by Lemma 11.8.5. Now use Lemma 11.8.27. □

We now consider also the function

$$y \mapsto \sigma(y) = \omega(z) + y^2 \colon\ K^\times \to K \qquad \text{where } z = -y^\dagger.$$

Note that $\sigma(y) = \sigma(-y)$ for all $y \in K^\times$. By Corollary 11.8.22, $\omega\big(\Lambda(K)\big) < \sigma\big(\Gamma(K)\big)$, and by embedding K into a Liouville closed H-field, this gives

$$\omega(K) \ < \ \sigma\big(\Gamma(K)\big).$$

LEMMA 11.8.29. *The restriction of σ to $\Gamma(K)$ is strictly increasing.*

PROOF. Let $y \in \Gamma(K)$, $z = -y^\dagger$ and $a > 1$ in K; we shall derive $\sigma(ay) > \sigma(y)$. We have $-(ay)^\dagger = z + b$ with $b := -a^\dagger$, so

$$\sigma(ay) - \sigma(y) \ = \ \omega(z+b) - \omega(z) + (a^2 - 1)y^2.$$

We have $a \in C$ iff $b = 0$, and in this case $\sigma(ay) = \sigma(y) + (a^2 - 1)y^2 > \sigma(y)$ as required. Suppose $a \notin C$. Then by Lemma 5.2.1,

$$\sigma(ay) - \sigma(y) \ = \ -b\big(2(b/y)^\dagger + b\big) + (a^2 - 1)y^2.$$

We now distinguish three cases:

CASE 1: $a \asymp 1$. Then $\Psi < va' = vb$, in particular $y \succ b$ and $(b/y)^\dagger \succ b$, hence $-b\big(2(b/y)^\dagger + b\big) \sim -2b(b/y)^\dagger$. From $vy \in \Psi < v(b/y)' = v(b/y) + v(b/y)^\dagger$, we obtain $v(y^2) < vb + v(b/y)^\dagger$. If $a \not\sim 1$, this gives $-b\big(2(b/y)^\dagger + b\big) \prec y^2 \asymp (a^2-1)y^2$, so $\sigma(ay) - \sigma(y) \sim (a^2 - 1)y^2 > 0$. If $a \sim 1$, then $a = 1 + \varepsilon$ with $0 < \varepsilon \prec 1$, so $b = -\varepsilon'/(1+\varepsilon) > 0$, hence $0 < b/y \prec 1$, which gives $-b\big(2(b/y)^\dagger + b\big) \sim -2b(b/y)^\dagger > 0$, and thus $\sigma(ay) - \sigma(y) > 0$.

CASE 2: $a \succ 1$ *and* $y \prec b$. Then $vy \in \Psi < v(y/b)' = v(y/b) + v(y/b)^\dagger$, so $vb < v(y/b)^\dagger = v(b/y)^\dagger$ and thus $-b\big(2(b/y)^\dagger + b\big) \sim -b^2$. Also, $v(1/a)' > \Psi$, in particular $vb = v(1/a)^\dagger = va + v(1/a)' > v(ay)$ and hence $b^2 \prec (ay)^2 \sim (a^2-1)y^2$. This gives $\sigma(ay) - \sigma(y) \sim (a^2 - 1)y^2 > 0$.

CASE 3: $a \succ 1$ *and* $y \succcurlyeq b$. Then $(b/y)^\dagger \prec y$ by Lemma 6.5.4(i), so

$$-b\big(2(b/y)^\dagger + b\big) \ \preccurlyeq \ by \ \preccurlyeq \ y^2 \ \prec \ (a^2 - 1)y^2,$$

and thus once again, $\sigma(ay) - \sigma(y) \sim (a^2 - 1)y^2 > 0$. □

As any pre-H-field embeds into some Liouville closed H-field, Lemmas 11.8.19 and 11.8.29 give that σ is strictly increasing on $\Gamma(K)^\uparrow$, even without assuming that K itself is Liouville closed. By earlier comments, if K has asymptotic integration, then the sequence $(\upgamma_\rho)$ in $\Gamma(K)$ is strictly decreasing and coinitial in $\Gamma(K)$.

COROLLARY 11.8.30. *Assume that K has asymptotic integration. Then the sequence $\big(\sigma(\upgamma_\rho)\big) = (\upomega_\rho + \upgamma_\rho^2)$ is strictly decreasing and coinitial in $\sigma\big(\Gamma(K)\big)$. Thus for any $\upomega$ in any pre-H-field extension of K we have the equivalence*

$$\upomega_\rho \rightsquigarrow \upomega \iff \omega\big(\Lambda(K)\big) < \upomega < \sigma\big(\Gamma(K)\big).$$

As to the behavior of σ on the complement $K^{>} \setminus \Gamma(K)$, we note:

LEMMA 11.8.31. *Suppose K is Liouville closed. Then*

$$\begin{aligned}\sigma\big(K^{>} \setminus \Gamma(K)\big) &\subseteq \omega(K)^{\downarrow}, \text{ so}\\ \sigma(K^{\times}) &= \sigma\big(K^{>} \setminus \Gamma(K)\big) \cup \sigma\big(\Gamma(K)\big) \quad \text{with } \sigma\big(K^{>} \setminus \Gamma(K)\big) < \sigma\big(\Gamma(K)\big).\end{aligned}$$

PROOF. Let $s \in K^{>} \setminus \Gamma(K)$. Using Lemma 11.8.19, take $a \in K^{\succ 1}$ with $s = (1/a)'$ and set $y := a^{\dagger} \in \Gamma(K)$, $z := -y^{\dagger} \in \Lambda(K)$. Then $s = -a'/a^2 = -y/a$, hence $-s^{\dagger} = -y^{\dagger} + a^{\dagger} = z + y$ and thus, using Corollary 5.2.2,

$$\sigma(s) \;=\; \omega(z+y) + (y/a)^2 \;=\; \omega(z) - y^2 + (y/a)^2 \;<\; \omega(z). \qquad \square$$

COROLLARY 11.8.32. *Suppose K is Liouville closed, and for every $a \in K$ the operator $\partial^2 - a \in K[\partial]$ splits over $K[i]$. Then $K \setminus \omega(K)^{\downarrow} = \sigma\big(\Gamma(K)\big)$. In particular, $\sigma\big(\Gamma(K)\big)$ is upward closed and K is ω-free.*

PROOF. From (5.2.1) we get $K = \omega(K) \cup \sigma(K^{\times})$, so

$$K \setminus \omega(K)^{\downarrow} \;=\; \sigma(K^{\times}) \setminus \omega(K)^{\downarrow} \;=\; \sigma\big(\Gamma(K)\big)$$

by Lemma 11.8.31. The rest now follows, using Corollary 11.8.30. $\square$

In Section 14.2 we consider additional restrictions on K ensuring that $\omega(K)$ is downward closed. Figure 11.2 is a sketch of the functions ω on $\Lambda(\mathbb{T})$ and σ on $\Gamma(\mathbb{T})$.

COROLLARY 11.8.33. *The following are equivalent for a Liouville closed H-field K:*

(i) $K = \omega\big(\Lambda(K)\big) \cup \sigma\big(\Gamma(K)\big)$;

(ii) *K is ω-free, $\omega\big(\Lambda(K)\big)$ is downward closed, and $\sigma\big(\Gamma(K)\big)$ is upward closed;*

(iii) *for every $a \in K$ the operator $\partial^2 - a \in K[\partial]$ splits over $K[i]$, and $\omega(K)$ is downward closed.*

PROOF. From (5.2.1) and $\omega(K) < \sigma\big(\Gamma(K)\big)$ we get (i) $\Rightarrow$ (iii). The implication (iii) $\Rightarrow$ (ii) follows from Corollary 11.8.32, and (ii) $\Rightarrow$ (i) from Corollary 11.8.30. $\square$

We say that a pre-H-field K is **Schwarz closed** if K is Liouville closed and satisfies the equivalent conditions in the previous corollary. This terminology is motivated by the role of the functions ω and σ in the Schwarzian derivative; see Section 5.2.

LEMMA 11.8.34. *If K is Schwarz closed, then so is K^{ϕ} for all $\phi \in K^{>}$.*

PROOF. Let K be Schwarz closed and $\phi \in K^{>}$. Then K^{ϕ} is Liouville closed, and with $\delta = \phi^{-1}\partial$, every operator $\delta^2 - a \in K^{\phi}[\delta]$ $(a \in K^{\phi})$ splits over $K^{\phi}[i] = K[i]^{\phi}$. Let ω^{ϕ} be the map $z \mapsto -\big(2\delta(z) + z^2\big) \colon K^{\phi} \to K^{\phi}$. We saw in Section 11.7 that it plays the role of ω in the differential field K^{ϕ}. The computations there give

$$\omega^{\phi}(K^{\phi}) \;=\; \phi^{-2}\big(\omega(K) - \omega(-\phi^{\dagger})\big),$$

hence $\omega^{\phi}(K^{\phi})$ is downward closed. $\square$

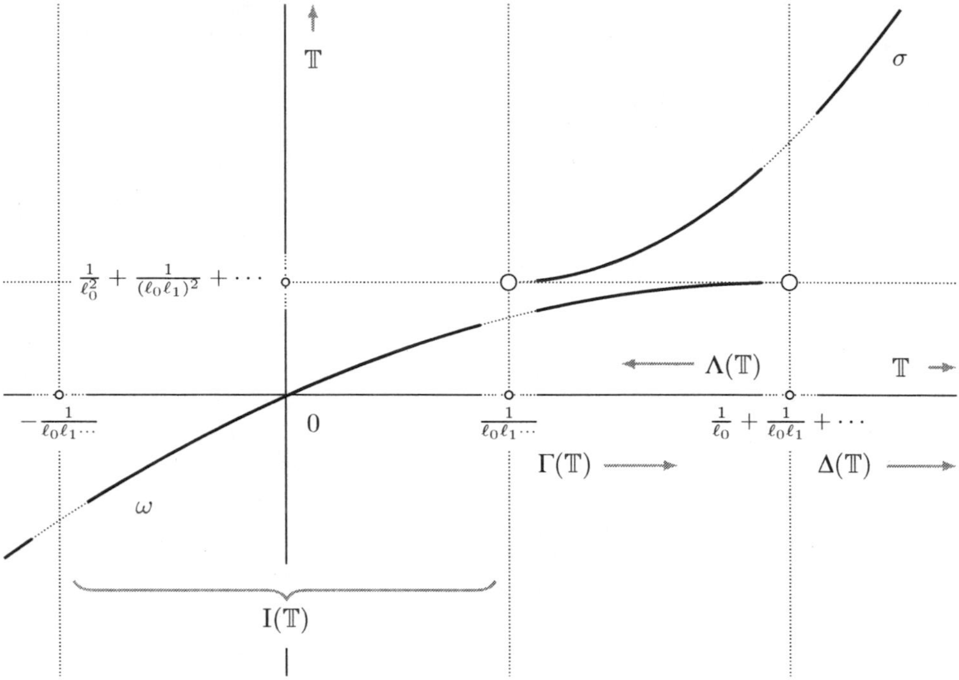

Figure 11.2: The functions ω and σ on $\mathbb{T}$.

LEMMA 11.8.35. *Let K and L be Schwarz closed H-subfields of an H-field M such that $C_L = C_M$. Then $K \cap L$ is a Schwarz closed H-subfield of M.*

PROOF. First, $K \cap L$ is a Liouville closed H-subfield of M by Lemma 10.6.1. It remains to show that $\omega\big(\Lambda(K)\big) \cap L \subseteq \omega\big(\Lambda(K \cap L)\big)$ and $\sigma\big(\Gamma(K)\big) \cap L \subseteq \sigma\big(\Gamma(K \cap L)\big)$. Let $f \in \omega\big(\Lambda(K)\big) \cap L$. Then $f = \omega(z_1)$ with $z_1 \in \Lambda(K)$. Since L is Schwarz closed,

$$f \in \omega\big(\Lambda(M)\big) \cap L \subseteq \omega\big(\Lambda(L)\big),$$

so $f = \omega(z_2)$ with $z_2 \in \Lambda(L)$. Now $\omega(z_1) = \omega(z_2)$ with $z_1, z_2 \in \Lambda(M)$, so

$$z := z_1 = z_2 \in K \cap L \cap \Lambda(K) = \Lambda(K \cap L),$$

where the last step uses Corollary 11.8.6. Hence $f = \omega(z) \in \omega\big(\Lambda(K \cap L)\big)$, as required. The other inclusion follows in the same way. □

LEMMA 11.8.36. *Suppose K is Schwarz closed and $(K_i)_{i \in I}$, $I \neq \emptyset$, is a family of Schwarz closed H-subfields of K, each with the same constant field as K. Then $F := \bigcap_{i \in I} K_i$ is a Schwarz closed H-subfield of K.*

PROOF. By Lemma 10.6.2, F is a Liouville closed H-subfield of K. We need to show that $F = \omega(\Lambda(F)) \cup \sigma(\Gamma(F))$, and for this it is enough to show that $F \cap \omega(\Lambda(K)) \subseteq \omega(\Lambda(F))$ and $F \cap \sigma(\Gamma(K)) \subseteq \sigma(\Gamma(F))$. Let $f \in F \cap \omega(\Lambda(K))$; then $f \notin \sigma(\Gamma(K))$. Hence for each $i \in I$ we have $f \in K_i \setminus \sigma(\Gamma(K_i)) = \omega(\Lambda(K_i))$, so we may take $z_i \in \Lambda(K_i)$ with $\omega(z_i) = f$. Since the map $z \mapsto \omega(z)\colon \Lambda(K) \to K$ is injective, we have $z_i = z_j$ for all $i, j \in I$. Denoting the common value of z_i $(i \in I)$ by z, we obtain $z \in F \cap \Lambda(K) = \Lambda(F)$ by Corollary 11.8.6, hence $f = \omega(z) \in \omega(\Lambda(F))$ as required. Similarly one shows that $F \cap \sigma(\Gamma(K)) \subseteq \sigma(\Gamma(F))$. □

Chapter Twelve

Triangular Automorphisms

Throughout this chapter K is a commutative ring containing $\mathbb{Q}$ as a subring.

When K carries also a derivation and $\phi \in K^{\times}$, then we saw in Section 5.7 that compositional conjugation by ϕ induces an automorphism of the K-algebra

$$K[Y_0, Y_1, Y_2, \dots],$$

where $Y_n := Y^{(n)}$ and Y is a differential indeterminate. These automorphisms are triangular (as defined below) and this facilitates their finer analysis by Lie techniques, as we show in the present chapter. An endomorphism σ of this K-algebra is uniquely determined by the sequence of images $\sigma(Y_0)$, $\sigma(Y_1)$, $\sigma(Y_2), \dots$, and is said to be *triangular* if for each n there are $\sigma_{0n}, \dots, \sigma_{nn} \in K$ with

$$\sigma(Y_n) \;=\; \sigma_{0n}Y_0 + \sigma_{1n}Y_1 + \cdots + \sigma_{nn}Y_n.$$

In Sections 12.2–12.7 of this chapter we introduce a formalism to analyze triangular automorphisms of such a polynomial algebra by means of their *logarithms*, the triangular derivations. In Section 12.8 we apply this to compositional conjugation in differential polynomial rings. From this chapter, only Corollaries 12.7.18 and 12.7.19 as well as Section 12.8 will be needed later.

12.1 FILTERED MODULES AND ALGEBRAS

In this preliminary section we collect some definitions and simple facts about filtered modules, filtered algebras, and graded algebras.

Filtered modules

Let V be a K-module. A **filtration** of V is a family $(V^i)_{i\in\mathbb{Z}}$ of K-submodules of V such that

(1) $\bigcup_i V^i = V$,

(2) $\bigcap_i V^i = \{0\}$, and

(3) $V^i \supseteq V^{i+1}$ for all i.

We call a filtration (V^i) of V **nonnegative** if $V^0 = V$; equivalently, $V^i = V$ for every $i \leqslant 0$. In this case we also call (V^n) a nonnegative filtration. The **trivial** filtration of V is the nonnegative filtration (V^n) of V with $V^n = \{0\}$ for $n > 0$. If (V^i) is a filtration of V and W is a submodule of V, then $(V^i \cap W)$ is a filtration of W, called the filtration of W **induced** by the filtration (V^i).

Let (V^i) be a filtration of V. To keep notations simple we denote V with this filtration also by V; the combined object is called a **filtered K-module**, and a submodule of V equipped with the filtration induced by (V^i) is called a **filtered submodule** of V. The filtration (V^i) is a fundamental system of neighborhoods of 0 for a unique topology on V making the additive group of V a topological group. For $x \in V$, let $|x| \in \mathbb{R}^{\geqslant 0}$ be given by

$$|x| := 2^{-i} \text{ if } x \in V^i \setminus V^{i+1}, \quad |0| := 0,$$

so $|x| \leqslant 2^{-i} \Leftrightarrow x \in V^i$. Thus $|\cdot|$ is an ultrametric norm on V (and a bit more): for all $x, y \in V$ and $a \in K$ and nonzero $k \in \mathbb{Z}$,

$$|x| = 0 \iff x = 0, \quad |ax| \leqslant |x|, \quad |kx| = |x|, \quad |x + y| \leqslant \max(|x|, |y|).$$

The topology defined above is induced by the metric $(x, y) \mapsto |x - y|$ on V. We call the filtered K-module V **complete** if V is complete with respect to this metric.

A family $(v_\lambda)_{\lambda\in\Lambda}$ in V is said to be **summable** if there is a (necessarily unique) $v \in V$ with the following property: for every $\varepsilon > 0$ there is a finite $I(\varepsilon) \subseteq \Lambda$ such that $\left|v - \sum_{\lambda\in I} v_\lambda\right| < \varepsilon$ for all finite $I \subseteq \Lambda$ with $I \supseteq I(\varepsilon)$; in that case $v_\lambda \neq 0$ for only countably many $\lambda \in \Lambda$, we call this v the **sum** of the family (v_λ), and denote it by $\sum_{\lambda\in\Lambda} v_\lambda$, or by $\sum v_\lambda$ if the index set Λ is understood from the context. If (u_λ) and (v_λ) are summable families in V with the same index set and $a \in K$, then $(u_\lambda + v_\lambda)$ and (av_λ) are summable, with

$$\sum u_\lambda + v_\lambda = \sum u_\lambda + \sum v_\lambda, \qquad \sum av_\lambda = a \sum v_\lambda.$$

Suppose V is complete and $(v_\lambda)_{\lambda\in\Lambda}$ is a family in V. Then (v_λ) is summable iff for each $\varepsilon > 0$ we have $|v_\lambda| < \varepsilon$ for all but finitely many λ. In particular, a sequence (v_n) in V is summable iff $v_n \to 0$ as $n \to \infty$. Suppose also that Λ is the disjoint union of $\Lambda(1)$ and $\Lambda(2)$; then $(v_\lambda)_{\lambda\in\Lambda}$ is summable iff $(v_\lambda)_{\lambda\in\Lambda(1)}$ and $(v_\lambda)_{\lambda\in\Lambda(2)}$ are summable; moreover, if $(v_\lambda)_{\lambda\in\Lambda}$ is summable, then

$$\sum_{\lambda\in\Lambda} v_\lambda = \sum_{\lambda\in\Lambda(1)} v_\lambda + \sum_{\lambda\in\Lambda(2)} v_\lambda.$$

We leave it to the reader to state and prove a similar statement for $\Lambda = \bigcup_{i\in I} \Lambda(i)$ where I is any index set with $\Lambda(i) \cap \Lambda(j) = \emptyset$ for all $i \neq j$.

Let W be a filtered K-module, with filtration (W^j), and let $\Phi\colon V \to W$ be K-linear. Then Φ is continuous iff for each j there is an i such that $\Phi(V^i) \subseteq W^j$. Thus if Φ is continuous and V and W are complete, then Φ is "strongly additive" in the following sense: if (v_λ) is a summable family in V, then $\big(\Phi(v_\lambda)\big)$ is summable in W, and

$\Phi\left(\sum v_\lambda\right) = \sum \Phi(v_\lambda)$. For $d \in \mathbb{Z}$ we say that Φ is **of rank** d if $\Phi(V^i) \subseteq W^{i+d}$ for each $i \in \mathbb{Z}$, and we call Φ **ranked** if Φ is of rank d for some d. (For example, if V is a filtered submodule of W, then the natural inclusion $V \to W$ is of rank 0.) If Φ is ranked, then Φ is continuous. If Φ is of rank $d \geqslant 0$, then Φ is of rank 0. If $\Phi\colon V \to V$ is of rank d, then Φ^n is of rank nd; here and below, Φ^n denotes the n-fold composition of Φ with itself: $\Phi^0 = \mathrm{id}$, $\Phi^{n+1} = \Phi \circ \Phi^n$ for each n.

In the rest of this subsection we assume that V is complete. We have the K-algebra $\mathrm{End}(V)$ of endomorphisms of V, whose elements are the K-linear maps $V \to V$, with multiplication given by composition of maps. The endomorphisms of V of rank 0 form a subalgebra of $\mathrm{End}(V)$. A family (Φ_λ) of endomorphisms of V is said to be **summable** if for each $v \in V$ the family $\big(\Phi_\lambda(v)\big)$ is summable in V; then the sum of (Φ_λ) is the endomorphism $\sum \Phi_\lambda$ of V given by

$$\left(\sum \Phi_\lambda\right)(v) := \sum \Phi_\lambda(v).$$

Below Φ ranges over $\mathrm{End}(V)$. We say that Φ is **topologically nilpotent** if for each j we have $\Phi^n(V) \subseteq V^j$ for all sufficiently large n. Note that if the filtration of V is nonnegative and Φ is of positive rank, then Φ is topologically nilpotent. We say that Φ is **weakly nilpotent** if for each $v \in V$ we have $\Phi^n(v) \to 0$ as $n \to \infty$. For example, if Φ is of positive rank or topologically nilpotent, then Φ is weakly nilpotent. If Φ is weakly nilpotent, then for each formal power series $f = \sum f_n z^n \in K[[z]]$ ($f_n \in K$ for all n), the sequence $(f_n \Phi^n)$ of endomorphisms of V is summable, and we set

$$f(\Phi) := \sum f_n \Phi^n \in \mathrm{End}(V).$$

We equip the K-module $K[[z]]$ with the nonnegative filtration $\big(z^n K[[z]]\big)$. Then $K[[z]]$ is complete. If Φ is weakly nilpotent and the family (f_λ) in $K[[z]]$ is summable, then the family $\big(f_\lambda(\Phi)\big)$ in $\mathrm{End}(V)$ is summable, and

$$\left(\sum f_\lambda\right)(\Phi) = \sum f_\lambda(\Phi).$$

In Section 12.2 we use the following.

LEMMA 12.1.1. *Suppose Φ is continuous and weakly nilpotent. Then:*

(i) *the map $f \mapsto f(\Phi)\colon K[[z]] \to \mathrm{End}(V)$ is a morphism of K-algebras;*

(ii) *if Φ is of rank 0, then $f(\Phi) \in \mathrm{End}(V)$ is of rank 0, for all $f \in K[[z]]$;*

(iii) *for $g \in zK[[z]]$, the endomorphism $g(\Phi)$ of V is weakly nilpotent and*

$$(f \circ g)(\Phi) = f\big(g(\Phi)\big) \text{ for all } f \in K[[z]].$$

PROOF. It is easy to check that $f \mapsto f(\Phi)$ is K-linear. Let $f, g \in K[[z]]$, $f = \sum_m f_m z^m$, $g = \sum_n g_n z^n$ with $f_m, g_n \in K$ for all m, n. Then for each $v \in V$,

$$\begin{aligned}\big(f(\Phi)\circ g(\Phi)\big)(v) &= \sum_m f_m\,\Phi^m\left(\sum_n g_n\,\Phi^n(v)\right)\\ &= \sum_m\left(\sum_n f_m g_n\,\Phi^{m+n}(v)\right)\\ &= \sum_{k=0}^{\infty}\left(\sum_{m+n=k} f_m g_n\right)\Phi^k(v) \;=\; (f\cdot g)(\Phi)(v),\end{aligned}$$

where we used continuity of Φ^m for the second equality. This shows (i).

To get (ii), just use that each V^i is closed. For (iii), let $g = \sum_{i\geqslant 1} g_i z^i$ where $g_i \in K$ for all $i \geqslant 1$. Note that $g(\Phi)^n = g^n(\Phi)$ by (i), and

$$g^n \;=\; g_1^n z^n + \text{higher degree terms in } z.$$

Let $v \in V$ and $j \geqslant 1$, and take $n_1 \geqslant 1$ such that $\Phi^n(v) \in V^j$ for all $n \geqslant n_1$. Then for all $n \geqslant n_1$ we have $g^n(\Phi)(v) = g_1^n \Phi^n(v) + \cdots \in V^j$, and so $g(\Phi)^n(v) \in V^j$. Thus $g(\Phi)$ is weakly nilpotent. Let also $f = \sum f_n z^n$ where $f_n \in K$ for all n. Then the family $(f_n g^n)$ in $K[[z]]$ is summable, and $f \circ g = \sum f_n g^n$ in $K[[z]]$, so

$$(f\circ g)(\Phi) \;=\; \left(\sum f_n g^n\right)(\Phi) \;=\; \sum f_n g^n(\Phi) \;=\; \sum f_n g(\Phi)^n \;=\; f\big(g(\Phi)\big)$$

as claimed. □

Thus if Φ is continuous and weakly nilpotent, and $f, g \in K[[z]]$ are such that $fg = 1$, then $f(\Phi)$ is a K-module automorphism of V with inverse $g(\Phi)$.

Filtered algebras

Throughout the rest of this section A is a (not necessarily commutative) K-algebra with $1 \neq 0$. As usual, this includes the requirement that

$$\lambda(xy) = (\lambda x)y = x(\lambda y) \qquad (\lambda \in K,\ x, y \in A).$$

So $\lambda \mapsto \lambda 1\colon K \to A$ is ring morphism taking values in center(A); via this morphism we identify $\mathbb{Q}$ with a subring of center(A). A **filtration** of the K-algebra A is a filtration $(A^i)_{i\in\mathbb{Z}}$ of the K-module A with $1 \in A^0$ and $A^i A^j \subseteq A^{i+j}$ for all i, j. Thus A^0 is a K-subalgebra of A, and the A^i are (left and right) A^0-submodules of A. Given a nonnegative filtration (A^n) of A, each A^n is a two-sided ideal of A.

Let (A^i) be a filtration of the K-algebra A. To keep notations simple we denote A with this filtration also by A; the combined object is called a **filtered K-algebra**. A subalgebra of A with the induced filtration is then also a filtered K-algebra. The norm on the additive group of A obtained from the filtration satisfies $|xy| \leqslant |x|\cdot|y|$ for all

$x, y \in A$, so A with the topology given by the filtration is a topological ring. If $(u_\lambda)_{\lambda\in\Lambda}$ and $(v_\omega)_{\omega\in\Omega}$ are summable families in A (with respect to this filtration), then so is the family $(u_\lambda v_\omega)$ indexed by $\Lambda \times \Omega$ and

$$\sum u_\lambda v_\omega = \sum u_\lambda \cdot \sum v_\omega.$$

EXAMPLE. Let R be a K-algebra with $1_R \neq 0$, and let $A := R[[z]]$ be the K-algebra of power series in one commuting indeterminate z with coefficients in R. (As usual we identify R with a subring of A via $r \mapsto rz^0$.) Then $(z^n A)$ is a nonnegative filtration of A making A a complete filtered K-algebra.

Logarithms

Let A be a complete filtered K-algebra with respect to the filtration (A^i). Then every sequence (a_n) in A with $a_n \in A^n$ for each n is summable in A. In particular, given a formal power series $f = \sum_n f_n z^n \in K[[z]]$ ($f_n \in K$ for each n) and $a \in A^1$, the sequence $(f_n a^n)$ is summable in A, and we set $f(a) := \sum f_n a^n$. Here is an analogue of Lemma 12.1.1:

LEMMA 12.1.2. *Let $a \in A^1$. Then the map $f \mapsto f(a)\colon K[[z]] \to A$ is a K-algebra morphism of rank 0. If $f \in K[[z]]$, $g \in zK[[z]]$, then the element $f \circ g \in K[[z]]$ satisfies $(f \circ g)(a) = f\big(g(a)\big)$. Also $f(bab^{-1}) = bf(a)b^{-1}$ for $f \in K[[z]]$, $b \in A^\times$.*

As a consequence, each element of $1 + A^1$ is a unit in A: if $a \in A^1$, then $1 - a$ has a multiplicative (two-sided) inverse given by $(1-a)^{-1} = \sum_{i=0}^\infty a^i$. Thus the multiplicative group $1 + A^1$ is a subgroup of the group $A^\times$ of units of A. Applying Lemma 12.1.2 to the series $\exp(z) = \sum z^n/n!$ and $\log(1+z) = \sum_{n=1}^\infty (-1)^{n+1} z^n/n$ we see that the maps

$$\begin{aligned} A^1 &\to 1 + A^1 : a \mapsto \exp(a) := \sum \frac{a^n}{n!}, \\ 1 + A^1 &\to A^1 : a \mapsto \log(a) := \sum_{n=1}^\infty (-1)^{n+1} \frac{(a-1)^n}{n} \end{aligned}$$

are each other's inverse, and thus bijective. Note that $\exp(0) = 1$.

LEMMA 12.1.3. *Let $a, b \in A$ and $ab = ba$. Then:*

(i) $\exp(a+b) = \exp(a)\exp(b)$ *for $a, b \in A^1$;*
(ii) $\log(ab) = \log(a) + \log b$ *for $a, b \in 1 + A^1$;*
(iii) $(\log a)(\log b) = (\log b)(\log a)$ *for $a, b \in 1 + A^1$.*

PROOF. We leave (i) and (ii) as routine verifications. As to (iii), this follows from the (almost obvious) fact that for $f, g \in K[[z]]$ and $s, t \in A^1$ with $st = ts$ we have $f(s)g(t) = g(t)f(s)$: for $a, b \in 1 + A^1$ we have $a = 1 + s$, $b = 1 + t$ with $s, t \in A^1$, and then $ab = ba$ gives $st = ts$, and so $\log(1+s)\log(1+t) = \log(1+t)\log(1+s)$. □

The mutually inverse nature of exp and log gives

$$\exp(A^n) = 1 + A^n, \qquad \log(1 + A^n) = A^n \qquad \text{for } n \geqslant 1.$$

For $r \in K$ and $a \in 1 + A^1$ we set $a^r := \exp(r \log a) \in 1 + A^1$. Thus for $a \in 1 + A^1$ and $r, s \in K$ we have $\log(a^r) = r \log a$ and

$$a^0 = 1, \qquad a^{r+s} = a^r \cdot a^s, \qquad a^{-r} = (a^r)^{-1}, \qquad a^n = \underbrace{a \cdots a}_{n \text{ times}}.$$

Also, $\exp(ra) = \exp(a)^r$ for $a \in A^1$ and $r \in K$, and

$$(ab)^r = a^r b^r \qquad \text{for } a, b \in 1 + A^1 \text{ with } ab = ba \text{ and } r \in K.$$

We also write e^a instead of $\exp(a)$ when $a \in A^1$. Thus $\mathrm{e}^z = \exp(z) = \sum z^n/n!$ in the filtered K-algebra $K[[z]]$ of the example above.

Lie algebras and filtered Lie algebras

Recall that a Lie algebra over K is a K-module L equipped with a K-bilinear operation $[\ ,\ ]\colon L \times L \to L$ (the Lie bracket of L) satisfying $[x, x] = 0$ for all $x \in L$, as well as the Jacobi Identity:

$$[x, [y, z]] + [y, [z, x]] + [z, [x, y]] = 0 \qquad \text{for all } x, y, z \in L.$$

(Thus $[x, y] = -[y, x]$ for $x, y \in L$, in view of $[x + y, x + y] = 0$.) A Lie algebra L over K is said to be **abelian** if $[x, y] = 0$ for all $x, y \in L$. Given a Lie algebra L over K, a **Lie subalgebra** of L is a K-submodule M of L such that $[x, y] \in M$ for all $x, y \in M$, and an **ideal** of L is a K-submodule M of L such that $[x, y] \in M$ for all $x \in L$ and $y \in M$. A **filtration** of a Lie algebra L over K is a filtration $(L^i)_{i\in\mathbb{Z}}$ of L as a K-module such that additionally $[L^i, L^j] \subseteq L^{i+j}$ for all indices i, j.

Let $(L^i)_{i\in\mathbb{Z}}$ be a filtration of the Lie algebra L over K. If $i \geqslant 0$, then $[L^i, L^i] \subseteq L^i$, so L^i is a Lie subalgebra of L. If the filtration (L^i) of L is nonnegative, i.e., if $L^0 = L$, then each L^i is even an ideal of L, since then $[L, L^i] = [L^0, L^i] \subseteq L^i$. The norm on A defined by the filtration satisfies $|[x, y]| \leqslant |x| \cdot |y|$ for all $x, y \in L$, and so the Lie bracket operation $[\ ,\ ]\colon L \times L \to L$ is continuous with respect to the topology on L given by the filtration (with the product topology on $L \times L$).

EXAMPLE. The binary operation on A given by

$$[a, b] := ab - ba \qquad (a, b \in A)$$

turns A into a Lie algebra A_{Lie} over K. Every filtration of the K-algebra A is also a filtration of A_{Lie}.

A **filtered Lie algebra over** K is a Lie algebra L over K together with a filtration of L. Let (L^i) be the filtration of a filtered Lie algebra L over K. Given a Lie subalgebra M of L, $(L^i \cap M)$ is a filtration of M, called the filtration of M **induced** by (L^i). A Lie subalgebra of L equipped with the filtration induced by (L^i) is called a **filtered Lie subalgebra** of L.

Derivations

A K-**derivation** on A is a K-linear map $\Delta\colon A \to A$ such that

$$\Delta(xy) = \Delta(x)y + x\Delta(y) \qquad \text{for all } x, y \in A$$

(and thus $\Delta(\lambda \cdot 1) = 0$ for $\lambda \in K$).

EXAMPLE. Given $a \in A$, the adjoint $x \mapsto \mathrm{ad}_a(x) := [a, x] = ax - xa$ is a K-derivation on A. The K-derivations ad_a on A (where $a \in A$) are called **internal.** If A is a filtered K-algebra and $a \in A^d$, then ad_a is of rank d.

We denote the set of all K-derivations on A by $\mathrm{der}_K(A)$. If $\Delta_1, \Delta_2 \in \mathrm{der}_K(A)$ and $r \in K$, then $\Delta_1 + \Delta_2 \in \mathrm{der}_K(A)$ and $r\Delta_1 \in \mathrm{der}_K(A)$, and so $\mathrm{der}_K(A)$ is naturally a left K-module. One also verifies easily that if $\Delta_1, \Delta_2 \in \mathrm{der}_K(A)$, then

$$[\Delta_1, \Delta_2] := \Delta_1\Delta_2 - \Delta_2\Delta_1 \in \mathrm{der}_K(A).$$

With this operation $[\ ,\ ]$, the K-module $\mathrm{der}_K(A)$ is a Lie algebra over K, in fact, a Lie subalgebra of the Lie algebra $\mathrm{End}(A)_{\mathrm{Lie}}$ over K, where $\mathrm{End}(A)$ is the K-algebra of endomorphisms of A as a K-module. If σ is an automorphism of the K-algebra A, then for every $\Delta \in \mathrm{der}_K(A)$ we have $\sigma\Delta\sigma^{-1} \in \mathrm{der}_K(A)$, and $\Delta \mapsto \sigma\Delta\sigma^{-1}$ is an automorphism of the Lie algebra $\mathrm{der}_K(A)$ over K, with inverse $\Delta \mapsto \sigma^{-1}\Delta\sigma$.

Given $\Delta \in \mathrm{der}_K(A)$, we have for all $x, y \in A$,

$$(12.1.1) \qquad \Delta^n(xy) = \sum_{i+j=n} \binom{n}{i} \Delta^i(x)\Delta^j(y) \qquad \text{(Leibniz rule)},$$

and more generally, for $m \geqslant 1$ and $x_1, \dots, x_m \in A$,

$$\Delta^n(x_1 \cdots x_m) = \sum_{i_1+\cdots+i_m=n} \frac{n!}{i_1! \cdots i_m!} \Delta^{i_1}(x_1) \cdots \Delta^{i_m}(x_m).$$

We shall often use the following facts, the second of which follows from (12.1.1) and the remark after Lemma 12.1.1:

LEMMA 12.1.4. *Let A be a complete filtered K-algebra with respect to the filtration (A^i). Suppose $\Delta \in \mathrm{der}_K(A)$ is continuous. Then*

(i) *if A is commutative and $a \in A^1$, then $\Delta(\mathrm{e}^a) = \mathrm{e}^a\,\Delta(a)$;*

(ii) *if Δ is weakly nilpotent, then the K-module endomorphism $\mathrm{e}^\Delta := \exp(\Delta)$ of A is a K-algebra automorphism of A with inverse $\mathrm{e}^{-\Delta}$.*

Graded algebras

A **grading** of A is a family $(A_i)_{i\in\mathbb{Z}}$ of K-submodules of A such that the following two conditions are satisfied:

(1) $A = \bigoplus_{i\in\mathbb{Z}} A_i$ (internal direct sum of K-submodules of A);

(2) $A_i A_j \subseteq A_{i+j}$ for all $i, j \in \mathbb{Z}$.

A little argument shows that then $1 \in A_0$, so A_0 is a K-subalgebra of A, and each A_i is a left-and-right A_0-submodule of A. A **graded K-algebra** is a K-algebra A together with a grading of A. Let A be a graded K-algebra and let (A_i) be its grading. The elements of A_i are said to be **homogeneous of degree** i. For every $a \in A$ there is a unique family (a_i) with $a_i \in A_i$ for each i and $a_i = 0$ for all but finitely many i such that $a = \sum_i a_i$. For this family (a_i) we call a_i the **homogeneous part of a of degree** i. For $a \in A$, $a \neq 0$, we define the **degree of** a as the largest i such that $a_i \neq 0$, denoted by $\mathrm{d}(a)$, or by $\mathrm{d}_{(A_i)}(a)$ if we want to indicate the dependence on (A_i). We also set $\mathrm{d}(0) := -\infty < \mathbb{Z}$. The grading (A_i) of A is said to be **nonnegative** if $A_i = \{0\}$ for all $i < 0$. From the grading (A_i) we obtain the filtration (A^i) of the K-algebra A by

$$A^i := \bigoplus_{j\geqslant i} A_j,$$

the filtration of A **associated** to (A_i). Clearly (A_i) is nonnegative iff (A^i) is.

Gradings of polynomial algebras

In the rest of this section $A = K[Y_0, Y_1, \dots]$ is the (commutative) K-algebra of polynomials in the distinct indeterminates Y_n, $n = 0, 1, 2, \dots$, and $\boldsymbol{i}$ ranges over the set $\mathbb{N}^{(\mathbb{N})}$ of sequences $\boldsymbol{i} = (i_0, i_1, \dots) \in \mathbb{N}^{\mathbb{N}}$ such that $i_n = 0$ for all but finitely many n. For each $\boldsymbol{i}$ we set

$$Y^{\boldsymbol{i}} := Y_0^{i_0} Y_1^{i_1} \cdots Y_n^{i_n} \cdots \in A.$$

Given $P \in A$ we have a unique family $(P_{\boldsymbol{i}})$ in K such that $P = \sum_{\boldsymbol{i}} P_{\boldsymbol{i}} Y^{\boldsymbol{i}}$ with $P_{\boldsymbol{i}} = 0$ for all but finitely many $\boldsymbol{i}$. Below, $Y^{\diamond} := \{Y^{\boldsymbol{i}} : \boldsymbol{i} \in \mathbb{N}^{(\mathbb{N})}\}$ is the multiplicative monoid of monomials.

DEFINITION 12.1.5. A **degree function on** A is a function $\mathrm{d}\colon Y^{\diamond} \to \mathbb{Z}$ such that $\mathrm{d}(1) = 0$ and $\mathrm{d}(s \cdot t) = \mathrm{d}(s) + \mathrm{d}(t)$ for all $s, t \in Y^{\diamond}$.

EXAMPLE. Given a sequence (d_n) of integers, define

$$\mathrm{d}(Y^{\boldsymbol{i}}) := \sum_n d_n i_n \in \mathbb{Z} \qquad \text{for each } \boldsymbol{i}.$$

Then d is a degree function on A. Any degree function d on A arises from a sequence (d_n) of integers in this manner, by setting $d_n := \mathrm{d}(Y_n)$ for each n.

Any degree function d on A yields a grading $(A_d)_{d\in\mathbb{Z}}$ of A,

$$A_d := \{P \in A :\ P_{\boldsymbol{i}} = 0 \text{ if } \mathrm{d}(Y^{\boldsymbol{i}}) \neq d\} = \sum_{\mathrm{d}(Y^{\boldsymbol{i}})=d} KY^{\boldsymbol{i}}.$$

Suppose (A_i) is a grading of A for which each indeterminate Y_n is homogeneous. This grading is induced by a degree function on A as above, namely the restriction of $\mathrm{d}_{(A_i)}$ to $Y^\diamond$. To simplify notation denote $\mathrm{d}_{(A_i)}$ by d. Then

$$\mathrm{d}(P) \;=\; \max\{\mathrm{d}(Y^{\boldsymbol{i}}) : P_{\boldsymbol{i}} \neq 0\} \in \mathbb{Z} \text{ if } P \neq 0, \qquad \mathrm{d}(0) = -\infty < \mathbb{Z},$$

and the elements of A_i are said to be **d-homogeneous of degree** i. Given $P \in A$ and $i \in \mathbb{Z}$, the homogeneous part of P of degree i with respect to (A_i) is called the **d-homogeneous part of** P **of degree** i. Given $d \in \mathbb{Z}$, a K-linear map $\Phi\colon A \to A$ is said to be **d-homogeneous of degree** d if $\Phi(A_i) \subseteq A_{i+d}$ for all i; given also a d-homogeneous K-linear map $\Psi\colon A \to A$ of degree $e \in \mathbb{Z}$, the K-linear map $\Psi \circ \Phi\colon A \to A$ is d-homogeneous of degree $d + e$.

Clearly the grading induced by a degree function d on A is nonnegative if and only if $\mathrm{d}(Y_n) \geqslant 0$ for each n; in this case we say that d is **nonnegative**. We now define two important nonnegative degree functions on A:

EXAMPLE. The usual (total) degree

$$\deg(Y^{\boldsymbol{i}}) \;:=\; |\boldsymbol{i}| \;=\; i_0 + i_1 + i_2 + \cdots + i_n + \cdots \in \mathbb{N}$$

yields the degree function deg on A. For $P \in A$ we have

$$\deg(P) \;=\; \max\{|\boldsymbol{i}| : P_{\boldsymbol{i}} \neq 0\} \in \mathbb{N} \text{ if } P \neq 0, \qquad \deg(0) = -\infty < \mathbb{Z}.$$

We denote the grading associated to deg by (A_d). In the rest of this chapter, the term **homogeneous** (no mention of a degree function or grading) is synonymous with deg-*homogeneous*. For $P \in A$ and $d \in \mathbb{N}$ we let

$$P_d \;=\; \sum_{|\boldsymbol{i}|=d} P_{\boldsymbol{i}} Y^{\boldsymbol{i}} \in A_d$$

be the homogeneous part of P of degree d. Thus $A_d = \{P \in A : P = P_d\}$ for each $d \in \mathbb{N}$. For example, $A_0 = K$, $A_1 = \bigoplus_n K Y_n$, $A_2 = \bigoplus_{m \leqslant n} K Y_m Y_n$.

EXAMPLE. Setting

$$\mathrm{wt}(Y^{\boldsymbol{i}}) \;:=\; \|\boldsymbol{i}\| \;=\; i_1 + 2i_2 + \cdots + n i_n + \cdots \in \mathbb{N}$$

gives a degree function wt on A. Note that $\mathrm{wt}(Y_i) = i$ for each $i \in \mathbb{N}$. For $P \in A$ we call $\mathrm{wt}(P)$ the **weight** of P:

$$\mathrm{wt}(P) \;=\; \max\{\|\boldsymbol{i}\| : P_{\boldsymbol{i}} \neq 0\} \in \mathbb{N} \text{ if } P \neq 0, \qquad \mathrm{wt}(0) = -\infty < \mathbb{Z}.$$

We denote the grading associated to wt by $(A_{[w]})$. In the rest of this chapter, **isobaric** is synonymous with wt-*homogeneous*. For $P \in A$ and $w \in \mathbb{N}$ we let

$$P_{[w]} \;=\; \sum_{\|\boldsymbol{i}\|=w} P_{\boldsymbol{i}} Y^{\boldsymbol{i}} \in A_{[w]}$$

be the **isobaric part of** P **of weight** w. Thus

$$A_{[w]} \;=\; \{P \in A : P = P_{[w]}\} \qquad \text{for each } w \in \mathbb{N},$$

so $A_{[0]} = K[Y_0]$, $A_{[1]} = K[Y_0]\,Y_1$, $A_{[2]} = K[Y_0]Y_1^2 + K[Y_0]Y_2$.

If K is equipped with a derivation, Y is a differential indeterminate over K, and the K-algebra $K\{Y\} = K[Y, Y', \dots]$ is identified with $K[Y_0, Y_1, \dots]$ by setting $Y_n = Y^{(n)}$, then these notions of degree and weight agree with the ones for differential polynomials introduced in Section 4.2.

12.2 TRIANGULAR LINEAR MAPS

In this section V is a K-module. We equip V with the trivial filtration. This makes V a complete filtered K-module. Recall from Section 12.1 that $\operatorname{End}(V)$ is the K-algebra of endomorphisms of V. Thus a family $(\Phi_i)_{i\in I}$ of endomorphisms of V is summable (as defined in Section 12.1) iff for each $v \in V$ we have $\Phi_i(v) = 0$ for all but finitely many i; recall that then $\sum \Phi_i$ is the endomorphism of V given by

$$\left(\sum \Phi_i\right)(v) = \sum \Phi_i(v).$$

If $(\Phi_i)_{i\in I}$ and $(\Psi_j)_{j\in J}$ are summable families of endomorphisms of V, then so is $(\Phi_i\Psi_j)_{(i,j)\in I\times J}$ with

$$\left(\sum_i \Phi_i\right)\left(\sum_j \Psi_j\right) = \sum_{i,j} \Phi_i\Psi_j.$$

Throughout this section $\Phi, \Psi \in \operatorname{End}(V)$.

Locally nilpotent and locally unipotent endomorphisms

We call Φ **locally nilpotent** if for every $v \in V$ there is some n such that $\Phi^n(v) = 0$. Note that Φ is locally nilpotent iff Φ is weakly nilpotent with respect to the trivial filtration on V. If Φ is locally nilpotent, then so is every power Φ^n with $n \geqslant 1$ and every scalar multiple $\lambda\Phi$, and for each family (λ_n) of scalars the family $(\lambda_n\Phi^n)$ of endomorphisms is summable, and $\sum_{n\geqslant 1} \lambda_n\Phi^n$ is locally nilpotent. We call Φ **locally unipotent** if $\Phi - 1$ is locally nilpotent. If σ is an automorphism of the K-module V, and Φ is locally nilpotent, respectively locally unipotent, then so is $\sigma\Phi\sigma^{-1}$.

In the rest of this subsection we assume that Φ is locally nilpotent. Then $(\frac{1}{n!}\Phi^n)$ is summable, so we can define the endomorphism $\exp\Phi$ of V by

$$\exp\Phi := \sum_{n=0}^{\infty} \frac{1}{n!}\Phi^n = 1 + \Phi + \frac{1}{2}\Phi^2 + \cdots + \frac{1}{n!}\Phi^n + \cdots.$$

Since $\exp\Phi - 1 = \sum_{n=1}^{\infty} \frac{1}{n!}\Phi^n$ is locally nilpotent, $\exp\Phi$ is locally unipotent.

Conversely, assume Ψ is locally unipotent. Then we define the locally nilpotent endomorphism $\log\Psi$ of V by

$$\log\Psi := \sum_{n=1}^{\infty} \frac{(-1)^{n+1}}{n}(\Psi - 1)^n = (\Psi - 1) - \frac{1}{2}(\Psi - 1)^2 + \frac{1}{3}(\Psi - 1)^3 - \cdots.$$

We have $\exp(\log \Psi) = \Psi$ and $\log(\exp \Phi) = \Phi$. If Φ_1, Φ_2 are commuting locally nilpotent endomorphisms of V, then $\Phi_1\Phi_2$ and $\Phi_1 + \Phi_2$ are locally nilpotent and

$$\exp(\Phi_1)\exp(\Phi_2) = \exp(\Phi_1 + \Phi_2).$$

Thus $\exp \Phi$ is an automorphism of the K-module V with inverse $\exp(-\Phi)$, and

$$(\exp \Phi)^k = \exp(k\Phi) \qquad (k \in \mathbb{Z}).$$

If σ is an automorphism of the K-module V, then

$$\exp(\sigma\Phi\sigma^{-1}) = \sigma\exp(\Phi)\sigma^{-1}, \qquad \log(\sigma\Psi\sigma^{-1}) = \sigma\log(\Psi)\sigma^{-1}.$$

We also note that for $v \in V$ we have

(12.2.1) $$\Phi(v) = 0 \iff (\exp \Phi)(v) = v.$$

Triangular matrices

We construe $K^{\mathbb{N}\times\mathbb{N}}$ as a K-module with the componentwise addition and scalar multiplication. The elements $M = (M_{ij})_{i,j\in\mathbb{N}}$ of $K^{\mathbb{N}\times\mathbb{N}}$ may be visualized as infinite square matrices with entries in K:

$$M = \begin{pmatrix} M_{00} & M_{01} & M_{02} & \cdots \\ M_{10} & M_{11} & M_{12} & \cdots \\ M_{20} & M_{21} & M_{22} & \cdots \\ \vdots & \vdots & \vdots & \ddots \end{pmatrix}.$$

We say that $M = (M_{ij}) \in K^{\mathbb{N}\times\mathbb{N}}$ is **column-finite** if for each j there are only finitely many i with $M_{ij} \neq 0$. Given column-finite matrices $M = (M_{ij})$ and $N = (N_{ij})$ we can define their matrix product $MN \in K^{\mathbb{N}\times\mathbb{N}}$ by

$$(MN)_{ij} := \sum_k M_{ik}N_{kj}.$$

Then MN is again column-finite. With this product operation, the K-submodule of $K^{\mathbb{N}\times\mathbb{N}}$ consisting of all column-finite matrices is a K-algebra with multiplicative identity 1 given by the identity matrix. We identify K with a subring of this K-algebra via $\lambda \mapsto \lambda \cdot 1$.

We say that $M = (M_{ij}) \in K^{\mathbb{N}\times\mathbb{N}}$ is (upper) **triangular** if $M_{ij} = 0$ for all i, j with $i > j$. The set $\mathfrak{tr}_K$ of triangular matrices is a subalgebra of the K-algebra of column-finite matrices. For every n we set

$$\mathfrak{tr}_K^n := \{M \in \mathfrak{tr}_K : M_{ij} = 0 \text{ for all } i, j \text{ with } j < i + n\},$$

so $\mathfrak{tr}_K^0 = \mathfrak{tr}_K$, and $\mathfrak{tr}_K^1 = \{M \in \mathfrak{tr}_K : M_{ii} = 0 \text{ for all } i\}$. It is easily verified that $(\mathfrak{tr}_K^n)$ is a complete nonnegative filtration of the K-algebra $\mathfrak{tr}_K$; in particular, each $\mathfrak{tr}_K^n$ is an ideal of the Lie algebra $(\mathfrak{tr}_K)_{\text{Lie}}$. We say that $M \in K^{\mathbb{N}\times\mathbb{N}}$ is **diagonal** if $M_{ij} = 0$ for all $i \neq j$. Then

$$D_K := \{M \in K^{\mathbb{N}\times\mathbb{N}} : M \text{ is diagonal}\}$$

is a (commutative) subalgebra of the K-algebra $\mathfrak{tr}_K$. For $M \in \mathfrak{tr}_K$ we define the matrix $M_0 \in D_K$ by $(M_0)_{ii} = M_{ii}$. Then $M \mapsto M_0 \colon \mathfrak{tr}_K \to D_K$ is a K-algebra morphism that is the identity on D_K. The multiplicative group of units of D_K is

$$D_K^\times = \{M \in D_K : M_{ii} \in K^\times \text{ for all } i\}.$$

The group morphism $M \mapsto M_0 \colon \mathfrak{tr}_K^\times \to D_K^\times$ from the group $\mathfrak{tr}_K^\times$ of units of $\mathfrak{tr}_K$ onto $D_K^\times$ is the identity on $D_K^\times$ and has kernel $1+\mathfrak{tr}_K^1$, so $1+\mathfrak{tr}_K^1$ is a normal subgroup of $\mathfrak{tr}_K^\times$ with $(1+\mathfrak{tr}_K^1)D_K^\times = \mathfrak{tr}_K^\times$ and $(1+\mathfrak{tr}_K^1) \cap D_K^\times = \{1\}$. Thus $\mathfrak{tr}_K^\times$ is the internal semidirect product of $(1+\mathfrak{tr}_K^1)$ with $D_K^\times$. We set $\mathcal{U} := 1+\mathfrak{tr}_K^1$ (also denoted by $\mathcal{U}_K$ if we need to indicate the dependence on K) and call its elements **unitriangular** matrices. A **unitriangular** group over K is a subgroup of $\mathcal{U}$.

Triangular linear maps

We now assume that V is a free K-module on the basis (Y_n). In particular,

$$V = \bigoplus_n K\,Y_n \qquad \text{(internal direct sum of } K\text{-submodules of } V).$$

The set of K-linear maps $V \to V$, under (pointwise) addition and composition, forms the K-algebra $\operatorname{End}(V)$. We say that Φ is **triangular** if

$$\Phi(Y_j) = \Phi_{0j}Y_0 + \Phi_{1j}Y_1 + \cdots + \Phi_{jj}Y_j \qquad \text{where } \Phi_{ij} \in K \text{ for } i,j \in \mathbb{N},\ i \leqslant j.$$

Triangular endomorphisms of the K-module V may be conveniently represented by triangular bi-infinite matrices with entries in K: for every triangular Φ define

$$M_\Phi := (\Phi_{ij})_{i,j\in\mathbb{N}} = \begin{pmatrix} \Phi_{00} & \Phi_{01} & \Phi_{02} & \Phi_{03} & \cdots \\ & \Phi_{11} & \Phi_{12} & \Phi_{13} & \cdots \\ & & \Phi_{22} & \Phi_{23} & \cdots \\ & & & \Phi_{33} & \cdots \\ & & & & \ddots \end{pmatrix} \in \mathfrak{tr}_K.$$

Here and below, given triangular Φ we set $\Phi_{ij} := 0$ for all $i,j \in \mathbb{N}$ with $i > j$. It is easily verified that if Φ and Ψ are triangular, then the composition $\Phi\Psi$ and the sum $\Phi+\Psi$ are also triangular, with $M_{\Phi\Psi} = M_\Phi \cdot M_\Psi$ and $M_{\Phi+\Psi} = M_\Phi + M_\Psi$. Hence the triangular endomorphisms of the K-module V form a K-subalgebra $\operatorname{tr}_K(V)$ of $\operatorname{End}(V)$, isomorphic to the K-algebra $\mathfrak{tr}_K$ via the isomorphism $\Phi \mapsto M_\Phi$. If the K-module V and the basis (Y_n) are clear from the context, we also abbreviate $\operatorname{tr}_K(V)$ as tr_K. For each n we set

$$\operatorname{tr}_K^n := \{\Phi \in \operatorname{tr}_K : M_\Phi \in \mathfrak{tr}_K^n\}.$$

Then (tr_K^n) is a complete nonnegative filtration of the K-algebra tr_K.

LEMMA 12.2.1. *If the sequence (Φ_n) in tr_K is summable in the sense of the filtration (tr_K^n), then (Φ_n) is summable as defined in the beginning of this section, and the sum $\sum \Phi_n \in \mathrm{tr}_K$ in the sense of the filtration (tr_K^n) equals $\sum \Phi_n$ as defined in the beginning of this section.*

We shall use this fact tacitly in what follows.

Suppose Φ is triangular and $\Phi_{ii} = 0$ for all i. Then $\Phi^n \in \mathrm{tr}_K^n$ for all n, so Φ is locally nilpotent. Moreover, $\exp \Phi$ is triangular with $(\exp \Phi)_{ii} = 1$ for all i, and $M_{\exp \Phi} = \exp M_\Phi$. We can reverse this as follows. Suppose Ψ is triangular and $\Psi_{ii} = 1$ for all i. Then $(\Psi - 1)^n \in \mathrm{tr}_K^n$ for all n, so Ψ is locally unipotent. Moreover, $\log \Psi$ is triangular with $(\log \Psi)_{ii} = 0$ for all i, and $M_{\log \Psi} = \log M_\Psi$.

Diagonals

Let $M \in \mathrm{tr}_K$. For each n we call the triangular matrix

$$M_n = \begin{pmatrix} 0 & \cdots & \cdots & \cdots & 0 & M_{0n} & 0 & \cdots & & \\ & 0 & \cdots & \cdots & \cdots & 0 & M_{1,n+1} & 0 & \cdots & \\ & & 0 & \cdots & \cdots & \cdots & 0 & M_{2,n+2} & 0 & \cdots \\ & & & \ddots & & & & \ddots & & \ddots \quad \ddots \end{pmatrix} \in \mathrm{tr}_K^n$$

the n-**diagonal** of M; i.e.,

$$(M_n)_{ij} = 0 \text{ if } j \neq i+n, \qquad (M_n)_{i,i+n} = M_{i,i+n}.$$

In the metric given by the filtration (tr_K^n) we have $\sum_{i=0}^n M_i \to M$ as $n \to \infty$, so (M_n) is summable with $M = M_0 + M_1 + \cdots + M_n + \cdots$. We say that M is n-**diagonal** if $M = M_n$. Thus M is diagonal as defined earlier iff M is 0-diagonal.

NOTATION. For a sequence $a = (a_i) \in K^{\mathbb{N}}$, define $\mathrm{diag}_n\, a \in \mathrm{tr}_K^n$ by

$$(\mathrm{diag}_n\, a)_{i,i+n} = a_i, \qquad (\mathrm{diag}_n\, a)_{ij} = 0 \ \text{ for } j \neq i+n.$$

Thus the n-diagonal matrices are precisely the matrices $\mathrm{diag}_n\, a$ with $a \in K^{\mathbb{N}}$. We also abbreviate $\mathrm{diag}_0\, a$ as $\mathrm{diag}\, a$.

The sum of two n-diagonal matrices is n-diagonal. As for products, we have:

LEMMA 12.2.2. *Let $M = \mathrm{diag}_m\, a$ be m-diagonal and $N = \mathrm{diag}_n\, b$ be n-diagonal, where $a = (a_i), b = (b_i) \in K^{\mathbb{N}}$. Then MN is $(m+n)$-diagonal, in fact*

$$MN = \mathrm{diag}_{m+n}(a_i \cdot b_{i+m})_i.$$

Therefore $[M, N]$ is $(m+n)$-diagonal, with

$$[M, N] = \mathrm{diag}_{m+n}(a_i \cdot b_{i+m} - b_i \cdot a_{i+n})_i,$$

and for each $k \in \mathbb{N}$, M^k is km-diagonal, with

$$M^k = \mathrm{diag}_{km}(a_i \cdot a_{i+m} \cdots a_{i+(k-1)m})_i.$$

COROLLARY 12.2.3. *Let $M = \operatorname{diag}_1 a$ where $a = (a_i) \in K^{\mathbb{N}}$. Then*

$$(\exp M)_{ij} \;=\; \frac{1}{(j-i)!}\, a_i \cdot a_{i+1} \cdots a_{j-1} \qquad \textit{for all } i \leqslant j.$$

Suppose $\Phi \in \operatorname{tr}_K$. For each n we call the triangular endomorphism $\Phi_n \in \operatorname{tr}_K^n$ of V with associated matrix $M_{\Phi_n} = (M_\Phi)_n$ the **n-diagonal** of Φ. We also say that Φ is **n-diagonal** if $\Phi = \Phi_n$ and **diagonal** if $\Phi = \Phi_0$. Thus (Φ_n) is summable with

$$\Phi \;=\; \Phi_0 + \Phi_1 + \cdots + \Phi_n + \cdots,$$

and $\Phi \in \operatorname{tr}_K^m$ if and only if $\Phi_0 = \cdots = \Phi_{m-1} = 0$. For $\Psi \in \operatorname{tr}_K$ we have

$$(\Phi\Psi)_n \;=\; \sum_{i+j=n} \Phi_i \Psi_j. \tag{12.2.2}$$

Hence if Ψ is a diagonal automorphism of V then

$$(\Psi\Phi\Psi^{-1})_n \;=\; \Psi\Phi_n\Psi^{-1}.$$

The identity (12.2.2) also implies, for $k \in \mathbb{N}$:

$$(\Phi^k)_n \;=\; \sum_{i_1+\cdots+i_k=n} \Phi_{i_1}\Phi_{i_2}\cdots\Phi_{i_k};$$

thus if $\Phi \in \operatorname{tr}_K^m$ and $n < mk$, then $(\Phi^k)_n = 0$. This immediately yields:

LEMMA 12.2.4. *Let $\Phi, \Psi \in \operatorname{tr}_K^1$ with $\Phi = \log(1+\Psi)$, and $m \geqslant 1$. Then $\Phi \in \operatorname{tr}_K^m$ if and only if $\Psi \in \operatorname{tr}_K^m$. Also, for all $n \geqslant 1$:*

$$\Phi_n \;=\; \sum_{i_1,\dots,i_k} \frac{(-1)^{k+1}}{k}\Psi_{i_1}\cdots\Psi_{i_k}, \qquad \Psi_n \;=\; \sum_{i_1,\dots,i_k} \frac{1}{k!}\Phi_{i_1}\cdots\Phi_{i_k},$$

both summed over the $(i_1,\dots,i_k)$ with $k \geqslant 1$, $i_1,\dots,i_k \geqslant 1$ and $i_1+\cdots+i_k = n$.

EXAMPLE. Let $\Phi, \Psi \in \operatorname{tr}_K^1$ with $\Phi = \log(1+\Psi)$. Then

$$\Phi_1 = \Psi_1, \quad \Phi_2 = \Psi_2 - \tfrac{1}{2}(\Psi_1)^2, \quad \Phi_3 = \Psi_3 - \tfrac{1}{2}(\Psi_1\Psi_2 + \Psi_2\Psi_1) + \tfrac{1}{3}(\Psi_1)^3, \quad \ldots.$$

12.3 THE LIE ALGEBRA OF AN ALGEBRAIC UNITRIANGULAR GROUP

In this section K is an integral domain. A K-algebra A is said to be **nontrivial** if $1_A \neq 0$. Given a morphism $\phi\colon A \to B$ of nontrivial commutative K-algebras, we extend ϕ to a morphism $\phi\colon \operatorname{tr}_A \to \operatorname{tr}_B$ of K-algebras by $\phi(M) := \big(\phi(M_{ij})\big)$. This extended ϕ maps tr_A^n to tr_B^n, restricts to a group morphism $\mathcal{U}_A \to \mathcal{U}_B$, and

$$\phi\big(f(M)\big) \;=\; f\big(\phi(M)\big) \quad \text{for } f \in K[[z]],\ M \in \operatorname{tr}_A^1.$$

Let $X = (X_{ij})_{i,j\in\mathbb{N}}$ be a family of distinct indeterminates. A unitriangular group $\mathcal{G}$ over K is **algebraic** if for some family (P_α) of polynomials $P_\alpha \in K[X]$:

(1) $\mathcal{G} = \{G = (G_{ij}) \in K^{\mathbb{N}\times\mathbb{N}} : P_\alpha(G) = 0 \text{ for all } \alpha\}$,

(2) for each nontrivial commutative K-algebra A, the set

$$\mathcal{G}_A := \{G = (G_{ij}) \in A^{\mathbb{N}\times\mathbb{N}} : P_\alpha(G) = 0 \text{ for all } \alpha\}$$

of common zeros of the P_α in A is a subgroup of $\mathcal{U}_A = 1 + \mathfrak{tr}^1_A$.

In particular, $\mathcal{U} = \mathcal{U}_K$ is algebraic. Below $\mathcal{G}$ is such an algebraic unitriangular group over K and (P_α) is a family of polynomials $P_\alpha \in K[X]$ as above. Thus a morphism $\phi\colon A \to B$ of nontrivial commutative K-algebras induces a group morphism

$$\mathcal{G}_A \to \mathcal{G}_B\colon \quad G \mapsto \phi(G).$$

We also let t be an indeterminate, and let ε be the image of t under the natural map $K[t] \to K[t]/(t^2)$, so $K[t]/(t^2) = K[\varepsilon] = K \oplus K\varepsilon$ with $\varepsilon^2 = 0$ is the K-algebra of dual numbers over K. Thus for $M \in \mathfrak{tr}^1_{K[\varepsilon]}$ we have $\exp(\varepsilon M) = 1 + \varepsilon M$ in $\mathfrak{tr}_{K[\varepsilon]}$.

LEMMA 12.3.1. *Let $M \in \mathfrak{tr}^1_K$. Then $\exp(tM) \in \mathcal{G}_{K[t]} \Longleftrightarrow 1 + \varepsilon M \in \mathcal{G}_{K[\varepsilon]}$.*

PROOF. The forward direction is clear by applying the K-algebra morphism

$$K[t] \to K[\varepsilon]\colon \quad t \mapsto \varepsilon.$$

For the converse suppose $1 + \varepsilon M \in \mathcal{G}_{K[\varepsilon]}$. Let $n \geqslant 1$ and let $t_1, \dots, t_n$ be distinct indeterminates with respective images $\varepsilon_1, \dots, \varepsilon_n$ under the natural map

$$K[t_1, \dots, t_n] \to K[t_1, \dots, t_n]/(t_1^2, \dots, t_n^2).$$

Then $\varepsilon_i^2 = 0$ for $i = 1, \dots, n$, and with

$$R_n := K[t_1, \dots, t_n]/(t_1^2, \dots, t_n^2) = K[\varepsilon_1, \dots, \varepsilon_n],$$

the kernel of the K-algebra morphism $K[t] \to R_n$ with $t \mapsto \tau := \varepsilon_1 + \dots + \varepsilon_n$ is generated as an ideal by t^{n+1}. Moreover, the image of $\exp(tM)$ under this K-algebra morphism is

$$\begin{aligned} 1 + \tau M + \frac{\tau^2}{2!}M^2 + \dots + \frac{\tau^n}{n!}M^n &= 1 + (\varepsilon_1 + \dots + \varepsilon_n)M + \dots + (\varepsilon_1 \cdots \varepsilon_n)M^n \\ &= (1 + \varepsilon_1 M) \cdots (1 + \varepsilon_n M) \in \mathcal{G}_{K[\tau]}. \end{aligned}$$

Since this holds for all $n \geqslant 1$, we obtain

$$\exp(tM) = 1 + tM + \frac{t^2}{2!}M^2 + \dots + \frac{t^n}{n!}M^n + \dots \in \mathcal{G}_{K[t]}$$

as required. □

LEMMA 12.3.2. *The set*

$$\mathfrak{g} := \{M \in \mathfrak{tr}^1_K : \exp(tM) \in \mathcal{G}_{K[t]}\} = \{M \in \mathfrak{tr}^1_K : 1 + \varepsilon M \in \mathcal{G}_{K[\varepsilon]}\}$$

is a Lie subalgebra of $\mathfrak{tr}^1_K$.

PROOF. Let $M \in \mathfrak{tr}_K^1$. Given a polynomial $P \in K[X]$ with $X = (X_{ij})$ as before, we have in $K[\varepsilon]$, by Taylor expansion and $\varepsilon^2 = 0$,

$$P(1 + \varepsilon M) \;=\; P(1) + \varepsilon \sum_{i,j} \frac{\partial P}{\partial X_{ij}}(1) \cdot M_{ij}.$$

Note that $\frac{\partial P}{\partial X_{ij}} \neq 0$ for only finitely many $(i, j) \in \mathbb{N}^2$. Now $1 \in \mathcal{G}$ gives $P_\alpha(1) = 0$ for each α, and so

$$M \in \mathfrak{g} \quad\Longleftrightarrow\quad 1 + \varepsilon M \in \mathcal{G}_{K[\varepsilon]} \quad\Longleftrightarrow\quad \sum_{i,j} \frac{\partial P_\alpha}{\partial X_{ij}}(1) \cdot M_{ij} = 0 \text{ for all } \alpha.$$

Thus $\mathfrak{g}$ is a submodule of the K-module $\mathfrak{tr}_K^1$. Let $M, N \in \mathfrak{g}$; to show that $[M, N] \in \mathfrak{g}$, we let $R := R_2 = K[\varepsilon_1, \varepsilon_2]$ be as in the proof of the previous lemma, so $G := 1 + \varepsilon_1 M \in \mathcal{G}_{K[\varepsilon_1]} \subseteq \mathcal{G}_R$ and $H := 1 + \varepsilon_2 N \in \mathcal{G}_{K[\varepsilon_2]} \subseteq \mathcal{G}_R$. Then in $\mathfrak{tr}_R^1$ we have

$$GH = 1 + \varepsilon_1 M + \varepsilon_2 N + \varepsilon_1\varepsilon_2 MN, \quad HG = 1 + \varepsilon_1 M + \varepsilon_2 N + \varepsilon_1\varepsilon_2 NM,$$

so $GH = HG\big(1 + \varepsilon_1\varepsilon_2[M, N]\big)$. Then $GH, HG \in \mathcal{G}_R$ gives $1 + \varepsilon_1\varepsilon_2[M, N] \in \mathcal{G}_{K[\varepsilon_1\varepsilon_2]}$. Applying the K-algebra isomorphism $K[\varepsilon] \to K[\varepsilon_1\varepsilon_2]$ with $\varepsilon \mapsto \varepsilon_1\varepsilon_2$ then yields $[M, N] \in \mathfrak{g}$. Hence $\mathfrak{g}$ is a Lie subalgebra of $\mathfrak{tr}_K^1$. □

The next lemma shows that $\mathfrak{g}$ depends only on $\mathcal{G}$, not on the particular family (P_α). We call $\mathfrak{g}$ the **Lie algebra** of $\mathcal{G}$, and consider it as a Lie subalgebra of $\mathfrak{tr}_K^1$. Note that if $M \in \mathfrak{g}$ then $\exp(tM) \in \mathcal{G}_{K[t]}$, and substitution of $t = 1$ yields $\exp(M) \in \mathcal{G}$. Thus $\exp(\mathfrak{g}) \subseteq \mathcal{G}$. Here is how $\mathcal{G}$ and $\mathfrak{g}$ determine each other:

LEMMA 12.3.3. $\exp(\mathfrak{g}) = \mathcal{G}$ *and* $\log(\mathcal{G}) = \mathfrak{g}$.

PROOF. Let $G \in \mathcal{G}$; it suffices to show that then $\log(G) \in \mathfrak{g}$, i.e., $\exp\big(t\log(G)\big) \in \mathcal{G}_{K[t]}$. Now for each α, the polynomial $P_\alpha\big(\exp(t\log(G))\big) \in K[t]$ vanishes upon substitution of integers for t, since $\exp(k\log(G)) = \exp(\log(G^k)) = G^k \in \mathcal{G}$ for each $k \in \mathbb{Z}$. Since K is assumed to be an integral domain, we therefore have $P_\alpha\big(\exp(t\log(G))\big) = 0$ for each α and so $\exp\big(t\log(G)\big) \in \mathcal{G}_{K[t]}$. □

The Lie algebra of the algebraic unitriangular group $\mathcal{U}$ over K is $\mathfrak{tr}_K^1$; we denote this Lie algebra also by $\mathfrak{u}$. We equip the K-module $\mathfrak{u}$ with the complete nonnegative filtration $(\mathfrak{u}^n)$ given by $\mathfrak{u}^n := \mathfrak{tr}_K^{n+1}$ for each n. This makes $\mathfrak{u}$ a *filtered* Lie algebra over K. The Lie subalgebra $\mathfrak{g}$ of $\mathfrak{u}$ is made into a filtered Lie algebra over K by giving it the filtration induced by $(\mathfrak{u}^n)$.

LEMMA 12.3.4. *$\mathcal{G}$ and $\mathfrak{g}$ are closed in $\mathfrak{tr}_K$. In particular, the filtered Lie algebra $\mathfrak{g}$ over K is complete.*

PROOF. For each $P \in K[X]$, the function $G \mapsto P(G)\colon \mathfrak{tr}_K \to K$ is clearly locally constant, and so its zero set is closed (and open) in $\mathfrak{tr}_K$. Thus $\mathcal{G}$ is closed in $\mathfrak{tr}_K$, and so is $\mathfrak{g}$ in view of the equivalence

$$M \in \mathfrak{g} \iff \sum_{i,j} \frac{\partial P_\alpha}{\partial X_{ij}}(1) \cdot M_{ij} = 0 \text{ for all } \alpha$$

from the proof of Lemma 12.3.2. □

Section 12.5 below is devoted to the investigation of the Lie algebra of a certain algebraic unitriangular group over $\mathbb{Q}$ which plays an important role in the study of triangular automorphisms of differential polynomial rings in Section 12.8.

Notes and comments

The results in this section are analogues of well-known facts about algebraic groups of (finite-size) matrices. The proof of Lemma 12.3.2 follows [400, Theorem I.5].

12.4 DERIVATIONS ON THE RING OF COLUMN-FINITE MATRICES

In this section we investigate two kinds of derivations on the ring of column-finite matrices: those induced by derivations on K and the internal K-derivations.

First, let a derivation ∂ on K be given. This is a $\mathbb{Q}$-derivation in the sense of Section 12.1. For $M = (M_{ij}) \in K^{\mathbb{N}\times\mathbb{N}}$ we set $\partial(M) := \big(\partial(M_{ij})\big)$. If M is column-finite, then so is $\partial(M)$, and $M \mapsto \partial(M)$ is a derivation on the $\mathbb{Q}$-algebra of column-finite matrices which restricts to a derivation on its $\mathbb{Q}$-subalgebra $\mathfrak{tr}_K$ consisting of the triangular matrices over K.

Now let t be an indeterminate. Then $\mathfrak{tr}_K$ is a K-subalgebra of $\mathfrak{tr}_{K[t]}$. We equip $K[t]$ with the derivation $\frac{d}{dt}$, and accordingly we define $\frac{dM}{dt} \in \mathfrak{tr}_{K[t]}$ for $M \in \mathfrak{tr}_{K[t]}$ as just explained. This gives a K-derivation $M \mapsto \frac{dM}{dt}$ of rank 0 on the K-algebra $\mathfrak{tr}_{K[t]}$ equipped with the filtration $(\mathfrak{tr}^n_{K[t]})$. The following two lemmas about this K-derivation are used in Section 12.5.

LEMMA 12.4.1. *Let $M \in \mathfrak{tr}^1_K$. Then*

$$\frac{d}{dt}\,\mathrm{e}^{tM} = \mathrm{e}^{tM}\,M.$$

PROOF. We have $(tM)^n = t^n M^n$ for every n, hence

$$\mathrm{e}^{tM} = \sum \frac{(tM)^n}{n!} = \sum \frac{t^n M^n}{n!}$$

and thus

$$\frac{d}{dt}\,\mathrm{e}^{tM} = \sum \frac{d}{dt}\left(\frac{t^n M^n}{n!}\right) = \sum_{n=1}^{\infty} \frac{t^{n-1} M^n}{(n-1)!} = \mathrm{e}^{tM}\,M. \qquad \square$$

The following lemma is a familiar fact about systems of linear differential equations with constant coefficients:

LEMMA 12.4.2. *Let $M \in \mathfrak{tr}_K^1$ and $Y \in \mathfrak{tr}_{K[t]}$, so $Y(0) \in \mathfrak{tr}_K$. Then*

$$\frac{dY}{dt} = YM \qquad \Longleftrightarrow \qquad Y = Y(0)\,\mathrm{e}^{tM}.$$

PROOF. Lemma 12.4.1 shows that if $Y = Y(0)\,\mathrm{e}^{tM}$, then $\frac{dY}{dt} = YM$. Conversely, suppose $\frac{dY}{dt} = YM$. Replacing Y by $Y - Y(0)\,\mathrm{e}^{tM}$ we arrange $Y(0) = 0$; we need to show that then $Y = 0$. Towards a contradiction, suppose $Y \neq 0$. Now $Y(0) = 0$, so for all i, j with $Y_{ij} \neq 0$ we have $Y_{ij} = t^{n_{ij}} Z_{ij}$ with $n_{ij} \in \mathbb{N}$, $n_{ij} \geqslant 1$, and $Z_{ij} \in K[t]$, $Z_{ij}(0) \neq 0$. Pick i, j with minimal n_{ij}. Then $\frac{dY}{dt} - YM$ gives

$$n_{ij} t^{n_{ij}-1} Z_{ij} + t^{n_{ij}} \frac{dZ_{ij}}{dt} = \frac{dY_{ij}}{dt} = \sum_k Y_{ik} M_{kj} = \sum_{Y_{ik} \neq 0} t^{n_{ik}} Z_{ik} M_{kj}.$$

Here the right-hand side is divisible in $K[t]$ by $t^{n_{ij}}$ while the left-hand side is not, a contradiction. □

Secondly, given a column-finite matrix $A \in K^{\mathbb{N}\times\mathbb{N}}$, the adjoint $M \mapsto \operatorname{ad}_A M = [A, M]$ of A is a K-derivation on the K-algebra of column-finite matrices. In the rest of this section we employ the derivation ad_A, for a particular choice of A, to establish commutator identities for certain diagonal matrices used in later sections.

DEFINITION 12.4.3. The (column-finite) **shift matrix** $S \in K^{\mathbb{N}\times\mathbb{N}}$ is given by

$$S_{j+1,j} = 1, \qquad S_{i,j} = 0 \text{ for } i \neq j+1.$$

A column-finite matrix $M \in K^{\mathbb{N}\times\mathbb{N}}$ has **derivative** $M' \in K^{\mathbb{N}\times\mathbb{N}}$ given by

$$M' := \operatorname{ad}_{-S} M = MS - SM$$

and so M' is also column-finite.

Multiplying a column-finite matrix $M \in K^{\mathbb{N}\times\mathbb{N}}$ by S has the following effect:

$(MS)_{ij} = M_{i,j+1}$: shifts M one column to the left, cancels the leftmost column,
$(SM)_{ij} = M_{i-1,j}$: shifts M one row downwards, adds a top row of zeros.

Here and below $M_{ij} := 0$ if $i < 0$ or $j < 0$. In particular,

$$(M')_{ij} = M_{i,j+1} - M_{i-1,j}.$$

EXAMPLE. Let $n \geqslant 1$, $a = (a_i) \in K^{\mathbb{N}}$, and $M = \operatorname{diag}_n a$. Then

$$(\operatorname{diag}_n a)' = \operatorname{diag}_{n-1}(a_0, a_1 - a_0, a_2 - a_1, \ldots).$$

The next lemma lists some properties of the derivation $M \mapsto M'$.

LEMMA 12.4.4. *Let* $M \in \mathfrak{tr}_K$. *Then*

$$M' = 0 \quad \Longleftrightarrow \quad M \in K,$$

and for $M \in \mathfrak{tr}_K^\times$, *we have*

$$(M^{-1})' = -M^{-1}M'M^{-1}.$$

Suppose now that $M \in \mathfrak{tr}_K^1$. *Then* $M' \in \mathfrak{tr}_K$, *and for* $n \geqslant 1$,

$$(M_n)' = (M')_{n-1}, \qquad M \in \mathfrak{tr}_K^n \implies M' \in \mathfrak{tr}_K^{n-1}.$$

Below, elements of $K^{\mathbb{N}}$ are column vectors, and $a = (a_i) \in K^{\mathbb{N}}$ is called **finite** if $a_i = 0$ for all but finitely many i. So $e := (1, 0, 0, \dots)^{\mathrm{t}} \in K^{\mathbb{N}}$ is finite. If $M \in K^{\mathbb{N}\times\mathbb{N}}$ is column-finite and $a \in K^{\mathbb{N}}$ is finite, then $Ma \in K^{\mathbb{N}}$ (defined in the obvious way) is finite, and in particular, Me is the leftmost column of M.

LEMMA 12.4.5. *Suppose* $a \in K^{\mathbb{N}}$ *is finite,* $B \in \mathfrak{tr}_K$, *and* $A, C \in K^{\mathbb{N}\times\mathbb{N}}$ *are column-finite. Then there is a unique column-finite matrix* $X \in K^{\mathbb{N}\times\mathbb{N}}$ *such that*

$$Xe = a, \tag{12.4.1}$$
$$X' = AXB + C. \tag{12.4.2}$$

If also $A, C \in \mathfrak{tr}_K$ *and* $a_i = 0$ *for all* $i \geqslant 1$, *then* $X \in \mathfrak{tr}_K$.

PROOF. Suppose $X \in K^{\mathbb{N}\times\mathbb{N}}$ is column-finite and satisfies (12.4.1) and (12.4.2). Then the leftmost column of X is a. Let j be given. Then for each i,

$$X_{i,j+1} = (XS)_{ij} = (X' + SX)_{ij} = (AXB)_{ij} + C_{ij} + X_{i-1,j}.$$

Since B is triangular, the sum $(AXB)_{ij} = \sum_{k,l} A_{ik}X_{kl}B_{lj}$ only involves entries of X from its columns with indices $l = 0, \dots, j$. Thus the column $(X_{i,j+1})_i$ of X with index $j+1$ is determined by its columns with lower index, and so there is at most one X as claimed. If in addition $A, C \in \mathfrak{tr}_K$ and $a_i = 0$ for all $i \geqslant 1$, then an induction on j shows that $X_{ij} = 0$ for all $i > j$.

Reversing these considerations we construct an X as claimed. □

For the next lemmas and corollaries (used in Sections 12.5 and 12.6), recall that

$$\binom{X}{n} := \frac{X(X-1)\cdots(X-n+1)}{n!} \in \mathbb{Q}[X]$$

is a polynomial of degree n, with $\binom{X}{0} = 1$. For $k \in \mathbb{Z}$ we let $\binom{k}{n}$ be its value at k; so $\binom{k}{n} = 0$ if $0 \leqslant k < n$. It is also convenient to set $\binom{n}{-1} := 0$. Consider now

$$A(n) := \operatorname{diag}_n \binom{i+n}{n},$$

so $A(0) = 1$ and $A(1) = \mathrm{diag}_1(1, 2, 3, \dots)$. Note that for all n,

$$A(n+1)' = A(n), \tag{12.4.3}$$

by the familiar recurrence relations for binomial coefficients. It is easy to verify, using Lemma 12.2.2, that $\big[A(m), A(n)\big] = 0$ for all m, n. We also set

$$B(n) := \mathrm{diag}_n \binom{i+n}{n+1},$$

so $B(0) = \mathrm{diag}(0, 1, 2, 3, \dots)$. For all n,

$$B(n+1)' = B(n), \tag{12.4.4}$$

again by the recurrence relation for binomial coefficients.

LEMMA 12.4.6. *For all m, n,*

$$\big[B(m), B(n)\big] = \left(\binom{m+n}{m-1} - \binom{m+n}{m+1}\right) B(m+n) \tag{12.4.5}$$

and

$$\big[A(m), B(n)\big] = \binom{m+n}{n+1} A(m+n). \tag{12.4.6}$$

PROOF. A simple computation yields $[B(0), B(n)] = -nB(n)$, and likewise we have $[B(m), B(0)] = -[B(0), B(m)] = mB(m)$ and $[A(0), B(n)] = 0$, $[A(m), B(0)] = mA(m)$. Thus (12.4.5) and (12.4.6) hold if $m = 0$ or $n = 0$. Let $m \geqslant 1$ and $n \geqslant 1$. Then by (12.4.4),

$$\begin{aligned}\big[B(m), B(n)\big]' &= \big[B(m)', B(n)\big] + \big[B(m), B(n)'\big] \\ &= \big[B(m-1), B(n)\big] + \big[B(m), B(n-1)\big].\end{aligned}$$

Inductively we can assume that the last sum equals the sum of

$$\begin{aligned}&\left(\binom{m+n-1}{m-2} - \binom{m+n-1}{m}\right) B(m+n-1) \text{ and} \\ &\left(\binom{m+n-1}{m-1} - \binom{m+n-1}{m+1}\right) B(m+n-1),\end{aligned}$$

so $\big[B(m), B(n)\big]' = \left(\binom{m+n}{m-1} - \binom{m+n}{m+1}\right) B(m+n)'$ by (12.4.4). Now (12.4.5) follows from Lemma 12.4.4. Similarly for (12.4.6), we have by (12.4.4) and (12.4.3):

$$\begin{aligned}\big[A(m), B(n)\big]' &= \big[A(m)', B(n)\big] + \big[A(m), B(n)'\big] \\ &= \big[A(m-1), B(n)\big] + \big[A(m), B(n-1)\big].\end{aligned}$$

Inductively, we can assume that the last sum equals

$$\binom{m+n-1}{n+1} A(m+n-1) + \binom{m+n-1}{n} A(m+n-1),$$

so $\big[A(m), B(n)\big]' = \binom{m+n}{n+1} A(m+n)'$ by (12.4.3). Now use Lemma 12.4.4. □

From (12.4.5) we obtain:

COROLLARY 12.4.7. *Let $c_1, c_2 \in K$, and define the sequence $(C(n))_{n\geqslant 1}$ in $\mathfrak{tr}_K$ by*

$$\begin{cases} C(n) = c_n B(n) & \text{for } n = 1, 2, \\ C(n+1) = \big[C(1), C(n)\big] & \text{for } n \geqslant 2. \end{cases}$$

Then $C(n) = c_n B(n)$ for all $n \geqslant 1$, where $c_{n+1} = \big(1 - \binom{n+1}{2}\big) c_1 c_n$ for $n \geqslant 2$.

REMARK. An easy induction on $n \geqslant 2$ shows that the terms $c_2, c_3, \dots$ of the sequence $(c_n)_{n\geqslant 1}$ in the previous lemma are explicitly given by

$$c_n = (-c_1)^{n-2} c_2 \cdot \frac{(n-2)!\,(n+1)!}{3 \cdot 2^{n-1}} \qquad \text{for } n \geqslant 2.$$

Notes and comments

The derivative of a column-finite matrix is defined in [215], and used there to give simple proofs for combinatorial identities.

12.5 ITERATION MATRICES

In this section we introduce special types of triangular matrices, called iteration matrices, and study their matrix logarithms.

Bell polynomials

Let $x, y_1, y_2, y_3, \dots, z$ be distinct indeterminates, and set

$$R := \mathbb{Q}[x, y_1, y_2, y_3, \dots], \qquad A := R[[z]].$$

We view the power series ring A as a complete filtered A-algebra with respect to the nonnegative filtration (A^n) given by $A^n = z^n A$. Set

$$y := \sum_{n=1}^{\infty} y_n \frac{z^n}{n!} \in zR[[z]] = A^1,$$

so $xy \in A^1$, and $\exp(xy) \in 1 + A^1$. Here is an explicit formula for $\exp(xy)$, where for $\boldsymbol{k} = (k_1, \dots, k_d) \in \mathbb{N}^d$, $d \geqslant 1$, we set

$$|\boldsymbol{k}| := k_1 + \cdots + k_d, \qquad \|\boldsymbol{k}\| := k_1 + 2k_2 + \cdots + dk_d.$$

PROPOSITION 12.5.1. *In A we have the identity*

$$\exp(xy) = \sum_{j=0}^{\infty} \left(\sum_{i=0}^{j} B_{ij} x^i \right) \frac{z^j}{j!}, \quad \textit{where}$$

$$B_{ij} = \sum_{\substack{\boldsymbol{k}=(k_1,\dots,k_d)\in\mathbb{N}^d \\ |\boldsymbol{k}|=i,\, \|\boldsymbol{k}\|=j}} \frac{j!}{k_1! k_2! \cdots k_d! \cdot (1!)^{k_1} (2!)^{k_2} \cdots (d!)^{k_d}} \, y_1^{k_1} y_2^{k_2} \cdots y_d^{k_d},$$

for $i \leqslant j$ *and* $d = j - i + 1$. *In particular, for such* i, j, d *we have* $B_{ij} \in \mathbb{Q}[y_1, \ldots, y_d]$, *and the coefficients of* B_{ij} *are in* $\mathbb{N}$.

PROOF. With $k_1, k_2, \ldots$ ranging over $\mathbb{N}$, the multinomial identity gives:

$$\begin{aligned}
\exp(x \cdot y) &= \sum_{i=0}^{\infty} \frac{x^i}{i!} \left(\sum_{n=1}^{\infty} y_n \frac{z^n}{n!} \right)^i \\
&= \sum_{i=0}^{\infty} \frac{x^i}{i!} \left(\sum_{k_1+k_2+\cdots=i} \frac{i!}{k_1!k_2!\cdots} \left(\frac{y_1 z^1}{1!}\right)^{k_1} \left(\frac{y_2 z^2}{2!}\right)^{k_2} \cdots \right) \\
&= \sum_{j=0}^{\infty} \left(\sum_{i=0}^{j} x^i \left(\sum_{\substack{k_1+k_2+\cdots=i \\ k_1+2k_2+\cdots=j}} \frac{j!}{k_1!k_2!\cdots(1!)^{k_1}(2!)^{k_2}\cdots} y_1^{k_1} y_2^{k_2} \cdots \right) \right) \frac{z^j}{j!}.
\end{aligned}$$

The condition on the infinite sequence $k_1, k_2, \ldots$ in the innermost sum forces $k_n = 0$ for $n > d := j - i + 1$, which yields the displayed formula for B_{ij}. By the following lemma, the coefficients of B_{ij} are in $\mathbb{N}$. □

LEMMA 12.5.2. *Let* $\boldsymbol{k} = (k_1, \ldots, k_d) \in \mathbb{N}^d$, $d \geqslant 1$, *and set* $n = \|\boldsymbol{k}\|$. *The number of partitions of an* n*-element set into exactly* k_1 *sets of cardinality* 1, k_2 *sets of cardinality* 2, *etc., is* $\dfrac{n!}{k_1! \cdots k_d! \cdot (1!)^{k_1} \cdots (d!)^{k_d}}$.

PROOF. Set $[n] = \{1, \ldots, n\}$. Consider first the special case $n = kd$ with $k, d \geqslant 1$. We claim that the number of partitions of $[n]$ into k sets of size d is $\frac{n!}{k!(d!)^k}$. This claim is obviously true for $k = 1$. Let $k > 1$ and assume inductively that the claim holds with $k - 1$ instead of k. Then the claim follows by the inductive assumption and the fact that there are $\binom{n-1}{d-1}$ subsets of $[n]$ of size d containing 1. As to the general case, the lemma clearly holds for $d = 1$, so let $d > 1$, and assume the lemma holds for $d - 1$ instead of d. This inductive assumption takes care of the case $k_d = 0$, so let $k_d \geqslant 1$. Then use that there are $\binom{n}{k_d d}$ subsets of $[n]$ of size $k_d d$, apply the claim above and the inductive assumption, and perform a routine computation. □

COROLLARY 12.5.3. *Let* $b, c_1, c_2, c_3, \ldots \in K$ *and set*

$$c := \sum_{n=1}^{\infty} c_n \frac{z^n}{n!} \in zK[[z]].$$

Then $e^{bc} = \sum_{j=0}^{\infty} \left(\sum_{i=0}^{j} B_{ij}(c_1, \ldots, c_{j-i+1}) b^i \right) \cdot \frac{z^j}{j!}$ *in* $K[[z]]$.

The $B_{ij} = B_{ij}(y_1, y_2, \ldots) \in \mathbb{Q}[y_1, y_2, \ldots]$ are the (partial) **Bell polynomials.** We also set $B_{ij} := 0$ for $i > j$, so the family $(B_{ij} x^i z^j / j!)_{i,j \in \mathbb{N}}$ is summable in A, with

$$\sum_{i \in \mathbb{N}} \frac{x^i y^i}{i!} = \sum_{i,j \in \mathbb{N}} B_{ij} x^i \frac{z^j}{j!}. \tag{12.5.1}$$

Note that $B_{0j} = 0$ and $B_{1j} = y_j$ for $j \geqslant 1$, and $B_{jj} = y_1^j$ for all j. Next we establish the following identity in $\mathbb{Q}[y_1, y_2, \dots][[z]]$, holding for all $i \in \mathbb{N}$:

$$\frac{y^i}{i!} = \sum_{j=0}^{\infty} B_{ij} \frac{z^j}{j!}. \tag{12.5.2}$$

To see why this identity holds we try to view both sides in (12.5.1) as power series in x and then compare the coefficients of x^i. To justify this idea, we note that R is a subring of $S := \mathbb{Q}[y_1, y_2, \dots][[x]]$, and accordingly, $A = R[[z]]$ is a subring of

$$B := S[[z]] = \mathbb{Q}[y_1, y_2, \dots][[x, z]] = \big(\mathbb{Q}[y_1, y_2, \dots][[z]]\big)[[x]],$$

a complete filtered B-algebra with respect to the nonnegative filtration (B^n) given by $B^n = (x, z)^n B$. Then $A^n \subseteq B^n$ for all n and $xy \in B^2$, so $\sum (x^i y^i)/i!$ takes the value in B that it has in A. Likewise, the right-hand sum in (12.5.1) is defined in B, and takes the value in B it has in A. Now (12.5.2) follows as indicated above.

The next recursion formula facilitates the computation of the Bell polynomials:

LEMMA 12.5.4. *Suppose $i_1 \leqslant i \leqslant j$. Then in $\mathbb{Q}[y_1, y_2, \dots]$ we have*

$$B_{ij} = \frac{1}{\binom{i}{i_1}} \sum_{j_1=0}^{j} \binom{j}{j_1} B_{i_1, j_1} B_{i-i_1, j-j_1}.$$

PROOF. By (12.5.2),

$$\frac{1}{i_1!} y^{i_1} = \sum_{j=0}^{\infty} B_{i_1,j} \frac{z^j}{j!}, \qquad \frac{1}{(i-i_1)!} y^{i-i_1} = \sum_{j=0}^{\infty} B_{i-i_1,j} \frac{z^j}{j!},$$

hence in $\mathbb{Q}[y_1, y_2, \dots][[z]]$ we have

$$\begin{aligned}\binom{i}{i_1} \cdot \frac{1}{i!} y^i = \frac{1}{i_1!} y^{i_1} \cdot \frac{1}{(i-i_1)!} y^{i-i_1} &= \left(\sum_{j=0}^{\infty} B_{i_1,j} \frac{z^j}{j!}\right) \cdot \left(\sum_{j=0}^{\infty} B_{i-i_1,j} \frac{z^j}{j!}\right) \\ &= \sum_{j=0}^{\infty} \left(\sum_{j_1=0}^{j} \binom{j}{j_1} B_{i_1,j_1} B_{i-i_1,j-j_1}\right) \frac{z^j}{j!},\end{aligned}$$

and the lemma follows. □

Using this with $i_1 = 1$ we get easily

$$\begin{aligned} B_{23} &= 3y_1y_2, & B_{24} &= 4y_1y_3 + 3y_2^2, & B_{25} &= 5y_1y_4 + 10y_2y_3, \\ B_{34} &= 6y_1^2y_2, & B_{35} &= 10y_1^2y_3 + 15y_1y_2^2, & B_{45} &= 10y_1^3y_2. \end{aligned}$$

Iteration matrices

Given a power series $f \in zK[[z]]$,

$$f = \sum_{n\geqslant 1} f_n \frac{z^n}{n!} \qquad (f_n \in K \text{ for each } n \geqslant 1), \tag{12.5.3}$$

we introduce the triangular matrix

$$[\![f]\!] := \big([\![f]\!]_{ij}\big)_{i,j\in\mathbb{N}} = \big(B_{ij}(f_1, f_2, \dots, f_{j-i+1})\big)_{i,j\in\mathbb{N}} =$$
$$\begin{pmatrix} 1 & 0 & 0 & 0 & 0 & 0 & \cdots \\ & f_1 & f_2 & f_3 & f_4 & f_5 & \cdots \\ & & f_1^2 & 3f_1f_2 & 4f_1f_3+3f_2^2 & 5f_1f_4+10f_2f_3 & \cdots \\ & & & f_1^3 & 6f_1^2f_2 & 10f_1^2f_3+15f_1f_2^2 & \cdots \\ & & & & f_1^4 & 10f_1^3f_2 & \cdots \\ & & & & & f_1^5 & \cdots \\ & & & & & & \ddots \end{pmatrix} \in \mathfrak{tr}_K\,.$$

Note that (12.5.2) gives $\frac{f^i}{i!} = \sum_{j=0}^{\infty} [\![f]\!]_{ij} \frac{z^j}{j!}$ in $K[[z]]$. As we see from the second row of the display, the map $f \mapsto [\![f]\!]\colon zK[[z]] \to \mathfrak{tr}_K$ is injective. It is also easy to check that $[\![z]\!] = 1$. The matrix $[\![f]\!]$ is called the **iteration matrix of** f, since $f \mapsto [\![f]\!]$ converts composition of power series into matrix multiplication:

LEMMA 12.5.5. *Let $f, g \in zK[[z]]$. Then* $[\![f \circ g]\!] = [\![f]\!] \cdot [\![g]\!]$.

PROOF. The above identity for powers of elements in $zK[[z]]$ gives

$$\sum_{k=0}^{\infty} [\![f\circ g]\!]_{ik} \frac{z^k}{k!} = \frac{1}{i!}(f\circ g)^i = \frac{1}{i!}\, f^i \circ g = \sum_{j=0}^{\infty} [\![f]\!]_{ij} \frac{g^j}{j!}$$
$$= \sum_{j=0}^{\infty} \left([\![f]\!]_{ij} \sum_{k=0}^{\infty} [\![g]\!]_{jk} \frac{z^k}{k!} \right) = \sum_{k=0}^{\infty} \left(\sum_{j=0}^{\infty} [\![f]\!]_{ij} [\![g]\!]_{jk} \right) \frac{z^k}{k!}.$$

Now compare the coefficients of $z^k/k!$ in the first and the last sum. □

The subset $zK^\times + z^2K[[z]]$ of $zK[[z]]$ is a group under formal composition with identity element z, and it admits a group embedding

$$f \mapsto [\![f]\!] \quad : \quad zK^\times + z^2K[[z]] \to \mathfrak{tr}_K^\times$$

into the group $\mathfrak{tr}_K^\times$ of units of $\mathfrak{tr}_K$. We say that f as in (12.5.3) is **unitary** if $f_1 = 1$. The set $z + z^2K[[z]]$ of unitary power series in $K[[z]]$ is a subgroup of $zK^\times + z^2K[[z]]$ under composition, whose image under $f \mapsto [\![f]\!]$ is a subgroup of $\mathcal{U} = 1 + \mathfrak{tr}_K^1$ which we denote by $\mathcal{I}$ and call the **group of iteration matrices over** K. It is easy to see that for $f \in zK[[z]]$ and $n \geqslant 1$, we have

$$f \in z + z^{n+1}K[[z]] \iff [\![f]\!] \in 1 + \mathfrak{tr}_K^n\,.$$

The Lie algebra of the group of iteration matrices

In the rest of this section K is an integral domain. It is easy to check that then the unitriangular group $\mathcal{I}$ over K is algebraic: use the way that the entries of an arbitrary element are given by polynomials in the entries of its second row. Thus $\mathcal{I}$ has an associated Lie algebra by Lemma 12.3.3 and the remarks preceding it. Our next goal is to give an explicit description of this Lie algebra.

DEFINITION 12.5.6. Let $h = \sum_{n=1}^{\infty} h_n \frac{z^n}{n!} \in zK[[z]]$, $h_n \in K$ for $n \geqslant 1$. The **infinitesimal iteration matrix** of h is the triangular matrix

$$\langle\!\langle h \rangle\!\rangle = (\langle\!\langle h \rangle\!\rangle_{ij}) = \begin{pmatrix} 0 & 0 & 0 & 0 & 0 & \cdots \\ & h_1 & h_2 & h_3 & h_4 & \cdots \\ & & 2h_1 & 3h_2 & 4h_3 & \cdots \\ & & & 3h_1 & 6h_2 & \cdots \\ & & & & 4h_1 & \cdots \\ & & & & & \ddots \end{pmatrix} \in \mathfrak{tr}_K$$

where $\langle\!\langle h \rangle\!\rangle_{ij} = \binom{j}{j-i+1} h_{j-i+1}$ for $i \leqslant j$.

Thus $h \mapsto \langle\!\langle h \rangle\!\rangle \colon zK[[z]] \to \mathfrak{tr}_K$ is an injective continuous K-linear map, and

$$h \in z^{n+1}K[[z]] \Longleftrightarrow \langle\!\langle h \rangle\!\rangle \in \mathfrak{tr}_K^n\,.$$

We introduce the K-submodule

$$\mathfrak{i} := \{\langle\!\langle h \rangle\!\rangle : h \in z^2K[[z]]\}$$

of $\mathfrak{u} = \mathfrak{tr}_K^1$. For each n, the matrix

$$\left\langle\!\!\left\langle \tfrac{z^{n+1}}{(n+1)!} \right\rangle\!\!\right\rangle = \operatorname{diag}_n \binom{i+n}{n+1} \in \mathfrak{tr}_K^n$$

is n-diagonal. In the notation from Section 12.4, we have $\left\langle\!\left\langle \frac{z^{n+1}}{(n+1)!} \right\rangle\!\right\rangle = B(n)$. Note that for $h \in zK[[z]]$ we have $\langle\!\langle h \rangle\!\rangle = \sum_{n=1}^{\infty} h_n \langle\!\langle z^n/n! \rangle\!\rangle$. The map

$$h \mapsto \langle\!\langle h \rangle\!\rangle \quad : \quad z^2K[[z]] \to \mathfrak{i}$$

is a K-module isomorphism. Lemma 12.4.6 implies that $\mathfrak{i}$ is a Lie subalgebra of the Lie algebra $\mathfrak{u}$ over K. We equip $\mathfrak{i}$ with the filtration $(\mathfrak{i}^n)$ induced by the filtration of $\mathfrak{u}$: $\mathfrak{i}^n = \mathfrak{i} \cap \mathfrak{u}^n = \mathfrak{i} \cap \mathfrak{tr}_K^{n+1}$. Then the isomorphism above and its inverse are both of rank 0 with respect to the filtration $(z^{n+2}K[[z]])$ of the K-module $z^2K[[z]]$ and the filtration $(\mathfrak{i}^n)$ of $\mathfrak{i}$. Here is the main result of this subsection:

THEOREM 12.5.7. $\exp(\mathfrak{i}) = \mathcal{I}$; *in other words,* $\mathfrak{i}$ *is the Lie algebra of* $\mathcal{I}$.

We give the proof of this theorem after some preparation. Let t be an indeterminate and set $K^* = K[t]$. We extend the K-derivation $\frac{d}{dt}$ of K^* to the K-derivation, also denoted by $\frac{d}{dt}$, of the power series ring $K^*[[z]]$ by

$$\frac{d}{dt}\left(\sum f_n z^n\right) = \sum \frac{df_n}{dt} z^n \qquad (\text{all } f_n \in K^*).$$

Of course, we also have the usual K^*-derivation $\frac{\partial}{\partial z}$ on $K^*[[z]]$.

LEMMA 12.5.8. *Let $f \in zK^*[[z]]$ and $h \in zK[[z]]$. Then*

$$\frac{df}{dt} = \frac{\partial f}{\partial z} h \text{ in } K^*[[z]] \implies \frac{d}{dt}[\![f]\!] = [\![f]\!]\langle\!\langle h\rangle\!\rangle \text{ in } \mathrm{tr}_{K^*}.$$

PROOF. Assume $\frac{df}{dt} = \frac{\partial f}{\partial z} h$; we need to show that for all i, j,

$$\frac{d}{dt}[\![f]\!]_{ij} = \left([\![f]\!]\langle\!\langle h\rangle\!\rangle\right)_{ij}.$$

For $i = 0$, both sides are 0, so let $i \geqslant 1$. By the formula for f^i,

$$\frac{1}{i!}\frac{\partial f^i}{\partial z} = \sum_{k=1}^{\infty} [\![f]\!]_{ik} \frac{z^{k-1}}{(k-1)!}$$

and hence with $h = \sum_{n=1}^{\infty} h_n z^n/n!$ (all $h_n \in K$),

$$\begin{aligned}\frac{1}{i!}\frac{\partial f^i}{\partial z} h &= \sum_{j=1}^{\infty}\left(\sum_{k=1}^{j} \frac{[\![f]\!]_{ik}\cdot h_{j-k+1}}{(k-1)!(j-k+1)!}\right) z^j \\ &= \sum_{j=1}^{\infty}\left(\sum_{k=1}^{j} [\![f]\!]_{ik}\langle\!\langle h\rangle\!\rangle_{kj}\right)\frac{z^j}{j!} = \sum_{j=1}^{\infty}\left([\![f]\!]\langle\!\langle h\rangle\!\rangle\right)_{ij}\frac{z^j}{j!}.\end{aligned}$$

Moreover

$$\frac{1}{i!}\frac{df^i}{dt} = \sum_{j=1}^{\infty} \frac{d}{dt}[\![f]\!]_{ij}\frac{z^j}{j!}.$$

By the hypothesis of the lemma

$$\frac{df^i}{dt} = if^{i-1}\frac{df}{dt} = if^{i-1}\frac{\partial f}{\partial z} h = \frac{\partial f^i}{\partial z} h,$$

hence $\frac{d}{dt}[\![f]\!]_{ij} = \left([\![f]\!]\langle\!\langle h\rangle\!\rangle\right)_{ij}$ for all j, as required. □

We can now prove the following important fact:

PROPOSITION 12.5.9. *Let $n \geqslant 1$, $h \in z^{n+1}K[[z]]$, so*

$$t\langle\!\langle h\rangle\!\rangle \in \mathrm{tr}^n_{K[t]}, \qquad \mathrm{e}^{t\langle\!\langle h\rangle\!\rangle} = 1 + t\langle\!\langle h\rangle\!\rangle + \cdots \in 1 + \mathrm{tr}^n_{K[t]}.$$

Set $f^{[t]} := \sum_{j=1}^{\infty} \left(\mathrm{e}^{t\langle\!\langle h\rangle\!\rangle}\right)_{1j} \frac{z^j}{j!} \in K^*[[z]]$. Then

$$f^{[t]} \in z + z^{n+1}tK^*[[z]], \qquad \frac{df^{[t]}}{dt} = \frac{\partial f^{[t]}}{\partial z} h, \qquad [\![f^{[t]}]\!] = \mathrm{e}^{t\langle\!\langle h\rangle\!\rangle}.$$

PROOF. That $f^{[t]} \in z + z^{n+1}tK^*[[z]]$ is an easy verification, and gives $[\![f^{[t]}]\!](0) = 1$. Let $h = \sum_{k=1}^{\infty} h_k z^k/k!$ with all $h_k \in K$. Using Lemma 12.4.1, we get

$$\begin{aligned}
\frac{df^{[t]}}{dt} &= \sum_{j=1}^{\infty} \left(\frac{d}{dt}\mathrm{e}^{t\langle\!\langle h\rangle\!\rangle}\right)_{1j} \cdot \frac{z^j}{j!} \\
&= \sum_{j=1}^{\infty} \left(\mathrm{e}^{t\langle\!\langle h\rangle\!\rangle}\langle\!\langle h\rangle\!\rangle\right)_{1j} \cdot \frac{z^j}{j!} \\
&= \sum_{j=1}^{\infty} \left(\sum_{i=1}^{j} (\mathrm{e}^{t\langle\!\langle h\rangle\!\rangle})_{1i}\langle\!\langle h\rangle\!\rangle_{ij}\frac{1}{j!}\right) z^j \\
&= \sum_{j=1}^{\infty} \left(\sum_{i=1}^{j} \frac{(\mathrm{e}^{t\langle\!\langle h\rangle\!\rangle})_{1i}}{(i-1)!}\frac{h_{j-i+1}}{(j-i+1)!}\right) z^j \;=\; \frac{\partial f^{[t]}}{\partial z}h.
\end{aligned}$$

So $\frac{d}{dt}[\![f^{[t]}]\!] = [\![f^{[t]}]\!]\langle\!\langle h\rangle\!\rangle$ by Lemma 12.5.8. Thus both $[\![f^{[t]}]\!]$ and $\mathrm{e}^{t\langle\!\langle h\rangle\!\rangle}$ satisfy $\frac{dY}{dt} = Y\langle\!\langle h\rangle\!\rangle$ and $Y(0) = 1$. Hence $[\![f^{[t]}]\!] = \mathrm{e}^{t\langle\!\langle h\rangle\!\rangle}$ by Lemma 12.4.2. □

For later use we give another description of the power series $f^{[t]}$ from the previous proposition. Let $h \in z^2K[[z]]$. Then we have the K^*-derivations $\Delta := h\frac{\partial}{\partial z}$ and $t\Delta$ on $K^*[[z]]$. These are both of rank 1, and so give rise to the K^*-algebra automorphisms e^{Δ} and $\mathrm{e}^{t\Delta}$ of $K^*[[z]]$, by Lemma 12.1.4. With these notations:

LEMMA 12.5.10. $\mathrm{e}^{t\Delta}(z) \in z + tz^2K^*[[z]], \qquad [\![\mathrm{e}^{t\Delta}(z)]\!] = \mathrm{e}^{t\langle\!\langle h\rangle\!\rangle}$.

PROOF. Set $f := \mathrm{e}^{t\Delta}(z) = \sum_{n=0}^{\infty} \frac{t^n}{n!}\Delta^n(z)$. Then

$$\frac{df}{dt} = \sum_{n=1}^{\infty} \frac{t^{n-1}}{(n-1)!}\Delta^n(z) = \Delta\left(\sum_{n=1}^{\infty} \frac{t^{n-1}}{(n-1)!}\Delta^{n-1}(z)\right) = \Delta(f) = h\frac{\partial f}{\partial z}.$$

As in the proof of Proposition 12.5.9 it now follows that $[\![f]\!] = \mathrm{e}^{t\langle\!\langle h\rangle\!\rangle}$. □

The following corollary, with $h \in z^2K[[z]]$ and Δ as before, is obtained by setting $t = 1$ in the last identity of Proposition 12.5.9 and in Lemma 12.5.10 above. It shows in particular that $\exp(\mathrm{i}) \subseteq \mathcal{I}$:

COROLLARY 12.5.11. *Set* $f := \sum_{j=1}^{\infty} (\mathrm{e}^{\langle\!\langle h\rangle\!\rangle})_{1j} \frac{z^j}{j!} \in z + z^2K[[z]]$. *Then*

$$\mathrm{e}^{\langle\!\langle h\rangle\!\rangle} = [\![f]\!], \qquad \mathrm{e}^{\Delta}(z) = f.$$

Recall that $B(k) = \langle\!\langle \frac{z^{k+1}}{(k+1)!}\rangle\!\rangle$ for $k \in \mathbb{N}$. Given $k_1, \dots, k_n \in \mathbb{N}$ and $k = k_1 + \cdots + k_n$, we have

$$B(k_1)\cdots B(k_n) = \mathrm{diag}_k\left(\binom{i+k_1}{k_1+1}\binom{i+k_1+k_2}{k_2+1}\cdots\binom{i+k_1+\cdots+k_n}{k_n+1}\right)_{i\geqslant 0}$$

by Lemma 12.2.2. Now let $h = \sum_{k=0}^{\infty} h_{k+1} \frac{z^{k+1}}{(k+1)!} \in zK[[z]]$ (all $h_{k+1} \in K$). Then

$$M := \langle\!\langle h \rangle\!\rangle = h_1 \langle\!\langle z \rangle\!\rangle + h_2 \langle\!\langle \tfrac{z^2}{2} \rangle\!\rangle + \cdots + h_{k+1} \langle\!\langle \tfrac{z^{k+1}}{(k+1)!} \rangle\!\rangle + \cdots \in \mathfrak{tr}_K,$$

and hence

$$M^n = \sum_{k_1,\dots,k_n \in \mathbb{N}} h_{k_1+1} \cdots h_{k_n+1} \langle\!\langle \tfrac{z^{k_1+1}}{(k_1+1)!} \rangle\!\rangle \cdots \langle\!\langle \tfrac{z^{k_n+1}}{(k_n+1)!} \rangle\!\rangle,$$

and so for $i, j \in \mathbb{N}$ and $n \geqslant 1$:

$$(12.5.4) \quad (M^n)_{ij} = \sum_{\substack{k_1,\dots,k_n \in \mathbb{N} \\ k_1+\cdots+k_n = j-i}} h_{k_1+1} \cdots h_{k_n+1} \tbinom{i+k_1}{k_1+1} \tbinom{i+k_1+k_2}{k_2+1} \cdots \tbinom{i+k_1+\cdots+k_n}{k_n+1}.$$

This observation leads to:

LEMMA 12.5.12. *Let $n \geqslant 1$. Then*

$$(M^n)_{11} = h_1^n, \qquad (M^n)_{1j} = \frac{j^n - 1}{j-1} h_1^{n-1} h_j + P_{nj}(h_1, \dots, h_{j-1}) \quad \textit{for } j \geqslant 2,$$

where $P_{nj}(Y_0, \dots, Y_{j-2}) \in \mathbb{Q}[Y_0, \dots, Y_{j-2}]$ has all its coefficients in $\mathbb{Z}$, is homogeneous of degree n, isobaric of weight $j-1$, and independent of h.

PROOF. Set $i = 1$ in (12.5.4). Then the only terms involving h_j in this sum are those of the form $h_1^{n-1} h_j \, j^{n-m}$ where $m \in \{1, \dots, n\}$. This yields the lemma. □

For example, $P_{24}(Y_0, Y_1, Y_2) = 10 Y_1 Y_2$ and $P_{34}(Y_0, Y_1, Y_2) = 76 Y_0 Y_1 Y_2 + 18 Y_1^3$. With h and $M = \langle\!\langle h \rangle\!\rangle$ as above, we get:

COROLLARY 12.5.13. *Suppose $h \in z^2 K[[z]]$. Then $M \in \mathfrak{tr}_K^1$, and for $j \geqslant 2$,*

$$(\mathrm{e}^M)_{1j} = h_j + P_j(h_2, \dots, h_{j-1})$$

where $P_j(Y_1, \dots, Y_{j-2}) \in \mathbb{Q}[Y_1, \dots, Y_{j-2}]$ is independent of h. Moreover, $P_2 = 0$, and for $j > 2$, P_j has degree at most $j-1$ and is isobaric of weight $j-1$.

PROOF. Let $j \geqslant 2$. We have $h_1 = 0$, so by Lemma 12.5.12,

$$(M^n)_{1j} = \begin{cases} h_j & \text{if } n = 1, \\ P_{nj}(h_1, \dots, h_{j-1}) & \text{if } 1 < n < j, \\ 0 & \text{if } n \geqslant j. \end{cases}$$

Hence

$$(\mathrm{e}^M)_{1j} = \sum_{n=1}^{j-1} \frac{1}{n!} (M^n)_{1j} = h_j + \sum_{n=2}^{j-1} \frac{1}{n!} P_{nj}(h_1, \dots, h_{j-1}).$$

Thus in view of $h_1 = 0$,

$$P_j(Y_1, \dots, Y_{j-2}) := \sum_{n=2}^{j-1} \frac{1}{n!} P_{nj}(0, Y_1, \dots, Y_{j-2})$$

has the right properties. □

Theorem 12.5.7 now follows immediately from Corollary 12.5.11 and the following:

PROPOSITION 12.5.14. *Let $f \in zK[[z]]$ be unitary. Then* $\log [\![f]\!] \in \mathfrak{i}$.

PROOF. We have $f = \sum_{j \geqslant 1} f_j \frac{z^j}{j!}$ (all $f_j \in K$). We define recursively the sequence $(h_j)_{j \geqslant 1}$ in K by $h_1 := 0$, and $h_{j+1} := f_{j+1} - P_{j+1}(h_2, \dots, h_j)$ for $j \geqslant 1$. Then $h := \sum_{j=1}^{\infty} h_j \frac{z^j}{j!} \in z^2 K[[z]]$, and by Corollary 12.5.13 we have $\left(\mathrm{e}^{\langle\!\langle h \rangle\!\rangle}\right)_{1j} = f_j$ for $j \geqslant 1$. Corollary 12.5.11 now yields $\mathrm{e}^{\langle\!\langle h \rangle\!\rangle} = [\![f]\!]$, hence $\log [\![f]\!] = \langle\!\langle h \rangle\!\rangle \in \mathfrak{i}$. □

Note: if $f \in z + z^{n+1} K[[z]]$, $n \geqslant 1$, then $[\![f]\!] \in 1 + \mathrm{tr}_K^n$, and so $\log [\![f]\!] \in \mathfrak{i}^{n-1}$.

The iterative logarithm

In this subsection we fix a unitary power series

$$f = z + \sum_{n=2}^{\infty} f_n \frac{z^n}{n!} \in z + z^2 K[[z]] \qquad \text{(all } f_n \in K\text{)}.$$

By Theorem 12.5.7 there is a (unique) power series $h \in z^2 K[[z]]$ with $\log [\![f]\!] = \langle\!\langle h \rangle\!\rangle$; we call this h the **iterative logarithm of** f and denote it by $\mathrm{itlog}(f)$. The proof of Proposition 12.5.14 gives $\mathrm{itlog}(f) = \sum_{n=2}^{\infty} h_n \frac{z^n}{n!}$ where $h_n = H_n(f_2, \dots, f_n)$ and $H_n \in \mathbb{Q}[Y_1, \dots, Y_{n-1}]$ is isobaric of weight $n-1$, for all $n \geqslant 2$: the "isobaric" statement follows inductively from the recursion

$$H_{n+1}(Y_1, \dots, Y_n) = Y_n - P_{n+1}\big(H_2(Y_1), \dots, H_n(Y_1, \dots, Y_{n-1})\big) \quad (n \geqslant 2).$$

The H_n for $n = 2, 3, 4$ are easily determined:

$$H_2 = Y_1, \quad H_3 = Y_2 - \tfrac{3}{2} Y_1^2, \quad H_4 = Y_3 - 5 Y_1 Y_2 + \tfrac{9}{2} Y_1^3, \quad \text{and so}$$

$$h_2 = f_2, \quad h_3 = f_3 - \tfrac{3}{2} f_2^2, \quad h_4 = f_4 - 5 f_2 f_3 + \tfrac{9}{2} f_2^2, \quad \text{and thus} \tag{12.5.5}$$

$$\langle\!\langle h \rangle\!\rangle = \log [\![f]\!] = \begin{pmatrix} 0 & 0 & 0 & 0 & 0 & \cdots \\ & 0 & f_2 & f_3 - \frac{3}{2} f_2^2 & f_4 - 5 f_2 f_3 + \frac{9}{2} f_2^3 & \cdots \\ & & 0 & 3 f_2 & 4 f_3 - 6 f_2^2 & \cdots \\ & & & 0 & 6 f_2 & \cdots \\ & & & & & \ddots \end{pmatrix}.$$

Here is another way to obtain $\mathrm{itlog}(f)$:

LEMMA 12.5.15. *Define* $f[n] \in z^{n+1}K[[z]]$ *recursively by*

$$f[0] = z, \qquad f[n+1] = f[n] \circ f - f[n].$$

Then

$$\operatorname{itlog}(f) = \sum_{n=1}^{\infty} \frac{(-1)^{n-1}}{n} f[n].$$

PROOF. Set $h := \operatorname{itlog}(f) = \sum_{j=1}^{\infty} h_j z^j/j!$ (all $h_j \in K$), and let $j \geqslant 1$. Then

$$h_j = \langle\!\langle h \rangle\!\rangle_{1j} = \sum_{n \geqslant 1} \frac{(-1)^{n-1}}{n} \left((\llbracket f \rrbracket - 1)^n \right)_{1j}.$$

By Lemma 12.5.5,

$$(\llbracket f \rrbracket - 1)^n = \sum_{k=0}^{n} (-1)^{n-k} \binom{n}{k} \llbracket f \rrbracket^k = \sum_{k=0}^{n} (-1)^{n-k} \binom{n}{k} \llbracket f^{[k]} \rrbracket$$

where $f^{[k]}$ denotes the kth compositional iterate of f, defined recursively by $f^{[0]} = z$ and $f^{[k+1]} = f^{[k]} \circ f$. Hence

$$\left((\llbracket f \rrbracket - 1)^n \right)_{1j} = \left(\sum_{k=0}^{n} (-1)^{n-k} \binom{n}{k} \llbracket f^{[k]} \rrbracket \right)_{1j} = \left\llbracket \sum_{k=0}^{n} (-1)^{n-k} \binom{n}{k} f^{[k]} \right\rrbracket_{1j}.$$

An easy induction gives $f[n] = \sum_{k=0}^{n} (-1)^{n-k} \binom{n}{k} f^{[k]}$. The lemma follows. □

COROLLARY 12.5.16. *Let* Δ *be the* K*-derivation* $\operatorname{itlog}(f)\frac{d}{dz}$ *on* $K[[z]]$ *of rank* 1. *Then the* K*-algebra automorphism* e^{Δ} *of* $K[[z]]$ *satisfies*

$$g \circ f = e^{\Delta}(g), \text{ for all } g \in K[[z]].$$

PROOF. This follows from Corollary 12.5.11 and the continuity of e^{Δ}. □

Set $h := \operatorname{itlog}(f)$, let A be a commutative ring extension of K, and $a \in A$. Then $e^{a\langle\!\langle h \rangle\!\rangle} = 1 + a\langle\!\langle h \rangle\!\rangle + \cdots \in 1 + a\,\mathrm{tr}_A^1$ and we set

$$f^{[a]} := \sum_{j=1}^{\infty} \left(e^{a\langle\!\langle h \rangle\!\rangle} \right)_{1j} \frac{z^j}{j!} \in z + z^2 aA[[z]].$$

(For $A = K^*$ and $a = t$ this is just the $f^{[t]}$ from Proposition 12.5.9.) The K-algebra morphism $g = g(t) \mapsto g(a)\colon K^* \to A$ sending t to a extends to the K-algebra morphism $M \mapsto M(a)\colon \mathrm{tr}_{K^*} \to \mathrm{tr}_A$ by substituting a for t in each entry, and it extends also to the K-algebra morphism

$$K^*[[z]] \to A[[z]], \qquad g = \sum g_n z^n \mapsto g\big|_{t=a} := \sum g_n(a) z^n \quad (\text{all } g_n \in K^*).$$

It is easy to check that then

$$e^{a\langle\!\langle h\rangle\!\rangle} = e^{t\langle\!\langle h\rangle\!\rangle}(a), \qquad f^{[a]} = f^{[t]}\big|_{t=a}.$$

Then Lemma 12.5.5 and Proposition 12.5.9 give

$$(12.5.6) \qquad f^{[0]} = z, \qquad f^{[1]} = f, \qquad f^{[a+b]} = f^{[a]} \circ f^{[b]} \quad (a, b \in A).$$

Thus the power series $f^{[a]}$ with $a \in K$ form a subgroup of $z + z^2K[[z]]$ under composition which contains f; they may be thought of as fractional iterates of f. In fact, $f^{[t]}$ is *unique* in the sense that if $g \in K^*[[z]]$ and $g\big|_{t=n} = f^{[n]}$ for all n, then $g = f^{[t]}$. (Use that K is an integral domain.)

COROLLARY 12.5.17.

$$(12.5.7) \qquad \operatorname{itlog}(f) = \left.\frac{df^{[t]}}{dt}\right|_{t=0}$$

and for $a \in K$,

$$(12.5.8) \qquad \operatorname{itlog}(f^{[a]}) = a \cdot \operatorname{itlog}(f).$$

If $g \in z + z^2K[[z]]$ *is unitary with* $f \circ g = g \circ f$, *then*

$$\operatorname{itlog}(f \circ g) = \operatorname{itlog}(f) + \operatorname{itlog}(g).$$

PROOF. Proposition 12.5.9 yields (12.5.7). Let $a \in K$. Then $f^{[at]}\big|_{t=n} = f^{[an]}$ for all n, as is easily checked, and $f^{[an]} = (f^{[a]})^{[n]}$ for all n by (12.5.6). Thus $f^{[at]} = (f^{[a]})^{[t]}$ by the uniqueness property of $(f^{[a]})^{[t]}$. Together with (12.5.7) this gives (12.5.8). The last statement follows from Lemma 12.1.3(ii). □

PROPOSITION 12.5.18 (Aczél and Jabotinsky). *In* $K^*[[z]]$ *we have*

$$(12.5.9) \qquad \operatorname{itlog}(f) \cdot \frac{\partial f^{[t]}}{\partial z} = \frac{df^{[t]}}{dt} = \operatorname{itlog}(f) \circ f^{[t]}$$

and by evaluating at $t = 1$ *we get in* $K[[z]]$,

$$(12.5.10) \qquad \operatorname{itlog}(f) \cdot f' = \operatorname{itlog}(f) \circ f.$$

PROOF. The first equality in (12.5.9) is from Proposition 12.5.9. To get the second one, let s be a new indeterminate, distinct from t and z. In $(K[s,t])[[z]]$ we have

$$f^{[s+t]} = f^{[s]} \circ f^{[t]}.$$

The K-derivations $\partial/\partial s$ and $\partial/\partial t$ on $K[s,t]$ extend to continuous K-derivations on $K[s,t][[z]]$ with $\partial z/\partial s = \partial z/\partial t = 0$. Then in $K[s,t][[z]]$,

$$\frac{\partial f^{[s+t]}}{\partial s} = \frac{\partial (f^{[s]} \circ f^{[t]})}{\partial s} = \frac{\partial f^{[s]}}{\partial s} \circ f^{[t]}.$$

At $s = 0$ the left-hand side becomes $\frac{df^{[t]}}{dt}$, and the right-hand side $\operatorname{itlog}(f) \circ f^{[t]}$. This gives the second equality of (12.5.9). □

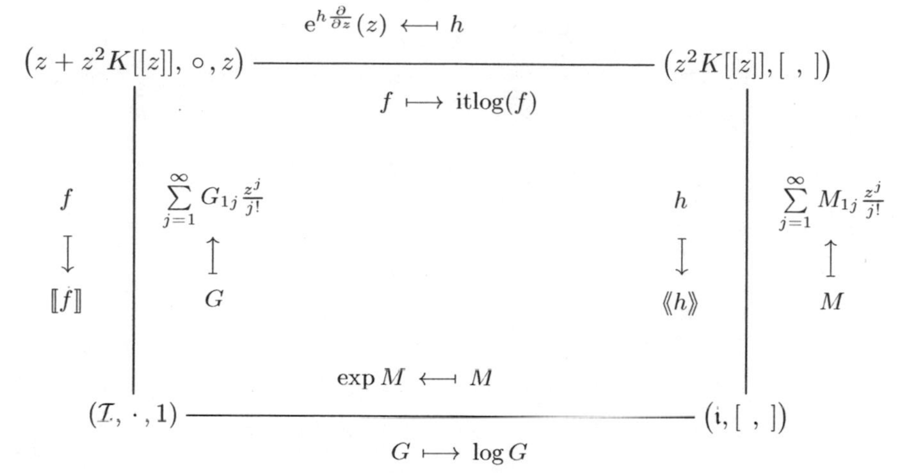

Figure 12.1: Relationship between power series and (infinitesimal) iteration matrices.

Notes and comments

A general reference for properties of the Bell polynomials (named after E. T. Bell [42]) is [90]. (Our notation differs slightly from that in [90]: our B_{ij} here is $\mathbf{B}_{ji}$ there.) Lemma 12.5.5 is from Jabotinsky [200, 201]. Theorem 12.5.7 for $K = \mathbb{C}$ is in [386]; the proof given here follows [16]. The Lie subalgebra $\bigoplus_{n\geqslant 1} K\langle\!\langle z^{n+1}\rangle\!\rangle$ of $\mathfrak{i}_K$, with $\left[\langle\!\langle z^{m+1}\rangle\!\rangle, \langle\!\langle z^{n+1}\rangle\!\rangle\right] = (m-n)\langle\!\langle z^{m+n+1}\rangle\!\rangle$ for all $m, n \geqslant 1$ (by (12.4.5)), is a variant of the "Witt algebra" [8, pp. 206–212]. The terminology "iterative logarithm" was coined in [118]. The construction of a family of power series $(f^{[a]})_{a\in K}$ satisfying (12.5.6) goes back to [130]. The functional equation (12.5.10) satisfied by the iterative logarithm is known as Julia's equation in iteration theory. (See [231, §8.5A].) It was found by Aczél [5] and Jabotinsky [201], although [157] suggests that G. Frege was already aware of it.

12.6 RIORDAN MATRICES

In this section K is an integral domain. We enlarge the group of iteration matrices to the group of so-called Riordan matrices.

The Riordan group

A **Riordan pair** over K is a pair (f, g) where $f \in zK[[z]]$ and $g \in K[[z]]$. Let (f, g) be a Riordan pair, $f = \sum f_n \frac{z^n}{n!}$ and $g = \sum g_n \frac{z^n}{n!}$ with $f_n, g_n \in K$ for all n, and $f_0 = 0$.

Then by (12.5.2), with i, j ranging over $\mathbb{N}$,

$$\frac{1}{i!} f^i g = \sum_{j \geqslant i} R_{ij} \frac{z^j}{j!}$$

where for $i \leqslant j$,

$$R_{ij} = \sum_{k=i}^{j} \binom{j}{k} B_{ik}(f_1, \dots, f_{k-i+1}) \cdot g_{j-k}.$$

We also set $R_{ij} = 0$ for $i > j$. We call the triangular matrix

$$[\![f, g]\!] := (R_{ij}) \in \mathfrak{tr}_K$$

the **Riordan matrix** of (f, g). Note that

$$R_{0j} = g_j \text{ for all } j, \qquad R_{1j} = \sum_{k=1}^{j} \binom{j}{k} f_k g_{j-k} \text{ for } j \geqslant 1.$$

Clearly $[\![f, 1]\!]$ is the iteration matrix $[\![f]\!]$ of f as introduced in Section 12.5. We have $R_{ii} = f_1^i g_0$ for each i, hence $[\![f, g]\!] \in 1 + \mathfrak{tr}_K^1$ iff $f_1 = g_0 = 1$. We say that the Riordan pair (f, g) is **unitary** if $f_1 = g_0 = 1$.

In the next lemma we identify a power series $a = \sum a_n \frac{z^n}{n!} \in K[[z]]$, $a_n \in K$ for each n, with the element (a_n) of $K^{\mathbb{N}}$, viewed as a row vector. With this convention:

LEMMA 12.6.1. *Let (f, g) be a Riordan pair over K and $a, b \in K[[z]]$. Then*

$$a \cdot [\![f, g]\!] = b \text{ in } K^{\mathbb{N}} \iff (a \circ f) \cdot g = b \text{ in } K[[z]].$$

PROOF. This is immediate from

$$(a \circ f) \cdot g = \left(\sum a_i \frac{f^i}{i!} \right) \cdot g = \sum_i a_i \left(\sum_{j \geqslant i} R_{ij} \frac{z^j}{j!} \right) = \sum_j \left(\sum_{i \leqslant j} a_i R_{ij} \right) \frac{z^j}{j!}. \quad \square$$

As a consequence the set of Riordan matrices of unitary Riordan pairs over K is a subgroup of the unitriangular group $\mathcal{U} = 1 + \mathfrak{tr}_K^1$:

COROLLARY 12.6.2. *Let (f, g) and (f^*, g^*) be Riordan pairs over K. Then*

$$\big(f \circ f^*, (g \circ f^*) \cdot g^*\big)$$

is a Riordan pair over K, and

$$[\![f, g]\!] \cdot [\![f^*, g^*]\!] = [\![f \circ f^*, (g \circ f^*) \cdot g^*]\!].$$

If (f, g), (f^*, g^*) are unitary, then so is $\big(f \circ f^*, (g \circ f^*) \cdot g^*\big)$. Moreover, if (f, g) is unitary, then $\big(f^{[-1]}, 1/(g \circ f^{[-1]})\big)$ is a unitary Riordan pair over K, and for $(f^*, g^*) = \big(f^{[-1]}, 1/(g \circ f^{[-1]})\big)$ we have

$$[\![f, g]\!] \cdot [\![f^*, g^*]\!] = [\![f^*, g^*]\!] \cdot [\![f, g]\!] = [\![z, 1]\!] = 1.$$

PROOF. The row with index i of $[\![f,g]\!]$ is $\frac{1}{i!}f^i g$, hence by the previous lemma, the row with index i of $[\![f,g]\!]\cdot[\![f^*,g^*]\!]$ is $\left((\frac{1}{i!}f^i g)\circ f^*\right)\cdot g^* = \frac{1}{i!}(f\circ f^*)^i\cdot\left((g\circ f^*)\cdot g^*\right)$. □

We call the subgroup $\mathcal{R}$ of $\mathcal{U}$ consisting of the Riordan matrices of unitary Riordan pairs over K the **Riordan group** over K. Note that the unitriangular group $\mathcal{R}$ over K is algebraic. The group $\mathcal{I}$ of iteration matrices is a subgroup of $\mathcal{R}$.

The Appell group and its Lie algebra

For $g=\sum g_n\frac{z^n}{n!}$ ($g_n\in K$ for all n), we set $[g] := [\![z,g]\!]$, that is,

$$[g] \;=\; \begin{pmatrix} g_0 & g_1 & g_2 & g_3 & g_4 & \cdots \\ & g_0 & 2g_1 & 3g_2 & 4g_3 & \cdots \\ & & g_0 & 3g_1 & 6g_2 & \cdots \\ & & & g_0 & 4g_1 & \cdots \\ & & & & g_0 & \cdots \\ & & & & & \ddots \end{pmatrix} \qquad \text{where } [g]_{ij}=\binom{j}{i}g_{j-i} \text{ for } i\leqslant j.$$

Note that $[g]\cdot[\![f]\!]=[\![f,g\circ f]\!]$ for $f\in zK[[z]]$, $g\in K[[z]]$. By Corollary 12.6.2, the map $g\mapsto[g]\colon K[[z]]\to\mathfrak{tr}_K$ is an embedding of K-algebras. Moreover,

$$g\in z^nK[[z]] \iff [g]\in\mathfrak{tr}_K^n \qquad (g\in K[[z]]).$$

Hence the image of this embedding, with the filtration induced by $(\mathfrak{tr}_K^n)$, is a complete filtered subalgebra of $\mathfrak{tr}_K$, and $g\mapsto[g]\colon K[[z]]\to\mathfrak{tr}_K$ is continuous. In particular, for $g\in zK[[z]]$ and $h\in 1+zK[[z]]$ we have

$$(12.6.1) \qquad [\exp g] \;=\; \exp\,[g], \qquad [\log h] \;=\; \log\,[h].$$

The embedding $g\mapsto[g]\colon K[[z]]\to\mathfrak{tr}_K$ maps the subgroup $1+zK[[z]]$ of $K[[z]]^\times$ onto a commutative normal subgroup of $\mathcal{R}$, called the **Appell group** over K and denoted here by $\mathcal{A}$. Now $\mathcal{A}\cap\mathcal{I}=\{1\}$, and from

$$[\![f,g]\!] \;=\; [\![f]\!]\cdot[g] \text{ for } [\![f,g]\!]\in\mathcal{R},$$

we get $\mathcal{R}=\mathcal{I}\cdot\mathcal{A}=\mathcal{A}\cdot\mathcal{I}$, so the group $\mathcal{R}$ is the internal semidirect product of its normal subgroup $\mathcal{A}$ with its subgroup $\mathcal{I}$. Now set

$$\mathfrak{a} \;:=\; \{[g] : g\in zK[[z]]\},$$

an abelian Lie subalgebra of the Lie algebra $\mathfrak{u}=\mathfrak{tr}_K^1$ over K; with the filtration induced by $\mathfrak{u}$, the filtered Lie algebra $\mathfrak{a}$ over K is complete. For $n\geqslant 1$, the matrix

$$\left[\tfrac{z^n}{n!}\right] \;=\; \operatorname{diag}_n\binom{i+n}{n} \;=\; A(n)\in\mathfrak{a}^{n-1}$$

is n-diagonal, and for $g=\sum_{n=1}^\infty g_n\frac{z^n}{n!}\in zK[[z]]$, where $g_n\in K$ for $n\geqslant 1$, we have

$$[g] \;=\; \sum_{n=1}^{\infty} g_n\left[\tfrac{z^n}{n!}\right] \;=\; \textstyle\sum_{n=1}^\infty g_nA(n) \text{ in } \mathfrak{a}.$$

By (12.6.1) we have $\exp(\mathfrak{a}) = \mathcal{A}$ and $\log(\mathcal{A}) = \mathfrak{a}$, so $\mathfrak{a}$ is the Lie algebra of the algebraic unitriangular group $\mathcal{A}$ over K.

For each unitary Riordan pair (f, g) and $\phi \in K^\times$, the pair $\big(\phi^{-1}f(\phi z), g(\phi z)\big)$ is also a unitary Riordan pair, and the diagonal matrix $D = \operatorname{diag}(\phi^i) = [\![\phi z, 1]\!]$ satisfies

$$D^{-1}[\![f, g]\!]D \;=\; [\![\phi^{-1}f(\phi z), g(\phi z)]\!].$$

Hence for all such D,

$$D\mathcal{R}D^{-1} = \mathcal{R}, \quad D\mathcal{I}D^{-1} = \mathcal{I}, \quad D\mathcal{A}D^{-1} = \mathcal{A}.$$

Some identities involving Riordan matrices

Recall from Section 12.4 the shift matrix $S \in K^{\mathbb{N}\times\mathbb{N}}$ and the derivative M' of a column-finite matrix $M \in K^{\mathbb{N}\times\mathbb{N}}$.

LEMMA 12.6.3. *Let (f, g) be a Riordan pair over K. Then*

$$[\![f, g]\!] \cdot S \;=\; S \cdot [\![f, f'g]\!] + [\![f, g']\!].$$

In particular, $[\![f]\!] \cdot S = S \cdot [\![f, f']\!]$ and $[g]' = [g]$.

PROOF. Let $i \in \mathbb{N}$. Then by definition of $[\![f, g]\!]$ we have

$$\frac{1}{i!}f^i g \;=\; \sum_{j \geqslant i} [\![f, g]\!]_{ij}\frac{z^j}{j!}.$$

Differentiating both sides with respect to z yields for $i \geqslant 1$,

$$\frac{1}{(i-1)!}f^{i-1}f'g + \frac{1}{i!}f^i g' \;=\; \sum_{j \geqslant i-1} [\![f, g]\!]_{i,j+1}\frac{z^j}{j!}.$$

The definition of $[\![f, f'g]\!]$ and $[\![f, g']\!]$ gives for $i \geqslant 1$,

$$\frac{1}{(i-1)!}f^{i-1}f'g + \frac{1}{i!}f^i g' \;=\; \left(\sum_{j \geqslant i-1} [\![f, f'g]\!]_{i-1,j}\frac{z^j}{j!}\right) + \left(\sum_{j \geqslant i} [\![f, g']\!]_{i,j}\frac{z^j}{j!}\right).$$

These last two identities together give for $i \geqslant 1$ and $j \geqslant i-1$:

$$\big([\![f, g]\!] \cdot S\big)_{ij} \;=\; [\![f, g]\!]_{i,j+1} \;=\; [\![f, f'g]\!]_{i-1,j} + [\![f, g']\!]_{ij} \;=\; \big(S \cdot [\![f, f'g]\!] + [\![f, g']\!]\big)_{ij}.$$

These equalities actually hold for all $i, j \in \mathbb{N}$, as is easily verified using $B_{0j} = 0$ for $j \geqslant 1$, and thus $[\![f, g]\!] \cdot S = S \cdot [\![f, f'g]\!] + [\![f, g']\!]$. Taking $g = 1$ yields $[\![f]\!] \cdot S = S \cdot [\![f, f']\!]$. Taking $f = z$ gives $[g]' = [g']$. □

COROLLARY 12.6.4. *Let $f \in z + z^2K[[z]]$ and $g \in K[[z]]$. Then*

$$\begin{aligned} f' &\in 1 + zK[[z]] \subseteq K[[z]]^\times, \\ [\![f,g]\!]' &= [\![f,g]\!] \cdot S \cdot [1 - (1/f')] + [\![f, g'/f']\!], \quad \textit{in particular,} \\ [\![f]\!]' &= [\![f]\!] \cdot S \cdot [1 - (1/f')]. \end{aligned}$$

PROOF. By the previous lemma,

$$[\![f,g]\!] \cdot S = (S \cdot [\![f,g]\!] + [\![f, g'/f']\!]) \cdot [f'].$$

Note that $f' \in 1 + zK[[z]]$, so $[f'] \in 1 + \mathfrak{tr}_K^1 = \mathcal{U}$, and hence

$$S \cdot [\![f,g]\!] = [\![f,g]\!] \cdot S \cdot [f']^{-1} - [\![f, g'/f']\!],$$

and thus

$$\begin{aligned} [\![f,g]\!]' &= [\![f,g]\!] \cdot S - S \cdot [\![f,g]\!] \\ &= [\![f,g]\!] \cdot S \cdot (1 - [f']^{-1}) + [\![f, g'/f']\!] \\ &= [\![f,g]\!] \cdot S \cdot [1 - (1/f')] + [\![f, g'/f']\!] \end{aligned}$$

as claimed. □

The Lie algebra of $\mathcal{R}$

The material in this subsection is not used later.

LEMMA 12.6.5. *Let $g \in zK[[z]]$ and $h \in z^2K[[z]]$. Then* $\big[[g], \langle\!\langle h \rangle\!\rangle\big] = [g'h]$.

PROOF. Let $m, n \geqslant 1$. With the notation from Section 12.4,

$$\left[\tfrac{z^m}{m!}\right] = A(m), \qquad \left\langle\!\!\left\langle \tfrac{z^n}{n!} \right\rangle\!\!\right\rangle = B(n-1),$$

and so by Lemma 12.4.6,

$$\begin{aligned} \left[\left[\tfrac{z^m}{m!}\right], \left\langle\!\!\left\langle \tfrac{z^n}{n!} \right\rangle\!\!\right\rangle\right] &= [A(m), B(n-1)] = \binom{m+n-1}{n} A(m+n-1) \\ &= \binom{m+n-1}{n} \left[\tfrac{z^{m+n-1}}{(m+n-1)!}\right]. \end{aligned}$$

The general case now follows from this special case and the continuity of the Lie bracket operation. □

Clearly $\mathfrak{a} \cap \mathfrak{i} = \{0\}$, where $\mathfrak{i}$ is the Lie algebra of $\mathcal{I}$. Let $\mathfrak{r}$ be the Lie algebra of the algebraic unitriangular group $\mathcal{R}$ over K. From the previous lemma, it follows that $\mathfrak{a} \oplus \mathfrak{i}$ is a Lie subalgebra of $\mathfrak{r}$ and $\mathfrak{a}$ is an ideal of $\mathfrak{a} \oplus \mathfrak{i}$. In fact, $\mathfrak{r} = \mathfrak{a} \oplus \mathfrak{i}$, but we will not prove this here.

Notes and comments

The Riordan group (introduced in [403] and named in honor of J. Riordan for his work [340] on combinatorial identities) is connected to Rota's "umbral calculus" dealing with sequences of polynomials. In this calculus, an Appell sequence (over K) is a sequence $(P_j)_{j\geqslant 0}$ of polynomials $P_j \in K[z]$ with P_0 of degree 0 and $P_j' = jP_{j-1}$ for $j > 0$ (see [358, Theorem 2.5.6]). The subgroup $\mathcal{A}$ of the Riordan group is called the Appell group because for $R = (R_{ij}) \in \mathcal{A}$, the sequence $\big(\sum_{i=0}^{j} R_{ij} z^i\big)_{j\geqslant 0}$ is an Appell sequence. The subgroup $\mathcal{I}$ of $\mathcal{R}$ is also called the "associated subgroup" in the combinatorics literature.

12.7 DERIVATIONS ON POLYNOMIAL RINGS

In this section A is a commutative ring containing K (and thus $\mathbb{Q}$) as a subring. Then a K-derivation on A as defined in Section 12.1 is the same as a derivation on A whose ring of constants contains K. Recall that $\operatorname{der}_K(A)$ is the Lie algebra over K consisting of the K-derivations on A, with Lie bracket

$$[\Delta, \Lambda] = \Delta\Lambda - \Lambda\Delta \qquad (\Delta, \Lambda \in \operatorname{der}_K(A)).$$

Since A is commutative, $a\Delta \in \operatorname{der}_K(A)$ for all $a \in A$ and $\Delta \in \operatorname{der}_K(A)$, and so $\operatorname{der}_K(A)$ is naturally a left A-module. For $\partial \in \operatorname{der}_K(A)$, let A^∂ be the ring of constants of ∂, so A^∂ is a K-subalgebra of A. Let $\Delta, \Lambda \in \operatorname{der}_K(A)$. Then we set $A^{\Delta,\Lambda} := A^\Delta \cap A^\Lambda$ and, with $\operatorname{Lie}(\Delta, \Lambda)$ the Lie subalgebra of $\operatorname{der}_K(A)$ over K generated by Δ, Λ, we have

(12.7.1)
$$A^{\Delta,\Lambda} = \bigcap_{\partial \in \operatorname{Lie}(\Delta,\Lambda)} A^\partial.$$

Exponential automorphisms

If $\Delta \in \operatorname{der}_K(A)$ is locally nilpotent, then the (locally unipotent) automorphism $\exp\Delta$ of the K-module A, with inverse $\exp(-\Delta)$, is a K-algebra automorphism (Lemma 12.1.4). The K-automorphisms of A of the form $\exp\Delta$ with locally nilpotent $\Delta \in \operatorname{der}_K(A)$ are said to be **exponential.** The above can be reversed:

PROPOSITION 12.7.1. *Let σ be a locally unipotent K-algebra endomorphism. Then $\log\sigma$ is a locally nilpotent K-derivation on A, and so $\sigma = \exp(\log\sigma)$ is an exponential automorphism.*

PROOF. Set $\Delta := \log\sigma$, so Δ is a locally nilpotent endomorphism of the K-module A, and $\sigma = \exp\Delta$. Let $a, b \in A$ and take $N \in \mathbb{N}^{\geqslant 1}$ with $\Delta^i(a) = \Delta^i(b) = \Delta^i(ab) = 0$ for all $i > N/2$. Then $\sigma(ab) = \sigma(a)\sigma(b)$ gives

$$\sum_{k=0}^{N} \frac{1}{k!}\Delta^k(ab) = \sum_{k=0}^{N} \sum_{i+j=k} \frac{1}{i!}\frac{1}{j!}\Delta^i(a)\Delta^j(b).$$

Replacing σ by σ^n changes Δ to $n\Delta$, and the above equality remains valid for Δ replaced by $n\Delta$, that is, $\sum_{k=0}^{N} c_k n^k = 0$ with

$$c_k := \frac{1}{k!}\Delta^k(ab) - \sum_{i+j=k} \frac{1}{i!}\frac{1}{j!}\Delta^i(a)\Delta^j(b) \in A.$$

The $(N+1) \times (N+1)$ Vandermonde matrix with rows $(n^0, n^1, \dots, n^N)$ for $n = 0, \dots, N$ is invertible, but annihilates the column vector $(c_0, \dots, c_N)^{\mathrm{t}}$, so $c_k = 0$ for $k = 0, \dots, N$. For $k = 1$ this yields $\Delta(ab) = \Delta(a)b + a\Delta(b)$. □

If $\Delta \in \mathrm{der}_K(A)$ is locally nilpotent and $a \in A^{\Delta}$ then $a\Delta$ is locally nilpotent. Using the Leibniz rule one also shows:

LEMMA 12.7.2. *If $S \subseteq A$ generates the K-algebra A, and $\Delta \in \mathrm{der}_K(A)$ is such that for every $s \in S$ there is n with $\Delta^n(s) = 0$, then Δ is locally nilpotent.*

If $\Delta, \Lambda \in \mathrm{der}_K(A)$ are locally nilpotent and $[\Delta, \Lambda] = 0$, then

$$\exp(\Delta)\exp(\Lambda) = \exp(\Delta + \Lambda)$$

is an exponential automorphism of A. In particular, for each locally nilpotent $\Delta \in \mathrm{der}_K(A)$ and $k \in \mathbb{Z}$, $\exp(\Delta)^k$ is an exponential automorphism of A with $\exp(\Delta)^k = \exp(k\Delta)$.

REMARK. The composition of two exponential automorphisms is not in general exponential. For example, take $A = K[Y_0, Y_1]$. Then the K-derivations $\Delta = Y_0\frac{\partial}{\partial Y_1}$, $\Lambda = Y_1\frac{\partial}{\partial Y_0}$ on A are locally nilpotent. A computation shows that

$$\Phi := \exp(\Delta)\exp(\Lambda) - 1$$

satisfies $\Phi(Y_0) = Y_0 + Y_1$, $\Phi(Y_1) = Y_0$ and hence $\Phi^n(Y_0) \neq 0$ for all n. So Φ is not locally nilpotent, and the K-algebra automorphism $\exp(\Delta)\exp(\Lambda)$ of A is not locally unipotent, and thus not exponential.

In the next subsection we study a class of exponential automorphisms of polynomial algebras over K which do form a group under composition.

Triangular derivations

In the rest of this section $A = K[Y_0, Y_1, Y_2, \dots]$ where (Y_n) is a sequence of distinct indeterminates. For each n we also put

$$Y_{<n} = (Y_0, \dots, Y_{n-1}), \text{ so}$$
$$K[Y_{<m}] \subseteq K[Y_{<n}] \subseteq K[Y] = A \qquad \text{for } m \leqslant n, \text{ and}$$
$$A = \bigcup_n K[Y_{<n}].$$

We abbreviate $\mathrm{der}_K(A)$ by der_K, and we denote by ∂_m the K-derivation $\frac{\partial}{\partial Y_m}$ of A as well as any restriction of this derivation to a K-subalgebra $K[Y_{<n}]$ where $m < n$. The following well-known fact has a routine proof:

LEMMA 12.7.3. *$\operatorname{der}_K(K[Y_{<n}])$ is a free $K[Y_{<n}]$-module with basis $\partial_0, \dots, \partial_{n-1}$, and $[\partial_i, \partial_j] = 0$ for $0 \leqslant i, j < n$. For each $\Delta \in \operatorname{der}_K(K[Y_{<n}])$ we have*

$$\Delta = \Delta(Y_0)\partial_0 + \cdots + \Delta(Y_{n-1})\partial_{n-1}.$$

In a similar vein: for each sequence (f_n) in A the sequence of endomorphisms $\left(f_n \frac{\partial}{\partial Y_n}\right)$ of the K-module A is summable (with respect to the trivial filtration on the K-module A as in Section 12.2), and $\Delta := \sum_{n=0}^{\infty} f_n \frac{\partial}{\partial Y_n}$ is the unique K-derivation of A with $\Delta(Y_n) = f_n$ for all n.

If $\Delta \in \operatorname{der}_K$ and $\Delta(Y_{n-1}) \in K[Y_{<n}]$ for all $n \geqslant 1$, then $\Delta(K[Y_{<n}]) \subseteq K[Y_{<n}]$ for all n. Note also that each K-derivation $\partial_n \in \operatorname{der}_K$ is locally nilpotent. More generally, every $\Delta \in \operatorname{der}_K$ with $\Delta(Y_n) \in K[Y_{<n}]$ for all n is locally nilpotent, by the following:

LEMMA 12.7.4. *Let Δ be a K-derivation of $K[Y_{<n}]$ with $\Delta(Y_j) \in K[Y_{<j}]$ for $j = 0, \dots, n-1$. Then Δ is locally nilpotent.*

PROOF. We proceed by induction on n. The case $n = 0$ is clear, so let $n > 0$. Then the derivation Δ restricts to a K-derivation of $K[Y_{<n-1}]$, and so inductively we can assume this restriction to be locally nilpotent. Now $\Delta(Y_{n-1}) \in K[Y_{<n-1}]$ and hence $\Delta^{m+1}(Y_{n-1}) = \Delta^m(\Delta(Y_{n-1})) = 0$ for some m. Thus Δ is locally nilpotent, by Lemma 12.7.2. □

DEFINITION 12.7.5. Let Δ be a K-derivation of A. Then we call Δ **triangular** (relative to $Y_0, Y_1, \dots$) if Δ restricts to a triangular endomorphism of the K-submodule $A_1 = \bigoplus_j K Y_j$ of A with respect to its basis $Y_0, Y_1, \dots$; that is,

$$\Delta(Y_j) = \Delta_{0j}Y_0 + \Delta_{1j}Y_1 + \cdots + \Delta_{jj}Y_j$$

where $\Delta_{ij} \in K$ for $i \leqslant j$. If also $\Delta_{jj} = 0$ for all j, then Δ is **strictly triangular**. Every triangular endomorphism of the K-submodule A_1 with respect to the basis $Y_0, Y_1, \dots$ extends uniquely to a (necessarily triangular) K-derivation of A.

Let $\operatorname{trder}_K = \operatorname{trder}_K(A)$ be the set of triangular K-derivations of A. The Lie bracket $[\Delta, \Lambda]$ of triangular K-derivations Δ, Λ of A is also triangular; hence trder_K is a Lie subalgebra (over K) of der_K. For $\Delta \in \operatorname{trder}_K$ with the Δ_{ij} as above, put

$$M_\Delta := M_{\Delta|A_1} = (\Delta_{ij})_{i,j \in \mathbb{N}} \in \mathfrak{tr}_K,$$

with $\Delta_{ij} := 0$ for $i > j$, by convention. We obtain a complete filtration $(\operatorname{trder}_K^n)_{n \geqslant 0}$ of the Lie algebra trder_K over K by setting

$$\operatorname{trder}_K^n := \{\Delta \in \operatorname{trder}_K : M_\Delta \in \mathfrak{tr}_K^n\}.$$

(So trder_K^1 consists of all strictly triangular K-derivations.) We have a commuting diagram of isomorphisms of Lie algebras over K:

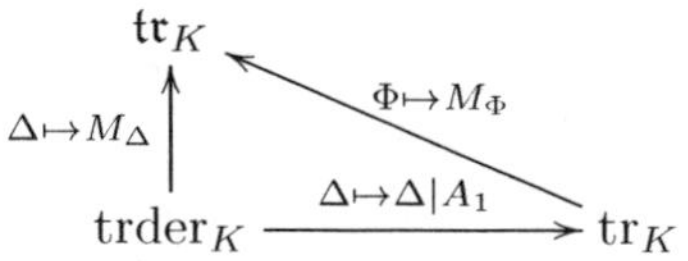

Every triangular K-derivation of $A = K[Y]$ restricts to a derivation of $K[Y_{<n}]$, for each n. Hence by Lemma 12.7.4, every strictly triangular K-derivation of A is locally nilpotent. Note that if $\Delta \in \operatorname{trder}_K^n$ and $P \in K[Y_{<n}]$, then $\Delta(P) = 0$.

Diagonals

In the next lemma d is a degree function on A and $\Delta \in \operatorname{der}_K$.

LEMMA 12.7.6. *Let $d \in \mathbb{Z}$ be such that for each n, $\Delta(Y_n)$ is* d*-homogeneous of degree* $\mathrm{d}(Y_n) + d$. *Then Δ is* d*-homogeneous of degree d.*

PROOF. Let $\boldsymbol{i} \in \mathbb{N}^{(\mathbb{N})}$ with $\mathrm{d}(Y^{\boldsymbol{i}}) = i$. It is enough to show that then $\Delta(Y^{\boldsymbol{i}})$ is d-homogeneous of degree $i + d$. For such $\boldsymbol{i}$ we have

$$\Delta(Y^{\boldsymbol{i}}) = \sum_{n \geqslant 0} \Delta(Y_n)\partial_n(Y^{\boldsymbol{i}})$$

with $\Delta(Y_n)$ d-homogeneous of degree $d_n + d$ and $\partial_n(Y^{\boldsymbol{i}})$ d-homogeneous of degree $i - d_n$, where $d_n := \mathrm{d}(Y_n)$. □

Thus if $d \in \mathbb{Z}$ and $\Delta(Y_n) \in A_{d+1}$ for every n, then Δ is homogeneous of degree d. If $w \in \mathbb{Z}$ and $\Delta(Y_n) \in A_{[n+w]}$ for every n, then Δ is isobaric of weight w.

DEFINITION 12.7.7. Let $\Delta \in \operatorname{trder}_K$. We call the triangular K-derivation Δ_n of A with associated matrix $M_{\Delta_n} = (M_\Delta)_n$ the nth **diagonal** or the n**-diagonal** of Δ. We also say that Δ is n**-diagonal** if $\Delta = \Delta_n$ and **diagonal** if $\Delta = \Delta_0$.

In the metric on trder_K given by the filtration $(\operatorname{trder}_K^n)$ we have $\sum_{i=0}^n \Delta_i \to \Delta$ as $n \to \infty$; more suggestively,

$$\Delta = \Delta_0 + \Delta_1 + \cdots + \Delta_n + \cdots .$$

Also, $\Delta \in \operatorname{trder}_K^m$ if and only if $\Delta_0 = \cdots = \Delta_{m-1} = 0$. Each n-diagonal derivation is homogeneous of degree 0 and isobaric of weight $-n$, by Lemma 12.7.6. In particular, for all $P \in A$ and $\Delta \in \operatorname{trder}_K^m$ we have

$$\deg\big(\Delta(P)\big) \leqslant \deg(P), \qquad \operatorname{wt}\big(\Delta(P)\big) \leqslant \operatorname{wt}(P) - m.$$

From Lemma 12.2.2 we obtain:

LEMMA 12.7.8. *Let $\Delta, \Lambda \in \operatorname{trder}_K$. If Δ is m-diagonal and Λ is n-diagonal, then $[\Delta, \Lambda]$ is $(m+n)$-diagonal. For each $k \in \mathbb{N}$ we have*

$$[\Delta, \Lambda]_k = \sum_{m+n=k} [\Delta_m, \Lambda_n].$$

Also, if $\Delta \in \operatorname{trder}_K$ is n-diagonal, then $A^\Delta \supseteq K[Y_{<n}]$ and

(12.7.2) $$\Delta = \Delta_{0n} Y_0 \partial_n + \Delta_{1,n+1} Y_1 \partial_{n+1} + \Delta_{2,n+2} Y_2 \partial_{n+2} + \cdots .$$

Triangular algebra endomorphisms

These are defined as follows.

DEFINITION 12.7.9. An endomorphism σ of the K-algebra A is called **triangular** (relative to $Y_0, Y_1, \dots$) if σ restricts to a triangular endomorphism of the K-submodule $A_1 = \bigoplus_j K\, Y_j$ of A with respect to its basis $Y_0, Y_1, \dots$; that is,

$$\sigma(Y_j) \;=\; \sigma_{0j}Y_0 + \sigma_{1j}Y_1 + \cdots + \sigma_{jj}Y_j$$

where $\sigma_{ij} \in K$ for $i \leqslant j$. If in addition $\sigma_{jj} = 1$ for all j, then σ is **unitriangular**. If σ is triangular with $\sigma_{ij} = 0$ for all $i < j$, then σ is called **diagonal.**

Given a triangular K-algebra endomorphism σ of A, set

$$M_\sigma \;:=\; M_{\sigma|A_1} \;=\; (\sigma_{ij})_{i,j\in\mathbb{N}} \in \mathfrak{tr}_K,$$

with $\sigma_{ij} := 0$ for $i > j$, by convention. The following is easy to verify:

LEMMA 12.7.10. *Let σ be a triangular K-algebra endomorphism of A. Then*

(i) *σ is bijective $\Longleftrightarrow \sigma_{jj} \in K^\times$ for all $j \Longleftrightarrow M_\sigma \in \mathfrak{tr}_K^\times$;*

(ii) *σ is unitriangular $\Longleftrightarrow M_\sigma \in 1 + \mathfrak{tr}_K^1$;*

(iii) *σ is bijective and diagonal $\Longleftrightarrow M_\sigma \in D_K^\times$.*

The triangular K-algebra automorphisms of A form a subgroup TrAut_K of the group of all automorphisms of the K-algebra A. The map

$$\sigma \mapsto M_\sigma\colon\ \operatorname{TrAut}_K \to \mathfrak{tr}_K^\times$$

is a group isomorphism. Let $m \geqslant 1$. Define TrAut_K^m to be the subgroup of TrAut_K consisting of the triangular K-algebra automorphisms σ of A with $M_\sigma \in 1 + \mathfrak{tr}_K^m$. For each $\Delta \in \operatorname{trder}_K^m$ we have $\exp\Delta \in \operatorname{TrAut}_K^m$ and $\exp M_\Delta = M_{\exp\Delta}$. So we have a commuting diagram

$$\begin{array}{ccc}
\operatorname{trder}_K^m & \xrightarrow{\ \Delta\mapsto M_\Delta\ } & \mathfrak{tr}_K^m \\
\big\downarrow{\scriptstyle\exp} & & \big\downarrow{\scriptstyle\exp} \\
\operatorname{TrAut}_K^m & \xrightarrow{\ \sigma\mapsto M_\sigma\ } & 1 + \mathfrak{tr}_K^m
\end{array}$$

where the horizontal arrows are group isomorphisms and the right (and hence also the left) vertical arrow is a bijection. If $\sigma \in \operatorname{TrAut}_K^m$, then $\log\sigma \in \operatorname{trder}_K^m$, $\log M_\sigma = M_{\log\sigma}$, and the automorphism $\sigma = \exp(\log\sigma)$ is exponential.

Let σ be a triangular K-automorphism of A. Then for every triangular K-derivation Δ on A, the K-derivation $\sigma\Delta\sigma^{-1}$ of A is triangular, the map $\Delta \mapsto \sigma\Delta\sigma^{-1}$ is an automorphism of the Lie algebra trder_K over K, and

$$M_{\sigma\Delta\sigma^{-1}} \;=\; M_\sigma\, M_\Delta\, (M_\sigma)^{-1}.$$

If $m \geqslant 1$, $\Delta \in \operatorname{trder}_K^m$, then $\sigma\Delta\sigma^{-1} \in \operatorname{trder}_K^m$ and

$$\exp(\sigma\Delta\sigma^{-1}) \;=\; \sigma\exp(\Delta)\sigma^{-1}.$$

Companion derivations

Let $\Delta \in \operatorname{trder}^1_K$ and $\sigma = \exp\Delta$. Then σ is homogeneous of degree 0; in fact $\deg\sigma(P) = \deg P$ for every $P \in A$. Next, we investigate the isobaric parts of the images of polynomials under Δ. The following lemma for P of the form $P = Y_n$ is already implicit in Lemma 12.2.4:

LEMMA 12.7.11. *Let $\Delta \in \operatorname{trder}^1_K$ and put $\sigma := \exp(\Delta)$. Suppose also that $P \in A_{[w]}$, $P \neq 0$, with $w \in \mathbb{N}$. Then $\operatorname{wt}\sigma(P) = w$, and*

$$\begin{aligned}
\sigma(P)_{[w]} &= P,\\
\sigma(P)_{[w-1]} &= \Delta_1(P),\\
\sigma(P)_{[w-2]} &= \Delta_2(P) + \tfrac{1}{2}(\Delta_1)^2(P),\\
\sigma(P)_{[w-3]} &= \Delta_3(P) + \tfrac{1}{2}(\Delta_1\Delta_2 + \Delta_2\Delta_1)(P) + \tfrac{1}{6}(\Delta_1)^3(P)\\
&\ \vdots\\
\sigma(P)_{[w-n]} &= \sum_{i_1,\dots,i_k} \frac{1}{k!}(\Delta_{i_1}\cdots\Delta_{i_k})(P) \qquad (n \geqslant 1),
\end{aligned}$$

summed over the $(i_1,\dots,i_k)$ with $k \geqslant 1$, $i_1,\dots,i_k \geqslant 1$ and $i_1 + \cdots + i_k = n$.

PROOF. Note that $\Delta^k(P) = 0$ for all big enough $k \in \mathbb{N}$. For all i,

$$\sigma(P)_{[i]} = P_{[i]} + \Delta(P)_{[i]} + \frac{1}{2}\Delta^2(P)_{[i]} + \cdots + \frac{1}{k!}\Delta^k(P)_{[i]} + \cdots.$$

Now $\operatorname{wt}(\Delta^k(P)) \leqslant \operatorname{wt}(P) - k < \operatorname{wt} P = w$ for $k \geqslant 1$, hence $\operatorname{wt}(\sigma(P)) = w$ and $\sigma(P)_{[w]} = P$. The family $(\Delta_i)_{i\geqslant 1}$ of endomorphisms of the K-module A is summable, with $\Delta = \sum_{i\geqslant 1}\Delta_i$. Let $k \geqslant 1$. Then

$$\Delta^k = \sum_{i_1,\dots,i_k} \Delta_{i_1}\cdots\Delta_{i_k}$$

summed over the $(i_1,\dots,i_k)$ with $i_1,\dots,i_k \geqslant 1$. Each operator $\Delta_{i_1}\cdots\Delta_{i_k}$ in this sum is isobaric of weight $-(i_1 + \cdots + i_k)$, so for all n,

$$\Delta^k(P)_{[w-n]} = \sum_{i_1,\dots,i_k} (\Delta_{i_1}\cdots\Delta_{i_k})(P),$$

summed over the $(i_1,\dots,i_k)$ with $i_1,\dots,i_k \geqslant 1$ and $i_1 + \cdots + i_k = n$. This yields the lemma. Note also that $\big(\Delta^k(P)\big)_{[w-n]} = 0$ if $k > n$. □

As before, $\Delta \in \operatorname{trder}^1_K$ and $\sigma = \exp(\Delta)$. We say that an element P of A is σ-**invariant** if $\sigma(P) = P$. We already observed that $P \in A$ is σ-invariant iff $\Delta(P) = 0$; see (12.2.1). We obtain a refinement of this fact for isobaric polynomials:

COROLLARY 12.7.12. *Let $P \in A_{[w]}$ where $w \in \mathbb{N}$. Then for $n = 1, \dots, w$*

$$\sigma(P)_{[w-1]} = \cdots = \sigma(P)_{[w-n]} = 0 \qquad \Longleftrightarrow \qquad \Delta_1(P) = \cdots = \Delta_n(P) = 0.$$

In particular: $\sigma(P) = P \iff \Delta_n(P) = 0$ for $n = 1, \dots, w$.

PROOF. From the previous lemma we have, for $n = 1, \dots, w$:

$$\begin{aligned}\sigma(P)_{[w-n]} &= \Delta_n(P) + \mathbb{Q}\text{-linear combination of terms } (\Delta_{i_1} \cdots \Delta_{i_k})(P) \\ &\qquad \text{with } k > 1 \text{ and } 1 \leqslant i_1, \dots, i_k < n.\end{aligned}$$

This yields the corollary. □

Thus if $P \in A$ is isobaric and σ-invariant, then

$$\Delta_1(P) = \Delta_2(P) = 0. \tag{12.7.3}$$

The polynomials $P \in A$, not necessarily isobaric, that satisfy (12.7.3), comprise the K-subalgebra A^{Δ_1,Δ_2} of A. Note that if $P \in A^{\Delta_1,\Delta_2}$ then $\Lambda(P) = 0$ for all Λ in the Lie subalgebra $\operatorname{Lie}(\Delta_1, \Delta_2)$ (over K) of trder_K generated by Δ_1, Δ_2; cf. (12.7.1). This Lie algebra contains in particular the so-called **companion** derivations

$$\begin{aligned}
\Delta_1^{\mathrm{c}} &= \Delta_1 \\
\Delta_2^{\mathrm{c}} &= \Delta_2 \\
\Delta_3^{\mathrm{c}} &= [\Delta_1, \Delta_2] = \operatorname{ad}_{\Delta_1}(\Delta_2) \\
\Delta_4^{\mathrm{c}} &= [\Delta_1, \Delta_3^{\mathrm{c}}] = [\Delta_1, \Delta_1, \Delta_2] = \operatorname{ad}_{\Delta_1}^2(\Delta_2) \\
\Delta_5^{\mathrm{c}} &= [\Delta_1, \Delta_4^{\mathrm{c}}] = [\Delta_1, \Delta_1, \Delta_1, \Delta_2] = \operatorname{ad}_{\Delta_1}^3(\Delta_2) \\
&\ \vdots \\
\Delta_n^{\mathrm{c}} &= [\Delta_1, \Delta_{n-1}^{\mathrm{c}}] = [\underbrace{\Delta_1, \dots, \Delta_1}_{n-2 \text{ times}}, \Delta_2] = \operatorname{ad}_{\Delta_1}^{n-2}(\Delta_2) \qquad \text{for } n \geqslant 2.
\end{aligned}$$

Each Δ_n^{c} is n-diagonal. We combine the Δ_n^{c} into the strictly triangular K-derivation

$$\Delta^{\mathrm{c}} := \Delta_1^{\mathrm{c}} + \Delta_2^{\mathrm{c}} + \cdots + \Delta_n^{\mathrm{c}} + \cdots,$$

which we call the **companion derivation** of Δ. Note that $\Delta_{0n}^{\mathrm{c}} = (\Delta_n^{\mathrm{c}})_{0n}$ for $n \geqslant 1$, in particular, $\Delta_{01}^{\mathrm{c}} = \Delta_{01}$ and $\Delta_{02}^{\mathrm{c}} = \Delta_{02}$.

PROPOSITION 12.7.13. *Suppose $\Delta_{0n}^{\mathrm{c}} \neq 0$ for all $n \geqslant 1$. Then $A^{\Delta_1,\Delta_2} = K[Y_0]$.*

PROOF. In connection with (12.7.2) we already observed that $K[Y_0] \subseteq A^{\Delta_1,\Delta_2}$. For the reverse inclusion, let $P \in A^{\Delta_1,\Delta_2}$. Take n minimal with $P \in K[Y_0, \dots, Y_n]$. Towards a contradiction, suppose that $n \geqslant 1$. By (12.7.2) and the hypothesis of the proposition, we have

$$\Delta_n^{\mathrm{c}} = \Delta_{0,n}^{\mathrm{c}} Y_0 \,\partial_n + \Delta_{1,n+1}^{\mathrm{c}} Y_1 \,\partial_{n+1} + \cdots, \qquad \Delta_{0,n}^{\mathrm{c}} \neq 0.$$

Thus $\Delta_n^{\mathrm{c}}(P) = \Delta_{0,n}^{\mathrm{c}} Y_0 \frac{\partial P}{\partial Y_n} \neq 0$, a contradiction. □

COROLLARY 12.7.14. *Suppose* $\Delta_{01} = \Delta_{02} = 0$ *and* $\Delta^{c}_{1,n} \neq 0$ *for all* $n \geqslant 2$. *Then* $A^{\Delta_1,\Delta_2} = K[Y_0, Y_1]$, *so all isobaric* σ*-invariants belong to* $K[Y_0, Y_1]$.

PROOF. Put $K^* := K[Y_0]$ and view A as the K^*-algebra of polynomials in the indeterminates $Y_1, Y_2, \dots$ over K^*. Since $\Delta_{01} = \Delta_{02} = 0$, Δ_1 and Δ_2 are triangular K^*-derivations on A relative to $Y_1, Y_2, \dots$, and the claim now follows from Proposition 12.7.13 and (12.7.3). □

The Stirling automorphism

In the rest of this section K *is an integral domain.* The Stirling automorphism Υ of A is the unitriangular automorphism of A whose matrix $M_\Upsilon \in \mathfrak{tr}_{\mathbb{Q}}$ has the signed Stirling numbers of the first kind as its entries:

$$M_\Upsilon = (\Upsilon_{ij}) := \begin{pmatrix} 1 & 0 & 0 & 0 & 0 & 0 & \cdots \\ & 1 & -1 & 2 & -6 & 24 & \cdots \\ & & 1 & -3 & 11 & -50 & \cdots \\ & & & 1 & -6 & 35 & \cdots \\ & & & & 1 & -10 & \cdots \\ & & & & & 1 & \cdots \\ & & & & & & \ddots \end{pmatrix} \quad \text{where } \Upsilon_{ij} = (-1)^{j-i}\begin{bmatrix} j \\ i \end{bmatrix}.$$

By the recurrence relation (5.7.3) for Stirling numbers of the first kind,

$$(12.7.4) \qquad M'_\Upsilon = -M_\Upsilon D \qquad \text{where } D = \operatorname{diag}(0,1,2,3,\dots).$$

The matrix M_Υ is an iteration matrix:

LEMMA 12.7.15. $M_\Upsilon = [\![\log(1+z)]\!]$.

PROOF. Let $f := \log(1+z) \in z + z^2\mathbb{Q}[[z]]$. Then

$$[1-(1/f')] = [-z] = -\operatorname{diag}_1(1,2,3,\dots)$$

and thus

$$S \cdot [1-(1/f')] = -\operatorname{diag}(0,1,2,3,\dots) = -D,$$

where D is as in (12.7.4). So by Corollary 12.6.4 and by (12.7.4), respectively, both $X = [\![f]\!]$ and $X = M_\Upsilon$ satisfy the differential equation $X' = -XD$, and have the same leftmost column. Thus $[\![f]\!] = M_\Upsilon$ by Lemma 12.4.5. □

Since $\mathrm{e}^z - 1$ is the formal compositional inverse of $\log(1+z)$ in $z + z^2K[[z]]$, we get

$$M_{\Upsilon^{-1}} = (M_\Upsilon)^{-1} = [\![\mathrm{e}^z - 1]\!].$$

By Lemma 12.4.4 and (12.7.4), its derivative satisfies the identity

$$M'_{\Upsilon^{-1}} = DM_{\Upsilon^{-1}}, \qquad D := \operatorname{diag}(0,1,2,3,\dots),$$

which gives a recurrence relation for the entries of $M_{\Upsilon^{-1}} = (\Upsilon^{-1}_{ij})$: for all i, j

$$\Upsilon^{-1}_{i,j+1} = \Upsilon^{-1}_{i-1,j} + i\Upsilon^{-1}_{ij} \qquad (\Upsilon_{-1,j} := 0 \text{ by convention})$$

with side conditions $\Upsilon^{-1}_{00} = 1$ and $\Upsilon^{-1}_{i0} = \Upsilon^{-1}_{0j} = 0$ for $i, j > 0$. Thus

$$M_{\Upsilon^{-1}} = \begin{pmatrix} 1 & 0 & 0 & 0 & 0 & 0 & \cdots \\ & 1 & 1 & 1 & 1 & 1 & \cdots \\ & & 1 & 3 & 7 & 15 & \cdots \\ & & & 1 & 6 & 25 & \cdots \\ & & & & 1 & 10 & \cdots \\ & & & & & 1 & \cdots \\ & & & & & & \ddots \end{pmatrix}.$$

From the recurrence relations we get $\Upsilon^{-1}_{ij} = \left\{ {j \atop i} \right\}$, where $\left\{ {j \atop i} \right\}$ is by definition the number of equivalence relations on a j-element set with exactly i equivalence classes. These numbers are called Stirling numbers of the second kind. (See [90, §§5.1, 5.3], [151, §6.1], [433, p. 8].)

The **Stirling derivation** $\nabla := \log \Upsilon \in \operatorname{trder}^1_K$ of A has matrix

$$M_\nabla = \begin{pmatrix} 0 & 0 & 0 & 0 & 0 & 0 & 0 & \cdots \\ & 0 & -1 & \frac{1}{2} & -\frac{1}{2} & \frac{2}{3} & -\frac{11}{12} & \cdots \\ & & 0 & -3 & 2 & -\frac{5}{2} & 4 & \cdots \\ & & & 0 & -6 & 5 & -\frac{15}{2} & \cdots \\ & & & & 0 & -10 & 10 & \cdots \\ & & & & & 0 & -15 & \cdots \\ & & & & & & 0 & \cdots \\ & & & & & & & \ddots \end{pmatrix}.$$

Since

$$M_\nabla = \log M_\Upsilon = \log [\![\log(1+z)]\!] = \langle\!\langle \operatorname{itlog}\big(\log(1+z)\big) \rangle\!\rangle,$$

this gives

$$\operatorname{itlog}\big(\log(1+z)\big) = -\frac{z^2}{2!} + \frac{1}{2}\frac{z^3}{3!} - \frac{1}{2}\frac{z^4}{4!} + \frac{2}{3}\frac{z^5}{5!} - \frac{11}{12}\frac{z^6}{6!} + \cdots.$$

Note that by (12.5.8) in Corollary 12.5.17 we have

$$\operatorname{itlog}\big(\log(1+z)\big) = -\operatorname{itlog}(\mathrm{e}^z - 1).$$

In view of Definition 12.5.6 and setting

$$h := \operatorname{itlog}\big(\log(1+z)\big) = \sum_{n=1}^{\infty} h_n \frac{z^n}{n!} \qquad \text{where } h_n \in \mathbb{Q},\ h_1 = 0,$$

the coefficients h_n determine the diagonals of ∇ as follows:

LEMMA 12.7.16. $M_{\nabla_n} = h_{n+1}\operatorname{diag}_n\binom{i+n}{n+1}$ *for each* n. *In particular,*

$$M_{\nabla_1} = -\operatorname{diag}_1\binom{i+1}{2}, \qquad M_{\nabla_2} = \frac{1}{2}\operatorname{diag}_2\binom{i+2}{3}.$$

The companion derivation ∇^{c} of ∇ has matrix

$$M_{\nabla^{\mathrm{c}}} = \begin{pmatrix} 0 & 0 & 0 & 0 & 0 & 0 & 0 & \cdots \\ & 0 & -1 & \frac{1}{2} & 1 & 5 & 45 & \cdots \\ & & 0 & -3 & 2 & 5 & 30 & \cdots \\ & & & 0 & -6 & 5 & 15 & \cdots \\ & & & & 0 & -10 & 10 & \cdots \\ & & & & & 0 & -15 & \cdots \\ & & & & & & 0 & \cdots \\ & & & & & & & \ddots \end{pmatrix}.$$

Here is an explicit formula for the entries of the matrix $M_{\nabla^{\mathrm{c}}}$:

PROPOSITION 12.7.17. *For* $n \geqslant 1$ *we have*

$$M_{\nabla^{\mathrm{c}}_n} = \operatorname{diag}_n c_n\binom{i+n}{n+1} \tag{12.7.5}$$

where $c_1 = -1$ *and* $c_n = \frac{(n-2)!(n+1)!}{3\cdot 2^n}$ *for* $n \geqslant 2$, *so* $c_2 = \frac{1}{2}$. *In particular, we have* $\nabla^{\mathrm{c}}_{0,n} = 0$ *for all* n, *and* $\nabla^{\mathrm{c}}_{1,n} = c_{n-1} \neq 0$ *for all* $n \geqslant 2$.

PROOF. Lemma 12.7.16 gives (12.7.5) for $n = 1, 2$, and then Corollary 12.4.7 and the subsequent remark gives (12.7.5) for $n > 2$. □

From Proposition 12.7.17 and Corollary 12.7.14 we obtain:

COROLLARY 12.7.18. *Suppose* $P \in A$ *is isobaric. Then*

$$\Upsilon(P) - P \iff \nabla_1(P) = \nabla_2(P) = 0 \iff P \in K[Y_0, Y_1].$$

Since $(\nabla_2)_{i2} = 0$ for all i (see the matrix of ∇), we have $\nabla_2(Y_2) = 0$, and so $K[Y_0, Y_1, Y_2] \subseteq A^{\nabla_2}$. Thus by Corollary 12.7.18:

COROLLARY 12.7.19. *Suppose* $P \in K[Y_0, Y_1, Y_2] \subseteq A$ *is isobaric. Then*

$$\nabla_1(P) = 0 \iff P \in K[Y_0, Y_1].$$

The algebra of partial differential operators

The K-derivations $\partial_n = \frac{\partial}{\partial Y_n}$ of A satisfy $[\partial_m, \partial_n] = 0$ for all m, n, so they generate a commutative subalgebra $K[\partial] := K[\partial_0, \partial_1, \ldots]$ of the K-algebra $\operatorname{End}(A)$ of endomorphisms of the K-module A. Let $\boldsymbol{i}$ range over the set $\mathbb{N}^{(\mathbb{N})}$ of sequences

$\boldsymbol{i} = (i_0, i_1, \dots) \in \mathbb{N}^{\mathbb{N}}$ such that $i_n = 0$ for all but finitely many n, and likewise with $\boldsymbol{j}$ and $\boldsymbol{k}$. For such $\boldsymbol{i}$, set

$$\partial^{\boldsymbol{i}} := \partial_0^{i_0}\partial_1^{i_1}\cdots\partial_n^{i_n}\cdots \in K[\partial],\ |\boldsymbol{i}| := i_0 + i_1 + \cdots,\ \boldsymbol{i}! := i_0!i_1!\cdots i_n!\cdots \in \mathbb{N}^{\geqslant 1}.$$

The $\partial^{\boldsymbol{i}}$ generate the K-module $K[\partial]$. We have a K-bilinear map

$$(12.7.6) \qquad (\Delta, P) \mapsto \langle \Delta, P\rangle := (\Delta P)\big|_{Y_0 = Y_1 = \cdots = Y_n = \cdots = 0} : \quad K[\partial] \times A \to K.$$

It is easy to check that $\langle \partial^{\boldsymbol{i}}, Y^{\boldsymbol{j}}\rangle = \boldsymbol{i}!$ if $\boldsymbol{i} = \boldsymbol{j}$ and $\langle \partial^{\boldsymbol{i}}, Y^{\boldsymbol{j}}\rangle = 0$ otherwise. Thus the pairing (12.7.6) is non-degenerate, and $(\partial^{\boldsymbol{i}})$ is a basis for the K-module $K[\partial]$. So for each $\Delta \in K[\partial]$ there is a unique family $(a_{\boldsymbol{i}})$ in K such that $a_{\boldsymbol{i}} = 0$ for all but finitely many $\boldsymbol{i}$ and $\Delta = \sum_{\boldsymbol{i}} a_{\boldsymbol{i}}\partial^{\boldsymbol{i}}$. Put

$$\begin{aligned} K[\partial_{<n}] &= K[\partial_0, \dots, \partial_{n-1}], \text{ so} \\ K[\partial_{<m}] \subseteq K[\partial_{<n}] &\subseteq K[\partial] \qquad \text{for } m \leqslant n, \text{ and} \\ K[\partial] &= \bigcup_n K[\partial_{<n}]. \end{aligned}$$

Note that if $\Delta \in K[\partial]$, then $\Delta\big(K[Y_{<n}]\big) \subseteq K[Y_{<n}]$, and the K-linear map

$$\Delta \mapsto \Delta|K[Y_{<n}] : K[\partial] \to \operatorname{End}\big(K[Y_{<n}]\big)$$

is injective on $K[\partial_{<n}]$.

For each family $(a_{\boldsymbol{i}})$ in K, the family $(a_{\boldsymbol{i}}\partial^{\boldsymbol{i}})$ of endomorphisms of the K-module A is summable with respect to the trivial filtration on the K-module A, so we have an endomorphism $\Delta := \sum_{\boldsymbol{i}} a_{\boldsymbol{i}}\partial^{\boldsymbol{i}} \in \operatorname{End}(A)$. We let $K[[\partial]]$ be the K-submodule of $\operatorname{End}(A)$ consisting of the endomorphisms $\sum_{\boldsymbol{i}} a_{\boldsymbol{i}}\partial^{\boldsymbol{i}}$ where $(a_{\boldsymbol{i}})$ is a family in K. By the properties of the pairing (12.7.6) there is for each $\Delta \in K[[\partial]]$ a unique family $(a_{\boldsymbol{i}})$ in K with $\Delta = \sum_{\boldsymbol{i}} a_{\boldsymbol{i}}\partial^{\boldsymbol{i}}$. For any families $(a_{\boldsymbol{i}})$ and $(b_{\boldsymbol{j}})$ in K,

$$\left(\sum_{\boldsymbol{i}} a_{\boldsymbol{i}}\partial^{\boldsymbol{i}}\right)\left(\sum_{\boldsymbol{j}} b_{\boldsymbol{j}}\partial^{\boldsymbol{j}}\right) = \sum_{\boldsymbol{k}}\left(\sum_{\boldsymbol{i}+\boldsymbol{j}=\boldsymbol{k}} a_{\boldsymbol{i}}b_{\boldsymbol{j}}\right)\partial^{\boldsymbol{k}}.$$

Thus $K[[\partial]]$ is a commutative K-subalgebra of $\operatorname{End}(A)$ which contains $K[\partial]$ as a subalgebra. We call the elements of $K[[\partial]]$ **partial differential operators** on A. We say that $\Delta = \sum_{\boldsymbol{i}} a_{\boldsymbol{i}}\partial^{\boldsymbol{i}} \in K[[\partial]]$ as above is **homogeneous of order** $r \in \mathbb{N}$ if $a_{\boldsymbol{i}} = 0$ whenever $|\boldsymbol{i}| \neq r$. Note that if $P \in A$ has degree $\leqslant d$ and $\Delta \in K[[\partial]]$ is homogeneous of order r, then $\deg \Delta(P) \leqslant d - r$; so if $d < r$, then $\Delta(P) = 0$.

Let a sequence $\Phi = (\Phi_n)$ in $K[[\partial]]$ be given; put

$$\Phi^{\boldsymbol{i}} := \Phi_0^{i_0}\Phi_1^{i_1}\cdots\Phi_n^{i_n}\cdots \in K[[\partial]].$$

For $\Delta = \sum_{\boldsymbol{i}} a_{\boldsymbol{i}}\partial^{\boldsymbol{i}} \in K[\partial]$ (all $a_{\boldsymbol{i}} \in K$, and $a_{\boldsymbol{i}} = 0$ for all but finitely many $\boldsymbol{i}$), set

$$\Delta(\Phi) := \sum_{\boldsymbol{i}} a_{\boldsymbol{i}}\Phi^{\boldsymbol{i}} \in K[[\partial]].$$

Routine arguments show:

LEMMA 12.7.20. *The map $\Delta \mapsto \Delta(\Phi)$ is the unique K-algebra morphism from $K[\partial]$ into $K[[\partial]]$ with $\partial_n(\Phi) = \Phi_n$ for each n. If $\Delta \in K[\partial]$ is homogeneous of order r and each Φ_n is homogeneous of order s, then $\Delta(\Phi)$ is homogeneous of order rs.*

Notes and comments

General references on locally nilpotent derivations and exponential automorphisms of *finite-dimensional* polynomial rings are [132, 143, 307]. Our usage of "triangular" differs from these sources, where a K-derivation Δ of $K[Y_{<n}]$ is called triangular if it satisfies the hypothesis of Lemma 12.7.4. Proposition 12.7.1 is [132, Proposition 2.1.3], with a different proof. In connection with Proposition 12.7.13, it follows from a theorem of Maurer [290] and Weitzenböck [463] that if K is a field and Δ is a strictly triangular K-derivation of A, then for each n the K-algebra $A^{\Delta} \cap K[Y_{<n}]$ is finitely generated; see also [307, Theorem 6.2.1].

The power series itlog$(\mathrm{e}^z - 1) \in \mathbb{Q}[[z]] \subseteq \mathbb{C}((z))$ is d-transcendental over the differential subfield of $\mathbb{C}((z))$ consisting of the convergent Laurent series, with d/dz as the derivation on $\mathbb{C}((z))$; see [17].

12.8 APPLICATION TO DIFFERENTIAL POLYNOMIALS

In this section K is a differential field and Y is a differential indeterminate over K. We apply the material in Section 12.7 to the K-algebra $A = K\{Y\} = K[Y_0, Y_1, \dots]$ where $Y_n = Y^{(n)}$ for each n. *Throughout this section $P \in A$ and $\phi \in K^{\times}$.* As in Section 5.7, let $\delta = \phi^{-1}\partial$ be the derivation of the compositional conjugate K^{ϕ} of K. Consider the unitriangular K-automorphism Υ_{ϕ} of A defined by

$$(12.8.1)\quad M_{\Upsilon_\phi} = \left(\phi^{-j} F_i^j(\phi)\right)_{i,j} = \begin{pmatrix} 1 & 0 & 0 & 0 & 0 & \cdots \\ & 1 & \phi'/\phi^2 & \phi''/\phi^3 & \phi^{(3)}/\phi^4 & \cdots \\ & & 1 & 3\phi'/\phi^2 & 4\phi''/\phi^3 + 3(\phi')^2/\phi^4 & \cdots \\ & & & 1 & 6\phi'/\phi^2 & \cdots \\ & & & & 1 & \cdots \\ & & & & & \ddots \end{pmatrix}.$$

Expressed in terms of the $\delta^n(\phi)$, this is

$$M_{\Upsilon_\phi} = \left(\phi^{-j} G_i^j(\phi)\right)_{i,j} = \begin{pmatrix} 1 & 0 & 0 & 0 & \cdots \\ & 1 & \delta(\phi)/\phi & \delta^2(\phi)/\phi + \delta(\phi)^2/\phi^2 & \cdots \\ & & 1 & 3\delta(\phi)/\phi & \cdots \\ & & & 1 & \cdots \\ & & & & \ddots \end{pmatrix}.$$

We also define the diagonal K-automorphism Ξ_ϕ of A by

$$M_{\Xi_\phi} = \operatorname{diag}(\phi^i) = \begin{pmatrix} 1 & & & & \\ & \phi & & & \\ & & \phi^2 & & \\ & & & \phi^3 & \\ & & & & \ddots \end{pmatrix}.$$

Thus $P^\phi = (\Upsilon_\phi \circ \Xi_\phi)(P)$. If P is isobaric of weight $w \in \mathbb{N}$, then $\Xi_\phi(P) = \phi^w P$ and hence $P^\phi = \phi^w\, \Upsilon_\phi(P)$. We set

$$\nabla_\phi := \log \Upsilon_\phi \in \operatorname{trder}^1_K.$$

Then by Lemma 12.7.11 we have:

LEMMA 12.8.1. *Suppose P is isobaric of weight $w \in \mathbb{N}$. Then $(P^\phi)_{[w]} = \phi^w P$ and for $n \geqslant 1$:*

$$(P^\phi)_{[w-n]} = \phi^w \sum_{i_1,\dots,i_k} \frac{1}{k!}(\nabla_{\phi,i_1} \cdots \nabla_{\phi,i_k})(P),$$

summed over the $(i_1,\dots,i_k)$ with $k \geqslant 1$, $i_1,\dots,i_k \geqslant 1$ and $i_1 + \cdots + i_k = n$. So

$$(P^\phi)_{[w-1]} = \phi^w \nabla_{\phi,1}(P), \qquad (P^\phi)_{[w-2]} = \phi^w\big(\nabla_{\phi,2} + \tfrac{1}{2}(\nabla_{\phi,1})^2\big)(P).$$

In the next lemma $x \in K$ satisfies $x' = 1$, and $t := \frac{1}{x}$, so $\Upsilon_t = \Upsilon$ by (5.7.4), and thus $\nabla_t = \nabla$, and $\nabla_{t,1} = \nabla_1$ as well as $\nabla_{t,2} = \nabla_2$.

LEMMA 12.8.2. *Suppose P is isobaric of weight $w \in \mathbb{N}$. Then the following conditions are equivalent:*

(i) $P \in K[Y, Y']$,

(ii) $P \in K[Y] \cdot (Y')^w$,

(iii) P^ϕ *is isobaric of weight w, for every ϕ,*

(iv) P^t *is isobaric of weight w,*

(v) $P^t = t^w P$.

In (v), $t^w P$ lies in the differential ring $K\{Y\}$ and P^t in its compositional conjugate $K^t\{Y\}$, but the equality makes sense as these differential rings have the same underlying ring $A = K[Y, Y', \dots]$.

PROOF. (i) $\Rightarrow$ (ii) $\Rightarrow$ (iii) $\Rightarrow$ (iv) is clear, and (iv) $\Rightarrow$ (v) follows from Lemma 12.8.1. To show (v) $\Rightarrow$ (i), suppose $P^t = t^w P$. Then $\Upsilon(P) = \Upsilon_t(P) = t^{-w} P^t = P$ and hence $P \in K[Y, Y']$ by Corollary 12.7.18. □

By Lemma 12.7.15, the matrix M_Υ is the iteration matrix of the formal power series $\log(1+z)$. More generally, the matrix M_{Υ_ϕ} from (12.8.1) is an iteration matrix as a consequence of the next proposition:

PROPOSITION 12.8.3. *Set*

$$f_\phi := \sum_{n=1}^{\infty} \phi^{(n-1)} \frac{z^n}{n!} = \phi z + \phi' \frac{z^2}{2!} + \phi'' \frac{z^3}{3!} + \cdots \in zK[[z]]$$

and

$$F_\phi := \big(F_i^j(\phi)\big)_{i,j\in\mathbb{N}} = \begin{pmatrix} 1 & 0 & 0 & 0 & 0 & \cdots \\ & \phi & \phi' & \phi'' & \phi^{(3)} & \cdots \\ & & \phi^2 & 3\phi\phi' & 4\phi\phi'' + 3(\phi')^2 & \cdots \\ & & & \phi^3 & 6\phi^2\phi' & \cdots \\ & & & & \phi^4 & \cdots \\ & & & & & \ddots \end{pmatrix} \in \operatorname{tr}_K .$$

Then $[\![f_\phi]\!] = F_\phi$.

PROOF. This amounts to showing that for $i \leqslant j$,

$$B_{ij}\big(\phi, \phi', \dots, \phi^{(j-i)}\big) = F_i^j(\phi)$$

where $B_{ij}(y_1, \dots, y_{j-i+1}) \in \mathbb{Q}[y_1, \dots, y_{j-i+1}]$ is the Bell polynomial defined in Section 12.5. Let x, z be distinct indeterminates, and note that $\frac{\partial}{\partial x}$ and $\partial := \frac{\partial}{\partial z}$ are commuting K-derivations on $K[[x,z]]$. With $R := K[x]$, these two derivations map the subring $R[[z]]$ of $K[[x,z]]$ into $R[[z]]$. With the nonnegative filtration $(z^n R[[z]])$ on $R[[z]]$, the restriction of ∂ to $R[[z]]$ is a continuous R-derivation on $R[[z]]$. Set $a := \mathrm{e}^{x f_\phi} \in 1 + zR[[z]]$, so

$$a = \sum_{j=0}^{\infty} \left(\sum_{i=0}^{j} B_{ij}(\phi, \phi', \dots) x^i \right) \frac{z^j}{j!} \quad \text{in } R[[z]] \tag{12.8.2}$$

by Corollary 12.5.3. Also $\frac{\partial a}{\partial x} = f_\phi a$. Below we set

$$\theta := \partial(f_\phi) = \phi + \phi' z + \phi'' \frac{z^2}{2!} + \cdots \in K[[z]]^\times \subseteq R[[z]]^\times,$$

so $\partial(a) = x\theta a$. Hence with δ denoting the derivation $\theta^{-1}\partial$ of $R[[z]]$, we have $\delta(a) = \theta^{-1}\partial(a) = xa$, so $\delta^n(a) = x^n a$ for all n and thus, by the definition of the differential polynomials F_n^j:

$$a^{-1}\partial^j(a) = \sum_{n=0}^{j} F_n^j(\theta)\, x^n. \tag{12.8.3}$$

Let $i \leqslant j$. Applying $\frac{\partial^i}{\partial x^i}$ to the left-hand side of (12.8.3) and using $\frac{\partial a}{\partial x} = f_\phi a$ yields

$$\begin{aligned} \frac{\partial^i}{\partial x^i}\big(a^{-1}\partial^j(a)\big) &= \frac{\partial^i}{\partial x^i}\left(a^{-1}\frac{\partial^j a}{\partial z^j}\right) = \sum_{k=0}^{i} \binom{i}{k} \frac{\partial^k a^{-1}}{\partial x^k} \cdot \frac{\partial^{i+j-k} a}{\partial x^{i-k} \partial z^j} \\ &= \sum_{k=0}^{i} \binom{i}{k} (-1)^k f_\phi^k a^{-1} \cdot \frac{\partial^{i+j-k} a}{\partial x^{i-k}\partial z^j} \end{aligned}$$

in $K[[x,z]]$, and so by (12.8.2),

$$\text{(12.8.4)} \qquad \frac{1}{i!}\frac{\partial^i}{\partial x^i}\left(a^{-1}\partial^j(a)\right)\bigg|_{x=z=0} = \frac{1}{i!}\frac{\partial^{i+j} a}{\partial x^i \partial z^j}\bigg|_{x=z=0} = B_{ij}(\phi,\phi',\dots).$$

Applying $\frac{1}{i!}\frac{\partial^i}{\partial x^i}$ to the right-hand side of (12.8.3) yields

$$\frac{1}{i!}\frac{\partial^i}{\partial x^i}\left(\sum_{n=0}^{j} F_n^j(\theta)\,x^n\right) = F_i^j(\theta) + \text{terms of positive degree in } x\text{, in } K[[x,z]],$$

hence

$$\frac{1}{i!}\frac{\partial^i}{\partial x^i}\left(\sum_{n=0}^{j} F_n^j(\theta)\,x^n\right)\bigg|_{x=0} = F_i^j(\theta).$$

Also, for $k \in \mathbb{N}$,

$$\partial^k\theta = \partial^{k+1}(f_\phi) = \phi^{(k)} + \phi^{(k+1)}z + \phi^{(k+2)}\frac{z^2}{2!} + \cdots,$$

hence $\partial^k\theta\big|_{z=0} = \phi^{(k)}$, and thus

$$\text{(12.8.5)} \qquad \frac{1}{i!}\frac{\partial^i}{\partial x^i}\left(\sum_{n=0}^{j} F_n^j(\theta)\,x^n\right)\bigg|_{x=z=0} = F_i^j(\theta)\big|_{z=0} = F_i^j(\phi).$$

Equality (12.8.3) together with (12.8.4) and (12.8.5) now yields the claim. □

As desired, we can now interpret M_{Υ_ϕ} as an iteration matrix. Recall from Section 5.8 the definition of the Riccati polynomials R_n.

COROLLARY 12.8.4. *Let*

$$g_\phi := \sum_{n=1}^{\infty}\left(\frac{\phi^{(n-1)}}{\phi^n}\right)\frac{z^n}{n!} = \sum_{n\geqslant 1}\left(\frac{R_{n-1}(\phi^\dagger)}{\phi^{n-1}}\right)\frac{z^n}{n!} \in z + z^2K[[z]].$$

Then $[\![g_\phi]\!] = M_{\Upsilon_\phi}$.

PROOF. Let f_ϕ and F_ϕ be as in Proposition 12.8.3. Then we have $g_\phi = f_\phi \circ (\phi^{-1}z)$, hence $[\![g_\phi]\!] = [\![f_\phi]\!]\cdot[\![\phi^{-1}z]\!] = F_\phi\cdot\operatorname{diag}(\phi^{-i}) = M_{\Upsilon_\phi}$, as claimed. □

The next proposition shows that the derivation $\nabla_{\phi,n}$ is a certain scalar multiple of ∇_n for $n = 1,2$. This fact will be very useful in Chapter 13:

PROPOSITION 12.8.5. *For $n = 1,2$ there is a polynomial $G_n \in \mathbb{Q}[Y_1,\dots,Y_n]$, isobaric of weight n and independent of ϕ and K, such that*

$$\nabla_{\phi,n} = \phi^{-n}G_n\big(R_1(\phi^\dagger),\dots,R_n(\phi^\dagger)\big)\nabla_n.$$

In more detail, with $\lambda := -\phi^\dagger$ and $\omega := -(2\lambda' + \lambda^2) = 2(\phi^\dagger)' - (\phi^\dagger)^2$, we have

$$\nabla_{\phi,1} = \phi^{-1}\big(-R_1(\phi^\dagger)\big)\nabla_1 = (\lambda/\phi)\nabla_1,$$
$$\nabla_{\phi,2} = \phi^{-2}\big(2R_2(\phi^\dagger) - 3R_1(\phi^\dagger)^2\big)\nabla_2 = (\omega/\phi^2)\nabla_2.$$

The proof below shows: if $n \geqslant 1$ and the coefficient of z^{n+1} in $\operatorname{itlog}(\log(1+z))$ is not zero, then there is a G_n as in the first sentence of the lemma.

PROOF. Let $g = g_\phi = \sum_{n=1}^{\infty} g_n \frac{z^n}{n!}$, where $g_n = \phi^{(n-1)}/\phi^n$ for $n \geqslant 1$, be the unitary power series from Corollary 12.8.4, and let $h = \sum_{n=2}^{\infty} h_n \frac{z^n}{n!}$ ($h_n \in K$ for $n \geqslant 2$) be the iterative logarithm of g. Then $M_{\nabla_\phi} = \log [\![g]\!] = \langle\!\langle h \rangle\!\rangle$. Let $n \geqslant 1$. Then

$$M_{\nabla_{\phi,n}} = h_{n+1} \operatorname{diag}_n \binom{i+n}{n+1}.$$

Recall that we have an isobaric polynomial $H_{n+1} \in \mathbb{Q}[Y_1, \dots, Y_n]$ of weight n with

$$h_{n+1} = H_{n+1}(g_2, \dots, g_{n+1}) = \phi^{-n} H_{n+1}\big(R_1(\phi^\dagger), \dots, R_n(\phi^\dagger)\big).$$

It remains to use Lemma 12.7.16 and (12.5.5). □

REMARK. It is sometimes convenient to express the transformation factors in the previous proposition in terms of the derivation $\delta = \phi^{-1}\partial$ of K:

$$\phi^{-1}\lambda = -\phi^{-1}\delta(\phi) \qquad \phi^{-2}\omega = \phi^{-2}\big(2\phi\delta^2(\phi) - \delta(\phi)^2\big).$$

From the previous proposition and Lemma 12.8.1 we obtain:

COROLLARY 12.8.6. *Suppose P is nonzero and set $w = \operatorname{wt}(P)$ and $Q := P^\phi$. Then with $\lambda = -\phi^\dagger$ and $\omega = -(2\lambda' + \lambda^2)$,*

$$\begin{aligned} Q_{[w]} &= \phi^w P_{[w]}, \\ Q_{[w-1]} &= \phi^{w-1}\big[P_{[w-1]} + \lambda \nabla_1(P_{[w]})\big], \\ Q_{[w-2]} &= \phi^{w-2}\big[P_{[w-2]} + \lambda\nabla_1(P_{[w-1]}) + \big(\omega\nabla_2 + \tfrac{1}{2}\lambda^2\nabla_1^2\big)(P_{[w]})\big]. \end{aligned}$$

Additive and multiplicative conjugates of partial differential operators

This is for use in Section 14.4 below and concerns partial differential operators on the K-algebra $A = K\{Y\}$, using the notions defined at the end of Section 12.7.

LEMMA 12.8.7. *Let $\Delta \in K[[\partial]]$ and $h \in K$. Then*

$$(\Delta P)_{+h} = \Delta(P_{+h}).$$

PROOF. For $\Delta = \partial_i$, this is (4.3.3). It extends easily, first to products $\Delta = a\partial_0^{i_0} \cdots \partial_n^{i_n}$ with $a \in K$, and next to Δ as an infinite sum of such products. □

As to multiplicative conjugation, let Δ range over $K[\partial]$ rather than over $K[[\partial]]$.

LEMMA 12.8.8. *Suppose $h \in K^\times$. There is a unique K-algebra morphism*

$$\Delta \mapsto \Delta_{\times h} : K[\partial] \to K[[\partial]]$$

such that for all Δ, P we have

$$(12.8.6) \qquad (\Delta_{\times h} P)_{\times h} = \Delta(P_{\times h}).$$

If Δ is homogeneous of order r, then so is $\Delta_{\times h}$.

PROOF. If $\Delta \mapsto \Delta_{\times h}\colon K[\partial] \to K[[\partial]]$ is a K-algebra morphism satisfying (12.8.6) for all Δ, P, then $\Delta_{\times h}P = \big(\Delta(P_{\times h})\big)_{\times h^{-1}}$ for all Δ, P, so there is at most one such K-algebra morphism. For existence, let $\Delta \mapsto \Delta_{\times h}$ be the unique K-algebra morphism from $K[\partial]$ to $K[[\partial]]$ sending each ∂_i to

$$(\partial_i)_{\times h} := \sum_{j \geqslant i} \binom{j}{i} h^{(j-i)} \partial_j.$$

Then (12.8.6) holds for all Δ, P by the identity (4.3.4). □

For $h = 1$ we have $\Delta_{\times 1}P = \Delta P$, and for $h \in K^{\times}$,

$$\deg \Delta_{\times h}(P) = \deg \Delta(P_{\times h}), \qquad \operatorname{wt} \Delta_{\times h}(P) = \operatorname{wt} \Delta(P_{\times h}).$$

Notes and comments

Bank [35, Lemma 13] proves Lemma 12.8.2 for $P \in C\{Y\}$, $C = \mathbb{C}$, by analytic techniques. Babakhanian [31, Theorem 8.3] has an algebraic proof, different from ours, for $P \in C\{Y\}$. Similar results play a role in the Newton diagram method for differential polynomials developed by Bank [35] and Strodt [437, 438]. It might be interesting to relate the Bank-Strodt method to our Newton diagram method from Chapters 13 and 14.

Chapter Thirteen

The Newton Polynomial

In this chapter K is a d*-valued field of H-type with asymptotic integration and small derivation. We also assume that K is equipped with a monomial group $\mathfrak{M}$*, and let $\mathfrak{m}$, $\mathfrak{n}$ range over $\mathfrak{M}$. As usual, (Γ, ψ) is the asymptotic couple of K. The unique element of $\Gamma^{>}$ fixed by ψ is denoted by 1, so $\Psi < 1 + 1$. We let γ range over Γ, and ϕ over the active elements of $\mathfrak{M}$ in K, so K^{ϕ} inherits the properties we imposed on K. Throughout, $P \in K\{Y\}^{\neq}$. We now present an overview of the main results to be established in this chapter.

Since K is d-valued, we have an isomorphism

$$c \mapsto \overline{c} = c + \scriptstyle{\mathcal{O}}\displaystyle : \ C \to \boldsymbol{k} = \mathcal{O}/\scriptstyle{\mathcal{O}}$$

from its constant field C (with its trivial derivation) onto the differential residue field $\boldsymbol{k}$ of K, and below we identify C with $\boldsymbol{k}$ via this isomorphism. We also extend the residue map $a \mapsto \overline{a} \colon \mathcal{O} \to C$ to the differential ring morphism

$$Q \mapsto \overline{Q} \colon \ \mathcal{O}\{Y\} \to C\{Y\}$$

that sends $Y^{(n)}$ to $Y^{(n)}$ for each n. The constant field C of K is also the constant field of K^{ϕ}, and so $C\{Y\}$ is a common differential subring of all $K^{\phi}\{Y\}$. As we did with K, we identify C with the differential residue field of K^{ϕ}, and extend the residue map $a \mapsto \overline{a} \colon \mathcal{O}^{\phi} \to C$ to the differential ring morphism

$$Q \mapsto \overline{Q} \colon \ \mathcal{O}^{\phi}\{Y\} \to C\{Y\},$$

where $\mathcal{O}^{\phi}$ is the valuation ring $\mathcal{O}$ of K^{ϕ} viewed as a differential subring of K^{ϕ}.

We now use $\mathfrak{M}$ to associate to $P \in K\{Y\}^{\neq}$ its dominant monomial $\mathfrak{d}_P \in \mathfrak{M}$ and its dominant part $D_P \in C\{Y\}$:

$$\mathfrak{d}_P \in \mathfrak{M} \text{ with } \mathfrak{d}_P \asymp P, \qquad D_P = \overline{\mathfrak{d}_P^{-1} P} \in C\{Y\}.$$

(This is in agreement with Section 6.6. Another choice of $\mathfrak{M}$ would multiply D_P by a factor in $C^{\times}$.) Likewise, $\mathfrak{d}_{P^{\phi}} \in \mathfrak{M}$ and $D_{P^{\phi}} \in C\{Y\}$ for each ϕ. As in Chapter 11 a condition $S(\phi)$ on elements ϕ is said to hold *eventually* if there is an active ϕ_0 in K such that $S(\phi)$ holds for all $\phi \preccurlyeq \phi_0$.

PROPOSITION 13.0.1. *Given any P there exists a differential polynomial $N \in C\{Y\}$ such that eventually $D_{P^{\phi}} = N$.*

This fact is derived in Section 13.1. We define the **Newton polynomial** of P to be the differential polynomial N in Proposition 13.0.1, and denote it by N_P. Section 13.2 contains some elementary results about these Newton polynomials. In Section 13.3 we establish a key consequence of ω-freeness:

THEOREM 13.0.2. *Suppose K is ω-free. Then $N_P \in C[Y](Y')^{\mathbb{N}}$ for all P.*

For ω-free K this allows us to describe explicitly the behavior of $P(y)$ near the constant field, not just in K, but in any d-valued field extension of K of H-type.

Another key consequence of ω-freeness proved in this chapter concerns *eventual equalizers.* To explain this, let Γ be divisible, and let $P, Q \in K\{Y\}^{\neq}$ be homogeneous of degrees $d > e$. By the Equalizer Theorem from Chapter 6 there is for each ϕ an $a \in K^{\times}$ such that $P^{\phi}_{\times a} \asymp Q^{\phi}_{\times a}$. We show in Section 13.5 that for sufficiently high $v\phi$ we can take such a independent of ϕ, provided K is ω-free:

THEOREM 13.0.3. *If K is ω-free, Γ is divisible, and $P, Q \in K\{Y\}^{\neq}$ are homogeneous of degrees $d > e$, then there exists $a \in K^{\times}$ such that, eventually, $P^{\phi}_{\times a} \asymp Q^{\phi}_{\times a}$.*

In Section 13.6 we consider more generally any ungrounded H-asymptotic field E with $\Gamma_E \neq \{0\}$ and prove for such E:

THEOREM 13.0.4. *If E is ω-free and F is a* d*-algebraic* d*-valued field extension of E of H-type, then $\Gamma^{<}_{E}$ is cofinal in $\Gamma^{<}_{F}$, and F is ω-free.*

In Section 13.6 we also extend the Eventual Equalizer Theorem 13.0.3 to ω-free E. Section 13.7 contains the construction, for any E that is λ-free but not ω-free, of a canonical extension $E\langle\gamma\rangle$ generated over E by a solution γ of a certain second-order differential equation. Section 13.8 on unraveling asymptotic equations is technical, but crucial in the next chapter. The last section describes some concrete H-fields $\mathbb{R}\langle\omega\rangle \subseteq \mathbb{R}\langle\lambda\rangle \subseteq \mathbb{R}\langle\gamma\rangle$ with interesting generic features.

13.1 REVISITING THE DOMINANT PART

We derived some basic facts on the dominant part of P in Section 6.6, and here we add to this in the more special setting of this chapter. Recall that in this setting D_P is defined using our monomial group $\mathfrak{M}$ of K. Thus $\mathfrak{d}_{\mathfrak{m}P} = \mathfrak{m}\mathfrak{d}_P$ and $D_{\mathfrak{m}P} = D_P$. If $a \in \mathcal{O}$ and $P(a) = 0$, then $D_P(\overline{a}) = 0$.

Elementary facts on the dominant part

We represent P as

$$P = \mathfrak{d}_P D_P + R_P \quad \text{with } R_P \in K\{Y\},\ R_P \prec P.$$

Note that if $R_P \neq 0$, then

$$\operatorname{wm}(P) \leqslant \operatorname{wm}(R_P) \leqslant \operatorname{wt}(R_P) \leqslant \operatorname{wt}(P).$$

It will also be convenient to define $\mathfrak{d}_Q$ and D_Q for $Q = 0 \in K\{Y\}$ by $\mathfrak{d}_0 := 0 \in K$ and $D_0 := 0 \in C\{Y\}$. If $Q \in K\{Y\}$, then $\mathfrak{d}_{PQ} = \mathfrak{d}_P\mathfrak{d}_Q$ and $D_{PQ} = D_P D_Q$.

LEMMA 13.1.1. *Let $Q \in C\{Y\}$, $Q \notin C$. Then $P(Q) \neq 0$, and*

$$\mathfrak{d}_{P(Q)} = \mathfrak{d}_P, \qquad D_{P(Q)} = D_P(Q), \qquad R_{P(Q)} = R_P(Q).$$

(In particular, $D_{P_{+c}} = (D_P)_{+c}$ for $c \in C$ and $D_{P_{\times c}} = (D_P)_{\times c}$ for $c \in C^{\times}$.)

PROOF. We have

$$P(Q) = \mathfrak{d}_P D_P(Q) + R_P(Q)$$

where $R_P(Q) \prec P$ and $D_P(Q) \in C\{Y\}^{\neq}$ by Lemma 4.3.12. So $P(Q) \asymp \mathfrak{d}_P$, hence $\mathfrak{d}_{P(Q)} = \mathfrak{d}_P$, and thus $D_{P(Q)} = D_P(Q)$, and $R_{P(Q)} = R_P(Q)$. □

LEMMA 13.1.2. *Suppose P is homogeneous. Then D_P is homogeneous and*

$$D_{\mathrm{Ri}(P)} = \mathrm{Ri}(D_P) \text{ in } C\{Z\}.$$

PROOF. Clearly D_P is homogeneous. From $P = \mathfrak{d}_P D_P + R$, we get

$$\mathrm{Ri}(P) = \mathfrak{d}_P \mathrm{Ri}(D_P) + \mathrm{Ri}(R),$$

with $\mathrm{Ri}(D_P) \in C\{Z\}$, $\mathrm{Ri}(D_P) \neq 0$, and $v(\mathrm{Ri}(R)) = v(R) > v(P) = v(\mathfrak{d}_P)$. This gives the desired result. □

The case that $D_P \in C[Y](Y')^{\mathbb{N}}$ is important later in this chapter. The next lemma concerns the even more special case $D_P \in C[Y]$, except that we use Z rather than Y as the indeterminate, since the lemma will be applied to differential polynomials obtained from Riccati transforms.

LEMMA 13.1.3. *Assume $Q \in K\{Z\}^{\neq}$ and $D_Q \in C[Z]$. Then $D_{Q^\phi} = D_Q$ for $\phi \preccurlyeq 1$.*

PROOF. We have $Q = \mathfrak{d}_Q D_Q + R_Q$, with $Q \succ R_Q$. Then for $\phi \preccurlyeq 1$ we have $Q^\phi = \mathfrak{d}_Q D_Q + (R_Q)^\phi$, with $(R_Q)^\phi \preccurlyeq R_Q \prec \mathfrak{d}_Q$ by Lemma 11.1.1, so $D_{Q^\phi} = D_Q$. □

From the dominant part to the Newton polynomial

If $\phi_0 \in \mathfrak{M}$ is active in K and $\phi \preccurlyeq \phi_0$ is such that $w = \mathrm{dwt}(P^\phi) = \mathrm{dwm}(P^{\phi_0})$, then

$$\mathfrak{d}_{P^\phi} = (\phi/\phi_0)^w \mathfrak{d}_{P^{\phi_0}}, \tag{13.1.1}$$

by Corollary 11.1.11(i), with K^{ϕ_0} in the role of K. Moreover:

LEMMA 13.1.4. *Suppose $\phi \preccurlyeq 1$ and $\mathrm{dwm}(P^\phi) = \mathrm{dwt}(P) = w$. Then $\mathfrak{d}_{P^\phi} = \phi^w \mathfrak{d}_P$, and D_P is isobaric of weight w with $D_{P^\phi} = D_P$.*

PROOF. If $\phi = 1$, then the lemma holds trivially, so assume $\phi \prec 1$. Then $\mathfrak{d}_{P^\phi} = \phi^w \mathfrak{d}_P$ by Corollary 11.1.11(ii) and (13.1.1). Also, by Corollary 11.1.11(iii),

$$\begin{aligned} D_{P^\phi} &= \sum_{v((P^\phi)_{[\sigma]}) = v(P^\phi)} \overline{\left((P^\phi)_{[\sigma]}/\mathfrak{d}_{P^\phi}\right)}\, Y^{[\sigma]} = \sum_{v(P_{[\sigma]}) = v(P)} \overline{\phi^w P_{[\sigma]}/\phi^w \mathfrak{d}_P}\, Y^{[\sigma]} \\ &= \sum_{v(P_{[\sigma]}) = v(P)} \overline{P_{[\sigma]}/\mathfrak{d}_P}\, Y^{[\sigma]} = D_P, \end{aligned}$$

and D_P is isobaric of weight w. □

Recall from Section 11.1 the definition of the Newton weight $\operatorname{nwt}(P)$ of P:

$$\operatorname{nwt}(P) = \operatorname{dwt}(P^{\phi}) = \operatorname{dwm}(P^{\phi}), \quad \text{eventually},$$

and so Lemma 13.1.4 yields a differential polynomial $N \in C\{Y\}$ such that

$$D_{P^{\phi}} = N, \quad \text{eventually}.$$

DEFINITION 13.1.5. The **Newton polynomial** of P is the unique $N_P \in C\{Y\}$ such that eventually $D_{P^{\phi}} = N_P$. By convention, $N_Q := 0 \in C\{Y\}$ for $Q = 0 \in K\{Y\}$.

Clearly N_P is isobaric of weight $\operatorname{nwt}(P)$, and if $\operatorname{dwt}(P) = \operatorname{nwt}(P)$, then $D_P = N_P$. In particular, if $P \in K[Y, Y']$, then $N_P \in C[Y](Y')^{\mathbb{N}}$. Our N_P depends on the monomial group $\mathfrak{M}$ of K, but the Newton polynomial of P obtained with another choice of $\mathfrak{M}$ equals cN_P for some $c \in C^{\times}$. If we want to stress the dependence of N_P on K equipped with its monomial group $\mathfrak{M}$, we write N_P^K for N_P.

EXAMPLE. Suppose $P = Q(Y) \cdot (Y')^w$ where $Q \in K[Y]^{\neq}$ and $w \in \mathbb{N}$. Then $P^{\phi} = \phi^w P$ for every ϕ, hence $\operatorname{nwt}(P) = w$ and $N_P = D_Q \cdot (Y')^w$.

EXAMPLE. Suppose $D_P \in C[Y]$. Then $D_{P^{\phi}} = D_P$ for every $\phi \preccurlyeq 1$, by Lemma 13.1.3, hence $N_P = D_P$.

The Newton polynomials $N_{P_{\times \mathfrak{m}}}$ of the multiplicative conjugates $P_{\times \mathfrak{m}}$ play a role in detecting zeros of P. To explain this, let $f \in K^{\times}$, and let $(c, \mathfrak{m})$ be the unique pair with $c \in C^{\times}$ such that $f \sim c\mathfrak{m}$. With these notations:

LEMMA 13.1.6. *Suppose $P(f) = 0$. Then $N_{P_{\times \mathfrak{m}}}(c) = 0$, and thus $\operatorname{nwt}(P_{\times \mathfrak{m}}) \geqslant 1$ or $N_{P_{\times \mathfrak{m}}}$ is not homogeneous.*

PROOF. We can reduce to the case that $v(f) = 0$, $v(P) = 0$ and $N_P = D_P$, so $\mathfrak{m} = 1$ and $f = c + \varepsilon$, $\varepsilon \prec 1$, hence

$$0 = P(f) = N_P(c + \varepsilon) + R_P(f) = N_P(c) + g, \quad \text{with } g \prec 1,$$

so $N_P(c) = 0$. If $\operatorname{wt}(N_P) = 0$, then $N_P \in C[Y]$ and N_P is not homogeneous. □

Motivated by this lemma we define a monomial $\mathfrak{m}$ to be a **starting monomial for** P if $\operatorname{nwt}(P_{\times \mathfrak{m}}) \geqslant 1$ or $N_{P_{\times \mathfrak{m}}}$ is not homogeneous; equivalently, $N_{P_{\times \mathfrak{m}}} \notin CY^{\mathbb{N}}$. We call $\mathfrak{m}$ an **algebraic starting monomial for** P if $N_{P_{\times \mathfrak{m}}}$ is not homogeneous. Note: if $\mathfrak{m}$ is a starting monomial for P, then $\operatorname{ndeg} P_{\times \mathfrak{m}} \geqslant 1$. Also, $\mathfrak{m}$ is an algebraic starting monomial for P iff $\mathfrak{m}/\mathfrak{n}$ is an algebraic starting monomial for $P_{\times \mathfrak{n}}$. By Corollary 11.2.5, P has at most $\deg P - \operatorname{mul} P$ *algebraic* starting monomials. But some P have infinitely many starting monomials:

EXAMPLE 13.1.7. Let K be the H-subfield $\mathbb{R}(\mathrm{e}^{\mathbb{R}x}, \ell_0^{\mathbb{R}}, \ell_1^{\mathbb{R}}, \dots)$ of $\mathbb{T}$, with monomial group $\mathfrak{M} = \bigcup_n \mathrm{e}^{\mathbb{R}x} \ell_0^{\mathbb{R}} \cdots \ell_n^{\mathbb{R}}$. Then K is ω-free, and for $P := Y''Y - (Y')^2$ and every $r \in \mathbb{R}$ we have $P(\mathrm{e}^{rx}) = 0$, so e^{rx} is a starting monomial for P.

Call $f \in K^\times$ an **approximate zero** of P if $N_{P_{\times\mathfrak{m}}}(c) = 0$, where $(c, \mathfrak{m})$ is the unique pair in $C^\times \times \mathfrak{M}$ with $f \sim c\mathfrak{m}$; the **multiplicity** of f as an approximate zero of P is then by definition the multiplicity of $N_{P_{\times\mathfrak{m}}}$ at c as defined just before Lemma 4.3.3. If $P(f) = 0$, then f is an approximate zero of P by Lemma 13.1.6.

In the next section we derive various useful properties of these Newton polynomials. We now continue with technicalities about dominant parts as needed later.

Decomposing P^ϕ

The identities (13.1.2) below provide a useful decomposition of P^ϕ for isobaric P. Accordingly, the asymptotic equivalence from Lemma 11.1.8 will be improved in Lemma 13.1.8. For $0 \leqslant k \leqslant n$ we define $\varepsilon_k^n(\phi) \in K$ by

$$\begin{aligned} \varepsilon_0^n(\phi) &= \varepsilon_n^n(\phi) = 0, \quad \varepsilon_k^n(\phi) = 0 \text{ if } \phi' = 0, \\ F_k^n(\phi) &= \begin{bmatrix} n \\ k \end{bmatrix} \phi^k (\phi^\dagger)^{n-k} \big(1 + \varepsilon_k^n(\phi)\big), \end{aligned}$$

so $\varepsilon_k^n(\phi) \prec 1$ if $\phi^\dagger \succ \phi$. Given $\boldsymbol{\tau} = \tau_1 \cdots \tau_d \geqslant \boldsymbol{\sigma} = \sigma_1 \cdots \sigma_d$ we put

$$\begin{aligned} \varepsilon_{\boldsymbol{\sigma}}^{\boldsymbol{\tau}}(\phi) &:= -1 + \prod_{i=1}^{d} \big(1 + \varepsilon_{\sigma_i}^{\tau_i}(\phi)\big), \quad \text{so} \\ F_{\boldsymbol{\sigma}}^{\boldsymbol{\tau}}(\phi) &= \begin{bmatrix} \boldsymbol{\tau} \\ \boldsymbol{\sigma} \end{bmatrix} \phi^{\|\boldsymbol{\sigma}\|} (\phi^\dagger)^{\|\boldsymbol{\tau}\| - \|\boldsymbol{\sigma}\|} \big(1 + \varepsilon_{\boldsymbol{\sigma}}^{\boldsymbol{\tau}}(\phi)\big), \end{aligned}$$

and if $\phi^\dagger \succ \phi$, then $\varepsilon_{\boldsymbol{\sigma}}^{\boldsymbol{\tau}}(\phi) \prec 1$. By Lemma 5.7.4,

$$(P^\phi)_{[\boldsymbol{\sigma}]} = \phi^{\|\boldsymbol{\sigma}\|} \sum_{\boldsymbol{\tau} \geqslant \boldsymbol{\sigma}} \begin{bmatrix} \boldsymbol{\tau} \\ \boldsymbol{\sigma} \end{bmatrix} (\phi^\dagger)^{\|\boldsymbol{\tau}\| - \|\boldsymbol{\sigma}\|} \big(1 + \varepsilon_{\boldsymbol{\sigma}}^{\boldsymbol{\tau}}(\phi)\big) P_{[\boldsymbol{\tau}]}.$$

For $i \in \mathbb{N}$ we define

$$P^{\phi,i} := \sum_{\|\boldsymbol{\sigma}\| = i} \left(\sum_{\boldsymbol{\tau} \geqslant \boldsymbol{\sigma}} \begin{bmatrix} \boldsymbol{\tau} \\ \boldsymbol{\sigma} \end{bmatrix} \big(1 + \varepsilon_{\boldsymbol{\sigma}}^{\boldsymbol{\tau}}(\phi)\big) P_{[\boldsymbol{\tau}]} \right) Y^{[\boldsymbol{\sigma}]} \in K\{Y\},$$

so $P^{\phi,i}$ is isobaric of weight i. Note also that $P^{\phi,0} = P_{[0]}$ is the isobaric part of P of weight 0. If P is isobaric of weight w, then $P^{\phi,w} = P$ and

$$\text{(13.1.2)} \quad P^\phi = \sum_{i=0}^{w} \phi^i (\phi^\dagger)^{w-i} P^{\phi,i}, \quad (P^\phi)_{[i]} = \phi^i (\phi^\dagger)^{w-i} P^{\phi,i} \text{ for } i = 0, \dots, w.$$

Suppose $x \in K$ satisfies $x \succ 1$ and $x' = 1$, and set $t := 1/x$. Then $t = x^\dagger$ and

$$\text{(13.1.3)} \qquad P^{t,i} = \sum_{\|\boldsymbol{\sigma}\| = i} \left(\sum_{\boldsymbol{\tau} \geqslant \boldsymbol{\sigma}} \begin{bmatrix} \boldsymbol{\tau} \\ \boldsymbol{\sigma} \end{bmatrix} P_{[\boldsymbol{\tau}]} \right) Y^{[\boldsymbol{\sigma}]}.$$

(To see this, note that Lemma 5.7.6 gives $\varepsilon^n_k(t) = 0$ for all n and $k = 0, \dots, n$.) The following lemma compares $P^{\phi,i}$ and $P^{t,i}$.

LEMMA 13.1.8. *The $P^{\phi,i}$ have the following properties:*

(i) *if $\mathrm{wt}(P) = w > 0$, then $P^{\phi,w-1} = P^{t,w-1}$;*

(ii) *if P is isobaric of weight $w > 0$, then $P^{\phi,w-1} = -\nabla_1(P)$;*

(iii) *if $\phi^\dagger \succ \phi$, then for all $i \in \mathbb{N}$,*

$$v\big(P^{\phi,i} - P^{t,i}\big) \geqslant v(P) + \psi\big(\psi(v\phi) - v\phi\big) - \psi(v\phi) > v(P).$$

PROOF. For (i), Example 5.7.7 shows that for $\boldsymbol{\sigma} \leqslant \boldsymbol{\tau}$ with $\|\boldsymbol{\tau}\| \leqslant \|\boldsymbol{\sigma}\| + 1$,

$$F^{\boldsymbol{\tau}}_{\boldsymbol{\sigma}}(\phi) = \begin{bmatrix} \boldsymbol{\tau} \\ \boldsymbol{\sigma} \end{bmatrix} \phi^{\|\boldsymbol{\sigma}\|} (\phi^\dagger)^{\|\boldsymbol{\tau}\| - \|\boldsymbol{\sigma}\|}.$$

For (ii), use (i), the second identity in (13.1.2) for $\phi = t$ and $i = w - 1$, and Lemma 12.8.1 for $\phi = t$, taking into account that $\nabla_{t,1} = \nabla_1$. For (iii), use the remark following Lemma 11.1.6 and the proof of Lemma 11.1.7. □

Even if K does not contain an element $x \succ 1$ with $x' = 1$, we define $P^{t,i} \in K\{Y\}$ by (13.1.3), and then Lemma 13.1.8 goes through. To see this, use that by Proposition 10.2.7 there is an $x \succ 1$ with $x' = 1$ in an immediate asymptotic extension of K.

More on the case that $D_P \in C[Y](Y')^{\mathbb{N}}$ and $R_P \prec^\flat P$

We recall from the subsection on flattening in Section 9.4 the convention on using $\prec^\flat$, etcetera, to denote, not only a certain binary relation on K, but also its extension to $K\langle Y \rangle$.

LEMMA 13.1.9. *Assume $D_P \in C[Y](Y')^{\mathbb{N}}$, $R_P \prec^\flat P$, and $\phi \preccurlyeq 1$. Then*

$$D_{P^\phi} = D_P = N_P, \qquad (R_P)^\phi = R_{P^\phi}, \qquad R_{P^\phi} \prec^\flat P^\phi.$$

PROOF. From $P = \mathfrak{d}_P D_P + R_P$ with $D_P \in C[Y](Y')^n$ we obtain

$$P^\phi = \phi^n \mathfrak{d}_P D_P + (R_P)^\phi \quad \text{in } K^\phi\{Y\}.$$

From $0 \leqslant v\phi < 1 + 1$ we get $\phi \asymp^\flat 1$. By Lemma 11.1.1 with $v^\flat$ in place of v,

$$(R_P)^\phi \asymp^\flat R_P \prec^\flat P \asymp^\flat \phi^n \mathfrak{d}_P \asymp^\flat P^\phi,$$

so $D_{P^\phi} = D_P$. As this holds for all $\phi \preccurlyeq 1$, we get $D_P = N_P$. □

Here is a slight variant, with almost the same proof:

LEMMA 13.1.10. *Assume $D_P \in C[Y](Y')^{\mathbb{N}}$, $\phi \preccurlyeq 1$, and $R_P \prec \phi^n P$ for all n. Then*

$$D_{P^\phi} = D_P, \qquad (R_P)^\phi = R_{P^\phi}, \qquad R_{P^\phi} \prec \phi^n P^\phi \text{ for all } n.$$

PROOF. From $P = \mathfrak{d}_P D_P + R_P$ with $D_P \in C[Y](Y')^m$ we obtain

$$P^\phi = \phi^m \mathfrak{d}_P D_P + (R_P)^\phi \quad \text{in } K^\phi\{Y\}.$$

By Lemma 11.1.1 we have

$$(R_P)^\phi \preccurlyeq R_P \prec \phi^n P \text{ for all } n,$$

hence $(R_P)^\phi \prec \phi^m \mathfrak{d}_P \asymp P^\phi$, so $D_{P^\phi} = D_P$. □

The next lemma and its corollary will only be needed in Section 14.4. For $i \in \mathbb{N}$ and $Q \in K\{Y\}$, let $\partial_i Q := \frac{\partial Q}{\partial Y^{(i)}}$ denote the partial derivative of Q with respect to $Y^{(i)}$. With this notation, we have:

LEMMA 13.1.11. *Suppose $D_P \in C[Y](Y')^{\mathbb{N}}$ and $R_P \prec^\flat P$. Let $i \in \{0,1\}$ be such that $\partial_i D_P \neq 0$. Then $\partial_i P \asymp P$ and*

$$D_{\partial_i P} = \partial_i D_P \in C[Y](Y')^{\mathbb{N}}, \quad R_{\partial_i P} \prec^\flat \partial_i P.$$

PROOF. From $\partial_i P - \mathfrak{d}_P\, \partial_i D_P + \partial_i R_P$ and $\partial_i R_P \preccurlyeq R_P \prec^\flat P$, we obtain $\mathfrak{d}_{\partial_i P} = \mathfrak{d}_P$, $D_{\partial_i P} = \partial_i D_P$, and $R_{\partial_i P} = \partial_i R_P \prec^\flat \partial_i P$. □

COROLLARY 13.1.12. *Suppose $D_P \in C[Y](Y')^{\mathbb{N}}$ and $R_P \prec^\flat P$. Let $k, l \in \mathbb{N}$ be such that $\frac{\partial^{k+l} D_P}{\partial Y^k \partial (Y')^l} \neq 0$. Then for $Q := \frac{\partial^{k+l} P}{\partial Y^k \partial (Y')^l}$ we have $Q \asymp P$ and*

$$\frac{\partial^{k+l} D_P}{\partial Y^k \partial (Y')^l} = D_Q = N_Q, \qquad R_Q \prec^\flat Q.$$

In the next lemma we use notation introduced in the preceding subsection.

LEMMA 13.1.13. *Suppose $\phi \preccurlyeq 1$ and* $\operatorname{dwm}(P^\phi) = \operatorname{dwt}(P)$. *Then*

$$R_{P^\phi} = (R_P)^\phi + \sum_{i=0}^{w-1} \mathfrak{d}_P \cdot \phi^i (\phi^\dagger)^{w-i} \cdot (D_P)^{\phi,i} \qquad \textit{where } w := \operatorname{dwm}(P).$$

PROOF. By (13.1.2) we have

$$(D_P)^\phi = \phi^w D_P + \sum_{i=0}^{w-1} \phi^i (\phi^\dagger)^{w-i} \cdot (D_P)^{\phi,i}.$$

Using (13.1.1) we obtain

$$\begin{aligned} P^\phi &= \mathfrak{d}_P (D_P)^\phi + (R_P)^\phi \\ &= \mathfrak{d}_{P^\phi} D_P + (R_P)^\phi + \sum_{i=0}^{w-1} \mathfrak{d}_P \cdot \phi^i (\phi^\dagger)^{w-i} \cdot (D_P)^{\phi,i}. \end{aligned}$$

This yields the displayed formula for R_{P^ϕ}, since $D_{P^\phi} = D_P$ by Lemma 13.1.4. □

Recall from Section 9.4 that $\prec^{\flat}_{\phi}$ refers to the flattening $v^{\flat}_{\phi}$ of the valuation v of K^{ϕ}.

LEMMA 13.1.14. *Suppose* $\phi \preccurlyeq 1$, $R_P \prec^{\flat}_{\phi} P$ *and* $\operatorname{dwm}(P^{\phi}) = \operatorname{dwt}(P)$. *Then*

$$(R_{P^{\phi}})_{[w]} \prec^{\flat}_{\phi} P^{\phi} \qquad \textit{for all } w \geqslant \operatorname{dwm}(P).$$

PROOF. Let $w \geqslant \operatorname{dwm}(P)$. By the last lemma $(R_{P^{\phi}})_{[w]} = \big((R_P)^{\phi}\big)_{[w]}$. Hence

$$v\big((R_{P^{\phi}})_{[w]}\big) \geqslant wv\phi + v(R_P) \geqslant v(P^{\phi}) + v(R_P) - v(P)$$

by Lemma 11.1.10 and Corollary 11.1.11. By assumption $v^{\flat}_{\phi}(R_P) > v^{\flat}_{\phi}(P)$, so

$$v^{\flat}_{\phi}\big((R_{P^{\phi}})_{[w]}\big) \geqslant v^{\flat}_{\phi}(P^{\phi}) + v^{\flat}_{\phi}(R_P) - v^{\flat}_{\phi}(P) > v^{\flat}_{\phi}(P^{\phi}).$$ □

Behavior of $P(y)$

If $D_P \in C[Y](Y')^{\mathbb{N}}$ and $R_P \prec^{\flat} P$, then we have a good description of $vP(y)$ in the region $1 \not\asymp y \preccurlyeq^{\flat} 1$, even in suitable extensions of K:

LEMMA 13.1.15. *Suppose* $D_P \in C[Y](Y')^{\mathbb{N}}$ *and* $R_P \prec^{\flat} P$. *Then we have for every* d*-valued field extension* L *of* H*-type of* K *and all* $y \in L$,

$$\begin{aligned} 1 \prec y \preccurlyeq^{\flat} 1 &\implies v\big(P(y)\big) = v(P) + \operatorname{ddeg}(P)\, vy + \operatorname{dwt}(P)\, \psi_L(vy), \\ 1 \succ y \preccurlyeq^{\flat} 1 &\implies v\big(P(y)\big) = v(P) + \operatorname{dmul}(P)\, vy + \operatorname{dwt}(P)\, \psi_L(vy). \end{aligned}$$

Moreover, if K *is equipped with an ordering making* K *an* H*-field, then there are* $\sigma, \tau \in \{-1, +1\}$ *such that for all* H*-field extensions* L *of* K *and all* $y \in L^{>}$,

$$\begin{aligned} 1 \prec y \preccurlyeq^{\flat} 1 &\implies \operatorname{sign} P(y) = \sigma, \\ 1 \succ y \preccurlyeq^{\flat} 1 &\implies \operatorname{sign} P(y) = \tau. \end{aligned}$$

PROOF. After dividing P by $\mathfrak{d}_P$ we may assume that $v(P) = 0$, so

$$P = D_P + R_P \qquad \text{where } v^{\flat}(R_P) > 0.$$

We have $D_P = D(Y) \cdot (Y')^w$ where $D \in C[Y]$, $w = \operatorname{wt}(D_P)$. Let L be a d-valued field extension of H-type of K. Now $1 \in \Gamma^{\flat} \subseteq \Gamma^{\flat}_L$, so $\psi_L\big((\Gamma^{\flat}_L)^{\neq}\big) \subseteq \Gamma^{\flat}_L$. Let $y \in L$. By these facts about L, if $y \preccurlyeq^{\flat} 1$, then $R_P(y) \prec^{\flat} 1$. Also $\operatorname{ddeg} P = \deg D + w$ and $\operatorname{dmul} P = \operatorname{mul} D + w$. If $y \succ 1$, then

$$v\big(D_P(y)\big) = v(D(y)) + w\, v(y') = (\deg D)\, vy + w\, \big(vy + \psi_L(vy)\big),$$

and for $y \prec 1$ these equalities hold with $\operatorname{mul} D$ instead of $\deg D$. Suppose now that $1 \not\asymp y \preccurlyeq^{\flat} 1$. Then $vy \in (\Gamma^{\flat}_L)^{\neq}$, and so $\psi_L(vy) \in \Gamma^{\flat}_L$, hence

$$D_P(y) \asymp^{\flat} 1, \qquad P(y) \sim^{\flat} D_P(y).$$

Suppose now that L is an H-field and $y > 0$. If $1 \prec y \preccurlyeq^{\flat} 1$, then $y' > 0$, so $\operatorname{sign} P(y) = \operatorname{sign} D(y)$, which equals the sign of the coefficient of the highest degree term of D. If $1 \succ y \preccurlyeq^{\flat} 1$, then $y' < 0$, so $\operatorname{sign} P(y) = (-1)^w \operatorname{sign} D(y)$, and $\operatorname{sign} D(y)$ equals the sign of the coefficient of the lowest degree term of D. □

REMARK. With the same assumptions as in Lemma 13.1.15, its proof also shows that if $\operatorname{ddeg} P > 0$ and L is a d-valued field extension of H-type of K, then

$$u, y \in L,\ u \asymp 1 \prec y \asymp^\flat 1 \implies P(u) \prec P(y),$$

whereas if $\operatorname{ddeg} P = 0$ and L is a d-valued field extension of H-type of K, then

$$y \in L,\ y \preccurlyeq^\flat 1 \implies P(y) \sim P(0) \sim P.$$

Notes and comments

In connection with Example 13.1.7 we mention that for P of order 1 there are only finitely many starting monomials; we omit the proof, since we do not use this fact later. In Section 14.2 we show that if K is ω-free and P has degree 1, then P has only finitely many starting monomials.

13.2 ELEMENTARY PROPERTIES OF THE NEWTON POLYNOMIAL

Note that from the definition of N_P we get

$$P^\phi \;=\; \mathfrak{d}_{P^\phi} N_P + R_{P^\phi}, \qquad \text{eventually.}$$

It is clear that $N_{P^\phi} = N_P$. We get $N_{PQ} = N_P N_Q$ for $Q \in K\{Y\}^{\neq}$ from the corresponding properties of dominant parts. In particular,

$$N_{\mathfrak{m}P} \;=\; N_P, \qquad N_{uP} \;=\; \overline{u}\, N_P \;\text{ for } u \in K,\ u \asymp 1.$$

Below we prove some other basic facts about Newton polynomials.

Lemmas on Newton polynomials

We begin with an easy consequence of the definitions and Corollary 11.1.11.

LEMMA 13.2.1. *Let $w = \operatorname{nwt}(P)$. Then*

$$D_{P^\phi} = N_P \textit{ for all } \phi \preccurlyeq 1 \iff D_P = N_P \iff \operatorname{dwt}(P) = w.$$

LEMMA 13.2.2. *Let $P = Q + R$ where $Q, R \in K\{Y\}$ and $R \prec^\flat P$. Then $N_P = N_Q$. If also $D_P = N_P$, then $D_{Q^\phi} = N_P$ for all $\phi \preccurlyeq 1$.*

PROOF. Note that $Q \neq 0$ and $v(P) = v(Q)$. Let $\phi \preccurlyeq 1$. Then $0 \leqslant v\phi < 1 + 1$, so

$$v(Q^\phi) \;\leqslant\; v(Q) + \operatorname{wt}(Q) v\phi \;<\; v(R) \;\leqslant\; v(R^\phi).$$

Since $P^\phi = Q^\phi + R^\phi$, this gives $v(P^\phi) = v(Q^\phi) < v(R^\phi)$, and so $D_{P^\phi} = D_{Q^\phi}$. This holds for all $\phi \preccurlyeq 1$, so $N_P = N_Q$. If in addition $D_P = N_P$, then by Lemma 13.2.1 we obtain that $D_{Q^\phi} = N_P$ for all $\phi \preccurlyeq 1$. □

LEMMA 13.2.3. *Suppose* $\mathfrak{m} \prec\!\!\prec \mathfrak{n} \succ^\flat 1$ *and* $P = Q + R$ *where* $R \prec_\mathfrak{n} P$. *Then*

$$N_{P_{\times \mathfrak{m}}} = N_{Q_{\times \mathfrak{m}}}.$$

PROOF. For $\phi \preccurlyeq 1$ we have $R^\phi \asymp_\mathfrak{n} R \prec_\mathfrak{n} Q \asymp_\mathfrak{n} Q^\phi$, by Corollary 11.1.13. Hence replacing K, P, Q, R by $K^\phi, P^\phi, Q^\phi, R^\phi$, respectively, for suitable $\phi \preccurlyeq 1$, we arrange that $D_{P_{\times \mathfrak{m}}} = N_{P_{\times \mathfrak{m}}}$ and $D_{Q_{\times \mathfrak{m}}} = N_{Q_{\times \mathfrak{m}}}$. Then by Corollary 9.4.21:

$$R_{\times \mathfrak{m}} \asymp_\mathfrak{n} R \prec_\mathfrak{n} Q \asymp_\mathfrak{n} Q_{\times \mathfrak{m}},$$

so $R_{\times \mathfrak{m}} \prec Q_{\times \mathfrak{m}}$ and hence $N_{P_{\times \mathfrak{m}}} = D_{P_{\times \mathfrak{m}}} = D_{Q_{\times \mathfrak{m}}} = N_{Q_{\times \mathfrak{m}}}$. □

COROLLARY 13.2.4. *Suppose* $\mathfrak{n} \succ 1$ *and* $\operatorname{ndeg} P = \operatorname{ndeg} P_{\times \mathfrak{n}} = d$. *Set* $Q := P_{\leqslant d}$. *Then for all* $\mathfrak{m} \prec\!\!\prec \mathfrak{n}$ *and all* $g \preccurlyeq 1$ *in* K *we have*

$$N_{P_{+g, \times \mathfrak{m}}} = N_{Q_{+g, \times \mathfrak{m}}}.$$

PROOF. After replacing K, P by K^ϕ, P^ϕ, respectively, for suitable ϕ, we may assume that $D_P = N_P$, $D_{P_{\times \mathfrak{n}}} = N_{P_{\times \mathfrak{n}}}$, and $\mathfrak{n} \succ^\flat 1$. Let $R := P - Q = P_{>d}$. Then by Corollary 9.4.20 we have $R \preccurlyeq_\mathfrak{n} \mathfrak{n}^{-1} P \prec_\mathfrak{n} P$. Thus given $g \in K^{\preccurlyeq 1}$, Lemma 13.2.3 applies to P_{+g}, Q_{+g}, R_{+g} in place of P, Q, R, respectively. □

Recall that $P_{|i|'}$ is the subhomogeneous part of P of subdegree i (see Section 4.2). By Corollary 5.7.5 we have $(P^\phi)_{|i|'} = (P_{|i|'})^\phi$, and $P^\phi_{|i|'}$ denotes either of these without ambiguity. When studying N_P, the following lemma sometimes allows us to reduce to the case where P is homogeneous or subhomogeneous.

LEMMA 13.2.5.

$$N_P = \sum_i N_{P_{|i|'}} = \sum_i N_{P_i},$$

where the first sum ranges over all $i \in \mathbb{N}$ *such that* $P^\phi \asymp P^\phi_{|i|'}$ *eventually, and the second sum ranges over all* $i \in \mathbb{N}$ *such that* $P^\phi \asymp P^\phi_i$ *eventually.*

PROOF. We will only prove $N_P = \sum_i N_{P_{|i|'}}$. (To show $N_P = \sum_i N_{P_i}$ one argues in an analogous way.) Below i ranges over elements of $\mathbb{N}$ with $P_{|i|'} \neq 0$, and likewise with j. First, after replacing P by P^ϕ for suitable $\phi \preccurlyeq 1$, we may assume that for all $\phi \preccurlyeq 1$ we have $v(P^\phi_{|i|'}) = v(P_{|i|'}) + \operatorname{nwt}(P_{|i|'}) v\phi$ and $N_{P_{|i|'}} = D_{P^\phi_{|i|'}}$. Therefore, for all i, j with $\operatorname{nwt}(P_{|i|'}) \neq \operatorname{nwt}(P_{|j|'})$, either $P^\phi_{|i|'} \prec P^\phi_{|j|'}$ eventually, or $P^\phi_{|i|'} \succ P^\phi_{|j|'}$ eventually. Hence, after replacing P by P^ϕ for suitable $\phi \preccurlyeq 1$, we may assume that for all i, j and all $\phi \preccurlyeq 1$ we have $P_{|i|'} \prec P_{|j|'}$ iff $P^\phi_{|i|'} \prec P^\phi_{|j|'}$. So for all i and $\phi \preccurlyeq 1$, $P \asymp P_{|i|'}$ iff $P^\phi \asymp P^\phi_{|i|'}$, hence for all i, $P \asymp P_{|i|'}$ iff $P^\phi \asymp P^\phi_{|i|'}$ eventually. Thus for each $\phi \preccurlyeq 1$,

$$D_{P^\phi} = \sum_{P^\phi_{|i|'} \asymp P^\phi} D_{P^\phi_{|i|'}} = \sum_{P \asymp P_{|i|'}} N_{P_{|i|'}},$$

and this yields the claim. □

LEMMA 13.2.6. *Let $P, Q \in K\{Y\}^{\neq}$ be homogeneous of different degrees. Then $N_{P+Q} \in \{N_P, N_Q, N_P + N_Q\}$. Also, $N_{(P+Q)_{\times \mathfrak{m}}}$ is homogeneous for every monomial $\mathfrak{m}$, with at most one exception.*

PROOF. Take ϕ such that

$$N_P = D_{P^\phi}, \quad N_Q = D_{Q^\phi}, \quad N_{P+Q} = D_{P^\phi + Q^\phi}.$$

If $v(P^\phi) < v(Q^\phi)$, then $N_{P+Q} = D_{P^\phi} = N_P$, and if $v(P^\phi) > v(Q^\phi)$, then $N_{P+Q} = D_{Q^\phi} = N_Q$. If $v(P^\phi) = v(Q^\phi)$, then

$$N_{P+Q} = D_{P^\phi + Q^\phi} = D_{P^\phi} + D_{Q^\phi} = N_P + N_Q,$$

since P^ϕ and Q^ϕ are homogeneous of different degrees.

For the second claim of the lemma, assume $\deg P = d < \deg Q = e$. Suppose $N_{(P+Q)_{\times \mathfrak{m}}}$ is not homogeneous. It suffices to show that then $N_{(P+Q)_{\times g}} = N_{P_{\times g}}$ for all nonzero $g \prec \mathfrak{m}$ in K. Towards proving this, we can arrange $\mathfrak{m} = 1$, so by the argument above we have $v(P^\phi) = v(Q^\phi)$ for ϕ as above. Let $g \in K^\times$, $g \prec 1$ and set $\gamma := vg$, so $\gamma > 0$. Take ϕ as above such that in addition

$$N_{P_{\times g}} = D_{P^\phi_{\times g}}, \qquad N_{Q_{\times g}} = D_{Q^\phi_{\times g}}, \qquad N_{(P+Q)_{\times g}} = D_{P^\phi_{\times g} + Q^\phi_{\times g}}.$$

By Corollary 6.1.3 we have

$$v(P^\phi_{\times g}) = v(P^\phi) + d\gamma + o(\gamma) < v(Q^\phi_{\times g}) = v(Q^\phi) + e\gamma + o(\gamma),$$

so $N_{(P+Q)_{\times g}} = N_{P_{\times g}}$ by the proof of the first claim of the lemma. □

Let J be the finite nonempty set of $j \in \mathbb{N}$ such that $P_j \neq 0$; then $P = \sum_{j \in J} P_j$.

COROLLARY 13.2.7. *There is a unique set $I \subseteq J$ such that $N_P = \sum_{i \in I} N_{P_i}$. This set I is determined by the condition that for all $i \in J$,*

$$i \in I \iff v(P_i^\phi) \leqslant v(P_j^\phi), \text{ eventually, for each } j \in J.$$

PROOF. Use that for all $i, j \in J$, either $v(P_i^\phi) < v(P_j^\phi)$, eventually, or $v(P_i^\phi) = v(P_j^\phi)$, eventually, or $v(P_i^\phi) > v(P_j^\phi)$, eventually. □

COROLLARY 13.2.8. *For all but finitely many $\mathfrak{m}$ there is $i \in J$ such that*

$$N_{P_{\times \mathfrak{m}}} = N_{P_{i,\times \mathfrak{m}}}.$$

PROOF. Let $\mathfrak{m}$ be such that for all distinct $i, j \in J$ the Newton polynomial

$$N_{(P_i + P_j)_{\times \mathfrak{m}}}$$

is homogeneous. Then there is a (necessarily unique) $i \in J$ such that $v(P^\phi_{i,\times \mathfrak{m}}) < v(P^\phi_{j,\times \mathfrak{m}})$, eventually, for every $j \in J \setminus \{i\}$. For this i we have $N_{P_{\times \mathfrak{m}}} = N_{P_{i,\times \mathfrak{m}}}$. □

Cleanness

In Section 13.1, having both $D_P \in C[Y](Y')^{\mathbb{N}}$ and $R_P \prec^\flat P$ turned out to be very strong. The next lemma shows that for divisible Γ the eventual form of the first condition implies the eventual form of the second condition.

LEMMA 13.2.9. *Suppose Γ is divisible and $N_P \in C[Y](Y')^{\mathbb{N}}$. Then*

$$R_{P^\phi} \prec^\flat_\phi P^\phi, \qquad \textit{eventually.}$$

PROOF. Set $w := \operatorname{nwt}(P)$. After replacing P by P^ϕ for suitable $\phi \preccurlyeq 1$, we may assume $w = \operatorname{dwm}(P) = \operatorname{dwm}(P^\phi) = \operatorname{dwt}(P)$ for all $\phi \preccurlyeq 1$, and thus for all such ϕ,

$$D_P = D_{P^\phi} = N_P, \qquad v(P^\phi) = v(P) + w\,v\phi.$$

In particular, $(D_P)^{\phi,i} = 0$ for $i = 0,\dots,w-1$ for $\phi \preccurlyeq 1$, so $(R_P)^\phi = R_{P^\phi}$ for $\phi \preccurlyeq 1$ by Lemma 13.1.13. If $R_P = 0$, this gives $R_{P^\phi} = 0$ for $\phi \preccurlyeq 1$. So assume $R_P \neq 0$. Replacing P by P^ϕ for suitable $\phi \preccurlyeq 1$, we arrange in addition:

$$v(R_{P^\phi}) = v(R_P) + \operatorname{nwt}(R_P)v\phi \quad \text{for all } \phi \preccurlyeq 1.$$

We need to show that $\psi^\phi\big(v(R_{P^\phi}) - v(P^\phi)\big) \leqslant 0$, eventually. For this, we distinguish three cases. Suppose first that $\operatorname{nwt}(R_P) > w$. Then we have for $\phi \prec 1$,

$$0 < v\phi \leqslant v(R_P) - v(P) + \big(\operatorname{nwt}(R_P) - w\big)v\phi = v(R_{P^\phi}) - v(P^\phi).$$

Hence, if $v\phi \geqslant 1$, then

$$0 \geqslant 1 - v\phi = \psi^\phi(v\phi) \geqslant \psi^\phi\big(v(R_{P^\phi}) - v(P^\phi)\big).$$

Now assume $\operatorname{nwt}(R_P) < w$. Let $\alpha := \frac{v(R_P)-v(P)}{w-\operatorname{nwt}(R_P)} \in \Gamma^>$, and take $\beta \in \Gamma^{\neq}$ such that $\beta + \psi(\beta) = \alpha$. If $v\phi \geqslant \psi(\beta)$, then by Lemma 9.2.2,

$$\psi^\phi\big(v(R_{P^\phi}) - v(P^\phi)\big) = \psi(\alpha - v\phi) - v\phi \leqslant 0.$$

Finally, if $\operatorname{nwt}(R_P) = w$ and $v\phi \geqslant \psi\big(v(R_P) - v(P)\big)$, then clearly

$$\psi^\phi\big(v(R_{P^\phi}) - v(P^\phi)\big) \leqslant 0.$$

□

COROLLARY 13.2.10. *If Γ is divisible and $P \in K[Y,Y']$, then*

$$N_P \in C[Y](Y')^{\mathbb{N}}, \textit{ and eventually } R_{P^\phi} \prec^\flat_\phi P^\phi.$$

Let L be a d-valued field extension of H-type of K with asymptotic integration, furnished with a monomial group $\mathfrak{M}_L \supseteq \mathfrak{M}$. Note that then L has small derivation, and each ϕ is active in L. If in addition $\Gamma^>$ is coinitial in $\Gamma_L^>$, then $N_P^L = N_P$. This fact yields a variant of Lemma 13.2.9:

LEMMA 13.2.11. *If K has rational asymptotic integration and $N_P \in C[Y](Y')^{\mathbb{N}}$, then $R_{P^\phi} \prec^\flat_\phi P^\phi$, eventually.*

PROOF. The algebraic closure L of K has by Lemma 3.3.33 a monomial group $\mathfrak{M}_L \supseteq \mathfrak{M}$, and $\Gamma^>$ is coinitial in $\Gamma_L^> = (\mathbb{Q}\Gamma)^>$. Thus if L has asymptotic integration, then $N_P^L = N_P \in C[Y](Y')^{\mathbb{N}}$, and Lemma 13.2.9 applies to L instead of K. $\square$

In Section 13.3 we shall prove: K is ω-free $\iff N_P \in C[Y](Y')^{\mathbb{N}}$ for all P.

COROLLARY 13.2.12. *Assume $N_P \in C[Y](Y')^{\mathbb{N}}$ and eventually $R_{P^\phi} \prec^\flat_\phi P^\phi$. Then there exists ϕ such that for every* d*-valued field extension L of H-type of K,*

$$\begin{aligned} y \in L,\ 1 \prec y \preceq^\flat_\phi 1 &\implies v\big(P(y)\big) \ = \ v^{\mathrm{e}}(P) + \operatorname{ndeg}(P)\, vy + \operatorname{nwt}(P)\,\psi_L(vy), \\ y \in L,\ 1 \succ y \preceq^\flat_\phi 1 &\implies v\big(P(y)\big) \ = \ v^{\mathrm{e}}(P) + \operatorname{nmul}(P)\, vy + \operatorname{nwt}(P)\,\psi_L(vy). \end{aligned}$$

If K is equipped with an ordering making K an H-field, then there are ϕ and $\sigma, \tau \in \{-1, +1\}$ such that for all H-field extensions L of K and all $y \in L^>$,

$$\begin{aligned} 1 \prec y \preceq^\flat_\phi 1 &\implies \operatorname{sign} P(y) = \sigma, \\ 1 \succ y \preceq^\flat_\phi 1 &\implies \operatorname{sign} P(y) = \tau. \end{aligned}$$

PROOF. This follows from Lemma 13.1.15 and Corollary 11.1.15, since for $y \not\asymp 1$ in any d-valued field extension L of H-type of K we have

$$v(P^\phi) + \operatorname{dwt}(P^\phi)\psi_L^\phi(vy) \ = \ v^{\mathrm{e}}(P) + \operatorname{nwt}(P)\,\psi_L(vy), \ \text{ eventually.} \qquad \square$$

DEFINITION 13.2.13. We say that K is **clean** if for every P we have

$$N_P \in C[Y](Y')^{\mathbb{N}}, \text{ and eventually } R_{P^\phi} \prec^\flat_\phi P^\phi.$$

Note that if K is clean, then so is each compositional conjugate K^ϕ.

Behavior under additive and multiplicative conjugation

LEMMA 13.2.14. *Let $c \in C$ and $\varepsilon \in K$, $\varepsilon \prec 1$. Then*

$$N_{P_{+c}} = (N_P)_{+c}, \qquad N_{P_{+\varepsilon}} = N_P.$$

PROOF. Eventually $N_{P_{+c}} = D_{(P_{+c})^\phi}$. Since $(P_{+c})^\phi = (P^\phi)_{+c}$, we have $N_{P_{+c}} = D_{(P^\phi)_{+c}}$, eventually. By Lemma 13.1.1, $D_{(P^\phi)_{+c}} = \big(D_{P^\phi}\big)_{+c}$, and eventually we have $\big(D_{P^\phi}\big)_{+c} = (N_P)_{+c}$. The other displayed item follows likewise from part (iii) of Lemma 6.6.5. $\square$

COROLLARY 13.2.15. *Let $f, g, h \in K$ and $f - g \prec h$. Then $N_{P_{+f,\times h}} = N_{P_{+g,\times h}}$.*

PROOF. Use that $P_{+g,\times h} = P_{+f,\times h,+\varepsilon}$ for $\varepsilon := \frac{g-f}{h} \prec 1$. $\square$

For use in Sections 13.5 and 14.4 we need:

LEMMA 13.2.16. *Suppose that $D_P \in C[Y](Y')^{\mathbb{N}}$, and suppose that $\gamma > 0$ is such that $v(R_P) > v(P) + m\gamma + n\gamma'$ for all m, n. Then for $g \in K$ with $vg = \gamma$ we have*

$$N_{P_{\times g}} \in C^\times \cdot Y^\mu, \quad \mu := \operatorname{mul} D_P.$$

PROOF. We have $D_P = D(Y) \cdot (Y')^j$ where

$$D \in C[Y], \qquad D = cY^i + \text{terms of higher degree}, c \in C^\times.$$

We can arrange $v(P) = 0$, so $P = D_P + R$, $R = R_P$. Let $g \in K$, $vg = \gamma$. Then

$$P_{\times g} = D(gY) \cdot g^j (g^\dagger Y + Y')^j + R_{\times g}.$$

Now $D(gY) = g^i c Y^i (1 + E)$ with $E \in YK[Y]$, $v(E) \geqslant \gamma$. Hence

$$P_{\times g} = g^{i+j} c Y^i (1 + E)(g^\dagger Y + Y')^j + R_{\times g}.$$

Let ϕ be such that $v\phi > \gamma^\dagger = v(g^\dagger)$. Then, in view of $E^\phi = E$,

$$\begin{aligned} P^\phi_{\times g} &= g^{i+j} c Y^i (1 + E)(g^\dagger Y + \phi Y')^j + R^\phi_{\times g}, \\ (g^\dagger Y + \phi Y')^j &= (g^\dagger Y)^j + F, \quad v(F) > j\gamma^\dagger, \text{ so} \\ P^\phi_{\times g} &= g^{i+j} (g^\dagger)^j c Y^{i+j} + G, \quad v(G) > (i+j)\gamma + j\gamma^\dagger = i\gamma + j\gamma'. \end{aligned}$$

Since $i + j = \operatorname{mul} D_P$, this gives the desired result. □

COROLLARY 13.2.17. *Suppose that $D_P \in C[Y](Y')^{\mathbb{N}}$ and $R_P \prec^\flat P$. Let $f, g \in K$ be such that $f \prec^\flat 1$ and $g \prec 1$, $g \asymp^\flat 1$. Then*

$$N_{P_{\times g}}, N_{P_{+f,\times g}} \in C^\times \cdot Y^\mu, \quad \textit{where } \mu := \operatorname{mul} D_P.$$

PROOF. From $g \prec 1$ and $g \asymp^\flat 1$ we get $N_{P_{\times g}} \in C^\times \cdot Y^\mu$ by Lemma 13.2.16. Lemma 4.5.1(i) and $f \prec^\flat 1$ yield $(D_P)_{+f} \sim^\flat D_P$ and $(R_P)_{+f} \asymp^\flat R_P$, and so in view of $P_{+f} = \mathfrak{d}_P (D_P)_{+f} + (R_P)_{+f}$ we have $\mathfrak{d}_{P_{+f}} = \mathfrak{d}_P$, $D_{P_{+f}} = D_P \in C[Y](Y')^{\mathbb{N}}$, and $R_{P_{+f}} \prec^\flat P_{+f}$. Hence $N_{P_{+f,\times g}} \in C^\times \cdot Y^\mu$ by Lemma 13.2.16. □

13.3 THE SHAPE OF THE NEWTON POLYNOMIAL

In this section we combine the material from the previous two sections with results from Chapters 11 and 12.

Statement of results

Recall from Section 12.7 that ∇_1 and ∇_2 are the $1,2$-diagonals of the triangular (Stirling) derivation ∇ of the polynomial K-algebra $K\{Y\} = K[Y, Y', Y'', \dots]$. By Corollary 12.7.18 and Lemma 12.8.2 we have

$$N_P \in C[Y](Y')^{\mathbb{N}} \iff \nabla_1(N_P) = \nabla_2(N_P) = 0.$$

The main goal of this section is to prove the following two results:

THEOREM 13.3.1. *If K is λ-free, then $\nabla_1(N_P) = 0$. Conversely, if $\nabla_1(N_Q) = 0$ for each homogeneous $Q \in K\{Y\}^{\neq}$ of degree 1, then K is λ-free.*

THEOREM 13.3.2. *If K is ω-free, then $\nabla_1(N_P) = \nabla_2(N_P) = 0$. Conversely, if $\nabla_1(N_Q) = \nabla_2(N_Q) = 0$ for each homogeneous $Q \in K\{Y\}^{\neq}$ of degree 2, then K is ω-free.*

Before we get to the proofs, let us first deduce some consequences. Theorem 13.3.2 and Lemma 13.2.11 immediately yield a characterization of cleanness showing that K being clean does not depend on the choice of monomial group $\mathfrak{M}$:

COROLLARY 13.3.3. *K is clean iff K is ω-free.*

The previous corollary in conjunction with Corollary 11.7.15 gives:

COROLLARY 13.3.4. *If K is a union of asymptotic subfields, each with a smallest comparability class, then K is clean.*

Another consequence: if K is spherically complete, then K is not clean.

The λ-free case

We precede the proof of Theorem 13.3.1 with a lemma. We identify the ordered group $\mathbb{Z}$ with the ordered subgroup $\mathbb{Z} \cdot 1$ of Γ via $k \mapsto k \cdot 1$.

LEMMA 13.3.5. *Suppose* $\operatorname{dwm}(P^\phi) = \operatorname{dwt}(P)$ *and* $v\phi > 1$. *Then* $D_{P^\phi} = D_{Q^\phi}$, *where* $Q := P_{[0]} + \cdots + P_{[w]}$, $w := \operatorname{dwt}(P)$.

PROOF. Replacing P by $\mathfrak{d}_P^{-1}P$, we may assume $\mathfrak{d}_P = 1$, so $\mathfrak{d}_{P^\phi} = \phi^w$. From the identity $(P-Q)^\phi = \sum_{i>w}(P_{[i]})^\phi$ and (13.1.2) we obtain

$$(P-Q)^\phi = \left(\sum_{i>w\geqslant j} \phi^j(\phi^\dagger)^{i-j}(P_{[i]})^{\phi,j}\right) + \left(\sum_{i\geqslant j>w} \phi^j(\phi^\dagger)^{i-j}(P_{[i]})^{\phi,j}\right).$$

It is clear that $v(P_{[i]})^{\phi,j} \geqslant v(P_{[i]}) \geqslant v(P) = 0$ for all i, j. Moreover, if $i > w \geqslant j$ then $i - j > (w-j)v\phi$ since $v\phi = 1 + o(1)$ by Corollary 6.5.5. Hence each term $\phi^j(\phi^\dagger)^{i-j}(P_{[i]})^{\phi,j}$ $(i > w \geqslant j)$ in the first sum has valuation

$$jv\phi + i - j + v(P_{[i]})^{\phi,j} \geqslant jv\phi + i - j > wv\phi.$$

Clearly each term $\phi^j(\phi^\dagger)^{i-j}(P_{[i]})^{\phi,j}$ $(i \geqslant j > w)$ in the second sum has valuation $> wv\phi$. Hence $v(P-Q)^\phi > wv\phi = vP^\phi$ and thus $D_{P^\phi} = D_{Q^\phi}$. □

Let us also recall the transformation formulas deduced in Corollary 12.8.6: with $w = \operatorname{wt}(P)$, $\lambda = -\phi^\dagger$ and $\omega = -(2\lambda' + \lambda^2)$,

$$(13.3.1)\qquad \begin{cases} (P^\phi)_{[w]} = \phi^w P_{[w]}, \\ (P^\phi)_{[w-1]} = \phi^{w-1}\big[P_{[w-1]} + \lambda\nabla_1(P_{[w]})\big], \\ (P^\phi)_{[w-2]} = \phi^{w-2}\big[P_{[w-2]} + \lambda\nabla_1(P_{[w-1]}) + \big(\omega\nabla_2 + \tfrac{1}{2}\lambda^2\nabla_1^2\big)(P_{[w]})\big]. \end{cases}$$

The next proposition and its corollary yield Theorem 13.3.1:

PROPOSITION 13.3.6. *Suppose K is λ-free. Then $\nabla_1(N_P) = 0$.*

PROOF. Let $w := \operatorname{nwt}(P)$. Then $N_P \in C\{Y\}$ is isobaric of weight w; if $w = 0$, then $N_P \in C[Y]$ and hence $\nabla_1(N_P) = \nabla_2(N_P) = 0$, so we can assume that $w > 0$. Replacing K and P by K^ϕ and P^ϕ for suitable $\phi \preccurlyeq 1$, we arrange $D_P = N_P$. (Here we use the invariance of λ-freeness under compositional conjugation of K.) Then Lemma 13.2.1 gives $D_{P^\phi} = N_P$ and $\operatorname{dwm}(P^\phi) = \operatorname{dwt}(P^\phi) = w$ for all $\phi \preccurlyeq 1$. By Lemma 13.3.5 we can further reduce to the case that $\operatorname{wt}(P) = w$, and by Lemma 13.1.14 we arrange $(R_P)_{[w]} \prec^\flat P$. From $P = \mathfrak{d}_P N_P + R_P$ we get $P_{[w]} = \mathfrak{d}_P N_P + (R_P)_{[w]}$. Multiplying P by $\mathfrak{d}_P^{-1}$, we obtain in addition $P \asymp 1$ and $P_{[w]} = N_P + R$ with $R \in K\{Y\}$, $R \prec^\flat 1$.

Suppose towards a contradiction that $\nabla_1 N_P \neq 0$. Then

$$\nabla_1 P_{[w]} = \nabla_1 N_P + \nabla_1 R, \qquad \nabla_1 N_P \in C\{Y\}^{\neq}, \quad \nabla_1 R \prec^\flat 1.$$

Recall the pc-sequence (λ_ρ) of width $\{\gamma \in \Gamma_\infty : \gamma > \Psi\}$ introduced in Section 11.5, and take a pseudolimit λ of (λ_ρ) in some immediate asymptotic field extension of K (Theorem 11.4.1). Since (λ_ρ) has no pseudolimit in K, there is no $a \in K$ with $v(a + \lambda) > \Psi$, which gives $\phi \preccurlyeq 1$ with $P_{[w-1]} + \lambda \nabla_1 P_{[w]} \succ \phi$. With $\lambda = -\phi^\dagger$ we have $\lambda - \lambda \prec \phi$ by Lemma 11.5.6, so

$$P_{[w-1]} + \lambda \nabla_1 P_{[w]} \sim P_{[w-1]} + \lambda \nabla_1 P_{[w]} \succ \phi.$$

By Proposition 11.1.4 we have $P^\phi \asymp \phi^w$, but the second identity of (13.3.1) gives

$$\begin{aligned}(P^\phi)_{[w-1]} &= \phi^{w-1}\big(P_{[w-1]} + \lambda \nabla_1 P_{[w]}\big) \\ &\sim \phi^{w-1}\big(P_{[w-1]} + \lambda \nabla_1 P_{[w]}\big) \succ \phi^w,\end{aligned}$$

a contradiction. □

REMARK. For later use we record a variant of Proposition 13.3.6:

Let K be an immediate extension of its valued differential subfield E, and assume E is λ-free, $\mathfrak{M} \subseteq E^\times$, and $P \in E\{Y\}^{\neq}$. Then $\nabla_1 N_P = 0$.

We did not assume here that K is λ-free, and while $C_E \subseteq C$ and $N_P \in C\{Y\}$, it does not follow from $P \in E\{Y\}$ that $N_P \in C_E\{Y\}$. Nevertheless, the proof of this variant is the same as that of Proposition 13.3.6, apart from minor changes.

The above remark and Corollary 12.7.19 give a result to be used in Section 13.7:

COROLLARY 13.3.7. *Let K be an immediate extension of its valued differential subfield E, and assume E is λ-free, $\mathfrak{M} \subseteq E^\times$, and $P \in E\{Y\}^{\neq}$ has order at most 2. Then $N_P \in C[Y](Y')^{\mathbb{N}}$.*

Next we relate λ-freeness to properties of homogeneous P of degree 1:

COROLLARY 13.3.8. *The following are equivalent:*

(i) *K is λ-free;*

(ii) *$\nabla_1(N_Q) = 0$ for every homogeneous $Q \in K\{Y\}^{\neq}$ of degree 1;*

(iii) *for every homogeneous $Q \in K\{Y\}^{\neq}$ of degree 1 there is $c \in C^{\times}$ such that*

$$N_Q = cY \text{ or } N_Q = cY';$$

(iv) *for every $a \in K$ and $Q(Y) = aY' + Y''$, we have $\operatorname{nwt}(Q) \leqslant 1$.*

PROOF. Proposition 13.3.6 gives (i) $\Rightarrow$ (ii). Assume (ii), and let $Q \in K\{Y\}^{\neq}$ be homogeneous of degree 1. Then $N_Q = cY^{(w)}$, $c \in C^{\times}$, $w = \operatorname{nwt}(Q)$. If $w > 1$, then

$$\nabla_1(N_Q) = -c \cdot \binom{w}{2} Y^{(w-1)} \neq 0$$

by Lemma 12.7.16, contradicting (ii). Thus $w = 0$ or $w = 1$. This shows (ii) $\Rightarrow$ (iii), and (iii) $\Rightarrow$ (iv) is obvious. To show the contrapositive of (iv) $\Rightarrow$ (i), suppose $\lambda \in K$ is a pseudolimit of (λ_ρ). Consider

$$Q(Y) = \lambda Y' + Y'' \in K\{Y\}.$$

Then $Q^{\phi} = (\phi' + \phi\lambda)Y' + \phi^2 Y''$, and by Lemma 11.5.6,

$$v(\phi' + \phi\lambda) = v\phi + v(\lambda + \phi^{\dagger}) > v(\phi^2),$$

hence $\operatorname{nwt}(Q) = 2$. □

The following variant is also useful:

COROLLARY 13.3.9. *Suppose K is an immediate extension of its valued differential subfield E, and E is λ-free and $\mathfrak{M} \subseteq E^{\times}$. Let $Q \in E\{Y\}^{\neq}$ be homogeneous of degree 1. Then $N_Q = cY$ or $N_Q = cY'$ for some $c \in C^{\times}$.*

PROOF. Since N_Q is homogeneous of degree 1 and isobaric of weight $w := \operatorname{nwt}(Q)$, we have $N_Q = cY^{(w)}$ with $c \in C^{\times}$. We have $\nabla_1 Q = 0$ by the remark after the proof of Proposition 13.3.6. Thus $w = 0$ or $w = 1$ as in the proof of Corollary 13.3.8. □

The next lemma shows that λ-freeness imposes restrictions on the shape of N_P for any P. Its corollary shows this to be decisive for $\operatorname{nwt}(P) \leqslant 3$.

LEMMA 13.3.10. *Suppose K is λ-free, and $\operatorname{nwt}(P) > 1$. Then $(N_P)_{|1|'} = 0$. Also $(N_P)^{\phi,1} = 0$ for every ϕ and $(N_P)^{t,1} = 0$.*

PROOF. Set $w := \operatorname{nwt}(P) > 1$ and $N := N_P$. Suppose towards a contradiction that $N_{|1|'} \neq 0$. Then $P_{|1|'} \neq 0$ and $N_{|1|'} = N_{P_{|1|'}}$ by Lemma 13.2.5, so after replacing P by $P_{|1|'}$ we may assume that P is subhomogeneous of subdegree 1. Using Lemma 13.2.5 we may further reduce to the case that P is homogeneous. Then

$P = Y^d Q$ where $d \in \mathbb{N}$ and $Q \in K\{Y'\} \subseteq K\{Y\}$ is homogeneous of degree 1. Now by Corollary 13.3.8, $N = Y^d N_Q = cY^d Y'$, $c \in C^\times$; hence $\operatorname{nwt}(P) = \operatorname{wt}(N) = 1$, a contradiction. Thus $N_{|1|'} = 0$.

Recall from Section 11.1 that $\left[\begin{smallmatrix}\tau\\ \sigma\end{smallmatrix}\right] = 0$ for $\boldsymbol{\tau} \geqslant \boldsymbol{\sigma}$ with $\operatorname{supp}\boldsymbol{\tau} \neq \operatorname{supp}\boldsymbol{\sigma}$. Also, if $\|\boldsymbol{\sigma}\| = 1$, $\boldsymbol{\tau} \geqslant \boldsymbol{\sigma}$ and $\operatorname{supp}\boldsymbol{\tau} = \operatorname{supp}\boldsymbol{\sigma}$, then $N_{[\boldsymbol{\tau}]} = 0$ (since $N_{|1|'} = 0$). Hence

$$N^{\phi,1} = \sum_{\|\boldsymbol{\sigma}\|=1} \left(\sum_{\boldsymbol{\tau} \geqslant \boldsymbol{\sigma}} \begin{bmatrix}\boldsymbol{\tau}\\ \boldsymbol{\sigma}\end{bmatrix} N_{[\boldsymbol{\tau}]} \big(1 + \varepsilon_{\boldsymbol{\sigma}}^{\boldsymbol{\tau}}(\phi)\big) \right) Y^{[\boldsymbol{\sigma}]} = 0,$$

and

$$N^{t,1} = \sum_{\|\boldsymbol{\sigma}\|=1} \left(\sum_{\boldsymbol{\tau} \geqslant \boldsymbol{\sigma}} \begin{bmatrix}\boldsymbol{\tau}\\ \boldsymbol{\sigma}\end{bmatrix} N_{[\boldsymbol{\tau}]} \right) Y^{[\boldsymbol{\sigma}]} = 0. \qquad \square$$

COROLLARY 13.3.11. *Suppose K is λ-free and* $\operatorname{nwt}(P) \leqslant 3$. *Then* $N_P \in C[Y](Y')^{\mathbb{N}}$.

PROOF. Assume $\operatorname{nwt}(P) = 3$, and set $N := N_P$. By (13.1.2),

$$N^\phi = (\phi^\dagger)^3 N^{\phi,0} + \phi(\phi^\dagger)^2 N^{\phi,1} + \phi^2 \phi^\dagger N^{\phi,2} + \phi^3 N^{\phi,3}.$$

Now $N^{\phi,0}$ is the isobaric part of N of weight 0, so $N^{\phi,0} = 0$. Next, $N^{\phi,1} = 0$ by Lemma 13.3.10. Also, $N^{\phi,2} = 0$ by Lemma 13.1.8(ii) and Proposition 13.3.6, and $N^{\phi,3} = N$. This gives $N^\phi = \phi^3 N$. Take $x \succ 1$ in an immediate asymptotic extension of K with $x' = 1$ and set $t = 1/x$. Replacing ϕ by t in the above arguments still gives us $N^t = t^3 N$, even if this immediate extension is not λ-free. Apply Lemma 12.8.2 to get $N \in C[Y](Y')^3$. The case $\operatorname{nwt}(P) \leqslant 2$ is similar. $\square$

The ω-free case

Towards proving Theorem 13.3.2 we need:

LEMMA 13.3.12. *Suppose* $\operatorname{wt}(P) = \operatorname{wt}(N_P) = w$, $D_P = N_P$, $P \asymp 1$, *and*

$$P_{[w]} = N_P + R, \quad \nabla_1 N_P = 0, \quad v(R) > \Psi.$$

Then there exists $\phi \preccurlyeq 1$ such that for $Q := P^\phi - (P^\phi)_{[w-1]}$ we have: $D_{Q^\theta} = N_P$ for all $\theta \in \mathfrak{M}^{\preccurlyeq 1}$ active in K^ϕ; note that $Q_{[w-1]} = 0$ and $N_P = N_Q$ for such Q.

PROOF. For all $\phi \preccurlyeq 1$ we have $P^\phi \sim (P^\phi)_{[w]}$ and by (13.3.1),

$$\begin{aligned}(P^\phi)_{[w]} &= \phi^w P_{[w]},\\ (P^\phi)_{[w-1]} &= \phi^{w-1} \cdot \big(P_{[w-1]} - \phi^\dagger \nabla_1 P_{[w]}\big) = \phi^{w-1} \cdot \big(P_{[w-1]} - \phi^\dagger \nabla_1 R\big),\end{aligned}$$

with $v(\nabla_1 R) > \Psi$, so $v(\phi^\dagger \nabla_1 R) > \Psi$. We claim that $v(P_{[w-1]}) > \Psi$. Assuming towards a contradiction that this claim is false, we get ϕ with $\phi \preccurlyeq P_{[w-1]} \preccurlyeq 1$, so

$$P^\phi \sim (P^\phi)_{[w]} \asymp \phi^w \preccurlyeq (P^\phi)_{[w-1]}$$

by the identities above. But $N_P = D_{P^\phi}$ is isobaric of weight w, so $(P^\phi)_{[w-1]} \prec P^\phi$. This contradiction proves the claim. Take $\gamma > \Psi$ with $v(P_{[w-1]}) > \gamma$ and $v(R) > \gamma$, and then take β with $\Psi < \beta < \gamma$. Then $\alpha := \gamma - \beta > 0$, and for all $\phi \preccurlyeq 1$,

$$v\big((P^\phi)_{[w-1]}\big) - v(P^\phi) \,>\, \gamma - v\phi \,>\, \alpha$$

by the identities above. Thus we can take $\phi \preccurlyeq 1$ such that $(P^\phi)_{[w-1]} \prec^\flat_\phi P^\phi$. Then $Q := P^\phi - (P^\phi)_{[w-1]}$ has by Lemma 13.2.2 applied to P^ϕ the desired property. □

The next proposition and its corollary imply Theorem 13.3.2.

PROPOSITION 13.3.13. *Suppose K is ω-free. Then* $\nabla_1(N_P) = \nabla_2(N_P) = 0$.

PROOF. As in the proof of Proposition 13.3.6 we first reduce to the case that $w := \operatorname{nwt}(P) = \operatorname{wt}(P) > 0$ and $D_P = N_P$, as well as

$$P \asymp 1, \qquad P_{[w]} \,=\, N_P + R, \quad R \prec^\flat 1.$$

Since K is ω-free, K is λ-free, so by Proposition 13.3.6 we have $\nabla_1 N_P = 0$. If $w = 1$, then $N_P \in K[Y, Y']$, so $\nabla_2 N_P = 0$ by Corollary 12.7.18. Assume $w > 1$ in the rest of the proof. Lemma 13.3.12 provides a further reduction to the case $P_{[w-1]} = 0$: this involves replacing K for a suitable $\phi \preccurlyeq 1$ by K^ϕ and P by $\phi^{-w}Q$ with $Q := P^\phi - (P^\phi)_{[w-1]}$.

Assume towards a contradiction that $\nabla_2 N_P \neq 0$. Then

$$\nabla_2 P_{[w]} \,=\, \nabla_2 N_P + \nabla_2 R, \qquad \nabla_2 N_P \in C\{Y\}^{\neq}, \quad \nabla_2 R \prec^\flat 1.$$

Recall the pc-sequence (ω_ρ) of width $\{\gamma \in \Gamma_\infty : \gamma > 2\Psi\}$ introduced in Section 11.5. Take a pseudolimit ω of (ω_ρ) in an immediate asymptotic field extension of K. Since (ω_ρ) has no pseudolimit in K, there is no $b \in K$ with $b + \omega \prec a^2$ for all active $a \in K$. This gives $\phi \preccurlyeq 1$ with $P_{[w-2]} + \omega\nabla_2 P_{[w]} \succ \phi^2$. From $P_{[w-1]} = 0$, $\nabla_1 P_{[w]} = \nabla_1 R \prec^\flat 1$, and (13.3.1) we obtain, with $\lambda = -\phi^\dagger, \omega = -(2\lambda' + \lambda^2)$:

$$\begin{aligned} (P^\phi)_{[w]} \,&=\, \phi^w P_{[w]}, \qquad (P^\phi)_{[w-1]} \prec^\flat 1, \\ (P^\phi)_{[w-2]} \,&=\, \phi^{w-2}\big[P_{[w-2]} + \omega\nabla_2 P_{[w]} + S\big], \quad S \prec^\flat 1. \end{aligned}$$

By Lemma 11.7.7 we have $\omega - \omega \prec \phi^2$, so

$$P_{[w-2]} + \omega\nabla_2 P_{[w]} + S \,\sim\, P_{[w-2]} + \omega\nabla_2 P_{[w]} \,\succ\, \phi^2.$$

By Proposition 11.1.4 we have $P^\phi \asymp \phi^w$, but the above also gives

$$(P^\phi)_{[w-2]} \,=\, \phi^{w-2}\big[P_{[w-2]} + \omega\nabla_2 P_{[w]} + S\big] \,\succ\, \phi^w,$$

a contradiction. □

REMARK. For later use we record a variant of Proposition 13.3.13:

If K is an immediate extension of its valued differential subfield E, and E is ω-free, $\mathfrak{M} \subseteq E^\times$, and $P \in E\{Y\}^\times$, then $\nabla_1 N_P = \nabla_2 N_P = 0$, so $N_P \in C[Y](Y')^{\mathbb{N}}$.

Taking into account the remark following the proof of Proposition 13.3.6, the proof of this variant is the same as that of Proposition 13.3.13, apart from routine changes.

COROLLARY 13.3.14. *The following are equivalent:*

(i) *K is ω-free;*

(ii) *$\nabla_1(Q) = \nabla_2(Q) = 0$ for every homogeneous $Q \in K\{Y\}^{\neq}$ of degree 2;*

(iii) *for every homogeneous $Q \in K\{Y\}^{\neq}$ of degree 2 we have:*

$$N_Q = cY^i(Y')^j \text{ for some } c \in C^\times \text{ and some } i, j \in \mathbb{N} \text{ with } i + j = 2;$$

(iv) *for every $a \in K$ and $Q(Y) = a(Y')^2 + 2Y'Y^{(3)} - 3(Y'')^2$, we have $\operatorname{nwt}(Q) \leqslant 3$.*

PROOF. Proposition 13.3.13 gives (i) $\Rightarrow$ (ii). Suppose (ii) holds, and let $Q \in K\{Y\}^{\neq}$ be homogeneous of degree 2. Then N_Q is homogeneous of degree 2, and isobaric, so $N_Q \in C[Y](Y')^{\mathbb{N}}$ by Corollary 12.7.18 and Lemma 12.8.2. Thus $N_Q = cY^i(Y')^j$ with $c \in C^\times$, $i + j = 2$. This shows (ii) $\Rightarrow$ (iii), and (iii) $\Rightarrow$ (iv) is obvious. To show the contrapositive of (iv) $\Rightarrow$ (i), suppose $\omega \in K$ is a pseudolimit of (ω_ρ). Let

$$N(Y) = 2Y'Y^{(3)} - 3(Y'')^2 \in \mathbb{Q}\{Y'\} \subseteq K\{Y'\} \subseteq K\{Y\}.$$

Note that N is homogeneous of degree 2, isobaric of weight 4, and

$$N^\phi = \phi^4 N + (2\phi\phi'' - 3(\phi')^2)(Y')^2 \in K^\phi\{Y'\}.$$

Consider the differential polynomial

$$Q := -\omega \cdot (Y')^2 + N \in K\{Y'\}.$$

We have

$$Q^\phi = (2\phi\phi'' - 3(\phi')^2 - \omega\phi^2) \cdot (Y')^2 + \phi^4 N.$$

Setting $\lambda := -\phi^\dagger$ and $\omega := -(2\lambda' + \lambda^2)$, we get

$$(\phi')^2 = \lambda^2\phi^2, \qquad \phi\phi'' = -\phi(\lambda\phi)' = (-\lambda' + \lambda^2)\phi^2,$$

so by Lemma 11.7.7,

$$2\phi\phi'' - 3(\phi')^2 - \omega\phi^2 = (-2\lambda' + 2\lambda^2 - 3\lambda^2 - \omega)\phi^2 = (\omega - \omega)\phi^2 \prec \phi^4.$$

Thus $D_{Q^\phi} = N$ for all ϕ, and hence $\operatorname{nwt}(Q) = \operatorname{wt}(N) = 4$. $\square$

Suppose K is ω-free. Then $N_{P_{\times\mathfrak{m}}} = A(Y)(Y')^j$ with $A \in C[Y]$ and $j \in \mathbb{N}$. Let $y \sim c\mathfrak{m}$ ($c \in C^\times$) be an approximate zero of P. Then the multiplicity of y as an approximate zero of P is $i+j$, where i is the multiplicity of A at c; we call i the **algebraic multiplicity** of the approximate zero y of P. We say that y is an **algebraic approximate zero of** P if $A(c)=0$, equivalently, its algebraic multiplicity is $\geqslant 1$.

LEMMA 13.3.15. *Suppose K is ω-free. Let P and $\mathfrak{m}$ be given. Then there are distinct $c_1,\dots,c_n \in C^\times$ such that $c_1\mathfrak{m},\dots,c_n\mathfrak{m}$ are algebraic approximate zeros of P with respective algebraic multiplicities $\mu_1,\dots,\mu_n$, and such that for any algebraic approximate zero $y \asymp \mathfrak{m}$ of P we have $y \sim c_k\mathfrak{m}$ for some k. These properties uniquely determine the set $\{(c_1,\mu_1),\dots,(c_n,\mu_n)\}$. If C is algebraically closed, then*

$$\mu_1+\cdots+\mu_n \;=\; \operatorname{ndeg} P_{\times\mathfrak{m}} - \operatorname{nmul} P_{\times\mathfrak{m}},$$

and if C is real closed, then

$$\mu_1+\cdots+\mu_n \;\equiv\; \operatorname{ndeg} P_{\times\mathfrak{m}} - \operatorname{nmul} P_{\times\mathfrak{m}} \mod 2.$$

PROOF. Take $A \in C[Y]$ without zeros in C, $i,j \in \mathbb{N}$, distinct $c_1,\dots,c_n \in C^\times$, and $\mu_1,\dots,\mu_n \in \mathbb{N}^{\geqslant 1}$ such that

$$N_{P_{\times\mathfrak{m}}} \;=\; A(Y)\cdot Y^i \cdot \left(\prod_{k=1}^{n}(Y-c_k)^{\mu_k}\right)\cdot (Y')^j.$$

Then $c_1,\dots,c_n,\mu_1,\dots,\mu_n$ have the desired property. If C is algebraically closed, then $\deg A = 0$. If C is real closed, then $\deg A$ is even. □

Newton polynomials and upward shift

This subsection assumes familiarity with Appendix A. Consider $\mathbb{T}$ equipped with its usual ordering, valuation and derivation $\partial = \frac{d}{dx}$. Then $\mathbb{T}$ is an ω-free H-field with small derivation, and constant field $\mathbb{R}$. The group of transmonomials is a monomial group of $\mathbb{T}$, and we equip $\mathbb{T}$ with this monomial group. We have the usual logarithmic sequence (ℓ_n) for $\mathbb{T}$ given by $\ell_0 = x$ and $\ell_{n+1} = \log(\ell_n)$, with corresponding $\gamma_n := \frac{1}{\ell_0\ell_1\cdots\ell_n}$. It is easy to check that the map $f \mapsto f{\uparrow}\colon \mathbb{T} \to \mathbb{T}$ is for each n an isomorphism $\mathbb{T}^{\gamma_n} \to \mathbb{T}^{\gamma_{n-1}}$ of H-fields, where $\gamma_{-1} := 1$. Defining $f{\uparrow}^0 := f$ and $f{\uparrow}^{n+1} := (f{\uparrow}^n){\uparrow}$ for $f \in \mathbb{T}$, we obtain for each n an isomorphism $f \mapsto f{\uparrow}^n\colon \mathbb{T}^{\gamma_{n-1}} \to \mathbb{T}$ of H-fields.

Let $P \in \mathbb{T}\{Y\}$. Recall that $P{\uparrow}$ denotes the differential polynomial in $\mathbb{T}\{Y\}$ obtained by applying $f \mapsto f{\uparrow}$ to the coefficients of $P^{1/x} \in \mathbb{T}^{1/x}\{Y\}$, and that $P(y){\uparrow} = P{\uparrow}(y{\uparrow})$ for $y \in \mathbb{T}$. Defining inductively $P{\uparrow}^0 := P$ and $P{\uparrow}^{n+1} := (P{\uparrow}^n){\uparrow}$, we have $P(y){\uparrow}^n = P{\uparrow}^n(y{\uparrow}^n)$ for all n and $y \in \mathbb{T}$.

LEMMA 13.3.16. *Let $P \in \mathbb{T}\{Y\}$. Then $P{\uparrow}^n \in \mathbb{T}\{Y\}$ is obtained by applying the operation $f \mapsto f{\uparrow}^n$ to the coefficients of $P^{\gamma_{n-1}} \in \mathbb{T}^{\gamma_{n-1}}\{Y\}$, and $D_{P{\uparrow}^n} = D_{P^{\gamma_{n-1}}}$.*

PROOF. Let $Q \in \mathbb{T}\{Y\}$ be obtained by applying $f \mapsto f{\uparrow}^n$ to the coefficients of $P^{\gamma_{n-1}}$. By Lemma 4.2.1, it suffices to show that $P(y){\uparrow}^n = Q(y{\uparrow}^n)$ for all $y \in \mathbb{T}$. For this we may assume $P = Y^{(i)}$, $i \geqslant 1$, so $P^{\gamma_{n-1}} = \sum_{j=1}^{i} F_j^i(\gamma_{n-1})Y^{(j)}$. With δ the derivation of $\mathbb{T}^{\gamma_{n-1}}$, we then have $\partial^j(y\uparrow^n) = \delta^j(y)\uparrow^n$ for all $j \in \mathbb{N}$, so

$$\begin{aligned} Q(y\uparrow^n) &= \sum_{j=1}^{i} F_j^i(\gamma_{n-1})\uparrow^n \partial^j(y\uparrow^n) = \sum_{j=1}^{i} F_j^i(\gamma_{n-1})\uparrow^n \delta^j(y)\uparrow^n \\ &= P^{\gamma_{n-1}}(y)\uparrow^n = P(y)\uparrow^n \end{aligned}$$

for $y \in \mathbb{T}$, as required. To prove the identity, assume $P \neq 0$, so

$$P^{\gamma_{n-1}} = \mathfrak{d} D_{P^{\gamma_{n-1}}} + R_n, \quad \mathfrak{d} := \mathfrak{d}_{P^{\gamma_{n-1}}}, \ R_n \in \mathbb{T}^{\gamma_{n-1}}\{Y\}, \ R_n \prec P^{\gamma_{n-1}},$$

so $P\uparrow^n = (\mathfrak{d}\uparrow^n) \cdot D_{P^{\gamma_{n-1}}} + S_n$ with $S_n \in \mathbb{T}\{Y\}$, $S_n \prec P\uparrow^n$. It remains to note that $\mathfrak{d}\uparrow^n$ is a transmonomial. □

Since $\mathbb{T}$ is ω-free, we may conclude:

COROLLARY 13.3.17. *Let $P \in \mathbb{T}\{Y\}^{\neq}$. Then there is an $n_0 \in \mathbb{N}$ such that*

$$D_{P\uparrow^n} = N_P \in \mathbb{R}[Y](Y')^{\mathbb{N}} \qquad \textit{for all } n \geqslant n_0.$$

Notes and comments

Corollary 13.3.17 is a translation to our "compositional conjugation" setting of [194, Section 8.3.1]. Some of the other basic properties of the Newton polynomial established in Section 13.2 (for example, Lemma 13.2.14) were first shown for $K = \mathbb{T}_{\mathrm{g}}$ in [194, Section 8.3]. We do not need this, but it is worth mentioning that in Corollary 13.3.17 one can take $n_0 = 2\,\mathrm{dwm}(P)$.

13.4 REALIZING CUTS IN THE VALUE GROUP

Throughout this section K is ω-free (and thus clean, by Corollary 13.3.3). For later use we study realizations of the following three cuts in the value group Γ of K:

$$\Gamma^{<}, \quad \Gamma^{\leqslant}, \quad (\Gamma^{<})' = \Psi^{\downarrow}.$$

Given any element y in any d-valued field extension L of H-type of K such that vy realizes one of these cuts, we shall derive some informative results about the H-asymptotic field extension $K\langle y\rangle$ of K. If vy realizes the cut $\Gamma^{<}$ or the cut $\Gamma^{\leqslant}$ (that is, $\Gamma^{<} < vy < 0$ or $0 < vy < \Gamma^{>}$), then this is fairly straightforward, based on earlier results in this chapter. The case where vy realizes $\Psi^{\downarrow}$ (that is, $\Psi < vy < (\Gamma^{>})'$) requires some facts about the operation $Q \mapsto Q^{\times\phi}$ on differential polynomials.

The cut $\Gamma^{<}$

Model-theoretic compactness yields a d-valued field extension L of H-type of K with an element $y \in L$ such that $\Gamma^{<} < vy < 0$. Let such L and $y \in L$ be given. We determine the asymptotic couple of $K\langle y\rangle$:

LEMMA 13.4.1. *We have* $v\big(P(y)\big) = v^{\mathrm{e}}(P) + \operatorname{ndeg}(P)\,vy + \operatorname{nwt}(P)\,\psi_L(vy)$. *Moreover,* $\Gamma_{K\langle y\rangle} = \Gamma \oplus \mathbb{Z}vy \oplus \mathbb{Z}\psi_L(vy)$ *(internal direct sum) with* $\max \Psi_{K\langle y\rangle} = \psi_L(vy)$.

PROOF. The first identity follows from Corollary 13.2.12. As this holds for every P, we get $\Gamma_{K\langle y\rangle} = \Gamma + \mathbb{Z}vy + \mathbb{Z}\psi_L(vy)$. Set $\alpha := vy$, $\beta := \psi_L(vy)$, and $\Gamma_1 = \Gamma + \mathbb{Z}\beta$. Since K has rational asymptotic integration by Corollary 11.6.8, and $\Psi < \beta < (\Gamma^{>})'$, we get from Corollary 9.8.6 that $k\beta \notin \Gamma$ for all nonzero $k \in \mathbb{Z}$, and $[\Gamma_1] = [\Gamma]$, so $\psi_L(\Gamma_1^{\neq}) = \Psi$. With ψ_1 the restriction of ψ_L to $\Gamma_1^{\neq}$ we get an H-asymptotic couple (Γ_1, ψ_1) with gap β. Since $0 < n|\alpha| < \Gamma_1^{>}$ for all $n \geqslant 1$, and $\psi_L\big(|\alpha|\big) = \beta$, this gives the desired result in view of $[\Gamma_1 + \mathbb{Z}\alpha] = [\Gamma] \cup \big\{[\alpha]\big\}$. □

The extension $K\langle y\rangle$ of K is determined up to isomorphism:

LEMMA 13.4.2. *Let y^* also be an element of a* d*-valued field extension L^* of H-type of K such that $\Gamma^{<} < vy^* < 0$. Then there is a unique valued differential field embedding $K\langle y\rangle \to L^*$ that is the identity on K and sends y to y^*.*

PROOF. Note that $P(y) \neq 0$ for all P by the first part of Lemma 13.4.1, and likewise for y^*, that is, y and y^* are d-transcendental over K. The proof of that lemma also gives an ordered group isomorphism $\Gamma_{K\langle y\rangle} \to \Gamma_{K\langle y^*\rangle}$ that is the identity on Γ, sends vy to vy^*, and $\psi_L(vy)$ to $\psi_{L^*}(vy^*)$. This yields the desired result. □

The cut $\Gamma^{\leqslant}$

There also exists a d-valued field extension L of H-type of K with an element $y \in L$ such that $0 < vy < \Gamma^{>}$. Let such L and $y \in L$ be given. Then the following analogues of the lemmas above hold, with similar proofs:

LEMMA 13.4.3. *We have* $v\big(P(y)\big) = v^{\mathrm{e}}(P) + \operatorname{nmul}(P)\,vy + \operatorname{nwt}(P)\,\psi_L(vy)$. *Moreover,* $\Gamma_{K\langle y\rangle} = \Gamma \oplus \mathbb{Z}vy \oplus \mathbb{Z}\psi_L(vy)$ *(internal direct sum) with* $\max \Psi_{K\langle y\rangle} = \psi_L(vy)$. *If y^* is an element of a* d*-valued field extension L^* of H-type of K such that $0 < vy^* < \Gamma^{>}$, then there is a unique valued differential field embedding $K\langle y\rangle \to L^*$ that is the identity on K and sends y to y^*.*

The residue field of $K\langle y\rangle$

Let L be a d-valued field extension of H-type of K and let $y \in L$.

LEMMA 13.4.4. *Suppose $0 < |vy| < \Gamma^{>}$. Then* $\operatorname{res} K\langle y\rangle = \operatorname{res} K$, *and so $K\langle y\rangle$ is* d*-valued with $C_{K\langle y\rangle} = C$.*

PROOF. Let $f \in K\langle y\rangle^{\times}$, so $f = P(y)/Q(y)$ where $P, Q \in K\{Y\}^{\neq}$. We have $N_P = D(Y)\cdot(Y')^w$ with $D(Y) \in C[Y]^{\neq}$ and $w = \operatorname{nwt}(P)$. The proof of Lemma 13.1.15 shows that eventually, with $\delta = \phi^{-1}\partial$,

$$P(y) = P^{\phi}(y) \sim \mathfrak{d}_{P^{\phi}} D(y)\delta(y)^w = \mathfrak{d}_{P^{\phi}} D(y)(y'/\phi)^w =: p(\phi) \in K(y,y')^{\times},$$

and likewise, $Q(y) \sim q(\phi) \in K(y,y')^\times$, eventually, so $f \sim p(\phi)/q(\phi) \in K(y,y')^\times$. Thus $\operatorname{res} K\langle y\rangle = \operatorname{res} K(y,y')$. By Lemmas 13.4.1 and 13.4.3 we have

$$vK(y,y')^\times \ = \ \Gamma \oplus \mathbb{Z}vy \oplus \mathbb{Z}\psi_L(vy),$$

so $\operatorname{res} K(y,y') \ = \ \operatorname{res} K$ by Lemma 3.1.30 applied to the valued field extensions $K(y)|K$ and $K(y,y')|K(y)$. □

Before we study the extension $K\langle y\rangle$ of K in the case where $\Psi < vy < (\Gamma^>)'$, we turn our attention to the operation $Q \mapsto Q^{\times\phi} = Q^\phi_{\times\phi}$.

Combining multiplicative and compositional conjugation

In Section 5.7 we defined $P^{\times\phi} := P^\phi_{\times\phi}$, so $P^{\times\phi}(Y') = P(Y')^\phi$. Hence $D_{P^{\times\phi}}(Y') = D_{P(Y')^\phi}$, and so $D_{P^{\times\phi}}(Y') = N_{P(Y')} \in C\{Y'\} \subseteq C\{Y\}$, eventually. Also $N_{P(Y')} \in C[Y](Y')^{\mathbb{N}}$, so $N_{P(Y')} = c(Y')^w$ with $c \in C^\times$ and $w \in \mathbb{N}$. We set $N_P^\times := cY^w$, so $N_P^\times(Y') = N_{P(Y')}$, so $D_{P^{\times\phi}} = N_P^\times$, eventually. We call $N_P^\times$ the **eventual dominant part of** $P^{\times\phi}$. For the above w we set

$$\operatorname{nwt}^\times(P) \ := \ w \ = \ \operatorname{nwt}\big(P(Y')\big).$$

If P is homogeneous of degree d, then so is each $P^{\times\phi}$, hence $\operatorname{nwt}^\times(P) = d$. Set

$$v^{\times\mathrm{e}}(P) \ := \ v^{\mathrm{e}}\big(P(Y')\big),$$

so in view of $v(P^{\times\phi}) = v\big(P(Y')^\phi\big)$ we get

$$v(P^{\times\phi}) \ = \ v^{\times\mathrm{e}}(P) + \operatorname{nwt}^\times(P)v\phi, \quad \text{eventually.}$$

Next an analogue of Corollary 13.2.12 for $v\big(P(z)\big)$ for active z.

COROLLARY 13.4.5. *There is an active $\phi_0 = \phi_0(P)$ in K such that for every* d*-valued field extension L of H-type of K and every active $z \prec \phi_0$ in L,*

$$vP(z) \ = \ v^{\times\mathrm{e}}(P) + \operatorname{nwt}^\times(P)\, vz.$$

PROOF. Take active ϕ_0 in K such that

$$D_{P^{\times\phi_0}} = N_P^\times, \quad R_{P(Y')^{\phi_0}} \prec^\flat_{\phi_0} P(Y')^{\phi_0}, \quad v(P^{\times\phi_0}) = v^{\times\mathrm{e}}(P) + \operatorname{nwt}^\times(P)v(\phi_0).$$

Using Lemma 13.1.1, the second relation gives $R_{P^{\times\phi_0}} \prec^\flat_{\phi_0} P^{\times\phi_0}$. Let L be a d-valued field extension of H-type of K and z an active element of L with $z \prec \phi_0$. Then $y := z\phi_0^{-1} \prec 1$ is active in L^{ϕ_0}, so $y \preceq^\flat_{\phi_0} 1$ and $P(z) = P^{\times\phi_0}(y)$. Hence by Lemma 13.1.15 and using $\operatorname{wt}(N_{P^\times}) = 0$:

$$\begin{aligned} v\big(P(z)\big) \ &= \ v(P^{\times\phi_0}) + \operatorname{dmul}(P^{\times\phi_0})\, v(y) + \operatorname{dwt}(P^{\times\phi_0})\, \psi_{L^{\phi_0}}(vy) \\ &= \ v(P^{\times\phi_0}) + \operatorname{nwt}^\times(P)\, v(y) \\ &= \ v^{\times\mathrm{e}}(P) + \operatorname{nwt}^\times(P)\, vz. \end{aligned}$$

□

Here is an immediate consequence:

COROLLARY 13.4.6. *There exists $y_0 \in K^\times$ with $y_0 \prec 1$ such that for every* d*-valued field extension L of H-type of K and any $y \in L$,*

$$0 < |vy| < v(y_0) \implies v\big(P(y^\dagger)\big) = v^{\times\mathrm{e}}(P) + \mathrm{nwt}^\times(P)\psi_L(vy).$$

COROLLARY 13.4.7. *Suppose P is homogeneous. Then $\mathrm{Ri}(N_P) = N^\times_{\mathrm{Ri}(P)}$, and*

$$\mathrm{nwt}^\times(P) = \deg(P), \quad \mathrm{nwt}(P) = \mathrm{nwt}^\times\big(\mathrm{Ri}(P)\big), \quad v^{\mathrm{e}}(P) = v^{\times\mathrm{e}}\big(\mathrm{Ri}(P)\big).$$

PROOF. By Lemma 5.8.6 we have $\mathrm{Ri}(P^\phi) = \mathrm{Ri}(P)^{\times\phi}$, and by Lemma 13.1.2 we have $D_{\mathrm{Ri}(P^\phi)} = \mathrm{Ri}(D_{P^\phi})$, hence, eventually,

$$\mathrm{Ri}(N_P) = \mathrm{Ri}(D_{P^\phi}) = D_{\mathrm{Ri}(P^\phi)} = D_{\mathrm{Ri}(P)^{\times\phi}} = N^\times_{\mathrm{Ri}(P)}.$$

Set $d := \deg(P)$ and $R := \mathrm{Ri}(P) \in K\{Z\}$. Since P is homogeneous, we have $\mathrm{nwt}^\times(P) = d$. Moreover, $N_P \in C^\times Y^{d-w}(Y')^w$ where $w = \mathrm{nwt}(P)$, so $N_R^\times = \mathrm{Ri}(N_P) \in C^\times Z^w$, and thus $\mathrm{nwt}^\times(R) = w$. To get $v^{\mathrm{e}}(P) = v^{\times\mathrm{e}}(R)$ we apply Corollary 13.4.6 to R in place of P. This gives $y_0 \in K^\times$ with $y_0 \prec 1$ such that for all $y \in K$ with $y_0 \prec y \prec 1$ we have

$$v\big(R(y^\dagger)\big) = v^{\times\mathrm{e}}(R) + \mathrm{nwt}^\times(R)\psi(vy).$$

By Corollary 13.2.12 we can arrange that in addition, for all $y \in K$ with $y_0 \prec y \prec 1$,

$$v\big(P(y)\big) = v^{\mathrm{e}}(P) + d\,vy + \mathrm{nwt}(P)\psi(vy).$$

Since $R(y^\dagger) = P(y)/y^d$ and $\mathrm{nwt}^\times(R) = \mathrm{nwt}(P)$, we get $v^{\times\mathrm{e}}(R) = v^{\mathrm{e}}(P)$. □

We also have a version of Corollary 13.4.5 for the behavior of $v\big(P(z)\big)$ for $z \in K$ as vz approaches Ψ from above:

COROLLARY 13.4.8. *There exists $\delta_0 \in (\Gamma^>)'$ such that for every* d*-valued field extension L of H-type of K and every $z \in L$ with $\Psi < vz \leqslant \delta_0$,*

$$v\big(P(z)\big) = v^{\times\mathrm{e}}(P) + \mathrm{nwt}^\times(P)\,vz.$$

PROOF. Take ϕ_0 as in the proof of Corollary 13.4.5. By Lemma 9.2.11 and the remark following it, we can take $\gamma_0 \in \Psi^{>v\phi_0}$ and $\delta_0 \in (\Gamma^>)'$ such that the map

$$\gamma \mapsto \psi_L(\gamma - v\phi_0)$$

is constant on the set $[\gamma_0, \delta_0]_{\Gamma_L}$, for each d-valued field extension L of H-type of K. Given such an L and $z \in L$ with $\Psi < vz \leqslant \delta_0$, set $y := z\phi_0^{-1} \prec 1$. Then $\psi_L(vy) = \psi_L(vz - v\phi_0) = \psi(\gamma_0 - v\phi_0)$, and the latter is $> v\phi_0$ by Lemma 6.5.4(i). Thus $y \preceq^\flat_{\phi_0} 1$. Since $P(z) = P^{\times\phi_0}(y)$, the claims now follow as in the proof of Corollary 13.4.5. □

Here is an easy consequence of Corollaries 13.4.5 and 13.4.8:

COROLLARY 13.4.9. *There exists $y_0 \in K^\times$ with $y_0 \prec 1$ such that for every* d*-valued field extension L of H-type of K and all $y \in L$,*

$$0 < |vy| < v(y_0) \implies v\big(P(y')\big) = v^{\times\mathrm{e}}(P) + \mathrm{nwt}^\times(P)v(y').$$

The cut $\Psi^{\downarrow}$

We now return to the setting of the beginning of this section. Model-theoretic compactness gives a d-valued field extension L of H-type of K with an element $z \in L$ such that $\Psi < vz < (\Gamma^{>})'$. Let such an L and $z \in L$ be given. Then $v\big(P(z)\big) = v^{\times\mathrm{e}}(P) + \operatorname{nwt}^{\times}(P)\,vz$ by Corollary 13.4.8.

LEMMA 13.4.10. *We have* $\Gamma_{K\langle z\rangle} = \Gamma \oplus \mathbb{Z}vz$ *(internally) with* $[\Gamma_{K\langle z\rangle}] = [\Gamma]$.

PROOF. Clearly $\Gamma_{K\langle z\rangle} = \Gamma + \mathbb{Z}vz$. From Corollary 9.8.6 we get $\Gamma + \mathbb{Z}vz = \Gamma \oplus \mathbb{Z}vz$ and $[\Gamma + \mathbb{Z}vz] = [\Gamma]$. □

Corollary 9.8.6 and Lemma 13.4.10 give an analogue of Lemma 13.4.2:

LEMMA 13.4.11. *Let* z^* *also be an element of a* d*-valued field extension* L^* *of* H*-type of* K *such that* $\Psi < vz^* < (\Gamma^{>})'$*. Then there is a unique valued differential field embedding* $K\langle z\rangle \to L^*$ *that is the identity on* K *and sends* z *to* z^**.*

COROLLARY 13.4.12. *The extension* $K\langle z\rangle$ *is* d*-valued with* $C_{K\langle z\rangle} = C$*.*

PROOF. Let $\phi_0 \in K$ be as in the proof of Corollary 13.4.5. So $D_{P^{\times\phi_0}} = N_P^{\times} = cY^w$ with $c \in C^{\times}$ and $w = \operatorname{nwt}^{\times}(P)$. For $y := z\phi_0^{-1}$ we have $P(z) = P^{\times\phi_0}(y)$. The proofs of Lemmas 13.1.15 and Corollary 13.4.8 show that then

$$P(z) \;=\; P^{\times\phi_0}(y) \;\sim\; \mathfrak{d}_{P^{\times\phi_0}}cy^w \in K(z)^{\times}.$$

Thus $\operatorname{res} K\langle z\rangle = \operatorname{res} K(z)$, and as $vK(z)^{\times} = \Gamma \oplus \mathbb{Z}vz$ by Lemma 13.4.10, we have $\operatorname{res} K(z) = \operatorname{res} K$ by Lemma 3.1.30. It remains to appeal to Lemma 9.1.2. □

Notes and comments

The following is worth mentioning. There is a $\delta_0 \in (\Gamma^{>})'$ such that $N_{P_{\times\mathfrak{m}}} = N_P^{\times}$ for all $\mathfrak{m}$ with $\Psi < v\mathfrak{m} < \delta_0$. Thus $\operatorname{ndeg}_{\mathcal{E}}(P) = \operatorname{nwt}^{\times}(P)$ for $\mathcal{E} := \{y \in K^{\times} : vy > \Psi\}$. (This will not be used in the present volume.)

13.5 EVENTUAL EQUALIZERS

Throughout this section K *is* ω*-free.* The goal of this section is to prove the Eventual Equalizer Theorem 13.0.3 and to derive from it a kind of Newton diagram for our differential polynomial $P \in K\{Y\}^{\neq}$: Proposition 13.5.7.

More general multiplicative-compositional conjugations

In the proof of the Eventual Equalizer Theorem we need to consider more general combinations of multiplicative and compositional conjugation than in Section 13.4. To define these we assume in this subsection that K is algebraically closed, and we also fix an algebraic closure $F = K\langle Y\rangle^{\mathrm{a}}$ of $K\langle Y\rangle$. We equip F with the unique extension of the derivation ∂ of $K\langle Y\rangle$ to a derivation, also denoted by ∂, of F. Next, extend the (gaussian) valuation of $K\langle Y\rangle$ to a valuation of F. This makes F a valued differential field, and its derivation is small by the remarks at the beginning of Section 6.3 and by Proposition 6.2.1. Also by Section 6.3, the image y of $Y \in \mathcal{O}_F$ in the differential residue field $\boldsymbol{k}_F$ of F is d-transcendental over the (trivial) differential residue field $\boldsymbol{k}$ of K, and $\boldsymbol{k}\langle y\rangle$ is the differential residue field of $K\langle Y\rangle$. Thus $\boldsymbol{k}_F = \boldsymbol{k}\langle y\rangle^{\mathrm{a}}$. Since $\boldsymbol{k}$ is algebraically closed, Lemma 5.7.10 gives:

LEMMA 13.5.1. *If $Q \in \mathcal{O}_F$ and $\overline{Q} \notin \boldsymbol{k}$, then $v\big(P(Q)\big) = v(P)$.*

Next, let any ϕ be given. Then F is also an algebraic closure of $K^\phi\langle Y\rangle$, and the unique extension of the derivation $\delta = \phi^{-1}\partial$ of $K^\phi\langle Y\rangle$ to a derivation δ of F is again small with respect to the given valuation of F, by the same arguments as before. Note that K^ϕ has the same (trivial) differential residue field $\boldsymbol{k}$ as K, and that the lemma above goes through as follows:

LEMMA 13.5.2. *If $Q \in \mathcal{O}_F$ and $\overline{Q} \notin \boldsymbol{k}$, then $v\big(P^\phi(Q)\big) = v(P^\phi)$.*

As in Section 5.7 we extend the ring isomorphism $P \mapsto P^\phi \colon K\{Y\} \to K^\phi\{Y\}$ to an automorphism $R \mapsto R^\phi$ of the field F, and choose a map

$$(R, q) \mapsto R^q \colon F^\times \times \mathbb{Q} \to F^\times$$

extending the usual map $(R, k) \mapsto R^k \colon F^\times \times \mathbb{Z} \to F^\times$, subject to $(R^q)^\dagger = qR^\dagger$ for $R \in F^\times$, $q \in \mathbb{Q}$. Thus $a^q \in K^\times$ for $a \in K^\times$. In that same section we defined

$$P^{\times q,\phi} := P^\phi_{\times \phi^q} \in K^\phi\{Y\} \quad (q \in \mathbb{Q}), \qquad P^{\times\phi} := P^{\times 1,\phi} = P^\phi_{\times\phi}.$$

PROPOSITION 13.5.3. *Suppose P is homogeneous, $d = \deg(P)$, and $q \in \mathbb{Q}^\times$. Then there exists an $\alpha \in \Gamma$ such that*

$$v(P^{\times q,\phi}) = \alpha + dq\, v\phi, \qquad \textit{eventually.}$$

PROOF. Lemma 5.7.12 gives homogeneous $E \in K\{Y\}^{\neq}$ of degree $w := \mathrm{wt}(P)$, with

$$P^{\times q,\phi}\big((Y')^q\big) =_{\mathrm{c}} (\phi Y')^{dq-w} \cdot E^{\times\phi}(Y') \qquad \text{for each } \phi.$$

Set $\alpha := v^{\times\mathrm{e}}(E)$. By Corollary 13.4.7 we have $\mathrm{nwt}^\times(E) = \deg(E) = w$. It is obvious that $v\big(E^{\times\phi}(Y')\big) = v(E^{\times\phi})$ for all ϕ, hence

$$v\big(E^{\times\phi}(Y')\big) = v^{\times\mathrm{e}}(E) + \mathrm{nwt}^\times(E)v\phi = \alpha + w\, v\phi, \qquad \text{eventually.}$$

By Lemma 13.5.2 we also have $v\big(P^{\times q,\phi}\big((Y')^q\big)\big) = v(P^{\times q,\phi})$. Hence

$$\begin{aligned} v\big(P^{\times q,\phi}\big((Y')^q\big)\big) &= v\big((\phi Y')^{dq-w}\big) + v\big(E^{\times\phi}(Y')\big) \\ &= (dq - w)\, v\phi + \big(\alpha + w\, v\phi\big) = \alpha + dq\, v\phi, \end{aligned}$$

eventually. □

Proof of the Eventual Equalizer Theorem

Complementing the "eventual" terminology, we say that a property $S(\phi)$ of elements ϕ holds **cofinally** if for every active ϕ_0 in K there is a $\phi \preccurlyeq \phi_0$ such that $S(\phi)$ holds.

PROOF OF THEOREM 13.0.3. We assume that Γ is divisible and $P, Q \in K\{Y\}^{\neq}$ are homogeneous of degrees $d > e$. Our job is to show that there exists $a \in K^\times$ such that, eventually, $P^\phi_{\times a} \asymp Q^\phi_{\times a}$. Note that $v(P^\phi_{\times a})$ depends only on va for $a \in K^\times$, where P and ϕ are given. Thus by passing to the algebraic closure of K we can arrange that K is algebraically closed. We extend the usual map $(k, a) \mapsto a^k : \mathbb{Z} \times K^\times \to K^\times$ to a map $(q, a) \mapsto a^q : \mathbb{Q} \times K^\times \to K^\times$ such that $(a^q)^\dagger = qa^\dagger$ for all $a \in K^\times$ and $q \in \mathbb{Q}$.

Take an elementary extension K_* of K with an active z in K_* such that $z \prec \phi$ for every ϕ. Then $\Gamma_{K\langle z\rangle} = \Gamma \oplus \mathbb{Z}v(z)$, by Lemma 13.4.10. Let $L := K\langle z\rangle^{\mathrm{a}}$ be the algebraic closure of $K\langle z\rangle$ in K_*. Then $\Gamma_L = \mathbb{Q}\Gamma_{K\langle z\rangle} = \Gamma \oplus \mathbb{Q}v(z)$. Now, working in the compositional conjugate L^z, Theorem 6.0.1 gives an element $\alpha + q\,v(z)$ in its value group ($\alpha \in \Gamma$, $q \in \mathbb{Q}$) such that

$$v_{P^z}\big(\alpha + q\,v(z)\big) = v_{Q^z}\big(\alpha + q\,v(z)\big),$$

so $v_{P^\phi}\big(\alpha + q\,v(\phi)\big) = v_{Q^\phi}\big(\alpha + q\,v(\phi)\big)$, cofinally. Take $a \in K^\times$ with $va = \alpha$. Then

$$P^\phi_{\times a\phi^q} \asymp Q^\phi_{\times a\phi^q}, \qquad \text{cofinally.}$$

We claim that $q = 0$. Towards a contradiction, assume $q \neq 0$. Renaming $P_{\times a}$ and $Q_{\times a}$ as P and Q, respectively, we get $P^{\times q,\phi} \asymp Q^{\times q,\phi}$, cofinally. However, Proposition 13.5.3 gives $\beta, \gamma \in \Gamma$ such that eventually

$$v(P^{\times q,\phi}) = \beta + dq\,v\phi, \qquad v(Q^{\times q,\phi}) = \gamma + eq\,v\phi,$$

so $\beta - \gamma = (e-d)q\,v\phi$, cofinally, contradicting $q \neq 0$. Hence $q = 0$. Thus $P^\phi_{\times a} \asymp Q^\phi_{\times a}$, cofinally, that is,

$$v^{\mathrm{e}}(P_{\times a}) + \operatorname{nwt}(P_{\times a})v\phi = v^{\mathrm{e}}(Q_{\times a}) + \operatorname{nwt}(Q_{\times a})v\phi, \qquad \text{cofinally.}$$

Thus $v^{\mathrm{e}}(P_{\times a}) = v^{\mathrm{e}}(Q_{\times a})$ and $\operatorname{nwt}(P_{\times a}) = \operatorname{nwt}(Q_{\times a})$, so the above holds not only cofinally, but even eventually, and we are done. □

The Eventual Equalizer Theorem is a key to the Newton diagram of P, but for this we also need Corollary 13.5.5 from the next subsection.

Transition from ndeg **to** nmul

Such a transition is given by the next lemma. Its proof uses that K is clean. Let f, g range over K. For $f \preccurlyeq \mathfrak{m}$ we let $f_{\mathfrak{m}}$ be the unique element of C such that $f = f_{\mathfrak{m}}\mathfrak{m} + g$ with $g \prec \mathfrak{m}$, so $f_{\mathfrak{m}} = 0$ if $f \prec \mathfrak{m}$.

LEMMA 13.5.4. *Suppose that $f \preccurlyeq \mathfrak{m}$. Then with $c := f_{\mathfrak{m}}$,*

$$\operatorname{ndeg}_{\prec\mathfrak{m}} P_{+f} = \operatorname{mul}(N_{P_{\times\mathfrak{m}}})_{+c}.$$

PROOF. For $\mathfrak{n} \prec \mathfrak{m}$ and $\mathfrak{n} = \mathfrak{e}\mathfrak{m}$ we have

$$P_{+f,\times\mathfrak{n}} = P_{\times\mathfrak{m},+\mathfrak{m}^{-1}f,\times\mathfrak{e}},$$

so replacing P by $P_{\times\mathfrak{m}}$ and f by $\mathfrak{m}^{-1}f$ we arrange $\mathfrak{m} = 1$. Set $Q := P_{+f}$, so by Lemma 13.2.14, $N_Q = (N_P)_{+c}$. Set $\mu := \operatorname{mul}(N_P)_{+c} = \operatorname{mul} N_Q \in \mathbb{N}$. Then for $\mathfrak{n} \prec 1$ and $i > \mu$ we have eventually $v(Q_i^\phi) \geqslant v(Q_\mu^\phi)$, hence eventually

$$v\big((Q^\phi)_{\times\mathfrak{n},i}\big) = v(Q_i^\phi) + iv\mathfrak{n} + o(v\mathfrak{n}) > v\big((Q^\phi)_{\times\mathfrak{n},\mu}\big) = v(Q_\mu^\phi) + \mu v\mathfrak{n} + o(v\mathfrak{n}),$$

so $\deg N_{Q_{\times\mathfrak{n}}} \leqslant \mu$. Take $\phi \preccurlyeq 1$ such that $N_Q = D_{Q^\phi}$ and $R_{Q^\phi} \prec_\phi^\flat Q^\phi$. Next, take $\mathfrak{n} \prec 1$ with $\psi^\phi(v\mathfrak{n}) = v\mathfrak{n}$, so $v_\phi^\flat(\mathfrak{n}) = 0$. Then by Lemma 13.2.16,

$$N_{Q_{\times\mathfrak{n}}} = N_{Q^\phi_{\times\mathfrak{n}}} \in C^\times \cdot Y^\mu,$$

in particular, $\deg N_{Q_{\times\mathfrak{n}}} = \mu$. □

COROLLARY 13.5.5. $\operatorname{ndeg}_{\prec\mathfrak{m}} P = \operatorname{nmul} P_{\times\mathfrak{m}}$.

The Newton diagram of P

In this subsection we assume that Γ *is divisible.* We begin with a reformulation of the Eventual Equalizer Theorem:

COROLLARY 13.5.6. *Assume $P, Q \in K\{Y\}^{\neq}$ are homogeneous of different degrees. Then there is a unique monomial $\mathfrak{m}$ such that $N_{(P+Q)_{\times\mathfrak{m}}}$ is not homogeneous.*

PROOF. There is at most one such monomial by Lemma 13.2.6. The proof of that lemma shows that if $\mathfrak{m}$ is such that eventually $P^\phi_{\times\mathfrak{m}} \asymp Q^\phi_{\times\mathfrak{m}}$, then $N_{(P+Q)_{\times\mathfrak{m}}} = N_{P_{\times\mathfrak{m}}} + N_{Q_{\times\mathfrak{m}}}$ is not homogeneous. Thus it remains to appeal to the Eventual Equalizer Theorem for the existence of such an $\mathfrak{m}$. □

For homogeneous $P, Q \in K\{Y\}^{\neq}$ of different degrees we call the unique $\mathfrak{m}$ as in Corollary 13.5.6 the **eventual equalizer** for P, Q and denote it by $\mathfrak{e}(P,Q)$.

We now focus on P, and let J be the finite nonempty set of $j \in \mathbb{N}$ with $P_j \neq 0$. Thus $\operatorname{ndeg} P_{\times\mathfrak{m}} \in J$ for all $\mathfrak{m}$. For distinct $i, j \in J$ we let $\mathfrak{e}(P,i,j)$ be the eventual equalizer for P_i, P_j. Therefore any algebraic starting monomial for P is of the form $\mathfrak{e}(P,i,j)$ with distinct $i, j \in J$. Note that $\mathfrak{M}$ is (totally) ordered by $\preccurlyeq$. Below we fix a $\preccurlyeq$-closed set $\mathcal{E} \subseteq K^\times$.

PROPOSITION 13.5.7. *There are $i_0, \dots, i_n \in J$ and eventual equalizers*

$$\mathfrak{e}(P,i_0,i_1) \prec \mathfrak{e}(P,i_1,i_2) \prec \cdots \prec \mathfrak{e}(P,i_{n-1},i_n)$$

with $\operatorname{mul} P = i_0 < \cdots < i_n = \operatorname{ndeg}_{\mathcal{E}} P$, *such that:*

(i) *the algebraic starting monomials for P in $\mathcal{E}$ are the $\mathfrak{e}(P,i_m,i_{m+1})$, $m < n$;*

(ii) *for $\mathfrak{m} = \mathfrak{e}(P,i_m,i_{m+1})$, $m < n$, we have* $\operatorname{mul} N_{P_{\times\mathfrak{m}}} = i_m$, $\deg N_{P_{\times\mathfrak{m}}} = i_{m+1}$.

PROOF. Let i, j range over J, and set $d := \operatorname{ndeg}_{\mathcal{E}} P$, so $\operatorname{mul} P \leqslant d \leqslant \deg P$. We proceed by induction on $d - \operatorname{mul} P$. If $d = \operatorname{mul} P$, then all $N_{P_{\times \mathfrak{m}}}$ with $\mathfrak{m} \in \mathcal{E}$ are homogeneous of degree d, and so there are no algebraic starting monomials of P in $\mathcal{E}$. Assume that $d > \operatorname{mul} P$, and take $i < d$ such that the eventual equalizer $\mathfrak{e} = \mathfrak{e}(P, i, d)$ is maximal with respect to $\preccurlyeq$. We claim that then $\mathfrak{e} \in \mathcal{E}$. Towards proving this, let $\mathfrak{n} \prec \mathfrak{e}$. Since $P^{\phi}_{i,\times\mathfrak{e}} \asymp P^{\phi}_{d,\times\mathfrak{e}}$, eventually, this gives $P^{\phi}_{i,\times\mathfrak{n}} \succ P^{\phi}_{d,\times\mathfrak{n}}$, eventually, and thus $\operatorname{ndeg} P_{\times\mathfrak{n}} \neq d$. Taking $\mathfrak{n} \in \mathcal{E}$ such that $\operatorname{ndeg} P_{\times\mathfrak{n}} = d$ therefore gives $\mathfrak{e} \preccurlyeq \mathfrak{n}$, and so we get $\mathfrak{e} \in \mathcal{E}$, as claimed. Also $\deg N_{P_{\times\mathfrak{e}}} = d$: otherwise $\deg N_{P_{\times\mathfrak{e}}} = j < d$, so $P^{\phi}_{j,\times\mathfrak{e}} \succ P^{\phi}_{d,\times\mathfrak{e}}$ eventually, hence $\mathfrak{e}(P, j, d) \succ \mathfrak{e}$, contradicting the maximality of $\mathfrak{e}$. Since $i < d$ and $(N_{P_{\times\mathfrak{e}}})_i = N_{P_{i,\times\mathfrak{e}}} \neq 0$, it follows that $\mathfrak{e}$ is an algebraic starting monomial for P.

CLAIM: *$\mathfrak{e}$ is the largest algebraic starting monomial for P in $\mathcal{E}$.*

To see this, suppose towards a contradiction that $\mathfrak{n} \succ \mathfrak{e}$ is an algebraic starting monomial for P in $\mathcal{E}$. Then $\operatorname{ndeg} P_{\times\mathfrak{e}} \leqslant \operatorname{ndeg} P_{\times\mathfrak{n}}$, so $\operatorname{ndeg} P_{\times\mathfrak{n}} = d$ by the maximality property of $\mathcal{E}$. So $\mathfrak{n} = \mathfrak{e}(P, j, d)$ with $j < d$, but this contradicts the maximality of $\mathfrak{e}$, and so proves the claim.

By decreasing i if necessary we arrange $i = \operatorname{mul} N_{P_{\times\mathfrak{e}}}$. Then $\operatorname{ndeg}_{\prec\mathfrak{e}} P = i$ by Corollary 13.5.5. It remains to apply the inductive assumption with $\mathcal{E}$ replaced by the set $\{g \in K^{\times} : g \prec \mathfrak{e}\}$. □

Let $(i_0, \dots, i_n)$ be as in Proposition 13.5.7. This tuple is uniquely determined by the data K, P, $\mathcal{E}$. If $\operatorname{mul} P = \operatorname{ndeg}_{\mathcal{E}} P$, then $n = 0$ and this tuple is just $(\operatorname{mul} P)$. To simplify notation, set $\mathfrak{e}_m := \mathfrak{e}(P, i_{m-1}, i_m)$ for $1 \leqslant m \leqslant n$. We now have a complete description of the behavior of $\operatorname{nmul} P_{\times g}$ and $\operatorname{ndeg} P_{\times g}$ for $g \in \mathcal{E}$:

COROLLARY 13.5.8. *Assume $\operatorname{mul} P \neq \operatorname{ndeg}_{\mathcal{E}} P$, so $n \geqslant 1$. Let g range over $\mathcal{E}$. Then $\operatorname{nmul} P_{\times g}$ and $\operatorname{ndeg} P_{\times g}$ lie in the set $\{i_0, \dots, i_n\}$, and we have:*

$$\begin{aligned}
\operatorname{nmul} P_{\times g} &= i_0 \iff g \preccurlyeq \mathfrak{e}_1; \\
\operatorname{ndeg} P_{\times g} &= i_0 \iff g \prec \mathfrak{e}_1; \\
\operatorname{nmul} P_{\times g} &= i_m \iff \mathfrak{e}_m \prec g \preccurlyeq \mathfrak{e}_{m+1}, \qquad (1 \leqslant m < n); \\
\operatorname{ndeg} P_{\times g} &= i_m \iff \mathfrak{e}_m \preccurlyeq g \prec \mathfrak{e}_{m+1}, \qquad (1 \leqslant m < n); \\
\operatorname{nmul} P_{\times g} &= i_n \iff \mathfrak{e}_n \prec g; \\
\operatorname{ndeg} P_{\times g} &= i_n \iff \mathfrak{e}_n \preccurlyeq g.
\end{aligned}$$

PROOF. Let $1 \leqslant m < n$, and set $\mathfrak{e} := \mathfrak{e}(P, i_{m-1}, i_m)$ and $\mathfrak{e}^* := \mathfrak{e}(P, i_m, i_{m+1})$. For $\mathfrak{e} \prec g \prec \mathfrak{e}^*$ we obtain from Proposition 13.5.7 and Corollary 11.2.5 that

$$i_m = \operatorname{ndeg} P_{\times\mathfrak{e}} \leqslant \operatorname{nmul} P_{\times g} = \operatorname{ndeg} P_{\times g} \leqslant \operatorname{nmul} P_{\times\mathfrak{e}^*} = i_m,$$

where $\operatorname{nmul} P_{\times g} = \operatorname{ndeg} P_{\times g}$ because P has no algebraic starting monomial $\mathfrak{m}$ with $\mathfrak{e} \prec \mathfrak{m} \prec \mathfrak{e}^*$. Thus $i_m = \operatorname{nmul} P_{\times g} = \operatorname{ndeg} P_{\times g}$, from which we obtain the "middle" equivalences. The "end" equivalences are derived in the same way. □

Of course, if $\operatorname{mul} P = \operatorname{ndeg}_{\mathcal{E}} P$, then $\operatorname{nmul} P_{\times g} = \operatorname{ndeg} P_{\times g} = \operatorname{mul} P$ for all $g \in \mathcal{E}$, and P has no algebraic starting monomials in $\mathcal{E}$.

Suppose now that $\operatorname{mul} P \neq \operatorname{ndeg}_{\mathcal{E}} P$, and let n be as above. So n is by Proposition 13.5.7 the number of algebraic starting monomials in $\mathcal{E}$ for P, and $1 \leqslant n \leqslant \operatorname{ndeg}_{\mathcal{E}} P - \operatorname{mul} P$. Thus we can take algebraic approximate zeros $y_1, \dots, y_N$ of P in $\mathcal{E}$, $N \in \mathbb{N}$, such that each algebraic approximate zero y of P in $\mathcal{E}$ satisfies $y \sim y_i$ for exactly one $i \in \{1, \dots, N\}$. Combining Proposition 13.5.7 with Lemma 13.3.15 yields:

COROLLARY 13.5.9. *For $i = 1, \dots, N$, let μ_i be the algebraic multiplicity of the approximate zero y_i of P, and set $\mu := \mu_1 + \cdots + \mu_N$. Then $\mu \leqslant \operatorname{ndeg}_{\mathcal{E}} P - \operatorname{mul} P$. If C is algebraically closed, then $\mu = \operatorname{ndeg}_{\mathcal{E}} P - \operatorname{mul} P$. If C is real closed, then $\mu \equiv \operatorname{ndeg}_{\mathcal{E}} P - \operatorname{mul} P \bmod 2$.*

Notes and comments

Corollary 13.5.6 and Proposition 13.5.7 for $K = \mathbb{T}_{\mathrm{g}}$ are Proposition 8.14(c) and Proposition 8.17, respectively, in [194]. Corollary 13.5.9 for $K = \mathbb{T}_{\mathrm{g}}[i]$ is [194, Exercise 8.14].

13.6 FURTHER CONSEQUENCES OF ω-FREENESS

In this section E is an ungrounded H-asymptotic field with $\Gamma_E \neq \{0\}$. Thus E is pre-d-valued but not necessarily d-valued. Also in contrast to K, we do not assume the derivation of E small, or that the valued field E has a monomial group. Let (Γ_E, ψ_E) be the asymptotic couple of E, so $\Psi_E \neq \emptyset$ and Ψ_E has no largest element. We fix a logarithmic sequence for E and corresponding sequences (λ_ρ) and (ω_ρ) in E as in Section 11.5 (with E instead of K).

THEOREM 13.6.1. *Suppose that E is ω-free, and that F is a pre-*d*-valued field extension of E of H-type which is* d*-algebraic over E. Then:*

(i) *there is no $y \in F$ with $0 < |vy| < \Gamma_E^{>}$;*

(ii) *there is no $z \in F$ with $\Psi_E < vz < (\Gamma_E^{>})'$; and*

(iii) *F is ω-free.*

In particular, if E is ω-free, then so is its differential-valued hull $\operatorname{dv}(E)$. (See Section 10.3 for the basic facts on differential-valued hulls.)

PROOF. If (i) holds, then $\Gamma_E^{<}$ is cofinal in $\Gamma_F^{<}$, and thus the logarithmic sequence for E can also serve as such for F. This remark helps to justify the reduction steps we make in the proof, especially in connection with (iii).

First we take some active $a \in E$ and replace E and F by their compositional conjugates E^a and F^a. In this way we arrange that the derivations of E and F are small. Next, replacing F by $\operatorname{dv}(F)$ we arrange that F is d-valued of H-type. Finally, replacing E and F by their algebraic closures, we can also assume that E and F are algebraically closed. This last replacement doesn't change Ψ_E, so we can keep the same logarithmic sequence for E.

Lemma 3.3.32 gives a monomial group $\mathfrak{M}$ for E. Since E is ω-free, it has asymptotic integration, so $K := \mathrm{dv}(E)$ is an immediate extension of E. We equip K with the monomial group $\mathfrak{M}$, so K satisfies our standing assumptions at the beginning of this chapter, and $\Gamma := \Gamma_K = \Gamma_E$ is divisible. Take the unique valued differential field embedding of K into F that is the identity on E, and identify K with a valued differential subfield of F via this embedding. Then F is a d-valued field extension of H-type of K. Now let $y \in F$ and take $P \in E\{Y\}^{\neq}$ with $P(y) = 0$. By the remark following the proof of Proposition 13.3.13 we have $N_P \in C[Y](Y')^{\mathbb{N}}$, and thus by Lemma 13.2.9 and Corollary 13.2.12 we cannot have $0 < |vy| < \Gamma^>$. This proves (i).

For (ii), suppose towards a contradiction that $z \in F$ and $\Psi < vz < (\Gamma^>)'$. Corollary 10.4.7 provides a d-algebraic d-valued field extension L of H-type of F with an element $y \in L^\times$ such that $y^\dagger = z$. If $y \asymp 1$, then $y = c + y_1$ with $c \in C_L^\times$ and $v(y_1) \geqslant \gamma$ for some $\gamma \in \Gamma^>$, by (i) applied to L in the role of F, so $v(z) = v(y_1') \geqslant \gamma'$, contradicting the assumption on z. Thus $y \not\asymp 1$, but then the assumption on z gives $0 < |vy| < \Gamma^>$, contradicting (i). This proves (ii). Therefore, F has asymptotic integration. For (iii), we first show that F is λ-free. Suppose it is not. Since we arranged Γ_F to be divisible, F has a gap creator s by Lemma 11.5.14. Proposition 10.4.1 gives a d-valued field extension $F(f)$ of H-type of F with $f \neq 0$ and $f^\dagger = s$, but then $\Psi < vf < (\Gamma^>)'$ by the remark preceding Corollary 11.6.1, and this contradicts (ii). Thus F is indeed λ-free, and this holds not just for the present F, which is algebraically closed, etcetera, but for any F satisfying the assumptions in the theorem. To show that our present F is even ω-free, assume towards a contradiction that $\omega_\rho \rightsquigarrow \omega \in F$. Then F has by Corollary 11.7.10 an immediate asymptotic extension that is d-algebraic over F (and thus over E) but not λ-free. This contradicts what we just proved, namely, that any F as in the hypothesis of the theorem is λ-free. □

Corollary 13.6.2. *Let E be an ω-free H-field. Then E has exactly one Liouville closure, up to isomorphism over E.*

Proof. Every Liouville H-field extension of E is d-algebraic over E, and thus no such extension has a gap, by Theorem 13.6.1. This gives the desired conclusion in view of the remarks following Theorem 10.6.12. □

Corollary 13.6.3. *Suppose E is ω-free. Then the pc-sequences (λ_ρ) and (ω_ρ) are of* d*-transcendental type over E.*

Proof. If (λ_ρ) is of d-algebraic type over E, then (λ_ρ) pseudoconverges in some immediate asymptotic extension of E that is d-algebraic over E, by Corollary 11.4.13 and Lemma 11.4.8, but this is impossible by Theorem 13.6.1. Likewise, (ω_ρ) is of d-transcendental type over E. □

We now combine this with Lemma 11.7.5 to get:

Proposition 13.6.4. *Suppose E is ω-free, and ω is an element in an H-asymptotic field extension of E such that $\omega_\rho \rightsquigarrow \omega$. Then $E\langle\omega\rangle$ is λ-free.*

PROOF. Since (ω_ρ) is of d-transcendental type over E, the asymptotic extension $F := E\langle\omega\rangle$ of E is immediate and ω is d-transcendental over E. Moreover, (λ_ρ) is a divergent pc-sequence in E. Towards a contradiction, suppose $P, Q \in E\{Y\}^{\neq}$ are such that

$$\lambda_\rho \rightsquigarrow \lambda := \frac{P(\omega)}{Q(\omega)} \in F \setminus E.$$

Put $R := P/Q \in E\langle Y\rangle \setminus E$; so $R(\omega) = \lambda$. Set

$$G(Y) := P(Y)Q(\omega) - Q(Y)P(\omega) \in F\{Y\}.$$

From $R \notin F$ we get $G \neq 0$. In view of $G(\omega) = 0$, this gives $G \notin F$. Lemma 11.7.5 then yields $G(\omega_\rho) \rightsquigarrow 0$. Also $Q(\omega_\rho) \not\rightsquigarrow 0$, and thus $Q(\omega_\rho) \asymp Q(\omega)$, eventually, by Lemma 11.7.5. Take ρ_0 such that $Q(\omega_\rho) \neq 0$ for $\rho > \rho_0$. Then

$$Q(\omega)Q(\omega_\rho)\big(R(\omega_\rho) - \lambda\big) = G(\omega_\rho) \qquad (\rho > \rho_0),$$

so for $\alpha := -2v\big(Q(\omega)\big) \in \Gamma_E$ we get

$$v\big(R(\omega_\rho) - \lambda\big) = \alpha + v\big(G(\omega_\rho)\big), \text{ eventually.} \tag{13.6.1}$$

In particular, $R(\omega_\rho) \rightsquigarrow \lambda$ $(\rho > \rho_0)$. If $R(\omega_\rho) \rightsquigarrow b \in E$ $(\rho > \rho_0)$, then

$$P(\omega_\rho) - bQ(\omega_\rho) = Q(\omega_\rho)\big(R(\omega_\rho) - b\big) \rightsquigarrow 0,$$

contradicting that (ω_ρ) is of d-transcendental type over E. Thus the pc-sequence $\big(R(\omega_\rho)\big)_{\rho>\rho_0}$ in E is divergent. Then by Corollary 2.2.20, the pc-sequences (λ_ρ) and $\big(R(\omega_\rho)\big)_{\rho>\rho_0}$ in E are equivalent, and so the latter has width $\{\gamma \in \Gamma_\infty : \gamma > \Psi\}$. Thus for any big enough ρ_0 the set $\big\{v\big(R(\omega_\rho) - \lambda\big) : \rho > \rho_0\big\}$ is a cofinal subset of $\Psi_E^{\downarrow}$. However, taking $a \in E^\times$ with $va = \alpha$ and setting $H := aG \in F\{Y\}$, it follows from (13.6.1) that for big enough ρ_0 this set equals $\big\{v\big(H(\omega_\rho)\big) : \rho > \rho_0\big\}$. As $H(\omega) = 0$, this contradicts Lemma 11.7.5. □

Towards extending Lemmas 13.4.1 and 13.4.2 to the present setting, we first note:

LEMMA 13.6.5. *Suppose E is ω-free. Let $P \in E\{Y\}^{\neq}$. Then there exists an active $\mathfrak{a}$ in E such that for every pre-d-valued field extension F of E of H-type,*

$$y \in F,\ 1 \prec y \preceq^{\flat}_{\mathfrak{a}} 1 \implies v\big(P(y)\big) = v^{\mathrm{e}}(P) + \operatorname{ndeg}(P)vy + \operatorname{nwt}(P)\psi_F(vy).$$

PROOF. In view of Lemma 11.1.17 we can arrange by compositional conjugation that E has small derivation. Next, replacing E by $\operatorname{dv}(E)$, and any F by $\operatorname{dv}(F)$, we also arrange that E is differential-valued, and then passing to algebraic closures we can even assume that E (and any F) is algebraically closed. Then E has a monomial group $\mathfrak{M}$, so E satisfies the assumptions made on K in this chapter. As E is ω-free, we can apply Corollary 13.2.12 to get the desired result. □

LEMMA 13.6.6. *Suppose E is an ω-free H-field. Let $P \in E\{Y\}^{\neq}$. Then there are active a in E and $\sigma \in \{-1,+1\}$ such that for every H-field extension F of E,*

$$y \in F^{>},\ 1 \prec y \preccurlyeq^{\flat}_{a} 1 \implies \operatorname{sign} P(y) = \sigma.$$

PROOF. As Lemma 13.6.5 follows from the first part of Corollary 13.2.12, this follows from its second part. Instead of algebraic closures, take real closures. □

PROPOSITION 13.6.7. *Suppose E is ω-free, F is a pre-*d*-valued field extension of E of H-type, and $y \in F$ satisfies $\Gamma_E^{<} < vy < 0$. Then for all $P \in E\{Y\}^{\neq}$,*

$$v\big(P(y)\big) = v^{\mathrm{e}}(P) + \operatorname{ndeg}(P)vy + \operatorname{nwt}(P)\psi_F(vy).$$

*Moreover, $\Gamma_{E\langle y\rangle} = \Gamma_E \oplus \mathbb{Z}vy \oplus \mathbb{Z}\psi_F(vy)$ (internal direct sum) and $\max \Psi_{E\langle y\rangle} = \psi_F(vy)$. For any y^* in any pre-*d*-valued extension F^* of E of H-type satisfying $\Gamma_E^{<} < vy^* < 0$, there is a unique valued differential field embedding $E\langle y\rangle \to F^*$ over E sending y to y^*.*

PROOF. The first part is immediate from Lemma 13.6.5, and the proof of the rest is like that of Lemmas 13.4.1 and 13.4.2. □

COROLLARY 13.6.8. *Let E be an ω-free H-field. Let F and F^* be H-field extensions of E with elements $y \in F^{>}$ and $y^* \in F^{*>}$ such that $\Gamma_E^{<} < vy < 0$ and $\Gamma_E^{<} < vy^* < 0$. Then there is a unique pre-H-field embedding $E\langle y\rangle \to F^*$ over E sending y to y^*.*

PROOF. This follows from Proposition 13.6.7 in view of Lemma 13.6.6. □

Next we extend the Eventual Equalizer Theorem to the present setting:

COROLLARY 13.6.9. *Assume E is ω-free, $P, Q \in E\{Y\}^{\neq}$ are homogeneous of degrees $d > e$, and $(d-e)\Gamma_E = \Gamma_E$. Then for some $a \in E^{\times}$ and active f_0 in E,*

$$P^f_{\times a} \asymp Q^f_{\times a} \text{ for all active } f \preccurlyeq f_0 \text{ in } E.$$

PROOF. Note: for active $f, g \in E$ with $f \asymp g$ and $a, b \in E^{\times}$ with $a \asymp b$ we have

$$P^f_{\times a} \asymp P^f_{\times b} \asymp P^g_{\times b}, \text{ and likewise } Q^f_{\times a} \asymp Q^g_{\times b}.$$

An initial compositional conjugation by an active element of E arranges that the derivation of E is small. Since $\operatorname{dv}(E)$ is still ω-free, by Theorem 13.6.1, and is an immediate extension of E, we can also arrange that E is d-valued of H-type. Then the algebraic closure K of E is d-valued of H-type and ω-free, and has a monomial group $\mathfrak{M}$, and so K with $\mathfrak{M}$ satisfies the conditions imposed at the beginning of this chapter, and in addition $\Gamma := \Gamma_K$ is divisible. By the Eventual Equalizer Theorem we can take $a \in K^{\times}$ and active ϕ_0 in K such that for all active $\phi \preccurlyeq \phi_0$ we have $P^{\phi}_{\times a} \asymp Q^{\phi}_{\times a}$, where, as always in this chapter, ϕ ranges over $\mathfrak{M}$. Now let $f \in E$ be active with $f \preccurlyeq \phi_0$. Taking ϕ such that $\phi \asymp f$ we note that $P^f_{\times a} \asymp Q^f_{\times a}$. Since $P^f, Q^f \in E^f\{Y\}^{\neq}$ are homogeneous of degrees $d > e$, and $(d-e)\Gamma_E = \Gamma_E$, we have by the Equalizer Theorem of Chapter 6 a unique $\alpha \in \Gamma_E$ such that $v_{P^f}(\alpha) = v_{Q^f}(\alpha)$. It follows that $\alpha = va$. Thus with a replaced by any element of E with valuation α, and with f_0 any active element of E with $f_0 \preccurlyeq \phi_0$, we have the desired conclusion. □

For $d = 1$ and $e = 0$ this means:

COROLLARY 13.6.10. *If E is ω-free and $P \in E\{Y\}^{\neq}$ is homogeneous of degree 1, then for each $a \in E^{\times}$ there are elements $g \in E^{\times}$ and active f_0 in E such that:*

$$P^f_{\times g} \asymp a \text{ for all active } f \preccurlyeq f_0 \text{ in } E.$$

The results in Section 13.3 yield in the present setting initially the following:

LEMMA 13.6.11. *Assume E is ω-free and $P \in E\{Y\}^{\neq}$ has Newton weight w. Then there are $A \in E[Y]^{\neq}$ with $A \asymp 1$, $w + \operatorname{mul} A = \operatorname{nmul} P$, and $w + \deg A = \operatorname{ndeg} P$, an active $\mathfrak{e}$ in E, an $a \in E^{\times}$, and an $R \in E^{\mathfrak{e}}\{Y\}$, such that*

$$P^{\mathfrak{e}} = a \cdot A \cdot (Y')^w + R, \qquad a \asymp P^{\mathfrak{e}}, \quad R \prec^{\flat}_{\mathfrak{e}} P^{\mathfrak{e}}.$$

PROOF. We have $\operatorname{order}(R) \leqslant \operatorname{order}(P)$ and $\deg R \leqslant \deg P$ for R as above, so we are dealing with an elementary statement about E and the tuple of coefficients of P. Hence we may pass to an elementary extension of E and arrange in this way that E is $\aleph_1$-saturated, and thus has a monomial group $\mathfrak{M}$, by Lemma 3.3.39. Moreover, an initial compositional conjugation by an active element of E arranges that the derivation of E is small. Let $K := \operatorname{dv}(E)$, so K is an immediate extension of E and K is an ω-free d-valued field of H-type with monomial group $\mathfrak{M} \subseteq E^{\times}$. This leads to the Newton polynomial $N_P \in C[Y](Y')^w$, so $N_P = B \cdot (Y')^w$, with $B \in C[Y]$, so $w + \deg B = \operatorname{ndeg} P$. Take $A \in E[Y]$ with $A \sim B$, $\operatorname{mul} A = \operatorname{mul} B$, $\deg A = \deg B$. Take $\gamma > 0$ in $\Gamma = \Gamma_E$ such that $v(A - B) > \gamma$. Take active $\mathfrak{e} \preccurlyeq 1$ in $\mathfrak{M}$ such that

$$\Gamma^{\flat}_{\mathfrak{e}} < \gamma, \quad P^{\mathfrak{e}} = \mathfrak{d}_{P^{\mathfrak{e}}} N_P + R_{P^{\mathfrak{e}}}, \qquad R_{P^{\mathfrak{e}}} \prec^{\flat}_{\mathfrak{e}} P^{\mathfrak{e}}.$$

Since $N_P = B \cdot (Y')^w = A \cdot (Y')^w + G$ with $vG > \gamma$, we obtain

$$P^{\mathfrak{e}} = \mathfrak{d}_{P^{\mathfrak{e}}} \cdot A \cdot (Y')^w + \mathfrak{d}_{P^{\mathfrak{e}}} G + R_{P^{\mathfrak{e}}},$$

so the desired conclusion holds with $a := \mathfrak{d}_{P^{\mathfrak{e}}}$ and $R := \mathfrak{d}_{P^{\mathfrak{e}}} G + R_{P^{\mathfrak{e}}}$. □

In the next two corollaries of Lemma 13.6.11 we let F range over H-asymptotic field extensions of E. Recall that for such F, if $\mathfrak{e} \in E$ is active in E, then $\mathfrak{e}$ remains active in F, and if $f \in F$ is active, then F^f has small derivation.

COROLLARY 13.6.12. *Assume E is ω-free and $P \in E\{Y\}^{\neq}$ has Newton weight w. Let $A \in E[Y]^{\neq}$ and $\mathfrak{e} \in E$ be as in Lemma 13.6.11. Then for all F and active $f \preccurlyeq \mathfrak{e}$ in F there are $a_f \in F^{\times}$ and $R_f \in F^f\{Y\}$ such that*

$$P^f = a_f \cdot A \cdot (Y')^w + R_f, \qquad a_f \asymp P^f, \quad R_f \prec^{\flat}_{\mathfrak{e}} P^f.$$

PROOF. To simplify notation we replace E, P by $E^{\mathfrak{e}}$, $P^{\mathfrak{e}}$ (and each F by $F^{\mathfrak{e}}$), and rename accordingly, so that $\mathfrak{e} = 1$. With a and R as in Lemma 13.6.11 we have

$$P = a \cdot A \cdot (Y')^w + R, \quad a \asymp P, \quad A \asymp 1, \quad R \prec^{\flat} P \quad (\text{in } E\{Y\}).$$

Note that $R \prec^\flat P$ remains true in $F\{Y\}$. An easy computation as in the proof of Lemma 13.1.9 now shows that if $f \preccurlyeq 1$ is active in F, then

$$P^f = af^w \cdot A \cdot (Y')^w + R^f, \qquad R^f \prec^\flat af^w \asymp^\flat P^f,$$

so $a_f := af^w$ and $R_f := R^f$ works. □

COROLLARY 13.6.13. *Suppose E is ω-free. Let $P \in E\{Y\}^{\neq}$. Then there is an active $e \in E$ such that for all F and active $f \preccurlyeq e$ in F,*

$$\operatorname{ddeg} P^f = \operatorname{ndeg} P, \quad \operatorname{dmul} P^f = \operatorname{nmul} P, \quad \operatorname{dwt} P^f = \operatorname{nwt} P.$$

Independent of whether E is ω-free, when we have a decomposition of P as in Lemma 13.6.11, the following is relevant:

LEMMA 13.6.14. *Assume the derivation of E is small, and $P \in E\{Y\}^{\neq}$,*

$$P = a \cdot A \cdot (Y')^w + R, \qquad a \in E^\times,\ A \in E[Y]^{\neq},\ w \in \mathbb{N},\ R \in E\{Y\},$$

with $a \asymp P$, $A \asymp 1$, $R \prec^\flat P$. Let $\overline{A}$ be the image of A in $\boldsymbol{k}_E[Y]$. Then $\operatorname{nmul} P = \operatorname{mul} \overline{A} + w$, $\operatorname{ndeg} P = \deg \overline{A} + w$, and there exists $\gamma \in \Gamma_E^>$ such that for all $g \in E$,

$$\begin{aligned} 0 < vg < \gamma &\Longrightarrow \operatorname{nmul}(P) = \operatorname{nmul}(P_{\times g}) = \operatorname{ndeg}(P_{\times g}), \\ -\gamma < vg < 0 &\Longrightarrow \operatorname{ndeg}(P) = \operatorname{ndeg}(P_{\times g}) = \operatorname{nmul}(P_{\times g}). \end{aligned}$$

PROOF. Let $f \prec 1$ be active in E. Then $vf \in \Gamma^\flat$ and

$$P^f = af^w \cdot A \cdot (Y')^w + R^f$$

with $v(R^f) \geqslant v(R) > v(af^w)$, so $\operatorname{dmul} P^f = \operatorname{mul} \overline{A} + w$ and $\operatorname{ddeg} P^f = \deg \overline{A} + w$. This proves the claim about $\operatorname{nmul} P$ and $\operatorname{ndeg} P$. Take $B, G \in E[Y]$ such that $A = B + G$, all nonzero coefficients of B are $\asymp 1$, and $G \prec 1$, so $B = b_m Y^m + \cdots + b_n Y^n$ with $m = \operatorname{mul} \overline{A}$ and $n = \deg \overline{A}$, $b_m, \dots, b_n \in E$, $b_m, b_n \neq 0$. Next, let $g \in E^\times$ be such that $g \not\asymp 1$. Then in $E^f\{Y\}$ we have

$$P^f_{\times g} = af^w \cdot A_{\times g} \cdot (Y'_{\times g})^w + R^f_{\times g} = af^w \cdot B_{\times g} \cdot (Y'_{\times g})^w + af^w \cdot G_{\times g} \cdot (Y'_{\times g})^w + R^f_{\times g}.$$

If $f \prec g^\dagger$, then $Y'_{\times g} = f^{-1} g' Y + g Y' \sim f^{-1} g' Y$ in $E^f\{Y\}$. To derive the first displayed implication, assume $0 < v(g) \in \Gamma^\flat$ and $m\, v(g) < v(G)$. Then $v(g') \in \Gamma^\flat$ and $B_{\times g} \sim b_m g^m Y^m \succ G_{\times g}$. Therefore, eventually with respect to f,

$$P^f_{\times g} \sim ab_m g^m (g')^w \cdot Y^{m+w},$$

and so $\operatorname{nmul}(P_{\times g}) = \operatorname{ndeg}(P_{\times g}) = m + w = \operatorname{nmul}(P)$. For the second implication, let g satisfy $vg < 0$, $vg \in \Gamma^\flat$, and $v(g^n) = nv(g) < v(G_{\times g})$. Then $v(g') \in \Gamma^\flat$, $B_{\times g} \sim b_n g^n Y^n \succ G_{\times g}$, so eventually with respect to f,

$$P^f_{\times g} \sim ab_n g^n (g')^w \cdot Y^{n+w},$$

and thus $\operatorname{ndeg}(P_{\times g}) = \operatorname{nmul}(P_{\times g}) = n + w = \operatorname{ndeg}(P)$. □

COROLLARY 13.6.15. *Suppose E is ω-free, and $P \in E\{Y\}^{\neq}$. Then there exists $\gamma \in \Gamma_E^>$ such that for all $g \in E$,*

$$0 < vg < \gamma \implies \operatorname{nmul}(P) = \operatorname{nmul}(P_{\times g}) = \operatorname{ndeg}(P_{\times g}),$$
$$-\gamma < vg < 0 \implies \operatorname{ndeg}(P) = \operatorname{ndeg}(P_{\times g}) = \operatorname{nmul}(P_{\times g}).$$

PROOF. Use Lemma 13.6.12 and compositional conjugation by some active element of E to arrange that E has small derivation and P has a decomposition as in the hypothesis of Lemma 13.6.14. Now apply Lemma 13.6.14. □

Note also that $\deg A = \operatorname{ddeg} A$ for A as in Lemma 13.6.11. The device in the proof of passing to an elementary extension with a monomial group and then to the differential-valued hull can also be used in other situations. For example, the results on the Newton diagram of P in Section 13.5 extend in a similar way to the present setting. Here we state what is needed later:

COROLLARY 13.6.16. *Assume E is ω-free, Γ_E is divisible, and $P \in E\{Y\}^{\neq}$ is not homogeneous. Then there are $i_0 < \cdots < i_n$ in $\{i \in \mathbb{N} : P_i \neq 0\}$ with $i_0 = \operatorname{mul} P$ and $i_n = \deg P$ and elements $\mathfrak{e}_1 \prec \cdots \prec \mathfrak{e}_n$ in $E^\times$ such that for all $g \in E^\times$:*

$$\begin{aligned}
\operatorname{nmul} P_{\times g} &= i_0 \iff g \preccurlyeq \mathfrak{e}_1; \\
\operatorname{ndeg} P_{\times g} &= i_0 \iff g \prec \mathfrak{e}_1; \\
\operatorname{nmul} P_{\times g} &= i_m \iff \mathfrak{e}_m \prec g \preccurlyeq \mathfrak{e}_{m+1}, \qquad (1 \leqslant m < n); \\
\operatorname{ndeg} P_{\times g} &= i_m \iff \mathfrak{e}_m \preccurlyeq g \prec \mathfrak{e}_{m+1}, \qquad (1 \leqslant m < n); \\
\operatorname{nmul} P_{\times g} &= i_n \iff \mathfrak{e}_n \prec g; \\
\operatorname{ndeg} P_{\times g} &= i_n \iff \mathfrak{e}_n \preccurlyeq g.
\end{aligned}$$

This follows from Corollary 13.5.8 for $\mathcal{E} = K^\times$.

LEMMA 13.6.17. *Assume E is ω-free, Γ_E is divisible, and (g_ρ) is a pc-sequence in E with $g_\rho \rightsquigarrow 0$. Let $\boldsymbol{g} := c_E(g_\rho)$ be the corresponding cut in E and*

$$\mathcal{E} := \{g \in E^\times : g \prec g_\rho, \text{ eventually}\}.$$

Let $P \in E\{Y\}^{\neq}$. If $\mathcal{E} \neq \emptyset$, then $\operatorname{ndeg}_{\boldsymbol{g}} P = \operatorname{ndeg}_{\mathcal{E}} P$. If $\mathcal{E} = \emptyset$, then (g_ρ) is a c-sequence in E, and $\operatorname{ndeg}_{\boldsymbol{g}} P = \operatorname{mul} P$.

PROOF. Set $\gamma_\rho = v(g_{\rho+1} - g_\rho) \in \Gamma_\infty$. Removing some initial terms we arrange that (γ_ρ) is strictly increasing, and $\gamma_\rho = v(g_\rho) \in \Gamma$ for each ρ. Then for all ρ,

$$\operatorname{ndeg}_{\geqslant \gamma_\rho} P_{+g_\rho} = \operatorname{ndeg}_{\geqslant \gamma_\rho} P = \operatorname{ndeg} P_{\times g_\rho}.$$

If $\mathcal{E} \neq \emptyset$, the desired result follows easily from Corollary 13.6.16: the critical case is when $v(\mathcal{E})$ has a least element of the form $v(\mathfrak{e}_m)$ using notation from that corollary. If $\mathcal{E} = \emptyset$, use Lemma 11.2.2. □

COROLLARY 13.6.18. *Suppose that E is ω-free and Γ_E is divisible. Let (h_ρ) be a pc-sequence in E with pseudolimit $h \in E$. Let $\boldsymbol{h} := c_E(h_\rho)$ and*

$$\mathcal{E} := \{g \in E^\times : g \prec h_\rho - h, \text{ eventually}\}.$$

Let $P \in E\{Y\}^{\neq}$. If $\mathcal{E} \neq \emptyset$, then $\operatorname{ndeg}_{\boldsymbol{h}} P = \operatorname{ndeg}_{\mathcal{E}} P_{+h}$, and if $\mathcal{E} = \emptyset$, then $\operatorname{ndeg}_{\boldsymbol{h}} P = \operatorname{mul} P_{+h}$.

PROOF. Put $g_\rho := h_\rho - h$. Then (g_ρ) is a pc-sequence in E with $g_\rho \rightsquigarrow 0$. By Lemma 11.2.12(iii) we have $\operatorname{ndeg}_{\boldsymbol{h}} P = \operatorname{ndeg}_{\boldsymbol{g}+h} P = \operatorname{ndeg}_{\boldsymbol{g}} P_{+h}$, with $\boldsymbol{g} = c_E(g_\rho)$. It remains to apply Lemma 13.6.17 to P_{+h} in place of P. □

Notes and comments

The conclusion of Corollary 13.6.9 goes through for λ-free E and homogeneous $P, Q \in E\{Y\}^{\neq}$ of order $\leqslant 1$ and degrees $d > e$ with $(d-e)\Gamma_E = \Gamma_E$. This fact will not be used in this volume, so we omit the proof.

13.7 FURTHER CONSEQUENCES OF λ-FREENESS

In this section E is a λ-free H-asymptotic field (and thus pre-d-valued with rational asymptotic integration). Our main goal here is to establish Proposition 13.7.1. The basic tool is Lemma 13.7.2, which rests on Corollary 13.3.7.

Recall from Section 5.2 that for elements $y \neq 0$ and z in a differential field, we let $\omega(z) = -(2z' + z^2)$ and $\sigma(y) = \omega(-y^\dagger) + y^2$. If F is a differential field, $f \in F$, and y is a nonzero solution in F of the equation $\sigma(y) = f$, then multiplication by y^2 yields $2y''y = 3(y')^2 - y^4 + fy^2$, so y satisfies the differential equation $A(y)y'' = B(y)$, with $A := 2Y$ and $B := 3(Y')^2 - Y^4 + fY^2$. This is why we consider such equations, for example in Corollary 13.7.5 below.

PROPOSITION 13.7.1. *Suppose $\omega_\rho \rightsquigarrow \omega \in E$. Then there is an element γ of a pre-d-valued field extension of E of H-type such that $\Psi_E < v\gamma < (\Gamma_E^>)'$ and $\sigma(\gamma) = \omega$. For any such elements γ and γ^*, in possibly different extensions, there is an isomorphism $E\langle\gamma\rangle \to E\langle\gamma^*\rangle$ over E sending γ to γ^*.*

Throughout this section we assume $P \in E\{Y\}^{\neq}$.

LEMMA 13.7.2. *Suppose P has order at most 2 and $\operatorname{nwt} P = w$. There are $a \in E$, $A \in E[Y]^{\neq}$ with $A \asymp 1$ and $w + \deg A = \operatorname{ndeg} P$, and active $\mathfrak{e}$ in E such that*

$$P^{\mathfrak{e}} = a \cdot A \cdot (Y')^w + R, \qquad a \asymp P^{\mathfrak{e}},\ R \in E^{\mathfrak{e}}\{Y\},\ R \prec^\flat_{\mathfrak{e}} P^{\mathfrak{e}}.$$

Given such $a, A, \mathfrak{e}$ and any active $\mathfrak{f} \preccurlyeq \mathfrak{e}$ in E there is an $a_{\mathfrak{f}} \in E$ such that

$$P^{\mathfrak{f}} = a_{\mathfrak{f}} \cdot A \cdot (Y')^w + R_{\mathfrak{f}}, \qquad a_{\mathfrak{f}} \asymp P^{\mathfrak{f}},\quad R_{\mathfrak{f}} \in E^{\mathfrak{f}}\{Y\},\ R_{\mathfrak{f}} \prec^\flat_{\mathfrak{e}} P^{\mathfrak{f}}.$$

PROOF. For the first part, follow the proof of Lemma 13.6.11 to reduce to the case that E has a monomial group $\mathfrak{M}$ and small derivation. Then $K := \operatorname{dv}(E)$ is d-valued of H-type and an immediate extension of E. Whether or not K is λ-free, Corollary 13.3.7 yields $N_P \in C[Y](Y')^{\mathbb{N}}$ where $C = C_K$. The rest of the proof of Lemma 13.6.11 can be copied verbatim, except that to get $R_{P^{\mathrm{e}}} \prec^{\flat}_{\mathrm{e}} P^{\mathrm{e}}$ as in that proof, we appeal to Lemma 13.2.11, using that K has rational asymptotic integration. For the second part, use the proof of Corollary 13.6.12. □

Next we establish variants of some results in Section 13.4. Recall from Section 5.7 that for $f \in E^\times$ we defined $P^{\times f} := P^f_{\times f} \in E^f\{Y\}$, so $P^{\times f}(Y') = P(Y')^f$. Let us generalize notation introduced in Section 13.4, and put

$$\operatorname{nwt}^\times(P) := \operatorname{nwt}\big(P(Y')\big), \qquad v^{\times\mathrm{e}}(P) := v^{\mathrm{e}}\big(P(Y')\big).$$

Since $v(P^{\times f}) = v\big(P(Y')^f\big)$, there is an active f_0 in E such that

$$v(P^{\times f}) = v^{\times\mathrm{e}}(P) + \operatorname{nwt}^\times(P)vf \quad \text{for active } f \preccurlyeq f_0 \text{ in } E.$$

Here is a variant of Corollary 13.4.5:

COROLLARY 13.7.3. *Suppose* $\operatorname{order}(P) \leqslant 1$. *Then there exists an active f in E such that for every H-asymptotic field extension F of E and active $z \prec f$ in F,*

$$v\big(P(z)\big) = v^{\times\mathrm{e}}(P) + \operatorname{nwt}^\times(P)vz.$$

PROOF. Let $w = \operatorname{nwt}^\times(P)$. Apply Lemma 13.7.2 to $P(Y')$ in place of P to get active f in E, $a \in E^\times$, and $R \in E^f\{Y\}$ such that

$$P^{\times f} = a\cdot(Y^w + R), \quad va = v(P^{\times f}) = v^{\times\mathrm{e}}(P) + w\,vf, \quad R \prec^{\flat}_f 1.$$

Let F be an H-asymptotic field extension of E and z an active element of F with $z \prec f$. Then $y := zf^{-1} \prec 1$ is active in F^f, so $y \asymp^{\flat}_f 1$. Now F^f has small derivation, so $R(y) \prec^{\flat}_f 1 \asymp^{\flat}_f y^w$. Thus $P(z) = P^{\times f}(y) \sim^{\flat}_f ay^w$, and hence $v\big(P(z)\big) = v(a\cdot y^w) = v^{\times\mathrm{e}}(P) + w\,vz$ as claimed. □

In a similar way we obtain an analogue of Corollary 13.4.8:

COROLLARY 13.7.4. *Suppose* $\operatorname{order}(P) \leqslant 1$. *Then there exists $\delta_0 \in (\Gamma_E^>)'$ such that for every H-asymptotic field extension F of E and $z \in F$ with $\Psi_E < vz \leqslant \delta_0$,*

$$v\big(P(z)\big) = v^{\times\mathrm{e}}(P) + \operatorname{nwt}^\times(P)vz.$$

PROOF. Let w, f, a and R be as in the proof of Corollary 13.7.3. By Lemma 9.2.11 and the remark following it, we can take $\gamma_0 \in \Gamma_E^{>vf}$ and $\delta_0 \in (\Gamma_E^>)'$ such that the map $\gamma \mapsto \psi_F(\gamma - vf)\colon [\gamma_0, \delta_0]_{\Gamma_F} \to \Gamma_F$ is constant, for each H-asymptotic field extension F of E. Given such an F and $z \in F$ with $\Psi_E < vz \leqslant \delta_0$, set $y := zf^{-1} \prec 1$. Then $y \asymp^{\flat}_f 1$. (See proof of Corollary 13.4.8.) Hence we get $P(z) = P^{\times f}(y) \sim^{\flat}_f ay^w = af^{-w}z^w$ as in the proof of Corollary 13.7.3. □

REMARK. Suppose $\operatorname{order}(P) \leqslant 1$, and E is equipped with an ordering making E a pre-H-field. The proofs of Corollaries 13.7.3 and 13.7.4 show that there are $\gamma_0 \in (\Gamma_E^{<})'$, $\delta_0 \in (\Gamma_E^{>})'$, and $\sigma \in \{-1, +1\}$, such that for every pre-H-field extension F of E and $z \in F^{>}$ with $\gamma_0 \leqslant vz \leqslant \delta_0$, we have $\operatorname{sign} P(z) = \sigma$.

Note that by Corollary 13.7.4, if z is an element in an H-asymptotic field extension of E with $\Psi_E < vz < (\Gamma_E^{>})'$, then z, z' are algebraically independent over E.

COROLLARY 13.7.5. *Let $A, B \in E\{Y\}$ be of order at most 1, $A \neq 0$, and let y, z be elements of (possibly different) H-asymptotic field extensions of E such that*

$$\Psi_E < vy < (\Gamma_E^{>})', \quad A(y)y'' = B(y), \tag{13.7.1}$$

$$\Psi_E < vz < (\Gamma_E^{>})', \quad A(z)z'' = B(z). \tag{13.7.2}$$

Then there is an isomorphism $E\langle y\rangle \to E\langle z\rangle$ over E sending y to z.

PROOF. By the remark preceding the corollary, y, y' are algebraically independent over E, and so are z, z'. Thus $A(y) \neq 0$, $A(z) \neq 0$, and so $E\langle y\rangle = E(y, y')$ and $E\langle z\rangle = E(z, z')$ as fields, and we have an isomorphism $h\colon E\langle y\rangle \to E\langle z\rangle$ of differential fields over E sending y to z. If P has order $\leqslant 1$, then

$$v\big(P(y)\big) \;=\; v^{\times \mathrm{e}}(P) + \operatorname{nwt}^{\times}(P)vy$$

by Corollary 13.7.4, and similarly with y replaced by z. In view of Corollary 9.8.6 it follows that h is also a valued field isomorphism. □

REMARK. Let A and B be as in Corollary 13.7.5, and suppose additionally that E is equipped with an ordering making E a pre-H-field. Let $y > 0$ and $z > 0$ be elements of pre-H-field extensions of E satisfying (13.7.1) and (13.7.2), respectively. The isomorphism $E\langle y\rangle \to E\langle z\rangle$ of that corollary is also an isomorphism of ordered fields, in view of the remark following the proof of Corollary 13.7.4.

LEMMA 13.7.6. *Let z be an element of an H-asymptotic field extension of E with $\Psi_E < vz < (\Gamma_E^{>})'$, and consider the valued field extension $F := E(z, z')$ of E. Then $\Gamma_F = \Gamma_E \oplus \mathbb{Z}vz$ (internally), $[\Gamma_F] = [\Gamma_E]$, and $\operatorname{res}(F) = \operatorname{res}(E)$.*

PROOF. Clearly $\Gamma_F = \Gamma_E + \mathbb{Z}vz$ by Corollary 13.7.4. Since E is $\boldsymbol{\lambda}$-free, the algebraic closure E^{a} of E (inside the algebraic closure of $E\langle z\rangle$) has asymptotic integration. Since $\Psi_{E^{\mathrm{a}}} = \Psi_E < vz < (\Gamma_{E^{\mathrm{a}}}^{>})'$, we have $vz \notin \Gamma_{E^{\mathrm{a}}} = \mathbb{Q}\Gamma_E$. Thus $\Gamma_F = \Gamma_E \oplus \mathbb{Z}vz$. From Lemma 2.4.17 applied to $G := \mathbb{Q}\Gamma_E$ and $b := vz$ we obtain $[\Gamma_F] = [\Gamma_E]$. To show $\operatorname{res}(F) = \operatorname{res}(E)$, suppose P has order $\leqslant 1$. Taking w, f, a, and R as in the proof of Corollary 13.7.3, that proof yields $P(z) \sim af^{-w}z^w$. Now let $g \in F^{\times}$. Then the above gives $e \in E^{\times}$ and $k \in \mathbb{Z}$ with $g \sim ez^k$. If $g \asymp 1$ this forces $k = 0$, and so $g \sim e \in E^{\times}$. Thus $\operatorname{res}(F) = \operatorname{res}(E)$. □

We can now prove Proposition 13.7.1.

PROOF OF PROPOSITION 13.7.1. The uniqueness statement is a consequence of Corollary 13.7.5, so it suffices to prove the existence of γ with the desired properties. We arrange that Γ_E is divisible, for example by passing to the algebraic closure of E. Corollary 11.7.13 yields an immediate asymptotic extension $E(\lambda)$ of E with $\lambda_\rho \rightsquigarrow \lambda$ and $\omega(\lambda) = \omega$. Note that λ is transcendental over E. By Lemma 11.5.14, $-\lambda$ creates a gap over $E(\lambda)$, and the proof of that lemma yields an element $\gamma \neq 0$ in an H-asymptotic field extension of $E(\lambda)$ such that $v\gamma \notin \Gamma_E$, $\gamma^\dagger = -\lambda$, $\Psi_{E(\lambda,\gamma)} = \Psi_E$ and $v\gamma$ is a gap in $E(\lambda,\gamma)$. Note that then γ is transcendental over $E(\lambda)$, and

$$\Gamma_{E(\lambda,\gamma)} = \Gamma_E \oplus \mathbb{Z}v\gamma, \qquad \operatorname{res} E(\lambda,\gamma) = \operatorname{res} E, \qquad \sigma(\gamma) = \omega + \gamma^2.$$

In the field extension $E(\lambda,\gamma) = E(\gamma,\gamma')$ of E the elements λ, γ are algebraically independent over E. Using Corollary 1.9.4, let E_γ be the valued differential field extension of E with $E(\lambda,\gamma)$ as its underlying valued field, and derivation ∂_γ given by

$$\partial_\gamma(\lambda) = \frac{1}{2}(\gamma^2 - \lambda^2 - \omega), \quad \partial_\gamma(\gamma) = -\gamma \cdot \lambda.$$

Since $-2\lambda' - \lambda^2 = \omega$, the first equality amounts to $\partial_\gamma(\lambda) = \lambda' + \frac{1}{2}\gamma^2$. The second equality just says that $\partial_\gamma(\gamma) = \gamma'$. Let ω_γ, σ_γ be the analogues of ω, σ in the differential field E_γ. Then $\omega_\gamma(z) = -2\partial_\gamma(z) - z^2$ for $z \in E_\gamma$ and

$$\sigma_\gamma(y) = \omega_\gamma\big(-\partial_\gamma(y)/y\big) + y^2 \qquad (y \in E_\gamma^\times),$$

and thus $\sigma_\gamma(\gamma) = \omega$, as is easily checked. To show that E_γ is pre-d-valued, we use Lemma 10.1.19 with E, E_γ in place of K, L, and with the multiplicative subgroup $T := E(\lambda)^\times \cdot \gamma^{\mathbb{Z}}$ of $E_\gamma^\times$. In order to apply that lemma, we need:

CLAIM: *If $s, t \in T$ and $s \preccurlyeq 1$, $t \prec 1$, then $\partial_\gamma(s) \prec \partial_\gamma(t)/t$ in E_γ.*

To prove this claim, let $t \in T$. Take $a \in E(\lambda)^\times$ and $k \in \mathbb{Z}$ such that $t = a\gamma^k$. Take $A \in E(Y)^\times$ with $A(\lambda) = a$. Let $R \mapsto R^\partial$ be the derivation on the field $E(Y)$ extending that of E with $Y^\partial = 0$. Then by Corollary 4.1.4,

$$a' = A^\partial(\lambda) + (\partial A/\partial Y)(\lambda)\cdot\lambda', \qquad \partial_\gamma(a) = A^\partial(\lambda) + (\partial A/\partial Y)(\lambda)\cdot\partial_\gamma(\lambda), \text{ so}$$

$$\frac{\partial_\gamma(t)}{t} - \frac{t'}{t} = \frac{\partial_\gamma(a)}{a} - \frac{a'}{a} = \left(\frac{\partial A/\partial Y}{A}\right)(\lambda)\cdot\big(\partial_\gamma(\lambda) - \lambda'\big) = \left(\frac{\partial A/\partial Y}{A}\right)(\lambda)\cdot\tfrac{1}{2}\gamma^2.$$

Then Corollary 11.6.10 gives

$$v\left(\frac{\partial_\gamma(t)}{t} - \frac{t'}{t}\right) > v\gamma > \Psi_E = \Psi_{E(\lambda,\gamma)}.$$

If $t \not\asymp 1$, this gives $\partial_\gamma(t)/t \sim t'/t$, and so $\partial_\gamma(t) \sim t'$. Suppose now that $t \asymp 1$. Then $k = 0$ and so $A(\lambda) = a = t \asymp 1$, hence

$$\partial_\gamma(t) - t' = (\partial A/\partial Y)(\lambda)\cdot\tfrac{1}{2}\gamma^2 \asymp \left(\frac{\partial A/\partial Y}{A}\right)(\lambda)\cdot\tfrac{1}{2}\gamma^2.$$

As before this yields $v\big(\partial_\gamma(t) - t'\big) > \Psi_E = \Psi_{E(\lambda,\gamma)}$, and thus $v\big(\partial_\gamma(t)\big) > \Psi_{E\langle\lambda,\gamma\rangle}$. The claim now follows. □

REMARKS. Let E and $E\langle\gamma\rangle$ be as in Proposition 13.7.1.

(1) By Lemma 13.7.6 we have $\operatorname{res} E\langle\gamma\rangle = \operatorname{res} E$. In view of Lemma 9.1.2 it follows that if E is d-valued, then so is $E\langle\gamma\rangle$ with $C_{E\langle\gamma\rangle} = C_E$.

(2) Lemma 13.7.6 also gives $[\Gamma_E] = [\Gamma_{E\langle\gamma\rangle}]$, and so $v\gamma$ is a gap in $E\langle\gamma\rangle$.

(3) Suppose $b \asymp 1$ in $E\langle\gamma\rangle$; then $b' \prec \gamma$. This is because $\operatorname{res} E\langle\gamma\rangle = \operatorname{res} E$ gives $u \asymp 1$ in E with $b \sim u$, so $u' \prec \gamma$ and $b' - u' \prec \gamma$, and thus $b' \prec \gamma$.

LEMMA 13.7.7. *Let E and γ be as in Proposition 13.7.1 and let E be equipped with an ordering making it a pre-H-field. Then $E\langle\gamma\rangle$ has a unique field ordering extending that of E in which $\gamma > 0$ and $\mathcal{O}_{E\langle\gamma\rangle}$ is convex. Moreover, $E\langle\gamma\rangle$ with this ordering is a pre-H-field in which $\mathcal{O}_{E\langle\gamma\rangle}$ is the convex hull of $\mathcal{O}_E$.*

PROOF. Let $f \in E\langle\gamma\rangle^\times$. As $E\langle\gamma\rangle = E(\gamma, \gamma')$, Lemma 13.7.6 gives

$$f = g\gamma^k(1+a), \quad \text{with } g \in E^\times,\ k \in \mathbb{Z},\ a \in E\langle\gamma\rangle,\ a \prec 1.$$

Thus for any field ordering on $E\langle\gamma\rangle$ extending that of E in which $\gamma > 0$ and the valuation ring of $E\langle\gamma\rangle$ is convex we have: $f > 0 \Longleftrightarrow g > 0$. Thus at most one such ordering on $E\langle\gamma\rangle$ exists. For any such ordering the valuation ring of $E\langle\gamma\rangle$ is the convex hull of $\mathcal{O}$, since $\operatorname{res}(E) = \operatorname{res}(E\langle\gamma\rangle)$ by Lemma 13.7.6.

It remains only to show that $E\langle\gamma\rangle$ has a field ordering making it a pre-H-field extension of E in which $\gamma > 0$. For this we use the construction of $E\langle\gamma\rangle$ in the proof of Proposition 13.7.1. In detail, we first arrange that Γ_E is divisible, for example, by passing to the real closure of E. Then we take an immediate asymptotic extension $E(\lambda)$ of E with $\lambda_\rho \rightsquigarrow \lambda$ and $\omega(\lambda) = \omega$, and use Lemma 10.5.8 to make it a pre-H-field extension of E. Next we consider a Liouville closed H-field extension of $E(\lambda)$, and take an element $\gamma > 0$ in this extension such that $\gamma^\dagger = -\lambda$. Then $E\langle\gamma\rangle = E(\lambda, \gamma)$ is a pre-H-field extension of $E(\lambda)$, and $v\gamma$ is a gap in $E\langle\gamma\rangle$ by the remark following the proof of Lemma 11.5.14. It now follows from Lemma 13.7.6 that $E(\lambda, \gamma)$ as a pre-d-valued field extension of E is exactly as constructed in the proof of Proposition 13.7.1, and in what follows we define ∂_γ and E_γ as in that proof, and use its notations. We also consider E_γ to be equipped with the field ordering of $E(\lambda, \gamma)$. It is enough to show that E_γ is a pre-H-field. Recall that $T := E(\lambda)^\times \cdot \gamma^{\mathbb{Z}}$. By Lemma 10.5.5 it suffices to show:

CLAIM: *Suppose $t \in T$ and $t \succ 1$; then $\partial_\gamma(t)/t > 0$.*

To establish this claim, note that $\partial_\gamma(t)/t \sim t'/t$ by the proof of Proposition 13.7.1. Since $E(\lambda, \gamma)$ is a pre-H-field, we have $t'/t > 0$, and the claim follows. □

REMARK. If in addition to the hypotheses of Lemma 13.7.7, E is an H-field, then $E\langle\gamma\rangle$ ordered as in that lemma is also an H-field, with $C_{E\langle\gamma\rangle} = C_E$.

Cases of low complexity

For use in the next chapter we derive here results for P of low complexity, under our standing assumption that E is λ-free.

COROLLARY 13.7.8. *If* order $P \leqslant 2$, *then there is* $\gamma \in \Gamma_E^>$ *such that for all* $g \in E$,

$$\begin{aligned} 0 < vg < \gamma &\Longrightarrow \operatorname{nmul}(P) = \operatorname{nmul}(P_{\times g}) = \operatorname{ndeg}(P_{\times g}), \\ -\gamma < vg < 0 &\Longrightarrow \operatorname{ndeg}(P) = \operatorname{ndeg}(P_{\times g}) = \operatorname{nmul}(P_{\times g}). \end{aligned}$$

PROOF. Suppose order $P \leqslant 2$. Use Lemma 13.7.2 and compositional conjugation by some active element of E to arrange that E has small derivation and P is as in the hypothesis of Lemma 13.6.14. Now apply Lemma 13.6.14. □

LEMMA 13.7.9. *If* $\operatorname{ndeg} P = 0$ *and* f_0 *is active in* E, *then for some active* $f \preccurlyeq f_0$ *in* E *we have* $P^f \sim_f^\flat a$ *in* $E^f\{Y\}$ *with* $a \in E$.

We omit the proof: it is like that of the next lemma but shorter.

LEMMA 13.7.10. *Suppose* $\operatorname{ndeg} P = 1$ *and* f_0 *is active in* E. *Then for some active* $f \preccurlyeq f_0$ *in* E *we have either* $P^f \sim_f^\flat a + bY$ *in* $E^f\{Y\}$ *with* $a \preccurlyeq b$ *in* E, *or* $P^f \sim_f^\flat bY'$ *in* $E^f\{Y\}$ *with* $b \in E$. *In particular,* $\operatorname{nwt}(P) \leqslant 1$.

PROOF. By compositional conjugation we arrange that E has small derivation. By passing to an elementary extension of E we further arrange that E has a monomial group $\mathfrak{M}$. Then $K := \operatorname{dv}(E)$ is an immediate extension of E. With $\mathfrak{M}$ as its monomial group K satisfies the assumptions at the beginning of this chapter. Then $\operatorname{ndeg} P = 1$ gives $P_1 \neq 0$, and either $N_P = N_{P_0} + N_{P_1}$ or $N_P = N_{P_1}$. Applying Corollary 13.3.9 to $Q := P_1$ gives $N_{P_1} = cY$ or $N_{P_1} = cY'$ with $c \in C^\times$. Since N_P is isobaric, this gives $N_P = c_0 + cY$ or $N_P = cY'$, with $c_0, c \in C$, $c \neq 0$. In particular, $N_P \in C[Y](Y')^{\mathbb{N}}$. We have $\mathfrak{d}_{P^\phi} \in \mathfrak{M} \subseteq E^\times$, and $P^\phi \sim_\phi^\flat \mathfrak{d}_{P^\phi} N_P$, eventually, by Lemma 13.2.11. We just do the case $N_P = c_0 + cY$. (The other case is similar.) Take $e_0 \preccurlyeq e_1 \asymp 1$ in E such that $c_0 - e_0 \prec 1$ and $c_1 - e_1 \prec 1$ in K. Recall that ϕ ranges over the elements of $\mathfrak{M}$ active in K, but $\mathfrak{M} \subseteq E^\times$, so $\phi \in E$, and eventually

$$P^\phi \sim_\phi^\flat \mathfrak{d}_{P^\phi} N_P = \mathfrak{d}_{P^\phi}(e_0 + e_1 Y) + \mathfrak{d}_{P^\phi}\big((c_0 - e_0) + (c_1 - e_1)Y\big).$$

Also $(c_0 - e_0) + (c_1 - e_1)Y \prec_\phi^\flat 1$, eventually, so $P^\phi \sim_\phi^\flat \mathfrak{d}_{P^\phi}(e_0 + e_1 Y)$, eventually. □

COROLLARY 13.7.11. *If* $\operatorname{ndeg} P \leqslant 1$, *then there is* $\gamma \in \Gamma_E^>$ *such that for all* $g \in E$,

$$\begin{aligned} 0 < vg < \gamma &\Longrightarrow \operatorname{nmul}(P) = \operatorname{nmul}(P_{\times g}) = \operatorname{ndeg}(P_{\times g}), \\ -\gamma < vg < 0 &\Longrightarrow \operatorname{ndeg}(P) = \operatorname{ndeg}(P_{\times g}) = \operatorname{nmul}(P_{\times g}). \end{aligned}$$

PROOF. Assume $\operatorname{ndeg} P \leqslant 1$. By Lemmas 13.7.9 and 13.7.10 and compositional conjugation we arrange that E has small derivation and P is as in the hypothesis of Lemma 13.6.14, with $w = 0$ or $w = 1$. Now apply Lemma 13.6.14. □

LEMMA 13.7.12. *If* $\operatorname{nmul} P = 1$, *then there is* $\gamma \in \Gamma_E^>$ *such that for all* $g \in E^\times$,

$$0 < vg < \gamma \implies \operatorname{nmul} P_{\times g} = \operatorname{ndeg} P_{\times g} = 1.$$

PROOF. Assume $\operatorname{nmul} P = 1$. As in the proof of Lemma 13.7.10 we arrange that E has small derivation, a monomial group, and an immediate (d-valued) extension K. With $P = P_0 + P_1 + R$, where $R := \sum_{d \geqslant 2} P_d$, we have $P_0 \prec P_1^\phi \succcurlyeq R^\phi$ in $K^\phi\{Y\}$, eventually, and $v(P_1^\phi) = v^{\mathrm{e}}(P) + \operatorname{nwt}(P_1)v\phi$, eventually. By Corollary 13.3.9 we have $\operatorname{nwt}(P_1) \leqslant 1$, and as E has asymptotic integration, this yields $\alpha \in \Gamma_E^>$ such that $v(P_0) > \alpha + v^{\mathrm{e}}(P) + \operatorname{nwt}(P_1)v\phi$ for all ϕ. Then for $\gamma \in \Gamma_E^>$ with $2\gamma < \alpha$ we have for all $g \in E^\times$,

$$0 < vg < \gamma \implies v(P_0) > v(P_{1,\times g}^\phi) < v(R_{\times g}^\phi), \text{ eventually,}$$

so any such γ has the desired property. □

Notes and comments

Corollary 13.7.4 improves on [20, Proposition 12.12].

13.8 ASYMPTOTIC EQUATIONS

This section is somewhat technical, but indispensable. The main facts we establish here are Proposition 13.8.8 and Lemma 13.8.13. These are needed in Chapter 14 in proving Theorems 14.0.2 and 14.3.5, which are of independent interest. Our quantifier elimination for $\mathbb{T}$ depends essentially on Theorem 14.0.2.

Recall the assumptions on K imposed in the beginning of this chapter. *In this section we assume in addition that K is ω-free,* so that we can use results of Section 13.5. Also, as before, P ranges over $K\{Y\}^{\neq}$ and ϕ over the elements of $\mathfrak{M}$ that are active in K. Let $\mathcal{E} \subseteq K^\times$ be $\preccurlyeq$-closed, and consider the asymptotic equation over K given by

$$\text{(E)} \qquad P(Y) = 0, \qquad Y \in \mathcal{E}.$$

An **approximate solution** of (E) is an approximate zero y of P such that $y \in \mathcal{E}$, and the **multiplicity** of an approximate solution y of (E) is its multiplicity as an approximate zero of P. An **algebraic approximate solution** of (E) is an algebraic approximate zero y of P, as defined at the end of Section 13.3, such that $y \in \mathcal{E}$. Let $y \in \mathcal{E}$ with $y \sim c\mathfrak{m}$ ($c \in C^\times$). Then by Lemma 13.5.4,

$$y \text{ is an approximate solution of (E)} \iff N_{P_{\times \mathfrak{m}}}(c) = 0 \iff \operatorname{ndeg}_{\prec \mathfrak{m}} P_{+y} \geqslant 1,$$

and if y is an approximate solution of (E), then its multiplicity equals $\operatorname{ndeg}_{\prec \mathfrak{m}} P_{+y}$, which is $\leqslant \operatorname{ndeg}_{\mathcal{E}} P$. If y is an approximate solution of (E), then so is every $z \in K^\times$ with $z \sim y$, with the same multiplicity. By Lemma 13.1.6, if y is a solution of (E), then y is an approximate solution of (E). For each ϕ, the asymptotic equation

$$(\mathrm{E}^\phi) \qquad P^\phi(Y) = 0, \qquad Y \in \mathcal{E}$$

has the same solutions and the same approximate solutions as (E), with the same multiplicities.

A **starting monomial** for (E) is a starting monomial $\mathfrak{m}$ for P with $\mathfrak{m} \in \mathcal{E}$; we define "algebraic starting monomial for (E)" likewise. If $y \sim c\mathfrak{m}$ ($c \in C^{\times}$) is an approximate solution of (E), then $\mathfrak{m}$ is a starting monomial for (E). Hence if (E) has no starting monomial (in particular, if $\operatorname{ndeg}_{\mathcal{E}} P = 0$), then (E) has no approximate solution. By Proposition 13.5.7, if Γ is divisible and $\operatorname{mul} P < \operatorname{ndeg}_{\mathcal{E}} P$, then there is an algebraic starting monomial for (E), and $\operatorname{ndeg} P_{\times \mathfrak{e}} = \operatorname{ndeg}_{\mathcal{E}} P$, $\mathfrak{e} :=$ largest algebraic starting monomial for (E).

Let $\mathcal{E}' \subseteq \mathcal{E}$ be $\preccurlyeq$-closed and let $f \in \mathcal{E} \cup \{0\}$. We call the asymptotic equation

$$(\mathrm{E}') \qquad P_{+f}(Y) = 0, \qquad Y \in \mathcal{E}'$$

a **refinement** of (E). By Lemma 11.2.7 we have $\operatorname{ndeg}_{\mathcal{E}'} P_{+f} \leqslant \operatorname{ndeg}_{\mathcal{E}} P$. Note that if y is a solution of (E$'$) and $f + y \neq 0$, then $f + y$ is a solution of (E). Moreover:

LEMMA 13.8.1. *Let $y \not\sim -f$ be an approximate solution of* (E$'$) *of multiplicity μ. Then $f + y$ is an approximate solution of* (E) *of multiplicity $\geqslant \mu$.*

PROOF. Since $y \not\sim -f$ we have $y \preccurlyeq f + y$ and thus

$$1 \leqslant \mu = \operatorname{ndeg}_{\prec y}(P_{+f})_{+y} = \operatorname{ndeg}_{\prec y} P_{+(f+y)} \leqslant \operatorname{ndeg}_{\prec f+y} P_{+(f+y)},$$

hence $f + y$ is an approximate solution of (E) of multiplicity $\geqslant \mu$. □

LEMMA 13.8.2. *Suppose $f \neq 0$, $\mathcal{E}' \subseteq K^{\prec f}$ and $\operatorname{ndeg}_{\mathcal{E}} P = \operatorname{ndeg}_{\mathcal{E}'} P_{+f} \geqslant 1$. Then f is an approximate solution of* (E).

PROOF. Using Lemma 11.2.7 we have

$$\operatorname{ndeg}_{\mathcal{E}'} P_{+f} \leqslant \operatorname{ndeg}_{\prec f} P_{+f} \leqslant \operatorname{ndeg}_{\preccurlyeq f} P_{+f} = \operatorname{ndeg}_{\preccurlyeq f} P \leqslant \operatorname{ndeg}_{\mathcal{E}} P,$$

and hence $\operatorname{ndeg}_{\prec f} P_{+f} = \operatorname{ndeg}_{\mathcal{E}} P \geqslant 1$. Thus f is an approximate solution of (E) by the equivalence displayed earlier in this section. □

Let an asymptotic equation (E) be given, with Newton degree $d = \operatorname{ndeg}_{\mathcal{E}} P$. Then $\operatorname{ndeg}_{\prec f} P_{+f} \leqslant d$ for all $f \in \mathcal{E}$. Moreover:

LEMMA 13.8.3. *Suppose $d \geqslant 1$. Then the following are equivalent:*

(i) $\operatorname{ndeg}_{\prec f} P_{+f} < d$ *for all* $f \in \mathcal{E}$;

(ii) $\operatorname{ndeg}_{\prec f} P_{+f} < d$ *for all* $f \in \mathcal{E}$ *with* $\operatorname{ndeg} P_{\times f} = d$;

(iii) *there is no approximate solution of* (E) *of multiplicity d.*

PROOF. Let $f \in \mathcal{E}$ and suppose $\operatorname{ndeg} P_{\times f} < d$. Then

$$\operatorname{ndeg}_{\prec f} P_{+f} \leqslant \operatorname{ndeg}_{\preccurlyeq f} P_{+f} = \operatorname{ndeg}_{\preccurlyeq f} P = \operatorname{ndeg} P_{\times f} < d.$$

Thus (i) $\Longleftrightarrow$ (ii). An earlier equivalence in this section gives (i) $\Longleftrightarrow$ (iii). □

We say that (E) is **unraveled** if $d \geqslant 1$ and one of the equivalent conditions in Lemma 13.8.3 holds. So if $d \geqslant 1$ and (E) does not have an approximate solution, then (E) is unraveled. If (E) is unraveled and has an approximate solution, then $d \geqslant 2$ by condition (iii) of Lemma 13.8.3.

EXAMPLE. Suppose $P \in K[Y](Y')^{\mathbb{N}}$. Then $d = \operatorname{ndeg}_{\mathcal{E}} P = \operatorname{ddeg}_{\mathcal{E}} P$. Assume that $d \geqslant 1$. If $P \in K[Y]$, then (E) is not unraveled if and only if there are $\mathfrak{m} \in \mathcal{E}$ and $a, b \in C^{\times}$ such that $D_{P_{\times \mathfrak{m}}} = a \cdot (Y - b)^d$. Suppose $\operatorname{wt} P \geqslant 1$. Then (E) is not unraveled if and only if $1 \in \mathcal{E}$ and $D_P \in C[Y](Y')^{\mathbb{N}}$ has a zero in $C^{\times}$ of multiplicity d.

If (E) is unraveled, then so is (E^{ϕ}) for each ϕ; moreover:

LEMMA 13.8.4. *Suppose the asymptotic equation* (E) *is unraveled. Let $f \in K$ be such that $f \preccurlyeq \mathfrak{e}$ for some algebraic starting monomial $\mathfrak{e}$ for* (E). *Then*

$$P_{+f}(Y) = 0, \qquad Y \in \mathcal{E}$$

is also unraveled.

PROOF. We have $\operatorname{ndeg}_{\mathcal{E}} P_{+f} = d \geqslant 1$. Let $g \in \mathcal{E}$; we need to show that

$$\operatorname{ndeg}_{\prec g} P_{+(f+g)} < d.$$

If $g \succ f$, then

$$\operatorname{ndeg}_{\prec g} P_{+(f+g)} = \operatorname{ndeg}_{\prec g} P_{+g} < d$$

since (E) is unraveled. If $g \prec f$, then similarly

$$\operatorname{ndeg}_{\prec g} P_{+(f+g)} \leqslant \operatorname{ndeg}_{\prec f} P_{+(f+g)} = \operatorname{ndeg}_{\prec f} P_{+f} < d.$$

Next, suppose that $g \asymp f$. If $f + g \asymp g$, then

$$\operatorname{ndeg}_{\prec g} P_{+(f+g)} = \operatorname{ndeg}_{\prec f+g} P_{+(f+g)} < d,$$

and if $f + g \prec g$, then we take an algebraic starting monomial $\mathfrak{e}$ for (E) with $f \preccurlyeq \mathfrak{e}$, and get from Corollary 13.5.5 that

$$\operatorname{ndeg}_{\prec g} P_{+(f+g)} = \operatorname{ndeg}_{\prec g} P = \operatorname{nmul} P_{\times f} \leqslant \operatorname{nmul} P_{\times \mathfrak{e}} < \operatorname{ndeg} P_{\times \mathfrak{e}} \leqslant d,$$

so $\operatorname{ndeg}_{\prec g} P_{+(f+g)} < d$. □

Lemma 13.8.4 has a converse of sorts:

LEMMA 13.8.5. *Suppose Γ is divisible and $\operatorname{ndeg}_{\mathcal{E}} P > \operatorname{mul} P$. Let $\mathfrak{e}$ be the largest algebraic starting monomial for* (E). *Suppose $\mathcal{E}_1 \subseteq K^{\times}$ is $\preccurlyeq$-closed and $f \in \mathcal{E}_1$ is such that $\operatorname{ndeg}_{\mathcal{E}_1} P = \operatorname{ndeg}_{\mathcal{E}} P$ and $P_{+f}(Y) = 0, Y \in \mathcal{E}_1$, is unraveled. Then $f \preccurlyeq \mathfrak{e}$.*

PROOF. We have $d := \operatorname{ndeg}_{\mathcal{E}} P = \operatorname{ndeg}_{\mathcal{E}_1} P_{+f}$. If $f \succ \mathfrak{e}$, then

$$d = \operatorname{ndeg} P_{\times \mathfrak{e}} \leqslant \operatorname{ndeg}_{\prec f} P = \operatorname{ndeg}_{\prec f} (P_{+f})_{+(-f)} < d,$$

a contradiction. □

Assume $\mathrm{ndeg}_{\mathcal{E}} P = d \geqslant 1$. Let $f \in \mathcal{E} \cup \{0\}$ and let $\mathcal{E}' \subseteq \mathcal{E}$ be $\preccurlyeq$-closed. We call the pair $(f, \mathcal{E}')$ a **partial unraveler** for (E) if $\mathrm{ndeg}_{\mathcal{E}'} P_{+f} = d$. Thus $(f, \mathcal{E})$ is a partial unraveler for (E). If $(f, \mathcal{E}')$ is a partial unraveler for (E) and $(f_1, \mathcal{E}_1)$ is a partial unraveler for (E′), then $(f + f_1, \mathcal{E}_1)$ is a partial unraveler for (E). Moreover, if $(f, \mathcal{E}')$ is a partial unraveler for (E), then $(f, \mathcal{E}')$ is a partial unraveler for (E^{ϕ}). An **unraveler** for (E) is a partial unraveler $(f, \mathcal{E}')$ for (E) with unraveled (E′). In the next easy lemma we continue assuming $\mathrm{ndeg}_{\mathcal{E}} P \geqslant 1$.

LEMMA 13.8.6. *Let $a \in K^{\times}$ and $a\mathcal{E} := \{ay \in K^{\times} : y \in \mathcal{E}\}$, and consider*

$$(a\mathrm{E}) \qquad\qquad P_{\times a^{-1}}(Y) = 0, \qquad Y \in a\mathcal{E}.$$

The Newton degree of (aE) *equals that of* (E). *If $(f, \mathcal{E}')$ is a partial unraveler for* (E), *then $(af, a\mathcal{E}')$ is a partial unraveler for* (aE); *similarly with* unraveler *instead of* partial unraveler. *If $a \in \mathfrak{M}$, then the algebraic starting monomials for* (aE) *are exactly the $a\mathfrak{e}$ with $\mathfrak{e}$ an algebraic starting monomial for* (E).

We also note an easy consequence of Lemma 13.8.4:

COROLLARY 13.8.7. *Assume* $\mathrm{ndeg}_{\mathcal{E}} P \geqslant 1$. *Let $(f, \mathcal{E}')$ be an unraveler for* (E), *and let $g \in K$ be such that $f - g \preccurlyeq \mathfrak{e}$ for some algebraic starting monomial $\mathfrak{e}$ of the refinement* (E′) *of* (E). *Then $(g, \mathcal{E}')$ is also an unraveler for* (E).

Recall that throughout this section K is ω-free. In the next result we use the notion *asymptotically* d*-algebraically maximal* defined in Section 9.1.

PROPOSITION 13.8.8. *Suppose K is asymptotically* d*-algebraically maximal, Γ is divisible, $d :=$* $\mathrm{ndeg}_{\mathcal{E}} P \geqslant 1$, *but there is no $f \in \mathcal{E} \cup \{0\}$ with* $\mathrm{mul}\, P_{+f} = d$. *Then there exists an unraveler for* (E).

PROOF. Let $\big((f_\lambda, \mathcal{E}_\lambda)\big)_{\lambda<\rho}$ be a sequence of partial unravelers for (E), indexed by the ordinals less than an ordinal $\rho > 0$, such that $(f_0, \mathcal{E}_0) = (0, \mathcal{E})$ and

(1) $\mathcal{E}_\lambda \supseteq \mathcal{E}_\mu$ for all $\lambda < \mu < \rho$,

(2) $f_\mu - f_\lambda \succ f_\nu - f_\mu$ for all $\lambda < \mu < \nu < \rho$,

(3) $f_{\lambda+1} - f_\lambda \in \mathcal{E}_\lambda \setminus \mathcal{E}_{\lambda+1}$ for all λ with $\lambda + 1 < \rho$.

Note that for $\rho = 1$ we have such a sequence. By (2) we have $f_\lambda - f_\mu \asymp f_\lambda - f_{\lambda+1}$ for $\lambda < \mu < \rho$. Take $\mathfrak{m}_\lambda \in \mathfrak{M}$ with $\mathfrak{m}_\lambda \asymp f_{\lambda+1} - f_\lambda$ for $\lambda + 1 < \rho$. Then by (3),

$$\begin{aligned} d = \mathrm{ndeg}_{\mathcal{E}_{\lambda+1}} P_{+f_{\lambda+1}} &\leqslant \mathrm{ndeg}_{\preccurlyeq \mathfrak{m}_\lambda} P_{+f_{\lambda+1}} \\ &= \mathrm{ndeg}_{\preccurlyeq \mathfrak{m}_\lambda} (P_{+f_\lambda})_{+(f_{\lambda+1}-f_\lambda)} \\ &= \mathrm{ndeg}_{\preccurlyeq \mathfrak{m}_\lambda} P_{+f_\lambda} \\ &\leqslant \mathrm{ndeg}_{\mathcal{E}_\lambda} P_{+f_\lambda} = d, \end{aligned}$$

and thus $\mathrm{ndeg}_{\preccurlyeq \mathfrak{m}_\lambda} P_{+f_\lambda} = d$, for all ordinals λ with $\lambda + 1 < \rho$.

Suppose first that ρ is a successor ordinal, $\rho = \sigma + 1$. Consider the refinement

$$(\mathrm{E}_\sigma) \qquad P_{+f_\sigma}(Y) = 0, \qquad Y \in \mathcal{E}_\sigma$$

of (E). If (E_σ) is unraveled, then $(f_\sigma, \mathcal{E}_\sigma)$ is an unraveler of (E), and we are done. Assume (E_σ) is not unraveled, and take $f \in \mathcal{E}_\sigma$ such that $\operatorname{ndeg}_{\prec f}(P_{+f_\sigma})_{+f} = d$. The subset

$$\mathcal{E}_\rho := \{y \in K^\times : y \prec f\}$$

of $\mathcal{E}_\sigma$ is $\preccurlyeq$-closed, with $\operatorname{ndeg}_{\mathcal{E}_\rho}(P_{+f_\sigma})_{+f} = d$. Hence $(f, \mathcal{E}_\rho)$ is a partial unraveler for (E_σ), so $(f_\rho, \mathcal{E}_\rho)$, where $f_\rho := f_\sigma + f$, is a partial unraveler for (E). Then $\big((f_\lambda, \mathcal{E}_\lambda)\big)_{\lambda<\rho+1}$ satisfies (1)–(3) with $\rho+1$ instead of ρ.

Now suppose ρ is a limit ordinal. Then $(f_\lambda)_{\lambda<\rho}$ is a pc-sequence in K. Let $\boldsymbol{f} = c_K(f_\lambda)$ be the corresponding cut in K. Then $\operatorname{ndeg}_{\boldsymbol{f}} P = d$. The d-valued field K of H-type is asymptotically d-algebraically maximal and has rational asymptotic integration, so we can take a pseudolimit f_ρ of $(f_\lambda)_{\lambda<\rho}$ in K, by Lemmas 11.4.8 and 11.4.12. Consider the subset

$$\mathcal{E}_\rho := \bigcap_{\lambda<\rho} \mathcal{E}_\lambda = \{y \in K^\times : y \prec \mathfrak{m}_\lambda \text{ for all } \lambda < \rho\}$$

of $K^\times$. If $\mathcal{E}_\rho = \emptyset$, then $\operatorname{mul} P_{+f_\rho} = \operatorname{ndeg}_{\boldsymbol{f}} P = d$ by Corollary 13.6.18, contradicting the hypothesis of the proposition. Thus $\mathcal{E}_\rho \neq \emptyset$, and so $\mathcal{E}_\rho$ is $\preccurlyeq$-closed. We have $\operatorname{ndeg}_{\mathcal{E}_\rho} P_{+f_\rho} = \operatorname{ndeg}_{\boldsymbol{f}} P = d$ by Corollary 13.6.18, hence $(f_\rho, \mathcal{E}_\rho)$ is a partial unraveler for (E), and $\big((f_\lambda, \mathcal{E}_\lambda)\big)_{\lambda<\rho+1}$ satisfies (1)–(3) with $\rho+1$ instead of ρ.

This building process must end in producing an unraveler for (E). □

Behavior of unravelers under immediate extensions

In this subsection we fix an ω*-free immediate* H*-asymptotic extension* L *of* K. By Lemma 9.1.2, L is d-valued; our monomial group $\mathfrak{M}$ for K continues to serve as a monomial group for L. Let $\mathcal{E} \subseteq K^\times$ be $\preccurlyeq$-closed. Then the subset

$$\mathcal{E}_L := \{y \in L^\times : vy \in v\mathcal{E}\}$$

of $L^\times$ is $\preccurlyeq$-closed with $\mathcal{E}_L \cap K = \mathcal{E}$. We consider the asymptotic equation

$$(\mathrm{E}_L) \qquad P(Y) = 0, \qquad Y \in \mathcal{E}_L$$

over L, which has the same Newton degree $\operatorname{ndeg}_{\mathcal{E}_L} P = \operatorname{ndeg}_{\mathcal{E}} P$ as (E). An element y of K is an approximate solution of (E) if and only if y is an approximate solution of (E_L), and in this case the multiplicity of y as an approximate solution of (E) agrees with the multiplicity of y as an approximate solution of (E_L). Hence (E) is unraveled if and only if (E_L) is. This yields:

LEMMA 13.8.9. *Assume* $\operatorname{ndeg}_{\mathcal{E}} P \geqslant 1$. *Let* $f \in \mathcal{E} \cup \{0\}$ *and let* $\mathcal{E}' \subseteq \mathcal{E}$ *be* $\preccurlyeq$*-closed. Then* $(f, \mathcal{E}')$ *is a partial unraveler for* (E) *if and only if* $(f, \mathcal{E}'_L)$ *is a partial unraveler for* (E_L)*; similarly with* unraveler *instead of* partial unraveler.

Let (a_ρ) be a divergent pc-sequence in K with minimal d-polynomial P over K, and suppose $a_\rho \rightsquigarrow \ell \in L$.

LEMMA 13.8.10. $\operatorname{mul}(P_{+\ell}) \leqslant 1$.

PROOF. By Lemma 11.3.8 we have $Q(\ell) \neq 0$ for all $Q \in K\{Y\}^{\neq}$ of smaller complexity than P. Thus $S_P(\ell) \neq 0$, and so $\operatorname{mul}(P_{+\ell}) \leqslant 1$. □

Let $\boldsymbol{a} = c_K(a_\rho)$ be the cut defined by (a_ρ) in K, and let $a \in K$, $\mathfrak{v} \in K^\times$ be such that $a - \ell \prec \mathfrak{v}$ and $\operatorname{ndeg}_{\prec\mathfrak{v}} P_{+a} = \operatorname{ndeg}_{\boldsymbol{a}} P$. Consider the asymptotic equation

$$P_{+a}(Y) = 0, \qquad Y \prec \mathfrak{v} \tag{13.8.1}$$

over K. In the next lemma we also consider it as an asymptotic equation over L. The following consequence of Proposition 13.8.8 is needed in Section 14.5 below:

LEMMA 13.8.11. *Suppose that Γ is divisible, L is asymptotically* d-*algebraically maximal, and* $\operatorname{ndeg}_{\boldsymbol{a}} P \geqslant 2$. *Then there exists an unraveler $(f, \mathcal{E})$ for* (13.8.1) *over L such that $f \neq 0$, $\operatorname{ndeg}_{\prec f} P_{+(a+f)} = \operatorname{ndeg}_{\boldsymbol{a}} P$, and $a_\rho \rightsquigarrow a + f + z$ for all $z \in \mathcal{E} \cup \{0\}$.*

PROOF. Take $g \in K$ such that $a - \ell \sim -g$. Then $0 \neq g \prec \mathfrak{v}$, so

$$\operatorname{ndeg}_{\boldsymbol{a}} P \;=\; \operatorname{ndeg}_{\prec\mathfrak{v}} P_{+a} \;=\; \operatorname{ndeg}_{\prec\mathfrak{v}} P_{+(a+g)} \;\geqslant\; \operatorname{ndeg}_{\prec g} P_{+(a+g)}.$$

Also $(a+g) - \ell \prec g$, so $\operatorname{ndeg}_{\boldsymbol{a}} P \leqslant \operatorname{ndeg}_{\prec g} P_{+(a+g)}$ by Lemma 11.4.12, and thus $\operatorname{ndeg}_{\boldsymbol{a}} P = \operatorname{ndeg}_{\prec g} P_{+(a+g)}$. Now $P_{+(a+g)}$ is a minimal differential polynomial for $\big(a_\rho - (a+g)\big)$ over K and $\operatorname{ndeg}_{\boldsymbol{a}-(a+g)} P_{+(a+g)} = \operatorname{ndeg}_{\boldsymbol{a}} P$. Suppose $\mathcal{E} \subseteq L^{\times,\prec g}$ is $\preccurlyeq$-closed in the sense of L, and $(h, \mathcal{E})$ is an unraveler for the asymptotic equation

$$P_{+(a+g)}(Y) \;=\; 0, \qquad Y \prec g$$

over L with $a_\rho - (a+g) \rightsquigarrow h + z$ for all $z \in \mathcal{E} \cup \{0\}$. Then $(f, \mathcal{E})$, $f := g + h \neq 0$, is an unraveler for (13.8.1) in L with $\operatorname{ndeg}_{\prec f} P_{+(a+f)} = \operatorname{ndeg}_{\boldsymbol{a}} P$ and $a_\rho \rightsquigarrow a + f + z$ for all $z \in \mathcal{E} \cup \{0\}$. Thus, after replacing P, (a_ρ), ℓ, $\mathfrak{v}$ by $P_{+(a+g)}$, $\big(a_\rho - (a+g)\big)$, $\ell - (a+g)$, g, respectively, we may assume $a = 0$, and only need to show the existence of an unraveler $(f, \mathcal{E})$ for (13.8.1) in L with $a_\rho \rightsquigarrow f + z$ for all $z \in \mathcal{E} \cup \{0\}$. For this, consider the subset

$$\mathcal{Z} \;:=\; \{z \in L^\times \,:\, z \prec a_\rho - \ell, \text{ eventually}\}$$

of $L^\times$. By Lemma 13.8.10 and since $\operatorname{ndeg}_{\boldsymbol{a}} P \geqslant 2$, there is no $z \in \mathcal{Z} \cup \{0\}$ such that $\operatorname{mul}(P_{+(\ell+z)}) = \operatorname{ndeg}_{\boldsymbol{a}} P$. So $\mathcal{Z} \neq \emptyset$, and thus $\mathcal{Z}$ is $\preccurlyeq$-closed, and $\operatorname{ndeg}_{\mathcal{Z}} P_{+\ell} = \operatorname{ndeg}_{\boldsymbol{a}} P$, by Corollary 13.6.18. Proposition 13.8.8 now provides us with an unraveler $(g, \mathcal{E})$ for the asymptotic equation

$$P_{+\ell}(Y) \;=\; 0, \qquad Y \in \mathcal{Z}$$

over L. Then $(f, \mathcal{E})$, $f := \ell + g$, is an unraveler for (13.8.1) (with $a = 0$), and $a_\rho \rightsquigarrow f + z$ for all $z \in \mathcal{Z} \cup \{0\}$. □

Neglecting terms of high degree

In this subsection we assume Γ *is divisible and* $d := \operatorname{ndeg}_{\mathcal{E}} P \geqslant 1$. Let $(f, \mathcal{E}')$ be an unraveler for (E), and set $\mathfrak{f} := \mathfrak{d}_f$. Suppose also that $d > \operatorname{mul}(P_{+f})$. Then (E′) has an algebraic starting monomial by Proposition 13.5.7, and we let $\mathfrak{e}$ be the largest algebraic starting monomial for (E′). Let $g \in K$, $\mathfrak{g} := \mathfrak{d}_g$, and suppose $\mathfrak{e} \prec \mathfrak{g} \prec \mathfrak{f}$. Put $\widetilde{f} := f - g$ (so $\widetilde{f} \sim f$), and consider the refinement

$$(\widetilde{\mathrm{E}}) \qquad P_{+\widetilde{f}}(Y) = 0, \qquad Y \preccurlyeq \mathfrak{g}$$

of (E). Set $\widetilde{\mathcal{E}}' := \{y \in \mathcal{E}' : y \prec \mathfrak{g}\}$, so $\mathfrak{e} \in \widetilde{\mathcal{E}}'$.

LEMMA 13.8.12. *The asymptotic equation* $(\widetilde{\mathrm{E}})$ *has Newton degree d, and* $(g, \widetilde{\mathcal{E}}')$ *is an unraveler for* $(\widetilde{\mathrm{E}})$.

PROOF. We have

$$d = \operatorname{ndeg}_{\preccurlyeq \mathfrak{e}} P_{+f} \leqslant \operatorname{ndeg}_{\preccurlyeq \mathfrak{g}} P_{+f} = \operatorname{ndeg}_{\preccurlyeq \mathfrak{g}} P_{+\widetilde{f}} \leqslant \operatorname{ndeg}_{\mathcal{E}} P_{+\widetilde{f}} = \operatorname{ndeg}_{\mathcal{E}} P = d$$

and hence $(\widetilde{\mathrm{E}})$ has Newton degree d. Also

$$d = \operatorname{ndeg}_{\preccurlyeq \mathfrak{e}} P_{+f} \leqslant \operatorname{ndeg}_{\widetilde{\mathcal{E}}'} P_{+f} \leqslant \operatorname{ndeg}_{\mathcal{E}} P_{+f} = \operatorname{ndeg}_{\mathcal{E}} P = d.$$

Hence the asymptotic equation

$$P_{+f}(Y) = 0, \qquad Y \in \widetilde{\mathcal{E}}',$$

which is a refinement of both (E′) and $(\widetilde{\mathrm{E}})$, has Newton degree d, and as (E′) is unraveled, the pair $(g, \widetilde{\mathcal{E}}')$ is an unraveler for $(\widetilde{\mathrm{E}})$. □

Recall that for $F \in K\{Y\}$ and $e \in \mathbb{N}$, we defined $F_{\leqslant e} := F_0 + F_1 + \cdots + F_e$. Note that if $e \geqslant \operatorname{ndeg} F$, then $N_F = N_{F_{\leqslant e}}$ by Corollary 13.2.7 and its proof. Set $F := P_{+\widetilde{f}}$. Then $d \geqslant \operatorname{ndeg} F_{\times \mathfrak{m}}$ for all $\mathfrak{m} \preccurlyeq \mathfrak{g}$. Consider the "truncation"

$$(\widetilde{\mathrm{E}}_{\leqslant d}) \qquad F_{\leqslant d}(Y) = 0, \qquad Y \preccurlyeq \mathfrak{g}$$

of $(\widetilde{\mathrm{E}})$ as an asymptotic equation over K. We have

$$N_{F_{\times \mathfrak{m}}} = N_{(F_{\times \mathfrak{m}})_{\leqslant d}} = N_{(F_{\leqslant d})_{\times \mathfrak{m}}} \qquad \text{for } \mathfrak{m} \preccurlyeq \mathfrak{g},$$

so $(\widetilde{\mathrm{E}}_{\leqslant d})$ has the same algebraic starting monomials and the same Newton degree d as $(\widetilde{\mathrm{E}})$. In the next lemma we show that under suitable conditions the unraveler $(g, \widetilde{\mathcal{E}}')$ for $(\widetilde{\mathrm{E}})$ is also an unraveler for $(\widetilde{\mathrm{E}}_{\leqslant d})$. This will be crucial in Section 14.4.

LEMMA 13.8.13. *Suppose* $(\mathfrak{e}/\mathfrak{g}) \prec\!\!\prec (\mathfrak{g}/\mathfrak{f})$. *Then* $(g, \widetilde{\mathcal{E}}')$ *is an unraveler for* $(\widetilde{\mathrm{E}}_{\leqslant d})$, *and* $\mathfrak{e}$ *is the largest algebraic starting monomial of the unraveled asymptotic equation*

$$(\widetilde{\mathrm{E}}'_{\leqslant d}) \qquad (F_{\leqslant d})_{+g}(Y) = 0, \qquad Y \in \widetilde{\mathcal{E}}'.$$

PROOF. For $(g, \widetilde{\mathcal{E}}')$ to be an unraveler for $(\widetilde{\mathrm{E}}_{\leqslant d})$ it is enough to show:

(1) $\operatorname{ndeg}_{\widetilde{\mathcal{E}}'}(F_{\leqslant d})_{+g} = d$;

(2) $\operatorname{ndeg}_{\prec h}(F_{\leqslant d})_{+(g+h)} < d$ for all $h \in \widetilde{\mathcal{E}}'$.

Until further notice we assume $\mathfrak{g} = 1$, so $\mathfrak{e} \prec 1 \prec \mathfrak{f}$, $\mathfrak{e} \prec\!\!\prec \mathfrak{f}$. At the end we reduce to this special case. We have $\operatorname{ndeg} F = \operatorname{ndeg}_{\preccurlyeq \mathfrak{g}} F = d$ by an equality from the proof of Lemma 13.8.12, so $d \leqslant \operatorname{ndeg} F_{\times \mathfrak{f}} \leqslant \operatorname{ndeg}_{\mathcal{E}} F = d$, and thus $\operatorname{ndeg} F = \operatorname{ndeg} F_{\times \mathfrak{f}} = d$. If $\mathfrak{m} \prec\!\!\prec \mathfrak{f}$, then by Corollary 13.2.4 with $F, \mathfrak{f}$ in the role of $P, \mathfrak{n}$ we have

$$N_{P_{+f,\times \mathfrak{m}}} = N_{F_{+g,\times \mathfrak{m}}} = N_{(F_{\leqslant d})_{+g,\times \mathfrak{m}}}.$$

In particular, if $\mathfrak{e} \preccurlyeq \mathfrak{m} \prec 1$, then $N_{P_{+f,\times \mathfrak{m}}} = N_{(F_{\leqslant d})_{+g,\times \mathfrak{m}}}$, and so $\mathfrak{e}$ is the largest algebraic starting monomial of $(\widetilde{\mathrm{E}}'_{\leqslant d})$. Also,

$$\mathfrak{e} \preccurlyeq \mathfrak{m} \in \widetilde{\mathcal{E}}' \implies \operatorname{ndeg}(F_{\leqslant d})_{+g,\times \mathfrak{m}} = \operatorname{ndeg} P_{+f,\times \mathfrak{m}} = d,$$

so (1) holds. For (2), let $h \in \widetilde{\mathcal{E}}'$, so $h \in \mathcal{E}'$, $h \prec 1$, and $h \sim c\mathfrak{h}$ with $c \in C^\times$, $\mathfrak{h} := \mathfrak{d}_h$. Applying Lemma 13.5.4 twice gives

$$\begin{aligned} \operatorname{ndeg}_{\prec h}(F_{\leqslant d})_{+(g+h)} &= \operatorname{mul}\big(N_{(F_{\leqslant d})_{+g,\times \mathfrak{h}}}\big)_{+c} \\ \operatorname{ndeg}_{\prec h} P_{+(f+h)} &= \operatorname{mul}\big(N_{P_{+f,\times \mathfrak{h}}}\big)_{+c}. \end{aligned}$$

Now (E') is unraveled, so if $\mathfrak{e} \preccurlyeq \mathfrak{h}$, then $\operatorname{ndeg}_{\prec h} P_{+(f+h)} < d$, and thus

$$\operatorname{ndeg}_{\prec h}(F_{\leqslant d})_{+(g+h)} < d$$

by combining various equalities above. If $\mathfrak{e}^2 \preccurlyeq \mathfrak{h} \prec \mathfrak{e}$, then $\mathfrak{h}^\dagger \asymp \mathfrak{e}^\dagger$, so $\mathfrak{h} \prec\!\!\prec \mathfrak{f}$, hence

$$\operatorname{ndeg}\,(F_{\leqslant d})_{+g,\times \mathfrak{h}} = \operatorname{ndeg}\, P_{+f,\times \mathfrak{h}} < \operatorname{ndeg}\, P_{+f,\times \mathfrak{e}} = d,$$

as $\mathfrak{e}$ is the largest algebraic starting monomial for (E'). Thus if $\mathfrak{h} \prec \mathfrak{e}$, then

$$\operatorname{ndeg}_{\prec h}(F_{\leqslant d})_{+(g+h)} = \operatorname{mul}\big(N_{(F_{\leqslant d})_{+g,\times \mathfrak{h}}}\big)_{+c} \leqslant \operatorname{ndeg}\,(F_{\leqslant d})_{+g,\times \mathfrak{h}} < d.$$

This gives (2) when $\mathfrak{g} = 1$. To reduce to the case $\mathfrak{g} = 1$, replace $P, f, g, \mathcal{E}, \mathcal{E}'$ by $P_{\times \mathfrak{g}}$, $f/\mathfrak{g}, g/\mathfrak{g}, \mathfrak{g}^{-1}\mathcal{E}, \mathfrak{g}^{-1}\mathcal{E}'$, respectively, and use Lemma 13.8.6. □

13.9 SOME SPECIAL H-FIELDS

This section will not be used later in this volume but is included for its intrinsic interest. We assume familiarity with Appendix A. We construct here H-fields,

$$\mathbb{R}\langle\omega\rangle \subseteq \mathbb{R}\langle\lambda\rangle \subseteq \mathbb{R}\langle\gamma\rangle \text{ (with } H\text{-field inclusions)},$$

each generated as a differential field over their common constant field $\mathbb{R}$ by a single element, where $\mathbb{R}\langle\omega\rangle$ is λ-free but not ω-free, $\mathbb{R}\langle\lambda\rangle$ is not λ-free but has rational asymptotic integration, and $\mathbb{R}\langle\gamma\rangle$ has a gap $v\gamma$. These H-fields and their asymptotic couples have certain canonical features that are worth documenting. Moreover, they can be realized as Hardy fields as we show at the end of this section.

The ambient H-field $\mathbb{L}$

Let $\mathfrak{L}_n := \ell_0^{\mathbb{R}} \cdots \ell_n^{\mathbb{R}}$ be the subgroup of the ordered multiplicative group $\mathbb{T}^{>}$ generated by the real powers of the iterated logarithms ℓ_i for $i = 0, \dots, n$. This yields ordered group inclusions

$$\mathfrak{L}_0 \subseteq \mathfrak{L}_1 \subseteq \mathfrak{L}_2 \subseteq \cdots \subseteq \bigcup_n \mathfrak{L}_n =: \mathfrak{L} \subseteq G^{\text{LE}}.$$

We view $\mathbb{L} := \mathbb{R}[[\mathfrak{L}]]$ in the natural way as an ordered subfield of the ordered Hahn field $\mathbb{R}[[G^{\text{LE}}]]$. The latter also contains $\mathbb{T}$ as an ordered subfield, with

$$\mathbb{L} \cap \mathbb{T} = \bigcup_n \mathbb{R}[[\mathfrak{L}_n]] = \mathbb{T}_{\log}.$$

We equip $\mathbb{L}$ with the unique strongly additive $\mathbb{R}$-linear derivation such that

$$(\ell_0^r)' = r\,\ell_0^{r-1}, \qquad (\ell_{n+1}^r)' = r\,\ell_{n+1}^{r-1}(\ell_0 \cdots \ell_n)^{-1} \qquad (r \in \mathbb{R}).$$

This makes $\mathbb{L}$ a spherically complete immediate real closed H-field extension of $\mathbb{T}_{\log}$, with constant field $\mathbb{R}$. Thus $\mathbb{L}$ and $\mathbb{T}_{\log}$ have the same asymptotic couple $(v(\mathfrak{L}), \psi)$. Moreover, $v(\mathfrak{L})$ is an ordered vector space over $\mathbb{R}$. We set $e_n := v(\ell_n)$, so $e_n < 0$, $[e_n] > [e_{n+1}]$, and $v(\mathfrak{L}) = \bigoplus_n \mathbb{R}\, e_n$ (internal direct sum) with

$$[v(\mathfrak{L})^{\neq}] = \{[e_n] : n = 0, 1, 2, \dots\}, \qquad e_n^{\dagger} = -(e_0 + e_1 + \cdots + e_n).$$

Note that $\Psi := \psi(v(\mathfrak{L})^{\neq})$ has no supremum in $v(\mathfrak{L})$, so the H-asymptotic couple $(v(\mathfrak{L}), \psi)$ has rational asymptotic integration.

The elements λ and ω of $\mathbb{L}$

Let $\mathfrak{M}$ be the subgroup of $\mathfrak{L}$ generated by $\ell_0, \ell_1, \dots$, so $\mathfrak{M} = \bigcup_n \ell_0^{\mathbb{Z}} \cdots \ell_n^{\mathbb{Z}}$. The ordered subfield $\mathbb{R}[[\mathfrak{M}]]$ of $\mathbb{L}$ is closed under the derivation of $\mathbb{L}$, which makes it an H-subfield of $\mathbb{L}$. Now $\mathbb{R}[[\mathfrak{M}]]$ has special elements

$$\sum_{n=1}^{\infty} \ell_n, \quad \lambda := \left(\sum_{n=1}^{\infty} \ell_n\right)' = \sum_{n=0}^{\infty} (\ell_0 \cdots \ell_n)^{-1}, \qquad \omega := \omega(\lambda) = \sum_{n=0}^{\infty} (\ell_0 \cdots \ell_n)^{-2},$$

none lying in $\mathbb{T}_{\log}$. This gives the H-subfields $\mathbb{R}\langle\omega\rangle$ and $\mathbb{R}\langle\lambda\rangle$ of $\mathbb{R}[[\mathfrak{M}]]$.

PROPOSITION 13.9.1. *$\mathbb{R}[[\mathfrak{M}]]$ is an immediate extension of $\mathbb{R}\langle\omega\rangle$ and of $\mathbb{R}\langle\lambda\rangle$. Also, $\mathbb{R}\langle\omega\rangle$ is λ-free and not ω-free.*

Towards the proof, first note that the asymptotic couple $(v(\mathfrak{L}), \psi)$ of $\mathbb{L}$ extends the asymptotic couple $(v(\mathfrak{M}), \psi)$ of $\mathbb{R}[[\mathfrak{M}]]$, with $v(\mathfrak{M}) = \bigoplus_n \mathbb{Z}\, e_n$. It follows that $[v(\mathfrak{M})] = [v(\mathfrak{L})]$, the two asymptotic couples have the same Ψ-set, namely Ψ, and the H-asymptotic couple $(v(\mathfrak{M}), \psi)$ has rational asymptotic integration.

LEMMA 13.9.2. *Let $G \neq \{0\}$ be an ordered subgroup of $v(\mathfrak{M})$ such that $\psi(G^{\neq}) \subseteq G$ and $G^{<}$ is coinitial in $v(\mathfrak{M})^{<}$. Then $G = v(\mathfrak{M})$.*

PROOF. Note: if $[e_n] \in [G]$, then $e_n^\dagger \in G$. Suppose $m < n$ and $[e_m], [e_n] \in [G]$. Then $-e_m^\dagger = e_0 + \cdots + e_m \in G$ and $-e_n^\dagger = e_0 + \cdots + e_n \in G$, hence

$$e_{m+1} + \cdots + e_n = e_m^\dagger - e_n^\dagger \in G,$$

so $[e_{m+1}] = [e_{m+1} + \cdots + e_n] \in [G]$, and thus $e_{m+1}^\dagger \in G$. Inductively it follows that G contains $e_m^\dagger, e_{m+1}^\dagger, \dots, e_n^\dagger$ and hence G contains

$$e_{m+1} = e_m^\dagger - e_{m+1}^\dagger,\ e_{m+2} = e_{m+1}^\dagger - e_{m+2}^\dagger,\ \dots,\ e_n = e_{n-1}^\dagger - e_n^\dagger.$$

Take m with $[e_m] \in [G]$. Then $-\psi(e_m^\dagger) = e_0 \in G$. Thus $e_n \in G$ for all n. □

Since $\mathbb{T}_{\log}$ is ω-free, it follows from Proposition 13.6.4 that the H-subfield $\mathbb{T}_{\log}\langle\omega\rangle$ of $\mathbb{L}$ is λ-free. In order to conclude that $\mathbb{R}\langle\omega\rangle$ is λ-free, it is clearly enough to get $v(\mathbb{R}\langle\omega\rangle^\times) = v(\mathfrak{M})$, and that is part of Corollary 13.9.4 below. The H-field $\mathbb{L}$ has asymptotic integration, with divisible value group, and $\lambda_n \rightsquigarrow \lambda$, so $-\lambda$ creates a gap over $\mathbb{L}$, by Lemma 11.5.14. We take some Liouville closed H-field extension L of $\mathbb{L}$ and take $\gamma \in L^{>}$ with $\gamma^\dagger = -\lambda$. Then $v\gamma$ is a gap in $\mathbb{L}\langle\gamma\rangle$ by the remark following the proof of Lemma 11.5.14. We use this gap to prove:

LEMMA 13.9.3. $v(\mathbb{Q}\langle\lambda\rangle^\times) = v(\mathfrak{M})$, *and* $\mathbb{R}[[\mathfrak{M}]]$ *is an immediate extension of* $\mathbb{R}\langle\lambda\rangle$.

PROOF. Take $z \in L$ with $z' = \gamma$. Subtracting a constant in L from z we arrange $z \not\asymp 1$, and then $v(z^\dagger) > \Psi$. Let Δ be the value group of $\mathbb{Q}\langle z\rangle$. Then $\psi_L(\Delta^{\neq}) \subseteq \Delta$. We apply Lemma 9.2.19 to Δ as a subgroup of the value group of L. From $\lambda \in \mathbb{Q}\langle z\rangle$ we get $v(1/\lambda) - v(\ell_0) \in \Delta^{<}$, and as $v(\ell_0^\dagger) < v(z^\dagger) \in \psi_L(\Delta^{\neq})$, that lemma yields

$$v(\ell_1) \;=\; \chi\big(v(\ell_0)\big) \in \Delta.$$

As $v(\ell_1^\dagger) < v(z^\dagger)$, we likewise get $v(\ell_2) \in \Delta$. Continuing this way we get $v(\ell_n) \in \Delta$ for all $n \geqslant 1$. Hence $[\Delta]$ is infinite, and since $\operatorname{trdeg}\big(\mathbb{Q}\langle z\rangle|\mathbb{Q}\langle\lambda\rangle\big) \leqslant 2$, also $\big[v(\mathbb{Q}\langle\lambda\rangle^\times)\big]$ is infinite. Thus $v\big(\mathbb{Q}\langle\lambda\rangle^\times\big) = v(\mathfrak{M})$ by Lemma 13.9.2. □

COROLLARY 13.9.4. *Let E be a differential subfield of $\mathbb{R}\langle\lambda\rangle$ not contained in $\mathbb{R}$. Then $v(E^\times) = v(\mathfrak{M})$, and E as a pre-H-subfield of $\mathbb{R}\langle\lambda\rangle$ is not ω-free.*

PROOF. Take $a \in E$, $a \notin \mathbb{R}$. Then λ is d-algebraic over $\mathbb{R}\langle a\rangle$ by Lemma 4.1.5. By Lemma 13.9.3, the set $\big[v(\mathbb{R}\langle\lambda\rangle^\times)\big]$ is infinite, so $\big[v(\mathbb{R}\langle a\rangle^\times)\big]$ is infinite, hence $\big[v(\mathbb{Q}\langle a\rangle^\times)\big]$ is infinite by Corollary 10.5.17 and Lemma 2.4.4, and thus $v(E^\times) = v(\mathfrak{M})$ by Lemma 13.9.2. Since $\mathbb{R}\langle\lambda\rangle$ is not λ-free, it is not ω-free, and as $\mathbb{R}\langle\lambda\rangle$ is d-algebraic over E, Theorem 13.6.1 yields that E is not ω-free. □

Applying this to $E = \mathbb{R}\langle\omega\rangle$ yields Proposition 13.9.1.

Properties of $\mathbb{R}\langle\gamma\rangle$

Just before Lemma 13.9.3 we introduced a pre-H-field extension $\mathbb{L}\langle\gamma\rangle$ of $\mathbb{L}$. It is generated as a differential field over $\mathbb{L}$ by an element $\gamma > 0$ with $\gamma^\dagger = -\lambda$. Since $\mathbb{L}\langle\gamma\rangle = \mathbb{L}(\gamma)$, as fields, and $v\gamma \notin v(\mathfrak{L})$, we have $v\big(\mathbb{L}(\gamma)^\times\big) = v(\mathfrak{L}) \oplus \mathbb{Z}v\gamma$ (internal direct sum) by Lemma 3.1.30. It follows that $\mathbb{L}(\gamma)$ has the same residue field as $\mathbb{L}$, and so it is an H-field with the same constant field $\mathbb{R}$ as $\mathbb{L}$. We think of γ informally as an infinite product,

$$\gamma = \exp\left(-\sum_{n=1}^{\infty} \ell_n\right) = 1/\ell_0\ell_1\ell_2\cdots,$$

which suggests $v\gamma = -(e_0+e_1+e_2+\cdots)$. At this point we attach no formal meaning to these identities, but they suggest other identities that do have meaning and that are easy to prove. For example, let $\alpha = \big(\sum_{i=0}^{\infty} r_i e_i\big) + kv\gamma \in v\big(\mathbb{L}(\gamma)^\times\big)$, $\alpha \neq 0$, where $r_i \in \mathbb{R}$ for $i = 0, 1, \dots$, $r_i = 0$ for all but finitely many i, and $k \in \mathbb{Z}$. Then in the asymptotic couple of $\mathbb{L}\langle\gamma\rangle$ we have

$$\text{(13.9.1)} \qquad \alpha^\dagger = e_m^\dagger = -(e_0 + \cdots + e_m) \quad \text{where } m = \min\{i \in \mathbb{N} : r_i \neq k\}.$$

This is because in $\mathbb{L}\langle\gamma\rangle$ we have:

$$\left(\gamma^k \prod_{i=0}^{\infty} \ell_i^{r_i}\right)^\dagger = k\gamma^\dagger + \sum_{i=0}^{\infty} r_i \ell_i^\dagger = \sum_{i=0}^{\infty} (r_i - k)\frac{1}{\ell_0\ell_1\cdots\ell_i}.$$

We now turn to the H-subfield $\mathbb{R}\langle\gamma\rangle = \mathbb{R}\langle\lambda\rangle(\gamma)$ of $\mathbb{L}\langle\gamma\rangle$. Since $v\gamma$ is a gap in $\mathbb{L}\langle\gamma\rangle$, it is a gap in $\mathbb{R}\langle\gamma\rangle$. The asymptotic couple $\big(v(\mathfrak{M}), \psi\big)$ of $\mathbb{R}\langle\lambda\rangle$ has rational asymptotic integration, so $v\big(\mathbb{R}\langle\gamma\rangle^\times\big) = v(\mathfrak{M}) \oplus \mathbb{Z}v\gamma$ (internal direct sum) by Corollaries 3.1.11 and 9.8.6, and $\alpha^\dagger$ for $0 \neq \alpha \in v\big(\mathbb{R}\langle\gamma\rangle^\times\big)$ is given by (13.9.1).

Realizing $\mathbb{R}\langle\gamma\rangle$ as a Hardy field

The H-field $\mathbb{R}\langle\gamma\rangle$ is isomorphic over $\mathbb{R}$ to a Hardy field extension of $\mathbb{R}$, and thus the same holds for the H-subfields $\mathbb{R}\langle\omega\rangle$ and $\mathbb{R}\langle\lambda\rangle$ of $\mathbb{R}\langle\gamma\rangle$. To see this, recall that $\mathcal{G}$ is the ring of germs at $+\infty$ of one-variable real-valued functions defined on half-lines $(a, +\infty)$, $a \in \mathbb{R}$. Define $l_n, e_n \in \mathcal{G}$ by recursion on n such that $l_0(t) = e_0(t) = t$, $l_{n+1}(t) = \log l_n(t)$, and $e_{n+1}(t) = \exp e_n(t)$, eventually. Every Hardy field extends to one that contains all l_n, e_n, and all real numbers. Boshernitzan [58] constructs a Hardy field with an element e_ω such that for every n, eventually $e_\omega(t) > e_n(t)$. In particular, e_ω is eventually strictly increasing and $e_\omega(t) \to +\infty$ as $t \to +\infty$. Let l_ω be the inverse of e_ω: the germ in $\mathcal{G}$ such that $l_\omega\big(e_\omega(t)\big) = t$ eventually. By [378] (see also [21, Theorem 1.7]), l_ω lies in a Hardy field extension of $\mathbb{R}$, and for each n and $r \in \mathbb{R}$ we have eventually $r < l_\omega(t) < l_n(t)$. Then $g := l_\omega^\dagger$ lies in the same Hardy field extension of $\mathbb{R}$.

LEMMA 13.9.5. *There is an isomorphism $\mathbb{R}\langle\gamma\rangle \to \mathbb{R}\langle g\rangle$ of ordered differential fields which is the identity on $\mathbb{R}$ and sends γ to g. (Here $\mathbb{R}\langle g\rangle$ is a Hardy field.)*

PROOF. Consider the H-fields

$$E := \mathbb{R}(\ell_0, \ell_1, \dots) \subseteq \mathbb{R}[[\mathfrak{M}]], \qquad F := \mathbb{R}(l_0, l_1, \dots),$$

the latter a Hardy field extension of $\mathbb{R}$. Lemma 10.2.3 and the remarks following it yield an isomorphism $h\colon E \to F$ of ordered differential fields that is the identity on $\mathbb{R}$ and sends ℓ_n to l_n, for each n. Since E is d-valued of H-type with small derivation and ω-free, and has monomial group $\mathfrak{M}$, both E and F may be used in place of K in Section 13.4. Since $\Psi_E < v\gamma < (\Gamma_E^>)'$ and $\Psi_F < vg < (\Gamma_F^>)'$, it follows from Lemma 13.4.11 that h extends to an isomorphism $E\langle\gamma\rangle \to F\langle g\rangle$ of valued differential fields sending γ to g. Since $\gamma > 0$ and $g > 0$, the proof of Corollary 13.4.12 shows that this isomorphism is also order preserving. □

Notes and comments

Rosenlicht [368, p. 831] states that he has no example of a Hardy field extension $K \subseteq K(u)$ such that $u \neq 0$, $u^\dagger \in K$, and $v(u^\dagger)$ is a gap in K, except when $K \subset \mathbb{R}$. We note here that $K = \mathbb{R}\langle g\rangle$ as in Lemma 13.9.5 and $u = e^{\int g}$ in a Hardy field extension of K furnish such an example.

Chapter Fourteen

Newtonian Differential Fields

In this chapter K is an ungrounded H-asymptotic field with $\Gamma := v(K^\times) \neq \{0\}$. So the subset Ψ of Γ is nonempty and has no largest element, and thus K is pre-d-valued by Corollary 10.1.3. We let ϕ range over the active elements of K. Since Ψ^ϕ contains positive elements, the differential residue field $\boldsymbol{k}^\phi$ of K^ϕ is just the residue field $\boldsymbol{k}$ of the valued field K with the trivial derivation. As $\boldsymbol{k}^\phi$ does not depend on ϕ, we let $\boldsymbol{k}$ stand for $\boldsymbol{k}^\phi$. We also fix a "monomial" set $\mathfrak{M} \subseteq K^\times$ that is mapped bijectively by v onto Γ and which gives us the dominant monomial $\mathfrak{d}_P \in \mathfrak{M} \cup \{0\}$ and the dominant part $D_P \in \boldsymbol{k}\{Y\}$ of any $P \in K^\phi\{Y\}$. (We do not require $\mathfrak{M}$ to be a monomial group of K; such a group might not even exist.)

By an *extension* of K we mean an H-asymptotic field extension of K.

Our main interest is in the case that K has asymptotic integration, but then K cannot be d-henselian in the sense of Chapter 7, by Corollary 9.4.10. The correct notion in that situation, called *newtonian,* is an eventual variant of d-*henselian.* We define $P \in K\{Y\}$ to be **quasilinear** if $\operatorname{ndeg} P = 1$, and we define K to be **newtonian** if every quasilinear $P \in K\{Y\}$ has a zero in the valuation ring $\mathcal{O}$ of K. If K is newtonian, then K is henselian as a valued field, by Lemma 3.3.10, and K^f is newtonian for every $f \in K^\times$. In Section 14.1 we show that for λ-free K,

$$K \text{ is newtonian} \iff \text{for each } \phi \text{ the flattening of } K^\phi \text{ is d-henselian,}$$

and derive some consequences of this link between newtonianity and differential-henselianity. In Section 14.2 we consider weak forms of newtonianity and apply this to $P \in K\{Y\}$ of low complexity. Among results needed later we show there that if K is a newtonian Liouville closed H-field, then the subset $\omega(K)$ of K is downward closed, and the subset $\sigma\big(\Gamma(K)\big)$ of K is upward closed.

In Section 14.3 we prove newtonian versions of d-henselian results in Chapter 7, leading to the following important analogue of Theorem 7.0.1:

THEOREM 14.0.1. *If K is λ-free and asymptotically* d-*algebraically maximal, then K is ω-free and newtonian.*

One (minor) part of this is an immediate consequence of Corollary 11.7.10: if K is λ-free and asymptotically d-algebraically maximal, then K is ω-free. Note also that by Zorn any ω-free K has an immediate asymptotically d-algebraically maximal d-algebraic extension, and that by Section 13.6 any such extension is also ω-free, and thus newtonian by Theorem 14.0.1.

The main result of this chapter is almost a converse to Theorem 14.0.1:

THEOREM 14.0.2. *If K is ω-free and newtonian with divisible value group, then K is asymptotically* d*-algebraically maximal.*

This is key to eliminating quantifiers for $\mathbb{T}$ in Chapter 16. After a rather technical Section 14.4 on unraveling, we prove Theorem 14.0.2 in Section 14.5.

14.1 RELATION TO DIFFERENTIAL-HENSELIANITY

We let a, b, y range over K, and P over $K\{Y\}^{\neq}$. Recall that throughout this chapter ϕ ranges over the active elements of K.

LEMMA 14.1.1. *Suppose $\Psi^{>0} \neq \emptyset$, P has order $\leqslant r$, $v(P_1) = 0$ and $v(P_i) > r\Psi$ for all $i \neq 1$. Then $D_{P^\phi} = D_{P_1^\phi}$ for all $\phi \preccurlyeq 1$, and thus* $\operatorname{ndeg} P = 1$.

PROOF. For $\phi \preccurlyeq 1$ in K we have $v(P_1^\phi) \leqslant rv\phi < v(P_i^\phi)$ for $i \neq 1$. □

If $\Psi^{>0} \neq \emptyset$, then we let 1 denote the unique element of $\Gamma^>$ with $\psi(1) = 1$, and we identify $\mathbb{Z}$ with a subgroup of Γ via $k \mapsto k \cdot 1$.

LEMMA 14.1.2. *Suppose K is newtonian with $\Psi^{>0} \neq \emptyset$ and Δ is a convex subgroup of Γ with $1 \in \Delta$. Then (K, v_Δ) is* d*-henselian.*

PROOF. Let $\dot{\mathcal{O}}$ be the valuation ring of $\dot{v} = v_\Delta$. Let $P \in \dot{\mathcal{O}}\{Y\}$ of order $\leqslant r$ be such that $P_1 \mathrel{\dot{\asymp}} 1$ and $P_i \mathrel{\dot{\prec}} 1$ for $i \geqslant 2$; by Lemma 7.2.1 it is enough to show that P has a zero in $\dot{\mathcal{O}}$. We can arrange $P_1 \asymp 1$. Take $\gamma < 0$ in Δ such that $(\gamma/2)+(r+1) < v(P_0)$. Take $g \in K^\times$ with $vg = \gamma$. Then $P_{\times g} = P_0 + L + R$ with $L = P_{1,\times g}$ and $R \mathrel{\dot{\prec}} 1$, so $vL = \gamma + o(\gamma)$. Take a with $va = -v(L)$. Then $Q := aP_{\times g} = aP_0 + aL + aR$ with

$$v(aP_0) = -\gamma + o(\gamma) + v(P_0) > r + 1 > r\Psi, \quad v(aL) = 0, \quad aR \mathrel{\dot{\prec}} 1,$$

so $\operatorname{ndeg} Q = 1$ by Lemma 14.1.1. Since K is newtonian, Q has a zero $y \in \mathcal{O}$, and then gy is a zero of P in $\dot{\mathcal{O}}$. □

Lemma 14.1.2 and its proof go through if the assumption that K is newtonian is replaced by: every $P \in K\{Y\}^{\neq}$ with $\operatorname{ndeg} P = 1$ has a zero in the valuation ring $\dot{\mathcal{O}}$ of v_Δ. Let $(K^\phi, v_\phi^\flat)$ be the differential field K^ϕ with valuation $v_\phi^\flat$. Then:

LEMMA 14.1.3. *Let $\phi \preccurlyeq \theta \in K$ (so θ is active). If $(K^\phi, v_\phi^\flat)$ is* d*-henselian, then so is $(K^\theta, v_\theta^\flat)$.*

PROOF. First, $(K^\phi, v_\theta^\flat)$ is isomorphic to a coarsening of $(K^\phi, v_\phi^\flat)$. Now $K^\theta = (K^\phi)^u$ with $u = \theta/\phi \asymp_\theta^\flat 1$, by Lemma 6.5.4(i). It remains to appeal to Lemma 7.3.4 and the subsection on compositional conjugation immediately preceding it. □

Thus the following two conditions on K are equivalent:

(1) for every active θ in K there is a $\phi \preccurlyeq \theta$ such that $(K^\phi, v_\phi^\flat)$ is d-henselian;

(2) $(K^\phi, v_\phi^\flat)$ is d-henselian, for every ϕ.

We already saw that these conditions are satisfied if K is newtonian. If K is λ-free we can reverse this implication:

LEMMA 14.1.4. *Suppose K is λ-free and $(K^\phi, v^\flat_\phi)$ is* d*-henselian, for every ϕ. Then K is newtonian.*

PROOF. Let $P \in K\{Y\}^{\neq}$ and $\operatorname{ndeg} P = 1$; we need to show that then P has a zero in $\mathcal{O}$. By Corollary 9.5.8 we can take $\gamma > 0$ such that P has no zero in the region $-\gamma < vy < 0$. By Lemma 13.7.10 we can take ϕ such that $\Gamma^\flat_\phi \subseteq (-\gamma, \gamma)$, and either $P^\phi \sim^\flat_\phi a + bY$ in $K^\phi\{Y\}$, $a \preccurlyeq b$ in K, or $P^\phi \sim^\flat_\phi bY'$ in $K^\phi\{Y\}$, $b \in K$. Then P^ϕ has a zero $y \preccurlyeq^\flat_\phi 1$ by Lemma 7.1.1, and the choice of γ and ϕ gives $y \in \mathcal{O}$. □

COROLLARY 14.1.5. *Suppose K is λ-free and newtonian, and* $\operatorname{cf}(\Gamma) = \omega$. *Then the completion K^c of K is also λ-free and newtonian.*

PROOF. Let a flattening $(K^\phi, v^\flat_\phi)$ of K be given with $v\phi \in \Psi$. It is d-henselian. By Lemma 3.4.4 it has as a completion the flattening $\big((K^c)^\phi, v^\flat_\phi\big)$, so the latter is d-henselian by Proposition 7.2.15. Now K^c is λ-free by Lemma 11.6.5, so the desired conclusion follows from Lemma 14.1.4 applied to K^c instead of K. □

In Chapter 16 we show that the ω-free newtonian Liouville closed H-fields are exactly the existentially closed H-fields; see Appendix B for the concept *existentially closed.* This makes the next result interesting, in particular for $K = \mathbb{T}$.

COROLLARY 14.1.6. *If K is an ω-free newtonian Liouville closed H-field with* $\operatorname{cf}(\Gamma) = \omega$, *then K^c is also an ω-free newtonian Liouville closed H-field.*

This is immediate from Lemmas 10.6.5 and 11.7.20, and Corollary 14.1.5.

Preparing for newtonization

We begin with an analogue of Lemma 7.5.5:

LEMMA 14.1.7. *Let $r \geqslant 1$, suppose K is newtonian, and let $G \in K\{Y\} \setminus K$ have order $\leqslant r$. Then there do not exist $y_0, \dots, y_{r+1} \in K$ such that*

(i) $y_{i-1} - y_i \succ y_i - y_{i+1}$ *for all* $i \in \{1, \dots, r\}$, *and* $y_r \neq y_{r+1}$;

(ii) $G(y_0) = \cdots = G(y_{r+1}) = 0$;

(iii) $\operatorname{ndeg} G_{+y_{r+1}, \times g} = 1$ *and* $y_0 - y_{r+1} \preccurlyeq g$ *for some* $g \in K^\times$.

PROOF. Towards a contradiction, suppose $y_0, \dots, y_{r+1} \in K$ satisfy (i), (ii), (iii). By taking ϕ with sufficiently high $v\phi$ we arrange that $y_{i-1} - y_i \succ^\flat_\phi y_i - y_{i+1}$ for all $i \in \{1, \dots, r\}$ and $\operatorname{ddeg} G^\phi_{+y_{r+1}, \times g} = 1$, where g witnesses (iii). But $(K^\phi, v^\flat_\phi)$ is d-henselian, and so we contradict Lemma 7.5.5. □

This yields also the newtonian version of Proposition 7.5.6:

LEMMA 14.1.8. *Suppose K is newtonian and $G \in K\{Y\}$ satisfies* $\operatorname{ndeg} G = 1$. *Let E be an immediate extension of K. Then G has the same zeros in $\mathcal{O}$ as in $\mathcal{O}_E$.*

PROOF. Note first that $\operatorname{ndeg} G_{+y} = 1$ for all $y \in \mathcal{O}_E$. Towards a contradiction, suppose $G(\ell) = 0$ with $\ell \in \mathcal{O}_E \setminus \mathcal{O}$. Now $\ell \preccurlyeq 1$ gives $\operatorname{ndeg} G_{+\ell} = 1$, and from $G(\ell) = 0$ it follows easily that $\operatorname{ndeg} G_{+\ell,\times g} = 1$ for all $g \preccurlyeq 1$ in K.

CLAIM: *Let $\gamma \in v(\ell - K)$, $\gamma \geqslant 0$. Then $G(y) = 0$ for some $y \in \mathcal{O}$ with $v(\ell - y) \geqslant \gamma$.*

To prove this claim, take $a \in K$ and $g \in K^{\times}$ such that $v(\ell - a) = v(g) = \gamma$. Then by Corollary 11.2.4 and the observation preceding the claim,

$$\operatorname{ndeg} G_{+a,\times g} = \operatorname{ndeg} G_{+\ell,\times g} = 1,$$

so we get $b \in \mathcal{O}$ such that $G(a + gb) = 0$, so $y := a + gb$ satisfies the claim. Taking $r \geqslant 1$ with G of order $\leqslant r$, the claim yields $y_0, \dots, y_r, y_{r+1} \in \mathcal{O}$ such that

$$\ell - y_0 \succ \ell - y_1 \succ \cdots \succ \ell - y_{r+1}, \quad G(y_0) = G(y_1) = \cdots = G(y_{r+1}) = 0,$$

contradicting Lemma 14.1.7: take $g = 1$ in (iii). □

COROLLARY 14.1.9. *If K has a newtonian immediate extension, then K has a newtonian immediate extension L such that:*

(i) *L is* d*-algebraic over K;*

(ii) *no proper differential subfield of L containing K is newtonian.*

PROOF. Suppose F is a newtonian immediate extension of K. Let L be the intersection inside F of the collection of all differential subfields E of F that contain K and are newtonian. Applying Lemma 14.1.8 to these extensions $E \subseteq F$ shows that L is newtonian. That (i) holds is because the differential subfield $E \supseteq K$ of F consisting of all $y \in F$ that are d-algebraic over F is newtonian. It is obvious that (ii) holds. □

The condition that an extension L of K is newtonian includes L being ungrounded (which is automatic if $L|K$ is immediate). A **newtonization** of K is a newtonian extension of K that embeds over K into every newtonian extension of K. At this stage this is just a definition. In Section 14.5 we can say more.

Strong newtonianity

This notion will be useful in Section 14.5.

LEMMA 14.1.10. *Let K be newtonian, (a_ρ) a pc-sequence in K, $G \in K\{Y\}^{\neq}$,* $\operatorname{ndeg}_{\boldsymbol{a}} G = 1$, *and $\boldsymbol{a} := c_K(a_\rho)$. Then $G(a) = 0$ and $a_\rho \rightsquigarrow a$ for some $a \in K$.*

PROOF. We can assume that we have a strictly increasing sequence (γ_ρ) in Γ such that $v(a_\sigma - a_\rho) = \gamma_\rho$ for all $\sigma > \rho$. For each ρ take $g_\rho \in K$ with $v(g_\rho) = \gamma_\rho$. We can further assume that $\operatorname{ndeg} G_{+a_\rho,\times g_\rho} = 1$ for all ρ. Then we get for each ρ an element $z_\rho \in K$ with $G(z_\rho) = 0$ and $z_\rho - a_\rho \preccurlyeq g_\rho$. Let

$$B_\rho := \{z \in K : v(z - a_\rho) \geqslant \gamma_\rho\}$$

be the closed ball in K centered at a_ρ with valuation radius γ_ρ, so $z_\rho \in B_\rho$. If $\rho < \sigma$ and $z \in B_\sigma$, then $v(z - a_\sigma) \geqslant \gamma_\sigma > \gamma_\rho = v(a_\sigma - a_\rho)$ and so $v(z - a_\rho) = \gamma_\rho$; in particular $B_\rho \supseteq B_\sigma$ whenever $\rho \leqslant \sigma$.

CLAIM: *There is an index ρ_0 such that $z_{\rho_0} \in B_\rho$ for all $\rho \geqslant \rho_0$.*

Suppose not. Then we get an infinite sequence $\rho_0 < \rho_1 < \cdots$ of indices such that $y_m := z_{\rho_m} \notin B_{\rho_n}$ for all $m < n$. Taking each y_m as center of B_{ρ_m} we see that

$$v(y_m - y_{m+1}) < v(y_{m+1} - y_{m+2})$$

for all m. Take $r \geqslant 1$ such that G has order $\leqslant r$. Then conditions (i) and (ii) of Lemma 14.1.7 are satisfied. Set $g := g_{\rho_0}$. Then also $y_0 - y_{r+1} \preccurlyeq g$ and $a_{\rho_0} - y_{r+1} \preccurlyeq g$. In view of Corollary 11.2.4 the latter gives $\operatorname{ndeg} G_{+y_{r+1},\times g} = \operatorname{ndeg} G_{+a_{\rho_0},\times g} = 1$, so condition (iii) in Lemma 14.1.7 also holds, which contradicts that lemma.

This proves the claim. Let ρ_0 be an index as in the claim; then $v(z_{\rho_0} - a_\rho) = \gamma_\rho$ for all $\rho \geqslant \rho_0$ and thus $a_\rho \rightsquigarrow z_{\rho_0}$, and so $a := z_{\rho_0}$ has the required property. ☐

Define K to be **strongly newtonian** if it is newtonian and for every divergent pc-sequence (a_ρ) in K with minimal d-polynomial $G(Y)$ over K we have $\operatorname{ndeg}_{\boldsymbol{a}} G = 1$, where $\boldsymbol{a} = c_K(a_\rho)$. For this notion we have an analogue of Theorem 7.0.3:

LEMMA 14.1.11. *Suppose K has rational asymptotic integration and is strongly newtonian. Then K is asymptotically* d*-algebraically maximal.*

PROOF. Towards a contradiction, assume K has a proper immediate d-algebraic extension. Then Lemma 11.3.8 yields a divergent pc-sequence (a_ρ) in K with a minimal d-polynomial $G(Y)$ over K. This contradicts Lemma 14.1.10. ☐

Quasilinear asymptotic equations

In this subsection $\mathcal{E} \subseteq K^\times$ is $\preccurlyeq$-closed.

LEMMA 14.1.12. *If K is newtonian and $\operatorname{ndeg}_{\mathcal{E}} P = 1$, then P has a zero in $\mathcal{E} \cup \{0\}$.*

PROOF. Assume K is newtonian and $\operatorname{ndeg}_{\mathcal{E}} P = 1$. Take $g \in \mathcal{E}$ with $\operatorname{ndeg} P_{\times g} = 1$. Then $P_{\times g}$ has a zero in $\mathcal{O}$, and thus P has a zero in $g\mathcal{O} \subseteq \mathcal{E} \cup \{0\}$. ☐

Next we consider an asymptotic equation

(E) $$P(Y) = 0, \qquad Y \in \mathcal{E}$$

over K. We say that (E) is **quasilinear** if $\operatorname{ndeg}_{\mathcal{E}} P = 1$. Recall from Section 11.2 the notion of a solution y of (E) best approximating a given element f of a valued differential field extension of K.

LEMMA 14.1.13. *Suppose K is newtonian, and* (E) *is quasilinear and has a solution. Let f be an element of a valued differential field extension of K. Then f is best approximated by some solution of* (E).

PROOF. We may assume that $f \not\preccurlyeq \mathcal{E}$. Take $\mathfrak{m} \in \mathcal{E}$ such that $f \preccurlyeq \mathfrak{m}$, $\operatorname{ndeg} P_{\times\mathfrak{m}} = \operatorname{ndeg}_{\mathcal{E}} P = 1$, and (E) has a solution $y \preccurlyeq \mathfrak{m}$. By Lemma 11.2.10 we may replace P by $P_{\times\mathfrak{m}}$ and $\mathcal{E}$ by $\mathcal{O}^{\neq}$, and thus assume $\mathcal{E} = \mathcal{O}^{\neq}$. Suppose f is not best approximated by any solution of (E). Then we get an infinite sequence $y_0, y_1, y_2, \dots$ of solutions of (E) with $y_0 - f \succ y_1 - f \succ y_2 - f \succ \cdots$. For each i we have $\operatorname{ndeg} P_{+y_i} = \operatorname{ndeg} P = 1$, and this leads to a contradiction with Lemma 14.1.7. □

LEMMA 14.1.14. *Suppose K is newtonian,* (E) *is quasilinear, and $f \in \mathcal{E}$ satisfies $\operatorname{ndeg}_{\prec f} P_{+f} \geqslant 1$. Then* (E) *has a solution $y \sim f$, and every solution y of* (E) *that best approximates f satisfies $y \sim f$.*

PROOF. Since (E) is quasilinear and $f \in \mathcal{E}$, we have

$$\operatorname{ndeg}_{\prec f} P_{+f} \leqslant \operatorname{ndeg}_{\mathcal{E}} P_{+f} = \operatorname{ndeg}_{\mathcal{E}} P = 1,$$

so $\operatorname{ndeg}_{\prec f} P_{+f} = 1$. By Lemma 14.1.12 we get $z \prec f$ in K with $P(f+z) = 0$, so $f + z$ is a solution of (E) with $f + z \sim f$. Given any solution y of (E) that best approximates f, we get $y - f \preccurlyeq (f+z) - f = z \prec f$, so $y \sim f$. □

14.2 CASES OF LOW COMPLEXITY

In this section we consider weak forms of newtonianity and differential polynomials of low complexity: degree 1 or order at most 2. At the end we apply this to the differential polynomial function ω and to the related function σ. Throughout this section, r ranges over $\mathbb{N}$, a, b, y over K, and γ over Γ.

Define K to be **r-newtonian** if every quasilinear $P \in K\{Y\}$ of order $\leqslant r$ has a zero in $\mathcal{O}$. Define K to be **$(1,1)$-newtonian** if every quasilinear $P \in K[Y, Y']$ of degree $\leqslant 1$ in Y' has a zero in $\mathcal{O}$. Define K to be **r-linearly newtonian** if every quasilinear $P \in K\{Y\}$ with $\deg P = 1$ and $\operatorname{order}(P) \leqslant r$ has a zero in $\mathcal{O}$. Define K to be **linearly newtonian** if K is r-linearly newtonian for every r. Each of these conditions on K is clearly invariant under compositional conjugation by any $f \in K^{\times}$. These notions are mainly used for $r = 1$ and $r = 2$. Thus for K we have:

$$\begin{array}{ccccccccc}
\text{newtonian} & \Longrightarrow & \text{2-newtonian} & \Rightarrow & \text{1-newtonian} & \Rightarrow & (1,1)\text{-newtonian} & \Rightarrow & \text{henselian} \\
\Downarrow & & \Downarrow & & & & \Downarrow & & \\
\text{linearly newtonian} & \Rightarrow & \text{2-linearly newtonian} & & \Longrightarrow & & \text{1-linearly newtonian} & &
\end{array}$$

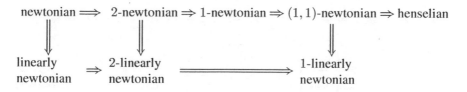

Linear newtonianity

Recall from Section 7.2 the notion of a valuation ring of a valued differential field with small derivation being linearly surjective. The following result relates this notion to linear newtonianity:

LEMMA 14.2.1. *Suppose K is r-linearly newtonian with $\Psi^{>0} \neq \emptyset$ and Δ is a convex subgroup of Γ with $1 \in \Delta$. Then the valuation ring of (K, v_Δ) is r-linearly surjective.*

PROOF. Like that of Lemma 14.1.2, but simpler since here $R = 0$. □

The conclusion of Lemma 14.2.1 implies in particular the r-linear surjectivity of K.

COROLLARY 14.2.2. *If K is r-linearly newtonian, then K is r-linearly surjective.*

PROOF. Apply the above to any compositional conjugate K^ϕ of K. □

In combination with Proposition 11.6.17 this yields:

COROLLARY 14.2.3. *If K has asymptotic integration and is 1-linearly newtonian, then K is λ-free.*

Asymptotic integrability plus 1-linear newtonianity has further nice consequences:

LEMMA 14.2.4. *If K has asymptotic integration, the following are equivalent:*

(i) *K is 1-linearly newtonian;*

(ii) *every $P \in K\{Y\}$ with $\operatorname{nmul} P = \deg P = 1$ and $\operatorname{order} P \leqslant 1$ has a zero in $\smallo$.*

PROOF. Assume K has asymptotic integration. Then $\Gamma^{>}$ has no least element.

To prove (i) ⇒ (ii), let K be 1-linearly newtonian and let $P \in K\{Y\}$ be such that $\operatorname{nmul} P = \deg P = 1$, $\operatorname{order} P \leqslant 1$. By the remark following the proof of Proposition 11.2.14 we have $g_0 \prec 1$ in K such that $\operatorname{nmul} P_{\times g} = \operatorname{ndeg} P_{\times g} = 1$ for all $g \in K$ with $g_0 \prec g \prec 1$. For such g we have a zero of P in $g\mathcal{O} \subseteq \smallo$.

Next, assume (ii), and let $P \in K\{Y\}$, $\operatorname{ndeg} P = \deg P = 1$, $\operatorname{order} P \leqslant 1$; our job is to show that P has a zero in $\mathcal{O}$. By the remark following the proof of Proposition 11.2.14 we have $f_0 \succ 1$ in K such that $\operatorname{nmul} P_{\times f} = \operatorname{ndeg} P_{\times f} = 1$ for all $f \in K$ with $f_0 \succ f \succ 1$. By Corollary 9.5.8 we can take f_0 such that also $P(y) \neq 0$ for all y with $f_0 \succ y \succ 1$. Take any $f \in K$ with $f_0 \succ f \succ 1$, and take a zero $y \in \smallo$ of $P_{\times f}$. Then $P(fy) = 0$, and so $fy \in \mathcal{O}$, since otherwise $f_0 \succ fy \succ 1$. □

In Section 11.8 we defined the set $\mathrm{I}(E) \subseteq E$ for pre-H-fields E. We now do this for any asymptotic field E by

$$\mathrm{I}(E) := \{f \in E : f \preccurlyeq g' \text{ for some } g \in \mathcal{O}_E\}.$$

Then $\mathrm{I}(E)$ is an $\mathcal{O}_E$-submodule of E with $\partial\mathcal{O}_E \subseteq \mathrm{I}(E)$ and $(\mathcal{O}_E^\times)^\dagger \subseteq \mathrm{I}(E)$.

LEMMA 14.2.5. *Assume K has asymptotic integration and is 1-linearly newtonian. Then K is* d*-valued and $\partial\smallo = \mathrm{I}(K) = (1+\smallo)^\dagger$.*

PROOF. Let $a \in \mathrm{I}(K)$, and $P := Y' - a$. Then $P^{\phi} = \phi Y' - a$, so $\operatorname{nmul} P = 1$, and thus P has a zero $y \in \mathcal{o}$ by Lemma 14.2.4. This gives $\partial\mathcal{o} = \mathrm{I}(K)$, and so K is d-valued. Next, take $Q := Y' - (a + aY)$. Then $Q^{\phi} = \phi Y' - (a + aY)$, so $\operatorname{nmul} Q = 1$, and thus Q has a zero $y \in \mathcal{o}$ by Lemma 14.2.4, and then $a = y'/(1+y) = (1+y)^{\dagger}$. This proves $\mathrm{I}(K) = (1+\mathcal{o})^{\dagger}$. □

Here is a generalization of Lemma 14.2.4:

LEMMA 14.2.6. *Suppose K has asymptotic integration and $r \geqslant 1$. Then the two conditions below are equivalent:*

(i) *K is r-linearly newtonian;*

(ii) *every $P \in K\{Y\}$ with* $\operatorname{nmul} P = \deg P = 1$ *and* $\operatorname{order} P \leqslant r$ *has a zero in $\mathcal{o}$.*

PROOF. By Corollary 14.2.3, (i) implies K is λ-free, and using also Lemma 14.2.4, so does (ii). Now follow the proof of Lemma 14.2.4, using Corollary 13.7.11 instead of the remark following the proof of Proposition 11.2.14. □

Application to linear differential equations

We determine here the dimension of the kernel of a linear differential operator over K for linearly newtonian d-valued K. The results in this subsection will not be used in this volume.

Let $A = a_0 + a_1\partial + \cdots + a_r\partial^r \in K[\partial]^{\neq}$ where $a_0, \dots, a_r \in K$, $a_r \neq 0$. In Section 11.1 we defined the set

$$\mathscr{E}^{\mathrm{e}}(A) \;=\; \{\gamma : \operatorname{nwt}_A(\gamma) \geqslant 1\} \;=\; \bigcap_{\phi} \mathscr{E}(A^{\phi})$$

of eventual exceptional values of A. If K is λ-free, then by Lemma 13.7.10 we have $\operatorname{nwt}_A(\gamma) = 1$ for all $\gamma \in \mathscr{E}^{\mathrm{e}}(A)$. In Section 11.1 we also defined the map $v^{\mathrm{e}}_A : \Gamma \to \Gamma$, and mentioned the fact that

$$v_{A^{\phi}}(\gamma) \;=\; v^{\mathrm{e}}_A(\gamma) + \operatorname{nwt}_A(\gamma)v\phi, \quad \text{eventually.} \tag{14.2.1}$$

Thus if $\gamma \notin \mathscr{E}^{\mathrm{e}}(A)$, then eventually $v_{A^{\phi}}(\gamma) = v^{\mathrm{e}}_A(\gamma)$.

LEMMA 14.2.7. *The map $\gamma \mapsto v^{\mathrm{e}}_A(\gamma) : \Gamma \setminus \mathscr{E}^{\mathrm{e}}(A) \to \Gamma$ is strictly increasing, and if K is ω-free, then this map is surjective.*

PROOF. The first part holds because $v_{A^{\phi}}$ is strictly increasing for each ϕ. Suppose K is ω-free and let $\alpha \in \Gamma$. Corollary 13.6.10 yields γ such that $v_{A^{\phi}}(\gamma) = \alpha$ eventually. Then $\operatorname{nwt}_A(\gamma) = 0$ by (14.2.1), so $\gamma \notin \mathscr{E}^{\mathrm{e}}(A)$ and $v^{\mathrm{e}}_A(\gamma) = \alpha$. □

We have $v(\ker^{\neq} A) \subseteq \mathscr{E}^{\mathrm{e}}(A)$; moreover:

PROPOSITION 14.2.8. *Suppose K is r-linearly newtonian and K has asymptotic integration. Then $v(\ker^{\neq} A) = \mathscr{E}^{\mathrm{e}}(A)$.*

PROOF. The case $r = 0$ is trivial, so assume $r \geqslant 1$. Let $\gamma \in \mathscr{E}^{\mathrm{e}}(A)$, and take $g \in K^{\times}$ with $vg = \gamma$; our job is to find $y \in \ker A$ with $y \asymp g$. Replacing A by Ag we arrange $\gamma = 0$ and $g = 1$. Since K is λ-free, we have $\mathrm{nwt}(A) = 1$. Put $P := a_0Y + a_1Y' + \cdots + a_rY^{(r)} \in K\{Y\}$. Then $D_{P^{\phi}} \in \boldsymbol{k}^{\times} \cdot Y'$, so by Lemma 6.6.5(i),

$$D_{P^{\phi}_{+1}} \in \boldsymbol{k}^{\times} \cdot (D_{P^{\phi}})_{+1} \;=\; \boldsymbol{k}^{\times} \cdot Y',$$

and hence $\mathrm{nmul}\, P_{+1} = 1$. By Lemma 14.2.6 we get $z \in \mathcal{o}$ with $P(1+z) = 0$, so we can take $y := 1 + z$. □

Corollary 14.2.2, Proposition 14.2.8, and Lemma 5.6.6 yield:

COROLLARY 14.2.9. *If K is* d*-valued and r-linearly newtonian, then*

$$\dim_C \ker A \;=\; |\mathscr{E}^{\mathrm{e}}(A)|.$$

Suppose K is d-valued. If K has an immediate r-linearly newtonian extension, then $|\mathscr{E}^{\mathrm{e}}(A)| \leqslant r$, by Corollary 14.2.9. Thus $|\mathscr{E}^{\mathrm{e}}(A)| \leqslant r$ if K is ω-free, by the remark following Theorem 14.0.1 (to be proved in Section 14.3). Also, if K is r-linearly newtonian and L is an immediate extension of K, then $\ker_L A = \ker A$.

PROPOSITION 14.2.10. *Suppose K is ω-free and r-linearly newtonian. Then for every $a \neq 0$ there is $y \neq 0$ such that $A(y) = a$, $vy \notin \mathscr{E}^{\mathrm{e}}(A)$, and $v^{\mathrm{e}}_A(vy) = va$.*

PROOF. It is enough to do the case $a = 1$. Lemma 14.2.7 gives $g \in K^{\times}$ with $vg \notin \mathscr{E}^{\mathrm{e}}(A)$ and $v^{\mathrm{e}}_A(vg) = 0$. Replacing A by Ag we arrange $0 \notin \mathscr{E}^{\mathrm{e}}(A)$ and $v^{\mathrm{e}}(A) = 0$; our job is now to find $y \asymp 1$ with $A(y) = 1$. Eventually, $\mathrm{dwt}(A^{\phi}) = 0$ and $v(A^{\phi}) = 0$. Hence eventually $A^{\phi}(1) \asymp 1$ and thus $a_0 = A(1) = A^{\phi}(1) \asymp 1$. Put $P := a_0Y + a_1Y' + \cdots + a_rY^{(r)} \in K\{Y\}$. Eventually $P^{\phi} \sim a_0Y$ in $K^{\phi}\{Y\}$, so eventually $P^{\phi}_{+(1/a_0)} \sim a_0Y + 1$. Then for $Q := (-1+P)_{+(1/a_0)}$ we have $Q^{\phi} = -1 + P^{\phi}_{+(1/a_0)} \sim a_0Y$, eventually, so $\mathrm{nmul}\, Q = 1$. Lemma 14.2.6 gives $z \in \mathcal{o}$ with $(-1+P)\big((1/a_0)+z\big) = Q(z) = 0$, and thus $y := (1/a_0) + z$ works. □

Newtonianity of order r

Lemmas 14.2.4 and 14.2.6 extend as follows:

LEMMA 14.2.11. *Suppose K has asymptotic integration and $r \geqslant 1$. Then the two conditions on K, r below are equivalent:*

(i) *K is r-newtonian;*

(ii) *every $P \in K\{Y\}$ with* $\mathrm{nmul}\, P = 1$ *and* $\mathrm{order}\, P \leqslant r$ *has a zero in $\mathcal{o}$.*

PROOF. Follow the proof of Lemma 14.2.6, but for the direction (i) ⇒ (ii), appeal to Lemma 13.7.12 instead of Corollary 13.7.11. □

COROLLARY 14.2.12. *Let K have asymptotic integration and be r-newtonian, $r \geqslant 1$, and let $P \in K\{Y\}$, $u \in \mathcal{O}$, and $A \in \boldsymbol{k}[Y]$ be such that* $\mathrm{order}\, P \leqslant r$, $A(\overline{u}) = 0$, $A'(\overline{u}) \neq 0$, *and $D_{P^{\phi}} \in \boldsymbol{k}^{\times} \cdot A$, eventually. Then P has a zero in $u + \mathcal{o}$.*

PROOF. By Lemma 6.6.5(i) we have $D_{P^{\phi}_{+u}} \in \boldsymbol{k}^{\times} \cdot \big(D_{P^{\phi}}\big)_{+\overline{u}}$, hence $\operatorname{nmul} P_{+u} = 1$. It remains to apply Lemma 14.2.11. □

COROLLARY 14.2.13. *Suppose K has small derivation and asymptotic integration and is 1-newtonian. Let $Q \in K[Y, Y']$ satisfy $vQ > \Psi$. Then there is a unique $y \in \smallo$ such that $y' = Q(y)$.*

PROOF. With $P = Y' - Q$ we have $P^{\phi} \sim \phi Y'$ for $\phi \preccurlyeq 1$, so $\operatorname{nmul} P = 1$. This gives $y \in \smallo$ with $y' = Q(y)$ by Lemma 14.2.11. Suppose $z \in \smallo$, $y \neq z$, and $z' = Q(z)$. With $\varepsilon := z - y$ we get by Taylor expansion

$$\begin{aligned} \varepsilon' &= Q(z) - Q(y) = \sum_{i+j \geqslant 1} Q_{(i,j)}(y)\varepsilon^i(\varepsilon')^j, \text{ so} \\ \varepsilon^{\dagger} &= \sum_{(i,j)} Q_{(i+1,j)}(y)\varepsilon^i(\varepsilon')^j + \sum_{j \geqslant 1} Q_{(0,j)}(y)(\varepsilon')^{j-1}\varepsilon^{\dagger}. \end{aligned}$$

The valuation of the first sum on the last line is $> \Psi$, and the valuation of the second term is $> v(\varepsilon^{\dagger}) \in \Psi$, and we have a contradiction. □

For our application to $\omega(K)$ it is enough to consider $(1,1)$-newtonianity, where the proof of Lemma 14.2.4 gives the following:

LEMMA 14.2.14. *If K has asymptotic integration, the following are equivalent:*

(i) *K is $(1,1)$-newtonian;*

(ii) *every $P \in K[Y, Y']$ with $\deg_{Y'} P \leqslant 1$ and $\operatorname{nmul} P = 1$ has a zero in $\smallo$.*

COROLLARY 14.2.15. *Assume K has asymptotic integration and is $(1,1)$-newtonian. Let $P \in K[Y, Y']$ have degree $\leqslant 1$ in Y', let $u \in \mathcal{O}$, and let $A \in \boldsymbol{k}[Y]$ be such that $A(\overline{u}) = 0$, $A'(\overline{u}) \neq 0$, and $D_{P^{\phi}} \in \boldsymbol{k}^{\times} \cdot A$, eventually. Then P has a zero in $u + \smallo$.*

PROOF. Like that of Corollary 14.2.12; use Lemma 14.2.14 instead of 14.2.11. □

Application to ω

Let $f \in K$. When is $f \in \omega(K)$, that is, when is there y such that $\omega(y) = f$, equivalently, when does $P(Y) := Y^2 + 2Y' + f$ have a zero in K? Before we give a partial answer, let $b \neq 0$ and note that

$$\begin{aligned} P_{+a} &= Y^2 + 2aY + 2Y' + P(a), \text{ so} \\ P_{+a,\times b} &= b^2Y^2 + (2ab + 2b')Y + 2bY' + P(a), \text{ and thus} \\ b^{-1}P^{\phi}_{+a,\times b} &= bY^2 + (2a + 2b^{\dagger})Y + 2\phi Y' + P(a)/b. \end{aligned}$$

This leads to the following result:

LEMMA 14.2.16. *Suppose K is a $(1,1)$-newtonian real closed H-field. Then the subset $\omega\big(\Lambda(K)^{\downarrow}\big)$ of K is downward closed.*

PROOF. Corollary 14.2.2 for $r = 1$ shows K to be 1-linearly surjective. As K is also d-valued, it has asymptotic integration. Thus K is λ-free by Corollary 14.2.3.

Let $f \in \omega\big(\Lambda(K)\big)^{\downarrow}$; we need to show that $f \in \omega\big(\Lambda(K)^{\downarrow}\big)$. By Corollary 11.8.21, we may take ρ so that $f < \omega_\rho$. Then $f - \omega_\rho \preccurlyeq f - \omega_{\rho'}$ for $\rho < \rho'$, so by increasing ρ and using Lemma 11.7.1 we arrange that $f - \omega_\rho \succ \gamma_\rho^2$. Set $a := \lambda_\rho = -\gamma_\rho^\dagger$. Take $b \in K^>$ such that $f - \omega_\rho = -b^2$. Then with P as above,

$$b^{-1}P^{\phi}_{+a,\times b} \;=\; bY^2 + 2(b/\gamma_\rho)^\dagger Y + 2\phi Y' - b.$$

We have $b \succ \gamma_\rho$, hence b is active and $b \succ (b/\gamma_\rho)^\dagger$, and eventually $(b/\gamma_\rho)^\dagger \succ \phi$. It follows that the hypothesis of Corollary 14.2.15 holds with $P_{+a,\times b}$ in the role of P and $u = -1$, $A = Y^2 - 1$. Then that corollary provides a zero $g \in -1 + o$ of $P_{+a,\times b}$, and so $\omega(a + bg) = f$. It remains to note that $a + bg < a \in \Lambda(K)$. □

COROLLARY 14.2.17. *If K is a $(1,1)$-newtonian Liouville closed H-field, then the subset $\omega(K) = \Omega(K)$ of K is downward closed.*

Application to the function σ

In this subsection we assume that $f \in K$. Solving the equation $\sigma(y) = f$ in $K^\times$ means finding a zero of

$$S(Y) \;:=\; \sigma(Y) - f \;=\; \omega(-Y^\dagger) + Y^2 - f \in K\langle Y\rangle$$

in $K^\times$. Let $b \in K^\times$. Extending the automorphism $P \mapsto P_{\times b}$ of the ring $K\{Y\}$ to an automorphism $R \mapsto R_{\times b}$ of its fraction field $K\langle Y\rangle$, we get

$$S_{\times b} \;=\; \omega(-b^\dagger - Y^\dagger) + b^2Y^2 - f.$$

Lemma 5.2.1 applied to $w = -b^\dagger - Y^\dagger$, $z = -b^\dagger$, gives

$$\begin{aligned}\omega(-b^\dagger - Y^\dagger) \;&=\; \omega(-b^\dagger) + Y^\dagger\cdot\big(2(Y^{\dagger\dagger} - b^\dagger - Y^\dagger) + Y^\dagger\big)\\ &=\; \omega(-b^\dagger) + Y^\dagger\cdot\big(2(Y')^\dagger - 2b^\dagger - 3Y^\dagger\big)\end{aligned}$$

and so, with $Q := Y^2 S_{\times b} \in K\{Y\}$, we have

$$\begin{aligned}Q \;&=\; b^2Y^4 + \big(\omega(-b^\dagger) - f\big)Y^2 + Y'Y\cdot\big(2(Y')^\dagger - 2b^\dagger - 3Y^\dagger\big)\\ &=\; b^2Y^4 + \big(\omega(-b^\dagger) - f\big)Y^2 + R, \quad\text{where}\\ R \;&:=\; 2Y''Y - 2b^\dagger Y'Y - 3(Y')^2 \in K\{Y\}.\end{aligned}$$

Now $(Y'')^\phi = \phi^2 Y'' + \phi' Y'$ and thus

$$\begin{aligned}R^\phi \;&=\; 2(\phi^2Y'' + \phi'Y')Y - 2b^\dagger\phi Y'Y - 3\phi^2(Y')^2\\ &=\; \phi\cdot\big(2\phi Y'' + 2(\phi/b)^\dagger Y'Y - 3\phi(Y')^2\big).\end{aligned}$$

We use this computation in the proof of the next result:

PROPOSITION 14.2.18. *Let K be a 2-newtonian real closed H-field with asymptotic integration. Then the subset $\sigma\big(\Gamma(K)^{\dagger}\big)$ of K is upward closed.*

PROOF. Assume $f > \sigma(\gamma_\rho)$; by Corollary 11.8.30 it suffices to show that then $f \in \sigma\big(\Gamma(K)^{\dagger}\big)$. Since $\sigma(\gamma_\rho) > \sigma(\gamma_{\rho+1})$ and $\sigma(\gamma_\rho) - \sigma(\gamma_{\rho+1}) \sim \gamma_\rho^2$, we can increase ρ and arrange that $f - \sigma(\gamma_\rho) \succ \gamma_\rho^2$. Take $b \in K^>$ with $b^2 = f - \sigma(\gamma_\rho)$; then $b \succ \gamma_\rho$, so b is active. Moreover, using Lemma 11.7.6 in the last step in the next line,

$$\sigma(b) - f \;=\; \sigma(b) - \sigma(\gamma_\rho) - b^2 \;=\; \omega(-b^{\dagger}) - \omega(-\gamma_\rho^{\dagger}) - \gamma_\rho^2 \;\prec\; b^2,$$

and so $\omega(-b^{\dagger}) - f \sim -b^2$. Eventually $\phi \prec b$, so $(\phi/b)^{\dagger} \prec b$, hence $R^{\phi} \prec b^2$, and thus $Q^{\phi} \sim b^2Y^2(Y^2-1)$. Corollary 14.2.12 gives $u \in 1 + o$ with $Q(u) = 0$. Then $\sigma(bu) = f$ and $bu \in \Gamma(K)^{\dagger}$. □

COROLLARY 14.2.19. *If K is a 2-newtonian Liouville closed H-field, then the subset $\sigma\big(\Gamma(K)\big)$ of K is upward closed.*

In combination with Corollaries 11.8.33 and 14.2.17 this yields:

COROLLARY 14.2.20. *If K is a 2-newtonian ω-free Liouville closed H-field, then K is Schwarz closed.*

Notes and comments

All newtonian K known to us are ω-free. Note that if K has asymptotic integration and is newtonian, then K is λ-free by Corollary 14.2.3. This leaves open whether there are newtonian K without asymptotic integration, that is, with a gap, and whether there are λ-free newtonian K that are not ω-free. Corollary 14.2.13 for $K = \mathbb{T}$ and $Q \in K[Y]$ is in [235, Corollary 63].

The intersection E of all maximal Hardy fields is clearly a Hardy field that contains $\mathbb{R}$ and is Liouville closed. By Boshernitzan [57], E is d-algebraic over $\mathbb{R}$, and by [59, Proposition 3.7] there is no $y \in E$ with $y'' + y = \mathrm{e}^{x^2}$. Hence E is not 2-linearly surjective and thus (by Corollary 14.2.2) E is not newtonian.

14.3 SOLVING QUASILINEAR EQUATIONS

In this section K is ω-free, a, b, y range over K, and P over $K\{Y\}^{\neq}$.

Newton position, and proof of Theorem 14.0.1

Suppose $\operatorname{nmul} P = 1$ with $P_0 \neq 0$. By Corollary 13.6.10 (which assumes ω-freeness) we can take $g \in K^{\times}$ such that eventually $P_0 \asymp P^{\phi}_{1,\times g}$. Since $P_0 \prec P_1^{\phi}$, eventually, we have $g \prec 1$. Let $i \geqslant 2$. Since $P_1^{\phi} \succcurlyeq P_i^{\phi}$, eventually, we get $P^{\phi}_{1,\times g} \succ P^{\phi}_{i,\times g}$, eventually. Thus $\operatorname{ndeg} P_{\times g} = 1$, and so if K is newtonian, then P has a zero in $g\mathcal{O}$, but P has no zero in $g o$.

Define P to be **in newton position at** a if $\operatorname{nmul} P_{+a} = 1$. Suppose P is in newton position at a and set $Q := P_{+a}$, so $Q(0) = P(a)$. If $P(a) \neq 0$, then by the above there is $g \in K^\times$ such that eventually $P(a) = Q(0) \asymp Q^\phi_{1,\times g}$, and as vg does not depend on the choice of such g, we set $v^e(P,a) := vg$. If $P(a) = 0$ we set $v^e(P,a) = \infty \in \Gamma_\infty$.

LEMMA 14.3.1. *Suppose K is newtonian and P is in newton position at a. Then $P(b) = 0$ and $v(a-b) \geqslant v^e(P,a)$ for some b; any such b satisfies $v(a-b) = v^e(P,a)$.*

PROOF. If $P(a) = 0$, then we must take $b = a$. Assume $P(a) \neq 0$, set $\gamma := v^e(P,a) \in \Gamma$, take $g \in K$ with $vg = \gamma$. With $Q := P_{+a}$ we have $P(a+gY) = Q_{\times g}$ and $\operatorname{ndeg} Q_{\times g} = 1$, so we get $y \asymp 1$ with $Q(gy) = 0$, and so for $b := a + gy$ we have $P(b) = 0$ and $v(a-b) = v^e(P,a)$. Conversely, if $v(a-b) \geqslant \gamma$ and $P(b) = 0$, then $b = a + gy$ with $y \preccurlyeq 1$, so $Q(gy) = 0$, hence $y \asymp 1$, and thus $v(a-b) = \gamma$. □

Without assuming K is newtonian, we get:

LEMMA 14.3.2. *Suppose P is in newton position at a and $P(a) \neq 0$. Then there exists b with the following properties:*

(i) *P is in newton position at b, $v(a-b) = v^e(P,a)$, and $P(b) \prec P(a)$;*

(ii) *for all $b^* \in K$ with $v(a-b^*) \geqslant v^e(P,a)$: $P(b^*) \prec P(a) \Leftrightarrow a - b \sim a - b^*$;*

(iii) *for all $b^* \in K$, if $a - b \sim a - b^*$, then P is at newton position at b^* and $v^e(P,b^*) > v^e(P,a)$.*

PROOF. With $Q = P_{+a}$, $\gamma = v^e(P,a)$, and $g \in K$ with $vg = \gamma$, we have

$$P(a) \asymp Q^\phi_{1,\times g} \succ Q^\phi_{i,\times g} \quad \text{for } i \geqslant 2\text{, eventually.}$$

Thus $D_{Q^\phi_{\times g}}$ is isobaric of weight 0, eventually, so we get $d \in K$ with $d \asymp P(a)$ such that eventually we have

$$Q^\phi_{\times g} = P(a) + dY + R_\phi, \quad R_\phi \in K^\phi\{Y\}, \quad R_\phi \prec P(a).$$

Taking $y \sim -P(a)/d$ gives $y \asymp 1$ and $Q(gy) \prec P(a)$, so with $b := a + gy$ we obtain $P(b) \prec P(a)$ and $v(a-b) = \gamma$. Now $P_{+b} = P_{+a+gy}$ with $gy \prec 1$, so $\operatorname{nmul} P_{+b} = \operatorname{nmul} P_{+a} = 1$, and thus P is in newton position at b. Conversely, if $b^* \in K$, $v(a-b^*) \geqslant \gamma$ and $P(b^*) \prec P(a)$, then $b^* = a + gy^*$ with $y^* \sim -P(a)/d$.

With y and b as above it remains to show that $v^e(P,b) > v^e(P,a)$. Let P have order $\leqslant r$, let $\boldsymbol{i}$ and $\boldsymbol{j}$ range over $\mathbb{N}^{1+r}$, and recall that $P_{(\boldsymbol{i})} = \frac{P^{(\boldsymbol{i})}}{\boldsymbol{i}!} \in K\{Y\}$, so

$$P^\phi_{+a}(Y) = P^\phi(a) + \sum_{|\boldsymbol{i}|\geqslant 1} (P^\phi)_{(\boldsymbol{i})}(a) Y^{\boldsymbol{i}}, \qquad (P^\phi_{+a})_1 = \sum_{|\boldsymbol{i}|=1} (P^\phi)_{(\boldsymbol{i})}(a) Y^{(\boldsymbol{i})},$$

and likewise with b instead of a. Taylor expanding $(P^\phi)_{(\boldsymbol{i})}$ at a for $|\boldsymbol{i}| = 1$ gives

$$(P^\phi)_{(\boldsymbol{i})}(b) = (P^\phi)_{(\boldsymbol{i})}(a) + \sum_{|\boldsymbol{j}|\geqslant 1} (P^\phi)_{(\boldsymbol{i})(\boldsymbol{j})}(a) \cdot (gy)^{\boldsymbol{j}} \quad \text{in } K^\phi.$$

As $(P^{\phi})_{(i)(j)}(a) = \binom{i+j}{j}(P^{\phi})_{(i+j)}(a)$, we get for $A_{\phi} := \left(P^{\phi}_{+a}\right)_1$ and $B_{\phi} := \left(P^{\phi}_{+b}\right)_1$ that $B_{\phi} = A_{\phi} + E_{\phi}$, where eventually $v(E_{\phi}) \geqslant v(A_{\phi}) + \gamma + o(\gamma)$. Together with $v_{A_{\phi}}(\gamma) = v(A_{\phi}) + \gamma + o(\gamma)$, we get

$$v_{E_{\phi}}(\gamma) \;=\; v(E_{\phi}) + \gamma + o(\gamma) \;\geqslant\; v(A_{\phi}) + 2\gamma + o(\gamma) \;>\; v_{A_{\phi}}(\gamma), \text{ eventually.}$$

Hence $v_{B_{\phi}}(\gamma) = v_{A_{\phi}}(\gamma)$, and so $P(b) \prec P(a)$ forces $v^{\mathrm{e}}(P,b) > v^{\mathrm{e}}(P,a)$. □

LEMMA 14.3.3. *Suppose P is in newton position at a and there is no b with $P(b) = 0$ and $v(a-b) = v^{\mathrm{e}}(P,a)$. Then there exists a divergent pc-sequence (a_{ρ}) in K such that $P(a_{\rho}) \rightsquigarrow 0$.*

PROOF. Let $(a_{\rho})_{\rho<\lambda}$ be a sequence in K with λ an ordinal > 0, $a_0 = a$, and

(1) P is in newton position at a_{ρ}, for all $\rho < \lambda$,

(2) $v(a_{\rho'} - a_{\rho}) = v^{\mathrm{e}}(P, a_{\rho})$ whenever $\rho < \rho' < \lambda$,

(3) $P(a_{\rho'}) \prec P(a_{\rho})$ and $v^{\mathrm{e}}(P, a_{\rho'}) > v^{\mathrm{e}}(P, a_{\rho})$ whenever $\rho < \rho' < \lambda$.

Note that there is such a sequence if $\lambda = 1$. Suppose $\lambda = \mu + 1$ is a successor ordinal. Then Lemma 14.3.2 yields $a_{\lambda} \in K$ such that $v(a_{\lambda} - a_{\mu}) = v^{\mathrm{e}}(P, a_{\mu})$, $P(a_{\lambda}) \prec P(a_{\mu})$ and $v^{\mathrm{e}}(P, a_{\lambda}) > v^{\mathrm{e}}(P, a_{\mu})$. Then the extended sequence $(a_{\rho})_{\rho<\lambda+1}$ has the above properties with $\lambda + 1$ instead of λ.

Suppose λ is a limit ordinal. Then (a_{ρ}) is a pc-sequence and $P(a_{\rho}) \rightsquigarrow 0$. If (a_{ρ}) has no pseudolimit in K we are done. Assume otherwise, and take a pseudolimit $a_{\lambda} \in K$ of (a_{ρ}). The extended sequence $(a_{\rho})_{\rho<\lambda+1}$ clearly satisfies condition (2) with $\lambda + 1$ instead of λ. Applying Lemma 14.3.2 to a_{ρ}, $a_{\rho+1}$ and a_{λ} in the place of a, b and b^*, where $\rho < \lambda$, we see that conditions (1) and (3) are also satisfied with $\lambda + 1$ instead of λ. This building process must come to an end. □

PROOF OF THEOREM 14.0.1. Assume that K has no proper immediate d-algebraic extension. In order for K to be newtonian, it suffices by Lemma 14.2.11 to show that every P with $\operatorname{nmul} P = 1$ has a zero in $\mathcal{O}$. So let $\operatorname{nmul} P = 1$ and suppose towards a contradiction that P has no zero in $\mathcal{O}$. Then P is in newton position at 0, and so by Lemma 14.3.3 there exists a divergent pc-sequence (a_{ρ}) in K with $P(a_{\rho}) \rightsquigarrow 0$. Then K has a proper immediate d-algebraic extension by Section 11.4, a contradiction. This concludes the proof of Theorem 14.0.1. □

Application to solving asymptotic equations

In this subsection K is d-*valued with small derivation, and $\mathfrak{M}$ is a monomial group for K.*

LEMMA 14.3.4. *Suppose K is newtonian. Let $g \in K^{\times}$ be an approximate zero of P such that* $\operatorname{ndeg} P_{\times g} = 1$. *Then there exists $y \sim g$ in K such that $P(y) = 0$.*

PROOF. Take $c \in C^{\times}$ and $\mathfrak{m} \in \mathfrak{M}$ with $g \sim c\mathfrak{m}$. Then $N_{P_{\times\mathfrak{m}}}(c) = 0$, so

$$\operatorname{nmul} P_{\times\mathfrak{m},+c} = \operatorname{mul} N_{P_{\times\mathfrak{m},+c}} = \operatorname{mul}(N_{P_{\times\mathfrak{m}}})_{+c} \geqslant 1.$$

Now $\operatorname{nmul} P_{\times\mathfrak{m},+c} \leqslant \operatorname{ndeg} P_{\times\mathfrak{m},+c} = \operatorname{ndeg} P_{\times\mathfrak{m}} = 1$, so $\operatorname{nmul} P_{\times\mathfrak{m},+c} = 1$, giving $z \in \mathcal{O}$ with $P_{\times\mathfrak{m},+c}(z) = 0$ by Lemma 14.2.11. Thus $y \sim g$ and $P(y) = 0$ for $y := (c+z)\mathfrak{m}$. □

Next we consider an asymptotic equation

$$\text{(E)} \qquad P(Y) = 0, \qquad Y \in \mathcal{E}$$

where $P \in K\{Y\}^{\neq}$ and $\mathcal{E} \subseteq K^{\times}$ is $\preccurlyeq$-closed.

In the next theorem and its corollaries 14.3.6 and 14.3.7, C is algebraically closed, Γ is divisible, and K is asymptotically d*-algebraically maximal.*

THEOREM 14.3.5. *If* $\operatorname{ndeg}_{\mathcal{E}} P > \operatorname{mul} P = 0$, *then* (E) *has a solution.*

PROOF. We proceed by induction on $d = \operatorname{ndeg}_{\mathcal{E}} P$. If $d = 1$ and $\operatorname{mul} P = 0$, then we take $g \in \mathcal{E}$ with $\operatorname{ndeg} P_{\times g} = 1$, so Theorem 14.0.1 gives $y \preccurlyeq 1$ with $P(gy) = 0$, and then gy is a solution of (E). Suppose $d > 1$, $\operatorname{mul} P = 0$, and (E) does not have a solution; we shall derive a contradiction. Proposition 13.8.8 gives an unraveler $(f, \mathcal{E}')$ for (E). The corresponding refinement (E′) of (E) has Newton degree d and is unraveled, with $\operatorname{mul} P_{+f} = 0$. Replacing P, $\mathcal{E}$ by P_{+f}, $\mathcal{E}'$, we may therefore assume that (E) is unraveled. As C is algebraically closed, we get from Corollary 13.5.9 an algebraic approximate solution f of (E). Since (E) is unraveled, we have $\operatorname{mul} P_{+f} = 0 < \operatorname{ndeg}_{\prec f} P_{+f} < d$, so by the inductive hypothesis, the refinement

$$P_{+f}(Y) = 0, \qquad Y \prec f$$

of (E), and hence (E) itself, has a solution. This is the desired contradiction. □

COROLLARY 14.3.6. *K is weakly differentially closed.*

PROOF. Let $P \in K\{Y\} \setminus K$. If $\operatorname{mul} P > 0$, then $P(0) = 0$. Otherwise we have $\deg P > \operatorname{mul} P = 0$. It remains to apply Theorem 14.3.5 with $\mathcal{E} = K^{\times}$. □

COROLLARY 14.3.7. *Suppose $g \in K^{\times}$ is an approximate zero of P. Then there exists $y \sim g$ such that $P(y) = 0$.*

PROOF. An equivalence at the beginning of Section 13.8 gives $\operatorname{ndeg}_{\prec g} P_{+g} \geqslant 1$, so

$$P_{+g}(Y) = 0, \qquad Y \prec g$$

has positive Newton degree. If $\operatorname{mul} P_{+g} \geqslant 1$, then $P(g) = 0$, so $y := g$ works. Suppose $\operatorname{mul} P_{+g} = 0$. Then Theorem 14.3.5 yields $z \prec g$ in $K^{\times}$ with $P_{+g}(z) = 0$, and so $P(y) = 0$ and $y \sim g$ for $y := g + z$. □

In the next theorem and its corollary we assume that C is real closed, Γ is divisible, and K is asymptotically d*-algebraically maximal.*

THEOREM 14.3.8. *If* $\operatorname{mul} P = 0$ *and* $\operatorname{ndeg}_{\mathcal{E}} P$ *is odd, then* (E) *has a solution.*

PROOF. We argue by induction on $d = \operatorname{ndeg}_{\mathcal{E}} P$ as in the proof of Theorem 14.3.5. The case $d = 1$ is dealt with as in the proof of that theorem. Suppose $d > 1$ is odd, $\operatorname{mul} P = 0$; assume towards a contradiction that (E) does not have a solution. As in the proof of Theorem 14.3.5 we arrange that (E) is unraveled.

Suppose $f = c\mathfrak{m}$ ($c \in C^\times$, $\mathfrak{m} \in \mathcal{E}$) is an approximate solution of (E) of odd multiplicity. This multiplicity equals $\mu + w$ where μ is the algebraic multiplicity of f as an approximate zero of P and $w := \operatorname{nwt} P_{\times\mathfrak{m}}$. We have $\operatorname{mul} P_{+f} = 0$. Also $\operatorname{ndeg}_{\prec f} P_{+f} = \operatorname{mul}(N_{P_{\times\mathfrak{m}}})_{+c} = \mu + w < d$ by Lemma 13.5.4 and (E) being unraveled. Then by the inductive hypothesis, the refinement

$$P_{+f}(Y) = 0, \qquad Y \prec f$$

of (E), and hence (E) itself, has a solution, and we have a contradiction. So it is enough to find such an f. In order to do so, we distinguish two cases.

CASE 1: *There exists* $\mathfrak{m} \in \mathcal{E}$ *such that* $w := \operatorname{nwt} P_{\times\mathfrak{m}}$ *is odd.* Take such $\mathfrak{m}$ and take $A \in C[Y]$ with $N_{P_{\times\mathfrak{m}}} = A(Y) \cdot (Y')^w$; pick $c \in C^\times$ with $A(c) \neq 0$. Then the approximate solution $f = c\mathfrak{m}$ of (E) has odd multiplicity w.

CASE 2: $\operatorname{nwt} P_{\times\mathfrak{m}}$ *is even for all* $\mathfrak{m} \in \mathcal{E}$. Corollary 13.5.9 gives an approximate zero $f \in \mathcal{E}$ of P with odd algebraic multiplicity, and thus with odd multiplicity. □

As Theorem 14.3.5 gives Corollary 14.3.6, so we get from Theorem 14.3.8:

COROLLARY 14.3.9. *If* $\deg P$ *is odd, then* P *has a zero in* K.

Recall from Section 13.1 that $\mathfrak{m} \in \mathfrak{M}$ is a starting monomial for P iff $\operatorname{nwt}(P_{\times\mathfrak{m}}) \geqslant 1$ or $N_{P_{\times\mathfrak{m}}}$ is not homogeneous; equivalently, $N_{P_{\times\mathfrak{m}}} \notin CY^{\mathbb{N}}$.

COROLLARY 14.3.10. *Let* $\mathfrak{m} \in \mathfrak{M}$. *Then* $\mathfrak{m}$ *is a starting monomial for* P *if and only if* $P(f) = 0$ *and* $f \asymp \mathfrak{m}$ *for some* f *in some* d*-valued extension of* K.

PROOF. Take an algebraic closure K^{a} of the d-valued field K. Use Lemma 3.3.33 to equip K^{a} with a monomial group $\mathfrak{M}^{\mathrm{a}}$ containing $\mathfrak{M}$. By Zorn we have an immediate asymptotically d-algebraically maximal d-algebraic extension L of K^{a}. Then L is d-valued, ω-free, C_L is an algebraic closure of C, $\Gamma_L = \mathbb{Q}\Gamma$, and $\mathfrak{M}^{\mathrm{a}}$ remains a monomial group for L. If $\mathfrak{m}$ is a starting monomial for P, then we can take $c \in C_L^\times$ such that $g := c\mathfrak{m}$ is an approximate zero of P, hence by Corollary 14.3.7 applied to L in place of K there is $f \sim g$ in L such that $P(f) = 0$. Conversely, suppose L is a d-valued extension of K and $f \in L$, $P(f) = 0$, and $f \asymp \mathfrak{m}$. Take $c \in C_L^\times$ with $f \sim c\mathfrak{m}$. As in the proof of Lemma 13.1.6 we get $N_{P_{\times\mathfrak{m}}}(c) = 0$, and so $\mathfrak{m}$ is a starting monomial for P. □

COROLLARY 14.3.11. *Let* $\mathfrak{m} \in \mathfrak{M}$. *Then there are* $\mathfrak{m}_0, \mathfrak{m}_1 \in \mathfrak{M}$ *with* $\mathfrak{m}_0 \prec \mathfrak{m} \prec \mathfrak{m}_1$ *such that there is no starting monomial* $\mathfrak{n}$ *for* P *with* $\mathfrak{m}_0 \prec \mathfrak{n} \prec \mathfrak{m}_1$ *and* $\mathfrak{n} \neq \mathfrak{m}$.

PROOF. Applying Corollary 13.2.12 to $P_{\times\mathfrak{m}}$ we get $\mathfrak{m}_0, \mathfrak{m}_1 \in \mathfrak{M}$ with $\mathfrak{m}_0 \prec \mathfrak{m} \prec \mathfrak{m}_1$ such that $P(f) \neq 0$ for all f in all d-valued extensions of K with $\mathfrak{m}_0 \prec f \prec \mathfrak{m}_1$ and $f \not\asymp \mathfrak{m}$. Then $\mathfrak{m}_0$, $\mathfrak{m}_1$ have the desired property by Corollary 14.3.10. □

Recall the concept of *newtonization* defined in Section 14.1.

COROLLARY 14.3.12. *If L is a newtonization of K, then L is an immediate* d*-algebraic extension of K, no proper differential subfield of L containing K is newtonian, and any newtonization of K is K-isomorphic to L.*

PROOF. By Zorn there exist immediate asymptotically d-algebraically maximal d-algebraic extensions of K. By Theorem 14.0.1 any such extension is newtonian. It remains to appeal to Corollary 14.1.9. □

In the rest of this subsection L is an ω-free immediate extension of K. To the asymptotic equation (E) above corresponds the asymptotic equation

$$(\mathrm{E}_L) \qquad P(Y) = 0, \qquad Y \in \mathcal{E}_L$$

over L. If (E) is quasilinear, then so is (E_L). Moreover, if K is newtonian and (E) is quasilinear, then every solution of (E_L) is a solution of (E), by Lemma 14.1.8. We say that $f \in L$ is **quasilinear** over K if $Q(f) = 0$ for some $Q \in K\{Y\}^{\neq}$ with $\operatorname{ndeg} Q_{\times f} = 1$. Every element of L that is linear over K is quasilinear over K. If K is newtonian, then no element of $L \setminus K$ is quasilinear over K.

LEMMA 14.3.13. *Suppose K is newtonian,* (E) *is quasilinear. Let $\mathcal{E}' \subseteq \mathcal{E}$ be $\preccurlyeq$-closed and let $f \in \mathcal{E}_L$ be such that the refinement*

$$(\mathrm{E}'_L) \qquad P_{+f}(Y) = 0, \qquad Y \in \mathcal{E}'_L$$

of (E_L) *is also quasilinear. Let $y \preccurlyeq f$ be a solution of* (E) *that best approximates f. Then $f - y \in \mathcal{E}'_L \cup \{0\}$.*

PROOF. Suppose $f \neq y$, and set $\mathfrak{m} := \mathfrak{d}_{f-y}$. Towards a contradiction, suppose $f - y \notin \mathcal{E}'_L$. Then $\mathcal{E}'_L \prec \mathfrak{m} \in \mathcal{E}$, so by quasilinearity of (E):

$$1 = \operatorname{ndeg}_{\mathcal{E}'_L} P_{+f} \leqslant \operatorname{ndeg}_{\preccurlyeq\mathfrak{m}} P_{+f} = \operatorname{ndeg}_{\preccurlyeq\mathfrak{m}} P_{+y} \leqslant \operatorname{ndeg}_{\mathcal{E}} P_{+y} = \operatorname{ndeg}_{\mathcal{E}} P = 1.$$

Hence the asymptotic equation

$$(14.3.1) \qquad P_{+y}(Y) = 0, \qquad Y \preccurlyeq \mathfrak{m}$$

over K is quasilinear. By quasilinearity of (E'_L) we have

$$\operatorname{ndeg}_{\prec\mathfrak{m}} (P_{+y})_{+(f-y)} = \operatorname{ndeg}_{\prec\mathfrak{m}} P_{+f} \geqslant \operatorname{ndeg}_{\mathcal{E}'_L} P_{+f} = 1,$$

so $f - y$ is an approximate solution of (14.3.1) over L, by an equivalence at the beginning of Section 13.8. Take $g \in K^{\times}$ with $g \sim f - y \asymp \mathfrak{m}$. Then g is an approximate solution of (14.3.1), and $\operatorname{ndeg} P_{+y,\times g} = \operatorname{ndeg}_{\preccurlyeq\mathfrak{m}} P_{+y} = 1$. Then Lemma 14.3.4 gives $z \sim f - y$ in K such that $P(y + z) = P_{+y}(z) = 0$. Using $y \preccurlyeq f$ here for the first time, we get $y + z \neq 0$, and thus $y + z$ is a better approximation to f by a solution of (E) than y, a contradiction. □

Notes and comments

Theorem 14.0.1, together with Zorn, is one source of ω-free newtonian H-asymptotic fields. The next chapter provides another source, and a more constructive procedure to build such objects.

14.4 UNRAVELERS

This section is somewhat technical. Our aim is Proposition 14.4.1 below. In the next section we derive Theorem 14.0.2 from Proposition 14.4.1.

In this section K is ω-free, d*-valued, with divisible value group* Γ *and small derivation, and* $\mathfrak{M}$ *is a monomial* group *of* K. We let $\mathfrak{m}$, $\mathfrak{n}$ range over $\mathfrak{M}$, and ϕ over the elements of $\mathfrak{M}$ that are active in K. Also, $\mathcal{E} \subseteq K^\times$ is $\preccurlyeq$-closed, and $P \in K\{Y\}^{\neq}$. Consider the asymptotic equation

$$\text{(E)} \qquad P(Y) = 0, \qquad Y \in \mathcal{E}$$

over K, and assume $d := \operatorname{ndeg}_{\mathcal{E}} P \geqslant 2$. In addition we fix an ω-free newtonian immediate extension $\widehat{K}$ of K, and use $\mathfrak{M}$ as a monomial group for $\widehat{K}$. Set

$$\widehat{\mathcal{E}} := \mathcal{E}_{\widehat{K}} = \{y \in \widehat{K}^\times : vy \in v\mathcal{E}\}.$$

Associated to (E) we have the asymptotic equation

$$(\widehat{\text{E}}) \qquad P(Y) = 0, \qquad Y \in \widehat{\mathcal{E}}$$

over $\widehat{K}$, with the same Newton degree d as (E). We assume there is given an unraveler $(\widehat{f}, \widehat{\mathcal{E}}')$ for $(\widehat{\text{E}})$ with $\widehat{f} \neq 0$, $\operatorname{ndeg}_{\prec \widehat{f}} P_{+\widehat{f}} = d$. Then $\widehat{f}$ is an approximate solution of $(\widehat{\text{E}})$ of multiplicity d, by an equivalence at the beginning of Section 13.8. Note also that $\widehat{\mathcal{E}}' = \mathcal{E}'_{\widehat{K}}$ for some $\preccurlyeq$-closed subset $\mathcal{E}'$ of $\mathcal{E}$, and $\operatorname{ndeg}_{\widehat{\mathcal{E}}'} P_{+\widehat{f}} = d$. It follows that $(\widehat{\text{E}})$ is not unraveled, and so (E) is not unraveled.

Finally, we assume $\operatorname{mul} P_{+\widehat{f}} < d$. Then by Proposition 13.5.7, the refinement

$$(\widehat{\text{E}}') \qquad P_{+\widehat{f}}(Y) = 0, \qquad Y \in \widehat{\mathcal{E}}'$$

of $(\widehat{\text{E}})$ has an algebraic starting monomial. Let $\mathfrak{e}$ be the largest algebraic starting monomial for $(\widehat{\text{E}}')$. The goal of this section is to prove the following:

PROPOSITION 14.4.1. *There exists $f \in \widehat{K}$ such that* (i) *or* (ii) *below holds:*

(i) $\widehat{f} - f \preccurlyeq \mathfrak{e}$ *and* $A(f) = 0$ *for some* $A \in K\{Y\}$ *with* $\mathrm{c}(A) < \mathrm{c}(P)$ *and* $\deg A = 1$*;*

(ii) $\widehat{f} \sim f$, $\widehat{f} - a \preccurlyeq f - a$ *for all* $a \in K$, *and* $A(f) = 0$ *for some* $A \in K\{Y\}$ *with* $\mathrm{c}(A) < \mathrm{c}(P)$ *and* $\operatorname{ndeg} A_{\times f} = 1$.

Below we first prove Proposition 14.4.1 in the special case $d = \deg P$, and then show how to reduce the general case to this special case via Lemma 13.8.13.

Note that for each $f \in \widehat{K}$ with $\widehat{f} - f \preccurlyeq \mathfrak{e}$, the pair $(f, \widehat{\mathcal{E}}')$ is an unraveler for $(\widehat{\mathrm{E}})$, by Corollary 13.8.7. So if $f \in K$ and $\widehat{f} - f \preccurlyeq \mathfrak{e}$, then $(f, \mathcal{E}')$ with $\mathcal{E}' := \widehat{\mathcal{E}}' \cap K^\times$ is an unraveler for (E), by Lemma 13.8.9.

Note also that towards proving Proposition 14.4.1 we may, for any given ϕ, replace $K, P, \widehat{K}$ by $K^\phi, P^\phi, \widehat{K}^\phi$, without changing $\mathcal{E}, d, \widehat{f}, \widehat{\mathcal{E}}', \mathfrak{e}$.

A special case

Set $G := P_{+\widehat{f},\times\mathfrak{e}} \in \widehat{K}\{Y\}$. Then $\operatorname{ndeg} G = d$ and for $w := \operatorname{nwt}(G)$ we have $N_G \in C[Y](Y')^w$. As N_G is not homogeneous, we have $w \leqslant d-1$. Towards proving Proposition 14.4.1 we may compositionally conjugate by an active element of $\mathfrak{M}$, and thus arrange $D_G = N_G$ and $R_G \prec^\flat G$. Let ∂_0, ∂_1 be the $\widehat{K}$-derivations $\partial/\partial Y$ and $\partial/\partial Y'$ on $\widehat{K}\{Y\}$. Using notations from the end of Section 12.8 we set

$$\Delta := \left(\partial_0^{d-1-w}\partial_1^w\right)_{\times\mathfrak{e}} \in \widehat{K}[[\partial]], \qquad Q := \Delta P \in K\{Y\}.$$

Then by Lemmas 12.8.7 and 12.8.8 we get

$$Q_{+\widehat{f},\times\mathfrak{e}} = \partial_0^{d-1-w}\partial_1^w G \neq 0,$$

hence by Corollary 13.1.12,

$$\partial_0^{d-1-w}\partial_1^w N_G = D_{Q_{+\widehat{f},\times\mathfrak{e}}} = N_{Q_{+\widehat{f},\times\mathfrak{e}}}, \qquad R_{Q_{+\widehat{f},\times\mathfrak{e}}} \prec^\flat Q_{+\widehat{f},\times\mathfrak{e}},$$

so $N_{Q_{+\widehat{f},\times\mathfrak{e}}} \in C[Y]$ has degree 1, and thus $\operatorname{ndeg}_{\preccurlyeq\mathfrak{e}} Q_{+\widehat{f}} = \operatorname{ndeg} Q_{+\widehat{f},\times\mathfrak{e}} = 1$. Then the asymptotic equation

$$Q_{+\widehat{f}}(Y) = 0, \qquad Y \preccurlyeq \mathfrak{e}$$

over $\widehat{K}$ is quasilinear. Note that

$$1 = \operatorname{ndeg}_{\preccurlyeq\mathfrak{e}} Q_{+\widehat{f}} \leqslant \operatorname{ndeg}_{\widehat{\mathcal{E}}} Q_{+\widehat{f}} = \operatorname{ndeg}_{\widehat{\mathcal{E}}} Q = \operatorname{ndeg}_{\mathcal{E}} Q.$$

LEMMA 14.4.2. *Suppose $\mathfrak{e} \prec \widehat{f}$ and the asymptotic equation*

$$(14.4.1) \qquad Q(Y) = 0, \qquad Y \in \widehat{\mathcal{E}}$$

over $\widehat{K}$ is quasilinear. Then (14.4.1) *has a solution $y \sim \widehat{f}$, and if f is any solution of* (14.4.1) *that best approximates $\widehat{f}$, then $f - \widehat{f} \preccurlyeq \mathfrak{e}$.*

PROOF. We have $\operatorname{ndeg}_{\prec\widehat{f}} Q_{+\widehat{f}} \leqslant \operatorname{ndeg}_{\widehat{\mathcal{E}}} Q_{+\widehat{f}} = \operatorname{ndeg}_{\widehat{\mathcal{E}}} Q = 1$, and from $\mathfrak{e} \prec \widehat{f}$ we get $1 = \operatorname{ndeg}_{\preccurlyeq\mathfrak{e}} Q_{+\widehat{f}} \leqslant \operatorname{ndeg}_{\prec\widehat{f}} Q_{+\widehat{f}}$, and so $\operatorname{ndeg}_{\prec\widehat{f}} Q_{+\widehat{f}} = 1$. Thus (14.4.1) has a solution $y \sim \widehat{f}$. It remains to apply Lemma 14.3.13 with $\widehat{K}$ in the role of both L and K, and $Q, \widehat{f}, f$ in the role of P, f, y in that lemma. □

If $\deg P = d$, then $\deg Q = 1$ and so (14.4.1) is automatically quasilinear. It follows that we are in case (i) of Proposition 14.4.1 when $\deg P = d$:

COROLLARY 14.4.3. *Suppose* $\deg P = d$. *Then there exist* $f \in \widehat{K}$ *and* $A \in K\{Y\}$ *such that* $\widehat{f} - f \preccurlyeq \mathfrak{e}$, $A(f) = 0$, $\mathrm{c}(A) < \mathrm{c}(P)$, *and* $\deg A = 1$.

PROOF. If $\widehat{f} \preccurlyeq \mathfrak{e}$, then we can take $f = 0$ and $A = Y$. Assume $\mathfrak{e} \prec \widehat{f}$. Then Lemmas 14.4.2 and 14.1.13 give a solution f of (14.4.1) with $f - \widehat{f} \preccurlyeq \mathfrak{e}$. Now $Q(f) = 0$, so $A := Q$ works. (Recall that $d \geqslant 2$.) □

Strictly speaking, Corollary 14.4.3 has been derived assuming that $D_G = N_G$ and $R_G \prec^\flat G$, but it holds without this assumption, using a compositional conjugation as indicated in the beginning of this subsection.

Tschirnhaus refinements

Let $\mathfrak{f} := \mathfrak{d}_{\widehat{f}}$, put $H := P_{\times \mathfrak{f}} \in K\{Y\}$ and $w := \operatorname{nwt}(H)$, so $w \leqslant d$. If $\mathfrak{e} \succcurlyeq \mathfrak{f}$, then case (i) of Proposition 14.4.1 holds for $f = 0$, with $A := Y$. For the rest of the proof of this proposition we assume $\mathfrak{e} \preccurlyeq \mathfrak{f}$. Then

$$d = \operatorname{ndeg}_{\preccurlyeq \mathfrak{e}} P_{+\widehat{f}} \leqslant \operatorname{ndeg}_{\preccurlyeq \mathfrak{f}} P_{+\widehat{f}} = \operatorname{ndeg}_{\preccurlyeq \mathfrak{f}} P \leqslant d,$$

so $\operatorname{ndeg} H = d$. Towards proving Proposition 14.4.1 we may compositionally conjugate by an active element of $\mathfrak{M}$ to arrange that $D_H = N_H \in C[Y](Y')^w$ and $R_H \prec^\flat H$. Then $D_{H^\phi} = N_H$ and $R_{H^\phi} \prec^\flat H^\phi$ for $\phi \preccurlyeq 1$, by Lemma 13.1.9. Let ∂_0 and ∂_1 be the K-derivations $\partial/\partial Y$ and $\partial/\partial Y'$ on $K\{Y\}$ and define $\Delta \in K[[\partial]]$ by

$$\Delta := \begin{cases} \left(\partial_0^{d-1-w}\partial_1^w\right)_{\times \mathfrak{f}} & \text{if } w \leqslant d-1, \\ \left(\partial_1^{d-1}\right)_{\times \mathfrak{f}} & \text{if } w = d, \end{cases}$$

and set $Q := \Delta P \in K\{Y\}$. If $w \leqslant d-1$, then $Q_{\times \mathfrak{f}} = \partial_0^{d-1-w}\partial_1^w H$, while if $w = d$, then $Q_{\times \mathfrak{f}} = \partial_1^{d-1} H$. Thus by Corollary 13.1.12,

$$N_{Q_{\times \mathfrak{f}}} = D_{Q_{\times \mathfrak{f}}}, \qquad R_{Q_{\times \mathfrak{f}}} \prec^\flat Q_{\times \mathfrak{f}}, \qquad \operatorname{ndeg} Q_{\times \mathfrak{f}} = \operatorname{ddeg} Q_{\times \mathfrak{f}} = 1,$$

so $\operatorname{ddeg} Q^\phi_{\times \mathfrak{f}} = 1$ for all $\phi \preccurlyeq 1$, and the asymptotic equation

$$(14.4.2) \qquad Q(Y) = 0, \qquad Y \preccurlyeq \mathfrak{f}$$

over $\widehat{K}$ is quasilinear. Thus by Corollary 6.6.11:

LEMMA 14.4.4. *Suppose* $f \in \widehat{K}$ *is a solution of* (14.4.2). *Then for all* $\mathfrak{g} \in \mathfrak{M}$ *with* $\mathfrak{g} \preccurlyeq \mathfrak{f}$ *we have* $\operatorname{mul} Q_{+f,\times \mathfrak{g}} = 1$ *and* $N_{Q_{+f,\times \mathfrak{g}}} = D_{Q_{+f,\times \mathfrak{g}}} \in C^\times Y \cup C^\times Y'$. *Thus* Q_{+f} *has no algebraic starting monomials* $\preccurlyeq \mathfrak{f}$, *and if* $\mathfrak{g} \in \mathfrak{M}$ *with* $\mathfrak{g} \preccurlyeq \mathfrak{f}$ *is a starting monomial for* Q_{+f}, *then each* $\widehat{g} \asymp \mathfrak{g}$ *in* $\widehat{K}$ *is an approximate zero of* Q_{+f}.

LEMMA 14.4.5. *The element* $\widehat{f}$ *of* $\widehat{K}$ *is an approximate solution of* (14.4.2).

PROOF. Take $c \in C^\times$ such that $\widehat{f} \sim c\mathfrak{f}$. Since $\widehat{f}$ is an approximate zero of P of multiplicity $d = \operatorname{ndeg} P_{\times\mathfrak{f}}$, Corollary 4.3.4 gives $N_H = a\,(Y-c)^{d-w}\,(Y')^w$ where $a \in C^\times$. If $w \leqslant d-1$, then $N_{Q_{\times\mathfrak{f}}} = a\,(d-w)!w!\,(Y-c)$ by Corollary 13.1.12, so $N_{Q_{\times\mathfrak{f}}}(c) = 0$. If $w = d$, then $N_{Q_{\times\mathfrak{f}}} = a\,d!\,Y'$, and again $N_{Q_{\times\mathfrak{f}}}(c) = 0$. □

Let $f \in \widehat{K}$ with $f \sim \widehat{f}$; then $\operatorname{ndeg}_{\prec\mathfrak{f}} P_{+f} = \operatorname{ndeg}_{\prec\mathfrak{f}} P_{+\widehat{f}} = d$, that is, the refinement

(T) $$P_{+f}(Y) = 0, \qquad Y \prec \mathfrak{f}$$

of $(\widehat{\mathrm{E}})$ still has Newton degree d. Moreover, the refinement

(ΔT) $$Q_{+f}(Y) = 0, \qquad Y \prec \mathfrak{f}$$

of (14.4.2) is also quasilinear: use Lemmas 14.1.13 and 14.4.5, and Corollary 14.1.14 to get a solution $f_0 \in \widehat{K}$ of (14.4.2) that best approximates $\widehat{f}$; then $f_0 \sim \widehat{f} \sim f$, and thus $\operatorname{ndeg}_{\prec\mathfrak{f}} Q_{+f} = \operatorname{ndeg}_{\prec\mathfrak{f}} Q_{+f_0} = 1$ by Lemma 14.4.4, as claimed.

DEFINITION 14.4.6. A **Tschirnhaus refinement** of $(\widehat{\mathrm{E}})$ is an asymptotic equation (T) over $\widehat{K}$ with $\widehat{f} \sim f \in \widehat{K}$ such that some solution $f_0 \in \widehat{K}$ of (14.4.2) (taken over $\widehat{K}$) best approximates $\widehat{f}$ and satisfies $f_0 - \widehat{f} \sim f - \widehat{f}$. Given $f, \widehat{g} \in \widehat{K}$ and $\mathfrak{m}$ with $\mathfrak{m} \prec f - \widehat{f} \preccurlyeq \widehat{g} \prec \mathfrak{f}$ (so $f \sim \widehat{f}$), the refinement

(TC) $$P_{+(f+\widehat{g})}(Y) = 0, \qquad Y \preccurlyeq \mathfrak{m}$$

of (T) is said to be **compatible** with (T) if it has Newton degree d and $\widehat{g}$ is not an approximate solution of (ΔT). In this compatibility definition we do not require (T) to be a Tschirnhaus refinement of $(\widehat{\mathrm{E}})$.

LEMMA 14.4.7. *Let $f, f_0, \widehat{g} \in \widehat{K}$ and $\mathfrak{m}$ be such that $\mathfrak{m} \prec f_0 - \widehat{f} \sim f - \widehat{f} \preccurlyeq \widehat{g} \prec \mathfrak{f}$, and* (TC) *has Newton degree d. Then $\widehat{g}$ is an approximate solution of* (T) *and of*

(T_0) $$P_{+f_0}(Y) = 0, \qquad Y \prec \mathfrak{f}.$$

PROOF. Since $\operatorname{ndeg}_{\preccurlyeq\mathfrak{m}} P_{+(f+\widehat{g})} = d = \operatorname{ndeg}_{\prec\mathfrak{f}} P_{+f}$, Lemma 13.8.2 yields that $\widehat{g}$ is an approximate solution of (T). Since $f_0 - f \prec \widehat{f} - f \preccurlyeq \widehat{g}$, we have $\operatorname{ndeg}_{\prec\widehat{g}} P_{+(f_0+\widehat{g})} = \operatorname{ndeg}_{\prec\widehat{g}} P_{+(f+\widehat{g})} \geqslant 1$. Hence $\widehat{g}$ is also an approximate solution of (T_0). □

LEMMA 14.4.8. *Let $f, f_0, \widehat{g} \in \widehat{K}$ satisfy $f_0 - \widehat{f} \sim f - \widehat{f} \preccurlyeq \widehat{g} \prec \mathfrak{f}$. Then $\widehat{g}$ is an approximate solution of* (ΔT) *if and only if $\widehat{g}$ is an approximate solution of*

($\Delta\mathrm{T}_0$) $$Q_{+f_0}(Y) = 0, \qquad Y \prec \mathfrak{f}.$$

PROOF. We have $\operatorname{ndeg}_{\prec\widehat{g}} Q_{+(f_0+\widehat{g})} = \operatorname{ndeg}_{\prec\widehat{g}} Q_{+(f+\widehat{g})}$, since $f_0 - f \prec \widehat{f} - f \preccurlyeq \widehat{g}$. □

In the rest of this subsection we fix a Tschirnhaus refinement (T) of $(\widehat{\mathrm{E}})$. If $\mathfrak{e} \prec f - \widehat{f}$, then compatible refinements of (T) are easy to come by:

LEMMA 14.4.9. *Suppose* $\mathfrak{e} \prec f - \widehat{f}$. *Then taking* $\widehat{g} := \widehat{f} - f$ *and* $\mathfrak{m} := \mathfrak{e}$, *the refinement* (TC) *of* (T) *is compatible.*

PROOF. (TC) has Newton degree d, since $\operatorname{ndeg}_{\preccurlyeq\mathfrak{e}} P_{+(f+\widehat{g})} = \operatorname{ndeg}_{\preccurlyeq\mathfrak{e}} P_{+\widehat{f}} = d$. Take a solution $f_0 \in \widehat{K}$ of (14.4.2) that best approximates $\widehat{f}$ with $f - \widehat{f} \sim f_0 - \widehat{f}$. By Lemma 14.4.8, if $\widehat{g}$ is an approximate solution of (ΔT), then $\widehat{g}$ is also an approximate solution of ($\Delta\mathrm{T}_0$). But ($\Delta\mathrm{T}_0$) has no approximate solution $\sim \widehat{f} - f_0$ in $\widehat{K}$: if it had one, then by Lemma 14.1.14 it has a solution $y \sim \widehat{f} - f_0$, so $Q(f_0 + y) = 0$, and $f_0 + y$ would be a solution of (14.4.2) with $f_0 + y - \widehat{f} = y - (\widehat{f} - f_0) \prec \widehat{f} - f_0$, contradicting that f_0 is a best approximation. Thus (TC) is compatible. □

We now describe the effect of multiplicatively conjugating by $\mathfrak{f}$ on the above:

REMARK 14.4.10. Consider the asymptotic equation

$$(\mathfrak{f}^{-1}\mathrm{E}) \qquad P_{\times\mathfrak{f}}(Y) = 0, \qquad Y \in \mathfrak{f}^{-1}\mathcal{E}$$

over K. By Lemma 13.8.6, $(\mathfrak{f}^{-1}\widehat{f}, \mathfrak{f}^{-1}\widehat{\mathcal{E}}')$ is an unraveler for

$$(\mathfrak{f}^{-1}\widehat{\mathrm{E}}) \qquad P_{\times\mathfrak{f}}(Y) = 0, \qquad Y \in \mathfrak{f}^{-1}\widehat{\mathcal{E}}$$

over $\widehat{K}$, and $\operatorname{ndeg}_{\prec 1}(P_{\times\mathfrak{f}})_{+\mathfrak{f}^{-1}\widehat{f}} = \operatorname{ndeg}_{\mathfrak{f}^{-1}\widehat{\mathcal{E}}} P_{\times\mathfrak{f}} = d$. Moreover,

$$(\mathfrak{f}^{-1}\mathrm{T}) \qquad (P_{\times\mathfrak{f}})_{+\mathfrak{f}^{-1}f}(Y) = 0, \qquad Y \prec 1$$

is a Tschirnhaus refinement of $(\mathfrak{f}^{-1}\widehat{\mathrm{E}})$, and if (TC) is a compatible refinement of (T), then

$$(\mathfrak{f}^{-1}\mathrm{TC}) \qquad (P_{\times\mathfrak{f}})_{+\mathfrak{f}^{-1}f+\mathfrak{f}^{-1}\widehat{g}}(Y) = 0, \qquad Y \preccurlyeq \mathfrak{f}^{-1}\mathfrak{m}$$

is a compatible refinement of $(\mathfrak{f}^{-1}\mathrm{T})$.

The Slowdown Lemma

Let (T) be a Tschirnhaus refinement of $(\widehat{\mathrm{E}})$ and (TC) a compatible refinement of (T), and set $\mathfrak{g} = \mathfrak{d}_{\widehat{g}}$. The goal of this subsection is the proof of the following lemma to the effect that the step from (E) to (T) is much larger than the step from (T) to (TC):

LEMMA 14.4.11 (Slowdown). $\dfrac{\mathfrak{m}}{\mathfrak{g}} \lll \dfrac{\mathfrak{g}}{\mathfrak{f}}$, *where* $\mathfrak{m}$ *is the monomial appearing in* (TC).

We first establish an auxiliary result used in the proof of Lemma 14.4.11:

LEMMA 14.4.12. *Suppose* $\mathfrak{f} = 1$. *Then* $Q_{+f}(\widehat{g}) \asymp_{\mathfrak{g}} \mathfrak{g}\, Q_{+f}$.

PROOF. Let $f_0 \in \widehat{K}$ be a solution of (14.4.2) that best approximates $\widehat{f}$ and satisfies $f - \widehat{f} \sim f_0 - \widehat{f}$. Then $f_0 \sim f \sim \widehat{f} \asymp \mathfrak{f} = 1$. Take $c \in C^{\times}$ such that $f_0 \sim c$.

CLAIM 1: $f_0 - c \prec^{\flat} 1$.

Suppose otherwise. Then $f_0 - c \succeq^\flat 1$. By Corollary 13.1.12,

$$Q = \mathfrak{d}_Q D_Q + R_Q \qquad \text{where } D_Q \in C^\times(Y-c) \cup C^\times Y', \text{ and } R_Q \prec^\flat Q,$$

hence $Q_{+c} = \mathfrak{d}_Q (D_Q)_{+c} + (R_Q)_{+c}$ with $(R_Q)_{+c} \asymp R_Q \prec^\flat Q \asymp Q_{+c}$, so

$$\mathfrak{d}_{Q_{+c}} = \mathfrak{d}_Q, \quad D_{Q_{+c}} = (D_Q)_{+c} \in C^\times Y \cup C^\times Y', \quad R_{Q_{+c}} = (R_Q)_{+c} \prec^\flat Q_{+c}.$$

Therefore, with $\mathfrak{d} = \mathfrak{d}_{f_0 - c}$ we have $N_{Q_{+c,\times\mathfrak{d}}} \in C^\times Y$ by Lemma 13.2.16, so $\mathfrak{d}$ is not a starting monomial for Q_{+c}, contradicting $Q_{+c}(f_0 - c) = Q(f_0) = 0$.

CLAIM 2: $\mathfrak{g} \prec^\flat 1$.

Suppose not. Then $\mathfrak{g} \succeq^\flat 1$. As in the proof of Claim 1 we get $D_{P_{+c}} \in C[Y](Y')^{\mathbb{N}}$ and $R_{P_{+c}} \prec^\flat P_{+c}$, so by Claim 1 and Corollary 13.2.17,

$$N_{P_{+f_0,\times\mathfrak{g}}} = N_{P_{+c,+(f_0-c),\times\mathfrak{g}}} =_{\mathrm{c}} N_{P_{+c,\times\mathfrak{g}}} \in C^\times Y^{\mathbb{N}}.$$

Therefore $\widehat{g}$ is not an approximate zero of P_{+f_0}, contradicting Lemma 14.4.7.

CLAIM 3: $Q_{+f}(\widehat{g}) \asymp_{\mathfrak{g}} Q_{+f,\times\mathfrak{g}}$.

Since $\widehat{g}$ is not an approximate solution of (ΔT), it is not an approximate solution of ($\Delta \mathrm{T}_0$) by Lemma 14.4.8, and therefore $N_{Q_{+f,\times\mathfrak{g}}} = N_{Q_{+f_0,\times\mathfrak{g}}} \in C^\times Y$ by Corollary 13.2.15 and Lemma 14.4.4. Take $\phi \preccurlyeq 1$ such that $D_{Q^\phi_{+f,\times\mathfrak{g}}} = N_{Q_{+f,\times\mathfrak{g}}}$. Then

$$Q^\phi_{+f,\times\mathfrak{g}} = aY + R \qquad \text{where } a \in K^\times, R \in \widehat{K}^\phi\{Y\}, R \prec a,$$

so with $\widehat{g} = u\mathfrak{g}$, $u \asymp 1$ in $\widehat{K}$,

$$Q_{+f}(\widehat{g}) = Q_{+f,\times\mathfrak{g}}(u) = Q^\phi_{+f,\times\mathfrak{g}}(u) = au + R(u) \sim au$$

and hence $Q_{+f}(\widehat{g}) \asymp a \asymp Q^\phi_{+f,\times\mathfrak{g}}$. We have $Q^\phi_{+f,\times\mathfrak{g}} \asymp_{\mathfrak{g}} Q_{+f,\times\mathfrak{g}}$ by Claim 2 and Corollary 11.1.13.

By Claim 3 and Lemma 4.5.1(i) we have

$$Q_{+f}(\widehat{g}) \asymp_{\mathfrak{g}} Q_{+f,\times\mathfrak{g}} = Q_{\times\mathfrak{g},+f/\mathfrak{g}} \sim Q_{\times\mathfrak{g},+f_0/\mathfrak{g}} = Q_{+f_0,\times\mathfrak{g}}.$$

Corollary 9.4.19 and $\operatorname{mul}(Q_{+f_0}) = \operatorname{ddeg}(Q_{+f_0}) = 1$ give $Q_{+f_0,\times\mathfrak{g}} \asymp_{\mathfrak{g}} \mathfrak{g}\, Q_{+f_0}$. Putting everything together, we get $Q_{+f}(\widehat{g}) \asymp_{\mathfrak{g}} \mathfrak{g}\, Q_{+f_0} \sim \mathfrak{g}\, Q_{+f}$. □

PROOF OF LEMMA 14.4.11. Until further notice we assume $\mathfrak{f} = 1$. Put $F := P_{+f}$ and $G := Q_{+f}$. Note: $\operatorname{ddeg} F_{+\widehat{g}} = \operatorname{ddeg} F = \operatorname{ddeg} P = d$, by Lemma 6.6.5.

CLAIM 1: $\mathfrak{g}\,(F_{+\widehat{g}})_d \preccurlyeq_{\mathfrak{g}} (F_{+\widehat{g}})_{d-1}$.

We have $G(\widehat{g}) \asymp_{\mathfrak{g}} \mathfrak{g}\, G$ by Lemma 14.4.12. If $w \leqslant d-1$, then by (4.3.3),

$$G(\widehat{g}) = Q(f + \widehat{g}) = \left(\partial_0^{d-1-w} \partial_1^w P\right)(f + \widehat{g})$$

is, up to a factor from $\mathbb{Q}^\times$, the coefficient of $Y^{d-1-w}(Y')^w$ in $P_{+(f+\widehat{g})} = F_{+\widehat{g}}$, and if $w = d$, then $G(\widehat{g})$ is likewise, up to a factor from $\mathbb{Q}^\times$, the coefficient of $(Y')^{d-1}$ in $F_{+\widehat{g}}$. Thus $\mathfrak{g}\, G \preccurlyeq_{\mathfrak{g}} (F_{+\widehat{g}})_{d-1}$. Since $G = Q_{+f} \asymp Q \asymp P$ and $F_{+\widehat{g}} \sim F \asymp P$ with $\operatorname{ddeg} F_{+\widehat{g}} = d$, we get $G \asymp F_{+\widehat{g}} \asymp (F_{+\widehat{g}})_d$, and the claim follows.

CLAIM 2: $\mathfrak{n} \prec_{\mathfrak{g}} \mathfrak{g} \implies (F_{+\widehat{g},\times\mathfrak{n}})_d \prec_{\mathfrak{n}} (F_{+\widehat{g},\times\mathfrak{n}})_{d-1}$.

Suppose $\mathfrak{n} \prec_{\mathfrak{g}} \mathfrak{g}$. Then $\mathfrak{n} \prec_{\mathfrak{n}} \mathfrak{g}$ by Lemma 9.4.15. Also $(F_{+\widehat{g}})_d \neq 0$, hence

$$\mathfrak{n}\,(F_{+\widehat{g}})_d \prec_{\mathfrak{n}} \mathfrak{g}\,(F_{+\widehat{g}})_d.$$

From $\mathfrak{n} \preccurlyeq \mathfrak{g} \prec 1$ we get $\mathfrak{g} \preccurlyeq\!\!\!\prec \mathfrak{n}$, and so by Claim 1 and Lemma 9.4.16,

$$\mathfrak{g}\,(F_{+\widehat{g}})_d \preccurlyeq_{\mathfrak{n}} (F_{+\widehat{g}})_{d-1}.$$

This, together with the previous display, yields $\mathfrak{n}\,(F_{+\widehat{g}})_d \prec_{\mathfrak{n}} (F_{+\widehat{g}})_{d-1}$.

By Lemma 9.4.18 we have for all $i \in \mathbb{N}$,

$$(F_{+\widehat{g},\times\mathfrak{n}})_i = \big((F_{+\widehat{g}})_i\big)_{\times\mathfrak{n}} \asymp_{\mathfrak{n}} \mathfrak{n}^i\,(F_{+\widehat{g}})_i.$$

Using this for $i = d-1$ and $i = d$ yields the claim.

CLAIM 3: $\mathfrak{n} \prec_{\mathfrak{g}} \mathfrak{g} \implies \operatorname{ndeg} F_{+\widehat{g},\times\mathfrak{n}} \leqslant d-1$.

Assume $\mathfrak{n} \prec_{\mathfrak{g}} \mathfrak{g}$. First we note that $\operatorname{ndeg} F_{+\widehat{g},\times\mathfrak{n}} \leqslant d$, since

$$\operatorname{ndeg} F_{+\widehat{g},\times\mathfrak{n}} \leqslant \operatorname{ndeg} F_{+\widehat{g}} = \operatorname{ndeg} F = \operatorname{ndeg} P_{+f} = \operatorname{ndeg} P = d.$$

From $\mathfrak{g} \prec^{\flat} 1$ by Claim 2 in the proof of Lemma 14.4.12, we get $\mathfrak{n} \prec^{\flat} 1$. It remains to use Corollary 11.1.13 and Claim 2 in the present proof.

Now $\operatorname{ndeg} F_{+\widehat{g},\times\mathfrak{m}} = d$, and so $\mathfrak{g} \preccurlyeq_{\mathfrak{g}} \mathfrak{m}$ by Claim 3, hence $\mathfrak{m}/\mathfrak{g} \asymp_{\mathfrak{g}} 1$, and thus $\mathfrak{m}/\mathfrak{g} \prec\!\!\!\prec \mathfrak{g}$, which is the content of the Slowdown Lemma for $\mathfrak{f} = 1$. The general case is reduced to this special case by multiplicative conjugation as in Remark 14.4.10. □

One more lemma

First an immediate consequence of Lemmas 14.4.9 and 14.4.11:

COROLLARY 14.4.13. *Suppose* (T) *is a Tschirnhaus refinement of* ($\widehat{\mathrm{E}}$). *Then*

$$\mathfrak{e} \prec \widehat{f} - f \implies \frac{\mathfrak{e}}{\widehat{f}-f} \prec\!\!\!\prec \frac{\widehat{f}-f}{\widehat{f}}.$$

LEMMA 14.4.14. *Suppose* (T) *is a Tschirnhaus refinement of* ($\widehat{\mathrm{E}}$) *and* $\mathfrak{e} \prec \widehat{f} - f$. *Then, with* $F := P_{+f}$, $\widehat{g} := \widehat{f} - f$, *and* $\mathfrak{g} := \mathfrak{d}_{\widehat{g}}$, *the asymptotic equation*

$$(\widehat{\mathrm{E}}_{\leqslant d}) \qquad F_{\leqslant d}(Y) = 0, \quad Y \preccurlyeq \mathfrak{g}$$

has Newton degree d. Moreover, $(\widehat{g}, \widehat{\mathcal{E}}'')$, *where* $\widehat{\mathcal{E}}'' := \{y \in \widehat{\mathcal{E}}' : y \prec \mathfrak{g}\}$, *is an unraveler for* $(\widehat{\mathrm{E}}_{\leqslant d})$, *and* $\mathfrak{e}$ *is the largest algebraic starting monomial for the unraveled asymptotic equation*

$$(\widehat{\mathrm{E}}'_{\leqslant d}) \qquad (F_{\leqslant d})_{+\widehat{g}}(Y) = 0, \qquad Y \in \widehat{\mathcal{E}}''$$

over $\widehat{K}$.

PROOF. Using Corollary 14.4.13, apply Lemma 13.8.13 to $\widehat{K}, \widehat{f}, f, \widehat{g}, \widehat{\mathcal{E}}, \widehat{\mathcal{E}}'$ in the role of $K, f, \widetilde{f}, g, \mathcal{E}, \mathcal{E}'$, respectively. □

We are now ready to turn to the proof of Proposition 14.4.1.

Proof of Proposition 14.4.1

By Lemmas 14.4.5, 14.1.14, and 14.1.13 we have a solution $f_0 \sim \widehat{f}$ in $\widehat{K}$ of (14.4.2) that best approximates $\widehat{f}$. If $\widehat{f} - a \preccurlyeq f_0 - a$ for all $a \in K$, then (ii) holds for $f := f_0$, witnessed by $A := Q$. Otherwise, take $f = a \in K$ such that $\widehat{f} - f \succ f_0 - f$, that is, $f_0 - \widehat{f} \sim f - \widehat{f}$. Then $f \sim \widehat{f}$, since $f_0 \sim \widehat{f}$, and so (T) is a Tschirnhaus refinement of $(\widehat{\mathrm{E}})$. If $\widehat{f} - f \preccurlyeq \mathfrak{e}$, then we are in case (i), as witnessed by $A := Y - f$. So assume from now on that $\mathfrak{e} \prec \widehat{f} - f$, and set $\widehat{g} := \widehat{f} - f$, $\mathfrak{g} := \mathfrak{d}_{\widehat{g}}$. Consider the asymptotic equation $(\widehat{\mathrm{E}}_{\leqslant d})$ introduced in Lemma 14.4.14. By that lemma, $(\widehat{\mathrm{E}}_{\leqslant d})$ has Newton degree d, and $(\widehat{g}, \widehat{\mathcal{E}}'')$, where $\widehat{\mathcal{E}}'' = \{y \in \widehat{\mathcal{E}}' : y \prec \mathfrak{g}\}$, is an unraveler for $(\widehat{\mathrm{E}}_{\leqslant d})$, and thus $\operatorname{ndeg}_{\prec \widehat{g}}(F_{\leqslant d})_{+\widehat{g}} = d$. Moreover, $\mathfrak{e}$ is the largest algebraic starting monomial for $(\widehat{\mathrm{E}}'_{\leqslant d})$. Since $f \in K$, we can view $(\widehat{\mathrm{E}}_{\leqslant d})$ as an asymptotic equation over K. Thus Corollary 14.4.3 applies to $(\widehat{\mathrm{E}}_{\leqslant d})$ viewed as an asymptotic equation over K in place of (E) and with $(\widehat{g}, \widehat{\mathcal{E}}'')$ in place of $(\widehat{f}, \widehat{\mathcal{E}}')$. So we get $g \in \widehat{K}$ and $B \in K\{Y\}$ such that $\widehat{g} - g \preccurlyeq \mathfrak{e}$, $B(g) = 0$, $\mathrm{c}(B) < \mathrm{c}(F_{\leqslant d})$, and $\deg B = 1$. Then $f + g$ has the property stated for f in (i) as witnessed by $A := B_{+(-f)}$. □

Easy consequences

We continue in the setting introduced at the beginning of this section. Here are some easy consequences of Proposition 14.4.1:

COROLLARY 14.4.15. *Suppose* (a_ρ) *is a divergent pc-sequence in* K *with pseudolimit* $\widehat{f}$ *and minimal* d*-polynomial* P *over* K. *Then there exist* $f \in \widehat{K}$ *and* $A \in K\{Y\}$ *such that* $\widehat{f} - f \preccurlyeq \mathfrak{e}$, $A(f) = 0$, $\mathrm{c}(A) < \mathrm{c}(P)$, *and* $\deg A = 1$.

PROOF. Suppose not. Then Proposition 14.4.1 gives $f \in \widehat{K}$ and $A \in K\{Y\}^{\neq}$ such that $\widehat{f} - a \preccurlyeq f - a$ for all $a \in K$, $A(f) = 0$, and $\mathrm{c}(A) < \mathrm{c}(P)$. Then $f \notin K$, since $\widehat{f} \notin K$. Take a divergent pc-sequence (b_σ) in K such that $b_\sigma \rightsquigarrow f$. From $\widehat{f} - b_\sigma \preccurlyeq f - b_\sigma$ for all σ, we get $b_\sigma \rightsquigarrow \widehat{f}$, hence (a_ρ) and (b_σ) are equivalent, so $a_\rho \rightsquigarrow f$, and thus $\widehat{f} - a \asymp f - a$ for all $a \in K$. Hence $Z(K, f) = Z(K, \widehat{f})$, using a notion from Section 11.4. Then $A \in Z(K, \widehat{f})$, by Lemma 11.4.3, hence $\mathrm{c}(A) \geqslant \mathrm{c}(P)$ by Corollary 11.4.13, a contradiction. □

COROLLARY 14.4.16. *If K is newtonian, then there is an $f \in K$ with $\widehat{f} - f \preccurlyeq \mathfrak{e}$.*

PROOF. Assume K is newtonian. Take $f \in \widehat{K}$ as in Proposition 14.4.1. Then $f \in K$ by Lemma 14.1.8. In case (ii) of Proposition 14.4.1, this gives $\widehat{f} = f$. □

14.5 NEWTONIZATION

In this section we assume K is ω-free and d*-valued with divisible value group* Γ. We do not assume here the very special setting introduced at the beginning of Section 14.4, but rather reduce to it in establishing Proposition 14.5.1 below. This will then be used to derive Theorem 14.0.2 and related results.

PROPOSITION 14.5.1. *Suppose (a_ρ) is a divergent pc-sequence in K, with minimal differential polynomial G over K, and set $\boldsymbol{a} := c_K(a_\rho)$. Then $\operatorname{ndeg}_{\boldsymbol{a}} G = 1$.*

PROOF. We first arrange by compositional conjugation that K has small derivation. Set $d := \operatorname{ndeg}_{\boldsymbol{a}} G$. Zorn gives an asymptotically d-algebraically maximal immediate d-algebraic extension $\widehat{K}$ of K. Then $\widehat{K}$ is ω-free by Theorem 13.6.1. Now forget how we got $\widehat{K}$: we only use below that it is an ω-free asymptotically d-algebraically maximal immediate extension of K. Note that then $\widehat{K}$ is newtonian by Theorem 14.0.1.

Take $\ell \in \widehat{K}$ such that $a_\rho \rightsquigarrow \ell$. Note that G is an element of $Z(K, \ell)$ of minimal complexity, by Corollary 11.4.13. Lemma 11.4.12 yields $d \geqslant 1$, and provides $a \in K$ and $\mathfrak{v} \in K^\times$ such that $a - \ell \prec \mathfrak{v}$ and $\operatorname{ndeg}_{\prec \mathfrak{v}} G_{+a} = d$.

Suppose towards a contradiction that $d \geqslant 2$. Then Lemma 13.8.11 provides an unraveler $(\widehat{f}, \mathcal{E})$ with $\widehat{f} \neq 0$ for the asymptotic equation

$$G_{+a}(Y) = 0, \qquad Y \prec \mathfrak{v}$$

over $\widehat{K}$ such that $\operatorname{ndeg}_{\prec \widehat{f}} G_{+(a+\widehat{f})} = d$ and $a_\rho \rightsquigarrow a + \widehat{f} + g$ for all $g \in \mathcal{E} \cup \{0\}$. Then $\operatorname{mul} G_{+(a+\widehat{f})} < d$ by Lemma 13.8.10.

Assume temporarily that K has a monomial group. Then we are in the set-up of Section 14.4 with $P := G_{+a}$, and can apply Corollary 14.4.15 to the divergent pc-sequence $(a_\rho - a)$ and $P := G_{+a}$. This yields an $f \in \widehat{K}$ and an $A \in K\{Y\}^{\neq}$ such that $f - \widehat{f} \in \mathcal{E} \cup \{0\}$, $A(f) = 0$, and $\mathrm{c}(A) < \mathrm{c}(P)$. Then $a_\rho - a \rightsquigarrow f$. But P is a minimal d-polynomial of $(a_\rho - a)$ over K, and so we contradict Section 11.4.

To reduce to the case that K has a monomial group, consider the valued differential field $\widehat{K}$ with K as a distinguished subset. This structure has the first-order property that $H \notin Z(K, \ell)$ for all $H \in K\{Y\}$ with $\mathrm{c}(H) < \mathrm{c}(G)$. Passing to an $\aleph_1$-saturated elementary extension of this structure and using Lemma 3.3.39 we arrange that K has a monomial group while preserving the first-order property just mentioned and other relevant first-order properties, but some minor issues arise:

(1) the updated $\widehat{K}$ might not be asymptotically d-algebraically maximal (though it remains an ω-free immediate extension of K);

(2) the old pc-sequence (a_ρ) might acquire a pseudolimit in the updated K.

To deal with (1), replace $\widehat{K}$ by an asymptotically d-algebraically maximal immediate d-algebraic extension of $\widehat{K}$. To deal with (2), replace (a_ρ) by some divergent pc-sequence in K with pseudolimit ℓ. This updated pc-sequence still has minimal differential polynomial G over K, and the Newton degree of G in the cut defined by this pc-sequence remains d, by Corollary 11.4.13 and Lemma 11.4.12. □

This proposition is the analogue of the henselian and d-henselian configuration results 3.3.20 and 7.4.1. Below we derive similar consequences from it.

COROLLARY 14.5.2. *The following are equivalent:*

(i) *K is newtonian;*

(ii) *K is strongly newtonian;*

(iii) *K is asymptotically* d-*algebraically maximal.*

PROOF. The implication (i) ⇒ (ii) follows from Proposition 14.5.1, (ii) ⇒ (iii) holds by Lemma 14.1.11, and (iii) ⇒ (i) by Theorem 14.0.1. □

This result contains Theorem 14.0.2: by Lemma 14.2.5 the standing assumption of this section that K is d-valued is satisfied under the hypotheses of Theorem 14.0.2.

COROLLARY 14.5.3. *Suppose K is newtonian. If C is algebraically closed, then K is weakly differentially closed. If C is real closed, then every $P \in K\{Y\}$ of odd degree has a zero in K.*

PROOF. Combine Corollaries 14.3.6 and 14.3.9 with Corollary 14.5.2. □

COROLLARY 14.5.4. *Let L be a newtonian* d-*algebraic immediate extension of K. Then L is a newtonization of K, as defined in Section 14.1.*

PROOF. Let E be a newtonian extension of K. To embed L over K into E we can assume $K \neq L$; it suffices to show that then $K\langle a\rangle$ can be embedded over K into E for some $a \in L \setminus K$. Since L is d-algebraic over K we can take a divergent pc-sequence (a_ρ) in K with a minimal differential polynomial G over K, by Lemmas 11.4.7, 11.4.8 and Corollary 11.4.13. Then $\operatorname{ndeg}_{\boldsymbol{a}} G = 1$ by Proposition 14.5.1, where $\boldsymbol{a} := c_K(a_\rho)$. Then Lemma 14.1.10 yields $a \in L$ such that $G(a) = 0$ and $a_\rho \rightsquigarrow a$. Since K is ω-free, we also have $\operatorname{ndeg}_{\boldsymbol{a}_E} G = 1$, where $\boldsymbol{a}_E := c_E(a_\rho)$, by Corollary 13.6.13. Then Lemma 14.1.10 gives $b \in E$ with $G(b) = 0$ and $a_\rho \rightsquigarrow b$. The results in Section 11.4 now yield an embedding $K\langle a\rangle \to E$ over K sending a to b. □

Thus K has a newtonization, and any two newtonizations of K are isomorphic over K, by Corollary 14.3.12; this permits us to speak of *the* newtonization of K.

Suppose L is an immediate newtonian extension of K. Then any K-embedding of the newtonization of K into L has image

$$\{f \in L : f \text{ is d-algebraic over } K\},$$

and so we refer to this image as the *newtonization of K in L.*

COROLLARY 14.5.5. *Let L be an immediate ω-free newtonian extension of K. Then L embeds over K into any $|\Gamma|^+$-saturated newtonian extension of K.*

PROOF. By passing to the newtonization of K in L we arrange that K is newtonian. Let K^* be a $|\Gamma|^+$-saturated newtonian extension of K, and let $y \in L \setminus K$; it is enough to show that then $K\langle y\rangle$ can be embedded over K into K^*. Take a divergent pc-sequence (a_ρ) in K with $a_\rho \rightsquigarrow y$. Since K is asymptotically d-algebraically maximal, it follows from Section 11.4 that (a_ρ) is of d-transcendental type over K. The saturation assumption on K^* gives $z \in K^*$ such that $a_\rho \rightsquigarrow z$. Then Section 11.4 gives a valued differential field embedding $K\langle y\rangle \to K^*$ over K sending y to z. □

Preservation of newtonianity

The results of this subsection are not used later, but are included for their intrinsic interest. First a descent property of newtonianity. A special case of it says that K is newtonian if $K[i]$ is newtonian.

COROLLARY 14.5.6. *Let L be an algebraic extension of K such that $L = K(C_L)$. If L is newtonian, then so is K.*

PROOF. Suppose M is an immediate d-algebraic extension of K. Then M is d-valued with constant field $C_M = C$. Using Corollary 4.6.8 we arrange that L and M are common differential subfields of some differential field. Then $M(C_L)$ is a differential field, and by Corollary 4.6.20, the constant field of $M(C_L)$ is C_L. By Lemma 4.6.16, M and C_L are linearly disjoint over $C_M = C$. We now use Proposition 10.5.15: it gives us a valuation on the field $M(C_L)$ that extends the valuation of M with the same value group Γ as M and trivial on C_L, and makes $M(C_L)$ a d-valued field extension of H-type of M. Applying Proposition 10.5.15 to L, in particular its uniqueness part, we see that $M(C_L)$ contains $L = K(C_L)$ as a d-valued subfield. Thus $M(C_L)$ is an immediate d-algebraic extension of L.

Assume L is newtonian. Then $M(C_L) = L = K(C_L)$ by Corollary 14.5.2. Since L is algebraic over K, we have $K(C_L) = K[C_L]$ and $M(C_L) = M[C_L]$ and so $K[C_L] = M[C_L]$. Since M is linearly disjoint from C_L over C, this gives $K = M$. Therefore K is asymptotically d-algebraically maximal, and thus newtonian. □

The next result is obtained by a reduction to Corollary 7.6.8.

PROPOSITION 14.5.7. *If K is newtonian, then so is any algebraic extension of K.*

PROOF. Assume K is newtonian. Recall that in the present chapter ϕ ranges over the active elements of K. Note that each flattening $(K^\phi, v^\flat_\phi)$ is d-henselian.

CLAIM: *Each $(K^\phi, v^\flat_\phi)$ is* d*-algebraically maximal in the sense of Chapter 7.*

To show this, assume towards a contradiction that $(K^\phi, v^\flat_\phi)$ is not d-algebraically maximal. Then $(K^\phi, v^\flat_\phi)$ has a proper immediate d-algebraic extension in the sense of Chapter 7. Such an extension is not required to be asymptotic, but has small derivation, and so is actually H-asymptotic in view of Lemmas 9.4.2 and 9.4.5 applied to $(K^\phi, v^\flat_\phi)$. By uncoarsening (Lemma 9.6.5 and the considerations preceding it), this yields a proper immediate d-algebraic H-asymptotic extension of K^ϕ, and so K^ϕ would not be asymptotically d-algebraically maximal, contradicting Corollary 14.5.2. This proves our claim.

Let L be an algebraic extension of K. Since Γ is assumed to be divisible, we have $\Gamma_L = \Gamma$. Let ϕ be given. The flattening $(L^\phi, v^\flat_\phi)$ of L^ϕ is an extension of $(K^\phi, v^\flat_\phi)$ in the sense of Chapter 7, and is also an algebraic extension. Thus by our claim and Corollary 7.6.8, $(L^\phi, v^\flat_\phi)$ is d-henselian. Since ϕ is arbitrary, Lemma 14.1.4 and the equivalence preceding it (both applied to L instead of K) yield that L is newtonian. □

COROLLARY 14.5.8. *Suppose K is newtonian and real closed. Then $K[i]$, where $i^2 = -1$, is linearly closed and so every monic $A \in K[\partial]^{\neq}$ is a product of monic irreducible operators in $K[\partial]$ of order* 1 *and order* 2.

PROOF. By Proposition 14.5.7 the algebraic closure $K[i]$ of K is newtonian. So by Corollary 14.5.3, $K[i]$ is weakly differentially closed and hence linearly closed by Lemma 5.8.9. The rest now follows from Lemma 5.1.35. □

Newton-Liouville closure

Besides assuming that K is ω-free and d*-valued with divisible value group, we also assume in this subsection that K comes equipped with a field ordering making K an H-field.* By Corollary 14.3.12, the newtonization K^{nt} of K is an immediate extension of K. Hence by Lemma 10.5.8, K^{nt} has a unique field ordering extending that of K in which the valuation ring of K^{nt} is convex; below we consider K^{nt} equipped with this ordering. Then K^{nt} is a newtonian H-field extension of K, and every embedding of K into a newtonian pre-H-field L extends to an embedding $K^{\mathrm{nt}} \to L$ of pre-H-fields.

LEMMA 14.5.9. *There exists a newtonian Liouville closed H-field extension K^{nl} of K which embeds over K into every newtonian Liouville closed H-field extension of K. Any such K^{nl} is* d*-algebraic over K, hence ω-free, and its constant field is a real closure of C.*

PROOF. We define inductively an infinite tower $K_0 \subseteq K_1 \subseteq K_2 \subseteq \cdots$ of ω-free H-field extensions of K with divisible value group as follows. Set $K_0 := K$, and assume inductively that K_n is an ω-free H-field extension of K with divisible value group. For even n we let K_{n+1} be the Liouville closure of K_n. For odd n we let $K_{n+1} := (K_n)^{\mathrm{nt}}$ be the newtonization of K_n. In both cases, K_{n+1} has divisible value group, and K_{n+1} is d-algebraic over K_n and thus remains ω-free, by Theorem 13.6.1. Thus $K^{\text{nl}} := \bigcup_n K_n$ is a newtonian Liouville closed H-field extension of K with the desired semiuniversal property. The second part of the lemma follows easily from this semiuniversal property. □

Let E be an ω-free H-field. We extend Lemma 14.5.9 to E by applying it to the ω-free real closed H-field extension $K := E^{\text{rc}}$ of E:

COROLLARY 14.5.10. *There is a newtonian Liouville closed H-field extension E^{nl} of E which embeds over E into every newtonian Liouville closed H-field extension of E. Any such E^{nl} is* d-*algebraic over E, thus ω-free, and its constant field is a real closure of C_E.*

COROLLARY 14.5.11. *Every pre-H-field extends to an ω-free newtonian Liouville closed H-field.*

PROOF. Any pre-H-field extends to an ω-free H-field by Corollary 11.7.18. Now apply Corollary 14.5.10. □

A **Newton-Liouville closure** of E is by definition a newtonian Liouville closed H-field extension E^{nl} of E with the embedding property stated in Corollary 14.5.10. Thus E has a Newton-Liouville closure. In Section 16.1 below we show that E has up to isomorphism over E a unique Newton-Liouville closure.

Schwarz closure

Let E be an ω-free H-field.

PROPOSITION 14.5.12. *There exists a Schwarz closed H-field extension of E that embeds over E into any Schwarz closed H-field extension of E.*

PROOF. Let E^{nl} be a Newton-Liouville closure of E. Then E^{nl} is Schwarz closed by Corollary 14.2.20, and its constant field is a real closure of C_E. Let E^{s} be the intersection of all Schwarz closed H-subfields of E^{nl} that contain E. Then E^{s} is a Schwarz closed H-field extension of E by Lemma 11.8.36. We show that it has the desired semiuniversal property. Let F be a Schwarz closed H-field extension of E. Take a Newton-Liouville closure F^{nl} of F. Then we have an H-field embedding $i\colon E^{\text{nl}} \to F^{\text{nl}}$ over E, so $i(E^{\text{s}})$ and F are Schwarz closed H-subfields of F^{nl}. Now F and F^{nl} have the same constant field, so $i(E^{\text{s}}) \cap F$ is a Schwarz closed H-subfield of $i(E^{\text{s}})$ containing E, by Lemma 11.8.35. By the minimality property of E^{s} this gives $i(E^{\text{s}}) \subseteq F$. □

Define a **Schwarz closure** of E to be a Schwarz closed H-field extension of E that embeds over E into every Schwarz closed H-field extension of E. So E has a Schwarz closure by Proposition 14.5.12, and because of the obvious minimality property of the Schwarz closure constructed in its proof, any two Schwarz closures of E are isomorphic over E. This allows us to speak of *the* Schwarz closure E^{s} of E. Thus by the proof of 14.5.12:

(1) E^{s} is d-algebraic over E,

(2) the constant field of E^{s} is a real closure of C_E,

(3) E^{s} has no proper Schwarz closed H-subfield containing E.

Chapter Fifteen

Newtonianity of Directed Unions

In this brief chapter we prove an analogue of Hensel's Lemma for ω-free d-valued fields of H-type: Theorem 15.0.1. *Throughout this chapter K is an H-asymptotic field with asymptotic couple (Γ, ψ), and γ ranges over Γ.*

THEOREM 15.0.1. *If K is* d*-valued with $\partial K = K$, and K is a directed union of spherically complete grounded* d*-valued subfields, then K is newtonian.*

Note that by Corollary 11.7.15, any K as in the hypothesis of this theorem is ω-free. The proof of the theorem is given in Section 15.4. As special cases we shall obtain at the end of that section:

COROLLARY 15.0.2. *The ω-free H-fields $\mathbb{T}$ and $\mathbb{T}_{\log}$ are newtonian.*

15.1 FINITELY MANY EXCEPTIONAL VALUES

In this section we assume $\sup \Psi = 0$, *and we let y range over $K^\times$.* Let $A \in K[\partial]^{\neq}$. Recall from Section 5.6 that

$$\mathscr{E}(A) = \{vy : A(y) \prec Ay\} = \{vy : \operatorname{dwm}(Ay) > 0\}$$

is the set of exceptional values for A. We have $v(\ker^{\neq} A) \subseteq \mathscr{E}(A)$, so knowing $\mathscr{E}(A)$ helps in locating the solutions in K of the differential equation $A(y) = 0$. If K is d-valued, then

$$\mathscr{E}(A) = \{\gamma : \operatorname{dwm}_A(\gamma) > 0\},$$

and $\dim_C \ker A \leqslant |\mathscr{E}(A)|$ by Lemma 5.6.6.

If K extends to a d-henselian asymptotic field, then $|\mathscr{E}(A)| \leqslant \operatorname{order} A$ for all $A \in K[\partial]^{\neq}$, by Lemma 7.5.2. Using this fact we prove:

LEMMA 15.1.1. *If K is pre-*d*-valued, then $|\mathscr{E}(A)| \leqslant$* order *$A$ for all $A \in K[\partial]^{\neq}$.*

PROOF. Assume K is pre-d-valued. Consider first the case that K is grounded (so $\max \Psi = 0$). Then K has an ω-free d-valued extension L of H-type, by Corollary 11.7.18. In view of Theorem 14.0.1 and the subsequent remarks we can pass to an immediate extension of L and arrange in this way that L is also newtonian. Then the flattening $(L, v^\flat)$ of L is d-henselian by Lemma 14.1.2, and the inclusion map $K \to L$ is actually an embedding of K into this flattening. If K is ungrounded, then $0 \notin \Psi$, and this case is taken care of by Corollary 10.1.14. □

15.2 INTEGRATION AND THE EXTENSION $K(x)$

In this section K is d*-valued with* $\sup \Psi = 0$. Then the equation $y' = 1$ has no solution in K, and so we adjoin a solution:

LEMMA 15.2.1. *Let $K(x)$ be a field extension of K with x transcendental over K. Then there is a unique pair consisting of a derivation of $K(x)$ and a valuation ring of $K(x)$ that makes $K(x)$ a* d*-valued extension of H-type of K with $x' = 1$ and $x \succ 1$. This extension $K(x)$ has the same constant field as K, has value group $\Gamma + \mathbb{Z}vx$ with $\Gamma^{<} < nvx < 0$ for all $n \geqslant 1$, and has small derivation.*

PROOF. If $\Psi < 0$, use Lemma 10.2.2 and subsequent remarks. If $\max \Psi = 0$, use Lemma 10.2.3 and the remarks following its proof. $\square$

Note that the extension $K(x)$ of K described in Lemma 15.2.1 is grounded with $\max \Psi_{K(x)} = -vx > 0$, has $K[x]$ as a differential subring, and that

$$\mathbb{Z}vx \;=\; \{\alpha \in \Gamma_{K(x)} : \psi(\alpha) > 0\} \;=\; \Gamma^{\flat}_{K(x)}.$$

LEMMA 15.2.2. *Assume K is spherically complete. Then $K = \partial K + C = \partial(K + Cx)$.*

PROOF. By Lemma 10.2.4 and the remarks following it we get $\mathcal{O} = \partial\mathcal{O}$. Now let $s \in K$; we need to show $s \in \partial K + C$. For this we may assume that $s \notin \partial K$ and hence $s - a' \succcurlyeq 1$ for all $a \in K$, since $\mathcal{O} = \partial\mathcal{O}$. From Lemma 10.2.6 it follows that

$$S := \{v(s - a') : a \in K\} \subseteq \Gamma^{\leqslant}$$

has a largest element β, and so $\beta = 0$ by Lemma 10.2.5. Take $a \in K$ with $v(s - a') = 0$, next take $c \in C^{\times}$, $\varepsilon \in \mathcal{O}$ such that $s - a' = c(1 + \varepsilon)$, and then take $b \in \mathcal{O}$ such that $\varepsilon = b'$. Then $s = (a + cb)' + c \in \partial K + C$. $\square$

COROLLARY 15.2.3. *Suppose K is spherically complete. Then $K[x] = \partial\big(K[x]\big)$.*

PROOF. Let $a \in K$; we show that $ax^n = f'$ for some $f \in K[x]$. For $n = 0$, use Lemma 15.2.2. Let $n \geqslant 1$ and take $b \in K$, $c \in C$ with $a = b' + c$, and inductively take $g \in K[x]$ with $g' = bx^{n-1}$. Set $f := \frac{c}{n+1}x^{n+1} + bx^n - ng$. Then $f' = ax^n$. $\square$

COROLLARY 15.2.4. *Let E be a spherically complete* d*-valued field of H-type with an element ϕ such that $v\phi = \max \Psi_E$. Let F be a differential field extension of E such that $\phi \in \partial F$. Then $E \subseteq \partial F$.*

PROOF. Take $x \in F$ with $\partial x = \phi$. Let $\delta := \phi^{-1}\partial$ be the derivation of F^{ϕ}. Then the standing assumption of this section is satisfied for $K := E^{\phi}$, since $\max \Psi_K = 0$. Also $\delta x = 1$, hence by Lemma 15.2.2 with δ instead of ∂, $K \subseteq \delta(K) + C = \delta(K + Cx) \subseteq \delta(F^{\phi})$, that is, $E \subseteq \phi^{-1}\partial F$, and thus $E = \phi E \subseteq \partial F$. $\square$

Notes and comments

In the next volume we extend Corollary 15.2.3 as follows: if K is spherically complete and $A \in K[\partial]^{\neq}$, then $K[x] = A\big(K[x]\big)$.

15.3 APPROXIMATING ZEROS OF DIFFERENTIAL POLYNOMIALS

In this section K is d*-valued with* $\sup\Psi = 0$*, we fix a* $P \in K\{Y\}^{\neq}$ *of order* $\leqslant r$*, and we let a be an element of $\mathcal{O}$ such that* $\operatorname{dmul} P_{+a} = 1$.

We also consider the condition $\operatorname{ddeg} P_{+a} = 1$, equivalent to $P_{+a} \sim (P_{+a})_1$ under our standing assumption that $\operatorname{dmul} P_{+a} = 1$.

We have $P(a) \prec (P_{+a})_1 \asymp P_{+a} \asymp P$. Take $\mathfrak{d} \in K^\times$ with $\mathfrak{d} \asymp P$. Then $\mathfrak{d}^{-1}P \in \mathcal{O}\{Y\}$ is in dh-position at a as defined in Section 7.1. We have elements $a_0, \dots, a_r \in K$ with $(P_{+a})_1 = a_0Y + a_1Y' + \cdots + a_rY^{(r)}$, so

$$A := L_{P_{+a}} = a_0 + a_1\partial + \cdots + a_r\partial^r \in K[\partial]^{\neq}.$$

As in Section 7.2 where $P \asymp 1$, let $v(P,a)$ be the unique $\alpha \in \Gamma_\infty$ with $v_A(\alpha) = v(P(a))$; thus $v(P,a) = v(\mathfrak{d}^{-1}P, a)$, and $v(P,a) = \infty$ iff $P(a) = 0$. We use $v(P,a)$ as a measure of how close a is to a potential zero of P.

Throughout the rest of this section we assume $P(a) \neq 0$. Thus $v(P,a) \in \Gamma^>$.

LEMMA 15.3.1. *Let* $b \in K$, $v(a-b) \geqslant v(P,a)$, $P(b) \prec P(a)$, *and* $B := L_{P_{+b}}$. *Then:*

(i) $\operatorname{dmul} P_{+b} = 1$;

(ii) $v(a-b) = v(P,a)$ *and* $v(P,b) > v(P,a)$;

(iii) *for all* $y \in K^\times$, *if* $vy = O\big(v(P,a)\big)$ *and* $A(y) \prec Ay$, *then* $B(y) \prec By$;

(iv) $\big\{\alpha \in \mathscr{E}(A) : \alpha = O\big(v(P,a)\big)\big\} \subseteq \mathscr{E}(B)$;

(v) *if* $\operatorname{ddeg} P_{+a} = 1$, *then* $\operatorname{ddeg} P_{+b} = 1$.

PROOF. For (i)–(iv), apply Lemma 7.2.4 to $\mathfrak{d}^{-1}P$; for (v), use Lemma 6.6.5. □

In Propositions 15.3.3 and 15.3.6 below we indicate how to improve $\gamma := v(P,a)$.

LEMMA 15.3.2. *If* $\gamma \notin \mathscr{E}(A)$, *then* $P(b) \prec P(a)$ *for some* $b \in K$ *with* $v(a-b) = \gamma$.

PROOF. We adapt an argument from the proofs of Lemmas 7.2.2 and 7.2.3. Take $g \in K^\times$ with $vg = \gamma$, and set $Q := P_{+a}$, $L := P(a)^{-1}Q_{1,\times g} \in K\{Y\}$. Then $g \prec 1$ and $L \asymp 1$, so $L = \sum_{i=0}^r b_iY^{(i)}$ with $b_0, \dots, b_r \in \mathcal{O}$ and $b_i \asymp 1$ for some $i \in \{0, \dots, r\}$. As in the proof of Lemma 7.2.2, we get

$$P(a+gY) = P(a)\cdot\big(1 + L(Y) + R(Y)\big) \quad \text{where } \operatorname{mul} R \geqslant 2 \text{ and } R \prec 1.$$

Now assume that $\gamma \notin \mathscr{E}(A)$. Then $\operatorname{dwm}(Q_{1,\times g}) = \operatorname{dwm}(Ag) = 0$, so $b_0 \asymp 1$. Take $c \in C^\times$ with $b_0c \sim -1$. Then $1 + L(c) = 1 + b_0c \prec 1$ and so

$$P(a+gc) = P(a)\cdot\big(1 + L(c) + R(c)\big) \prec P(a).$$

Hence $b := a + gc$ has the desired properties. □

PROPOSITION 15.3.3. *Suppose K is spherically complete, $v(P,a) \notin \mathscr{E}(L_{P_{+a}})$, and there is no $b \in K$ with $v(a-b) = v(P,a)$ and $P(b) = 0$. Then for some $b \in \mathcal{O}$ we have $v(a-b) = v(P,a)$, $P(b) \prec P(a)$, and $v(P,b) \in \mathscr{E}(L_{P_{+b}})$.*

PROOF. Let $(a_\rho)_{\rho<\lambda}$ be a sequence in $\mathcal{O}$ with λ an ordinal > 0, $a_0 = a$, and

(1) $\operatorname{dmul} P_{+a_\rho} = 1$, for all $\rho < \lambda$,

(2) $v(a_{\rho'} - a_\rho) = v(P, a_\rho)$ whenever $\rho < \rho' < \lambda$,

(3) $P(a_{\rho'}) \prec P(a_\rho)$ and $v(P, a_{\rho'}) > v(P, a_\rho)$ whenever $\rho < \rho' < \lambda$, and

(4) $v(P, a_\rho) \notin \mathscr{E}(L_{P_{+a_\rho}})$ for all $\rho < \lambda$.

Note that there is such a sequence for $\lambda = 1$. We now keep extending this sequence as in the proof of Lemma 7.2.5 while preserving (1)–(4), using Lemma 15.3.2 instead of Lemma 7.2.3. This extension procedure cannot go on indefinitely, and it must end in a violation of clause (4), which gives an element b as required. □

In what follows $K(x)$ is the d-valued extension of H-type of K from Lemma 15.2.1, so x is transcendental over K, $x' = 1$, and $x \succ 1$. We also assume that a d-valued extension F of $K(x)$ of H-type is given such that Ψ_F has a supremum in Γ_F. We use the flattening $v^\flat\colon F^\times \to \Gamma_F^\sharp = \Gamma_F/\Gamma_F^\flat$ of the valuation $v\colon F^\times \to \Gamma_F$ of F, with the associated dominance relation $\preccurlyeq^\flat$. Here, as usual,

$$\Gamma_F^\flat = \{\alpha \in \Gamma_F : \psi_F(\alpha) > 0\},$$

a convex subgroup of Γ_F. Note that $x \asymp^\flat 1$. Also $\Gamma_F^\flat \cap \Gamma = \{0\}$, so for $b \in K$ we have $b \prec 1 \Longleftrightarrow b \prec^\flat 1$. In the next lemma we set $n := \operatorname{dwm}_A(\gamma)$.

LEMMA 15.3.4. *For some $b \in K[x]$ we have $v(a-b) = \gamma + nvx$ and $P(b) \prec^\flat P(a)$.*

PROOF. With the notations from the proof of Lemma 15.3.2 we have $\operatorname{dwm}(Q_{1,\times g}) = \operatorname{dwm}(Ag) = n$, hence $b_0, \dots, b_{n-1} \prec 1 \asymp b_n$. Take $c \in C^\times$ with $b_n c \sim -1$ and set $y := (c/n!)x^n \in K[x]$. Then $y \asymp^\flat 1$, so $R(y) \preccurlyeq^\flat R \prec^\flat 1$, and thus

$$\begin{aligned} P(a+gy) &= P(a)\cdot\big(1 + L(y) + R(y)\big) \\ &= P(a)\cdot\left(1 + \sum_{i=0}^{n-1} b_i\big(c/(n-i)!\big)x^{n-i} + b_n c + R(y)\right) \prec^\flat P(a), \end{aligned}$$

so that $b := a + gy$ has the desired property. □

Let $(F, v^\flat)$ be the valued differential field whose underlying differential field is that of F and whose valuation is $v^\flat$. Let $\mathcal{O}^\flat := \mathcal{O}_F^\flat$ be the valuation ring of $(F, v^\flat)$. Note that $(F, v^\flat)$ is an H-asymptotic field extension of K with $\Gamma_{(F,v^\flat)} = \Gamma_F^\sharp$ and $\sup \Psi_{(F,v^\flat)} = 0$. Thus $\partial^{-1}P$ remains in dh-position at a with respect to $(F, v^\flat)$.

Suppose $\partial^{-1}P$ is in dh-position at $b \in \mathcal{O}^\flat$, with respect to $(F, v^\flat)$. Then $v^\flat(P,b)$ is the unique $\beta \in (\Gamma_F^\sharp)_\infty$ with $v_B^\flat(\beta) = v^\flat\big(P(b)\big)$, where $B := L_{P_{+b}}$. An argument in the proof of Lemma 7.3.4 gives that P is in dh-configuration at b with respect to the valuation v of F, and so $v(P,b) \in (\Gamma_F)_\infty$ is defined. By Lemma 4.5.5 we have $v^\flat(P,b) = v(P,b) + \Gamma_F^\flat$ if $P(b) \neq 0$. Thus by Lemma 15.3.4, and by Lemma 7.2.4 applied to $(F, v^\flat)$ in the role of K, we obtain:

COROLLARY 15.3.5. *There is a $b \in K[x]$ with $v^\flat(a-b) = v^\flat(P,a)$ and $P(b) \prec^\flat P(a)$. For any $b \in F$ with $v^\flat(a-b) \geqslant v^\flat(P,a)$ and $P(b) \prec^\flat P(a)$, and any $y \in F^\times$:*

(i) *$\partial^{-1}P$ is in dh-position at b with respect to $(F, v^\flat)$;*

(ii) *$v^\flat(a-b) = v^\flat(P,a)$ and $v^\flat(P,b) > v^\flat(P,a)$;*

(iii) *if $vy = O\big(v^\flat(P,a)\big)$, and $A(y) \prec^\flat Ay$, then $B(y) \prec^\flat By$ for $B := L_{P_{+b}}$.*

Note also that if $b \subset F$ and $v^\flat(a-b) \geqslant v^\flat(P,a)$, then $b \preccurlyeq 1$. Next, take $\phi \in F$ with $v\phi = \sup \Psi_F$. We have $\phi \preccurlyeq 1/x \prec 1$, $\phi \asymp^\flat 1$, and F^ϕ is again (like K) a d-valued field of H-type with $\sup \Psi_{F^\phi} = 0$. We can now state and prove another key fact:

PROPOSITION 15.3.6. *Assume* $\operatorname{ddeg} P_{+a} = 1$. *Then for some $b \in K[x]$ with $b \preccurlyeq 1$,*

(i) *$P^\phi_{+b} \sim (P^\phi_{+b})_1$ in $F^\phi\{Y\}$, and $v(P^\phi, b) > v(P,a)$;*

(ii) *$\big\{\alpha \in \mathscr{E}_K(A) : \alpha = O\big(v(P,a)\big)\big\} \subseteq \mathscr{E}_{F^\phi}(B^\phi)$ where $B := L_{P_{+b}}$.*

PROOF. We have $\operatorname{dmul} P_{+a} = \operatorname{ddeg} P_{+a} = 1$ with respect to K and thus with respect to the valued field extension $(F, v^\flat)$ of K. Take $b \in K[x] \subseteq F$ as in Corollary 15.3.5. Then $b \preccurlyeq 1$ and $v^\flat(a-b) > 0$, so $\operatorname{dmul} P_{+b} = \operatorname{ddeg} P_{+b} = 1$ with respect to $(F, v^\flat)$, by Lemma 6.6.5, and so for $H := (P_{+b})_1 \in F\{Y\}$ we have

$$P_{+b} \;=\; H + S \quad \text{where } S \prec^\flat H.$$

By Proposition 11.1.4, $v(H^\phi) \equiv v(H) \bmod \Gamma_F^\flat$ and $v(S^\phi) \equiv v(S) \bmod \Gamma_F^\flat$, so

$$P^\phi_{+b} \;=\; H^\phi + S^\phi \quad \text{where } S^\phi \prec^\flat H^\phi = (P^\phi_{+b})_1,$$

and thus $P^\phi_{+b} \sim (P^\phi_{+b})_1$. Also $v_{H^\phi}\big(v(P^\phi,b)\big) = v\big(P(b)\big) = v_H\big(v(P,b)\big)$, hence

$$v_{H^\phi}\big(v(P^\phi,b)\big) + \Gamma_F^\flat \;=\; v^\flat\big(P(b)\big) \;=\; v_H\big(v(P,b)\big) + \Gamma_F^\flat \;=\; v_{H^\phi}\big(v(P,b)\big) + \Gamma_F^\flat,$$

using Proposition 11.1.4 for the last equality. Applying Lemma 4.5.5 to F^ϕ in the role of K, using the injectivity of $v^\flat_{H^\phi}$, and using Corollary 15.3.5, we get

$$v(P^\phi,b) + \Gamma_F^\flat \;=\; v(P,b) + \Gamma_F^\flat \;=\; v^\flat(P,b) \;>\; v^\flat(P,a) \;=\; v(P,a) + \Gamma_F^\flat$$

and hence $v(P^\phi, b) > v(P,a)$. We have now established (i).

For (ii), let $\alpha \in \mathscr{E}_K(A)$ and $\alpha = O\big(v(P,a)\big)$. Take $y \in K^\times$ such that $vy = \alpha$ and $A(y) \prec Ay$. Then $A(y) \prec^\flat Ay$ with respect to $(F, v^\flat)$, and so $B(y) \prec^\flat By$ by (iii) of Corollary 15.3.5. In the common underlying valued subfield of $(F, v^\flat)$ and $(F, v^\flat)^\phi = (F^\phi, v^\flat)$ we have $B(y) = B^\phi(y)$, and by Proposition 11.1.4 we have $By \asymp^\flat (By)^\phi = B^\phi y$. Hence $B^\phi(y) \prec^\flat B^\phi y$, so $B^\phi(y) \prec B^\phi y$ with respect to F^ϕ, and thus $\alpha = vy \in \mathscr{E}_{F^\phi}(B^\phi)$. This yields (ii). □

For the proof of Theorem 15.4.1 below it is also relevant that for $B = L_{P_{+b}}$ as in (ii) of Proposition 15.3.6 we have $B^\phi = L_{P^\phi_{+b}}$, by Corollary 5.7.5.

15.4 PROOF OF NEWTONIANITY

We begin with a result of independent interest and with more constructive content than Theorem 15.0.1.

THEOREM 15.4.1. *Let* $K_0 \subseteq K_1 \subseteq \cdots \subseteq K_r$ *be a tower of spherically complete* d-*valued fields of* H*-type. For* $i = 0, \dots, r$, *let* $\phi_i \in K_i^\times$ *be such that* $v\phi_i = \max \Psi_{K_i}$. *Assume that* $\phi_i \in \partial K_{i+1}$ *for* $i = 0, \dots, r-1$. *Let* $P \in K_0\{Y\}^{\neq}$ *have order* $\leqslant r$, *and let* $a \in \mathcal{O}_{K_0}$ *be such that* $P_{+a}^{\phi_0} \sim (P_{+a}^{\phi_0})_1$. *Then* $P(b) = 0$ *for some* $b \in \mathcal{O}_{K_r}$.

PROOF. Towards a contradiction, assume there is no such b (so $r \geqslant 1$). We shall build a sequence $a_0, b_0, a_1, b_1, \dots, a_r, b_r$ with $a_i, b_i \in \mathcal{O}_{K_i}$ and $P_{+a_i}^{\phi_i} \sim (P_{+a_i}^{\phi_i})_1$ and $P_{+b_i}^{\phi_i} \sim (P_{+b_i}^{\phi_i})_1$ for $i = 0, \dots, r$. This sequence will be shown to have certain further properties that lead to a contradiction with Lemma 15.1.1.

We set $a_0 := a$, and $K := K_0^{\phi_0}$. If $v(P^{\phi_0}, a_0) \in \mathscr{E}_K(L_{P_{+a_0}^{\phi_0}})$, take $b_0 = a_0$; otherwise, use Proposition 15.3.3 to get $b_0 \in \mathcal{O}_K = \mathcal{O}_{K_0}$ such that $v(a_0 - b_0) = v(P^{\phi_0}, a_0)$, $P(b_0) \prec P(a_0)$, and $v(P^{\phi_0}, b_0) \in \mathscr{E}_K(L_{P_{+b_0}^{\phi_0}})$. By Lemma 15.3.1 we have $\operatorname{dmul} P_{+b_0}^{\phi_0} = \operatorname{ddeg} P_{+b_0}^{\phi_0} = 1$, and so $P_{+b_0}^{\phi_0} \sim (P_{+b_0}^{\phi_0})_1$.

To get a_1, first take $x \in K_1$ with $x' = \phi_0$, that is, $\partial_0 x = 1$ for the derivation $\partial_0 := \phi_0^{-1}\partial$ of $K_r^{\phi_0}$ that extends the derivation of K. Then $x \succ 1$, and x is transcendental over K by Lemma 4.6.10. By Lemma 10.2.3 and a subsequent remark, $K(x)$ is a d-valued field extension of K of H-type. Hence $K(x)$ is as described in Lemma 15.2.1. Applying Proposition 15.3.6 to $F := K_1^{\phi_0}$ with P^{ϕ_0} and b_0 in the roles of P and a yields an $a_1 \in \mathcal{O}_{K_1}$ such that $P_{+a_1}^{\phi_1} \sim (P_{+a_1}^{\phi_1})_1$, $v(P^{\phi_1}, a_1) > v(P^{\phi_0}, b_0)$, and every $\alpha \in \mathscr{E}_{K_0^{\phi_0}}(L_{P_{+b_0}^{\phi_0}})$ with $0 \leqslant \alpha \leqslant v(P^{\phi_0}, b_0)$ lies in $\mathscr{E}_{K_1^{\phi_1}}(L_{P_{+a_1}^{\phi_1}})$.

With K_1, a_1 as a new starting point instead of K_0, a_0, we repeat the above construction, and obtain by iteration elements $a_i, b_i \in \mathcal{O}_{K_i}$ $(i = 0, \dots, r)$ with the properties listed in the beginning of the proof, such that moreover for $i = 0, \dots, r$:

(1) $v(P^{\phi_i}, b_i) \geqslant v(P^{\phi_i}, a_i)$;

(2) $v(P^{\phi_i}, b_i) \in \mathscr{E}_{K_i^{\phi_i}}(L_{P_{+b_i}^{\phi_i}})$;

(3) $v(P^{\phi_{i+1}}, a_{i+1}) > v(P^{\phi_i}, b_i)$, if $i < r$;

(4) all $\alpha \in \mathscr{E}_{K_i^{\phi_i}}(L_{P_{+a_i}^{\phi_i}})$ with $0 \leqslant \alpha \leqslant v(P^{\phi_i}, a_i)$ lie in $\mathscr{E}_{K_i^{\phi_i}}(L_{P_{+b_i}^{\phi_i}})$;

(5) all $\alpha \in \mathscr{E}_{K_i^{\phi_i}}(L_{P_{+b_i}^{\phi_i}})$ with $0 \leqslant \alpha \leqslant v(P^{\phi_i}, b_i)$ lie in $\mathscr{E}_{K_{i+1}^{\phi_{i+1}}}(L_{P_{+a_{i+1}}^{\phi_{i+1}}})$, if $i < r$.

It follows that $v(P^{\phi_0}, b_0) < v(P^{\phi_1}, b_1) < \cdots < v(P^{\phi_r}, b_r)$ are $r+1$ distinct elements of $\mathscr{E}_{K_r^{\phi_r}}(L_{P_{+b_r}^{\phi_r}})$, contradicting Lemma 15.1.1. □

The proof shows that the conclusion of Theorem 15.4.1 can be strengthened to:

$P(b) = 0$ *for some* $b \in \mathcal{O}_{K_r}$ *with* $v(a - b) > 0$.

EXAMPLE. Let K_n be the spherically complete H-subfield $\mathbb{R}[[\ell_0^{\mathbb{Z}} \cdots \ell_n^{\mathbb{Z}}]]$ of $\mathbb{T}_{\log}$. Then $\max \Psi_{K_n} = v(\ell_n^\dagger)$, and $\phi_n := \ell_n^\dagger = \frac{1}{\ell_0 \ell_1 \cdots \ell_n} = \ell_{n+1}' \in \partial K_{n+1}$. Therefore Theorem 15.4.1 applies to each tower $K_n \subseteq K_{n+1} \subseteq \cdots \subseteq K_{n+r}$: if $P \in K_n\{Y\}^{\neq}$ has order $\leqslant r$ and $D_{P^{\phi_n}}$ is homogeneous of degree 1, then P has a zero in $\mathcal{O}_{K_{n+r}}$. All this goes through if $\ell_n^{\mathbb{Z}}$ is replaced by $\ell_n^{\mathbb{R}}$ for each n.

PROOF OF THEOREM 15.0.1. Let K be d-valued such that $\partial K = K$, and K is a directed union of spherically complete grounded d-valued subfields. Let $P \in K\{Y\}^{\neq}$ be quasilinear, $r := \operatorname{order}(P)$; we need to show that P has a zero in $\mathcal{O}$. By Corollary 11.7.15, K is ω-free. Hence by Corollary 13.6.12 and after replacing K, P by suitable compositional conjugates we can assume that K has small derivation, and that one of the following holds:

(1) for all active $\phi \preccurlyeq 1$ in K there is an $f \in K^\times$ such that $P^\phi \sim fY'$,

(2) for all active $\phi \preccurlyeq 1$ in K there is an $f \in K^\times$ such that $P^\phi \sim fY$,

(3) for all active $\phi \preccurlyeq 1$ in K there are $f, g \in K^\times$ with $P^\phi \sim f + gY$ and $f \asymp g$.

Take spherically complete d-valued subfields $K_0 \subseteq K_1 \subseteq \cdots \subseteq K_r$ of K with elements $\phi_i \in K_i^\times$ such that $P \in K_0\{Y\}$, $v\phi_i = \max \Psi_{K_i}$ $(i = 0, \dots, r)$, and $\phi_i \in \partial K_{i+1}$ for $i = 0, \dots, r-1$. Then $\phi_0 \preccurlyeq 1$ and ϕ_0 is active. It follows that either $P^{\phi_0} \sim fY'$ for some $f \in K_0^\times$, or $P^{\phi_0} \sim fY$ for some $f \in K_0^\times$, or $P^{\phi_0} \sim f + gY$ for some $f, g \in K_0^\times$ with $f \asymp g$. In the first two cases the hypothesis of Theorem 15.4.1 is satisfied for $a = 0$, and in the last case it is satisfied for $a = -f/g$. Thus $P(b) = 0$ for some $b \in \mathcal{O}_{K_r}$. □

In concrete cases the hypothesis $K = \partial K$ in the theorem can often be verified by means of Corollary 15.2.4. Taking in the example above $\ell_n^{\mathbb{R}}$ in place of $\ell_n^{\mathbb{Z}}$ we get $\mathbb{T}_{\log}$ as the directed union of the spherically complete K_n, with $K_n \subseteq \partial K_{n+1}$ by Corollary 15.2.4, and therefore $\mathbb{T}_{\log} = \partial \mathbb{T}_{\log}$. Thus $\mathbb{T}_{\log}$ is newtonian.

For $\mathbb{T}$, use the increasing union representation $\mathbb{T} = \bigcup_n E_{2n}{\downarrow}_n$ from Appendix A. We show there that each $E_{2n}{\downarrow}_n$ is a spherically complete grounded H-subfield of $\mathbb{T}$ with $\max \Psi_{E_{2n}\downarrow_n} = v(\ell_n^\dagger)$. Thus $\mathbb{T}$ is newtonian.

Notes and comments

Berarducci and Mantova [43] construct a derivation on Conway's field **No** of surreal numbers with $\omega' = 1$ that makes it a Liouville closed H-field with constant field $\mathbb{R}$. They show moreover that this derivation is simplest possible in a certain sense. They ask whether **No** with this derivation is elementarily equivalent to the differential field $\mathbb{T}$. This question has a positive answer, based on Theorem 15.0.1 and the next chapter; see [24].

Chapter Sixteen

Quantifier Elimination

We are now close to establishing the main result of this volume: the theory T^{nl} of ω-free newtonian Liouville closed H-fields eliminates quantifiers in a certain natural language. This theory has two completions: $T^{\mathrm{nl}}_{\mathrm{small}}$, whose models are the models of T^{nl} with small derivation, and $T^{\mathrm{nl}}_{\mathrm{large}}$, in whose models the derivation is not small. One can move from models of $T^{\mathrm{nl}}_{\mathrm{small}}$ to models of $T^{\mathrm{nl}}_{\mathrm{large}}$ by compositional conjugation. These two "sides" of T^{nl} reflect in a way the gap phenomenon, and we do not wish to obscure this by restriction to the "small derivation" case.

This chapter does not depend on the previous one, where the H-field $\mathbb{T}$ is shown to be a model of $T^{\mathrm{nl}}_{\mathrm{small}}$, but our quantifier elimination should of course be viewed in light of that fact about $\mathbb{T}$.

To state the results of the present chapter with full precision, we introduce some first-order languages. Recall from Section 4.7 that $\mathcal{L}_\partial = \{0, 1, -, +, \cdot, \partial\}$ is the language of differential rings. We augment it here with binary relation symbols $\leqslant$ and $\preccurlyeq$ to obtain the language

$$\mathcal{L} := \{0,\ 1,\ +,\ -,\ \cdot\ ,\ \partial,\ \leqslant,\ \preccurlyeq\}$$

of ordered valued differential rings. Each ordered valued differential field is viewed as an $\mathcal{L}$-structure in the natural way, interpreting $\leqslant$ as the ordering and $\preccurlyeq$ as the dominance relation as suggested by these symbols. It is clear that the ω-free newtonian Liouville closed H-fields are exactly the models of an $\mathcal{L}$-theory T^{nl}, but T^{nl} does not eliminate quantifiers in this language, as we shall see in Section 16.5. To achieve quantifier elimination we consider a certain extension $T^{\mathrm{nl},\iota}_{\Lambda\Omega}$ by definitions of T^{nl}, in a language $\mathcal{L}^{\iota}_{\Lambda\Omega}$ that augments $\mathcal{L}$ by a new unary function symbol ι and new unary relation symbols I, Λ and Ω. We obtain defining axioms in $T^{\mathrm{nl},\iota}_{\Lambda\Omega}$ for these new symbols by requiring that every model K of T^{nl} expands uniquely to a model $K^{\iota}_{\Lambda\Omega}$ of $T^{\mathrm{nl},\iota}_{\Lambda\Omega}$ such that for all $a \in K$,

$$\begin{aligned}
\iota(a) &= a^{-1} \text{ if } a \neq 0, \qquad \iota(0) = 0,\\
\mathrm{I}(a) &\iff a = y' \text{ for some } y \prec 1 \text{ in } K,\\
\Lambda(a) &\iff a = -y^{\dagger\dagger} \text{ for some } y \succ 1 \text{ in } K,\\
\Omega(a) &\iff 4y'' + ay = 0 \text{ for some } y \in K^{\times}.
\end{aligned}$$

Note that I, Λ, and Ω get interpreted in the expansion $K^{\iota}_{\Lambda\Omega}$ of a model K of T^{nl} as the convex additive subgroup $\mathrm{I}(K)$ of K and as the downward closed subsets $\Lambda(K)$

and $\Omega(K)$ of K that were introduced in Sections 11.8 and 5.2. We can now state our main result, proved in Section 16.6:

THEOREM 16.0.1. *The theory $T^{\mathrm{nl},\iota}_{\Lambda\Omega}$ eliminates quantifiers.*

This fails if either Λ or Ω is dropped from the language; see Section 16.5. On the other hand, the predicate I is only included for convenience, to simplify some later proofs and formulations: in Theorem 16.0.1 we can drop I from the language, since for K as above and $a \in K^\times$ we have by Lemmas 11.8.5 and 16.3.10:

$$\mathrm{I}(a) \iff \neg\Lambda(-a^\dagger), \qquad \mathrm{I}(a) \iff \Omega\big(\sigma(a)\big).$$

In Section 16.6 we also derive some consequences of Theorem 16.0.1:

COROLLARY 16.0.2. *Let K be an ω-free newtonian Liouville closed H-field. Then:*

(i) *K has* NIP*;*

(ii) *a subset of C^n is definable in K if and only if it is semialgebraic in the sense of the real closed field C;*

(iii) *K is o-minimal at infinity: if $X \subseteq K$ is definable in K, then there exists $a \in K$ such that $(a, +\infty) \subseteq X$ or $(a, +\infty) \cap X = \emptyset$.*

We indicate briefly the ideas behind the proof of Theorem 16.0.1, and some steps that still need to be taken. As is well-known, a theory eliminates quantifiers if and only if its models and their substructures have certain embedding properties. It is for this reason that we have built an arsenal of embedding and extension results in the previous chapters: the universal property of the H-field hull of a pre-H-field in Section 10.5, the algebraic and Liouville extensions of Sections 10.5 and 10.6, the construction of immediate extensions in Section 11.4 and of F_ω in Section 11.7, Proposition 13.6.7 and Corollary 13.6.8, and, crucially, the Newton-Liouville closure of Section 14.5 (the latter requiring a lengthy development over several chapters). An important consequence of Theorem 14.0.2 is that it enables us to deal with immediate extensions. It is worth noting that immediate extensions typically require the most attention in analogous situations—going back to the work by Ax-Kochen [28] and Eršov [131] in the 1960s—and such extensions indeed preoccupied us in several chapters. In contrast to prior model-theoretic work on (enriched) valued fields, however, the models of our theory T^{nl} are never spherically complete: both Liouville closedness and ω-freeness prevent that.

Constant field extensions are taken care of by Propositions 10.5.15 and 10.5.16. It remains to deal with extensions that are completely controlled by the corresponding extension of asymptotic couples. We handle such extensions in Section 16.1, where it leads to a result of independent interest:

THEOREM 16.0.3. *If K is an ω-free newtonian Liouville closed H-field, then K has no proper* d*-algebraic H-field extension with the same constant field.*

In Section 16.2 we use this to get uniqueness-up-to-isomorphism of Newton-Liouville closures of ω-free H-fields, and then employ our arsenal to prove an embedding result that has the model completeness of T^{nl} as a consequence.

Going beyond model completeness to quantifier elimination requires attention to the substructures of models of $T^{\mathrm{nl},\iota}_{\Lambda\Omega}$ rather than of T^{nl}, and this involves the extra predicates I, Λ, Ω. Therefore we first determine in Section 16.3 the substructures of models of $T^{\mathrm{nl},\iota}_{\Lambda\Omega}$: they are the expanded pre-H-fields (K, I, Λ, Ω) where (I, Λ, Ω) is a ΛΩ-cut in the pre-H-field K as defined in that section. It turns out that a given pre-H-field K has either exactly one ΛΩ-cut or exactly two ΛΩ-cuts. Moreover, an ω-free pre-H-field K has just one ΛΩ-cut, and any (K, I, Λ, Ω) has an extension $(K^*, I^*, \Lambda^*, \Omega^*)$ with K^* an ω-free H-field, such that $(K^*, I^*, \Lambda^*, \Omega^*)$ embeds over K into any model of $T^{\mathrm{nl},\iota}_{\Lambda\Omega}$ extending (K, I, Λ, Ω): Proposition 16.4.1. These two facts allow us to focus henceforth for embedding purposes on ω-free H-fields, and forget about ΛΩ-cuts, and this is taken care of by the results in Section 16.2.

Notes and comments

The first suggestion that the primitives $\Lambda(\mathbb{T})$ and $\Omega(\mathbb{T})$ might be enough to eliminate quantifiers for $\mathbb{T}$, in addition to the usual primitives for ordered valued differential fields, is in [23]. The very end of that paper indicates a possible further obstruction to QE that would require also a certain partial inverse of the function ω as an extra primitive. Fortunately, we were able to get rid of this obstruction: this ultimately rests on observing Corollary 12.7.19, which gives Corollary 13.3.7 needed in Section 13.7, which in turn is used in the proofs of Lemmas 16.3.16 and 16.4.6 below.

A minor issue is whether to include field inversion among the primitives to get QE. We found it convenient to do so. If we drop the symbol ι for inversion from the language, we would probably loose QE, but we would certainly regain it upon replacing the unary relation symbols Λ and Ω by binary relation symbols Λ_2 and Ω_2, to be interpreted in $\mathbb{T}$ according to

$$\Lambda_2(a,b) \Leftrightarrow a \in b \cdot \Lambda(\mathbb{T}), \qquad \Omega_2(a,b) \Leftrightarrow a \in b \cdot \Omega(\mathbb{T}).$$

16.1 EXTENSIONS CONTROLLED BY ASYMPTOTIC COUPLES

In this section we deal with a kind of H-field extension that is in some sense controlled by the corresponding extension of asymptotic couples. At this point Proposition 9.9.2 about closed H-asymptotic couples becomes relevant, and we use it to obtain the following:

LEMMA 16.1.1. *Let K be an ω-free newtonian Liouville closed H-field and L an H-field extension with $C_L = C$, and let $f \in L \setminus K$. Suppose K is maximal in L in the sense that there is no $y \in L \setminus K$ for which $K\langle y\rangle$ is an immediate extension of K. Then the vector space $\mathbb{Q}\Gamma_{K\langle f\rangle}/\Gamma$ over $\mathbb{Q}$ is infinite-dimensional.*

PROOF. We claim that there is no divergent pc-sequence in K with a pseudolimit in L. To see this, let (y_ρ) be a divergent pc-sequence in K. It cannot be of d-algebraic type, since K is asymptotically d-algebraically maximal. So it is of d-transcendental type, and if it had a pseudolimit $y \in L$, then $K\langle y\rangle$ would be an immediate extension of K. This proves our claim. Thus for each $y \in L \setminus K$ the set $v(y - K) \subseteq \Gamma_L$ has a largest element: otherwise there would be a divergent pc-sequence in K with pseudolimit y. Given $y \in L \setminus K$, a *best approximation in K to y* is by definition an element $y_0 \in K$ such that $v(y - y_0) = \max v(y - K)$; note that then $v(y - y_0) \notin \Gamma$, since $C_L = C$. For convenience we set $L = K\langle f\rangle$ below.

Pick a best approximation b_0 in K to $f_0 := f$, and set $f_1 := (f_0 - b_0)^\dagger \in L$. Then $f_1 \notin K$, since K is Liouville closed and $C = C_L$. Thus we can take a best approximation b_1 in K to f_1, and continuing this way, we obtain a sequence (f_n) in $L \setminus K$ and a sequence (b_n) in K, such that b_n is a best approximation in K to f_n and $f_{n+1} = (f_n - b_n)^\dagger$ for all n. Thus $v(f_n - b_n) \notin \Gamma$ for all n.

CLAIM: $v(f_0 - b_0), v(f_1 - b_1), v(f_2 - b_2), \dots$ *are $\mathbb{Q}$-linearly independent over* Γ.

To prove this claim, take $a_n \in K^\times$ with $a_n^\dagger = b_n$ for $n \geqslant 1$. Then

$$f_n - b_n = (f_{n-1} - b_{n-1})^\dagger - a_n^\dagger = \left(\frac{f_{n-1} - b_{n-1}}{a_n}\right)^\dagger \qquad (n \geqslant 1).$$

With $\psi := \psi_L$ and $\alpha_n = va_n \in \Gamma$ for $n \geqslant 1$, we get

$$\begin{aligned} v(f_n - b_n) &= \psi\big(v(f_{n-1} - b_{n-1}) - \alpha_n\big), \quad \text{so by an easy induction on } n,\\ v(f_n - b_n) &= \psi_{\alpha_1,\dots,\alpha_n}\big(v(f_0 - b_0)\big), \qquad (n \geqslant 1). \end{aligned}$$

Suppose towards a contradiction that $v(f_0 - b_0), \dots, v(f_n - b_n)$ are $\mathbb{Q}$-linearly dependent over Γ. Then we have $m < n$ and $q_1, \dots, q_{n-m} \in \mathbb{Q}$ such that

$$v(f_m - b_m) + q_1 v(f_{m+1} - b_{m+1}) + \cdots + q_{n-m} v(f_n - b_n) \in \Gamma.$$

For $\gamma := v(f_m - b_m) \in \Gamma_L \setminus \Gamma$ this gives

$$\gamma + q_1 \psi_{\alpha_{m+1}}(\gamma) + \cdots + q_{n-m} \psi_{\alpha_{m+1},\dots,\alpha_n}(\gamma) \in \Gamma,$$

but this contradicts Proposition 9.9.2. □

PROOF OF THEOREM 16.0.3. Let K be an ω-free newtonian Liouville closed H-field, and assume towards a contradiction that L is a proper d-algebraic H-field extension of K with $C_L = C$. Since K is asymptotically d-algebraically maximal, there is no $y \in L \setminus K$ for which $K\langle y\rangle$ is an immediate extension of K. Taking any f in $L \setminus K$, the transcendence degree of $K\langle f\rangle$ over K is finite, but this contradicts Lemma 16.1.1 in view of the Zariski-Abhyankar Inequality (Corollary 3.1.11). □

Description of $K\langle f\rangle$

Let K, L, and f be as in Lemma 16.1.1. Elaborating on the proof of this lemma we shall obtain a complete description of $K\langle f\rangle$ as an H-field extension of K generated by f. For this we use the notations in that proof, and set $\beta_n := v(f_n - b_n) - \alpha_{n+1} \in \Gamma_{K\langle f\rangle}$. Thus $\beta_0, \beta_1, \beta_2, \dots$ are $\mathbb{Q}$-linearly independent over Γ, by the proof of Lemma 16.1.1.

LEMMA 16.1.2. *The asymptotic couple of $K\langle f\rangle$ has the following properties:*

(i) $\Gamma_{K\langle f\rangle} = \Gamma \oplus \bigoplus_n \mathbb{Z}\beta_n$ *(internal direct sum);*

(ii) $\beta_n^\dagger \notin \Gamma$ *for all* n, *and* $\beta_m^\dagger \neq \beta_n^\dagger$ *for all* $m \neq n$;

(iii) $\psi\big(\Gamma_{K\langle f\rangle}^{\neq}\big) = \Psi \cup \{\beta_n^\dagger : n = 0, 1, 2, \dots\}$;

(iv) $\big[\Gamma_{K\langle f\rangle}\big] = [\Gamma] \cup \big\{[\beta_n] : n = 0, 1, 2, \dots\big\}$;

(v) $\Gamma^<$ *is cofinal in* $\Gamma_{K\langle f\rangle}^<$, *and* $\beta_0^\dagger < \beta_1^\dagger < \beta_2^\dagger < \cdots$.

PROOF. Consider the "monomials" $\mathfrak{m}_n := \frac{f_n - b_n}{a_{n+1}}$ with $v(\mathfrak{m}_n) = \beta_n$. Then

$$\begin{aligned}\mathfrak{m}_{n+1} &= \frac{f_{n+1} - b_{n+1}}{a_{n+2}} = \frac{(f_n - b_n)^\dagger - b_{n+1}}{a_{n+2}} \\ &= \frac{(a_{n+1}\mathfrak{m}_n)^\dagger - b_{n+1}}{a_{n+2}} = \frac{a_{n+1}^\dagger + \mathfrak{m}_n^\dagger - b_{n+1}}{a_{n+2}} = \frac{\mathfrak{m}_n^\dagger}{a_{n+2}},\end{aligned}$$

and so $\mathfrak{m}_n' = a_{n+2}\mathfrak{m}_n\mathfrak{m}_{n+1}$. Thus $f = b_0 + a_1\mathfrak{m}_0$ gives $f' = b_0' + a_1'\mathfrak{m}_0 + a_1a_2\mathfrak{m}_0\mathfrak{m}_1$, and continuing by induction on n gives

$$f^{(n)} = F_n(\mathfrak{m}_0, \dots, \mathfrak{m}_n), \quad F_n(Y_0, \dots, Y_n) \in K[Y_0, \dots, Y_n], \quad \deg F_n \leqslant n + 1.$$

Thus for $P \in K\{Y\}^{\neq}$ of order $\leqslant r \in \mathbb{N}$ we have

$$P(f) = \sum_{\boldsymbol{i} \in I} a_{\boldsymbol{i}}\, \mathfrak{m}_0^{i_0} \cdots \mathfrak{m}_r^{i_r}$$

where the sum is over a finite nonempty set I of tuples $\boldsymbol{i} = (i_0, \dots, i_r) \in \mathbb{N}^{1+r}$, and $a_{\boldsymbol{i}} \in K^\times$ for all $\boldsymbol{i} \in I$. Since $v(\mathfrak{m}_0) = \beta_0, v(\mathfrak{m}_1) = \beta_1, \dots$ are $\mathbb{Q}$-linearly independent over Γ, we obtain $v\big(P(f)\big) \in \Gamma + \sum_n \mathbb{N}\beta_n$, which proves (i).

We have $\beta_n^\dagger \notin \Gamma$ because by the proof of Lemma 16.1.1,

$$\beta_n^\dagger = \psi\big(v(f_n - b_n) - \alpha_{n+1}\big) = v(f_{n+1} - b_{n+1}) = \beta_{n+1} + \alpha_{n+2} \notin \Gamma.$$

Since $\beta_1, \beta_2, \beta_3, \beta_4, \dots$ are $\mathbb{Q}$-linearly independent, so are $\beta_0^\dagger, \beta_1^\dagger, \beta_2^\dagger, \beta_3^\dagger, \dots$ by these equalities. This proves (ii), which in view of (i) yields (iii). From (ii) we get $[\beta_n] \notin [\Gamma]$ for all n, and $[\beta_m] \neq [\beta_n]$ for all $m \neq n$. Again by (i), this gives (iv).

To get (v), assume towards a contradiction that $\Gamma^<$ is not cofinal in $\Gamma_{K\langle f\rangle}^<$. Then by (iv) we get n with $[\beta_n] < [\alpha]$ for all $\alpha \in \Gamma^{\neq}$, hence $\Psi < \beta_n^\dagger < (\Gamma^>)'$. Then $[\beta_n^\dagger - \alpha] \in [\Gamma]$ for all $\alpha \in \Gamma$, by Corollary 9.8.6. For $\alpha := \alpha_{n+2}$ this means $[\beta_n^\dagger - \alpha_{n+2}] = [\beta_{n+1}] \in [\Gamma]$, contradicting (ii). Thus $\Gamma^<$ is indeed cofinal in $\Gamma_{K\langle f\rangle}^<$. For any n we can therefore take $\alpha \in \Gamma^{\neq}$ with $[\alpha] < [\beta_n]$. Also $[\beta_{n+1}] \notin [\Gamma]$ and $\beta_n^\dagger - \alpha^\dagger \in (\Gamma + \mathbb{Z}\beta_{n+1}) \setminus \Gamma$, and thus by Lemmas 2.4.4 and 6.5.4,

$$[\beta_{n+1}] \leqslant [\beta_n^\dagger - \alpha^\dagger] < [\beta_n - \alpha] = [\beta_n].$$

So we have a strictly decreasing sequence $[\beta_0] > [\beta_1] > [\beta_2] > \cdots$ in $[\Gamma_{K\langle f\rangle}]$, and thus a strictly increasing sequence $\beta_0^\dagger < \beta_1^\dagger < \beta_2^\dagger < \cdots$ in view of (ii). $\square$

The following consequence is not needed later but worth pointing out:

COROLLARY 16.1.3. *$K\langle f\rangle$ is ω-free.*

PROOF. Assume towards a contradiction that $K\langle f\rangle$ is not ω-free. Since $\Gamma^<$ is cofinal in $\Gamma^<_{K\langle f\rangle}$ this gives an element $\omega \in K\langle f\rangle$ such that $\omega_\rho \rightsquigarrow \omega$, where (ω_ρ) is the sequence in K obtained in the usual way from a logarithmic sequence in K. Now (ω_ρ) is of d-transcendental type over K, so $K\langle\omega\rangle \subseteq K\langle f\rangle$ is an immediate extension of K. Since $\omega \notin K$, this contradicts Lemma 16.1.1. $\square$

LEMMA 16.1.4. *Suppose g in some H-field extension M of K realizes the same cut in the ordered set K as f does. Then $v(g-b_0) = \max v(g-K) \notin \Gamma$, and $g_1 := (g-b_0)^\dagger$ realizes the same cut in the ordered set K as $f_1 = (f-b_0)^\dagger$.*

PROOF. Let $\alpha \in \Gamma$ and $b \in K$. We claim that

$$v(f-b) < \alpha \iff v(g-b) < \alpha, \quad v(f-b) > \alpha \iff v(g-b) > \alpha.$$

To prove this claim, take $a \in K^>$ with $va = \alpha$. Consider first the case that $v(f-b) < \alpha$. Then $|f-b| > a$, so $|g-b| > a$, and thus $v(g-b) \leqslant \alpha$. Since $\Gamma^<$ is cofinal in $\Gamma^<_{K\langle f\rangle}$ we have $\alpha_1 \in \Gamma$ such that $v(f-b) < \alpha_1 < \alpha$. Take $a_1 \in K^>$ such that $va_1 = \alpha_1$. Then the argument above with a_1 instead of a gives $v(g-b) \leqslant \alpha_1 < \alpha$. In a similar way we show: $v(f-b) > \alpha \Rightarrow v(g-b) > \alpha$. Finally, assume $v(f-b) = \alpha$. Then $C = C_{K\langle f\rangle}$ gives $c \in C^\times$ with $f-b \sim ca$, so $\frac{|c|}{2}a < |f-b| < 2|c|a$, hence $\frac{|c|}{2}a < |g-b| < 2|c|a$, and thus $v(g-b) = va = \alpha$. This finishes the proof of the claim.

It follows from this claim and $v(f-b_0) \notin \Gamma$ that $v(g-b_0) \notin \Gamma$. This gives $v(g-b_0) = \max v(g-K)$: otherwise we have $b \in K$ with $v(g-b_0) < v(g-b)$, and so $v(g-b_0) = v(b-b_0) \in \Gamma$, a contradiction. Next we get $(g-b_0)^\dagger \notin K$: otherwise, $(g-b_0)^\dagger = a^\dagger$ with $a \in K^\times$, so $g-b_0 = ca$ for some $c \in C_M^\times$, and thus $v(g-b_0) = va \in \Gamma$, a contradiction.

Next, we show that $(g-b_0)^\dagger$ realizes the same cut in K as $(f-b_0)^\dagger$. If $f < b_0$, then we can replace f, g, b_0 by $-f, -g, -b_0$ to reduce to the case $f > b_0$. So we assume $f > b_0$, which gives $g > b_0$. Suppose towards a contradiction that $h \in K$ is such that $(f-b_0)^\dagger < h$ in $K\langle f\rangle$ and $h < (g-b_0)^\dagger$ in M. Take $\phi \in K^>$ such that $h = \phi^\dagger$. Then $s := (f-b_0)/\phi > 0$ and $s^\dagger < 0$, so $s' < 0$, and thus $s = c + \varepsilon$ with $c \in C^\geqslant$ and $0 < \varepsilon \prec 1$ in $K\langle f\rangle$. If $c \neq 0$, then $v(f-b_0) = v\phi \in \Gamma$ contradicts $v(f-b_0) \notin \Gamma$. So $c = 0$ and thus $f = b_0 + \phi\varepsilon$. Likewise, $h < (g-b_0)^\dagger$ gives $t := (g-b_0)/\phi > 0$ and $t^\dagger > 0$, so $t' > 0$, and thus either $t = c^* - \varepsilon^*$ with $c^* \in C_M^>$ and $0 < \varepsilon^* \prec 1$ in M, or $t > C_M$. The first case would give $v(g-b_0) = v\phi$, a contradiction, so we get $t > C_M$, but that would give

$$f = b_0 + \phi\varepsilon < b_0 + \phi < b_0 + \phi t = g,$$

with $b_0 + \phi \in K$, contradicting that f and g realize the same cut in K.

The other way that $(g - b_0)^\dagger$ does not realize the same cut in K as $(f - b_0)^\dagger$ is that we have $h \in K$ such that $(g - b_0)^\dagger < h$ in M and $h < (f - b_0)^\dagger$ in $K\langle f \rangle$. Taking as before $\phi \in K^>$ with $h = \phi^\dagger$, a similar argument gives us $g < b_0 + \phi < f$, and we have again a contradiction. □

PROPOSITION 16.1.5. *Suppose g in some H-field extension M of K realizes the same cut in the ordered set K as f does. Then there is an embedding $K\langle f \rangle \to M$ of H-fields over K sending f to g.*

PROOF. By Lemma 16.1.4 we can recursively define $g_n \in M \setminus K$ to realize the same cut in K as f_n by $g_0 := g$, and $g_{n+1} := (g_n - b_n)^\dagger$. Then $v(g_n - b_n) \notin \Gamma$ for all n. The same argument as in the proof of Lemma 16.1.1 shows:

$$v(g_0 - b_0),\ v(g_1 - b_1),\ v(g_2 - b_2), \dots \text{ are } \mathbb{Q}\text{-linearly independent over } \Gamma.$$

Set $\beta_n^* := v(g_n - b_n) - \alpha_{n+1}$, and $\mathfrak{m}_n^* := \frac{g_n - b_n}{a_{n+1}}$. Then $\beta_0^*, \beta_1^*, \beta_2^*, \dots$ are $\mathbb{Q}$-linearly independent over Γ. With $F_n \in K[Y_0, \dots, Y_n]$ as in the proof of Lemma 16.1.2 we get $g^{(n)} = F_n(\mathfrak{m}_0^*, \dots, \mathfrak{m}_n^*)$, and so for $P \in K\{Y\}^{\neq}$ of order $\leqslant r \in \mathbb{N}$ we get

$$P(g) \ = \ \sum_{\boldsymbol{i} \in I} a_{\boldsymbol{i}}\, \mathfrak{m}_0^{*\,i_0} \cdots \mathfrak{m}_r^{*\,i_r}$$

where the sum is over the same finite nonempty set I of tuples $\boldsymbol{i} = (i_0, \dots, i_r) \in \mathbb{N}^{1+r}$ as in the proof of Lemma 16.1.2, and with the same coefficients $a_{\boldsymbol{i}} \in K^\times$ as in that proof. Since $v(\mathfrak{m}_0^*) = \beta_0^*, v(\mathfrak{m}_1^*) = \beta_1^*, \dots$ are $\mathbb{Q}$-linearly independent over Γ, we obtain $v\big(P(g)\big) \in \Gamma + \sum_n \mathbb{N}\beta_n^*$. The rest of that proof then shows that Lemma 16.1.2 goes through with f replaced by g and each β_n by β_n^*. In particular,

$$[\beta_0^*] \ > \ [\beta_1^*] \ > \ [\beta_2^*] \ > \ \cdots .$$

Next, $\mathfrak{m}_n \in K\langle f \rangle$ realizes the same cut in K as $\mathfrak{m}_n^*$, and so β_n realizes the same cut in the ordered set Γ as β_n^*. It follows that we have an ordered abelian group isomorphism $j\colon \Gamma_{K\langle f \rangle} \to \Gamma_{K\langle g \rangle}$ over Γ sending β_n to β_n^* for each n. Using the expressions above for $P(f)$ and $P(g)$ it follows that $j\big(v(P(f))\big) = v(P(g))$ for all $P \in K\{Y\}^{\neq}$, so we have a valued differential field embedding $K\langle f \rangle \to M$ over K sending f to g. This is even an embedding of ordered valued differential fields, since the facts stated about the $\mathfrak{m}_n$, β_n and $\mathfrak{m}_n^*$, β_n^* yield: $P(f) > 0 \iff P(g) > 0$, for all $P \in K\{Y\}^{\neq}$. □

Notes and comments

Lemma 16.1.2 is a version for H-fields of Proposition 5.3 in [18]. That proposition analyzes "simple extensions of H-triples of type (V)."

In applying Theorem 16.0.3 one should watch out: replacing *H-field extension* in its statement by *pre-H-field extension* results in a false statement.

16.2 MODEL COMPLETENESS

We first show the uniqueness-up-to-isomorphism of Newton-Liouville closures of ω-free H-fields, although existence of such closures is enough for model completeness.

Uniqueness of Newton-Liouville closure

Let E be an ω-free H-field. We saw in Section 14.5 that E has a Newton-Liouville closure.

LEMMA 16.2.1. *Let E^{nl} be any Newton-Liouville closure of E and $i\colon E^{\text{nl}} \to L$ an embedding into an H-field L with $C_L \subseteq i(E^{\text{nl}})$. Then*

$$i(E^{\text{nl}}) = \{f \in L : f \text{ is d-algebraic over } i(E)\}.$$

PROOF. Since E^{nl} is d-algebraic over E, every element of $i(E^{\text{nl}})$ is d-algebraic over $i(E)$. Also, $i(E^{\text{nl}})$ is a newtonian Liouville closed H-subfield of L with the same constants as L, so every $f \in L$ that is d-algebraic over $i(E^{\text{nl}})$ lies in $i(E^{\text{nl}})$ by Theorem 16.0.3. □

COROLLARY 16.2.2. *Any two Newton-Liouville closures of E are isomorphic over E. If E^{nl} is a Newton-Liouville closure of E, then E^{nl} does not have any proper newtonian Liouville closed H-subfield containing E.*

PROOF. Let E^{nl} and L be Newton-Liouville closures of E. Then there exists an embedding $E^{\text{nl}} \to L$ over E, and any such embedding is necessarily surjective by Lemma 16.2.1. This proves the first part. The minimality property of E^{nl} also follows from Lemma 16.2.1 by considering embeddings $E^{\text{nl}} \to E^{\text{nl}}$ over E. □

Model completeness of T^{nl}

As usual, $|S|$ denotes the cardinality of a set S, and κ^+ the next cardinal after the cardinal κ. Here is the decisive embedding property:

PROPOSITION 16.2.3. *Let E be an ω-free H-subfield of an ω-free newtonian Liouville closed H-field K with $C_E = C$, let $i\colon E \to L$ be an embedding into an ω-free newtonian Liouville closed H-field L. Assume $\text{cf}(\Gamma_L^<) > |\Gamma|$ and the underlying ordered set of L is $|K|^+$-saturated. Then i extends to an embedding $K \to L$.*

PROOF. Note that every differential subfield of K containing E is an H-subfield of K. Assume $E \neq K$; it is enough to show that then i can be extended to an embedding $F \to L$ for some ω-free H-subfield F of K properly containing E.

Consider first the case that $\Gamma_E^<$ is not cofinal in $\Gamma^<$. Then we have $y \in K^>$ such that $\Gamma_E^< < vy < 0$. By the cofinality assumption on $\Gamma_L^<$ we also have $y^* \in L^>$ such that $\Gamma_{iE}^< < vy^* < 0$. Then Corollary 13.6.8 yields an extension of i to an embedding $E\langle y\rangle \to L$ sending y to y^*. Now $E\langle y\rangle$ is a grounded H-field by Proposition 13.6.7. Then by Lemma 11.7.17 we can extend this embedding $E\langle y\rangle \to L$ to an embedding $F := E\langle y\rangle_\omega \to L$, and for the same reason we can identify the extension F of $E\langle y\rangle$

over $E\langle y\rangle$ with an ω-free H-subfield of K. This achieves our goal of extending i to an embedding $F \to L$.

We are left with the case that $\Gamma_E^{<}$ is cofinal in $\Gamma^{<}$. Then every differential subfield of K containing E is an ω-free H-subfield of K.

SUBCASE 1: *E is not a newtonian Liouville closed H-field.* Then we can extend i to an embedding $F \to L$ where

$$F := \{f \in K : f \text{ is d-algebraic over } E\}$$

is the Newton-Liouville closure of E inside K, by Lemma 16.2.1.

SUBCASE 2: *E is newtonian and Liouville closed, and $E\langle y\rangle$ is an immediate extension of E for some $y \in K \setminus E$.* For such y we take a divergent pc-sequence (a_ρ) in E such that $a_\rho \rightsquigarrow y$. Since E is asymptotically d-algebraically maximal, (a_ρ) is of d-transcendental type over E. The saturation assumption on L gives $z \in L$ such that $i(a_\rho) \rightsquigarrow z$, by Lemma 2.4.2. Then Section 11.4 yields a valued differential field embedding $F := E\langle y\rangle \to L$ that extends i and sends y to z. By Lemma 10.5.8 this embedding $F \to L$ is an embedding of ordered valued differential fields.

SUBCASE 3: *E is newtonian and Liouville closed, and there is no $y \in K \setminus E$ such that $E\langle y\rangle$ is an immediate extension of E.* Then we take any $f \in K \setminus E$, and take some $g \in L$ such that for all $a \in E$ we have: $a < f \Rightarrow i(a) < g$, and $a > f \Rightarrow i(a) > g$. Then Proposition 16.1.5 yields an extension of i to an embedding $E\langle f\rangle \to L$ sending f to g. □

When $C_E \neq C$, we require an extra hypothesis:

COROLLARY 16.2.4. *Let E, K, L, and i be as in Proposition 16.2.3, except that we drop the assumption $C_E = C$. Assume in addition that the underlying ordered set of C_L is $|C|^+$-saturated. Then i extends to an embedding $K \to L$.*

PROOF. The real closed constant field C_L is $|C|^+$-saturated, so the ordered field embedding $i|C_E \colon C_E \to C_L$ extends to an ordered field embedding $j\colon C \to C_L$. Then Propositions 10.5.15 and 10.5.16 yield an extension of i to an embedding $E(C) \to L$ that agrees with j on C. The H-subfield $E(C)$ of K is d-algebraic over E, so is ω-free. Now apply Proposition 16.2.3 with $E(C)$ in place of E. □

The cofinality and saturation hypotheses in Proposition 16.2.3 and Corollary 16.2.4 are of course satisfied if L (as an ordered valued differential field) is $|K|^+$-saturated, but the weaker assumption of Proposition 16.2.3 is useful in [24]. In view of the next result we recall from Chapter 15 that $\mathbb{T}$ is a model of T^{nl}.

COROLLARY 16.2.5. *The $\mathcal{L}$-theory T^{nl} is model complete. Thus T^{nl} is the model companion of the $\mathcal{L}$-theory of H-fields and of the $\mathcal{L}$-theory of pre-H-fields.*

PROOF. Apply Corollary 16.2.4 to the models E of T^{nl} and use B.10.4. □

Tournant dangereux

Recall that the valuation ring $\mathcal{O}_{\mathbb{T}}$ of $\mathbb{T}$ is existentially definable without parameters in the differential field $\mathbb{T}$, that is, in $\mathbb{T}$ construed as an $\mathcal{L}_\partial$-structure. On the other hand:

COROLLARY 16.2.6. *The valuation ring $\mathcal{O}_{\mathbb{T}}$ of $\mathbb{T}$ is not universally definable in the differential field $\mathbb{T}$, even if we allow parameters.*

PROOF. Take an elementary extension $\mathbb{T}_1$ of the H-field $\mathbb{T}$ with an element α in the value group Γ_1 of $\mathbb{T}_1$ such that $0 < \alpha < \Gamma_{\mathbb{T}}^{>}$. Let $\mathbb{T}_2$ be the Δ-coarsening of $\mathbb{T}_1$, for $\Delta := \{\gamma \in \Gamma_1 : |\gamma| < \Gamma_{\mathbb{T}}^{>}\}$. It follows easily from Proposition 10.1.5 that $\mathbb{T}_2$ is a pre-H-field extension of the H-field $\mathbb{T}$. Extend the pre-H-field $\mathbb{T}_2$ to an ω-free newtonian Liouville closed H-field K. Thus $\mathbb{T} \preccurlyeq K$, as H-fields.

Suppose $\phi(y)$ is a universal formula in the language $\mathcal{L}_\partial$ augmented by names for elements of $\mathbb{T}$ such that $\phi(y)$ defines $\mathcal{O}_{\mathbb{T}}$ in $\mathbb{T}$. Then it defines $\mathcal{O}_{\mathbb{T}_1}$ in $\mathbb{T}_1$ and $\mathcal{O}_K$ in K, since $\mathbb{T}_1$ and K are elementary extensions of $\mathbb{T}$. Take $b \in \mathbb{T}_1$ such that $v(b) = -\alpha$. Then $b \notin \mathcal{O}_{\mathbb{T}_1}$, so $\mathbb{T}_1 \models \neg\phi(b)$, hence $K \models \neg\phi(b)$, since $\mathbb{T} \subseteq \mathbb{T}_1 \subseteq K$ as differential fields. Hence $b \notin \mathcal{O}_K$. But $b \in \mathcal{O}_{\mathbb{T}_2}$, and so $b \in \mathcal{O}_K$, a contradiction. □

In particular, the theory of $\mathbb{T}$ as a differential field is not model complete.

Notes and comments

In the next volume we expect to pay more attention to various natural H-subfields of $\mathbb{T}$ such as the field $\mathbb{T}_{\mathrm{g}}$ of grid-based transseries. (The *Notes and comments* to Appendix A sketch how to build $\mathbb{T}_{\mathrm{g}}$ inside $\mathbb{T}$.) The construction of $\mathbb{T}_{\mathrm{g}}$ easily yields that it is ω-free and Liouville closed. It is newtonian by [194], especially Chapter 5. Thus $\mathbb{T}_{\mathrm{g}} \preccurlyeq \mathbb{T}$ by Corollary 16.2.5. Since $\mathbb{R} \subseteq \mathbb{T}_{\mathrm{g}}$, we can combine these facts with Theorem 16.0.3 to conclude: *every $f \in \mathbb{T} \setminus \mathbb{T}_{\mathrm{g}}$ is* d*-transcendental over* $\mathbb{T}_{\mathrm{g}}$. For example, given any nonzero real numbers $c_1, c_2, c_3, \dots$, the transseries

$$c_1 + c_2 2^{-x} + c_3 3^{-x} + \cdots = \sum_{n=1}^{\infty} c_n \, \mathrm{e}^{-x \log n}$$

is d-transcendental over $\mathbb{T}_{\mathrm{g}}$. Thus the series $\zeta(x) = 1 + 2^{-x} + 3^{-x} + \cdots$ for the zeta function is d-transcendental over $\mathbb{T}_{\mathrm{g}}$; this improves the classical fact that $\zeta(x)$ is d-transcendental, stated by Hilbert in his 1900 ICM address [183, p. 287] and proved in [312, 429]; see [373] for the history, and [223, 432] for other d-transcendence results about Dirichlet series that can be strenghtened likewise.

Corollary 16.2.6 corrects an error in [20]: Lemma 14.1 there is false. The definition of "existentially closed H-field" in that paper is therefore not equivalent to the usual definition. The mistake is in the equivalence "$z \succ 1 \iff \cdots$" claimed in the alleged proof of that lemma. This also led to an incorrect model completeness conjecture on p. 279 of [22]: we should have included the valuation ring of $\mathbb{T}$ among the primitives. (This is the "minor change" referred to in our Preface.)

In our treatment the proof of model completeness of T^{nl} is a key step towards QE (quantifier elimination) in a suitably extended language, rather than model completeness being obtained as a consequence of QE, as often happens. This gives hope that in possible extensions of our work, for example to $\mathbb{T}_{\log}$, model completeness, rather than QE, might be a realistic first aim.

16.3 $\Lambda\Omega$-CUTS AND $\Lambda\Omega$-FIELDS

Throughout this section K is a pre-H-field. The reader needs to be familiar with Sections 10.5, 10.6 and 11.8. We shall determine the sets $I, \Lambda, \Omega \subseteq K$ for which (K, I, Λ, Ω) can be embedded into $(L, \mathrm{I}(L), \Lambda(L), \Omega(L))$ for some ω-free newtonian Liouville closed H-field L. We first consider for any given K just the possibilities for I, next we determine the possibilities for (I, Λ), and finally, the possibilities for (I, Λ, Ω). This section is independent of the previous two.

I-sets

By an **I-set** in K we mean an $\mathcal{O}$-submodule I of K such that $\partial\mathcal{O} \subseteq I$ and $a^\dagger \notin I$ for all $a \succ 1$ in K. An I-set in K is in particular a convex additive subgroup of the ordered additive group of K. Note that

$$\mathrm{I}(K) \;=\; \{y \in K : y \preccurlyeq f' \text{ for some } f \in \mathcal{O}\}$$

is the smallest I-set in K. If K has no gap, then $\mathrm{I}(K)$ is the only I-set in K. If K has a gap β and $v(b') = \beta$ for some $b \asymp 1$ in K, then $\mathrm{I}(K)$ is the only I-set in K.

LEMMA 16.3.1. *Suppose K has gap β and $v(b') \neq \beta$ for all $b \asymp 1$ in K. Then K has exactly two I-sets, namely $\mathrm{I}(K) = \{a \in K : va > \beta\}$ and $\{a \in K : va \geqslant \beta\}$.*

Clearly, if K is an H-field with gap β, then the hypothesis of Lemma 16.3.1 holds.

LEMMA 16.3.2. *If L is a pre-H-field extension of K and J is an I-set in L, then $J \cap K$ is an I-set in K. As a strong converse, if I is an I-set in K, then K has a Liouville closed H-field extension L such that $I = \mathrm{I}(L) \cap K$.*

PROOF. The first part of the lemma is obvious. For the second part, first note that K has a Liouville closed H-field extension L, and so the desired conclusion follows from the first part if K has just one I-set. It remains to consider the case that K has a gap β and $v(b') \neq \beta$ for all $b \asymp 1$ in K. Take $s \in K$ with $vs = \beta$.

We first deal with $I = \{a \in K : va \geqslant \beta\}$. Lemma 10.2.1 and subsequent remarks, combined with Corollary 10.5.10, give a pre-H-field extension $K(y)$ of K such that $y' = s$, $y \prec 1$, and $K(y)$ has no gap. Then $s \in \mathrm{I}\big(K(y)\big)$, so $\mathrm{I}\big(K(y)\big) \cap K = I$ by Lemma 16.3.1. Taking a Liouville closed H-field extension L of $K(y)$, we get $\mathrm{I}(L) \cap K(y) = \mathrm{I}\big(K(y)\big)$, and thus $\mathrm{I}(L) \cap K = I$.

Next, let $I = \mathrm{I}(K) = \{a \in K : va > \beta\}$. The H-field hull $H(K)$ of K is an immediate extension of K by Corollary 10.3.2. So β is still a gap in $H(K)$ and $v(b') \neq \beta$ for all $b \asymp 1$ in $H(K)$, and thus $\mathrm{I}\big(H(K)\big) \cap K = I$. Since $H(K)$ is d-valued, we can apply Corollary 10.5.11 to give an H-field extension $F := H(K)(y)$ such that $y' = s$, $y \succ 1$, and F has no gap. Then $s \notin \mathrm{I}(F)$, and so $\mathrm{I}(F) \cap K = I$ by Lemma 16.3.1. Taking a Liouville closed H-field extension L of F, we get $\mathrm{I}(L) \cap F = \mathrm{I}(F)$, and thus $\mathrm{I}(L) \cap K = I$. □

Λ-cuts

To motivate the notion of a Λ-cut, suppose K is a Liouville closed H-field. Various definitions and results in Section 11.8 show that then the subsets $I := \mathrm{I}(K)$ and $\Lambda := \Lambda(K) = -(K^{\succ 1})^{\dagger\dagger}$ of K have the following universal properties:

(ΛΩ1) I is an I-set in K;

(ΛΩ2) $\Lambda(K) \subseteq \Lambda$;

(ΛΩ3) for all $a \in K^\times$: $a \in I \iff -a^\dagger \notin \Lambda$;

(ΛΩ4) $\Lambda + I \subseteq \Lambda$;

(ΛΩ5) Λ is downward closed;

(ΛΩ6) for all $a, \phi \in K$, if $a, \phi > 0$ and $a \succcurlyeq 1$, then $\phi a - \phi^\dagger \notin \Lambda$.

We define a **Λ-cut** in K to be a pair (I, Λ) of sets $I, \Lambda \subseteq K$ satisfying conditions (ΛΩ1)–(ΛΩ6) above. Only (ΛΩ5) and (ΛΩ6) involve the ordering of K. Although I is determined by Λ via (ΛΩ3), it is convenient to make it part of a Λ-cut.

Suppose that (I, Λ) is a Λ-cut in K. It easily follows from (ΛΩ1) and (ΛΩ3) that $\Lambda \cap \Delta(K) = \emptyset$, with $\Delta(K)$ as defined in Section 11.8. For any differential subfield E of K we have a Λ-cut $(I \cap E, \Lambda \cap E)$ in the pre-H-subfield E of K. Thus every pre-H-field, having a Liouville closed H-field extension, has a Λ-cut. It is easy to check that for $\phi \in K^>$ the pair

$$(I, \Lambda)^\phi := \big(\phi^{-1} I, \phi^{-1}(\Lambda + \phi^\dagger)\big)$$

is a Λ-cut in K^ϕ.

LEMMA 16.3.3. *Suppose K is grounded. Then K has a unique Λ-cut.*

PROOF. By compositional conjugation we arrange $\max \Psi = 0$. Then $\mathrm{I}(K) = \smallo$. Assume $(\mathrm{I}(K), \Lambda)$ is a Λ-cut in K. Since $1 \notin \mathrm{I}(K)$, we have $-1^\dagger = 0 \in \Lambda$ by (ΛΩ3), so $\smallo^\downarrow \subseteq \Lambda$ by (ΛΩ4) and (ΛΩ5). If $a \in K^>$ and $a \succcurlyeq 1$, then taking $\phi = 1$ in (ΛΩ6) shows that $a \notin \Lambda$. Thus $\Lambda = \smallo^\downarrow$. □

LEMMA 16.3.4. *Suppose K has a gap β and $v(b') = \beta$ for some $b \asymp 1$ in K. Then K has a unique Λ-cut.*

PROOF. We arrange by compositional conjugation that $\beta = 0$. Let (I, Λ) be a Λ-cut in K. Then $I = \mathrm{I}(K) = \mathcal{O}$, so $1 \in I$, hence $0 \notin \Lambda$ by (ΛΩ3), and thus

$$\Lambda \subseteq \{a \in K : a < I\}$$

by (ΛΩ4) and (ΛΩ5). Suppose $a \in K$ and $a < I$. Then $-a^\dagger \in \Lambda$ by (ΛΩ3), so $-a^\dagger < I$, hence $a^\dagger \succ 1$. Since the derivation of K is small, we get $v(a^\dagger) = o(va)$ by Lemma 9.2.10(iv), and so $a < -a^\dagger < 0$, hence $a \in \Lambda$ by (ΛΩ5). This yields $\Lambda = \{a \in K : a < I\}$. □

LEMMA 16.3.5. *Suppose K has a gap β and $v(b') \neq \beta$ for all $b \asymp 1$ in K. Then there are exactly two Λ-cuts (I, Λ) in K, one with $I = \mathrm{I}(K)$, and the other with $I = \{y \in K : vy \geqslant \beta\}$.*

PROOF. Lemma 16.3.2 provides for each of the two I-sets I in K a Liouville closed H-field extension L of K with $\mathrm{I}(L) \cap K = I$, giving rise to a Λ-cut $\big(I, \Lambda(L) \cap K\big)$ in K. To get uniqueness, arrange $\beta = 0$ by compositional conjugation. Then $\mathrm{I}(K) = \smallo$ and $\{y \in K : vy \geqslant \beta\} = \mathcal{O}$. If $(\smallo, \Lambda)$ is a Λ-cut in K, then we get $\Lambda = \smallo^\downarrow$ as in the proof of Lemma 16.3.3. If $(\mathcal{O}, \Lambda)$ is a Λ-cut in K, then we get $\Lambda = \{a \in K : a < \mathcal{O}\}$ as in the proof of Lemma 16.3.4. □

Thus if K is an H-field with a gap β, then there are exactly two Λ-cuts (I_1, Λ_1) and (I_2, Λ_2) in K, with $I_1 = \mathrm{I}(K) = \{y \in K : vy > \beta\}$ and $I_2 = \{y \in K : vy \geqslant \beta\}$.

LEMMA 16.3.6. *If K is λ-free, then $\big(\mathrm{I}(K), \Lambda(K)^\downarrow\big)$ is the only Λ-cut in K.*

PROOF. Let K be λ-free. Then K has asymptotic integration, so $\mathrm{I}(K)$ is the only I-set in K. Let $\big(\mathrm{I}(K), \Lambda\big)$ be a Λ-cut in K. Then $\Lambda(K)^\downarrow \subseteq \Lambda \subseteq K \setminus \Delta(K)^\uparrow$, and so it remains to note that by λ-freeness we have $\Lambda(K)^\downarrow = K \setminus \Delta(K)^\uparrow$. □

LEMMA 16.3.7. *Suppose K has asymptotic integration and is not λ-free. Then there are exactly two Λ-cuts in K: $\big(\mathrm{I}(K), \Lambda(K)^\downarrow\big)$, and $\big(\mathrm{I}(K), K \setminus \Delta(K)^\uparrow\big)$.*

PROOF. Let (I, Λ) be a Λ-cut in K. Then $I = \mathrm{I}(K)$, and we claim that $\Lambda = \Lambda(K)^\downarrow$ or $\Lambda = K \setminus \Delta(K)^\uparrow$. To see this, first note that $\Lambda(K)^\downarrow \subseteq \Lambda \subseteq K \setminus \Delta(K)^\uparrow$. Suppose $\Lambda \neq \Lambda(K)^\downarrow$, and take $a \in \Lambda \setminus \Lambda(K)^\downarrow$. Then $\Lambda(K) < a < \Delta(K)$, and for every $b \in K$ with $\Lambda(K) < b < \Delta(K)$ we have $v(b - a) > \Psi$ by Corollary 11.8.16, so $b - a \in \mathrm{I}(K)$, and hence $b \in \Lambda + \mathrm{I}(K) \subseteq \Lambda$. Thus $\Lambda = K \setminus \Delta(K)^\uparrow$, which proves our claim. It remains to show that there is more than one Λ-cut in K. Let $E := H(K)^{\mathrm{rc}}$ be the real closure of the H-field hull of K. Now $H(K)$ is an immediate extension of K by Corollary 10.3.2, so E is an H-field extension of K with divisible value group $\Gamma_E = \mathbb{Q}\Gamma$ and so $\Psi_E = \Psi$. We now distinguish two cases:

CASE 1: *E has a gap.* Then we can take $s \in E^\times$ and $n \geqslant 1$ such that vs is a gap in E and $s^n \in K$. Note that then $s^\dagger \in K$. Now E is an H-field, so Lemma 16.3.5 yields Λ-cuts (I_1, Λ_1) and (I_2, Λ_2) in E, with $I_1 = \mathrm{I}(E) = \{y \in E : y \prec s\}$ and $I_2 = \{y \in E : y \preccurlyeq s\}$. Now $s \notin I_1$, so $-s^\dagger \in \Lambda_1$ by (ΛΩ3), and $s \in I_2$, so $-s^\dagger \notin \Lambda_2$, also by (ΛΩ3). Hence $-s^\dagger \in \Lambda_1 \cap K$ and $-s^\dagger \notin \Lambda_2 \cap K$. So we have two distinct Λ-cuts in K, namely

$$(I_1 \cap K, \Lambda_1 \cap K) = \big(\mathrm{I}(K), K \setminus \Delta(K)^\uparrow\big), \quad (I_2 \cap K, \Lambda_2 \cap K) = \big(\mathrm{I}(K), \Lambda(K)^\downarrow\big).$$

CASE 2: *E has no gap.* Then E has asymptotic integration, and the sequence (λ_ρ) for K also serves for E. Take $\lambda \in K$ such that $\lambda_\rho \rightsquigarrow \lambda$. Then $-\lambda$ creates a gap over E by Lemma 11.5.14. Take an element $f \neq 0$ in some Liouville closed H-field extension of E such that $f^\dagger = -\lambda$. Then vf is a gap in $E(f)$ by the remark following the proof of Lemma 11.5.14. Using Lemma 9.1.2 it follows that the pre-H-field $E(f)$ is actually an H-field. Hence by Lemma 16.3.5, $E(f)$ has Λ-cuts (I_1, Λ_1) and (I_2, Λ_2), with $I_1 = \mathrm{I}(E(f))$ and $I_2 = \{a \in E(f) : a \preccurlyeq f\}$. Now $f \notin I_1$, so $-f^\dagger = \lambda \in \Lambda_1$ by ($\Lambda\Omega$3), and $f \in I_2$, so $\lambda \notin \Lambda_2$, again by ($\Lambda\Omega$3). Thus

$$(I_1 \cap K, \Lambda_1 \cap K) = \big(\mathrm{I}(K), K \setminus \Delta(K)^\dagger\big), \quad (I_2 \cap K, \Lambda_2 \cap K) = \big(\mathrm{I}(K), \Lambda(K)^\downarrow\big)$$

are two distinct Λ-cuts in K. □

The proofs above have the following byproduct:

COROLLARY 16.3.8. *Let (I, Λ) be a Λ-cut in K. Then K has a Liouville closed H-field extension L such that $(I, \Lambda) = \big(\mathrm{I}(L) \cap K, \Lambda(L) \cap K\big)$.*

PROOF. If K has a unique Λ-cut, then the conclusion of the corollary holds for any Liouville closed H-field extension L of K. The case that K has a gap β and $v(b') \neq \beta$ for all $b \asymp 1$ in K is treated in the proof of Lemma 16.3.5. The case that K has asymptotic integration and is not λ-free reduces to the previous cases by extending K to E or $E(f)$ as in the proof of Lemma 16.3.7. □

This corollary, when combined with Proposition 11.8.20 and Lemma 11.8.29, has the following consequence:

COROLLARY 16.3.9. *Let (I, Λ) be a Λ-cut in K. Then the functions*

$$z \mapsto \omega(z) \colon \Lambda \to K, \qquad y \mapsto \sigma(y) \colon K^{>} \setminus I \to K$$

are strictly increasing.

$\Lambda\Omega$-cuts

By various results in Section 11.8 the sets $I := \mathrm{I}(K)$, $\Lambda := \Lambda(K)$, and $\Omega := \Omega(K)$ of any Schwarz closed K have the following universal properties:

($\Lambda\Omega$7) $\omega(K) \subseteq \Omega$ (and so $0 \in \Omega$);

($\Lambda\Omega$8) for all $f, g \in K^{>}$, if $-\frac{1}{2}f^\dagger \in \Lambda$ and $f \asymp g$, then $\omega(-\frac{1}{2}f^\dagger) + g \notin \Omega$;

($\Lambda\Omega$9) for all $f \in K^\times$, if $-\frac{1}{2}f^\dagger \notin \Lambda$, then $\Omega + f \subseteq \Omega$;

($\Lambda\Omega$10) Ω is downward closed.

Accordingly we define a **$\Lambda\Omega$-cut** in K to be a triple (I, Λ, Ω) of sets $I, \Lambda, \Omega \subseteq K$ such that (I, Λ) is a Λ-cut in K and conditions ($\Lambda\Omega$7)–($\Lambda\Omega$10) above are satisfied. Let us record some easy consequences of the axioms above:

LEMMA 16.3.10. *Let (I, Λ, Ω) be a $\Lambda\Omega$-cut in K. Then*

($\Lambda\Omega$11) *for all $g, h \in I$, we have $\Omega + gh \subseteq \Omega$;*

($\Lambda\Omega$12) *for all $f \in K^\times$: $f \in I \Longleftrightarrow \sigma(f) \in \Omega$.*

PROOF. To show ($\Lambda\Omega$11), let $0 \neq g, h \in I$, and set $f := gh$. By ($\Lambda\Omega$9), it suffices to prove $-\frac{1}{2}f^\dagger \notin \Lambda$. We may assume $g = ha$, $a \preccurlyeq 1$. So $f = h^2 a$ and hence $-\frac{1}{2}f^\dagger = -h^\dagger - \frac{1}{2}a^\dagger$, and $-h^\dagger \notin \Lambda$ by ($\Lambda\Omega$3). If $a \asymp 1$, then ($\Lambda\Omega$4) and $a^\dagger \in \mathrm{I}(K) \subseteq I$ give $-\frac{1}{2}f^\dagger \notin \Lambda$. If $a \prec 1$, then $-\frac{1}{2}a^\dagger > 0$ and hence $-\frac{1}{2}f^\dagger > -h^\dagger$ and so $-\frac{1}{2}f^\dagger \notin \Lambda$.

For ($\Lambda\Omega$12), let $f \in K^\times$. If $f \in I$, then $\sigma(f) = \omega(-f^\dagger) + f^2 \in \Omega$ by ($\Lambda\Omega$11). Suppose $f \notin I$. Then $-\frac{1}{2}(f^2)^\dagger = -f^\dagger \in \Lambda$ by ($\Lambda\Omega$3), so $\sigma(f) = \omega(-f^\dagger) + f^2 = \omega(-\frac{1}{2}(f^2)^\dagger) + f^2 \notin \Omega$ by ($\Lambda\Omega$8). □

Suppose (I, Λ, Ω) is a $\Lambda\Omega$-cut in K. It follows easily from ($\Lambda\Omega$10) and ($\Lambda\Omega$12) that if $a \in K \setminus I$, then $\Omega < \sigma(a)$. In particular, we have $\Omega < \sigma\big(\Gamma(K)\big)$, so

$$\omega(K)^\downarrow \subseteq \Omega \subseteq K \setminus \sigma\big(\Gamma(K)\big)^\uparrow.$$

For any differential subfield E of K we have a $\Lambda\Omega$-cut $(I \cap E, \Lambda \cap E, \Omega \cap E)$ in the pre-H-subfield E of K. Thus every pre-H-field, having a Schwarz closed H-field extension by Corollary 11.7.18 and Proposition 14.5.12, has a $\Lambda\Omega$-cut. Moreover, any Λ-cut (I, Λ) in K is part of a $\Lambda\Omega$-cut (I, Λ, Ω) in K. (Proof: let (I, Λ) be a Λ-cut in K. Take a Liouville closed extension L of K with $\mathrm{I}(L) \cap K = I$ and $\Lambda(L) \cap K = \Lambda$, and next a Schwarz closed extension M of L. Then $\Omega = \Omega(M) \cap K$ gives a $\Lambda\Omega$-cut (I, Λ, Ω) in K.) For $\phi \in K^>$ one can use the identities at the end of Section 11.7 to show that the triple $(I, \Lambda, \Omega)^\phi = (I^\phi, \Lambda^\phi, \Omega^\phi)$ with

$$(16.3.1) \qquad I^\phi := \phi^{-1} I, \quad \Lambda^\phi := \phi^{-1}(\Lambda + \phi^\dagger), \quad \Omega^\phi := \phi^{-2}\big(\Omega - \omega(-\phi^\dagger)\big)$$

is a $\Lambda\Omega$-cut in K^ϕ.

LEMMA 16.3.11. *Suppose K is grounded. Then K has a unique $\Lambda\Omega$-cut.*

PROOF. By compositional conjugation we arrange $\max \Psi = 0$. Let (I, Λ, Ω) be a $\Lambda\Omega$-cut in K. Then $I = o$ and $\Lambda = o^\downarrow$ by the proof of Lemma 16.3.3. Let $u \asymp 1$ in K. Then $u^\dagger \prec 1$ and $(u^\dagger)' \prec 1$, so $\sigma(u) = \omega(-u^\dagger) + u^2 \sim u^2$. Also $u \notin I$, so $\Omega < \sigma(u)$. Hence $\Omega < 2u^2$ for all units u of $\mathcal{O}$, and so $\Omega \subseteq o^\downarrow$. We claim that $\Omega = o^\downarrow$. By ($\Lambda\Omega$10) it is enough to show $o \subseteq \Omega$. We distinguish two cases:

CASE 1: *Γ has no least positive element.* Then every element of o is a product gh with $g, h \in o = I$, and as $0 \in \Omega$, we get $o \subseteq \Omega$ from ($\Lambda\Omega$11), and so our claim holds.

CASE 2: *Γ has a least positive element α.* Take $a \in K$ with $va = \alpha$ and set $b := -2a$. From $\alpha^\dagger = 0$ we get $v(a') = \alpha$ and so $\omega(a) \sim -2a' = b'$, hence $v(\omega(a) - b') \geqslant 2\alpha$, and then $b' \in \Omega$ by ($\Lambda\Omega$7) and ($\Lambda\Omega$11). Thus $b' \in \Omega$ for each $b \in K$ with $vb = \alpha$. For such b and $u \asymp 1$ in K we have $v\big((ub)' - ub'\big) = v(u'b) \geqslant 2\alpha$, and thus $ub' \in \Omega$ by ($\Lambda\Omega$11). Fixing b and varying u we see that all elements of K of valuation α belong to Ω, and thus $o \subseteq \Omega$, as desired. □

LEMMA 16.3.12. *Suppose K has a gap β and $v(b') = \beta$ for some $b \asymp 1$ in K. Then K has a unique $\Lambda\Omega$-cut.*

PROOF. We arrange by compositional conjugation that $\beta = 0$. Let (I, Λ, Ω) be a $\Lambda\Omega$-cut in K. Then $I = \mathcal{O}$ and $\Lambda = \{a \in K : a < \mathcal{O}\}$ by the proof of Lemma 16.3.4. From $0 \in \Omega$ and ($\Lambda\Omega$10) and ($\Lambda\Omega$11) we obtain $\mathcal{O}^{\downarrow} \subseteq \Omega$. We claim that $\Omega = \mathcal{O}^{\downarrow}$. This is clear if $\Gamma = \{0\}$, so suppose $\Gamma \neq \{0\}$. Then $\Gamma^{<}$ does not have a largest element. For $a \in K^{>}$ with $a \succ 1$ we have $\sigma(a) \sim a^2$, as well as $a \notin I$, so $\sigma(a) > \Omega$ by ($\Lambda\Omega$7) and ($\Lambda\Omega$12), and thus $\Omega < 2a^2$. This yields the claim. □

LEMMA 16.3.13. *Suppose K has a gap β and $v(b') \neq \beta$ for all $b \asymp 1$ in K. Then there are exactly two $\Lambda\Omega$-cuts (I, Λ, Ω) in K, one with $I = \mathrm{I}(K)$, and the other with $I = \{y \in K : vy \geqslant \beta\}$.*

PROOF. Lemma 16.3.2 provides for each of the two I-sets I in K a Liouville closed H-field extension L of K with $\mathrm{I}(L) \cap K = I$, and by further extending we can arrange L to be even Schwarz closed, giving rise to a $\Lambda\Omega$-cut $\big(I, \Lambda(L) \cap K, \omega(L) \cap K\big)$ in K. To get uniqueness, arrange $\beta = 0$ by compositional conjugation. Then by the proof of Lemma 16.3.5 we have a $\Lambda\Omega$-cut $(o, o^{\downarrow}, \Omega)$ in K. The same arguments as in the proof of Lemma 16.3.11 show that for each such $\Lambda\Omega$-cut we have $\Omega = o^{\downarrow}$. We also have a $\Lambda\Omega$-cut $(\mathcal{O}, \Lambda, \Omega)$ with $\Lambda = \{a \in K : a < \mathcal{O}\}$. The same argument as in the proof of Lemma 16.3.12 shows that $\Omega = \mathcal{O}^{\downarrow}$ for each such $\Lambda\Omega$-cut. □

In the next lemmas we treat the case where K has asymptotic integration. Recall Lemma 16.3.6: if K is λ-free, then $\big(\mathrm{I}(K), \Lambda(K)^{\downarrow}\big)$ is the only Λ-cut in K.

LEMMA 16.3.14. *Suppose K is ω-free. Then the only $\Lambda\Omega$-cut in K is*

$$\big(\mathrm{I}(K), \Lambda(K)^{\downarrow}, \omega(K)^{\downarrow}\big).$$

PROOF. Let (I, Λ, Ω) be a $\Lambda\Omega$-cut in K, so $I = \mathrm{I}(K)$ and $\Lambda = \Lambda(K)^{\downarrow}$. Also, $\omega(K)^{\downarrow} \subseteq \Omega \subseteq K \setminus \sigma\big(\Gamma(K)\big)^{\uparrow}$, and so it remains to note that by ω-freeness we have $\omega(K)^{\downarrow} = K \setminus \sigma\big(\Gamma(K)\big)^{\uparrow}$. □

LEMMA 16.3.15. *Suppose K has asymptotic integration and the set 2Ψ does not have a supremum in Γ. Then for each $\Lambda\Omega$-cut (I, Λ, Ω) in K we have*

$$\Omega = \omega\big(\Lambda(K)\big)^{\downarrow} = \omega(K)^{\downarrow} \quad \textit{or} \quad \Omega = K \setminus \sigma\big(\Gamma(K)\big)^{\uparrow}.$$

PROOF. Let (I, Λ, Ω) be a $\Lambda\Omega$-cut in K. Recall that

$$\omega\big(\Lambda(K)\big)^{\downarrow} \subseteq \Omega \subseteq K \setminus \sigma\big(\Gamma(K)\big)^{\uparrow}.$$

Note that if $\Omega = \omega\big(\Lambda(K)\big)^{\downarrow}$, then $\Omega = \omega(K)^{\downarrow}$ by ($\Lambda\Omega$7). Suppose $\Omega \neq \omega\big(\Lambda(K)\big)^{\downarrow}$, and take $a \in \Omega \setminus \omega\big(\Lambda(K)\big)^{\downarrow}$. Then

$$\omega\big(\Lambda(K)\big) < a < \sigma\big(\Gamma(K)\big),$$

and for every $b \in K$ with $\omega\big(\Lambda(K)\big) < b < \sigma\big(\Gamma(K)\big)$ we have $v(b - a) > 2\Psi$ by Corollaries 11.7.2 and 11.8.30, so by Lemma 9.2.17 there are $g, h \in \mathrm{I}(K)$ with $b - a = gh$, and thus $b \in \Omega + gh \subseteq \Omega$ by ($\Lambda\Omega$11). Thus $\Omega = K \setminus \sigma\big(\Gamma(K)\big)^{\uparrow}$. □

LEMMA 16.3.16. *Suppose K is λ-free, but not ω-free. Then there are exactly two ΛΩ-cuts in K:* $\big(\mathrm{I}(K), \Lambda(K)^{\downarrow}, \omega(K)^{\downarrow}\big)$, *and* $\big(\mathrm{I}(K), \Lambda(K)^{\downarrow}, K \setminus \sigma(\Gamma(K))^{\uparrow}\big)$.

PROOF. Since K is λ-free, K has rational asymptotic integration, and so 2Ψ does not have a supremum in Γ. Let (I, Λ, Ω) be a ΛΩ-cut in K. Then $I = \mathrm{I}(K)$, $\Lambda = \Lambda(K)^{\downarrow}$, and either $\Omega = \omega(K)^{\downarrow}$ or $\Omega = K \setminus \sigma\big(\Gamma(K)\big)^{\uparrow}$ by Lemma 16.3.15.

It remains to show that there is more than one ΛΩ-cut in K. Take $\omega \in K$ such that $\omega_\rho \rightsquigarrow \omega$. Then Corollary 11.7.13 and Lemma 10.5.8 yield an immediate pre-H-field extension $K(\lambda)$ of K with $\lambda_\rho \rightsquigarrow \lambda$ and $\omega(\lambda) = \omega$. Now $K(\lambda)$ has a ΛΩ-cut $(I_\lambda, \Lambda_\lambda, \Omega_\lambda)$. Then $\omega \in \Omega_\lambda$, and so intersecting with K gives a ΛΩ-cut (I, Λ, Ω) in K with $\omega \in \Omega$. It remains to find such a ΛΩ-cut in K with $\omega \notin \Omega$. For that we take a pre-H-field extension $K\langle\gamma\rangle$ as in Lemma 13.7.7 in which $\sigma(\gamma) = \omega$ and $v\gamma$ is a gap. By remark (3) following the proof of Proposition 13.7.1 there is no $b \asymp 1$ in $K\langle\gamma\rangle$ with $b' \asymp \gamma$. Then by Lemma 16.3.5 we have a Λ-cut $(I_\gamma, \Lambda_\gamma)$ in $K\langle\gamma\rangle$ with $\gamma \notin I_\gamma$. Take a ΛΩ-cut $(I_\gamma, \Lambda_\gamma, \Omega_\gamma)$ in $K\langle\gamma\rangle$. Then $\omega = \sigma(\gamma) \notin \Omega_\gamma$ by (ΛΩ12). Intersecting with K gives a ΛΩ-cut (I, Λ, Ω) in K with $\omega \notin \Omega$. □

LEMMA 16.3.17. *Suppose K has asymptotic integration and is not λ-free, and the set 2Ψ does not have a supremum in Γ. Then there are exactly two ΛΩ-cuts in K:*

$$\big(\mathrm{I}(K), \Lambda(K)^{\downarrow}, K \setminus \sigma\big(\Gamma(K)\big)^{\uparrow}\big) \text{ and } \big(\mathrm{I}(K), K \setminus \Delta(K)^{\uparrow}, K \setminus \sigma\big(\Gamma(K)\big)^{\uparrow}\big).$$

PROOF. Let (I, Λ, Ω) be a ΛΩ-cut in K. Let $\lambda_\rho \rightsquigarrow \lambda \in K$. Then $\omega_\rho \rightsquigarrow \omega := \omega(\lambda)$ by Corollary 11.7.3 and $\omega \notin \omega\big(\Lambda(K)\big)^{\downarrow}$ by Corollary 11.8.30, but $\omega \in \omega(K) \subseteq \Omega$. Hence $\Omega = K \setminus \sigma\big(\Gamma(K)\big)^{\uparrow}$ by Lemma 16.3.15. It remains to use Lemma 16.3.7. □

It follows from Lemma 9.2.17 that in Lemma 16.3.17 we can drop the condition that 2Ψ has no supremum in Γ if Γ is divisible.

LEMMA 16.3.18. *Suppose K has asymptotic integration and $vf = \sup 2\Psi$, $f \in K^{>}$. Then there are exactly two ΛΩ-cuts in K, namely*

$$\big(\mathrm{I}(K), \Lambda(K)^{\downarrow}, K \setminus \sigma\big(\Gamma(K)\big)^{\uparrow}\big) \text{ and } \big(\mathrm{I}(K), K \setminus \Delta(K)^{\uparrow}, \Omega_f\big), \text{ where}$$
$$\Omega_f := \big\{a \in K : a \leqslant \omega(-\tfrac{1}{2}f^{\dagger}) + g \text{ for some } g \prec f \text{ in } K\big\}.$$

Moreover, $\Lambda(K)^{\downarrow} \neq K \setminus \Delta(K)^{\uparrow}$ *and* $K \setminus \sigma\big(\Gamma(K)\big)^{\uparrow} \neq \Omega_f$.

PROOF. Let $\sqrt{f}$ be a positive element of the real closure of K with $(\sqrt{f})^2 = f$. The pre-H-field extension $L = K(\sqrt{f})$ of K has gap $\beta = v(\sqrt{f}) = \frac{1}{2}vf$. Hence $\Lambda(K) < -\frac{1}{2}f^{\dagger} = -(\sqrt{f})^{\dagger} < \Delta(K)$ by Corollaries 11.5.7 and 11.8.16. Thus K is not λ-free, and hence by Lemma 16.3.7, K has exactly two Λ-cuts, $\big(\mathrm{I}(K), \Lambda(K)^{\downarrow}\big)$ and $\big(\mathrm{I}(K), K \setminus \Delta(K)^{\uparrow}\big)$. By Lemma 10.3.3 there is no $b \asymp 1$ in L with $b' \asymp \sqrt{f}$, so by Lemma 16.3.5, L also has exactly two Λ-cuts, (I_1, Λ_1) and (I_2, Λ_2), with

$$I_1 = \mathrm{I}(L) = \{y \in L : vy > \beta\}, \qquad I_2 = \{y \in L : vy \geqslant \beta\}.$$

We have $\sqrt{f} \notin I_1$ and $\sqrt{f} \in I_2$, hence $-\frac{1}{2}f^\dagger = -(\sqrt{f})^\dagger \in \Lambda_1 \cap K$ and $-\frac{1}{2}f^\dagger \notin \Lambda_2 \cap K$ by (ΛΩ3). Therefore

$$I_1 \cap K = I_2 \cap K = \mathrm{I}(K), \quad \Lambda_1 \cap K = K \setminus \Delta(K)^\uparrow, \quad \Lambda_2 \cap K = \Lambda(K)^\downarrow.$$

Note that $\sigma(\sqrt{f}) = \omega(-\frac{1}{2}f^\dagger) + f$ lies in K. Also, by Lemma 16.3.13 we have just one ΛΩ-cut $(I_1, \Lambda_1, \Omega_1)$ in L, and just one ΛΩ-cut $(I_2, \Lambda_2, \Omega_2)$ in L. This yields distinct ΛΩ-cuts $(I_1 \cap K, \Lambda_1 \cap K, \Omega_1 \cap K)$ and $(I_2 \cap K, \Lambda_2 \cap K, \Omega_2 \cap K)$ in K, since $\Lambda_1 \cap K \neq \Lambda_2 \cap K$; we claim that these are the only ΛΩ-cuts in K, and that $\Omega_f = \Omega_1 \cap K \neq \Omega_2 \cap K = K \setminus \sigma(\Gamma(K))^\uparrow$.

For this, let (I, Λ, Ω) be a ΛΩ-cut in K. Note that if $g \in K^\times$ and $g \prec f$, then Lemma 9.2.17 gives $a, b \in \mathrm{I}(K) = I$ with $g = ab$, and so by (ΛΩ11) we obtain $\Omega + g \subseteq \Omega$. In particular, using (ΛΩ10) we get $\Omega_f \subseteq \Omega$. Also as in the proof of Lemma 16.3.17 we obtain $\omega(-\frac{1}{2}f^\dagger) \in \Omega \setminus \omega(\Lambda(K))^\downarrow$.

CASE 1: $\Lambda = \Lambda(K)^\downarrow$. Then $-\frac{1}{2}f^\dagger \notin \Lambda_2 \cap K = \Lambda$, and thus $\Omega + f \subseteq \Omega$ by (ΛΩ9). Replacing the role of f by uf for any $u \in K^>$ with $u \asymp 1$ we get $\Omega + g \subseteq \Omega$ for all $g \in K^\times$ with $vg > 2\Psi$. Then the proof of Lemma 16.3.15 yields $\Omega = K \setminus \sigma(\Gamma(K))^\uparrow$. In particular, $\Omega = \Omega_2 \cap K$ and $\sigma(\sqrt{f}) = \omega(-\frac{1}{2}f^\dagger) + f \in \Omega$.

CASE 2: $\Lambda = K \setminus \Delta(K)^\uparrow$. Then $-\frac{1}{2}f^\dagger \in \Lambda_1 \cap K = \Lambda$, and $\sqrt{f} \notin I_1 = \mathrm{I}(L)$ gives $\sigma(\sqrt{f}) \notin \Omega_1 \cap K$ by (ΛΩ12). We have $\Omega_f \subseteq \Omega$, and (ΛΩ8) and (ΛΩ10) yield $\Omega \subseteq \Omega_f$, so $\Omega = \Omega_f = \Omega_1 \cap K$. □

We have now covered all possibilities, and conclude that K has either exactly one ΛΩ-cut, or exactly two. Moreover:

COROLLARY 16.3.19. *The pre-H-field K has a unique ΛΩ-cut if and only if*

(i) *K is grounded, or*

(ii) *there exists $b \asymp 1$ in K such that $v(b')$ is a gap in K, or*

(iii) *K is ω-free.*

The proofs above have the following byproduct:

COROLLARY 16.3.20. *Let (I, Λ, Ω) be a ΛΩ-cut in K. Then K has a Schwarz closed H-field extension L such that $(I, \Lambda, \Omega) = (\mathrm{I}(L) \cap K, \Lambda(L) \cap K, \Omega(L) \cap K)$.*

PROOF. If there is a unique ΛΩ-cut in K, then the conclusion of the corollary holds for any Schwarz closed H-field extension L of K. The case that K has a gap β with $v(b') \neq \beta$ for all $b \asymp 1$ in K is treated in the proof of Lemma 16.3.13. Suppose K has asymptotic integration but is not λ-free. Then by Lemmas 16.3.17 and 16.3.18, there are exactly two ΛΩ-cuts

$$(I_1, \Lambda_1, \Omega_1) = (\mathrm{I}(K), \Lambda(K)^\downarrow, \dots), \quad (I_2, \Lambda_2, \Omega_2) = (\mathrm{I}(K), K \setminus \Delta(K)^\downarrow, \dots)$$

in K. By Corollary 16.3.8 we can take for $i = 1, 2$ a Liouville closed extension L_i of K such that $(\mathrm{I}(L_i) \cap K, \Lambda(L_i) \cap K) = (I_i, \Lambda_i)$. Extending L_i if necessary, we can even arrange that each L_i is Schwarz closed. Then

$$\big(\mathrm{I}(L_i) \cap K, \Lambda(L_i) \cap K, \Omega(L_i) \cap K\big) = (I_i, \Lambda_i, \Omega_i) \quad \text{for } i = 1, 2.$$

Finally, the case that K is λ-free, but not ω-free, reduces to the case that K is not λ-free or K has a gap by extending K to $K(\lambda)$ and to $K\langle\gamma\rangle$ as in the proof of Lemma 16.3.16.

□

Let (I, Λ, Ω) be a $\Lambda\Omega$-cut in K, let L be as in Corollary 16.3.20, and take a Newton-Liouville closure L^{nl} of L. Then the conclusion of that corollary remains valid for L^{nl} in place of L, since $\mathrm{I}(L^{\mathrm{nl}}) \cap L = \mathrm{I}(L)$, $\Lambda(L^{\mathrm{nl}}) \cap L = \Lambda(L)$, $\Omega(L^{\mathrm{nl}}) \cap L = \Omega(L)$.

$\Lambda\Omega$-fields

For model-theoretic use we rephrase some of the results above in the terminology of $\Lambda\Omega$-fields:

DEFINITION 16.3.21. A **pre-$\Lambda\Omega$-field** is a quadruple (K, I, Λ, Ω) where K is a pre-H-field (as throughout this section) and (I, Λ, Ω) is a $\Lambda\Omega$-cut in K. If in addition K is an H-field, then we call (K, I, Λ, Ω) a **$\Lambda\Omega$-field**.

Since our pre-H-field K has a $\Lambda\Omega$-cut (I, Λ, Ω), we can turn K into a pre-$\Lambda\Omega$-field (K, I, Λ, Ω). If $\boldsymbol{K} = (K, I, \Lambda, \Omega)$ is a pre-$\Lambda\Omega$-field and $\phi \in K^>$, then $\boldsymbol{K}^\phi = (K^\phi, I^\phi, \Lambda^\phi, \Omega^\phi)$, where I^ϕ, Λ^ϕ, Ω^ϕ are as in (16.3.1), is also a pre-$\Lambda\Omega$-field. Below, any qualifier that applies to pre-H-fields, such as *has asymptotic integration*, when applied to a pre-$\Lambda\Omega$-field $(K, \dots)$ means that the underlying pre-H-field K has the property in question.

Let $\boldsymbol{K} = (K, I, \Lambda, \Omega)$ and $\boldsymbol{L} = (L, I_L, \Lambda_L, \Omega_L)$ be pre-$\Lambda\Omega$-fields. An **embedding** $h\colon \boldsymbol{K} \to \boldsymbol{L}$ of pre-$\Lambda\Omega$-fields is an embedding $h\colon K \to L$ of pre-H-fields such that

$$h(I) = h(K) \cap I_L, \qquad h(\Lambda) = h(K) \cap \Lambda_L, \qquad h(\Omega) = h(K) \cap \Omega_L.$$

We also say that $\boldsymbol{L}$ is an **extension** of $\boldsymbol{K}$ if L is a pre-H-field extension of K with $I = I_L \cap K$, $\Lambda = \Lambda_L \cap K$, and $\Omega = \Omega_L \cap K$ (so the natural inclusion $K \to L$ is an embedding $\boldsymbol{K} \to \boldsymbol{L}$ of pre-$\Lambda\Omega$-fields); notation: $\boldsymbol{K} \subseteq \boldsymbol{L}$. If $(L, I_L, \Lambda_L, \Omega_L)$ is a pre-$\Lambda\Omega$-field and K is a pre-H-subfield of L, then

$$(K, I_L \cap K, \Lambda_L \cap K, \Omega_L \cap K) \subseteq (L, I_L, \Lambda_L, \Omega_L).$$

From Corollary 16.3.19 we see that K has a unique expansion (K, I, Λ, Ω) to a pre-$\Lambda\Omega$-field if and only if one of the following conditions is satisfied:

(i) K is grounded;

(ii) there exists $b \asymp 1$ in K such that $v(b')$ is a gap in K;

(iii) K is ω-free.

From this equivalence we obtain:

COROLLARY 16.3.22. *Let* $\boldsymbol{K} = (K, \dots)$ *and* $\boldsymbol{L} = (L, \dots)$ *be pre-*$\Lambda\Omega$*-fields where* K *satisfies one of the conditions* (i), (ii), (iii) *above. Then any embedding* $K \to L$ *of pre-*H*-fields is also an embedding of pre-*$\Lambda\Omega$*-fields* $\boldsymbol{K} \to \boldsymbol{L}$.

By the remark following the proof of Corollary 16.3.20, every pre-$\Lambda\Omega$-field extends to some ω-free newtonian Liouville closed $\Lambda\Omega$-field. In the next section we establish a more precise result of this kind.

Notes and comments

For Liouville closed K the set $\mathrm{I}(K)$ lives in some sense in the asymptotic couple of K, and reflects the extra predicate needed to get QE for such asymptotic couples in [18]. Such an "equivalence" to a definable set in the asymptotic couple no longer exists for $\Lambda(K)$ and $\Omega(K)$.

16.4 EMBEDDING PRE-$\Lambda\Omega$-FIELDS INTO ω-FREE $\Lambda\Omega$-FIELDS

In this section we fix a pre-$\Lambda\Omega$-field $\boldsymbol{K} = (K, I, \Lambda, \Omega)$, and construct an ω-free $\Lambda\Omega$-field extension of $\boldsymbol{K}$ with a useful semiuniversal property:

PROPOSITION 16.4.1. *There exists an* ω*-free* $\Lambda\Omega$*-field extension* $\boldsymbol{K}^*$ *of* $\boldsymbol{K}$ *such that* $\operatorname{res} \boldsymbol{K}^*$ *is algebraic over* $\operatorname{res} \boldsymbol{K}$ *and any embedding of* $\boldsymbol{K}$ *into a Schwarz closed* $\Lambda\Omega$*-field* $\boldsymbol{L}$ *extends to an embedding of* $\boldsymbol{K}^*$ *into* $\boldsymbol{L}$.

This result is contained in the next lemmas with their corollaries.

LEMMA 16.4.2. *Suppose* K *is grounded, or there exists* $b \asymp 1$ *in* K *such that* $v(b')$ *is a gap in* K. *Then* $\boldsymbol{K}$ *has an* ω*-free* $\Lambda\Omega$*-field extension* $\boldsymbol{K}^*$ *such that any embedding of* $\boldsymbol{K}$ *into a* $\Lambda\Omega$*-field* $\boldsymbol{L}$ *closed under logarithms extends to an embedding* $\boldsymbol{K}^* \to \boldsymbol{L}$.

PROOF. The H-field hull $F := H(K)$ of K is grounded by Corollary 10.3.2(i)(iii). Take $K^* = F_\omega$ and apply Lemma 11.7.17 and Corollary 16.3.22. □

LEMMA 16.4.3. *Suppose* K *has gap* β *and* $v(b') \neq \beta$ *for all* $b \asymp 1$ *in* K. *Then there exists a grounded pre-*$\Lambda\Omega$*-field extension* $\boldsymbol{K}_1$ *of* $\boldsymbol{K}$ *such that any embedding of* $\boldsymbol{K}$ *into a* $\Lambda\Omega$*-field* $\boldsymbol{L}$ *closed under integration extends to an embedding* $\boldsymbol{K}_1 \to \boldsymbol{L}$.

PROOF. Take $s \in K$ with $vs = \beta$. We distinguish two cases:

CASE 1: $s \notin I$. Then $I = \mathrm{I}(K) = \{a \in K : va > \beta\}$ by Lemma 16.3.13. By Corollary 10.3.2, $H(K)$ is an immediate extension of K, so β remains a gap in $H(K)$. Since $H(K)$ is d-valued of H-type, Corollary 10.5.11 yields an H-field extension $K_1 := H(K)(y)$ of K such that $y' = s$ and $y \succ 1$. Then K_1 is grounded, and thus admits a unique expansion $\boldsymbol{K}_1 = (K_1, I_1, \Lambda_1, \Omega_1)$ to a pre-$\Lambda\Omega$-field. From $I_1 = \mathrm{I}(K_1)$ we get $s \notin I_1$, so $\boldsymbol{K}_1 \supseteq \boldsymbol{K}$ by Lemma 16.3.13. Let $\boldsymbol{L} \supseteq \boldsymbol{K}$ be a $\Lambda\Omega$-field which is closed under integration. Take $z \in L$ with $z' = s$. Then $z \succ 1$ since $s \notin I$. The universal property of $H(K)$ and Corollary 10.5.11 give a unique embedding

$K_1 = H(K)(y) \to L$ of pre-H-fields over K sending y to z, and by Corollary 16.3.22 this is an embedding $\boldsymbol{K}_1 \to \boldsymbol{L}$ of pre-$\Lambda\Omega$-fields.

CASE 2: $s \in I$. Then $I = \{a \in K : va \geqslant \beta\}$ by Lemma 16.3.13. Let $K_1 = K(y)$ be a pre-H-field extension of K with $y' = s$ and $y \prec 1$, as in Corollary 10.5.10 and the subsequent remarks. Then K_1 is grounded, and thus admits a unique expansion $\boldsymbol{K}_1 = (K_1, I_1, \Lambda_1, \Omega_1)$ to a pre-$\Lambda\Omega$-field. From $s \in \mathrm{I}(K_1) = I_1$ and Lemma 16.3.13 we get $\boldsymbol{K}_1 \supseteq \boldsymbol{K}$. Let $\boldsymbol{L} \supseteq \boldsymbol{K}$ be a $\Lambda\Omega$-field closed under integration. Take $z \in L$ with $z' = s$. From $s \in I$ we get $z \preccurlyeq 1$, and by subtracting a constant from z we arrange $z \prec 1$. Then Corollary 10.5.10 yields a unique embedding $K_1 = K(y) \to L$ of pre-H-fields over K sending y to z. By Corollary 16.3.22 this is an embedding $\boldsymbol{K}_1 \to \boldsymbol{L}$ of pre-$\Lambda\Omega$-fields. □

These two lemmas and the constructions in their proofs yield the following:

COROLLARY 16.4.4. *Suppose K does not have asymptotic integration. Then $\boldsymbol{K}$ has an ω-free $\Lambda\Omega$-field extension $\boldsymbol{K}^*$ such that* $\operatorname{res}\boldsymbol{K}^* = \operatorname{res}\boldsymbol{K}$ *and any embedding of $\boldsymbol{K}$ into a $\Lambda\Omega$-field $\boldsymbol{L}$ closed under integration extends to an embedding $\boldsymbol{K}^* \to \boldsymbol{L}$.*

The next two lemmas deal with the case where K has asymptotic integration.

LEMMA 16.4.5. *Assume K has asymptotic integration and is not λ-free. Then $\boldsymbol{K}$ extends to an ω-free $\Lambda\Omega$-field $\boldsymbol{K}^*$ such that* $\operatorname{res}\boldsymbol{K}^* = (\operatorname{res}\boldsymbol{K})^{\mathrm{rc}}$ *and any embedding of $\boldsymbol{K}$ into a Liouville closed $\Lambda\Omega$-field $\boldsymbol{L}$ extends to an embedding $\boldsymbol{K}^* \to \boldsymbol{L}$.*

PROOF. By Corollary 16.4.4 it is enough to show: $\boldsymbol{K}$ has a $\Lambda\Omega$-field extension $\boldsymbol{K}_1$ with a gap such that $\operatorname{res}\boldsymbol{K}_1 = (\operatorname{res}\boldsymbol{K})^{\mathrm{rc}}$ and any embedding of $\boldsymbol{K}$ into a Liouville closed $\Lambda\Omega$-field $\boldsymbol{L}$ extends to an embedding $\boldsymbol{K}_1 \to \boldsymbol{L}$. By Lemmas 16.3.17 and 16.3.18, we have $\Lambda(K)^{\downarrow} \neq K \setminus \Delta(K)^{\uparrow}$, and the pre-$H$-field K has precisely two $\Lambda\Omega$-cuts, $\big(\mathrm{I}(K), \Lambda(K)^{\downarrow}, \dots\big)$ and $\big(\mathrm{I}(K), K \setminus \Delta(K)^{\uparrow}, \dots\big)$. Let $E := H(K)^{\mathrm{rc}}$. By Corollary 10.3.2(i) we have $\Gamma_{H(K)} = \Gamma$, so $\Gamma_E = \mathbb{Q}\Gamma$. We distinguish two cases:

CASE 1: *E has a gap.* Take $s \in E^{\times}$ and $n \geqslant 1$ such that vs is a gap in E and $s^n \in K$. Then $s^{\dagger} = \frac{1}{n}(s^n)^{\dagger} \in K$. By Lemma 16.3.13, E has exactly two $\Lambda\Omega$-cuts $(I_1, \Lambda_1, \Omega_1)$ and $(I_2, \Lambda_2, \Omega_2)$, with $I_1 = \mathrm{I}(E) = \{y \in E : y \prec s\}$ and $I_2 = \{y \in E : y \preccurlyeq s\}$. We have $s \notin I_1$, so $-s^{\dagger} \in \Lambda_1 \cap K$, and $s \in I_2$, so $-s^{\dagger} \notin \Lambda_2 \cap K$. If $-s^{\dagger} \in \Lambda$, then we set $\boldsymbol{K}_1 := (E, I_1, \Lambda_1, \Omega_1)$, and if $-s^{\dagger} \notin \Lambda$, then we set $\boldsymbol{K}_1 := (E, I_2, \Lambda_2, \Omega_2)$. Then $\boldsymbol{K}_1$ is an extension of the pre-$\Lambda\Omega$-field $\boldsymbol{K}$. Given an embedding $i \colon \boldsymbol{K} \to \boldsymbol{L}$ into a Liouville closed $\Lambda\Omega$-field $\boldsymbol{L}$, there is a unique embedding $j \colon E \to L$ of H-fields such that $j(a) = i(a)$ for all $a \in K$, and it is easy to check that j is an embedding $\boldsymbol{K}_1 \to \boldsymbol{L}$ of $\Lambda\Omega$-fields.

CASE 2: *E has no gap.* Then E has asymptotic integration, and the sequence (λ_{ρ}) for K also serves for E. Take $\lambda \in K$ such that $\lambda_{\rho} \rightsquigarrow \lambda$. Then $-\lambda$ creates a gap over E by Lemma 11.5.14. Take an element $f \neq 0$ in some Liouville closed H-field extension of E such that $f^{\dagger} = -\lambda$. Then vf is a gap in $E(f)$ by the remark following the proof of Lemma 11.5.14 with $\operatorname{res}E(f) = \operatorname{res}E$. Using Lemma 9.1.2 it follows that the pre-H-field $E(f)$ is actually an H-field. Therefore, by Lemma 16.3.13, $E(f)$ has exactly two $\Lambda\Omega$-cuts $(I_1, \Lambda_1, \Omega_1)$ and $(I_2, \Lambda_2, \Omega_2)$, with

$$I_1 = \mathrm{I}(E(f)) = \{y \in E(f) : y \prec f\}, \qquad I_2 = \{y \in E(f) : y \preccurlyeq f\}.$$

We have $f \notin I_1$, so $\lambda = -f^\dagger \in \Lambda_1 \cap K$, and $f \in I_2$, so $\lambda = -f^\dagger \notin \Lambda_2 \cap K$. If $\lambda \in \Lambda$, then we set $\boldsymbol{K}_1 := \big(E(f), I_1, \Lambda_1, \Omega_1\big)$, and if $\lambda \notin \Lambda$, then we set $\boldsymbol{K}_1 := \big(E(f), I_2, \Lambda_2, \Omega_2\big)$. In any case, we have $\boldsymbol{K}_1 \supseteq \boldsymbol{K}$. Let $i\colon \boldsymbol{K} \to \boldsymbol{L}$ be an embedding into a Liouville closed $\Lambda\Omega$-field $\boldsymbol{L}$. By Lemma 11.5.13 the set

$$S := \big\{v(\lambda + a^\dagger) : a \in E^\times\big\}$$

is a cofinal subset of $\Psi_E^\downarrow$ and f is transcendental over E. Then Lemmas 10.4.5 and 10.5.19 provide an embedding $j\colon E(f) \to L$ of H-fields with $j(a) = i(a)$ for all $a \in K$, and any such embedding is an embedding $\boldsymbol{K}_1 \to \boldsymbol{L}$ of $\Lambda\Omega$-fields. □

LEMMA 16.4.6. *Suppose K is λ-free but not ω-free. Then $\boldsymbol{K}$ has an ω-free $\Lambda\Omega$-field extension $\boldsymbol{K}^*$ such that* $\operatorname{res} \boldsymbol{K}^*$ *is algebraic over* $\operatorname{res} \boldsymbol{K}$ *and any embedding of $\boldsymbol{K}$ into a Schwarz closed $\Lambda\Omega$ field $\boldsymbol{L}$ extends to an embedding of $\boldsymbol{K}^*$ into $\boldsymbol{L}$.*

PROOF. Take $\omega \in K$ such that $\omega_\rho \rightsquigarrow \omega$, so $\omega\big(\Lambda(K)\big)^\downarrow < \omega < \sigma\big(\Gamma(K)\big)^\uparrow$. By Lemma 16.3.16, there are exactly two $\Lambda\Omega$-cuts in K:

$$\big(\mathrm{I}(K), \Lambda(K)^\downarrow, \omega(K)^\downarrow\big), \qquad \big(\mathrm{I}(K), \Lambda(K)^\downarrow, K \setminus \sigma\big(\Gamma(K)\big)^\uparrow\big).$$

Since $\omega \in K\setminus\sigma\big(\Gamma(K)\big)^\uparrow$, it follows from the proof of Lemma 16.3.16 that $\omega \notin \omega(K)^\downarrow$. We distinguish two cases:

CASE 1: $\Omega = \omega(K)^\downarrow$. Lemma 13.7.7 yields a pre-H-field extension $K_\gamma := K\langle\gamma\rangle$ of K such that $\sigma(\gamma) = \omega$, $\gamma > 0$, $v\gamma$ is gap in K_γ, and $\operatorname{res}(K_\gamma) = \operatorname{res}(K)$. By remark (3) following the proof of Proposition 13.7.1 there is no $b \asymp 1$ in K_γ with $b' \asymp \gamma$, so by Lemma 16.3.13 we have exactly two $\Lambda\Omega$-cuts $(I_1, \Lambda_1, \Omega_1)$, $(I_2, \Lambda_2, \Omega_2)$ in K_γ, where

$$I_1 = \mathrm{I}(K_\gamma) = \{y \in K_\gamma : y \prec \gamma\}, \qquad I_2 = \{y \in K_\gamma : y \preccurlyeq \gamma\}.$$

Put $\boldsymbol{K}_\gamma := (K_\gamma, I_1, \Lambda_1, \Omega_1)$. We have $\gamma \notin I_1$, so $\omega = \sigma(\gamma) \notin \Omega_1$ by ($\Lambda\Omega$12), and thus $\boldsymbol{K}_\gamma \supseteq \boldsymbol{K}$. Let $\boldsymbol{K}^*$ be an ω-free $\Lambda\Omega$-field extension of $\boldsymbol{K}_\gamma$ obtained by applying Corollary 16.4.4 to $\boldsymbol{K}_\gamma$ instead of $\boldsymbol{K}$. Let $\boldsymbol{L}$ be a Schwarz closed $\Lambda\Omega$-field extension of $\boldsymbol{K}$; we claim that there is an embedding $\boldsymbol{K}^* \to \boldsymbol{L}$ over K. Now $\omega \notin \Omega = \Omega(L) \cap K$, hence $\omega \in \sigma\big(\Gamma(L)\big)$. By Lemma 11.8.29, the restriction of σ to $\Gamma(L)$ is strictly increasing. Let γ^* be the unique element of $\Gamma(L)$ such that $\sigma(\gamma^*) = \omega$. From $\Gamma(L) = L^> \setminus \mathrm{I}(L)$ we get $\gamma^* > 0$ and $\gamma^* \notin \mathrm{I}(L)$, and thus $v\gamma^* < (\Gamma^>)'$. Also $\sigma(\gamma^*) = \omega < \sigma\big(\Gamma(K)\big)$ gives $0 < \gamma^* < \Gamma(K)$, and so $\Psi < v\gamma^*$. Thus Proposition 13.7.1 and Lemma 13.7.7 yield an embedding $h\colon K_\gamma \to L$ of pre-H-fields over K with $h(\gamma) = \gamma^*$. Since $\gamma \notin I_1$ and $\gamma^* \notin \mathrm{I}(L)$, h is an embedding $\boldsymbol{K}_\gamma \to \boldsymbol{L}$ of pre-$\Lambda\Omega$-fields. By Corollary 16.4.4 we can extend h to an embedding $\boldsymbol{K}^* \to \boldsymbol{L}$.

CASE 2: $\Omega = K\setminus\sigma\big(\Gamma(K)\big)^\uparrow$. Corollary 11.7.13 and Lemma 10.5.8 yield an immediate pre-H-field extension $K_\lambda := K(\lambda)$ of K with $\lambda_\rho \rightsquigarrow \lambda$ and $\omega(\lambda) = \omega$. Then K_λ has rational asymptotic integration and is not λ-free, and $\Lambda(K_\lambda) < \lambda < \Delta(K_\lambda)$, so by Lemma 16.3.17 there are exactly two $\Lambda\Omega$-cuts in K_λ:

$$\big(\mathrm{I}(K_\lambda), \Lambda(K_\lambda)^{\downarrow}, K_\lambda \setminus \sigma\big(\Gamma(K_\lambda)\big)^{\uparrow}\big) \text{ and } \big(\mathrm{I}(K_\lambda), K_\lambda \setminus \Delta(K_\lambda)^{\uparrow}, K_\lambda \setminus \sigma\big(\Gamma(K_\lambda)\big)^{\uparrow}\big).$$

Note that $\omega \in \Omega$ as well as $\omega \in \omega(K_\lambda) \subseteq K_\lambda \setminus \sigma\big(\Gamma(K_\lambda)\big)^{\uparrow}$. Therefore, setting

$$\boldsymbol{K}_\lambda := \big(K_\lambda, \mathrm{I}(K_\lambda), K_\lambda \setminus \Delta(K_\lambda)^{\uparrow}, K_\lambda \setminus \sigma\big(\Gamma(K_\lambda)\big)^{\uparrow}\big),$$

we get $\boldsymbol{K}_\lambda \supseteq \boldsymbol{K}$. Let $\boldsymbol{K}^*$ be an ω-free $\Lambda\Omega$-field extension of $\boldsymbol{K}_\lambda$ obtained by applying Lemma 16.4.5 to $\boldsymbol{K}_\lambda$ instead of $\boldsymbol{K}$. Let $\boldsymbol{L}$ be a Schwarz closed $\Lambda\Omega$-field extension of $\boldsymbol{K}$; we claim that there is an embedding $\boldsymbol{K}^* \to \boldsymbol{L}$ over K. Recall from Corollary 11.8.13 and Proposition 11.8.20 that ω is strictly increasing on $\Lambda(L)$ and $\Lambda(L) < \Delta(L)$. Since $\omega \in \Omega = \omega\big(\Lambda(L)\big) \cap K$, we get a unique $\lambda^* \in \Lambda(L)$ such that $\omega(\lambda^*) = \omega$. Then $\Lambda(K) < \lambda^* < \Delta(K)$, so $\lambda_\rho \rightsquigarrow \lambda^*$ and Corollary 11.7.13 yields an embedding $h\colon K_\lambda \to L$ of pre-H-fields over K with $h(\lambda) = \lambda^*$. Since $\lambda \notin \Lambda(K_\lambda)^{\downarrow}$ and $\lambda^* \in \Lambda(L)$, h is an embedding $\boldsymbol{K}_\lambda \to \boldsymbol{L}$ of pre-$\Lambda\Omega$-fields, and so by Lemma 16.4.5, h extends to an embedding $\boldsymbol{K}^* \to \boldsymbol{L}$. □

LEMMA 16.4.7. *If K is ω-free, then $\boldsymbol{K}$ has an ω-free $\Lambda\Omega$-field extension $\boldsymbol{K}^*$ such that any embedding of $\boldsymbol{K}$ into a $\Lambda\Omega$-field $\boldsymbol{L}$ extends to an embedding of $\boldsymbol{K}^*$ into $\boldsymbol{L}$.*

PROOF. Assume K is ω-free. Then $H(K)$ is ω-free by Theorem 13.6.1. Let $\boldsymbol{K}^*$ be an expansion of $H(K)$ to a $\Lambda\Omega$-field. Then $\boldsymbol{K} \subseteq \boldsymbol{K}^*$ by Corollary 16.3.19. It remains to use the universal property of $H(K)$ and Corollary 16.3.22. □

Corollary 16.4.4 and Lemmas 16.4.5, 16.4.6, and 16.4.7 now have Proposition 16.4.1 as an immediate consequence. Note: this proof yields an extension $\boldsymbol{K}^*$ of $\boldsymbol{K}$ as in Proposition 16.4.1 that is d-algebraic over $\boldsymbol{K}$.

The Newton-Liouville closure of a pre-$\Lambda\Omega$-field

Here we extend the results on Newton-Liouville closures of ω-free H-fields to pre-$\Lambda\Omega$-fields.

PROPOSITION 16.4.8. *Let $\boldsymbol{K} = (K, I, \Lambda, \Omega)$ be a pre-$\Lambda\Omega$-field. Then $\boldsymbol{K}$ has an ω-free newtonian Liouville closed $\Lambda\Omega$-field extension $\boldsymbol{K}^{\mathrm{nl}}$ that embeds over $\boldsymbol{K}$ into any ω-free newtonian Liouville closed $\Lambda\Omega$-field extension of $\boldsymbol{K}$.*

PROOF. First we take an ω-free $\Lambda\Omega$-field extension $\boldsymbol{K}^* = (K^*, \dots)$ of $\boldsymbol{K}$ as in Proposition 16.4.1. Next we take the Newton-Liouville closure E^{nl} of the ω-free H-field $E := K^*$. Then the unique expansion of E^{nl} to a $\Lambda\Omega$-field is an extension $\boldsymbol{K}^{\mathrm{nl}}$ of $\boldsymbol{K}$ as claimed. □

We define a **Newton-Liouville closure** of a pre-$\Lambda\Omega$-field $\boldsymbol{K}$ to be an extension $\boldsymbol{K}^{\mathrm{nl}}$ as in Proposition 16.4.8. Thus every pre-$\Lambda\Omega$-field has a Newton-Liouville closure.

PROPOSITION 16.4.9. *Let $\boldsymbol{K}$ be a pre-$\Lambda\Omega$-field. Any two Newton-Liouville closures of $\boldsymbol{K}$ are isomorphic over $\boldsymbol{K}$. If $\boldsymbol{K}^{\mathrm{nl}}$ is a Newton-Liouville closure of $\boldsymbol{K}$, then $\boldsymbol{K}^{\mathrm{nl}}$ does not have any proper newtonian ω-free Liouville closed $\Lambda\Omega$-subfield containing $\boldsymbol{K}$ as a substructure.*

PROOF. Let $\boldsymbol{K}^{\mathrm{nl}}$ be the Newton-Liouville closure of $\boldsymbol{K}$ constructed in the proof of Proposition 16.4.8. Then $\boldsymbol{K}^{\mathrm{nl}}$ is d-algebraic over $\boldsymbol{K}$ and the residue field of $\boldsymbol{K}^{\mathrm{nl}}$ is a real closure of $\operatorname{res} \boldsymbol{K}$. Let $\boldsymbol{L}$ be any Newton-Liouville closure of $\boldsymbol{K}$. Embedding $\boldsymbol{L}$ into $\boldsymbol{K}^{\mathrm{nl}}$ over $\boldsymbol{K}$, we see that $\boldsymbol{L}$ is d-algebraic over $\boldsymbol{K}$ and its residue field is a real closure of $\operatorname{res} \boldsymbol{K}$. Consider any embedding $i\colon \boldsymbol{K}^{\mathrm{nl}} \to \boldsymbol{L}$ over $\boldsymbol{K}$. Then $i(\boldsymbol{K}^{\mathrm{nl}}) = \boldsymbol{L}$ by Theorem 16.0.3. This proves the first part, and the minimality property of $\boldsymbol{K}^{\mathrm{nl}}$ is likewise a consequence of Theorem 16.0.3. □

16.5 THE LANGUAGE OF $\Lambda\Omega$-FIELDS

In the introduction to this chapter we specified the language

$$\mathcal{L} := \{0,\ 1,\ +,\ -,\ \cdot\ ,\ \partial,\ \leqslant,\ \preccurlyeq\}$$

of ordered valued differential rings. Each ordered valued differential field is viewed as an $\mathcal{L}$-structure in the natural way. In this section we show that Theorem 16.0.1 fails if we drop either the symbol Λ or the symbol Ω from the language $\mathcal{L}^{\iota}_{\Lambda\Omega}$ of $\Lambda\Omega$-fields. (We prove somewhat sharper versions of this fact.)

Throughout this section K is a pre-H-field. In Section 11.8 we introduced the special subsets

$$\mathrm{I}(K), \quad \Gamma(K), \quad \Lambda(K), \quad \Delta(K), \quad \omega(K), \quad \sigma\big(\Gamma(K)\big)$$

of K. If K is Schwarz closed, then each of these sets is clearly existentially definable as well as universally definable in the $\mathcal{L}$-structure K, both forms of definability holding without parameters from K and witnessed by $\mathcal{L}$-formulas independent of K. In this section we successively investigate the quantifier-free definability of these sets in a Schwarz closed K, in the language $\mathcal{L}$ and some extensions of this language with predicates for some of these sets. Our first result shows that no ω-free real closed H-field eliminates quantifiers in $\mathcal{L}$:

PROPOSITION 16.5.1. *Suppose K is an ω-free real closed H-field. Then the subset $\mathrm{I}(K)$ of K is not quantifier-free definable (with parameters) in the $\mathcal{L}$-structure K.*

PROOF. Take an element ℓ in an elementary extension K^* of K with $\ell > 0$ and $1 \prec \ell \prec \ell_\rho$ for all ρ, and set $\gamma := \ell^\dagger$, $\lambda := -\gamma^\dagger$. Then λ and $\lambda + \gamma$ are pseudolimits of (λ_ρ), by Corollary 11.5.7, and (λ_ρ) is of d-transcendental type over K, by Corollary 13.6.3. Hence by Lemma 11.4.7 and Corollary 11.4.13, the pre-H-subfields $K\langle\lambda\rangle$ and $K\langle\lambda+\gamma\rangle$ of K^* are immediate extensions of K (so they are H-fields), and we have an isomorphism $K\langle\lambda\rangle \to K\langle\lambda+\gamma\rangle$ of H-fields over K sending λ to $\lambda+\gamma$. By Lemma 11.5.14, the element $-\lambda$ creates a gap over $K\langle\lambda\rangle$. Likewise, $-(\lambda+\gamma)$ creates a gap over $K\langle\lambda+\gamma\rangle$. Let $f := (1/\ell)^\dagger = -\gamma$ and $g := (1/\ell)' = -\gamma/\ell$, so $f < 0$ and

$g < 0$ (using $\ell > 0$). Then $f^\dagger = -\lambda$ and $g^\dagger = -(\lambda + \gamma)$, so the above isomorphism $K\langle\lambda\rangle \to K\langle\lambda+\gamma\rangle$ extends by Lemma 11.5.13 and the uniqueness parts of Lemmas 10.4.5 and 10.5.19 to an isomorphism $K\langle\lambda, f\rangle \to K\langle\lambda+\gamma, g\rangle$ of $\mathcal{L}$-structures sending f to g. Now, if $\mathrm{I}(K)$ were defined in K by a quantifier-free formula $\varphi(y)$ in the language $\mathcal{L}$ augmented by names for the elements of K, then we would have $K^* \models \neg\varphi(f)$ and $K^* \models \varphi(g)$, and so $K\langle\lambda, f\rangle \models \neg\varphi(f)$ and $K\langle\lambda+\gamma, g\rangle \models \varphi(g)$, violating the above isomorphism between $K\langle\lambda, f\rangle$ and $K\langle\lambda+\gamma, g\rangle$. □

We extend $\mathcal{L}$ by a single unary function symbol ι to the language $\mathcal{L}^\iota$. Any ordered valued differential field F will be construed as an $\mathcal{L}^\iota$-structure by interpreting this new function symbol as the function $F \to F$ that agrees with $f \mapsto f^{-1}$ on $F^\times$ and sends 0 to 0. Thus the underlying ring of an $\mathcal{L}^\iota$-substructure of an ordered valued differential field is a field. Passing from $\mathcal{L}$ to $\mathcal{L}^\iota$ does not increase what we can express quantifier-free in ordered valued differential fields:

COROLLARY 16.5.2. *Given any quantifier-free $\mathcal{L}^\iota$-formula $\varphi^\iota(x_1,\dots,x_n)$, there is a quantifier-free $\mathcal{L}$-formula $\varphi(x_1,\dots,x_n)$ which in every ordered valued differential field F defines the same subset of F^n as $\varphi^\iota(x_1,\dots,x_n)$.*

PROOF. Let OVD^ι be the $\mathcal{L}^\iota$-theory of ordered valued differential fields. By B.11.5 it is enough to show that OVD^ι has closures of $\mathcal{L}$-substructures. Let $E, F \models \mathrm{OVD}^\iota$ have a common $\mathcal{L}$-substructure D. Thus D is an ordered subring of both E and F with a derivation on it that agrees on D with the derivations of E and F such that for all $f, g \in D$ we have $f \preccurlyeq_E g \Longleftrightarrow f \preccurlyeq_D g \Longleftrightarrow f \preccurlyeq_F g$, where $\preccurlyeq_D, \preccurlyeq_E, \preccurlyeq_F$ are the interpretations of the symbol $\preccurlyeq$ of $\mathcal{L}$ in D, E, F, respectively. Let K_E and K_F be the fraction fields of the integral domain D in E and F, respectively. Then K_E is the underlying ring of an $\mathcal{L}^\iota$-substructure of E, to be denoted also by K_E. Likewise, K_F denotes the corresponding $\mathcal{L}^\iota$-substructure of F. The unique field isomorphism $K_E \to K_F$ over D is clearly an $\mathcal{L}^\iota$-isomorphism. □

Thus if K is an ω-free real closed H-field, then $\mathrm{I}(K)$ is not quantifier-free definable (with parameters) in the $\mathcal{L}^\iota$-structure K.

Let $\mathcal{L}^\iota_\Lambda$ be the language $\mathcal{L}^\iota$ augmented by unary predicate symbols I and Λ. Given a Λ-cut (I, Λ) in K, we have the $\mathcal{L}^\iota_\Lambda$-structure (K, I, Λ): interpret I and Λ by I and Λ. Recall that if K is λ-free, then there is only one Λ-cut in K. By Lemma 11.8.5, if K is Liouville closed, then $\mathrm{I}(K)$ is quantifier-free definable in $(K, \Lambda(K))$, with K construed as an $\mathcal{L}^\iota$-structure; nevertheless, we include the symbol I in $\mathcal{L}^\iota_\Lambda$. Note that if K is Liouville closed, then $\Delta(K) = K\setminus\Lambda(K)$ is quantifier-free definable in the $\mathcal{L}^\iota_\Lambda$-structure $(K, \mathrm{I}(K), \Lambda(K))$ as well. However:

PROPOSITION 16.5.3. *Suppose K is an ω-free real closed H-field. Then the subsets $\omega(K)$ and $\omega(K)^\downarrow$ of K are not quantifier-free definable (even allowing parameters) in the $\mathcal{L}^\iota_\Lambda$-structure $(K, \mathrm{I}(K), \Lambda(K)^\downarrow)$.*

PROOF. Take K^* and ℓ as in the proof of Proposition 16.5.1, and set $\gamma := \ell^\dagger$, $\lambda := -\gamma^\dagger$, and $\omega := \omega(\lambda)$. Then $\lambda_\rho \rightsquigarrow \lambda$ by Corollary 11.5.7 and hence $\omega_\rho \rightsquigarrow \omega$ by Corollary 11.7.3. In view of Corollary 11.7.2 and $v(\gamma^2) > 2\Psi$, the pc-sequences (ω_ρ)

and $(\omega_\rho + \gamma_\rho^2)$ in K are equivalent, and $\sigma(\gamma_\rho) = \omega_\rho + \gamma_\rho^2 \rightsquigarrow \sigma(\gamma) = \omega + \gamma^2$. By Corollary 13.6.3, (ω_ρ) is of d-transcendental type over K. Hence by Lemma 11.4.7 and Corollary 11.4.13, the pre-H-subfields $K\langle\omega\rangle$ and $K\langle\omega+\gamma^2\rangle$ of K^* are immediate extensions of K (so they are H-fields) and we have an isomorphism

$$K\langle\omega\rangle \to K\langle\omega+\gamma^2\rangle$$

of H-fields over K sending ω to $\omega + \gamma^2$. By Proposition 13.6.4, $K\langle\omega\rangle$ is λ-free, and hence by Lemma 16.3.6, $\big(\mathrm{I}(K\langle\omega\rangle), \Lambda(K\langle\omega\rangle)^{\downarrow}\big)$ is the unique Λ-cut in $K\langle\omega\rangle$, and likewise for $K\langle\omega+\gamma^2\rangle$ instead of $K\langle\omega\rangle$. Thus

$$\Lambda(K^*)^{\downarrow} \cap K\langle\omega\rangle = \Lambda\big(K\langle\omega\rangle\big)^{\downarrow}, \qquad \Lambda(K^*)^{\downarrow} \cap K\langle\omega+\gamma^2\rangle = \Lambda\big(K\langle\omega+\gamma^2\rangle\big)^{\downarrow},$$

so our isomorphism $K\langle\omega\rangle \to K\langle\omega+\gamma^2\rangle$ is an isomorphism between $\mathcal{L}^{\iota}_{\Lambda}$-substructures of $(K^*, \mathrm{I}(K^*), \Lambda(K^*)^{\downarrow})$. Now $\omega \in \omega(K^*)$ and $\omega + \gamma^2 \in \sigma\big(\Gamma(K^*)\big)$, so $\omega + \gamma^2 \notin \omega(K^*)$, and thus $\omega(K)$ is not quantifier-free definable (with parameters) in the $\mathcal{L}^{\iota}_{\Lambda}$-structure $(K, \mathrm{I}(K), \Lambda(K)^{\downarrow})$. Likewise with $\omega(K)^{\downarrow}$ instead of $\omega(K)$. □

Let $\mathcal{L}^{\iota}_{\Omega}$ be the language $\mathcal{L}^{\iota}$ augmented by the unary predicate symbols I and Ω. Then we have the following analogue of Proposition 16.5.3:

PROPOSITION 16.5.4. *Suppose K is an ω-free real closed H-field. Then the subsets $\Lambda(K)$ and $\Lambda(K)^{\downarrow}$ of K are not quantifier-free definable (even allowing parameters) in the $\mathcal{L}^{\iota}_{\Omega}$-structure $(K, \mathrm{I}(K), \omega(K)^{\downarrow})$.*

PROOF. Again, take K^* and ℓ as in the proof of Proposition 16.5.1, and set $\gamma := \ell^{\dagger}$ and $\lambda := -\gamma^{\dagger}$. Then $K\langle\lambda\rangle$ and $K\langle\lambda + \gamma\rangle$ are immediate H-field extensions of K, and we have an isomorphism $K\langle\lambda\rangle \to K\langle\lambda + \gamma\rangle$ of H-fields over K which sends λ to $\lambda+\gamma$. Since K has asymptotic integration with divisible value group, so does $K\langle\lambda\rangle$. As $K\langle\lambda\rangle$ is not λ-free, there are by Lemma 16.3.17 exactly two $\Lambda\Omega$-cuts in $K\langle\lambda\rangle$,

$$(I, \Lambda_1, \Omega) \quad \text{and} \quad (I, \Lambda_2, \Omega),$$

the key point being that these two $\Lambda\Omega$-cuts have the same first component I and same third component Ω. In particular,

$$\mathrm{I}(K^*) \cap K\langle\lambda\rangle = I, \qquad \omega(K^*)^{\downarrow} \cap K\langle\lambda\rangle = \Omega.$$

Likewise with $K\langle\lambda+\gamma\rangle$ in place of $K\langle\lambda\rangle$. Hence our isomorphism $K\langle\lambda\rangle \to K\langle\lambda+\gamma\rangle$ is an isomorphism between $\mathcal{L}^{\iota}_{\Omega}$-substructures of $\big(K^*, \mathrm{I}(K^*), \omega(K^*)^{\downarrow}\big)$. But $\lambda = -\ell^{\dagger\dagger} \in \Lambda(K^*)$ and $\lambda + \gamma = -(1/\ell)'^{\dagger} \in \Delta(K^*)$, so $\lambda + \gamma \notin \Lambda(K^*)$, and thus $\Lambda(K)$ is not quantifier-free definable (even allowing parameters) in the $\mathcal{L}^{\iota}_{\Omega}$-structure $(K, \mathrm{I}(K), \omega(K)^{\downarrow})$. Similarly with $\Lambda(K)^{\downarrow}$ in place of $\Lambda(K)$. □

Thus ω-free newtonian Liouville closed H-fields do not eliminate quantifiers when viewed in the usual way either as $\mathcal{L}^{\iota}_{\Lambda}$-structures or as $\mathcal{L}^{\iota}_{\Omega}$-structures. This goes to explain our choice of language $\mathcal{L}^{\iota}_{\Lambda\Omega}$ (see the introduction to this chapter). Could a mild "algebraic" extension of $\mathcal{L}^{\iota}$, for example by a square root function, allow us to drop one of Λ, Ω? To eliminate this possibility and make our choice of language more compelling, we now indicate some stronger negative results. Towards this end we specify a language $\mathcal{L}^{\mathrm{a}}$ that serves as a more robust version of $\mathcal{L}^{\iota}$.

To define $\mathcal{L}^{\mathrm{a}}$, note that $\mathcal{L}$ has the language of ordered rings as a sublanguage. We consider $\mathbb{R}$ as a structure for the language of ordered rings in the usual way. A function $\mathbb{R}^n \to \mathbb{R}$ is said to be **$\mathbb{Q}$-semialgebraic** if its graph is defined in the structure $\mathbb{R}$ by a (quantifier-free) formula in the language of ordered rings; we do not allow names for arbitrary real numbers in such formulas. We extend $\mathcal{L}$ to the language $\mathcal{L}^{\mathrm{a}}$ by adding for each $\mathbb{Q}$-semialgebraic function $f\colon \mathbb{R}^n \to \mathbb{R}$ an n-ary function symbol f. We construe any real closed valued differential field E as an $\mathcal{L}^{\mathrm{a}}$-structure by interpreting such f as the function $E^n \to E$ whose graph is defined in E by any formula in the language of ordered rings that defines the graph of f in $\mathbb{R}$. For example, the function $a \mapsto a^{-1}\colon E \to E$ (with $0^{-1} := 0$ by convention) is named by a function symbol ι of $\mathcal{L}^{\mathrm{a}}$. For each integer $d \geqslant 1$, the function $y \mapsto y^{1/d}\colon E \to E$, taking the value 0 for $y \leqslant 0$ by convention, is also named by a function symbol. With the richer language $\mathcal{L}^{\mathrm{a}}$ replacing $\mathcal{L}^{\iota}$ the above results go through. For example, Corollary 16.5.2 extends as follows:

COROLLARY 16.5.5. *Given any quantifier-free $\mathcal{L}^{\mathrm{a}}$-formula $\varphi^{\mathrm{a}}(x_1,\dots,x_n)$, there is a quantifier-free $\mathcal{L}$-formula $\varphi(x_1,\dots,x_n)$ which defines in every real closed K the same subset of K^n as $\varphi^{\mathrm{a}}(x_1,\dots,x_n)$. (Recall that K ranges over pre-H-fields.)*

(In the proof of Corollary 16.5.2, replace OVD^{ι} by the $\mathcal{L}^{\mathrm{a}}$-theory of real closed ordered valued differential fields whose valuation ring is convex, use Corollary 3.5.18, and take real closures of fraction fields instead of just fraction fields.)

Thus if K is any ω-free real closed H-field, then $\mathrm{I}(K)$ is still not quantifier-free definable (with parameters) in the $\mathcal{L}^{\mathrm{a}}$-structure K.

Let $\mathcal{L}^{\mathrm{a}}_{\Lambda}$ be the language $\mathcal{L}^{\mathrm{a}}$ augmented by unary relation symbols I and Λ. Then Proposition 16.5.3 goes through with $\mathcal{L}^{\mathrm{a}}_{\Lambda}$ instead of $\mathcal{L}^{\iota}_{\Lambda}$: replace in the proof of that proposition $K\langle\omega\rangle$ and $K\langle\omega+\gamma^2\rangle$ by their real closures in K^*, and use that these real closures are immediate extensions of K and $\mathcal{L}^{\mathrm{a}}$-substructures of K^*.

Likewise, let $\mathcal{L}^{\mathrm{a}}_{\Omega}$ be the language $\mathcal{L}^{\mathrm{a}}$ augmented by unary relation symbols I and Ω. Then Proposition 16.5.4 goes through with $\mathcal{L}^{\mathrm{a}}_{\Omega}$ instead of $\mathcal{L}^{\iota}_{\Omega}$: replace in the proof of that proposition $K\langle\lambda\rangle$ and $K\langle\lambda+\gamma\rangle$ by their real closures in K^*.

Notes and comments

For $K = \mathbb{T}$, Proposition 16.5.1 and the $\mathcal{L}^{\mathrm{a}}_{\Lambda}$-variant of Proposition 16.5.3 are in [23], with slightly different notation: $\mathcal{L}'$ instead of our $\mathcal{L}^{\mathrm{a}}$. Corollary 16.5.5 also occurs there as Proposition 5.5, but with a defective proof.

16.6 ELIMINATION OF QUANTIFIERS WITH APPLICATIONS

In the introduction to this chapter we defined the theory $T^{\mathrm{nl},\iota}_{\Lambda\Omega}$. Its models are exactly the ω-free newtonian Liouville closed ΛΩ-fields; we defined ΛΩ-fields at the end of

Section 16.3. As noted at the end of that section, the substructures of models of $T^{\mathrm{nl},\iota}_{\Lambda\Omega}$ are exactly the pre-$\Lambda\Omega$-fields. Thus by the embedding criterion B.11.9 for QE, Theorem 16.0.1 is a consequence of the following:

THEOREM 16.6.1. *Let $\boldsymbol{K}$ and $\boldsymbol{L}$ be ω-free newtonian Liouville closed $\Lambda\Omega$-fields such that $\boldsymbol{L}$ is κ^+-saturated, where κ is the cardinality of the underlying set of $\boldsymbol{K}$. Let $\boldsymbol{E}$ be a substructure of $\boldsymbol{K}$ and let $i\colon \boldsymbol{E} \to \boldsymbol{L}$ be an embedding. Then i can be extended to an embedding $\boldsymbol{K} \to \boldsymbol{L}$.*

$$\begin{array}{ccc} \boldsymbol{K} & \dashrightarrow & \boldsymbol{L} \\ {\scriptstyle\subseteq}\uparrow & \nearrow_{i} & \\ \boldsymbol{E} & & \end{array}$$

As to the proof of Theorem 16.6.1, note that by Proposition 16.4.1 we can reduce to the case that $\boldsymbol{E}$ is an ω-free $\Lambda\Omega$-field. In view of Corollary 16.3.22 this case is taken care of by Corollary 16.2.4.

COROLLARY 16.6.2. *$T^{\mathrm{nl},\iota}_{\Lambda\Omega}$ is the model completion of the $\mathcal{L}^{\mathrm{nl},\iota}_{\Lambda\Omega}$-theory of $\Lambda\Omega$-fields.*

Theorem 16.0.1 has some immediate logical consequences for T^{nl}:

COROLLARY 16.6.3. *The completions of T^{nl} are the $\mathcal{L}$-theories $T^{\mathrm{nl}}_{\mathrm{small}}$ and $T^{\mathrm{nl}}_{\mathrm{large}}$. These two theories as well as T^{nl} itself are decidable.*

PROOF. Consider the Hardy field $E = \mathbb{Q}(x)$ ($x > \mathbb{Q}$, $x' = 1$). Any Liouville closed H-field K with small derivation has an element $f > C$ with $f' = 1$, and this yields an embedding $E \to K$ sending x to f. Note also that E is grounded. In view of Corollary 16.3.22, Theorem 16.0.1 and a well-known completeness criterion (Corollary B.11.7) the completeness of $T^{\mathrm{nl}}_{\mathrm{small}}$ then follows.

Next, set $a := x^{-2}$ and consider the compositional conjugate E^a of E. Its derivation $\delta = x^2\partial$ is not small, since $x^{-1} \prec 1$ and $\delta(x^{-1}) = -1 \asymp 1$. Let K be any Liouville closed H-field whose derivation is not small. Take $f \in K$ such that $f' = -1$. By subtracting a constant from f we arrange that $f \not\asymp 1$, and thus $f \prec 1$ and $f > 0$. This yields an embedding $E^a \to K$ sending x^{-1} to f. As with $T^{\mathrm{nl}}_{\mathrm{small}}$, we derive from this the completeness of $T^{\mathrm{nl}}_{\mathrm{large}}$.

Decidability of these theories then follows since we can effectively enumerate a set of first-order axioms for T^{nl}; see B.6. □

Let $E = \mathbb{Q}(x)$ be as in the proof of Corollary 16.6.3. Then E is a grounded H-field with constant field $\mathbb{Q}$, and so E has a unique $\Lambda\Omega$-cut. Therefore E has by Proposition 16.4.8 a Newton-Liouville closure E^{nl} in the sense that E^{nl} is an ω-free newtonian Liouville closed H-field extension of E that embeds over E into any ω-free newtonian Liouville closed H-field extension of E. Thus E^{nl} is a model of $T^{\mathrm{nl}}_{\mathrm{small}}$ that embeds into any model of $T^{\mathrm{nl}}_{\mathrm{small}}$. As $T^{\mathrm{nl}}_{\mathrm{small}}$ is model complete, this means that E^{nl} is a so-called *prime model* of $T^{\mathrm{nl}}_{\mathrm{small}}$, as defined in B.10.

The E^{nl} obtained above is d-algebraic over E, hence over $\mathbb{Q}$, and the constant field of E^{nl} is a real closure of the constant field $\mathbb{Q}$ of E. Thus by Theorem 16.0.3 any prime model of $T^{\mathrm{nl}}_{\mathrm{small}}$ is isomorphic to E^{nl}.

We also wish to call attention to the H-subfield $\mathbb{T}^{\mathrm{da}}$ of $\mathbb{T}$ given by

$$\mathbb{T}^{\mathrm{da}} := \{f \in \mathbb{T} : f \text{ is d-algebraic (over } \mathbb{Q})\}.$$

Note that $\mathbb{T}^{\mathrm{da}}$ contains the ω-free H-subfield $\mathbb{R}(\ell_0, \ell_1, \ell_2, \dots)$ of $\mathbb{T}$. It follows from Lemma 16.2.1 that $\mathbb{T}^{\mathrm{da}}$ is actually a Newton-Liouville closure of $\mathbb{R}(\ell_0, \ell_1, \ell_2, \dots)$. In particular, we have $\mathbb{T}^{\mathrm{da}} \preccurlyeq \mathbb{T}$.

Eliminating quantifiers is contingent on a good choice of primitives, but a reasonable QE should have consequences of a more intrinsic nature that would be hard to obtain otherwise. Below we derive such consequences.

A further reduction

We can eliminate the primitives $\preccurlyeq$, Λ, Ω, ι by introducing some "ideal" elements. In this way we reduce quantifier-free formulas to a very simple form, to be used in proving the results in the present section.

More precisely, let $y = (y_1, \dots, y_n)$ be a tuple of distinct syntactic variables, and $Y = (Y_1, \dots, Y_n)$ a corresponding tuple of distinct differential indeterminates. A routine induction on terms shows that for any $\mathcal{L}^{\iota}$-term $t(y)$, there are quantifier-free formulas $\phi_1(y), \dots, \phi_m(y)$ $(m \geqslant 1)$ in the language of differential rings, and differential polynomials $F_1(Y), G_1(Y), \dots, F_m(Y), G_m(Y) \in \mathbb{Q}\{Y\}$, such that for all differential fields K and $a \in K^n$,

$$K \models \phi_1(a) \vee \cdots \vee \phi_m(a), \text{ and for } i = 1, \dots, m,$$
$$\text{if } K \models \phi_i(a), \text{ then } G_i(a) \neq 0 \text{ and } t(a) = \frac{F_i(a)}{G_i(a)}.$$

Let K be an H-field. Then we give K its order topology and K^n the corresponding product topology. (Note that by Lemma 2.4.1 the order topology on K equals its valuation topology if the valuation is nontrivial.)

COROLLARY 16.6.4. *Suppose K is an ω-free newtonian Liouville closed H-field and $X \subseteq K^n$ is definable in K. Then X has empty interior in K^n if and only if $X \subseteq \{a \in K^n : P(a) = 0\}$ for some $P \in K\{Y\}^{\neq}$.*

PROOF. Note that Lemma 4.4.10 goes through for differential polynomials over K in n indeterminates, by induction on n. Next, observe that the sets $\Lambda(K)$ and $\Omega(K)$ are open and closed in K. Now use the remarks above and Theorem 16.0.1. □

Now let K be an ω-free newtonian Liouville closed H-field. Take an immediate H-field extension L of K with an element λ such that $\Lambda(K) < \lambda < \Delta(K)$, and set $\omega := \omega(\lambda)$. Then $\Omega(K) < \omega < K \setminus \Omega(K)$, and for $f, g \in K$ with $g > 0$,

$$\frac{f}{g} \in \Lambda(K) \iff f < \lambda g, \qquad \frac{f}{g} \in \Omega(K) \iff f < \omega g.$$

Thus for any $\mathcal{L}^{\iota}$-term $t(y)$ as before the atomic formula $\Lambda(t(y))$ is equivalent, in K and for $y_1, \ldots, y_n$ ranging over K, to a boolean combination of formulas, each of which has one of the following forms:

$$F(y) < \lambda G(y), \quad G(y) > 0, \quad G(y) = 0 \qquad (F, G \in \mathbb{Q}\{Y\}).$$

Likewise for the atomic formula $\Omega(t(y))$ with ω instead of λ.

We can also eliminate occurrences of $\preccurlyeq$, but for this we take a further H-field extension L^* of L with an element c^* such that $C < c^* < a$ for all $a \in K^{>C}$. Then for all $f_1, g_1, f_2, g_2 \in K$ with $g_1, g_2 \neq 0$,

$$\frac{f_1}{g_1} \preccurlyeq \frac{f_2}{g_2} \iff |f_1 g_2| \leqslant c^* |f_2 g_1|.$$

Thus for any $\mathcal{L}^{\iota}$-terms $t_1(y)$ and $t_2(y)$ the atomic formula $t_1(y) \preccurlyeq t_2(y)$ is equivalent, in K and for $y_1, \ldots, y_n$ ranging over K, to a boolean combination of formulas, each of which has one of the following forms:

$$F(y) \leqslant c^* G(y), \quad G(y) > 0, \quad G(y) = 0 \qquad (F, G \in \mathbb{Q}\{Y\}).$$

To summarize some of the above in a single lemma, let z_{ij} ($i = 1, \ldots, n$; $j \in \mathbb{N}$) and v_λ, v_ω, v_{c^*}, be distinct syntactic variables, and set $z_j := (z_{1j}, \ldots, z_{nj})$. For $a = (a_1, \ldots, a_n) \in K^n$ we set $a^{(i)} := (a_1^{(i)}, \ldots, a_n^{(i)})$, and as usual $a' = a^{(1)}$. Recall from Section 3.6 that $\mathcal{L}_{\text{OR}} := \{0, 1, -, +, \cdot, \leqslant\}$ is the language of ordered rings.

LEMMA 16.6.5. *Let $X \subseteq K^n$ be definable without parameters in K. Then there is a quantifier-free $\mathcal{L}_{\text{OR}}$-formula $\varphi(z_0, z_1, \ldots, z_r, v_\lambda, v_\omega, v_{c^*})$, for some $r \in \mathbb{N}$, such that*

$$X = \{a \in K^n : L^* \models \varphi(a, a', \ldots, a^{(r)}, \lambda, \omega, c^*)\}.$$

NIP

We refer to B.13 for a definition and discussion of the very robust but highly restrictive property NIP that some model theoretic structures enjoy. In this subsection we establish part (i) of Corollary 16.0.2:

PROPOSITION 16.6.6. *Every ω-free newtonian Liouville closed H-field has* NIP.

PROOF. Let K be an ω-free newtonian Liouville closed H-field. Assume towards a contradiction that the relation $R \subseteq K^m \times K^n$ is definable without parameters in K and independent. We just do the case $m = n = 1$; the general case only involves more notation. Thus for every $N \geqslant 1$ there are $a_1, \ldots, a_N \in K$ and $b_I \in K$ ($I \subseteq \{1, \ldots, N\}$), such that for $i = 1, \ldots, N$ and all $I \subseteq \{1, \ldots, N\}$,

$$R(a_i, b_I) \iff i \in I.$$

Let L^* be an H-field extension of K as at the beginning of this section, containing λ, ω, c^*. By Lemma 16.6.5 we can take a quantifier-free $\mathcal{L}_{\text{OR}}$-formula

$$\varphi(x_0, x_1, \ldots, x_r, y_0, y_1, \ldots, y_r, v_\lambda, v_\omega, v_{c^*}),$$

such that for all $a, b \in K$:

$$R(a,b) \iff L^* \models \varphi(a, a', \dots, a^{(r)}, b, b', \dots, b^{(r)}, \lambda, \omega, c^*).$$

Thus the relation $R^* \subseteq (L^*)^{r+1} \times (L^*)^{r+4}$ given by

$$R^*(a_0, \dots, a_r, b_0, \dots, b_{r+3}) \iff L^* \models \varphi(a_0, \dots, a_r, b_0, \dots, b_{r+3})$$

is independent and quantifier-free definable in the $\mathcal{L}_{\mathrm{OR}}$-structure L^*, that is, in the ordered field L^*. This contradicts B.13.8. □

The induced structure on the constant field

The goal of this subsection is to establish the following:

PROPOSITION 16.6.7. *Let K be an ω-free newtonian Liouville closed H-field, and let $X \subseteq K^n$ be definable in K. Then $X \cap C^n$ is semialgebraic in the sense of C.*

The proof goes by reduction to Proposition 3.6.13, using our QE and the fact that a real closed H-field K with constant field $C \neq K$ yields a tame pair (K, C) as defined in Section 3.6.

PROOF OF PROPOSITION 16.6.7. Take an immediate real closed H-field extension L of K and $\lambda, \omega \in L$ as earlier in this section. As in the proof of Corollary 4.7.4, our QE reduces the problem to showing for polynomials $p, q \in K[Y_1, \dots, Y_n]$ that the following subsets of C^n are semialgebraic in the sense of C:

$$\{c \in C^n : p(c) = 0\}, \qquad \{c \in C^n : p(c) > 0\}, \qquad \{c \in C^n : p(c) \preccurlyeq q(c)\},$$
$$\{c \in C^n : p(c) < \lambda q(c)\}, \quad \{c \in C^n : p(c) < \omega q(c)\}.$$

This holds for the first three sets by a direct application of Proposition 3.6.13 to the tame pair (K, C). The two sets involving λ and ω are semialgebraic in the sense of C by applying Proposition 3.6.13 to the tame pair (L, C). □

O-minimality at infinity

By this we mean the following:

PROPOSITION 16.6.8. *Let K be an ω-free newtonian Liouville closed H-field, and let $X \subseteq K$ be definable in K. Then there exists an element $a \in K$ such that*

$$(a, +\infty) \subseteq X \ \text{ or } \ (a, +\infty) \cap X = \emptyset.$$

This is proved by logarithmic decomposition of differential polynomials; we refer to Section 4.2 for how these decompositions are defined. First some lemmas.

LEMMA 16.6.9. *Let K be a Liouville closed H-field and $K < a$ where a lies in some H-field extension of K. Then $K < (a^\dagger)^m < a$ for all $m \geqslant 1$.*

PROOF. Since $(K^\times)^\dagger = K$, the set $\Psi \subseteq \Gamma$ is downward closed. Set $\alpha := va$. Then $\alpha < \Gamma$, and so $\alpha^\dagger < \Gamma$: otherwise we have $\gamma \in \Gamma^<$ such that $\alpha^\dagger > \gamma^\dagger$, and so $\alpha > \gamma$, contradicting $\alpha < \Gamma$. Also, $\alpha^\dagger = o(\alpha)$ by Lemma 9.2.10; to apply this lemma, first shift by an element of Γ to reduce to the small derivation case. It remains to note that $a^\dagger > 0$ and $v(a^\dagger) = \alpha^\dagger$. □

LEMMA 16.6.10. *With K and a as in Lemma 16.6.9, a is* d*-transcendental over K, $C_{K\langle a\rangle} = C$, and $K\langle a\rangle$ is an H-field extension of K whose value group*

$$v(K\langle a\rangle^\times) = \Gamma \oplus \bigoplus_n \mathbb{Z}v(a^{\langle n\rangle}) \qquad \textit{(internal direct sum)}$$

contains Γ as a convex subgroup. If K is ω-free, then so is $K\langle a\rangle$.

PROOF. By induction and Lemma 16.6.9 we have

$$K < (a^{\langle n+1\rangle})^m < a^{\langle n\rangle} \qquad (n = 0, 1, 2, \dots,\ m = 1, 2, \dots).$$

Let $P \in K\{Y\}^{\neq}$ of order $\leqslant r$ have logarithmic decomposition

$$P = \sum_{\boldsymbol{i}} P_{\langle \boldsymbol{i}\rangle} Y^{\langle \boldsymbol{i}\rangle}$$

with $\boldsymbol{i}$ ranging over $\mathbb{N}^{1+r}$, all $P_{\langle \boldsymbol{i}\rangle} \in K$, and $P_{\langle \boldsymbol{i}\rangle} \neq 0$ for only finitely many $\boldsymbol{i}$. Take $\boldsymbol{j} \in \mathbb{N}^{1+r}$ lexicographically maximal with $P_{\langle \boldsymbol{j}\rangle} \neq 0$. It follows from the above that

$$P(a) \sim P_{\langle \boldsymbol{j}\rangle} \cdot a^{\langle \boldsymbol{j}\rangle}, \text{ and thus}$$

$$P(a) \neq 0, \qquad \operatorname{sign} P(a) = \operatorname{sign} P_{\langle \boldsymbol{j}\rangle}, \qquad v\big(P(a)\big) = vP_{\langle \boldsymbol{j}\rangle} + \sum_{n=0}^{r} j_n v(a^{\langle n\rangle}).$$

Thus a is d-transcendental over K, and for any $f \in K\langle a\rangle^{\neq}$ there are $g \in K^\times$ and $k_0, \dots, k_r \in \mathbb{Z}$ such that $f \sim g \cdot \big(a^{\langle 0\rangle}\big)^{k_0} \cdots \big(a^{\langle r\rangle}\big)^{k_r}$. Therefore $\operatorname{res} K\langle a\rangle = \operatorname{res} K$, and so $K\langle a\rangle$ is an H-field extension of K with $C = C_{K\langle a\rangle}$. The statement about the value group of $K\langle a\rangle$ also follows easily. Suppose now that K is ω-free, and $K\langle a\rangle$ is not; it remains to derive a contradiction from this assumption. Since $\Gamma^<$ is cofinal in $\Gamma^<_{K\langle a\rangle}$ this gives an element $\omega \in K\langle a\rangle$ such that $\omega_\rho \rightsquigarrow \omega$, where (ω_ρ) is the sequence in K obtained in the usual way from a logarithmic sequence in K. Now (ω_ρ) is of d-transcendental type over K, so $K\langle\omega\rangle$ is an immediate extension of K, and ω is d-transcendental over K. Now $\omega = P(a)/Q(a)$ with $P, Q \in K\{Y\}^{\neq}$, so a is a zero of the differential polynomial $\omega Q(Y) - P(Y) \in K\langle\omega\rangle\{Y\}^{\neq}$, and thus $K\langle a\rangle$ is d-algebraic over $K\langle\omega\rangle$. It follows that $K\langle a\rangle$ has finite trancendence degree over $K\langle\omega\rangle$, and so $\mathbb{Q}v\big(K\langle a\rangle^\times\big)/\Gamma$ has finite dimension as a vector space over $\mathbb{Q}$, contradicting the above structure of $v\big(K\langle a\rangle^\times\big)$. □

PROOF OF PROPOSITION 16.6.8. By a routine translation into model-theoretic terms it is enough to show the following:

CLAIM: *Let L be an elementary extension of K with elements $a, b > K$. Then there is a pre-H-field isomorphism $i\colon K\langle a\rangle \to K\langle b\rangle$ over K with $i(a) = b$ such that also*

$$i\big(\Lambda(L) \cap K\langle a\rangle\big) = \Lambda(L) \cap K\langle b\rangle, \qquad i\big(\Omega(L) \cap K\langle a\rangle\big) = \Omega(L) \cap K\langle b\rangle.$$

A pre-H-field isomorphism $i\colon K\langle a\rangle \to K\langle b\rangle$ over K with $i(a) = b$ is obtained from Lemma 16.6.10 and its proof, in particular, the equalities for sign $P(a)$ and $v\big(P(a)\big)$ in that proof. Since K is ω-free, so are $K\langle a\rangle$ and $K\langle b\rangle$ by the same lemma, and so the additional property claimed for i is now a consequence of Corollary 16.3.19. □

Using fractional linear transformations we get analogous behavior of any definable set $X \subseteq K$ to the left as well as to the right of any point in K. In other words, K is locally o-minimal in the sense of Marker and Steinhorn; see [449]. Thus:

COROLLARY 16.6.11. *Let K be an ω-free newtonian Liouville closed H-field, and let $X \subseteq K$ be definable in K. Then X is the disjoint union of an open definable subset of K and a discrete definable subset of K. Moreover, X is discrete in K iff $X \subseteq \big\{y \in K : P(y) = 0\big\}$ for some $P \in K\{Y\}^{\neq}$.*

PROOF. The interior of X in K is definable, and so X with its interior removed is discrete, by local o-minimality. For the second part, use Corollary 16.6.4. □

O-minimality at $C^{\downarrow}$

We also have o-minimality at another important cut:

PROPOSITION 16.6.12. *Let K be an ω-free newtonian Liouville closed H-field, and let $X \subseteq K$ be definable in K. Then there exists an element $a > C$ in K such that*

$$\{f \in K : C < f < a\} \subseteq X \ \textit{or}\ \{f \in K : C < f < a\} \cap X = \emptyset.$$

PROOF. By a routine translation into model-theoretic terms it suffices to show:

CLAIM: *Let L be an elementary extension of K and $f, g \in L$ be such that $C_L < f < a$ and $C_L < g < a$ for all $a > C$ in K. Then there is a pre-H-field isomorphism $i\colon K\langle f\rangle \to K\langle g\rangle$ over K with $i(f) = g$ such that also*

$$i\big(\Lambda(L) \cap K\langle f\rangle\big) = \Lambda(L) \cap K\langle g\rangle, \qquad i\big(\Omega(L) \cap K\langle f\rangle\big) = \Omega(L) \cap K\langle g\rangle.$$

To prove this claim, note first that $\Gamma^{<} < vf < 0$ and $\Gamma^{<} < vg < 0$. So Corollary 13.6.8 yields a pre-H-field isomorphism $K\langle f\rangle \to K\langle g\rangle$ over K sending f to g. Also, $K\langle f\rangle$ and $K\langle g\rangle$ have a smallest comparability class by Proposition 13.6.7, and thus $K\langle f\rangle$ and $K\langle g\rangle$ have unique $\Lambda\Omega$-cuts by Corollary 16.3.19. This yields the claim. □

Notes and comments

The uniqueness-up-to-isomorphism of prime models of $T^{\text{nl}}_{\text{small}}$ holds also on general model-theoretic grounds: [284, Corollary 4.2.16].

Some of the arguments in the subsections on the induced structure on the constant field and o-minimality at infinity were already used in Section 5 of [23] to prove quantifier-free versions of Propositions 16.6.7 and 16.6.8 for $K = \mathbb{T}$.

Shelah [410] considers a strengthening of NIP, called *dp-minimality*; see [103] for basic facts about this notion. Algebraically closed valued fields, the field of p-adic numbers, and o-minimal structures are dp-minimal. Simon [415, Theorem 3.6] proved that if an expansion $(G; \leqslant, 0, +, \dots)$ of a divisible ordered abelian group is dp-minimal, then all infinite definable subsets of G have nonempty interior. Thus if K is any pre-H-field and $K \neq C$, then K is *not* dp-minimal, since the definable set $C \subseteq K$ has empty interior. Another strengthening of NIP is the notion of *distality*, due to Simon [416]. O-minimal structures and the field of p-adic numbers are distal, but algebraically closed valued fields are not. We intend to show elsewhere that $\mathbb{T}$ is distal.

In view of [105, §2.25], Corollary 16.6.4 yields a natural notion of dimension for definable sets $X \subseteq \mathbb{T}^n$; for details, see [25]. In connection with Corollary 16.6.11, the standard example of an infinite discrete definable subset of $\mathbb{T}$ is of course the set $\mathbb{R}$ of constants. The question arises if this is the source of all discreteness: is every discrete definable subset of $\mathbb{T}^n$ the image of some semialgebraic set $S \subseteq \mathbb{R}^m$ under a definable map $S \to \mathbb{T}^n$? It turns out that the answer is negative; see [25].

Appendix A

Transseries

We assume here familiarity with well-based series and Hahn fields as exposed in Section 3.1. We begin by adding some items to this material. Our construction of $\mathbb{T}$ is self-contained as to concepts and definitions, but for proofs of some key properties we refer to [112], where $\mathbb{T}$ is denoted by $\mathbb{R}((x^{-1}))^{\mathrm{LE}}$, or $\mathbb{R}((t))^{\mathrm{LE}}$ with $t = x^{-1}$, and called the *field of logarithmic-exponential series* (in x over $\mathbb{R}$). The construction is also very similar to the treatment in Schmeling's thesis [388].

The reader should be aware that notations and terminology concerning Hahn fields and transseries vary considerably across the literature, even in our own earlier works. (For example, the $\mathbb{T}$ in [22] is not the $\mathbb{T}$ constructed here.) In the present volume we have systematized things by adopting many notations from [194].

Summability in Hahn fields

In what follows, $\mathfrak{M}$ is a multiplicative (totally) ordered abelian group, ordered by $\preccurlyeq$. Also C will be a (coefficient) field, so that we have the Hahn field $C[[\mathfrak{M}]]$, with the internal direct sum decomposition

$$C[[\mathfrak{M}]] \;=\; C[[\mathfrak{M}^{\succ 1}]] \oplus C \oplus C[[\mathfrak{M}^{\prec 1}]]$$

into C-linear subspaces. Note that $C[[\mathfrak{M}^{\prec 1}]] = C[[\mathfrak{M}]]^{\prec 1}$.

A family $(f_\lambda)_{\lambda \in \Lambda}$ in $C[[\mathfrak{M}]]$ is said to be **summable** if $\bigcup_\lambda \operatorname{supp} f_\lambda$ is well-based and for each $\mathfrak{m} \in \mathfrak{M}$ there are only finitely many $\lambda \in \Lambda$ such that $f_{\lambda,\mathfrak{m}} \neq 0$; in that case we define its sum $\sum_\lambda f_\lambda$ to be the series $f \in C[[\mathfrak{M}]]$ such that $f_\mathfrak{m} = \sum_\lambda f_{\lambda,\mathfrak{m}}$ for each $\mathfrak{m} \in \mathfrak{M}$. (This agrees with the usual notation for elements of $C[[\mathfrak{M}]]$: for a series $f = \sum_\mathfrak{m} f_\mathfrak{m}\mathfrak{m} \in C[[\mathfrak{M}]]$ the family $f_\mathfrak{m}\mathfrak{m}$ is indeed summable with sum f.)

Let $t = (t_1, \dots, t_n)$ be a tuple of distinct variables and let

$$F \;=\; F(t) \;=\; \sum_\nu c_\nu t^\nu \in C[[t]] \;:=\; C[[t_1, \dots, t_n]]$$

be a formal power series over C; here the sum ranges over all multiindices $\nu = (\nu_1, \dots, \nu_n) \in \mathbb{N}^n$, and $c_\nu \in C$, $t^\nu := t_1^{\nu_1} \cdots t_n^{\nu_n}$. For any tuple $\varepsilon = (\varepsilon_1, \dots, \varepsilon_n)$ of elements of $C[[\mathfrak{M}]]^{\prec 1}$ the family $(c_\nu \varepsilon^\nu)$ is summable, where $\varepsilon^\nu := \varepsilon_1^{\nu_1} \cdots \varepsilon_n^{\nu_n}$. Put

$$F(\varepsilon) := \sum_\nu c_\nu \varepsilon^\nu \in C[[\mathfrak{M}]]^{\preccurlyeq 1} \;=\; C[[\mathfrak{M}^{\preccurlyeq 1}]].$$

For example, if C has characteristic zero, then for $n = 1$ and $t = t_1$ the formal series $\exp(t) = \sum_{\nu=0}^{\infty} t^\nu/\nu!$ yields a partial exponential function

$$\varepsilon \mapsto \exp(\varepsilon) = \sum_{\nu=0}^{\infty} \varepsilon^\nu/\nu! \, : \, C[[\mathfrak{M}]]^{\prec 1} \to 1 + C[[\mathfrak{M}]]^{\prec 1},$$

an isomorphism of the additive subgroup $C[[\mathfrak{M}]]^{\prec 1}$ of $C[[\mathfrak{M}]]$ onto the multiplicative subgroup $1 + C[[\mathfrak{M}]]^{\prec 1}$ of $C[[\mathfrak{M}]]^\times$, with inverse

$$1+\delta \mapsto \log(1+\delta) := \sum_{\nu=1}^{\infty}(-1)^{\nu-1}\delta^\nu/\nu \, : \, 1 + C[[\mathfrak{M}]]^{\prec 1} \to C[[\mathfrak{M}]]^{\prec 1}.$$

Let $\mathfrak{N}$ also be a multiplicative ordered abelian group. Then a map

$$\Phi \, : \, C[[\mathfrak{M}]] \to C[[\mathfrak{N}]]$$

is said to be **strongly additive** if for each summable family (f_λ) in $C[[\mathfrak{M}]]$ the family $\big(\Phi(f_\lambda)\big)$ in $C[[\mathfrak{N}]]$ is summable with $\Phi(\sum_\lambda f_\lambda) = \sum_\lambda \Phi(f_\lambda)$. Note that if Φ is strongly additive, then it is additive.

The case where $\mathfrak{M}$ is a product with convex factor $\mathfrak{G}$

Suppose now that $\mathfrak{G}$ and $\mathfrak{R}$ are ordered subgroups of $\mathfrak{M}$ such that

$$\mathfrak{G} \text{ is convex in } \mathfrak{M}, \qquad \mathfrak{G} \cap \mathfrak{R} = \{1\}, \quad \mathfrak{M} = \mathfrak{G}\mathfrak{R} := \{\mathfrak{g}\mathfrak{r} : \, \mathfrak{g} \in \mathfrak{G}, \, \mathfrak{r} \in \mathfrak{R}\}.$$

Then we have an isomorphism $C[[\mathfrak{M}]] \to C[[\mathfrak{G}]][[\mathfrak{R}]]$ of $C[[\mathfrak{G}]]$-algebras given by

$$f = \sum_{\mathfrak{m}} f_{\mathfrak{m}}\mathfrak{m} \mapsto \sum_{\mathfrak{r}\in\mathfrak{R}} \left(\sum_{\mathfrak{g}\in\mathfrak{G}} f_{\mathfrak{g}\mathfrak{r}}\mathfrak{g} \right) \mathfrak{r}.$$

For $f \in C[[\mathfrak{M}]]$ we have in fact $f = \sum_{\mathfrak{r}\in\mathfrak{R}} \left(\sum_{\mathfrak{g}\in\mathfrak{G}} f_{\mathfrak{g}\mathfrak{r}}\mathfrak{g}\right) \mathfrak{r}$ where the indicated sums exist in $C[[\mathfrak{M}]]$ according to the definition of summability. Whenever convenient we identify below $C[[\mathfrak{M}]]$ and $C[[\mathfrak{G}]][[\mathfrak{R}]]$ via the above isomorphism.

If in addition C is an ordered field, then $C[[\mathfrak{M}]]$ and $C[[\mathfrak{G}]]$ are ordered Hahn fields, and so is $C[[\mathfrak{G}]][[\mathfrak{R}]]$, and the above isomorphism is also an isomorphism of ordered fields. (In this remark and in what follows the reader is assumed to be familiar with Section 3.5.)

Directed unions of Hahn fields

A key feature of $\mathbb{T}$ will be its structure as a directed union of Hahn fields over its constant field $\mathbb{R}$. (A Hahn field over $\mathbb{R}$ with its natural valuation and ordering and any derivation is never a Liouville closed H-field, by [20, Corollary 7.2], and so cannot have the properties we expect of $\mathbb{T}$.) It is therefore useful to extend the notions above to such directed unions, and so we consider here a directed family $(\mathfrak{M})_{i\in I}$ with $I \neq \emptyset$, of ordered subgroups of the ordered multiplicative group $\mathfrak{M}$ such that $\mathfrak{M} = \bigcup_i \mathfrak{M}_i$. Here "directed" means that for all $i, j \in I$ there exists $k \in I$ with $\mathfrak{M}_i, \mathfrak{M}_j \subseteq \mathfrak{M}_k$. This leads to a directed union of Hahn fields over C, namely the valued subfield

$$K := \bigcup_i C[[\mathfrak{M}_i]]$$

of $C[[\mathfrak{M}]]$. Define a **K-subgroup** of $\mathfrak{M}$ to be an ordered subgroup $\mathfrak{G}$ of $\mathfrak{M}$ such that $C[[\mathfrak{G}]] \subseteq K$, inside the ambient $C[[\mathfrak{M}]]$; thus each $\mathfrak{M}_i$ is a K-subgroup of $\mathfrak{M}$. We say that the family $(\mathfrak{M}_i)$ is **healthy** if every K-subgroup of $\mathfrak{M}$ is contained in some $\mathfrak{M}_i$; this might depend on C. An easy diagonal argument shows: if I is countable (the relevant case for us), then $(\mathfrak{M}_i)$ is healthy. Also, by an easy cofinality argument, if every $\mathfrak{M}_i$ is convex in $\mathfrak{M}$, then $(\mathfrak{M}_i)$ is healthy.

Assume below that $(\mathfrak{M}_i)$ *is healthy*. A family (f_λ) in K is said to be **summable** if there exists a K-subgroup $\mathfrak{G}$ of $\mathfrak{M}$ such that all $f_\lambda \in C[[\mathfrak{G}]]$ and (f_λ) is summable as a family in $C[[\mathfrak{G}]]$; note that then $\sum_\lambda f_\lambda$ is defined as an element of K (lying in $C[[\mathfrak{G}]]$ for $\mathfrak{G}$ as above). Thus for $F = \sum_\nu c_\nu t^\nu \in C[[t]]$ with $t = (t_1, \dots, t_n)$ and $\varepsilon \in \mathcal{o}_K^n$, the family $(c_\nu \varepsilon^\nu)_{\nu\in\mathbb{N}^n}$ is summable, with $F(\varepsilon) := \sum_\nu c_\nu \varepsilon^\nu \in \mathcal{O}_K$.

Let $\mathfrak{N}$ also be an ordered abelian group, with $\mathfrak{N} = \bigcup_j \mathfrak{N}_j$ and $(\mathfrak{N}_j)$ a directed family of ordered subgroups of $\mathfrak{N}$. Let $L = \bigcup_j C[[\mathfrak{N}_j]] \subseteq C[[\mathfrak{N}]]$, and assume $(\mathfrak{N}_j)$ is healthy. Then a map $\Phi\colon K \to L$ is said to be **strongly additive** if for every summable family (f_λ) in K the family $(\Phi(f_\lambda))$ is summable in L, and $\Phi(\sum_\lambda f_\lambda) = \sum_\lambda \Phi(f_\lambda)$. A map $\Phi\colon K \to L$ is said to be **healthy** if for each K-subgroup $\mathfrak{G}$ of $\mathfrak{M}$ there exists an L-subgroup $\mathfrak{H}$ of $\mathfrak{N}$ such that $\Phi\big(C[[\mathfrak{G}]]\big) \subseteq C[[\mathfrak{H}]]$.

Exponential ordered fields

An **exponentiation** on a field E is a group morphism $\exp\colon E \to E^\times$ from the additive group of E into its multiplicative group. An **exponential ordered field** is an ordered field E equipped with a strictly increasing exponentiation on E (denoted by exp unless specified otherwise, necessarily taking values in the multiplicative subgroup $E^>$ of $E^\times$). A **logarithmic-exponential ordered field** is an exponential ordered field E with $\exp(E) = E^>$; the inverse of the ordered group isomorphism $\exp\colon E \to E^>$ is then an ordered group isomorphism $\log\colon E^> \to E$. Below we consider the ordered field $\mathbb{R}$ of real numbers as a logarithmic-exponential ordered field with exponentiation $r \mapsto \mathrm{e}^r$. For any logarithmic-exponential ordered field E we set $a^f := \exp(f \log a) \in E^>$ for $a \in E^>$ and $f \in E$, so $a^0 = 1$ and $a^1 = a$, and the usual identitites follow:

$$a^{f+g} = a^f a^g, \quad (ab)^f = a^f b^f, \quad a^{fg} = (a^f)^g \qquad (a, b \in E^>,\ f, g \in E).$$

Initially we shall construct $\mathbb{T}$ as a logarithmic-exponential ordered field extension of $\mathbb{R}$; the definition of the derivation on $\mathbb{T}$ comes later. This construction involves the following general procedure. We define a **pre-exponential ordered field** to be a tuple $(E, A, B, \exp)$ such that:

(1) E is an ordered field;

(2) A and B are additive subgroups of E with $E = A \oplus B$ and B convex in E;

(3) $\exp\colon B \to E^\times$ is a strictly increasing group morphism (so $\exp(B) \subseteq E^>$).

Let $(E, A, B, \exp)$ be a pre-exponential ordered field. We view A as the part of E where exponentiation is not yet defined, and accordingly we introduce a "bigger" pre-exponential ordered field $(E^*, A^*, B^*, \exp^*)$ as follows: Take a *multiplicative* copy $\exp^*(A)$ of the ordered additive group A with order-preserving isomorphism $\exp^*\colon A \to \exp^*(A)$, and put $E^* := E[[\exp^*(A)]]$. Viewing E^* as an ordered Hahn field over the ordered coefficient field E, we set

$$A^* := E[[\exp^*(A)^{\succ 1}]], \qquad B^* := (E^*)^{\preccurlyeq 1} = E \oplus (E^*)^{\prec 1} = A \oplus B \oplus (E^*)^{\prec 1}.$$

Note that $\exp^*(A)^{\succ 1} = \exp^*(A^>)$. Next we extend $\exp^*$ to $\exp^*\colon B^* \to (E^*)^\times$ by

$$\exp^*(a + b + \varepsilon) := \exp^*(a) \cdot \exp(b) \cdot \sum_{n=0}^{\infty} \frac{\varepsilon^n}{n!} \qquad (a \in A,\ b \in B,\ \varepsilon \in (E^*)^{\prec 1}).$$

Then $E \subseteq B^* = \operatorname{domain}(\exp^*)$, and $\exp^*$ extends $\exp$. Note that $E < (A^*)^>$ (but $\exp^*(E)$ is cofinal in E^* if $A \neq \{0\}$). In particular, for $a \in A^>$, we have

$$\exp^*(a) \in \exp^*(A^>) \subseteq (A^*)^>, \quad \text{so } \exp^*(a) > E.$$

Suppose now that $E = \mathbb{R}[[\mathfrak{N}]]$, where $\mathfrak{N}$ is a multiplicative ordered abelian group. We identify $\mathfrak{N}$ and $\exp(A)$ with subgroups of the product group $\mathfrak{N}^* = \mathfrak{N} \times \exp(A)$ via $\mathfrak{n} \mapsto (\mathfrak{n}, 1)$ and $e \mapsto (1, e)$ for $\mathfrak{n} \in \mathfrak{N}$ and $e \in \exp(A)$. Then $\mathfrak{N}^* = \mathfrak{N}\exp(A)$ and $\mathfrak{N} \cap \exp(A) = \{1\}$ (in $\mathfrak{N}^*$). We make $\mathfrak{N}^*$ into an ordered abelian group so that $\mathfrak{N}$ and $\exp(A)$ are ordered subgroups and $\mathfrak{N}$ is convex in $\mathfrak{N}^*$. The effect is that

$$E^* = \mathbb{R}[[\mathfrak{N}]]\,[[\exp(A)]] = \mathbb{R}[[\mathfrak{N}^*]],$$

after the natural identifications, so the inclusion $E \subseteq E^*$ is now the inclusion $\mathbb{R}[[\mathfrak{N}]] \subseteq \mathbb{R}[[\mathfrak{N}^*]]$ induced by $\mathfrak{N} \subseteq \mathfrak{N}^*$. Viewing E and E^* as Hahn fields over $\mathbb{R}$, we get: if all infinitesimals of E lie in B and $\exp(\varepsilon) = \sum_{\nu=0}^{\infty} \varepsilon^\nu/\nu!$ for all infinitesimal $\varepsilon \in E$, then all infinitesimals of E^* lie in B^* and $\exp^*(\varepsilon) = \sum_{\nu=0}^{\infty} \varepsilon^\nu/\nu!$ for all infinitesimal $\varepsilon \in E^*$.

Construction of $\mathbb{T}_{\exp}$

Starting with $E_0 := \mathbb{R}[[G_0]]$, with $G_0 = x^{\mathbb{R}}$, we construct the field $\mathbb{T}_{\exp} = \bigcup_m E_m$ of *exponential transseries* as the union of an increasing sequence of Hahn fields $E_m = \mathbb{R}[[G_m]]$. First we make the ordered Hahn field E_0 over $\mathbb{R}$ into the pre-exponential ordered field

$$(E_0, A_0, B_0, \exp_0), \quad A_0 := \mathbb{R}[[G_0^{\succ 1}]], \quad B_0 := E_0^{\preccurlyeq 1} = \mathbb{R} \oplus E_0^{\prec 1}, \text{ with}$$

$$\exp_0\colon B_0 \to E_0^\times \text{ given by } \exp_0(r + \varepsilon) := \mathrm{e}^r \sum_{n=0}^{\infty} \varepsilon^n/n! \qquad (r \in \mathbb{R},\ \varepsilon \in E_0^{\prec 1}).$$

Inductively, we assume given the pre-exponential ordered field $(E_m, A_m, B_m, \exp_m)$ with the ordered Hahn field $E_m = \mathbb{R}[[G_m]]$ over $\mathbb{R}$, and set

$$(E_{m+1}, A_{m+1}, B_{m+1}, \exp_{m+1}) \;:=\; (E_m^*, A_m^*, B_m^*, \exp_m^*), \text{ so}$$

$$E_m \;=\; \mathbb{R}[[G_m]] \;\subseteq\; E_{m+1} \;=\; \mathbb{R}[[G_{m+1}]] \text{ (inclusions of ordered Hahn fields)}$$

with G_m a convex ordered subgroup of $G_{m+1} = G_m \exp(A_m)$. We put

$$G^{\mathrm{E}} := \bigcup_m G_m, \qquad \mathbb{T}_{\exp} = \mathbb{R}[[x^{\mathbb{R}}]]^{\mathrm{E}} := \bigcup_m E_m,$$

with G^{E} construed as the multiplicative ordered abelian group having the G_m as ordered subgroups, and $\mathbb{T}_{\exp}$ as the ordered field with the E_m as ordered subfields. The elements of G^{E} are called **exponential transmonomials** (or E-monomials), and those of $\mathbb{T}_{\exp}$ are called **exponential transseries** (or E-series). The alternative notation $\mathbb{R}[[x^{\mathbb{R}}]]^{\mathrm{E}}$ for $\mathbb{T}_{\exp}$ highlights the role of the formal variable x and the initial Hahn field $\mathbb{R}[[x^{\mathbb{R}}]]$ in the construction of $\mathbb{T}_{\exp}$, with the superscript E indicating closure under exponentiation. Let $\exp\colon \mathbb{T}_{\exp} \to \mathbb{T}_{\exp}^{\times}$ be the common extension of the $\exp_m$. Then $\mathbb{T}_{\exp}$ with $\exp$ is an exponential ordered field extension of $\mathbb{R}$. The ordered Hahn field $\mathbb{R}[[G^{\mathrm{E}}]]$ gives an ordered field inclusion $\mathbb{T}_{\exp} \subseteq \mathbb{R}[[G^{\mathrm{E}}]]$. We think of any $f \in \mathbb{T}_{\exp}$ as a series $f(x) \in \mathbb{R}[[G^{\mathrm{E}}]]$ with $\operatorname{supp} f \subseteq G^{\mathrm{E}}$. Considering $\mathbb{T}_{\exp}$ also as a *valued* subfield of the Hahn field $\mathbb{R}[[G^{\mathrm{E}}]]$ we have

$$\exp(\varepsilon) \;=\; \sum_{\nu=0}^{\infty} \varepsilon^{\nu}/\nu! \qquad \text{for infinitesimal } \varepsilon \in \mathbb{T}_{\exp}.$$

Note: $\mathbb{T}_{\exp}$ is dense in this valued field $\mathbb{R}[[G^{\mathrm{E}}]]$, since every G_m is convex in G^{E}. In order to indicate an element of $\mathbb{R}[[G^{\mathrm{E}}]]$ outside $\mathbb{T}_{\exp}$, set $\exp_0(x) := x$ and $\exp_{n+1}(x) := \exp\big(\exp_n(x)\big)$; note that we abandon here the earlier meaning of $\exp_n\colon B_n \to E_n^{\times}$. Induction gives $\exp_{n+1}(x) \in \exp(A_n^{>})$, so $\exp_{n+1}(x) \in G_{n+1}$ and $\exp_{n+1}(x) > G_n$. Thus the series $\sum_{n=0}^{\infty} 1/\exp_n(x)$ lies in $\mathbb{R}[[G^{\mathrm{E}}]]$ but not in $\mathbb{T}_{\exp}$.

Straightforward inductions on m yield:

LEMMA A.1. *The G_m and A_m have the following basic properties:*

(i) $A_m \;=\; \{f \in \mathbb{T}_{\exp} : G_{m-1} \prec \operatorname{supp} f \subseteq G_m\}$, *with* $G_{-1} := \{1\}$;

(ii) $|a| > A_{m-1}$ *for all* $a \in A_m^{\neq}$, *with* $A_{-1} := \{0\}$;

(iii) $\big\{f \in \mathbb{R}[[G_m]] : \operatorname{supp} f \succ 1\big\} \;=\; A_0 \oplus \cdots \oplus A_m$;

(iv) $G_m = x^{\mathbb{R}} \cdot \exp(A_0 \oplus \cdots \oplus A_{m-1})$ *and* $x^{\mathbb{R}} \cap \exp(A_0 \oplus \cdots \oplus A_{m-1}) = \{1\}$.

COROLLARY A.2. $\{f \in \mathbb{T}_{\exp} : \operatorname{supp} f \succ 1\} \;=\; \bigoplus_{m=0}^{\infty} A_m$, *and*

$$G^{\mathrm{E}} \;=\; x^{\mathbb{R}} \cdot \exp\left(\bigoplus_{m=0}^{\infty} A_m\right), \qquad x^{\mathbb{R}} \cap \exp\left(\bigoplus_{m=0}^{\infty} A_m\right) = \{1\}.$$

LEMMA A.3. $\mathbb{T}^{>}_{\exp} = x^{\mathbb{R}} \cdot \exp(\mathbb{T}_{\exp})$.

PROOF. Let $f \in E^{>}_{m}$. By (iv) of Lemma A.1, $f = cx^r \exp(a)(1+\delta)$ with $c \in \mathbb{R}^{>}$, $r \in \mathbb{R}$, $a \in A_0 + \cdots + A_{m-1}$, and infinitesimal δ in the Hahn field $E_m = \mathbb{R}[[G_m]]$. Since $1+\delta = \exp\left(\sum_{\nu=1}^{\infty}(-1)^{\nu-1}\delta^{\nu}/\nu\right) \in \exp(E_m)$, we get $f \in x^{\mathbb{R}}\exp(E_m)$. □

It is easy to check that $x \notin \exp(\mathbb{T}_{\exp})$, so we are still missing $\log x$. Next we show that copying the above procedure with $\log x$ instead of x, and then with $\log\log x$, and so on, and taking a union, is enough to enlarge the exponential ordered field $\mathbb{T}_{\exp}$ to a logarithmic-exponential ordered field $\mathbb{T}$.

From $\mathbb{T}_{\exp}$ to $\mathbb{T}$

The idea is to use distinct symbols $\ell_0, \ell_1, \ell_2, \ldots$ in the role of $x, \log x, \log\log x, \ldots$. Replacing for any given n the formal variable x in $\mathbb{R}[[x^{\mathbb{R}}]]^{\mathrm{E}}$ by ℓ_n changes informally any E-series $f(x) \in \mathbb{R}[[x^{\mathbb{R}}]]^{\mathrm{E}}$ into a series $f(\ell_n) \in \mathbb{R}[[\ell_n^{\mathbb{R}}]]^{\mathrm{E}}$. Formally: take for each n an isomorphism

$$\mathfrak{m} \mapsto \mathfrak{m}{\downarrow_n} \; : \; G^{\mathrm{E}} \to G^{\mathrm{E},n}$$

of (multiplicative) ordered abelian groups, with $x^r{\downarrow_n} = \ell_n^r \in G^{\mathrm{E},n}$ for $r \in \mathbb{R}$. Given n, this isomorphism extends uniquely to a strongly additive $\mathbb{R}$-linear map

$$f \mapsto f{\downarrow_n} \; : \; \mathbb{R}[[G^{\mathrm{E}}]] \to \mathbb{R}[[G^{\mathrm{E},n}]].$$

This map is the identity on $\mathbb{R}$ and is an isomorphism of ordered (Hahn) fields; we denote the image of $\mathbb{R}[[x^{\mathbb{R}}]]^{\mathrm{E}}$ under this isomorphism by $\mathbb{R}[[\ell_n^{\mathbb{R}}]]^{\mathrm{E}}$, and make the ordered subfield $\mathbb{R}[[\ell_n^{\mathbb{R}}]]^{\mathrm{E}}$ of the ordered Hahn field $\mathbb{R}[[G^{\mathrm{E},n}]]$ into an exponential ordered field, with exponentiation denoted also by exp, in such a way that

$$f \mapsto f{\downarrow_n} \; : \; \mathbb{R}[[x^{\mathbb{R}}]]^{\mathrm{E}} \to \mathbb{R}[[\ell_n^{\mathbb{R}}]]^{\mathrm{E}}$$

is an isomorphism of exponential ordered fields. Also $\ell_n := \ell_n^1 \in \ell_n^{\mathbb{R}}$ by notational convention. For $n = 0$ we take $G^{\mathrm{E},0} = G^{\mathrm{E}}$, with $\mathfrak{m}{\downarrow_0} = \mathfrak{m}$ for $\mathfrak{m} \in G^{\mathrm{E}}$. Thus $x^r = \ell_0^r$ for $r \in \mathbb{R}$, and $f{\downarrow_0} = f$ for $f \in \mathbb{R}[[G^{\mathrm{E}}]]$. Given n, we have the increasing sequence $(G_m{\downarrow_n})_{m=0}^{\infty}$ of convex subgroups of $G^{\mathrm{E},n}$ with $G^{\mathrm{E},n} = \bigcup_m G_m{\downarrow_n}$, and likewise, $\mathbb{R}[[\ell_n^{\mathbb{R}}]]^{\mathrm{E}} = \bigcup_m \mathbb{R}[[G_m{\downarrow_n}]]$.

It is straightforward to define inductively a strongly additive $\mathbb{R}$-linear embedding

$$\mathbb{R}[[x^{\mathbb{R}}]]^{\mathrm{E}} \to \mathbb{R}[[x^{\mathbb{R}}]]^{\mathrm{E}}$$

of exponential ordered fields that sends x^r to $\exp(rx)$ for each $r \in \mathbb{R}$, and show that these properties define the embedding uniquely. (It maps G_m into G_{m+1} and E_m into E_{m+1}.) By transport to isomorphic copies of $\mathbb{R}[[x^{\mathbb{R}}]]^{\mathrm{E}}$ we have for each n a unique strongly additive $\mathbb{R}$-linear embedding $\mathbb{R}[[\ell_n^{\mathbb{R}}]]^{\mathrm{E}} \to \mathbb{R}[[\ell_{n+1}^{\mathbb{R}}]]^{\mathrm{E}}$ of exponential ordered fields that sends ℓ_n^r to $\exp(r\ell_{n+1})$ for each $r \in \mathbb{R}$; it maps $G_m{\downarrow_n}$ into $G_{m+1}{\downarrow_{n+1}}$ and thus $G^{\mathrm{E},n}$ into $G^{\mathrm{E},n+1}$. We identify $\mathbb{R}[[\ell_n^{\mathbb{R}}]]^{\mathrm{E}}$ with its image in $\mathbb{R}[[\ell_{n+1}^{\mathbb{R}}]]^{\mathrm{E}}$ under this embedding. So $\ell_n^r = \exp(r\ell_{n+1})$ for $r \in \mathbb{R}$, and $G_{m\downarrow n} \subseteq G_{m+1\downarrow n+1}$, and we have inclusions

$$\begin{aligned} \mathbb{T}_{\exp} &= \mathbb{R}[[\ell_0^{\mathbb{R}}]]^{\mathrm{E}} \subseteq \mathbb{R}[[\ell_1^{\mathbb{R}}]]^{\mathrm{E}} \subseteq \mathbb{R}[[\ell_2^{\mathbb{R}}]]^{\mathrm{E}} \subseteq \cdots, \\ G^{\mathrm{E}} &= G^{\mathrm{E},0} \subseteq G^{\mathrm{E},1} \subseteq G^{\mathrm{E},2} \subseteq \cdots \end{aligned}$$

of exponential ordered fields and ordered abelian groups. We now set

$$\mathbb{T} = \mathbb{R}[[x^{\mathbb{R}}]]^{\mathrm{LE}} := \bigcup_n \mathbb{R}[[\ell_n^{\mathbb{R}}]]^{\mathrm{E}}, \qquad G^{\mathrm{LE}} := \bigcup_n G^{\mathrm{E},n} \subseteq \mathbb{T},$$

with $\mathbb{T}$ construed as an exponential ordered field having the $\mathbb{R}[[\ell_n^{\mathbb{R}}]]^{\mathrm{E}}$ as exponential ordered subfields, and G^{LE} construed as a multiplicative ordered abelian group with the $G^{\mathrm{E},n}$ as ordered subgroups. In particular, we have the ordered Hahn field $\mathbb{R}[[G^{\mathrm{LE}}]]$ with the ordered field inclusion $\mathbb{T} \subseteq \mathbb{R}[[G^{\mathrm{LE}}]]$. We also consider $\mathbb{T}$ as a *valued* subfield of the Hahn field $\mathbb{R}[[G^{\mathrm{LE}}]]$. Continuing to denote the exponentiation of $\mathbb{T}$ by exp, we have $\exp(\varepsilon) = \sum_{\nu=0}^{\infty} \varepsilon^\nu/\nu!$ for infinitesimal $\varepsilon \in \mathbb{T}$. From $x^r = \exp(r\ell_1)$ for $r \in \mathbb{R}$, and Lemma A.3 we obtain

$$\left(\mathbb{R}[[x^{\mathbb{R}}]]^{\mathrm{E}}\right)^{>} \subseteq \exp\left(\mathbb{R}[[\ell_1^{\mathbb{R}}]]^{\mathrm{E}}\right),$$

and likewise $\left(\mathbb{R}[[\ell_n^{\mathbb{R}}]]^{\mathrm{E}}\right)^{>} \subseteq \exp\left(\mathbb{R}[[\ell_{n+1}^{\mathbb{R}}]]^{\mathrm{E}}\right)$ for all n. Thus $\mathbb{T}$ is a logarithmic-exponential ordered field. For the inverse $\log\colon \mathbb{T}^{>} \to \mathbb{T}$ of the exponentiation of $\mathbb{T}$ it is now literally true that $\ell_1 = \log x$, $\ell_2 = \log\log x$, and so on. As in any logarithmic-exponential field we set $a^f := \exp(f \log a)$ for $a \in \mathbb{T}^{>}$ and $f \in \mathbb{T}$. Thus for $f \in \mathbb{T}$ we have $\mathrm{e}^f = \exp(f)$, and so we use e^f as an alternative notation for $\exp(f)$. (The above identification $\ell_n^r = \exp(r\ell_{n+1})$ for real r agrees with this definition of powers.) The elements of G^{LE} are the **transmonomials** (or LE-monomials). For $f \in \mathbb{T}^\times$ we have the dominant monomial $\mathfrak{d}(f) \in G^{\mathrm{LE}}$.

For any $f \in \mathbb{T}$ and $S \subseteq G^{\mathrm{LE}}$ the subseries $f|_S := \sum_{\mathfrak{m}\in S} f_{\mathfrak{m}}\mathfrak{m} \in \mathbb{R}[[G^{\mathrm{LE}}]]$ also lies in $\mathbb{T}$: to see this, one first observes this holds for each E_m instead of $\mathbb{T}$, and thus for each $\mathbb{R}[[\ell_n^{\mathbb{R}}]]$. In particular, $\mathbb{T}$ is a truncation closed subfield of $\mathbb{R}[[G^{\mathrm{LE}}]]$.

The valuation of $\mathbb{T}$

Note that the valuation ring of $\mathbb{T}$ is $\{f \in \mathbb{T} : \operatorname{supp} f \preccurlyeq 1\}$, which is also the convex hull of $\mathbb{R}$ in the ordered field $\mathbb{T}$. We make the value group $\Gamma_{\mathbb{T}}$ of the valuation into an ordered vector space over $\mathbb{R}$ by $r\gamma := v(g^r)$ for $r \in \mathbb{R}$, $\gamma \in \Gamma_{\mathbb{T}}$ and $g \in \mathbb{T}^{>}$ with $vg = \gamma$. Setting $A := \{f \in \mathbb{T} : \operatorname{supp} f \succ 1\}$ we have the internal direct sum decomposition

$$\mathbb{T} = A \oplus \mathbb{R} \oplus o_{\mathbb{T}}$$

of $\mathbb{T}$ into $\mathbb{R}$-linear subspaces. The valuation has a concrete realization, based on the interesting fact that $\exp(A) = G^{\mathrm{LE}}$. (Proof of this fact: by Corollary A.2 we have

$$\exp(A^{\mathrm{E}}) = G^{\mathrm{E}}, \text{ where } A^{\mathrm{E}} := \left\{f \in \mathbb{R}[[x^{\mathbb{R}}]]^{\mathrm{E}} : \operatorname{supp} f \succ 1\right\} + \mathbb{R}\ell_1.$$

Use the natural analogues of this for $\mathbb{R}[[\ell_n^{\mathbb{R}}]]^{\mathrm{E}}$, $G^{\mathrm{E},n}$ in the role of $\mathbb{R}[[x^{\mathbb{R}}]]^{\mathrm{E}}$, G^{E}.) Thus the surjective map $f \mapsto -\log \mathfrak{d}(f)\colon \mathbb{T}^{\times} \to A$ can serve as the valuation of $\mathbb{T}$, with the ordered subgroup A of $\mathbb{T}$ as value group, its structure as $\mathbb{R}$-linear subspace of $\mathbb{T}$ agreeing with the earlier defined vector space structure on the value group. Note also that $\exp(A) = G^{\mathrm{LE}}$ gives that if $\mathfrak{m} \in G^{\mathrm{LE}}$ and $r \in \mathbb{R}$, then $\mathfrak{m}^r \in G^{\mathrm{LE}}$.

We saw in the construction of $\mathbb{T}_{\exp}$ that the sequence $x, \mathrm{e}^x, \mathrm{e}^{\mathrm{e}^x}, \mathrm{e}^{\mathrm{e}^{\mathrm{e}^x}}, \dots$ in $\mathbb{T}_{\exp}$ is strictly increasing and cofinal in it. Hence for each n the analogous sequence

$$\ell_n, \ell_{n-1}, \dots, \ell_1, x, \mathrm{e}^x, \mathrm{e}^{\mathrm{e}^x}, \dots$$

in $\mathbb{R}[[\ell_n^{\mathbb{R}}]]^{\mathrm{E}}$ is strictly increasing and cofinal in $\mathbb{R}[[\ell_n^{\mathbb{R}}]]^{\mathrm{E}}$. Thus the sequence

$$x, \mathrm{e}^x, \mathrm{e}^{\mathrm{e}^x}, \mathrm{e}^{\mathrm{e}^{\mathrm{e}^x}}, \dots$$

is even cofinal in $\mathbb{T}$. This argument also shows that the sequence $\ell_0, \ell_1, \ell_2, \ell_3, \dots$ in $\mathbb{T}^{>\mathbb{R}}$ is strictly decreasing. We claim that it is coinitial in $\mathbb{T}^{>\mathbb{R}}$. To see this, we note first that $[v(x)]$ is the smallest nontrivial archimedean class of the value group of $\mathbb{R}[[x^{\mathbb{R}}]]^{\mathrm{E}}$, with $[v(x)] < [v(\mathrm{e}^x)]$. Hence $[v(\ell_n)]$ is the smallest nontrivial archimedean class of the value group of $\mathbb{R}[[\ell_n^{\mathbb{R}}]]^{\mathrm{E}}$, and $[v(\ell_n)] < [v(\ell_{n-1}]$ when $n \geqslant 1$. In this way we get a sequence

$$[v(\ell_0)] > [v(\ell_1)] > [v(\ell_2)] > \cdots$$

of archimedean classes of the value group of $\mathbb{T}$, which is moreover coinitial in the set of nontrivial archimedean classes of this value group. It remains to note that we are dealing with a convex valuation on the ordered field $\mathbb{T}$.

Representing $\mathbb{T}$ as a directed union of Hahn fields

This was not done in [112] in the strong form we need for the hypothesis in Theorem 15.0.1 (in view of its Corollary 15.0.2), so we give full details here.

By construction, $\mathbb{T}$ is an increasing union of increasing unions of Hahn fields over $\mathbb{R}$. To represent it as as a directed (and also as an increasing) union of Hahn fields over $\mathbb{R}$, we recall that $G_m{\downarrow}_n \subseteq G_{m+1}{\downarrow}_{n+1}$. Also $G_m{\downarrow}_n \subseteq G_{m+1}{\downarrow}_n$, and from these two kinds of inclusions we easily obtain $G_m{\downarrow}_n \subseteq G_{2\nu}{\downarrow}_\nu$ for $\nu = \max(m,n)$; moreover, $G_{2m}{\downarrow}_m \subseteq G_{2n}{\downarrow}_n$ for all $m \leqslant n$. Thus the countable family $(G_m{\downarrow}_n)_{m,n}$ of ordered subgroups of G^{LE} is directed, with $G^{\mathrm{LE}} = \bigcup_{m,n} G_m{\downarrow}_n$. This family is in particular a healthy family (with respect to the coefficient field $\mathbb{R}$), and $\mathbb{T} = \bigcup_{m,n} \mathbb{R}[[G_m{\downarrow}_n]]$, and so the notion of a strongly additive map $\mathbb{T} \to \mathbb{T}$ makes sense. We even have an increasing sequence $(G_{2n}{\downarrow}_n)_n$ of ordered subgroups of G^{LE} with $\mathbb{T} = \bigcup_n \mathbb{R}[[G_{2n}{\downarrow}_n]]$. Note also that

$$\mathbb{R}[[G_m{\downarrow}_n]] = \mathbb{R}[[G_m]]{\downarrow}_n = E_m{\downarrow}_n.$$

The upward shift operator

A very useful automorphism of $\mathbb{T}$ is the **upward shift**, informally to be thought of as $f(x) \mapsto f(\mathrm{e}^x)$. Formally it is the unique strongly additive $\mathbb{R}$-linear automorphism $f \mapsto f{\uparrow}$ of the exponential ordered field $\mathbb{T}$ that sends x to e^x. (To construct it, take for each n the strongly additive $\mathbb{R}$-linear embedding $\mathbb{R}[[\ell_n^{\mathbb{R}}]]^{\mathrm{E}} \to \mathbb{R}[[\ell_n^{\mathbb{R}}]]^{\mathrm{E}}$ of exponential ordered fields that sends ℓ_n^r to $\exp(r\ell_n)$ for each $r \in \mathbb{R}$, and show that these maps have a common extension to a map $\mathbb{T} \to \mathbb{T}$; this common extension is the upward shift.) It is easy to check that

$$G^{\mathrm{E}}{\uparrow} \subseteq G^{\mathrm{E}} \subseteq G^{\mathrm{E},1}{\uparrow},$$

so the upward shift maps G^{LE} onto itself. The inverse of the upward shift operator is the downward shift operator $f \mapsto f{\downarrow}$. The nth iterate of the upward (respectively, downward) shift operator is $f \mapsto f{\uparrow^n}$, respectively, $f \mapsto f{\downarrow_n}$. Thus $x{\downarrow_n} = \ell_n$. If $f \in \mathbb{R}[[x^{\mathbb{R}}]]^{\mathrm{E}}$, then this $f{\downarrow_n}$ equals $f{\downarrow_n}$ as defined in the subsection "From $\mathbb{T}_{\exp}$ to $\mathbb{T}$." Thus $\big(\mathbb{R}[[\ell_n^{\mathbb{R}}]]^{\mathrm{E}}\big){\uparrow^n} = \mathbb{T}_{\exp}$.

LEMMA A.4. *If* $0 < g \in E_m{\downarrow_n}$, *then* $\log g \in E_{m+1}{\downarrow_{n+1}}$.

PROOF. Applying ${\uparrow^n}$, this reduces to the case $n = 0$. So let $0 < g \in E_m$; we have to show that $\log g \in E_{m+1}{\downarrow}$. By (iv) of Lemma A.1 we have $g = cx^r \exp(a)(1+\delta)$ with $c \in \mathbb{R}^{>}$, $r \in \mathbb{R}$, $a \in A_0 + \cdots + A_{m-1}$, and $\delta \prec 1$ in E_m. Then $\log g = \log c + r\ell_1 + a + \log(1+\delta)$; now use that $a, \log(1+\delta) \in E_m \subseteq E_{m+1}{\downarrow}$. □

Differentiating and integrating transseries

In the rest of this appendix we also assume familiarity with Section 10.5. It would be nice if transseries $f = f(x)$ could be differentiated so that the following rules hold:

(D1) $r' = 0$ for all $r \in \mathbb{R}$, and $x' = 1$;

(D2) $(\exp f)' = f' \exp f$ for all $f \in \mathbb{T}$ (and thus $(\log f)' = f^{\dagger}$ for all $f \in \mathbb{T}^{>}$);

(D3) $f \mapsto f' \colon \mathbb{T} \to \mathbb{T}$ is strongly additive.

The inductive construction of $\mathbb{T}$ makes it plausible that this can be done uniquely:

PROPOSITION A.5. *There is a unique derivation on* $\mathbb{T}$ *satisfying* (D1), (D2), (D3).

Section 3 of [112] constructs a derivation on $\mathbb{T}$, denoted by ∂ below, satisfying (D1), (D2), and (D3). (The paper cited denotes ∂ by d/dx, since ∂ is thought of as differentiation with respect to x.) These properties clearly determine ∂ uniquely, and below we consider $\mathbb{T}$ as a differential field whose derivation is ∂.

The map $\partial \colon \mathbb{T} \to \mathbb{T}$ is healthy. This is because $\partial(E_m{\downarrow_n}) \subseteq E_m{\downarrow_n}$ for all $m \geqslant n$. To prove these inclusions, note first that by the construction of ∂ each E_m is a differential subfield of $\mathbb{T}$ (so $\mathbb{T}_{\exp}$ is as well). Next, with $\exp_n = n$th iterate of exp,

$$\exp_n(x)' = \prod_{i=1}^{n} \exp_i(x), \qquad (f{\uparrow^n})' = f'{\uparrow^n} \exp_n(x)' \quad (f \in \mathbb{T}),$$

by [112]. In view of $\exp_n(x) \in E_m$ for $m \geqslant n$, this yields the desired inclusion.

A consequence of the strong additivity of the derivation is that for an ordinary power series $F \in \mathbb{R}[[t_1, \dots, t_n]]$ and $\varepsilon = (\varepsilon_1, \dots, \varepsilon_n) \in \sigma_{\mathbb{T}}^n$ we have

$$F(\varepsilon)' = \sum_{i=1}^{n} \frac{\partial F}{\partial t_i}(\varepsilon) \cdot \varepsilon_i'.$$

Here are some basic properties of $\mathbb{T}$ as an ordered valued differential field:

PROPOSITION A.6. *$C_{\mathbb{T}} = \mathbb{R}$, and $\mathbb{T}$ is a Liouville closed H-field.*

Theorem 3.9(4) in [112] gives $C_{\mathbb{T}} = \mathbb{R}$, which together with Proposition 4.3 in that paper makes $\mathbb{T}$ an H-field. The surjectivity of $\partial \colon \mathbb{T} \to \mathbb{T}$ is [112, Theorem 5.6]. Hence $\mathbb{T}$ is Liouville closed: given $a \in \mathbb{T}$, take $b \in \mathbb{T}$ with $b' = a$; then $(\mathrm{e}^b)^\dagger = a$.

The E_m and thus the $E_{2n}{\downarrow}_n$ are spherically complete H-subfields of $\mathbb{T}$. Moreover, $[v(x)]$ is the smallest element of $[\Gamma_{E_m}^{\neq}]$, so E_m is grounded with $\max \Psi_{E_m} = v(x^\dagger) = v(1/x)$. Likewise, $[v(\ell_n)]$ is the smallest elements of $[(E_m{\downarrow}_n)^{\neq}]$, so $E_{2n}{\downarrow}_n$ is grounded with $\max \Psi_{E_{2n}{\downarrow}_n} = v(\ell_n^\dagger)$. Thus $\mathbb{T} = \bigcup_n E_{2n}{\downarrow}_n$ represents $\mathbb{T}$ as an increasing union of spherically complete grounded H-subfields.

These facts also lead to an alternative proof that $\partial \colon \mathbb{T} \to \mathbb{T}$ is surjective: using Corollary 15.2.4 it suffices to note that $\ell_n^\dagger = \ell_{n+1}' \in \partial(E_{n+1}{\downarrow}_{n+1})$.

Composition

Let f range over $\mathbb{T}$ and g, h over $\mathbb{T}^{>\mathbb{R}}$. With this convention:

PROPOSITION A.7. *There is a unique operation*

$$(f, g) \mapsto f \circ g \; : \; \mathbb{T} \times \mathbb{T}^{>\mathbb{R}} \to \mathbb{T}$$

such that for each g the following conditions are satisfied:

(i) $x \circ g = g$;

(ii) $f \mapsto f \circ g \colon \mathbb{T} \to \mathbb{T}$ *is an $\mathbb{R}$-linear embedding of exponential ordered fields;*

(iii) $f \mapsto f \circ g \colon \mathbb{T} \to \mathbb{T}$ *is strongly additive.*

This is Theorem 6.2 in [112], except that the uniqueness is stated there with further conditions than just (i), (ii), (iii) above. It is rather easy to check, however, that there can be at most one operation as in the above proposition. Thus the other conditions of that Theorem 6.2, namely (iv), (v), (vi) below, are consequences:

(iv) $f \circ x = f$;

(v) $f{\uparrow} = f \circ \mathrm{e}^x$;

(vi) $(f \circ g) \circ h = f \circ (g \circ h)$;

(vii) $(f \circ g)' = (f' \circ g) \cdot g'$.

The last item here, the Chain Rule, is part of Proposition 6.3 in [112]. Note also that $\mathrm{e}^x \circ g = \mathrm{e}^g$, $\ell_1 \circ g = \log g$ and $f\downarrow = f \circ \ell_1$. If f and g are thought of as series $f(x)$ and $g(x)$, then $f \circ g$ can be thought of as $f\big(g(x)\big)$. For example, in Section 5.7 the identities $f\uparrow = f(\mathrm{e}^x)$ and $f\downarrow = f(\log x)$ use this suggestive notation.

PROPOSITION A.8. *Given g the map $f \mapsto f \circ g$ is healthy. More precisely, assume that $f \in E_m\downarrow_n$ and $g \in E_p\downarrow_q$, $p, q \in \mathbb{N}$. Then $f \circ g \in E_{m+n+p}\downarrow_{n+q}$.*

PROOF. We have $f \circ g = f\uparrow^n \circ \log^n g$, with $f\uparrow^n \in E_m$, and $\log^n g \in E_{p+n}\downarrow_{q+n}$ by Lemma A.4. It remains to use subsection 6.8 in [112]. □

The subfield $\mathbb{T}_{\log}$ of logarithmic transseries

This subfield is a particularly transparent part of $\mathbb{T}$, more so than $\mathbb{T}_{\exp}$. It also has much stronger algebraic closure properties than $\mathbb{T}_{\exp}$. Let $\mathfrak{L}_n := \ell_0^{\mathbb{R}} \cdots \ell_n^{\mathbb{R}}$ be the subgroup of G^{LE} generated by the real powers of the ℓ_i for $i = 0, \dots, n$. Then $\mathfrak{L}_n \subseteq G_n\downarrow_n$, and $\mathbb{R}[[\mathfrak{L}_n]]$ is an H-subfield of $\mathbb{R}[[G_n\downarrow_n]] = E_n\downarrow_n$. The ordered group inclusions

$$\mathfrak{L}_0 \subseteq \mathfrak{L}_1 \subseteq \mathfrak{L}_2 \subseteq \cdots \subseteq \mathfrak{L} := \bigcup_n \mathfrak{L}_n \subseteq G^{\mathrm{LE}}$$

induce H-field inclusions

$$\mathbb{R}[[x^{\mathbb{R}}]] = \mathbb{R}[[\mathfrak{L}_0]] \subseteq \mathbb{R}[[\mathfrak{L}_1]] \subseteq \mathbb{R}[[\mathfrak{L}_2]] \subseteq \cdots \subseteq \mathbb{T},$$

and we set

$$\mathbb{T}_{\log} := \bigcup_n \mathbb{R}[[\mathfrak{L}_n]].$$

So $\mathbb{T}_{\log}$ is an H-subfield of $\mathbb{T}$ with $G^{\mathrm{LE}} \cap \mathbb{T}_{\log} = \mathfrak{L}$. It is routine to check that for $f \in \mathbb{T}_{\log}^{>}$ we have $f^r \in \mathbb{T}_{\log}$ for all $r \in \mathbb{R}$, and $\log f \in \mathbb{T}_{\log}$.

Notes and comments

Analytic functions of infinitesimal arguments in a Hahn field were considered by Neumann [302, pp. 206–210]. For more on strongly additive maps between Hahn fields, see [191].

The exponential field of transseries was introduced by Dahn and Göring [96] as a natural candidate for an elementary extension of the exponential ordered field of real numbers, and independently in Écalle's work [120] on the Dulac Conjecture, where the derivation is also prominent. In connection with the construction of $\mathbb{T}$ as a directed union of Hahn fields, we note that by [237] no nontrivial Hahn field over $\mathbb{R}$ can be made into a logarithmic-exponential ordered field.

In this appendix we have followed closely [112], but the reader who consults that paper will note that it allows any logarithmic-exponential ordered field $\boldsymbol{k}$ as coefficient field in constructing $\boldsymbol{k}((x^{-1}))^{\mathrm{LE}}$ while here we stick to $\boldsymbol{k} = \mathbb{R}$. Apart from that, the exponential and logarithmic maps E and L there are denoted here by exp and log, the fields K_n and L_n there are our E_n and $\mathbb{R}[[x^{\mathbb{R}}]]^{\mathrm{E},n}$, and $G_{m,n}$ and $L_{m,n}$ there are our $G_n{\downarrow}_m$ and $E_n{\downarrow}_m$; the map Φ there is our upward shift operator $\uparrow$.

We wish to say a few words on the grid-based setting, which has been mentioned several times in earlier *Notes and comments.* This appendix rests on the notion of a *well-based* subset of an ordered abelian (multiplicative) group $\mathfrak{M}$. We now replace *well-based* by a much stronger restriction, namely *grid-based.* A **grid-based** subset of $\mathfrak{M}$ is one that is contained in $\mathfrak{m}\,\mathfrak{n}_1^{\mathbb{N}}\cdots\mathfrak{n}_n^{\mathbb{N}}$ for some $\mathfrak{m}\in\mathfrak{M}$ and $\mathfrak{n}_1,\dots,\mathfrak{n}_n\in\mathfrak{M}^{\prec 1}$. Let C be a field. Set

$$C[\![\mathfrak{M}]\!] := \big\{f\in C[[\mathfrak{M}]] : \operatorname{supp} f \text{ is grid-based}\big\}.$$

Then $C[\![\mathfrak{M}]\!]$ is a subfield of $C[[\mathfrak{M}]]$. For $\mathfrak{M}=x^{\mathbb{Q}}$ we obtain the field $C[\![x^{\mathbb{Q}}]\!]=\mathrm{P}(C)$ of Puiseux series over C (Example 3.3.23).

Except for the statements mentioning spherical completeness, the appendix goes through if everywhere *well-based* is replaced by *grid-based*, with the Hahn fields $C[[\mathfrak{M}]]$ accordingly replaced by their grid-based versions $C[\![\mathfrak{M}]\!]$. In particular, the starting point $\mathbb{R}[[x^{\mathbb{R}}]]$ in the construction of $\mathbb{T}$ is replaced by the subfield $\mathbb{R}[\![x^{\mathbb{R}}]\!]$ of $\mathbb{R}[[x^{\mathbb{R}}]]$. This leads to a logarithmic-exponential ordered subfield $\mathbb{T}_{\mathrm{g}}$ of $\mathbb{T}$ which is also an H-subfield of $\mathbb{T}$ and shares key algebraic closure properties with $\mathbb{T}$ like being Liouville closed. Every $f\in\mathbb{T}_{\mathrm{g}}$ has grid-based support.

Grid-based transseries were introduced in [120]. There $\mathbb{T}_{\mathrm{g}}$ is denoted by $\mathbb{R}[[[x]]]$, in [194] it is written as $\mathbb{T}$, and in [112] as $\mathbb{R}(\!(t)\!)^{\mathrm{LE,ft}}$; in this last paper *grid-based* was called *of finite type.*

Logarithmic transseries occur in the work of Loeb and Rota [264, 265] in connection with difference equations.

As already noted, our field $\mathbb{T}$ of transseries does not contain the well-based series $x+\log x+\log\log x+\cdots$. Van der Hoeven's thesis [190] shows how to go beyond $\mathbb{T}$ to include such series by building a strictly increasing ordinal sequence of fields of transseries, alternately closing off under well-based summation and exponentiation, and defining differentiation and composition in this setting. In Schmeling's thesis [388] this is put in an axiomatic framework for Hahn fields with a logarithm map, leading to fields T of transseries with additional structure, such as an *iterator* $\log_\omega\colon T^{>\mathbb{R}}\to T^{>\mathbb{R}}$ of the logarithm satisfying the identity

$$\log_\omega\log y = (\log_\omega y)-1 \qquad (y>\mathbb{R}).$$

This axiomatic setting for fields of transseries plays a role in the recent work by Berarducci and Mantova [43]. We plan to consider these extensions in more detail in our second volume in the light of the results of the present volume.

Appendix B

Basic Model Theory

This appendix is written for readers unfamiliar with model theory. It provides a rigorous treatment, with examples, of the model-theoretic tools used in our work: back-and-forth, compactness, types, model completeness, quantifier elimination, and (in a different vein) NIP. We adopt a many-sorted setting from the outset. In the first four sections we specify the notion of a *model-theoretic structure* and build up a formalism for handling these structures efficiently in the rest of the appendix.

B.1 STRUCTURES AND THEIR DEFINABLE SETS

The notion of structure we introduce in this section is very general and includes not only familiar (one-sorted) algebraic objects such as groups and rings, but also two-sorted structures such as group actions and topological spaces.

We associate to any structure its category of *definable sets and definable maps.* In the case of a field, the definable sets include the solution sets of finite systems of polynomial equations, but as we shall see, the notion of definable set goes beyond this by allowing boolean operations and projections.

Model-theoretic structures

A (model-theoretic) **structure** consists of:

(S1) a family $(M_s)_{s\in S}$ of nonempty sets M_s,

(S2) a family $(R_i)_{i\in I}$ of relations $R_i \subseteq M_{s_1} \times \cdots \times M_{s_m}$ on these sets, with the tuple $(s_1, \ldots, s_m) \in S^m$ determined by the index $i \in I$, and

(S3) a family $(f_j)_{j\in J}$ of functions $f_j \colon M_{s_1} \times \cdots \times M_{s_n} \to M_s$ with the tuple $(s_1, \ldots, s_n, s) \in S^{n+1}$ determined by the index $j \in J$.

Notation:

$$\boldsymbol{M} = \big((M_s); (R_i), (f_j)\big).$$

The elements of the index set S are called **sorts**, and M_s is the **underlying set** of sort s of $\boldsymbol{M}$. Elements of M_s are also called *elements of $\boldsymbol{M}$ of sort s.* The R_i and f_j are called the **primitives** of $\boldsymbol{M}$. For $m = 0$ the set S^m has the empty tuple as its only element, and the product $M_{s_1} \times \cdots \times M_{s_m}$ is then a singleton, that is, a one-element set. For $n = 0$ in (S3), the function f_j is then identified with its unique value in M_s

and is called a **constant.** The family $M = (M_s)$ is said to **underlie** the structure $\boldsymbol{M}$, or, abusing language, to be the **underlying set** of $\boldsymbol{M}$.

We call $\boldsymbol{M}$ as above an **S-sorted structure**. If S is finite of size n, then we also speak of an **n-sorted structure**. Many texts only consider one-sorted structures, so S doesn't need to be mentioned. But the extra generality adds useful flexibility and is natural, since mathematical structures are often many-sorted to begin with. In fact, virtually anything that mathematicians consider as a structure can be viewed as a structure in the above sense. We only mention a few examples here; more are discussed in B.2.2 below.

In specifying one-sorted structures we often take the liberty of using the same capital letter for the underlying set of the structure as for the structure itself.

EXAMPLES B.1.1.

(1) Any (additively written) abelian group M is construed naturally as a one-sorted structure $(M; 0, -, +)$ with the constant $0 \in M$ and the functions
$$a \mapsto -a\colon M \to M, \quad (a, b) \mapsto a + b\colon M \times M \to M.$$

(2) Every ring R is viewed naturally as a one-sorted structure $(R; 0, 1, -, +, \cdot)$ with constants 0, 1 and functions
$$r \mapsto -r\colon R \to R, \ (r, s) \mapsto r + s\colon R \times R \to R, \ (r, s) \mapsto r \cdot s\colon R \times R \to R.$$

(3) Let R be a ring and M an R-module. Then M can be viewed as a one-sorted structure, equipped with the functions in (1) and for each $r \in R$ the function $a \mapsto ra\colon M \to M$.

(4) Sometimes it is convenient to construe a module M over a ring R as a *two-sorted* structure, one sort with underlying set R and the same functions as in (2), the other sort with underlying set M and the same functions as in (1), as well as with a function $(r, a) \mapsto ra\colon R \times M \to M$ for the action of R on M.

(5) A nonempty ordered set in the sense of Section 2.1 is the same thing as a one-sorted structure $(M; \leqslant)$ where the binary relation $\leqslant$ on M is reflexive, antisymmetric, transitive, and total.

(6) An incidence geometry is a two-sorted structure consisting of a nonempty set P whose elements are called points, a nonempty set L whose elements are called lines, and a relation $R \subseteq P \times L$ between points and lines. For example, a nonempty topological space can be viewed as an incidence geometry, taking the underlying set of the space for P, the collection of open sets for L, and the membership relation between points and open sets for R.

Definable sets

Let $\boldsymbol{M} = \big(M; (R_i), (f_j)\big)$ be an S-sorted structure as above, where $M = (M_s)$. The main role of the primitives of $\boldsymbol{M}$ is to generate the *definable sets* of $\boldsymbol{M}$. Let $\boldsymbol{s} = s_1 \dots s_m$ and $\boldsymbol{t} = t_1 \dots t_n$ be elements of S^*, that is, words on the alphabet S. Then $\boldsymbol{st} = s_1 \dots s_m t_1 \dots t_n \in S^*$ is the concatenation of $\boldsymbol{s}$ and $\boldsymbol{t}$. We set $M_{\boldsymbol{s}} := M_{s_1} \times \cdots \times M_{s_m}$, and identify $M_{\boldsymbol{s}} \times M_{\boldsymbol{t}}$ with $M_{\boldsymbol{st}}$ in the obvious way.

There is no need to require $M_s \cap M_t = \emptyset$ for distinct $s, t \in S$. (This would be unnatural in Example (4) above, since R is naturally an R-module.) Instead we impose the convention that in referring to an element $a \in M_{\boldsymbol{s}}$, this is short for a reference to an ordered pair $(a, \boldsymbol{s})$ such that $a \in M_{\boldsymbol{s}}$. A similar convention is in force when we refer to a set $X \subseteq M_{\boldsymbol{s}}$. Given a map $f \colon X \to Y$, where $X \subseteq M_{\boldsymbol{s}}$, $Y \subseteq M_{\boldsymbol{t}}$, we let $\Gamma(f) \subseteq M_{\boldsymbol{st}}$ be the graph of f.

DEFINITION B.1.2. The 0-**definable** (or absolutely definable) sets of $\boldsymbol{M}$ are the relations $X \subseteq M_{\boldsymbol{s}}$ obtained recursively as follows:

(D1) the relations $R_i \subseteq M_{s_1 \dots s_m}$ and the graphs $\Gamma(f_j) \subseteq M_{s_1 \dots s_n s}$ are 0-definable;

(D2) if $X, Y \subseteq M_{\boldsymbol{s}}$ are 0-definable, then so are $X \cup Y \subseteq M_{\boldsymbol{s}}$ and $(M_{\boldsymbol{s}} \setminus X) \subseteq M_{\boldsymbol{s}}$;

(D3) if $X \subseteq M_{\boldsymbol{s}}$ and $Y \subseteq M_{\boldsymbol{t}}$ are 0-definable, then so is $X \times Y \subseteq M_{\boldsymbol{st}}$;

(D4) for all distinct $i, j \in \{1, \dots, m\}$ with $s_i = s_j$, the diagonal

$$\Delta_{ij} := \big\{(x_1, \dots, x_m) \in M_{\boldsymbol{s}} : x_i = x_j\big\}$$

is 0-definable;

(D5) if $X \subseteq M_{\boldsymbol{st}}$ is 0-definable, then so is $\pi(X) \subseteq M_{\boldsymbol{s}}$, where $\pi \colon M_{\boldsymbol{st}} \to M_{\boldsymbol{s}}$ is the obvious projection map.

We extend this notion to A-*definability*. Here A is a so-called **parameter set** in $\boldsymbol{M}$, that is, a family $A = (A_s)$ with $A_s \subseteq M_s$ for each s (shorthand: $A \subseteq M$). Then the structure $\boldsymbol{M}_A$ is obtained from $\boldsymbol{M}$ by adding for each $a \in A_s$ the constant $a \in M_s$ as a primitive. The A-**definable sets** of $\boldsymbol{M}$ are just the 0-definable sets of $\boldsymbol{M}_A$. The parameter set A with $A_s = \emptyset$ for each s is also denoted by 0, so $\boldsymbol{M}_0 = \boldsymbol{M}$ and the terminology 0-*definable* is unambiguous. Instead of M-*definable* we just write *definable*; so all finite sets $X \subseteq M_{\boldsymbol{s}}$ are definable.

For any family $(A_i)_{i \in I}$ of parameter sets in $\boldsymbol{M}$ we let $\bigcup_i A_i$ be the parameter set A in $\boldsymbol{M}$ such that $A_s = \bigcup_i A_{i,s}$ for every $s \in S$.

Describing definable sets

For constructing new definable sets from old ones, it is convenient to systematize the correspondence between sets and the conditions used to define them. Let x be a variable ranging over a set X, and let $\phi(x)$, $\psi(x)$ be formulas ("conditions" on x) defining the subsets

$$\Phi := \big\{x \in X : \phi(x) \text{ holds}\big\} \quad \text{and} \quad \Psi := \big\{x \in X : \psi(x) \text{ holds}\big\}$$

of X. (We will make this precise in Section B.4 below.) Various logical operations on formulas correspond to operations on the sets that these formulas define:

(C1) $\neg\phi(x)$ defines the complement $X \setminus \Phi$;

(C2) $\phi(x) \vee \psi(x)$ defines the union $\Phi \cup \Psi$;

(C3) $\phi(x) \wedge \psi(x)$, also written as $\phi(x) \,\&\, \psi(x)$, defines the intersection $\Phi \cap \Psi$; thus $\neg\big(\phi(x) \vee \psi(x)\big)$ and $\neg\phi(x) \wedge \neg\psi(x)$ define the same subset of X.

Moreover, if y is a variable ranging over a set Y and $\theta(x, y)$ is a formula defining the subset Θ of $X \times Y$, then

(C4) $\exists y\, \theta(x, y)$ defines $\pi(\Theta) \subseteq X$ where $\pi\colon X \times Y \to X$ is the natural projection;

(C5) $\forall y\, \theta(x, y)$ defines the set $\{x \in X : \{x\} \times Y \subseteq \Theta\}$; hence the formulas $\neg\exists y\, \theta(x, y)$ and $\forall y\, \neg\theta(x, y)$ define the same subset of X.

Thus, with $X = M_s$ and $Y = M_t$, if the sets Φ, Ψ defined by $\phi(x)$, $\psi(x)$ are A-definable, then so are the sets defined by $\neg\phi(x)$, $\phi(x) \vee \psi(x)$, $\phi(x) \wedge \psi(x)$, and if Θ is A-definable, then so are the sets defined by $\exists y\, \theta(x, y)$ and $\forall y\, \theta(x, y)$. The advantage of the logical formalism is that it is often more suggestive and transparent than traditional set-theoretic notation, in particular when quantifiers are involved:

EXAMPLE. Let $A \subseteq \mathbb{R}$, and let $\boldsymbol{M} = (\mathbb{R}; \dots)$ be a one-sorted structure with the usual ordering $\leqslant$ of the real line among the A-definable subsets of $\mathbb{R}^2$. Equip $\mathbb{R}^m$ with its usual product topology. Suppose $X \subseteq \mathbb{R}^m$ is A-definable (in $\boldsymbol{M}$). Then the interior $\operatorname{int}(X)$ of X in the ambient space $\mathbb{R}^m$ is A-definable. To see this, note that, with $\phi \to \psi$ short for $(\neg\phi) \vee \psi$, we have the equivalence

$$x \in \operatorname{int}(X) \iff \exists u\, \exists v\, \big[(u < x < v) \,\&\, \forall y\, (u < y < v \to y \in X)\big]$$

where u, v, x, y range over $\mathbb{R}^m$ and $u < x < v$ is short for

$$u_1 < x_1 < v_1 \,\&\, \cdots \,\&\, u_m < x_m < v_m.$$

Likewise, if $f\colon X \to \mathbb{R}$ is A-definable, that is, its graph $\Gamma(f)$ is A-definable, then the set $\{x \in X : f \text{ is continuous at } x\}$ is easily seen to be A-definable.

In general the 0-definable relations of a structure cannot be described in a way that is significantly more explicit than the recursive definition in B.1.2. For example, this is the case for the ring $\mathbb{Z}$ viewed as a one-sorted structure as in Example B.1.1(3); see [411, §7.5] for details. But in some cases, a more explicit description does exist. One example is the field $\mathbb{C}$ of complex numbers; here we construe $\mathbb{C}$ as a one-sorted structure as in B.1.1(2). (Including also as a primitive, say, $x \mapsto x^{-1}\colon \mathbb{C}^\times \to \mathbb{C}$, extended to a total function $\mathbb{C} \to \mathbb{C}$ by declaring $0^{-1} := 0$, wouldn't add to the 0-definable relations of $\mathbb{C}$.) Then by the Chevalley-Tarski Constructibility Theorem from Section B.12 the 0-definable subsets of $\mathbb{C}^n$ are just the finite unions of sets

$$\{a \in \mathbb{C}^n : P_1(a) = \cdots = P_m(a) = 0,\ Q(a) \neq 0\}$$

with $P_1, \dots, P_m, Q \in \mathbb{Q}[X_1, \dots, X_n]$. For a subfield A of $\mathbb{C}$ as a parameter set we get the same description of the A-definable subsets of $\mathbb{C}^n$ but now the polynomials have their coefficients in A. The above goes through for any algebraically closed field K instead of $\mathbb{C}$, with the prime field of K in place of $\mathbb{Q}$. So in this case the notion of "A-definable" is akin to Weil's notion of an algebraic variety defined over A. (Model-theoretic notions are often similar to foundational items in Weil's algebraic geometry.) Another prominent example, more relevant to our work, where a concise description of the definable sets is available is the field $\mathbb{R}$ of real numbers. Here, by the Tarski-Seidenberg Theorem in Section B.12, the 0-definable subsets of $\mathbb{R}^n$ are the $\mathbb{Q}$-*semialgebraic* subsets of $\mathbb{R}^n$, that is, finite unions of sets

$$\{a \in \mathbb{R}^n : P(a) = 0,\ Q_1(a) > 0, \dots, Q_m(a) > 0\}$$

with $P, Q_1, \dots, Q_m \in \mathbb{Q}[X_1, \dots, X_n]$. Note that the usual ordering relation on $\mathbb{R}$ is 0-definable in the field $\mathbb{R}$:

$$x \leqslant y \iff \exists z\,[z^2 = y - x].$$

In this appendix we develop some tools which allow us to identify benign structures, like $\mathbb{C}$ and $\mathbb{R}$, whose definable sets allow such explicit descriptions.

Conventions and notations

Let $M = (M_s)$ and $M' = (M'_s)$ be families of sets indexed by a set S. The **size** $|M|$ of M is defined to be the sum $\sum |M_s|$ of the cardinalities of the sets M_s, in other words, it is the cardinality of the disjoint union $\bigcup_s M_s \times \{s\}$. (We also use *size* as a synonym for *cardinality*.) Define $M \subseteq M' :\Leftrightarrow M_s \subseteq M'_s$ for all s. A map $h\colon M \to M'$ is a family $h = (h_s)$ of maps $h_s\colon M_s \to M'_s$. Let $h = (h_s)\colon M \to M'$ be such a map. We say that h is *injective* if each h_s is injective, and similarly with *surjective* or *bijective* in place of *injective*. Given $\boldsymbol{s} = s_1 \dots s_m \in S^*$ and $a = (a_1, \dots, a_m) \in M_{\boldsymbol{s}}$ we put $ha = h_{\boldsymbol{s}}a := (h_{s_1}a_1, \dots, h_{s_m}a_m) \in M'_{\boldsymbol{s}}$. For $A = (A_s) \subseteq M$ we put $h(A) := \big(h(A_s)\big) \subseteq M'$. Given also $M'' := (M''_s)$ and a map $h'\colon M' \to M''$, we let $h' \circ h$ denote the map $M \to M''$ given by $(h' \circ h)_s = h'_s \circ h_s$ for all s.

Notes and comments

The idea of "structure" emerged in the 19th century in algebra (Dedekind) and the foundations of mathematics [182, 393]. It led to the structural point of view in algebra, later extended by Bourbaki to other parts of mathematics; see [93]. In the early literature in mathematical logic, structures were called systems until the current terminology came into widespread use in the 1950s (see for example [345]), perhaps under Bourbaki's influence.

The underlying set M_s of sort s of a structure $\boldsymbol{M}$ is sometimes called the *universe* of sort s of $\boldsymbol{M}$, after de Morgan [296]. The notion of definable set can be traced back to Weyl [464], but became more familiar after Kuratowski and Tarski [246] spelled out the above correspondence between logical connectives and set-theoretic operations for routine use in *descriptive set theory*.

B.2 LANGUAGES

In this section we formalize the idea of *structures of the same kind* by defining the category of *$\mathcal{L}$-structures* for a given (model-theoretic) *language* $\mathcal{L}$. We also discuss various constructions with $\mathcal{L}$-structures: products, direct limits.

Languages

A **language** $\mathcal{L}$ is a triple $(S, \mathcal{L}^{\mathrm{r}}, \mathcal{L}^{\mathrm{f}})$ consisting of

(L1) a set S whose elements are the **sorts** of $\mathcal{L}$,

(L2) a set $\mathcal{L}^{\mathrm{r}}$ whose elements are the **relation symbols** of $\mathcal{L}$, and

(L3) a set $\mathcal{L}^{\mathrm{f}}$ whose elements are the **function symbols** of $\mathcal{L}$,

where $\mathcal{L}^{\mathrm{r}}$ and $\mathcal{L}^{\mathrm{f}}$ are disjoint, each $R \in \mathcal{L}^{\mathrm{r}}$ is equipped with a word $s_1 \dots s_m \in S^*$, called its **sort**, and each $f \in \mathcal{L}^{\mathrm{f}}$ is equipped with a **sort** $s_1 \dots s_n s \in S^*$. The elements of $\mathcal{L}^{\mathrm{r}} \cup \mathcal{L}^{\mathrm{f}}$ are called **nonlogical symbols** of $\mathcal{L}$. We also call $\mathcal{L}$ an **S-sorted language**. It is customary to present a language as a disjoint union $\mathcal{L} = \mathcal{L}^{\mathrm{r}} \cup \mathcal{L}^{\mathrm{f}}$ of its sets of relation and function symbols, while separately specifying the set S of sorts as well as the kind (relation symbol or function symbol) and sort of each nonlogical symbol of $\mathcal{L}$. If $R \in \mathcal{L}^{\mathrm{r}}$ has sort $s_1 \dots s_m \in S^*$, then we also call R an **m-ary** relation symbol of $\mathcal{L}$; similarly a function symbol of $\mathcal{L}$ of sort $s_1 \dots s_n s$ is called an n-ary function symbol of $\mathcal{L}$. Instead of 0-*ary*, 1-*ary*, 2-*ary* we say *nullary*, *unary*, *binary*, respectively. A **constant symbol** is a nullary function symbol. An m-ary relation symbol of sort $s \dots s$ is also said to be of sort s; an n-ary function symbol of sort $s \dots ss$ is said to be of sort s.

A language $\mathcal{L}' = (S', \mathcal{L}'^{\mathrm{r}}, \mathcal{L}'^{\mathrm{f}})$ is an **extension** of a language $\mathcal{L} = (S, \mathcal{L}^{\mathrm{r}}, \mathcal{L}^{\mathrm{f}})$ (and $\mathcal{L}$ is a **sublanguage** of $\mathcal{L}'$) if $S \subseteq S'$, $\mathcal{L}^{\mathrm{r}} \subseteq \mathcal{L}'^{\mathrm{r}}$, $\mathcal{L}^{\mathrm{f}} \subseteq \mathcal{L}'^{\mathrm{f}}$, and each nonlogical symbol of $\mathcal{L}$ has the same sort in $\mathcal{L}'$ as it has in $\mathcal{L}$; notation: $\mathcal{L} \subseteq \mathcal{L}'$.

EXAMPLES B.2.1.

(1) The one-sorted language $\mathcal{L}_{\mathrm{G}} = \{1, {}^{-1}, \cdot\}$ of groups has constant symbol 1, unary function symbol ${}^{-1}$, and binary function symbol $\cdot$.

(2) The one-sorted language $\mathcal{L}_{\mathrm{A}} = \{0, -, +\}$ of (additive) abelian groups has constant symbol 0, unary function symbol $-$, and binary function symbol $+$.

(3) The one-sorted language $\mathcal{L}_{\mathrm{R}} = \{0, 1, -, +, \cdot\}$ of rings is the extension of $\mathcal{L}_{\mathrm{A}}$ by a constant symbol 1 and a binary function symbol $\cdot$.

(4) The one-sorted language $\mathcal{L}_{\mathrm{O}} = \{\leqslant\}$ of ordered sets has just one binary relation symbol $\leqslant$.

(5) Combining (2) and (4) yields the one-sorted language $\mathcal{L}_{\mathrm{OA}} = \{\leqslant, 0, -, +\}$ of ordered abelian groups. Combining (3) and (4) yields the one-sorted language $\mathcal{L}_{\mathrm{OR}} = \{\leqslant, 0, 1, -, +, \cdot\}$ of ordered rings.

(6) Let R be a ring. The language $\mathcal{L}_{R\text{-mod}} = \mathcal{L}_{\mathrm{A}} \cup \{\lambda_r : r \in R\}$ of R-modules is the one-sorted language which extends $\mathcal{L}_{\mathrm{A}}$ by unary function symbols λ_r, one for each $r \in R$.

(7) The language

$$\mathcal{L}_{\mathrm{Mod}} = \{0_{\mathrm{R}}, 1_{\mathrm{R}}, -_{\mathrm{R}}, +_{\mathrm{R}}, \cdot_{\mathrm{R}}, 0_{\mathrm{M}}, -_{\mathrm{M}}, +_{\mathrm{M}}, \lambda\}$$

of modules (with unspecified scalar ring) is a two-sorted language with sorts R and M. The symbols 0_{R}, 1_{R}, 0_{M} are constant symbols, $-_{\mathrm{R}}$, $-_{\mathrm{M}}$ are unary function symbols, and $+_{\mathrm{R}}$, $\cdot_{\mathrm{R}}$, $+_{\mathrm{M}}$ are binary function symbols. Here the symbols indexed with a subscript R are of sort R, and similarly for M. The symbol λ is a function symbol of sort R M M.

In the rest of this appendix $\mathcal{L}$ is a language with S as its set of sorts, unless specified otherwise. We let s (possibly subscripted) range over S and $\boldsymbol{s} = s_1 \dots s_m$ over S^*. The **size** of $\mathcal{L}$ is the cardinal

$$|\mathcal{L}| := \max\{\aleph_0, |S|, |\mathcal{L}^{\mathrm{r}} \cup \mathcal{L}^{\mathrm{f}}|\},$$

and we say that $\mathcal{L}$ is **countable** if $|\mathcal{L}| = \aleph_0$, that is, S, $\mathcal{L}^{\mathrm{r}}$, and $\mathcal{L}^{\mathrm{f}}$ are countable.

$\mathcal{L}$-structures

An **$\mathcal{L}$-structure** is an S-sorted structure

$$\boldsymbol{M} = \big(M; (R^{\boldsymbol{M}})_{R \in \mathcal{L}^{\mathrm{r}}}, (f^{\boldsymbol{M}})_{f \in \mathcal{L}^{\mathrm{f}}}\big) \qquad \text{where } M = (M_s),$$

such that for every $R \in \mathcal{L}^{\mathrm{r}}$ of sort $s_1 \dots s_m$ its **interpretation** $R^{\boldsymbol{M}}$ in $\boldsymbol{M}$ is a subset of $M_{s_1 \dots s_m}$, and for $f \in \mathcal{L}^{\mathrm{f}}$ of sort $s_1 \dots s_n s$ its **interpretation** $f^{\boldsymbol{M}}$ in $\boldsymbol{M}$ is a function $M_{s_1 \dots s_n} \to M_s$. For a constant symbol c of $\mathcal{L}$ of sort s the corresponding M_s-valued function $c^{\boldsymbol{M}}$ is identified with its unique value in M_s, so $c^{\boldsymbol{M}} \in M_s$. If $\boldsymbol{M}$ is understood from the context we often omit the superscript $\boldsymbol{M}$ in denoting the interpretation in $\boldsymbol{M}$ of a nonlogical symbol of $\mathcal{L}$. The reader is supposed to keep in mind the distinction between symbols of $\mathcal{L}$ and their interpretation in an $\mathcal{L}$-structure, even if we use the same notation for both.

EXAMPLES B.2.2.

(1) Every group is considered as an $\mathcal{L}_{\mathrm{G}}$-structure by interpreting the symbols 1, ${}^{-1}$, and $\cdot$ as the identity element of the group, its group inverse, and its group multiplication, respectively.

(2) Let $\boldsymbol{A} = (A; 0, -, +)$ be an abelian group; here $0 \in A$ is the zero element of the group, and $-\colon A \to A$ and $+\colon A^2 \to A$ denote the group operations of $\boldsymbol{A}$. We consider $\boldsymbol{A}$ as an $\mathcal{L}_{\mathrm{A}}$-structure by taking as interpretations of the symbols 0, $-$ and $+$ of $\mathcal{L}_{\mathrm{A}}$ the group operations 0, $-$ and $+$ on A. (We took here the liberty of using the same notation for possibly entirely different things: $+$ is an element of the set $\mathcal{L}^{\mathrm{f}}_{\mathrm{A}}$, but also denotes in this context its interpretation as a binary operation

on the set A. Similarly with 0 and $-$.) In fact, any set A for which we single out an element of A, a unary operation on A, and a binary operation on A, is an $\mathcal{L}_{\mathrm{A}}$-structure if we choose to construe it that way.

(3) Likewise, any ring is construed as an $\mathcal{L}_{\mathrm{R}}$-structure in the obvious way.

(4) $(\mathbb{N}; \leqslant)$ is an $\mathcal{L}_{\mathrm{O}}$-structure where we interpret $\leqslant$ as the usual ordering relation on $\mathbb{N}$. Similarly for $(\mathbb{Z}; \leqslant)$, $(\mathbb{Q}; \leqslant)$ and $(\mathbb{R}; \leqslant)$. (Here we take even more notational liberties, by letting $\leqslant$ denote five different things: a symbol of $\mathcal{L}_{\mathrm{O}}$, and the usual orderings of $\mathbb{N}$, $\mathbb{Z}$, $\mathbb{Q}$, and $\mathbb{R}$, respectively.) Again, any nonempty set equipped with a binary relation on it can be viewed as an $\mathcal{L}_{\mathrm{O}}$-structure.

(5) Any ordered abelian group is an $\mathcal{L}_{\mathrm{OA}}$-structure.

(6) The ordered ring of integers and any ordered field are $\mathcal{L}_{\mathrm{OR}}$-structures.

(7) Let R be a ring and M be an R-module. Then M becomes an $\mathcal{L}_{R\text{-mod}}$-structure by interpreting the symbols of $\mathcal{L}_{\mathrm{A}}$ as in (2) and the function symbol λ_r $(r \in R)$ by the function $x \mapsto rx \colon M \to M$. We can also construe M as an $\mathcal{L}_{\mathrm{Mod}}$-structure whose underlying set of sort R is R and whose underlying set of sort M is M, and where the symbols 0_{R}, 1_{R}, $-_{\mathrm{R}}$, $+_{\mathrm{R}}$, $\cdot_{\mathrm{R}}$ are interpreted by the ring operations on R, the symbols 0_{M}, $-_{\mathrm{M}}$, $+_{\mathrm{M}}$ by the group operations on M, and λ by $(r, x) \mapsto rx \colon R \times M \to M$.

Let $\mathcal{L}'$ be an S'-sorted extension of $\mathcal{L}$ (so $S' \supseteq S$). An $\mathcal{L}'$-structure $\boldsymbol{M}' = (M'; \dots)$ is said to be an **$\mathcal{L}'$-expansion** of an $\mathcal{L}$-structure $\boldsymbol{M} = (M; \dots)$ (and $\boldsymbol{M}$ is said to be the **$\mathcal{L}$-reduct** of $\boldsymbol{M}'$) if $M_s = M'_s$ for all s, and each nonlogical symbol of $\mathcal{L}$ has the same interpretation in $\boldsymbol{M}$ as in $\boldsymbol{M}'$. Given such an $\mathcal{L}'$-expansion $\boldsymbol{M}'$ of $\boldsymbol{M}$ we sometimes abuse notation by letting a parameter set $A = (A_s)$ in $\boldsymbol{M}$ denote also the parameter set $(A_{s'})_{s' \in S'}$ in $\boldsymbol{M}'$, where $A_{s'} = \emptyset$ for $s' \in S' \setminus S$.

EXAMPLE. The ordered field of reals is an $\mathcal{L}_{\mathrm{OR}}$-expansion of the real line, where "the real line" stands for the $\mathcal{L}_{\mathrm{O}}$-structure $(\mathbb{R}; \leqslant)$.

Substructures

Let $\boldsymbol{M} = (M; \dots)$ and $\boldsymbol{N} = (N; \dots)$ be $\mathcal{L}$-structures. Then $\boldsymbol{M}$ is called a **substructure** of $\boldsymbol{N}$ (notation: $\boldsymbol{M} \subseteq \boldsymbol{N}$) if $M_s \subseteq N_s$ for all s, and

$$\begin{aligned} R^{\boldsymbol{N}} \cap M_{s_1 \dots s_m} &= R^{\boldsymbol{M}} && \text{whenever } R \in \mathcal{L}^{\mathrm{r}} \text{ has sort } s_1 \dots s_m, \\ f^{\boldsymbol{N}}|_{M_{s_1 \dots s_n}} &= f^{\boldsymbol{M}} \colon M_{s_1 \dots s_n} \to M_s && \text{whenever } f \in \mathcal{L}^{\mathrm{f}} \text{ has sort } s_1 \dots s_n s. \end{aligned}$$

We also say in this case that **$\boldsymbol{N}$ is an extension of $\boldsymbol{M}$** or that **$\boldsymbol{N}$ extends $\boldsymbol{M}$**. For groups, this is just the notion of subgroup, and for rings the notion of subring.

Morphisms

Let $\boldsymbol{M}$ and $\boldsymbol{N}$ be $\mathcal{L}$-structures. A **morphism** $h\colon \boldsymbol{M} \to \boldsymbol{N}$ is a map $h\colon M \to N$ that respects the primitives of $\boldsymbol{M}$, that is:

(M1) for every $R \in \mathcal{L}^{\mathrm{r}}$ of sort $s_1 \dots s_m$ and all $a \in M_{s_1 \dots s_m}$,

$$a \in R^{\boldsymbol{M}} \quad \Rightarrow \quad h_{s_1 \dots s_m} a \in R^{\boldsymbol{N}};$$

(M2) for every $f \in \mathcal{L}^{\mathrm{f}}$ of sort $s_1 \dots s_n s$ and all $a \in M_{s_1 \dots s_n}$,

$$h_s\big(f^{\boldsymbol{M}}(a)\big) \;=\; f^{\boldsymbol{N}}(h_{s_1 \dots s_n} a).$$

Replacing $\Rightarrow$ in (M1) by $\Longleftrightarrow$ and also requiring h to be injective yields the notion of an **embedding**. An **isomorphism** is a bijective embedding, and an **automorphism** of $\boldsymbol{M}$ is an isomorphism $\boldsymbol{M} \to \boldsymbol{M}$. If $\boldsymbol{M} \subseteq \boldsymbol{N}$, then the inclusions $a \mapsto a\colon M_s \to N_s$ yield an embedding $\boldsymbol{M} \to \boldsymbol{N}$, called the **natural inclusion** of $\boldsymbol{M}$ into $\boldsymbol{N}$. Conversely, a morphism $h\colon \boldsymbol{M} \to \boldsymbol{N}$ yields a substructure $h(\boldsymbol{M})$ of $\boldsymbol{N}$ whose underlying set of sort s is $h_s(M_s)$, and if h is an embedding we have an isomorphism $\boldsymbol{M} \to h(\boldsymbol{M})$ given by

$$a \mapsto h_s(a)\colon M_s \to h(M_s).$$

If $h\colon \boldsymbol{M} \to \boldsymbol{M}'$ and $h'\colon \boldsymbol{M}' \to \boldsymbol{M}''$ are morphisms (embeddings, isomorphisms, respectively), then so is $h' \circ h\colon \boldsymbol{M} \to \boldsymbol{M}''$. The automorphisms of $\boldsymbol{M}$ form a group $\mathrm{Aut}(\boldsymbol{M})$ under composition. A parameter set A in $\boldsymbol{M}$ yields the subgroup

$$\mathrm{Aut}(\boldsymbol{M}|A) \;:=\; \big\{f \in \mathrm{Aut}(\boldsymbol{M}) : f_s(a) = a \text{ for every } s \text{ and } a \in A_s\big\}$$

of $\mathrm{Aut}(\boldsymbol{M})$. We write $\boldsymbol{M} \cong \boldsymbol{N}$ if there is an isomorphism $\boldsymbol{M} \to \boldsymbol{N}$.

EXAMPLES. If $\boldsymbol{M}$ and $\boldsymbol{N}$ are abelian groups, construed as $\mathcal{L}_{\mathrm{A}}$-structures according to B.2.2(2), then a morphism $\boldsymbol{M} \to \boldsymbol{N}$ is exactly what in algebra is called a (homo)morphism from the group $\boldsymbol{M}$ into the group $\boldsymbol{N}$. Likewise with rings, and other kinds of algebraic structures.

Products

Now let $(\boldsymbol{N}_\lambda)_{\lambda \in \Lambda}$ be a family of $\mathcal{L}$-structures, where $\Lambda \neq \emptyset$, and let λ range over Λ. For each s put $N_s := \prod_\lambda (N_\lambda)_s$. (The Axiom of Choice guarantees that $N_s \neq \emptyset$.) We write the typical element of N_s as $a = \big(a(\lambda)\big)$. For $a = (a_1, \dots, a_m) \in N_{\boldsymbol{s}}$, put

$$a(\lambda) \;:=\; \big(a_1(\lambda), \dots, a_m(\lambda)\big) \in (N_\lambda)_{\boldsymbol{s}}.$$

The **product** $\prod_\lambda \boldsymbol{N}_\lambda$ of $(\boldsymbol{N}_\lambda)$ is defined to be the $\mathcal{L}$-structure $\boldsymbol{N}$ whose underlying set of sort s is N_s, and where the basic relations and functions are defined coordinatewise: for $R \in \mathcal{L}^{\mathrm{r}}$ of sort $s_1 \dots s_m$ and $a \in N_{s_1 \dots s_m}$,

$$a \in R^{\boldsymbol{N}} \quad :\Longleftrightarrow \quad a(\lambda) \in R^{\boldsymbol{N}_\lambda} \text{ for all } \lambda,$$

and for $f \in \mathcal{L}^{\mathrm{f}}$ of sort $s_1 \dots s_n s$ and $a \in N_{s_1 \dots s_n}$,

$$f^{\boldsymbol{N}}(a) := \big(f^{\boldsymbol{N}_\lambda}(a(\lambda))\big) \in N_s.$$

For each λ the projection to the λth factor is the morphism $\pi_\lambda = \pi_\lambda^{\boldsymbol{N}} : \boldsymbol{N} \to \boldsymbol{N}_\lambda$ given by $(\pi_\lambda)_s(a) = a(\lambda)$ for $a \in N_s$. This product construction makes it possible to combine several morphisms with a common domain into a single one: if for each λ we have a morphism $h_\lambda : \boldsymbol{M} \to \boldsymbol{N}_\lambda$, then we obtain a morphism $h : \boldsymbol{M} \to \boldsymbol{N}$ with $\pi_\lambda \circ h = h_\lambda$ for each λ. If each h_λ is an embedding, then so is h. This yields:

LEMMA B.2.3. *For each λ, let $h_\lambda : \boldsymbol{M}_\lambda \to \boldsymbol{N}_\lambda$ be a morphism. Then we have a morphism $h : \boldsymbol{M} := \prod_\lambda \boldsymbol{M}_\lambda \to \boldsymbol{N} = \prod_\lambda \boldsymbol{N}_\lambda$ such that $\pi_\lambda^{\boldsymbol{N}} \circ h = h_\lambda \circ \pi_\lambda^{\boldsymbol{M}}$ for all λ. If every h_λ is an embedding, then so is h.*

If $\boldsymbol{N}_\lambda = \boldsymbol{M}$ for all λ, then $\prod_\lambda \boldsymbol{N}_\lambda$ is denoted by $\boldsymbol{M}^\Lambda$; in this case we have an embedding $\Delta : \boldsymbol{M} \to \boldsymbol{M}^\Lambda$ with $\big(\Delta_s(a)\big)(\lambda) = a$ for all $a \in M_s$ and λ, called the **diagonal embedding** of $\boldsymbol{M}$ into $\boldsymbol{M}^\Lambda$.

Direct limits

Let Λ be a nonempty partially ordered set, and let $\lambda, \lambda', \lambda_1, \lambda_2, \dots$ range over Λ. Suppose that Λ is **directed**, that is, for all λ_1, λ_2 there exists λ with $\lambda_1, \lambda_2 \leqslant \lambda$. A **directed system** of $\mathcal{L}$-structures indexed by Λ consists of a family $(\boldsymbol{M}_\lambda)$ of $\mathcal{L}$-structures together with a family $(h_{\lambda\lambda'})_{\lambda \leqslant \lambda'}$ of morphisms $h_{\lambda\lambda'} : \boldsymbol{M}_\lambda \to \boldsymbol{M}_{\lambda'}$ such that $h_{\lambda\lambda} = \mathrm{id}_{\boldsymbol{M}_\lambda}$ for all λ and $h_{\lambda\lambda''} = h_{\lambda'\lambda''} \circ h_{\lambda\lambda'}$ whenever $\lambda \leqslant \lambda' \leqslant \lambda''$. Suppose that $\big((\boldsymbol{M}_\lambda), (h_{\lambda\lambda'})\big)$ is a directed system of $\mathcal{L}$-structures indexed by Λ. For this situation we have the following routine lemma.

LEMMA B.2.4. *There exists an $\mathcal{L}$-structure $\boldsymbol{M}$ and a family (h_λ) of morphisms $h_\lambda : \boldsymbol{M}_\lambda \to \boldsymbol{M}$ with the following properties:*

(i) *if $\lambda \leqslant \lambda'$, then $h_\lambda = h_{\lambda'} \circ h_{\lambda\lambda'}$;*

(ii) *if $\boldsymbol{N}$ is an $\mathcal{L}$-structure and (f_λ) is a family of morphisms $f_\lambda : \boldsymbol{M}_\lambda \to \boldsymbol{N}$ such that $f_\lambda = f_{\lambda'} \circ h_{\lambda\lambda'}$ for all $\lambda \leqslant \lambda'$, then there is a unique morphism $g : \boldsymbol{M} \to \boldsymbol{N}$ such that $f_\lambda = g \circ h_\lambda$ for all λ.*

We call an $\mathcal{L}$-structure $\boldsymbol{M}$ together with a family (h_λ) of morphisms $h_\lambda : \boldsymbol{M}_\lambda \to \boldsymbol{M}$ as in the previous lemma a **direct limit** of the directed system $\big((\boldsymbol{M}_\lambda), (h_{\lambda\lambda'})\big)$. If $\boldsymbol{M}$, (h_λ) and $\widetilde{\boldsymbol{M}}$, $(\widetilde{h}_\lambda)$ are two direct limits of $\big((\boldsymbol{M}_\lambda), (h_{\lambda\lambda'})\big)$, then the unique morphism $h : \boldsymbol{M} \to \widetilde{\boldsymbol{M}}$ such that $\widetilde{h}_\lambda = h \circ h_\lambda$ for all λ is actually an isomorphism. This allows us to speak of *the* direct limit of $\big((\boldsymbol{M}_\lambda), (h_{\lambda\lambda'})\big)$. One verifies easily that if all $h_{\lambda\lambda'}$ are embeddings, then so are all h_λ.

Now let $(\boldsymbol{M}_\lambda)$ be a family of $\mathcal{L}$-structures such that for $\lambda \leqslant \lambda'$, $\boldsymbol{M}_\lambda$ is a substructure of $\boldsymbol{M}_{\lambda'}$, with natural inclusion $h_{\lambda\lambda'} : \boldsymbol{M}_\lambda \hookrightarrow \boldsymbol{M}_{\lambda'}$. Then $\big((\boldsymbol{M}_\lambda), (h_{\lambda\lambda'})\big)$ is a directed system of $\mathcal{L}$-structures, and its direct limit $\bigcup_{\lambda \in \Lambda} \boldsymbol{M}_\lambda$, (h_λ) is called the **direct union** of $(\boldsymbol{M}_\lambda)$. In the following we always identify each $\boldsymbol{M}_\lambda$ with a substructure of $\bigcup_{\lambda \in \Lambda} \boldsymbol{M}_\lambda$ via the embedding h_λ, and we also simply speak of $\bigcup_{\lambda \in \Lambda} \boldsymbol{M}_\lambda$ as the direct union of $(\boldsymbol{M}_\lambda)$.

EXAMPLE B.2.5. View fields as $\mathcal{L}_{\mathrm{R}}$-structures in the natural way. The algebraic closure $\mathbb{F}_p^{\mathrm{a}}$ of the finite field $\mathbb{F}_p$ with p elements is the direct union of the family of finite subfields of $\mathbb{F}_p^{\mathrm{a}}$.

Notes and comments

What we call here a language is also known as a *signature*, or a *vocabulary*. Uncountable languages appear in Mal′cev [281], and many-sorted languages and structures in Herbrand [177] and Schmidt [389, 390]. Structures in mathematical practice are usually one-sorted or two-sorted, but the case of infinitely many sorts does naturally arise in model theory. For example, even if the structure $\boldsymbol{M}$ is one-sorted, the structure $\boldsymbol{M}^{\mathrm{eq}}$ associated to $\boldsymbol{M}$ by Shelah [409, III.6] is most naturally viewed as infinitely-sorted; see [186, Section 4.3].

B.3 VARIABLES AND TERMS

In the rest of this appendix $\boldsymbol{M} = (M; \dots)$ and $\boldsymbol{N} = (N; \dots)$ are $\mathcal{L}$-structures unless noted otherwise. In this section and the next we introduce the syntax (variables, terms, formulas) that helps to specify the definable sets of $\boldsymbol{M}$ along the lines of B.1, uniformly for all $\mathcal{L}$-structures $\boldsymbol{M}$.

Variables

We assume that for each sort s we have available infinitely many symbols, called **variables of sort** s, chosen so that if s and s' are different sorts, then no variable of sort s is a variable of sort s'. We also assume that no variable of any sort is a nonlogical symbol of any language. A **variable** of $\mathcal{L}$ is a variable of some sort s, and a **multivariable** of $\mathcal{L}$ is a tuple $x = (x_i)_{i\in I}$ of *distinct* variables of $\mathcal{L}$. The size of the index set I is called the size of x, and the x_i are called the variables in x. Often the index set I is finite, say $I = \{1, \dots, n\}$, so that $x = (x_1, \dots, x_n)$; in this case, if x_i is of sort s_i $(i = 1, \dots, n)$, we say that x is of sort $s_1 \dots s_n$. Instead of *x has finite size* we also say *x is finite.* Let $x = (x_i)_{i\in I}$ be a multivariable of $\mathcal{L}$, with x_i of sort s_i for $i \in I$. We define the **x-set** of $\boldsymbol{M}$ as

$$M_x := \prod_{i\in I} M_{s_i},$$

and we think of x as a variable running over M_x. When $I = \emptyset$, then x is said to be **trivial**, and M_x is a singleton. If $h\colon \boldsymbol{M} \to \boldsymbol{N}$ is a morphism, we obtain a map $(a_i) \mapsto \big(h_{s_i}(a_i)\big)\colon M_x \to N_x$ between x-sets, which we denote by h_x. For a parameter set $A = (A_s)$ in $\boldsymbol{M}$, we set

$$A_x := \prod_{i\in I} A_{s_i} \subseteq M_x.$$

Multivariables $x = (x_i)_{i\in I}$ and $y = (y_j)_{j\in J}$ of $\mathcal{L}$ are said to be **disjoint** if $x_i \neq y_j$ for all $i \in I$, $j \in J$, and in that case we put $M_{x,y} := M_x \times M_y$. *From now on x and y denote multivariables of $\mathcal{L}$, unless specified otherwise.*

Terms

We define **$\mathcal{L}$-terms** to be words on the alphabet consisting of the function symbols and variables of $\mathcal{L}$, obtained recursively as follows:

(T1) each variable of sort s, when viewed as a word of length 1, is an $\mathcal{L}$-term of sort s, and

(T2) if f is a function symbol of $\mathcal{L}$ of sort $s_1 \cdots s_n s$ and $t_1, \ldots, t_n$ are $\mathcal{L}$-terms of sort $s_1, \ldots, s_n$, respectively, then $ft_1 \cdots t_n$ is an $\mathcal{L}$-term of sort s.

Thus every constant symbol of $\mathcal{L}$ of sort s is an $\mathcal{L}$-term of sort s. Usually we write $f(t_1, \ldots, t_n)$ to denote $ft_1 \ldots t_n$, and shun prefix notation if dictated by tradition.

EXAMPLE. Let x, y, z be $\mathcal{L}_{\mathrm{R}}$-variables. Then the word $\cdot + x - yz$ is an $\mathcal{L}_{\mathrm{R}}$-term in the official prefix notation. For easier reading we indicate this term instead by $(x+(-y))\cdot z$ or even $(x-y)z$.

The following allows us to give definitions and proofs by *induction on terms.*

LEMMA B.3.1. *Every $\mathcal{L}$-term of sort s is either a variable of sort s, or equals $ft_1 \ldots t_n$ for a unique tuple $(f, t_1, \ldots, t_n)$ with f an n-ary function symbol of some sort $s_1 \ldots s_n s$, and t_i an $\mathcal{L}$-term of sort s_i for $i = 1, \ldots, n$.*

For now we shall assume this lemma without proof. At the end of this section we establish more general results which are also needed in the next section.

Let x be a multivariable. An $(\mathcal{L}, x)$-term is a pair (t, x) where t is an $\mathcal{L}$-term such that each variable in t is a variable in x. Such an $(\mathcal{L}, x)$-term is also written more suggestively as $t(x)$, and referred to as *the $\mathcal{L}$-term* $t(x)$. (It is not required that each variable in x actually occurs in t; this is like indicating a polynomial in the indeterminates $X_1, \ldots, X_n$ by $P(X_1, \ldots, X_n)$, where some of these indeterminates might not occur in P. Note that only finitely many of the variables in x can occur in any $\mathcal{L}$-term.) Given an $\mathcal{L}$-term $t(x)$ of sort s, we define a function

$$t^{\boldsymbol{M}} \colon M_x \to M_s$$

as follows, with $x = (x_i)$, and with $a = (a_i)$ ranging over M_x:

(1) if $t = x_i$, then $t^{\boldsymbol{M}}(a) := a_i$;

(2) if $t = ft_1 \cdots t_n$ where $f \in \mathcal{L}^{\mathrm{f}}$ is of sort $s_1 \ldots s_n s$ and $t_1, \ldots, t_n$ are $\mathcal{L}$-terms of sort $s_1, \ldots, s_n$, respectively, then

$$t^{\boldsymbol{M}}(a) := f^{\boldsymbol{M}}\big(t_1^{\boldsymbol{M}}(a), \ldots, t_n^{\boldsymbol{M}}(a)\big) \in M_s.$$

This inductive definition is justified by Lemma B.3.1.

EXAMPLE. Let R be a commutative ring viewed as an $\mathcal{L}_{\mathrm{R}}$-structure in the natural way. Let x, y, z be distinct $\mathcal{L}_{\mathrm{R}}$-variables, and let $t(x, y, z)$ be the $\mathcal{L}_{\mathrm{R}}$-term $(x - y)z$. Then the function $t^R \colon R^3 \to R$ is given by $t^R(a, b, c) = (a - b)c$. In fact, for each $\mathcal{L}_{\mathrm{R}}$-term $t(x_1, \dots, x_n)$ there is a (unique) polynomial $P^t(X_1, \dots, X_n)$ with integer coefficients such that for every commutative ring R we have $t^R(a) = P^t(a)$ for all $a \in R^n$. Conversely, there is for each polynomial $P \in \mathbb{Z}[X_1, \dots, X_n]$ an $\mathcal{L}_{\mathrm{R}}$-term $t(x_1, \dots, x_n)$ such that $t^R(a) = P(a)$ for every commutative ring R and all $a \in R^n$.

Let $t(x)$ be an $\mathcal{L}$-term of sort s and $a \in M_x$. If $h \colon \boldsymbol{M} \to \boldsymbol{N}$ is a morphism of $\mathcal{L}$-structures, then $h_s(t^{\boldsymbol{M}}(a)) = t^{\boldsymbol{N}}(h_x(a))$; so if $\boldsymbol{M} \subseteq \boldsymbol{N}$, then $t^{\boldsymbol{M}}(a) = t^{\boldsymbol{N}}(a)$. If $\mathcal{L}'$ is an extension of $\mathcal{L}$ and $\boldsymbol{M}'$ is an $\mathcal{L}'$-expansion of $\boldsymbol{M}$, then each $\mathcal{L}$-term $t(x)$ is also an $\mathcal{L}'$-term and $t^{\boldsymbol{M}} = t^{\boldsymbol{M}'} : M_x \to M_s$.

Variable-free terms

A term is said to be **variable-free** if no variables occur in it. Let t be a variable-free $\mathcal{L}$-term of sort s. Then the above gives a nullary function $t^{\boldsymbol{M}}$ with value in M_s, identified as usual with its value, so $t^{\boldsymbol{M}} \in M_s$. In particular, if t is a constant symbol c, then $t^{\boldsymbol{M}} = c^{\boldsymbol{M}} \in M_s$, where $c^{\boldsymbol{M}}$ is as in Section B.2, and if $t = ft_1 \dots t_n$ with $f \in \mathcal{L}^{\mathrm{f}}$ of sort $s_1 \dots s_n s$ and variable-free $\mathcal{L}$-terms $t_1, \dots, t_n$ of sorts $s_1, \dots, s_n$, respectively, then $t^{\boldsymbol{M}} = f^{\boldsymbol{M}}(t_1^{\boldsymbol{M}}, \dots, t_n^{\boldsymbol{M}}) \in M_s$.

Names

Let $A \subseteq M$ be a parameter set. We extend $\mathcal{L}$ to a language $\mathcal{L}_A$ by adding a constant symbol $\underline{a}$ of sort s for each $a \in A_s$, called the **name** of a. These names are symbols not in $\mathcal{L}$. Note that $|\mathcal{L}_A| = \max\{|A|, |\mathcal{L}|\}$. We make $\boldsymbol{M}$ into an $\mathcal{L}_A$-structure by interpreting each name $\underline{a}$ as the element $a \in A_s$. The $\mathcal{L}_A$-structure thus obtained is indicated by $\boldsymbol{M}_A$. (This is consistent with the notation introduced in Section B.1.) Hence for each variable-free $\mathcal{L}_A$-term t of sort s we have a corresponding element $t^{\boldsymbol{M}_A}$ of M_s, which for simplicity of notation we denote instead by $t^{\boldsymbol{M}}$. All this applies to the case $M = A$, where in $\mathcal{L}_M$ we have a name $\underline{a}$ for each element a of M_s. If $\boldsymbol{M} \subseteq \boldsymbol{N}$, then we consider $\mathcal{L}_M$ to be a sublanguage of $\mathcal{L}_N$ in such a way that each $a \in M$ has the same name in $\mathcal{L}_M$ as in $\mathcal{L}_N$. Then for each variable-free $\mathcal{L}_M$-term t we have $t^{\boldsymbol{M}} = t^{\boldsymbol{N}}$. Given a multivariable $x = (x_i)$ of $\mathcal{L}$ and $a = (a_i) \in M_x$, we set $\underline{a} := (\underline{a_i})$.

Substitution

Let α be an $\mathcal{L}_M$-term, $x = (x_i)_{i \in I}$ a multivariable, and $t = (t_i)_{i \in I}$ a family of $\mathcal{L}_M$-terms such that t_i is of the same sort as x_i, for each i. Then $\alpha(t/x)$ denotes the word obtained by replacing all occurrences of x_i in α by t_i, *simultaneously* for all i. If α is given in the form $\alpha(x)$, then $\alpha(t)$ is short for $\alpha(t/x)$ and for $a \in M_x$ we often write $\alpha(a)$ instead of $\alpha(\underline{a})$.

LEMMA B.3.2. *Let $\alpha(x)$ be an $\mathcal{L}_M$-term of sort s, and recall that α defines a map $\alpha^{\boldsymbol{M}} \colon M_x \to M_s$. Let $t = (t_i)_{i \in I}$ be a family of $\mathcal{L}_M$-terms, with each t_i of the same* *sort as* x_i. *Then* $\alpha(t)$ *is an* $\mathcal{L}_M$*-term of sort* s. *Moreover, if all* t_i *are variable-free, then so is* $\alpha(t)$, *and with* $a_i := t_i^{\boldsymbol{M}}$, $a := (a_i) \in M_x$ *we have*

$$\alpha(t)^{\boldsymbol{M}} = \alpha(a)^{\boldsymbol{M}} = \alpha^{\boldsymbol{M}}(a).$$

This follows by a straightforward induction on the length of α.

Generators

Assume $\mathcal{L}$ has for each s a constant symbol of sort s. Let $A \subseteq M$ be given. For each s we set

$$B_s := \{t^{\boldsymbol{M}}(a) : t(x) \text{ is an } \mathcal{L}\text{-term of sort } s, \text{ and } a \in A_x\} \subseteq M_s.$$

Then $B = (B_s)$ underlies a substructure $\boldsymbol{B} = (B; \dots)$ of $\boldsymbol{M}$, and clearly $\boldsymbol{B} \subseteq \boldsymbol{C}$ for all substructures $\boldsymbol{C}$ of $\boldsymbol{M}$ with $A \subseteq C$. We call $\boldsymbol{B}$ the substructure of $\boldsymbol{M}$ **generated** by A; notation: $\boldsymbol{B} = \langle A \rangle_{\boldsymbol{M}}$. Note that $|A| \leqslant |\langle A \rangle_{\boldsymbol{M}}| \leqslant \max\{|A|, |\mathcal{L}|\}$. If $\boldsymbol{M} = \langle A \rangle_{\boldsymbol{M}}$, then we say that $\boldsymbol{M}$ is generated by A. If $\boldsymbol{N}$ is an $\mathcal{L}$-structure, then each map $A \to N$ has clearly at most one extension to a morphism $\langle A \rangle_{\boldsymbol{M}} \to \boldsymbol{N}$.

Unique readability

We finish this section with the promised general result on unique readability. We let F be a set of symbols with a function $a \colon F \to \mathbb{N}$ (called the **arity** function). A symbol $f \in F$ is said to have arity n if $a(f) = n$. A word on F is said to be **admissible** if it can be obtained by applying the following rules:

(1) If $f \in F$ has arity 0, then f viewed as a word of length 1 is admissible.

(2) If $f \in F$ has arity $n \geqslant 1$ and $t_1, \dots, t_n$ are admissible words on F, then the concatenation $ft_1 \dots t_n$ is admissible.

Below we just write *admissible word* instead of *admissible word on* F. Note that the empty word is not admissible, and that the last symbol of an admissible word cannot be of arity $\geqslant 1$.

EXAMPLE. The $\mathcal{L}$-terms are admissible words on the alphabet consisting of the function symbols of $\mathcal{L}$ and the variables of $\mathcal{L}$, where each n-ary $f \in \mathcal{L}^{\mathrm{f}}$ has arity n and each variable has arity 0.

LEMMA B.3.3. *Let* $t_1, \dots, t_m$ *and* $u_1, \dots, u_n$ *be admissible words and* w *any word on* F *such that* $t_1 \dots t_m w = u_1 \dots u_n$. *Then* $m \leqslant n$, $t_i = u_i$ *for* $i = 1, \dots, m$, *and* $w = u_{m+1} \dots u_n$.

PROOF. By induction on the length ℓ of $u_1 \dots u_n$. If $\ell = 0$, then $m = n = 0$ and w is the empty word. Suppose $\ell > 0$, and assume the lemma holds for smaller lengths. Note that $n > 0$. If $m = 0$, then the conclusion of the lemma holds, so suppose $m > 0$. The first symbol of t_1 equals the first symbol of u_1. Say this first symbol is $h \in F$ with arity k. Then $t_1 = ha_1 \dots a_k$ and $u_1 = hb_1 \dots b_k$ where $a_1, \dots, a_k$ and $b_1, \dots, b_k$ are admissible words. Canceling the first symbol h gives

$$a_1 \dots a_k t_2 \dots t_m w = b_1 \dots b_k u_2 \dots u_n.$$

(Caution: any of $k, m-1, n-1$ could be 0.) We have $\mathrm{length}(b_1 \dots b_k u_2 \dots u_n) = \ell - 1$, so the induction hypothesis applies. It yields $k + m - 1 \leqslant k + n - 1$ (so $m \leqslant n$), $a_1 = b_1, \dots, a_k = b_k$ (so $t_1 = u_1$), $t_2 = u_2, \dots, t_m = u_m$, and $w = u_{m+1} \cdots u_n$. □

In particular, if $t_1, \dots, t_m$ and $u_1, \dots, u_n$ are admissible words such that $t_1 \dots t_m = u_1 \dots u_n$, then $m = n$ and $t_i = u_i$ for $i = 1, \dots, m$. Thus:

COROLLARY B.3.4 (unique readability). *Each admissible word equals $ft_1 \dots t_n$ for a unique tuple $(f, t_1, \dots, t_n)$ with $f \in F$ of arity n and $t_1, \dots, t_n$ admissible words.*

B.4 FORMULAS

We now fix once and for all the eight **logical symbols**:

$$\top \quad \bot \quad \neg \quad \vee \quad \wedge \quad = \quad \exists \quad \forall$$

to be thought of as *true, false, not, or, and, equals, there exists,* and *for all,* respectively. These are assumed to be distinct from all nonlogical symbols and variables of every language. The symbols $\neg$, $\vee$, $\wedge$ are called (logical) **connectives** and $\exists$, $\forall$ are called **quantifiers**. It will be convenient to fix once and for all a sequence $v_0, v_1, v_2, \dots$ of distinct symbols, called *unsorted* variables, and to declare that for each sort s the symbols v_n^s are the **quantifiable variables** of sort s. For any S-sorted language $\mathcal{L}$ and $s \in S$, these are among its variables of sort s, but $\mathcal{L}$ can have other (unquantifiable) variables of sort s. A multivariable is called quantifiable if each variable in it is quantifiable.

Formulas

The **atomic $\mathcal{L}$-formulas** are the following words on the alphabet

$$\mathcal{L}^{\mathrm{r}} \cup \mathcal{L}^{\mathrm{f}} \cup \{\text{variables of } \mathcal{L}\} \cup \{\top, \bot, =\} :$$

(A1) $\top$ and $\bot$ (as words of length 1);

(A2) the words $Rt_1 \dots t_m$ where $R \in \mathcal{L}^{\mathrm{r}}$ is of sort $s_1 \dots s_m$ and $t_1, \dots, t_m$ are $\mathcal{L}$-terms of sort $s_1, \dots, s_m$, respectively;

(A3) the words $=t_1 t_2$ where t_1, t_2 are $\mathcal{L}$-terms of the same sort.

The **$\mathcal{L}$-formulas** are the words on the alphabet

$$\mathcal{L}^{\mathrm{r}} \cup \mathcal{L}^{\mathrm{f}} \cup \{\text{variables of } \mathcal{L}\} \cup \{\top, \bot, \neg, \vee, \wedge, =, \exists, \forall\}$$

obtained as follows:

(F1) atomic $\mathcal{L}$-formulas are $\mathcal{L}$-formulas;

(F2) if φ, ψ are $\mathcal{L}$-formulas, then so are $\neg\varphi$, $\vee\varphi\psi$, and $\wedge\varphi\psi$;

(F3) if φ is an $\mathcal{L}$-formula and x is a quantifiable variable, then $\exists x\varphi$ and $\forall x\varphi$ are $\mathcal{L}$-formulas.

REMARKS.

(1) Having the connectives $\vee$ and $\wedge$ in front of the $\mathcal{L}$-formulas they "connect" rather than in between, is called *prefix notation* or *Polish notation.* This is theoretically elegant, but for the sake of readability we usually write $\varphi \vee \psi$ and $\varphi \wedge \psi$ to denote $\vee\varphi\psi$ and $\wedge\varphi\psi$, respectively, and we also use parentheses and brackets if this helps to clarify the structure of an $\mathcal{L}$-formula.

(2) All $\mathcal{L}$-formulas are admissible words on the alphabet consisting of the nonlogical symbols of $\mathcal{L}$, the variables of $\mathcal{L}$, and the eight logical symbols, where $\top$ and $\bot$ have arity 0, $\neg$ has arity 1, $\vee$, $\wedge$, $=$, $\exists$ and $\forall$ have arity 2, and the other symbols have the arities assigned to them earlier. Thus the results on unique readability are applicable to $\mathcal{L}$-formulas. (However, not all admissible words on this alphabet are $\mathcal{L}$-formulas: the word $\exists xx$ is admissible but not an $\mathcal{L}$-formula.)

(3) The reader should distinguish between different ways of using the symbol $=$. Sometimes it denotes one of the eight formal logical symbols, but we also use it to indicate equality of mathematical objects in the usual way. The context should always make it clear what our intention is in this respect without having to spell it out. To increase readability we usually write an atomic formula $=t_1t_2$ as $t_1 = t_2$ and its negation $\neg{=}t_1t_2$ as $t_1 \neq t_2$, where t_1, t_2 are $\mathcal{L}$-terms of the same sort.

We shall use the following notational conventions: $\varphi \to \psi$ denotes $\neg\varphi \vee \psi$, and $\varphi \leftrightarrow \psi$ denotes $(\varphi \to \psi) \wedge (\psi \to \varphi)$. We sometimes write $\varphi \,\&\, \psi$ instead of $\varphi \wedge \psi$.

DEFINITION B.4.1. Let φ be an $\mathcal{L}$-formula. Written as a word on the alphabet introduced above we have $\varphi = a_1 \dots a_m$. A **subformula** of φ is a subword of the form $a_i \dots a_k$ (where $1 \leqslant i \leqslant k \leqslant m$) which also happens to be an $\mathcal{L}$-formula. An occurrence of a variable x in φ at the jth place (that is, $a_j = x$) is said to be **bound** if φ has a subformula $a_i a_{i+1} \dots a_k$ with $i \leqslant j \leqslant k$ that is of the form $\exists x\psi$ or $\forall x\psi$. An occurrence which is not bound is said to be **free**. A **free variable** of φ is a variable that occurs free in φ.

EXAMPLE. In the $\mathcal{L}_{\mathrm{A}}$-formula $\big(\exists x(x = y)\big) \wedge x = 0$, where x and y are distinct variables, the first two occurrences of x are bound, the third is free, and the only occurrence of y is free. (Note: this formula is actually the string $\wedge\exists x = xy = x0$, and the occurrences of x and y are the occurrences in this string.)

An $(\mathcal{L}, x)$-formula is a pair (φ, x) with φ an $\mathcal{L}$-formula and x a multivariable such that all free variables of φ are in x. Such an $(\mathcal{L}, x)$-formula is written more suggestively as $\varphi(x)$, and referred to as *the $\mathcal{L}$-formula* $\varphi(x)$. Likewise, when referring to an $\mathcal{L}$-formula $\varphi(x,y)$ we really mean a triple (φ,x,y) consisting of an $\mathcal{L}$-formula φ and disjoint multivariables x and y such that all free variables of φ are in x or y. If $\mathcal{L}$ and x both have size $\leqslant \kappa$, then the set of $(\mathcal{L},x)$-formulas has size $\leqslant \kappa$.

An **$\mathcal{L}$-sentence** is an $\mathcal{L}$-formula without free variables. The set of all $\mathcal{L}$-sentences has size $|\mathcal{L}|$: that is why in (F3) we let x be quantifiable.

Substitution

Let φ be an $\mathcal{L}$-formula, let $x = (x_i)_{i\in I}$ be a multivariable, and let $t = (t_i)_{i\in I}$ be a family of $\mathcal{L}$-terms with t_i of the same sort as x_i for all i. Then $\varphi(t/x)$ denotes the word obtained by replacing all the free occurrences of x_i in φ by t_i, *simultaneously* for all i. If $\varphi = \varphi(x)$, then we also write $\varphi(t)$ instead of $\varphi(t/x)$. We have the following facts whose routine proofs are left to the reader.

LEMMA B.4.2. *The word $\varphi(t/x)$ is an $\mathcal{L}$-formula. If t is variable-free and $\varphi = \varphi(x)$, then $\varphi(t)$ is an $\mathcal{L}$-sentence.*

Given an $\mathcal{L}$-structure $\boldsymbol{M}$, an $\mathcal{L}_M$-formula $\varphi(x)$ and $a \in M_x$, we shall avoid many ugly expressions by writing the $\mathcal{L}_M$-sentence $\varphi(\underline{a})$ as just $\varphi(a)$.

Truth and definability

We can now define what it means for an $\mathcal{L}_M$-sentence σ to be **true** in the $\mathcal{L}$-structure $\boldsymbol{M}$ (notation: $\boldsymbol{M} \models \sigma$, also read as $\boldsymbol{M}$ **satisfies** σ, or σ **holds** in $\boldsymbol{M}$). First we consider atomic $\mathcal{L}_M$-sentences:

(T1) $\boldsymbol{M} \models \top$ and $\boldsymbol{M} \not\models \bot$;

(T2) $\boldsymbol{M} \models Rt_1 \dots t_m$ if and only if $(t_1^{\boldsymbol{M}}, \dots, t_m^{\boldsymbol{M}}) \in R^{\boldsymbol{M}}$, for $R \in \mathcal{L}^{\mathrm{r}}$ of sort $s_1 \dots s_m$ and variable-free $\mathcal{L}_M$-terms $t_1, \dots, t_m$ of sort $s_1, \dots, s_m$, respectively;

(T3) $\boldsymbol{M} \models t_1 = t_2$ if and only if $t_1^{\boldsymbol{M}} = t_2^{\boldsymbol{M}}$, for variable-free $\mathcal{L}_M$-terms t_1, t_2 of the same sort.

We extend the definition inductively to arbitrary $\mathcal{L}_M$-sentences as follows:

(T4) Suppose $\sigma = \neg\tau$; then $\boldsymbol{M} \models \sigma$ if and only if $\boldsymbol{M} \not\models \tau$.

(T5) Suppose $\sigma = \sigma_1 \vee \sigma_2$; then $\boldsymbol{M} \models \sigma$ if and only if $\boldsymbol{M} \models \sigma_1$ or $\boldsymbol{M} \models \sigma_2$.

(T6) Suppose $\sigma = \sigma_1 \wedge \sigma_2$; then $\boldsymbol{M} \models \sigma$ if and only if $\boldsymbol{M} \models \sigma_1$ and $\boldsymbol{M} \models \sigma_2$.

(T7) Suppose $\sigma = \exists x \varphi$ where x is a quantifiable variable and $\varphi(x)$ is an $\mathcal{L}_M$-formula; then $\boldsymbol{M} \models \sigma$ if and only if $\boldsymbol{M} \models \varphi(a)$ for some $a \in M_x$.

(T8) Suppose $\sigma = \forall x \varphi$ where x is a quantifiable variable and $\varphi(x)$ is an $\mathcal{L}_M$-formula; then $\boldsymbol{M} \models \sigma$ if and only if $\boldsymbol{M} \models \varphi(a)$ for all $a \in M_x$.

Even if we just want to define $\boldsymbol{M} \models \sigma$ for $\mathcal{L}$-sentences σ, one can see that if σ has the form considered in (T7) or (T8), the inductive definition above forces us to consider $\mathcal{L}_M$-sentences $\varphi(a)$. This is why we introduced names. ("Inductive" refers here to induction with respect to the number of logical symbols in σ.)

DEFINITION B.4.3. Given an $\mathcal{L}_M$-formula $\varphi(x)$ we let

$$\varphi^{\boldsymbol{M}} := \{a \in M_x : \boldsymbol{M} \models \varphi(a)\} \subseteq M_x.$$

The formula $\varphi(x)$ is said to **define** the set $\varphi^{\boldsymbol{M}}$ in $\boldsymbol{M}$. Given a parameter set $A \subseteq M$, a subset of M_x is said to be A-**definable** in $\boldsymbol{M}$ if it is of the form $\varphi^{\boldsymbol{M}}$ for some $\mathcal{L}_A$-formula $\varphi(x)$. A map $f\colon X \to M_y$, where $X \subseteq M_x$, is said to be A-definable if its graph $\Gamma(f) \subseteq M_{x,y}$ is. We use **definable** synonymously with M-definable.

EXAMPLES. Let $\boldsymbol{R} = (\mathbb{R}; \leqslant, 0, 1, +, -, \cdot)$.

(1) The set $\{r \in \mathbb{R} : r \leqslant \sqrt{2}\}$ is 0-definable in $\boldsymbol{R}$: it is defined by the formula $(x^2 \leqslant 1+1) \vee (x \leqslant 0)$. (Here x^2 abbreviates the term $x \cdot x$.)

(2) The set $\{r \in \mathbb{R} : r \leqslant \pi\}$ is definable in $\boldsymbol{R}$: it is defined by the formula $x \leqslant \pi$. (It takes more effort to show that it is not 0-definable; see Example B.12.16.)

It is easy to see that given $A \subseteq M$, a subset of M_s is A-definable in the sense of the previous definition if and only if it is A-definable in the sense of Definition B.1.2 (using the correspondences (C1)–(C5) between formulas and sets defined by them from Section B.1). If $\psi(y)$ is an $\mathcal{L}_A$-formula, then there is a finite multivariable x disjoint from y, an $\mathcal{L}$-formula $\varphi(x, y)$ and an $a \in A_x$, such that $\varphi(a, y) = \psi(y)$. Thus a set $Y \subseteq M_y$ is A-definable iff for some finite x disjoint from y, some $a \in A_x$ and some 0-definable $Z \subseteq M_{x,y}$ we have $Y = Z(a) := \{b \in M_y : (a, b) \in Z\}$. In this way *$A$-definability* reduces to *0-definability*.

We defined what it means for an $\mathcal{L}$-sentence to hold in $\boldsymbol{M}$. It is convenient to extend this to arbitrary $\mathcal{L}$-formulas. First, given $A \subseteq M$ we define an A-**instance** of an $\mathcal{L}_A$-formula $\varphi = \varphi(x)$ to be an $\mathcal{L}_A$-sentence of the form $\varphi(a)$ with $a \in A_x$. Of course φ can also be written as $\varphi(y)$ for another multivariable y. Thus for the above to count as a definition of *A-instance*, the reader should check that these different ways of specifying variables (including at least the free variables of φ) give the same A-instances.

DEFINITION B.4.4. An $\mathcal{L}$-formula φ is said to be **valid** in $\boldsymbol{M}$ (notation: $\boldsymbol{M} \models \varphi$) if all its M-instances are true in $\boldsymbol{M}$.

Suppose the multivariable $x = (x_1, \dots, x_m)$ is (finite and) quantifiable, and φ is an $\mathcal{L}$-formula. Then $\forall x \varphi$ denotes $\forall x_1 \cdots \forall x_m \varphi$, and likewise with $\forall$ replaced by $\exists$. The reader should check that if $\varphi = \varphi(x)$, then $\forall x \varphi$ is an $\mathcal{L}$-sentence and

$$\boldsymbol{M} \models \varphi \iff \boldsymbol{M} \models \forall x\, \varphi.$$

We define $\models \varphi$ to mean: $\boldsymbol{M} \models \varphi$ for all $\mathcal{L}$-structures $\boldsymbol{M}$. We call $\mathcal{L}$-formulas ψ, θ **equivalent** if $\models \psi \leftrightarrow \theta$. Thus $\neg\forall x \varphi$ and $\exists x \neg\varphi$ are equivalent.

Let $\varphi_1, \dots, \varphi_n$ be $\mathcal{L}$-formulas, where $n \geqslant 1$. We inductively define

$$\varphi_1 \wedge \cdots \wedge \varphi_n := \begin{cases} \varphi_1 & \text{if } n = 1, \\ \varphi_1 \wedge \varphi_2 & \text{if } n = 2, \\ (\varphi_1 \wedge \cdots \wedge \varphi_{n-1}) \wedge \varphi_n & \text{if } n \geqslant 3. \end{cases}$$

Similarly define $\varphi_1 \vee \cdots \vee \varphi_n$. For each permutation i of $\{1, \dots, n\}$, the $\mathcal{L}$-formulas $\varphi_1 \wedge \cdots \wedge \varphi_n$ and $\varphi_{i1} \wedge \cdots \wedge \varphi_{in}$ are equivalent; similarly with $\wedge$ replaced by $\vee$.

Formulas of a special form

In this subsection formula *means* $\mathcal{L}$-formula. We single out formulas by syntactical conditions with semantic counterparts in terms of behavior under embeddings, as explained in the next subsection. A formula is said to be **quantifier-free** if it has no occurrences of $\exists$ and no occurrences of $\forall$. A formula is said to be **existential** (or an **$\exists$-formula**) if it has the form $\exists x\, \rho$ with a finite multivariable x and a quantifier-free formula ρ, and **universal** (or a **$\forall$-formula**) if it has the form $\forall x\, \rho$ with finite x and quantifier-free ρ. If φ is a $\forall$-formula, then $\neg\varphi$ is equivalent to an $\exists$-formula, and similarly with "$\forall$" and "$\exists$" interchanged. A formula is said to be **universal-existential** (or a **$\forall\exists$-formula**) if it has the form $\forall x \exists y\, \rho$ with finite disjoint x, y and quantifier-free ρ.

LEMMA B.4.5. *If φ, ψ are $\exists$-formulas, then $\varphi \wedge \psi$ and $\varphi \vee \psi$ are equivalent to $\exists$-formulas; likewise with $\forall$ and with $\forall\exists$ in place of $\exists$.*

PROOF. The easy proofs of these facts use the device of "renaming variables" which we often use tacitly below. Let $\varphi = \exists x\, \rho$, $\psi = \exists y\, \theta$ where ρ, θ are quantifier-free; it is easy to check that if x, y are disjoint, then $\varphi \wedge \psi$ and $\exists x \exists y (\rho \wedge \theta)$ are equivalent. In the general case, first choose disjoint finite quantifiable multivariables x', y' of the same sort as x, y, respectively, such that no variable occurring in ρ, θ is in x' or y', and replace φ, ψ by the respective equivalent formulas $\exists x'\, \rho(x'/x)$, $\exists y'\, \theta(y'/y)$. □

Maps preserving formulas

In this subsection A is a parameter set in $\boldsymbol{M}$ and $h \colon A \to N$ is a map. For an $\mathcal{L}_A$-term t, let $h(t)$ be the $\mathcal{L}_N$-term obtained from t by replacing every occurrence of a name of an element $a \in A_s$ by the name of $h_s a \in N_s$. For an $\mathcal{L}_A$-formula φ, let $h(\varphi)$ be the $\mathcal{L}_N$-formula obtained from φ by replacing every occurrence of a name of an element $a \in A_s$ by the name of $h_s a$. If t is variable-free, then so is $h(t)$, and if φ is a sentence, then so is $h(\varphi)$. If $\varphi(x)$ is an $\mathcal{L}$-formula and $a \in A_x$, then $h\big(\varphi(a)\big) = \varphi(ha)$. We say that h **preserves atomic formulas** if for all atomic $\mathcal{L}$-formulas $\varphi(x)$ and $a \in A_x$ such that $\boldsymbol{M} \models \varphi(a)$ we have $\boldsymbol{N} \models \varphi(ha)$. In the same way we define *h preserves quantifier-free formulas* and *h preserves formulas.*

LEMMA B.4.6. *Suppose $A = M$. Then we have the following two equivalences:*

(i) *h is a morphism $\boldsymbol{M} \to \boldsymbol{N}$ iff h preserves atomic formulas;*

(ii) *h is an embedding $\boldsymbol{M} \to \boldsymbol{N}$ iff h preserves quantifier-free formulas.*

We leave the proofs of this routine lemma to the reader.

COROLLARY B.4.7. *Suppose $\boldsymbol{M} \subseteq \boldsymbol{N}$, and let σ be an $\mathcal{L}_M$-sentence.*

(i) *If σ is quantifier-free, then we have: $\boldsymbol{M} \models \sigma \iff \boldsymbol{N} \models \sigma$.*

(ii) *If σ is existential, then we have: $\boldsymbol{M} \models \sigma \Rightarrow \boldsymbol{N} \models \sigma$.*

(iii) *If σ is universal, then we have: $\boldsymbol{N} \models \sigma \Rightarrow \boldsymbol{M} \models \sigma$.*

PROOF. Part (i) is immediate from Lemma B.4.6(ii). Part (ii) follows easily from (i), and (iii) follows from (ii). □

In the next corollary we assume that $\mathcal{L}$ has for every s a constant symbol of sort s. This is to guarantee that $\langle A \rangle_{\boldsymbol{M}}$ is defined.

COROLLARY B.4.8. *The map h extends to a morphism $\langle A \rangle_{\boldsymbol{M}} \to \boldsymbol{N}$ iff h preserves atomic formulas. The map h extends to an embedding $\langle A \rangle_{\boldsymbol{M}} \to \boldsymbol{N}$ iff h preserves quantifier-free formulas.*

PROOF. The forward directions in both statements follow from Lemma B.4.6 and Corollary B.4.7(i). Suppose for every atomic $\mathcal{L}_A$-sentence σ true in $\boldsymbol{M}$, its image $h(\sigma)$ is true in $\boldsymbol{N}$. Now every element of sort s of $\langle A \rangle_{\boldsymbol{M}}$ has the form $t^{\boldsymbol{M}}(a)$ for some $\mathcal{L}$-term $t(x)$ of sort s and some $a \in A_x$. Moreover, if $t_1(x)$ and $t_2(y)$ are $\mathcal{L}$-terms of sort s and $t_1(a) = t_2(b)$, where $a \in A_x$ and $b \in A_y$, then $\boldsymbol{M} \models t_1(a) = t_2(b)$, and hence $\boldsymbol{N} \models t_1(h_x(a)) = t_2(h_y(b))$, and thus $t_1^{\boldsymbol{N}}(h_x(a)) = t_2^{\boldsymbol{N}}(h_y(b))$. These two facts easily yield the backward directions of the two equivalences. □

We say that $h\colon A \to N$ is **elementary** if it preserves formulas, that is, for all $\mathcal{L}$-formulas $\varphi(x)$ and all $a \in A_x$,

$$\boldsymbol{M} \models \varphi(a) \quad \iff \quad \boldsymbol{N} \models \varphi(ha).$$

By Lemmas B.4.6, every elementary map $M \to N$ is an embedding $\boldsymbol{M} \to \boldsymbol{N}$, and every isomorphism of $\mathcal{L}$-structures is elementary. Suppose the $\mathcal{L}_A$-formula $\varphi(x)$ defines the set $X \subseteq M_x$. If $h \in \operatorname{Aut}(\boldsymbol{M})$, then the $\mathcal{L}_{h(A)}$-formula $h(\varphi)(x)$ defines the set $h(X) \subseteq M_x$, so if $h \in \operatorname{Aut}(\boldsymbol{M}|A)$, then $X = h(X)$. This observation is often used to show that certain relations are not definable:

EXAMPLE. The usual ordering relation $\leqslant$ on the set of real numbers is not definable in the structure $(\mathbb{R}; 0, -, +)$: for any $r_1, \dots, r_n \in \mathbb{R}$ there is an automorphism σ of this structure such that $\sigma(r_i) = r_i$ for $i = 1, \dots, n$ and $\sigma(r) < 0 < r$ for some $r \in \mathbb{R}$.

More on substitution

The next lemma shows that substitution and evaluation in terms and formulas behave correctly. Let $x = (x_i)_{i \in I}$ and $y = (y_j)_{j \in J}$ be multivariables, and let $t = (t_i)$ be a family of $\mathcal{L}$-terms with $t_i = t_i(y)$ of the same sort as x_i, for all $i \in I$. Then for $a \in M_y$ we put $t^{\boldsymbol{M}}(a) := \big(t_i^{\boldsymbol{M}}(a)\big) \in M_x$.

LEMMA B.4.9. *Let $a \in M_y$, and let $\alpha(x)$ be an $\mathcal{L}_M$-term of sort s. Then we have $\alpha(t)^{\boldsymbol{M}}(a) = \alpha^{\boldsymbol{M}}(t^{\boldsymbol{M}}(a)) \in M_s$. Let $\varphi(x)$ be a quantifier-free $\mathcal{L}_M$-formula. Then*

$$\boldsymbol{M} \models \varphi(t)(a) \iff \boldsymbol{M} \models \varphi\big(t^{\boldsymbol{M}}(a)\big).$$

PROOF. The claim about α follows by induction on terms. Suppose φ is atomic, say $\varphi = R\alpha_1 \cdots \alpha_m$ with m-ary $R \in \mathcal{L}^{\mathrm{r}}$ and $\mathcal{L}_M$-terms $\alpha_1(x), \dots, \alpha_m(x)$. Then

$$\begin{aligned} \varphi(t) \;&=\; R\alpha_1(t) \cdots \alpha_m(t), \qquad \varphi(t^{\boldsymbol{M}}(a)) \;=\; R\alpha_1\big(t^{\boldsymbol{M}}(a)\big) \cdots \alpha_m\big(t^{\boldsymbol{M}}(a)\big), \text{ so} \\ \boldsymbol{M} \models \varphi(t)(a) \;&\iff\; \big(\alpha_1(t)^{\boldsymbol{M}}(a), \dots, \alpha_m(t)^{\boldsymbol{M}}(a)\big) \in R^{\boldsymbol{M}} \\ &\iff\; \big(\alpha_1^{\boldsymbol{M}}(t^{\boldsymbol{M}}(a)), \dots, \alpha_m^{\boldsymbol{M}}(t^{\boldsymbol{M}}(a))\big) \in R^{\boldsymbol{M}} \\ &\iff\; \boldsymbol{M} \models \varphi\big(t^{\boldsymbol{M}}(a)\big). \end{aligned}$$

The case that φ is $\alpha = \beta$ is handled the same way. The desired property is clearly inherited by disjunctions, conjunctions and negations. □

Notes and comments

The definition (T1)–(T8) of the satisfaction relation goes back to Tarski's paper [440], but in the form above is closer to Tarski-Vaught [446].

B.5 ELEMENTARY EQUIVALENCE AND ELEMENTARY SUBSTRUCTURES

The syntax (terms, formulas, sentences) and semantics (truth, definability) from the previous section will be used in this section to compare $\mathcal{L}$-structures.

Elementary equivalence

We say that $\boldsymbol{M}$ and $\boldsymbol{N}$ are **elementarily equivalent** (notation: $\boldsymbol{M} \equiv \boldsymbol{N}$) if they satisfy the same $\mathcal{L}$-sentences. So isomorphic $\mathcal{L}$-structures are elementarily equivalent. By B.5.6 below, however, the non-isomorphic ordered sets $(\mathbb{Q}; \leqslant)$ and $(\mathbb{R}; \leqslant)$ are elementarily equivalent as well. This uses the *back-and-forth method,* a general tool that we also relied on in Chapter 8.

DEFINITION B.5.1. A **partial isomorphism** from $\boldsymbol{M}$ to $\boldsymbol{N}$ is a bijection $\gamma\colon A \to B$ with $A \subseteq M$, $B \subseteq N$, such that

(1) for each $R \in \mathcal{L}^{\mathrm{r}}$ of sort $s_1 \dots s_m$ and $a \in A_{s_1 \dots s_m}$,

$$a \in R^{\boldsymbol{M}} \iff \gamma a \in R^{\boldsymbol{N}};$$

(2) for each $f \in \mathcal{L}^{\mathrm{f}}$ of sort $s_1 \ldots s_n s$ and $a \in A_{s_1 \ldots s_n}$, $b \in A_s$,

$$f^{\boldsymbol{M}}(a) = b \iff f^{\boldsymbol{N}}(\gamma a) = \gamma b.$$

Note that in this definition we do not assume that A and B are the underlying sets of substructures of $\boldsymbol{M}$ and $\boldsymbol{N}$, respectively. If A and B *are* the underlying sets of substructures $\boldsymbol{A}$ of $\boldsymbol{M}$ and $\boldsymbol{B}$ of $\boldsymbol{N}$, respectively, then a partial isomorphism $A \to B$ from $\boldsymbol{M}$ to $\boldsymbol{N}$ is the same thing as an isomorphism $\boldsymbol{A} \to \boldsymbol{B}$.

Given a partial isomorphism $\gamma\colon A \to B$ from $\boldsymbol{M}$ to $\boldsymbol{N}$, we set $\operatorname{domain}(\gamma_s) := A_s$ and $\operatorname{domain}(\gamma) := A$, and likewise $\operatorname{codomain}(\gamma_s) := B_s$ and $\operatorname{codomain}(\gamma) := B$. If $(\gamma_\lambda)_{\lambda \in \Lambda}$ is a family of partial isomorphisms indexed by a directed set $(\Lambda, \leqslant)$, such that $\gamma_{\lambda'}$ extends γ_λ for $\lambda \leqslant \lambda'$ in Λ, then there is a unique partial isomorphism $\gamma := \bigcup_{\lambda \in \Lambda} \gamma_\lambda$ from $\boldsymbol{M}$ to $\boldsymbol{N}$ with domain $\bigcup_{\lambda \in \Lambda} \operatorname{domain}(\gamma_\lambda)$ such that $\gamma(a) = \gamma_\lambda(a)$ for all $s \in S$, $\lambda \in \Lambda$, and $a \in \operatorname{domain}(\gamma_{\lambda,s})$.

EXAMPLE. Let $\boldsymbol{M} = (M; \leqslant)$, $\boldsymbol{N} = (N; \leqslant)$ be ordered sets, and let $a_1, \ldots, a_n \in M$ and $b_1, \ldots, b_n \in N$ with $a_1 < a_2 < \cdots < a_n$ and $b_1 < b_2 < \cdots < b_n$; then the map $a_i \mapsto b_i\colon \{a_1, \ldots, a_n\} \to \{b_1, \ldots, b_n\}$ is a partial isomorphism from $\boldsymbol{M}$ to $\boldsymbol{N}$.

DEFINITION B.5.2. A **back-and-forth system** from $\boldsymbol{M}$ to $\boldsymbol{N}$ is a collection Γ of partial isomorphisms from $\boldsymbol{M}$ to $\boldsymbol{N}$ such that $\Gamma \neq \emptyset$ and:

("Forth") for each $\gamma \in \Gamma$, $s \in S$, and $a \in M_s$ there is a $\gamma' \in \Gamma$ extending γ such that $a \in \operatorname{domain}(\gamma'_s)$;

("Back") for each $\gamma \in \Gamma$, $s \in S$, and $b \in N_s$ there is a $\gamma' \in \Gamma$ extending γ such that $b \in \operatorname{codomain}(\gamma'_s)$.

We say that $\boldsymbol{M}$ and $\boldsymbol{N}$ are **back-and-forth equivalent** (notation: $\boldsymbol{M} \equiv_{\mathrm{bf}} \boldsymbol{N}$) if there exists a back-and-forth system from $\boldsymbol{M}$ to $\boldsymbol{N}$.

PROPOSITION B.5.3. *If $\boldsymbol{M}$, $\boldsymbol{N}$ are countable and $\boldsymbol{M} \equiv_{\mathrm{bf}} \boldsymbol{N}$, then $\boldsymbol{M} \cong \boldsymbol{N}$.*

PROOF. Suppose Γ is a back-and-forth system from $\boldsymbol{M}$ to $\boldsymbol{N}$. Note first that for $\gamma \in \Gamma$ and $s \in S$ we have: $\operatorname{domain}(\gamma_s) = M_s$ iff $\operatorname{codomain}(\gamma_s) = N_s$. In view of this fact, and assuming M and N are countable, we can start with any $\gamma_0 \in \Gamma$ and build recursively a sequence (γ_n) in Γ such that γ_{n+1} extends γ_n for all n (going forth if n is even, and going back if n is odd) such that $\bigcup_n \operatorname{domain}(\gamma_n) = M$ and $\bigcup_n \operatorname{codomain}(\gamma_n) = N$. Then we have an isomorphism $\bigcup_n \gamma_n\colon \boldsymbol{M} \to \boldsymbol{N}$. □

In applying this proposition and the next one in a concrete situation, the key is to guess a back-and-forth system. That is where insight and imagination (and experience) come in. In the following result we do not assume countability.

PROPOSITION B.5.4. *If $\boldsymbol{M} \equiv_{\mathrm{bf}} \boldsymbol{N}$, then $\boldsymbol{M} \equiv \boldsymbol{N}$.*

Towards the proof, define an $\mathcal{L}$-formula to be **unnested** if every atomic subformula of it is either $\top$, or $\bot$, or has one of the following forms:

(1) $Rx_1 \ldots x_m$; here $R \in \mathcal{L}^{\mathrm{r}}$ and $x_1, \ldots, x_m$ are distinct variables;

(2) $x = y$; here x and y are distinct variables;

(3) $fx_1 \dots x_n = y$; here $f \in \mathcal{L}^{\mathrm{f}}$ and $x_1, \dots, x_n, y$ are distinct variables.

LEMMA B.5.5. *Each atomic $\mathcal{L}$-formula $\varphi(x)$ is equivalent to an unnested existential $\mathcal{L}$-formula $\varphi_\exists(x)$, and also to an unnested universal $\mathcal{L}$-formula $\varphi_\forall(x)$. Each $\mathcal{L}$-formula $\varphi(x)$ is equivalent to an unnested $\mathcal{L}$-formula $\varphi_{\mathrm{u}}(x)$.*

We leave the proof of this lemma to the reader.

PROOF OF PROPOSITION B.5.4. Let Γ be a back-and-forth system from $\boldsymbol{M}$ to $\boldsymbol{N}$, and let $\varphi(x)$ be an unnested $\mathcal{L}$-formula. By induction on the number of logical symbols in φ we show that for all $\gamma \in \Gamma$ with domain A and $a \in A_x$,

$$\boldsymbol{M} \models \varphi(a) \iff \boldsymbol{N} \models \varphi(\gamma a).$$

This yields $\boldsymbol{M} \equiv \boldsymbol{N}$, using Lemma B.5.5 for sentences. The stated equivalence follows from the definitions if φ is atomic, and its validity is preserved under $\wedge$, $\vee$, $\neg$. Suppose $\varphi = \exists y \psi$ where y is a quantifiable variable of sort s not occurring in x and $\psi(x, y)$ is unnested, and let $\gamma \in \Gamma$ with domain A and $a \in A_x$. If $\boldsymbol{M} \models \varphi(a)$, then we take $b \in M_s$ such that $\boldsymbol{M} \models \psi(a, b)$, and then $\gamma' \in \Gamma$ extending γ with $b \in \mathrm{domain}(\gamma'_s)$; by inductive hypothesis $\boldsymbol{N} \models \psi(\gamma a, \gamma' b)$ and hence $\boldsymbol{N} \models \varphi(\gamma a)$. Similarly one shows that $\boldsymbol{N} \models \varphi(\gamma a) \Rightarrow \boldsymbol{M} \models \varphi(a)$. □

EXAMPLE B.5.6. Let $\boldsymbol{M} = (M; \leqslant)$ and $\boldsymbol{N} = (N; \leqslant)$ be dense ordered sets without endpoints. Then the collection of all strictly increasing bijections $A \to B$, where $A \subseteq M$ and $B \subseteq N$ are finite, is a back-and-forth system from $\boldsymbol{M}$ to $\boldsymbol{N}$. Hence $\boldsymbol{M} \equiv \boldsymbol{N}$, and if M, N are countable, then $\boldsymbol{M} \cong \boldsymbol{N}$.

Elementary substructures

Let $\boldsymbol{M} \subseteq \boldsymbol{N}$. One says that $\boldsymbol{M}$ is an **elementary substructure** of $\boldsymbol{N}$ (and that the extension $\boldsymbol{M} \subseteq \boldsymbol{N}$ is **elementary**) if the natural inclusion $\boldsymbol{M} \hookrightarrow \boldsymbol{N}$ is elementary; notation: $\boldsymbol{M} \preccurlyeq \boldsymbol{N}$. We have

$$\boldsymbol{M} \preccurlyeq \boldsymbol{N} \iff \varphi^{\boldsymbol{M}} = \varphi^{\boldsymbol{N}} \cap M_x \text{ for each } \mathcal{L}_M\text{-formula } \varphi(x).$$

Also, $\boldsymbol{M} \preccurlyeq \boldsymbol{N} \iff \boldsymbol{M}_M \equiv \boldsymbol{N}_M$, so $\boldsymbol{M} \preccurlyeq \boldsymbol{N} \Rightarrow \boldsymbol{M} \equiv \boldsymbol{N}$.

EXAMPLE B.5.7. View groups as $\mathcal{L}_{\mathrm{G}}$-structures as in B.2.2(1). Let G, H be groups with $G \preccurlyeq H$, and suppose that H is simple. Then G is also simple: to see this, let $g, g' \in G$, $g \neq 1$; it suffices to show that g' is in the normal subgroup of G generated by g. Now $g' = h_1 g^{k_1} h_1^{-1} \cdots h_n g^{k_n} h_n^{-1}$ where $n \geqslant 1$, $h_1, \dots, h_n \in H$, $k_1, \dots, k_n \in \mathbb{Z}$. Therefore H satisfies the $\mathcal{L}_G$-sentence

$$\exists x_1 \cdots \exists x_n \big(g' = x_1 g^{k_1} x_1^{-1} \cdots x_n g^{k_n} x_n^{-1}\big),$$

and so does G, since $G \preccurlyeq H$.

We have the following useful criterion for a substructure to be elementary:

PROPOSITION B.5.8 (Tarski-Vaught test). *Let $A \subseteq N$. Suppose that for every $\mathcal{L}_A$-formula $\varphi(x)$ with x a quantifiable variable, if $\boldsymbol{N} \models \exists x \varphi(x)$, then $\boldsymbol{N} \models \varphi(a)$ for some $a \in A_x$. Then A underlies an elementary substructure of $\boldsymbol{N}$.*

PROOF. Note first that $N_s \neq \emptyset$ gives $A_s \neq \emptyset$, for all s. If f is a function symbol of $\mathcal{L}$ of sort $s_1 \dots s_n s$ and $a \in A_{s_1 \dots s_n}$, consider the $\mathcal{L}_A$-formula $\varphi(x)$ given by $f(a) = x$, with x a quantifiable variable of sort s; then $\boldsymbol{N} \models \exists x \varphi$ and hence $f^{\boldsymbol{N}}(a) \in A_x$. Thus A is the underlying set of a substructure $\boldsymbol{A}$ of $\boldsymbol{N}$. We show that for each $\mathcal{L}_A$-sentence σ we have $\boldsymbol{A} \models \sigma \iff \boldsymbol{N} \models \sigma$ by induction on the construction of σ. If σ is atomic, then this holds by Corollary B.4.7(i), and it is clear that the desired property is preserved under taking negations, conjunctions, and disjunctions. It remains to treat the case where $\sigma = \exists x \varphi$ with an $\mathcal{L}_A$-formula $\varphi(x)$. Then

$$\begin{aligned} \boldsymbol{A} \models \sigma \quad &\iff \quad \boldsymbol{A} \models \varphi(a) \text{ for some } a \in A_x \\ &\iff \quad \boldsymbol{N} \models \varphi(a) \text{ for some } a \in A_x \\ &\iff \quad \boldsymbol{N} \models \sigma, \end{aligned}$$

using the hypothesis in the proposition for the third equivalence. □

COROLLARY B.5.9. *If $\boldsymbol{M} \subseteq \boldsymbol{N}$ and for all finite $A \subseteq M$ and all $b \in N_s$ there exists $h \in \operatorname{Aut}(\boldsymbol{N}|A)$ with $h(b) \in M_s$, then $\boldsymbol{M} \preccurlyeq \boldsymbol{N}$.*

EXAMPLE. The previous corollary easily yields $(\mathbb{Q}; \leqslant) \preccurlyeq (\mathbb{R}; \leqslant)$.

The Tarski-Vaught test can be used to construct small elementary substructures:

PROPOSITION B.5.10 (Downward Löwenheim-Skolem). *Let $A \subseteq N$ and suppose κ is a cardinal such that* $\max\{|A|, |\mathcal{L}|\} \leqslant \kappa \leqslant |N|$. *Then $\boldsymbol{N}$ has an elementary substructure $\boldsymbol{M}$ of size κ with $A \subseteq M$.*

PROOF. After enlarging A to a parameter set in $\boldsymbol{N}$ of size κ, we may assume that $|A| = \kappa$. For every $B \subseteq N$ and s, let $\Phi_{B,s}$ be the set of all $\mathcal{L}_B$-formulas $\varphi(x)$ with x a quantifiable variable of sort s such that $\boldsymbol{N} \models \exists x \varphi$. For every $\varphi \in \Phi_{B,s}$ choose a $b_\varphi \in N_s$ with $\boldsymbol{N} \models \varphi(b_\varphi)$, and let $B'_s := \{b_\varphi : \varphi \in \Phi_{B,s}\}$. For each $b \in B_s$ and quantifiable variable x of sort s the $\mathcal{L}_B$-formula $x = b$ is in $\Phi_{B,s}$, so $b = b_{x=b} \in B'_s$, hence $B_s \subseteq B'_s$. Setting $B' := (B'_s)$, we have $|B| \leqslant |B'| \leqslant \max\{|B|, |\mathcal{L}|\}$. We now inductively define an increasing sequence $A_0 \subseteq A_1 \subseteq \cdots$ of parameter sets in $\boldsymbol{M}$ by $A_0 := A$ and $A_{n+1} := A'_n$, and put $M := \bigcup_n A_n$. Then $|A_n| = \kappa$ for each n, hence also $|M| = \kappa$. By Proposition B.5.8, M is the underlying set of an elementary substructure of $\boldsymbol{N}$. □

EXAMPLE B.5.11. Any infinite simple group H has a simple subgroup of any given infinite size $\leqslant |H|$, by Example B.5.7 and Proposition B.5.10.

Direct unions

Suppose $\boldsymbol{M}$ is the direct union of the family $(\boldsymbol{M}_\lambda)_{\lambda\in\Lambda}$ of substructures of $\boldsymbol{M}$. Let λ, λ' range over Λ. By (i) and (ii) of Corollary B.4.7, if σ is a $\forall\exists$-sentence such that $\boldsymbol{M}_\lambda \models \sigma$ for all λ, then $\boldsymbol{M} \models \sigma$.

EXAMPLE B.5.12. Let p be a prime number and σ a universal-existential $\mathcal{L}_{\mathrm{R}}$-sentence. If σ holds in all sufficiently large finite fields of characteristic p, then σ holds in the algebraic closure $\mathbb{F}_p^{\mathrm{a}}$ of $\mathbb{F}_p$; see also Example B.2.5.

LEMMA B.5.13. *Suppose $\boldsymbol{M}_\lambda \preccurlyeq \boldsymbol{M}_{\lambda'}$ for all $\lambda \leqslant \lambda'$. Then $\boldsymbol{M}_\lambda \preccurlyeq \boldsymbol{M}$ for all λ.*

PROOF. By induction on n we show that for each λ and each $\mathcal{L}_{M_\lambda}$-sentence σ of length n we have $\boldsymbol{M}_\lambda \models \sigma \iff \boldsymbol{M} \models \sigma$. This is clear if σ is atomic by Corollary B.4.7(i), and the desired property is preserved under taking negations, conjunctions, and disjunctions. So let $\varphi(x)$ be an $\mathcal{L}_{M_\lambda}$-formula, where x is a single quantifiable variable, and $\sigma = \exists x\varphi$. Suppose first that $\boldsymbol{M} \models \sigma$, and take $a \in M_x$ with $\boldsymbol{M} \models \varphi(a)$. Then we can take some $\lambda' \geqslant \lambda$ such that $a \in (M_{\lambda'})_x$. By inductive hypothesis $\boldsymbol{M}_{\lambda'} \models \varphi(a)$ and so $\boldsymbol{M}_{\lambda'} \models \sigma$, hence $\boldsymbol{M}_\lambda \models \sigma$ since $\boldsymbol{M}_\lambda \preccurlyeq \boldsymbol{M}_{\lambda'}$. Conversely, suppose $\boldsymbol{M}_\lambda \models \sigma$, and take $a \in (M_\lambda)_x$ with $\boldsymbol{M}_\lambda \models \varphi(a)$; then $\boldsymbol{M} \models \varphi(a)$ by inductive hypothesis, hence $\boldsymbol{M} \models \sigma$. $\square$

Algebraic closure and definable closure

Let A be a parameter set in $\boldsymbol{M}$ and $b \in M_{\boldsymbol{s}}$. The tuple b is said to be A-**definable** in $\boldsymbol{M}$ (or **definable over** A in $\boldsymbol{M}$) if $\{b\} \subseteq M_{\boldsymbol{s}}$ is A-definable, and is said to be A-**algebraic** in $\boldsymbol{M}$ (or **algebraic over** A in $\boldsymbol{M}$) if $b \in X$ for some finite A-definable set $X \subseteq M_{\boldsymbol{s}}$. If $\boldsymbol{M} \preccurlyeq \boldsymbol{N}$, then b is definable over A in $\boldsymbol{M}$ iff b is definable over A in $\boldsymbol{N}$, and similarly with *algebraic* in place of *definable.* In the above we omit "in $\boldsymbol{M}$" if $\boldsymbol{M}$ is clear from the context. Clearly A-*definable* implies A-*algebraic.*

LEMMA B.5.14. *The tuple b is A-definable if and only if $f(a) = b$ for some finite multivariable x, 0-definable $X \subseteq M_x$, $a \in X \cap A_x$, and 0-definable $f\colon X \to M_{\boldsymbol{s}}$.*

PROOF. The "if" direction is obvious. For the "only if" direction, suppose b is A-definable. Take a finite multivariable x, a 0-definable set $Y \subseteq M_x \times M_{\boldsymbol{s}}$, and an $a \in A_x$ such that $Y(a) = \{b\}$. Then the set

$$X \;:=\; \big\{a' \in M_x : |Y(a')| = 1\big\} \;\subseteq\; M_x$$

is 0-definable, and $f\colon X \to M_{\boldsymbol{s}}$ given by $Y(a') = \big\{f(a')\big\}$ does the job. $\square$

Let $\operatorname{dcl}(A)$ be the parameter set in $\boldsymbol{M}$ such that for every s,

$$\operatorname{dcl}(A)_s \;=\; \{b \in M_s : b \text{ is definable over } A\},$$

and define $\operatorname{acl}(A)$ likewise, with *algebraic* instead of *definable.* Call $\operatorname{dcl}(A)$ (respectively, $\operatorname{acl}(A)$) the **definable closure** of A in $\boldsymbol{M}$ (respectively, the **algebraic closure** of A in $\boldsymbol{M}$). We say that A is **definably closed** in $\boldsymbol{M}$ (respectively, **algebraically**

closed in $\boldsymbol{M}$) if $\mathrm{dcl}(A) = A$ (respectively, $\mathrm{acl}(A) = A$). It is easy to check that $\mathrm{dcl}(A)$ is definably closed in $\boldsymbol{M}$, and that $\mathrm{acl}(A)$ is algebraically closed in $\boldsymbol{M}$. Both $\mathrm{dcl}(A)$ and $\mathrm{acl}(A)$ do not change if we pass from $\boldsymbol{M}$ to an elementary extension. Clearly $\mathrm{dcl}(A) \subseteq \mathrm{acl}(A)$, and if A is algebraically closed in $\boldsymbol{M}$, then A is definably closed in $\boldsymbol{M}$. Note that if $\boldsymbol{M}$ is one-sorted and there exists an A-definable total ordering on the underlying set M of $\boldsymbol{M}$, then $\mathrm{dcl}(A) = \mathrm{acl}(A)$.

Suppose $\mathcal{L}$ has for each s a constant symbol of sort s. If A is definably closed in $\boldsymbol{M}$, then A underlies a substructure of $\boldsymbol{M}$. Hence both $\mathrm{dcl}(A)$ and $\mathrm{acl}(A)$ underlie substructures of $\boldsymbol{M}$, also denoted by $\mathrm{dcl}(A)$ and $\mathrm{acl}(A)$, respectively, with $\langle A \rangle_{\boldsymbol{M}} \subseteq \mathrm{dcl}(A) \subseteq \mathrm{acl}(A)$.

If $a \in M_s$ is A-definable, then a is A_0-definable for some finite subset A_0 of A, and similarly with *algebraic* in place of *definable*. If $A \subseteq B \subseteq M$, then $\mathrm{dcl}(A) \subseteq \mathrm{dcl}(B)$ and $\mathrm{acl}(A) \subseteq \mathrm{acl}(B)$.

LEMMA B.5.15. *Let $b \in M_s$ and set $X := \{\sigma(b) : \sigma \in \mathrm{Aut}(\boldsymbol{M}|A)\} \subseteq M_s$. If $b \in \mathrm{dcl}(A)_s$, then $X = \{b\}$, and if $b \in \mathrm{acl}(A)_s$, then X is finite.*

EXAMPLE. Let K be a field, let V be an infinite K-linear space viewed as an $\mathcal{L}_{K\text{-mod}}$-structure as in Example B.2.2(7), and let $A \subseteq V$. Then

$$\mathrm{dcl}(A) = \mathrm{acl}(A) = \text{the subspace of } V \text{ generated by } A.$$

To see this, let W be the subspace of V generated by A. Then $V \setminus W$ is infinite and is an orbit for the action of $\mathrm{Aut}(V|A)$ on V. Now use Lemma B.5.15.

Notes and comments

The notion of elementary equivalence is from [442]. The fact about countable dense ordered sets without endpoints from B.5.6 is due to Cantor [68], with the back-and-forth proof found by Huntington [196] and Hausdorff [172]. Proposition B.5.4 goes back to Ehrenfeucht [123] and Fraïssé [141]. (See [136, 318] for the history of *back-and-forth.*) Elementary extensions as well as Propositions B.5.8 and B.5.10 and Lemma B.5.13 are from [446]. Löwenheim [266] and Skolem [423] had shown Proposition B.5.10 in the case where $\mathcal{L}$ is countable; in the stated form it appears in [446]. Example B.5.11 is from [186, Section 3.1]. Model-theoretic algebraic closure was introduced by A. Robinson [354, p. 157] and gained prominence through the work of Morley [297] and Baldwin-Lachlan [34].

B.6 MODELS AND THE COMPACTNESS THEOREM

In the rest of this appendix, unless indicated otherwise, t is an $\mathcal{L}$-term, φ, ψ, and θ are $\mathcal{L}$-formulas, σ is an $\mathcal{L}$-sentence, and Σ is a set of $\mathcal{L}$-sentences. We drop the prefix $\mathcal{L}$ in $\mathcal{L}$-*term*, $\mathcal{L}$-*formula* and so on, unless this would cause confusion.

Models

We say that M is a **model of** Σ or Σ **holds in** M (notation: $M \models \Sigma$) if $M \models \sigma$ for all $\sigma \in \Sigma$.

To discuss examples it is convenient to introduce some notation. Suppose $\mathcal{L}$ contains (at least) the constant symbol 0 and the binary function symbol $+$. Given any terms $t_1, \dots, t_n$ we define the term $t_1 + \cdots + t_n$ inductively as follows: it is the term 0 if $n = 0$, the term t_1 if $n = 1$, and the term $(t_1 + \cdots + t_{n-1}) + t_n$ for $n \geqslant 2$. We write nt for the term $t + \cdots + t$ with n summands, in particular, $0t$ and $1t$ denote the terms 0 and t, respectively. Suppose $\mathcal{L}$ contains the constant symbol 1 and the binary function symbol $\cdot$ (the multiplication sign). Then we have similar notational conventions for $t_1 \cdot \ldots \cdot t_n$ and t^n; in particular, for $n = 0$ both stand for the term 1, and t^1 is just t.

EXAMPLES B.6.1. Fix three distinct quantifiable variables x, y, z.

(1) *Groups* are the $\mathcal{L}_{\mathrm{G}}$-structures that are models of

$$\mathrm{Gr} := \big\{\forall x(x \cdot 1 = x),\ \forall x\big(x \cdot x^{-1} = 1\big),\ \forall x \forall y \forall z\big((x \cdot y) \cdot z = x \cdot (y \cdot z)\big)\big\}.$$

(2) *Abelian groups* are the $\mathcal{L}_{\mathrm{A}}$-structures that are models of

$$\begin{aligned}\mathrm{Ab} := \big\{&\forall x(x + 0 = x),\ \forall x\big(x + (-x) = 0\big),\ \forall x \forall y(x + y = y + x),\\ &\forall x \forall y \forall z\big((x + y) + z = x + (y + z)\big)\big\}.\end{aligned}$$

(3) *Torsion-free abelian groups* are the $\mathcal{L}_{\mathrm{A}}$-structures that are models of

$$\mathrm{Tf} := \mathrm{Ab} \cup \big\{\forall x(nx = 0 \to x = 0) : n = 1, 2, 3, \dots\big\},$$

and *divisible abelian groups* are the $\mathcal{L}_{\mathrm{A}}$-structures that are models of

$$\mathrm{Div} := \mathrm{Ab} \cup \big\{\forall x \exists y(ny = x) : n = 1, 2, 3, \dots\big\}.$$

(4) *Ordered sets* are the $\mathcal{L}_{\mathrm{O}}$-structures that are models of

$$\begin{aligned}\mathrm{Or} := \big\{&\forall x(x \leqslant x),\ \forall x \forall y \forall z\big((x \leqslant y \wedge y \leqslant z) \to x \leqslant z\big),\\ &\forall x \forall y\big((x \leqslant y \wedge y \leqslant x) \to x = y\big)\big\}.\end{aligned}$$

Abbreviating $x \leqslant y \ \wedge \neg x = y$ by $x < y$, *dense ordered sets without endpoints* are the $\mathcal{L}_{\mathrm{O}}$-structures that are models of

$$\mathrm{DLO} := \mathrm{Or} \cup \big\{\forall x \forall y \exists z(x < y \to x < z \wedge z < y),\ \forall x \exists y \exists z(y < x \wedge x < z)\big\}.$$

(5) *Ordered abelian groups* are the $\mathcal{L}_{\mathrm{OA}}$-structures that are models of

$$\mathrm{OAb} := \mathrm{Or} \cup \mathrm{Ab} \cup \big\{\forall x \forall y \forall z(x \leqslant y \to x + z \leqslant y + z)\big\}.$$

(6) *Rings* are the $\mathcal{L}_{\rm R}$-structures that are models of

$$\mathrm{Ri} := \mathrm{Ab} \cup \{\forall x \forall y \forall z\big((x \cdot y) \cdot z = x \cdot (y \cdot z)\big),\ \forall x(x \cdot 1 = x \wedge 1 \cdot x = x), \\ \forall x \forall y \forall z\big(x \cdot (y+z) = x \cdot y + x \cdot z \wedge (x+y) \cdot z = x \cdot z + y \cdot z\big)\}.$$

(7) *Fields* are the $\mathcal{L}_{\rm R}$-structures that are models of

$$\mathrm{Fl} := \mathrm{Ri} \cup \{\forall x \forall y(x \cdot y = y \cdot x),\ 1 \neq 0,\ \forall x\big(x \neq 0 \rightarrow \exists y(x \cdot y = 1)\big)\}.$$

(8) *Ordered rings* are the $\mathcal{L}_{\rm OR}$-structures that are models of

$$\mathrm{ORi} := \mathrm{OAb} \cup \mathrm{Ri} \cup \{\forall x \forall y(0 \leqslant x \wedge 0 \leqslant y \rightarrow 0 \leqslant x \cdot y)\},$$

and *ordered fields* are the $\mathcal{L}_{\rm OR}$-structures that are models of $\mathrm{OFl} := \mathrm{ORi} \cup \mathrm{Fl}$.

(9) *Fields of characteristic* 0 are the $\mathcal{L}_{\rm R}$-structures that are models of

$$\mathrm{Fl}(0) := \mathrm{Fl} \cup \{n1 \neq 0 : n = 2, 3, 5, 7, 11, \ldots\},$$

and given a prime number p, *fields of characteristic* p are the $\mathcal{L}_{\rm R}$-structures that are models of $\mathrm{Fl}(p) := \mathrm{Fl} \cup \{p1 = 0\}$.

(10) *Algebraically closed fields* are the $\mathcal{L}_{\rm R}$-structures that are models of

$$\mathrm{ACF} := \mathrm{Fl} \cup \{\forall u_1 \cdots \forall u_n \exists x(x^n + u_1 x^{n-1} + \cdots + u_n = 0) : n \geqslant 2\}.$$

Here $u_1, u_2, u_3, \ldots$ is some fixed infinite sequence of distinct quantifiable variables, distinct also from x, and $u_i x^{n-i}$ abbreviates $u_i \cdot x^{n-i}$, for $i = 1, \ldots, n$.

(11) Given a prime number p or $p = 0$, *algebraically closed fields of characteristic* p are the $\mathcal{L}_{\rm R}$-structures that are models of $\mathrm{ACF}(p) := \mathrm{ACF} \cup \mathrm{Fl}(p)$.

Here is the important Compactness Theorem:

THEOREM B.6.2. *If every finite subset of* Σ *has a model, then* Σ *has a model.*

We give the proof of this theorem in the next section. The rest of this section contains some reformulations and simple but instructive applications of this theorem.

Logical consequence

We say that σ is a **logical consequence** of Σ (written $\Sigma \models \sigma$) if σ is true in every model of Σ. More generally, we say that a formula φ is a **logical consequence** of Σ (notation: $\Sigma \models \varphi$) if $\boldsymbol{M} \models \varphi$ for all models $\boldsymbol{M}$ of Σ. We also write $\sigma \models \varphi$ instead of $\{\sigma\} \models \varphi$. Note that $\Sigma \cup \{\sigma_1, \ldots, \sigma_n\} \models \varphi$ iff $\Sigma \models (\sigma_1 \wedge \cdots \wedge \sigma_n) \rightarrow \varphi$.

EXAMPLE. It is well-known that in any ring R we have $r \cdot 0 = 0$ for all $r \in R$. This can now be expressed as $\mathrm{Ri} \models x \cdot 0 = 0$.

Here is a version of the Compactness Theorem in terms of *logical consequence*:

THEOREM B.6.3. *If $\Sigma \models \sigma$, then $\Sigma_0 \models \sigma$ for some finite $\Sigma_0 \subseteq \Sigma$.*

PROOF. Suppose $\Sigma_0 \not\models \sigma$ for all finite $\Sigma_0 \subseteq \Sigma$. Then every finite subset of $\Sigma \cup \{\neg\sigma\}$ has a model, so by Theorem B.6.2, $\Sigma \cup \{\neg\sigma\}$ has a model, and thus $\Sigma \not\models \sigma$. □

A similar argument as in the proof of Theorem B.6.3 shows:

COROLLARY B.6.4. *Let $(\sigma_i)_{i\in I}$ and $(\tau_j)_{j\in J}$ be families of sentences such that*

$$\bigwedge_{i\in I} \sigma_i \models \bigvee_{j\in J} \tau_j,$$

that is, in each structure where all sentences σ_i are true, one of the sentences τ_j is true. Then there are $i_1, \dots, i_m \in I$ and $j_1, \dots, j_n \in J$ such that

$$\sigma_{i_1} \wedge \cdots \wedge \sigma_{i_m} \models \tau_{j_1} \vee \cdots \vee \tau_{j_n}.$$

PROOF. The hypothesis expresses that $\{\sigma_i : i \in I\} \cup \{\neg\tau_j : j \in J\}$ has no model. By the Compactness Theorem there are $i_1, \dots, i_m \in I$ and $j_1, \dots, j_n \in J$ such that $\{\sigma_{i_1}, \cdots, \sigma_{i_m}, \neg\tau_{j_1}, \dots, \neg\tau_{j_n}\}$ has no model, in other words, $\sigma_{i_1} \wedge \cdots \wedge \sigma_{i_m} \models \tau_{j_1} \vee \cdots \vee \tau_{j_n}$. □

Here is one of many routine applications of the Compactness Theorem:

COROLLARY B.6.5. *If the $\mathcal{L}_{\mathrm{R}}$-sentence σ is true in all fields of characteristic 0, then σ is true in all fields of sufficiently high prime characteristic.*

PROOF. If $\mathrm{Fl} \cup \{n1 \neq 0 : n \geqslant 1\} \models \sigma$, then compactness yields $N \in \mathbb{N}$ such that $\mathrm{Fl} \cup \{n1 \neq 0 : n = 1, \dots, N\} \models \sigma$, so σ holds in all fields of characteristic $p > N$. □

Note that $\mathrm{Fl}(0)$ is infinite; the previous corollary implies that there is no finite set of $\mathcal{L}_{\mathrm{R}}$-sentences whose models are exactly the fields of characteristic 0. Here is a typical application of compactness (via Corollary B.6.5) in algebra:

EXAMPLE B.6.6 (Noether-Ostrowski). Let $T = (T_1, \dots, T_n)$ and $P \in \mathbb{Z}[T]$. Given a prime number p, let $P \bmod p$ denote the image of P under the ring morphism $\mathbb{Z}[T] \to \mathbb{F}_p[T]$ which extends $a \mapsto a + p\mathbb{Z}\colon \mathbb{Z} \to \mathbb{Z}/p\mathbb{Z} = \mathbb{F}_p$ and sends T_i to T_i ($i = 1, \dots, n$). *Suppose P is irreducible over $\mathbb{C}$. Then there is some $N \in \mathbb{N}$ such that for all primes $p > N$ and every field F of characteristic p, the polynomial $P \bmod p$ is irreducible over F.* To see this note that every finitely generated field of characteristic zero can be embedded in $\mathbb{C}$, so the hypothesis implies that P is irreducible over every field of characteristic zero, and irreducibility of P over a given field can be expressed by a sentence: there exists an $\mathcal{L}_{\mathrm{Ri}}$-sentence σ_P such that for each field F, if F has characteristic zero then $F \models \sigma_P \iff P$ is irreducible over F, and if F has characteristic $p > 0$ then $F \models \sigma_P \iff P \bmod p$ is irreducible over F.

Completeness and compactness

We call Σ **complete** if Σ has a model and for all σ, either $\Sigma \models \sigma$ or $\Sigma \models \neg\sigma$; equivalently, Σ has a model and $\boldsymbol{M} \equiv \boldsymbol{N}$ for all models $\boldsymbol{M}$, $\boldsymbol{N}$ of Σ. Completeness is a strong property and it can be hard to show that a given set Σ is complete.

EXAMPLES. The set of $\mathcal{L}_{\mathrm{A}}$-sentences Ab (the set of axioms for abelian groups) is not complete: consider $\exists x(x \neq 0)$. The set of $\mathcal{L}_{\mathrm{O}}$-sentences DLO (the set of axioms for dense ordered sets without endpoints) is complete; see Example B.5.6.

An **$\mathcal{L}$-theory** is a set T of $\mathcal{L}$-sentences such that for all σ, if $T \models \sigma$, then $\sigma \in T$. In particular, we have for any Σ the $\mathcal{L}$-theory generated by Σ:

$$\mathrm{Th}(\Sigma) \;:=\; \{\sigma : \Sigma \models \sigma\}.$$

It has the same models as Σ. An **axiomatization** of an $\mathcal{L}$-theory T is a set Σ such that $\mathrm{Th}(\Sigma) = T$. We use the abbreviation Th also as follows: for any $\boldsymbol{M}$, the set

$$\mathrm{Th}(\boldsymbol{M}) \;:=\; \{\sigma : \boldsymbol{M} \models \sigma\}$$

is a complete $\mathcal{L}$-theory, called the **theory of $\boldsymbol{M}$**. (Thus if Σ has a model, then $\Sigma \subseteq T$ for some complete $\mathcal{L}$-theory T.) Given a class $\mathcal{C}$ of $\mathcal{L}$-structures, we set

$$\mathrm{Th}(\mathcal{C}) \;:=\; \bigcap_{\boldsymbol{M} \in \mathcal{C}} \mathrm{Th}(\boldsymbol{M}) \;=\; \{\sigma : \sigma \text{ is true in every } \boldsymbol{M} \in \mathcal{C}\}.$$

This is an $\mathcal{L}$-theory (not necessarily complete), called the **theory of $\mathcal{C}$**. Thus $\mathrm{Th}(\Sigma) = \mathrm{Th}(\mathcal{C})$, where $\mathcal{C}$ is the class of models of Σ.

A complete $\mathcal{L}$-theory that contains Σ is called a **completion** of Σ. Let $\mathrm{S}(\Sigma)$ be the set of completions of Σ; thus $|\mathrm{S}(\Sigma)| \leqslant 2^{|\mathcal{L}|}$. We set $\langle\sigma\rangle := \{T \in \mathrm{S}(\Sigma) : \sigma \in T\}$. One verifies easily that for sentences σ, τ we have $\langle\sigma\rangle = \langle\tau\rangle$ iff $\Sigma \models \sigma \leftrightarrow \tau$, and

$$\langle\sigma \wedge \tau\rangle = \langle\sigma\rangle \cap \langle\tau\rangle, \quad \langle\sigma \vee \tau\rangle = \langle\sigma\rangle \cup \langle\tau\rangle, \quad \langle\neg\sigma\rangle = \mathrm{S}(\Sigma) \setminus \langle\sigma\rangle.$$

The topology on $\mathrm{S}(\Sigma)$ with the sets $\langle\sigma\rangle$ as a basis is called the **Stone topology**. Its open sets are the unions $\bigcup_{\sigma\in\Delta}\langle\sigma\rangle$ with Δ a set of sentences. Note that the basic open sets $\langle\sigma\rangle$ are also closed, and so the Stone topology on $\mathrm{S}(\Sigma)$ is hausdorff. We have $\mathrm{S}(\Sigma) \neq \bigcup_{\sigma\in\Delta}\langle\sigma\rangle$ iff $\Sigma \cup \{\neg\sigma : \sigma \in \Delta\}$ has a model; hence Theorem B.6.2 also has the following reformulation, explaining the name *Compactness Theorem*.

THEOREM B.6.7. *The hausdorff space* $\mathrm{S}(\Sigma)$ *is compact.*

Let Σ' be a set of $\mathcal{L}$-sentences. Then Σ and Σ' are said to be **equivalent** if they have the same logical consequences. Thus Σ and Σ' are equivalent iff they have the same models, iff $\mathrm{S}(\Sigma) = \mathrm{S}(\Sigma')$. For example, Σ and $\mathrm{Th}(\Sigma)$ are equivalent.

Completeness and decidability

This subsection concerns the relation between completeness and (algorithmic) decidability, a logical issue that is hardly model-theoretic in nature. We just give an outline and refer to the literature for details, since decidability only makes an appearance in Corollary 16.6.3.

First, one should distinguish the notion of *logical consequence of* from that of *provable from.* To make the latter concept precise requires a *proof system*, which specifies logical axioms and inference rules for generating certain finite sequences of formulas, called (formal) proofs: a *proof of* φ *from* Σ is a sequence $\varphi_1, \dots, \varphi_n$ of formulas with $n \geqslant 1$ and $\varphi_n = \varphi$, such that for $k = 1, \dots, n$, either $\varphi_k \in \Sigma$ or φ_k is a logical axiom, or φ_k is "inferred" from some of the earlier formulas $\varphi_1, \dots, \varphi_{k-1}$ by applying an inference rule. (For example, among the logical axioms might be all formulas of the form $\varphi \vee \neg\varphi$, and among the inference rules is usually *Modus Ponens,* which allows one to infer ψ from ϕ and $\phi \to \psi$; see for example [411, Section 2.6] for an explicit proof system.) We call φ *provable from* Σ (in symbols: $\Sigma \vdash \varphi$) if there exists a proof of φ from Σ. The logical axioms and inference rules are chosen so that the logical axioms are valid in all $\mathcal{L}$-structures, and if φ is inferred from $\varphi_1, \dots, \varphi_k$ by an inference rule, then φ is valid in all $\mathcal{L}$-structures where $\varphi_1, \dots, \varphi_k$ are valid. Hence $\Sigma \vdash \varphi \Rightarrow \Sigma \models \varphi$. We now fix some traditional proof system for classical predicate logic. Then the converse also holds:

THEOREM B.6.8 (Gödel's Completeness Theorem). *If* $\Sigma \models \varphi$, *then* $\Sigma \vdash \varphi$.

See [411, Chapter 4] for a proof. "Completeness" here refers to the proof system, not to Σ. Theorem B.6.8 and its converse immediately yield version B.6.3 of the Compactness Theorem (of which we give an independent proof in the next section).

Suppose now that the language $\mathcal{L}$ has only finitely many nonlogical symbols. Then formulas can be made into inputs of computer programs, and the logical axioms and inference rules of the proof system hiding behind the notation $\vdash$ can also be effectively given. One says that Σ is *effectively enumerable* if there is an effective procedure that enumerates all elements of Σ. We say that an $\mathcal{L}$-theory T is *decidable* if there is an algorithm (program) that takes any sentence σ as input and decides whether or not $\sigma \in T$. If Σ is effectively enumerable, then so is the set $\mathrm{Th}(\Sigma)$ of all sentences provable from Σ, and we obtain:

COROLLARY B.6.9. *If* $\mathcal{L}$ *has only finitely many nonlogical symbols and* Σ *is complete and effectively enumerable, then* $\mathrm{Th}(\Sigma)$ *is decidable.*

Notes and comments

The notion of "model" goes back to the Hilbert school. (But it was Tarski [445] who first spoke of "model theory" as a subject in its own right.) Theorem B.6.8 was proved by Gödel [148] (1930) for countable languages and by Henkin [173] (1949) in general. The Compactness Theorem was shown independently by Mal′cev [281] (1936). The formulation in Theorem B.6.7 is due to Tarski [444]. Corollary B.6.5 is due to

A. Robinson [354]. Algebraic proofs of the statement in Example B.6.6 were given by Noether [305] and Ostrowski [311].

B.7 ULTRAPRODUCTS AND PROOF OF THE COMPACTNESS THEOREM

In this section Λ is a nonempty index set and λ, λ' range over Λ.

Filters and ultrafilters

A **proper filter** on Λ is a nonempty collection $\mathcal{F}$ of subsets of Λ such that $\emptyset \notin \mathcal{F}$ and for all $A, B \subseteq \Lambda$:

(Fi1) if $A, B \in \mathcal{F}$, then $A \cap B \in \mathcal{F}$;

(Fi2) if $A \subseteq B \subseteq \Lambda$ and $A \in \mathcal{F}$, then $B \in \mathcal{F}$.

Note that then $\Lambda \in \mathcal{F}$, and that for all $A \subseteq \Lambda$, either $A \notin \mathcal{F}$ or $\Lambda \setminus A \notin \mathcal{F}$.

EXAMPLE. If Λ is infinite, then the set of all cofinite subsets of Λ is a proper filter on Λ, called the **Fréchet filter** on Λ.

A proper filter on Λ that is maximal with respect to inclusion is called an **ultrafilter** on Λ. By Zorn, every proper filter on Λ is included in an ultrafilter on Λ.

LEMMA B.7.1. *Suppose $\mathcal{F}$ is a proper filter on Λ. Then the following are equivalent:*

(i) *$\mathcal{F}$ is an ultrafilter on Λ;*

(ii) *for all $A, B \subseteq \Lambda$ with $A \cup B \in \mathcal{F}$ we have $A \in \mathcal{F}$ or $B \in \mathcal{F}$;*

(iii) *for all $A \subseteq \Lambda$ we have $A \in \mathcal{F}$ or $\Lambda \setminus A \in \mathcal{F}$.*

PROOF. Suppose $\mathcal{F}$ is an ultrafilter on Λ and $A, B \subseteq \Lambda$, $A \cup B \in \mathcal{F}$. Then $A \cap F \neq \emptyset$ for all $F \in \mathcal{F}$, or $B \cap F \neq \emptyset$ for all $F \in \mathcal{F}$; we may assume that the first alternative holds. Then $\{C \subseteq \Lambda : C \supseteq A \cap F \text{ for some } F \in \mathcal{F}\}$ is a proper filter on Λ which includes $\mathcal{F} \cup \{A\}$, so $A \in \mathcal{F}$. This shows (i) $\Rightarrow$ (ii), and (ii) $\Rightarrow$ (iii) follows by taking $B = \Lambda \setminus A$. The direction (iii) $\Rightarrow$ (i) is obvious. □

A nonempty collection $\mathcal{F}$ of subsets of Λ has the **finite intersection property** (FIP) if for all $A_1, \dots, A_n \in \mathcal{F}$ with $n \geqslant 1$ we have $A_1 \cap \dots \cap A_n \neq \emptyset$. In particular, any proper filter on Λ has the FIP.

LEMMA B.7.2. *Suppose the nonempty collection $\mathcal{F}$ of subsets of Λ has the* FIP. *Then there exists an ultrafilter $\mathcal{U} \supseteq \mathcal{F}$ on Λ.*

PROOF. Let $\mathcal{F}^*$ be the collection of all subsets of Λ that contain a finite intersection $A_1 \cap \dots \cap A_n$ with $A_1, \dots, A_n \in \mathcal{F}$, $n \geqslant 1$. Then $\mathcal{F}^*$ is a proper filter and $\mathcal{F}^* \supseteq \mathcal{F}$. By Zorn there exists an ultrafilter $\mathcal{U} \supseteq \mathcal{F}^*$. □

Ultraproducts

Let $(\boldsymbol{M}_\lambda)$ be a family of $\mathcal{L}$-structures and $\boldsymbol{M} := \prod_\lambda \boldsymbol{M}_\lambda$ be its product. (See Section B.2.) Let σ be an $\mathcal{L}_M$-sentence. Take an $\mathcal{L}$-formula $\varphi(x)$ and $a \in M_x$ such that $\sigma = \varphi(a)$, and set

$$\|\sigma\| := \{\lambda : \boldsymbol{M}_\lambda \models \varphi(a(\lambda))\}.$$

This notation is justified since the $\mathcal{L}_{M_\lambda}$-sentence $\varphi(a(\lambda))$ depends only on σ and λ, not on the choice of $\varphi(x)$ and a such that $\sigma = \varphi(a)$. If σ, τ are $\mathcal{L}_M$-sentences, then

$$\|\sigma \mathbin{\&} \tau\| = \|\sigma\| \cap \|\tau\|, \quad \|\sigma \vee \tau\| = \|\sigma\| \cup \|\tau\|, \quad \|\neg\sigma\| = \Lambda \setminus \|\sigma\|, \tag{B.7.1}$$

$$\models \sigma \to \tau \quad \Rightarrow \quad \|\sigma\| \subseteq \|\tau\|. \tag{B.7.2}$$

Moreover:

LEMMA B.7.3. *Let $\psi(y)$ be an $\mathcal{L}_M$-formula, where y is a quantifiable variable of sort s. Then for all $b \in M_s$ we have $\|\psi(b)\| \subseteq \|\exists y\psi\|$, and there exists $b \in M_s$ such that equality holds.*

PROOF. For all $b \in M_s$ we have $\|\psi(b)\| \subseteq \|\exists y\psi\|$ by (B.7.2). To obtain equality, choose $b = (b(\lambda)) \in M_s$ as follows: Take an $\mathcal{L}$-formula $\varphi(x, y)$ and $a \in M_x$ such that $\psi(y) = \varphi(a, y)$. Then we have for any $b \in M_s$,

$$\|\psi(b)\| = \{\lambda : \boldsymbol{M}_\lambda \models \varphi(a(\lambda), b(\lambda))\}, \quad \|\exists y\psi\| = \{\lambda : \boldsymbol{M}_\lambda \models (\exists y\varphi)(a(\lambda))\}.$$

For $\lambda \in \|\exists y\psi\|$ we pick $b(\lambda) \in (M_\lambda)_s$ such that $\boldsymbol{M}_\lambda \models \varphi\big(a(\lambda), b(\lambda)\big)$, and for $\lambda \notin \|\exists y\psi\|$ we let $b(\lambda) \in (M_\lambda)_s$ be arbitrary. Then $\|\psi(b)\| \supseteq \|\exists y\psi\|$ as required. □

Let $\mathcal{F}$ be a proper filter on Λ. For $s \in S$ we define a binary relation $\sim_s$ on M_s by

$$a \sim_s b \quad :\Longleftrightarrow \quad \|a = b\| \in \mathcal{F} \quad \Longleftrightarrow \quad \{\lambda : a(\lambda) = b(\lambda)\} \in \mathcal{F}.$$

For $a = (a_1, \dots, a_m), b = (b_1, \dots, b_m) \in M_{s_1 \dots s_m}$ we set

$$a \sim_{s_1 \dots s_m} b \quad :\Longleftrightarrow \quad a_1 \sim_{s_1} b_1 \mathbin{\&} \cdots \mathbin{\&} a_m \sim_{s_m} b_m.$$

Using (B.7.2) one easily shows:

LEMMA B.7.4.

(i) *The relation $\sim_{s_1 \dots s_m}$ is an equivalence relation on $M_{s_1 \dots s_m}$.*

(ii) *If $R \in \mathcal{L}^{\mathrm{r}}$ has sort $s_1 \dots s_m$, and $a, b \in M_{s_1 \dots s_m}$, $a \sim_{s_1 \dots s_m} b$, then*

$$\|R\,a\| \in \mathcal{F} \quad \Longleftrightarrow \quad \|R\,b\| \in \mathcal{F}.$$

(iii) *If $f \in \mathcal{L}^{\mathrm{f}}$ has sort $s_1 \dots s_n s$ and $a, b \in M_{s_1 \dots s_n}$, $a \sim_{s_1 \dots s_n} b$, then*

$$f^{\boldsymbol{M}}(a) \sim_s f^{\boldsymbol{M}}(b).$$

For $a \in M_{s_1 \dots s_m}$ we let $a^\sim$ denote the equivalence class of a with respect to $\sim_{s_1 \dots s_m}$, and we let $M^\sim_{s_1 \dots s_m}$ be the set of equivalence classes of $\sim_{\boldsymbol{s}}$. We identify $M^\sim_{s_1 \dots s_m}$ with $M^\sim_{s_1} \times \cdots \times M^\sim_{s_m}$ in the natural way. We now define an $\mathcal{L}$-structure $\boldsymbol{M}^\sim$ whose underlying set of sort s is $M^\sim_s$: for $R \in \mathcal{L}^{\mathrm{r}}$ of sort $s_1 \dots s_m$ and $a \in M_{s_1 \dots s_m}$ we set

$$a^\sim \in R^{\boldsymbol{M}^\sim} \quad :\Longleftrightarrow \quad \|R\,a\| \in \mathcal{F},$$

and for $f \in \mathcal{L}^{\mathrm{f}}$ of sort $s_1 \dots s_n s$ and $a \in M_{s_1 \dots s_n}$ we put

$$f^{\boldsymbol{M}^\sim}(a^\sim) \ := \ f^{\boldsymbol{M}}(a)^\sim \in M_s.$$

The $\mathcal{L}$-structure $\boldsymbol{M}^\sim$ is called the **reduced product** of $(\boldsymbol{M}_\lambda)$ with respect to $\mathcal{F}$, and is also denoted by $\boldsymbol{M}/\mathcal{F} = (\prod_\lambda \boldsymbol{M}_\lambda)/\mathcal{F}$. If $\boldsymbol{M}_\lambda = \boldsymbol{N}$ for all λ, then $\boldsymbol{N}^\Lambda/\mathcal{F}$ is also called the **reduced power** of $\boldsymbol{N}$ with respect to $\mathcal{F}$. If $\mathcal{F}$ is an ultrafilter on Λ, then we speak of the **ultraproduct** $\boldsymbol{M}/\mathcal{F}$ of $(\boldsymbol{M}_\lambda)$ with respect to $\mathcal{F}$ and of the **ultrapower** $\boldsymbol{N}^\Lambda/\mathcal{F}$ of $\boldsymbol{N}$ with respect to $\mathcal{F}$. The maps $a \mapsto a^\sim \colon M_s \to M^\sim_s$ combine to a surjective morphism $\pi \colon \boldsymbol{M} \to \boldsymbol{M}/\mathcal{F}$.

LEMMA B.7.5. *Let x be finite and $a \in M_x$. Then for any $\mathcal{L}$-term $t(x)$ we have $t(a^\sim)^{\boldsymbol{M}/\mathcal{F}} = t^{\boldsymbol{M}/\mathcal{F}}(a^\sim) = t^{\boldsymbol{M}}(a)^\sim$ and for any atomic $\mathcal{L}$-formula $\varphi(x)$ we have*

$$\boldsymbol{M}/\mathcal{F} \models \varphi(a^\sim) \quad \Longleftrightarrow \quad \|\varphi(a)\| \in \mathcal{F}.$$

PROOF. The first equality about terms is part of Lemma B.3.2, and the second equality follows by an easy induction on terms. The statement about atomic formulas is a routine consequence of the equalities about terms. □

COROLLARY B.7.6. *Let $\Delta \colon \boldsymbol{N} \to \boldsymbol{N}^\Lambda$ be the diagonal embedding. Then the morphism $\pi \circ \Delta \colon \boldsymbol{N} \to \boldsymbol{N}^\Lambda/\mathcal{F}$ is an embedding, called the* **diagonal embedding** *of $\boldsymbol{N}$ into its reduced power $\boldsymbol{N}^\Lambda/\mathcal{F}$.*

Here is the main fact about ultraproducts:

THEOREM B.7.7 (Łoś). *Suppose $\mathcal{U}$ is an ultrafilter on Λ. Let $\varphi(x)$ be a formula with finite x, and let $a \in M_x$. Then $\boldsymbol{M}/\mathcal{U} \models \varphi(a^\sim) \Longleftrightarrow \|\varphi(a)\| \in \mathcal{U}$.*

PROOF. We proceed by induction on the construction of φ. The case where φ is atomic is covered by Lemma B.7.5. The cases $\varphi = \neg\psi$, $\varphi = \psi_1 \,\&\, \psi_2$, and $\varphi = \psi_1 \vee \psi_2$ with formulas ψ, ψ_1, ψ_2 follow by induction, (B.7.1), and Lemma B.7.1. Suppose $\varphi = \exists y \psi$ where $\psi(x, y)$ is a formula and y is of sort s; then

$$\begin{aligned} \boldsymbol{M}/\mathcal{U} \models \varphi(a^\sim) \quad &\Longleftrightarrow \quad \boldsymbol{M}/\mathcal{U} \models \psi(a^\sim, b^\sim) \text{ for some } b \in M_s \\ &\Longleftrightarrow \quad \|\psi(a, b)\| \in \mathcal{F} \text{ for some } b \in M_s, \end{aligned}$$

by inductive hypothesis. By Lemma B.7.3, $\|\psi(a, b)\| \in \mathcal{U}$ for some $b \in M_s$, if and only if $\|\varphi(a)\| \in \mathcal{U}$. The case $\varphi = \forall y \psi$ follows from $\models \forall y \psi \leftrightarrow \neg\exists y \neg\psi$. □

In particular, if $\mathcal{U}$ is an ultrafilter on Λ and $\boldsymbol{M}_\lambda \models \Sigma$ for all λ, then $\boldsymbol{M}/\mathcal{U} \models \Sigma$.

EXAMPLE. View abelian groups as $\mathcal{L}_{\mathrm{A}}$-structures in the natural way. Let $\mathcal{U}$ be an ultrafilter on $\mathbb{N}^{\geqslant 1}$ containing the Fréchet filter, set $G := \prod_{n \geqslant 1} (\mathbb{Z}/n\mathbb{Z})/\mathcal{U}$, and $g := (1 + n\mathbb{Z})^{\sim} \in G$. Then for every $m \geqslant 1$ we have $G \not\models mg = 0$ by Theorem B.7.7. So g has infinite order in the abelian group G. Thus there is no set of $\mathcal{L}_{\mathrm{A}}$-sentences whose models are the abelian torsion groups.

COROLLARY B.7.8. *If $\mathcal{U}$ is an ultrafilter on Λ, then the diagonal embedding of $\boldsymbol{N}$ into $\boldsymbol{N}^{\Lambda}/\mathcal{U}$ is elementary.*

PROOF. Let $\varphi(x)$ be a formula with finite x and $a \in N_x$ such that $\boldsymbol{N} \models \varphi(a)$. Then $\|\varphi(\Delta(a))\| = \Lambda \in \mathcal{U}$, so $\boldsymbol{N}^{\Lambda}/\mathcal{U} \models \varphi(\Delta(a)^{\sim})$ by Theorem B.7.7. □

Proof of the Compactness Theorem

We mean here Theorem B.6.2. Thus, suppose every finite subset of Σ has a model; we need to show that Σ has a model. We take Λ to be the set of all finite subsets of Σ; for each λ we take a model $\boldsymbol{M}_{\lambda}$ of λ and set $F(\lambda) := \{\lambda' : \lambda \subseteq \lambda'\} \subseteq \Lambda$. Then $F(\lambda_1) \cap F(\lambda_2) = F(\lambda_1 \cup \lambda_2)$ for all $\lambda_1, \lambda_2 \in \Lambda$, so $\mathcal{F} := \{F(\lambda) : \lambda \in \Lambda\}$ has the finite intersection property. By Lemma B.7.2 we have an ultrafilter $\mathcal{U}$ on Λ with $\mathcal{U} \supseteq \mathcal{F}$. Set $\boldsymbol{M} := \prod_{\lambda} \boldsymbol{M}_{\lambda}$. We claim that $\boldsymbol{M}/\mathcal{U} \models \Sigma$. Let $\sigma \in \Sigma$. Then $\{\sigma\} \in \Lambda$ and

$$F(\{\sigma\}) \subseteq \{\lambda : \boldsymbol{M}_{\lambda} \models \sigma\} = \|\sigma\|.$$

Now $F(\{\sigma\}) \in \mathcal{U}$, so $\|\sigma\| \in \mathcal{U}$, and thus $\boldsymbol{M}/\mathcal{U} \models \sigma$ by Theorem B.7.7. □

Functoriality of reduced products

Let $(\boldsymbol{M}_{\lambda})$ and $(\boldsymbol{N}_{\lambda})$ be families of $\mathcal{L}$-structures, and let $h_{\lambda} : \boldsymbol{M}_{\lambda} \to \boldsymbol{N}_{\lambda}$ be a morphism for each λ. Lemma B.2.3 gives a morphism $h : \boldsymbol{M} := \prod_{\lambda} \boldsymbol{M}_{\lambda} \to \boldsymbol{N} := \prod_{\lambda} \boldsymbol{N}_{\lambda}$ such that $\pi_{\lambda}^{\boldsymbol{N}} \circ h = h_{\lambda} \circ \pi_{\lambda}^{\boldsymbol{M}}$ for all λ. Let $\mathcal{F}$ be a proper filter on Λ. Then for all $a, b \in M_s$ we have $a \sim_s b \Rightarrow h_s a \sim_s h_s b$. It easily follows that we have a morphism $h/\mathcal{F} : \boldsymbol{M}/\mathcal{F} \to \boldsymbol{N}/\mathcal{F}$ making the diagram

$$\begin{array}{ccc} \boldsymbol{M} & \xrightarrow{\;h\;} & \boldsymbol{N} \\ \big\downarrow & & \big\downarrow \\ \boldsymbol{M}/\mathcal{F} & \xrightarrow{\;h/\mathcal{F}\;} & \boldsymbol{N}/\mathcal{F} \end{array}$$

commute. Here the vertical arrows are the morphisms $\boldsymbol{M} \to \boldsymbol{M}/\mathcal{F}$ and $\boldsymbol{N} \to \boldsymbol{N}/\mathcal{F}$ defined before Lemma B.7.5. If each h_{λ} is an embedding, then so is $h/\mathcal{F}$.

We can now prove a statement used in Section 9.5. Given $\mathcal{L}$-structures $\boldsymbol{M}$ and $\boldsymbol{N}$, we say that $\boldsymbol{M}$ is **existentially closed in** $\boldsymbol{N}$ if $\boldsymbol{M} \subseteq \boldsymbol{N}$ and every existential $\mathcal{L}_M$-sentence true in $\boldsymbol{N}$ is true in $\boldsymbol{M}$; equivalently, $\boldsymbol{M} \subseteq \boldsymbol{N}$ and every universal $\mathcal{L}_M$-sentence true in $\boldsymbol{M}$ is true in $\boldsymbol{N}$; notation: $\boldsymbol{M} \preccurlyeq_{\exists} \boldsymbol{N}$. Note that if $\boldsymbol{M} \preccurlyeq_{\exists} \boldsymbol{N}$, then also each universal-existential $\mathcal{L}_M$-sentence true in $\boldsymbol{N}$ is true in $\boldsymbol{M}$. Clearly $\boldsymbol{M} \preccurlyeq \boldsymbol{N} \Rightarrow \boldsymbol{M} \preccurlyeq_{\exists} \boldsymbol{N}$.

COROLLARY B.7.9. *Suppose $\boldsymbol{M}$ is the direct union of a directed family $(\boldsymbol{M}_\lambda)$ of models $\boldsymbol{M}_\lambda$ of Σ. Then $\boldsymbol{M}$ is existentially closed in some model of Σ.*

PROOF. Let $\mathcal{F}$ be the collection of sets $F_\lambda := \{\lambda' : \lambda' \geqslant \lambda\} \subseteq \Lambda$. Since $(\Lambda, \leqslant)$ is directed, $\mathcal{F}$ has the FIP. Then Lemma B.7.2 yields an ultrafilter $\mathcal{U} \supseteq \mathcal{F}$ on Λ, and so $\boldsymbol{M}^* := (\prod_\lambda \boldsymbol{M}_\lambda)/\mathcal{U} \models \Sigma$ by Theorem B.7.7. Let $\iota_\lambda \colon \boldsymbol{M}_\lambda \to \boldsymbol{M}$ be the natural inclusion, let $\iota/\mathcal{U} \colon \boldsymbol{M}^* \to \boldsymbol{M}^\Lambda/\mathcal{U}$ be the embedding obtained from the family of embeddings (ι_λ) as described before the corollary, and let $d\colon \boldsymbol{M} \to \boldsymbol{M}^\Lambda/\mathcal{U}$ be the diagonal embedding. Then $d(\boldsymbol{M}) \subseteq (\iota/\mathcal{U})(\boldsymbol{M}^*) \subseteq \boldsymbol{M}^\Lambda/\mathcal{U}$ and $d(\boldsymbol{M}) \preccurlyeq \boldsymbol{M}^\Lambda/\mathcal{U}$, hence $d(\boldsymbol{M}) \preccurlyeq_\exists (\iota/\mathcal{U})(\boldsymbol{M}^*)$. □

Notes and comments

Ultrafilters were introduced by H. Cartan [69, 70], and also appear in Stone [435]. The definition of the ultraproduct and Theorem B.7.7 are from [271], but versions of ultraproducts had been used already by Skolem [424], Hewitt [178], and Arrow [10]. Reduced products and the proof of the Compactness Theorem via ultraproducts given above are due to Frayne, Morel and Scott [142].

B.8 SOME USES OF COMPACTNESS

We first consider *diagrams,* which provide a way to construct embeddings using compactness. Next we study the relationship between *substructures* and *universal sentences.* We also prove the "upward" version of the Löwenheim-Skolem Theorem.

Diagrams

In this subsection we assume $A \subseteq M$. Let $\mathrm{Diag}(\boldsymbol{M})$ be the set of all quantifier-free sentences σ such that $\boldsymbol{M} \models \sigma$, and set $\mathrm{Diag}_A(\boldsymbol{M}) := \mathrm{Diag}(\boldsymbol{M}_A)$. We call $\mathrm{Diag}_M(\boldsymbol{M})$ the (quantifier-free) **diagram** of $\boldsymbol{M}$.

LEMMA B.8.1. *Let $h\colon A \to N$, and $\boldsymbol{N}_h$ the expansion of $\boldsymbol{N}$ to an $\mathcal{L}_A$-structure given by $\underline{a}^{\boldsymbol{N}_h} := h_s(a)$ for $a \in A_s$. Then:*

(i) *h preserves quantifier-free formulas iff $\boldsymbol{N}_h \models \mathrm{Diag}_A(\boldsymbol{M})$;*

(ii) *h is elementary iff $\boldsymbol{N}_h \models \mathrm{Th}(\boldsymbol{M}_A)$.*

This is rather obvious from the definitions, and yields:

LEMMA B.8.2 (Diagram Lemma).

(i) *There exists a map $A \to N$ preserving quantifier-free formulas if and only if some $\mathcal{L}_A$-expansion of $\boldsymbol{N}$ is a model of* $\mathrm{Diag}_A(\boldsymbol{M})$*;*

(ii) *there exists an elementary map $A \to N$ if and only if some $\mathcal{L}_A$-expansion of $\boldsymbol{N}$ is elementarily equivalent to $\boldsymbol{M}_A$.*

In particular, there exists an embedding $\boldsymbol{M} \to \boldsymbol{N}$ iff $\boldsymbol{N}$ can be expanded to a model of the diagram of $\boldsymbol{M}$, and there exists an elementary embedding $\boldsymbol{M} \to \boldsymbol{N}$ iff $\boldsymbol{N}$ can be expanded to a model of $\mathrm{Th}(\boldsymbol{M}_M)$. The Diagram Lemma acquires its power through the Compactness Theorem:

COROLLARY B.8.3. *The following conditions on a structure $\boldsymbol{M}$ are equivalent:*

(i) *$\boldsymbol{M}$ can be embedded into a model of Σ;*

(ii) *every finite subset of $\Sigma \cup \mathrm{Diag}_M(\boldsymbol{M})$ has a model.*

For one-sorted $\boldsymbol{M}$, these conditions are also equivalent to:

(iii) *every finitely generated substructure of $\boldsymbol{M}$ can be embedded into a model of Σ.*

Likewise, $\boldsymbol{M}$ has an elementary extension that is a model of Σ iff every finite subset of $\Sigma \cup \mathrm{Th}(\boldsymbol{M}_M)$ has a model.

PROOF. Use the Diagram Lemma and compactness to get (i) $\Leftrightarrow$ (ii). For one-sorted $\boldsymbol{M}$, use that every finite subset of $\Sigma \cup \mathrm{Diag}_M(\boldsymbol{M})$ is contained in a set of the form $\Sigma \cup \mathrm{Diag}_N(\boldsymbol{N})$ where $\boldsymbol{N}$ is a finitely generated substructure of $\boldsymbol{M}$. $\square$

COROLLARY B.8.4. *Let $\boldsymbol{M}$ and $\boldsymbol{N}$ be given. Then $\boldsymbol{M} \equiv \boldsymbol{N}$ if and only if there exists an $\mathcal{L}$-structure into which both $\boldsymbol{M}$ and $\boldsymbol{N}$ can be elementarily embedded.*

PROOF. Assume $\boldsymbol{M} \equiv \boldsymbol{N}$. Extend $\mathcal{L}$ to $\mathcal{L}'$ by adding names for the elements of $\boldsymbol{M}$ and $\boldsymbol{N}$ such that no name of any a in $\boldsymbol{M}$ is the name of any b in $\boldsymbol{N}$. We show that the set $\mathrm{Th}(\boldsymbol{M}_M) \cup \mathrm{Th}(\boldsymbol{N}_N)$ of $\mathcal{L}'$-sentences has a model; clearly, $\boldsymbol{M}$ as well as $\boldsymbol{N}$ admits an elementary embedding into the $\mathcal{L}$-reduct of such a model. By compactness (and replacing a finite subset of $\mathrm{Th}(\boldsymbol{M}_M)$ by the conjunction of the sentences in it) it suffices that $\{\varphi(a)\} \cup \mathrm{Th}(\boldsymbol{N}_N)$ has a model, for any formula $\varphi(x)$ with finite x and $a \in M_x$ such that $\boldsymbol{M} \models \varphi(a)$. For such $\varphi(x)$ and a we have $\boldsymbol{M} \models \exists x \varphi(x)$, so $\boldsymbol{N} \models \exists x \varphi(x)$, and thus $\{\varphi(a)\} \cup \mathrm{Th}(\boldsymbol{N}_N)$ has indeed a model.

The other direction is obvious. $\square$

In the same way one shows:

COROLLARY B.8.5. *Let $\boldsymbol{M} \subseteq \boldsymbol{N}$. Then $\boldsymbol{M} \preccurlyeq_{\exists} \boldsymbol{N}$ if and only if $\boldsymbol{N}$ embeds over $\boldsymbol{M}$ into some elementary extension of $\boldsymbol{M}$.*

Substructures and universal sentences

Call Σ **universal** if all sentences in Σ are universal. *In this subsection we assume that $\mathcal{L}$ has for each s a constant symbol of sort s.* Thus $\langle A \rangle_{\boldsymbol{M}}$ is defined for any parameter set A in $\boldsymbol{M}$.

PROPOSITION B.8.6. *Suppose Σ is universal, $x = (x_1, \dots, x_m)$, $y = (y_1, \dots, y_n)$ are disjoint, and $\varphi(x, y)$ is quantifier-free such that $\Sigma \models \forall x \exists y\, \varphi(x, y)$. Then there are n-tuples $t_1 = (t_{11}, \dots, t_{1n}), \dots, t_k = (t_{k1}, \dots, t_{kn})$ of terms $t_{ij}(x)$, $k \geqslant 1$, with*

$$\Sigma \models \varphi\big(x, t_1(x)\big) \vee \cdots \vee \varphi\big(x, t_k(x)\big).$$

The device we use in proving this is often applied: we extend the language $\mathcal{L}$ by new constant symbols and let them play the role of free parameters. Let $x_1, \ldots, x_m$ have sort $s_1, \ldots, s_m$, respectively. Let $\mathcal{L}_c := \mathcal{L} \cup \{c_1, \ldots, c_m\}$ where $c_1, \ldots, c_m$ are distinct new constant symbols of sort $s_1, \ldots, s_m$, respectively. An $\mathcal{L}_c$-structure $(\boldsymbol{M}, a)$ is just an $\mathcal{L}$-structure $\boldsymbol{M}$ together with any m-tuple $a = (a_1, \ldots, a_m) \in M_{s_1,\ldots,s_m}$. Thus, given any $\mathcal{L}$-formula $\psi(x_1, \ldots, x_m)$,

$$\Sigma \models \psi(c_1, \ldots, c_m) \text{ (relative to } \mathcal{L}_c) \iff \Sigma \models \psi(x_1, \ldots, x_m) \text{ (relative to } \mathcal{L}).$$

PROOF OF PROPOSITION B.8.6. Let $\boldsymbol{M} \models \Sigma$ and $a = (a_1, \ldots, a_m) \in M_x$. Since Σ is universal, $\boldsymbol{N} := \langle a_1, \ldots, a_m \rangle_{\boldsymbol{M}}$ is also a model of Σ. Hence $\boldsymbol{N} \models \forall x \exists y\, \varphi$ and so $\boldsymbol{N} \models \exists y\, \varphi(a, y)$. The elements of $\boldsymbol{N}$ of sort s are of the form $t(a)$ where $t(x)$ is an $\mathcal{L}$-term of sort s. Hence there is an n-tuple $t = (t_1, \ldots, t_n)$ of $\mathcal{L}$-terms with $t_i(x)$ of the same sort as y_i, such that $\boldsymbol{N} \models \varphi(a, t(a))$, and thus $\boldsymbol{M} \models \varphi(a, t(a))$. Hence every model of Σ, viewed as a set of $\mathcal{L}_c$-sentences, satisfies some $\mathcal{L}_c$-sentence $\varphi(c, t(c))$ with $t = (t_1, \ldots, t_n)$ an n-tuple of $\mathcal{L}$-terms with $t_i(x)$ of the same sort as y_i and $c = (c_1, \ldots, c_m)$. Now use B.6.4. □

The substructures of fields viewed as $\mathcal{L}_{\mathrm{R}}$-structures are exactly the integral domains, that is, the models of the set of universal $\mathcal{L}_{\mathrm{R}}$-sentences

$$\mathrm{Ri} \cup \big\{0 \neq 1,\ \forall x \forall y\, (xy = yx),\ \forall x \forall y\, (xy = 0 \to x = 0 \vee y = 0)\big\}.$$

This is an instance of the following general fact:

PROPOSITION B.8.7. *Let*

$$\Sigma_\forall := \big\{\sigma : \sigma \text{ is a universal sentence with } \Sigma \models \sigma\big\}$$

be the set of universal logical consequences of Σ. *Then for all* $\boldsymbol{M}$,

$$\boldsymbol{M} \models \Sigma_\forall \iff \boldsymbol{M} \text{ is a substructure of a model of } \Sigma.$$

PROOF. The direction $\Leftarrow$ is clear from Corollary B.4.7(iii). For $\Rightarrow$, suppose $\boldsymbol{M} \models \Sigma_\forall$. To show that $\boldsymbol{M}$ embeds into a model of Σ, let Δ be a finite subset of $\mathrm{Diag}_M(\boldsymbol{M})$; by Corollary B.8.3 it suffices to show that the set $\Sigma \cup \Delta$ of $\mathcal{L}_M$-sentences has a model. Replacing the sentences in Δ by their conjunction we arrange $\Delta = \big\{\varphi(a)\big\}$ where $\varphi(x)$ is a quantifier-free $\mathcal{L}$-formula, $x = (x_1, \ldots, x_m)$, and $a = (a_1, \ldots, a_m) \in M_x$ with distinct $a_1, \ldots, a_m$, such that $\boldsymbol{M} \models \varphi(a)$. If $\Sigma \cup \Delta$ has no model, then $\Sigma \models \neg\varphi(a)$ and hence $\Sigma \models \forall x\, \neg\varphi$, so $\forall x\, \neg\varphi \in \Sigma_\forall$, and thus $\boldsymbol{M} \models \forall x\, \neg\varphi$, contradicting $\boldsymbol{M} \models \varphi(a)$. □

COROLLARY B.8.8. *The following conditions on* Σ *are equivalent:*

(i) *every substructure of every model of* Σ *is a model of* Σ;

(ii) Σ *and* $\Sigma_\forall$ *are equivalent;*

(iii) Σ *is equivalent to some set of universal sentences.*

We say that φ, ψ are **Σ-equivalent** if $\Sigma \models \varphi \leftrightarrow \psi$. (So "$\emptyset$-equivalent" is the same as "equivalent" in the sense of Section B.4.) Note that Σ-equivalence is an equivalence relation on the collection of $\mathcal{L}$-formulas. If $\varphi(x)$, $\psi(x)$ are $(\mathcal{L}, x)$-formulas, then φ, ψ are Σ-equivalent iff $\varphi^{\boldsymbol{M}} = \psi^{\boldsymbol{M}}$ for all $\boldsymbol{M} \models \Sigma$.

COROLLARY B.8.9. *The following are equivalent for* $\Sigma, \varphi(x), x = (x_1, \dots, x_m)$*:*

(i) *for all* $\boldsymbol{M} \subseteq \boldsymbol{N}$ *and* $a \in M_x$*, if* $\boldsymbol{M}, \boldsymbol{N} \models \Sigma$ *and* $\boldsymbol{N} \models \varphi(a)$*, then* $\boldsymbol{M} \models \varphi(a)$*;*

(ii) $\varphi(x)$ *is* Σ*-equivalent to a universal formula* $\psi(x)$*.*

PROOF. Assume (i), and consider the set $\Sigma' := \Sigma \cup \{\varphi(c)\}$ of $\mathcal{L}_c$-sentences. By B.8.7 and (i) we have $(\Sigma')_\forall \cup \Sigma \models \varphi(c)$. By compactness, we can take a universal $\mathcal{L}$-formula $\psi(x)$ such that $\psi(c) \in (\Sigma')_\forall$ and $\{\psi(c)\} \cup \Sigma \models \varphi(c)$. Then $\Sigma \models \varphi \leftrightarrow \psi$. This shows (i) $\Rightarrow$ (ii). The reverse implication follows from B.4.7(iii). □

The Löwenheim-Skolem Theorem

In this subsection we assume that $\mathcal{L}$ *is one-sorted.* (We only use Theorem B.8.10 in this case, and we wish to keep formulations simple.) Call an $\mathcal{L}$-structure *infinite* if its underlying set is infinite.

First-order logic cannot limit the size of an infinite structure:

THEOREM B.8.10 (Upward Löwenheim-Skolem). *Suppose* $\boldsymbol{M}$ *is infinite and* κ *is a cardinal* $\geqslant \max\{|\mathcal{L}|, |M|\}$*. Then* $\boldsymbol{M}$ *has an elementary extension of size* κ*.*

PROOF. Let C be a set of new constant symbols with $|C| = \kappa$, and set $\mathcal{L}' := \mathcal{L} \cup C$. Every finite subset of the set

$$\Sigma' := \operatorname{Th}(\boldsymbol{M}_M) \cup \{c \neq d : c, d \in C \text{ are distinct}\}$$

of $\mathcal{L}'_M$-sentences has a model; in fact, as M is infinite, $\boldsymbol{M}_M$ itself can be expanded to a model by suitably interpreting the constants in C. By compactness, we obtain a model $\boldsymbol{N}'$ of Σ'; by the Diagram Lemma, there exists an elementary embedding of $\boldsymbol{M}$ into the $\mathcal{L}$-reduct of $\boldsymbol{N}'$. Thus we obtain an elementary extension $\boldsymbol{N}$ of $\boldsymbol{M}$ of size $\geqslant \kappa$. Proposition B.5.10 gives an elementary substructure $\boldsymbol{N}_1$ of $\boldsymbol{N}$ with $M \subseteq N_1$ and $|N_1| = \kappa$. Then $\boldsymbol{M} \preccurlyeq \boldsymbol{N}_1$. □

The downward and upward Löwenheim-Skolem Theorems (B.5.10 and B.8.10) yield:

COROLLARY B.8.11. *If* Σ *has an infinite model, then* Σ *has a model of any given size* $\kappa \geqslant |\mathcal{L}|$*.*

COROLLARY B.8.12 (Vaught's Test). *Suppose* $|\mathcal{L}| \leqslant \kappa$*,* Σ *has a model, all models of* Σ *are infinite, and all models of* Σ *of size* κ *are isomorphic. Then* Σ *is complete.*

PROOF. Assume σ is true in some model of Σ. Then σ is true in some model of Σ of size κ, by Corollary B.8.11. The isomorphism assumption then gives that σ is true in all models of Σ of size κ, and so again by Corollary B.8.11, σ is true in all models of Σ. It follows that Σ is complete. □

Vaught's Test is a test for completeness. Here is an application:

THEOREM B.8.13. *Let p be a prime number or $p = 0$. The set* $\mathrm{ACF}(p)$ *of axioms for algebraically closed fields of characteristic p is complete.*

PROOF. We use basic facts about transcendence bases; see [249, Chapter VIII]. Let K and L be algebraically closed fields of characteristic p of the same uncountable size; let $\boldsymbol{k} := \mathbb{F}_p$ if p is a prime and $\boldsymbol{k} := \mathbb{Q}$ if $p = 0$. View $\boldsymbol{k}$ as a subfield of K and of L as usual, and note that $|\boldsymbol{k}| \leqslant \aleph_0$. Let B be a transcendence basis of K over $\boldsymbol{k}$ and C a transcendence basis of L over $\boldsymbol{k}$. Then $|K| = |\boldsymbol{k}(B)| = |B|$ and likewise $|L| = |C|$. So there is a bijection $B \to C$, and this bijection extends to a field isomorphism $\boldsymbol{k}(B) \to \boldsymbol{k}(C)$ and then further to an isomorphism $K \to L$ between their algebraic closures. Thus $\mathrm{ACF}(p)$ is complete by Vaught's Test. □

Theorem B.8.13 and Corollary B.6.9 imply that $\mathrm{ACF}(p)$ is decidable. Applications of Theorem B.8.13 and another proof of this theorem are given in Section B.12.

Notes and comments

Diagrams and their role are explicit in A. Robinson [345]. Corollary B.8.3 is in Henkin [174]. A sharper version of Corollary B.8.4 is due to Keisler [214] and Shelah [404]: if $\boldsymbol{M} \equiv \boldsymbol{N}$ then there is an ultrafilter $\mathcal{U}$ on some nonempty set Λ with $\boldsymbol{M}^\Lambda/\mathcal{U} \cong \boldsymbol{N}^\Lambda/\mathcal{U}$. Proposition B.8.6 is a weak version of a theorem of Herbrand [177]. Corollary B.8.8 is due to Łoś [272] and Tarski [445], and Corollary B.8.12 to Łoś [270] and Vaught [455].

B.9 TYPES AND SATURATED STRUCTURES

Roughly speaking, a *type* is a set of formulas specifying a potential property of a tuple of elements in a structure, similar to a system of equations and inequalities that we wish to solve. A *saturated structure* realizes many types.

Types

Let $\Phi = \Phi(x)$ be a set of $(\mathcal{L}, x)$-formulas. We say that $a \in M_x$ **realizes Φ in $\boldsymbol{M}$** if $\boldsymbol{M} \models \varphi(a)$ for all $\varphi \in \Phi$. Clearly if $\boldsymbol{M} \preccurlyeq \boldsymbol{N}$ and $a \in M_x$, then a realizes Φ in $\boldsymbol{M}$ iff a realizes Φ in $\boldsymbol{N}$. We say that Φ **is realized in $\boldsymbol{M}$** if some $a \in M_x$ realizes Φ in $\boldsymbol{M}$. Assume $x = (x_i)_{i \in I}$ and take a tuple $c = (c_i)_{i \in I}$ of distinct new constant symbols with each c_i of the same sort as x_i. Let $\mathcal{L}_c$ be $\mathcal{L}$ augmented by these new constant symbols c_i. Then Φ is realized in some structure iff the set

$$\Phi(c) := \{\varphi(c) : \varphi \in \Phi\}$$

of $\mathcal{L}_c$-sentences has a model. Hence by compactness, Φ is realized in some structure iff every finite subset of Φ is realized in some structure. If this happens we call Φ an **x-type**. An x-type Φ is said to be **complete** if for each $\varphi(x)$, either $\varphi \in \Phi$ or $\neg\varphi \in \Phi$; equivalently, $\Phi(c)$ is a complete $\mathcal{L}_c$-theory. For $a \in M_x$ we let $\mathrm{tp}_x^{\boldsymbol{M}}(a)$ be the complete

x-type in $\boldsymbol{M}$ realized by a (and we leave out the superscript $\boldsymbol{M}$ or subscript x if $\boldsymbol{M}$ or x are clear from the context); that is, for each $\mathcal{L}$-formula $\varphi(x)$, we have $\varphi \in \mathrm{tp}_x^{\boldsymbol{M}}(a)$ iff $\boldsymbol{M} \models \varphi(a)$. Every x-type is contained in a complete one, namely one of the form $\mathrm{tp}_x^{\boldsymbol{M}}(a)$. If $\boldsymbol{M} \preccurlyeq \boldsymbol{N}$ and $a \in M_x$, then $\mathrm{tp}_x^{\boldsymbol{M}}(a) = \mathrm{tp}_x^{\boldsymbol{N}}(a)$.

DEFINITION B.9.1. We say that Φ is **Σ-realizable** if Φ is realized in some model of Σ; that is, if $\Sigma \cup \Phi(c)$, as a set of $\mathcal{L}_c$-sentences, has a model. The set of all complete Σ-realizable x-types is denoted by $\mathrm{S}_x(\Sigma)$.

Thus for a *complete* x-type Φ, we have: Φ is Σ-realizable iff $\Sigma \subseteq \Phi(c)$. A complete x-type is usually denoted by a letter like p or q.

Let Σ_c be Σ viewed as set of $\mathcal{L}_c$-sentences. Then for $p = p(x) \in \mathrm{S}_x(\Sigma)$ we have $p(c) \in \mathrm{S}(\Sigma_c)$, and the map $p \mapsto p(c) \colon \mathrm{S}_x(\Sigma) \to \mathrm{S}(\Sigma_c)$ is a bijection. The **Stone topology** on $\mathrm{S}_x(\Sigma)$ is the topology making this bijection a homeomorphism. That is, its basic open sets are the sets $\langle\varphi\rangle := \{p \in \mathrm{S}_x(\Sigma) : \varphi \in p\}$, with $\varphi = \varphi(x)$. Two formulas $\varphi(x)$ and $\psi(x)$ are Σ-equivalent iff $\langle\varphi\rangle = \langle\psi\rangle$. By Theorem B.6.7, the Stone topology makes $\mathrm{S}_x(\Sigma)$ a compact hausdorff space.

Separating types

In this subsection we fix a set $\Theta = \Theta(x)$ of $(\mathcal{L}, x)$-formulas such that $\top, \bot \in \Theta$, and for all $\theta_1, \theta_2 \in \Theta$, also $\theta_1 \wedge \theta_2 \in \Theta$ and $\theta_1 \vee \theta_2 \in \Theta$. For example, Θ could be the set of quantifier-free formulas $\theta(x)$. In Section B.11 we shall need the next lemma; its corollary was already used in Section 8.3.

LEMMA B.9.2. *Let a formula $\psi(x)$ be given. Then the following are equivalent:*

(i) *ψ is Σ-equivalent to some formula from Θ;*

(ii) *for all $p, q \in \mathrm{S}_x(\Sigma)$ with $\psi \in p$ and $\neg\psi \in q$ there exists $\theta \in \Theta$ such that $\theta \in p$ and $\neg\theta \in q$.*

PROOF. The direction (i) $\Rightarrow$ (ii) is clear. Conversely, assume (ii). Consider the open-and-closed subset $P := \langle\psi\rangle$ of $\mathrm{S}_x(\Sigma)$ and its complement $P^{\mathrm{c}} = \langle\neg\psi\rangle$. Let $p \in P$ be given. Then $P^{\mathrm{c}} \subseteq \bigcup_{\theta \in \Theta \cap p} \langle\neg\theta\rangle$ by (ii). Compactness of $\mathrm{S}_x(\Sigma)$ and $\Theta \cap p$ being closed under conjunction gives $\theta \in \Theta \cap p$ such that $P^{\mathrm{c}} \subseteq \langle\neg\theta\rangle$ and hence $p \in \langle\theta\rangle \subseteq P$. Since $p \in P$ was arbitrary, this yields $P = \bigcup_{\theta \in \Delta} \langle\theta\rangle$ for some $\Delta \subseteq \Theta$. By compactness of $\mathrm{S}_x(\Sigma)$ again and Θ being closed under disjunction, we obtain $\theta \in \Theta$ with $P = \langle\theta\rangle$, and then ψ is Σ-equivalent to θ. □

COROLLARY B.9.3. *Suppose $\neg\theta \in \Theta$ for all $\theta \in \Theta$. Then every formula $\psi(x)$ is Σ-equivalent to one in Θ iff $p \cap \Theta \neq q \cap \Theta$ for all $p \neq q$ in $\mathrm{S}_x(\Sigma)$.*

Types over a parameter set

In the rest of this section A is a parameter set in $\boldsymbol{M}$. An **x-type over A in $\boldsymbol{M}$** is a $\mathrm{Th}(\boldsymbol{M}_A)$-realizable x-type (in the language $\mathcal{L}_A$). Equivalently, a set of $(\mathcal{L}_A, x)$-formulas is an x-type over A in $\boldsymbol{M}$ if every finite subset of it is realized in $\boldsymbol{M}_A$.

For $b \in M_x$ we let $\mathrm{tp}_x^{\boldsymbol{M}}(b|A) := \mathrm{tp}_x^{\boldsymbol{M}_A}(b)$ be the complete x-type over A in $\boldsymbol{M}$ realized by b (and we leave out the superscript $\boldsymbol{M}$ or subscript x if $\boldsymbol{M}$ or x are clear from the context). If $\boldsymbol{M} \preccurlyeq \boldsymbol{N}$ and $b \in M_x$, then $\mathrm{tp}_x^{\boldsymbol{M}}(b|A) = \mathrm{tp}_x^{\boldsymbol{N}}(b|A)$. Let $\mathrm{S}_x^{\boldsymbol{M}}(A)$ (or $\mathrm{S}_x(A)$ if $\boldsymbol{M}$ is clear from the context) denote the space $\mathrm{S}_x\big(\mathrm{Th}(\boldsymbol{M}_A)\big)$ of complete x-types over A in $\boldsymbol{M}$. A basis for the Stone topology on $\mathrm{S}_x(A)$ is given by the sets $\langle\varphi\rangle = \{p \in \mathrm{S}_x(A) : \varphi \in p\}$ with $\varphi(x)$ an $\mathcal{L}_A$-formula. Note that if $\mathcal{L}$, x, A all have size $\leqslant \kappa$, then $|\mathrm{S}_x(A)| \leqslant 2^\kappa$.

Saturated structures

In this subsection κ is a cardinal > 0. We declare $\boldsymbol{M}$ to be **κ-saturated** if for all A of size $< \kappa$ and every variable v of $\mathcal{L}$, each complete v-type over A in $\boldsymbol{M}$ is realized in $\boldsymbol{M}$; equivalently, for all A of size $< \kappa$ and $s \in S$, each collection of A-definable subsets of M_s with the finite intersection property has a nonempty intersection.

EXAMPLE. Suppose the ordered abelian group $(G; +, -, 0, \leqslant)$ with $G \neq \{0\}$ is 2-saturated. Then $[G]$ has no largest element: given $a \in G^>$ a realization b of the v-type $\{v > na : n = 0, 1, 2, \dots\}$ over $\{a\}$ yields $[b] > [a]$. In particular, the ordered abelian group $(\mathbb{R}; +, -, 0, \leqslant)$ is not 2-saturated.

If M_s is finite for all s, then $\boldsymbol{M}$ is κ-saturated for all κ. If $\boldsymbol{M}$ is κ-saturated and M_s is infinite, then $|M_s| \geqslant \kappa$. (Take a variable v of sort s and consider the v-type $\{v \neq a : a \in M_s\}$.) If $\kappa \leqslant \kappa'$ and $\boldsymbol{M}$ is κ'-saturated, then $\boldsymbol{M}$ is κ-saturated. If $\boldsymbol{M}$ is κ-saturated, then so is any reduct of $\boldsymbol{M}$. If $\boldsymbol{M}$ is κ-saturated, κ is infinite, and $|A| < \kappa$, then $\boldsymbol{M}_A$ is κ-saturated.

The definition of "κ-saturated" ostensibly only concerns families of subsets of M_s for $s \in S$. It is a pleasant feature of model theory, however, that a one-variable property often yields a many-variable analogue, with some effort as in this case:

LEMMA B.9.4. *Suppose $\boldsymbol{M}$ is κ-saturated, κ is infinite, A has size $< \kappa$ and x has size $\leqslant \kappa$. Then every x-type over A in $\boldsymbol{M}$ is realized in $\boldsymbol{M}$.*

PROOF. Let $x = (x_i)_{i \in I}$, and let i, j range over I. Let $p \in \mathrm{S}_x^{\boldsymbol{M}}(A)$; it suffices to show that p is realized in $\boldsymbol{M}$. Take a well-ordering $\leqslant$ of I of order type $\leqslant |I|$. Then each proper downward closed subset of I has cardinality $< |I|$. For each j let $x_{\leqslant j} := (x_i)_{i \leqslant j}$ and let $p_{\leqslant j}$ be the set of all formulas in p with free variables in $x_{\leqslant j}$; then $p_{\leqslant j}$ is a complete $x_{\leqslant j}$-type over A in $\boldsymbol{M}$. Similarly, for each $j \in I$ we define $x_{<j}$ and the complete $x_{<j}$-type $p_{<j}$ over A in $\boldsymbol{M}$; then for each $\varphi \in p_{\leqslant j}$ we have $\exists x_j \varphi \in p_{<j}$. By recursion on i we construct a point $(b_i) \in M_x$ such that for each j, $(b_i)_{i \leqslant j}$ realizes $p_{\leqslant j}$. Suppose that for a certain j we have already a point $b = (b_i)_{i<j} \in M_{x_{<j}}$ realizing $p_{<j}$. Then $p_j := \big\{\varphi(b/x_{<j}) : \varphi \in p_{\leqslant j}\big\}$ is an x_j-type over $A \cup \{b_i : i < j\}$ in $\boldsymbol{M}$. Since $A \cup \{b_i : i < j\}$ has size $< \kappa$ and $\boldsymbol{M}$ is κ-saturated, we have $b_j \in M_{x_j}$ realizing p_j. Then $(b_i)_{i \leqslant j}$ realizes $p_{\leqslant j}$. □

In κ-saturated structures one can do things that otherwise would require passing to an elementary extension. As an example, here is a variant of Corollary B.8.4:

COROLLARY B.9.5. *Suppose that $\boldsymbol{M} \equiv \boldsymbol{N}$ and $\boldsymbol{N}$ is κ-saturated for some infinite $\kappa \geqslant |M|$. Then there exists an elementary embedding $\boldsymbol{M} \to \boldsymbol{N}$.*

PROOF. Assume for simplicity that our language $\mathcal{L}$ is one-sorted. Take a multivariable $x = (x_a)_{a \in M}$, and let $\vec{a} := (a)_{a \in M}$. Since $\boldsymbol{M} \equiv \boldsymbol{N}$, $\Phi := \{\varphi(x) : \boldsymbol{M} \models \varphi(\vec{a})\}$ is an x-type over $\emptyset$ in $\boldsymbol{N}$. By Lemma B.9.4, Φ is realized in $\boldsymbol{N}$. If $(a')_{a \in M}$ realizes Φ in $\boldsymbol{N}$, then $a \mapsto a'$ $(a \in M)$ is an elementary embedding $\boldsymbol{M} \to \boldsymbol{N}$. □

Every structure has a κ-saturated elementary extension:

PROPOSITION B.9.6. *Suppose $|\mathcal{L}| \leqslant \kappa$ and $|M| \leqslant 2^\kappa$. Then $\boldsymbol{M}$ has a κ^+-saturated elementary extension $\boldsymbol{N}$ with $|N| \leqslant 2^\kappa$.*

We first establish the following lemma, with the same assumptions on κ as in Proposition B.9.6.

LEMMA B.9.7. *There exists $\boldsymbol{M}' \succcurlyeq \boldsymbol{M}$ with $|M'| \leqslant 2^\kappa$ such that for all $A \subseteq M$ with $|A| \leqslant \kappa$ and any variable v of $\mathcal{L}$, each type in $\mathrm{S}_v^{\boldsymbol{M}}(A)$ is realized in $\boldsymbol{M}'$.*

PROOF. There are at most 2^κ many pairs (A, v) with $A \subseteq M$, $|A| \leqslant \kappa$, and v a quantifiable variable of $\mathcal{L}$. For such (A, v) we have $|\mathrm{S}_v^{\boldsymbol{M}}(A)| \leqslant 2^\kappa$; take for each $p \in \mathrm{S}_v^{\boldsymbol{M}}(A)$ a new constant symbol c_p of the same sort as v. Let $\mathcal{L}'$ be $\mathcal{L}$ augmented by these c_p. By compactness and Proposition B.5.10 applied to the set

$$\mathrm{Th}(\boldsymbol{M}_M) \cup \bigcup \{p(c_p) : p \in \mathrm{S}_v^{\boldsymbol{M}}(A),\ A \subseteq M,\ |A| \leqslant \kappa,\ v \text{ is quantifiable}\}$$

of $\mathcal{L}'_M$-sentences, we obtain an elementary extension $\boldsymbol{M}'$ of $\boldsymbol{M}$ as desired. □

Below we use the fact that the least cardinal κ^+ that is greater than κ is a regular ordinal as defined in Section 2.1; see [168, Theorem 3.1.11].

PROOF OF PROPOSITION B.9.6. Let α, β range over ordinals $< \kappa^+$. Recursion on α yields a sequence $(\boldsymbol{M}_\alpha)$ of structures of size $\leqslant 2^\kappa$ such that:

(1) $\boldsymbol{M}_0 = \boldsymbol{M}$ and $\boldsymbol{M}_\alpha \preccurlyeq \boldsymbol{M}_\beta$ if $\alpha \leqslant \beta$;

(2) if $\alpha = \beta + 1$ is a successor ordinal, then $\boldsymbol{M}_\beta$ is obtained from $\boldsymbol{M}_\alpha$ as $\boldsymbol{M}'$ is from $\boldsymbol{M}$ in the previous lemma;

(3) if $\alpha > 0$ is a limit ordinal, then $\boldsymbol{M}_\alpha = \bigcup_{\beta < \alpha} \boldsymbol{M}_\beta$.

Then $\boldsymbol{N} := \bigcup_\alpha \boldsymbol{M}_\alpha$ is an elementary extension of $\boldsymbol{M}$ with $|N| \leqslant 2^\kappa$. Suppose $B \subseteq N$ and $|B| < \kappa^+$. Since κ^+ is regular, we have β such that $B \subseteq M_\beta$. Then every $p \in \mathrm{S}_v^{\boldsymbol{N}}(B) = \mathrm{S}_v^{\boldsymbol{M}_\beta}(B)$ is realized in $\boldsymbol{M}_{\beta+1}$ and thus in $\boldsymbol{N}$. □

Notes and comments

The notion of κ-saturated structure goes back to the η_α-sets of Hausdorff [172, p. 181], but only appeared in model theory in the late 1950s in the work of Morley and Vaught [297, 298, 456], who also introduced types. See [136] for the history. A motivation for Hausdorff was du Bois-Reymond's "infinitary pantachy" [53] (the partially ordered set of germs at $+\infty$ of continuous real-valued functions): see [126, 431].

B.10 MODEL COMPLETENESS

If the primitives of an $\mathcal{L}$-structure $\boldsymbol{M}$ are computationally or topologically well-behaved, then the sets defined in $\boldsymbol{M}$ by quantifier-free formulas are often fairly tame as well. For example, the subsets of $\mathbb{R}$ definable by quantifier-free formulas in the ordered field of real numbers (viewed as an $\mathcal{L}_{\mathrm{OR}}$-structure) are finite unions of open intervals and singletons. In the ordered ring of integers, the subsets of $\mathbb{Z}$ definable by quantifier-free formulas are just the traces of the preceding sets in $\mathbb{Z}$. But taking arbitrary $\mathcal{L}_{\mathrm{OR}}$-formulas, one can define much more complicated sets in the ordered ring of integers: for example, the formula $\pi(x)$ given by

$$1+1 \leqslant x \,\&\, \forall u \forall v\big((1 \leqslant u \,\&\, 1 \leqslant v \,\&\, u \cdot v = x) \;\rightarrow\; u = 1 \;\vee\; v = 1\big)$$

defines the set of prime numbers. In general, the more quantifiers that occur in a formula $\varphi(x)$, the more complicated the set $\varphi^{\boldsymbol{M}}$ defined by $\varphi(x)$ in $\boldsymbol{M}$ can be. When this typical behavior does *not* occur, it is worth noting!

This is why in this section we consider *model completeness*. In the next section we study a sharper version of this, called *quantifier elimination*. A structure $\boldsymbol{M}$ is said to be **model complete** if for each formula $\varphi(x)$ there exists an existential formula $\psi(x)$ such that $\varphi^{\boldsymbol{M}} = \psi^{\boldsymbol{M}}$. For one-sorted model complete $\boldsymbol{M}$, every definable subset of M^m is, for some n, the image of a quantifier-free definable set in M^{m+n} under the natural projection map $M^{m+n} \to M^m$. Such projections often preserve many desirable topological properties, for example, having only finitely many connected components (in a suitable topological environment).

Model completeness also makes sense for a set of sentences:

DEFINITION B.10.1. Σ is said to be **model complete** if every formula is Σ-equivalent to an existential formula. (Note that then every formula $\varphi(x)$ is actually Σ-equivalent to an $\exists$-formula $\varphi'(x)$ with no more free variables than those of φ.)

Thus $\boldsymbol{M}$ is model complete iff $\mathrm{Th}(\boldsymbol{M})$ is model complete. The following lemma slightly reduces the job of proving model completeness:

LEMMA B.10.2. *Suppose every universal formula is Σ-equivalent to an existential formula. Then Σ is model complete.*

PROOF. We show by induction that every formula φ is Σ-equivalent to an $\exists$-formula. This is clear if φ is quantifier-free, and the conclusion is preserved under $\wedge$, $\vee$, and $\exists$. Suppose $\varphi = \neg\psi$. Assuming inductively that ψ is Σ-equivalent to an $\exists$-formula, φ is Σ-equivalent to a $\forall$-formula, and thus by hypothesis, φ is Σ-equivalent to an $\exists$-formula. The case where $\varphi = \forall y \psi$ with a single quantifiable variable y follows, since then φ is equivalent to $\neg\exists y \neg\psi$. □

Here is Robinson's model completeness test:

PROPOSITION B.10.3. *The following are equivalent:*

(i) *Σ is model complete;*

(ii) *for all models $\boldsymbol{M}$, $\boldsymbol{N}$ of Σ, if $\boldsymbol{M} \subseteq \boldsymbol{N}$, then $\boldsymbol{M} \preccurlyeq \boldsymbol{N}$;*

(iii) *for all models $\boldsymbol{M}$, $\boldsymbol{N}$ of Σ, if $\boldsymbol{M} \subseteq \boldsymbol{N}$, then $\boldsymbol{M} \preccurlyeq_{\exists} \boldsymbol{N}$.*

PROOF. Clearly (i) $\Rightarrow$ (ii) $\Rightarrow$ (iii). Assume (iii), and let $\varphi(x)$ be universal. Then for $\boldsymbol{M}, \boldsymbol{N} \models \Sigma$ with $\boldsymbol{M} \subseteq \boldsymbol{N}$ and $a \in M_x$ we have $\boldsymbol{M} \models \varphi(a) \Rightarrow \boldsymbol{N} \models \varphi(a)$. Hence φ is Σ-equivalent to an $\exists$-formula, by Corollary B.8.9. Thus Σ is model complete by the previous lemma. □

Condition (ii) in Proposition B.10.3 means that $\Sigma \cup \mathrm{Diag}_M(\boldsymbol{M})$ is complete, for all $\boldsymbol{M} \models \Sigma$; this explains the terminology *model complete.* Model completeness often entails completeness: A **prime model** of Σ is a model of Σ that embeds elementarily into every model of Σ. Note that if Σ has a prime model, then Σ is complete. If Σ is model complete and $\boldsymbol{M}$ is a model of Σ that embeds into every model of Σ, then $\boldsymbol{M}$ is a prime model of Σ.

Here is a variant of the above test for model completeness:

COROLLARY B.10.4. *The following are equivalent:*

(i) *Σ is model complete;*

(ii) *for all models $\boldsymbol{M}$, $\boldsymbol{N}$ of Σ with $\boldsymbol{M} \subseteq \boldsymbol{N}$ and every elementary extension $\boldsymbol{M}^*$ of $\boldsymbol{M}$ that is κ-saturated for some $\kappa > |N|$, there is an embedding $\boldsymbol{N} \to \boldsymbol{M}^*$ that extends the natural inclusion $\boldsymbol{M} \to \boldsymbol{M}^*$.*

PROOF. Suppose Σ is model complete, and $\boldsymbol{M}$, $\boldsymbol{N}$, $\boldsymbol{M}^*$ are as in the hypothesis of (ii). Then $\Sigma \cup \mathrm{Diag}_M(\boldsymbol{M})$ is complete, with models $\boldsymbol{M}^*_M$, $\boldsymbol{N}_M$. Since $\boldsymbol{M}^*_M$ is $|N|$-saturated, Corollary B.9.5 yields an embedding $\boldsymbol{N}_M \to \boldsymbol{M}^*_M$. This shows (i) $\Rightarrow$ (ii). The implication (ii) $\Rightarrow$ (i) follows from Corollary B.8.5, Proposition B.9.6 and the equivalence of (i) and (iii) in Theorem B.10.3. □

EXAMPLE B.10.5. Let $\mathcal{L} := \mathcal{L}_{\mathrm{A}}$ be the language of additive groups and

$$\Sigma := \mathrm{Tf} \cup \mathrm{Div} \cup \mathrm{Ab} \cup \big\{\exists x\,(x \neq 0)\big\}.$$

Then the models of Σ are the nontrivial divisible torsion-free abelian groups: see Example B.6.1(2),(3). A model of Σ may be viewed as a $\mathbb{Q}$-linear space in the natural way, and if the model is κ-saturated, then it has dimension $\geqslant \kappa$. Together with Corollary B.10.4 this yields model completeness of Σ.

Before trying to show that a certain theory is model complete, one better check that it has an axiomatization by $\forall\exists$-sentences:

PROPOSITION B.10.6. *Suppose Σ is model complete. Then Σ is equivalent to a set of $\forall\exists$-sentences.*

PROOF. Let $\Sigma_\forall$ be as in Proposition B.8.7 (so the models of $\Sigma_\forall$ are the substructures of models of Σ). Every sentence of the form $\forall x(\varphi \to \psi)$, where $\varphi(x)$ is universal and $\psi(x)$ is existential, is equivalent to a $\forall\exists$-sentence. Let $\Sigma_{\forall\exists}$ be the set of all

$\forall\exists$-sentences equivalent to a sentence $\forall x(\varphi \to \psi)$, where $\varphi(x)$ is universal, $\psi(x)$ is existential, and φ, ψ are Σ-equivalent. We claim that Σ and $\Sigma' := \Sigma_\forall \cup \Sigma_{\forall\exists}$ are equivalent. Note first that Σ' is also model complete: Let $\varphi(x)$ be universal, and take an $\exists$-formula $\psi(x)$ which is Σ-equivalent to φ; then φ is also Σ'-equivalent to ψ, since up to equivalence $\forall x(\psi \to \varphi)$ lies in $\Sigma_\forall$ and $\forall x(\varphi \to \psi)$ in $\Sigma_{\forall\exists}$. Let $\boldsymbol{M} \models \Sigma'$; we need to show $\boldsymbol{M} \models \Sigma$. Now $\boldsymbol{M} \models \Sigma_\forall$ yields some $\boldsymbol{N} \models \Sigma$ with $\boldsymbol{M} \subseteq \boldsymbol{N}$. Then $\boldsymbol{N} \models \Sigma'$ and so $\boldsymbol{M} \preccurlyeq \boldsymbol{N}$ by model completeness of Σ', thus $\boldsymbol{M} \models \Sigma$. □

Call Σ **inductive** if the direct union of any directed family of models of Σ is a model of Σ. If Σ is a set of $\forall\exists$-sentences, then Σ is inductive. Hence by B.10.6:

COROLLARY B.10.7. *If Σ is model complete, then Σ is inductive.*

Existentially closed models

An **existentially closed model of** Σ is a model $\boldsymbol{M}$ of Σ that is existentially closed in every extension $\boldsymbol{N} \models \Sigma$ of $\boldsymbol{M}$. Thus by Robinson's test, Σ is model complete iff every model of Σ is an existentially closed model of Σ. Inductive theories have existentially closed models:

LEMMA B.10.8. *Suppose Σ is inductive and $\boldsymbol{M} \models \Sigma$. Then $\boldsymbol{M}$ extends to an existentially closed model of Σ.*

PROOF. We first show that $\boldsymbol{M}$ has an extension $\boldsymbol{M}^* \models \Sigma$ which satisfies every existential $\mathcal{L}_M$-sentence that holds in some extension of $\boldsymbol{M}^*$ to a model of Σ.

Let $(\sigma_\lambda)_{\lambda<\kappa}$ be an enumeration of all $\mathcal{L}_M$-sentences, where κ is the cardinality of the set of $\mathcal{L}_M$-sentences. By recursion on $\mu < \kappa$ we construct a sequence $(\boldsymbol{M}_\mu)_{\mu\leqslant\kappa}$ of extensions of $\boldsymbol{M}$ to models of Σ such that $\boldsymbol{M}_\lambda \subseteq \boldsymbol{M}_{\lambda'}$ for all $\lambda \leqslant \lambda' \leqslant \kappa$ as follows: Set $\boldsymbol{M}_0 := \boldsymbol{M}$. Suppose $0 < \mu \leqslant \kappa$ and $(\boldsymbol{M}_\lambda)_{\lambda<\mu}$ is a sequence of extensions of $\boldsymbol{M}$ to models of Σ such that $\boldsymbol{M}_\lambda \subseteq \boldsymbol{M}_{\lambda'}$ for all $\lambda \leqslant \lambda' < \mu$. If μ is a limit ordinal, then set $\boldsymbol{M}_\mu := \bigcup_{\lambda<\mu} \boldsymbol{M}_\lambda$; since Σ is inductive, we have $\boldsymbol{M}_\mu \models \Sigma$. Suppose $\mu = \mu' + 1$; if some extension of $\boldsymbol{M}_{\mu'}$ is a model of $\Sigma \cup \{\sigma_{\mu'}\}$, then we let $\boldsymbol{M}_\mu$ be such an extension, and otherwise $\boldsymbol{M}_\mu := \boldsymbol{M}_{\mu'}$. Finally, put $\boldsymbol{M}^* := \boldsymbol{M}_\kappa$. It is routine to check that $\boldsymbol{M}^*$ has the desired property.

Now inductively define $\boldsymbol{M}^0 := \boldsymbol{M}$ and $\boldsymbol{M}^{n+1} := (\boldsymbol{M}^n)^*$. Then $\bigcup_n \boldsymbol{M}^n$ is an existentially closed model of Σ which extends $\boldsymbol{M}$. □

Model companions and model completions

In this subsection we assume that T is an inductive $\mathcal{L}$-theory T. Even though T may not be model complete, we may hope for the existence of a theory whose models are exactly the existentially closed models of T. This suggests the following notion:

DEFINITION B.10.9. A **model companion** of T is a model complete $\mathcal{L}$-theory $T^* \supseteq T$ such that every model of T embeds into a model of T^*.

LEMMA B.10.10. *Let T^* be an $\mathcal{L}$-theory. Then T^* is a model companion of T iff the models of T^* are exactly the existentially closed models of T.*

PROOF. Suppose T^* is a model companion of T. Then every model $\boldsymbol{M}^*$ of T^* is an existentially closed model of T: suppose $\boldsymbol{N} \models T$ extends $\boldsymbol{M}^* \models T^*$; extend $\boldsymbol{N}$ to a model $\boldsymbol{N}^*$ of T^*; then $\boldsymbol{M}^* \preccurlyeq \boldsymbol{N}^*$ by model completeness of T^* and thus $\boldsymbol{M}^* \preccurlyeq_\exists \boldsymbol{N}$ by Corollary B.4.7(ii). Conversely, suppose $\boldsymbol{M}$ is an existentially closed model of T. Take $\boldsymbol{M}^* \models T^*$ with $\boldsymbol{M} \subseteq \boldsymbol{M}^*$; then $\boldsymbol{M} \preccurlyeq_\exists \boldsymbol{M}^*$ and thus $\boldsymbol{M} \models T^*$ by Proposition B.10.6. This shows the "only if" direction of the lemma; the "if" direction follows from Theorem B.10.3 and Lemma B.10.8. □

COROLLARY B.10.11. *There is at most one model companion of T.*

Thus if a model companion of T exists, we may speak of *the* model companion of T. Clearly T is model complete iff it is its own model companion.

EXAMPLE B.10.12. The $\mathcal{L}_{\text{A}}$-theory T of torsion-free abelian groups is inductive. We let T^* be the $\mathcal{L}_{\text{A}}$-theory of nontrivial divisible torsion-free abelian groups. Then T^* is model complete (Example B.10.5), and every $A \models T$ extends to some $A^* \models T^*$: if $A = \{0\}$, take $A^* := \mathbb{Q}$, and if $A \neq \{0\}$, let $A^* := A \otimes_{\mathbb{Z}} \mathbb{Q}$ be the divisible hull of A. Hence T^* is the model companion of T.

DEFINITION B.10.13. A **model completion** of T is a model companion T^* of T such that in addition $T^* \cup \operatorname{Diag}_A(\boldsymbol{A})$ is complete, for all $\boldsymbol{A} \models T$.

We say that Σ **has AP** (short for: Σ **has the amalgamation property**) if for all models $\boldsymbol{A}$, $\boldsymbol{M}_1$, $\boldsymbol{M}_2$ of Σ and embeddings $f_i \colon \boldsymbol{A} \to \boldsymbol{M}_i$ $(i = 1,2)$ there exist a model $\boldsymbol{N}$ of Σ and embeddings $g_i \colon \boldsymbol{M}_i \to \boldsymbol{N}$ $(i = 1,2)$ making the diagram

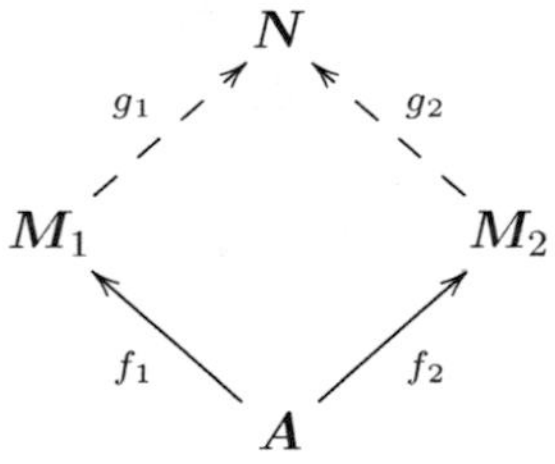

commute. Using the definitions and Corollary B.8.4 one easily shows:

LEMMA B.10.14. *Let T^* be an $\mathcal{L}$-theory. Then T^* is a model completion of T iff T^* is a model companion of T and T has* AP.

EXAMPLE B.10.15. Let T, T^* be as in Example B.10.12. Then T has AP. To see why, let embeddings $f_i \colon A \to M_i$ $(i = 1,2)$ of abelian groups be given. Set

$$F := \big\{(b,c) \in M_1 \times M_2 : \ nb = f_1(a),\ nc = -f_2(a) \text{ for some } n \geqslant 1 \text{ and } a \in A\big\},$$

a subgroup of the direct sum $M_1 \oplus M_2$ of M_1 and M_2. Put $N := (M_1 \oplus M_2)/F$; it is easy to check that N is torsion-free and that the natural morphisms $g_i \colon M_i \to N$ $(i = 1,2)$ are embeddings. Hence T^* is the model completion of T.

Notes and comments

The notion of model completeness and Proposition B.10.3 are due to A. Robinson [346, 349, 350, 354]. See [276] for a survey of classical model completeness results. In connection with B.10.6 and B.10.7 we should mention that Σ is inductive iff Σ is equivalent to a set of $\forall\exists$-sentences; this is due to Chang [72] and Łoś-Suszko [273]; see [186, Theorem 6.5.9] for a proof. Existentially closed models were introduced by Rabin [327], model companions by Barwise and Robinson [37], and the connection between the two notions is from [128]. By Cherlin [73], the $\mathcal{L}_{\mathrm{R}}$-theory of commutative rings has no model companion.

B.11 QUANTIFIER ELIMINATION

To keep formulations simple *we assume in this section that our language $\mathcal{L}$ has for each s a constant symbol of sort s.* Quantifier elimination is often the first key step to understanding the category of definable sets of a structure.

DEFINITION B.11.1. We say that Σ **admits quantifier elimination** (QE) if each formula $\varphi(x)$ with finite x is Σ-equivalent to a quantifier-free formula $\varphi'(x)$. We also express this by saying that Σ *has* QE or Σ *eliminates quantifiers.* We say that $\boldsymbol{M}$ admits QE if $\mathrm{Th}(\boldsymbol{M})$ admits QE.

(Of course, the restriction to finite x in this definition is superfluous.) Note that if Σ has QE, then Σ is model complete. In this section we give criteria for Σ to have QE. For a given structure $\boldsymbol{M}$ of interest, one tries to find a Σ that has $\boldsymbol{M}$ as a model and passes such a test. By Corollary B.9.3, Σ admits QE iff for every multivariable x, each type in $\mathrm{S}_x(\Sigma)$ is determined by its subset of quantifier-free formulas; this fact underlies the tests. To verify that Σ admits QE, it suffices to show that formulas of a special form are Σ-equivalent to quantifier-free formulas:

LEMMA B.11.2. *Suppose each formula $\exists y\, \theta(x,y)$ with quantifier-free θ, finite x, and a single variable y is Σ-equivalent to a quantifier-free formula. Then Σ has* QE.

PROOF. We proceed by induction on the number of logical symbols in a formula. To show that any formula $\varphi(x)$ is Σ-equivalent to a quantifier-free formula $\varphi'(x)$, we first note that this is obvious for quantifier-free φ, and that this property is preserved when taking disjunctions, conjunctions and negations. Suppose $\varphi = \exists y\psi$. We can assume inductively that $\psi(x,y)$ is Σ-equivalent to a quantifier-free formula $\psi'(x,y)$. The hypothesis of the lemma then yields a quantifier-free formula $\varphi'(x)$ that is Σ-equivalent to $\exists y\psi'(x,y)$, and hence to $\varphi(x)$. The case $\varphi = \forall y\psi$ reduces to the previous cases, since φ is equivalent to $\neg\exists y\neg\psi$. □

Before trying to show that a certain theory has QE, it might be a good idea to verify the following necessary condition, analogous to Proposition B.10.6:

COROLLARY B.11.3. *Suppose Σ has* QE. *Then Σ is equivalent to a set of sentences of the form $\forall x\exists y\theta(x,y)$ with finite x, a single variable y, and quantifier-free θ.*

PROOF. Let Σ' be the set of sentences $\forall x \exists y \theta(x, y)$ of the indicated form that are logical consequences of Σ. An easy exercise using Lemma B.11.2 shows that Σ' has QE. Let any sentence σ be given. Then we have a quantifier-free sentence σ' with $\Sigma' \models \sigma \leftrightarrow \sigma'$, and thus $\Sigma \models \sigma \leftrightarrow \sigma'$. Assume also that $\sigma \in \Sigma$. Then $\Sigma \models \sigma'$, hence $\Sigma' \models \sigma'$ (as σ' is quantifier-free) and thus $\Sigma' \models \sigma$. $\square$

A general elimination theorem

Suppose $\mathcal{L}$ is a sublanguage of the language $\mathcal{L}^*$. Let $S^* \supseteq S$ be the set of sorts for $\mathcal{L}^*$; assume that $\mathcal{L}^*$ has for each $s^* \in S^*$ a constant symbol of sort s^*. Let $\boldsymbol{M}^*$ and $\boldsymbol{N}^*$ range over $\mathcal{L}^*$-structures, and let $\boldsymbol{M}$ and $\boldsymbol{N}$ be their $\mathcal{L}$-reducts. Let Σ^* be a set of $\mathcal{L}^*$-sentences. Here is a criterion for an $\mathcal{L}^*$-formula to be Σ^*-equivalent to a quantifier-free $\mathcal{L}$-formula:

LEMMA B.11.4. *Let x be a multivariable of $\mathcal{L}$ and $\varphi^*(x)$ an $\mathcal{L}^*$-formula. Then $\varphi^*(x)$ is Σ^*-equivalent to a quantifier-free $\mathcal{L}$-formula $\varphi(x)$ iff for all $\boldsymbol{M}^*, \boldsymbol{N}^* \models \Sigma^*$, every common $\mathcal{L}$-substructure $\boldsymbol{A} = (A; \dots)$ of $\boldsymbol{M}$ and $\boldsymbol{N}$, and all $a \in A_x$:*

$$\text{(B.11.1)} \qquad \boldsymbol{M}^* \models \varphi^*(a) \iff \boldsymbol{N}^* \models \varphi^*(a).$$

PROOF. Forward direction: use Corollary B.4.7(i), and $\boldsymbol{M}^* \models \sigma$ iff $\boldsymbol{M} \models \sigma$, when σ is an $\mathcal{L}_M$-sentence. Conversely, suppose (B.11.1) holds for all $\boldsymbol{M}^*$, $\boldsymbol{N}^*$, $\boldsymbol{A}$ and a as specified in the lemma. Let $p, q \in \mathrm{S}_x(\Sigma^*)$ contain the same quantifier-free $\mathcal{L}$-formulas. By Lemma B.9.2 it is enough to show that then $\varphi^* \in p$ iff $\varphi^* \in q$. Take $\boldsymbol{M}^*, \boldsymbol{N}^* \models \Sigma^*$ and $a \in M_x$ realizing p in $\boldsymbol{M}^*$ and $b \in N_x$ realizing q in $\boldsymbol{N}^*$. Put $\boldsymbol{A} := \langle a \rangle_{\boldsymbol{M}}$ and $\boldsymbol{B} := \langle b \rangle_{\boldsymbol{N}}$. Since a and b realize the same quantifier-free $\mathcal{L}$-formulas in $\boldsymbol{M}$ and $\boldsymbol{N}$, respectively, we get an isomorphism $\boldsymbol{A} \to \boldsymbol{B}$ sending a to b. Then $\boldsymbol{M}^* \models \varphi^*(a)$ iff $\boldsymbol{N}^* \models \varphi^*(b)$, by (B.11.1), so $\varphi^* \in p$ iff $\varphi^* \in q$. $\square$

Typically, for $\mathcal{L}^* \neq \mathcal{L}$, the criterion in the lemma above gets used via its corollary below. To state that corollary, we define Σ^* to have **closures of $\mathcal{L}$-substructures** if for all $\boldsymbol{M}^*, \boldsymbol{N}^* \models \Sigma^*$ with a common $\mathcal{L}$-substructure $\boldsymbol{A} = (A; \dots)$ of $\boldsymbol{M}$ and $\boldsymbol{N}$, there is a (necessarily unique) isomorphism over A from the $\mathcal{L}^*$-substructure of $\boldsymbol{M}^*$ generated by A onto the $\mathcal{L}^*$-substructure of $\boldsymbol{N}^*$ generated by A.

COROLLARY B.11.5. *If Σ^* has closures of $\mathcal{L}$-substructures, then every quantifier-free $\mathcal{L}^*$-formula is Σ^*-equivalent to a quantifier-free $\mathcal{L}$-formula.*

EXAMPLE. Let $\mathcal{L} = \mathcal{L}_{\mathrm{R}}$ be the language of rings and let $\mathcal{L}^*$ be $\mathcal{L}$ augmented by a unary function symbol ι. Any field is naturally an $\mathcal{L}$-structure and expands to an $\mathcal{L}^*$-structure by $\iota(0) := 0$ and $\iota(a) := a^{-1}$ for nonzero a in the field. Note that

$$\Sigma^* := \mathrm{Fl} \cup \{\iota(0) = 0,\ \forall x\big(x \neq 0 \to x \cdot \iota(x) = 1\big)\}$$

has closures of $\mathcal{L}$-substructures, by the universal property of the fraction field of an integral domain. Hence every quantifier-free $\mathcal{L}^*$-formula is Σ^*-equivalent to a quantifier-free $\mathcal{L}$-formula. (This also follows easily by a simple direct argument; Corollaries 16.5.2 and 16.5.5 are more substantial applications.)

Application to QE

The expression $\boldsymbol{M} \equiv_A \boldsymbol{N}$ is short for $\boldsymbol{M}_A \equiv \boldsymbol{N}_A$; here A is a parameter set in both $\boldsymbol{M}$ and $\boldsymbol{N}$, that is, $A \subseteq M$ and $A \subseteq N$. Typically we use this notation when A underlies a common substructure $\boldsymbol{A}$ of $\boldsymbol{M}$ and $\boldsymbol{N}$. Taking $\mathcal{L}^* = \mathcal{L}$ in Lemma B.11.4 yields:

COROLLARY B.11.6. *Given Σ, the following conditions are equivalent:*

(i) *Σ has* QE*;*

(ii) *$\boldsymbol{M} \equiv_A \boldsymbol{N}$, for all $\boldsymbol{M}, \boldsymbol{N} \models \Sigma$ and any common substructure $\boldsymbol{A}$ of $\boldsymbol{M}$, $\boldsymbol{N}$;*

(iii) *$\Sigma \cup \mathrm{Diag}_A(\boldsymbol{A})$ is complete, for all $\boldsymbol{A} \models \Sigma_\forall$.*

COROLLARY B.11.7. *Suppose Σ admits* QE *and has a model. Then Σ is complete iff some $\mathcal{L}$-structure (not necessarily a model of Σ) embeds into every model of Σ.*

A test for quantifier elimination

We now come to our first QE test:

PROPOSITION B.11.8. *Let Σ be given. Then the following are equivalent:*

(i) *Σ has* QE*;*

(ii) *whenever $\boldsymbol{A}$ is a substructure of a model $\boldsymbol{M}$ of Σ, any embedding of $\boldsymbol{A}$ into any model $\boldsymbol{N}$ of Σ extends to an embedding of $\boldsymbol{M}$ into some $\boldsymbol{N}^* \succcurlyeq \boldsymbol{N}$.*

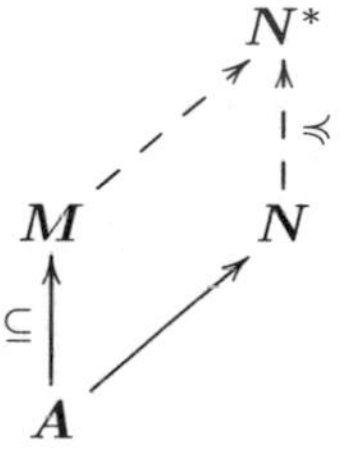

PROOF. Suppose Σ has QE, let $\boldsymbol{M}, \boldsymbol{N} \models \Sigma$, $\boldsymbol{A} \subseteq \boldsymbol{M}$, and let $i \colon \boldsymbol{A} \to \boldsymbol{N}$ be an embedding; we need to show that i extends to an embedding of $\boldsymbol{M}$ into some elementary extension of $\boldsymbol{N}$. Replacing $\boldsymbol{N}$ by an isomorphic structure we can assume that $\boldsymbol{A} \subseteq \boldsymbol{N}$ and i is the natural inclusion $\boldsymbol{A} \to \boldsymbol{N}$. Then by Corollary B.11.6 we have $\boldsymbol{M} \equiv_A \boldsymbol{N}$. Now use Corollary B.8.4 to obtain an extension of i to an (elementary) embedding of $\boldsymbol{M}$ into some elementary extension of $\boldsymbol{N}$.

For the converse, suppose (ii) holds. Let $\varphi(x)$ be an existential formula; by Lemma B.11.2 it suffices to show that $\varphi(x)$ is Σ-equivalent to a quantifier-free formula $\varphi'(x)$. For this we use Lemma B.11.4 in the case $\mathcal{L}^* = \mathcal{L}$. So let $\boldsymbol{M}, \boldsymbol{N} \models \Sigma$, let $\boldsymbol{A}$ be a common substructure of $\boldsymbol{M}$ and $\boldsymbol{N}$, and $a \in A_x$; by symmetry it is enough to show: $\boldsymbol{M} \models \varphi(a) \Rightarrow \boldsymbol{N} \models \varphi(a)$. Extend the natural inclusion $\boldsymbol{A} \to \boldsymbol{N}$ to an embedding of $\boldsymbol{M}$ into an elementary extension $\boldsymbol{N}^*$ of $\boldsymbol{N}$. Then

$$\boldsymbol{M} \models \varphi(a) \quad \Rightarrow \quad \boldsymbol{N}^* \models \varphi(a) \quad \Rightarrow \quad \boldsymbol{N} \models \varphi(a),$$

where we used Corollary B.4.7(ii) for the first implication. □

QE tests using saturation

Using saturated structures yields variants of the above test that are often easier to use:

COROLLARY B.11.9 (QE test, first variant). *Let Σ be given. Suppose that for all $\boldsymbol{M}, \boldsymbol{N} \models \Sigma$ with $|M|^+$-saturated $\boldsymbol{N}$ and all substructures $\boldsymbol{A}$ of $\boldsymbol{M}$, every embedding $\boldsymbol{A} \to \boldsymbol{N}$ extends to an embedding $\boldsymbol{M} \to \boldsymbol{N}$. Then Σ has* QE.

PROOF. Any structure has a κ-saturated elementary extension, for any $\kappa > 0$, so condition (ii) of Proposition B.11.8 is satisfied. □

The freedom to choose b in the next variant is sometimes convenient.

COROLLARY B.11.10 (QE test, second variant). *Let Σ be given. Suppose that for every $\boldsymbol{M} \models \Sigma$ and substructure $\boldsymbol{A}$ of $\boldsymbol{M}$ with $A \neq M$ and every embedding i of $\boldsymbol{A}$ into an $|A|^+$-saturated model $\boldsymbol{N}$ of Σ there exist $s \in S$ and $b \in M_s \setminus A_s$ such that i extends to an embedding $\boldsymbol{A}\langle b\rangle \to \boldsymbol{N}$. Then Σ has* QE.

PROOF. By Zorn the hypothesis of Corollary B.11.9 is satisfied. □

A **Σ-closure** of an $\mathcal{L}$-structure $\boldsymbol{A}$ is a model $\boldsymbol{A}_\Sigma$ of Σ with $\boldsymbol{A} \subseteq \boldsymbol{A}_\Sigma$ such that every embedding of $\boldsymbol{A}$ into a model $\boldsymbol{M}$ of Σ extends (not necessarily uniquely) to an embedding $\boldsymbol{A}_\Sigma \to \boldsymbol{M}$. Existence of Σ-closures simplifies the above test:

COROLLARY B.11.11 (QE test, third variant). *Let Σ be given. Assume:*

(i) *every substructure of every model of Σ has a Σ-closure; and*

(ii) *for all models $\boldsymbol{M}$, $\boldsymbol{N}$ of Σ with $\boldsymbol{M} \subseteq \boldsymbol{N}$ and $M \neq N$, and any $|M|^+$-saturated elementary extension $\boldsymbol{M}^*$ of $\boldsymbol{M}$ there exist $s \in S$, $b \in N_s \setminus M_s$, and an embedding of $\boldsymbol{M}\langle b\rangle$ into $\boldsymbol{M}^*$ over $\boldsymbol{M}$.*

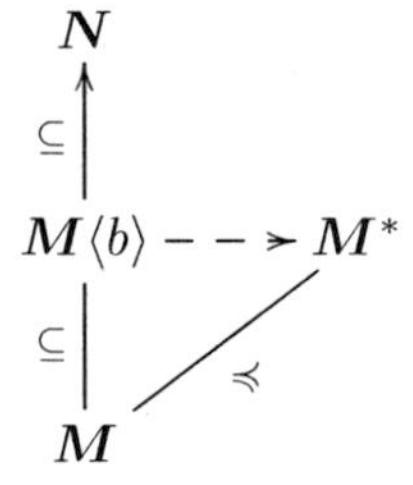

Then Σ has QE.

PROOF. Check that the hypothesis of B.11.10 is satisfied. □

EXAMPLE B.11.12. Suppose $\mathcal{L} = \mathcal{L}_{\mathrm{OA}}$ is the language of ordered abelian groups and $\Sigma = \mathrm{Div} \cup \mathrm{OAb} \cup \{\exists x\, x \neq 0\}$, so the models of Σ are the nontrivial divisible ordered abelian groups; see Example B.6.1(3),(5). Substructures of models of Σ are ordered abelian groups. Each ordered abelian group A has a Σ-closure A_Σ: if $A = \{0\}$, let A_Σ

be the ordered additive group $\mathbb{Q}$; if $A \neq \{0\}$, let $A_\Sigma := \mathbb{Q} \otimes_\mathbb{Z} A$ be the ordered divisible hull of A; see Section 2.4.

Let $M, N \models \Sigma$ with $M \subseteq N$, let M^* be an $|M|^+$-saturated elementary extension of M, and let $a \in N \setminus M$. Take an element $a^* \in M^*$ realizing the same cut in M as a. We have $M\langle a \rangle = M + \mathbb{Z}a = M \oplus \mathbb{Z}a$ and $M + \mathbb{Z}a^* = M \oplus \mathbb{Z}a^*$ (internal direct sums), so we get an embedding $M\langle a \rangle \to M^*$ over M sending a to a^*. Thus Σ has QE by B.11.11.

QE and definable closure

In this subsection we assume that Σ *is universal, and has* QE, *and* $\boldsymbol{M} \models \Sigma$. This has several useful consequences:

COROLLARY B.11.13. *Let* $A \subseteq M$. *Denoting the underlying set of the substructure* $\langle A \rangle_{\boldsymbol{M}}$ *of* $\boldsymbol{M}$ *by* $\langle A \rangle$, *we have* $\langle A \rangle = \operatorname{dcl}(A)$.

PROOF. This is because $\langle A \rangle \subseteq \operatorname{dcl}(A)$, and $\langle A \rangle_{\boldsymbol{M}} \preccurlyeq \boldsymbol{M}$. □

The next result is known as: $\boldsymbol{M}$ has definable Skolem functions.

COROLLARY B.11.14. *Let* $x = (x_1, \dots, x_m)$ *and* $y = (y_1, \dots, y_n)$ *be disjoint finite multivariables, and suppose the relation* $Z \subseteq M_x \times M_y$ *is* 0*-definable in* $\boldsymbol{M}$. *Let* $\pi \colon M_x \times M_y \to M_x$ *be the natural projection map. Then there is a* 0*-definable map* $f \colon M_x \to M_y$ *such that* $\big(a, f(a)\big) \in Z$ *for all* $a \in \pi(Z)$.

PROOF. Replacing Z by $Z \cup \big((M_x \setminus \pi(Z)) \times M_y\big)$ we arrange that $\pi(Z) = M_x$. Take a quantifier-free formula $\varphi(x, y)$ that defines Z in $\boldsymbol{M}$. Set

$$\Sigma(\boldsymbol{M}) := \Sigma \cup \{\sigma : \sigma \text{ is quantifier-free and } \boldsymbol{M} \models \sigma\}.$$

Then $\Sigma(\boldsymbol{M})$ is complete, so $\Sigma(\boldsymbol{M}) \models \forall x \exists y\, \varphi(x, y)$. Now use Proposition B.8.6. □

The proof above also shows that definable functions are piecewise given by terms:

COROLLARY B.11.15. *Suppose* $f \colon X \to M_s$ *with* $X \subseteq M_x$ *is* 0*-definable. Then there are terms* $t_1(x), \dots, t_k(x)$ *of sort* s, $k \in \mathbb{N}^{\geqslant 1}$, *such that for every* $a \in X$ *we have* $f(a) = t_i^{\boldsymbol{M}}(a)$ *for some* $i \in \{1, \dots, k\}$.

QE and model completeness

Let T be an $\mathcal{L}$-theory. By Corollary B.11.6, T has QE iff it is the model completion of $T_\forall$. Thus:

LEMMA B.11.16. *If* T *has a universal axiomatization and has a model completion, then this model completion has* QE.

EXAMPLE. Let $\mathcal{L} = \mathcal{L}_{\mathrm{OA}}$ and Σ be as in Example B.11.12: the models of Σ are the nontrivial divisible ordered abelian groups viewed as $\mathcal{L}$-structures. Then $\operatorname{Th}(\Sigma)$ is the model completion of the theory $\operatorname{Th}(\mathrm{OAb})$ of ordered abelian groups.

COROLLARY B.11.17. T *has* QE $\Longleftrightarrow$ T *is model complete and* $T_\forall$ *has* AP.

Notes and comments

QE goes back to Skolem [422, 423], Langford [250, 251], Presburger [321], and Tarski [439, 441, 443]. Corollary B.11.6 is from Sacks [377]. Corollary B.11.9 is close to Shoenfield [412, 413] and to [377, Theorem 17.2], the latter attributed there to L. Blum. Example B.11.12 is related to "Fourier-Motzkin elimination" [392, III, pp. 209–223]. See [110] for an application of Lemma B.11.15. Lemma B.11.16 is from [352] and Corollary B.11.17 from [128]. A survey of QE results from before 1984 is [462].

B.12 APPLICATION TO ALGEBRAICALLY CLOSED AND REAL CLOSED FIELDS

To demonstrate how the material of the previous sections is used in practice, we now apply it to *algebraically closed fields* and *real closed fields*.

Algebraically closed fields

We establish here the basic model-theoretic facts about algebraically closed fields: elimination of quantifiers, the Nullstellensatz, strong minimality, definably closed = perfect subfield, and definable functions.

In this subsection K, E, F are fields, and *algebraic over*, *algebraically closed*, and *algebraic closure* are taken in the sense of field theory; these notions will turn out to agree for algebraically closed ambient fields with the model-theoretic notions. By ACF we mean here the set of axioms for algebraically closed fields in the language of rings, and for p a prime number or $p = 0$ we let $\mathrm{ACF}(p)$ be the set of axioms for algebraically closed fields of characteristic p, as in Examples B.6.1(10),(11). Below p ranges over the set $\{0, 2, 3, 5, \dots\}$ of possible characteristics.

THEOREM B.12.1. ACF *has* QE.

PROOF. Every integral domain has an ACF-closure, namely the algebraic closure of its fraction field. Let E be algebraically closed, let K be a proper algebraically closed subfield of E, and let F be algebraically closed, $|K|^+$-saturated, also with K as a subfield. Take any $a \in E \setminus K$. Then $P(a) \neq 0$ for all $P(T) \in K[T]^{\neq}$, that is, a is transcendental over K. By saturation we can take $b \in F \setminus K$, so b is transcendental over K. Then the natural inclusion $K \to F$ extends to an embedding $K[a] \to F$ sending a to b. Thus ACF has QE by Corollary B.11.11. □

The substructures of algebraically closed fields are exactly the integral domains, so by the above, $\mathrm{Th}(\mathrm{ACF})$ is the model completion of the theory of integral domains. Here is a well-known manifestation (but not a special case) of Theorem B.12.1:

EXAMPLE. Let $X_1, \dots, X_n, Y$ be distinct indeterminates, $X = (X_1, \dots, X_n)$, and $P, Q \in \mathbb{Z}[X, Y]^{\neq}$ be monic in Y. Then there is a polynomial $R \in \mathbb{Z}[X]$ (the *resultant* of P and Q) such that for every algebraically closed E and $a \in E^n$: $R(a) = 0$ iff $P(a, Y), Q(a, Y) \in E[Y]$ have a common zero in E; see [249, IV, §8].

The following consequences of QE make up the Constructibility Theorem (Chevalley-Tarski) and the Nullstellensatz (Hilbert). Let $T = (T_1, \dots, T_n)$ be an n-tuple of distinct indeterminates.

COROLLARY B.12.2. *Let E be algebraically closed with subfield K.*

(i) *A subset of E^n is K-definable in E if and only if it is a boolean combination of zero sets $\{a \in E^n : P(a) = 0\}$ of polynomials $P \in K[T]$.*

(ii) *If $P_1, \dots, P_k, Q_1, \dots, Q_l \in K[T]$ and there is an overfield F of K with a point $a \in F^n$ such that*

$$P_1(a) = \cdots = P_k(a) = 0,\ Q_1(a) \neq 0, \dots, Q_l(a) \neq 0,$$

then there is such a point $a \in E^n$.

(iii) *Let $P_1, \dots, P_m \in K[T]$. Then $P_1, \dots, P_m$ have no common zero in E iff there are $Q_1, \dots, Q_m \in K[T]$ such that $P_1Q_1 + \cdots + P_mQ_m = 1$ (in $K[T]$).*

PROOF. Item (i) is immediate from QE. In (ii), extend F to be algebraically closed, and use that then $E \equiv_K F$ by QE. In (iii), suppose there are no $Q_1, \dots, Q_m \in K[T]$ such that $P_1Q_1 + \cdots + P_mQ_m = 1$. Then the ideal of $K[T]$ generated by $P_1, \dots, P_m$ is a proper ideal, and thus contained in a maximal ideal $\mathfrak{m}$ of $K[T]$. Put $t_i := T_i + \mathfrak{m}$ for $i = 1, \dots, n$ and $t := (t_1, \dots, t_n)$. Then $\mathfrak{m} \cap K = \{0\}$, so $K[T]/\mathfrak{m} = K[t]$ is a field extension of K, and $P(t) = P(T) + \mathfrak{m}$ for each $P \in K[T]$; in particular, $P_1, \dots, P_m$ have t as a common zero in an extension field of K, and thus $P_1, \dots, P_m$ have a common zero in E by (ii). □

The following is known as the *strong minimality* of algebraically closed fields. (This property alone has already many consequences.)

COROLLARY B.12.3. *Let E be algebraically closed. Then a set $X \subseteq E$ is definable in E iff X is finite or cofinite.*

PROOF. For a single indeterminate T and $P \in E[T]^{\neq}$ the set $\{a \in E : P(a) = 0\}$ is finite. Now use Corollary B.12.2(i) for $n = 1$. □

In Section B.8 we showed that ACF(p) is complete. This is also a consequence of Corollary B.11.7 and Theorem B.12.1, since for $p > 0$ the field $\mathbb{F}_p$ embeds into every model of ACF(p), and the ring $\mathbb{Z}$ embeds into every model of ACF(0). Thus an $\mathcal{L}_R$-sentence holds in *some* algebraically closed field of characteristic p iff it holds in *every* algebraically closed field of characteristic p. Moreover:

COROLLARY B.12.4. *Let σ be an $\mathcal{L}_R$-sentence. Then the following are equivalent:*

(i) *σ holds in some algebraically closed field of characteristic zero;*

(ii) *for all but finitely many prime numbers p, σ holds in all algebraically closed fields of characteristic p.*

PROOF. The direction (i) $\Rightarrow$ (ii) is clear from the remark preceding the corollary. Suppose conversely that $n_0 \in \mathbb{N}$ is such that σ holds in all algebraically closed fields of characteristic $p > n_0$. To get (i), it suffices to show that the set $\Sigma := \{\sigma\} \cup \mathrm{ACF}(0)$ of $\mathcal{L}_{\mathrm{R}}$-sentences has a model. Every finite $\Sigma_0 \subseteq \Sigma$ contains only finitely many sentences $p_1 1 \neq 0, \dots, p_n 1 \neq 0$ from $\mathrm{Fl}(0) \subseteq \mathrm{ACF}(0)$; since σ holds in every algebraically closed field of characteristic $p > \max\{n_0, p_1, \dots, p_n\}$, every such algebraically closed field is a model of Σ_0. Hence Σ has a model by compactness. This shows (ii) $\Rightarrow$ (i). □

Combining Example B.5.12 and Corollary B.12.4 yields:

COROLLARY B.12.5. *Let σ be a universal-existential $\mathcal{L}_{\mathrm{R}}$-sentence, and suppose that for all but finitely many prime numbers p, σ holds in all finite fields of characteristic p. Then σ holds in every algebraically closed field of characteristic zero.*

EXAMPLE B.12.6 (Serre). Suppose E is algebraically closed and $\mathrm{char}(E) \neq 2$. For $P = (P_1, \dots, P_n) \in E[T]^n$ and $a \in E^n$ set $P(a) := \big(P_1(a), \dots, P_n(a)\big) \in E^n$. Then *for any $P \in E[T]^n$ such that $P\big(P(a)\big) = a$ for all $a \in E^n$, there exists an $a \in E^n$ with $P(a) = a$.* To see this note that for fixed n and a bound d on the total degree of the P_i, the claim can be expressed as a universal-existential $\mathcal{L}_{\mathrm{R}}$-sentence $\sigma_{n,d}$. Obviously $\sigma_{n,d}$ holds in all finite fields of characteristic > 2, and so it holds in all algebraically closed fields of characteristic 0 by Corollary B.12.5, and in all algebraically closed fields of characteristic > 2 by Example B.5.12 and completeness of $\mathrm{ACF}(p)$, $p > 2$.

If E is algebraically closed with subset A, then the (model-theoretic) algebraic closure $\mathrm{acl}(A)$ of A in E contains obviously the field-theoretic algebraic closure in E of the subfield of E generated by A, and is in fact equal to (the underlying set of) this field-theoretic algebraic closure:

LEMMA B.12.7. *Let E be algebraically closed with algebraically closed subfield K. Then K is algebraically closed in E in the model theory sense.*

PROOF. Obvious from $K \preccurlyeq E$. □

If E is algebraically closed and $A \subseteq E$, then the definable closure $\mathrm{dcl}(A)$ of A in E contains at least the subfield of E generated by A, and equals (the underlying set of) this subfield when E has characteristic zero:

PROPOSITION B.12.8. *Let K be a subfield of the algebraically closed field E of characteristic zero. Then K is definably closed in E.*

PROOF. Let $a \in E \setminus K$; we claim that then $\sigma a \neq a$ for some $\sigma \in \mathrm{Aut}(E|K)$. (By Lemma B.5.15 the proposition follows from this claim.) If a is transcendental over K, take a transcendence basis B of E over K with $a \in B$, take the automorphism of $K(B)$ over K that sends each $b \in B$ to $b+1$, and then extend it to an automorphism of the algebraic closure E of $K(B)$. Suppose a is algebraic over K. Since $a \notin K$, the minimum polynomial of a over K is of degree > 1, so has a zero $b \in E$ with $b \neq a$ (here we use that E has characteristic zero). Take an automorphism σ of the algebraic closure K^{a} of K in E over K that sends a to b, take a transcendence basis B of E

over K^{a}, and extend σ to the automorphism of $K^{\mathrm{a}}(B)$ that is the identity on B, and then extend further to an automorphism of E. □

Characterizations of definable closures of this type lead to corresponding descriptions of definable functions. In this case definable functions are piecewise rational:

COROLLARY B.12.9. *Let E be algebraically closed of characteristic zero, with subfield K, and let $X \subseteq E^n$ and $f\colon X \to E$ be K-definable in E. Then there are $P_1, \dots, P_k, Q_1, \dots, Q_k \in K[T]$ such that for each $x \in X$ there is $i \in \{1, \dots, k\}$ with $Q_i(x) \neq 0$ and $f(x) = P_i(x)/Q_i(x)$.*

PROOF. Extending E if necessary we can assume E is $|K|^+$-saturated. Let $x \in X$. Then $f(x) \in \mathrm{dcl}(K \cup \{x\}) = K(x)$ by the proposition above, that is, $f(x) = P(x)/Q(x)$ with polynomials $P, Q \in K[T]$, $Q(x) \neq 0$. Now use saturation. □

Suppose E is algebraically closed of characteristic $p > 0$. Then we have a 0-definable automorphism $x \mapsto x^p$ of E, the *Frobenius map*, and the inverse $y \mapsto y^{1/p}$ of this map is not given piecewise by rational functions. The nth iterate $x \mapsto x^{p^n}$ of the Frobenius map has inverse $y \mapsto y^{1/p^n}$, and as we shall see, these inverse maps are the only obstructions in getting an analogue in positive characteristic of the above. Recall that a field K of characteristic $p > 0$ is said to be *perfect* if every element of K is a pth power x^p of some $x \in K$. So every finite field is perfect. For any subfield K of E there is a smallest perfect subfield of E containing K, namely

$$K^{1/p^\infty} := \bigcup_n K^{1/p^n} \quad \text{where } K^{1/p^n} := \{x^{1/p^n} : x \in K\} \subseteq E,$$

and by the next result K^{1/p^∞} is the definable closure of K in E.

PROPOSITION B.12.10. *Let E be algebraically closed of characteristic $p > 0$, with perfect subfield K. Then K is definably closed in E.*

The proof is identical to that of Proposition B.12.8, using the fact that an irreducible polynomial in one variable over a perfect field is separable.

COROLLARY B.12.11. *Let E be algebraically closed of characteristic $p > 0$, with perfect subfield K, and let $X \subseteq E^n$ and $f\colon X \to E$ be K-definable in E. Then there are $P_1, \dots, P_k, Q_1, \dots, Q_k \in K[T]$ and an m with the following property: for each $x \in X$ there is $i \in \{1, \dots, k\}$ such that*

$$Q_i(x^{1/p^m}) \neq 0 \text{ and } f(x) = P_i(x^{1/p^m})/Q_i(x^{1/p^m}),$$

where $x^{1/p^m} := (x_1^{1/p^m}, \dots, x_n^{1/p^m})$ for $x = (x_1, \dots, x_n) \in E^n$.

Here is an application, usually stated only for injective endomorphisms of algebraic varieties as a theorem of Ax:

COROLLARY B.12.12. *Let E be algebraically closed, and suppose $X \subseteq E^n$ and $f\colon X \to X$ are definable in E and f is injective. Then f is surjective.*

PROOF. Consider first the case that E is an algebraic closure of a finite field K of characteristic $p > 0$. After increasing K we can assume that X and f are definable over K. Now E is the union of the intermediate finite fields F with $K \subseteq F \subseteq E$, and all such F being perfect, it follows from Corollary B.12.11 that f maps $X(F) := X \cap F^n$ into $X(F)$, so $f\big(X(F)\big) = X(F)$ by injectivity of f and finiteness of F. Taking the union over all these F we get $f(X) = X$, so we are done for this particular E. The corollary is equivalent to certain $\mathcal{L}_{\mathrm{Ri}}$-sentences holding in all algebraically closed fields; we have shown these sentences hold in all algebraic closures of finite fields. Therefore they hold in all algebraically closed fields. □

Real closed fields

In this subsection we view ordered fields as structures in the language $\mathcal{L}_{\mathrm{OR}}$, and we let K, E and F denote ordered fields. We assume here that the reader is familiar with the basic *algebraic* facts concerning real closed fields as exposed in Section 3.5, partly based on [249, Chapter XI]. The models of

$$\begin{aligned}\mathrm{RCF} := \mathrm{OFl} \cup \big\{ &\forall x \exists y\, (x \geqslant 0 \to x = y^2),\\ &\forall u_1 \cdots \forall u_{2n+1} \exists x\, (x^{2n+1} + u_1 x^{2n} + \cdots + u_{2n+1} = 0) : n \geqslant 1 \big\},\end{aligned}$$

are the real closed ordered fields. We have an analogue of Theorem B.12.1:

THEOREM B.12.13. RCF *has* QE.

PROOF. We use Corollary B.11.11. Every ordered integral domain has an RCF-closure, namely the real closure of its ordered fraction field. Let E and F be real closed and K be a real closed ordered subfield of both E and F. Suppose $K \neq E$ and F is $|K|^+$-saturated. It is enough to show that some ordered subfield of E properly containing K embeds over K into F. Take any $a \in E \setminus K$. Then a is transcendental over K by Corollary 3.5.5, and realizes a certain cut A in K. Saturation gives an element $b \in F$ realizing the same cut A. Then b is transcendental over K, so we have a field embedding $K(a) \to F$ over K sending a to b. This embedding is also order preserving by Corollary 3.5.8. □

EXAMPLE. Let $\varphi(x_1, x_2) = \exists y\, (y^2 + x_1 y + x_2 = 0)$. Then φ is RCF-equivalent to the quantifier-free $\mathcal{L}_{\mathrm{OR}}$-formula $4x_2 \leqslant x_1^2$.

The substructures of real closed ordered fields are exactly the ordered integral domains, so $\mathrm{Th}(\mathrm{RCF})$ is the model completion of the theory of ordered integral domains. The ordered integral domain $\mathbb{Z}$ embeds into every model of RCF, hence RCF is complete and thus decidable.

REMARK. If E is real closed, then $E \models x \geqslant 0 \leftrightarrow \exists y(y^2 = x)$, so the ordering of E is definable by an existential $\mathcal{L}_{\mathrm{R}}$-formula. Let RCF' be a set of $\mathcal{L}_{\mathrm{R}}$-sentences whose models are the real closed fields. Then RCF' is model complete, but does not have QE: if $\{a \in \mathbb{R} : a \geqslant 0\}$ were definable in the field $\mathbb{R}$ by the quantifier-free $\mathcal{L}_{\mathrm{R}}$-formula $\varphi(x)$, then $\mathbb{R} \models \varphi(\sqrt{2})$, so $\mathbb{R} \models \varphi(-\sqrt{2})$, a contradiction.

COROLLARY B.12.14. *The following are equivalent, for an $\mathcal{L}_{\mathrm{OR}}$-sentence σ:*

(i) *σ holds in the ordered field $\mathbb{R}$;*

(ii) *σ holds in some real closed ordered field;*

(iii) *σ holds in every real closed ordered field.*

In the next three corollaries of Theorem B.12.13, E is real closed with ordered subfield K, and $T = (T_1, \dots, T_n)$ is an n-tuple of distinct indeterminates. First an analogue of part (i) of Corollary B.12.2 for RCF:

COROLLARY B.12.15. *A set $X \subseteq E^n$ is K-definable in E iff X is a boolean combination of sets of the form $\{a \in E^n : P(a) \geqslant 0\}$ where $P \in K[T]$. (In particular, if K is real closed, then the K-definable subsets of E are exactly the finite unions of singletons and intervals (a, b) where $a < b$ are in $K \cup \{-\infty, \infty\}$.)*

EXAMPLE B.12.16. The set $X = \{r \in \mathbb{R} : r \leqslant \pi\}$ is $\mathbb{R}$-definable in $\mathbb{R}$ but not 0-definable in $\mathbb{R}$. (Take $E = \mathbb{R}$ and $K = \mathbb{Q}^{\mathrm{rc}} \subseteq \mathbb{R}$ in the previous corollary.)

PROPOSITION B.12.17. *The definably closed subsets of E are exactly (the underlying sets of) the real closed subfields of E.*

PROOF. If K is a real closed subfield of E, then $K \preccurlyeq E$ by QE, hence K, as a subset of E, is definably closed in E. Conversely, suppose K is a subset of E and K is definably closed in E. Then K is (the underlying set of) a subfield of E. The 0-definable ordering on E guarantees that K is algebraically closed in E in the field-theoretic sense. Hence K is a real closed subfield of E, by Corollary 3.5.5. □

In the next corollary U is an indeterminate different from $T_1, \dots, T_n$.

COROLLARY B.12.18. *Let $X \subseteq E^n$ and $f\colon X \to E$ be K-definable in E. Then there exists $P \in K[T, U]^{\neq}$ such that $P(x, f(x)) = 0$ for all $x \in X$.*

PROOF. We can assume E is $|K|^+$-saturated. Let $x \in X$. Then $\mathrm{dcl}(K \cup \{x\})$ equals the real closure of $K(x)$ in E, by Proposition B.12.17, and contains $f(x)$. So $Q(x, f(x)) = 0$ for some $Q \in K[T, U]^{\neq}$. By saturation we get $Q_1, \dots, Q_m \in K[T, U]^{\neq}$ such that for each $x \in X$ we have $Q_i(x, f(x)) = 0$ for some $i \in \{1, \dots, m\}$; now put $P := Q_1 \cdots Q_m$. □

In the proof of the corollary above we tacitly used that for a real closed ordered field extension F of E, and $X \subseteq E^n$ and $f\colon X \to E$ that are K-definable in E, we have a K-definable set $X_F \subseteq F^n$ with $X_F \cap E^n = X$, and an extension of f to a K-definable function $f_F\colon X_F \to F$: define X_F in F by any formula defining X in E, and likewise with f_F, and note that X_F, f_F do not depend on the choice of these formulas. We finish this section with an application to the asymptotics of functions definable in the ordered field $\mathbb{R}$; in the proof we employ some facts about Hahn fields from Section 3.5.

COROLLARY B.12.19. *Let $f\colon \mathbb{R}^{>} \to \mathbb{R}$ be definable in the ordered field $\mathbb{R}$. Then either $f(r) = 0$ for all sufficiently large $r > 0$, or there are $c \in \mathbb{R}^{\times}$ and $q \in \mathbb{Q}$ such that $f(r)/cr^q \to 1$ as $r \to \infty$.*

PROOF. We can assume $f(r) \neq 0$ for arbitrarily large $r > 0$. Then $f(r) \neq 0$ for all sufficiently large $r > 0$ by Corollary B.12.15. Consider the real closed ordered field extension $E = \mathbb{R}[[x^{\mathbb{Q}}]]$ of $\mathbb{R}$. From $x > \mathbb{R}$ we get $f_E(x) \neq 0$, so

$$f_E(x) = cx^q(1+\delta) \qquad (c \in \mathbb{R}^\times,\ q \in \mathbb{Q},\ \delta \in E^{\prec 1}).$$

Hence for every $\varepsilon \in \mathbb{R}^>$ we have $|f_E(x)/cx^q - 1| < \varepsilon$, and thus $|f(r)/cr^q - 1| < \varepsilon$ for all sufficiently large $r \in \mathbb{R}^>$. □

EXAMPLE B.12.20 (Hörmander). Let $P \in \mathbb{R}[T]^{\neq}$ and define $f\colon \mathbb{R}^> \to \mathbb{R}$ by

$$f(r) := \min\big\{|P(a)| : a = (a_1, \dots, a_n) \in \mathbb{R}^n,\ |a_1| + \cdots + |a_n| = r\big\}.$$

Then there are $c \in \mathbb{R}^\times$ and $q \in \mathbb{Q}$ such that $f(r)/cr^q \to 1$ as $r \to \infty$.

Notes and comments

Theorem B.12.1 was known to Tarski (see [443, p. 54, Note 16]), but the first published proof was given by Seidenberg [395, p. 373]. Part (i) of Corollary B.12.2 is also due independently to Chevalley [79, Théorème 3, Corollaire]; see also [398]. A. Robinson [351] saw model-theoretic significance in Hilbert's Nullstellensatz [181]. This led him to the notion of model completeness. Corollary B.12.4 may be seen as a model-theoretic formulation of a heuristic principle in algebraic geometry named after Lefschetz [461, p. 242f]. See [397] for a discussion of this principle and [36, 127] for other attempts to make it precise. Example B.12.6 is from [401]. Corollary B.12.12 for endomorphisms of algebraic varieties X is due to Ax [26, 27], and, independently, to Grothendieck [158, §10.4.11]; Borel [54] gave a different proof of this result for $K = \mathbb{C}$.

Theorem B.12.13 is due to Tarski [439, 443]. The proof above was given by A. Robinson [350], who also used this theorem to derive in a few lines a solution of Hilbert's 17th Problem, originally solved by Artin [11]; see [347, 348]. Other proofs of Theorem B.12.13 were given by Łojasiewicz [269] and Seidenberg [395]. See [104] for the history of Tarski's theorem and some applications. A real Nullstellensatz (analogue of Corollary B.12.2(iii)) is in Dubois [113], Krivine [224], and Risler [341]; see [47, Section 4.1]. Corollary B.12.14 is sometimes called the *Tarski Principle.* Example B.12.20 is from [189, Lemma 3.9].

B.13 STRUCTURES WITHOUT THE INDEPENDENCE PROPERTY

In the previous sections we assembled a toolbox that helps in finding an intelligible description of the sets and maps definable in a structure M. Such a description often leads to useful geometric invariants (dimensions, Euler characteristics, ...) of these objects. But in this final section of the appendix we take another path by considering a robust dividing line in the realm of all structures discovered by Shelah. The tame side of this dividing line is called: *not having the independence property* (NIP). The NIP condition forbids certain combinatorial patterns in the definable binary relations of a structure. (As is often the case in this subject, a "tame" property is introduced here as the negation of a "wild" one.) Motivating NIP is a striking combinatorial dichotomy (Theorem B.13.1), which we prove first. This dichotomy leads to the definition of a measure of complexity for families of sets called *VC dimension* and a dual quantity known as *independence dimension.* After discussing these quantities we introduce the NIP property, and show that real closed fields have NIP. (This fact is used in our proof that $\mathbb{T}$ has NIP in Section 16.6.)

VC dimension

Given a set A we let 2^A be the power set of A and $\binom{A}{n}$ the set of n-element subsets of A. *Throughout this subsection X is an infinite set and $\mathcal{S}$ a collection of subsets of X.* For $A \subseteq X$ we set $\mathcal{S} \cap A := \{S \cap A : S \in \mathcal{S}\}$, and we say that A is **shattered** by $\mathcal{S}$ if $\mathcal{S} \cap A = 2^A$. We have a function $\pi_{\mathcal{S}} : \mathbb{N} \to \mathbb{N}$ given by

$$\pi_{\mathcal{S}}(n) \ := \ \max\left\{ |\mathcal{S} \cap A| : \ A \in \binom{X}{n} \right\}.$$

Thus $0 \leqslant \pi_{\mathcal{S}}(n) \leqslant 2^n$, and $\pi_{\mathcal{S}}(n) = 2^n$ means: some $A \in \binom{X}{n}$ is shattered by $\mathcal{S}$. If $\pi_{\mathcal{S}}(n) = 2^n$, then $\pi_{\mathcal{S}}(m) = 2^m$ for all $m \leqslant n$. Here is the promised dichotomy:

THEOREM B.13.1 (Sauer, Shelah). *Either $\pi_{\mathcal{S}}(n) = 2^n$ for all n, or else there exists $d \in \mathbb{N}$ such that $\pi_{\mathcal{S}}(n) \leqslant n^d$ for all sufficiently large n.*

Here is the result from finite combinatorics that underlies this dichotomy:

LEMMA B.13.2. *Suppose $|A| = n$, and $\mathcal{C} \subseteq 2^A$ and $d \in \{0, \dots, n\}$ are such that $|\mathcal{C}| > \sum_{i<d} \binom{n}{i}$. Then A has a subset B with $|B| = d$ that is shattered by $\mathcal{C}$.*

The hypothesis is sharp since the collection of subsets of A of size $< d$ has cardinality equal to the indicated sum of binomial coefficients, and this particular collection violates the conclusion of the lemma. Let us denote the indicated sum of binomial coefficients by $p_d(n)$. There is clearly a unique polynomial $p_d(X) \in \mathbb{Q}[X]$ (of degree $d-1$ if $d \geqslant 1$, and $p_0(X) = 0$), whose value at n is $p_d(n)$ for $n \geqslant d$. Note:

$$p_{d-1}(X-1) + p_d(X-1) \ = \ p_d(X) \qquad (d \geqslant 1).$$

PROOF OF LEMMA B.13.2. By induction on n. The desired result holds trivially for $d = 0$ and $d = n$, so let $0 < d < n$. Pick a point $a \in A$ and set $A' := A \setminus \{a\}$. Also put $C' := C \setminus \{a\}$ for $C \in \mathcal{C}$, and set $\mathcal{C}' := \{C' : C \in \mathcal{C}\}$. Under the map $C \mapsto C' : \mathcal{C} \to \mathcal{C}'$ a set $D \in \mathcal{C}'$ has either exactly one preimage or exactly two preimages; in the latter case these two preimages are D and $D \cup \{a\}$. So $\mathcal{C}' = \mathcal{C}_1 \cup \mathcal{C}_2$ (disjoint union) where $\mathcal{C}_1$ contains those $D \in \mathcal{C}'$ having one preimage in $\mathcal{C}$, and $\mathcal{C}_2$ those with two preimages. If $|\mathcal{C}'| > p_d(n-1)$, then by the inductive assumption applied to A' and $\mathcal{C}'$ there exists $B \subseteq A'$ of size d that is shattered by $\mathcal{C}'$ and thus by $\mathcal{C}$. So assume $|\mathcal{C}'| \leqslant p_d(n-1)$. Then

$$\begin{aligned} p_d(n-1) + p_{d-1}(n-1) \ &= \ p_d(n) \ < \ |\mathcal{C}| \ = \ |\mathcal{C}_1| + 2|\mathcal{C}_2| \ = \ \big(|\mathcal{C}_1| + |\mathcal{C}_2|\big) + |\mathcal{C}_2| \\ &= \ |\mathcal{C}'| + |\mathcal{C}_2|, \end{aligned}$$

hence $|\mathcal{C}_2| > p_{d-1}(n-1)$, so again by the inductive assumption applied to A' and $\mathcal{C}_2$ we get $B' \subseteq A'$ of size $d-1$ that is shattered by $\mathcal{C}_2$. Since for each $D \in \mathcal{C}_2$ we have $D \in \mathcal{C}$ and $D \cup \{a\} \in \mathcal{C}$, the set $B := B' \cup \{a\}$ is shattered by $\mathcal{C}$. □

This yields Theorem B.13.1 in a slightly stronger form:

THEOREM B.13.3. *If $d \in \mathbb{N}$ and $\pi_{\mathcal{S}}(d) < 2^d$, then $\pi_{\mathcal{S}}(n) \leqslant p_d(n)$ for all n.*

PROOF. Assume $\pi_{\mathcal{S}}(d) < 2^d$. Then the desired inequality holds for $n < d$, since $p_d(n) = 2^n$ for such n. Let $n \geqslant d$ and $A \in \binom{X}{n}$. If $|\mathcal{S} \cap A| > p_d(n)$, then the above lemma would give a set $B \subseteq A$ of size d shattered by $\mathcal{S}$, a contradiction. □

We define

$$\mathrm{VC}(\mathcal{S}) \ := \ \sup\left\{n : \pi_{\mathcal{S}}(n) = 2^n\right\} \in \mathbb{N} \cup \{-\infty, \infty\},$$

so $\mathrm{VC}(\mathcal{S}) := -\infty$ iff $\mathcal{S} = \emptyset$, and $\mathrm{VC}(\mathcal{S}) = \infty$ iff subsets of X of arbitrarily large finite size are shattered by $\mathcal{S}$. We call $\mathrm{VC}(\mathcal{S})$ the **VC dimension** of $\mathcal{S}$ and say that $\mathcal{S}$ is a **VC class** if $\mathrm{VC}(\mathcal{S}) < \infty$. The letters V and C here stand for the initials of Vapnik and Chervonenkis, the authors of [453].

Independence dimension

Let X be a nonempty set. Given $A_1, \dots, A_n \subseteq X$ we let $S(A_1, \dots, A_n)$ be the set of atoms of the boolean algebra of subsets of X that is generated by the A_j: the "nonempty fields in the Venn diagram of $A_1, \dots, A_n$"; that is, the elements of $S(A_1, \dots, A_n) \subseteq 2^X$ are the nonempty sets

$$\bigcap_{i \in I} A_i \cap \bigcap_{i \in [n] \setminus I} X \setminus A_i \qquad \text{where } I \subseteq [n] := \{1, \dots, n\}.$$

Thus $S(A_1, \dots, A_n) = \{X\}$ when $n = 0$, and always $S(A_1, \dots, A_n) \neq \emptyset$.

LEMMA B.13.4. *Let $A_1, \dots, A_m, B_1, \dots, B_n \subseteq X$. Then*

(i) $|S(A_1, \dots, A_m, B_1, \dots, B_n)| \leqslant |S(A_1, \dots, A_m)| \cdot |S(B_1, \dots, B_n)|$, *and*

(ii) *if $A_1, \dots, A_m$ are boolean combinations of $B_1, \dots, B_n$, then*

$$|S(A_1, \dots, A_m)| \ \leqslant \ |S(B_1, \dots, B_n)|.$$

PROOF. Part (i) is clear. For (ii) note that under the hypothesis of (ii), the boolean algebra of subsets of X generated by the B_j contains that generated by the A_i as a subalgebra, so every atom of the latter is a disjoint union of atoms of the former. □

We have $1 \leqslant |S(A_1, \dots, A_n)| \leqslant 2^n$, and we say that the sequence $A_1, \dots, A_n$ is **independent** (in X) if $|S(A_1, \dots, A_n)| = 2^n$, and call $A_1, \dots, A_n$ **dependent** (in X) otherwise. Next, let $\mathcal{S} \subseteq 2^X$, $\mathcal{S} \neq \emptyset$. Define

$$\pi^{\mathcal{S}}(n) \ := \ \max\left\{|S(A_1, \dots, A_n)| : A_1, \dots, A_n \in \mathcal{S}\right\}.$$

Note that $1 \leqslant \pi^{\mathcal{S}}(n) \leqslant 2^n$ for each n. We say that $\mathcal{S}$ is **independent** (in X) if $\pi^{\mathcal{S}}(n) = 2^n$ for every n, that is, if for every n there is an independent sequence of elements of $\mathcal{S}$ of length n. Otherwise, we say that $\mathcal{S}$ is **dependent** (in X). If $\mathcal{S}$ is dependent, we define the **independence dimension** of $\mathcal{S}$ as the largest n such that $\pi^{\mathcal{S}}(n) = 2^n$; notation: $n = \mathrm{IND}(\mathcal{S})$. If $\mathcal{S}$ is independent, we set $\mathrm{IND}(\mathcal{S}) = \infty$. If $\mathcal{S}$ is finite, then clearly $\mathrm{IND}(\mathcal{S}) \leqslant |\mathcal{S}|$.

VC duality

Often our collection $\mathcal{S}$ will be indexed by elements of a parameter space or index set, and then the set X and this index set play dual roles. To make this duality explicit, let X and Y be infinite sets and $\Phi \subseteq X \times Y$. For $x \in X, y \in Y$,

$$\Phi_x := \{y \in Y : (x,y) \in \Phi\}, \qquad \Phi_X := \{\Phi_x : x \in X\} \subseteq 2^Y$$
$$\Phi^y := \{x \in X : (x,y) \in \Phi\}, \qquad \Phi^Y := \{\Phi^y : y \in Y\} \subseteq 2^X.$$

One verifies easily that, given a finite set $A = \{x_1, \dots, x_n\} \subseteq X$, the map

$$B \mapsto \bigcap_{x \in B} \Phi_x \cap \bigcap_{x \in A \setminus B} Y \setminus \Phi_x : \Phi^Y \cap A \to S(\Phi_{x_1}, \dots, \Phi_{x_n})$$

is a bijection. It follows that $\pi_{\Phi^Y} = \pi^{\Phi_X}$. Thus $\mathrm{VC}(\Phi^Y) = \mathrm{IND}(\Phi_X)$. Reversing the role of X and Y also yields $\mathrm{VC}(\Phi_X) = \mathrm{IND}(\Phi^Y)$.

LEMMA B.13.5. *If* $\mathrm{VC}(\Phi^Y) < \infty$, *then* $\mathrm{VC}(\Phi_X) < 2^{1+\mathrm{VC}(\Phi^Y)}$.

PROOF. Let ε range over the set of functions $[n] \to \{-1, +1\}$. Suppose $\mathrm{VC}(\Phi_X) = \mathrm{IND}(\Phi^Y) \geqslant 2^n$. This yields an independent family $\left(\Phi^{y_\varepsilon}\right)_\varepsilon$ of 2^n subsets of X (with $y_\varepsilon \in Y$ for all ε). So for each $m \in [n]$ we have

$$\bigcap_{\varepsilon(m)=+1} \Phi^{y_\varepsilon} \cap \bigcap_{\varepsilon(m)=-1} (X \setminus \Phi^{y_\varepsilon}) \neq \emptyset,$$

so we can take an element $x(m)$ of the intersection on the left. Then for each ε and $m \in [n]$ we have: $x(m) \in \Phi_{y_\varepsilon} \longleftrightarrow \varepsilon(m) = +1$, so $x(1), \dots, x(n)$ are distinct and the set $\{x(1), \dots, x(n)\}$ is shattered by Φ^Y. Hence $\mathrm{VC}(\Phi^Y) \geqslant n$. □

We therefore have the equivalences

$$\Phi^Y \text{ is dependent} \Leftrightarrow \Phi_X \text{ is a VC class} \Leftrightarrow \Phi^Y \text{ is a VC class} \Leftrightarrow \Phi_X \text{ is dependent}.$$

We say that Φ is **dependent** if Φ^Y is dependent. We also set $\pi^\Phi := \pi^{\Phi^Y}$.

Let $\neg\Phi$ be the relative complement $(X \times Y) \setminus \Phi$ of Φ in $X \times Y$. It is clear that $\pi^{\neg\Phi} = \pi^\Phi$. For $\Phi, \Psi \subseteq X \times Y$ we have $\pi^{\Phi\cup\Psi}, \pi^{\Phi\cap\Psi} \leqslant \pi^\Phi \cdot \pi^\Psi$, by Lemma B.13.4. By Theorem B.13.1, this yields:

LEMMA B.13.6. *If* $\Phi, \Psi \subseteq X \times Y$ *are dependent, then so are* $\neg\Phi$, $\Phi \cup \Psi$, *and* $\Phi \cap \Psi$.

NIP

Let $\boldsymbol{M}$ be an $\mathcal{L}$-structure each of whose underlying sets M_s is infinite. Then we define $\boldsymbol{M}$ to have **NIP** (short for: **the non-independence property**) if every 0-definable relation $\Phi \subseteq M_x \times M_y$ with finite nonempty multivariables x, y is dependent. Note that if $\boldsymbol{M}$ has NIP, then so does $\boldsymbol{M}_A$ for each parameter set A in $\boldsymbol{M}$, and so does every reduct of $\boldsymbol{M}$. If $\boldsymbol{M}$ has NIP and $\boldsymbol{M} \equiv \boldsymbol{N}$, then $\boldsymbol{N}$ has NIP. Thus $\boldsymbol{M}$ having NIP is really a property of its theory $\mathrm{Th}(\boldsymbol{M})$.

Our goal in this subsection is to show that real closed fields have NIP. We obtain this as a consequence of the following:

PROPOSITION B.13.7. *Let $\mathcal{F}$ be an m-dimensional real vector space of real-valued functions on an infinite set X, and for each $f \in \mathcal{F}$, put*

$$\mathrm{pos}(f) := \{x \in X : f(x) > 0\}.$$

Then $\mathrm{pos}(\mathcal{F}) := \{\mathrm{pos}(f) : f \in \mathcal{F}\}$ *is a VC class of VC dimension m.*

PROOF. Let $A \in \binom{X}{m+1}$. The restriction map $f \mapsto f|A \colon \mathcal{F} \to \mathbb{R}^A$ is not surjective, since $\dim \mathbb{R}^A = m+1 > \dim \mathcal{F}$. Therefore we can take a nonzero $w \in \mathbb{R}^A$ that is orthogonal to all restrictions $f|A$ ($f \in \mathcal{F}$) with respect to the standard inner product $\langle u, v\rangle := \sum_{a\in A} u(a)\cdot v(a)$ on $\mathbb{R}^A$. Replacing w by $-w$ if necessary we can assume that $A^+ := \{a \in A : w(a) > 0\}$ is nonempty. If there were $f \in \mathcal{F}$ with $A^+ = A \cap \mathrm{pos}(f)$, then we would have $0 = \langle w, f|A\rangle = \sum_{a\in A} w(a)\cdot f(a) > 0$, a contradiction. Hence A is not shattered by $\mathrm{pos}(\mathcal{F})$. Thus $\mathrm{VC}(\mathrm{pos}(\mathcal{F})) \leqslant m$, and we leave the proof that equality holds as an exercise. □

The proposition above applies to the vector space of all real polynomial functions on $X = \mathbb{R}^n$ ($n \geqslant 1$) of degree $\leqslant d$. Together with B.12.13 and B.13.6 this yields at once:

COROLLARY B.13.8. *Real closed fields have* NIP.

Notes and comments

Theorem B.13.1 was found independently by Sauer [381] and Shelah [407]. The notion of VC class first arose in probability theory in the work of Vapnik and Chervonenkis [453]. The independence property was introduced into model theory by Shelah [404], who also proved in [405] by a curious set-theoretic argument the very useful fact that in the above definition of NIP it suffices to consider the case that x is a single variable. Laskowski [252] realized the connection between VC classes and NIP and gave a more elementary proof of the reduction to a single variable; see [107, Chapter 5] for an exposition. Proposition B.13.7 was shown in [114, Theorem 7.2] and used in [430] to prove that the ordered field $\mathbb{R}$ has NIP. More generally, all *o-minimal* structures have NIP [317], and so do many other structures: for example, *stable* structures (such as algebraically closed fields), ordered abelian groups [160], and the field $\mathbb{Q}_p$ of p-adic numbers [40]. For more about the significance of NIP in model theory, see [417].

Bibliography

[1] N. H. Abel, *Précis d'une théorie des fonctions elliptiques,* J. Reine Angew. Math. **4** (1829), 309–348.

[2] ———, *Sur les séries,* in: S. Lie, L. Sylow (eds.), *Œuvres Complètes de N. H. Abel,* nouvelle éd., vol. 2, pp. 197–205, Grøndahl, Christiania, 1881.

[3] S. Abhyankar, *On the valuations centered in a local domain,* Amer. J. Math. **78** (1956), 321–348.

[4] ———, *Historical ramblings in algebraic geometry and related algebra,* Amer. Math. Monthly **83** (1976), 409–448.

[5] J. Aczél, *Einige aus Funktionalgleichungen zweier Veränderlichen ableitbare Differentialgleichungen,* Acta Univ. Szeged. Sect. Sci. Math. **13** (1950), 179–189.

[6] K. Adjamagbo, *Sur l'effectivité du lemme du vecteur cyclique,* C. R. Acad. Sci. Paris Sér. I Math. **306** (1988), 543–546.

[7] Y. Akizuki, *Einige Bemerkungen über primäre Integritätsbereiche mit Teilerkettensatz,* Proc. Phys.-Math. Soc. Japan Ser. III **17** (1935), 327–336.

[8] R. Amayo, I. Stewart, *Infinite-Dimensional Lie Algebras,* Noordhoff International Publishing, Leyden, 1974.

[9] M. Anderson, *Solution Spaces for Linear Equations in Valued D Fields,* Ph. D. Thesis, University of California at Berkeley, 2011.

[10] K. J. Arrow, *A difficulty in the concept of social welfare,* J. Polit. Econ. **58** (1950), 328–346.

[11] E. Artin, *Über die Zerlegung definiter Funktionen in Quadrate,* Abh. Math. Sem. Univ. Hamburg **5** (1927), 100–115.

[12] E. Artin, O. Schreier, *Algebraische Konstruktion reeller Körper,* Abh. Math. Sem. Univ. Hamburg **5** (1926), 83–99.

[13] ———, *Eine Kennzeichnung der reell abgeschlossenen Körper,* Abh. Math. Sem. Univ. Hamburg **5** (1927), 225–231.

[14] M. Artin, *Algebraic approximation of structures over complete local rings,* Inst. Hautes Études Sci. Publ. Math. **36** (1969), 23–58.

[15] M. Aschenbrenner, *Some remarks about asymptotic couples,* in: F.-V. Kuhlmann et al. (eds.): *Valuation Theory and its Applications,* Vol. II, pp. 7–18, Fields Inst. Commun., vol. 33, Amer. Math. Soc., Providence, RI, 2003.

[16] ———, *Logarithms of iteration matrices, and proof of a conjecture by Shadrin and Zvonkine,* J. Combin. Theory Ser. A **119** (2012), 627–654.

[17] M. Aschenbrenner, W. Bergweiler, *Julia's Equation and differential transcendence,* Illinois J. Math. **59** (2015), 277–294.

[18] M. Aschenbrenner, L. van den Dries, *Closed asymptotic couples,* J. Algebra **225** (2000), 309–358.

[19] ———, *H-fields and their Liouville extensions,* Math. Z. **242** (2002), 543–588.

[20] ———, *Liouville closed H-fields,* J. Pure Appl. Algebra, **197** (2005), 83–139.

[21] ———, *Asymptotic differential algebra,* in: O. Costin, M. D. Kruskal, A. Macintyre (eds.), *Analyzable Functions and Applications,* pp. 49–85, Contemp. Math., vol. 373, Amer. Math. Soc., Providence, RI, 2005.

[22] M. Aschenbrenner, L. van den Dries, J. van der Hoeven, *Differentially algebraic gaps,* Selecta Math. **11** (2005), 247–280.

[23] ———, *Toward a model theory for transseries,* Notre Dame J. Form. Log. **54** (2013), 279–310.

[24] ———, *The surreal numbers as a universal H-field,* J. Eur. Math. Soc. (JEMS), to appear.

[25] ———, *Dimension in the realm of transseries,* Contemp. Math., to appear.

[26] J. Ax, *The elementary theory of finite fields,* Ann. of Math. **88** (1968), 239–271.

[27] ———, *Injective endomorphisms of varieties and schemes,* Pacific J. Math. **31** (1969), 1–7.

[28] J. Ax, S. Kochen, *Diophantine problems over local fields,* I, Amer. J. Math. **87** (1965), 605–630.

[29] ———, *Diophantine problems over local fields,* III, Ann. of Math. **83** (1966), 437–456.

[30] G. Azumaya, *On maximally central algebras,* Nagoya Math. J. **2** (1951), 119–150.

[31] A. Babakhanian, *On primitive elements in differentially algebraic extension fields,* Trans. Amer. Math. Soc. **134** (1968), 71–83.

[32] R. Baer, *Algebraische Theorie der differentiierbaren Funktionenkörper,* Sitzungsber. Heidelb. Akad. Wiss. (1927), 15–32.

[33] ______, *Dichte, Archimedizität und Starrheit geordneter Körper,* Math. Ann. **188** (1970), 165–205.

[34] J. Baldwin, A. Lachlan, *On strongly minimal sets,* J. Symbolic Logic **36** (1971), 79–96.

[35] S. Bank, *On the instability theory of differential polynomials,* Ann. Mat. Pura Appl. **74** (1966), 83–111.

[36] J. Barwise, P. Eklof, *Lefschetz's principle,* J. Algebra **13** (1969), 554–570.

[37] J. Barwise, A. Robinson, *Completing theories by forcing,* Ann. Math. Logic **2** (1970), 119–142.

[38] W. Baur, *On the elementary theory of pairs of real closed fields,* J. Symb. Logic **47** (1982), 669–679.

[39] T. Becker, *Real closed rings and ordered valuation rings,* Z. Math. Logik Grundlag. Math. **29** (1983), 417–425.

[40] L. Bélair, *Types dans les corps valués munis d'applications coefficients,* Illinois J. Math. **43** (1999), 410–425.

[41] L. Bélair, A. Macintyre, T. Scanlon, *Model theory of the Frobenius on the Witt vectors,* Amer. J. Math. **129** (2007), 665–721.

[42] E. T. Bell, *Exponential polynomials,* Ann. of Math. **35** (1934), 258–277.

[43] A. Berarducci, V. Mantova, *Surreal numbers, derivations, and transseries,* J. Eur. Math. Soc. (JEMS), to appear.

[44] J. Bertrand, *Règles sur la convergence des séries*, J. Math. Pures Appl. **7** (1842), 35–54.

[45] P. Blasiak, P. Flajolet, *Combinatorial models of creation-annihilation,* Sém. Lothar. Combin. **65** (2010/12), Art. B65c, 78 pp.

[46] L. Blum, *Differentially closed fields: a model-theoretic tour,* in: H. Bass, P. Cassidy, J. Kovacic (eds.), *Contributions to Algebra,* pp. 37–61, Academic Press, New York, 1977.

[47] J. Bochnak, M. Coste, M.-F. Roy, *Real Algebraic Geometry,* Ergeb. Math. Grenzgeb., vol. 36, Springer-Verlag, Berlin, 1998.

[48] P. du Bois-Reymond, *Sur la grandeur relative des infinis des fonctions,* Ann. Mat. Pura Appl. **4** (1871), 338–353.

[49] ———, *Théorie générale concernant la grandeur relative des infinis des fonctions et de leurs dérivées,* J. Reine Angew. Math. **74** (1872), 294–304.

[50] ———, *Eine neue Theorie der Convergenz und Divergenz von Reihen mit positiven Gliedern,* J. Reine Angew. Math. **76** (1873), 61–91.

[51] ———, *Ueber asymptotische Werthe, infinitäre Approximationen und infinitäre Auflösung von Gleichungen,* Math. Ann. **8** (1875), 362–414.

[52] ———, *Ueber die Paradoxen des Infinitärcalcüls,* Math. Ann. **11** (1877), 149–167.

[53] ———, *Die allgemeine Functionentheorie,* Verlag der H. Laupp'schen Buchhandlung, Tübingen, 1882.

[54] A. Borel, *Injective endomorphisms of algebraic varieties,* Arch. Math. **20** (1969), 531–537.

[55] É. Borel, *Mémoire sur les séries divergentes,* Ann. Sci. École Norm. Sup. **16** (1899), 9–131.

[56] M. Boshernitzan, *An extension of Hardy's class L of "orders of infinity,"* J. Analyse Math. **39** (1981), 235–255.

[57] ———, *New "orders of infinity,"* J. Analyse Math. **41** (1982), 130–167.

[58] ———, *Hardy fields and existence of transexponential functions,* Aequationes Math. **30** (1986), 258–280.

[59] ———, *Second order differential equations over Hardy fields,* J. London Math. Soc. **35** (1987), 109–120.

[60] N. Bourbaki, *Topologie Générale,* Chapitre II, *Structures Uniformes,* Hermann, Paris, 1961.

[61] ———, *Algèbre,* Chapitre IV, *Polynômes et Fractions Rationelles*; Chapitre V, *Corps Commutatifs,* Hermann, Paris, 1967.

[62] ———, *Fonctions d'une Variable Réelle,* Chapitre V, *Étude Locale des Fonctions,* Hermann, Paris, 1976.

[63] E. Brieskorn, H. Knörrer, *Plane Algebraic Curves,* Birkhäuser Verlag, Basel, 1986.

[64] Ch. Briot, J.-C. Bouquet, *Mémoire sur l'intégration des équations différentielles au moyen des fonctions elliptiques,* J. Éc. Poly. **21** (1856), 199–254.

[65] R. Brown, *Valued vector spaces of countable dimension,* Publ. Math. Debrecen **18** (1971), 149–151 (1972).

[66] A. Buium, P. Cassidy, *Differential algebraic geometry and differential algebraic groups: From algebraic differential equations to Diophantine geometry,* in: H. Bass et al. (eds.), *Selected Works of Ellis Kolchin with Commentary,* pp. 567–636, American Mathematical Society, Providence, RI, 1999.

[67] J. Cano, *On the series defined by differential equations, with an extension of the Puiseux polygon construction to these equations,* Analysis **13** (1993), 103–119.

[68] G. Cantor, *Beiträge zur Begründung der transfiniten Mengenlehre,* Math. Ann. **46** (1895), 481–512.

[69] H. Cartan, *Théorie des filtres,* C. R. Acad. Sci. Paris **205** (1937), 595–598.

[70] ———, *Filtres et ultrafiltres,* C. R. Acad. Sci. Paris **205** (1937), 777–779.

[71] P. Cartier, *Dérivations dans les corps,* in: *Séminaire H. Cartan et C. Chevalley, 8e année: 1955/1956,* exp. no. 13, Secrétariat mathématique, Paris, 1956.

[72] C. C. Chang, *On unions of chains of models,* Proc. Amer. Math. Soc. **10** (1959), 120–127.

[73] G. Cherlin, *Algebraically closed commutative rings,* J. Symbolic Logic **38** (1973), 493–499.

[74] ———, *Model Theoretic Algebra—Selected Topics,* Lecture Notes in Mathematics, vol. 521, Springer-Verlag, Berlin-New York, 1976.

[75] G. Cherlin, M. A. Dickmann, *Real closed rings,* II, Ann. Pure Appl. Logic **25** (1983), 213–231.

[76] C. Chevalley, *On the theory of local rings,* Ann. of Math. **44** (1943), 690–708.

[77] ———, *On the notion of the ring of quotients of a prime ideal,* Bull. Amer. Math. Soc. **50** (1944), 93–97.

[78] ———, *Introduction to the Theory of Algebraic Functions of One Variable,* Mathematical Surveys, vol. VI, American Mathematical Society, New York, 1951.

[79] ———, *Sur la théorie des variétés algébriques,* Nagoya Math. J. **8** (1955), 1–43.

[80] G. Chrystal, *Algebra,* Part II, 2nd ed., Adam and Charles Black, Edinburgh, 1900.

[81] R. C. Churchill, J. Kovacic, *Cyclic vectors,* in: Li Guo et al. (eds.), *Differential Algebra and Related Topics,* pp. 191–218, World Scientific Publishing Co., Inc., River Edge, NJ, 2002.

[82] I. S. Cohen, A. Seidenberg, *Prime ideals and integral dependence,* Bull. Amer. Math. Soc. **52** (1946), 252–261.

[83] L. W. Cohen, C. Goffman, *The topology of ordered Abelian groups,* Trans. Amer. Math. Soc. **67** (1949), 310–319.

[84] P. J. Cohen, *Decision procedures for real and p-adic fields,* Comm. Pure Appl. Math. **22** (1969), 131–151.

[85] P. M. Cohn, *Universal Algebra,* Harper & Row, New York-London, 1965.

[86] ______, *Free Ideal Rings and Localization in General Rings,* New Mathematical Monographs, vol. 3, Cambridge University Press, Cambridge, 2006.

[87] R. Cohn, *Solutions in the general solution,* in: H. Bass, P. Cassidy, J. Kovacic (eds.), *Contributions to Algebra,* pp. 117–128, Academic Press, New York, 1977.

[88] P. Colmez, J.-P. Serre (eds.), *Grothendieck-Serre Correspondence,* American Mathematical Society, Providence, RI, 2004.

[89] L. Comtet, *Une formule explicite pour les puissances successives de l'opérateur de dérivation de Lie,* C. R. Acad. Sci. Paris Sér. A–B **276** (1973), A165–A168.

[90] ______, *Advanced Combinatorics,* D. Reidel Publishing Co., Dordrecht, 1974.

[91] P. F. Conrad, *Embedding theorems for abelian groups with valuations,* Amer. J. Math. **75** (1953), 1–29.

[92] J. H. Conway, *On Numbers and Games,* London Mathematical Society Monographs, vol. 6, Academic Press, London-New York, 1976.

[93] L. Corry, *Modern Algebra and the Rise of Mathematical Structures,* 2nd ed., Birkhäuser Verlag, Basel, 2004.

[94] J. Cozzens, C. Faith, *Simple Noetherian Rings,* Cambridge Tracts in Mathematics, vol. 69, Cambridge University Press, Cambridge-New York-Melbourne, 1975.

[95] T. Crespo, Z. Hajto, M. van der Put, *Real and p-adic Picard-Vessiot fields,* Math. Ann. **365** (2016), 93–103.

[96] B. Dahn, P. Göring, *Notes on exponential-logarithmic terms,* Fund. Math. **127** (1987), 45–50.

[97] R. Dedekind, *Über den Zusammenhang zwischen der Theorie der Ideale und der Theorie der höheren Kongruenzen,* Abhandl. Kgl. Ges. Wiss. Göttingen **23** (1878), 1–23.

[98] F. Delon, *Types sur* $\mathbb{C}((X))$, in: *Groupe d'Étude Théories Stables* (*Bruno Poizat*), 2e année: 1978/79, Exp. No. 5, Secrétariat Math., Paris, 1981.

[99] J. Della Dora, Cl. Dicrescenzo, E. Tournier, *An algorithm to obtain formal solutions of a linear homogeneous differential equation at an irregular singular point,* in: J. Calmet (ed.), *Computer Algebra,* pp. 273–280, Lecture Notes in Computer Science, vol. 144, Springer-Verlag, Berlin-New York, 1982.

[100] J. Denef, *p-adic semi-algebraic sets and cell decomposition,* J. Reine Angew. Math. **369** (1986), 154–166.

[101] J. Denef, L. Lipshitz, *Power series solutions of algebraic differential equations,* Math. Ann. **267** (1984), 213–238.

[102] M. Deuring, *Verzweigungstheorie bewerteter Körper,* Math. Ann. **105** (1931), 277–307.

[103] A. Dolich, J. Goodrick, D. Lippel, *Dp-minimal theories: basic facts and examples,* Notre Dame J. Formal Logic **52** (2011), 267–288.

[104] L. van den Dries, *Alfred Tarski's elimination theory for real closed fields,* J. Symbolic Logic **53** (1988), 7–19.

[105] ———, *Dimension of definable sets, algebraic boundedness and Henselian fields,* Ann. Pure Appl. Logic **45** (1989), 189–209.

[106] ———, *T-convexity and tame extensions,* II, J. Symbolic Logic **62** (1997), 14–34.

[107] ———, *Tame Topology and O-minimal Structures,* London Mathematical Society Lecture Note Series, vol. 248, Cambridge University Press, Cambridge, 1998.

[108] ———, *Limit sets in o-minimal structures,* in: M. Edmundo et al. (eds.), *Proceedings of the RAAG Summer School Lisbon 2003: O-minimal Structures,* pp. 172–215, Lecture Notes in Real Algebraic and Analytic Geometry, Cuvillier Verlag, Göttingen, 2005.

[109] ———, *Truncation in Hahn fields,* in: A. Campillo et al. (eds.), *Valuation Theory in Interaction,* pp. 578–595, European Mathematical Society, 2014.

[110] L. van den Dries, A. Macintyre, D. Marker, *The elementary theory of restricted analytic fields with exponentiation,* Ann. of Math. **140** (1994), 183–205.

[111] ———, *Logarithmic-exponential power series,* J. London Math. Soc. **56** (1997), 417–434.

[112] ———, *Logarithmic-exponential series,* Ann. Pure Appl. Logic **111** (2001), 61–113.

[113] D. W. Dubois, *A Nullstellensatz for ordered fields,* Ark. Mat. **8** (1969), 111–114.

[114] R. M. Dudley, *Central limit theorems for empirical measures,* Ann. Probab. **6** (1978), 899–929.

[115] H. Dulac, *Sur les cycles limites,* Bull. Soc. Math. France **51** (1923), 45–188.

[116] G. Dumas, *Sur quelques cas d'irréductibilité des polynomes à coefficients rationnels,* J. Math. Pures Appl. **61** (1906), 191–258.

[117] G. Duval, *Valuations and Differential Galois Groups,* Mem. Amer. Math. Soc. **212** (2011), no. 998.

[118] J. Écalle, *Théorie itérative: introduction à la théorie des invariants holomorphes,* J. Math. Pures Appl. **54** (1975), 183–258.

[119] ______, *Finitude des cycles-limites et accéléro-sommation de l'application de retour,* in: J.-P. Françoise, R. Roussarie (eds.), *Bifurcations of Planar Vector Fields,* pp. 74–159, Lecture Notes in Mathematics, vol. 1455, Springer-Verlag, Berlin, 1990.

[120] ______, *Introduction aux Fonctions Analysables et Preuve Constructive de la Conjecture de Dulac,* Actualités Mathématiques, Hermann, Paris, 1992.

[121] ______, *Six lectures on transseries, analysable functions and the constructive proof of Dulac's conjecture,* in: D. Schlomiuk (ed.), *Bifurcations and Periodic Orbits of Vector Fields,* pp. 75–184, NATO Adv. Sci. Inst. Ser. C Math. Phys. Sci., vol. 408, Kluwer Acad. Publ., Dordrecht, 1993.

[122] G. A. Edgar, *Transseries for beginners,* Real Anal. Exchange **35** (2010), 253–309.

[123] A. Ehrenfeucht, *An application of games to the completeness problem for formalized theories,* Fund. Math. **49** (1960/1961), 129–141.

[124] P. Ehrlich, *Hahn's 'Über die nichtarchimedischen Grössensysteme' and the development of the modern theory of magnitudes and numbers to measure them,* in: J. Hintikka (ed.), *From Dedekind to Gödel* (*Boston, MA, 1992*), pp. 165–213, Synthese Lib., vol. 251, Kluwer Acad. Publ., Dordrecht, 1995.

[125] ______, *The rise of non-archimedean mathematics and the roots of a misconception,* I, Arch. Hist. Exact Sci. **60** (2006), 1–121.

[126] ______, *The absolute arithmetic continuum and the unification of all numbers great and small,* Bull. Symbolic Logic **18** (2012), 1–45.

[127] P. Eklof, *Lefschetz's principle and local functors,* Proc. Amer. Math. Soc. **37** (1973), 333–339.

[128] P. Eklof, G. Sabbagh, *Model-completions and modules,* Ann. Math. Logic **2** (1971), 251–295.

[129] A. J. Engler, A. Prestel, *Valued Fields,* Springer Monographs in Mathematics, Springer-Verlag, Berlin, 2005.

[130] P. Erdős, E. Jabotinsky, *On analytic iteration,* J. Analyse Math. **8** (1960/1961), 361–376.

[131] Ju. L. Eršov, *On the elementary theory of maximal normed fields*, Soviet Math. Dokl. **6** (1965), 1390–1393.

[132] A. van den Essen, *Polynomial Automorphisms and the Jacobian Conjecture,* Progress in Mathematics, vol. 190, Birkhäuser Verlag, Basel, 2000.

[133] L. Euler, *De integratione aequationum differentialium altiorum graduum,* Misc. Berol. **7** (1743), 193–242.

[134] ______, *De seriebus divergentibus,* Novi comm. acad. sci. Petrop. **5** (1760), 205–237.

[135] J.-M. Farto, *Multiplicity of the solutions of a differential polynomial,* J. Pure Appl. Algebra **108** (1996), 203–218.

[136] U. Felgner, *Die Hausdorffsche Theorie der η_α-Mengen und ihre Wirkungsgeschichte,* in: E. Brieskorn et al. (eds.), *Felix Hausdorff—gesammelte Werke,* vol. II, pp. 645–674, Springer-Verlag, Berlin, 2002.

[137] H. B. Fine, *On the functions defined by differential equations, with an extension of the Puiseux polygon construction to these equations,* Amer. J. Math. **11** (1889), 317–328.

[138] ______, *Singular solutions of ordinary differential equations,* Amer. J. Math. **12** (1890), 295–322.

[139] G. Fisher, *The infinite and infinitesimal quantities of du Bois-Reymond and their reception,* Arch. Hist. Exact Sci. **24** (1981), 101–163.

[140] I. Fleischer, *Maximality and ultracompleteness in normed modules,* Proc. Amer. Math. Soc. **9** (1958), 151–157.

[141] R. Fraïssé, *Application des γ-opérateurs au calcul logique du premier échelon,* Z. Math. Logik Grundlagen Math. **2** (1956), 76–92.

[142] T. Frayne, A. Morel, D. Scott, *Reduced direct products,* Fund. Math. **51** (1962), 195–228.

[143] G. Freudenburg, *Algebraic Theory of Locally Nilpotent Derivations,* Encyclopaedia of Mathematical Sciences, vol. 136, Invariant Theory and Algebraic Transformation Groups, VII, Springer-Verlag, Berlin, 2006.

[144] L. Fuchs, *Partially Ordered Algebraic Systems,* Pergamon Press, Oxford-London-New York-Paris; Addison-Wesley Publishing Co., Inc., Reading, Mass.-Palo Alto, Calif.-London, 1963.

[145] A. Gabrielov, *Projections of semianalytic sets,* Functional Anal. Appl. **2** (1968), 282–291.

[146] A. Gehret, *The asymptotic couple of the field of logarithmic transseries,* J. Algebra **470** (2017), 1–36.

[147] ______, *NIP for the asymptotic couple of the field of logarithmic transseries,* J. Symbolic Logic, to appear, arXiv:1503.06496.

[148] K. Gödel, *Die Vollständigkeit der Axiome des logischen Funktionenkalküls,* Monatsh. Math. Phys. **37** (1930), 349–360.

[149] A. W. Goldie, *The structure of prime rings under ascending chain conditions,* Proc. London Math. Soc. **8** (1958), 589–608.

[150] H. Gonshor, *An Introduction to the Theory of Surreal Numbers,* London Mathematical Society Lecture Note Series, vol. 110, Cambridge University Press, Cambridge, 1986.

[151] R. L. Graham, D. E. Knuth, O. Patashnik, *Concrete Mathematics,* 2nd ed., Addison-Wesley Publishing Company, Reading, MA, 1994.

[152] K. A. H. Gravett, *Valued linear spaces,* Quart. J. Math. **6** (1955), 309–315.

[153] ______, *Note on a result of Krull,* Proc. Cambridge Philos. Soc. **52** (1956), 379.

[154] H. Grell, *Beziehungen zwischen den Idealen verschiedener Ringe,* Math. Ann. **97** (1927), 490–523.

[155] D. Yu. Grigor′ev, M. F. Singer, *Solving ordinary differential equations in terms of series with real exponents,* Trans. Amer. Math. Soc. **327** (1991), 329–351.

[156] V. I. Gromak, I. Laine, S. Shimomura, *Painlevé Differential Equations in the Complex Plane,* de Gruyter Studies in Mathematics, vol. 28, Walter de Gruyter & Co., Berlin, 2002.

[157] D. Gronau, *Gottlob Frege, a pioneer in iteration theory,* in: L. Reich, J. Smítal, and G. Targonski, *Iteration Theory (ECIT 94),* pp. 105–119, Grazer Math. Ber., vol. 334, Karl-Franzens-Univ. Graz, Graz, 1997.

[158] A. Grothendieck, *Éléments de géométrie algébrique,* IV. *Étude locale des schémas et des morphismes de schémas, quatrième partie,* Inst. Hautes Études Sci. Publ. Math. **32** (1967), 5–361.

[159] A. Günaydın, P. Hieronymi, *Dependent pairs,* J. Symbolic Logic **76** (2011), 377–390.

[160] Y. Gurevich, P. H. Schmitt, *The theory of ordered abelian groups does not have the independence property,* Trans. Amer. Math. Soc. **284** (1984), 171–182.

[161] N. Guzy, F. Point, *Topological differential fields,* Ann. Pure Appl. Logic **161** (2010), 570–598.

[162] H. Hahn, *Über die nichtarchimedischen Größensysteme,* S.-B. Akad. Wiss. Wien, Math.-naturw. Kl. Abt. IIa **116** (1907), 601–655.

[163] G. H. Hardy, *Properties of logarithmico-exponential functions,* Proc. London Math. Soc. **10** (1911), 54–90.

[164] ______, *Some results concerning the behaviour at infinity of a real and continuous solution of an algebraic differential equation of the first order,* Proc. London Math. Soc. **10** (1912), 451–468.

[165] ______, *Orders of Infinity,* 2nd ed., Cambridge Univ. Press, Cambridge, 1924.

[166] W. A. Harris, Y. Sibuya, *The reciprocals of solutions of linear ordinary differential equations,* Adv. in Math. **58** (1985), 119–132.

[167] P. Hartman, *On the linear logarithmico-exponential differential equation of the second order,* Amer. J. Math. **70** (1948), 764–779.

[168] E. Harzheim, *Ordered Sets,* Advances in Mathematics, vol. 7, Springer, New York, 2005.

[169] H. Hasse, F. K. Schmidt, *Die Struktur diskret bewerteter Körper,* J. Reine Angew. Math. **170** (1933), 4–63.

[170] K. Hauschild, *Cauchyfolgen höheren Typus in angeordneten Körpern,* Z. Math. Logik Grundlagen Math. **13** (1967), 55–66.

[171] F. Hausdorff, *Die Graduierung nach dem Endverlauf,* Abh. Sächs. Akad. Wiss. Leipzig Math.-Natur. Kl. **31** (1909), 295–334.

[172] ______, *Grundzüge der Mengenlehre,* Veit & Comp., Leipzig, 1914.

[173] L. Henkin, *The completeness of the first-order functional calculus,* J. Symbolic Logic **14** (1949), 159–66.

[174] ______, *Some interconnections between modern algebra and mathematical logic,* Trans. Amer. Math. Soc. **74** (1953), 410–427.

[175] K. Hensel, *Neue Grundlagen der Arithmetik,* J. Reine Angew. Math. **127** (1904), 51–84.

[176] K. Hensel, G. Landsberg, *Theorie der algebraischen Funktionen einer Variabeln,* B. G. Teubner, Leipzig, 1902.

[177] J. Herbrand, *Recherches sur la théorie de la démonstration,* Trav. Soc. Sci. Lett. Varsovie Cl. III **33** (1930), 1–128.

[178] E. Hewitt, *Rings of real-valued continuous functions,* Trans. Amer. Math. Soc. **64** (1948), 45–99.

[179] G. Higman, *Ordering by divisibility in abstract algebras,* Proc. London Math. Soc. **2** (1952), 326–336.

[180] D. Hilbert, *Ueber die Theorie der algebraischen Formen,* Math. Ann. **36** (1890), 473–534.

[181] ______, *Über die vollen Invariantensysteme,* Math. Ann. **42** (1893), 313–373.

[182] ______, *Grundlagen der Geometrie,* B. G. Teubner, Leipzig, 1899.

[183] ______, *Mathematische Probleme,* Nachr. Königl. Gesell. Wiss. zu Göttingen, Math.-Phys. Kl. (1900), 253–297.

[184] E. Hille, *Nonoscillation theorems,* Trans. Amer. Math. Soc. **64** (1948), 234–252.

[185] ______, *Ordinary Differential Equations in the Complex Domain,* Pure and Applied Mathematics, Wiley-Interscience, New York-London-Sydney, 1976.

[186] W. Hodges, *Model Theory*, Encyclopedia of Mathematics and its Applications, vol. 42, Cambridge University Press, Cambridge, 1993.

[187] O. Hölder, *Zurückführung einer beliebigen algebraischen Gleichung auf eine Kette von Gleichungen,* Math. Ann. **34** (1889), 26–56.

[188] ______, *Die Axiome der Quantität und die Lehre vom Mass,* Berichte über die Verhandlungen der Sächsischen Akademie der Wissenschaften, Leipzig, Math.-Phys. K1. **53** (1901), 1–64.

[189] L. Hörmander, *On the theory of general partial differential operators,* Acta Math. **94** (1955), 161–248.

[190] J. van der Hoeven, *Asymptotique Automatique,* Thèse, École Polytechnique, Paris, 1997.

[191] ______, *Operators on generalized power series,* Illinois J. Math. **45** (2001), 1161–1190.

[192] ______, *Complex transseries solutions to algebraic differential equations,* Technical Report 2001-34, Université d'Orsay, 2001.

[193] ______, *A differential intermediate value theorem,* in: B. L. J. Braaksma et al. (eds.), *Differential Equations and the Stokes Phenomenon,* pp. 147–170, World Scientific Publishing Co., Inc., River Edge, NJ, 2002.

[194] ______, *Transseries and Real Differential Algebra,* Lecture Notes in Math., vol. 1888, Springer-Verlag, New York, 2006.

[195] ______, *Transserial Hardy fields,* Astérisque **323** (2009), 453–487.

[196] E. V. Huntington, *The Continuum and other Types of Serial Order, with an Introduction to Cantor's Transfinite Numbers,* Harvard University Press, Cambridge, MA, 1904.

[197] Yu. S. Il′yashenko, *Dulac's memoir "On limit cycles" and related questions of the local theory of differential equations,* Uspekhi Mat. Nauk **40** (1985), 41–78, 199.

[198] ———, *Finiteness Theorems for Limit Cycles,* Translations of Mathematical Monographs, vol. 94, American Mathematical Society, Providence, RI, 1991.

[199] Yu. S. Il′yashenko, S. Yakovenko, *Lectures on Analytic Differential Equations,* Graduate Studies in Mathematics, vol. 86, American Mathematical Society, Providence, RI, 2008.

[200] E. Jabotinsky, *Sur la représentation de la composition de fonctions par un produit de matrices. Application à l'itération de e^z et de $e^z - 1$,* C. R. Acad. Sci. Paris **224** (1947), 323–324.

[201] ———, *Analytic iteration,* Trans. Amer. Math. Soc. **108** (1963), 457–477.

[202] N. Jacobson, *Pseudo-linear transformations,* Ann. Math. **38** (1937), 484–507.

[203] ———, *A topology for the set of primitive ideals in an arbitrary ring,* Proc. Nat. Acad. Sci. U. S. A. **31** (1945), 333–338.

[204] J. Johnson, *Kähler differentials and differential algebra,* Ann. of Math. **89** (1969), 92–98.

[205] ———, *Systems of n partial differential equations in n unknown functions: the conjecture of M. Janet,* Trans. Amer. Math. Soc. **242** (1977), 329–334.

[206] C. Jordan, *Commentaire sur Galois,* Math. Ann. **1** (1869), 141–160.

[207] E. Kähler, *Algebra und Differentialrechnung,* in: *Bericht über die Mathematikertagung in Berlin vom 14.-18. 1. 1953,* pp. 58–163, Dt. Verl. d. Wissenschaften, Berlin, 1953.

[208] M. Kamensky, A. Pillay, *Interpretations and differential Galois extensions,* preprint (2014).

[209] I. Kaplansky, *Maximal fields with valuations,* Duke Math. J. **9** (1942), 303–321.

[210] ———, *An Introduction to Differential Algebra,* 2nd ed., Hermann, Paris, 1976.

[211] N. M. Katz, *On the calculation of some differential galois groups,* Invent. Math. **87** (1987), 13–61.

[212] K. Kedlaya, *The algebraic closure of the power series field in positive characteristic,* Proc. Amer. Math. Soc. **129** (2001), 3461–3470.

[213] ______, *On the algebraicity of generalized power series,* Beitr. Algebra Geom., to appear, arXiv:1508.01836, 2015.

[214] H. J. Keisler, *Ultraproducts and elementary classes,* Indag. Math. **23** (1961), 477–495.

[215] J. G. Kemeny, *Matrix representation for combinatorics,* J. Combin. Theory Ser. A **36** (1984), 279–306.

[216] M. Knebusch, C. Scheiderer, *Einführung in die reelle Algebra,* Vieweg Studium: Aufbaukurs Mathematik, vol. 63, Friedr. Vieweg & Sohn, Braunschweig, 1989.

[217] M. Knebusch, M. Wright, *Bewertungen mit reeller Henselisierung,* J. Reine Angew. Math. **286/287** (1976), 314–321.

[218] E. Kolchin, *Extensions of differential fields,* II, Ann. of Math. **45** (1944), 358–361.

[219] ______, *Algebraic matric groups and the Picard-Vessiot theory of homogeneous linear ordinary differential equations,* Ann. of Math. **49** (1948), 1–42.

[220] ______, *Galois theory of differential fields,* Amer. J. Math. **75** (1953), 753–824.

[221] ______, *Differential Algebra and Algebraic Groups,* Pure and Applied Mathematics, vol. 54, Academic Press, New York-London, 1973.

[222] ______, *Constrained extensions of differential fields,* Advances in Math. **12** (1974), 141–170.

[223] T. Komatsu, *On continued fraction expansions of Fibonacci and Lucas Dirichlet series,* Fibonacci Quart. **46/47** (2008/09), 268–278.

[224] J. L. Krivine, *Anneaux préordonnés,* J. Analyse Math. **21** (1964), 307–326.

[225] L. Kronecker, *Grundzüge einer arithmetischer Theorie der algebraischen Grössen,* J. Reine Angew. Math. **92** (1882), 1–122.

[226] W. Krull, *Primidealketten in allgemeinen Ringbereichen,* Sitzungsber. Heidelb. Akad. Wiss. Math.-Natur. Kl. (1928), 7. Abh.

[227] ______, *Idealtheorie in Ringen ohne Endlichkeitsbedingung,* Math. Ann. **101** (1929), 729–744.

[228] ______, *Über einen Hauptsatz der allgemeinen Idealtheorie,* Sitzungsber. Heidelb. Akad. Wiss. Math.-Natur. Kl. (1929), 2. Abh.

[229] ______, *Allgemeine Bewertungstheorie,* J. Reine Angew. Math. **167** (1932), 160–196.

[230] ______, *Dimensionstheorie in Stellenringen,* J. Reine Angew. Math. **179** (1938), 204–226.

[231] M. Kuczma, B. Choczewski, G. Roman, *Iterative Functional Equations,* Encyclopedia of Mathematics and its Applications, vol. 32, Cambridge University Press, Cambridge, 1990.

[232] F.-V. Kuhlmann, *Abelian groups with contractions,* I, Contemp. Math. **171** (1994), 217–241.

[233] ———, *Abelian groups with contractions,* II: *Weak o-minimality,* in: A. Facchini, C. Menini (eds.): *Abelian Groups and Modules,* pp. 323–342, Kluwer, Dordrecht, 1995.

[234] ———, *Approximation of elements in henselizations,* Manuscripta Math. **136** (2011), 461–474.

[235] ———, *Maps on ultrametric spaces, Hensel's Lemma, and differential equations over valued fields,* Comm. Algebra **39** (2011), 1730–1776.

[236] ———, book on valuation theory, in preparation.

[237] F.-V. Kuhlmann, S. Kuhlmann, S. Shelah, *Exponentiation in power series fields,* Proc. Amer. Math. Soc. **125** (1997), 3177–3183.

[238] S. Kuhlmann, *Valuation bases for extensions of valued vector spaces,* Forum Math. **8** (1996), 723–735.

[239] ———, *Ordered Exponential Fields,* Fields Institute Monographs, vol. 12, American Mathematical Society, Providence, RI, 2000.

[240] S. Kuhlmann, M. Matusinski, *Hardy type derivations on generalised series fields,* J. Algebra **351** (2012), 185–203.

[241] V. B. Kul′chinovskiĭ, *Ordered differential fields,* Comm. Algebra **26** (1998), 2491–2521.

[242] ———, *Ordered differential fields,* Siberian Math. J. **40** (1999), 326–340.

[243] E. E. Kummer, *Über die Zerlegung der aus Wurzeln der Einheit gebildeten complexen Zahlen in ihre Primfaktoren,* J. Reine Angew. Math. **35** (1847), 327–367.

[244] E. Kunz, *Kähler Differentials,* Advanced Lectures in Mathematics, Friedr. Vieweg & Sohn, Braunschweig, 1986.

[245] ———, *Why 'Kähler' differentials?,* in: R. Berndt, O. Riemenschneider (eds.), *Erich Kähler. Mathematische Werke,* pp. 848–853, Walter de Gruyter & Co., Berlin, 2003.

[246] C. Kuratowski, A. Tarski, *Les opérations logiques et les ensembles projectifs,* Fund. Math. **17** (1931), 240–248.

[247] J. Kürschak, *Über Limesbildung und allgemeine Körpertheorie,* J. Reine Angew. Math. **142** (1913), 211–253.

[248] E. Landau, *Ein Satz über die Zerlegung homogener linearer Differentialausdrücke in irreducible Factoren,* J. Reine Angew. Math. **24** (1902), 115–120.

[249] S. Lang, *Algebra,* 3rd ed., Addison-Wesley Publishing Company, Reading, MA, 1993.

[250] C. H. Langford, *Some theorems on deducibility,* Ann. Math. **28** (1927), 16–40.

[251] ———, *Theorems on deducibility (second paper),* Ann. Math. **28** (1927), 459–471.

[252] M. C. Laskowski, *Vapnik-Chervonenkis classes of definable sets*, J. London Math. Soc. **45** (1992), 377–384.

[253] D. Laugwitz, *Eine nichtarchimedische Erweiterung angeordneter Körper,* Math. Nachr. **37** (1968), 225–236.

[254] J. B. Leicht, *Zur Charakterisierung reell abgeschlossener Körper,* Monatsh. Math. **70** (1966), 452–453.

[255] A. H. M. Levelt, *Differential Galois theory and tensor products,* Indag. Math. **1** (1990), 439–449.

[256] F. W. Levi, *Arithmetische Gesetze im Gebiete diskreter Gruppen,* Rend. Circ. Math. Palermo **35** (1913), 225–236.

[257] T. Levi-Civita, *Sugli infiniti ed infinitesimi attuali quali elementi analitici,* Ist. Veneto Sci. Lett. Arti Atti Cl. Sci. Mat. Natur. **4** (1892-93), 1765–1815.

[258] ———, *Sui numeri transfiniti,* Atti Della R. Accademia Dei Lincei **7** (1898), 91–96, 113–121.

[259] A. H. Lightstone, A. Robinson, *Nonarchimedean Fields and Asymptotic Expansions,* North-Holland Mathematical Library, vol. 13, North-Holland Publishing Co., Amsterdam-Oxford; American Elsevier Publishing Co., Inc., New York, 1975.

[260] J. Liouville, *Mémoire sur la classification des transcendantes, et sur les racines de certaines équations en fonction finie explicite des coefficients,* J. Math. Pures et Appl. **2** (1837), 56–104.

[261] ———, *Suite du mémoire sur la classification des transcendantes, et sur les racines de certaines équations en fonction finie explicite des coefficients,* J. Math. Pures et Appl. **3** (1838), 523–546.

[262] ———, *Mémoire sur l'intégration d'une classe d'équations différentielles du second ordre en quantités finies explicites,* J. Math. Pures et Appl. **4** (1839), 423–456.

[263] ______, *Remarques nouvelles sur l'équation de Riccati,* J. Math. Pures et Appl. **6** (1841), 1–13.

[264] D. E. Loeb, *The iterated logarithmic algebra,* Adv. Math. **86** (1991), 155–234.

[265] D. E. Loeb, G.-C. Rota, *Formal power series of logarithmic type,* Adv. Math. **75** (1989), 1–118.

[266] L. Löwenheim, *Über Möglichkeiten im Relativkalkül,* Math. Ann. **76** (1915), 447–470.

[267] A. Loewy, *Über reduzible lineare homogene Differentialgleichungen,* Math. Ann. **56** (1903), 549–584.

[268] ______, *Über einen Fundamentalsatz für Matrizen oder lineare homogene Differentialsysteme,* Sitzungsber. Heidelb. Akad. Wiss. **5** (1918), 1–36.

[269] S. Łojasiewicz, *Ensembles semi-analytiques,* Inst. Hautes Études Sci., Bures-sur-Yvette, 1964.

[270] J. Łoś, *On the categoricity in power of elementary deductive systems and some related problems,* Colloq. Math. **3** (1954), 58–62.

[271] ______, *Quelques remarques, théorèmes et problèmes sur les classes définissables d'algèbres,* in: L. E. J. Brouwer et al. (eds.), *Mathematical Interpretation of Formal Systems*, pp. 98–113, Studies in Logic and the Foundations of Math., North-Holland, Amsterdam (1955).

[272] ______, *On extending of models,* I, Fund. Math. **42** (1955), 38–54.

[273] J. Łoś, R. Suszko, *On extending of models,* IV: *Infinite sums of models,* Fund. Math. **44** (1957), 52–60.

[274] A. Macintyre, *Classifying Pairs of Real Closed Fields,* Ph. D. Thesis, Stanford University, 1968.

[275] ______, *On definable subsets of p-adic fields,* J. Symbolic Logic **41** (1976), 605–610.

[276] ______, *Model completeness,* in: J. Barwise (ed.), *Handbook of Mathematical Logic,* pp. 139–180, Studies in Logic and the Foundations of Mathematics, vol. 90, North-Holland Publishing Co., Amsterdam-New York-Oxford, 1977.

[277] S. Mac Lane, *The uniqueness of the power series representation of certain fields with valuations,* Ann. of Math. **39** (1938), 370–382.

[278] ______, *The universality of formal power series fields,* Bull. Amer. Math. Soc. **45** (1939), 888–890.

[279] A. Magid, *The Picard-Vessiot anti-derivative closure,* J. Algebra **244** (2001), 1–18.

[280] E. Maillet, *Sur les fonctions hypertranscendantes,* C. R. Acad. Sci. Paris **142** (1906), 829–830.

[281] A. I. Mal′cev, *Untersuchungen aus dem Gebiete der mathematischen Logik,* Rec. Math. **1** (1936), 323–336.

[282] ———, *On the embedding of group algebras in division algebras,* Doklady Akad. Nauk SSSR **60** (1948), 1499–1501.

[283] B. Malgrange, *Sur les points singuliers des equations différentielles,* Enseignement Math. **20** (1974), 147–176.

[284] D. Marker, *Model Theory,* Graduate Texts in Mathematics, vol. 217, Springer-Verlag, New York, 2002.

[285] D. Marker, M. Messmer, A. Pillay, *Model Theory of Fields,* 2nd ed., Lecture Notes in Logic, vol. 5, A K Peters, Ltd., Wellesley, MA, 2006.

[286] D. Marker, C. Steinhorn, *Definable types in o-minimal theories*, J. Symbolic Logic **59** (1994), 185–198.

[287] C. Massaza, *Sulle valutazioni che inducono la topologia di un ordinamento non archimedeo,* Atti Accad. Sci. Torino Cl. Sci. Fis. Mat. Natur. **109** (1975), 343–359.

[288] H. Matsumura, *Commutative Ring Theory,* Cambridge Studies in Advanced Mathematics, vol. 8, Cambridge University Press, Cambridge, 1986.

[289] M. Matusinski, *On generalized series fields and exponential-logarithmic series fields with derivations,* in: A. Campillo et al. (eds.), *Valuation Theory in Interaction,* pp. 350–372, EMS Series of Congress Reports, European Mathematical Society (EMS), Zürich, 2014.

[290] L. Maurer, *Über die Endlichkeit der Invariantensysteme,* Sitzungsber. Math.-Phys. Kl. Kgl. Bayer. Akad. Wiss. München **29** (1899), 147–175.

[291] C. Michaux, *Differential Fields, Machines over the Real Numbers and Automata,* Thèse, Université de Mons-Hainaut, 1991.

[292] C. Miller, *Basics of o-minimality and Hardy fields,* in: C. Miller et al. (eds.), *Lecture Notes on O-minimal Structures and Real Analytic Geometry,* pp. 43–69, Fields Institute Communications, vol. 62, Springer, New York, 2012.

[293] J. S. Milne, *Étale Cohomology,* Princeton Mathematical Series, vol. 33, Princeton University Press, Princeton, N. J., 1980.

[294] E. H. Moore, *Concerning transcendentally transcendental functions,* Math. Ann. **48** (1896), 49–74.

[295] A. de Morgan, *The Differential and Integral Calculus,* Baldwin and Craddock, London, 1842.

[296] ———, *On the syllogism,* I. *On the structure of the syllogism,* Transactions of the Cambridge Philosophical Society **8** (1846), 379–408.

[297] M. Morley, *Categoricity in power,* Trans. Amer. Math. Soc. **114** (1965), 514–538.

[298] M. Morley, R. Vaught, *Homogeneous universal models,* Math. Scand. **11** (1962), 37–57.

[299] S. Morrison, *Continuous derivations,* J. Algebra **110** (1987), 468–479.

[300] M. Nagata, *On the theory of Henselian rings,* Nagoya Math. J. **5** (1953), 45–57.

[301] ———, *Local Rings,* Interscience Tracts in Pure and Applied Mathematics, vol. 13, John Wiley & Sons, New York-London, 1962.

[302] B. H. Neumann, *On ordered division rings,* Trans. Amer. Math. Soc. **66** (1949), 202–252.

[303] I. Newton, *Letter to Oldenburg dated Oct. 24, 1676,* in: *The Correspondence of Isaac Newton,* Vol. II (1676–1687), pp. 126–127, Cambridge University Press, New York-Cambridge, 1960.

[304] E. Noether, *Idealtheorie in Ringbereichen,* Math. Ann. **83** (1921), 24–66.

[305] ———, *Ein algebraisches Kriterium für absolute Irreduzibilität,* Math. Ann. **85** (1922), 26–33.

[306] ———, *Abstrakter Aufbau der Idealtheorie in algebraischen Zahl- und Funktionenkörpern,* Math. Ann. **96** (1927), 26–61.

[307] A. Nowicki, *Polynomial Derivations and their Rings of Constants*, Uniwersytet Mikołaja Kopernika, Toruń, 1994.

[308] F. W. J. Olver, *Asymptotics and Special Functions,* reprint of the 1974 original, AKP Classics, A K Peters, Ltd., Wellesley, MA, 1997.

[309] Ö. Ore, *Formale Theorie der linearen Differentialgleichungen,* I, J. Reine Angew. Math. **167** (1932), 221–234.

[310] ———, *Formale Theorie der linearen Differentialgleichungen,* II, J. Reine Angew. Math. **168** (1932), 233–252.

[311] A. Ostrowski, *Zur arithmetischen Theorie der algebraischen Größen,* Nachr. Akad. Wiss. Göttingen, math.-physik. Klasse (1919), 279–298.

[312] ———, *Über Dirichletsche Reihen und algebraische Differentialgleichungen,* Math. Z. **8** (1920), 241–298.

[313] ———, *Algebraische Funktionen von Dirichletschen Reihen,* Math. Z. **37** (1933), 98–133.

[314] ———, *Untersuchungen zur arithmetischen Theorie der Körper,* Math. Z. **39** (1934), 269–404.

[315] J. Pas, *Uniform p-adic cell decomposition and local zeta functions,* J. Reine Angew. Math. **399** (1989), 137–172.

[316] ———, *On the angular component map modulo P,* J. Symbolic Logic **55** (1990), 1125–1129.

[317] A. Pillay, C. Steinhorn, *Definable sets in ordered structures,* I, Trans. Amer. Math. Soc. **295** (1986), 565–592.

[318] J. M. Plotkin, *Who put the "back" in back-and-forth?*, in: J. N. Crossley et al. (eds.), *Logical Methods,* pp. 705–712, Progr. Comput. Sci. Appl. Logic, vol. 12, Birkhäuser, Boston, MA, 1993.

[319] H. Poincaré, *Mémoire sur les fonctions zétafuchsiennes,* Acta Math. **5** (1884), 209–278.

[320] E. G. C. Poole, *Introduction to the Theory of Linear Differential Equations,* Oxford Univ. Press, London, 1936.

[321] M. Presburger, *Über die Vollständigkeit eines gewissen Systems der Arithmetik ganzer Zahlen, in welchem die Addition als einzige Operation hervortritt,* in: *Comptes-rendus du I Congrés des Mathématiciens des Pays Slaves,* pp. 92–101, 395, Warsaw, 1930.

[322] A. Prestel, *Lectures on Formally Real Fields,* Lecture Notes in Math., vol. 1093, Springer-Verlag, Berlin, 1984.

[323] S. Prieß-Crampe, *Angeordnete Strukturen: Gruppen, Körper, projektive Ebenen,* Ergebnisse Math., vol. 98, Springer-Verlag, Berlin (1983).

[324] ———, *Der Banachsche Fixpunktsatz für ultrametrische Räume,* Results Math. **18** (1990), 178–186.

[325] H. Prüfer, *Untersuchungen über die Zerlegbarkeit der abzählbaren primären Abelschen Gruppen,* Math. Zeit. **17** (1923), 35–61.

[326] V. Puiseux, *Recherches sur les fonctions algébriques,* J. Math. Pures Appl. **15** (1850), 365–480.

[327] M. Rabin, *Non-standard models and independence of the induction axiom,* in: Y. Bar-Hillel et al. (eds.), *Essays on the Foundations of Mathematics,* pp. 287–99, North-Holland, Amsterdam, 1962.

[328] J.-P. Ramis, *Dévissage Gevrey,* Astérisque **59-60** (1978), 173–204.

[329] H. W. Raudenbush, *Differential fields and ideals of differential forms,* Ann. of Math. **34** (1933), 509–517.

[330] ______, *Ideal theory and algebraic differential equations,* Trans. Amer. Math. Soc. **36** (1934), 361–368.

[331] F. J. Rayner, *Relatively complete fields,* Proc. Edinburgh Math. Soc. **11** (1958/1959), 131–133.

[332] T. Rella, *Ordnungsbestimmungen in Integritätsbereichen und Newtonsche Polygone,* J. Reine Angew. Math. **158** (1927), 33–48.

[333] J.-P. Ressayre, *La théorie des modèles, et un petit problème de Hardy,* in: J.-P. Pier (ed.), *Development of Mathematics 1950–2000,* pp. 925–938, Birkhäuser Verlag, Basel, 2000.

[334] P. Ribenboim, *Corps maximaux et complets par des valuations de Krull,* Math. Z. **69** (1958), 466–479.

[335] ______, *Sur la théorie du prolongement des valuations de Krull,* Math. Z. **75** (1960/1961), 449–466.

[336] ______, *Théorie des Valuations,* Les Presses de l'Université de Montréal, Montréal, 1965.

[337] J. Riccati, *Animadversiones in aequationes differentiales secundi gradus,* Acta Eruditorum **8** (1724), 66–73.

[338] B. Riemann, *Über die Darstellbarkeit einer Function durch eine trigonometrische Reihe,* in: H. Weber (ed.), *Bernhard Riemann's gesammelte mathematische Werke und wissenschaftlicher Nachlass,* pp. 213–253, B. G. Teubner, Leipzig, 1876.

[339] D. S. Rim, *Relatively complete fields,* Duke Math. J. **24** (1957), 197–200.

[340] J. Riordan, *Combinatorial Identities,* John Wiley & Sons, Inc., New York-London-Sydney, 1968.

[341] J.-J. Risler, *Une caráctérisation des idéaux des variétés algébriques réelles,* C. R. Acad. Sci. Paris **271** (1970), 1171–1173.

[342] J. F. Ritt, *Differential Algebra,* Amer. Math. Soc. Colloquium Publications, vol. 33, Amer. Math. Soc., New York, N. Y., 1950.

[343] P. Robba, *Lemmes de Hensel pour les opérateurs différentiels. Application à la réduction formelle des équations différentielles,* Enseign. Math. **26** (1980), 279–311.

[344] S. Roberts, *Genius at Play: The Curious Mind of John Horton Conway,* Bloomsbury USA, New York, 2015.

[345] A. Robinson, *On the application of symbolic logic to algebra,* in: J. A. Todd (ed.), *Proceedings of the International Congress of Mathematicians, Cambridge, Mass., 1950,* vol. 1, pp. 686–694, Amer. Math. Soc., Providence, RI, 1952.

[346] ______, *Completeness and persistence in the theory of models,* Z. Math. Logik Grundlag. Math. **2** (1953), 15–26.

[347] ______, *On ordered fields and definite functions,* Math. Ann. **130** (1955), 257–271.

[348] ______, *Further remarks on ordered fields and definite functions,* Math. Ann. **130** (1955), 405–409.

[349] ______, *Ordered structures and related concepts,* in: L. E. J. Brouwer et al. (eds.), *Mathematical Interpretation of Formal Systems,* pp. 51–56, Studies in Logic and the Foundations of Math., North-Holland, Amsterdam, 1955.

[350] ______, *Complete Theories*, North-Holland Publishing Co., Amsterdam, 1956.

[351] ______, *Some problems of definability in the lower predicate calculus,* Fund. Math. **44** (1957), 309–329.

[352] ______, *Relative model-completeness and the elimination of quantifiers,* Dialectica **12** (1958), 394–407.

[353] ______, *On the concept of a differentially closed field,* Bull. Res. Council Israel Sect. F **8F** (1959), 113–128.

[354] ______, *Introduction to Model Theory and to the Metamathematics of Algebra,* 2nd ed., Studies in Logic and the Foundations of Math., North-Holland, Amsterdam, 1965.

[355] ______, *On the real closure of a Hardy field,* in: G. Asser et al. (eds.), *Theory of Sets and Topology,* Deutscher Verlag der Wissenschaften, Berlin, 1972.

[356] ______, *Function theory on some nonarchimedean fields,* Amer. Math. Monthly **80** (1973), 87–109.

[357] ______, *Ordered differential fields,* J. Comb. Theory **14** (1973), 324–333.

[358] S. Roman, *The Umbral Calculus,* Pure and Applied Mathematics, vol. 111, Academic Press, Inc., New York, 1984.

[359] P. Roquette, *On the prolongation of valuations,* Trans. Amer. Math. Soc. **88** (1958), 42–56.

[360] ______, *History of valuation theory,* I, in: F.-V. Kuhlmann et al. (eds.), *Valuation Theory and its Applications,* Vol. I, pp. 291–355, Fields Inst. Commun., vol. 32, Amer. Math. Soc., Providence, RI, 2002.

[361] M. Rosenlicht, *Integration in finite terms,* Amer. Math. Monthly **79** (1972), 963–972.

[362] ______, *The nonminimality of the differential closure,* Pacific J. Math. **52** (1974), 529–537.

[363] ______, *On the value group of a differential valuation,* Amer. J. Math. **101** (1979), 258–266.

[364] ______, *Differential valuations,* Pacific J. Math. **86** (1980), 301–319.

[365] ______, *On the value group of a differential valuation,* II, Amer. J. Math. **103** (1981), 977–996.

[366] ______, *Hardy fields,* J. Math. Analysis and Appl. **93** (1983), 297–311.

[367] ______, *The rank of a Hardy field,* Trans. Amer. Math. Soc. **280** (1983), 659–671.

[368] ______, *Rank change on adjoining real powers to Hardy fields,* Trans. Amer. Math. Soc. **284** (1984), 829–836.

[369] ______, *Growth properties of functions in Hardy fields,* Trans. Amer. Math. Soc. **299** (1987), 261–272.

[370] ______, *Asymptotic solutions of* $Y'' = F(x)Y$, J. Math. Anal. Appl. **189** (1995), 640–650.

[371] M. Rosenlicht, M. Singer, *On elementary, generalized elementary, and Liouvillian extension fields,* in: H. Bass, P. Cassidy, J. Kovacic (eds.), *Contributions to Algebra,* pp. 329–342, Academic Press, New York, 1977.

[372] L. A. Rubel, *An elimination theorem for systems of algebraic differential equations,* Houston J. Math. **8** (1982), 289–295.

[373] ______, *A survey of transcendentally transcendental functions,* Amer. Math. Monthly **96** (1989), 777–788.

[374] K. Rychlík, *Zur Bewertungstheorie der algebraischen Körper,* J. Reine Angew. Math. **153** (1924), 94–107.

[375] M. Saarimäki, P. Sorjonen, *Valued groups,* Math. Scand. **70** (1992), 265–280.

[376] G. E. Sacks, *The differential closure of a differential field,* Bull. Amer. Math. Soc. **78** (1972), 629–634.

[377] ______, *Saturated Model Theory,* Mathematics Lecture Note Series, W. A. Benjamin, Inc., Reading, MA, 1972.

[378] B. Salvy, J. Shackell, *Asymptotic expansions of functional inverses,* in: P. Wang (ed.), *Symbolic and Algebraic Computation. Proceedings of ISSAC '92,* pp. 130–137, ACM Press, New York, 1992.

[379] D. Sarason, *The product formula for Fredholm operators,* Amer. Math. Monthly **94** (1987), 68–70.

[380] H. Sarges, *Ein Beweis des Hilbertschen Basissatzes,* J. Reine Angew. Math. **283/284** (1976), 436–437.

[381] N. Sauer, *On the density of families of sets,* J. Combinatorial Theory Ser. A **13** (1972), 145–147.

[382] T. Scanlon, *A model complete theory of valued D-fields,* J. Symbolic Logic **65** (2000), 1758–1784.

[383] ______, *Quantifier elimination for the relative Frobenius,* in: F.-V. Kuhlmann et al. (eds.), *Valuation Theory and its Applications,* vol. II, pp. 323–352, Fields Inst. Commun., vol. 33, Amer. Math. Soc., Providence, RI, 2003.

[384] ______, *Differentially valued fields are not differentially closed,* in: Z. Chatzidakis et al. (eds.), *Model Theory with Applications to Algebra and Analysis,* Vol. 1, pp. 111–115, London Math. Soc. Lecture Note Ser., vol. 349, Cambridge Univ. Press, Cambridge, 2008.

[385] H. F. Scherk, *De evolvenda functione* $\frac{yd.yd.yd...yd\,X}{dx^n}$ *disquisitiones nonnullae analyticae,* Ph. D. thesis, Berlin, 1823.

[386] E. Schippers, *A power matrix approach to the Witt algebra and Loewner equations,* Comput. Methods Funct. Theory **10** (2010), 399–420.

[387] L. Schlesinger, *Handbuch der Theorie der linearen Differentialgleichungen,* vol. 2, part 1, B. G. Teubner, Leipzig, 1897.

[388] M. C. Schmeling, *Corps de Transséries*, Ph. D. thesis, Université Paris-VII, 2001.

[389] A. Schmidt, *Über deduktive Theorien mit mehreren Sorten von Grunddingen,* Math. Ann. **115** (1938), 485–506.

[390] ______, *Die Zulässigkeit der Behandlung mehrsortigen Theorien mittels der üblichen einsortigen Prädikatenlogik,* Math. Ann. **123** (1951), 187–200.

[391] O. Schreier, *Über den Jordan-Hölderschen Satz,* Abh. Math. Semin. Univ. Hambg. **6** (1928), 300–302.

[392] A. Schrijver, *Theory of Linear and Integer Programming,* Wiley-Interscience Series in Discrete Mathematics, John Wiley & Sons, Ltd., Chichester, 1986.

[393] E. Schröder, *Vorlesungen über die Algebra der Logik,* vol. 3, B. G. Teubner, Leipzig, 1895.

[394] D. Scott, *On completing ordered fields,* in: W. A. J. Luxemburg (ed.), *Applications of Model Theory to Algebra, Analysis, and Probability,* 274–278, Holt, Rinehart and Winston, New York, 1969.

[395] A. Seidenberg, *A new decision method for elementary algebra,* Ann. of Math. **60** (1954), 365–374.

[396] ______, *An elimination theory for differential algebra,* Univ. California Publ. Math. **3** (1956), 31–65.

[397] ______, *Comments on Lefschetz's principle,* Amer. Math. Monthly **65** (1958), 685–690.

[398] ______, *On k-constructable sets, k-elementary formulae, and elimination theory,* J. Reine Angew. Math. **239/240** (1969), 256–267.

[399] J.-P. Serre, *Faisceaux algébriques cohérents,* Ann. of Math. **61** (1955), 197–278.

[400] ______, *Lie Algebras and Lie Groups,* 2nd ed., Lecture Notes in Math., vol. 1500, Springer-Verlag, New York-Berlin, 1992.

[401] ______, *How to use finite fields for problems concerning infinite fields,* in: G. Lachaud et al. (eds.), *Arithmetic, Geometry, Cryptography and Coding Theory,* pp. 183–193, Contemp. Math., vol. 487, Amer. Math. Soc., Providence, RI, 2009.

[402] J. Shackell, *Symbolic Asymptotics,* Algorithms and Computation in Mathematics, vol. 12, Springer-Verlag, Berlin, 2004.

[403] L. W. Shapiro, S. Getu, W. J. Woan, L. C. Woodson, *The Riordan group,* Discrete Appl. Math. **34** (1991), 229–239.

[404] S. Shelah, *Every two elementarily equivalent models have isomorphic ultrapowers,* Israel J. Math. **10** (1971), 224–233.

[405] ______, *Stability, the f.c.p., and superstability; model theoretic properties of formulas in first-order theory,* Ann. Math. Logic **3** (1971), 271–362.

[406] ______, *Uniqueness and characterization of prime models over sets for totally transcendental first-order theories,* J. Symbolic Logic **37** (1972), 107–113.

[407] ______, *A combinatorial problem; stability and order for models and theories in infinitary languages,* Pacific J. Math. **41** (1972), 247–261.

[408] ______, *Differentially closed fields,* Israel J. Math. **16** (1973), 314–328.

[409] ______, *Classification Theory and the Number of Nonisomorphic Models,* Studies in Logic and the Foundations of Mathematics, North-Holland Publishing Co., Amsterdam, 1978.

[410] ______, *Strongly dependent theories,* Israel J. Math. **204** (2014), 1–83.

[411] J. R. Shoenfield, *Mathematical Logic,* Addison-Wesley, Reading, MA, 1967.

[412] ______, *A theorem on quantifier elimination,* in: *Symposia Math.* V (Istituto Nazionali di Alta Matematica), pp. 173–176, Academic Press, London, 1971.

[413] ______, *Quantifier elimination in fields,* in: A. I. Arruda et al. (eds.), *Non-classical Logics, Model Theory and Computability,* pp. 243–252, Stud. Logic Found. Math., vol. 89, North-Holland, Amsterdam, 1977.

[414] J. Silverman, *The Arithmetic of Elliptic Curves,* Graduate Texts in Mathematics, vol. 106, Springer-Verlag, New York, 1986.

[415] P. Simon, *On dp-minimal ordered structures,* J. Symbolic Logic **76** (2011), 448–460.

[416] ______, *Distal and non-distal NIP theories,* Ann. Pure Appl. Logic **164** (2013), 294–318.

[417] ______, *A Guide to NIP Theories,* Lecture Notes in Logic, vol. 44, Cambridge University Press, Cambridge, 2015.

[418] M. Singer, *The model theory of ordered differential fields,* J. Symbolic Logic **43** (1978), 82–91.

[419] ______, *A class of differential fields with minimal differential closures,* Proc. Amer. Math. Soc. **69** (1978), 319–322.

[420] ______, *Algebraic solutions of nth order linear differential equations,* in: P. Ribenboim (ed.), *Proceedings of the Queen's Number Theory Conference, 1979,* pp. 379–420, Queen's Papers in Pure and Appl. Math., vol. 54, Queen's University, Kingston, Ont., 1980.

[421] ______, *Testing reducibility of linear differential operators: a group-theoretic perspective,* Appl. Algebra Engrg. Comm. Comput. **7** (1996), 77–104.

[422] Th. Skolem, *Untersuchungen über die Axiome des Klassenkalküls und über Produktations- und Summationsprobleme, welche gewisse Klassen von Aussagen betreffen,* Videnskapsselskapets Skrifter, I. Matem.-naturv. klasse **4** (1919), 1–37.

[423] ______, *Logisch-kombinatorische Untersuchungen über die Erfüllbarkeit oder Beweisbarkeit mathematischer Sätze nebst einem Theorem über dichte Mengen,* Videnskapsselskapets Skrifter, I. Matem.-naturv. Klasse **4** (1920), 1–36.

[424] ______, *Über einige Satzfunktionen in der Arithmetik,* Skrifter utgitt av det Norske Videnskaps-Akademi i Oslo **7** (1931), 1–28.

[425] S. Smale, *Mathematical problems for the next century,* Math. Intelligencer **20** (1998), 7–15.

[426] A. L. Smirnov, *Torus schemes over a discrete valuation ring,* St. Petersburg Math. J. **8** (1997), 651–659.

[427] H. J. S. Smith, *On systems of linear indeterminate equations and congruences,* Phil. Trans. R. Soc. Lond. **151** (1861), 293–326.

[428] ______, *On the higher singularities of plane curves,* Proc. London Math. Soc. **6** (1875), 153–182.

[429] V. E. E. Stadigh, *Ein Satz ueber Funktionen die algebraische Differentialgleichungen befriedigen und ueber die Eigenschaft der Funktion $\zeta(s)$ keiner solchen Gleichung zu genügen,* Ph. D. thesis, Helsinki, 1902.

[430] G. Stengle, J. E. Yukich, *Some new Vapnik-Chervonenkis classes,* Ann. Statist. **17** (1989), 1441–1446.

[431] J. Steprāns, *History of the continuum in the 20th century,* in: D. M. Gabbay et al. (eds.), *Handbook of the History of Logic,* vol. 6, pp. 73–144, North-Holland, Amsterdam, 2012.

[432] J. Steuding, *The Fibonacci zeta-function is hypertranscendental,* Cubo **10** (2008), 133–136.

[433] J. Stirling, *Methodus Differentialis: sive Tractatus de Summatione et Interpolatione Serierum Infinitarum,* G. Strahan, London, 1730.

[434] O. Stolz, *Zur Geometrie der Alten, insbesondere über ein Axiom des Archimedes,* Math. Ann. **22** (1883), 504–519.

[435] M. H. Stone, *The theory of representation for Boolean algebras,* Trans. Amer. Math. Soc. **40** (1936), 37–111.

[436] ______, *Applications of the theory of Boolean rings to general topology,* Trans. Amer. Math. Soc. **41** (1937), 375–481.

[437] W. Strodt, *Contributions to the asymptotic theory of ordinary differential equations in the complex domain,* Mem. Amer. Math. Soc. **13** (1954).

[438] ______, *Principal solutions of ordinary differential equations in the complex domain,* Mem. Amer. Math. Soc. **26** (1957).

[439] A. Tarski, *Sur les ensembles définissables de nombres réels,* I, Fund. Math. **17** (1931), 210–239.

[440] ______, *Der Wahrheitsbegriff in den formalisierten Sprachen,* Studia Philos. **1** (1935), 261–405.

[441] ______, *Grundzüge des Systemenkalküls,* I, Fund. Math. **25** (1935), 503–526.

[442] ______, *Grundzüge des Systemenkalküls,* II, Fund. Math. **26** (1936), 283–301.

[443] ______, *A Decision Method for Elementary Algebra and Geometry,* 2nd ed., University of California Press, Berkeley and Los Angeles, Calif., 1951.

[444] ______, *Some notions and methods on the borderline of algebra and metamathematics,* in: J. A. Todd (ed.), *Proceedings of the International Congress of Mathematicians, Cambridge, Mass., 1950,* vol. 1, pp. 705–720, Amer. Math. Soc., Providence, RI, 1952.

[445] ______, *Contributions to the theory of models,* I, II, Indag. Math. **16** (1954), 572–588.

[446] A. Tarski, R. Vaught, *Arithmetical extensions of relational systems,* Compositio Math. **13** (1956), 81–102.

[447] A. I. Thaler, *On the Newton polytope,* Proc. Amer. Math. Soc. **15** (1964), 944–950.

[448] P. G. Todorov, *New explicit formulas for the nth derivative of composite functions,* Pacific J. Math. **92** (1981), 217–236.

[449] C. Toffalori, K. Vozoris, *Notes on local o-minimality,* Math. Log. Q. **55** (2009), 617–632.

[450] M. Tressl, *Pseudo completions and completions in stages of o-minimal structures,* Arch. Math. Logic **45** (2006), 983–1009.

[451] E. W. von Tschirnhaus, *Nova methodus auferendi omnes terminos intermedios ex data æquatione,* Acta Eruditorium **2** (1683), 204–207.

[452] A. I. Uzkov, *On rings of quotients of commutative rings,* Mat. Sbornik N. S. **22** (1948), 439–441.

[453] V. N. Vapnik, A. Ja. Chervonenkis, *On the uniform convergence of relative frequencies of events to their probabilities,* Theor. Probability Appl. **16** (1971), 264–280.

[454] V. S. Varadarajan, *Euler Through Time: A New Look at Old Themes,* American Mathematical Society, Providence, RI, 2006.

[455] R. Vaught, *Applications of the Löwenheim-Skolem-Tarski theorem to problems of completeness and decidability,* Indag. Math. **16** (1954), 467–472.

[456] ______, *Denumerable models of complete theories,* in: *Infinitistic Methods,* pp. 303–321, Pergamon Press, Oxford-London-New York-Paris; Państwowe Wydawnictwo Naukowe, Warsaw, 1961.

[457] G. Veronese, *Il continuo rettilineo e l'assioma V d'Archimede,* Atti Della R. Accademia Dei Lincei **6** (1889), 603–624.

[458] S. Warner, *Residual fields in valuation theory,* Math. Scand. **56** (1985), 203–221.

[459] H. Weber, *Die partiellen Differential-Gleichungen der mathematischen Physik,* vol. 2, 5th ed., Vieweg-Verlag, Braunschweig, 1912.

[460] J. H. M. Wedderburn, *Noncommutative domains of integrity,* J. Reine Angew. Math. **167** (1932), 129–141.

[461] A. Weil, *Foundations of Algebraic Geometry,* Amer. Math. Soc. Colloq. Publ., vol. 29, American Mathematical Society, Providence, R.I., 1946.

[462] V. Weispfenning, *Aspects of quantifier elimination in algebra,* in: P. Burmeister et al. (eds.), *Universal Algebra and its Links with Logic, Algebra, Combinatorics and Computer Science (Darmstadt, 1983),* pp. 85–105, Res. Exp. Math., vol. 4, Heldermann, Berlin, 1984.

[463] R. Weitzenböck, *Über die invarianten Gruppen,* Acta. Math. **58** (1932), 231–293.

[464] H. Weyl, *Über die Definitionen der mathematischen Grundbegriffe,* Math.-naturw. Bl. **7** (1910), 93–95 and 109–113.

[465] A. Wilkie, *Model completeness results for expansions of the ordered field of real numbers by restricted Pfaffian functions and the exponential function,* J. Amer. Math. Soc. **9** (1996), 1051–1094.

[466] O. Zariski, *The concept of a simple point of an abstract algebraic variety,* Trans. Amer. Math. Soc. **62** (1947), 1–52.

[467] ______, *The fundamental ideas of abstract algebraic geometry,* in: J. A. Todd (ed.), *Proceedings of the International Congress of Mathematicians, Cambridge, Mass., 1950,* vol. 2, pp. 77–89. Amer. Math. Soc., Providence, RI, 1952.

[468] O. Zariski, P. Samuel, *Commutative Algebra,* Vol. II, Graduate Texts in Mathematics, vol. 29, Springer-Verlag, New York-Heidelberg, 1975.

List of Symbols

Algebra

$[\Gamma' : \Gamma]$	index of the abelian group extension $\Gamma' \supseteq \Gamma$, 117
$[L : K]$	degree of the field extension $L \supseteq K$, 117
$\mathrm{Aut}(L\|K)$	group of automorphisms of L over K, 201
$\mathcal{F}_G$	monoid of Fredholm operators on G, 97
$\mathcal{S}_G$	monoid of surjective Fredholm operators on G, 97
coker A	cokernel of a linear map A, 95
$\mathrm{d}\,a$	differential of a (in $\Omega_{A\|K}$), 62
deg P	degree of P, 186
$\dim_C V$	dimension of the vector space V over the field C, 93
$\mathrm{End}(V)$	algebra of endomorphisms of a module V, 534
$\mathrm{Hom}_R(A, M)$	module of R-linear maps $A \to M$, 55
$\Im(f)$	imaginary part of $f \in K[\mathrm{i}]$, 256
i	an element of a field with $\mathrm{i}^2 = -1$, 172
index A	index of a Fredholm operator A, 95
ker A	kernel of a linear map A, 95
$\mu(M)$	minimal number of generators of the module M, 44
$\Omega_{A\|K}$	A-module of Kähler differentials of the K-algebra A, 62
$\mathrm{GL}_n(R)$	group of invertible $n \times n$ matrices over a ring R, 266
$\mathrm{nil}(R)$	nilradical of R, 30

$\operatorname{rank}_{\mathbb{Q}}(M)$	rational rank of the abelian group M, 58
$\operatorname{trdeg}(L\|K)$	transcendence degree of the field extension $L \supseteq K$, 117
$\mathbb{Q}G$	divisible hull $\mathbb{Q} \otimes_{\mathbb{Z}} G$ of an abelian group G, 98
$\Re(f)$	real part of $f \in K[i]$, 256
$\operatorname{Spec}(R)$	set of prime ideals of the commutative ring R, 29
$\operatorname{mul} P$	multiplicity of P at 0, 186
$A^{\neq}$	$A \setminus \{0\}$, for an additively written abelian group A, xv
$A_{\mathfrak{p}}$	localization of A with respect to its prime ideal $\mathfrak{p}$, 116
KL	compositum of the fields K and L, 29
M_{tor}	torsion submodule of M, 266
P_{+h}	additive conjugate $P(Y+h)$ of P by h, 186
$P_{\times h}$	multiplicative conjugate $P(hY)$ of P by h, 186
$R^{\times}$	group of units of a ring R, xv
$S^{-1}A$	localization of A at its multiplicative subset S, 46
$a\|\|b$	a totally divides b, 264
$(I : S)$	$\{r \in R : rS \subseteq I\}$, 30
$\sqrt{I}$	radical of an ideal I, 30

Model Theory

$\mathcal{L}_{\preccurlyeq}$	language of integral domains with a dominance relation, 179
$\mathcal{L}_{\mathrm{OR}}$	language of ordered rings, 182
$\mathcal{L}_{\mathrm{OR},\preccurlyeq}$	language of ordered integral domains with a dominance relation, 182
$\mathcal{L}_{\mathrm{tame}}$	language of tame pairs, 184
$\mathcal{L}_{\partial}$	language of differential rings, 237
$\mathcal{L}_{\Lambda\Omega}^{\iota}$	language of $\Lambda\Omega$-fields, 678

$\mathcal{L}_{\mathrm{R}}$	language of rings, 729
$\mathcal{L}_{\mathrm{O}}$	language of ordered sets, 729
$\mathcal{L}_{\mathrm{OA}}$	language of ordered abelian groups, 729
ACVF	theory of algebraically closed fields with nontrivial dominance relation in the language $\mathcal{L}_{\preccurlyeq}$, 179
RCVF	theory of real closed ordered fields with nontrivial convex dominance relation in the language $\mathcal{L}_{\mathrm{OR},\preccurlyeq}$, 182
$\mathrm{RCF}_{\mathrm{tame}}$	theory of nontrivial tame pairs in the language $\mathcal{L}_{\mathrm{tame}}$, 184
DCF	theory of differentially closed fields in the language $\mathcal{L}_{\partial}$, 237
T^{nl}	theory of ω-free newtonian Liouville closed H-fields in the language of ordered valued differential rings, 678
$T^{\mathrm{nl}}_{\mathrm{small}}$, $T^{\mathrm{nl}}_{\mathrm{large}}$	completions of T^{nl}, 678
ACF	theory of algebraically closed fields in the language $\mathcal{L}_{\mathrm{R}}$, 751
ACF(p)	theory of algebraically closed fields of characteristic p in the language $\mathcal{L}_{\mathrm{R}}$, 751
RCF	theory of real closed ordered fields in the language $\mathcal{L}_{\mathrm{OR}}$, 780
$T^{\mathrm{nl},\iota}_{\Lambda\Omega}$	theory of ω-free newtonian Liouville closed H-fields in the language $\mathcal{L}^{\iota}_{\Lambda\Omega}$, 678
$\boldsymbol{M}_A$	expansion of $\boldsymbol{M}$ by names for the elements of A, 736
$\boldsymbol{M} \subseteq \boldsymbol{N}$	$\boldsymbol{M}$ is a substructure of $\boldsymbol{N}$, 731
$\langle A\rangle_{\boldsymbol{M}}$	substructure of $\boldsymbol{M}$ generated by $A \subseteq M$, 737
$\boldsymbol{M} \preccurlyeq \boldsymbol{N}$	$\boldsymbol{M}$ is an elementary substructure of $\boldsymbol{N}$, 746
$\boldsymbol{M} \preccurlyeq_{\exists} \boldsymbol{N}$	$\boldsymbol{M}$ is existentially closed in $\boldsymbol{N}$, 758
$\boldsymbol{M} \cong \boldsymbol{N}$	$\boldsymbol{M}$, $\boldsymbol{N}$ are isomorphic, 732
$\boldsymbol{M} \equiv \boldsymbol{N}$	$\boldsymbol{M}$, $\boldsymbol{N}$ are elementarily equivalent, 744
$\boldsymbol{M} \equiv_{\mathrm{bf}} \boldsymbol{N}$	$\boldsymbol{M}$, $\boldsymbol{N}$ are back-and-forth equivalent, 745
$\boldsymbol{M} \equiv_A \boldsymbol{N}$	$\boldsymbol{M}_A$, $\boldsymbol{N}_A$ are elementarily equivalent, 773
$\mathrm{Aut}(\boldsymbol{M})$	group of automorphisms of $\boldsymbol{M}$, 732

$\mathrm{Aut}(\boldsymbol{M}\vert A)$	group of automorphisms of $\boldsymbol{M}$ over A, 732
$\boldsymbol{M} \models \sigma$	σ is true in $\boldsymbol{M}$, 740
$\boldsymbol{M} \models \Sigma$	$\boldsymbol{M}$ is a model of Σ, 750
$\Sigma \models \sigma$	σ is a logical consequence of Σ, 751
$\varphi^{\boldsymbol{M}}$	set defined by φ in $\boldsymbol{M}$, 741
$\mathrm{acl}(A)$	algebraic closure of A, 748
$\mathrm{dcl}(A)$	definable closure of A, 748
$\mathrm{Th}(\boldsymbol{M})$	theory of $\boldsymbol{M}$, 753
$\mathrm{Th}(\mathcal{C})$	theory of a class $\mathcal{C}$ of $\mathcal{L}$-structures, 753
$\mathrm{Th}(\Sigma)$	set of logical consequences of Σ, 753
$\mathrm{Diag}_M(\boldsymbol{M})$	diagram of $\boldsymbol{M}$, 759
$\mathrm{tp}_x^{\boldsymbol{M}}(b)$	complete x-type in $\boldsymbol{M}$ realized by $b \in M_x$, 763
$\mathrm{tp}_x^{\boldsymbol{M}}(b\vert A)$	complete x-type in $\boldsymbol{M}$ realized by $b \in M_x$ over A, 765
$\mathrm{S}_x^{\boldsymbol{M}}(A)$	space of complete x-types in $\boldsymbol{M}$ over A, 765

Combinatorics

S^*	set of (finite) words on a set S, 210
$\binom{\boldsymbol{j}}{\boldsymbol{i}}$	$\binom{j_0}{i_0}\cdots\binom{j_r}{i_r}$, for $\boldsymbol{i} = (i_0, \dots, i_r) \leqslant \boldsymbol{j} = (j_0, \dots, j_r)$ in $\mathbb{N}^{1+r}$, 210
$\boldsymbol{i}!$	$i_0! \cdots i_r!$ for $\boldsymbol{i} = (i_0, \dots, i_r) \in \mathbb{N}^{1+r}$, 210
$\boldsymbol{i} \leqslant \boldsymbol{j}$	partial ordering on $\mathbb{N}^{1+r}$, 210
$\vert\boldsymbol{i}\vert$	degree $i_0 + \cdots + i_r$ of $\boldsymbol{i} = (i_0, \dots, i_r) \in \mathbb{N}^{1+r}$, 209
$\vert\boldsymbol{i}\vert'$	subdegree $\vert\boldsymbol{i}\vert' = i_1 + \cdots + i_r$ of $\boldsymbol{i} = (i_0, \dots, i_r) \in \mathbb{N}^{1+r}$, 212
$\Vert\boldsymbol{i}\Vert$	weight $\Vert\boldsymbol{i}\Vert = i_1 + 2i_2 + \cdots + ri_r$ of $\boldsymbol{i} = (i_0, \dots, i_r) \in \mathbb{N}^{1+r}$, 212
$\binom{\boldsymbol{\tau}}{\boldsymbol{\sigma}}$	$\binom{\tau_1}{\sigma_1}\cdots\binom{\tau_d}{\sigma_d}$, for $\boldsymbol{\sigma} = \sigma_1 \cdots \sigma_d \leqslant \boldsymbol{\tau} = \tau_1 \cdots \tau_d$ in $\mathbb{N}^*$, 215
$\boldsymbol{\sigma} \leqslant \boldsymbol{\tau}$	partial ordering on $\mathbb{N}^*$, 215

$\lvert\boldsymbol{\sigma}\rvert$	length n of $\boldsymbol{\sigma} = \sigma_1 \cdots \sigma_n \in \mathbb{N}^*$, 210
$\lVert\boldsymbol{\sigma}\rVert$	weight $\lVert\boldsymbol{\sigma}\rVert = \sigma_1 + \cdots + \sigma_d$ of $\boldsymbol{\sigma} = \sigma_1 \cdots \sigma_d \in \mathbb{N}^*$, 212
$\operatorname{supp} \boldsymbol{\sigma}$	set of $i \in \{1, \dots, d\}$ with $\sigma_i \neq 0$, for $\boldsymbol{\sigma} = \sigma_1 \cdots \sigma_d \in \mathbb{N}^*$, 477
$\left[\begin{smallmatrix}\boldsymbol{\tau}\\ \boldsymbol{\sigma}\end{smallmatrix}\right]$	$\left[\begin{smallmatrix}\tau_1\\ \sigma_1\end{smallmatrix}\right] \cdots \left[\begin{smallmatrix}\tau_d\\ \sigma_d\end{smallmatrix}\right]$, for $\boldsymbol{\sigma} = \sigma_1 \cdots \sigma_d, \boldsymbol{\tau} = \tau_1 \cdots \tau_d \in \mathbb{N}^*$, 476
$\left[\begin{smallmatrix}n\\ k\end{smallmatrix}\right]$	unsigned Stirling numbers of the first kind, 294
$s(n, k)$	signed Stirling numbers of the first kind, 294
$\left\{\begin{smallmatrix}j\\ i\end{smallmatrix}\right\}$	Stirling numbers of the second kind, 576

Ordered Sets

$(a, b) = (a, b)_S$	interval in S, 71
$[a, b] = [a, b]_S$	convex hull of $\{a, b\}$ in S, 70
$\operatorname{cf}(S)$	the cofinality of an ordered set S, 72
$\operatorname{ci}(S)$	the coinitiality of an ordered set S, 72
$\operatorname{conv}(A)$	convex hull of A, 70
$\operatorname{ot}(S)$	order type of the ordered set S, 71
$A^{\downarrow}$	smallest downward closed subset containing A, 72
$A^{\uparrow}$	smallest upward closed subset containing A, 72

Valued Abelian Groups and Vector Spaces

(G, S, v)	valued abelian group, 75
$\overline{B}(s)$	$\overline{B}_0(s)$, 74
$\overline{B}_a(s)$	closed ball centered at a with radius s, 73
$\operatorname{supp} g$	support of an element $g \in \prod_s G_s$, 74
A^{-1}	distinguished right-inverse of A, 97
$a_\rho \rightsquigarrow a$	(a_ρ) pseudoconverges to a, 75

Ordered Abelian Groups and Vector Spaces

Valued Fields

G^{d}	decomposition group of a valuation ring, 122
K^{Δ}	maximal Δ-immediate extension of the valued field K, 164
K^{c}	completion of the valued field K, 130
K^{d}	completion of the ordered field K, 178
K^{h}	henselization of K, 147
K^{sc}	a step-completion of the valued field K, 162
K^{unr}	maximal unramified extension of K, 155
$K^{\prec 1}$	maximal ideal of the valuation ring of a valued field K, 112
$K^{\preccurlyeq 1}$	valuation ring of a valued field K, 112
$K^{\succ 1}$	the complement of $K^{\preccurlyeq 1}$ in a valued field K, 112
L^{d}	decomposition field of $L \supseteq K$, 122
$P_{\beta(\mathrm{E})}$	primary dominant part of (E), 190
$v = v_K$	valuation of a valued field K, 112

Differential Algebra

$\mathrm{Aut}_\partial(L\|K)$	group of automorphisms of a differential ring L over K, 201
∂	derivation on a ring, 56
δ	derivation $\delta = \phi^{-1}\partial$ of K^{ϕ}, 290
$\omega(z)$	$\omega(z) = -(2z' + z^2)$, 258
$\mathrm{trdeg}_\partial(L\|K)$	differential transcendence degree of the differential field extension $L \supseteq K$, 206
$\mathrm{S}(u)$	Schwarzian derivative of u, 260
$\mathrm{s}(y)$	$\omega(-y^\dagger) - y^2$, for $y \neq 0$, 260
$\sigma(y)$	$\omega(-y^\dagger) + y^2$, for $y \neq 0$, 262
$\mathrm{sol}_R(A)$	set of solutions of $y' = Ay$ over R, 276
$\Omega(K)$	set of $f \in K$ such that $4y'' + fy = 0$ for some $y \in K^\times$, 259

$\mathrm{Wr}(y_0, \dots, y_n)$	Wronskian matrix of $y_0, \dots, y_n$, 206
$\mathrm{wr}(y_0, \dots, y_n)$	Wronskian determinant $\det \mathrm{Wr}(y_0, \dots, y_n)$, 206
$a^\dagger$	logarithmic derivative $a^\dagger = a'/a$ of a unit a, 200
$a^{(n)}$	nth derivative $a^{(n)} = \partial^n(a)$ of a, 199
$a^{\langle n \rangle}$	nth iterated logarithmic derivative of a, 213
C_K	field of constants of the differential field K, 200
$f =_c g$	$f = c \cdot g$ for some constant $c \neq 0$, 296
$K[[\partial]]$	algebra of partial differential operators on $K\{Y\}$, 578
$K\langle Y_1, \dots, Y_m \rangle$	field of differential rational functions in $Y_1, \dots, Y_m$ with coefficients in the differential field K, 202
$K\langle y_1, \dots, y_m \rangle$	differential field generated by $y_1, \dots, y_m$ over K, 202
$K\{a\}$	differential ring generated by a over K, 199
$K\{Y\}$	ring of differential polynomials in Y over K, 201
$K\{Y_1, \dots, Y_m\}$	ring of differential polynomials in $Y_1, \dots, Y_m$ over K, 202
K^ϕ	compositional conjugate of K by $\phi \in K^\times$, 291
K^{dc}	differential closure of the differential field K, 239
$[S]$	differential ideal generated by S, 231

Differential Polynomials

$\mathrm{c}(P)$	complexity of P, 216
$\deg(P)$	degree of P, 201
$\lvert \boldsymbol{i} \rvert$	degree $\lvert \boldsymbol{i} \rvert = i_0 + \cdots + i_r$ of $\boldsymbol{i} = (i_0, \dots, i_r) \in \mathbb{N}^{1+r}$, 209
$\mathrm{m}(P)$	$\sum_{c \in C} \mathrm{mul}(P_{+c})$, for $P \in C[Y]$, 249
$\mathrm{sdeg}(P)$	subdegree of P, 213
$\mathrm{order}(P)$	order of P, 201
$\mathrm{Ri}(P)$	Riccati transform of P, 300

$\mathrm{mul}(P)$	multiplicity of P at 0, 209
$\mathrm{wt}(P)$	weight of P, 212
$\mathrm{wm}(P)$	weighted multiplicity of P, 212
F_k^n	used in expressing ∂^n in terms of δ, 290
G_k^n	used in expressing ∂^n in terms of δ, 292
I_P	initial of P, 216
L_P	linear part of P, 242
$P \circ Q = P(Q)$	composition of P with Q, 202
P^∂	result of applying the derivation ∂ to the coefficients of P, 67
P^ϕ	compositional conjugate of P by ϕ, 291
$P^{(\boldsymbol{i})}$	$\frac{\partial^{\lvert \boldsymbol{i} \rvert} P}{\partial^{i_0} Y \cdots \partial^{i_r} Y^{(r)}}$, for $\boldsymbol{i} = (i_0, \ldots, i_r) \in \mathbb{N}^{1+r}$, 210
$P^{[\boldsymbol{\sigma}]}$	$P^{(\sigma_1)} \ldots P^{(\sigma_n)}$, for $\boldsymbol{\sigma} = \sigma_1 \cdots \sigma_n \in \mathbb{N}^*$, 210
$P^{\boldsymbol{i}}$	$P^{i_0} (P')^{i_1} \cdots (P^{(r)})^{i_r}$, for $\boldsymbol{i} = (i_0, \ldots, i_r) \in \mathbb{N}^{1+r}$, 209
$P^{\times \phi}$	$P^{\times 1, \phi}$, 297
$P^{\times q, \phi}$	$P^{\phi}_{\times \phi^q}$, 298
$P^{\times q}$	$P((Y')^q)$, 298
$P^{\times}$	$P(Y')$, 297
P_d	homogeneous part of degree d of P, 209
$P_{(\boldsymbol{i})}$	$\frac{P^{(\boldsymbol{i})}}{\boldsymbol{i}!}$, for $\boldsymbol{i} \in \mathbb{N}^{1+r}$, 210
P_{+h}	additive conjugate $P(Y+h)$ of P by h, 214
$P_{[\boldsymbol{\sigma}]}$	coefficient of $Y^{[\boldsymbol{\sigma}]}$ in the decomposition along orders of P, 210
$P_{[w]}$	isobaric part of P of weight w, 212
$P_{\langle \boldsymbol{i} \rangle}$	coefficient of $Y^{\langle \boldsymbol{i} \rangle}$ in the logarithmic decomposition of P, 213
$P_{\boldsymbol{i}}$	coefficient of $Y^{\boldsymbol{i}}$ in the natural decomposition of P, 209
$P_{\lvert d \rvert'}$	subhomogeneous part of P of subdegree d, 212
$P_{\times h}$	multiplicative conjugate $P(hY)$ of P by h, 214

$P\!\uparrow$	upward shift of the differential polynomial P over $\mathbb{T}$, 293
R_k^n	G_{k+1}^{n+1}, 297
R_n	nth Riccati polynomial, 299
S_P	separant of P, 216

Linear Differential Operators

$\mathscr{E}(A)$	set of exceptional values of A, 287
$\mathscr{E}^{\mathrm{e}}(A)$	set of eventual exceptional values of A, 481
$\ker A$	kernel of A, 242
$\mathrm{m}(A)$	$\mathrm{m}(P)$ for $P \in C[Y]$ with $A = P(\partial) \in C[\partial]$, 249
$\mathrm{Ri}(A)$	Riccati transform of A, 301
A'	derivative of A, 243
A^*	adjoint of A, 246
A_L	companion matrix of L, 272
$A_{\ltimes a}$	twist of A by a, 243
$K[\partial]$	ring of linear differential operators over K, 241
M^*	dual of the $K[\partial]$-module M, 279
M_A	differential module associated to an $n \times n$ matrix A, 277
$y' = A^*y$	adjoint equation of $y' = Ay$, 280

Asymptotic Relations

$f \asymp g$	f and g are asymptotic, 113
$f \asymp\!\!\!\asymp g$	f and g are comparable, 383
$f \prec\!\!\!\prec g$	f is flatter than g, 383
$f \preceq\!\!\!\preceq g$	f is flatter than or comparable with g, 383
$f \prec 1$	f is infinitesimal, 114

$f \prec g$	f is strictly dominated by g, 113
$f \preccurlyeq 1$	f is bounded, 114
$f \preccurlyeq g$	f is dominated by g, 113
$f \sim g$	f and g are equivalent, 113
$f \succ 1$	f is infinite, 114
$\mathrm{Cl}(f)$	comparability class of f, 384
$\preccurlyeq^\flat$	dominance relation associated to $v^\flat$, 406
$\preccurlyeq^\flat_\phi$	dominance relation associated to $v^\flat_\phi$, 406
$\preccurlyeq_\Delta$ or $\dot{\preccurlyeq}$	dominance relation associated to $\dot{v} = v_\Delta$, 157
$\preccurlyeq_\Phi$	dominance relation $\preccurlyeq^\flat_{\Phi^\dagger}$, for $\Phi \in K^\times$, $\Phi \not\asymp 1$, 407

Asymptotic Couples

$(\dot{\Gamma}, \dot{\psi})$	coarsening of (Γ, ψ) by a convex subgroup, 396
(Γ, ψ)	asymptotic couple, 322
χ	contraction map of an H-asymptotic couple, 392
γ'	$\gamma' = \gamma + \psi(\gamma)$, for $\gamma \neq 0$, 381
$\gamma^\dagger$	$\gamma^\dagger = \psi(\gamma)$, for $\gamma \neq 0$, 381
Ψ	$\psi(\Gamma^{\neq})$, 381
$\psi(* - \alpha)$	value of $\psi(\gamma - \alpha)$ for all large enough $\gamma \in \Psi$, 389
∇	valuation on the value group, 320

Dominant Quantities

$\mathrm{ddeg}(P)$	dominant degree of P, 327
$\mathrm{ddeg}_{\mathcal{E}}(P)$	dominant degree of P on $\mathcal{E}$, 328
$\mathrm{dmul}(P)$	dominant multiplicity of P at 0, 327
$\mathrm{dwt}(P)$	dominant weight of P, 229

$\operatorname{dwt}_P(\gamma)$	dominant weight of $P_{\times g}$, where $vg = \gamma$, 230
$\operatorname{dwm}(P)$	dominant weighted multiplicity of P, 229
$\operatorname{dwm}_P(\gamma)$	dominant weighted multiplicity of $P_{\times g}$, where $vg = \gamma$, 230
$\mathfrak{d}_P$	dominant monomial of P, 325
$\mu(R)$	dominant multiplicity of R, 302
$\mu_R(g + o)$	$\mu(R_{+g})$, 303
$\nu(R)$	dominant weight of R, 302
$\nu_R(g + \mathcal{O})$	$\nu(R_{+g})$, 303
D_P	dominant part of P, 325
$v_P(\gamma)$	gaussian valuation of $P_{\times f}$, where $vf = \gamma$, 228

Eventual Quantities

$\mathfrak{e}(P, Q)$	eventual equalizer for P, Q, 613
$\operatorname{ndeg}(P)$	Newton degree of P, 480
$\operatorname{ndeg}_P(\gamma)$	Newton degree of $P_{\times g}$, where $vg = \gamma$, 482
$\operatorname{ndeg}_{\boldsymbol{a}}(P)$	Newton degree of P in the cut $\boldsymbol{a}$, 485
$\operatorname{ndeg}_{\mathcal{E}}(P)$	Newton degree of P on $\mathcal{E}$, 483
$\operatorname{ndeg}_{\geqslant\gamma}(P)$	Newton degree of P on $\{\mathfrak{n} : v\mathfrak{n} \geqslant \gamma\}$, 483
$\operatorname{ndeg}_{\prec\mathfrak{m}}(P)$	Newton degree of P on $\{\mathfrak{n} : \mathfrak{n} \prec \mathfrak{m}\}$, 483
$\operatorname{nmul}(P)$	Newton multiplicity of P, 480
$\operatorname{nmul}_P(\gamma)$	Newton multiplicity of $P_{\times g}$, where $vg = \gamma$, 482
$\operatorname{nwt}(P)$	Newton weight of P, 479
$\operatorname{nwt}^{\times}(P)$	Newton weight of $P(Y')$, 608
$\operatorname{nwt}_P(\gamma)$	Newton weight of $P_{\times g}$, where $vg = \gamma$, 480
$d(P, (a_\rho))$	eventual value of $\operatorname{ndeg}_{\geqslant\gamma_\rho} P_{+a_\rho}$, 484
$N_P^{\times}$	eventual dominant part of $P^{\times\phi}$, 608

Triangular Automorphisms

Υ_ϕ	triangular automorphism with matrix $\big(\phi^{-j}F_i^j(\phi)\big)_{i,j}$, 579
Ξ_ϕ	diagonal automorphism with matrix $\operatorname{diag}(\phi^i)$, 580
A^∂	algebra of constants of the derivation ∂ on A, 568
$A^{\Delta,\Lambda}$	algebra of common constants of Δ, Λ, 568
A_{Lie}	Lie algebra associated to the algebra A, 537
B_{ij}	Bell polynomials, 553
D_K	K-algebra of diagonal matrices over K, 543
$f^{[a]}$	fractional iterates of the power series f, 562
$K^{\mathbb{N}\times\mathbb{N}}$	K-module of matrices $M=(M_{ij})_{i,j\in\mathbb{N}}$ over K, 542

Asymptotic Fields

(ℓ_ρ)	logarithmic sequence, 499
$(\upgamma_\rho)$	$(\ell_\rho^\dagger)$, 500
$(\uplambda_\rho)$	$\big(-(\ell_\rho^{\dagger\dagger})\big)$, 500
$(\upomega_\rho)$	$\big(\omega\big((-\ell_\rho)^\dagger\big)\big)$, 511
$\mathcal{O}_\phi^\flat$	valuation ring of $v_\phi^\flat$, 407
$\Gamma^\flat$	$\{\gamma : \psi(\gamma) > 0\}$, 406
$\Gamma_\phi^\flat$	$\{\gamma : \psi(\gamma) > v\phi\}$, 406
$\Gamma^\sharp$	$\Gamma^\sharp = \Gamma/\Gamma^\flat$, 406
$\mathrm{I}(K)$	$\{y \in K : y \preccurlyeq f' \text{ for some } f \in \mathcal{O}\}$, 519
$\mathrm{dv}(K)$	differential-valued hull of K, 443
$\scriptstyle\mathcal{O}_\phi^\flat$	maximal ideal of the valuation ring of $v_\phi^\flat$, 407
$\Delta(K)$	$-(K^{\neq,\prec 1})'^\dagger$, 520
$\Gamma(K)$	$(K^{\succ 1})^\dagger$, 520
$\Lambda(K)$	$-(K^{\succ 1})^{\dagger\dagger}$, 520

Index